APPENDIX C, TABLE VI   *t* Distribution Values

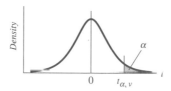

Degrees of Freedom ν	$t_{.10}$	$t_{.05}$	$t_{.025}$	$t_{.01}$	$t_{.005}$
1	3.078	6.314	12.706	31.821	63.657
2	1.886	2.920	4.303	6.965	9.925
3	1.638	2.353	3.182	4.541	5.841
4	1.533	2.132	2.776	3.747	4.604
5	1.476	2.015	2.571	3.365	4.032
6	1.440	1.943	2.447	3.143	3.707
7	1.415	1.895	2.365	2.998	3.499
8	1.397	1.860	2.306	2.896	3.355
9	1.383	1.833	2.262	2.821	3.250
10	1.372	1.812	2.228	2.764	3.169
11	1.363	1.796	2.201	2.718	3.106
12	1.356	1.782	2.179	2.681	3.055
13	1.350	1.771	2.160	2.650	3.012
14	1.345	1.761	2.145	2.624	2.977
15	1.341	1.753	2.131	2.602	2.947
16	1.337	1.746	2.120	2.583	2.921
17	1.333	1.740	2.110	2.567	2.898
18	1.330	1.734	2.101	2.552	2.878
19	1.328	1.729	2.093	2.539	2.861
20	1.325	1.725	2.086	2.528	2.845
21	1.323	1.721	2.080	2.518	2.831
22	1.321	1.717	2.074	2.508	2.819
23	1.319	1.714	2.069	2.500	2.807
24	1.318	1.711	2.064	2.492	2.797
25	1.316	1.708	2.060	2.485	2.787
26	1.315	1.706	2.056	2.479	2.779
27	1.314	1.703	2.052	2.473	2.771
28	1.313	1.701	2.048	2.467	2.763
29	1.311	1.699	2.045	2.462	2.756
30	1.310	1.697	2.042	2.457	2.750
40	1.303	1.684	2.021	2.423	2.704
60	1.296	1.671	2.000	2.390	2.660
120	1.289	1.658	1.980	2.358	2.617
∞	1.282	1.645	1.960	2.326	2.576

Note: For example, if $\alpha = .05$ and $\nu = 15$, then $t_{\alpha, \nu} = t_{.05, 15} = 1.753$.

QUANTITATIVE METHODS AND APPLIED STATISTICS SERIES

ALLYN AND BACON

Barry Render, Consulting Editor
Roy E. Crummer Graduate School of Business, Rollins College

STATISTICS
for Management and Economics

STATISTICS

for Management and Economics

COLLIN J. WATSON
University of Utah

PATRICK BILLINGSLEY
University of Chicago

D. JAMES CROFT
Mortgage Asset Research Institute

DAVID V. HUNTSBERGER
Iowa State University

ALLYN AND BACON

Boston   London   Toronto   Sydney   Tokyo   Singapore

Editor-in-Chief, Business and Economics: *Rich Wohl*
Editorial Assistant: *Dominique Vachon*
Cover Administrator: *Linda Dickinson*
Composition Buyer: *Linda Cox*
Manufacturing Buyer: *Megan Cochran*
Editorial-Production Service: *Sylvia Dovner, Technical Texts*
Text Designer: *Sylvia Dovner, Technical Texts*

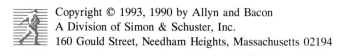

Copyright © 1993, 1990 by Allyn and Bacon
A Division of Simon & Schuster, Inc.
160 Gould Street, Needham Heights, Massachusetts 02194

Previous editions were published under the title *Statistical Inference for Management and Economics,* © 1986, 1980, 1975 by Allyn and Bacon.

In regard to the software disk accompanying this text, the publisher assumes no responsibility for damages, errors, or omissions, without any limitation, which may result from the use of the program or the text.

Library of Congress Cataloging-in-Publication Data
Statistics for management and economics / Collin J. Watson . . . [et
    al.]. — 5th ed.
        p. cm.
    Includes bibliographical references and index.
    ISBN 0-205-14094-7
    1. Social sciences—Statistical methods.   2. Statistics.
    3. Economics—Statistical methods.   I. Watson, Collin J.
    HA29.S78487   1993
    519.5′4--dc20   92-37332
                                                                        CIP

ISBN 0-205-14094-7

Printed in the United States of America
10 9 8 7 6 5 4 3 2   97 96 95 94

To L. Dee, Joyce, Nancy, Lauren, James, Stephen and Katie

Contents

CHAPTER 3   Descriptive Measures   71

CHAPTER 4   Probability   131

Discrete Probability Distributions   189   **CHAPTER 5**

Continuous Probability Distributions   237   **CHAPTER 6**

CHAPTER 7   Sampling Distributions   283

CHAPTER 8   Estimation   331

Tests of Hypotheses   383   **CHAPTER 9**

CHAPTER 10   Analysis of Variance   **469**

CHAPTER 11   Tests Using Categorical Data   **555**

CHAPTER 14   Time Series Analysis   747

Preface

Statistics for Management and Economics, Fifth Edition, is an introductory statistics text for students of management, business, and economics. It was written for a one- or two-term course. Our major objective for the fifth edition has been to present *principles* and *practices* of statistics in the context of substantive, engaging, and up-to-date *applications* in management, business, and economics. As we made changes in the text, we emphasized three themes: *clarity, accuracy,* and *applicability.*

 CHANGES IN THE FIFTH EDITION

The fifth edition of *Statistics for Management and Economics* includes many improvements in organization, content, and design.

Expanded Coverage of Statistical Process Control. We expanded and improved our discussion of *statistical process control,* which we now cover in Chapter 15. In this chapter, we discuss quality improvement, we present a complete and accurate discussion of process control, we discuss control charts and process capability, and we introduce Pareto charts.

Supplemental Sections. We reorganized most of the chapters so that *supplemental sections* are now placed at the ends of chapters. Instructors can omit the supplemental sections to streamline their courses, or they can include the supplemental sections to concentrate on the additional topics at their discretion.

Enhanced Content. Chapter 1 now includes important applications of statistics in management, business, and economics, along with pictorial graphics and more emphasis on time series or sequence plots for data generated by a process. Chapter 2 discusses data distributions, including histograms and stem-and-leaf diagrams. Chapter 6, on continuous probability distributions, emphasizes normal distributions; in addition, we have added a supplemental section on exponential probability distributions. For this edition, we have reintroduced a supplemental section on type II errors and power of hypothesis tests in Chapter 9. Analysis of residuals from regression analysis has been reorganized in Chapters 12 and 13. Chapter 14 includes a section on autocorrelation and autoregressive models for stationary series. In the nonparametric methods chapter (Chapter 18), we have included new sections on the runs test for randomness and the Kruskal–Wallis test, including multiple pairwise comparisons. We moved the discussion of index numbers to Chapter 19.

The fifth edition has been edited throughout to improve readability, to enhance clarity, and to improve our conversational and narrative style. Readers will find many new, color-enhanced graphics that contribute to the visual attractiveness of the book. Examples, problems, graphics, cases, data analyses, tables, data files, and references have been added, updated, and enhanced with color in every chapter.

Some material has been omitted or de-emphasized to keep the length of the book more manageable and to reflect increased emphasis on contemporary practices of statistics. For example, summaries are now shorter because they do not include material from the supplemental sections.

FEATURES

The fifth edition of *Statistics for Management and Economics* contains a wealth of special features.

Topics. The core topics that we included in the fifth edition reflect contemporary trends and widely applicable practices of statistics in management, business, and economics—from exploratory data analysis and statistical process control to estimation and hypothesis testing, and from analysis of variance and multiple regression to time series analysis.

Genuine Management and Business Examples and Cases. Throughout the text, we have used management and business examples and cases that are practical, relevant, and widely applicable. Students become involved as they think about and solve problems related to stock returns, yields of treasury bills, bankruptcy, predicting the catastrophic failure of the space shuttle *Challenger,* reducing risk with portfolio theory, and much more.

Abundant Assignment Problems. The fifth edition is unexcelled for the quality, abundance, and variety of its problems. Plentiful assignment problems, many with multiple parts, are located at the ends of sections and at the ends of chapters. Up-to-date problems are taken from all of the functional areas of management, business, and economics.

Unique Statistical Graphics. Well-designed, colorful graphics are used to present concepts, functions, and data. Graphics make statistical concepts physically perceptible, objectively real, and easier to comprehend. For graphical excellence in statistics, this text is unrivaled.

Actual Data. To enhance the realism of problems and cases, we have taken data from practical management and business sources such as *Business Statistics, Business Week, Forbes, Fortune, Statistical Abstract of the United States,* and *The Sporting News.* We have also used data from the COMPUSTAT files, business transactions, consulting engagements, surveys, experiments, processes, original research, and academic journals. A data disk containing files for larger data sets taken from the problems and cases is available to the instructor, and to the student in the student software workbooks.

Computer Integration. We have used computer applications to demonstrate data analysis throughout the book. Data analyses using MINITAB, SAS, and MYSTAT are presented in most of the chapters. We have annotated the computer exhibits with unique boxes that explain commands, program statement listings, and outputs.

ANCILLARIES FOR STUDENTS

Ancillaries available for students include the following items.

⬛ *Student Solutions Manual to Accompany Statistics for Management and Economics,* **Fifth Edition.** The student's Solutions Manual includes complete solutions for all odd-numbered problems in the text.

⬛ **Business MYSTAT with QC Tools Software.** Business MYSTAT is the educational version of SYSTAT, a highly rated, nationally recognized statistics package for personal computers. It is available packaged with the text on either 5 1/4″ or 3 1/2″ disks.

⬛ *Study Guide for Statistics for Management and Economics,* **Fifth Edition** (by Edward Mansfield). The study guide contains review tests and problems with worked out solutions corresponding to each chapter in the text.

⬛ *MINITAB Workbook for Statistics for Management and Economics,* **Fifth Edition** (by Leonard Presby). The workbook contains examples of data analysis with MINITAB using problems from the text. It is packaged with a data disk with over 50 files of data from the text, including the credit and ratio data sets.

⬛ *Business MYSTAT Workbook with QC Tools for Statistics for Management and Economics,* **Fifth Edition.** The workbook contains examples of data analyses from Business MYSTAT with QC Tools, using problems from the text. It is packaged with a data disk with over 50 files of data from the text, including the credit and ratio data sets.

 ## INSTRUCTOR'S ANCILLARIES

Available for the instructor are the following materials:

- *Instructor's Annotated Edition*
- *Instructor's Solutions Manual*
- Transparency masters
- Data disk
- Test bank
- Computerized test bank
- Video tapes

Detailed information about these ancillaries are available in the Instructor's Annotated Edition.

 ## ACKNOWLEDGMENTS

We have benefited from suggestions made by students, colleagues, and reviewers. Comments from the late Bruce Baird, Susan Chesteen, Michael Gardocki, Matthew Marlin, Steven Nahmias, Don Wardell, and Jerry Wiest have been beneficial. Brad Baird, Fred Cooper, and Winter Nie also provided useful comments. We thank Dean Wichern for insights provided in his book review in the *Journal of the American Statistical Association.*

For their helpful reviews and suggestions for the fifth edition, we thank Anthony Casey, University of Dayton; Angela Dean, The Ohio State University; George Dery, University of Lowell; Sharon Fitzgibbons, University of Nevada, Reno; John Haehl, Sonoma State University; Marilyn Hart, University of Wisconsin, Oshkosh; Gordon Johnson, California State University, Northridge; Clark E. Kristensen, Indiana University and Purdue University at Indianapolis; Darlene Lanier, Louisiana State University; Jamal Munshi, Sonoma State University; Dana Reneau, California

State University, Stanislaus; Peter Royce, University of New Hampshire; Richard Stockbridge, University of Kentucky; and Kathryn Szabat, LaSalle University.

We credit the following reviewers for many worthwhile suggestions over past editions: Rasoul Afifi, Northeastern Illinois University; Belva Cooley, Indiana State University; J. Devore, California Polytechnic Institute, San Luis Obispo; Robert Elrod, Georgia State University; Tom Foote, San Diego State University; Sam Graves, Boston College; John E. Hanke, Eastern Washington University; Rebecca Klemm, Temple University; Shu Jan Liang, Loyola University; Gordon H. Otto, University of Houston; Carol Pfrommer, University of Alabama; Neil Pelhemus, University of North Carolina, Chapel Hill; Leonard Presby, William Paterson College; Bill Seaver, University of Tennessee; James Sullivan, Bowling Green State University; Willbann Terpening, University of Notre Dame; Chipei Tseng, Northern Illinois University; and Thomas Yancey, University of Illinois, Urbana. We acknowledge contributions of the officers and staff of the publisher; the text's editor, Rich Wohl; Sylvia Dovner, and the officers and staff of the editorial and production services. We credit Edward Mansfield for preparing the Study Guide, Clark Kristensen for the test bank, and Leonard Presby for the Minitab Workbook.

For checking the solutions of the examples, problems, and case assignments, we thank Clark E. Kristensen, Indiana University and Purdue University at Indianapolis; Darlene Lanier, Louisiana State University; Peter Royce, University of New Hampshire; and Kathryn Szabat, LaSalle University.

We thank Wendy Ackerman, Michelle Barber, Kirk Doherty, Chris Lee, Sharon Lee, Karen Lindquist, Julia Smith, James Watson, and Lauren Watson for assistance in manuscript preparation. We thank the University of Utah Teaching Committee for a University Teaching Grant that supported the development of management and business applications and graphics that are included in the text. We thank Professor E. S. Pearson and the Biometrika Trustees for permission to use the material presented in Tables VII, VIII and IX.

REWARD

One theme for our text is accuracy. We work hard to be sure that our discussion of principles and practices of statistics, and our graphics, are completely accurate. We try to correct, in reprints, any misspelled words or mathematical errors that crop up in the text. To reward those willing to help correct any possible spelling or mathematical mistakes, we hereby offer $3 per misspelled word or mathematical error (not rounding differences, follow-through errors, and the like) in *Statistics for Management and Economics,* Fifth Edition, to the first instructor or student using our text who reports the mistake to us. Please submit reports of mistakes directly to the lead author's mail box or to the publisher.

We invite readers to send us comments or suggestions about how we should continue to improve the book. Please send us your ideas.

Introduction to Statistics

Basic concepts that are important for understanding and applying statistical methods to business and economic problems are introduced in this chapter. First, we discuss the modern meaning of *statistics* as it applies to business, managerial, and economic problems, and we consider some examples of today's important applications of statistics in the functional areas of business. Second, we discuss several types of business and economic statistical problems. Four areas into which statistical methods can be categorized, depending on the types of problems that these methods are used to solve, are discussed.

The four areas of statistical methods are *descriptive statistics* and *exploratory data analysis, probability, statistical inference,* and *special topics in statistics.* These areas do not stand alone. For example, statistical inference requires the use of descriptive statistics and probability, and the special topics in statistics involve the methods of statistical inference. We define a *population* and a *sample* in this chapter because they are important to understanding applications in all areas of statistics, especially in the area of statistical inference.

In this chapter, we also discuss *data* and *introductory statistical graphics.* Data are used to make rational business and economic decisions. Statistical graphics are used throughout this text to present statistical methods in a form that makes them easier to understand. In the remainder of the text, we will also see how to apply the statistical methods from each of the four areas to business and economic problems. Illustrating applications of statistical methods to aid business and economic decision making, providing an understanding of variability and risk, and demonstrating methods for analyzing processes are major objectives of this text.

1.1
WHAT IS STATISTICS?

The word *statistics* often calls up images of numbers piled upon numbers in vast tables. In common usage the word statistics is synonymous with the word *data*. The conception of statistics as data does not correspond to the science that carries the name statistics, nor is it descriptive of the activities of present-day statisticians. An up-to-date definition of **statistics** follows.

> **Definition**
>
> **Statistics** is the science that deals with collecting, analyzing, and interpreting data.

The science of business statistics aids business decision making. When there is uncertainty as to the conclusions that should be made during business decision making, concepts of probability are used to evaluate the reliability of conclusions based on data. For this reason, concepts and rules of probability play a fundamental role in the science and application of business statistics.

Although statistical science can be used in many fields of work and research, the following chapters concentrate on the use of statistical science in decision making by people in all areas of business and economics. Decision makers are interested in business statistics because the methods can help them make intelligent, reliable decisions based on scientific methods. Thus managers are most interested in the applications of statistics to their particular problems. To use applications of statistical science wisely, decision makers must have a good foundation in the principles and practices of probability and statistics in business and economics. The importance of business statistics in the functional areas of business is illustrated by actual business examples as follows.

EXAMPLE 1.1 *Accounting:* Many decisions that are made by accountants about the financial status, liquidity, and inventory of corporations are based on the analysis of financial ratios. Consequently, accountants use statistical methods with financial ratio data to analyze the financial condition of corporations. Statistical analyses have shown that financial ratios for some companies differ appreciably from or are outliers from those seen typically in the companies' industry groups. Managers, employees, and investors in these companies are very interested in these results because the companies that have atypical financial ratios may become bankrupt.

REFERENCE: Watson, Collin J. (1990), "Multivariate distributional properties, outliers and transformation of financial ratios," *The Accounting Review,* 65: 682–95.

EXAMPLE 1.2 *Economics:* Economists are often concerned with economic productivity, inflation, interest rates, and unemployment. They have used statistical methods to develop indexes such as the Consumer Price Index and Producer Price Index to measure inflation over time. Economists also commonly use time series analysis and forecasting to analyze and predict the future of the economy.

REFERENCE: U. S. Department of Labor, Bureau of Labor Statistics (1992), *Statistical Abstract of the United States* (Washington, D. C.: U. S. Government Printing Office).

EXAMPLE 1.3 *Finance:* Investors and managers generally invest in portfolios of securities, not just one security, because they do not like to put all of their eggs in one basket. Financial analysts and investors use portfolio theory to select combinations of such assets as stocks and bonds for their investment portfolios. Efficient portfolios are those that provide the highest expected return or expected profitability for any degree of risk, where risk is measured by the variability of possible returns. Statistical principles are used to measure expected returns, and statistical measures of variability or dispersion of returns are used to measure risk in portfolio theory. Harry Markowitz developed procedures that use statistical methods to find efficient portfolios. Markowitz was awarded the Nobel Prize in 1990 for his understanding and use of statistics in developing basic concepts of portfolio theory in finance.

REFERENCE: Brigham, E. F. (1991), *Financial Management,* 6th ed. (New York: Dryden).

EXAMPLE 1.4 *Management:* To compete effectively in the global marketplace for the production of goods and services, managers must constantly attempt to improve quality and productivity. Methods of statistical quality control are employed by modern companies to manage and constantly improve production processes. Deming (1986) and others advocate a total quality philosophy with on-going quality improvements. Motorola, Inc., was awarded the first annual Malcolm Baldridge National Quality Award (the United States counterpart to Japan's Deming Prize) for attaining preeminent leadership in using statistical quality control to improve company performance.

REFERENCE: Deming, W. Edwards (1986), *Out of the Crisis* (Cambridge, Mass.: MIT Press).

EXAMPLE 1.5 *Marketing:* Manufacturers of consumer products conduct *marketing research* to collect and analyze data relating to the marketing of goods and services. Marketing research often includes market potential and market share studies, product research, promotion research, and distribution research. Retailers want to know the answers to the following types of questions: How large is the market for a new product? What benefits does the new product provide? What are the personal characteristics, attitudes, interests, and opinions of consumers who are attracted to the product? Marketing research often uses questionnaires and surveys by mail, telephone, or personal interview to provide information that helps companies to decide whether and how they should market a product.

REFERENCE: Green, P. E., and D. S. Tull (1990), *Research for Marketing Decisions* (Englewood Cliffs, N. J.: Prentice-Hall).

Many entry level jobs that are available for business and economics students require the application of statistical science to the collection and analysis of data. Graduates will want to be prepared to use statistics effectively when they enter the business world. Furthermore, the understanding and application of business statistics are very important for managers to make rational decisions in today's global marketplace.

Managers who are knowledgeable about risk and variability and who think effectively under conditions of uncertainty can improve the profitability and performance of their organizations. Consequently, they can enhance their chances of achieving higher positions in organizations during their careers.

1.2
TYPES OF PROBLEMS IN STATISTICS

As we mentioned in the chapter introduction, statistical methods can be divided into four areas, depending on the type of problems they are used to solve. The first area is **descriptive statistics** and **exploratory data analysis.** Methods of organizing, summarizing, and presenting numerical data fall into this area of statistics. Chapters 2 and 3 cover the topics of descriptive statistics and exploratory data analysis. Chapter 2 deals with pictorial and graphical summary of data, and Chapter 3 deals primarily with numerical summaries.

The second area of statistical study is **probability.** Probability problems arise when a statistician takes a **sample** from a **population** and wishes to make statements about the likelihood of the sample's having certain characteristics.

> **Definition**
>
> A **population** is a set, or collection, of items of interest in a statistical study.
> A **sample** is a subset of items that have been selected from the population.

A typical probability problem is stated next:

> The manager of the Shop and Go retail store knows that 40% of the customers who enter the store will purchase one or more products and 60% will leave without making a purchase. What is the probability that in a sample of ten customers selected by random sampling, five will purchase one or more products?

Note that in this problem statement the population consists of all the company's customers. Also note that the purchaser/nonpurchaser breakdown in the population is given. The sample is the ten customers selected at random, and the question asks what the sample will look like. Chapters 4 through 6 deal with statistical methods that can be used to solve problems of this type—that is, probability problems.

The third area of statistics is called **statistical inference.** Problems involving statistical inference arise when a statistician takes a sample from a population and wishes to make statements about the population's characteristics from the information contained in the sample. A typical inference problem follows:

> The manager of employee benefits in the personnel department of CHF, Inc., questioned 100 people selected randomly from a work force of 7000 concerning their opinions of a proposed change in the company's medical insurance plan. The medical insurance salesperson claimed, "At least 80% of your work force will favor this change." However, the manager's sample of 100 showed that only 70 people in the sample favored the change. Is the insurance salesperson's claim true? If not, what proportion of the total work force favors the change?

In this problem the company's work force is the population, and the 100 people whose opinions were recorded form the sample. Here, however, we are given the sample results, and we ask questions about what the population looks like. This situation is directly opposite to the situation presented in probability problems. In a probability problem, the problem solver knows what the population looks like and raises questions about sample results that are likely to occur. In statistical inference situations, the problem solver knows what the sample result looks like and raises questions about the population from which the sample came. Statistical methods that can be used to solve inference problems are presented in Chapters 7 through 11.

The customers sampled at the Shop and Go retail store and the employees sampled by the personnel benefits department at CHF, Inc., in our examples were selected randomly. **Random sampling** from a finite population means that we use a sampling procedure whereby the items included in the sample are selected from the population so that every possible sample of the same size has an equal chance of being chosen. One requirement that must be met when samples are selected by random sampling is that each item in the finite population must have the same chance of being included in the sample. Procedures for selecting random samples are discussed in more detail in Chapter 7.

The fourth area of statistics covered in this text is difficult to label. We might call it **special topics in statistics.** Methods covered under this area are used to solve a wide range of statistical problems — from economic forecasting to deciding how large a production run a manager should order. Chapters 12 through 19 can be classified into the category of special topics.

These four areas of statistics build on one another. For instance, probability methods would not be understood by someone who did not know descriptive statistics. Also, problems in statistical inference cannot be solved by someone who does not understand probability. And the techniques discussed as special topics often use the methods of inference.

The objective of this book is to present *principles* and *practices* of statistics in the context of substantive, engaging, and up-to-date *applications* in management, business, and economics. Readers will learn enough about statistical techniques to be able to handle many common types of problems themselves. They will also learn to recognize those more complex problems that require consultation with a professional statistician.

Problems: Section 1.1 and 1.2

Answers to odd-numbered problems are given in the back of the text.

1. When we are dealing with problems in probability and statistical inference, which is usually the larger value in the problem, the population size or the sample size? Why?

2. State which type of statistical problem has the following assumption:
   a. You know what the population characteristics are.
   b. You know what the sample characteristics are.

3. When a firm test-markets a product, is the firm involved in a probability problem or a statistical inference problem? Why?

1.3 DATA, DATA TYPES, AND DATA SOURCES

Statistics is the science of collecting, analyzing, and interpreting data. Consequently, we discuss data, types of data, and sources of data in this section.

A **data set** is a collection of data. Data sets or data files are often collections of data that are organized to facilitate data analysis. Data are collected about entities known as **items.** An item may be a person, a company, a product, or any other entity of interest. Items are variously referred to in the practice of statistics as elements, objects, subjects, or units. A **variable** is a characteristic of interest for a set of items. A variable takes on different values or varies for different items. A **data value** is a *measurement* for a variable, and many data sets include several variables. The data values for a single item from all of the variables make up an **observation.** An observation is variously referred to in the practice of statistics as a case, a row, or a record.

=== **EXAMPLE 1.6**   Table 1.1 shows a *data set* for 55 large companies. An *item* in the data set listed in Table 1.1 is a company; consequently, we have 55 items in the data set. The items are listed in rows in Table 1.1, and the first item is the company Abbott Laboratories. The data set in Table 1.1 has three financial characteristics or *variables* arranged in columns, which are the companies' sales, profits, and margin. The *observation* for Abbott Laboratories is made up of the three *data values* 1,063.2, 160.6, and 15.1 for the sales, profits, and margin variables. ===

TABLE 1.1	*Company*	*Sales (in $millions)*	*Profits (in $millions)*	*Margin (%)*
Data Set for Large	Abbott Laboratories	$1,063.2	$160.6	15.1
Companies	Alcoa	1,160.0	165.3	14.3
	American Brands	2,197.2	117.4	5.3
	American Home Products	1,192.8	199.4	16.7
	Amoco	4,839.0	165.0	3.4
	AMR	1,498.6	6.6	0.4
	Atlantic Richfield	3,724.0	64.0	1.7
	Bellsouth	2,800.0	361.1	12.9
	Boeing	4,982.0	191.0	3.8
	Boise Cascade	969.3	32.0	3.3
	Borden	1,356.4	68.8	5.1
	Bristol-Myers	1,205.0	151.7	12.6
	Burlington Northern	1,609.8	75.6	4.7
	Caterpillar	1,775.0	−148.0	NM
	Champion International	1,058.0	64.5	6.1
	Chevron	5,900.0	−86.0	NM
	CSX	1,604.0	132.0	8.2
	Exxon	18,836.0	1,480.0	7.9
	General Electric	12,270.0	730.0	6.0
	Georgia-Pacific	1,781.0	83.0	4.7
	Hercules	640.5	38.9	6.1
	IBM	16,945.0	1,390.0	8.2
	Ingersoll-Rand	768.4	41.1	5.3
	Inland Steel Industries	817.5	39.7	4.9
	International Paper	1,900.0	113.0	5.9
	Kraft	2,414.5	117.1	4.9
	Martin Marietta	1,268.5	43.4	3.4
	Merck	1,149.1	171.8	15.0
	Merrill Lynch	2,764.8	197.6	7.1
	Mobil	12,009.0	201.0	1.7
	Monsanto	1,570.0	23.0	1.5
	Motorola	1,619.0	63.0	3.9
	NCR	1,575.1	134.1	8.5
	North American Philips	1,356.0	47.6	3.5
	Nynex	2,900.0	293.7	10.1

Company	Sales (in $millions)	Profits (in $millions)	Margin (%)*	
Pfizer	1,162.9	158.4	13.6	**TABLE 1.1** Continued
PPG Industries	1,200.8	72.9	6.1	
Procter & Gamble	4,255.0	190.0	4.5	**Data Set for Large**
Quaker Oats	1,083.6	30.1	2.8	**Companies**
Raytheon	2,041.2	99.6	4.9	
Reynolds Metals	863.9	−4.3	NM	
Santa Fe Southern Pacific	1,503.9	−384.7	NM	
Smithkline Beckman	1,043.3	147.4	14.1	
Standard Oil	2,264.0	51.0	2.7	
Tandy	1,196.0	103.8	8.7	
Teledyne	834.8	47.0	5.6	
Texas Instruments	1,335.8	26.6	2.0	
3M	2,112.0	185.0	8.8	
United Technologies	4,401.4	−252.6	NM	
USG	695.0	56.7	8.2	
USX	2,770.0	−1,167.0	NM	
Warner-Lambert	812.6	56.5	7.0	
Westinghouse Electric	2,869.1	203.2	7.1	
Weyerhaeuser	1,450.6	96.1	6.6	
Whirlpool	942.9	42.2	4.5	

*NM = not meaningful

SOURCE: *Business Week,* February 9, 1987, p. 31. Data from Standard & Poor's Compustat Services, Inc.

There are two types of data, broadly defined, known as **quantitative data** and **qualitative** or **categorical data.**

> **Definition**
>
> **Quantitative data** are data values that are measured on a numerical scale.

For example, the temperature measured in degrees Fahrenheit, the area in square feet, and the selling price of a condominium in dollars are quantitative data.*

> **Definition**
>
> **Qualitative** or **categorical data** are data values, each of which can be classified into a single category that belongs to a set of categories.

*In a finer classification of types of data, quantitative data can be subdivided into interval or ratio types of data. **Interval data** are quantitative data that have an equal and fixed distance between points on the measurement scale. For example, today's temperature measured in degrees Fahrenheit or Celcius is interval data. But the zero points on these scales are arbitrary. **Ratio data** are quantitative data that have a meaningful zero point and the ratio of two data values is meaningful. For example, data on prices measured in U. S. dollars are ratio data; for example, a condominium with a price of zero dollars is free and a condominium with a price of $80,000 costs twice as much as one with a price of $40,000.

Examples of qualitative data and some possible categories are gender (male or female), education (no college degree, undergraduate college degree, or graduate college degree), and marital status (married or not married).*

Statistical data often come from statistical studies, business transactions, surveys, databases, outputs of processes, trade associations, or government agencies. Statistical studies can be classified into two broad categories known as *observational studies* and *experimental studies.* In an **observational study,** the investigator examines variables of interest by using observed or historical data. The investigator does not directly control or determine which subjects or items receive treatments that are thought to affect the variables of interest in the study. For example, an investor in real estate may examine the relationship between the selling price and the area in square feet of condominiums. Since the investigator does not directly control the area of condominiums and since the data are collected from historical sales records, the study is an observational study.

In an **experimental study,** the investigator directly controls or determines which subjects or experimental items or material receive treatments that are thought to affect variables of interest. For example, in a water pollution study, an analyst may impartially or randomly select several containers of water (experimental items or material) that have been directly assigned to be filtered by one of four types of filters (treatments). The analyst then measures pollution counts (variable of interest) to determine which filter(s) result in the purest water. The analyst directly controls or determines the type of filter used to filter each container of water so the study is an experimental study. Random selection or assignment of subjects or experimental material so that each has an equal probability of receiving any of the treatments guards against bias.

Businesses maintain numerous data files that result from **business transactions.** Some of these include sales records, accounts receivable and payable records, cash receipts and disbursements, purchases, real assets, employee records, and inventory and production records. Business transactions data are often analyzed by using statistical methods presented in the following chapters.

Survey data is obtained from responses to questionnaires from personal interviews, telephone interviews, or mailings to respondents. Large **data bases** are available from organizations that provide business and economic data base services. For example, Standard & Poors Compustat Services, Inc., markets the COMPUSTAT files that contain financial data for companies listed on stock exchanges, and the Center for Research in Security Prices at the University of Chicago markets the CRSP Stock Files, which contain stock market data. Inputs to processes that are transformed into outputs over time generate **process data.** For example, a periodic inspection of memory chips produced by a manufacturing process at Digital Industries provides data on the proportion of defective chips being output by the process.

Federal, state, and local government agencies are the largest gatherers and publishers of data in the country. Business and trade organizations also gather and publish a great deal of data concerning their members' activities. Some of the most helpful sources of statistical information are the *Statistical Abstract of the United States,* the *Survey of Current Business,* and the *Encyclopedia of Associations.*

---

*In a finer classification of types of data, qualitative data can be subdivided into nominal and ordinal types of data. **Nominal data** are measurements that classify items into one category taken from a set of categories. For example, a person's gender—male or female—is nominal data. **Ordinal data** classify items into categories that can be ranked or ordered for the variable. For example, education can be measured on an ordinal scale where the categories can be ordered according to the amount of education—no college degree, undergraduate college degree, and graduate college degree.

Sources such as these contain hundreds of tables of statistics that describe the U. S. population and activities such as education, banking, energy, manufacturing, transportation, and so forth. They also provide lists of organizations and associations in business, health, cultural, athletic, and many other areas of interest.

Problems: Section 1.3

4. A data set collected by the credit department of National Retail Stores, Inc., contains data values for the gender or sex, marital status, age, and job income of applicants for credit. Classify the data as quantitative or qualitative for the different variables.

5. Three companies had corporate headquarters located in the states of New York, Pennsylvania, and Texas. What type of variable is the state for the location of a company's headquarters.

6. Consult the most recent issue of the *Survey of Current Business* to determine the following:

   a. The gross national product in current dollars for the fourth quarter of last year.
   b. Final sales of automobiles in current dollars for the latest quarter reported.

1.4
INTRODUCTION TO STATISTICAL GRAPHICS

Graphics can be used as an effective method of visual communication. Statistical graphics are beneficial for the presentation and analysis of data. The statistical graphics forms we discuss in this section are *line charts, bar* or *column charts, grouped bar charts, combination charts, pie charts,* and *pictorial charts.*

Line charts use lines between data points to depict the magnitudes of data for two variables or for one variable over time. The heights of the line allow the user to compare magnitudes easily. For example, the amounts of retail sales at Four Points Shopping Mall for the years 1989, 1990, 1991, and 1992 were $5, $6, $8, and $9 million. These data are presented in the form of a line chart in Figure 1.1. The magnitudes of the sales figures can be compared very easily in this chart. Data values for a variable over time are known as a *time series. Process data* are also collected over time. A line chart for a time series or for process data is known as a **time series plot,** or as a **sequence plot.**

FIGURE 1.1

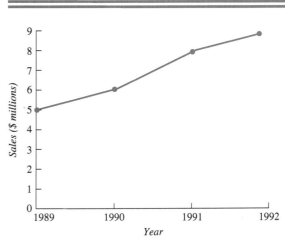

Line Chart Depicting Retail Sales at Four Points Shopping Mall over Time

FIGURE 1.2

**Bar Chart Depicting
Proposed Federal
Expenditures**

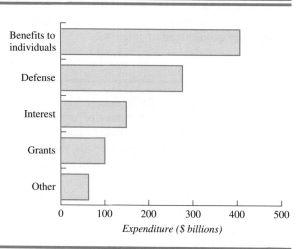

Bar charts are used to depict the magnitudes of data for different qualitative categories or over time. The lengths or heights of the bars allow the user to compare the magnitudes easily. For example, the expenditures proposed for a recent federal budget for benefit payments to individuals, defense, interest, grants to states and localities, and other operations as qualitative categories were $410, $280, $150, $100, and $60 billion. These data are presented in the form of a bar chart in Figure 1.2. The magnitudes of the proposed expenditures can be compared very easily in this chart.

For a second example, the amounts of retail sales at Four Points Shopping Mall given as a line chart in Figure 1.1 are depicted in the form of a bar or *column* chart in Figure 1.3. The magnitudes of the sales figures can be compared very easily in this chart.

Grouped bar charts can be used to depict the magnitudes of two or more grouped data values for different qualitative categories or over time. For example, the

FIGURE 1.3

**Bar Chart for Retail Sales
over Time**

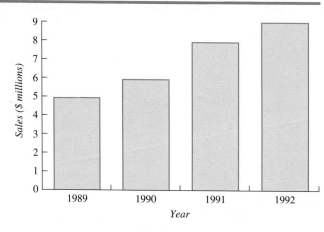

FIGURE 1.4

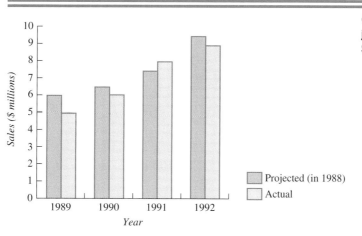

Grouped Bar Chart for Projected and Actual Retail Sales over Time

retail sales amounts at the mall were projected in 1988 to be $6, $6.5, $7.5, and $9.5 million for the years 1989, 1990, 1991, and 1992, respectively. The projected and actual sales amounts are depicted in a grouped bar chart in Figure 1.4. Comparisons of magnitudes of the projected and actual sales amounts over time can be made easily with this chart.

Combination charts use lines and bars to depict the magnitudes of two or more data values for different categories or for different times. For example, from 1988 through 1992 a software company had sales of $20, $28, $36, $64, and $73 thousand. A statistical method known as *regression analysis* was used to estimate sales for 1988 through 1995 to be $18, $29, $42, $56, $74, $94, $117, and $141 thousand. The actual and estimated sales amounts are depicted in a combination chart in Figure 1.5. The magnitudes of these values can be compared easily in this figure.

FIGURE 1.5

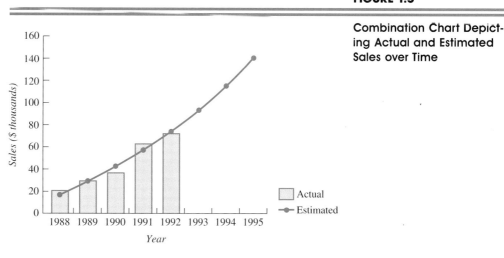

Combination Chart Depicting Actual and Estimated Sales over Time

FIGURE 1.6

Pie Charts for Percentages of Assets and Liabilities

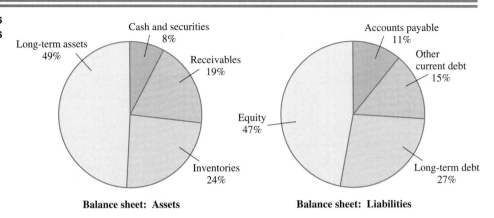

Balance sheet: Assets   Balance sheet: Liabilities

Pie charts can be used effectively to depict the proportions or percentages of a total quantity that correspond to several qualitative categories (usually five or fewer). Each category is depicted as a wedge of a circle, or a piece of a pie. The angle (in degrees) of each wedge is equal to the category's proportion multiplied by 360°. For example, a study of the balance sheets of 510 firms found that the average *proportions* of total assets for the cash and securities, receivables, inventories, and long-term assets accounts were 8%, 19%, 24%, and 49% (Stowe, Watson, and Robertson, 1980). The average *proportions* of total liabilities designated as accounts payable, other current debt, long-term debt, and equity accounts were 11%, 15%, 27%, and 47%. The *proportions* of total assets and total liabilities for the account categories are depicted in the pie charts of Figure 1.6.

Only five forms of statistical graphics have been presented in this section because our purpose in this section is simply to introduce statistical graphics. For a complete discussion of statistical graphics, see Schmid (1983) or Tufte (1983).

FIGURE 1.7

Misleading Graph of Sales Growth at Four Points Shopping Mall with Disproportionate Axes and No Zero Line at the Bottom

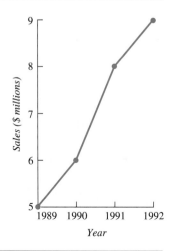

FIGURE 1.8

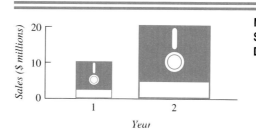

Some statistical graphics can be misused. The line chart in Figure 1.1 was used to show increasing sales over a four-year period at the Four Points Shopping Mall. However, if we wanted to overemphasize the growth in sales, say, to encourage First Bank to grant the mall a larger line of credit, we might *change the proportions of the axes* and *omit the zero line* at the bottom and present the bank with Figure 1.7. The figure now overemphasizes the growth in sales and may be misinterpreted as showing even more growth than actually took place.

Pictorial charts, which use pictorial symbols to represent data, are often used to gain attention, but they can be difficult to interpret and are also misused at times. For example, in its second year of operations, Softstat, Inc., doubled its sales of software from $10 million to $20 million dollars. The pictogram using the disk symbols in Figure 1.8 was intended to gain attention and depict the doubling of sales. However, human judgment often interprets areas during the comparison of two-dimensional symbols or volumes of three-dimensional symbols. Consequently, unsuspecting readers could judge that Softstat's sales had quadrupled since the area of the larger disk symbol is four times that of the smaller symbol. For a discussion of some misuses of statistical graphics see Huff (1965), Schmid (1983), and Tufte (1983).

We shall use statistical graphics frequently in the remainder of this text. Many of the figures presented in the text were originally plotted with the aid of a computer. As personal computers and spreadsheet or charting software become increasingly available, the use of statistical graphics such as line, bar, combination, and pie charts will undoubtedly increase.

Problems: Section 1.5

7. An automobile dealership had sales of $5, $13, $21, $49, and $68 million in 1988 to 1992. Present the data in a line chart.

8. The amounts of all federal revenues in a recent year were projected for individual income taxes, social insurance receipts, borrowing, corporation income taxes, and other revenues to be $390, $300, $140, $90, and $80 billion, respectively. Present the revenue data as a bar chart.

9. A franchise chain of restaurants for 1988 through 1992 had 4, 8, 16, 26, and 82 restaurants, respectively. Present the data in a bar (column) chart.

10. The populations (in thousands) of New York, Chicago, Los Angeles, Houston, and Salt Lake City were found in a study to be 7072, 3005, 2967, 1595, and 163. The number of individuals below the age of 20 for these same cities were found to be 2470, 900, 920, 610, and 55 (in thousands), respectively. Present the data in a grouped bar chart.

11. A franchise chain of restaurants for 1988 through 1992 had 4, 8, 16, 26, and 82 restaurants, respectively. A statistical model estimated the number of restaurants for 1988 through 1994 to be 4, 8, 16, 33, 140, 281, and 586. Present the data in a combination chart.

12. The proportions of all expenditures proposed for a recent federal budget for benefit payments to individuals, defense, interest, grants to states and localities, and other operations were .41, .28, .15, .10, and .06, respectively. The proportions of all federal revenues were projected for individual income taxes, social insurance receipts, borrowing, corporation income taxes, and other revenues to be .39, .30, .14, .09, and .08, respectively. Present the expenditure and revenue data as pie charts.

13. A pictorial chart showing the gasoline mileage attained by a Jeep Cherokee automobile before and after a light-weight, fuel-efficient engine was adopted for environmental reasons is shown in Figure 1.9. How might this pictorial chart be misleading?

FIGURE 1.9

Pictorial Chart of Gasoline
Mileage for a Jeep

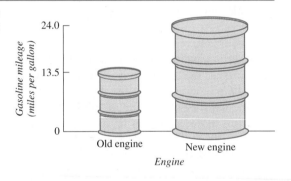

1.5
SUMMARY

The term *statistics* is often used as a synonym for data. However, in the modern sense of the word, *statistics* is the science of collecting, analyzing, and interpreting data. An important contribution of the science of statistics is that the reliability of conclusions based on the data may be evaluated objectively by means of probability statements.

Four areas of statistics are important in the application of statistics to aid business and economic decision making. These areas are *descriptive statistics* and *exploratory data analysis, probability, statistical inference,* and *special topics in statistics.* As discussed in this chapter, *populations* and *samples* are important elements of a statistical analysis. Statistical inference, which is a major topic for this book, involves taking a sample from a population, collecting data from the sample items, analyzing the data, and making inferences with respect to the population based on the sample results.

Data, types of data and *sources of statistical data,* and some *introductory statistical graphics* were discussed in the latter sections of this chapter. These concepts are important for modern managers who wish to apply statistical methods to aid them in making business and economic decisions.

Answers to odd-numbered problems are given in the back of the text.

14. You are assigned by your boss to examine each of last week's sales transactions, find their average, find the difference between the highest and lowest sales figures, and construct a chart showing the differences between charge account and cash customers. What type of statistical problem is this: descriptive statistics, probability, or inference?

15. Look at the problem statement that follows and determine whether it is a probability problem or a statistical inference problem.

    A quality control engineer took a sample of ten items from a production line and found that 20% of them were defective. What is the proportion of defectives in the population of all items coming off this production line?

16. Look at the next problem statement and determine whether it is a probability problem or a statistical inference problem.

    The manager of a hospital laboratory knows that 5% of the lab's tests have to be rerun. Her subordinates perform ten tests in one morning. What is the chance she will have to run none of them again?

17. Consult the *Statistical Abstract of the United States* to determine which three states have the lowest death rates. Also, use this source to determine any likely explanation of these states' low death rates from the distribution of people's ages in these states. [*Hint:* What proportion of each of the states' populations is over 65 years old? Compare this proportion with the national average.]

18. Consult the *Encyclopedia of Associations* to determine the location and founding date of the American Women's Society of CPAs.

19. A data set on the financial ratios for a sample of companies taken from the COMPUSTAT files contains data values for the Standard Industry Classification (SIC) Code that designates the industry in which the company operates and also gives ratios computed from financial data known as current assets–sales and net income–total assets. Classify the data as quantitative or qualitative for the different variables.

20. No one knows for sure how many heroin addicts there are in this country. However, the National Institute of Drug Abuse released a study in which they guessed that there were well over five hundred thousand. Researchers who conducted the study used 24 metropolitan areas and estimated the following number of addicts in each of these five cities: New York, 69,000; Los Angeles, 60,000; Chicago, 47,700; Detroit, 33,200; San Francisco, 28,600. If the head of the narcotics division of the New York City police force used these figures to argue that a higher proportion of the police department budget should go to this division, since New York has the worst problem in heroin addiction, would he be justified in doing so? [*Hint:* Use the *Statistical Abstract of the United States* population figures for these cities to adjust them to a per-capita basis.]

21. Consult the *Statistical Abstract of the United States* and use the table that lists numbers of physicians, dentists, and nurses in the population of each state to answer the following:

    a. Which area listed has the highest number of physicians per capita?
    b. Why does this location have a higher rate than the others listed?

22. Consult the "Current Business Statistics" supplement in the most recent *Survey of Current Business* to find the following:

    a. United States exports of goods and services (excluding military grant transfers), in millions of dollars.

b. Estimated total civilian labor force for the latest month on which data are available.

23. Consult the *Encyclopedia of Associations* and determine the following information about the National Retail Merchants Association:

   a. When was it founded, and where are its central offices?
   b. How many members does it currently have?
   c. How many publications does it distribute?

24. The annual return on capital for Statsoftware, Inc. was .08, .10, .13, .12, and .18 in 1988 through 1992. Present the data in a line chart.

25. An individual investor has a portfolio consisting of cash and money market funds of $20,000, bonds worth $50,000, stocks worth $70,000, and real estate worth $150,000. Present the data as a bar chart.

26. The populations (in thousands) of New York, Chicago, Los Angeles, Houston, and Salt Lake City are 7072, 3005, 2967, 1595, and 163, respectively. Present the data in a bar chart.

27. A study was conducted to assess the ability of humans who have different training in accounting (elementary, intermediate, or expert) to detect a change in the financial condition of the firm by using multidimensional graphics or by using tabular presentations (Stock and Watson, 1984). The resulting mean classification accuracies are given in the accompanying table.

   a. Present the data in a grouped bar chart.
   b. Does the graphic suggest that classification accuracy is higher for the multidimensional graphic format or for the tabular format?

Training	*Format*	*Mean Classification Accuracy*
Elementary	Graphic	.51
	Tabular	.47
Intermediate	Graphic	.50
	Tabular	.46
Expert	Graphic	.52
	Tabular	.43

28. A company has experienced the following financial results for earnings before interest and taxes (EBIT) and profits (EBIT and profits are in millions of dollars):

Year	1989	1990	1991	1992
EBIT	1.5	2.0	2.2	3.0
Profits	.08	1.0	1.2	1.6

   Present the data in a grouped bar chart.

 29. An automobile parts store had sales of $5, $13, $21, $49, and $68 thousand in 1988 through 1992. Estimated sales were $5, $12, $25, $44, $70, $100, $137, and $181 thousand for 1988 through 1995. Present the data in a combination line and bar chart.

Note: Problems with a computer symbol can be solved with a computer and data analysis software. The data disk symbol indicates that the data set for the problem is on the data disk available with this text.

30. A study of the specialties selected by 291 graduates of a medical school found that 68 of the physicians are practicing surgery, 99 are in family medicine, 20 are in obstetrics or gynecology, 22 are in psychiatry, and 82 are in other specialties (Watson and Croft, 1978). Present the data in a pie chart.

31. A study of the asset/liability structures of commercial banks found the proportions of assets for cash, liquid securities, investment securities, loans, and other assets to be .18, .03, .14, .59, and .06 (Simonson, Stowe, and Watson, 1983). The proportions of liabilities and capital for demand deposits, purchased funds, core deposits, other liabilities, and equity were .29, .40, .21, .05, and .06, respectively. Present these data in pie charts.

32. A study of the practice location of 272 physicians who had graduated from a public medical school found that 43 of the physicians were practicing in a rural location, 42 were practicing in a suburban location, 136 were practicing in an urban location, and 51 were practicing in a mega-urban location (Watson, 1980). Present the data in a pie chart.

33. The manager of a local area network of personal computers used the accompanying pictorial chart to depict the number of personal computers managed by the network during an annual performance review of her department. How might this pictorial chart be misleading to the information system manager who is conducting the annual review?

FIGURE 1.10

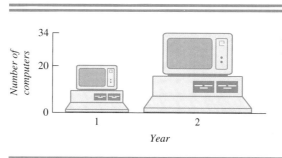

Pictorial Chart for Number of Personal Computers Connected to a Local Area Network

Refer to the 141 companies listed in the Ratio Data Set in Appendix A.

Ratio Data Set

Questions

34. Why is the term *data set* used with the ratio data located in Appendix A?
35. What is an item in the ratio data set in Appendix A?
36. How many items are in the ratio data set in Appendix A?
37. How many variables are included in the ratio data set in Appendix A?
38. What type of variables are *CNUM, DNUM* and *CONAME* in the ratio data set in Appendix A?
39. What type of variables are *CA/S, QA/S, CA/CL* and *NI/TA* in the ratio data set in Appendix A?

Refer to the 113 applicants for credit listed in the Credit Data Set in Appendix A.

Credit Data Set

Questions

40. What is an item in the credit data set in Appendix A?
41. How many items are in the credit data set in Appendix A?
42. How many variables are included in the credit data set in Appendix A?
43. What type of variables are *CLASS, SEX,* and *MSTATUS* in the credit data set in Appendix A?

44. What type of variables are *AGE, JOBYRS, ADDINC* and *TOTBAL* in the credit data set in Appendix A?

45. What types of applicants seem to have the value 99 listed for the variable JOBYRS?

46. Does it appear logical to you that a blank entry (indicated by 9999) and a 0 entry for the variable SPINC mean the same thing?

| **CASE 1.1** | **USING STATISTICAL GRAPHICS IN CORPORATE ANNUAL REPORTS** |

Corporations use annual reports to provide information to stockholders. The annual report generally describes the firm's operating results during the year and discusses company objectives. The report also presents financial statements, including the income statement, the balance sheet, the statement of returned earnings, and the statetment of changes in financial position. Statistical graphics are often used in annual reports to present information about the company.

FIGURE 1

Pie Chart Depicting Gulf & Western's Operating Income Contributions before Corporate Expenses ($ Millions) (SOURCE: Paramount Communications, Inc. (formerly Gulf & Western) 1987 *Annual Report.* Reprinted with permission.)

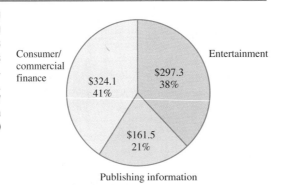

FIGURE 2

Bar Chart Depicting Gulf & Western's Operating Income 1983–1987 (SOURCE: Paramount Communications, Inc. (formerly Gulf & Western) 1987 *Annual Report.* Reprinted by permission.)

Gulf & Western, Inc. (now known as Paramount Communications, Inc.) is a broadly diversified, multi-industry company that ranked 31st on *Fortune* magazine's 1987 list of the nation's largest diversified service companies measured by sales. In fiscal 1987, 41% of Gulf & Western's operating income came from consumer and commercial finance, entertainment accounted for 38%, and publishing accounted for 21%. Gulf & Western depicted their fiscal 1987 operating income contributions from consumer and commercial finance, entertainment, and publishing as shown in the pie chart in Figure 1. Gulf & Western's operating income increased dramatically from 1983 through 1987, and Figure 2 shows a bar chart that depicts operating incomes in a format that allows the reader to compare the magnitudes easily.

a. Why is a pie chart appropriate for depicting operating income contributions?

b. Why is a bar chart appropriate for depicting operating incomes over time?

Case Assignment

EVALUATING THE EFFECTIVENESS OF TEACHING AIDS

CASE 1.2

Midcontinent Polytechnical Institute requires all students to take Statistics 101 in their freshman or sophomore years. Recently, an alumnus, Mr. William T. Kahi, donated a large sum of money that he wished to be used to provide teaching aids for students of statistics. Mr. Kahi had nearly flunked out of Midcontinent Poly in 1939 because of his difficulty in mastering this subject.

The faculty used the funds to convert two classrooms into a statistics laboratory. These rooms contained desks, lounge chairs, chalkboards, and a small library of statistics reference books. In addition, teaching assistants were on duty 14 hours a day. The teaching assistants helped students with their homework problems and checked out library materials to the students. The students could not only check out books from the statistics library but also obtain electronic calculators, personal computers, programmed learning aids, and a series of videotaped lectures on elementary statistics. The series of videotaped lectures was obtained only last term.

Dr. Walter Goss, chairman of the Department of Statistics, was interested in determining the results obtained by using these videotaped lectures and whether all statistics students would benefit by using the tapes. Since Dr. Goss taught a small section of Statistics 101 last term, he decided to compare the examination results of students who had used the tapes with those who had not. Thus, at the end of the term, he collected the sheets showing who had checked out the taped lectures. The results he found are as follows.

Name	Final Exam Score	Checked Out Tapes?	Name	Final Exam Score	Checked Out Tapes?	Name	Final Exam Score	Checked Out Tapes?
Michael B.	82	Yes	William S.	39	No	Susan D.	88	Yes
Harry O.	99	Yes	Harry T.	78	Yes	Bruce U.	100	No
Lucy McB.	77	No	Max W.	66	Yes	Earl S.	25	No
Ralph T.	45	No	Patty M.	86	No	Reed F.	95	Yes
Mary L.	99	No	Shauna T.	72	Yes	Edward B.	87	Yes
Jim C.	66	No	Terry C.	70	Yes	George T.	65	No
Bill R.	79	Yes	John B.	55	No	Juan A.	83	Yes
David H.	80	No	Bob G.	*	No	Janice N.	77	Yes
Roy T.	89	Yes	Betty S.	81	Yes	Kent Z.	94	Yes
Gary D.	90	No	Debbie B.	75	No			

*Dropped class.

  a. What is the population of interest to Dr. Goss?
  b. Do the students on the list constitute a sample?
  c. Why is the central problem for this case a problem of statistical inference?
  d. Is the final exam score a quantitative or qualitative variable?
  e. Is the variable on whether the tapes were checked out a quantitative or a qualitative variable?
  f. Is the study an observational or an experimental study?

CASE 1.3   **MAKING THE DECISION TO LAUNCH THE SPACE SHUTTLE *CHALLENGER***

The space shuttle *Challenger* was in launch position the morning of January 28, 1986, with seven astronauts and passengers on board. Just prior to the launch, sheets of ice clung hauntingly to the fuselage. Moments later, with national television coverage as it blasted into orbit, the shuttle disintegrated in a catastrophic explosion. The remains of the astronauts and passengers were never recovered.

Thiokol Corporation manufactures the two solid-fuel rocket motors that propelled the shuttle into space. The night before the catastrophe, executives of Thiokol and the National Aeronautics and Space Administration debated whether they should launch the shuttle according to schedule or postpone the mission. The weather report called for a temperature of 31°F at blast off.

From April 12, 1981, to January 12, 1986, prior to the catastrophe, the space shuttle had flown 24 successful missions. Six primary O-rings were used to seal the sections of the two solid-fuel rocket motors that were used to thrust the shuttle into space. On several flights, the motors had experienced O-ring erosion or gas blow-by incidents. On one flight, the motors were not recovered. The number of erosion or blow-by incidents and the temperature of the rocket joints for 23 successful flights prior to the catastrophe are given in the accompanying table. A plot that was obtained from a data analysis system of the number of O-ring incidents against the joint temperature is shown in Figure 1.

After examining the data, Thiokol executives recommended that the shuttle be launched on schedule because they felt that they did not have conclusive evidence that the

Mission	O-Ring Incidents	Temperature (°F)	Mission	O-Ring Incidents	Temperature (°F)
1	0	66	13	0	67
2	1	70	14	2	53
3	0	69	15	0	67
4	0	68	16	0	75
5	0	67	17	0	70
6	0	72	18	0	81
7	0	73	19	0	76
8	0	70	20	0	79
9	1	57	21	2	75
10	1	63	22	0	76
11	1	70	23	1	58
12	0	78			

FIGURE 1

Plot of Number of O-Ring
Incidents Against Joint
Temperature for the Space
Shuttle

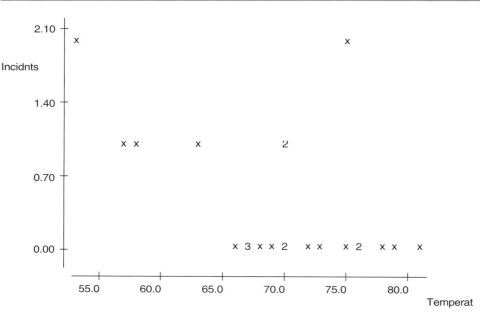

Note: The symbol 2 is for two points, and 3 is for three points.

low temperature would influence the capability of the solid-fuel rocket motors to thrust their payload into orbit. The next day, the shuttle exploded.

Case Assignment

a. If you were in charge of the launch decision, how would you go about deciding whether or not to launch the space shuttle?

b. How would the data influence your decision?

REFERENCES: Dalal, S, E. Fowlkes, and B. Hoadley (1989), "Risk Analysis of the Space Shuttle: Pre-*Challenger* Prediction of Failure," *Journal of the American Statistical Association,* 84: 945–957; and Presidential Commission on the Space Shuttle *Challenger* Accident (1986), *Report of the Presidential Commission on the Space Shuttle Challenger Accident,* vols. 1 and 2. Washington, D. C.

REFERENCES

Burek, Deborah M., ed. 1992 *Encylcopedia of Associations,* 26th ed. Detroit: Gale Research.

Campbell, Stephen K. 1974. *Flaws and Fallacies in Statistical Thinking.* Englewood Cliffs, N.J.: Prentice-Hall.

Cleveland, W. S. 1985. *The Elements of Graphing Data.* Monterey, Calif.: Wadsworth.

Huff, Darrell. 1965. *How to Lie with Statistics.* New York: Norton.

Microsoft Corporation. 1984. *Microsoft Chart.* Bellevue, Wash.: Microsoft Corporation.

Schmid, C. F. 1983. *Statistical Graphics.* New York: Wiley.

Simonson, D. G., J. D. Stowe, and C. J. Watson. 1983. "A Canonical Correlation Analysis of Commercial Bank Asset/Liability Structures." *Journal of Financial and Quantitative Analysis,* 18 (March): 125–40.

Stock, D., and C. J. Watson. 1984. "Human Judgment Accuracy, Multidimensional Graphics, and Humans versus Models." *Journal of Accounting Research,* 22 (Spring): 192–206.

Stowe, J. D., C. J. Watson, and T. Robertson. 1980. "Relationships Between the Two Sides of the Balance Sheet." *Journal of Finance,* 35 (September): 973–80.

Tanur, J. M., F. M. Mosteller, W. H. Kruskal, R. F. Link, R. S. Pieters, G. R. Rising, and E. L. Lehmann, eds. 1978. *Statistics: A Guide to the Unknown.* San Francisco: Holden-Day.

Tufte, E. R. 1983. *The Visual Display of Quantitative Information.* Cheshire, Conn.: Graphics Press.

U.S. Department of Labor, Bureau of Labor Statistics. 1992. *Statistical Abstract of the United States.* Washington, D.C.: U.S. Government Printing Office.

U.S. Department of Commerce. 1992. *Survey of Current Business, 1992.* Washington, D.C.: U.S. Goverment Printing Office.

Watson, C. J. 1980. "An Empirical Model of Physician Practice Location Decisions." *Computers and Biomedical Research* 13: 363–81.

Watson, C. J., and D. J. Croft. 1978. "A Multiple Discriminant Analysis of Physician Specialty Choice." *Computers and Biomedical Research* 11: 405–21.

**1.6
MISUSES OF
STATISTICS**

Another of the objectives of this book is worthy of special mention. When readers finish studying the statistical methods covered in this book, they should have a good feeling for the difference between appropriate and inappropriate uses of statistics. This book concentrates, naturally, on the appropriate uses. However, some of the examples, problems, and cases will point out potential statistical abuses of which the reader should be aware.

The misuses of statistics are so varied and broad that entire books have been written on them. One excellent example is Darrell Huff's delightful book entitled *How to Lie with Statistics,* which is referenced at the end of Chapter 1. A few of the more common problems that lead to misused statistics and incorrect conclusions are discussed next. For the most part these problems involve the way in which data are gathered and used in some sort of statistical analysis. Even the most careful applications of the statistical methods presented in this book are useless if some of the problems cited here are not avoided.

One of the simplest data problems to overcome in a statistical analysis is the *failure to adjust gross data to a per-item basis.*

EXAMPLE 1.7   A company president reports that the company made $30 million last year and $40 million this year. On the surface this result looks like a 33% improvement in the company's performance. However, if the company had 15 million shares of stock outstanding last year and 20 million shares outstanding this year, then the per-share performance is unchanged:

$$\text{Earnings per share last year} = \frac{\$30 \text{ million}}{15 \text{ million shares}} = \$2.00/\text{share}$$

$$\text{Earnings per share this year} = \frac{\$40 \text{ million}}{20 \text{ million shares}} = \$2.00/\text{share}$$

From this example we can see that a failure to adjust data to per-item figures can result in a misleading impression. For this reason, much of the financial and economic data reported in private and government publications are presented on a per-capita, per-share, per-household, or per-transaction basis. When one sees gross figures reported, one should always ask, Are these figures relevant as they stand or should they be presented on a per-something basis?

Another misleading impression can be produced if the data used in a statistical analysis are *not adjusted for inflation*. The high levels of inflation that economies all over the world have experienced in recent years make adjustments of this type very necessary if correct conclusions are to be drawn from financial and economic figures. For instance, a plant manager may proudly report that the average wages of workers in the plant have risen 60% in the past ten years. However, if the general price level of goods purchased by these workers has risen 80% during that same period, they are worse off in terms of what their wages will purchase than they were ten years ago.

Thus depreciation figures based on historical costs, which are not adjusted for inflation, create an impression that a company is more profitable than it really is in light of today's costs. This problem is so serious that the Securities and Exchange Commission now requires many large companies to report the impact that depreciation of assets costed at replacement values would have on their earnings figures. Methods for making these adjustments are discussed in Chapter 19, which deals with index numbers.

Another problem that can creep into statistical analysis is *induced bias*. When the data used in a statistical study are gathered through interviews with people, the people doing the study must take care that the interviewers do not in any way convey the responses they wish or expect to obtain. The average person is really rather nice. If this person is being interviewed, he or she may "read" the interviewer in an attempt to find out what the interviewer wants or likes to hear and adjust his or her responses accordingly. This adjustment, of course, produces a bias in the responses toward what the interviewer wanted to hear in the first place. Trained interviewers are very good at hiding their own opinions and expectations during an interview. Sometimes, however, no amount of training can eliminate an induced bias. For instance, if a black person interviews people about their attitude toward minority groups, that interviewer will likely get somewhat different responses than would a white interviewer.

Often statistical methods are applied to data that have been gathered from questionnaires. Such data obviously have no interviewer-induced bias. But the way in which questions are asked can induce bias in the results. Consider the example of a union official who wishes to know how members of his union feel about a wage settlement recently offered to a group of employees. He might ask the following question in a survey of their opinion:

> Should we accept or reject management's latest offer of a $1.05-per-hour increase over the next three years?

His responses to that question might be rather different from those he would receive if he asked his question in this manner:

> Should we accept or turn down flat our tight-fisted management's skimpy offer of a $1.05-per-hour increase over the next three years in light of the fact that the guys over at Universal Industries held out and got $1.30 per hour over the same period?

Methods of avoiding induced bias during interviews and in questionnaires will not be discussed in this book. But a little common sense and judgment will go a long way in eliminating this problem for those who are aware of it in advance.

A very serious problem, which is, unfortunately, rather common, involves *inappropriate comparisons of groups*. Statistical studies often involve the comparison of two or more groups with one another. Banks versus savings and loans, purchasers versus nonpurchasers of your product, married versus nonmarried credit applicants, and western versus midwestern versus eastern versus southern markets are examples of comparisons that are common. To compare two or more groups may be inappropriate for several reasons. We will discuss only two: *self-selection* and *hidden differences*.

Let us first deal with the problem of self-selection as it applies to the comparison of two different groups.

━━ **EXAMPLE 1.8**   A large national corporation offered a course in "Selling to Nonprofit Organizations" to any of its salespeople who were interested in enrolling. Approximately 35 of the company's 160 salespeople signed up and took the course. In an effort to show there had been an impact on the sales effort, the training director of the corporation reported to the vice president of personnel the following comparison:

> In the six months following the training course offered by this office, the people who took the course showed an average sales to nonprofit organizations of $1298 per salesperson per month. In the group that did not take the course, this average was only $706 per salesperson per month. More customer-oriented sales-training programs of this type would be of obvious benefit.

> The training director correctly adjusted the sales figures to a per-salesperson and per-month basis. However, the problem here is the self-selection of the training program's participants. Perhaps only the salespeople who had large nonprofit organizations in their territories signed up in the first place. Or perhaps the people who took the course already had substantial nonprofit organizations as accounts and merely wanted the training to see whether they could increase their already high sales level for these accounts. A better way to measure the impact of the training is to measure the participants' level of sales to nonprofit organizations before the training and then measure it after the training. The training director should make sure that comparable sales periods are measured both before and after the training and that any effects of price changes between the measured periods (inflation) are taken into account. ▀▀

The problems involved with comparing groups with hidden differences are similar to those involved with comparing groups that have self-selection. In fact, some of the self-selection situations, such as the one described previously, produce hidden differences in the groups being compared — such as the suggested differences in the sales potential to nonprofit organizations in the groups mentioned. However, consider a less obvious problem of comparing two groups that have hidden differences. Not long ago, a state official bragged that the climate in her state was obviously healthier than that in most states since the death rate for the population (expressed in deaths per 100,000 residents) was one of the lowest in the nation. On the surface, again, this evidence sounds like a convincing argument for this state's climate. However, a closer examination of the population living in this state revealed that a higher-than-average proportion of the people in the state were young people, and a much lower-than-average proportion were old. Thus one would expect a group of people with this type of age distribution to have a lower-than-average death rate, regardless of the climate.

The examples of misusing statistical data cited previously are only a very few that could be mentioned. Other examples of failure to convert data to a per-something basis, failure to adjust for inflation, induced bias, self-selection, and hidden differences will be presented for the readers to discover in the problems.

Problems: Section 1.6

47. In the past few years many corporations have made an effort to cut energy costs. The vice president of operations for a regional trucking company was recently shocked to learn that despite the company's efforts to increase its fleet's fuel economy through improved maintenance and purchase of more fuel-efficient trucks, the fuel costs for this year are running 12% ahead of those for last year.

    a. Can you suggest *two* possible explanations for the increase?

    b. Can you suggest a way to measure fuel use that would more accurately reflect the company's success or failure in reducing fuel use?

48. Enrollments in many large universities have remained stable in terms of numbers of students taking classes each quarter. However, many students are taking fewer classes

because they have to work to support themselves and to cover the high cost of their educations. Rather than expressing their costs in dollars per student enrolled each term, how should university administrators measure the cost of providing education so that the lighter student loads do not distort their figures?

49. During the past ten years a manager's income has doubled. Give two reasons why the purchasing power of her income has not doubled.

50. The manufacturer of a certain health food supplement showed in its advertising that a study of people who used its product had a lower incidence of heart attack than that found in the population as a whole. Why may such advertising not be justified?

51. Two islands are located off the East Coast of the United States. Both islands have 1000 residents. The first island, however, has 800 people whose ages are in the twenties and 200 people in their sixties. The second island has the opposite situation—800 people in their sixties and 200 in their twenties.

  a. Which island is likely to have the higher birthrate?
  b. Which island is likely to have the higher death rate?
  c. If you were the proprietor of the general store located on the second island, what kinds of products might you stock differently than you would if you were the proprietor of the store on the first island?

52. The National Anti-Union committee sent a survey to people across the nation asking their opinions on proposals to allow unionization of the armed forces. One of the questions was the following:

  > Do you want to have union bosses exercising tyrannical control over our troops when we are under attack by hordes of ruthless soldiers from Russia or Mainland China?   Yes _____   No _____

  a. Name the specific statistical problem that this question will cause.
  b. Change the statement of the question so that the essence of the question remains the same but the emotion is removed.

Data Distributions

In this chapter we begin our study of the first major topic area in the subject of statistics: descriptive statistics and exploratory data analysis. Statistical data obtained by means of census, sample survey, or experiment usually consist of raw, unorganized sets of numerical values. Data must be summarized before they can be used as a basis for making inferences about some phenomenon under investigation. Information must be extracted from the data before the data can be used as a basis for decision.

Our purpose in this chapter is to introduce several tabular and graphical formats that can be used for organizing, summarizing, exploring, and presenting numerical data. These tabular and graphical formats are known, in general, as *frequency distributions,* or since the data are observed, as *empirical frequency distributions.* We also discuss a *stem-and-leaf diagram* that is used to explore the features of a data set.

Tabular and graphical frequency distributions are often used by managers to describe or explore a set of data or to extract information from a set of data. Numerical measures for describing data sets are discussed in Chapter 3.

2.1
FREQUENCY
DISTRIBUTIONS

We must have information to make decisions or inferences. For example, an investor wants to know the amounts of the stock dividends that are paid by stocks included in her investment portfolio. Also, managers at a large Western Electronics Corporation manufacturing plant are concerned about employee absenteeism, and they want to develop information concerning the number of employees that are absent from day to day from a large set of absenteeism data. A staffing policy will be developed based on the information.

One type of information that may be desired of a set of data relates to the pattern or grouping into which the data fall. As a first step in organizing a set of data to provide this type of information, we could construct an array. An **array** is a list of the data with the numerical values ordered from low to high (or high to low). Arrays are often used to help make the overall pattern of the data apparent. For example, the stock dividends paid to an investor by five companies listed on the New York Stock Exchange are $5, 12, 8, 10 and 15. An *array* of the stock dividends paid by the five companies, arranged in increasing numerical order is

5   8   10   12   15

The smallest data value and the largest data value are easily found when the data have been ordered in an array. However, if the number of values is large, construction of the array may have to be done on a computer, and even then, the array may turn out to be so large that it is difficult to comprehend.

A more useful way to summarize a large set of data is to construct a frequency distribution. A frequency distribution is constructed by dividing the overall range of the values in our set of data into a number of classes. A **class** or *bin* is an interval of values within the overall range of values spanned by the data. Then we count the number of observations that fall into each of these classes. By looking at such a **frequency distribution** we can readily see the overall pattern of the data.

> **Definition**
>
> A **frequency distribution** is a tabular summary of a set of data that shows the frequency or number of items that fall in each of several distinct classes.

A frequency distribution is also known as a **frequency table.**

To construct a frequency distribution, we must do the following tasks:

1. Select the *number of classes.*
2. Choose the *class interval* or width of the classes.
3. Select the *class boundaries* or the values that form the interval for each class.
4. *Count* the number of values in the data set that fall in each class.

There are no universal rules for making the choices required to construct frequency distributions, because there are no simple rules that fit all situations. In this section, we consider procedures for constructing frequency distributions that are designed to summarize the data effectively, and then we will look at some alternative approaches.

TABLE 2.1

146	141	139	140	145	141	142	131	142	140
144	140	138	139	147	139	141	137	141	132
140	140	141	143	134	146	134	142	133	149
140	143	143	149	136	141	143	143	141	140
138	136	138	144	136	145	143	137	142	146
140	148	140	140	139	139	144	138	146	153
148	142	133	140	141	145	148	139	136	141
140	139	158	135	132	148	142	145	145	121
129	143	148	138	149	146	141	142	144	137
153	148	144	138	150	148	138	145	145	142
143	143	148	141	145	141				

Absences from Work Last 106 Days

We begin with the construction of a frequency distribution by using the data values given in Table 2.1 for the number of employees absent from work each day at a Western Electronics manufacturing plant over the last 106 working days.

The first step in constructing frequency distributions is to decide on the **number of classes** to use. Data analysts generally use between 5 and 20 classes for frequency distributions, and sometimes the number of classes is selected arbitrarily. However, it is often better to use a rule for selecting the number of classes that is designed to summarize the data effectively. One rule for effectively determining the *approximate number of classes* for a frequency distribution for quantitative data is given next.*

Rule for Determining the Approximate Number of Classes

$$\text{Approximate number of classes} = [(2)\,(\textit{Number of items in the data set})]^{.3333} \quad (2.1)$$

The *actual number of classes* is the integer value that just exceeds the number obtained from the rule for the approximate number of classes.

There are 106 values in Table 2.1 for the employees who are absent from work, so we determine the approximate number of classes for our frequency distribution as follows:

$$\textit{Approximate number of classes} = [(2)\,(106)]^{.3333} = 5.96$$

The integer value that just exceeds 5.96 is 6, so we will use 6 classes in this example. Table 2.2 gives the numbers of classes, or the integer values just exceeding the values given by the rule for determining the approximate number of classes for data sets of various sizes.

The next step in constructing frequency distributions is to determine the **class interval** or **width.** We use the highest data value and the lowest data value in the data

---

*The rule is practical for determining the number of classes, and it is designed to result in a frequency distribution that gives a nearly optimal approximation for a variety of underlying distributions (see Terrell and Scott, 1985).

TABLE 2.2	Number of Items in the Data Set	Number of Classes
	14– 32	4
Number of Classes for	33– 62	5
Various Sizes of Data Sets	63– 108	6
	109– 171	7
	172– 255	8
	256– 364	9
	365– 500	10
	501– 665	11
	666– 864	12
	865–1099	13
	1100–1373	14

set for the variable of interest to determine the approximate width for the class intervals according to the following equation.

Equation for Determining the Approximate Class Interval or Width

$$\frac{Approximate\ class}{interval\ or\ width} = \frac{Largest\ data\ value - Smallest\ data\ value}{Number\ of\ classes} \quad (2.2)$$

Once we determine the approximate class interval or width, we always *round up* or *increase* the approximate class interval to get the *actual class width*. However, there is no rule that tells us exactly how far we should round up. In some cases we may want to round up to the next highest integer, but in other cases we may want to round up to the next multiple of .1, or .5, or 10, or 100, and so forth. The precision with which the data are measured helps us make this decision.

For the data given in Table 2.1 on the number of employees who are absent from work at Western Electronic, the smallest data value is 121 and the largest is 158, so we determine the approximate class interval or width for our frequency distribution as follows:

$$\frac{Approximate\ class}{interval\ or\ width} = \frac{158 - 121}{6} = \frac{37}{6} = 6.17$$

To use a class interval of 6.17, which is not a whole-number multiple of the basic unit (an absence in this case), is very inconvenient, so we will use an interval of the next highest integer value of 7. Thus we can use exactly 6 classes, each of which is 7 units wide. Notice that all 6 classes will be 7 units wide. Classes for frequency distributions are typically *all the same width* to facilitate comparisons of frequencies among classes.

The next step in constructing frequency distributions is to specify the **class boundaries.** Class boundaries are selected so that the classes cover all of the data values and so that each data value falls in a distinct class. The *lower boundary* for the first class in the distribution, the class for the smaller values, will generally be an arbitrary but convenient value below the lowest data value (since the classes will generally overlap the actual data values). The *upper boundary* for the first class is then found by adding the class width to the lower boundary. The boundaries for the remaining classes are found by successively incrementing by the class width until all of the classes have been determined. The last class will generally have an upper boundary above the highest value for the data.

For the data given in Table 2.1 on absenteeism at Western Electronics, we have decided to use 6 classes with each class having a width of 7, so the classes will span (6)(7) or 42 values. Also, the span of the actual data is 37, from 121 to 158. The classes will overlap the span of the actual data values by 42 − 37, or 5 units.

Now we can find the lower boundary for the first class by using an arbitrary but convenient value somewhat below 121, the lowest value. In this case we should not select a lower boundary that is more than 5 units below 121. It is common to use a lower boundary that is about (1/2)(5) or about 2.5 units below 121. The value we select is 118.5. The upper boundary for the first class is the lower boundary 118.5 plus the class width, or 118.5 plus 7, which gives 125.5. The boundaries for the second, third, fourth, fifth, and sixth classes are found by successively incrementing by the class width of 7, resulting in lower and upper boundaries of 125.5 and 132.5, 132.5 and 139.5, 139.5 and 146.5, 146.5 and 153.5, and 153.5 and 160.5.

Notice that each data value in Table 2.1 falls into a distinct class for the Western Electronics absenteeism example and the classes cover all of the data values. Class boundaries are often set at values that are *not* included in the data set. This forces each actual data value to fall into a distinct class.

Now we must find the *frequency* for each class by **counting** the number of items that fall in each class. With these preliminary considerations taken care of, we can now construct the frequency distribution. The result is presented in Table 2.3. The tally shown in the central portion of the table is for convenience in counting only and would not be displayed in the completed distribution.

The frequency distribution given in Table 2.3 indicates that from 140 to 146 employees are commonly absent from work at Western Electronics. To have fewer than 119 or more than 160 employees absent from work would be a rare occurrence. Managers at Western Electronics can use this information in making staffing decisions. Our frequency distribution summarizes the information in the data set effectively. However, if the pattern of the data were not apparent, then we could construct an improved distribution by adjusting the number, width, or boundaries of the classes.

In some data sets the items have only distinct values. For example, the number of employees absent from work must be integers, or a data set may include the ages of

Number of Absences	Tally	Number of Items	
118.5–125.5	|	1	**TABLE 2.3**
125.5–132.5	||||	4	
132.5–139.5	ⵢ ⵢ ⵢ ⵢ ⵢ |	26	Distribution and Tally of
139.5–146.5	ⵢ ⵢ ⵢ ⵢ ⵢ ⵢ ⵢ ⵢ ⵢ ⵢ ⵢ ||||	59	Employees Absent at West-
146.5–153.5	ⵢ ⵢ ⵢ	15	ern Electronics in 106 Days
153.5–160.5	|	1	
	Total	106	

employees and the ages may be recorded to the nearest year. If the items can only take distinct values, then the classes in a frequency distribution can be defined by the lowest and highest *attainable* values in the class; these attainable values are called the **class limits.** For example, the class limits are 119 and 125 for the first class in the frequency distribution given in Table 2.3. Class boundaries are thus half way between the limits of the successive classes.

A convenient practice is to select one value from each class to serve as a representative of the class. The value commonly used is the **midpoint** of the class, or the **class mark,** which is the average of the two class boundaries. For the distribution of employee absences, the class mark for the first class is $(118.5 + 125.5)/2 = 122$.

The number of observations in any class is the **class frequency.** We sometimes wish to present the data in a **relative frequency distribution.** The relative class frequencies are found by dividing the class frequencies by the total number of items. A **percentage distribution** is determined by multiplying the relative class frequencies by 100 to convert them to percentages. For example, for the third class in Table 2.3, the class frequency is 26, so the relative frequency is $26/106$, or .245, and the percentage is $(.245)(100)$, or 24.5%.

Class limits, class marks, class relative frequencies, and class percentages are all illustrated in Table 2.4 for the Western Electronics employee absence data. In practice, not all the columns given in Table 2.4 are displayed in a frequency distribution.

The approach to constructing frequency distributions that we have described is summarized next.

Procedures for Constructing Frequency Distributions

1. Determine the **number of classes** as the integer that exceeds the number for the approximate number of classes found by using the following rule:

$$\textit{Approximate number of classes} = [(2)(\textit{Number of items in the data set})]^{.3333} \quad (2.1)$$

2. Determine the **class interval** or **width** as a larger value than the approximate width that is determined by using the following equation:

$$\frac{\textit{Approximate class}}{\textit{interval or width}} = \frac{\textit{Largest data value} - \textit{Smallest data value}}{\textit{Number of classes}} \quad (2.2)$$

3. Determine the **class boundaries.** Class boundaries must be selected so that the classes cover all of the actual data values and so that each data value falls in a distinct class. The lower boundary for the first class is an arbitrary but convenient value below the lowest data value. The upper boundary for the first class is then found by adding the class width to the lower boundary. The boundaries for the remaining classes are found by successively incrementing by the class width until all of the class boundaries have been determined.

4. Find the **frequency** for each class by *counting* the data values and present the results in a table.

Class Limits (number of employees absent)	Class Boundaries (number of employees absent)	Class Mark	Class Frequency	Relative Frequency	Percentage of Observations	
119–125	118.5–125.5	122	1	.009	0.9	**TABLE 2.4**
126–132	125.5–132.5	129	4	.038	3.8	
133–139	132.5–139.5	136	26	.245	24.5	**Distribution of Absences**
140–146	139.5–146.5	143	59	.557	55.7	**at Western Electronics in**
147–153	146.5–153.5	150	15	.142	14.2	**106 Days**
154–160	153.5–160.5	157	1	.009	0.9	
		Total	106	1.000	100.0	

We indicated earlier in this section that all the classes in a frequency distribution are usually of the same width. But sometimes it happens that our data contain a few **outlying** observations whose numerical values are much smaller or much larger than the rest. If we include these values in the ordinary way, we may find that a number of our class frequencies are equal to zero. This result is the case with the frequency distribution for the times between failures (in thousands of hours) of flexible disk drives produced by Digital Data Corporation given in Table 2.5. Because of the one extreme value, or **outlier,** in the data, five of the class frequencies shown in the "number of disk drives" column in Table 2.5 are equal to zero.

We can obtain a clearer picture of the distribution of the times to failure for disk drives without increasing the number of classes if we make use of an **open-ended interval,** which has no boundary on one end, for the last class. The resulting distribution is given in Table 2.6. Open-ended intervals permit the inclusion of a wide range of extreme values, but, unhappily, the actual numerical values are lost, and we

Times Between Failures (in thousands of hours)	Number of Disk Drives	
At least 0, but less than 5	41	**TABLE 2.5**
At least 5, but less than 10	94	
At least 10, but less than 15	172	**Distribution of Times Be-**
At least 15, but less than 20	238	**tween Failures for Flexible**
At least 20, but less than 25	290	**Disk Drives**
At least 25, but less than 30	277	
At least 30, but less than 35	157	
At least 35, but less than 40	105	
At least 40, but less than 45	0	
At least 45, but less than 50	0	
At least 50, but less than 55	0	
At least 55, but less than 60	0	
At least 60, but less than 65	0	
At least 65, but less than 70	1	
Total	1375	

TABLE 2.6	*Times Between Failures (in thousands of hours)*	*Number of Disk Drives*
Distribution of Times Between Failures for Flexible Disk Drives with an Open-Ended Interval	At least 0, but less than 5	41
	At least 5, but less than 10	94
	At least 10, but less than 15	172
	At least 15, but less than 20	238
	At least 20, but less than 25	290
	At least 25, but less than 30	277
	At least 30, but less than 35	157
	At least 35, but less than 40	105
	40 or more	1
	Total	1375

do not know how much larger or smaller the extremes actually are unless some indication is given in a footnote or elsewhere.

The class boundaries in Table 2.5 are defined somewhat differently than they were in the Western Electronics employee absences example. The employee absences were counts involving whole numbers, but the time between flexible disk drive failures is a continuous measure. As such, it may be a fractional value such as 5.4667 thousand hours. Thus the 47 observations in the second class of Table 2.5 could have values ranging from 5.0 to 9.99999 thousand hours—that is, from 5.0 up to but not including 10.0. Notice that each data value falls in a distinct class when class intervals are defined from the lower boundary up to but not including the upper boundary.

We have discussed procedures for constructing frequency distributions that summarize data effectively, and we have presented some alternative approaches. There is flexibility in selecting the number, width, and boundaries or limits of classes. Furthermore, if there are just a few distinct data values in a data set, then a frequency distribution may be constructed with a separate class for each value.

Problems: Section 2.1

Answers to odd-numbered problems are given in the back of the text.

1. How many classes would frequency distributions have if they contained the following number of data items?

   a. 50   **b.** 70   **c.** 250   **d.** 1200   **e.** 4000

2. The accompanying listing gives the stock dividends paid by 20 companies listed with the New York Stock Exchange.

   a. How many classes should be used in grouping these data?
   b. How wide should each class be, assuming that you want the class width rounded up to the nearest dollar?
   c. Construct the frequency distribution.

3	10	9	7	10	10	9	9	10	10
10	11	12	7	8	11	11	10	9	13

3. Assume that you are given the weights (in pounds) of 250 inventory items at a Stokely-Van Camp, Inc., warehouse. The smallest item weighs 8.8, and the largest item weighs 79.9.

    a. How many classes does the data's frequency distribution contain?
    b. What is the width of each class, assuming that you want an integer as the class width?
    c. How much overlap is there between the frequency distribution's classes and the actual data?
    d. Give the class limits of the first class, assuming that you split the overlap approximately in half.

4. The class limits in a frequency distribution are 32–36, 37–41, 42–46, and 47–51.

    a. What is the class width?
    b. What are the class boundaries?
    c. What are the class marks?

5. The accompanying listing gives the sales commissions paid to 48 part-time clerks in a Gap Stores, Inc., clothing store.

    a. How many classes should be used in grouping these data?
    b. How wide should each class be, assuming that you want the class width rounded up to the nearest dollar?
    c. Construct the frequency distribution.

181	182	184	193	125	70
168	172	161	149	77	123
115	114	135	136	115	97
112	109	104	108	184	117
80	64	128	115	132	92
123	120	75	131	118	161
84	100	106	71	56	148
111	128	83	114	112	135

2.2 CUMULATIVE FREQUENCY DISTRIBUTIONS

Frequency distributions as described in Section 2.1 are valuable aids for organizing and summarizing sets of data and for presenting data in such a way that the outstanding features are readily apparent. Sometimes, however, we require information on the number of observations whose numerical value is *less than* a given value. This information is contained in the **cumulative frequency distribution.** For instance, a stockbroker might be interested in the number of stocks on the New York Stock Exchange that have price-earnings ratios of less than 30. A salesperson might be interested in the number of companies in an industry that have sales under $10,000,000. A quality control engineer might be interested in the number of days for which the rejection rate on a production line was less than 1%. The cumulative frequency distribution is useful in all three cases.

   If we return to Table 2.3, we see that of the days observed, none had fewer than 119 absences, 1 had fewer than 126, 5 (1 + 4) had fewer than 133, 31 (1 + 4 + 26) had fewer than 140, and so on. To obtain the cumulative frequency of the number of absences less than the lower limit or boundary of a specified class, we add the class frequencies of all the preceding classes. The completed distribution is presented in Table 2.7. Note that this table uses the lower boundary of each class in order to avoid confusion about where values such as 119 and 126 should be counted.

	Number of Absences	Cumulative Number of Observations
TABLE 2.7		
Cumulative Distribution of	Less than 118.5	0
Employee Absences	Less than 125.5	1
	Less than 132.5	5
	Less than 139.5	31
	Less than 146.5	90
	Less than 153.5	105
	Less than 160.5	106

Table 2.7 contains further information. The number of days on which there were 140 or more absences is, of course, 106 − 31, or 75. And the number of days on which there were at least 140 but fewer than 154 absences is 105 − 31, or 74.

Cumulative distributions may be constructed for relative frequencies and percentages as well as for absolute frequencies. The procedures are identical except that we add the relative frequencies or percentages, as the case may be, instead of the absolute frequencies. Cumulative distributions that show the number of observations whose value exceeds a certain amount can also be constructed. Such distributions contain exactly the same information as the "less than" cumulative distributions discussed previously.

Problems: Section 2.2

6. The following data are the numbers of parts produced on a production line at Mark Industries, Inc., during 20 randomly selected days.
   a. How wide should each class be if you want the class width expressed to the nearest integer?
   b. Construct the frequency distribution.
   c. Find the cumulative frequency distribution.

2786	2426	2855	2220	2838
2832	3217	3346	3137	3010
2312	2524	2618	2978	2581
3218	2697	2427	2845	2714

7. Develop a cumulative distribution for the data on the sales commissions paid to 48 part-time clerks that were given in Problem 5.

8. The accompanying frequency distribution was prepared from the price-earnings ratios of several common stocks for companies listed on the American Stock Exchange. Construct a cumulative frequency distribution.

Class	Frequency
From 5 to less than 10	4
From 10 to less than 15	10
From 15 to less than 20	14
From 20 to less than 25	8
From 25 to less than 30	6

Although frequency distributions are effective in presenting the salient features of a set of data and are indispensable for computations, pictorial representation of the same information often makes the important characteristics more immediately apparent. Here we consider **histograms.**

> **Definition**
>
> A **histogram** is a graphic presentation of a frequency distribution and is constructed by erecting bars or rectangles on the class intervals.

Along the horizontal scale of the histogram we record the values of the variable concerned, marking off the *class boundaries.* Along the vertical scale we mark off frequencies. If we have equal class intervals, we erect over each class a rectangle whose height is proportional to the frequency of that class. For the distribution of employee absences that we divided into 6 classes (Table 2.4), we obtain the histogram of Figure 2.1. It shows very plainly that the number of employees absent each day tends to be in the vicinity of 140 and is generally within 10 absences of this central value.

The class interval used in constructing the frequency table has a marked effect on the appearance of the histogram. For example, Table 2.3 shows the employee absence data of Table 2.1 sorted into classes having a class interval of 7. If a class interval of 2 or of 15 is used, the distributions of Table 2.8 result. The differences among the histograms of these distributions are illustrated in Figure 2.2 (see page 39). Panels (a) and (c) of the figure come from the two distributions in Table 2.8, and the histogram in panel (b) is a repeat of Figure 2.1. The horizontal scale is the same in each case.

When raw data are grouped into classes, a certain amount of information is lost, since no distinction is made between items falling in the same class. The larger the

FIGURE 2.1

Histogram of Employee Absences at Western Electronics

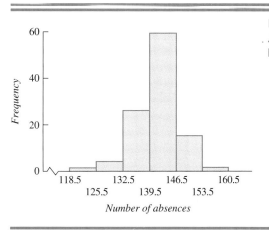

	Class Interval = 2		Class Interval = 15	
TABLE 2.8	*Number of Absences*	*Number of Observations*	*Number of Absences*	*Number of Observations*
Distributions for Employee Absences	120–121	1	117–131	3
	122–123	0	132–146	87
	124–125	0	147–161	16
	126–127	0	Total	106
	128–129	1		
	130–131	1		
	132–133	4		
	134–135	3		
	136–137	7		
	138–139	14		
	140–141	24		
	142–143	17		
	144–145	13		
	146–147	6		
	148–149	11		
	150–151	1		
	152–153	2		
	154–155	0		
	156–157	0		
	158–159	1		
	Total	106		

class interval or width is, the greater is the amount of information lost. For the employee absence data, a class interval of 15 is so large that the corresponding histogram gives very little idea of the shape of the distribution. A class interval of 2 is so small, in contrast, that it gives a ragged histogram. Little information has been lost in the histogram of Figure 2.2(c), but the presentation of information is somewhat misleading because the small irregularities in the histogram merely reflect the accidents of sampling. By using the rule for determining the approximate number of classes as given in Equation (2.1), we produced a histogram between those for the intervals of 2 and 15. Thus the rule for determining the approximate number of classes as given in Equation (2.1) will likely produce histograms between the extremes of giving too much detail or giving too little. However, the number of classes may need to be adjusted for some sets of data in order to obtain a clearer presentation of the information, particularly if the histogram is very uneven or asymmetrical.

Figure 2.3 shows the histogram for the times between failures of flexible disk drives given in Table 2.6. Observe that although the open class in this distribution cannot be represented by a bar in the histogram, we indicate by a note that there was an extreme value.

Statisticians sometimes use class relative frequencies or class percentages, rather than frequencies, for the heights of the rectangles in a histogram. **Frequency histograms, relative frequency histograms,** and **percentage histograms** are all commonly called *histograms.* The various histograms are identical in appearance if the same heights are used for frequencies, relative frequencies, or percentages.

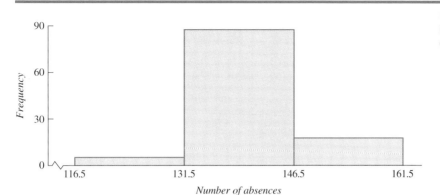

(a) **Classes = 3, class interval = 15, too few classes**

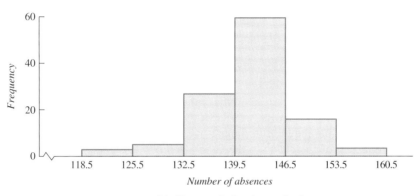

(b) **Classes = 6, class interval = 7**

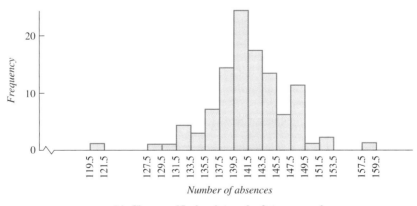

(c) **Classes = 15, class interval = 2, too many classes**

In the first section of this chapter, we mentioned that in constructing frequency distributions, we usually select all the classes to be of the same width. One of the advantages of equal class intervals is that the areas of the bars will be proportional to the heights of the bars. We therefore can draw the bars so that the height is propor-

FIGURE 2.3

Histogram for Times Be-
tween Failures for Flexible
Disk Drives

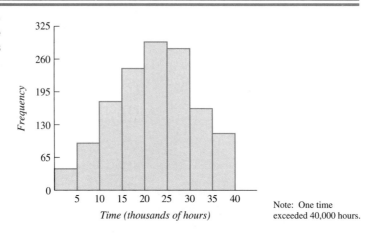

Note: One time
exceeded 40,000 hours.

tional to the frequency. If the class intervals are unequal, we must modify our approach somewhat.

EXAMPLE 2.1   If the last three classes of the distribution of employee absences given in Table 2.4 are combined, then we obtain a frequency distribution with *unequal class intervals,* as shown in Table 2.9. The rule for constructing a histogram is that the *area* of the bar over a class interval must be proportional to the frequency of the class. In the case of equal class intervals, this rule is the same as requiring that the height of the bar be proportional to the class frequency. But what if the class intervals are unequal?

Figure 2.4 shows the histogram for the new distribution with three combined classes in Table 2.9 drawn with height proportional to frequency. Here, because we have not followed the area rule, the 75 values in the range 140–160 are given an undue emphasis, and the histogram gives the impression that on a very large proportion of the days more than 139 employees were absent.

The correct representation is shown in Figure 2.5, where the areas of the bars are proportional to the frequencies. The height of the rightmost bar is one-third of what it was in Figure 2.4, since the width of the class is three times the width of the other classes in the histogram, or 21 absences. This drawing technique is correct because the eye naturally interprets size by area.

When large data sets are involved, data analysts often use a personal computer or a large computer system to obtain frequency distributions and his-

TABLE 2.9	*Number of Absences*	*Number of Observations*
Distribution of Employee	119–125	1
Absences with the Last	126–132	4
Three Classes Combined	133–139	26
	140–160	75

FIGURE 2.4

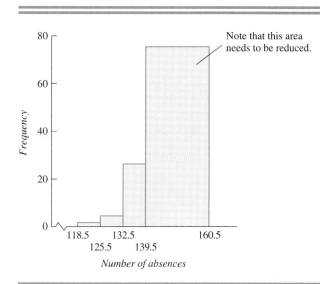

Incorrectly Drawn Histogram for Distribution with Unequal Class Intervals

Note that this area needs to be reduced.

Number of absences

tograms. Many data analysis software packages are available. The names, or acronyms, of five such products that are commonly used for data analysis are MINITAB, SAS, SPSS, SYSTAT (or Business MYSTAT), and BMDP. Users of these packages provide the appropriate program commands and the data that they want analyzed. In some modern software packages, analysts may provide commands to the computer by using graphical user interfaces or by navigating through menus and dialog boxes. Manuals are available for most data analysis software packages, and Appendix D gives a brief command reference for MINITAB as well as a users' manual for Business MYSTAT.

FIGURE 2.5

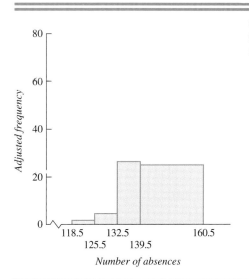

Correctly Drawn Histogram for Distribution with Unequal Class Intervals

Number of absences

Computer Exhibits 2.1A, 2.1B, and 2.1C show the histograms and command summaries from three data analysis software packages we used for the Western Electronics absenteeism data given in Table 2.1. We will follow the convention of summarizing the commands and presenting the output whenever we use a computer

---

MINITAB   **COMPUTER EXHIBIT 2.1A** Histograms for the Western Electronics Absenteeism Data

---

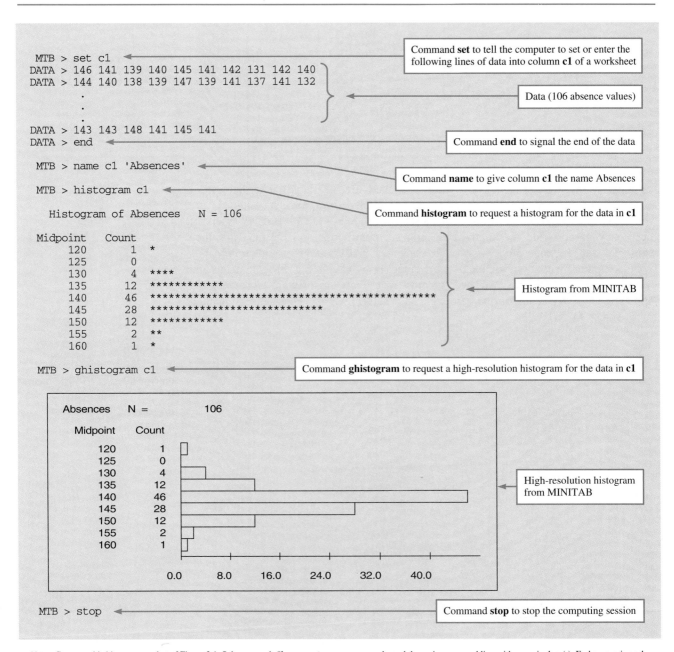

```
MTB > set c1                          Command set to tell the computer to set or enter the
DATA > 146 141 139 140 145 141 142 131 142 140     following lines of data into column c1 of a worksheet
DATA > 144 140 138 139 147 139 141 137 141 132
       .                               Data (106 absence values)
       .
       .
DATA > 143 143 148 141 145 141
DATA > end                            Command end to signal the end of the data

MTB > name c1 'Absences'              Command name to give column c1 the name Absences

MTB > histogram c1                    Command histogram to request a histogram for the data in c1

   Histogram of Absences    N = 106

Midpoint    Count
     120       1   *
     125       0
     130       4   ****
     135      12   ************
     140      46   **********************************************
     145      28   ****************************
     150      12   ************
     155       2   **
     160       1   *
                                       Histogram from MINITAB

MTB > ghistogram c1                   Command ghistogram to request a high-resolution histogram for the data in c1
```

```
 Absences   N =          106

    Midpoint    Count
       120       1
       125       0
       130       4
       135      12
       140      46                     High-resolution histogram
       145      28                     from MINITAB
       150      12
       155       2
       160       1

              0.0   8.0   16.0   24.0   32.0   40.0
```

```
MTB > stop                            Command stop to stop the computing session
```

Note: Compare this histogram to that of Figure 2.1. Subcommands "**increment 7;**" and "**start 122.**" could be used to obtain equivalent histograms. To use subcommands, end the main command line with a semicolon (;). End successive subcommands with semicolons (;), but end the last subcommand line with a period (.).

in the remainder of the text. The data analyses will be shown in the computer exhibits in several formats since the software packages differ in some respects. For instance, in constructing histograms, software packages often use different numbers, widths, and boundaries for classes; they may present class midpoints rather

COMPUTER EXHIBIT 2.1B Histograms for the Western Electronics Absenteeism Data

SAS

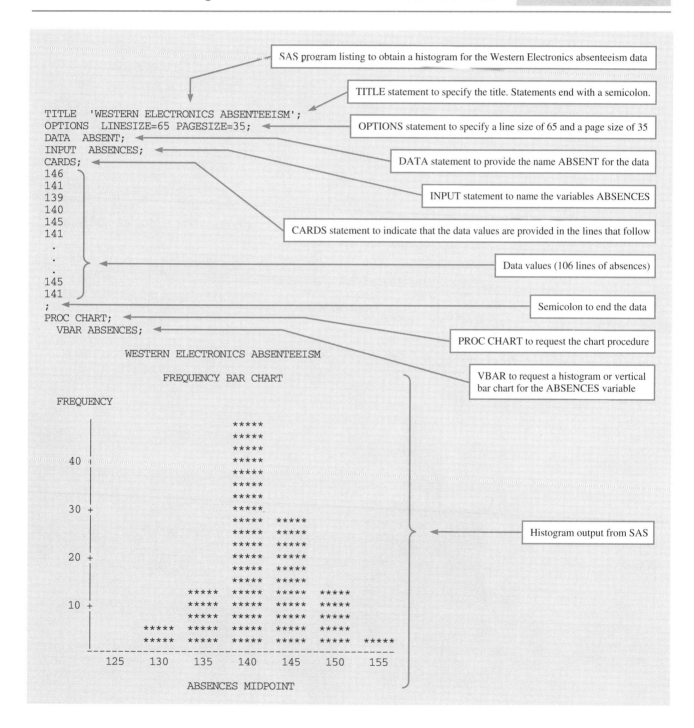

SAS program listing to obtain a histogram for the Western Electronics absenteeism data

TITLE statement to specify the title. Statements end with a semicolon.

```
TITLE   'WESTERN ELECTRONICS ABSENTEEISM';
OPTIONS  LINESIZE=65 PAGESIZE=35;
DATA  ABSENT;
INPUT  ABSENCES;
CARDS;
146
141
139
140
145
141
 .
 .
 .
145
141
;
PROC CHART;
   VBAR ABSENCES;
```

OPTIONS statement to specify a line size of 65 and a page size of 35

DATA statement to provide the name ABSENT for the data

INPUT statement to name the variables ABSENCES

CARDS statement to indicate that the data values are provided in the lines that follow

Data values (106 lines of absences)

Semicolon to end the data

PROC CHART to request the chart procedure

VBAR to request a histogram or vertical bar chart for the ABSENCES variable

```
         WESTERN ELECTRONICS ABSENTEEISM

              FREQUENCY BAR CHART

   FREQUENCY

                         *****
                         *****
                         *****
      40                 *****
                         *****
                         *****
                         *****
      30 +               *****
                         *****  *****
                         *****  *****
                         *****  *****
      20 +               *****  *****
                         *****  *****
                         *****  *****
                  *****  *****  *****  *****
      10 +        *****  *****  *****  *****
                  *****  *****  *****  *****
           *****  *****  *****  *****  *****
           *****  *****  *****  *****  *****  *****
         ----------------------------------------------
           125    130    135    140    145    150    155

                      ABSENCES MIDPOINT
```

Histogram output from SAS

than boundaries; and they may select classes that have convenient values for mid-points. Managers often use computers to provide statistical analyses, such as those presented in Computer Exhibit 2.1 and others given later in this chapter, to assist them in the decision making process.

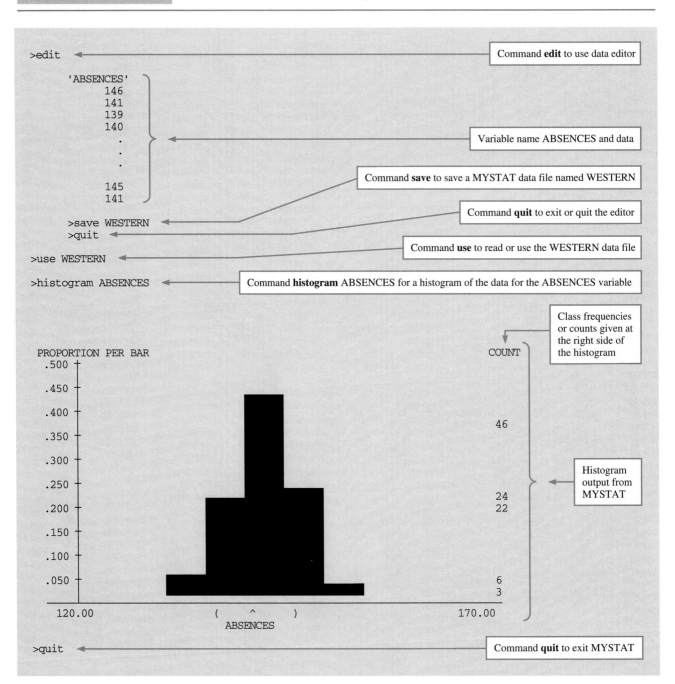

Problems: Section 2.3

9. The accompanying table gives the distribution of the ages (in days) of 100 accounts receivable at the Allred Veterinary Clinic.

Class	25–49	50–74	75–99	100–124	125–149
Frequency	15	25	30	20	10

   a. What is the width of each class?
   b. List the class boundaries.
   c. Draw the histogram.

10. As an experiment, toss four coins 50 times and construct the frequency distribution for the number of heads that appear on each toss. The values will be 0, 1, 2, 3, and 4. Draw the histogram.

11. All the classes in a frequency distribution were 3 units wide, except the last class, which was 9 units wide and contained 21 data items. How high should the bar for the last class be drawn in this distribution's histogram?

12. Three histograms are presented in Figure 2.6. One of the histograms shows the distribution for incomes for a set of executives of *Fortune* 500 companies; another histogram shows the distribution for the diameters of steel drive shafts produced at ACME Lathe and Mill by a process that is in control; and another histogram shows the distribution for the time that it takes people to pass through a border inspection station at Tiajuana.

   a. Is the distribution for incomes shown by the histogram in panel (a), (b), or (c)? Explain.
   b. Is the distribution for steel drive shaft diameters shown by the histogram in panel (a), (b), or (c)? Explain.
   c. Is the distribution for inspection times shown by the histogram in panel (a), (b), or (c)? Explain.

13. The accompanying listing gives the service times (in minutes) for a sample of customers who arrived at the drive-in service window of Second Mortgage Bank.

   a. How many classes should be used in grouping these data?
   b. How wide should each class be, assuming that you want the class width rounded up to the nearest tenth of a minute?
   c. Construct the frequency distribution.
   d. Draw the histogram.

7.4	4.9	0.8	1.7	0.5	0.1	2.5	0.9	0.8	2.9
4.7	0.9	2.9	1.6	1.9	2.6	1.9	5.1	0.8	1.4
0.2	1.2	1.0	0.4	3.7	2.3	3.5	2.9	2.6	0.8
0.9	0.9	2.1	3.6	6.5	0.5	0.2	0.1	2.2	0.4

Histograms for Incomes, Drive Shaft Diameters, and Inspection Times   FIGURE 2.6

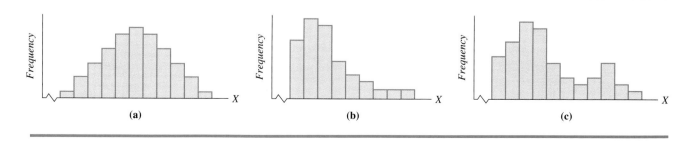

(a)   (b)   (c)

2.4
GRAPHIC PRESENTATION: OGIVE

Just as a frequency distribution can be represented graphically by a histogram, a cumulative frequency distribution is represented graphically by an **ogive.** To construct an ogive, we first lay out the class boundaries on the horizontal scale, just as for a histogram. Above each class boundary we next plot a point at a vertical distance proportional to the cumulative frequency—proportional, in other words, to the number of observations whose numerical value is *less than* that class boundary. Then we connect these points by straight lines.* By using the distribution in Table 2.7 in this way, we obtain the ogive shown in Figure 2.7(a).

We can interpolate graphically from an ogive. From the cumulative frequency distribution for employee absences, we may, for example, get an approximation for the number of observations whose numerical value is less than 150 by finding the height of the curve over that point. With the help of the dashed line in Figure 2.7(a), we find that on approximately 98 days fewer than 150 employees were absent.

If we convert the cumulative frequencies to cumulative relative frequencies by dividing each cumulative frequency by the number of items in the data set (there are 106 items in the absence data), then we have a **cumulative relative frequency distribution.** A cumulative relative frequency distribution is presented in Figure 2.7(b).

The primary use of a cumulative relative frequency distribution is to *approximate* percentiles. A **percentile** for a large set of data values that are not too repetitive is essentially a value below which a given proportion of the values in the data set fall. The 25th, 50th, and 75th percentiles for a set of data values are termed the first, second, and third **quartiles** (Q_1, Q_2, and Q_3). The three quartiles divide the data in an array into four parts, with each part containing about 25% of the data values. A **quantile** for a set of data values refers to any of the percentile or quartile values that measure positions in the set of data values. The 50th percentile, or second quartile

*If there are just a few distinct or discrete data values in a data set, then a cumulative frequency distribution can be constructed as a step function with a step showing the cumulative frequency less than or equal to each possible data value.

FIGURE 2.7   Ogives for Cumulative Employee Absences, Obtained from Table 2.7

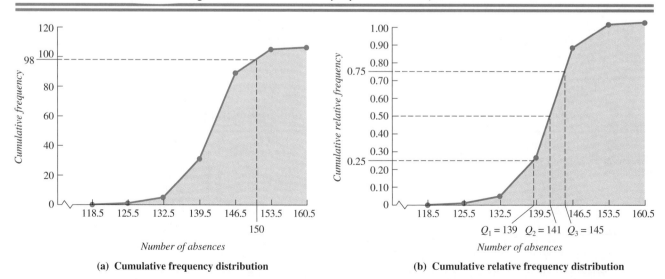

(a) **Cumulative frequency distribution**   (b) **Cumulative relative frequency distribution**

Q_2, split an ordered set of data values essentially in half. Consequently, the 50th percentile and second quartile Q_2 are measures of *central tendency* of the data.

━━ **EXAMPLE 2.2**   We want to find or approximate the 25th, 50th, and 75th *percentiles* for the Western Electronics absenteeism data, which are the same as the first, second, and third *quartiles* (Q_1, Q_2, and Q_3). We also want to find or approximate the .50 *quantile* for the Western Electronics absenteeism data. With the help of the dashed lines of Figure 2.7(b), our approximations of the 25th, 50th, and 75th percentiles are 139, 141, and 145. In other words, approximately 25 percent of all the absence values are less than 139. The approximations of the first, second, and third quartiles are also $Q_1 = 139$, $Q_2 = 141$, and $Q_3 = 145$. The .50 quantile is the same as the 50th percentile and the second quartile for a set of data values. Thus the approximation for the .50 quantile for the absenteeism data is 141. ━━

Problems: Section 2.4

14. The accompanying table gives the distribution of the ages of married women in the labor force of the eastern United States.

   a. Draw the histogram. Note that the class intervals are not equal.
   b. Draw the ogive to scale.
   c. From the ogive, find the age such that 50% are older; estimate the 75th percentile.

Age	Number of Married Women (thousands)	Age	Number of Married Women (thousands)
15–19	2212	40–44	3005
20–24	3503	45–54	5096
25–29	1313	55–64	2904
30–34	939	65–80	885
35–39	3490		

15. The accompanying frequency distribution was prepared by using data on the starting salary (in dollars per pay period) for several business school graduates.

   a. Construct an ogive.
   b. From the ogive, find the salary such that 50% earn more; estimate the 80th percentile.

Class	Frequency
From 1000 to less than 1200	8
From 1200 to less than 1400	14
From 1400 to less than 1600	16
From 1600 to less than 1800	6
From 1800 to less than 2000	4

16. The accompanying table gives the distribution for the number of defective items found in 100 lots of manufactured items at a Pittsburgh–Des Moines Corporation plant.

   a. What percentage of the lots contained more than five defective items?
   b. What is the relative frequency of the lots that contained two or fewer defectives?

c. What are the class limits? The class boundaries?

d. Draw the histogram and the ogive.

Number Defective	Frequency
0	23
1	25
2	19
3	14
4	11
5	5
6	2
7	1

2.5

EXPLORATORY DATA ANALYSIS: STEM-AND-LEAF GRAPHIC DISPLAY

Managers often wish to explore the important features of their data to obtain information that can be used to make decisions. **Exploratory data analysis** exposes patterns of data and drives us to see characteristics or features of data that we did not expect. Graphical displays such as histograms are often used for exploratory data analysis. Another graphical display that is useful for exploratory data analysis is known as a *stem-and-leaf diagram.*

Frequency distributions are effective for presenting the important features of data, and histograms are effective for presenting the same information pictorially. A **stem-and-leaf diagram** presents the data values in a pictorial display and thereby serves both purposes in a single diagram. To illustrate the construction of a stem-and-leaf display, consider the following data, which represent the ages (measured in years) of the 16 employees of American Department Stores, Inc., sales department:

18	21	22	19	32	33	40	41
56	57	64	28	29	29	38	39

To construct a stem-and-leaf display for the age data, we first write the leading digit (the *stem*) of each of the age values in ascending order to the left of a vertical line, as depicted in Figure 2.8(a). We then add the trailing digit (the *leaf*) for each employee's age on the appropriate row to the right of the vertical line. Figure 2.8(b) shows the trailing digit (a leaf) for the first age value (18 years) positioned to the right of its stem, and Figure 2.8(c) shows the finished stem-and-leaf diagram.

FIGURE 2.8

Stem-and-Leaf Display for Sales Employees' Ages

```
1 |                 1 | 8              1 | 8 9
2 |                 2 |                2 | 1 2 8 9 9
3 |                 3 |                3 | 2 3 8 9
4 |                 4 |                4 | 0 1
5 |                 5 |                5 | 6 7
6 |                 6 |                6 | 4
```

(a) Stem   (b) Stem with leaf for first   (c) Finished stem-and-
                      employee's age (18 years)   leaf display

The stem-and-leaf diagram shows that most of this department's employees are in their twenties and thirties, and it also shows the spread and symmetry of the data. Note that the actual age values are retained in the diagram, and they are presented as a type of horizontal histogram.

No strict rules are required for constructing stem-and-leaf displays. Each leaf in our example was added in ascending order, but that particular order is not required. The stem on each row of our example consisted of a separate single digit. However, a single-digit stem is not required in all displays. Sometimes two-digit stems or multiple-digit stems are used and other times stems are used over two lines of the display and the leaves for the stem are *stretched* over the two lines. Computer Exhibits 2.2A and 2.2B show stem-and-leaf displays for the sales department employee age data that were obtained by using data analysis systems. Note that the outputs presented in Computer Exhibits 2.2A and 2.2B have stems stretched across two lines.*

Stem-and-leaf displays and histograms are both useful for summarizing important features of a data set. One advantage of a stem-and-leaf display compared to a histogram is that the actual data values are presented in the display. Consequently, scanning the data to find unexpected characteristics or patterns in the data is rela-

---

When a stem is used on two lines, it is sometimes marked with "" on the first line and with "." on the second line. Then leaves (trailing digits) 0 through 4 are put on the "*" line and 5 through 9 are put on the "." line. Also, sometimes stems are used over five lines, in which case stems are marked with "*," "t," "f," "s," or ".". Then leaves 0 or 1 are put on the "*" line, 2 or 3 on the "t" line, 4 or 5 on the "f" line, 6 or 7 on the "s" line, and 8 or 9 on the "." line. Additional examples of stem-and-leaf displays can be found in the chapter references (see Hoaglin, Mosteller, and Tukey, 1983).

COMPUTER EXHIBIT 2.2A Stem-and-Leaf Display for the American Department Stores Age Data   **MINITAB**

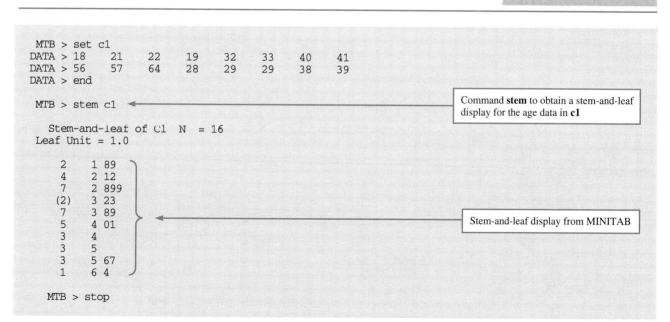

```
MTB > set c1
DATA > 18     21     22     19     32     33     40     41
DATA > 56     57     64     28     29     29     38     39
DATA > end

MTB > stem c1        ◄──────────  Command stem to obtain a stem-and-leaf
                                  display for the age data in c1

 Stem-and-leaf of C1   N  = 16
Leaf Unit = 1.0

    2    1 89  ⎫
    4    2 12  ⎪
    7    2 899 ⎪
   (2)   3 23  ⎪
    7    3 89  ⎬──────  Stem-and-leaf display from MINITAB
    5    4 01  ⎪
    3    4     ⎪
    3    5     ⎪
    3    5 67  ⎪
    1    6 4   ⎭

MTB > stop
```

Note: The first column of the display shows the **depth** for a stem or a count of the number of leaves on that stem or on stems closer to the nearer edge of the display. For example, there are 4 employees age 22 or younger and 5 employees age 40 or older. The stem that contains the leaf midway from each end shows a count of values (in parentheses) on that stem [(2) for the age data].

MYSTAT

COMPUTER EXHIBIT 2.2B Stem-and-Leaf Display for the American Department Stores Age Data

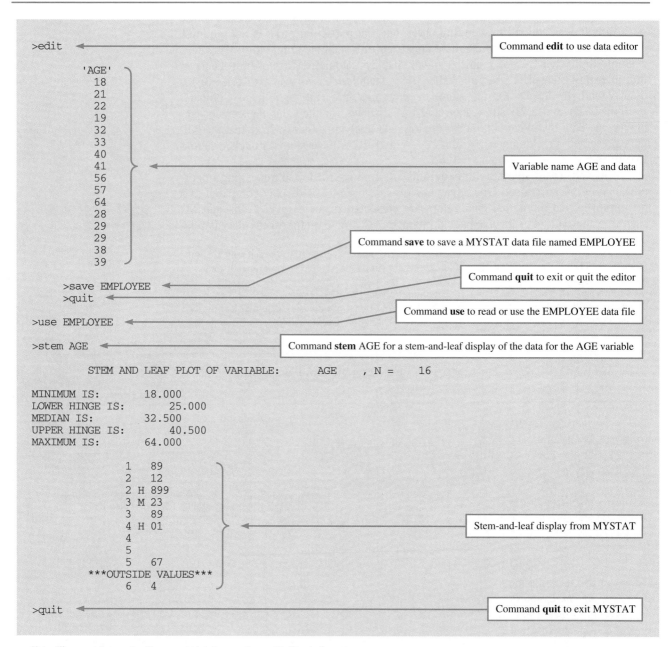

```
>edit                                                         ┌─────────────────────────────┐
                                                              │ Command edit to use data editor │
                                                              └─────────────────────────────┘
        'AGE'
         18
         21
         22
         19
         32
         33
         40                                                   ┌─────────────────────────────┐
         41                                                   │ Variable name AGE and data  │
         56                                                   └─────────────────────────────┘
         57
         64
         28
         29                              ┌──────────────────────────────────────────────────────┐
         29                              │ Command save to save a MYSTAT data file named EMPLOYEE │
         38                              └──────────────────────────────────────────────────────┘
         39                                        ┌──────────────────────────────────────┐
                                                   │ Command quit to exit or quit the editor │
     >save EMPLOYEE                                └──────────────────────────────────────┘
     >quit                                    ┌──────────────────────────────────────────────┐
                                              │ Command use to read or use the EMPLOYEE data file │
  >use EMPLOYEE                               └──────────────────────────────────────────────┘

  >stem AGE          ┌───────────────────────────────────────────────────────────────────────┐
                     │ Command stem AGE for a stem-and-leaf display of the data for the AGE variable │
                     └───────────────────────────────────────────────────────────────────────┘
        STEM AND LEAF PLOT OF VARIABLE:      AGE    , N =    16

MINIMUM IS:          18.000
LOWER HINGE IS:         25.000
MEDIAN IS:           32.500
UPPER HINGE IS:         40.500
MAXIMUM IS:          64.000

          1   89
          2   12
          2 H 899
          3 M 23
          3   89                                             ┌──────────────────────────────┐
          4 H 01                                             │ Stem-and-leaf display from MYSTAT │
          4                                                  └──────────────────────────────┘
          5
          5   67
     ***OUTSIDE VALUES***
          6   4
                                                             ┌──────────────────────────────┐
>quit                                                        │ Command quit to exit MYSTAT   │
                                                             └──────────────────────────────┘
```

Note: Hinges, medians, and outliers or outside values are discussed in Chapter 3.

tively simple. However, when we want to summarize the information from vast amounts of data, stem-and-leaf displays can become too large and histograms are generally preferred. We also have more flexibility in selecting class intervals for histograms than we have in selecting stems for stem-and-leaf displays. Overall, stem-and-leaf displays are generally used for purposes of exploratory data analysis, whereas frequency distributions and histograms are generally used to summarize the information in large sets of business data.

Problems: Section 2.5

17. Consider the data given in Problem 5 for the sales commissions paid to 48 part-time clerks. [*Hint:* Use the values 5, 6, 7, . . . , 19 as stems.]

    a. Construct a stem-and-leaf display for the data.
    b. Comment on an estimated value for the central tendency of the data.
    c. Comment on how spread out the values are, or the dispersion of the values.
    d. Comment on how symmetric the values are.
    e. Comment on any values that are far removed from the rest (called *outliers*).
    f. Comment on any gaps that exist in the data.
    g. Where do concentrations of the data exist?

18. Construct a stem-and-leaf display for the data given in Problem 6 and repeated below that represent the numbers of parts produced on a production line at Mark Industries, Inc., during 20 randomly selected days. Does the stem-and-leaf display have any advantage relative to the frequency distribution for the data? [*Hint:* Use the numbers 22, 23, 24, . . . , 33 for the stems, and use two digits for each leaf.]

2786	2426	2855	2220	2838
2832	3217	3346	3137	3010
2312	2524	2618	2978	2581
3218	2697	2427	2845	2714

2.6 SUMMARY

Statistical data are often in the form of raw, unorganized numerical values. Tabular and graphical formats are often used to organize, summarize, explore, present, and describe the data.

A *frequency distribution* is a tabular summary of a set of data that shows the frequency or number of data items that fall into each of several distinct classes. We discussed a procedure and several alternative approaches for constructing frequency distributions in this chapter. A *histogram* is a graphic presentation of a frequency distribution and is constructed by erecting bars or rectangles above the class boundaries. A *cumulative frequency distribution* and *ogive* are tabular and graphic presentations of the number of data values that are less than a given value. Managers use frequency distributions, relative frequency distributions, and percentage frequency distributions (or histograms) interchangeably. A *stem-and-leaf display* is a hybrid between a tabular display and a graphical display. It shows the numerical values so that the profile of the numbers looks like a histogram.

Tabular and graphical summaries of data sets are used to organize, summarize, explore, and present data in a form that provides information that is useful for making decisions or inferences. Statistical methods that accomplish these tasks fit into the category of descriptive statistics and exploratory data analysis.

The data distributions discussed in this chapter summarize observed or empirical data. The data values that are available are usually only a small part of a larger set. Conceptually, we could construct a frequency distribution for all of the data values in the larger set. Such conceptual distributions, known as theoretical or population distributions, are to be distinguished from the empirical data distributions of this chapter. Theoretical distributions are discussed in Chapters 5 and 6. An empirical data distribution estimates the corresponding theoretical distribution.

REVIEW PROBLEMS   *Answers to odd numbered problems are given in the back of the text.*

19. The accompanying listing gives the price-earnings ratios for a set of stocks whose prices are quoted by the National Association of Security Dealers Automated Quotations (NASDAQ).

    a. How many classes should be used in grouping these data?
    b. How wide should each class be, assuming that you want the class width rounded up to a whole number?
    c. Construct the frequency distribution.
    d. Draw the histogram.

4	35	20	25	19	21	28	21	15	18
10	23	25	37	29	31	29	30	33	16
21	32	27	34	26	34	18	22	28	24

20. Investors and managers often analyze and interpret financial statements by ratio analysis. A ratio that measures short-term liquidity is the current ratio, and it is defined as current assets divided by current liabilities (CA/CL). The values of the CA/CL ratio computed from the financial statements for a sample of 30 companies are given in the accompanying listing.

    a. How many classes should be used in grouping these data?
    b. How wide should each class be if you round the approximate class width up to a value with one digit to the right of the decimal?
    c. Construct the histogram.
    d. Comment on the central tendency of the histogram.
    e. Comment on how spread out the values are, or the dispersion of the values.
    f. Comment on the symmetry of the histogram.
    g. Comment on any values that are far removed from the rest.
    h. Comment on any gaps that exist in the data.
    i. Where do concentrations of the data exist?

5.32	1.41	3.32	2.69	2.33	2.36	1.54	2.33	2.05	1.83
2.28	1.59	5.50	1.12	1.55	3.14	5.00	3.19	1.78	3.79
4.70	1.32	3.11	3.17	2.92	5.45	1.07	1.67	5.21	9.98

21. Investors and managers often analyze and interpret financial statements by ratio analysis. A ratio that measures profitability is return on total assets, and it is defined as net income divided by total assets (NI/TA). The values of the NI/TA ratio computed from the financial statements for a sample of 30 companies are given in the accompanying listing.

    a. How many classes should be used in grouping these data?
    b. How wide should each class be if you round the approximate class width up to a value two digits to the right of the decimal?
    c. Construct the histogram.
    d. Comment on the central tendency of the histogram.
    e. Comment on how spread out the values are, or the dispersion of the values.
    f. Comment on the symmetry of the histogram.
    g. Comment on any values that are far removed from the rest.
    h. Comment on any gaps that exist in the data.
    i. Where do concentrations of the data exist?

.04	.12	.08	.11	.03	.05	.03	.03	−.13	.10
.09	.02	.06	.09	.02	.11	.09	.05	.07	.10
.10	.12	.01	.11	.11	.07	.09	.12	.07	.06

22. The accompanying relative frequency distribution shows the distribution of 50 Civil Service examination scores achieved by recent job applicants with the United States Postal Service. Find the original class frequencies.

Class	Relative Frequency
20–39	.12
40–59	.28
60–79	.36
80–99	.24

23. The accompanying data are the amounts, in parts per million, of pollutant materials found in 80 water samples taken at the confluence of the Monongahela and Allegheny rivers.

a. Construct a frequency distribution.
b. Construct the cumulative distribution.
c. Draw the histogram.
d. Construct a stem-and-leaf display.

3.64	5.08	3.87	3.52	3.05	4.98	3.30	3.64
3.20	2.55	4.40	4.61	3.74	4.42	2.76	3.20
3.30	3.63	5.05	2.87	4.50	4.44	4.40	4.74
3.64	3.81	4.61	4.04	3.40	4.74	3.52	5.06
2.76	3.87	3.39	4.39	5.50	3.52	3.86	4.74
3.40	3.05	2.63	3.08	4.48	5.74	3.64	4.08
4.54	3.72	4.50	3.98	3.96	2.74	3.74	2.76
3.30	5.51	3.07	4.42	4.62	4.10	4.30	2.98
2.76	3.22	2.53	2.98	6.06	3.40	4.50	3.42
5.38	3.73	3.06	5.42	3.98	3.66	4.18	6.08

24. The accompanying data are the percentages of employees at 96 firms who have had computer programming experience.

a. How many classes would Equation (2.1) suggest be used in this problem?
b. Construct a frequency distribution.
c. Construct the cumulative distribution.
d. Sketch the histogram.

7.78	6.02	9.68	6.14	7.28	6.23	4.64	7.41
6.75	4.51	8.42	7.92	6.84	5.81	4.67	7.38
5.43	11.28	7.96	6.38	5.92	4.80	10.72	6.17
6.17	9.54	7.82	5.33	5.79	4.76	9.78	5.84
6.00	5.23	6.48	6.53	5.46	4.83	9.51	6.82
7.26	10.90	10.82	9.81	10.73	10.46	10.63	7.79
6.54	9.72	9.74	10.62	4.87	5.23	4.68	7.19
5.28	10.62	7.24	8.38	5.05	4.98	8.62	6.76
5.03	8.41	8.65	8.37	8.46	7.93	7.19	6.53
4.29	8.62	8.50	8.83	7.78	7.64	5.39	7.21
5.31	7.76	6.42	5.13	6.49	6.31	7.72	4.15
4.97	6.13	5.29	6.34	7.82	6.68	6.44	4.68

25. The two histograms in Figure 2.9 were constructed for the rates of return from 200 financial investments of type A (e.g., bonds) and 200 investments of type B (e.g., stocks). A rate of return can be considered a measure of profitability.

a. Estimate the central tendency for the rates of return of type A and type B investments.
b. Comment on the dispersion or variability in rates of return for type A and type B investments.
c. Comment on the symmetry of each histogram.

FIGURE 2.9

Histograms for Rates of
Return

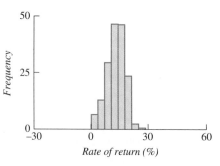

(a) Type A investments

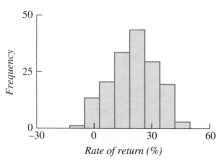

(b) Type B investments

26. The annual rates of return (as percentages) for a set of preferred stocks listed on the New York Stock Exchange are given in the accompanying schedule.

   a. How many classes should be used in grouping these data?
   b. How wide should each class be, assuming that you want the class width rounded up to the nearest dollar?
   c. Construct the frequency distribution.
   d. Draw the histogram.

−3	18	20	14	10	5	17	18	22	12
10	10	14	16	8	12	8	14	4	15
15	16	15	10	12	15	12	7	15	12
18	10	6	7	7	12	15	12	1	8
9	13	10	7	16	11	19	12	14	10
17	12	11	13	15	17	9	13	12	13
14	15	11	15						

27. A report on the profitability of stocks traded on the New York Stock Exchange, the American Stock Exchange, and the over-the-counter market listed the *best* 50 stocks for the year based on return to investors. The list is given in Table 2.10. Small but growing companies had the year's best returns.

   a. Use a computer to construct a histogram for return to investors for these stocks.
   b. Comment on an estimated value for the central tendency of returns.
   c. Comment on how spread out the values are, or the dispersion of the returns.
   d. Comment on the symmetry of the histogram for the returns.
   e. Comment on any values that are far removed from the rest.
   f. Comment on any gaps that exist in the data.
   g. Where do concentrations of the data exist?

28. As we noted in Problem 27, a report on the profitability of stocks traded on the New York Stock Exchange, the American Stock Exchange, and the over-the-counter market gave the *best* 50 stocks for a year based on return to investors (Table 2.10). Small but growing companies generally had the year's best returns. Computer Exhibit 2.3 (page 57) shows a histogram from MYSTAT of year 1 prices for these stocks.

   a. Comment on the central tendency of year 1 stock prices.
   b. Comment on how spread out the values are, or the dispersion of the prices.
   c. Comment on the symmetry of the histogram.
   d. Comment on any values that are far removed from the rest.
   e. Comment on any gaps that exist in the data.
   f. Where do concentrations of the data exist?

50 Best Stocks	Stock Price		Dividends per Share	Total Return to Investors (%)
	Year 1	Year 2		
1. Viratek (Costa Mesa, Calif.)	$10.625	$63.00	$0.00	493
2. Chambers Development (Pittsburgh)	$10.50	$16.875	$10.44	221
3. Control Resource Industries (Michigan City, Ind.)	$6.125	$18.75	$0.00	206
4. Crown Crafts (Calhoun, Ga.)	$8.25	$24.00	$0.00	191
5. Jefferson Smurfit (Alton, Ill.)	$13.625	$38.50	$0.50	188
6. COMB (Minneapolis)	$7.812	$22.25	$0.00	185
7. North American Holding (Hartford)	$5.125	$7.25	$6.38	183
8. Envirodyne (Chicago)	$8.125	$22.125	$0.00	172
9. Old National Bancorp (Spokane)	$15.568	$42.00	$0.20	171
10. Woodstream (Lititz, Pa.)	$9.00	$23.875	$0.40	171
11. Seagate Technology (Scotts Valley, Calif.)	$7.25	$19.125	$0.00	164
12. Olson Industries (Sherman Oaks, Calif.)	$11.75	$29.50	$0.00	151
13. Reebok International (Avon, Mass.)	$9.333	$23.375	$0.00	150
14. Hovnanian Enterprises (Red Bank, N.J.)	$7.444	$18.50	$0.00	149
15. Owens-Corning Fiberglas (Toledo)	$37.50	$13.75	$69.60	148
16. First Capital Holdings (Las Vegas)	$6.00	$14.875	$0.00	148
17. Nord Resources (Dayton)	$8.583	$21.25	$0.00	148
18. Circuit City Stores (Richmond)	$12.437	$30.625	$0.06	147
19. Genentech (South San Francisco)	$34.875	$85.00	$0.00	144
20. Ashton-Tate (Culver City, Calif.)	$18.375	$44.50	$0.00	142
21. Cherokee Group (North Hollywood, Calif.)	$9.50	$23.00	$0.00	142
22. Maxtor (San Jose, Calif.)	$7.187	$17.375	$0.00	142
23. HAL (Honolulu)	$9.00	$21.75	$0.00	142
24. Convenient Food Mart (Rosemont, Ill.)	$5.818	$14.00	$0.00	141
25. SPI Pharmaceuticals (Covina, Calif.)	$11.75	$28.00	$0.06	139
26. First Federal Bank FSB (Nashua, N.H.)	$17.75	$41.50	$0.40	137
27. Roper (Kankakee, Ill.)	$7.937	$18.375	$0.36	137
28. DST Systems (Kansas City, Mo.)	$11.25	$26.00	$0.20	133
29. Gap (San Bruno, Calif.)	$15.687	$35.75	$0.31	130
30. Westamerica Bancorp (San Rafael, Calif.)	$22.00	$49.00	$0.70	128
31. Bayly (Denver)	$6.00	$13.50	$0.12	127
32. Intermet (Atlanta)	$5.416	$12.125	$0.16	127
33. American Barrick Resources (Toronto)	$6.687	$15.125	$0.00	126

TABLE 2.10

Stock Data

(continues)

TABLE 2.10 Continued	50 Best Stocks	Stock Price		Dividends per Share	Total Return to Investors (%)
		Year 1	*Year 2*		
Stock Data	34. Ponderosa (Dayton)	$13.00	$28.75	$0.40	126
	35. Texas Air (Houston)	$15.00	$33.75	$0.00	125
	36. Presidential Life (Nyack, N.Y.)	$12.50	$28.00	$0.06	125
	37. Countrywide Credit (Pasadena, Calif.)	$5.637	$12.375	$0.25	124
	38. American Israeli Paper Mills (Hadera, Israel)	$7.875	$17.125	$0.30	123
	39. Riverside Group (Jacksonville, Fla.)	$5.00	$11.125	$0.00	123
	40. Duquesne Systems (Pittsburgh)	$14.375	$31.75	$0.00	121
	41. Ryan's Family Steak Houses (Greenville, S.C.)	$9.722	$21.375	$0.00	120
	42. Computer Factory (New York City)	$7.968	$17.50	$0.00	120
	43. For Better Living (San Juan Capistrano, Calif.)	$5.125	$11.00	$0.10	118
	44. Rollins Environmental Services (Wilmington, Del.)	$11.916	$25.875	$0.06	118
	45. View-Master Ideal Group (Beaverton, Ore.)	$8.333	$18.00	$0.00	116
	46. Puerto Rican Cement (San Juan)	$8.125	$17.375	$0.10	115
	47. Eastern Bancorp (Burlington, Vt.)	$11.00	$23.50	$0.10	115
	48. Micropolis (Chatsworth, Calif.)	$8.75	$18.75	$0.00	114
	49. Ero Industries (Chicago)	$6.833	$14.625	$0.00	114
	50. Laidlaw Industries (Hinsdale, Ill.)	$7.75	$16.375	$0.12	114

SOURCE: *Fortune*, February 2, 1987, p. 78. © 1987 Time Inc. All rights reserved.

29. Refer to the report on the profitability of the 50 best stocks traded on the New York Stock Exchange, the American Stock Exchange, and the over-the-counter market for a year based on return to investors given in Table 2.10. Small but growing companies generally had the year's best returns. Create a new variable, percentage annual increase in stock price, by using the following transformation: Percentage price increase = [100(Price 2 − Price 1)]/Price 1.

   a. Use a data analysis system to construct a histogram for percentage price increase for these stocks.
   b. Comment on the central tendency of percentage price increase.
   c. Comment on how spread out the values are, or the dispersion of the values.
   d. Comment on the symmetry of the histogram.
   e. Comment on any values that are far removed from the rest.
   f. Comment on any gaps that exist in the data.
   g. Where do concentrations of the data exist?

30. A report on fourth-quarter sales, profits, and profit margins for large companies listed the data in Table 2.11 (page 58). Cost-cutting moves helped some companies achieve higher profits as the dollar became stronger, but low oil prices resulted in lower prof-

COMPUTER EXHIBIT 2.3 Histogram for Stock Price Data **MYSTAT**

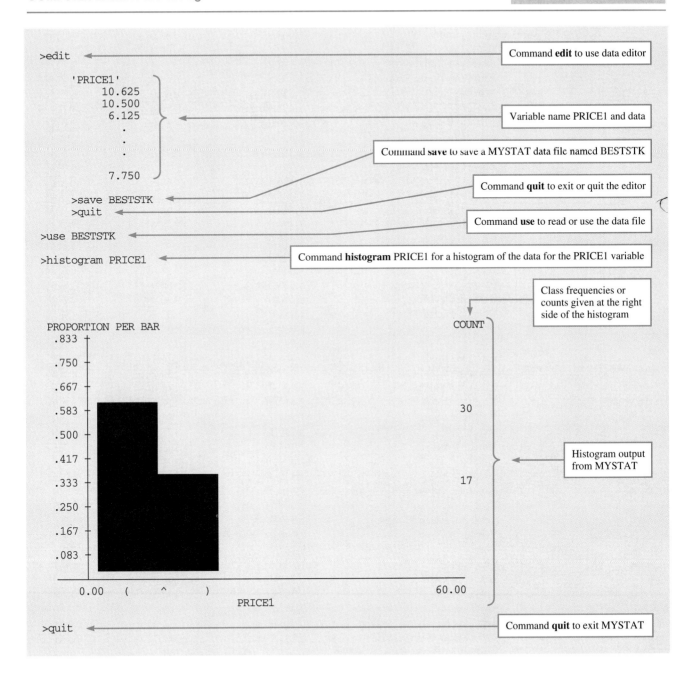

its or losses for some companies in the energy industry. A histogram for the margins of the companies with positive profits is presented in Computer Exhibit 2.4 (page 59).

a. Comment on an estimated value for the central tendency of margins.
b. Comment on how spread out the values are, or the dispersion of the margins.
c. Comment on the symmetry of the histogram for the margins.
d. Comment on any values that are far removed from the rest.
e. Comment on any gaps that exist in the data.
f. Where do concentrations of the data exist?

TABLE 2.11	*Company*	*Sales (in $millions)*	*Profits (in $millions)*	*Margin (%)**
Company Data	Abbott Laboratories	$1,063.2	$160.6	15.1
	Alcoa	1,160.0	165.3	14.3
	American Brands	2,197.2	117.4	5.3
	American Home Products	1,192.8	199.4	16.7
	Amoco	4,839.0	165.0	3.4
	AMR	1,498.6	6.6	0.4
	Atlantic Richfield	3,724.0	64.0	1.7
	Bellsouth	2,800.0	361.1	12.9
	Boeing	4,982.0	191.0	3.8
	Boise Cascade	969.3	32.0	3.3
	Borden	1,356.4	68.8	5.1
	Bristol-Myers	1,205.0	151.7	12.6
	Burlington Northern	1,609.8	75.6	4.7
	Caterpillar	1,775.0	−148.0	NM
	Champion International	1,058.0	64.5	6.1
	Chevron	5,900.0	−86.0	NM
	CSX	1,604.0	132.0	8.2
	Exxon	18,836.0	1,480.0	7.9
	General Electric	12,270.0	730.0	6.0
	Georgia-Pacific	1,781.0	83.0	4.7
	Hercules	640.5	38.9	6.1
	IBM	16,945.0	1,390.0	8.2
	Ingersoll-Rand	768.4	41.1	5.3
	Inland Steel Industries	817.5	39.7	4.9
	International Paper	1,900.0	113.0	5.9
	Kraft	2,414.5	117.1	4.9
	Martin Marietta	1,268.5	43.4	3.4
	Merck	1,149.1	171.8	15.0
	Merrill Lynch	2,764.8	197.6	7.1
	Mobil	12,009.0	201.0	1.7
	Monsanto	1,570.0	23.0	1.5
	Motorola	1,619.0	63.0	3.9
	NCR	1,575.1	134.1	8.5
	North American Philips	1,356.0	47.6	3.5
	Nynex	2,900.0	293.7	10.1
	Pfizer	1,162.9	158.4	13.6
	PPG Industries	1,200.8	72.9	6.1
	Procter & Gamble	4,255.0	190.0	4.5
	Quaker Oats	1,083.6	30.1	2.8
	Raytheon	2,041.2	99.6	4.9
	Reynolds Metals	863.9	−4.3	NM
	Santa Fe Southern Pacific	1,503.9	−384.7	NM
	Smithkline Beckman	1,043.3	147.4	14.1
	Standard Oil	2,264.0	51.0	2.7
	Tandy	1,196.0	103.8	8.7
	Teledyne	834.8	47.0	5.6
	Texas Instruments	1,335.8	26.6	2.0
	3M	2,112.0	185.0	8.8
	United Technologies	4,401.4	−252.6	NM
	USG	695.0	56.7	8.2
	USX	2,770.0	−1,167.0	NM
	Warner-Lambert	812.6	56.5	7.0

Company	Sales (in $millions)	Profits (in $millions)	Margin (%)*	**TABLE 2.11** Continued
				Company Data
Westinghouse Electric	2,869.1	203.2	7.1	
Weyerhaeuser	1,450.6	96.1	6.6	
Whirlpool	942.9	42.2	4.5	

*NM = not meaningful

SOURCE: *Business Week,* February 9, 1987, p. 31. Data from Standard & Poor's Compustat Services, Inc.

COMPUTER EXHIBIT 2.4 Histogram of the Profit Margins of 49 Companies   **SPSS/PC**

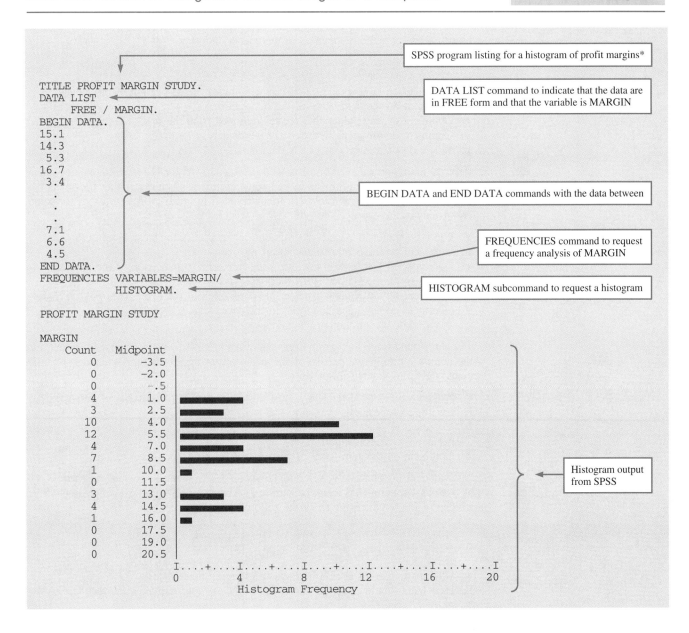

SPSS program listing for a histogram of profit margins*

```
TITLE PROFIT MARGIN STUDY.
DATA LIST
      FREE / MARGIN.
BEGIN DATA.
15.1
14.3
 5.3
16.7
 3.4
  .
  .
  .
 7.1
 6.6
 4.5
END DATA.
FREQUENCIES VARIABLES=MARGIN/
        HISTOGRAM.
```

DATA LIST command to indicate that the data are in FREE form and that the variable is MARGIN

BEGIN DATA and END DATA commands with the data between

FREQUENCIES command to request a frequency analysis of MARGIN

HISTOGRAM subcommand to request a histogram

```
PROFIT MARGIN STUDY

MARGIN
   Count   Midpoint
      0     -3.5
      0     -2.0
      0      -.5
      4      1.0
      3      2.5
     10      4.0
     12      5.5
      4      7.0
      7      8.5
      1     10.0
      0     11.5
      3     13.0
      4     14.5
      1     16.0
      0     17.5
      0     19.0
      0     20.5
         I....+....I....+....I....+....I....+....I....+....I
         0        4        8       12       16       20
                   Histogram Frequency
```

Histogram output from SPSS

31. Refer to the report on the profitability of the 50 best stocks traded on the New York Stock Exchange, the American Stock Exchange, and the over-the-counter market for a year based on return to investors. The list is given in Table 2.10. Small but growing companies generally had the year's best returns.

   a. Use a computer to construct a stem-and-leaf display of dividends per share for these stocks.
   b. Comment on the central tendency of dividends.
   c. Comment on how spread out the values are, or the dispersion of the values.
   d. Comment on the symmetry of the data.
   e. Comment on any values that are far removed from the rest.
   f. Comment on any gaps that exist in the data.
   g. Where do concentrations of the data exist?

32. Refer to the report on fourth-quarter sales, profits, and profit margins for large companies given in Table 2.11. Cost-cutting moves helped some companies achieve higher profits as the dollar became stronger, but low oil prices resulted in lower profits or losses for some companies in the energy industry.

   a. Use a computer to construct a histogram for sales for these companies.
   b. Comment on the central tendency of sales.
   c. Comment on how spread out the values are, or the dispersion of the values.
   d. Comment on the symmetry of the histogram.
   e. Comment on any values that are far removed from the rest.
   f. Comment on any gaps that exist in the data.
   g. Where do concentrations of the data exist?

33. Refer to the report on sales, profits, and profit margins for large companies given in Table 2.11. Cost-cutting moves helped some companies achieve higher profits as the dollar became stronger, but low oil prices resulted in lower profits or losses for some companies in the energy industry.

   a. Use a computer to construct a histogram for profits for these large companies.
   b. Comment on the central tendency of profits.
   c. Comment on how spread out the values are, or the dispersion of the values.
   d. Comment on the symmetry of the histogram.
   e. Comment on any values that are far removed from the rest.
   f. Comment on any gaps that exist in the data.
   g. Where do concentrations of the data exist?

34. The three percentage histograms presented in Figure 2.10 were created from data for the number of surgeries per day done in the operating rooms of Associates, Beneficient, and Consolidated hospitals.

   a. Estimate the central tendency, or a middle value, of the number of surgeries per day for each hospital.
   b. Comment on the day-to-day dispersion, or variation, in the number of surgeries per day for each hospital.
   c. Comment on the symmetry or asymmetry for each histogram.

35. The value of the Japanese yen in dollars has tended to increase over the past few years as Japan's economy has become stronger. The yen-per-dollar exchange rate for a sample of 20 months is given in the following listing:

   129   114   139   141   137   144   123   134   132   105
   118   134   140   139   129   124   131   120   142   137

   a. How many classes should be used in grouping these data?
   b. How wide should each class be if you round the approximate class width up to the nearest integer?

Histograms for Numbers of Surgeries per Day at Three Hospitals   **FIGURE 2.10**

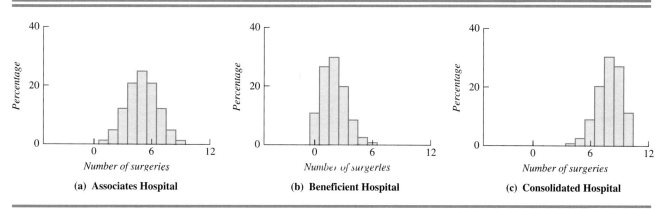

(a) **Associates Hospital**   (b) **Beneficient Hospital**   (c) **Consolidated Hospital**

    c. Construct the histogram.
    d. Comment on the central tendency of the data.
    e. Comment on how spread out the values are, or the dispersion of the values.
    f. Comment on the symmetry of the histogram.
    g. Explain how the histogram could be used by an investor for making investment decisions.

36. As we noted in Problem 35, the value of the Japanese yen in dollars has tended to increase over the past few years as Japan's economy has become stronger. The yen-per-dollar exchange rate for a sample of 20 months is repeated in the following listing:

| 129 | 114 | 139 | 141 | 137 | 144 | 123 | 134 | 132 | 105 |
| 118 | 134 | 140 | 139 | 129 | 124 | 131 | 120 | 142 | 137 |

    a. Construct a stem-and-leaf diagram.
    b. Comment on the central tendency of the data.
    c. Comment on how spread out the values are, or the dispersion of the values.

37. Investors in long-term bonds have tended to incur devaluations in their portfolios in the past few years. As corporations have been taken over or have undergone restructuring, the level of debt in corporate capital structures has been increased. The increased debt increases the risk that the companies will default on their long-term debt obligations, and the increased risk drives long-term bond values down. A measure of the risk from events such as mergers, leveraged buy-outs, or recapitalizations is the spread between the yield of long-term bonds and the yield of essentially riskless Treasury bonds. The spread is measured in basis points (100 basis points are equivalent to one percentage point). The spreads for a sample of 20 new bonds are given in the following listing:

| 98 | 85 | 106 | 96 | 80 | 90 | 93 | 97 | 98 | 102 |
| 93 | 96 | 110 | 111 | 86 | 100 | 109 | 118 | 88 | 121 |

    a. How many classes should be used in grouping these data in a frequency distribution?
    b. How wide should each class be if you round the approximate class width up to the nearest integer?
    c. Construct the histogram.
    d. Comment on the central tendency of the data.
    e. Comment on how spread out the values are, or the dispersion of the values.
    f. Comment on the symmetry of the histogram.
    g. Explain how the histogram could be used by an investor for making investment decisions.

38. Refer to Problem 37, which describes how investors in long-term bonds have tended to incur devaluations in their portfolios in the past few years. A measure of the risk to investors in bonds, from events such as mergers, leveraged buy-outs, or recapitalizations, is the spread between the yield of long-term bonds and the yield of essentially riskless Treasury bonds. The spread is measured in basis points (100 basis points are equivalent to one percentage point). The spreads for a sample of 20 new bonds are repeated in the following listing:

| 98 | 85 | 106 | 96 | 80 | 90 | 93 | 97 | 98 | 102 |
| 93 | 96 | 110 | 111 | 86 | 100 | 109 | 118 | 88 | 121 |

   a. Construct a stem-and-leaf diagram.
   b. Comment on the central tendency of the data.
   c. Comment on how spread out the values are, or the dispersion of the values.

39. Recently, the most expensive securities markets internationally have been stock exchanges in Japan. The price-earnings ratio is an indicator of the cost of a stock in terms of how much investors have been willing to pay per dollar of income statement profits. For instance, the price-earnings ratios for stocks traded on U.S. securities markets have tended to vary around a central value of about 13 in recent times. The price-earnings ratios for a sample of 35 stocks traded on a Japanese securities market are given in the accompanying listing.

   a. How many classes should be used in grouping these data in a frequency distribution?
   b. How wide should each class be if you round the approximate class width up to the nearest integer?
   c. Construct the histogram.
   d. Comment on the central tendency of the data.
   e. Comment on how spread out the values are, or the dispersion of the values.

65	61	68	64	60	65	61	67	60
55	66	59	69	61	59	58	59	65
61	62	56	53	54	61	54	54	62
62	54	63	64	60	61	60	63	

40. The most costly securities markets internationally in recent times have been Japanese stock exchanges, as described in Problem 39. The price-earnings ratio is an indicator of a stock's cost in terms of how much investors have been willing to pay per dollar of income statement profits. The price-earnings ratios for a sample of 35 stocks traded on a Japanese securities market are repeated in the accompanying listing.

   a. Construct a stem-and-leaf diagram.
   b. Comment on the central tendency of the data.
   c. Comment on how spread out the values are, or the dispersion of the values.

65	61	68	64	60	65	61	67	60
55	66	59	69	61	59	58	59	65
61	62	56	53	54	61	54	54	62
62	54	63	64	60	61	60	63	

41. Two histograms are shown in Figure 2.11. One histogram is for the amount of time spent waiting on hold on the telephone before receiving service at National Direct Mail Marketers; the other is for birth weights of infants whose parents buy baby clothing at Blueberry Baby Clothes, Inc.

   a. Is the histogram for birth weight shown in panel (a) or (b)? Explain.
   b. Is the histogram for waiting time shown in panel (a) or (b)? Explain.

Histograms for Birth Weight and Waiting Time **FIGURE 2.11**

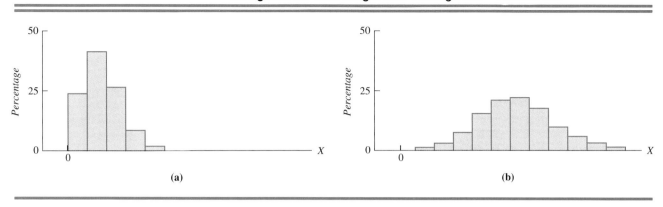

(a)   (b)

42. A financial analyst with First Metro Bancshares wishes to examine the yields on 90-day Treasury bills. The information will be used to ascertain the pattern of short-term interest rates. The yield data were collected monthly for two years and the results are presented in order (left to right from first row to last row) in the accompanying listing.

   a. Construct a histogram for the data.
   b. Comment on the central tendency for the data.
   c. Construct a time series plot for the data.
   d. Would the central tendency for the yields be a good approximation for the short-term interest rate in the month following the data collection period?

7.76	8.22	8.57	8.00	7.56	7.01	7.05	7.18
7.08	7.17	7.20	7.07	7.04	7.03	6.59	6.06
6.12	6.21	5.84	5.57	5.19	5.18	5.35	5.49

SOURCE: U.S. Department of Labor, Bureau of Labor Statistics (1986), *Business Statistics, 1986* (Washington, DC.: U.S. Government Printing Office).

43. An economist with First Metro Bancshares wishes to examine the discount rate at the New York Federal Reserve Bank. The information will be used to ascertain the pattern of interest rates for banks borrowing funds from the Federal Reserve. The discount data were collected monthly for two years and the results are presented in order (left to right from first row to last row) in the accompanying listing.

   a. Construct a histogram for the data.
   b. Comment on the central tendency for the data.
   c. Construct a time series plot for the data.
   d. Would the central tendency for the discounts be a good approximation for the interest rate for bank borrowing one month after the data were collected?

8.00	8.00	8.00	8.00	7.81	7.50	7.50	7.50
7.50	7.50	7.50	7.50	7.50	7.50	7.10	6.83
6.50	6.50	6.16	5.82	5.50	5.50	5.50	5.50

SOURCE: U.S. Department of Labor, Bureau of Labor Statistics (1986), *Business Statistics, 1986* (Washington, D.C.: U.S. Government Printing Office).

44. A bond trader with Stephen, James, and Roberts Investment Bankers wishes to examine the total amount of the gross federal debt. Data on the gross federal debt are given in Table 2.12. The bond trader is concerned that increasing federal debt will influence long-term bond yields, thereby increasing the riskiness of the firm's bond portfolio.

TABLE 2.12	Year	Total	Federal Government Account	The Public	Federal Reserve System	As Percent of GNP
			Held by—			
Gross Federal Debt	1965	322.3	61.5	260.8	39.1	47.9
(in billions of dollars)	1966	328.5	64.8	263.7	42.2	44.5
	1967	340.4	73.8	266.6	46.7	42.8
	1968	368.7	79.1	289.5	52.2	43.4
	1969	365.8	87.7	278.1	54.1	39.4
	1970	380.9	97.7	283.2	57.7	38.5
	1971	408.2	105.1	303.0	65.5	38.7
	1972	435.9	113.6	322.4	71.4	37.8
	1973	466.3	125.4	340.9	75.2	36.4
	1974	483.9	140.2	343.7	60.6	34.2
	1975	541.9	147.2	394.7	85.0	35.6
	1976	629.0	151.6	477.4	94.7	37.0
	1977	706.4	157.3	549.1	105.0	36.5
	1978	776.6	169.5	607.1	115.5	35.8
	1979	828.9	189.2	639.8	115.6	33.9
	1980	908.5	199.2	709.3	120.8	34.0
	1981	994.3	209.5	784.8	124.5	33.3
	1982	1,136.8	217.6	919.2	134.5	36.2
	1983	1,371.2	240.1	1,131.0	155.5	41.3
	1984	1,564.1	264.3	1,300.0	155.1	42.4
	1985	1,817.0	317.6	1,499.4	169.8	46.0
	1986	2,120.1	383.9	1,736.2	190.9	50.7
	1987	2,345.6	457.4	1,888.1	212.0	52.9
	1988	2,600.8	550.5	2,050.2	229.2	54.3
	1989	2,866.2	676.9	2,189.2	220.1	55.6
	1990	3,113.3	814.6	2,298.7	(NA)	56.7

NA = not available; GNP = gross national product

Source: U.S. Bureau of the Census (1991), *Statistical Abstract of the United States, 1991,* 111th ed. (Washington, D.C.: U.S. Government Printing Office).

a. Construct a histogram for the total amount of the gross federal debt data.

b. Comment on the central tendency of the data.

c. Construct a time series plot for the data. A time series plot is a plot of data values for a variable over time.

d. Would the central tendency of the histogram be a good approximation of the total amount of the gross federal debt for the year following the period of the data collection?

 45. An economist with the Federal Reserve Bank of Cleveland wishes to examine the total amount of national income. Data on national income are given in Table 2.13. The economist is concerned that the recession will result in a decrease in national income.

a. Construct a histogram for the total amount of national income data.

b. Comment on the central tendency of the data.

c. Construct a time series plot for the data. A time series plot is a plot of data values for a variable over time.

d. Would the central tendency of the histogram be a good approximation of the total amount of national income for the year following the period of the data collection?

		Compensation of Employees			
Year	National Income	Total	Wages and Salaries	Supplements to Wages and Salaries	**TABLE 2.13**
1959	410.1	281.2	259.8	21.4	National Income (in billions
1960	425.7	296.7	272.8	23.8	of dollars)
1961	440.5	305.6	280.5	25.1	
1962	474.5	327.4	299.3	28.1	
1963	501.5	345.5	314.8	30.7	
1964	539.1	371.0	337.7	33.2	
1965	586.9	399.8	363.7	36.1	
1966	643.7	443.0	400.3	42.7	
1967	679.9	475.5	428.9	46.6	
1968	741.0	524.7	471.9	52.8	
1969	798.6	578.4	518.3	60.1	
1970	833.5	618.3	551.5	66.8	
1971	899.5	659.4	584.5	74.9	
1972	992.9	726.2	638.7	87.6	
1973	1,119.5	812.8	708.6	104.2	
1974	1,198.8	891.3	772.2	119.1	
1975	1,285.3	948.7	814.7	134.0	
1976	1,435.5	1,058.3	899.6	158.7	
1977	1,609.1	1,177.3	994.0	183.3	
1978	1,829.8	1,333.0	1,120.9	212.1	
1979	2,038.9	1,496.4	1,255.3	241.1	
1980	2,198.2	1,644.4	1,376.6	267.8	
1981	2,432.5	1,815.5	1,515.6	299.8	
1982	2,522.5	1,916.0	1,593.3	322.7	
1983	2,720.8	2,029.4	1,684.2	345.2	
1984	3,058.3	2,226.9	1,850.0	376.9	
1985	3,268.4	2,382.8	1,986.3	396.5	
1986	3,437.9	2,523.8	2,105.4	418.4	
1987	3,692.3	2,698.7	2,261.2	437.4	
1988	4,002.6	2,921.3	2,443.0	478.3	
1989	4,244.7	3,101.3	2,585.8	515.5	
1990	4,459.6	3,290.3	2,738.9	551.4	

SOURCE: U.S. Bureau of Economic Analysis (1991), *Survey of Current Business, November* (Washington, D.C.: U.S. Government Printing Office).

Refer to the 141 companies listed in the Ratio Data Set in Appendix A.

Ratio Data Set

Questions

46. Locate the data for the current assets–sales ratio.

   a. Use a computer to obtain a histogram for this variable.
   b. How would this summary of the data be used for decision making?

47. Locate the data for the current assets–sales ratio.

   a. Use a computer to obtain a stem-and-leaf diagram for this variable.
   b. How would this summary of the data be used for decision making?

48. Locate the data for the current assets–current liabilities ratio.

   a. Use a computer to obtain a histogram for this variable.
   b. How would this summary of the data be used for decision making?

49. Locate the data for the current assets–current liabilities ratio.

    a. Use a computer to obtain a stem-and-leaf diagram for this variable.
    b. How would this summary of the data be used for decision making?

50. Locate the data for the net income–total assets ratio.

    a. Use a computer to obtain a histogram for this variable.
    b. How would this summary of the data be used for decision making?

51. Locate the data for the net income–total assets ratio.

    a. Use a computer to obtain a stem-and-leaf diagram for this variable.
    b. How would this summary of the data be used for decision making?

Credit Data Set   *Refer to the 113 applicants for credit listed in the Credit Data Set in Appendix A.*

Questions   52. If Equation (2.1) were used to construct a frequency distribution of JOBINC, how many classes would be used in the following distributions?

    a. All the applicants.
    b. All applicants who were granted credit.
    c. All applicants who were denied credit.

53. Use a data analysis system to construct a histogram for JOBINC of the applicants who were granted credit. Ignore the 6 applicants who did not list an income, and simply add a note to your histogram concerning them.

54. A three-dimensional histogram can be constructed by counting the number of items that fall in classes for two variables. A three-dimensional histogram of the CLASS and SEX variables from the credit data set in the data base appendix was constructed by using the Execustat package and the result is shown in Figure 2.12. What applicants for credit does the higher bar represent?

FIGURE 2.12

Three-Dimensional Histogram of the CLASS and SEX Variables for the Applicants for Credit

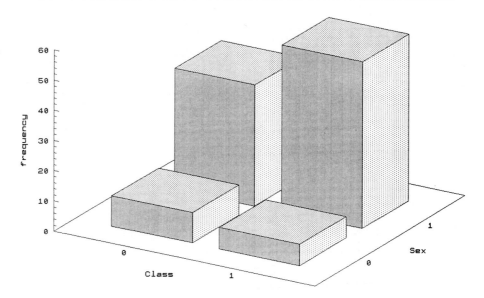

Note: Class = 1 indicates credit was granted; Class = 0 indicates credit was not granted; Sex = 1 indicates the applicant was male; Sex = 0 indicates the applicant was female.

USING HISTOGRAMS TO DEPICT INVESTMENT RISKS OF STOCKS AND BONDS

Variation in the expected future returns of an investment in a financial security is often considered to be a measure of the risk of the investment. The lower the variation in returns, the lower is the riskiness of the investment; and the higher the variation in returns or the more volatile the returns are, the higher is the riskiness of the investment.

The two histograms in Figure 1 were constructed for the rates of return obtained from 200 long-term corporate bonds and from 200 stocks. A rate of return can be considered to be a measure of profitability for an investment. The histograms show clearly that the variation in return for the stocks is greater than the variation in return for the bonds.

The riskiness of investments in stocks is generally considered to be greater than the riskiness of investments in bonds. Investors often try to minimize risk by selecting portfolios of stocks and bonds that are likely to reduce the variation in returns for the portfolio. Diversification is one strategy used by investors seeking to reduce their risk.

FIGURE 1

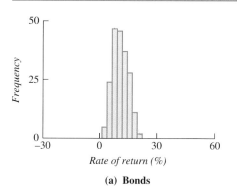
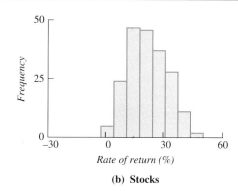

Histograms for Returns of Bonds and Stocks

(a) Bonds

(b) Stocks

a. Estimate the central tendency for the rates of return of the bonds and of the stocks shown in the histograms.

b. Comment on the dispersion or variation in rates of return for investments in bonds and investments in stocks.

c. Comment on the symmetry for each histogram.

Case Assignment

REFERENCE: Brigham E.F. (1988), *Financial Management Theory and Practice*, 4th ed. (New York: Dryden).

SUMMARIZING AGE DISTRIBUTIONS AT DIAGNOSIS FOR DISEASES OF THE BREAST

The cancer epidemic in America has dramatically influenced life and health insurers, health care management, health care providers, and consumers. The American Cancer Society now tells the public that one out of every nine American women will develop cancer of the breast during her lifetime. However, many abnormalities or lumps, such as cysts and fibroadenomas, that are noncancerous or benign can also develop.

The histograms in Figure 1 show the age distributions for women diagnosed with cysts, fibroadenomas, and cancer. However, be aware that the distributions may be chang-

FIGURE 1

Histograms of Age Distributions at Diagnosis for Diseases of the Breast

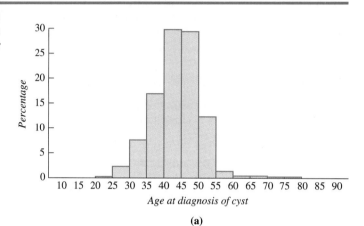

(a)

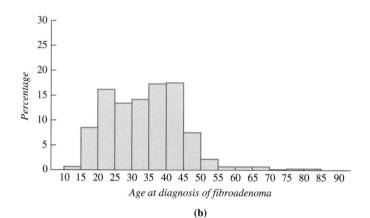

(b)

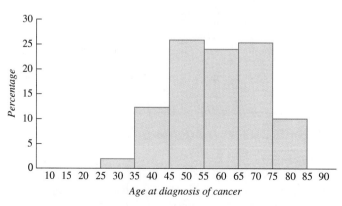

(c)

ing over time. The most common lumps are benign cysts. Figure 1(a) shows that cysts can occur at any age after adolescence, but they are most common during the later reproductive years and during menopause. Another benign condition is known as fibroadenoma. Fibroadenomas are firm, smooth, and freely movable masses. Figure 1(b) shows that fibroadenomas predominate in younger women after adolescence, can occur during reproductive years, and predominate again around menopause, after which the incidence declines. The primary sign of cancer is a principal mass with ill-defined margins. Figure 1(c) shows that cancer tends to occur at two peaks in the age distribution. Around the onset of menopause in the 45- to 49-year-old group, and after 60 years of age, possibly due to a postmenopausal hormonal stimulus. The histograms allow us to summarize and effectively present information on the age distributions for these diseases.

Case Assignment

  a. Based on the histograms, if an adolescent or young adult woman below the age of 25 has one of these conditions, which condition would you suspect?
  b. If a woman over 60 years of age has one of these conditions, which condition would you suspect?

REFERENCES: DeWaard, F., E. Baanders-Van Halewijn, and J. Huizinga (1964), *Cancer*, 17:141–51; and Haagensen, C.D. (1986), *Diseases of the Breast* (Philadelphia: W.B. Saunders Co.)

EVALUATING THE DISTRIBUTION OF ADDITIONAL COLLECTIONS AFTER A TAX AUDIT

CASE 2.3

Patrick J. Flagle founded Flagle Family Enterprises several years ago using money he had borrowed from his father and four of his six brothers. Despite the fact that Flagle paid back the initial loans from the family after only three years of operation, he continued to carry "family" as part of the company's name since he felt he owed much of the business' current success to these initial loans. The Flagle Family Enterprises owned several distributorships in the eastern states. Each distributorship was incorporated in the state in which it operated.

Flagle always believed in using to his advantage any information that might come his way. Thus he was particularly interested in a newspaper article that appeared in the local paper of the city where his largest distributorship is located.

The newspaper article was written by a reporter who had interviewed several past and current employees of the state tax commission in an effort to determine how tax returns are selected for audit and how much additional taxes are collected from those who are audited. In the article, the reporter gave several hints on how individuals might avoid audits of their personal returns. Flagle felt sure he could use information on the additional collections to help determine how aggressive he should be on deducting expenses on his return.

Regarding the collection of taxes, the reporter wrote the following:

It was not difficult to get the Tax Commission to show me data on the taxes they recover through their audits. They are very proud of the fact that this part of their agency is a "money maker." That is, they bring in more in additional tax assessments than the costs of the audits. For instance, their records show additional taxes collected through audits last year ran as follows:

Distribution of Tax Audit Results

Additional Collections		Returns
−$10,000.00 to	−$5,000.00	5
−4,999.99 to	−2,000.00	6
−1,999.99 to	−1,000.00	7
−999.99 to	Zero	90
.01 to	1,000.00	250
1,000.01 to	2,000.00	121
2,000.01 to	5,000.00	9
5,000.01 to	10,000.00	5
10,000.01 to	15,000.00	5
15,000.01 to	20,000.00	5

The negative figures in the table indicate that some lucky people who got audited received refunds.

Case Assignment

a. Construct a histogram for the "Additional Collection" data. Notice that the class intervals are not all equal.

b. How worried should Flagle be about additional collections if his tax return is audited?

REFERENCES

BMDP Statistical Software. 1985. *BMDP Statistical Software Manual.* Los Angeles: University of California Press.

Hoaglin, David C., Frederick Mosteller, and John W. Tukey, 1983. *Understanding Robust and Exploratory Data Analysis.* New York: Wiley.

Huff, Darell. 1965. *How to Lie with Statistics.* New York: Norton.

Miller, R. 1988. *MINITAB Handbook for Business and Economics.* Boston: Duxbury Press.

Ryan, T., B. Joiner, and B. Ryan. 1985. *MINITAB Handbook.* Boston: Duxbury Press.

SAS Institute, Inc. 1985. *SAS User's Guide: Basics, Version 5 Edition.* Cary, N.C.: SAS Institute, Inc.

Schmid, Calfin F. 1983. *Statistical Graphics.* New York: Wiley.

SPSS, Inc. 1986. *SPSS/PC+.* Chicago: SPSS, Inc.

Sturges, H.A. 1926. "The Choice of a Class Interval." *Journal of the American Statistical Association,* 21 (March): 65–66.

SYSTAT, Inc. 1990. *Business MYSTAT Manual.* Evanston, Ill.: SYSTAT, Inc.

Terrel, G.R., and D.W. Scott. 1985. "Oversmoothed Nonparametric Density Estimates." *Journal of the American Statistical Association,* 80 (March): 209–14.

Descriptive Measures

3

In the previous chapter we saw how tabular and graphical forms of presentation may be used to summarize and describe quantitative data. Although these techniques help us to sort out important features of the distribution of the data, statistical methods for the most part require *numerical* descriptions. These numerical descriptions are arrived at through arithmetic operations on the data, which yield *descriptive measures or descriptive statistics.* The basic descriptive statistics are the measures of central tendency or location and the measures of dispersion or scatter.

The mean, weighted mean, median, and mode are presented in this chapter as measures of central tendency. The range, mean deviation, variance, standard deviation, and coefficient of variation are presented as measures of dispersion. We cover what the standard deviation tells us in detail because it is an important measure in many of the remaining chapters of the text.

Descriptive measures are often determined with the aid of a data analysis system, and the results are used in exploratory data analyses and in written reports of statistical analyses. *Box plots* are presented as one method for exploring data and for graphically displaying descriptive measures.

3.1
SYMBOLS

If our data consist of measurements of some characteristic of a number of individuals or items—such as the annual incomes of some group of persons, the weights of a number of packages of cookies, or the weekly sales of several stores—we designate the variable by some letter or symbol, say, X. If we have measured two or more characteristics, we use different letters or symbols for each variable. If in addition to obtaining annual income, we also record the age of each person interviewed, we can represent income by X and age by Y. A third characteristic, such as educational level, we can represent by the letter Z, and so on.

To differentiate between the same kind of measurements made on different items or individuals, or between similar repeated measurements made on the same item or individual, we add a subscript to the corresponding symbol; thus X_1 stands for the income of the first person interviewed, X_2 for that of the second, X_{23} for that of the twenty-third, and so on.

In general, any arbitrary observed value is represented by X_i, where the subscript i is variable in the sense that it represents any one of the observed items and need only be replaced by the proper number in order to specify a particular observation. The income, age, and educational level of the ith, or general, individual are represented by X_i, Y_i, and Z_i, respectively.

3.2
MEASURES OF LOCATION

When we work with numerical data and their frequency distributions, we soon see that in most sets of data there is a tendency for the observed values to group themselves about some interior value; some central value seems to be characteristic of the data. This phenomenon, referred to as *central tendency,* may be used to describe the data in the sense that the central value locates the "middle" of the distribution. The statistics we calculate for this purpose are **measures of location,** also called **measures of central tendency.**

For a given set of data, the measure of location we use depends on what we mean by *middle.* Different definitions give rise to different measures. In the following sections we consider four such measures and their interpretations:

1. Mean
2. Weighted mean
3. Median
4. Mode

3.3
MEAN

All of us are familiar with the concept of the mean, or average value. We read and speak of batting averages, grade point averages, mean annual rainfall, the average weight of a boxcar of coal, and the like. In most cases the term *average* used in connection with a set of numbers refers to their *arithmetic mean.* For the sake of simplicity, we call it the **mean.**

For a given *population* of N values, $X_1, X_2, X_3, \ldots, X_N$, the **population mean** is denoted by μ (lowercase Greek mu) and defined as follows.

Definition and Equation for Population Mean μ

The mean for a population of N data values, $X_1, X_2, \ldots, X_N$, is their sum divided by N, or

$$\mu = \frac{1}{N} \sum_{i=1}^{N} X_i \qquad\qquad (3.1)$$

The symbol Σ (uppercase Greek sigma) is the summation symbol* and means

$$\sum_{i=1}^{N} X_i = X_1 + X_2 + X_3 + \cdots + X_N$$

EXAMPLE 3.1 If we receive three carloads of coal and their weights are 10, 13, and 16 tons, their mean weight is

$$\mu = \frac{1}{3}\left(10 + 13 + 16\right) = \frac{1}{3}(39) = 13 \text{ tons}$$

When we wish to find the mean of a set of numbers, and when those numbers form the population in which we are interested, then we use Equation (3.1) and call our mean μ. However, at times the numbers we have at our disposal are sample values, which we are using to estimate the mean of a larger population. For instance, suppose we are interested in estimating the mean number of employees absent per day for all days by using the *sample* of the number of employees absent each day for 106 days (see Table 2.1). Then we perform the same type of calculation as the one in Equation (3.1), but we divide by the sample size n (as opposed to the population size N), and we call our resulting **sample mean** $\overline{X}$.

Definition and Equation for Sample Mean $\overline{X}$

The mean for a sample of n data values, $X_1, X_2, \ldots, X_n$, is their sum divided by n, or

$$\overline{X} = \frac{1}{n} \sum_{i=1}^{n} X_i \qquad\qquad (3.1a)$$

---

*Discussion of descriptive measures requires some familiarity with the mathematical shorthand used to express them. Since we are concerned with large data sets and since the operation of addition plays a large role in our calculations, we need a way to express sums in compact and simple form. The summation notation meets this requirement. Appendix B, "Summation Notation," is included in the back of the text for students needing a review of summation notation.

FIGURE 3.1

Mean as a Center of
Gravity

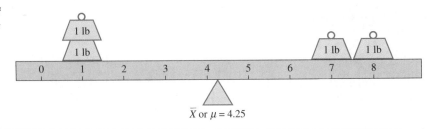

$\overline{X}$ or $\mu = 4.25$

EXAMPLE 3.2 The record of employee absences at a Western Electronics plant for 106 days in Table 2.1 has a sum equal to 15,005, and the mean number of employees absent per workday is

$$\overline{X} = \frac{1}{106}(146 + 141 + 139 + \cdots + 142) = \frac{1}{106}(15,005)$$

$$= 141.6 \text{ employees}$$

Notice that the unit for the mean is the same as the unit for the observations themselves (tons, employees, dollars, and so on). Notice also that even though the observations may be whole numbers (such as employees), the mean may have fractional or decimal parts (such as 141.6 employees).

When someone reviewing our work sees that our mean is denoted by $\overline{X}$, they know that the numbers we used to calculate the mean came from a sample. However, if they see that the mean is denoted by μ, they know that our mean was figured from numbers that form the entire population whose mean we seek. It is only the context of the situation or the problem statement that will allow the reader to determine if he or she is dealing with a sample and should call the mean $\overline{X}$ (or $\overline{Y}$ if we are calling our variable Y) or with a population and should call the mean μ.

The mean of a set of numbers is their "middle" in the sense that it is their center of gravity. Suppose we have four observations, 1, 8, 7, and 1, with a mean of $(1 + 8 + 7 + 1)/4$, or 4.25. Imagine a seesaw with a scale marked off along its edge, and imagine, for each of the four observations, a 1-pound weight positioned according to this scale, as shown in Figure 3.1. The mean 4.25 is the point at which the fulcrum of the seesaw must be placed in order to make it balance.

3.4
WEIGHTED MEAN

When we compute the simple arithmetic mean of a set of numbers, we assume that all the observed values are of equal importance, and we give them equal weight in our calculations. In situations where the numbers are not equally important, we can assign to each a weight that is proportional to its relative importance and calculate the **weighted mean.**

Definition and Equation for Weighted Mean

Let $X_1, X_2, X_3, \ldots, X_N$ be a set of N data values, and let $w_1, w_2, w_3, \ldots, w_N$ be the weights assigned to them. The **weighted mean** is found by dividing the sum of the products of the values and their weights by the sum of the weights:

$$\mu = \frac{w_1 X_1 + w_2 X_2 + w_3 X_3 + \cdots + w_N X_N}{w_1 + w_2 + w_3 + \cdots + w_N} \qquad (3.2)$$

or in summation notation

$$\mu = \frac{\sum w_i X_i}{\sum w_i} \qquad (3.2a)$$

In this case the summation symbol Σ is used without designating its index i or the values for the index. Implied here is that the summation is for all values of $w_i X_i$ in the numerator (or w_i in the denominator). Naturally, if the X values come from a sample, we denote the sample mean as $\overline{X}$ rather than μ and we use n rather than N.

If we compute the weighted mean for the classes from a frequency distribution by letting the value for each class be the class mark or midpoint and the weights for each class be the class frequency, then the mean for the distribution is its point of balance. If we draw a frequency distribution's histogram on some stiff material of uniform density, such as plywood or metal, cut it out, and balance it on a knife-edge arranged perpendicular to the horizontal scale, as shown in Figure 3.2, the point of balance is the sample mean $\overline{X}$.

Most students are familiar with the concept of the weighted mean, for the grade point average is such a measure. It is the mean of the numerical values of the letter grades weighted by the numbers of credit hours in which the various grades are earned.

FIGURE 3.2

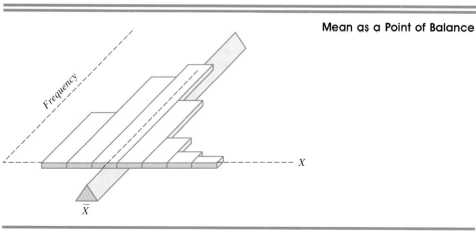

Mean as a Point of Balance

▄▄ **EXAMPLE 3.3**   If a student gets A's in two 3-credit courses, a B in a 5-credit course, a C in a 4-credit course, and a D in a 2-credit course, and if the numerical values of the letter grades are A = 4, B = 3, C = 2, and D = 1, the student's grade point average for the term is

$$\mu = \frac{(3)(4) + (3)(4) + (5)(3) + (4)(2) + (2)(1)}{3 + 3 + 5 + 4 + 2} = \frac{49}{17} = 2.882 \quad ▄▄$$

Failure to weight the values when one is combining data is a common error.

▄▄ **EXAMPLE 3.4**   A chemical compound produced by Federated Cyan Company is made up of two ingredients, chemical A and chemical B. If chemical A costs $5 per gallon and chemical B costs $10 per gallon, one might be tempted to say that the average cost of ingredients for the compound is ($5 + $10)/2 = $7.50 per gallon. This answer is not correct unless the compound is made up of equal parts of A and B. For example, if 1 gallon of the compound consists of .4 gallon of A and .6 gallon of B, then the true mean cost of ingredients per gallon is

$$\mu = \frac{(.4)(\$5) + (.6)(\$10)}{.4 + .6} = \$8$$

in accordance with Equation (3.2) for $N = 2$ weighted values.   ▄▄

3.5
MEDIAN

Another measure of central tendency, or location, of a set of data values is known as the median. Conventionally, the **median** *Md* is generally defined as follows.

Definition

A **median** is a number that splits an ordered set of data essentially in half. Specifically, when the values in a set of data have been arranged in numerical order:

1. If there are an *odd* number of values in the data set, then the **median** *Md* is the value in the *middle* position
2. If there are an *even* number of values in the data set, then the **median** *Md* is the *mean* of the *two values* in the central positions.

If there are *n* observed values and *n* is an *odd* number, the array is formed (the values are lined up in either increasing or decreasing order), and the median is by definition the observation in the middle position.

▄▄ **EXAMPLE 3.5**   The number of people being supervised by each of 11 department heads in an accounting firm is arrayed as follows:

1   1   2   3   3   8   11   14   19   19   20

The median is 8 people (there are five observations to its left and five to its right).   ▄▄

Histograms Showing Median and Modes   **FIGURE 3.3**

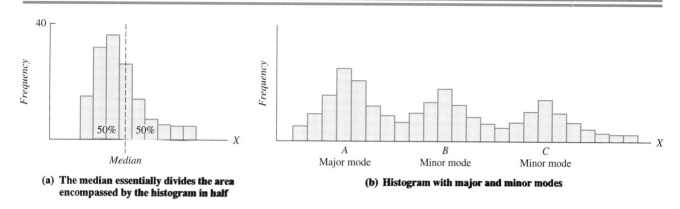

(a) **The median essentially divides the area encompassed by the histogram in half**

(b) **Histogram with major and minor modes**

If *n* is even, there is no one observation in the middle position. In this case, we take the median to be the average of the *pair* of observations occupying the two central positions.

EXAMPLE 3.6   In the next array we see the number of people supervised by 12 department heads in a department store:

   2   5   5   6   7   10   15   21   21   23   23   25

The observations 10 and 15 occupy the two center positions (there are five observations to the left of 10 and five to the right of 15), and so the median is (10 + 15)/2, or 12.5 people.

The methods for finding a median are appropriate for either samples of *n* data values or populations of *N* data values.

In summary, the median is a number that has essentially half of the arrayed data values below it and half above it. A histogram for a large set of data values that are not too repetitive has essentially 50% of its area to the left of the median and 50% to the right, as shown in Figure 3.3(a).

The **mode** is useful as a measure of location in cases where we desire to know which number comes up most often.

3.6
MODE

Definition

The **mode** *Mo* of a set of data values is the most frequently occurring value in the set.

EXAMPLE 3.7   A manager for Gulfbrook Land & Development looked at the winning bids for jobs on which he had submitted bids. He found that the winning bids, expressed as a percentage of his estimated costs, were as follows:

140   125   130   125   125   110   105   125   135   125   105

The winning bid most often seems to be 125% of his estimated costs (it occurs five times). Thus he might wish to enter a bid at 124% of his estimated costs on the next job he wants to get. Since the bid of 125% seems to be the winning bid most often, he will increase his chances of winning the contract with this knowledge of the mode.

The mode is generally not as useful a measure of location as the mean and the median. That is, when the data consist of only a few numbers, often each value occurs only once, and thus there is no single mode. If, however, one value does occur more than once, there is no guarantee that this value will in any way show the central tendency of the data if there are only a few values in the data set. For example, for the numbers

7   12   18   22   31   31

the mode is 31 since it appears twice and all other numbers appear only once. But 31 could hardly be considered a measure of central tendency for these data since it is at the extreme high end of the values. The concept for a mode as presented here is appropriate for sample data or population data.

When data are grouped in a frequency distribution, we often find that there is one class that has maximum frequency and that the frequencies of the other classes tend to fall away continually as we move away from the maximum class in either direction. This class is called the *modal class,* and we define the **mode** as the class mark of that class.

For example, for the distribution of employee absences given in Table 2.4, the class with the greatest frequency is 140–146, the frequency being 59. Moreover, as a glance at the histogram in Figure 2.1 shows, the frequencies taper off on either side of this one. This class is therefore the modal class, and its midpoint or class mark 143 is the mode *Mo*:

$$Mo = 143$$

If there is such a class with maximum frequency, then the distribution is said to be *unimodal.* If there is no such class, the mode is undefined or each value can be taken as a mode. In a histogram like that in Figure 3.3(b), the distribution is *multimodal; A* is the major mode, and *B* and *C* are minor modes. The concentration of values is greatest around *A,* and there are secondary concentrations around *B* and *C.* This discussion of the mode applies to both population and sample data.

Problems: Sections 3.3–3.6

Answers to odd-numbered problems are given in the back of the text.

  1. The percentage yields for a sample of money market mutual funds are listed as follows:

8   3   6   4   4

a. Determine the mean.   **b.** Determine the median.
c. Find the mode (or modes) if it exists.

Note: Problems with a hand symbol can be solved without a calculator.

2. The waiting times in minutes at a Shop and Go grocery store checkout counter for a sample of shoppers are given in the following listing:

    8   0   2   2   3

    a. Determine the mean.   **b.** Determine the median.
    c. Find the mode (or modes) if it exists.

3. The losses at financial services companies from bad loans are plaguing banks, savings and loan associations, and insurance companies. The following data are a sample of loan losses (in millions of dollars) suffered by a bank:

    7   2   3   6   3   3

    a. Determine the mean.   **b.** Determine the median.
    c. Find the mode (or modes) if it exists.

4. An auditor for Abbott & Kasteler found that the billing errors given in the accompanying table occurred on invoices. Positive errors indicate the customer was billed too much; whereas negative errors indicate the customer was billed too little. For these sample data, find the following:

    a. The mean error.   **b.** The median error.

Customer	Date	Size of Error
Jense	Mar. 4	$32
Lea	Jan. 8	−45
Stew	Dec. 12	66
Mill	June 7	2
Johns	May 23	−8
Blodge	Sept. 9	−51
Seman	Feb. 5	12
Glen	Oct. 30	18

5. The weights (in tons) of a population of boxcars coming into a Nocaff-Cola Company warehouse were 30.6, 26.2, 44.3, 21.9, and 27.0. For these figures, find the following:

    a. The mean.   **b.** The median.

6. A student has accumulated 20 credits with the grade of A, 25 credits with B's, 10 credits with C's, and 2 credits with D's. The school uses the grading scale in which A = 4 grade points, B = 3, C = 2, and D = 1. Determine the student's grade point average.

7. Three departments had mean sales of $15, $20, and $24 on the basis of 30, 35, and 50 items. What is the mean sales if these three departments are combined?

The measures of location discussed in the preceding sections describe one characteristic of a set of numbers—the typical or central value. However, we often wish to know about another characteristic of a number set, the variation or scatter among the values. For instance, a student about to graduate from college has accepted a job with a company. She has her choice of working at the company's offices in city A or city B. Since she likes outdoor activities, she decides to check on the climate in each of the cities. Her investigations tell her that both cities have a mean daily high temperature of 70°F. This figure alone, however, may be misleading. Further investigation reveals that city A's high temperature varies from 65°F in the winter to 75°F in

3.7
MEASURES OF
VARIATION

the summer. But city B's high varies from 20°F in the winter to 120°F in the summer. This information may have a rather significant effect on the student's location selection.

The statistics we calculate to measure the variation or scatter in a set of numbers are referred to as **measures of variation** or **measures of dispersion.** Since several sets of data could have the same—or nearly the same—mean, median, and mode but could vary considerably in the extent to which the individual observations differ from one another, a more complete description of the data results when we evaluate one of the measures of variation in addition to one or more of the measures of location. Four such measures are considered in the following sections: range, mean deviation, variance, and standard deviation.

3.8
RANGE

One measure of the variation or dispersion of a set of numbers is the **range.**

Definition and Equation for the Range

The **range** is the difference between the largest and smallest values in the data, or

$$Range = X_{\text{largest}} - X_{\text{smallest}}$$

In Table 2.1 we presented data concerning the number of employees absent from their work at a large Western Electronics manufacturing plant over a 106-day period. The smallest number of absences on any of the 106 days was 121 and the largest was 158; hence the range is $158 - 121$, or 37, employees.

Although the range is easy to calculate and is commonly used as a rough-and-ready measure of variability, it is generally *not* a satisfactory measure of variation, for several reasons. First, its calculation involves only two of the observed values regardless of the number of observations available. Therefore it utilizes only a fraction of the available information concerning variation in the data, and it reveals nothing with respect to the way in which the bulk of the observations are dispersed within the interval bounded by the smallest and largest values. Second, as the number of observations is increased, the range generally tends to become larger. Therefore, to use the ranges to compare the variation in two sets of data is not proper unless they contain the same numbers of values. Finally, the range is the least stable of our measures of variation for all but the smallest samples sizes; that is, in repeated samples taken from the same source, the ranges will exhibit more variation from sample to sample than will the other measures.

The range differs from most of our statistical measures in that it is a relatively good measure of variation for small numbers of observations, but it becomes less and less reliable as the sample size increases. Because it is easy to calculate and is reasonably stable in small samples, it is commonly used in statistical quality control where samples of four or five observations are often sufficient. The definition of the range applies to both sample and population data.

Since the disadvantages of the range limit its usefulness, we need to consider other measures of variation. Suppose we have a population of N numbers, $X_1, X_2, X_3, \ldots, X_N$, whose mean is μ. If we were to plot these values as dots in a **dot diagram** and μ on the X axis, as in Figure 3.4 (only five values are shown), and measure the distance of each of the X's from the mean μ, then we might be tempted to think that the average, or mean, of these distances should provide a measure of variation. These distances, or **deviations from the mean,** are equal to $(X_1 - \mu)$, $(X_2 - \mu), \ldots, (X_N - \mu)$. The general term $X_i - \mu$ is the amount by which X_i, the ith value of X, differs from the population mean. If $X_i - \mu$ is 5, then the value of X_i is 5 higher than the mean. If, in contrast, $X_i - \mu$ is -8, then the value of X_i is 8 below the mean. Thus the deviations from the mean will be either positive or negative depending on whether X_i is above or below the mean.

To find the mean of these deviations, take their sum and divide it by the number of deviations. This value is

$$\frac{1}{N} \sum_{i=1}^{N} (X_i - \mu)$$

But this quantity is always equal to zero, since the algebraic sum of the deviations of a set of numbers from its mean is always equal to zero.* That is,

$$\sum_{i=1}^{N} (X_i - \mu) = 0 \qquad (3.3)$$

*This statement can be shown as follows: By Rules B.1 and B.3 in Appendix B,

$$\sum (X_i - \mu) = \sum X_i - \sum \mu = \sum X_i - N\mu$$

since μ is a constant. But

$$\mu = \frac{1}{N} \sum X_i \quad \text{or} \quad N\mu = \sum X_i$$

Therefore,

$$\sum (X_i - \mu) = \sum X_i - \sum X_i = 0$$

FIGURE 3.4

Deviations from the Population Mean

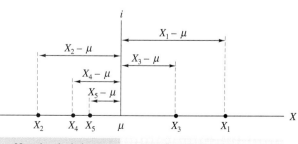

The same result holds true for deviations around the sample mean $\overline{X}$. That is,

$$\sum_{i=1}^{n}(X_i - \overline{X}) = 0$$

The reason is that we have used *signed distances* rather than unsigned distances of the X_i to the mean μ. We can eliminate the algebraic signs on these values in two ways: We may simply ignore them, or we may square them. If we ignore the signs on these distances, we then have the **mean deviation**

$$\frac{1}{N}\sum_{i=1}^{N}|X_i - \mu| \tag{3.4}$$

where $|X_i - \mu|$, the *absolute value* of $X_i - \mu$, is just $X_i - \mu$ with the sign converted to $+$ (positive) if it happens to be $-$ (negative).

The mean deviation, though it appears to be relatively simple, is not of great interest. The absolute values make the mean deviation difficult to use in the mathematical operations we wish to perform in later chapters. Since the mean deviation does not have the mathematical properties we desire, the better procedure is to eliminate the signs on the deviations around the mean by squaring them.

3.10
POPULATION VARIANCE AND STANDARD DEVIATION

Suppose we do use the square of the deviations instead of the absolute value as a measure of deviation. In the squaring process the negative signs will disappear; hence the sum of the squares of the deviations from the mean will always be a positive number greater than zero (unless all the observations have the same value, and then the sum of squares of the deviations from the mean will be equal to zero). This quantity, for simplicity, will be referred to as the **sum of squares.**

Equation for Sum of Squares

The **sum of squares** for a population of N data values, $X_1, X_2, \ldots, X_N$, is the sum of the squared deviations for the values from their mean μ, or

$$Sum\ of\ squares = \sum_{i=1}^{N}(X_i - \mu)^2 \tag{3.5}$$

Hereafter, whenever the phrase *sum of squares* appears, it should be taken to mean the *sum of squares of deviations from the mean*. To avoid ambiguity, if the sum of squares of the observations, or the sum of squares for any other quantities (not deviations), is meant, the phrase will be qualified accordingly.

The sum of squares provides a *measure of dispersion*. If all the observed values are identical, the sum of squares is equal to zero. If the values tend to be close together, the sum of squares will be small; but if they scatter over a wide range, the sum of squares will be correspondingly large.

EXAMPLE 3.8 To illustrate this characteristic of sums of squares and, incidentally, to show how the sum of squares is calculated according to the definition given in Formula (3.5), we use the numbers that follow. They show the sales (in thousands) of ten salespeople at Pop Boys—Minny Mike and Jock's retail automobile dealership for last week:

$$12 \quad 6 \quad 15 \quad 3 \quad 12 \quad 6 \quad 21 \quad 15 \quad 18 \quad 12$$

Their mean is

$$\mu = \frac{1}{N} \sum X_i = \frac{1}{10}(120) = 12$$

Notice that we denoted the mean as μ rather than $\overline{X}$. This symbol indicates that we are interested in *this* set of values and do not consider it to be a sample. Thus these salespeople are all the salespeople in the company and not just a sample. If they were a sample, we would have denoted their mean as $\overline{X}$. To find the deviations from the mean, we subtract $\mu = 12$ from each of the X values. We get the values shown in the second column of the accompanying table. We can use the fact that the sum of the deviations from the mean is equal to zero, as shown at the bottom of the second column of the table, to check our arithmetic.

To complete our calculations, we square each of the deviations and sum them. The squared deviations and their sum

$$\sum_{i=1}^{10}(X_i - \mu)^2 = 288$$

are shown in the third column of the following table:

X_i	$X_i - \mu$	$(X_i - \mu)^2$
12	0	0
6	−6	36
15	3	9
3	−9	81
12	0	0
6	−6	36
21	9	81
15	3	9
18	6	36
12	0	0
120	0	288

$$N = 10$$
$$\mu = 12$$

EXAMPLE 3.9 In contrast to the preceding example, which displays a fair amount of variation, suppose we have the numbers

$$12 \quad 10 \quad 12 \quad 14 \quad 10 \quad 13 \quad 12 \quad 11 \quad 14 \quad 12$$

for the sales (in thousands) of ten salespeople at Pop Boys—Minny Mike and Jock's dealership for a *different* week. We find the *sum of squares* as before (see the next table). As could be expected, since the variation in the second set of numbers is less than in the first set, the sum of squares, 18, is considerably smaller than that calculated for the first set. Notice that the mean is the same as in the preceding example; the sum of squares measures spread *from* the mean.

X_i	$X_i - \mu$	$(X_i - \mu)^2$
12	0	0
10	-2	4
12	0	0
14	2	4
10	-2	4
13	1	1
12	0	0
11	-1	1
14	2	4
12	0	0
120	0	18

$$N = 10$$
$$\mu = 12$$

To compare the variability of the sales for the ten salespeople at Pop Boys, we can graph the deviations of the values for each of the two weeks as in Figure 3.5. A comparison of the deviations around the means and the sums of squares for the sales in the two weeks (288 contrasted with 18) shows clearly that the variation in the second week is less than the variation in the first week.

If we have a large number of values, or if the mean is not a whole number, calculating the sum of squares by finding the deviations and squaring them could become tedious. Fortunately, there is another method for computing the sum of squares that does not necessitate finding the individual deviations from the mean. The sum of squares of deviations from the mean can be written as given next.*

*Algebraically we know that $(a - b)^2 = a^2 - 2ab + b^2$. Therefore we can rewrite the sum of squares as

$$\sum(X_i - \mu)^2 = \sum(X_i^2 - 2\mu X_i + \mu^2)$$

If we now apply Rules B.1, B.2, and B.3 in Appendix B, we get

$$\sum(X_i - \mu)^2 = \sum X_i^2 - 2\mu\sum X_i + N\mu^2$$

But

$$\mu = \frac{1}{N}\sum X_i \quad \text{and} \quad \mu^2 = \frac{1}{N^2}\left(\sum X_i\right)^2$$

Therefore,

$$\sum(X_i - \mu)^2 = \sum X_i^2 - 2\frac{1}{N}\left(\sum X_i\right)^2 + \frac{1}{N}\left(\sum X_i\right)^2$$

so that combining the two right-hand terms gives

$$\sum(X_i - \mu)^2 = \sum X_i^2 - \frac{1}{N}\left(\sum X_i\right)^2$$

FIGURE 3.5

Deviations Around the
Mean and Sums of Squares
for Two Data Sets

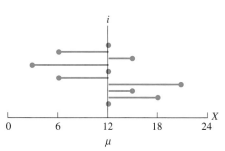

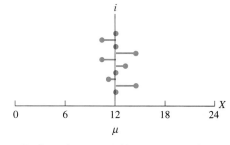

(a) Sum of squares is 288, larger variation
around the mean

(b) Sum of squares is 18, smaller variation
around the mean

Computational Formula for Sum of Squares

$$\sum(X_i - \mu)^2 = \sum X_i^2 - \frac{1}{N}\left(\sum X_i\right)^2 \qquad (3.6)$$

The expression on the right-hand side of Equation (3.6) provides an alternative procedure for computing a sum of squares that is especially useful when a calculator is available. For this reason it is often referred to as the *computational formula* for sums of squares, and it should be used for all but the simplest sets of data. To apply the formula, we obtain the sum of the squares of the observed values as well as their sum, square the sum, divide it by N (the number of observations), and subtract the result from the sum of squares of the observed values.

Although the sum of squares is a measure of variation, it is usually more convenient to use the mean of the squared deviations as a dispersion measure. This mean of squared deviations for population data is called the **population variance** σ^2 (lowercase Greek sigma squared).

Definition and Equation for Population Variance σ^2

The variance for a population of N data values, $X_1, X_2, \ldots, X_N$, is the sum of the squared deviations for the values from their mean μ divided by N, or

$$\sigma^2 = \frac{1}{N}\sum_{i=1}^{N}(X_i - \mu)^2 \qquad (3.7)$$

Thus the variance is the mean value of the squares of the deviations. It can also be found when the computational formula is used to obtain the sum of squares.

> **Computational Formula for Population Variance**
>
> $$\sigma^2 = \frac{1}{N}\left[\sum X_i^2 - \frac{1}{N}\left(\sum X_i\right)^2\right] \qquad (3.8)$$

The **population standard deviation** σ, or root-mean-squared deviation, of the numbers in a population of size N is the square root of the variance.

> **Definition and Equation for Population Standard Deviation σ**
>
> The standard deviation for a population of N data values, $X_1, X_2, \ldots, X_N$, is the square root of the population variance, or
>
> $$\sigma = \sqrt{\sigma^2} = \sqrt{\frac{1}{N}\sum(X_i - \mu)^2} \qquad (3.9)$$

EXAMPLE 3.10   Let us work out an illustrative calculation of the population variance and standard deviation by using the following population data on the number of defectives found by a quality control inspector during nine successive inspection hours at NEDCOR, Inc.:

    3   6   2   5   3   8   6   7   5

Step One   Find the sum of squares. First, find the mean:

$$\sum X_i = 45$$

$$\mu = \frac{1}{N}\sum X_i = \frac{45}{9} = 5$$

Thus the mean value is 5 defects per hour. Now we have the following table for intermediate computations:

X_i	$X_i - \mu$	$(X_i - \mu)^2$
3	-2	4
6	1	1
2	-3	9
5	0	0
3	-2	4
8	3	9
6	1	1
7	2	4
5	0	0
45	0	32

Thus the sum of squares, $\Sigma(X_i - \mu)^2$, is 32.

Step Two   To get the population variance, divide the sum of squares by N:

$$\sigma^2 = \frac{1}{N}\Sigma(X_i - \mu)^2$$

$$= \frac{1}{9}(32) = 3.56$$

Thus the variance has a value of 3.56 measured in units of (defects)^2.

Step Three   To obtain the standard deviation, take the square root of the variance:

$$\sigma = \sqrt{\sigma^2} = \sqrt{3.56 \text{ (defects)}^2} = 1.89 \text{ defects}$$

Thus the standard deviation is 1.89 defects and has the same units of measure as the individual data values and the mean.

If we were to use the computational formula (3.6) for the sum of squares in Step 1 of Example 3.10, we would find

$$\Sigma X_i = 45$$

$$\Sigma X_i^2 = 9 + 36 + 4 + 25 + 9 + 64 + 36 + 49 + 25 = 257$$

$$\Sigma X_i^2 - \frac{1}{N}\left(\Sigma X_i\right)^2 = 257 - \frac{1}{9}(45)^2$$

$$= 257 - 225 = 32$$

Since finding the standard deviation involves summing the squared deviations, dividing by N, and taking a square root, the standard deviation is sometimes called the *root-mean-squared deviation,* as we indicated earlier.

3.11 SAMPLE VARIANCE AND STANDARD DEVIATION

Recall that in Section 3.3 we made a distinction between μ, the mean of a population, and $\overline{X}$, the mean of a sample. While the methods for computing these means were shown to be the same, we used the different notations to indicate whether our data set was a population or a sample chosen to represent a population.

The same type of distinction is made between the population standard deviation σ and the sample standard deviation s. If the numbers in our data set form the population of interest, then we use the following equation to find the standard deviation:

$$\sigma = \sqrt{\frac{1}{N}\Sigma(X_i - \mu)^2} \tag{3.9}$$

Also, we can use the computational formula of Equation (3.8) to obtain the expression under the square root sign.

However, if the numbers in our data set form a *sample* and we wish to estimate a population variance or standard deviation, we can compute the sample mean $\overline{X} = (1/n)\Sigma X$ and then find the **sample variance s^2.**

Definition and Equation for Sample Variance s^2

The variance for a sample of n data values, $X_1, X_2, \ldots, X_n$, is the sum of the squared deviations for the values from their mean $\overline{X}$ divided by $n - 1$, or

$$s^2 = \frac{1}{n-1} \sum (X_i - \overline{X})^2 \tag{3.10}$$

When the computational formula is used to obtain the sum of squares, the sample variance is computed from the following equation.

Computational Formula for Sample Variance s^2

$$s^2 = \frac{1}{n-1} \left[\sum X_i^2 - \frac{1}{n} \left(\sum X_i \right)^2 \right] \tag{3.10a}$$

The standard deviation is always the square root of the variance. So we define the **sample standard deviation** s as follows.

Definition and Equation for Sample Standard Deviation s

The standard deviation for a sample of n data values, $X_1, X_2, \ldots, X_n$, is the square root of the sample variance, or

$$s = \sqrt{s^2} = \sqrt{\frac{1}{n-1} \sum (X_i - \overline{X})^2} \tag{3.11}$$

The only differences between the formulas for the population and sample standard deviations are as follows:

1. In the population standard deviation formula, we call the mean μ, and in the sample standard deviation formula we call it $\overline{X}$.
2. In the population formula for σ we divide the sum of squared deviations by N, but in the sample formula we divide that sum by $n - 1$.

The first difference is only a notational one. We use μ and $\overline{X}$ to denote whether we are working with population or sample data, but the mathematical computations are the same. We also use N for the size of the population and n for the sample size. The second difference, however, is more substantial. Dividing by $n - 1$ when we calculate the standard deviation from sample data produces a mathematically different result from what we obtain when using population data and dividing by N.

An intuitive reason why we divide the sample sum of squares by $n - 1$ rather than by n is shown in Figure 3.6. We must remember that the standard deviation is a

FIGURE 3.6

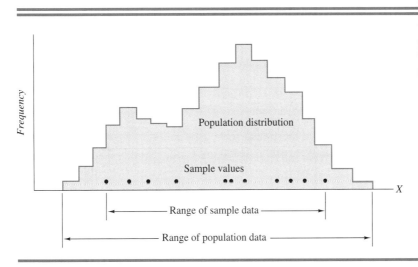

Variation in the Sample Data Smaller Than the Variation in the Population Data

measure of dispersion, spread, or scatter of a set of values around the mean. Figure 3.6 pictures a population and the range over which its values are spread. A typical set of randomly selected values from the population is depicted by the dots (●'s) falling along the X axis in a **dot diagram.** Note that the range over which the sample values are spread is smaller than that over which the population is spread. It would be a rather unusual sample that included both the largest and the smallest value in the population. Thus we can see intuitively that the spread of values in a sample will typically be less than the spread in the population. If we are using sample data to estimate the population standard deviation, then we must adjust our calculations somehow to make up for the smaller spread in the sample. Dividing the sum of squares by $n - 1$ rather than by n adjusts the sample standard deviation up to where it better estimates the population standard deviation.*

One might logically ask, "Why not divide by $n - 2$ or by $n - 3$?" Dividing the sample sum of squares by $n - 1$ gives a better estimate of the population variance, and thus standard deviation, than if any other divisor is used. We will not present a mathematical proof of this fact. However, it does involve the concept of *degrees of freedom.*

For any sum of squared deviations for a sample of data values, there is an expression or number known as its **degrees of freedom.** The degrees of freedom can be explained as the number of squared deviations included in the sum minus the number of independent restrictions imposed on the deviations involved.[†] We will give the degrees of freedom associated with any new sum of squared deviations whenever it is important to do so.

---

*Actually, the division of the sample sum of squares by $n - 1$ makes the sample variance s^2 a better estimator of the population variance σ^2 in the sense that it is not biased downward under the situation discussed in Section 8.1. Since the standard deviation is the (positive) square root of the variance, we proceed using s as the estimator of σ.

†For a sample of n data values there are n squared deviations from the mean. However, only $n - 1$ of the deviations are independent due to the restriction that the sum of the deviations from the mean must be zero. Phrased differently, when $n - 1$ of the deviations are freely specified, the nth one is not free to vary. The sum of squares of deviations of the n values from their mean has, therefore, $n - 1$ degrees of freedom.

PRACTICE PROBLEM 3.1

The real estate multiple-listing service in Las Vegas would like to know how long it takes homes that it lists to sell. Thus the director of the service took a sample of $n = 10$ homes listed last year and found they sold in the following number of weeks (rounded to the nearest whole week): 4, 8, 22, 1, 12, 7, 9, 5, 20, 6. Find the estimates of the mean and the standard deviation of the time to sale for the population of all homes listed by the service last year.

Solution: Since these data form a sample and we are interested in the population, we will find $\overline{X}$, the sample mean, and s, the sample data's estimate of the population standard deviation. Computations are given in the following table.

X_i	$X_i - \overline{X}$	$(X_i - \overline{X})^2$
4	-5.4	29.16
8	-1.4	1.96
22	12.6	158.76
1	-8.4	70.56
12	2.6	6.76
7	-2.4	5.76
9	-0.4	0.16
5	-4.4	19.36
20	10.6	112.36
6	-3.4	11.56
94		416.40

Using the computations given in the table, we find

$$\overline{X} = \frac{1}{n}\sum_{i=1}^{10}X_i = \frac{1}{10}(94) = 9.4 \text{ weeks}$$

and

$$s = \sqrt{\frac{1}{n-1}\sum_{i=1}^{10}(X_i - \overline{X})^2} = \sqrt{\frac{1}{9}(416.40)} = 6.8 \text{ weeks}$$

EXAMPLE 3.11   In the previous section in Example 3.10, we calculated the population mean and standard deviation of the number of defects found by a quality control inspector during nine successive inspection hours. We found that the mean and the standard deviation of those nine values were $\mu = 5$ defects and $\sigma = 1.89$ defects.

In a quality control situation, however, an inspector typically takes a sample of roughly this size and tries to estimate the population mean and standard deviation from the sample. If we view the nine values in that example as a sample, then we find that the sample mean, $\overline{X} = 5$ defects, remains unchanged, except that the notation of $\overline{X}$ rather than μ is used. We find the sample standard deviation s as

$$s = \sqrt{\frac{1}{8}(32)} = \sqrt{4} = 2 \text{ defects}$$

The reader may wonder how he or she is to determine whether to use the formula of σ, the population standard deviation, or s, the sample standard deviation. The answer is that most statistical problems involve sampling. Thus the formula for s, using $n - 1$ as the divisor, is the most commonly used formula. But in situations where the reader is unsure, he or she should ask this question: "Am I interested in these particular values, or are they just representative of some larger group that I would examine if I had the time, resources, and patience?" If the numbers you have are the values you are interested in, then calculate μ and σ, using N as the divisor, to describe them. If you have a sample and are interested in some larger group or population, then calculate the sample mean $\overline{X}$ using n as the divisor, and calculate the sample standard deviation s using $n - 1$ as the divisor.

Problems: Sections 3.8–3.11

8. The waiting times (in minutes) at a Shop and Go grocery store checkout counter for a sample of shoppers are given in the accompanying listing (see Problem 2):

    8   0   2   2   3

    a. Determine the range.   **b.** Determine the variance.
    c. Find the standard deviation.

9. The losses at financial services companies from bad loans are troubling banks, savings and loan associations, and insurance companies. The accompanying data are a sample of loan losses (in millions of dollars) suffered by a bank (see Problem 3):

    7   2   3   6   3   3

    a. Determine the range.   **b.** Determine the variance.
    c. Find the standard deviation.

10. An auditor for Abbott & Kasteler found that the billing errors given in the accompanying table occurred on invoices. Positive errors indicate the customer was billed too much, and negative errors indicate the customer was billed too little. For these sample data, find the following (see Problem 4):

    a. The range.   **b.** The mean deviation.
    c. The variance.   **d.** The standard deviation.

Customer	Date	Size of Error
Jense	Mar. 4	$32
Lea	Jan. 8	−45
Stew	Dec. 12	66
Mill	June 7	2
John	May 23	−8
Blodge	Sept. 9	−51
Seman	Feb. 5	12
Glen	Oct. 30	18

11. The weights (in tons) of boxcars coming into a Nocaff-Cola Company warehouse were found to be 30.6, 26.2, 44.3, 21.9, and 27.0. For these figures, find the following (see Problem 5):

    a. The range.
    b. The variance (assuming these values form the population).
    c. The variance (assuming these values form a sample).
    d. The standard deviations from parts b and c.

12. A sample of 28 cattle that have just arrived at the MS Feed Exchange feedlot has been weighed, and their beginning weights (in pounds) are presented in the accompanying table.
    a. Determine the variance.
    b. Determine the standard deviation of the beginning weights.
    c. Determine the range.

375	329	450	357	387	316	323
312	396	303	437	401	359	341
433	430	412	450	378	369	377
343	438	440	349	444	355	301

3.12
WHAT THE STANDARD DEVIATION TELLS US

So far we have concentrated on how to *calculate* the standard deviation of a set of numbers. We have said very little about what the standard deviation *means,* other than to say that it is a measure of *spread* or *dispersion* in the data set. A data set with a large standard deviation has much dispersion with values widely scattered around its mean, and a data set with a small standard deviation has little dispersion with the values tightly clustered about its mean.

The range of a data set, the largest minus the smallest value, is a rough measure of this spread and it is very understandable. The standard deviation, in contrast, has little intuitive appeal. It is the mathematical properties of the standard deviation, which we will use in subsequent chapters, that justify its wide use in statistics. Since we will be using this measure often, we will spend some time at this point to relate more fully what the standard deviation tells us about a set of numbers. First, we will present two histograms that represent data sets that have the same mean but different standard deviations. Second, we will present two rules that help us understand what the standard deviation tells us.

Figure 3.7 depicts two histograms that have the same mean but different standard deviations. The standard deviation of the data depicted in Figure 3.7(a) is about one-third as large as the standard deviation of the data depicted in Figure 3.7(b). Compare the two histograms, and note this difference in variability.

The following general rule, which was developed both from theory and empirical observations is helpful for the use of the *standard deviation.*

General Empirical Rule for Using Standard Deviations

For any set of data values, essentially 90 percent or more of the values lie within plus or minus 3 standard deviations from the mean.

Let us examine what this rule tells us. Assume that a set of data values for stock prices has a mean of $50 and a standard deviation of $5. Then the rule tells us that at least essentially 90 percent of the stocks in the data set have prices that lie between the limits $50 \pm (3)(5)$ or between 50 ± 15, or from $35 to $65.*

In the previous section we presented a practice problem in which the estimated mean and standard deviation of time to sale of homes listed with a multiple-listing service were $\overline{X} = 9.4$ weeks and $s = 6.8$ weeks. If we form an interval that is three

---

*The general empirical rule is based in part on Chebyshev's rule, which states that for any set of data values, at least $100(1 - 1/k^2)\%$ of the values must lie within plus or minus k standard deviations of the mean where $k > 1$. This rule is also known as the Bienaymé–Chebyshev rule in honor of its developers.

FIGURE 3.7

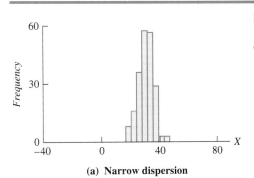

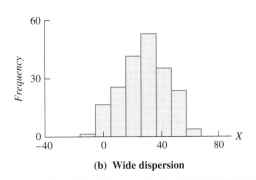

Distributions with Equal Means and Unequal Variability

standard deviations on either side of 9.4 weeks, we have $9.4 \pm (3)(6.8) = 9.4 \pm 20.4$, or -11.0 to 29.8 weeks. Thus the director of this service can say that essentially 90% or more of the homes listed with his service sell within 0 to 29.8 weeks of their listing. (It would be silly to apply any meaning to the lower end of the interval since -11.0 weeks is meaningless.)

The quality control inspector of Example 3.9 of the previous section estimated the population mean and standard deviation for the number of hourly defectives to be $\overline{X} = 5$ defects and $s = 2$ defects. If she were to form an interval of three standard deviations on either side of the mean, she would obtain $5 \pm (3)(2) = 5 \pm 6$, or -1 to 11. Then she could say that essentially 90% or more of the hours of operation for her process produce between 0 and 11 defects.

A more exact rule about the proportion of a data set that lies in an interval can be made if we have a data set whose histogram shows that the numbers tend to follow a *bell-shaped curve.*

Empirical Rule for Bell-Shaped Data Distributions

If the histogram for a set of data values is *shaped like a bell,* then the interval bounded by:

$\mu \pm 1\sigma$ contains about 68%, or about two-thirds of the values

$\mu \pm 2\sigma$ contains about 95% of the values

$\mu \pm 3\sigma$ contains about 99.7%, or virtually all of the values

FIGURE 3.8

Data with a Bell-Shaped Histogram

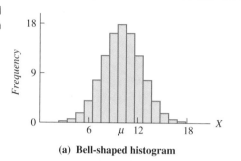

(a) Bell-shaped histogram

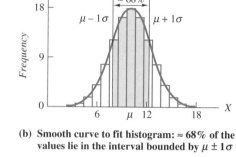

(b) Smooth curve to fit histogram: ≈ 68% of the values lie in the interval bounded by $\mu \pm 1\sigma$

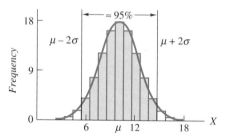

(c) Smooth curve to fit histogram: ≈ 95% of the values lie in the interval bounded by $\mu \pm 2\sigma$

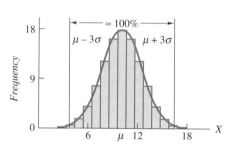

(d) Smooth curve to fit histogram: ≈100% of the values lie in the interval bounded by $\mu \pm 3\sigma$

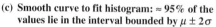

Three things should be noted about this rule for bell-shaped data distributions. The first is the restriction that the rule can be used only when there is reason to believe that the data are symmetrical about their mean and have a histogram that looks bell-shaped, or similar to the one shown in Figure 3.8(a). The empirical rule for bell-shaped data is shown in Figures 3.8(b), (c), and (d). Typically, physical measurements such as weights, lengths, and pressures fit a bell-shaped curve. Figure 3.8 also shows that if the number of values in a data set is large, a histogram can be approximated by a smooth curve.

The second thing to note is that this rule is consistent with the general rule stated earlier. The general rule holds for *any* set of data, while this rule holds only for bell-shaped data. But bell-shaped data qualify as *any* data set and the "virtually all" statement in the rule qualifies as "essentially 90% or more" that we get from the general rule for the interval three standard deviations on either side of the mean.

The third thing to note about the empirical rule for bell-shaped data is that the range of the values, the highest minus the lowest, is about six standard deviations (plus three and minus three standard deviations gives an interval that is six standard deviations wide).

EXAMPLE 3.12   Trucks traveling on the I-70 interstate highway weigh an average of 12.5 tons and have a standard deviation of 2.2 tons. What does this standard deviation tell you?

Since weights often follow a bell-shaped histogram when plotted, you can estimate that about 68% of the trucks on that highway weigh between 12.5 − 2.2 = 10.3 tons and 12.5 + 2.2 = 14.7 tons. In addition, you can estimate that about 95% of the trucks weigh between 8.1 and 16.9 tons (plus two and minus two standard deviations). Virtually all of the trucks weigh between 5.9 and 19.1 tons.

We should not apply this bell-shaped data rule to the previous examples involving the time to sell homes and the number of defects each hour on a production line. We have no evidence that those data follow bell-shaped histograms when they are plotted, and they are discrete counts, not continuous physical measurements such as the weights of trucks.

Problems: Section 3.12

13. Zero-coupon bonds (called CATS, TIGRS, or STRIPS by some stock brokerage firms) sell at a deep discount from face value and do not pay any interest or principal until maturity. As market interest rates change, the resale value and yield for zeros experience greater fluctuations than do conventional bonds. Two histograms are depicted in Figure 3.9. One of the histograms shows the yields (as percentages) over time for zero-coupon bonds, and the other shows the yields over time for conventional bonds. Is the histogram for zero-coupon bonds shown in Figure 3.9(a) or in Figure 3.9(b)? Explain.

Histograms Showing Yields for Zero-Coupon Bonds or Conventional Bonds   FIGURE 3.9

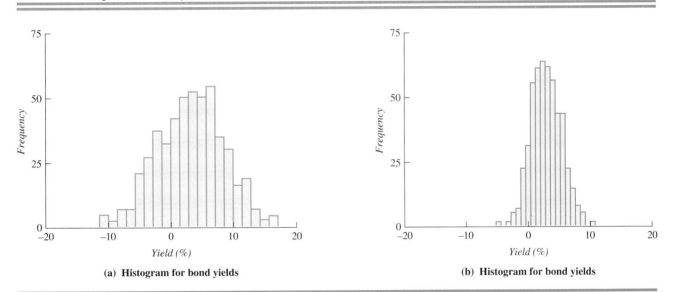

(a) Histogram for bond yields

(b) Histogram for bond yields

14. A set of stocks traded on the New York Stock Exchange has a mean price of $50 and a standard deviation of $3. What is the interval within which you are sure that at least 90% of the prices lie?

15. The wieghts of packages of candy coming off a production line at Valley Candies have a mean of 10.20 pounds and a standard deviation of 0.10 pound.

   a. Which rule can be applied to these measurements: the general rule, the bell-shaped data rule, or both?

b. What proportion of the packages have weights between 10.10 and 10.30 pounds?

c. What proportion weigh between 10.00 and 10.40 pounds?

d. If the packages are labeled as containing 10.0 pounds, about what proportion of the packages will be underweight?

16. Use the data given in the statement of Problem 15 to estimate the proportion of packages with weights between 10.10 and 10.40 pounds.

3.13
SELECTING MEASURES OF LOCATION AND DISPERSION

We have presented four measures of location in this chapter: mean, weighted mean, median, and mode. However, we have not indicated when using the different measures of location is appropriate. That is the purpose of this section.

When we are deciding which measure of location to report for a set of data, a primary consideration is the use to which the results are to be put. In addition, we need to know the advantages and disadvantages of each measure of location regarding its calculation and interpretation.

The *shape* of a frequency distribution or histogram influences the relationship among the measures of central tendency. If the distribution is *symmetric* and *unimodal,* then the mean $\overline{X}$, the median *Md*, and the mode *Mo* will all coincide. But some frequency distributions are asymmetrical, or **skewed.** A frequency distribution is *skewed to the right* (or skewed positively) if it extends farther to the right of the mode than it does to the left. It is *skewed to the left* (or skewed negatively) if the opposite is true. As the distribution becomes more and more skewed, the differences among the measures of central tendency become greater. The relationships among the mean $\overline{X}$, the median *Md*, and the mode *Mo* for some typical symmetric and skewed distributions are depicted in Figure 3.10.

The mean is sensitive to extreme values. If in a small town the average annual income of the 100 heads of household is reported as $9990, this figure is correct but very misleading if one head of household is a multimillionaire with an income of $900,000 and the remaining 99 are paupers with incomes of $1000. A few extreme values have little or no effect on the median and the mode. For the numbers

$$1 \quad 3 \quad 4 \quad 6 \quad 6 \quad 9 \quad 13$$

we have $\overline{X} = 6$, $Md = 6$, and $Mo = 6$. If we include the number 70 in this set of numbers, we have the values

$$1 \quad 3 \quad 4 \quad 6 \quad 6 \quad 9 \quad 13 \quad 70$$

and the mean will be equal to 14, a shift of 8 units, but the median and the mode will be unchanged. For this reason, when the data are skewed or contain extreme values, the median or the mode may be more characteristic and therefore provide a better description.

One reason why the median is a better measure of central tendency than the mean for some data sets is that the median is resistant to extreme values and the mean is not. **Resistance** means that the measure is insensitive to outliers or to changes in a few data values.

If we want to combine measures for several sets of data, the algebraic properties of the mean give it a distinct advantage. We have seen that we can use the weighted mean for this purpose. The median and the mode are not subject to this type of algebraic treatment.

FIGURE 3.10

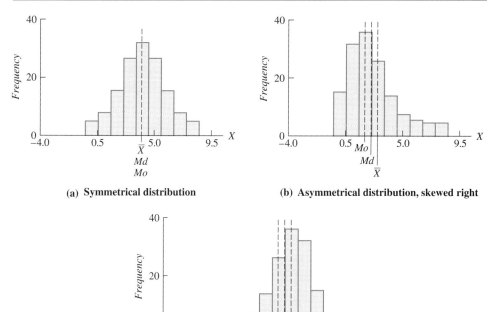

(a) Symmetrical distribution

(b) Asymmetrical distribution, skewed right

(c) Asymmetrical distribution, skewed left

Mean, Median, and Mode for Symmetrical and Asymmetrical Distributions

If the evaluation of a measure of location is a first step toward making inferences about the source of the data, the mathematical and distributional properties of the mean give it a distinct advantage. In the realm of statistical inference a primary consideration is statistical stability. If a large number of sets of data are taken from the same source and all three measures are calculated for each set, there will be less variation among the means than among the medians or the modes; hence the mean is more stable. This result, coupled with the fact that the mean is more amenable to mathematical and theoretical treatment, makes the mean an almost universal choice for all but purely descriptive or exploratory purposes. The three measures of location are usually ranked according to their overall desirability in the following order: mean, median, and mode.

We have presented four measures of variability or dispersion: the range, mean deviation, variance, and standard deviation. We have indicated that due to mathematical properties possessed by the variance and standard deviation (which are virtually the same measure of variation), we most often use them to describe the variability or dispersion in a set of data values. Since it is the difference between the largest and smallest data values, the range is sensitive to outliers or it is not resistant. The mean deviation, variance, and standard deviation can also be influenced appreciably by a few data values, even though the variance and standard deviation possess desirable mathematical properties.

3.14
COEFFICIENT OF VARIATION

A descriptive statistic that combines the standard deviation and the mean is called the *coefficient of variation*. The following calculation will show why the coefficient of variation is a useful measure of relative dispersion when the means are positive.

EXAMPLE 3.13   Consider two stocks, A and B. If we take a random sample of the daily closing prices of these stocks, we might find that the respective standard deviations of these closing prices are

$$s_A = \$.50 \quad \text{and} \quad s_B = \$5.00$$

Since the standard deviation measures variation in a set of numbers, we might conclude that the second stock's closing prices vary much more than those of the first stock. In fact, we might wish to avoid investment in such a volatile stock and choose to put our funds into stock A.

Before we call our broker, however, we might be wise to note that

$$\overline{X}_A = \$1.00 \quad \text{and} \quad \overline{X}_B = \$100.00$$

Now the situation looks rather different. Stock A is really the one with volatile price movements because its standard deviation of price closings is 50% of its mean value. Stock B's standard deviation of price closings is only 5% of its mean value.

To convert the standard deviation to a value that can be *compared* between two number sets of rather different magnitudes, we compute the **coefficient of variation, CV.**

Definition and Equation for Population or Sample Coefficient of Variation

The **coefficient of variation** is a unitless figure that expresses the standard deviation as a percentage of the mean:

$$CV = \left(\frac{\sigma}{\mu}\right) \times 100 \qquad \mu > 0$$

$$CV = \left(\frac{s}{\overline{X}}\right) \times 100 \qquad \overline{X} > 0$$

(3.12)

EXAMPLE 3.14   We intuitively found the coefficients of variation for the two stocks in Example 3.11:

$$CV_A = \left(\frac{\$.50}{\$1.00}\right) \times 100 = 50\%$$

and

$$CV_B = \left(\frac{\$5.00}{\$100.00}\right) \times 100 = 5\%$$

Financial managers sometimes use the coefficient of variation to measure and compare the riskiness of competing portfolios of investments. Those portfolios with higher coefficients of variation go through wider fluctuations in their market value from period to period than do those portfolios with smaller coefficients of variation.

Problems: Sections 3.13-3.14

17. The sample of 28 cattle described in Problem 12 has now been fed at the MS Feed Exchange feedlot for 100 days. The accompanying data give the weights of the cattle at the end of this feeding period. Use these data and the results you obtained for Problem 12 to determine which set of weights, the beginning weights or the ending weights, of the cattle in the MS Feed Exchange have more *relative* variation. Discuss why your result makes sense. That is, why does the group you find with more relative variation logically have more variation?

552	487	739	599	729	617	589
498	556	430	687	613	596	612
720	701	645	738	492	637	712
451	578	689	654	721	640	Died

18. Which of the following statements are true?
    a. For symmetrically distributed data the mean, the median, and the mode all have the same numerical value.
    b. The range of a set of numbers is usually about three times the standard deviation.
    c. If a set of numbers is skewed to the right (has a few values quite a bit larger than the rest of the numbers), then the mean will exceed the median.
    d. If all the numbers in a data set have the same value, then the variance and the standard deviation are zero.

FIGURE 3.11

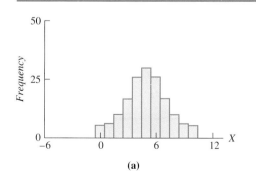 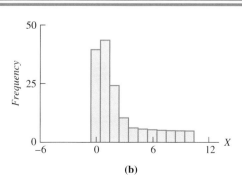

Histograms for Symmetric and Skewed Distributions

19. An investor in financial securities has several type A (such as bonds) and several type B (such as stocks) investments. The type A investments have a mean rate of return of 10% and a standard deviation of 4%. The type B investments have a mean rate of return of 20% and a standard deviation of 10%. The rate of return can be considered to be a measure of profitability. In relative terms, which type of investment has the larger variation in rate of return?

20. Specify the relative locations of the mean, the median, and the mode for each of the histograms in Figure 3.11. Comment on the skewness for each histogram.

3.15

EXPLORATORY DATA ANALYSIS WITH COMPUTERS: DESCRIPTIVE MEASURES AND BOX PLOTS

The important features of a set of data often need to be explored to obtain information that can be used to aid decision makers. **Exploratory data analysis** exposes patterns of data and forces us to see characteristics or features of data that we may not have expected to see. In this section we comment on the use of calculators for obtaining descriptive measures, and we use a computer to obtain descriptive measures that are helpful in exploratory data analyses or in reporting results of statistical studies. *Box plots* for summarizing key descriptive measures graphically are introduced, and a comprehensive descriptive summary from a computer program is presented.

3.15.1 Descriptive Measures

The calculation of descriptive measures for a large set of data can be rather involved and require a vast number of computations. Calculators can cut the time required for obtaining descriptive measures in a large data set. Some calculators have automatic function keys that allow the user to find the mean, the standard deviation, and so forth, of a data set by merely entering the data and pushing one or two keys. Readers are urged to work several problems at the end of this chapter without using such automatic functions. They may find that they understand the descriptive measures better if they work several problems by using hand calculations. If one does use a calculator with automatic function keys, one should determine whether the automatic function for computing the standard deviation divides the sum of squared deviations by N or by $n - 1$.

When large data sets are involved, decision makers often use a personal computer or a large computer to explore the important features of their data and to obtain descriptive measures.

EXAMPLE 3.15   A computer was used to obtain descriptive measures for the sample data on the number of employees absent at Western Electronics, presented earlier in Table 2.1. The results of the data analyses are presented in Computer Exhibits 3.1A and 3.1B. Most of the computer programs that perform statistical analyses present results similar to those shown in Computer Exhibit 3.1, and it is usually very easy to adapt to different outputs if you have access to a different software package. Compare the descriptive measures in Computer Exhibit 3.1 with the histograms of Computer Exhibit 2.1.

3.15.2 Box Plots

Graphical displays, including **box plots,** are often used to give effective visual summaries of key descriptive measures of sets of data.

COMPUTER EXHIBIT 3.1A Exploratory Data Analysis for the Western Electronics Absenteeism Data

MINITAB

```
MTB > set c1
DATA > 146 141 139 140 145 141 142 131 142 140
DATA > 144 140 138 139 147 139 141 137 141 132
        .
        .
        .
DATA > 143 143 148 141 145 141
DATA > end

MTB > name c1 'Absences'

MTB > describe c1  ◄───────────────────────────────
```
> Command **describe** to obtain descriptive measures for **c1**

```
                  N      MEAN    MEDIAN    TRMEAN    STDEV    SEMEAN
Absences        106    141.56    141.00    141.64     5.26      0.51

                MIN       MAX        Q1        Q3
Absences     121.00    158.00    139.00    145.00

MTB > stop
```

Note: The value labeled TRMEAN is a **trimmed mean** that was computed for the middle 90% of the values after the largest and smallest 5% of the values were omitted. SEMEAN is a label for the **standard error of the mean** (to be discussed in Chapter 7). The value for SEMEAN was found by dividing the standard deviation by the square root of the sample size.

The first and third quartiles for the sample data are Q_1 and Q_3. Q_1 and Q_3 are in positions $(n + 1)/4$ and $3(n + 1)/4$ of the data arrayed in ascending order. For example, consider the three ordered values 1, 2, and 4. Q_1 and Q_3 are in positions $(3 + 1)/4 = 1$ and $3(3 + 1)/4 = 3$, so $Q_1 = 1$ and $Q_3 = 4$. If the position values for Q_1 and Q_3 have fractional parts, then the quartile values will be weighted means of two contiguous ordered values (with one minus the fractional part as the weight for the smaller value and the fractional part as the weight for the larger value). For example, consider the four ordered values 1, 2, 4, and 6. Q_1 is in position $(4 + 1)/4 = 1.25$, so it is a weighted mean of the first and second ordered values, or $Q_1 = (1 - .25)(1) + (.25)(2) = 1.25$. Q_3 is in position $3(4 + 1)/4 = 3.75$, so it is the weighted mean of the third and fourth ordered values, or $Q_3 = (1 - .75)(4) + (.75)(6) = 5.5$.

Quartiles and trimmed means are generally resistant to outliers or to changes in a few data values.

Definition

Box plots are graphical displays that summarize the main features of a set of data including the central tendency, dispersion, symmetry, and distances (or tail lengths) to the minimum and maximum values.

Although there are several versions of box plots, Figure 3.12 shows a simple box plot and the descriptive measures needed to draw the plot. The edges of a box are drawn at the first quartile (Q_1) and the third quartile (Q_3) of the data set so that the box covers the middle half of the values. A line is placed in the box at the median. The lines emanating from the box are known as **whiskers** and they extend to cover the values from the box down to the minimum value and up to the maximum value. The values of *Minimum* = 121, Q_1 = 139, *Median* = 141, Q_3 = 145, and *Maximum* = 158 for the Western Electronics absence data are displayed graphically by the box plot and are known as a **5-number-summary** in exploratory data analysis.

MYSTAT

COMPUTER EXHIBIT 3.1B Exploratory Data Analysis for the Western
Electronics Absenteeism Data

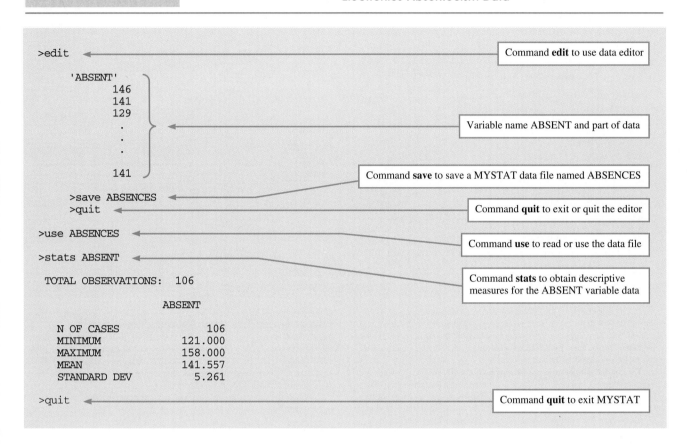

```
>edit                                                    Command edit to use data editor

     'ABSENT'
        146
        141
        129                                              Variable name ABSENT and part of data
          .
          .
          .
        141                                              Command save to save a MYSTAT data file named ABSENCES

       >save ABSENCES
       >quit                                             Command quit to exit or quit the editor

>use ABSENCES                                            Command use to read or use the data file

>stats ABSENT                                            Command stats to obtain descriptive
                                                         measures for the ABSENT variable data
  TOTAL OBSERVATIONS:   106

                     ABSENT

   N OF CASES               106
   MINIMUM              121.000
   MAXIMUM              158.000
   MEAN                 141.557
   STANDARD DEV           5.261

>quit                                                    Command quit to exit MYSTAT
```

Computers are typically used to obtain box plots, and some programs draw different versions of box plots that are helpful in identifying outliers. **Outliers** are
surprising values that stick out or stand apart from the other values. Outliers may
occur if an item is in error and needs to be corrected or if it is an unusual item and
needs to be discarded or given special attention. The whiskers of a box plot are
often drawn so they emanate at most the distance $1.5(Q_3 - Q_1)$ from the box and

FIGURE 3.12

**Box Plot of the Western
Electronics Absenteeism
Data**

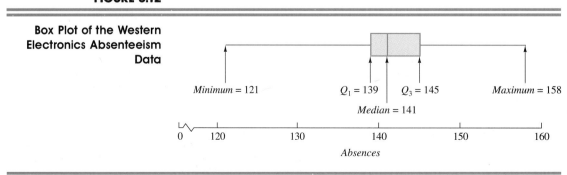

$Q_3 - Q_1$ is known as the **interquartile range, IQR.** The interquartile range is a resistant measure of variation for a set of data values; it is generally not influenced by outliers or changes in a few data values. Data values in the interval from $1.5(Q_3 - Q_1)$ to $3.0(Q_3 - Q_1)$ away from the box are marked with a symbol, such as an asterisk, and are treated as possible outliers; values that are more extreme are marked with another symbol, such as a letter "O," and are treated as probable outliers. The following example illustrates the use of box plots to help identify outliers and compare the important features of several data sets across time, populations, or variables.

EXAMPLE 3.16

Investors and managers often analyze and interpret financial statements by using ratio analyses. A ratio that measures a company's profitability is return on assets, and it is defined as net income divided by total assets (NI/TA). Descriptive measures for samples of 400 companies for three different years are summarized by the box plots shown in Computer Exhibit 3.2. The box plots show that distributions for NI/TA did not change too much during the three years, that the central tendency for return on assets was about .05, or 5%, and that the middle half of the companies had returns of from 2.5% to 10%.

Perhaps the most important feature of the box plots of NI/TA are the possible and probable outliers, marked with *'s and O's, respectively. Practically all of the probable outliers have negative returns on assets, and these companies may be

---

COMPUTER EXHIBIT 3.2 Box Plots of Returns on Assets for 400 Companies
Across Three Years

MINITAB

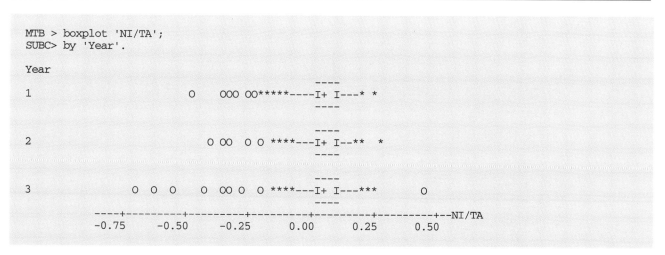

Note: The edges of MINITAB box plots (the I's) are located at values known as **hinges**. Hinges are variations of quartiles, and often they are equal to quartiles. Hinges are found by their *depth*, or the number of ordered values the hinge value is from the nearer edge of a data array. The depths of hinges are $depth_{hinges} =$ int($depth_{Md}$ + 1)/2, where int is the integer part function and $depth_{Md}$ is the depth of the median. The depth of the median is $depth_{Md} = (n + 1)/2$.

For example, consider the five ordered values 1, 2, 4, 6, and 7. The depth of the median for the five values is $depth_{Md} = (5 + 1)/2 = 3$ (the median is then the third value from either edge of the ordered data, so the median is 4). The depths of the hinges for the five values are $depth_{hinges} =$ int(3 + 1)/2 = 4/2 = 2. The hinges are then found by locating the second value from either edge of the

ordered data, so the values for the hinges are 2 and 6.

If the depths of the hinges have a fractional part, then the value of a hinge will be a weighted mean of two contiguous ordered values. For example, consider the four ordered values 1, 2, 4, and 6. The depth of the median for the four values is $depth_{Md} = (4 + 1)/2 = 2.5$; and the depths of the hinges are $depth_{hinges} =$ int(2.5 + 1)/2 = int(3.5)/2 = 3/2 = 1.5. The value of the lower hinge is the weighted mean of the first and second ordered values from the left of the array (with the fractional part and one minus the fractional part as weights); so the lower hinge is [(.5)(1) + (1 − .5)(2)]/1 = 1.5. Likewise, the upper hinge is the weighted mean of the first and second ordered values from the right of the array, or [(.5)(6) + (1 − .5)(4)]/1 = 5.

experiencing financial difficulties. Managers need to turn these companies around, investors should be wary of these companies, and analysts need to be wary of outliers in financial ratio data.

EXAMPLE 3.17 A computer was used to obtain a comprehensive data summary of the Western Electronics absenteeism data. Computer Exhibit 3.3 provides

SAS

COMPUTER EXHIBIT 3.3 Comprehensive Data Summary of the Western Electronics Absenteeism Data

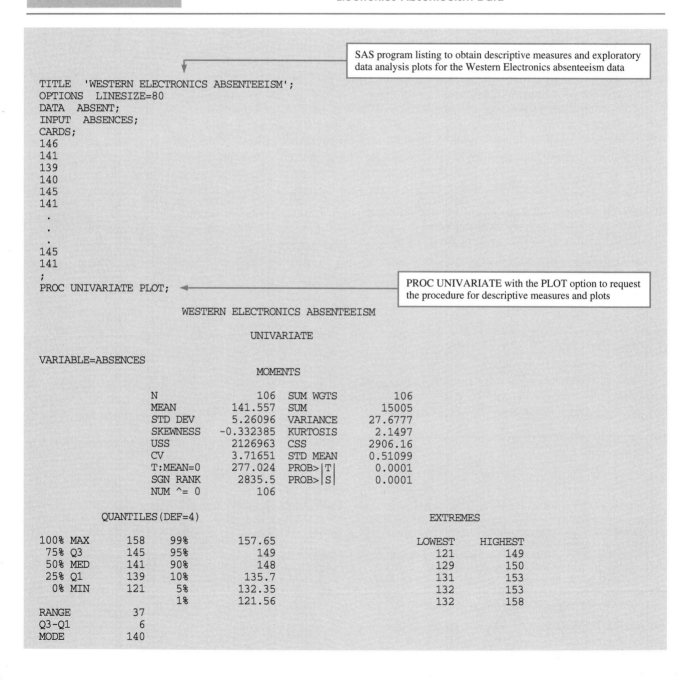

SAS program listing to obtain descriptive measures and exploratory data analysis plots for the Western Electronics absenteeism data

```
TITLE  'WESTERN ELECTRONICS ABSENTEEISM';
OPTIONS  LINESIZE=80
DATA  ABSENT;
INPUT  ABSENCES;
CARDS;
146
141
139
140
145
141
 .
 .
 .
145
141
;
PROC UNIVARIATE PLOT;
```

PROC UNIVARIATE with the PLOT option to request the procedure for descriptive measures and plots

```
                      WESTERN ELECTRONICS ABSENTEEISM

                              UNIVARIATE

VARIABLE=ABSENCES
                                    MOMENTS

                N              106  SUM WGTS           106
                MEAN       141.557  SUM              15005
                STD DEV    5.26096  VARIANCE       27.6777
                SKEWNESS -0.332385  KURTOSIS        2.1497
                USS        2126963  CSS            2906.16
                CV         3.71651  STD MEAN       0.51099
                T:MEAN=0   277.024  PROB>|T|        0.0001
                SGN RANK    2835.5  PROB>|S|        0.0001
                NUM ^= 0       106

        QUANTILES(DEF=4)                              EXTREMES

100% MAX    158    99%   157.65           LOWEST    HIGHEST
 75% Q3     145    95%      149              121        149
 50% MED    141    90%      148              129        150
 25% Q1     139    10%    135.7              131        153
  0% MIN    121     5%   132.35              132        153
                    1%   121.56              132        158
RANGE        37
Q3-Q1         6
MODE        140
```

SAS		

COMPUTER EXHIBIT 3.3 Comprehensive Data Summary of the Western Electronics Absenteeism Data (Continued)

```
STEM LEAF                                          #              BOXPLOT
 158 0                                             1                 0
 156
 154
 152 00                                            2
 150 0                                             1                 |
 148 00000000000                                  11
 146 000000                                        6                 |
 144 0000000000000                                13              +-----+
 142 00000000000000000                            17              |     |
 140 000000000000000000000000                     24              *--+--*
 138 00000000000000                               14              +-----+
 136 0000000                                       7                 |
 134 000                                           3
 132 0000                                          4                 |
 130 0                                             1                 |
 128 0                                             1                 0
 126
 124
 122
 120 0                                             1                 0
     ----+----+----+----+----
```

Note: A sample **skewness** coefficient, such as $g_1 = [\Sigma(X_i - \overline{X})^3/n]/[s(n-1)/n]^3$, measures the direction and magnitude of asymmetry about the mean for a set of data. Skewness is zero for symmetric sample distributions, positive for distributions skewed or stretched to the right, and negative for distributions skewed to the left.

A sample **kurtosis** coefficient, such as $g_2 = [\Sigma(X_i - \overline{X})^4/n]/[s(n-1)/n]^4 - 3$, is generally considered to be a measure of a distribution's tail heaviness and peakedness near the center. Generally, kurtosis is zero for symmetric, bell-shaped distributions; positive for distributions with larger tail frequencies and sharper peakedness near the center than bell-shaped distributions; and negative for distri-butions with smaller tail frequencies and flatter peakedness near the center than bell-shaped distributions. Skewness and kurtosis measures for small samples can be unreliable, especially if the data contain outliers.

The plots are stem-and-leaf and box plots. Possible outliers are marked with a zero on the box plot; and if any values were probable outliers, they would be marked with asterisks. A third plot, known as a normal probability plot, was omitted since it is beyond the scope of our discussion.

a host of descriptive measures, percentiles or quantiles, a stem-and-leaf display, and a box plot for the absence data. ═══

Managers often use computers and data analysis software to provide descriptive measures and exploratory data analyses like those presented in Computer Exhibits 3.1 through 3.3 to assist them in the decision-making process.

Problems: Section 3.15

21. Lenders of home mortgage money often employ flexible interest rates that begin at a specific interest rate and then either increase or decrease as market interest rates change over the years. A sample of beginning mortgage interest rates (given as percentages) for flexible rate loans is presented here. Use a computer to compute descriptive measures for the following sample of interest rates (as percentages):

12.07	12.40	11.61	11.55	12.37	11.90	11.81	12.13	12.29
12.33	12.42	11.70	12.35	12.09	12.02	12.41	11.66	11.90
11.58	12.33	12.23	12.02	12.43	12.18	12.17		

22. A manager for Demobank Corporation is concerned about the time that the bank's customers spend in the waiting line prior to receiving service at a teller's window. A sample of waiting times (in minutes) is presented in the accompanying data. The

bank manager has established the policy that the normal waiting time should be about .5 minute and that customers should rarely be required to wait in line for more than 1 minute. Use a computer to compute appropriate statistics for the sample, and use the box plot from MYSTAT in Computer Exhibit 3.4 to comment on the degree to which the bank's operations are meeting the manager's intended policy.

.08	.73	.77	.91	.19	.91	.74	.26	.50
.63	.99	.13	.57	.54	.30	.17	.79	.53
.39	.94	.59	.91	.31	.10	.80		

 23. A manager of human resources at Transpacific Realty Investors is concerned about employee job satisfaction. Several employees participate in a study and respond to a questionnaire designed to measure job satisfaction. The following numbers give the sample of job satisfaction scores. Higher scores indicate a greater degree of satisfaction. Use a computer to determine appropriate descriptive measures and a box plot for the degree of job satisfaction of the employees.

42.9	49.1	59.9	40.6	73.2	47.8	36.5	64.5	32.8
56.4	84.2	58.3	82.9	44.0	43.3	49.1	63.5	
77.1	43.8	78.6	75.7	87.8	85.5	30.2	60.0	

MYSTAT   **COMPUTER EXHIBIT 3.4** Box Plot of Waiting Times

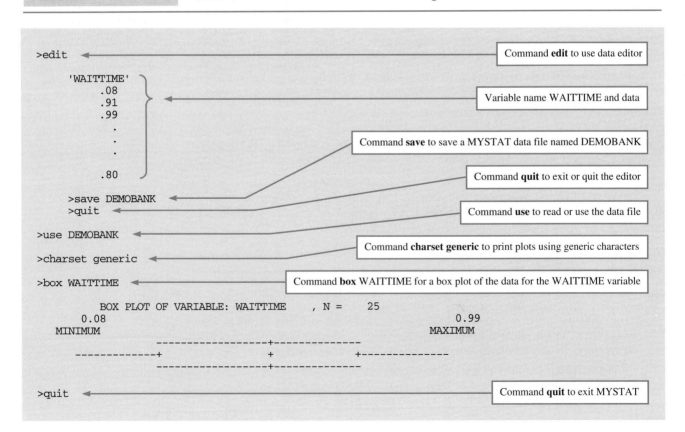

In this chapter we considered several descriptive measures for both population data and sample data. *Descriptive measures* give numerical values that allow the decision maker to describe or measure the central tendency (or location) and also the dispersion (or scatter) of the data.

We differentiated descriptive measures for population data and sample data because the sample measures (or sample statistics) will be used to estimate unknown population measures (or parameters) under the topic of statistical inference. The

3.16
SUMMARY

TABLE 3.1   Summary of Formulas for Descriptive Measures

Descriptive Measure	*Formula*	
Population mean	$\mu = \dfrac{1}{N} \displaystyle\sum_{i=1}^{N} X_i$	(3.1)
Sample mean	$\overline{X} = \dfrac{1}{n} \displaystyle\sum_{i=1}^{n} X_i$	(3.1a)
Weighted mean	$\mu = \dfrac{\sum w_i X_i}{\sum w_i}$	(3.2a)
Population variance	$\sigma^2 = \dfrac{1}{N} \sum (X_i - \mu)^2$	(3.7)
Computational formula for population variance	$\sigma^2 = \dfrac{1}{N}\left[\sum X_i^2 - \dfrac{1}{N}\left(\sum X_i\right)^2\right]$	(3.8)
Population standard deviation	$\sigma = \sqrt{\sigma^2} = \sqrt{\dfrac{1}{N}\sum (X_i - \mu)^2}$	(3.9)
Sample variance	$s^2 = \dfrac{1}{n-1}\sum (X_i - \overline{X})^2$	(3.10)
Computational formula for sample variance	$s^2 = \dfrac{1}{n-1}\left[\sum X_i^2 - \dfrac{1}{n}\left(\sum X_i\right)^2\right]$	(3.10a)
Sample standard deviation	$s = \sqrt{s^2} = \sqrt{\dfrac{1}{n-1}\sum (X_i - \overline{X})^2}$	(3.11)
Population or sample coefficient of variation	$CV = \left(\dfrac{\sigma}{\mu}\right) \times 100 \quad \mu > 0$ $CV = \left(\dfrac{s}{\overline{X}}\right) \times 100 \quad \overline{X} > 0$	(3.12)

mean, weighted mean, median, and *mode* are measures of central tendency. The *range, mean deviation, variance, standard deviation,* and *coefficient of variation* are measures of dispersion. We examined what the standard deviation tells us because this measure of variation is typically new to statistics students and because it is an important measure in many of the remaining chapters of the text.

This chapter presented computational formulas for variances and standard deviations as tools to lessen the computational burden. Some of the important formulas for descriptive measures are summarized in Table 3.1. Properties of the mean, the median, and the mode relative to *resistance* to outliers or to changes in a few data values and to the shape (symmetry or skewness) of the histogram for the data were also discussed.

Many students have access to a computer that will provide descriptive measures, exploratory analyses, and *box plots* similiar to those presented in the printouts we provided in the chapter.

REVIEW PROBLEMS

Answers to odd-numbered problems are given in the back of the text.

24. The losses at financial services companies from bad loans are plaguing banks, savings and loan associations, and insurance companies. The following data are a sample of loan losses (in millions of dollars) suffered by a savings and loan association:

    2   5   9   11   13

    a. Determine the mean.   **b.** Determine the median.
    c. Find the mode (or modes) if it exists. **d.** Determine the range.
    e. Determine the variance.   **f.** Find the standard deviation.

25. The losses at financial services companies from bad loans are plaguing banks, savings and loan associations, and insurance companies. The following data are a sample of loan losses (in millions of dollars) suffered by an insurance company:

    1   3   3   5   6   6

    a. Determine the mean.   **b.** Determine the median.
    c. Find the mode (or modes) if it exists. **d.** Determine the range.
    e. Determine the variance.   **f.** Find the standard deviation.

26. The following data are a sample of the changes in the Dow–Jones Industrial Average over successive days:

    0   0   0   0

    a. Determine the mean.   **b.** Determine the median.
    c. Find the mode (or modes) if it exists. **d.** Determine the range.
    e. Determine the variance.   **f.** Find the standard deviation.

27. The following data are a sample of the three-month Treasury bill percentage yields over time:

    3   6   2   5   3   8   6   7   5

    a. Determine the mean.   **b.** Determine the median.
    c. Find the mode (or modes) if it exists. **d.** Determine the range.
    e. Determine the variance.   **f.** Find the standard deviation.

28. The following data are a sample of the four-month percentage rates of return for several stocks:

    −4   2   −6   0   −4   6   2   4   0

a. Determine the mean. **b.** Determine the median.
c. Find the mode (or modes) if it exists. **d.** Determine the range.
e. Determine the variance. **f.** Find the standard deviation.

29. The following data are a sample of the number of defects found by inspectors for automobiles exiting an assembly line:

    1 2 3 4 5

a. Determine the mean. **b.** Determine the median.
c. Find the mode (or modes) if it exists. **d.** Determine the range.
e. Determine the variance. **f.** Find the standard deviation.

30. The following data are a sample of the weights of infants that use disposable diapers produced by Baggy No-Leak Environmental Disposables:

    16 2 22 8 6 20 24 14

a. Determine the mean. **b.** Determine the median.
c. Find the mode (or modes) if it exists. **d.** Determine the range.
e. Determine the variance. **f.** Find the standard deviation.

31. The following data are a sample of the percentage rates of return experienced by six-month Eurodollar deposit purchasers:

    −1 2 −2 1 −1 4 2 3 1

a. Determine the mean. **b.** Determine the median.
c. Find the mode (or modes) if it exists. **d.** Determine the range.
e. Determine the variance. **f.** Find the standard deviation.

32. The amounts of precipitation acidity (pH) as measured in a sample of seven containers of rainfall collected at Acid Rain Lake are as follows:

    4.2 4.0 3.8 3.0 6.0 4.0 4.0

a. Determine the mean. **b.** Determine the median.
c. Find the mode (or modes) if it exists. **d.** Determine the range.
e. Determine the variance. **f.** Find the standard deviation.

33. The accompanying data represent the weights of 30 cattle (in pounds) after they have been in a Midland Industries feedlot for 60 days. For these sample data, find the following.
a. The mean. **b.** The median. **c.** The range.
d. The variance. **e.** The standard deviation.
f. The standard deviation if you consider the data to be a population.

982	1205	258	927	620	1023
395	1406	1012	762	840	960
1056	793	713	736	1582	895
1384	862	1152	1230	1261	624
862	1650	368	358	956	1425

34. The following data are the daily sales (in thousands of dollars) achieved during the previous day by a sample of people who sell to industrial firms for National Chemical Company:

    2.3 3.4 2.2 3.5 4.5 5.7

a. Determine the mean. **b.** Determine the variance.
c. Determine the standard deviation. **d.** Determine the median.
e. Determine the range.

35. The yield to maturity of a set of tax-free municipal bonds that an individual investor has in his bond portfolio is given here. The yield to maturity can be considered to be a type of profitability measure. The following data represent the yields (in percentages) for all of the bonds of this type in the investor's portfolio, and the investor wants a numerical description of the yield of just these bonds:

    8.8   9.0   8.6   8.4   8.7   9.3   9.5

    a. Is the set of bonds a sample or a population?
    b. Determine the mean, the variance, the standard deviation, the median, and the range.

36. A sample of patients at the Eastern Clinic has incurred the following charges (in hundreds of dollars):

    2.3   3.2   2.8   4.2   8.3   5.6

    a. Determine the mean.   **b.** Determine the variance.
    c. Determine the standard deviation.   **d.** Determine the median.
    e. Determine the range.

37. The amounts of property taxes (in thousands of dollars) assessed by Potawatomie County for a sample of residential properties are given here:

    8   7   3   3   6   4   4

    a. Determine the mean.   **b.** Determine the variance.
    c. Determine the standard deviation.   **d.** Determine the median.
    e. Determine the range.
    f. Determine an interval within which essentially 90% or more of the values in this data set will fall according to the empirical rule.

38. For fixed-income securities *duration* is the weighted maturity of a bond. The duration measures for a sample of 7 bonds traded on the over-the-counter market are presented here:

    7   1   6   3   3   4   4

    a. Determine the mean.   **b.** Determine the variance.
    c. Determine the standard deviation.   **d.** Determine the median.
    e. Determine the range.

39. The number of customers waiting in line for service at the drive-in window at a Burger Inn restaurant for 7 different observations taken at different times are presented here:

    5   0   0   0   3   3   3

    a. Determine the mean.   **b.** Determine the variance.
    c. Determine the standard deviation.   **d.** Determine the median.
    e. Determine the range.

40. A store manager for Fern's Candies found that the store's mean daily receipts totaled $550, with a standard deviation of $200. What can this manager say about the proportion of days on which receipts lie between $0 and $1150?

41. The balances in the escrow account for Grand Western Financial's mortgage portfolio have a mean of $305 and a standard deviation of $85.

    a. According to the general empirical rule defining the meaning of a $\pm$ three-standard deviation interval, essentially, at least what proportion of the accounts have balances between $50.00 and $560.00?
    b. According to the empirical rule for bell-shaped data, about what proportion of the accounts have balances between $220.00 and $390.00?

42. Two histograms are presented in Figure 3.13, and both are labeled with the variable X. One of these histograms represents the amount of overfill (in ounces) that are being placed in flour sacks by an automatic filling machine. The other histogram represents the monthly salaries (in thousands of dollars) of a sample of managers.

   a. Does Figure 3.13(a) or 3.13(b) match the amount of overfill being placed in the flour sacks? Explain.

   b. Which histogram is appropriate for the salaries per month of managers? Explain.

   c. Determine the relative locations of the mean, the median, and the mode for each of the histograms. Comment on the skewness of each histogram.

FIGURE 3.13

Histograms for Problem 42

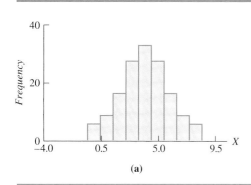
(a)

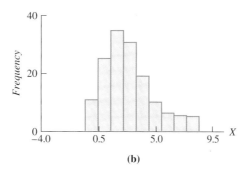
(b)

43. During a recent maintenance check the maximum thrust of the engines of 40 DC9's was measured. The following data give the differences between the thrusts of the port and starboard engines, measured in pounds. Find the mean, the variance, and the standard deviation for the following sample:

70	50	480	−270	−50	−40	−100	−30	10	20
0	−40	190	60	−150	200	220	−230	−240	−450
310	−20	−330	−400	440	80	30	90	90	110
−340	−90	360	−20	−180	60	210	130	−80	−200

44. A report on the profitability of stocks traded on the New York Stock Exchange, the American Stock Exchange, and the over-the-counter market listed the *best* 50 stocks for the year based on return to investors. The list, assumed to be a sample here, is given in Table 3.2. Small but growing companies had the year's best returns.

   a. Use a computer to find the descriptive statistics for return to investors for these stocks with the best returns.

   b. Why is the mean return not equal to the median return?

   c. If an individual investor owned a high-performing stock for the year, then in what interval would the return for the stock fall?

   d. Determine an interval symmetric about the mean within which essentially 90% or more of the returns must fall.

   e. Can we say that about 68% of the returns for the best stocks will fall in the interval from one standard deviation below the mean to one standard deviation above the mean? Explain.

TABLE 3.2	*50 Best Stocks*	Stock Price		*Dividends per Share*	*Total Return to Investors (%)*
		Year 1	*Year 2*		
Stock Data for Problem 44	1. Viratek (Costa Mesa, Calif.)	$10.625	$63.00	$0.00	493
	2. Chambers Development (Pittsburgh)	$10.50	$16.875	$10.44	221
	3. Control Resource Industries (Michigan City, Ind.)	$6.125	$18.75	$0.00	206
	4. Crown Crafts (Calhoun, Ga.)	$8.25	$24.00	$0.00	191
	5. Jefferson Smurfit (Alton, Ill.)	$13.625	$38.50	$0.50	188
	6. COMB (Minneapolis)	$7.812	$22.25	$0.00	185
	7. North American Holding (Hartford)	$5.125	$7.25	$6.38	183
	8. Envirodyne (Chicago)	$8.125	$22.125	$0.00	172
	9. Old National Bancorp (Spokane)	$15.568	$42.00	$0.20	171
	10. Woodstream (Lititz, Pa.)	$9.00	$23.875	$0.40	171
	11. Seagate Technology (Scotts Valley, Calif.)	$7.25	$19.125	$0.00	164
	12. Olson Industries (Sherman Oaks, Calif.)	$11.75	$29.50	$0.00	151
	13. Reebok International (Avon, Mass.)	$9.333	$23.375	$0.00	150
	14. Hovnanian Enterprises (Red Bank, N. J.)	$7.444	$18.50	$0.00	149
	15. Owens-Corning Fiberglas (Toledo)	$37.50	$13.75	$69.60	148
	16. First Capital Holdings (Las Vegas)	$6.00	$14.875	$0.00	148
	17. Nord Resources (Dayton)	$8.583	$21.25	$0.00	148
	18. Circuit City Stores (Richmond)	$12.437	$30.625	$0.06	147
	19. Genentech (South San Francisco)	$34.875	$85.00	$0.00	144
	20. Ashton-Tate (Culver City, Calif.)	$18.375	$44.50	$0.00	142
	21. Cherokee Group (North Hollywood, Calif.)	$9.50	$23.00	$0.00	142
	22. Maxtor (San Jose, Calif.)	$7.187	$17.375	$0.00	142
	23. HAL (Honolulu)	$9.00	$21.75	$0.00	142
	24. Convenient Food Mart (Rosemont, Ill.)	$5.818	$14.00	$0.00	141
	25. SPI Pharmaceuticals (Covina, Calif.)	$11.75	$28.00	$0.06	139
	26. First Federal Bank FSB (Nashua, N. H.)	$17.75	$41.50	$0.40	137
	27. Roper (Kankakee, Ill.)	$7.937	$18.375	$0.36	137
	28. DST Systems (Kansas City, Mo.)	$11.25	$26.00	$0.20	133
	29. Gap (San Bruno, Calif.)	$15.687	$35.75	$0.31	130
	30. Westamerica Bancorp (San Rafael, Calif.)	$22.00	$49.00	$0.70	128
	31. Bayly (Denver)	$6.00	$13.50	$0.12	127
	32. Intermet (Atlanta)	$5.416	$12.125	$0.16	127

	Stock Price		Dividends per Share	Total Return to Investors (%)	
50 Best Stocks	Year 1	Year 2			**TABLE 3.2** Continued

Stock Data for Problem 44

50 Best Stocks	Year 1	Year 2	Dividends per Share	Total Return to Investors (%)
33. American Barrick Resources (Toronto)	$6.687	$15.125	$0.00	126
34. Ponderosa (Dayton)	$13.00	$28.75	$0.40	126
35. Texas Air (Houston)	$15.00	$33.75	$0.00	125
36. Presidential Life (Nyack, N.Y.)	$12.50	$28.00	$0.06	125
37. Countrywide Credit (Pasadena, Calif.)	$5.637	$12.375	$0.25	124
38. American Israeli Paper Mills (Hadera, Israel)	$7.875	$17.125	$0.30	123
39. Riverside Group (Jacksonville, Fla.)	$5.00	$11.125	$0.00	123
40. Duquesne Systems (Pittsburgh)	$14.375	$31.75	$0.00	121
41. Ryan's Family Steak Houses (Greenville, S.C.)	$9.722	$21.375	$0.00	120
42. Computer Factory (New York City)	$7.968	$17.50	$0.00	120
43. For Better Living (San Juan Capistrano, Calif.)	$5.125	$11.00	$0.10	118
44. Rollins Environmental Services (Wilmington, Del.)	$11.916	$25.875	$0.06	118
45. View-Master Ideal Group (Beaverton, Ore.)	$8.333	$18.00	$0.00	116
46. Puerto Rican Cement (San Juan)	$8.125	$17.375	$0.10	115
47. Eastern Bancorp (Burlington, Vt.)	$11.00	$23.50	$0.10	115
48. Micropolis (Chatsworth, Calif.)	$8.75	$18.75	$0.00	114
49. Ero Industries (Chicago)	$6.833	$14.625	$0.00	114
50. Laidlaw Industries (Hinsdale, Ill.)	$7.75	$16.375	$0.12	114

SOURCE: *Fortune*, February 2, 1987, p. 78. © 1987 Time Inc. All rights reserved.

45. A report on fourth-quarter sales, profits, and profit margins for large companies, assumed to be a sample here, listed the data shown in Table 3.3. Cost-cutting moves helped some companies achieve higher profits as the dollar became stronger, but low oil prices resulted in lower profits or losses for some companies in the energy industry.

   a. Use a computer to find the descriptive statistics for profit margins for these companies.
   b. Why is the mean profit margin not equal to the median margin?
   c. If a manager believes that the profit margin for her corporation is comparable to those reported here, then in what interval would the margin for her corporation fall?
   d. Determine an interval within which essentially 90% or more of the margins will fall.
   e. Can we say that about 95% of the margins for the best stocks will fall in the interval from two standard deviations below the mean to two standard deviations above the mean? Explain.

TABLE 3.3	Company	Sales (in $millions)	Profits (in $millions)	Margin (%)
Company Data for	Abbott Laboratories	$1,063.2	$160.6	15.1
Problem 45	Alcoa	1,160.0	165.3	14.3
	American Brands	2,197.2	117.4	5.3
	American Home Products	1,192.8	199.4	16.7
	Amoco	4,839.0	165.0	3.4
	AMR	1,498.6	6.6	0.4
	Atlantic Richfield	3,724.0	64.0	1.7
	Bellsouth	2,800.0	361.1	12.9
	Boeing	4,982.0	191.0	3.8
	Boise Cascade	969.3	32.0	3.3
	Borden	1,356.4	68.8	5.1
	Bristol-Myers	1,205.0	151.7	12.6
	Burlington Northern	1,609.8	75.6	4.7
	Caterpillar	1,775.0	−148.0	NM
	Champion International	1,058.0	64.5	6.1
	Chevron	5,900.0	−86.0	NM
	CSX	1,604.0	132.0	8.2
	Exxon	18,836.0	1,480.0	7.9
	General Electric	12,270.0	730.0	6.0
	Georgia-Pacific	1,781.0	83.0	4.7
	Hercules	640.5	38.9	6.1
	IBM	16,945.0	1,390.0	8.2
	Ingersoll-Rand	768.4	41.1	5.3
	Inland Steel Industries	817.5	39.7	4.9
	International Paper	1,900.0	113.0	5.9
	Kraft	2,414.5	117.1	4.9
	Martin Marietta	1,268.5	43.4	3.4
	Merck	1,149.1	171.8	15.0
	Merrill Lynch	2,764.8	197.6	7.1
	Mobil	12,009.0	201.0	1.7
	Monsanto	1,570.0	23.0	1.5
	Motorola	1,619.0	63.0	3.9
	NCR	1,575.1	134.1	8.5
	North American Philips	1,356.0	47.6	3.5
	Nynex	2,900.0	293.7	10.1
	Pfizer	1,162.9	158.4	13.6
	PPG Industries	1,200.8	72.9	6.1
	Procter & Gamble	4,255.0	190.0	4.5
	Quaker Oats	1,083.6	30.1	2.8
	Raytheon	2,041.2	99.6	4.9
	Reynolds Metals	863.9	−4.3	NM
	Santa Fe Southern Pacific	1,503.9	−384.7	NM
	Smithkline Beckman	1,043.3	147.4	14.1
	Standard Oil	2,264.0	51.0	2.7
	Tandy	1,196.0	103.8	8.7
	Teledyne	834.8	47.0	5.6
	Texas Instruments	1,335.8	26.6	2.0
	3M	2,112.0	185.0	8.8

Company	Sales (in $millions)	Profits (in $millions)	Margin (%)	
				TABLE 3.3 Continued
United Technologies	4,401.4	−252.6	NM	**Company Data for**
USG	695.0	56.7	8.2	**Problem 45**
USX	2,770.0	−1,167.0	NM	
Warner-Lambert	812.6	56.5	7.0	
Westinghouse Electric	2,869.1	203.2	7.1	
Weyerhaeuser	1,450.6	96.1	6.6	
Whirlpool	942.9	42.2	4.5	

NM = not meaningful

SOURCE: *Business Week,* February 9, 1987, p. 31. Data from Standard & Poor's Compustat Services, Inc.

46. For the following percentages of defective items in eight samples, find the percentage of defective items when the samples are combined into one large sample:

Sample Size	Percentage Defective
50	6.0
20	5.0
35	20.0
150	0.0
100	1.0
75	4.0
40	2.5
200	0.5

47. The value of the Japanese yen in dollars has tended to increase over the past few years as Japan's economy has become stronger. The yen-per-dollar exchange rate for a sample of 20 months is given in the accompanying listing. A currency trader wants to measure the central tendency for the data.

a. Find the mean.   b. Find the median.

c. What advantage does the median have as a measure of central tendency for the data?

129	114	139	141	137	144	123	134	132	105
118	134	140	139	129	124	131	120	142	137

48. As we noted in Problem 47, the value of the Japanese yen in dollars has tended to increase over the past few years as Japan's economy has become stronger. The yen-per-dollar exchange rate for a sample of 20 months is given in the accompanying listing. A currency trader wants to measure the variability of the exchange rate.

a. Find the range.   b. Find the variance.

c. Find the standard deviation.

129	114	139	141	137	144	123	134	132	105
118	134	140	139	129	124	131	120	142	137

49. Investors in long-term bonds have tended to incur devaluations in their portfolios in the past few years. As corporations have been taken over or have undergone restructuring, the level of debt in corporate capital structures has been increased. The increased debt increases the risk that the companies will default on their long-term debt obligations, and the

increased risk drives long-term bond values down. A measure of the risk from events such as mergers, leveraged buy-outs, or recapitalizations is the spread between the yield of long-term bonds and the yield of essentially riskless Treasury bonds. The spread is measured in basis points (100 basis points are equivalent to 1 percentage point). The spreads for a sample of 20 new bonds are given in the accompanying listing. A financial analyst wants to know the central tendency for the spreads.

a. Find the mean.   **b.** Find the median.

c. Is the mean or the median the more resistant measure of central tendency?

98	85	106	96	80	90	93	97	98	102
93	96	110	111	86	100	109	118	88	121

50. Refer to Problem 49, which described how investors in long-term bonds have tended to incur devaluations in their portfolios in the past few years. A measure of the risk to investors in bonds from events such as mergers, leveraged buy-outs, or recapitalizations is the spread between the yield of long-term bonds and the yield of essentially riskless Treasury bonds. The spread is measured in basis points (100 basis points are equivalent to 1 percentage point). The spreads for a sample of 20 new bonds are given in the accompanying listing. A financial analyst wants to describe the variation in the data.

a. Find the range.   **b.** Find the variance.   **c.** Find the standard deviation.

d. Find an interval around the mean in which about 95% of the data values should fall. Assume the data have a bell-shaped distribution.

98	85	106	96	80	90	93	97	98	102
93	96	110	111	86	100	109	118	88	121

51. Recently, the most expensive securities markets internationally have been stock exchanges in Japan. The price-earnings ratio is an indicator of the cost of a stock in terms of how much investors have been willing to pay per dollar of income statement profits. For instance, the price-earnings ratios for stocks traded on U.S. securities markets have tended to vary around a central value of about 13 in recent times. The price-earnings ratios for a sample of 35 stocks traded on a Japanese securities market are given in the accompanying listing. A global investor wishes to describe the central tendency of the price-earnings ratio for Japan.

a. Find the mean.   **b.** Find the median.

65	61	68	64	60	65	61	67	60
55	66	59	69	61	59	58	59	65
61	62	56	53	54	61	54	54	62
62	54	63	64	60	61	60	63	

52. As we described in Problem 51, the most costly securities markets internationally in recent times have been Japanese stock exchanges. The price-earnings ratio is an indicator of a stock's cost in terms of how much investors have been willing to pay per dollar of income statement profits. The price-earnings ratios for a sample of 35 stocks traded on a Japanese securities market are given in the accompanying listing. A global investor wishes to describe the variation in the price-earnings ratio in Japanese markets.

a. Find the range.   **b.** Find the variance.   **c.** Find the standard deviation.

d. Find an interval around the mean in which about 95% of the data values should fall. Assume the data have a bell-shaped distribution.

65	61	68	64	60	65	61	67	60
55	66	59	69	61	59	58	59	65
61	62	56	53	54	61	54	54	62
62	54	63	64	60	61	60	63	

 53. A financial analyst with First Metro Bancshares wishes to examine the yields on 90-day Treasury bills. The information will be used to ascertain the pattern of short-term interest

rates. The yield data were collected monthly for two years, and the results are presented in order (left to right from first row to last row) in the accompanying listing.

a. Use a computer to obtain descriptive measures for the data.
b. Use a computer to obtain a box plot for the data.
c. To help explain the variability of short-term interest rates to your bank's treasurer, find a symmetric interval around the mean in which you can be sure that 90% or more of the yields lie.

7.76	8.22	8.57	8.00	7.56	7.01	7.05	7.18
7.08	7.17	7.20	7.07	7.04	7.03	6.59	6.06
6.12	6.21	5.84	5.57	5.19	5.18	5.35	5.49

SOURCE: U.S. Department of Labor, Bureau of Labor Statistics (1986), *Business Statistics, 1986* (Washington, D.C.: U.S. Government Printing Office).

54. An economist with First Metro Bancshares wishes to examine the discount rate at the New York Federal Reserve Bank. The information will be used to ascertain the pattern of interest rates for banks borrowing funds from the Federal Reserve. The discount data were collected monthly for two years, and the results are presented in order (left to right from first row to last row) in the accompanying listing.

a. Use a computer to obtain descriptive measures for the data.
b. Use a computer to obtain a box plot for the data.
c. To help explain the variability of the discount rate to your bank's treasurer, find a symmetric interval around the mean in which you can be sure that 90% or more of the discount rates lie.

8.00	8.00	8.00	8.00	7.81	7.50	7.50	7.50
7.50	7.50	7.50	7.50	7.50	7.50	7.10	6.83
6.50	6.50	6.16	5.82	5.50	5.50	5.50	5.50

SOURCE: U.S. Department of Labor, Bureau of Labor Statistics (1986), *Business Statistics, 1986* (Washington, D.C.: U.S. Government Printing Office).

55. An economist with the Federal Reserve Bank of Cleveland wishes to examine the total amount of personal income. Data on personal income are given in Table 3.4. The economist is concerned that a recession will result in a decrease in personal income.

a. Find the mean. b. Find the median.
c. Why is the median an advantageous measure of the central tendency for the data?
d. Find the range. e. Find the variance. f. Find the standard deviation.

Refer to the 145 companies listed in the Ratio Data Set in Appendix A.

Ratio Data Set

Questions

56. Locate the current-assets–to–sales ratio data.

a. Use a computer to obtain descriptive measures for the data.
b. Use a computer to obtain a box plot for the data.
c. What is notable about this summary of the data for decision making?

57. Locate the quick-assets–to–sales ratio data.

a. Use a computer to obtain descriptive measures for the data.
b. Use a computer to obtain a box plot for the data.
c. Explain how the box plot is resistant to outliers or to changes in a few data values.

58. Locate the net-income–to–total-assets ratio data.

a. Use a computer to obtain descriptive measures for the data.
b. Use a computer to obtain a box plot for the data.
c. Explain how the box plot is resistant to outliers or to changes in a few data values.

TABLE 3.4	Year	Personal Income	Less: Personal Tax and Nontax Pay- ments	Equals: DPI*	Less: Personal Outlays	Equals: Personal Savings	Saving as Per- centage of DPI	DPI in Constant (1987) Dollars
Personal Income	1959	391.2	44.5	346.7	324.7	22.0	6.4	1,284.9
	1960	409.2	48.7	360.5	339.9	20.6	5.7	1,313.0
	1961	426.5	50.3	376.2	351.3	24.9	6.6	1,356.4
	1962	453.4	54.8	398.7	372.8	25.9	6.5	1,414.8
	1963	476.4	58.0	418.4	393.7	24.6	5.9	1,461.1
	1964	510.7	56.0	454.7	423.1	31.6	6.9	1,562.2
	1965	552.9	61.9	491.0	456.4	34.6	7.0	1,653.5
	1966	601.7	71.0	530.7	494.3	36.4	6.9	1,734.3
	1967	646.5	77.9	568.6	522.8	45.9	8.1	1,811.4
	1968	709.9	92.1	617.8	573.9	43.9	7.1	1,886.8
	1969	773.7	109.9	663.8	620.4	43.4	6.5	1,947.4
	1970	831.0	109.0	722.0	664.4	57.6	8.0	2,025.3
	1971	893.5	108.7	784.9	719.3	65.5	8.3	2,099.9
	1972	980.5	132.0	848.5	788.6	59.9	7.1	2,186.2
	1973	1,098.7	140.6	958.1	871.9	86.2	9.0	2,334.1
	1974	1,205.7	159.1	1,046.5	953.0	93.5	8.9	2,317.0
	1975	1,307.3	156.4	1,150.9	1,050.4	100.4	8.7	2,355.4
	1976	1,446.3	182.3	1,264.0	1,170.7	93.2	7.4	2,440.9
	1977	1,601.3	210.0	1,391.3	1,303.1	88.1	6.3	2,512.6
	1978	1,807.9	240.1	1,567.8	1,459.6	108.1	6.9	2,638.4
	1979	2,033.1	280.2	1,753.0	1,629.3	123.7	7.1	2,710.1
	1980	2,265.4	312.4	1,952.9	1,798.6	154.3	7.9	2,733.6
	1981	2,534.7	360.2	2,174.5	1,982.1	192.4	8.8	2,795.8
	1982	2,690.9	371.4	2,319.6	2,119.6	200.0	8.6	2,820.4
	1983	2,862.5	368.8	2,493.7	2,324.7	169.1	6.8	2,893.6
	1984	3,154.6	395.1	2,759.5	2,537.2	222.3	8.1	3,080.1
	1985	3,379.8	436.8	2,943.0	2,753.2	189.8	6.4	3,162.1
	1986	3,590.4	459.0	3,131.5	2,943.6	187.8	6.0	3,261.9
	1987	3,802.0	512.5	3,289.5	3,146.9	142.6	4.3	3,289.6
	1988	4,075.9	527.7	3,548.2	3,392.0	156.2	4.4	3,404.3
	1989	4,380.2	591.7	3,788.6	3,621.6	166.9	4.4	3,471.2
	1990	4,679.8	621.0	4,058.8	3,852.2	206.6	5.1	3,538.3

*Disposable personal income.

SOURCE: *Survey of Current Business* (November, 1991), U.S. Bureau of Economic Analysis, Washington, D.C.

Credit Data Set *Refer to the 113 applicants for credit listed in the Credit Data Set in Appendix A.*

Questions 59. Find the mean and the standard deviation of JOBYRS for all applicants who were employed for wages. That is, leave out applicants who appear to have been retired, nonwork-
ing students, or housewives. [*Hint: n = 107.*]

60. Find the mean and the standard deviation of ADDINC for all applicants who were denied
credit.

ANALYZING THE FAIRNESS OF THE 1970 DRAFT LOTTERY

On December 1, 1969, the United States Selective Service System conducted the 1970 draft lottery for draft-eligible 19- to 25-year-old men. The lottery was supposed to determine a random selection sequence that was fair and impartial by which eligible men would be inducted into the military. Registrants who received lower-priority draft numbers were inducted into the armed forces first, and those who received higher draft numbers were not drafted. The draft lottery was extremely important because the United States was in the midst of a controversial war in Vietnam, and many men who were drafted ended up in combat. There were more than fifty thousand casualties during the Vietnam War.

Prior to conducting a televised drawing, Selective Service officials placed 366 dates, January 1 through December 31, in capsules. The dates were placed in the capsules in

						Month							
Day	Jan	Feb	Mar	Apr	May	Jun	Jul	Aug	Sep	Oct	Nov	Dec
1	305	086	108	032	330	249	093	111	225	359	019	129
2	159	144	029	271	298	228	350	045	161	125	034	328
3	251	297	267	083	040	301	115	261	049	244	348	157
4	215	210	275	081	276	020	279	145	232	202	266	165
5	101	214	293	269	364	028	188	054	082	024	310	056
6	224	347	139	253	155	110	327	114	006	087	076	010
7	306	091	122	147	035	085	050	168	008	234	051	012
8	199	181	213	312	321	366	013	048	184	283	097	105
9	194	338	317	219	197	335	277	106	263	342	080	043
10	325	216	323	218	065	206	284	021	071	220	282	041
11	329	150	136	014	037	134	248	324	158	237	046	039
12	221	068	300	346	133	272	015	142	242	072	066	314
13	318	152	259	124	295	069	042	307	175	138	126	163
14	238	004	354	231	178	356	331	198	001	294	127	026
15	017	089	169	273	130	180	322	102	113	171	131	320
16	121	212	166	148	055	274	120	044	207	254	107	096
17	235	189	033	260	112	073	098	154	255	288	143	304
18	140	292	332	090	278	341	190	141	246	005	146	128
19	058	025	200	336	075	104	227	311	177	241	203	240
20	280	302	239	345	183	360	187	344	063	192	185	135
21	186	363	334	062	250	060	027	291	204	243	156	070
22	337	290	265	316	326	247	153	339	160	117	009	053
23	118	057	256	252	319	109	172	116	119	201	182	162
24	059	236	258	002	031	358	023	036	195	196	230	095
25	052	179	343	351	361	137	067	286	149	176	132	084
26	092	365	170	340	357	022	303	245	018	007	309	173
27	355	205	268	074	296	064	289	352	233	264	047	078
28	077	299	223	262	308	222	088	167	257	094	281	123
29	349	285	362	191	226	353	270	061	151	229	099	016
30	164		217	208	103	209	287	333	315	038	174	003
31	211		030		313		193	011		079		100

TABLE 1

1970 Draft Numbers by Day of Birth

FIGURE 1

**Box Plots of 1970 Draft
Numbers by Month**

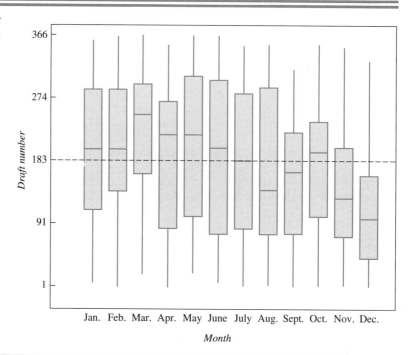

chronological order and then the capsules were placed in a box, one month at a time. Just before the drawing the capsules were poured from the box into a bowl. The capsules were generally drawn from the top of the bowl, and draft sequence numbers were assigned to the eligible men whose birth dates matched the capsule dates according to the order in which the capsules were drawn. Some mixing of the capsules occurred during this process, but whether the mixing was thorough enough to make the lottery fair and impartial is questionable.

The results for the 1970 draft lottery are presented in Table 1. If the lottery had been conducted fairly and impartially so that all eligible draftees were treated equally, then we would expect the distribution of draft numbers to be similar for each month. Box plots of the 1970 draft numbers by month are presented in Figure 1. The box plots suggest that draft-eligible men having birthdays later in the year, especially in the months of November and December, tended to receive lower draft numbers than men who had birthdays earlier in the year. The pattern that can be seen with the aid of the box plots suggests that the capsules were not mixed thoroughly enough to obtain a fair and impartial sequence for induction into the armed services. The box plots indicate that the draft numbers reflect the order in which the capsules were placed in the box and then in the bowl.

To conduct a fair and impartial draft lottery, the Selective Service System should have achieved a thorough mixing of the capsules. Unfortunately, a thorough mixing of capsules in a large bowl is difficult to achieve. Apparently, the capsules were not mixed thoroughly, and all registrants were not treated equally in the 1970 draft lottery.

In response to criticisms of the 1970 lottery, the Selective Service System used a different procedure for the 1971 lottery. Two different drums were used for drawing birth dates and draft numbers in the 1971 lottery. Furthermore, birth dates were placed in capsules in a fair and impartial order, and draft numbers were also placed in capsules in an impartial order. Both the drum containing the birth dates and the drum containing the draft numbers were rotated to ensure that the capsules were mixed thoroughly. The mixing that was achieved in the 1971 draft lottery made the system of selection fair and impartial, and all registrants were treated equally. Figure 3.18 shows that box plots can be used effectively to explore the features of sets of data and to compare distributions of data across time.

Use a computer to obtain descriptive measures of the draft numbers in Table 1 by month.

Case Assignment

REFERENCE: Fienberg, S.E. (1971), "Randomization and Social Affairs: The 1970 Draft Lottery," *Science,* 171: 255–61.

EXPLORING COMPUSTAT DATA FILES

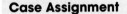

CASE 3.2

Standard & Poor's COMPUSTAT Services, Inc., provides files of financial, statistical, and market data for several thousand companies. Most of the companies listed on the New York and American stock exchanges are included in the COMPUSTAT files. The COMPUSTAT files list key balance sheet, income statement, and market items that can be used by managers, security analysts, investors, and lenders in financial analyses. For example, COMPUSTAT users can compute financial ratios, calculate growth rates, analyze investment portfolios, and perform exploratory data analyses.

A file of financial ratio data is given in Appendix A. The ratio data file contains 141 companies; their company numbers *CNUM;* industry numbers *DNUM;* current assets/sales (*CA/S*), quick assets/sales (*QA/S*), current assets/current liabilities (*CA/CL*), and net income/total assets (*NI/TA*) accounting ratios; and the company name. The ratios were computed from COMPUSTAT data, and the companies were sampled from the COMPUSTAT files. The ratios measure financial dimensions of inventory intensiveness, receivables intensiveness, short-term liquidity, and return on assets. The first five lines of the file are reproduced in the accompanying listing.

CNUM	DNUM	CA/S	QA/S	CA/CL	NI/TA	CONAME
449842	800	0.854	0.854	3.611	0.123	IP TIMBERLANDS –LP
000776	1040	22.770	22.770	39.895	−0.005	ABM GOLD CORP –CL A
422704	1040	0.361	0.221	3.216	0.061	HECLA MINING CO
053435	1311	1.304	1.305	1.661	0.022	AVALON CORP
730448	1311	0.411	0.384	0.876	−0.001	POGO PRODUCING CO

The COMPUSTAT files are often used for exploring the characteristics of companies and industries and for examining the relationships among variables for companies and industries. A computer was used to obtain descriptive measures and a box plot for the *CA/CL* ratio, and the results are presented in Figure 1. In addition to displaying key descriptive measures, including the first and third quartiles and the median, the box plot shows that several of the companies are outliers and have very large *CA/CL* ratios relative to the bulk of the remaining companies. These companies may be maintaining more short-term liquidity than is needed, or perhaps some financial difficulty may exist. Managers, investors, or lenders may need to do more detailed examinations of these companies.

FIGURE 1

Descriptive Measures and
Box Plots for *CA/CL* from
Ratio Data Set of
Appendix A Using MINITAB

	N	MEAN	MEDIAN	TRMEAN	STDEV	SEMEAN
CA/CL	141	2.699	1.866	2.146	3.966	0.334

	MIN	MAX	Q1	Q3
CA/CL	0.345	39.895	1.310	2.953

```
            ---
          --+ I--** O      O              O                        O
            ---
          --+---------+---------+---------+---------+---------+----CA/CL
          0.0       8.0      16.0      24.0      32.0      40.0
```

Case Assignment

Use the ratio data set from Appendix B for the 141 companies and a computer to solve the following problems.

a. Find descriptive measures for the *CA/S, QA/S,* and *NI/TA* ratios.

b. Obtain box plots for the *CA/S, QA/S,* and *NI/TA* ratios.

c. Recode or transform the SIC code data according to the nine industries listed in the Ratio Data Set of Appendix A. Find descriptive measures for the *NI/TA* ratio data by the nine industries.

d. Obtain box plots for the *NI/TA* ratio by the nine industries.

REFERENCE: Watson, Collin J. (1990), "Multivariate Distributional Properties, Outliers, and Transformation of Financial Ratios," *The Accounting Review,* 65: 682–695.

REFERENCES

Hoel, Paul G. 1971. *Elementary Statistics.* 3rd ed. New York: Wiley.

Huff, Darrell. 1965. *How to Lie with Statistics.* New York: Norton.

Ryan, B., B. Joiner, and T. Ryan. 1985. *Minitab Handbook.* 2nd ed. Boston: Duxbury Press.

Snedecor, George W., and William G. Cochran. 1989. *Statistical Methods.* 8th ed. Ames: Iowa State University Press.

Velleman, P. F., and D. C. Hoaglin. 1981. *Applications, Basics, and Computing of Exploratory Data Analysis.* Boston: Duxbury Press.

There are times when we want to find the descriptive statistics—mean, median, mode, and standard deviation—of a data set that has been grouped into a frequency distribution. Maybe someone has already grouped the data for us, and we do not have access to the raw data to make these calculations as they have been presented in the previous sections of this chapter. Or sometimes, when we are faced with finding the mean for a very large number of observed values and do not have a data analysis system available, we can materially decrease the labor involved by grouping the data into a frequency distribution and then finding the statistics for the grouped data. However, keep in mind that by grouping the data, we have lost information, and the statistics obtained from the grouped data will only approximate those of the ungrouped data. If the number of observations is large and the class intervals are small, the approximation will be very good.

To illustrate the calculation of the mean for grouped data, we apply the procedures to the population data of the annual earnings of 300 part-time employees at NuFlor, Inc., that are summarized in Table 3.4. To find the mean of this distribution, we operate as if each observed value in a given class were equal to the class mark for that class. Now the observations are $X_1, X_2, \ldots, X_N$, where $N = 300$ for this set of data, and these observations are grouped into k classes, where $k = 10$ in this case. Let v_j denote the class mark for the jth class, and let f_j denote the class frequency for that class. Now if each observation X_i falling into the jth class had a value of exactly v_j instead of approximately v_j, the sum of all the observations in the jth class would be exactly $v_j f_j$. In Table 3.4 the class mark of the fourth class (1100 but less than 1300) is $v_4 = 1200$, and the class frequency is $f_4 = 70$. If each employee in the fourth class made exactly \$1200, the sum of the earnings for these 70 employees would be $v_4 f_4 = 1200 \times 70$, or \$84,000. This value will *approximate* their total earnings even if their individual earnings are spread out over the range \$1100–\$1300. The total earnings for all 300 employees is approximated by summing the individual class totals:

$$\sum_{j=1}^{k} v_j f_j = v_1 f_1 + v_2 f_2 + \cdots + v_k f_k$$

Notice that since each of the N observations goes into exactly one of the k classes, the class frequencies f_j must add to the total number of observations N:

$$\sum_{j=1}^{k} f_j = f_1 + f_2 + \cdots + f_k = N$$

	Annual Earnings	*Class Mark,* v_j	*Number of Employees,* f_j
TABLE 3.4			

Annual Earnings	*Class Mark,* v_j	*Number of Employees,* f_j
$ 500 but less than 700	$ 600	12
700 but less than 900	800	21
900 but less than 1100	1000	52
1100 but less than 1300	1200	70
1300 but less than 1500	1400	68
1500 but less than 1700	1600	36
1700 but less than 1900	1800	16
1900 but less than 2100	2000	11
2100 but less than 2300	2200	9
2300 but less than 2500	2400	5
	Total	300

Annual Earnings of 300 Part-Time Employees

The **mean for grouped data** is a weighted mean with the class midpoints as data values and the frequencies as weights. It can be found as follows.

Equation for Mean for Grouped Data

$$\mu = \frac{\sum v_j f_j}{\sum f_j} = \frac{1}{N} \sum_{j=1}^{k} v_j f_j \qquad (3.13)$$

For the data of Table 3.4, which we will view to be a population of 300 earnings figures, we calculate intermediate values for finding the mean as follows:

Class Mark, v_j	*Frequency,* f_j	*Product,* $v_j f_j$
$ 600	12	7,200
800	21	16,800
1000	52	52,000
1200	70	84,000
1400	68	95,200
1600	36	57,600
1800	16	28,800
2000	11	22,000
2200	9	19,800
2400	5	12,000
	$N = \sum f_j = 300$	$\sum v_j f_j = 395,400$

Using the intermediate values in the table, the mean is

$$\mu = \frac{395,400}{300} = \$1318$$

If the data with which we are working form a sample, we find the mean just as is shown in Equation (3.13), but we divide by the sample size n, rather than the population size N, and we call it $\overline{X}$ rather than μ.

Since we often have data in the form of a grouped frequency distribution, we need a way of estimating the median for grouped data. Recall that when raw data are grouped into classes, information is lost because the distinction between the various observations in each individual class is lost. The following method assumes that the observations in each class are more or less evenly spread over that class, so that this loss of information is of little importance.

We found the 50th percentile, or median, in Chapter 2 by using the ogive. There is an arithmetic procedure equivalent to the graphical procedure with the ogive; it is less convenient if the ogive has been constructed, but it does not require the ogive. The equation for finding the **median for grouped data** is given next.

Equation for Median for Grouped Data

$$Md = LB_{Md} + \frac{n/2 - Cum.\ freq.}{f_{Md}} \times Width \qquad (3.14)$$

where: Md = median

LB_{Md} = lower boundary of class containing median

$Width$ = width of class containing median

f_{Md} = frequency of class containing median

$n/2 - Cum.\ freq.$ = $n/2$ minus the cumulative frequency of the class preceding the class containing the median, n being the total number of observations (if the values are a population of data, then use N in place of n)

=== **EXAMPLE 3.18** Consider the population cumulative frequency distribution in Table 3.5 for the earnings of part-time employees given in Table 3.4.

Since 150 (half of 300, the number of employees) is between 85 and 155, the median must fall in the fourth class. The lower boundary for that class is 1100, the width of the class is 200, its frequency is 70, $N/2$ is 150, and the cumulative frequency at the lower boundary of the fourth class is 85. Thus

$$LB_{Md} = 1100 \qquad Width = 200 \qquad f_{Md} = 70$$
$$N/2 - Cum.\ freq. = 150 - 85 = 65$$

The formula gives

$$Md = 1100 + \frac{65}{70} \times 200 = 1100 + 185.7 = 1285.7$$

So the median earnings is about $1286. ==

Annual Earnings	Number of Employees	Cumulative Frequency	**TABLE 3.5**
$ 500 but less than 700	12	12	**Cumulative Frequency**
700 but less than 900	21	33	**Distribution for Data**
900 but less than 1100	52	85	**in Table 3.4**
1100 but less than 1300	70	155	
1300 but less than 1500	68	223	
1500 but less than 1700	36	259	
1700 but less than 1900	16	275	
1900 but less than 2100	11	286	
2100 but less than 2300	9	295	
2300 but less than 2500	5	300	
Total	300		

So far in this section we have identified methods for finding the mean and the median of data that have been grouped into a frequency distribution. These measures, however, are measures of location—measures of the typical value in the data set. We often want to measure the variation in a set of data values too. Since the variance and the standard deviation are the most commonly used measures of variation in statistics, we will only consider methods of finding these two measures in grouped data.

When we want to calculate the variance and the standard deviation for data that have been grouped to form a frequency table, we follow the three-step procedure outlined previously, but the calculation of the sum of squares requires a slight modification. As when computing the mean μ, we proceed as though each value in a given class were equal to the class mark. To find the sum of squares, we calculate the mean for the distribution by the procedure of Equation (3.13) and then find the deviations of the class marks v_j from the mean. The square of the deviation of a class mark from the mean $(v_j - \mu)^2$, is the amount each value in the class contributes to the sum of squares. If there are f_j items in the jth class, the total contribution of the class is $(v_j - \mu)^2 f_j$. Therefore, the sum of squares is the sum of the contributions of all of the classes,

$$\sum (v_j - \mu)^2 f_j$$

The **variance of grouped data** is given next.

Equation for Population Variance of Grouped Data

$$\sigma^2 = \frac{1}{N} \sum_{j=1}^{k} (v_j - \mu)^2 f_j \qquad\qquad (3.15)$$

Of course, if the grouped data we are using form a sample, then we use $\overline{X}$ to represent the mean of the sample, and the **sample variance of grouped data** is calculated as follows.

Equation for Sample Variance of Grouped Data

$$s^2 = \frac{1}{n-1} \sum_{j=1}^{k} (v_j - \overline{X})^2 f_j \qquad\qquad (3.16)$$

where $n = \sum_{j=1}^{k} f_j$ is the total number of observations in the sample.

EXAMPLE 3.19   Let us use the numbers in a frequency distribution to go through an illustrative calculation. The following table shows the age distribution in a *sample* of 101 viewers who regularly watch the advertising on a television game show:

Age	Class Mark, v_j	Frequency, f_j
0–4	2	30
5–9	7	51
10–14	12	10
15–19	17	10
Total		101

The intermediate calculations required to obtain the variance and the standard deviation for this distribution are displayed in the next table:

v_j	f_j	$v_j f_j$	$v_j - \overline{X}$	$(v_j - \overline{X})^2$	$(v_j - \overline{X})^2 f_j$
2	30	60	−5	25	750
7	51	357	0	0	0
12	10	120	5	25	250
17	10	170	10	100	1000
	101	707			2000

Using the intermediate calculations shown in the table, the sample mean is

$$\overline{X} = \frac{1}{n} \sum v_j f_j = \frac{1}{101}(707) = 7 \text{ years}$$

and the sample variance and standard deviation are

$$s^2 = \frac{1}{n-1}\sum (v_j - \overline{X})^2 f_j = \frac{1}{100}(2000) = 20\,(\text{years})^2$$
$$s = \sqrt{s^2} = \sqrt{20} = 4.47 \text{ years}$$

Note that in Example 3.19 the deviations of the class marks from the mean do *not* sum to zero. For grouped data, it is the sum of the products of the frequencies times the deviations of the class marks from the mean that is equal to zero.

Computational Formula for Sum of Squares of Grouped Data

$$\sum v_j^2 f_j - \frac{1}{n}\left(\sum v_j f_j\right)^2 \tag{3.17}$$

where v_j is the class mark.

EXAMPLE 3.20 If we use the computational formula to find the sum of squares for the preceding table of ages of regular viewers of the advertising on a game show, the calculations are as follows:

v_j	v_j^2	f_j	$v_j f_j$	$v_j^2 f_j$
2	4	30	60	120
7	49	51	357	2499
12	144	10	120	1440
17	289	10	170	2890
		101	707	6949

The sum of squares is thus

$$\sum v_j^2 f_j - \frac{1}{n}\left(\sum v_j f_j\right)^2 = 6949 - \frac{1}{.101}(707)^2$$
$$= 6949 - 4949 = 2000$$

and the sample variance and standard deviation are

$$s^2 = \frac{1}{100}(2000) = 20 \,(\text{years})^2$$

$$s = \sqrt{20} = 4.47 \text{ years}$$

Because of the information lost in grouping, the variance and standard deviation are, in general, not exact when calculated from the grouped data. They are good approximations, however; and the smaller the class interval, the better the approximation is. Neither the mean nor the variance can be calculated when the distribution contains open-ended classes.

Problems: Section 3.17

61. The accompanying distribution gives the number of defectives found in a sample of 404 lots of manufactured items. Find the mean, the median, the mode, the variance, and the standard deviation for the number of defectives. [*Hint:* Use the formulas for grouped data as though the numbers 1, 2,..., 12 were class marks v_j.]

Number of Defective Items	Number of Lots	Number of Defective Items	Number of Lots
0	53	7	12
1	110	8	9
2	82	9	3
3	58	10	1
4	35	11	2
5	20	12	1
6	18		

62. The following frequency distribution shows the ages of new-car purchasers at a large midwestern automobile dealership. Find the mean, the variance, and the standard deviation of this distribution. (Assume it is a population.)

Class	Frequency
33–37	10
38–42	10
43–47	51
48–52	30

63. The following frequency distribution shows the ages of used-car purchasers at the same large midwestern automobile dealership mentioned in Problem 62. Find the mean, the variance, and the standard deviation of this distribution. Do the new-car and used-car departments appear to be selling to different age groups? What implications does this information have for the management of this dealership?

Class	Frequency
16–21	15
22–27	16
28–33	5
34–39	5

64. Find the sample mean and median for the following relative frequency distribution, which represents the number of cabinets that are produced with zero defects at a local cabinet shop per day.

Class	Relative Frequency
20 but less than 40	.12
40 but less than 60	.28
60 but less than 80	.36
80 but less than 100	.24

65. Find the sample mean and median for the following relative frequency distribution, which represents the number of defective items assembled per day in an electronics equipment assembly plant.

Class	Relative Frequency
5 but less than 10	.2
10 but less than 15	.5
15 but less than 20	.3

66. The following frequency distribution shows the ages of home buyers in a Chicago neighborhood. Find the mean, the variance, and the standard deviation for this sample distribution.

Class	Frequency
18–32	5
33–37	10
38–42	10
43–47	30
48–52	35
53–67	10

67. If a set of grouped data has an open-ended class for either the first or the last class, which of the following *cannot* be found from the grouped distribution?

   a. The class marks of all the classes.   **b.** The boundaries of all the classes.
   c. The mean.   **d.** The range.
   e. The standard deviation.   **f.** The median.

68. For the estimated numbers of married women in the labor force of a region of the eastern United States, as given in the accompanying frequency table, find the following.

   a. The mean age.   **b.** The median age.
   c. The variance.   **d.** The standard deviation.
   e. The coefficient of variation.

Age	Number of Married Women (thousands)
15–19	2212
20–24	3503
25–29	1313
30–34	939
35–39	3490
40–44	3005
45–54	5096
55–64	2904
65–80	885

69. In an investigation of electronics equipment reliability, a record of the time between equipment malfunctions was kept for nearly two years. Forty pieces of equipment were involved and gave rise to a sample of 850 observations on times between malfunctions.

    a. Find the median and the mode.
    b. Find the mean, the variance, and the standard deviation for this sample.

Time Between Malfunctions (hours)	Number of Observations
$0 \leq X < 100$	20
$100 \leq X < 200$	43
$200 \leq X < 300$	60
$300 \leq X < 400$	75
$400 \leq X < 500$	95
$500 \leq X < 600$	138
$600 \leq X < 700$	240
$700 \leq X < 800$	97
$800 \leq X < 900$	62
$900 \leq X < 1000$	20
Total	850

Probability

4

In this chapter we begin our study of the second major topic area in the subject of statistics: probability. As we saw in Chapter 1, probability theory provides a basis for evaluating the reliability of the conclusions we reach and the inferences we make when we apply statistical techniques to the collection, analysis, and interpretation of data. Since probability plays so important a role in statistics, we need to be familiar with the elements of the subject.

In this chapter we consider briefly some of the basic ideas of probability. Some of the concepts are illustrated by straightforward and familiar examples involving cards and dice to show the historical development of probability. Connections between probability and decision making in management and economics are also made in this chapter.

We begin this chapter by discussing the *meaning of probability* in terms of random experiments, outcomes, sample spaces, and events. *Relative frequency, classical,* and *subjective methods of assigning probabilities* to events are also considered. Then we cover *properties* of probabilities and *formulas for computing probabilities* for several compound events. *Probability tree diagrams* are used to illustrate some of the probability concepts. Probability distributions for discrete and continuous random variables are covered in the next two chapters.

4.1
PROBABILITY CONCEPTS

Consider an operation that can result in any one of a definite set of possible observations known as **outcomes**. Further, the operation must be governed by chance—so the actual outcome cannot be predicted with complete certainty—and it must be repeatable under stable conditions. We have in mind, for example, the drawing of a card from a well-shuffled deck, the rolling of a pair of dice, the selection of an invoice during an audit, or the inspection of a part coming off a production line. A repeatable operation such as this is known as a **random experiment.**

> **Definition**
>
> A **random experiment** or **random trial** is an operation that is repeatable under stable conditions and that results in any one of a set of outcomes; furthermore, the actual outcome cannot be predicted with certainty.

If the random experiment consists of drawing a card from an ordinary deck, there are 52 possible outcomes: the 52 cards in the deck. This set of all possible outcomes, or results, of the experiment is called the **sample space,** denoted by S.

> **Definition**
>
> A **sample space S** is the set of all possible outcomes of an experiment.

EXAMPLE 4.1   For the roll of a single die the sample space—the set of 6 possible outcomes—is exhibited in Figure 4.1. An accounting firm in an auditing engagement may have 100 invoices in which 95 are correctly written and 5 are not. If an auditor for the firm randomly selects only one invoice, then the sample space consists of 100 possible outcomes, which can be separated into two subsets: correct, which has 95 elements, and error, which has 5 elements.

FIGURE 4.1

Simple Space for Rolling a Single Die

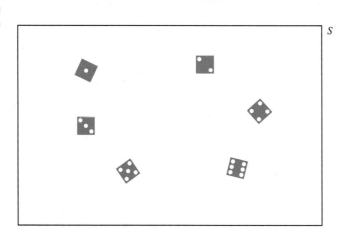

In each of the operations in the previous examples, there is a definite set of possible outcomes (the 52 cards, the 6 outcomes for the rolled die, or the 100 invoices), known as the sample space. Within the sample space, there is usually some smaller set of outcomes that are of interest (the spades, the even numbers of dots on a die, or the invoices that are correctly written). This smaller set is called an **event**, or a **subset**, and the outcomes in the subset are called the outcomes *favorable* to the event.

Definition

An **event** is a subset of one or more outcomes of a random experiment.

An event that is a single possible outcome of an experiment that cannot be decomposed into simpler outcomes, such as any one of the 6 possible outcomes of a rolled die, is called a *simple event* or a *sample point*.

We want to assign to each event a numerical quantity, called the **probability of the event,** that measures the chance that the event will occur; furthermore, the probability must be a number between 0 and 1. We denote events by A, B, C, and so on, and their probabilites by $P(A)$, $P(B)$, $P(C)$, and so on.

There are several different ways to assign probabilities to events. The *relative frequency method,* the *classical method,* and the *subjective method* of assigning probabilities to events are discussed in the next three subsections.

4.1.1 Relative Frequency Method of Assigning Probabilities to Events

Suppose that we repeatedly perform an experiment, such as inspecting parts coming off a Unitel, Inc., production line; and we keep track of an event such as *part rejected.* At each repetition of the experiment, we compute the relative frequency with which the event has occurred in the sequence of trials up to that point. Quite generally, over a very long series of repetitions, the relative frequency will stabilize at some limit or it will approach some limit, which we take to be the probability of the event. This probability is to be regarded as a natural characteristic of the experiment and of the event itself.

For example, experience shows that if a coin is tossed repeatedly, the relative frequency of heads tends to stabilize at a definite value, which we take to be the probability of heads. Figure 4.2 shows the results of such an experiment. The horizontal scale in this figure represents the number of tosses of the coin, and the height of the curve over any given point on the horizontal scale represents the relative frequency of heads up to that point. In this experiment the coin was tossed 600 times in all. Since the relative frequency seems to be settling down at or approaching 1/2, we can say that the probability of a head is 1/2 and the coin is well balanced.

Figure 4.3 shows the same effect for a different coin. Here the relative frequency is converging to approximately .3, and thus the coin is unbalanced or bent.

By now we can see that probability can be assigned to an event in terms of its **relative frequency probability** in the long run.

Relative Frequency Method of Assigning Probabilities to Events

The **relative frequency probability** that an event will be the result of a random experiment is assigned as the *proportion* of times that the event occurs as the outcome of the experiment *in the long run*.

FIGURE 4.2

**Relative Frequencies of
Heads for a Balanced Coin**

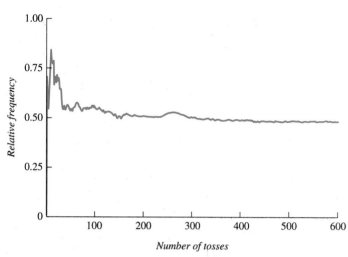

If we undertake a large number of repetitions of a random experiment, we can compute the approximate probability of an event by using Equation (4.1).

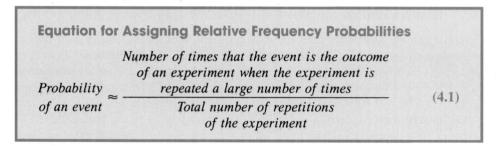

Equation for Assigning Relative Frequency Probabilities

$$\frac{Probability}{of\ an\ event} \approx \frac{\begin{array}{c}Number\ of\ times\ that\ the\ event\ is\ the\ outcome\\ of\ an\ experiment\ when\ the\ experiment\ is\\ repeated\ a\ large\ number\ of\ times\end{array}}{\begin{array}{c}Total\ number\ of\ repetitions\\ of\ the\ experiment\end{array}} \qquad (4.1)$$

FIGURE 4.3

**Relative Frequencies of
Heads for an Unbalanced
Coin**

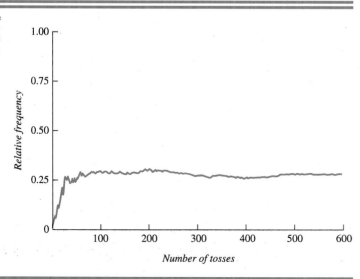

4.1.2 Classical Method of Assigning Probabilities to Events

We need not perform all experiments repeatedly and observe the results in order to determine the probabilities of events. For instance, we do not have to roll an ordinary die for a long time in order to determine that the probability of rolling a 1 (for one dot) is 1/6. If the die is fair (not loaded), everyone expects the outcomes depicted in Figure 4.1 to be **equally likely.** In addition, all outcomes depicted in Figure 4.1 are **mutually exclusive** in that all of the outcomes such as 1 and 3 are distinctly different and none is a subset of or is contained in the others.

> **Definition**
>
> Two sets of outcomes are **mutually exclusive** if there is no outcome that be-longs to both sets.

If the outcomes of an experiment are *mutually exclusive* and *equally likely,* then we can assign probabilities to events by the following method.

> **Classical Method of Assigning Probabilities to Mutually Exclusive and Equally Likely Events**
>
> If the outcomes of an experiment are *mutually exclusive* and *equally likely,* then the **classical probability** that an event will be the result of a random experiment is the number of outcomes of an experiment that are favorable to the event divided by the total number of outcomes of the experiment.

We can assign a probability to an event by the classical method by using the follow-ing equation.

> **Equation for Computing Probabilities for Mutually Exclusive and Equally Likely Events**
>
> $$\frac{\text{Probability}}{\text{of an event}} = \frac{\substack{\textit{Number of outcomes of an experiment} \\ \textit{that are favorable to the event}}}{\substack{\textit{Total number of outcomes of} \\ \textit{the experiment}}} \qquad (4.2)$$

For the experiment of rolling a single die, consider the event of interest to be that the *number of dots showing is an even number.* Of the 6 possible outcomes depicted in Figure 4.1, three of the outcomes are favorable to the event (namely, 2, 4, and 6), and we compute the probability of rolling an even number to be 3/6. The chance of an auditor for an accounting firm selecting a *correctly written invoice* from our well-shuffled 100 invoices is 95/100, or .95. The chance of selecting one containing an *error* is 5/100, or .05.

We are able to compute these probabilities without tossing a single die or drawing an invoice. In other words, we define probabilities in terms of the physical structure of the experiment. Thus we know that a die has six sides, and from this knowledge we can calculate the theoretical probability of various outcomes for rolling a fair die. But we can also determine the same probabilities by using the relative frequency method and conducting experiments with a large number of repetitions.

4.1.3 Subjective Method of Assigning Probabilities to Events

The probabilities discussed to this point have all been objective in the sense that we have taken the probability of an event as being a property of the event itself. Statistics is sometimes placed in the framework of a personal belief of probability or **subjective probability.**

Subjective Method of Assigning Probabilities to Events

A **subjective probability** is an individual's degree of belief in the occurrence of an event.

For example, L and M Enterprises had a project loan with Sumitomo Bank for the building of a domed arena that had never been built before. The project loan agreement required that the arena be completed within 360 days. Let C denote the event that the arena would be completed within 360 days. People in the company had varying degrees of belief in C, saying things such as "It is fairly likely that we will complete the project in 360 days or less" or "It is improbable that we can finish the project in 360 days." Ultimately, the arena was completed on schedule.

Adherents of the subjective method of assigning probabilities hold that one can assign to C a probability that represents numerically the degree of a person's belief in C. The idea is that the probability will be different for different people, because they have different information about C and assess it in different ways. Subjective probability does not require ideas of repeated trials and the stabilizing of relative frequencies or of equally likely outcomes. Subjective probabilities should be rational, or in other words, they should conform to the properties of probabilities that we will discuss in the next section.

Problems: Section 4.1

Answers to odd-numbered problems are given in the back of the text.

1. Electronic Digital Devices, a microprocessor producer, wants to know the probability that a randomly selected microprocessor is defective. A batch of 400 microprocessors was selected, and each microprocessor was tested to see whether it was defective. Figure 4.4 shows the percentage of microprocessors that were found to be defective as each was tested.
   a. What is the probability that a microprocessor selected at random is defective?
   b. What concept of probability is applicable for this problem?

2. List the different possible outcomes when three coins are tossed. How many are there? How many result in two heads? Two or more heads?

3. Construction Enterprises has a project financing agreement with Citifirst Bank, and the loan agreement requires the building of a one-of-a-kind office building. Citifirst will receive payments on the financing from the cash flows generated by events that will be held in the arena. The project loan agreement requires that the project be com-

FIGURE 4.4

Percentages of Defective Microprocessors

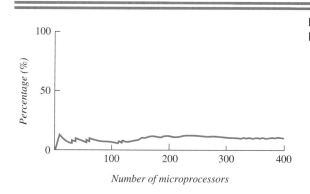

pleted within 460 days. Let A denote the event that the project will be completed within 460 days. People in the company will certainly have varying degrees of belief in A and will say things such as, "It is fairly likely that we will complete the project in 460 days or less" or "It is improbable that we can finish the project in 460 days."

a. What method of assigning probabilities to events is being used by the people in the company?
b. Why can't other methods of assigning probabilities to events be used in this situation?

REFERENCE: Brigham, E. F. (1991), *Financial Management,* 6th ed. (New York: Prentice-Hall).

We turn now to three general properties to which probabilities must conform regardless of the method used to assign probabilities.

4.2
PROPERTIES OF PROBABILITIES

Property 1

If A is an event, then

$$0 \le P(A) \le 1 \qquad (4.3)$$

In other words, the probability of an event may be 0 or 1 or a value between 0 and 1. The next property relates to an experiment with sample space S.

Property 2

The sample space contains all possible outcomes of the experiment. Thus

$$P(S) = 1 \qquad (4.4)$$

In other words, the outcome of an experiment will be in the sample space.

Another property deals with adding probabilities to determine the probability of the event "*A* or *B*." Now the event "*A* or *B*" consists of all the outcomes favorable to event *A*, or favorable to event *B*, or favorable to both events *A* and *B*. It is known as the **union** of event *A* with event *B*. As was discussed in Section 4.1, two events are *mutually exclusive,* or *disjoint,* if they have no outcomes in common. In this case the number of outcomes in the event "*A* or *B*" is the number in *A* plus the number in *B* (since there are no outcomes in both *A* and *B*), and the following equation holds.

Property 3

If event *A* is *mutually exclusive* of event *B*, then

$$P(A \text{ or } B) = P(A) + P(B) \tag{4.5}$$

In other words, this property tells us to add the probabilities for mutually exclusive events *A* and *B* in order to determine $P(A \text{ or } B)$. The property of adding probabilities for the union of mutually exclusive events also extends to the summation of probabilities for three or more mutually exclusive events.

4.3
COMPUTING PROBABILITIES

We now consider some rules for computing probabilities for different types of compound events. Business applications of probability concepts require the use of these rules.

4.3.1 Addition Rules for Computing Probabilities

The following example introduces one rule that calls for the addition of probabilities.

A single ordinary die is rolled: *A* is the event that the number of dots showing is *even*, and *B* is the event of rolling a number *less than or equal to* 4. We want to find $P(A \text{ or } B)$. Recall that the event "*A* or *B*" is known as the *union* of the event *A* with the event *B*. Here "*A* or *B*" is the event of rolling an *even* number or a number *less than or equal to* 4—the word *or* does not exclude the possibility that both events occur. Event "*A* and *B*" is the event that the die shows an *even* number and that the number is *less than or equal to* 4, and it is known as the **intersection** of event *A* and event *B*. The probability of "*A* and *B*" is denoted $P(A \text{ and } B)$ and is called the **joint probability** of events *A* and *B*.

Figure 4.5 shows that the number of outcomes in *A* is 3, the number in *B* is 4, and the number in "*A* and *B*" is 2. Notice, therefore, that *A* is not mutually exclusive of *B* because two of the outcomes (rolling a 2 or rolling a 4) are included in both *A* and *B*. If event *A* and event *B* are not mutually exclusive, as is the case in this example, the property expressed by Equation (4.5) does not apply, but the following equation does apply.

General Rule for Addition of Probabilities

If *A* and *B* are events, then

$$P(A \text{ or } B) = P(A) + P(B) - P(A \text{ and } B) \tag{4.6}$$

FIGURE 4.5

Sample Space for Rolling a Single Die

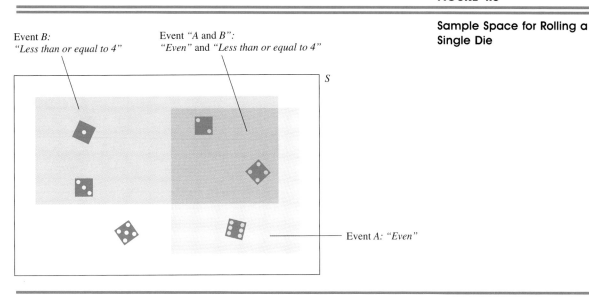

Event *B:*
"*Less than or equal to 4*"

Event "*A and B*":
"*Even*" and "*Less than or equal to 4*"

S

Event *A:* "*Even*"

Let us continue with our example, where we want to find the probability of rolling an *even* number (*A*) or a number *less than or equal to* 4 (*B*). By Equation (4.6), the probability we seek is

$$P(A \text{ or } B) = P(A) + P(B) - P(A \text{ and } B) = \frac{3}{6} + \frac{4}{6} - \frac{2}{6} = \frac{5}{6}$$

Each of the two terms $P(A)$ and $P(B)$ on the right in Equation (4.6) includes the outcomes (rolling a 2 or rolling a 4) that are in *A* and in *B*, so these outcomes have been counted twice; subtracting the term $P(A \text{ and } B)$ once compensates for this double counting.

A diagram known as a **Venn diagram** (after the English mathematician, John Venn) illustrates the concept behind Equation (4.6). The sample space *S* is represented by the area in the rectangle in Figure 4.6. For the case when event *A* and event *B* are *not* mutually exclusive, the Venn diagram of Figure 4.6(a) shows event *A* as the area in a circle, event *B* as the area in a circle, and event "*A and B*" (or the intersection of *A* with *B*) as the overlapped, color area. The Venn diagram in Figure 4.6(b) shows

FIGURE 4.6

Events "*A and B*" and "*A or B*" When *A* and *B* Are *Not* Mutually Exclusive

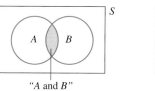

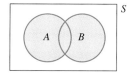

(a) Event "*A and B*"   **(b) Event "*A or B*"**

event "*A* or *B*" as the color area. Figures 4.6(a) and 4.6(b) demonstrate that the sum *P(A)* plus *P(B)* includes *P(A* and *B)* twice, and thus in Equation (4.6) we subtract *P(A* and *B)* once to compensate when *A* and *B* are *not* mutually exclusive events.

Applications of probability concepts often use tables of combined frequency counts for two or more variables. Tables of frequency counts are known as **contingency tables** or **cross-tabulation tables.**

Definition

Contingency tables or **cross-tabulation tables** present frequency counts for combinations of two or more variables.

The use of contingency tables or cross-tabulation tables is illustrated in the following probability problem.

 PRACTICE PROBLEM 4.1

Omnicare, Inc., has 245 customers that are classified in the accompanying contingency table by the frequency with which they place a regular or an irregular order and by their payment terms, cash or credit. We would like to find the probability that a randomly selected customer is either a *regular* or a *credit* customer.

	Payment Terms		
Order Type	Cash	Credit	Total
Regular	10	15	25
Irregular	20	200	220
Total	30	215	245

Solution Designating the selection of a *regular* order customer as event *A* and the selection of a *credit* payment term customer as event *B*, we desire *P(A* or *B)*. If the selection of the customer is random, each of the 245 customers has an equal opportunity of being selected. The number of customers in event *A* is 25, and the number in event *B* is 215. The number in "*A* and *B*" is the number that are both regular and credit, and this number is 15. Thus by Equation (4.7) the probability of randomly selecting a customer that is either a *regular* or a *credit* customer is

$$P(A \text{ or } B) = P(A) + P(B) - P(A \text{ and } B)$$

$$= \frac{25}{245} + \frac{215}{245} - \frac{15}{245} = \frac{225}{245}$$

Note that the 25 regular customers in the first term of this equation include 15 credit customers. But these same 15 credit customers are part of the 215 credit customers in the second term. Since these 15 are double-counted in the first two terms of the sum, they are subtraced once in the third term.

As a special case, Equation (4.6) contains the property of probabilities expressed by Equation (4.5), which applies when event A and event B are mutually exclusive. If A and B are mutually exclusive events, then the event "A and B" cannot happen and hence has a probability of zero. If we set $P(A$ and $B)$ equal to zero in Equation (4.6), we get Equation (4.5). Thus if A and B are *mutually exclusive* events, then we use Equation (4.5) as a rule of addition for computing $P(A$ or $B)$.

Rule for Addition of Probabilities of Mutually Exclusive Events

If event A and event B are *mutually exclusive,* then

$$P(A \text{ or } B) = P(A) + P(B) \tag{4.5}$$

EXAMPLE 4.1   For a roll of a single die the sample space consists of the 6 simple outcomes in Figure 4.1. If A is the event that the number rolled is *even*, we know that A contains three outcomes:

$$A = \{2, 4, 6\}$$

Event B, that the number rolled is 5, is a simple event, or the outcome

$$B = \{5\}$$

Event A and event B are *mutually exclusive* because none of the outcomes included in A is included in B; that is, if the number rolled is even, it cannot also be 5.

Now "A or B" stands for the event that the number rolled is either *even* or 5. The event "A or B" contains four outcomes, the three of A or $\{2, 4, 6\}$ and the one of B or $\{5\}$:

$$(A \text{ or } B) = \{2, 4, 5, 6\}$$

By Equation (4.5) for computing probabilities of mutually exclusive events, we see that $P(A$ or $B)$ must be equal to 4/6 and can be found as follows:

$$P(A \text{ or } B) = P(A) + P(B) = \frac{3}{6} + \frac{1}{6} = \frac{4}{6}$$

Event A may be said to be *mutually exclusive* of event B if both A and B cannot happen at the same time. Understanding this point will remove a common source of confusion. The sample space represents the possible outcomes of a single trial of the experiment, and A and B are mutually exclusive if they cannot both occur in one trial. The roll of a single die in Example 4.1 is a trial, and the outcome of the trial cannot be both an even number and 5 on one role of a die; but a single die can be even on one roll and 5 on another roll.

The *Venn diagram* of Figure 4.7 illustrates the concept of mutually exclusive events and the application of Equation (4.5). In this figure event A is the set of outcomes enclosed in the circle A and event B is the set of outcomes enclosed in the circle B. Event A and event B are mutually exclusive because they do not overlap.

FIGURE 4.7

Mutually Exclusive Events

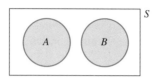

Event "*A* or *B*" is represented by the shaded area in *A* plus the shaded area in *B*, and $P(A \text{ or } B) = P(A) + P(B)$, according to Equation (4.5).

Equation (4.5) can be extended to three mutually exclusive events. If *A*, *B*, and *C* are *mutually exclusive* in the sense that no two can happen at once, then

$$P(A \text{ or } B \text{ or } C) = P(A) + P(B) + P(C) \qquad (4.7)$$

The same principle holds for four or more mutually exclusive events, as illustrated in the next example.

EXAMPLE 4.2   A group of employees at Toys Unlimited, Inc., consists of 5 production workers, 7 clerks, 4 secretaries, 6 staff trainees, 3 managers, and 2 executives, for a total of 27 people. The probability that a randomly selected person is a *secretary,* a *staff* trainee, a *manager,* or an *executive* is

$$P(Secretary \text{ or } Staff \text{ or } Manager \text{ or } Executive)$$

$$= P(Secretary) + P(Staff) + P(Manager) + P(Executive)$$

$$= \frac{4}{27} + \frac{6}{27} + \frac{3}{27} + \frac{2}{27} = \frac{15}{27} = \frac{5}{9}$$

Notice that when a situation can be phrased in terms of "either" event *A* "or" event *B* "or" event *C*, and so on, then Equations (4.5) through (4.7), which involve addition, should be considered. That is, the either-or sequence implies that probabilities should be *added* (with the possible need to subtract the probabilities that the events occur together).

Now we turn to another event, the complement event. The **complement** of an event *A* is the event that happens exactly when *A* does not. We denote the complement of event *A* by $\overline{A}$. The probabilities of event *A* and its complement $\overline{A}$ are related according to Equation (4.8).

Rule for Complementary Events

If $\overline{A}$ is the complement of event *A*, then

$$P(\overline{A}) = 1 - P(A) \qquad (4.8)$$

Figure 4.8 is a Venn diagram showing complementary events.

FIGURE 4.8

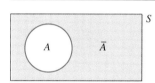

Complementary Events

EXAMPLE 4.3 In a stack of invoices being audited by an accountant with Price Waterhouse, 25 invoices are for single items, 30 are for two items, and 20 are for three or more items. We want to find the probability of selecting an *invoice with two or more items*. Notice that selecting an invoice with two or more items, event $\overline{A}$, is the complement of selecting an invoice with a single item, event A.

We can determine the probability that a randomly selected invoice is for two or more items, event $\overline{A}$, simply by subtracting the probability of selecting a single item invoice, event A, from 1. Thus the probability that a randomly selected invoice is for two or more items is

$$P(\overline{A}) = 1 - P(A) = 1 - \frac{25}{75} = \frac{50}{75} = \frac{2}{3}$$

We could also use Equation (4.5) to find the probability of selecting an invoice with *two or more items* as follows:

$$P(\textit{Two or more items}) = P(\textit{Two items or Three or more items})$$

$$= P(\textit{Two items}) + P(\textit{Three or more items})$$

$$= \frac{30}{75} + \frac{20}{75} = \frac{50}{75} = \frac{2}{3}$$

The answer is the same regardless of the way the problem is worked, but in many problems finding $P(\overline{A}) = 1 - P(A)$ is the simpler method.

4.3.2 Conditional Probability

Next, we will consider a situation that requires a *conditional probability*—the probability of an event given that another event has occurred. After we have discussed conditional probability, we will develop a general equation that uses multiplication to compute probabilities for *joint* events.

Let us introduce conditional probability with an example. Suppose we roll a single die and we define event B as rolling an *odd* number and event A as rolling a number *less than or equal to* 3. We are interested in the probability that event B, rolling an *odd* number, will occur *given* that event A, rolling a number *less than or equal to* 3, has occurred or is certain to occur.

FIGURE 4.9

Sample Space and Re-
duced Sample Space for
Rolling a Single Die

Event *A:*
"Less than or equal to 3"

Event *"A* and *B":*
*"Less than or equal
to 3"* and *"Odd"*

Event *B: "Odd"*

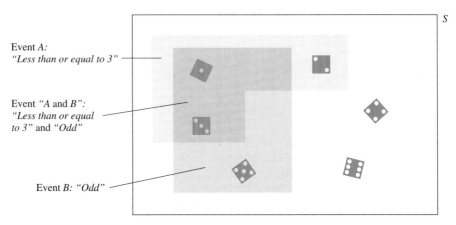

(a) **Sample space for rolling a single die**

Event *A*

Event *"A* and *B"*

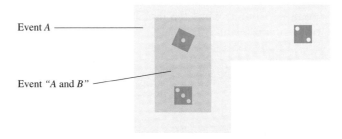

(b) **Reduced sample space given event *A***

We call the probability of event *B given* event *A* a **conditional probability,** and
we denote it as $P(B \mid A)$. Figure 4.9(a) shows the sample space for rolling a single
die. However, *given* that we know that event *A*, the number rolled is *less than or
equal to* 3, must occur, we need only consider the shaded outcomes labeled event *A*
in Figure 4.9(a). In other words, we can *reduce* the sample space *given* that event *A*
has occurred or must occur. The **reduced sample space** that results given event *A* is
depicted in Figure 4.9(b), and it shows that there are three equally likely outcomes in
the reduced sample space (1, 2, and 3) and that two of these outcomes (1 and 3) correspond to event *B*, that the number rolled is *odd*. So by Equation (4.2) we see that
$P(B \mid A) = 2/3$ from the reduced sample space.

By comparing the outcomes in the reduced sample space of Figure 4.9(b) with
the outcomes in the full sample space of Figure 4.9(a), we see that the outcomes in
the reduced sample space that correspond to event *B*, rolling an odd number, are the
outcomes that correspond to the event *"A* and *B"* in the full sample space of Figure 4.9(a). So if we find $P(A \text{ and } B) = 2/6$ and $P(A) = 3/6$ from the full sample
space of Figure 4.9(a), we see that we can compute $P(B \mid A)$ as $P(A \text{ and } B)/P(A) =
(2/6)/(3/6) = 2/3$. This result corresponds to the following definition for **condi-
tional probabilities.**

> **Definition**
>
> The **conditional probability** of event B given event A is denoted $P(B\,|\,A)$ and is defined as
>
> $$P(B\,|\,A) = \frac{P(A \text{ and } B)}{P(A)} \qquad (4.9)$$
>
> where $P(A) > 0$.

The ordinary probability $P(B)$ is often referred to as a **simple probability,** a **marginal probability,** or an **unconditional probability,** to distinguish it from the conditional probability $P(B\,|\,A)$. Conditional probabilities have the same properties as simple probabilities.

We presented the concept of conditional probability with the example of rolling a single die. The concept of conditional probability can be explained in general with the Venn diagram of Figure 4.10(a), which shows events A, B, and "A and B" as areas in a sample space. If we are given the fact that event A must occur, then we can reduce the sample space that we must consider to those outcomes in A, which gives us the denominator of Equation (4.9). Now we can see in Figure 4.10(b) that the outcomes that are favorable to event B in the reduced sample space must be in the event "A and B" of the full sample space, so we have the numerator of Equation (4.9). Now putting the numerator and denominator together gives us Equation (4.9) for the conditional probability of event B given event A.

At times, we know $P(A)$ and $P(A$ and $B)$ in advance, and we put them into Equation (4.9) to find the value of $P(B\,|\,A)$. The following practice problems demonstrate how Equation (4.9) can be used to find conditional probabilities.

 PRACTICE PROBLEM 4.2

The accompanying contingency table gives frequencies for a classification of the equipment used in a Fruehauf Corporation manufacturing plant. Find the probability

FIGURE 4.10

Sample Space and Reduced Sample Space

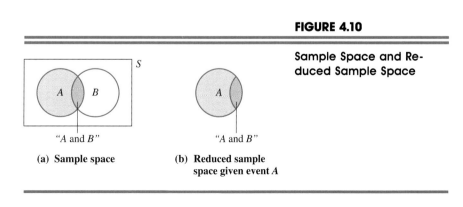

(a) Sample space (b) Reduced sample space given event A

that a randomly selected piece of equipment is a *high-use* item *given* that it is *in working order*.

Working Status	Equipment Use			Total
	Low	*Moderate*	*High*	
In working order	10	18	12	40
Under repair	2	6	8	16
Total	12	24	20	56

Solution Using Equation (4.9), we find

$$P(High\ use\ |\ In\ working\ order) = \frac{P(High\ use\ \text{and}\ In\ working\ order)}{P(In\ working\ order)}$$

$$= \frac{12/56}{40/56} = \frac{12}{40} = \frac{3}{10}$$

Note that in the double fraction of Practice Problem 4.2 for the Fruehauf Corporation data, the value 56 is in the denominator of both fractions. When this value is divided, we are left with 12/40. This result leads us to the general conclusion that when we have a table such as the one in Practice Problem 4.2, the "given" part of the conditional probability restricts or *reduces* our attention to one single row or column of the table, or reduces our sample space. For example, we were "given" that the piece of equipment selected was in working order. That information restricted our attention to the first row of the table, where there were 40 pieces of equipment. Given these 40, there were only 12 that were high use, and thus $P(High\ use\ |In\ working\ order) = 12/40$, or 3/10. Consider the equivalent approach in the following problem.

PRACTICE PROBLEM 4.3

Use the contingency table for Fruehauf Corporation in Practice Problem 4.2 to find the probability that a randomly selected piece of equipment is *under repair,* given that it is a *moderate-use* item.

Solution We can use Equation (4.9) as follows:

$$P(Under\ repair\ |\ Moderate\ use) = \frac{P(Under\ repair\ \text{and}\ Moderate\ use)}{P(Moderate\ use)}$$

$$= \frac{6/56}{24/56} = \frac{6}{24} = \frac{1}{4}$$

However, we can also consider the fact that the "given moderate use" statement restricts our attention to the second column of the table, where there are 24 pieces of equipment. Of these 24, only 6 are under repair. Thus $P(Under\ repair\ |Moderate\ use) = 6/24$, or 1/4.

4.3.3 General Rule for Multiplication of Probabilities

When any two of the three terms in Equation (4.9) are known, we can solve for the third. For instance, multiplying the formula by $P(A)$ gives the following multiplication rule, which allows us to compute **joint probabilities.**

General Rule for Multiplication of Probabilities

If A and B are events, then

$$P(A \text{ and } B) = P(A) \times P(B|A) \qquad (4.10)$$

Since we can change the positions of A and B in Equation (4.10), we can express the equation as $P(B \text{ and } A) = P(B)P(A|B)$; but since $P(B \text{ and } A) = P(A \text{ and } B)$, we can also say that $P(A \text{ and } B) = P(B)P(A|B)$.

Sometimes, we know $P(A)$ and $P(B|A)$ in advance, and we put them into Equation (4.10) to find the value of $P(A \text{ and } B)$. The following examples demonstrate how Equation (4.10) can be used to find joint probabilities.

EXAMPLE 4.4 The experiment for this example consists of drawing two cards in succession from a deck, *without replacing* the card after the first drawing. Let A be the event that the first card drawn is a *spade,* and let B be the event that the second card drawn is a *diamond.* The chance of A, $P(A)$, is 13/52. Now if the first card drawn is indeed a spade, then just prior to the second drawing the deck consists of 51 cards, since we did not replace the first card; and of the 51 remaining cards, 13 are diamonds. Thus the chance of a *diamond* on the second draw given that we drew a *spade* on the first draw without replacement must be (13/51), or $P(B|A) = 13/51$. By Equation (4.10) the chance of a *spade* and then a *diamond* is

$$P(A \text{ and } B) = P(A) \times P(B|A) = \frac{13}{52} \times \frac{13}{51} = \frac{169}{2652}$$

In this example interpreting $P(B|A)$ is accomplished by imagining yourself between the two draws, knowing that the first card was a spade but ignorant of what the second will be.

EXAMPLE 4.5 Consider again the 245 customers of Omnicare, Inc., in Practice Problem 4.1. The accompanying cross-tabulation table is repeated here from that problem.

	Payment Terms		
Order Type	Cash	Credit	Total
Regular	10	15	25
Irregular	20	200	220
Total	30	215	245

We can use Equation (4.10) to find the probability that a randomly selected customer has both of any two characteristics. Suppose we desire the probability that a randomly selected customer is both a *credit* customer (event *A*) and *orders irregularly* (event *B*). From the table we see that *P(A)* equals 215/245 and that $P(B \mid A)$ equals 200/215. Thus by Equation (4.10) the probability that a randomly selected customer both is a *credit* customer and orders *irregularly* is

$$P(A \text{ and } B) = P(A) \times P(B \mid A) = \frac{215}{245} \times \frac{200}{215} = \frac{200}{245}$$

Of course, if one had the contingency table from Omnicare's customers, then there would not be any need for calculating these probabilities. The number of favorable events is immediately apparent from the table and need only be divided by 245 to give the appropriate probabilities. The calculations have been made here only to show that the relationships in various equations are true.   ▆▆

Note that when the situation can be phrased in terms of the words *both* event *A and* event *B*, then the *multiplication* of Equation (4.10) applies.

4.3.4 Independent Events

The concept of *independent events* is the key topic of the following discussion and examples. Consider the experiment of drawing a single card from a deck of 52 cards. If we let *A* be the event *spade* and let *B* be the event *face card*, then the event "*A* and *B*" consists of the three face cards that are also spades. Since *P(A)* equals 13/52 and *P(A* and *B)* equals 3/52, we can use Equation (4.9) to determine the probability of a face card, *given* that the card is known to be a spade. We obtain

$$P(B \mid A) = \frac{P(A \text{ and } B)}{P(A)} = \frac{3/52}{13/52} = \frac{3}{13}$$

As before, this result makes sense. If someone draws a card from the deck depicted in Figure 4.11(a) and tells you it is a spade, as far as you are concerned, the 13 spades are now the possible outcomes, or in other words, they become the reduced sample space as shown in Figure 4.11(b). Of the cards in the reduced sample space, 3 favor the event *face card,* which gives $P(B \mid A) = 3/13$ just as did the use of Equation (4.9). Notice that the three outcomes that are favorable to the event *B* (*face card*) in the reduced sample space are the outcomes of event "*A* and *B*" in the complete sample space.

The important point to notice about this illustration is that since $P(B) = 12/52 = 3/13 = P(B \mid A)$, the *conditional* and *unconditional* probabilities are *equal,* which is not true in Practice Problems 4.2, 4.3, and 4.4. In other words, our knowledge of the card's suit does not influence the probability that it is a face card: All suits have the same proportion (3/13) of face cards so $P(B) = P(B \mid A)$. In this case we call *A* and *B independent events.*

Rephrasing the results from the illustration, *A* and *B* are independent events if

$$P(B \mid A) = P(B) \tag{4.11}$$

FIGURE 4.11

Sample Space and Reduced Sample Space

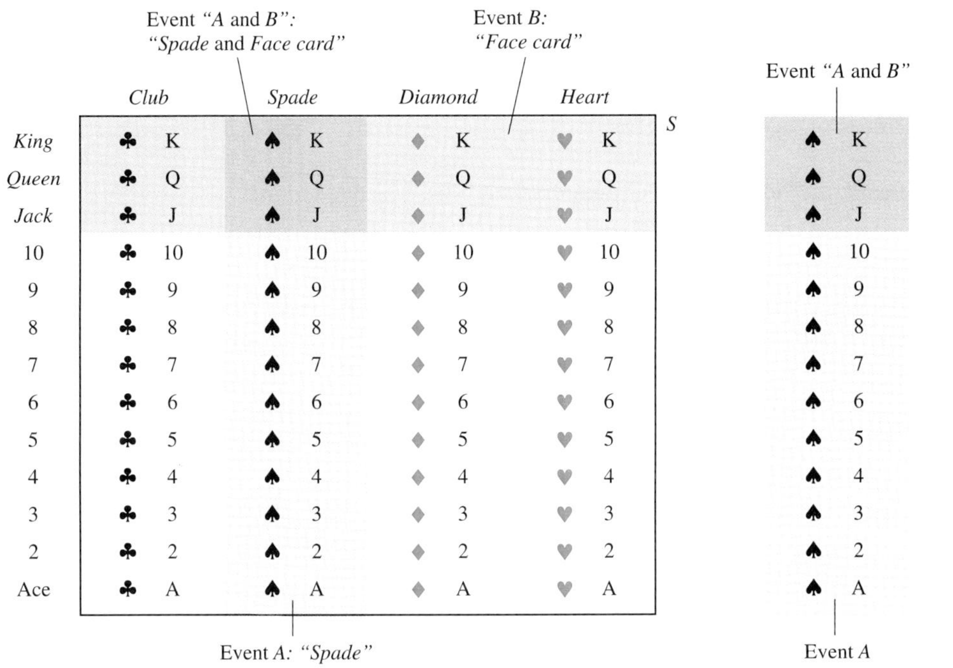

Event "A and B":
"Spade and Face card"

Event B:
"Face card"

Event "A and B"

Event A: "Spade"

Event A

(a) Sample space for drawing a card

(b) Reduced sample space given event A

In words, Equation (4.11) tells us that A and B are independent events if A gives no information about the probability of B. We also say that events A and B are independent if the conditional probability of B, given A, is the same as the unconditional probability of B. Still another way of stating the concept of independence is as follows: If knowing that one event has already occurred does not change the probability that the other event has also occurred, then the two events are independent.

EXAMPLE 4.6   A coin is to be tossed twice and a bet is made that a head will occur on the second toss. The chance of winning the bet is $1/2$. Now assume that an observer states that "A tail came up on the first toss." This information does not change the chance of winning since the probability of a head on the second toss remains $1/2$. That is, the tosses of a coin produce *independent* results: Knowing the outcome of one toss does not change the probabilities of what will happen on subsequent tosses.

An important aspect of Equation (4.11) is that it leads to a definition of independent events. If $P(B\,|\,A) = P(B)$, then events A and B are independent, and we can replace $P(B\,|\,A)$ with $P(B)$ in Equation (4.9), multiply both sides of the result by $P(A)$, and rearrange terms. Accordingly, we have the following definition of **independent events.**

> **Definition**
>
> *A* and *B* are **independent events** if and only if
>
> $$P(A \text{ and } B) = P(A) \times P(B) \tag{4.12}$$

If events are *not* independent then we say they are **dependent**.

4.3.5 Rule for Multiplication of Probabilities for Independent Events

An important aspect of the definition of independent events is the following multiplication rule that can be used to find $P(A \text{ and } B)$ if *A* and *B* are independent events.

> **Rule for Multiplication of Probabilities for Independent Events**
>
> If *A* and *B* are **independent events,** then
>
> $$P(A \text{ and } B) = P(A) \times P(B) \tag{4.12}$$

Equation (4.12) can be extended to find the joint probability of three or more independent events as the product of their simple probabilities.

 PRACTICE PROBLEM 4.4

The following contingency table classifies a college's accounting graduates who have taken out student loans at Gibraltar Savings & Loan by their gender and whether they have defaulted on their loans.

Find the probability that a randomly selected graduate is a *man* given that the graduate's loan is *not in default*. Also, find the probability that a randomly selected graduate is a *man*. Comment on the dependence or independence of the event being a *man* and *not being in default* on a student loan for these data.

	Default Status		
Gender	*In Default*	*Not in Default*	*Total*
Man	18	42	60
Woman	12	28	40
Total	30	70	100

Solution If we use Equation (4.9), we find

$$P(Man \mid Not \text{ in } default) = \frac{P(Man \text{ and } Not \text{ in } default)}{P(Not \text{ in } default)}$$

$$= \frac{42/100}{70/100} = \frac{42}{70} = \frac{3}{5}$$

and by Equation (4.2), we find

$$P(Man) = \frac{60}{100} = \frac{3}{5}$$

Since $P(Man \mid Not\ in\ default) = P(Man) = 3/5$, the events *man* and *not in default* are independent. Thus the probability of selecting a person who is a *man* and who is *not in default* could be computed by Equation (4.12) as follows:

$$P(Man \mid Not\ in\ default) = P(Man) \times P(Not\ in\ default)$$

$$= \frac{60}{100} \times \frac{70}{100} = \frac{42}{100}$$

Notice that we were required to establish that the events were independent before we used Equation (4.12) for this problem. This requirement is generally true for probability problems in management and economics; but in some problems, such as rolling dice or flipping coins, we can ascertain independence by the structure of the problem.

EXAMPLE 4.7　The Fruehauf Corporation of Practice Problem 4.2 had the following contingency table for working status and usage of manufacturing equipment:

Working Status	Equipment Use			Total
	Low	*Moderate*	*High*	
In working order	10	18	12	40
Under repair	2	6	8	16
Total	12	24	20	56

A piece of equipment has been selected randomly, and you think that it is in working order (as you might well believe if you have been boasting to a customer about the company's ability to handle any work that it is given). Your unconditional probability that the equipment is *in working order* is 40/56. However, a friend now tells you, "The equipment selected is a low-use item." Now your probability of the equipment's being *in working order* given that it is a *low-use* item increases to 10/12. It increases because the use of the equipment and its repair status are not independent. That is, knowing the level of equipment use tells us something about whether a piece of equipment will be in working order or under repair.

Equation (4.12) defines independence. Notice, however, that Equation (4.12) holds whenever Equation (4.11) holds and *A* and *B* are independent events if $P(B \mid A) = P(B)$. In addition, we could divide both sides of Equation (4.12) by $P(B)$, which leads to Equation (4.13) as follows:

$$P(A \mid B) = P(A) \tag{4.13}$$

Here the occurrence of *B* neither increases nor decreases the probability of *A*. Thus, when Equations (4.11), (4.12), and (4.13) hold, *A* has no influence on *B* if and only if B has no influence on A, and A is independent of B if and only if B is independent of A.

The contingency or cross-tabulation tables in this chapter, such as the Fruehauf Corporation contingency table shown in Example 4.7, have all been tabulated for us. Counting the frequencies for a contingency table can be tedious and is often done with a data analysis system when the categories for the variables have been coded. For example, the categories for equipment use at Fruehauf Corporation were coded as *Low* = 1, *Moderate* = 2, and *High* = 3; and the categories for the working status of the equipment were coded as *In working order* = 1 and *Under repair* = 2. The data are presented in the following listing:

Item	Status	Use	Item	Status	Use		Status	Use	Item	Status	Use
1	1	3	15	1	2	29	1	2	43	2	3
2	1	3	16	1	2	30	1	2	44	2	3
3	1	3	17	1	2	31	1	1	45	2	3
4	1	3	18	1	2	32	1	1	46	2	3
5	1	3	19	1	2	33	1	1	47	2	3
6	1	3	20	1	2	34	1	1	48	2	3
7	1	3	21	1	2	35	1	1	49	2	2
8	1	3	22	1	2	36	1	1	50	2	2
9	1	3	23	1	2	37	1	1	51	2	2
10	1	3	24	1	2	38	1	1	52	2	2
11	1	3	25	1	2	39	1	1	53	2	2
12	1	3	26	1	2	40	1	1	54	2	2
13	1	2	27	1	2	41	2	3	55	2	1
14	1	2	28	1	2	42	2	3	56	2	1

Computer programs were used to arrange the Fruehauf Corporation data into a contingency table, and the results are presented in Computer Exhibits 4.1A and 4.1B.

Problems: Sections 4.2–4.3

4. Give an example of two events that are mutually exclusive (cannot occur together).

5. Give an example of two events that are independent of each other (knowing that one has occurred does not change the probability that the other has occurred).

6. Humane West Hospital has two emergency sources of power. Historical records show that there is a .97 chance that the one source will operate during a total power failure and that there is a .92 chance that the other source will operate. Find the probability that in a total power failure neither emergency power source will operate. If one power source fails, it does not influence the probability that the second power source will fail.

7. There are 30 employees in one department, and only 10 of them are enrolled in the Minnesota Power and Light savings bond program. If 2 employees from this department are selected at random to serve on a benefits committee, what is the probability that both of them are enrolled in the program?

8. A national automobile rental firm with a very large number of customers claims that 95% of its customers are satisfied with the service they receive. If you were to interview two randomly selected customers, find the probability that both of them would be dissatisfied with the service.

COMPUTER EXHIBIT 4.1A Contingency Table for Fruehauf Corporation Equipment   **MINITAB**

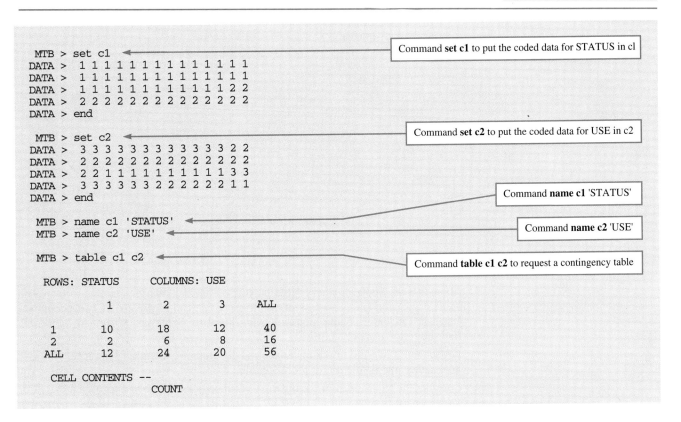

```
MTB > set c1                              Command set c1 to put the coded data for STATUS in c1
DATA > 1 1 1 1 1 1 1 1 1 1 1 1 1 1 1
DATA > 1 1 1 1 1 1 1 1 1 1 1 1 1 1 1
DATA > 1 1 1 1 1 1 1 1 1 1 1 1 1 2 2
DATA > 2 2 2 2 2 2 2 2 2 2 2 2 2 2
DATA > end
MTB > set c2                              Command set c2 to put the coded data for USE in c2
DATA > 3 3 3 3 3 3 3 3 3 3 3 3 3 2 2
DATA > 2 2 2 2 2 2 2 2 2 2 2 2 2 2
DATA > 2 2 1 1 1 1 1 1 1 1 1 1 3 3
DATA > 3 3 3 3 3 3 2 2 2 2 2 2 1 1
DATA > end
MTB > name c1 'STATUS'                    Command name c1 'STATUS'
MTB > name c2 'USE'                       Command name c2 'USE'

MTB > table c1 c2                         Command table c1 c2 to request a contingency table

  ROWS: STATUS      COLUMNS: USE

            1         2         3       ALL

    1       10        18        12       40
    2        2         6         8       16
  ALL       12        24        20       56

     CELL CONTENTS --
                   COUNT
```

COMPUTER EXHIBIT 4.1B Contingency Table for Freuhauf Corporation Equipment   **MYSTAT**

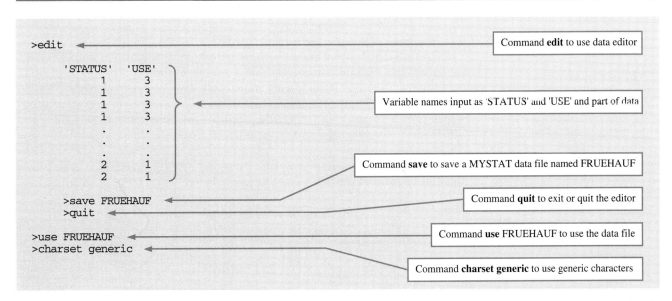

```
>edit                                     Command edit to use data editor

      'STATUS'    'USE'
          1         3
          1         3
          1         3                     Variable names input as 'STATUS' and 'USE' and part of data
          1         3
          .         .
          .         .
          .         .
          2         1
          2         1
     >save FRUEHAUF                        Command save to save a MYSTAT data file named FRUEHAUF
     >quit                                 Command quit to exit or quit the editor
>use FRUEHAUF                             Command use FRUEHAUF to use the data file
>charset generic                          Command charset generic to use generic characters
```

(continues)

MYSTAT	**COMPUTER EXHIBIT 4.1B** Contingency Table for Freuhauf Corporation Equipment (Continued)

```
>tabulate STATUS*USE
```
Command **tabulate** STATUS*USE to request contingency table

```
TABLE OF   STATUS   (ROWS) BY      USE     (COLUMNS)
FREQUENCIES
            1.000    2.000    3.000    TOTAL
          ---------------------------------
   1.000 |    10       18       12   |    40
   2.000 |     2        6        8   |    16
          ---------------------------------
   TOTAL      12       24       20        56

   .
   .
   .
>quit
```
Command **quit** to exit MYSTAT

9. About 8% of people in a very large population have an allergic reaction to materials used in the production process at Southwest Chemical Company. If two people selected at random from the population are hired by Southwest Chemical tomorrow, find the following probabilities.

   a. Neither is allergic to the materials.   **b.** Both are allergic to the materials.
   c. Exactly one is allergic to the materials.

10. Consider the experiment of rolling a pair of dice (a white die and a colored die). The sample space is shown in Figure 4.12. We define event B as the *white die shows one dot* and event A as the *sum of the dots on the two dice is less than or equal to 5*.

    a. List the outcomes favorable to event B.   **b.** Determine $P(B)$.
    c. List the outcomes favorable to event A.   **d.** Determine $P(A)$.
    e. List the outcomes favorable to event "A and B."   **f.** Determine $P(A \text{ and } B)$.
    g. List the outcomes in the reduced sample space given that event A is certain to occur.
    h. List the outcomes favorable to event B given that event A has occurred.
    i. Determine $P(B \mid A)$ by the methods used to define classical probabilities.
    j. Determine $P(B \mid A)$ by the definitional formula for conditional probabilities.
    k. Are events A and B independent? Show why or why not.

11. Seventy-five percent of the graduates of a business education program end up in management. About 25% of the graduates have both management jobs and the good public-speaking skills that such a job usually requires. Find the conditional probability that graduates of this program will have good public-speaking skills, given that they have management jobs.

12. Consider the accompanying table, which classifies Mortgage Investors' Trust's sales markets by national location (east or west) and by population (large or small). A mar-

	Population		
Location	*Large*	*Small*	*Total*
East	56	14	70
West	4	26	30
Total	60	40	100

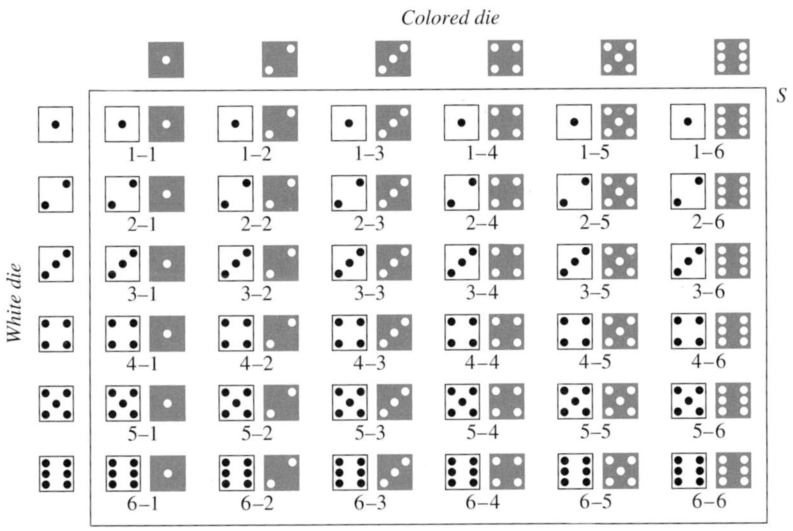

FIGURE 4.12

Sample Space for Rolling a
Pair of Dice

keting vice president will be assigned at random to develop a marketing plan for a
sales market.

a. Find $P(Large)$.
b. Find $P(Large \mid East)$.
c. Use the numbers in the table to find $P(East$ and $Large)$.
d. Find $P(East)$.
e. Use a multiplication formula to find $P(East$ and $Large)$.
f. Are the events *Large* and *East* independent? Show why or why not.

We can illustrate many of the concepts presented in the previous section by using
probability tree diagrams. These diagrams are useful in presenting a visual picture
of a probability situation.

4.4
PROBABILITY TREE
DIAGRAMS

Definition

Probability tree diagrams depict events or sequences of events as branches
of a tree.

Each branch of the tree is labeled to indicate which event it represents. Along
each branch is given the probability of the event's occurrence, given that the sequence of events has reached that point in the tree. Thus all the branches that emanate
from one point must be (1) mutually exclusive and (2) collectively exhaustive. That
is, they must represent distinct events, and they must account for all possible events
that can occur at that point.

EXAMPLE 4.8 Consider again the breakdown of Omnicare's 245 customers from Practice Problem 4.1. The table associated with those customers is presented again here:

| | Payment Terms | | |
Order Type	Cash	Credit	Total
Regular	10	15	25
Irregular	20	200	220
Total	30	215	245

The probability tree diagram that describes the random selection of a customer is shown in Figure 4.13. This diagram illustrates several features of probability trees. First, all the branches emanating from one point have *probabilities that add to unity,* because the events represented by the branches are *mutually exclusive* and form a sample space or are *collectively exhaustive.* Second, the probabilities associated with all the branches after the first set are *conditional probabilities.* That is, the 10/25 probability that someone is a cash customer on the top branch is the conditional probability that someone is a cash customer *given* that he or she is a regular customer—that is, given that we are on the branch labeled *Regular.* Third, the probabilities at the ends of the branches are *joint probabilities* and can be found easily by multiplying all the probabilities along the branches leading to the endpoint. That is, the 200/245 probability found at the end of the bottom set of branches is the joint probability that a customer is both irregular and credit. We find this value by multiplying the two probabilities along these branches: (220/245) × (200/220) = 200/245.

There is nothing about the construction of probability tree diagrams that says we had to draw the tree in this example by using the regular/irregular classification first. We could just as easily have drawn a cash/credit branch first and then followed with regular/irregular branches at the ends. The order in which we are given the probability information or the order in which we will learn about

FIGURE 4.13

Probability Tree Diagram Describing the Random Selection of a Customer

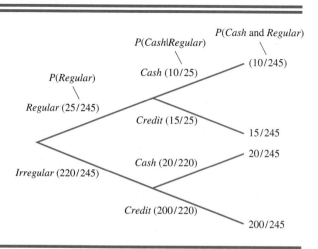

the occurrence of events, however, often dictates which branches are first. (See Practice Problem 4.5.)

Tree diagrams can be used to find joint probabilities rather easily, since, as was illustrated previously, one merely has to multiply all the probabilities along the branches leading to the joint event to obtain the correct value. Unconditional probabilities are also easy to calculate. For instance, the unconditional probability of selecting a regular customer is found by adding all the probabilities at the ends of the branches that were reached by going down a branch labeled *Regular.* Thus $P(Regular) = (10/245) + (15/245) = 25/245$. This result does not seem too profound since the *Regular* branch itself has a probability of 25/245.

However, consider the more complex problem of finding $P(Cash)$ from this diagram. We merely add the probabilities at the ends of the branches that were reached by going down any *Cash* branch. Thus $P(Cash) = P(Regular$ and $Cash) + P(Irregular$ and $Cash) = (10/245) + (20/245) = 30/245$.

 PRACTICE PROBLEM 4.5

The personnel manager of a Database Systems, Inc., operation classifies job applicants as qualified or unqualified for the jobs they seek. The manager says that only 25% of the job applicants are qualified, and of those that are qualified, 20% list high school as their highest level of education. But 30% of the qualified applicants list trade school, and 50% list college. The situation is different among the unqualified applicants in that 40% of them list high school as their highest level of education, another 40% list trade school, and only 20% list college. Draw a probability tree diagram of this situation, find the joint probability that an applicant is both *qualified* and a *college graduate,* and find the unconditional probability that an applicant comes from a *trade school* background.

Solution Since the problem states "of those that are qualified," the probabilities that follow this statement are conditional—they assume the condition that the person is qualified to begin with. Thus we are constrained to drawing a probability tree with the qualified/not-qualified branches as our first branching. We can list the other branches and associated probabilities by noting carefully the description of the situation in the problem statement.

The tree is presented in Figure 4.14 with the probabilities given in the problem statement. To find all of the joint probabilities, multiply the probabilities along the branches. Thus the problem asks us to find the joint probability $P(Qualified$ and *College*). To find this probability, go up the first branch and down the third branch that starts there. Thus $P(Qualified$ and $College) = .25 \times .50 = .125$. We could have obtained this same result by using Equation (4.10), which, adapted to this situation, would state $P(Qualified$ and $College) = P(Qualified) \times P(College|Qualified) = .25 \times .50 = .125$.

The problem also asks us to find the unconditional probability that an applicant comes from a trade school. To do so, we merely compute and add all the joint probabilities found at the ends of any branches that are reached by going through *trade school.* There are two such branches, and the resulting computations are $(.25 \times .30) + (.75 \times .40) = .075 + .300 = .375$. If we had used Equations (4.5) and (4.10), we could have obtained the same result, but in a much more cumbersome

FIGURE 4.14

Probability Tree Diagram
Describing the Random Se-
lection of a Job Applicant

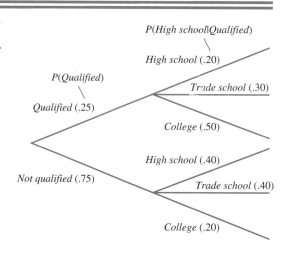

fashion. Let T represent trade school, Q represent qualified, and NQ represent not qualified. Then

$$P(T) = P[(T \text{ and } Q) \text{ or } (T \text{ and } NQ)]$$
$$= P(Q) \times P(T \mid Q) + P(NQ) \times P(T \mid NQ)$$
$$= (.25 \times .30) + (.75 \times .40) = .075 + .300 = .375$$

which is the same answer we obtained by using the tree diagram.

Tracing through the semantics of the probability formulas can be rather difficult. So probability tree diagrams such as the one presented for Database Systems of Practice Problem 4.5 can be very useful in simplifying the computations of joint and unconditional probabilities.

Tree diagrams can also be used to demonstrate the concept of statistical independence. Consider Figure 4.15. The tree diagram of Figure 4.15(a) demonstrates that the

FIGURE 4.15

Probability Tree Diagrams
Depicting Independent
and Dependent Events

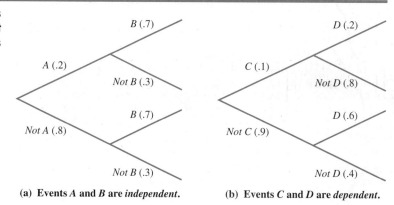

(a) Events A and B are *independent*.　　(b) Events C and D are *dependent*.

events A and B are *independent*. Regardless of whether A or *not A* occurs first, the chance that event B will occur remains at .7 and the chance of *not B* remains at .3. That is, that the probabilities assigned to events B and *not B* at the end of the A and *not A* branches are the same indicates that knowing about event A has no influence on the probabilities we assign to event B.

The tree diagram of Figure 4.15(b) demonstrates that events C and D are *dependent*. Note that the probabilities for events D and *not D* are .2 and .8, respectively, if event C has already occurred. However, if C has *not* occurred, the chances of D and *not D* are .6 and .4. Thus knowing about event C's outcome changes the probabilities assigned to event D.

In general, then, two events A and B are independent if the pair of probabilities assigned to B and *not B* at the end of the A branch are the same as those assigned to B and *not B* at the end of the *not A* branch. The two events are dependent if the pair of probabilities assigned to B and *not B* at the end of the A branch differ from the pair of probabilities for B and *not B* at the end of the *not A* branch.

Problems: Section 4.4

13. With the aid of a tree diagram, list the different possible outcomes, with respect to the gender of each child in succession, when a family has 3 children.

   a. How many outcomes are there?
   b. How many result in two girls?
   c. How many result in 2 or more girls?

14. A Southwest Energy Company pipeline has 3 safety shutoff valves in case the line springs a leak. These valves are designed to operate independently of one another. There is only a 7% chance that valve 1 will fail at any one time, a 10% failure chance for valve 2, and a 5% failure chance for value 3. If there is a leak, find the following probabilities [*Hint:* You might want to draw a tree diagram.]:

   a. That all 3 valves operate correctly.
   b. That all 3 valves fail.
   c. That only 1 valve operates correctly.
   d. That at least 1 valve operates correctly.

15. Thor Systems Company has 800 employees. Twenty percent of the employees have college degrees, but half of those people are in nonmanagement positions. Thirty percent of the nondegree people are in management positions.

   a. How many managers does the company have?
   b. What is the conditional probability of being a college graduate, given that a person is a manager?
   c. What is the conditional probability of being a manager, given that a person has a college degree?

16. In Problem 15, find the probability of randomly selecting someone who is either a college graduate or a manager.

17. Company records at Informatics Realty Group show that 80 employees have been reprimanded more than once, and 300 have never been reprimanded. The company has 400 employees. The people who have zero or one reprimand on their records are both split evenly between men and women. The people who have more than one reprimand are 75% men. Draw a probability tree to help you determine the probability of randomly selecting a person who is the following:

   a. A woman with one reprimand.
   b. A man.
   c. One with no reprimand, given that the employee is a man.

4.5
BAYES' THEOREM

We have analyzed several problems using probabilities of events, or prior probabilities. Many prior probabilities have been subjective estimates. It is often advisable to obtain additional information in the form of samples, surveys, tests, or experiments that can be used to revise probabilities. For example, a consultant may be retained to provide information. Or tests such as blood tests, credit tests, or tests of raw materials may be used to obtain sample information that can be used to revise probabilities.

The revision of prior probabilities by using sample information is done with Bayes' theorem (named in honor of the English mathematician and reverend, Sir Thomas Bayes 1702–1761). The sample information provides conditional probabilities or likelihoods and Bayes' theorem combines the *prior probabilities* and the *conditional probabiities* to give *revised or posterior probabilities* that reflect the sample information. The revised probabilities can enrich the analysis of probability problems and eventually lead to improved decisions when decisions are based on probability information.

EXAMPLE 4.9 Hughes Tool Company has a contract requiring it to build a piece of equipment that it has never built before. The contract requires that the project be completed within 90 days. Let A denote the hypothesis that the project will be completed within 90 days, and let $\overline{A}$ denote the opposite hypothesis that the project will take longer than 90 days to complete. One can assign the probability $P(A)$ to A and the probability $P(\overline{A})$ to $\overline{A}$.

Now Hughes hires a work-methods engineer to provide information about how long it will take to complete the project, and she says that it can be completed within the specified time. The new information, call it B, will alter our attitude toward the hypothesis A. Still, we know that the methods engineer is not always correct. After asking some questions, we determine that 80% of the projects that are completed on time have been correctly forecast as being completed on time by this engineer. That is, $P(B\,|\,A) = .80$. We also know that for projects that were not completed on time, the engineer forecasted completion by the deadline 10% of the time; thus $P(B\,|\,\overline{A}) = .10$.

We are to compute a *new* probability $P(A\,|\,B)$ of the hypothesis, given the information from the engineer.

To arrive at Bayes' theorem, we proceed as follows. By Equation (4.9) for conditional probabilities, $P(A\,|\,B)$ is equal to $P(A \text{ and } B)/P(B)$, and applying the multiplication of Equation (4.10) to find the joint probability of the numerator gives

$$P(A\,|\,B) = \frac{P(A) \times P(B\,|\,A)}{P(B)}$$

Now B can happen in two ways: if "A and B" happens or if "$\overline{A}$ and B" happens. Consequently, by adding probabilities for the mutually exclusive events, we get $P(B) = P(A \text{ and } B) + P(\overline{A} \text{ and } B)$. Using Equation (4.10) on each of these last two terms now gives the denominator as $P(B) = P(A) \times P(B\,|\,A) + P(\overline{A}) \times P(B\,|\,\overline{A})$. Substituting this result for the denominator in the preceding formula gives the answer. The equation for $P(A\,|\,B)$ is known as **Bayes' theorem.**

Bayes' Theorem

$$P(A\,|\,B) = \frac{P(A) \times P(B\,|\,A)}{P(A) \times P(B\,|\,A) + P(\overline{A}) \times P(B\,|\,\overline{A})} \qquad (4.14)$$

Given the prior probabilities $P(A)$ and $P(\overline{A}) = 1 - P(A)$ and the conditional probabilities, or likelihoods $P(B\,|\,A)$ and $P(B\,|\,\overline{A})$, of observing information B if A holds or if $\overline{A}$ holds, we can use Bayes' theorem to compute the *posterior probability* $P(A\,|\,B)$, which is the revised probability for A after the information in B has been taken into account.

The sample space and reduced sample space for problems addressed by Bayes' theorem are depicted in Figure 4.16. Figure 4.16(b) shows the reduced sample space given that event B must occur and the events that lead to Bayes' theorem.

In Example 4.9, $P(B\,|\,A) = .8$ and $P(B\,|\,\overline{A}) = .1$, so the formula gives

$$P(A\,|\,B) = \frac{P(A) \times .8}{P(A) \times .8 + P(\overline{A}) \times .1}$$

If $P(A) = .3$, say, so that $P(\overline{A}) = .7$, then

$$P(A\,|\,B) = \frac{.3 \times .8}{(.3 \times .8) + (.7 \times .1)} = \frac{.24}{.24 + .07} = .77$$

Notice that $P(A\,|\,B)$ exceeds $P(A)$ here; the information from B or observing B has increased our probability of A because A "explains" B better than $\overline{A}$ does. In statistical notation, in this example, $P(B\,|\,A) > P(B\,|\,\overline{A})$.

Bayes' theorem is somewhat difficult to comprehend at first. However, the ideas may be simpler to understand if we use the probability tree diagrams discussed in the

Venn Diagrams for Bayes' Theorem **FIGURE 4.16**

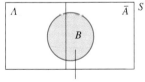

Probability of event B:

$$\begin{aligned} P(B) &= P[(A \text{ and } B) \text{ or } (\overline{A} \text{ and } B)] \\ &= P(A \text{ and } B) + (\overline{A} \text{ and } B) \\ &= P(A) \times P(B|A) + P(\overline{A}) \times P(B|\overline{A}) \end{aligned}$$

Probability of event "A and B":

$$P(A \text{ and } B) = P(A) \times P(B|A)$$

Thus

$$P(A|B) = \frac{P(A \text{ and } B)}{P(B)}$$

$$= \frac{P(A) \times P(B|A)}{P(A) \times P(B|A) + P(\overline{A}) \times P(B|\overline{A})}$$

Probability of event "$\overline{A}$ and B":

$$P(\overline{A} \text{ and } B) = P(\overline{A}) \times P(B|\overline{A})$$

(a) **Sample space**

(b) **Reduced sample space**

FIGURE 4.17

Probability Tree Diagram of
the Bayes' Theorem Com-
putations for the Equip-
ment Contract

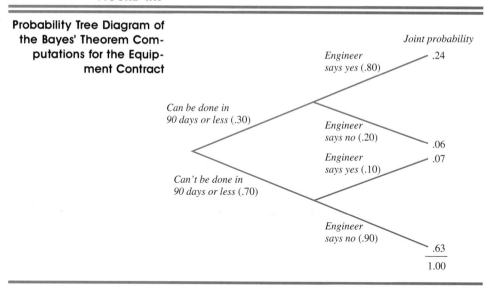

previous section. The example of trying to complete the project in 90 days or less is diagrammed in Figure 4.17. The joint probabilities at the ends of the branches were found by multiplying the probabilities along the branches. For instance, .06 is the probability that the project can be done in 90 days or less and that the engineer will not predict that it can be done in 90 days or less.

In the statement of this situation, however, we desire to know $P(A \mid B) = P(Project\ complete\ in\ 90\ days\ or\ less \mid Engineer\ says\ it\ can\ be\ done)$. Since the engineer says the project can be done by the deadline, then we must be at the end of the first or third branches in our diagram. The chance of our being at either one of these points is $.24 + .07 = .31$. The fact remains, however, that we *must be* at the end of one of those branches—that is, it has been *given* that the engineer says, "Yes the project can be done by the deadline." Thus if we want to know if the project can actually be done by that time in the face of our original prior estimate of $P(Project\ can\ be\ done\ in\ 90\ days\ or\ less) = .30$ and in light of the engineer's prediction, then we must have come down the *Can be done in 90 or less* branch with a probability that is proportional to the .24 contribution that this branch makes to the two situations where the *Engineer says yes*. Thus,

$$P(Can\ be\ done\ in\ 90\ days\ or\ less \mid Engineer\ says\ yes) = \frac{.24}{.24 + .07}$$

$$= .77$$

This calculation is a direct application of Bayes' theorem.

We have developed Bayes' theorem for the case where the event of interest has just two partitions A and $\overline{A}$. Nevertheless, Bayes' theorem extends in the intuitive way if there are k partitions on the event, say, $A_1, A_2, \ldots, A_k$. In this situation, if we want to determine the probability $P(A_i \mid B)$ for the ith partition A_i given B, we simply extend the summation in the denominator to k terms and $P(A_i \mid B) = P(A_i) \times P(B \mid A_i)/[P(A_1) \times P(B \mid A_1) + P(A_2) \times P(B \mid A_2) + \cdots + P(A_k) \times P(B \mid A_k)]$.

PRACTICE PROBLEM 4.6

Screening for a Rare Disease: Screening health care consumers for a rare disease is fraught with difficulties when screening tests are not perfect. Many supposedly routine tests used for purposes of screening are far from perfect. For instance, the chance that someone has a somewhat rare disease is .02. A test that identifies this disease is imperfect, even though it appears to be fairly reliable, in that it gives a positive reaction in 97% of the people who have the disease. It also gives a positive reaction in 5% of the people who do not have the disease. Given that an individual just received a positive reaction from this test, what is the probability that the individual has the disease?

Solution The situation is diagrammed in Figure 4.18. Since it was given that the individual received a positive test, this person is out at the end of the top branch or at the end of the third branch down. The chance of a positive test is just the sum of the probabilities at the ends of these two branches. Thus, *P(Positive test)* = .0194 + .0490 = .0684. But the chance that the individual has the disease or actually came down the disease branch to get the positive test is just the top branch's share of this sum, or .0194. Thus, the probability that the individual has the disease given the positive test results is found as follows:

$$P(Disease \mid Positive\ test) = \frac{.0194}{.0194 + .0490} = \frac{.0194}{.0684} = .2836$$

A direct application of Bayes' theorem yields the same numerical results as follows:

$$P(Disease \mid Test\ positive)$$

$$= \frac{P(Disease)P(Test\ positive \mid Disease)}{P(Disease)P(Test\ positive \mid Disease) + P(No\ disease)P(Test\ positive \mid No\ disease)}$$

$$= \frac{.02 \times .97}{(.02 \times .97) + (.98 \times .05)} = .2836$$

FIGURE 4.18

Probability Tree Diagram of the Bayes' Theorem Computations for the Disease Test

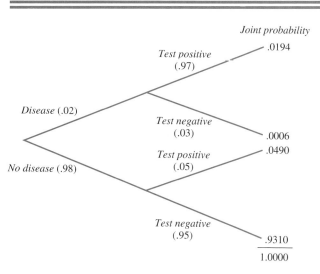

Notice that the probability that the individual has the disease, given that the test was positive, has increased from .02 to .2836. However, notice also, in this case, that although about 28% of health care consumers who receive a positive test actually have the somewhat rare disease, 72% who receive a positive test actually do *not* have the disease.

Health care consumers who undergo imperfect screening tests for rare diseases should thoroughly understand the phenomenon that, of the individuals who receive a positive test for a rare disease, substantially more will *not* have the disease than will have the disease. Before consenting to expensive or invasive and risky treatments, individuals should be aware that it is more likely that screening results are erroneous and that they are actually disease free. The hazard of supposedly routine screening tests for rare diseases is that even though an occasional individual may benefit, many more may be subjected to anxiety, perilous surgery, or invasive treatments based on erroneous test results. Similar difficulties are encountered in screening of other rare occurrences such as illegal drug usage, bankruptcy prediction, cancer screening, routine tests given during physical examinations, lie detector tests, and so forth.

Problems: Section 4.5

18. A human resources manager at Construction Machinery & Equipment, Inc., knows from experience that 80% of applicants for a job will be able to perform satisfactorily on the job. Of those applicants who do satisfactory work on the job, 90% pass an abilities test. Of those applicants who do not perform satisfactorily on the job, 60% do not pass an abilities test. If an applicant selected at random passes the test, find the following probabilities:

    a. That the applicant will perform satisfactorily.
    b. That the applicant will not perform satisfactorily.

19. Twenty percent of the home buyers who have mortgages with Second Financial Corporation default on their mortgage payments. Forty percent of those who default on their payments received the score of "good credit risk" on the qualifying credit check. Eighty percent of those who did not default on their mortgage payments scored "good credit risk" on the qualifying credit check.

    a. What is the probability that a randomly selected home buyer who has a mortgage with the savings and loan will default on his or her mortgage payments, given that the person scored "good credit risk" on the credit check?
    b. Comment on the value to the savings and loan of the credit check for the home buyers who have mortgages with the institution.

20. A type of cathode ray tube, A, and its ruggedized version, B, which are produced by Electrical Equipment and Supply, Inc., are installed at random in single-tube units. Thirty percent of the tubes are of type B. The probability that type B will fail in the first hundred hours of continuous operation is .1; the probability that type A will fail is .3. If a particular unit fails, what is the probability it had a type A tube installed? What is the probability that a type B tube was installed? Use Bayes' theorem.

21. Mega Memory Devices, a firm that assembles memory boards for personal computers, buys 60% of its memory chips from supplier A and the remainder from supplier B. Supplier A produces memory chips that are 5% defective, and B produces 10% defective. A memory chip is selected at random from the inventory. A test of the chip shows that it is defective.

    a. What is the probability that the chip was supplied by A?
    b. What is the probability that the chip was supplied by B?

We considered several probability concepts in this chapter. These concepts were illustrated with a variety of straightforward examples that use cards and dice and with decision-making problems in business and economics. Probability provides a basis for evaluating the reliability of the conclusions we reach and the inferences we make under conditions of uncertainty. Thus probability is an important topic in the study of statistics.

The meaning of probability was explored by a discussion of the relative frequency, classical, and subjective concepts of probability. Random experiments, outcomes, the sample space, and events are important concepts for understanding probability.

The main equations that are used for computing probabilities according to the definitions, properties, and rules for probabilities that are given in this chapter are summarized in Table 4.1.

We may more easily determine probabilities involving two or more events by remembering that when events are related by the word *or,* the probabilities of the individual events are *added* and the probability that the events occur together is subtracted from this sum. When events are related by the word *and,* the probabilities of the events are *multiplied* as in rules 5 and 6 of Table 4.1.

Probability tree diagrams can be used to illustrate many of the concepts presented in this chapter.

TABLE 4.1   Summary of Equations for Computing Probabilities

Methods

1. If we undertake a large number of repetitions of a random experiment, then according to the relative frequency method of assigning probabilities, we can approximate the probability of an event as follows:

$$\text{Probability of an event} \approx \frac{\begin{array}{c}\textit{Number of times that the event is the outcome}\\ \textit{of an experiment when the experiment is}\\ \textit{repeated a large number of times}\end{array}}{\begin{array}{c}\textit{Total number of repetitions}\\ \textit{of the experiment}\end{array}} \tag{4.1}$$

2. If the outcomes of an experiment are *equally likely* and *mutually exclusive,* then according to the classical method of assigning probabilities, we can compute the probability of an event as follows:

$$\text{Probability of an event} = \frac{\begin{array}{c}\textit{Number of outcomes of an experiment that are}\\ \textit{favorable to the event}\end{array}}{\begin{array}{c}\textit{Total number of outcomes of}\\ \textit{the experiment}\end{array}} \tag{4.2}$$

(continues)

TABLE 4.1   Summary of Equations for Computing Probabilities (Continued)

Rules

3. If A and B can occur together, then

$$P(A \text{ or } B) = P(A) + P(B) - P(A \text{ and } B) \tag{4.6}$$

But if A and B are *mutually exclusive* (if they cannot both occur at the same time), then

$$P(A \text{ or } B) = P(A) + P(B) \tag{4.5}$$

4. If A and $\overline{A}$ are *complementary* (if the occurrence of $\overline{A}$ is the same thing as the nonoccurrence of A), then

$$P(\overline{A}) = 1 - P(A) \tag{4.8}$$

5. The *conditional probability* of B, given A, is defined by

$$P(B \mid A) = \frac{P(A \text{ and } B)}{P(A)} \qquad \text{where} \quad P(A) > 0 \tag{4.9}$$

This equation is sometimes used to find $P(A \text{ and } B)$ as

$$P(A \text{ and } B) = P(A) \times P(B \mid A) \tag{4.10}$$

6. The events A and B are defined to be *independent* (to relect the idea that the occurrence of one event does not alter the probability of occurrence of the other) if and only if

$$P(A \text{ and } B) = P(A) \times P(B) \tag{4.12}$$

The independence condition is the same as

$$P(B \mid A) = P(B) \qquad \text{and} \qquad P(A \mid B) = P(A) \tag{4.11}$$

7. Bayes' theorem is used to compute *posterior* probabilities from *prior* and *observed* probabilities according to the formula

$$P(A \mid B) = \frac{P(A) \times P(B \mid A)}{P(A) \times P(B \mid A) + P(\overline{A}) \times P(B \mid \overline{A})} \tag{4.14}$$

REVIEW PROBLEMS   *Answers to odd-numbered problems are given in the back of the text.*

22. International Farm Insurance Company wants to know the probability that an insured owner of a new home, selected at random, will not file a claim in the first year of the policy. A file on the insurance history of 600 new houses in similar locations is

FIGURE 4.19

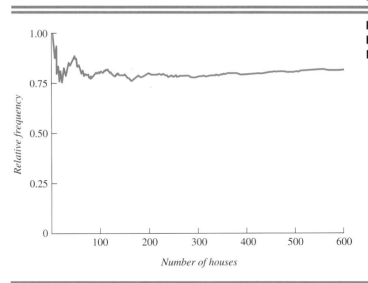

Relative Frequencies of No Insurance Claims on New Homes

Number of houses

checked, and Figure 4.19 shows the proportion of new homes that did not have an insurance claim filed during the first year of the policy.

a. From the figure, what is the probability that an insured owner of a new house will not file an insurance claim in the first year?

b. What method of assigning probabilities to events is applicable for this problem?

23. List the different possible outcomes when a pair of dice, one white and one colored die, are tossed. How many are there? How many outcomes have an even number for the total number of dots showing? How many are greater than 4?

24. Journey's End Medical Clinic serves 4000 regular patients. Clinic records show that 2500 of these patients are covered by private medical insurance; 1000 are on Medicare or welfare programs, which cover their medical costs; and 500 do not have any medical coverage. If a patient is selected at random, find the following probabilities:

a. That the patient is covered by private insurance.

b. That the patient is covered by some third-party private insurance or other program.

c. That the payment for services will not involve filing government forms.

25. A box of price tags at Cedar Home Furnishings contains ten red tags numbered from 1 through 10 and ten white tags numbered 1 through 10. If one tag is selected at random, find the following probabilities:

a. That it is red.　　　　　**b.** That it is an even number.

c. That it is red and even.　　**d.** That it is red or even.

26. A parts bin at Consolidated Electric of New York contains 100 parts, 5 of which are defective and 7 of which are used. All the used parts are good. Find the probability that a randomly selected part is the following:

a. Neither defective nor used.　　**b.** Neither used nor good.

c. Both defective and new.　　　　**d.** Both defective and used.

27. Deal Drilling & Exploration, Inc., is drilling test wells both in Alaska and off the coast of Texas. The managers of the company feel that the chance of oil in Alaska is .6 and the chance off the coast of Texas is .8. Assuming independence, find the following probabilities:

a. That both wells find oil.　　**b.** That exactly one well finds oil.

c. That neither well finds oil.

28. The payments for hospital bills at the Midstate Valley Hospital are either private payments (made by the patients) or third-party payments (made by insurance companies or government agencies). The payments made at Midstate Valley are divided between private and third-party in such a way that the probability of a third-party payment is .75. Two-thirds of those third-party payments are for surgical bills, and one-third are for nonsurgical bills. The private payments, however, are split 50–50 between payments for surgery and nonsurgery items.

   a. Find the probability that a randomly selected payment is both private and for a surgical item.
   b. Find the probability that a randomly selected payment both is third-party and does not involve a surgical item.
   c. Find the probability that a randomly selected payment is for a surgical item.

29. The accompanying table classifies the stock price change of Faux Leasing, Inc., today as up or down (down includes no change) and yesterday's price change (up or down) for 100 days. Assume that stock markets are "efficient," or that stock prices follow a *random walk,* as suggested by some financial theorists. That is, the current price of a stock fully reflects available information concerning a stock's expected future returns. Then, among other things, the price change today for a stock must be independent of yesterday's price change. For example, consider a randomly selected stock price change.

   a. Find *P(Up today).*   b. Find *P(Up yesterday).*
   c. Use the numbers in the table to find *P(Up today | Up yesterday).*
   d. Use the numbers in the table to find *P(Up yesterday and Up today).*
   e. Use the general rule for multiplication of probabilities to find *P(Up yesterday and Up today).*
   f. Are the events *"Up today"* and *"Up yesterday"* independent? Show why or why not.
   g. Is your answer in part f consistent with the theory that stock price changes follow a random walk?

Today's Stock Price Change	Yesterday's Stock Price Change		Total
	Up	Down	
Up	49	21	70
Down	21	9	30
Total	70	30	100

REFERENCE: Malkiel, B. (1975), *A Random Walk Down Wall Street* (New York: Norton).

30. There are three automatic-control boxes on a production line at a Solarcell Corporation plant. Two were replaced last year. The third one needs to be replaced this year, but no one can remember which box needs replacement. If two boxes are randomly selected for replacement, what is the probability that the oldest box (one not replaced last year) will get replaced?

31. Cedar Home Furnishings has found in its historical records that 30% of its customers are urban and 70% are suburban. It also classifies customers as "good" (60%), "borderline" (30%), and "poor" (10%). Of the good customers, 20% are urban and 80% are suburban. But borderline are split 40% to 60% in the urban-suburban breakdown.

   a. Find *P*(both *Good* and *Suburban*).   b. Find *P*(both *Poor* and *Urban*).
   c. Find *P*(either *Suburban* or *Good*).   d. Find *P*(*Borderline*).
   e. Find *P*(*Suburban | Poor*).

32. Two tennis balls are drawn blindly from a box by a customer at an Associated Corporation variety store. The box contains three white and seven black balls that are thoroughly mixed. What is the probability of obtaining two white balls for the following conditions?

    a. The first is replaced before the second is drawn.
    b. The first is *not* replaced before the second is drawn.

33. Consider the experiment of drawing a card from a well-shuffled, ordinary deck. The sample space is shown in Figure 4.20. We define event A as an *ace* and event B as a *black suit.*

    a. List the outcomes favorable to event A.
    b. Determine $P(A)$.
    c. List the outcomes favorable to event B.
    d. Determine $P(B)$.
    e. List the outcomes favorable to event "A and B."
    f. Determine $P(A \text{ and } B)$.
    g. List the outcomes in the reduced sample space given that event B is certain to occur.
    h. List the outcomes favorable to event A given that event B has occurred.
    i. Determine $P(A \mid B)$ by the methods used to define classical probabilities.
    j. Determine $P(A \mid B)$ by the definitional formula for conditional probabilities.
    k. Are events A and B independent? Show why or why not.

FIGURE 4.20

Sample Space for Drawing A Card

	Club	Spade	Diamond	Heart
King	♣ K	♠ K	♦ K	♥ K
Queen	♣ Q	♠ Q	♦ Q	♥ Q
Jack	♣ J	♠ J	♦ J	♥ J
10	♣ 10	♠ 10	♦ 10	♥ 10
9	♣ 9	♠ 9	♦ 9	♥ 9
8	♣ 8	♠ 8	♦ 8	♥ 8
7	♣ 7	♠ 7	♦ 7	♥ 7
6	♣ 6	♠ 6	♦ 6	♥ 6
5	♣ 5	♠ 5	♦ 5	♥ 5
4	♣ 4	♠ 4	♦ 4	♥ 4
3	♣ 3	♠ 3	♦ 3	♥ 3
2	♣ 2	♠ 2	♦ 2	♥ 2
Ace	♣ A	♠ A	♦ A	♥ A

S

34. Consider the list of companies shown in Table 4.2, which gives annual changes in market values (as percentages). Some of the companies had increases in market values, and some had decreases. Also, some of the companies are in the oil service and supply industry, and some are in the building materials industry. An individual investor is going to select one of the companies at random and then purchase stock in that company. The investor is interested in examining several characteristics of the

		Market Value		Market Value
TABLE 4.2	*Oil Service and Supply*	*Change (%)*	*Building Materials*	*Change (%)*
Market Value Data	Baker International	−39	CalMat	41
	CBI Industries	27	CertainTeed	12
	Cameron Iron Works	−31	De Soto	14
	Dresser Industries	7	Jim Walter	64
	Gearhart Industries	−79	Lafarge	18
	Halliburton	−12	Lone Star Industries	19
	Hughes Tool	−39	Masco	39
	NL Industries	−64	Moore McCormack	
	Ocean Drilling &		Resources	−26
	Exploration	−42	Owens-Corning Fiberglas	−72
	Schlumberger	−12	Sherwin-Williams	22
	Smith International	−48	Texas Industries	−15
	Western Co. of North		USG	69
	America	−70		

SOURCE: *Business Week,* December 29, 1986, p. 141.

companies. For the randomly selected company, determine the following.

a. Find P(*Increase in company's market value last year*).
b. Find P(*Building materials company*).
c. Find P(*Increase in market value* | *Building materials company*).
d. Find P(*Increase in market value* and *Building materials company*).
e. Use the general rule for multiplication of probabilities to find P(*Increase in market value* and *Building materials company*).
f. Are the events *Increase in market value* and *Building materials company* independent? Show why or why not.
g. Should investors be aware of the industries in which companies operate?

35. Providers of services, including public accounting firms, face the threat of litigation and the costs of court-imposed judgments or out-of-court settlements. Consider the following table, which classifies Big Five and Non–Big Five accounting firms according to the resolution of legal cases involving audit services. The litigations result in payments by the audit firms, or the cases may be dismissed or have no payments made by the firms. An investor believes that he sustained losses because sampling techniques used during an audit were faulty. If a legal case involving audit services is selected at random, find the probabilities for the following events:

a. Big Five.   **b.** Payment.
c. Payment if it is Big Five.   **d.** Non–Big Five or Payment.

	Resolution of Litigation		
Firm	*No Payment*	*Payment*	*Total*
Big Five	60	90	150
Non–Big Five	10	20	30
Total	70	110	180

REFERENCE: Palmrose, Z. (1988), "An Analysis of Auditor Litigation and Audit Service Quality," *Accounting Review,* LXIII: 55–73.

36. As a defense against unwanted takeovers, many corporations have adopted antitakeover strategies. Incentive packages, known as golden parachutes, have been adopted

by many large corporations to provide compensation for corporate executive officers in the event of a takeover. Consider the accompanying table, which classifies corporations by whether they have adopted golden parachutes and whether they have been targets in takeover attempts. If a corporation is selected at random, find the probabilities for the following events:

a. Golden parachute.   **b.** Takeover target.
c. Takeover target if it has golden parachutes.
d. Takeover target or golden parachute.

| Takeover | Golden Parachute | | Total |
	Parachute	No Parachute	
Target	3	7	10
Not a target	17	73	90
Total	20	80	100

REFERENCE: Singh, H., and F. Harianto (1989), "Management-Board Relationships, Takeover Risk and the Adoption of Golden Parachutes," *Academy of Management Journal,* 32: 7–24.

37. An investor has enough capital to purchase just two of four available stocks. After the stocks have been well shuffled, the investor selects stocks by a blind draw without replacement. The stocks retain their values, three of the four available stocks will pay the investor a dividend of $1.00, but the fourth stock requires that the investor pay an additional dollar to the company.

   a. Without applying any of the rules for computing probabilities given in the chapter, does it seem more likely that the investor will increase her wealth or that she will not increase her wealth?
   b. Find the probability that the investor will receive $2.00 in total dividend payments.
   c. Find the probability that the total of the dividend payment and the additional payment that may be required by the investor will be equal to zero dollars.

   REFERENCE: Tarasov, Lev (1988), *The World is Built on Probability* (Moscow: Mir Publishers).

38. Certified public accountants engaged in audits of corporations often use models to decide whether audit opinions for company financial statements should be classified as qualified or as nonqualified. A study of statistical models known as least squares or logit models that were used to predict audit opinion decisions for 100 companies resulted in the accompanying table. The table classifies the prediction as correct or incorrect for the least squares and logit models. If an audit qualification prediction is selected at random, find the probabilities for the following prediction events.

   a. Least squares correct   **b.** Logit correct
   c. Logit correct given it is least squares correct
   d. Least squares correct or logit correct.

| Least Squares | Logit | | Total |
	Correct	Incorrect	
Correct	66	5	71
Incorrect	7	22	29
Total	73	27	100

REFERENCE: Stone, M., and J. Rasp (1991), "Tradeoffs in the Choice Between Logit and OLS for Accounting Choice Studies," *The Accounting Review,* 66: 170–87.

39. A numerically controlled milling machine at AAA Rivet Company is designated as being out of control if a sample of two milled parts has any that are defective. If 5% of the parts from the machine are defective and the process is stable, what is the probability that the milling machine will be designated as out of control after a given sample is checked?

40. American Policy and Casualty, Inc., an insurer, has a very large number of policyholders. In the past, American has made payments to 10% of the beneficiaries of its policyholders on claims for property damages. For two insurance policies selected at random, find the following probabilities:

    a. That no claims are paid.   **b.** That one claim is paid.
    c. That two claims are paid.

41. A management information system has three components that operate in series, as shown in the accompanying diagram. Since the components operate in series, all three components must function properly if the management information system is to function properly. The probabilities of failure for the three components during one period of operation are .01, .02, and .04. The component failures are statistically independent.

    input ⟶ ① ⟶ ② ⟶ ③ ⟶ output

    a. Find the probability that the management information system functions properly.
    b. Find the probability that the management information system fails during one period of operation.

42. A management information system has two components that operate in parallel, as shown in the accompanying diagram. The components convert raw data into machine readable files (1) by using an optical scanner and (2) by keying the data with a keyboard. Since the components operate in parallel, either of the components must function properly if the management information system is to function properly. The probabilities of failure for the components during one period of operation are .20 and .03. The component failures are statistically independent.

    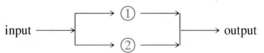

    a. Find the probability that the management information system functions properly.
    b. Find the probability that the management information system fails during one period of operation.

43. Canyonlands Petroleum Exploration Corporation is analyzing the riskiness of drilling for oil in the canyonlands oil basin. Exploration for oil in canyonlands is risky because well productivity in the region is highly variable. Furthermore, the productivity of a geologic structure in the basin is not influenced by or is independent of the productivity of other structures in the basin. The possible levels of productivity (in barrels) that might occur over the life of a well and their associated probabilities are given in the accompanying table.

    a. What is the probability that an exploratory well will discover oil?
    b. What is the probability that the productivity of an exploratory well will exceed 250,000 barrels?
    c. If two exploratory wells are drilled in the basin, then what is the probability that at least one well will discover oil?

Oil Well Productivity	Probability
0 (dry well)	.980
250,000	.010
500,000	.009
1,000,000	.001

44. Overthrust Drilling & Exploration has leased a tract of land located in the overthrust belt and plans to drill for oil. From a great deal of experience the manager of the oil company believes that the probability of obtaining a productive well on the site during a drilling operation is .20. A geologist for the company indicates that the probability of a known geologic structure (*KGS*) beneath similar tracts of land, given that the tracts resulted in productive oil wells, is .80. The probability of a *KGS*, given that the tracts do not result in productive oil wells, is .30.

   a. What is the probability that a well drilled at the site will be productive?
   b. After some research the geologist reports that there is a *KGS* beneath the tract. With this information, what is the probability that a well drilled on the site will be productive?

45. BSX Corporation has two suppliers that sell it batches of a chemical. The only way to determine whether the chemical is good is to use it in the company's production process. Vendor X supplies 40% of the batches, and vendor Y supplies 60%. Eighty percent of the batches from vendor X are good, and 85% of the batches from vendor Y are good. If one of the company's production runs is spoiled owing to a bad chemical batch, find the probabilities that the batch came from vendor X and vendor Y, respectively.

46. Cervical cancer accounts for around 4% of all cancers diagnosed in women and is responsible for 3.5% of cancer deaths. The American Cancer Society recommends that women over the age of 20 without symptoms be screened at least every three years by cervical cytology or cervical smears using Papanicolaou's method. However, the annual incidence rate of this cancer and the accuracy of screening do not seem to be well known, and the effectiveness of cervical cytology has never been tested in an experimental study. The annual incidence rate of this cancer is somewhat rare and has been reported to be about 1.1 in 1000 patients (Jones, Wentz, and Burdette, 1988) or the probability that a patient selected randomly for screening has this type of cancer is $P(Cancer) = .0011$. Unfortunately, screening tests for cancer are not perfect. For example, the sample may become air dried, and all cytologists will not reach the same conclusions from the same slides. Consequently, cervical cytology can result in a positive test or in a recommendation for further diagnosis or treatment given that there is no cancer. The probability of a positive test given that there is no cancer has been estimated to be about .006 (Soot et al., 1991); although be aware that the probabilities can vary among populations, providers, or labs. Furthermore, adequate cells may not be included in a sample, and a sample may contain several hundred thousand cells that must be inspected individually. Consequently, cytology can result in a negative test or no recommendation for further diagnosis given that there actually is cancer present. The probability of a negative test given that the individual has cancer has been estimated to be .30 (ACS, 1980).

   a. Find the probability of having this cancer given a positive test result.
   b. Find the probability of *not* having this cancer given a negative test.

   REFERENCES: "ACS Report on the Cancer-Related Health Checkup" (1980), *CA,* 30: 194–240; Jones, H., A. Wentz, and L. Burdette (1988), *Novak's Textbook of Gynecology* (Baltimore: Williams and Wilkins); Soot, H., et al. (1991), "The Validation of Cervical Cytology," *Acta Cytologica,* 35: 8–14; and *Wall Street Journal* (1987), "The Pap Test Misses Much Cervical Cancer Through Labs' Errors" (January 2): A1, Col. 5.

47. Cancer of the prostate is the number two cancer killer of American men. About one in 10 men will be victimized by this cancer in his lifetime. A blood test has been developed for screening for early detection of this cancer and the National Cancer Institute plans an experimental study of the effectiveness of screening. The annual incidence rate of this cancer is somewhat rare and has been reported to be about 1 in 1000 patients or the probability that an asymptotic patient selected randomly for screening has this type of cancer is $P(Cancer) = .001$. Unfortunately, screening tests for cancer are not perfect. The protein that is screened by the test is not generated in

a uniform fashion among all patients. The probability of a positive test given that there is no cancer has been estimated to be about .16. The probability of a negative test given that the individual has cancer has been estimated to be .20.

a. Find the probability of having this cancer given a positive test result.
b. Find the probability of *not* having this cancer given a negative test.

REFERENCE: *Wall Street Journal* (1991), "Prostate Cancer Detection May Be Aided by Test" (April 24): B1, Col. 6.

Ratio Data Set   *Refer to the 141 companies listed in the Ratio Data Set in Appendix A.*

Questions   48. Locate the data for the net income–total assets ratio (*NI/TA*) as a measure of profitability.

    a. Use a computer to obtain a tabulation for the companies that have zero or positive values and the companies that have negative values for this variable.
    b. If a company is selected at random from the list, what is the probability that the company has a negative value for this ratio?
    c. If two companies are selected at random from the list without replacement, what is the probability that both companies have negative values for this ratio?

49. Locate the data for the net income–total assets ratio as a measure of profitability and the Standard Industry Classification Code for classifying the companies by industry. Use a computer to obtain a two-row by two-column cross-tabulation for the companies that have zero or positive values and those that have negative values for the *NI/TA* variable and the companies that are services (SIC codes 7000–8999) or non-services companies.

50. From the cross-tabulation table constructed in the previous question, find the probability that a company selected randomly from the data set will have negative profits (negative *NI/TA* values) given that the company is in the services industry.

51. From the cross-tabulation table constructed in Question 49, find the probability that a company selected randomly from the data set will have negative profits (negative *NI/TA* values) and that the company is not in the services industry.

52. From the cross-tabulation table constructed in Question 49, find the probability that a company selected randomly from the data set will have negative profits (negative *NI/TA* values) or that the company is in the services industry.

53. From the cross-tabulation table constructed in Question 49, find the probability that a company selected randomly from the data set will have zero or positive profits (zero or positive *NI/TA* values) given that the company is not in the services industry.

Credit Data Set   *Refer to the 113 applicants for credit listed in the Credit Data Set in Appendix A.*

Questions   54. Construct a classification table in which the rows are married and unmarried, and the columns are credit granted and credit denied. Find the number of people who fall into each of this table's four cells.

55. Use the table constructed in Question 54 to find the following probabilities if an applicant is selected at random:

    a. *P*(Credit granted | Married).
    b. *P*(Credit granted | Unmarried).

56. Construct a probability tree in which the first branches deal with marital status and the second set of branches deals with whether credit was granted or denied.

EVALUATING CANCER SCREENING TESTS

Cancer of the breast is the most frequent cancer and the leading cause of cancer death in women. The American Cancer Society tells us that about one in nine women will be victimized by this cancer over her lifetime and recommends that women over the age of 50 without symptoms be screened annually by X-ray mammography. However, the annual incidence rate of this cancer, the accuracy of screening, and the statistical evaluation of the effectiveness of screening do not seem to be well known by either health care consumers or health care providers.

The annual incidence rate of this cancer is somewhat rare and has been reported to be about 2 in 1000 women (Moskowitz, 1983). We can thus take the probability that a patient selected randomly for screening has this type of cancer to be

$$P(Cancer) = .002$$

However, be aware that the probabilities of acquiring cancer are not known exactly, and they may change over time or across populations.

Unfortunately, mammographic screening tests for cancer are not perfect. Not all radiologists will reach the same conclusion from the same mammogram and some forms of benign abnormalities, such as cysts, fibroadenomas, and so forth, may simulate cancer. Consequently, mammography can result in a positive test or in a recommendation for further diagnosis or treatment given that there is no cancer. The probability of a positive test or mammogram given that there is no cancer has been estimated to be about .04 (Moskowitz, 1983), so we use the following probability:

$$P(Positive\ test\,|\,No\ cancer) = .04$$

Furthermore, some cancers are not found by mammography owing to poor quality in the radiographic image, questionable skills of the mammogram reader, minute size or subtle details of the lesions and so forth. Consequently, mammography can result in a negative test or no further diagnosis given that there actually is cancer present. The probability of a negative test, or mammogram that does not lead to further diagnosis or treatment, given that the individual has cancer has been estimated to be .36 (Goldberg and Wittes, 1981) so we use the following probability:

$$P(Negative\ test\,|\,Cancer) = .36$$

The probabilities can be summarized as shown in the probability tree diagram presented in Figure 1.

To statistically evaluate mammographic screening for cancer, one probability we want to know about is the probability of having cancer given a positive test or mammogram (the predictive value of a positive test or mammogram). Using the probability information presented in Figure 1, the probability that a randomly selected patient has *cancer* given that the individual received a *positive test* from mammography is determined by using Bayes' theorem as follows:

$$
\begin{aligned}
P(Cancer\,|\,Positive\ test) &= \frac{P(Cancer\ \text{and}\ Positive\ test)}{P(Positive\ test)} \\
&= \frac{P(Cancer)P(Positive\ test\,|\,Cancer)}{\begin{array}{l}P(Cancer)P(Positive\ test\,|\,Cancer)\ + \\ P(No\ cancer)P(Positive\ test\,|\,No\ cancer)\end{array}} \\
&= \frac{(.002 \times .64)}{(.002 \times .64) + (.998 \times .04)} \\
&= \frac{.00128}{.00128 + .03992} = \frac{.00128}{.0412} = .031
\end{aligned}
$$

FIGURE 1

**Probability Tree Diagram
for Mammography**

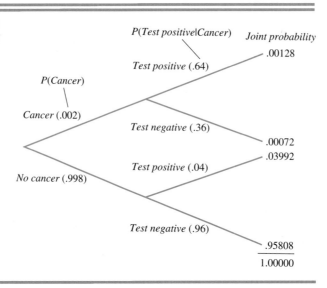

Consequently, the probability that the individual has cancer given that the test or mammogram was positive has increased from .002 to .031; but, of those receiving a positive mammogram, the probability of having cancer still seems low (.031). Unfortunately, this result also means that 96.9% or about than 97 out of 100 patients receiving a positive mammogram do *not* have cancer. Based on these probabilities, of the individuals who receive a positive test or mammogram for cancer, substantially more will *not* have the disease (about 97 out of 100) than will have the disease (about 3 in 100). The hazard to the overwhelming majority of patients who do not have cancer even though they have received an erroneous positive test or mammogram is that they will undergo time-consuming, costly, invasive, potentially life-threatening diagnostic procedures, such as excisional biopsy. Some patients may be subjected to unnecessary treatments, hospitalization, and anesthesia. They will also suffer a great deal of anxiety believing that they may have cancer.

Now, we also want to find the probability of *not* having cancer given a negative test or mammogram (the predictive value of a negative test). Using the probability information presented in Figure 1, the probability that a randomly selected patient does *not have cancer* given that the individual received a *negative test* or mammogram is determined by using Bayes' theorem as follows:

$$P(No\ cancer \mid Negative\ test) = \frac{P(No\ cancer\ \text{and}\ Negative\ test)}{P(Negative\ test)}$$

$$= \frac{P(No\ cancer)P(Negative\ test \mid No\ cancer)}{P(No\ cancer)P(Negative\ test \mid No\ cancer) + P(Cancer)P(Negative\ test \mid Cancer)}$$

$$= \frac{(.998)(.96)}{(.998)(.96) + (.002)(.36)}$$

$$= \frac{.95808}{.95808 + .00072} = \frac{.95808}{.9588} = .99924006$$

Consequently, the probability that the individual does not have cancer given that the test was negative has increased just slightly from .998 to .99924006. This also means that the probability of an individual actually having cancer given a negative mammogram is only .00075994 (or $1 - .99924006$). The hazard to these very few individuals is that they will not be diagnosed as actually having cancer as soon as they would be with an accurate screening test, and the delay in treatment may decrease their chances of survival.

Overall, the overwhelming majority of patients who receive positive mammograms actually do *not* have cancer. A patient who receives a negative mammogram also very likely does not have cancer. Statistical evaluation of mammographic screening for cancer suggests that it would be in the best interests of health care consumers if the accuracy of the mammographic technique could be improved or if other more accurate cancer screening methods could be developed.

Case Assignment

a. What risk does a consumer take if the individual foregoes screening for cancer and the disease is actually present?

b. What risk does a consumer take if the individual undergoes screening for cancer and the disease is not actually present?

REFERENCES: Goldberg, J., and J. Wittes (1981), "The Evaluation of Medical Screening Procedures," *The American Statistician*, 35: 4–11; and Moskowitz, M. (1983), "Minimal Breast Cancer Redux," *Radiologic Clinics of North America*, 21: 93–113.

DETERMINING THE RISK OF AIDS FROM BLOOD TRANSFUSIONS

CASE 4.2

A dramatic increase in life expectancy has occurred over the past century. However, the influenza virus in 1918 and the epidemics of lung cancer and heart disease in the past fifty years have tended to decrease life expectancies. The acquired immune deficiency syndrome (AIDS) epidemic is also certain to decrease life expectancies. Furthermore, the AIDS epidemic will dramatically influence the health care system, including costs of health care, insurers, and health care management.

Prior to March 1985 several hundred people were victimized by AIDS when they received transfusions of blood that contained the human immune-deficiency virus that results in AIDS. At the time about 4 in 10,000 donor units of blood were infected with the AIDS virus, or the probability that a randomly selected unit of blood was infected with the AIDS virus was

$$P(Donor\ blood\ was\ infected\ with\ AIDS\ virus\ prior\ to\ 1985) = .0004$$

Routine testing of blood donations for antibodies of the AIDS virus began in March 1985. Since testing was started, health care administrators have said that the public need not worry about getting AIDS from blood transfusions.

One desirable effect of screening potential blood donations for the AIDS virus has been that the proportion of infected donors has decreased. The numbers are not known exactly; however, it is believed that with testing for AIDS the number of blood donors that are infected with the AIDS virus has decreased to about 1 in 10,000. Thus since March 1985 the probability that a randomly selected donor is infected with the AIDS virus is conjectured to have decreased to about

$$P(Donor\ blood\ is\ infected\ with\ AIDS\ virus\ after\ 1985) = .0001$$

Unfortunately, blood tests for the AIDS virus are not error-free. If a donor is infected with AIDS and the donor's immune system has not yet recognized the virus, then the immune system will not develop the antibodies that are screened by the test. It has been estimated that about 1 in 100,000 current blood donations test negative, given the blood is virus-positive. Note that the numbers are not known exactly and that improvements in the

FIGURE 1

Probability Tree Diagram for the AIDS Test for Blood Donors

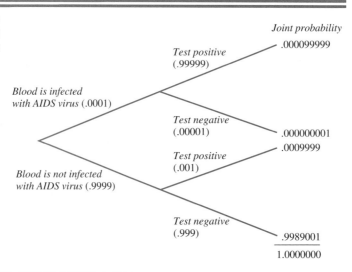

Joint probability

Blood is infected with AIDS virus (.0001) → Test positive (.99999)	.000099999
Blood is infected with AIDS virus (.0001) → Test negative (.00001)	.000000001
Blood is not infected with AIDS virus (.9999) → Test positive (.001)	.0009999
Blood is not infected with AIDS virus (.9999) → Test negative (.999)	.9989001
	1.0000000

tests are being made. We take the probability that the test for AIDS is negative given that the blood is infected with the virus for a randomly selected unit of blood to be

$$P(\textit{Test is negative} \mid \textit{Blood is infected with the AIDS virus}) = 1/100{,}000$$

$$= .00001$$

If about 1 in 1000 units of blood test positive when the blood does not have the virus, then the probability that the test is negative given that the blood is not infected with the virus is about

$$P(\textit{Test is negative} \mid \textit{Blood is not infected with the AIDS virus}) = .999$$

After March 1985, when screening began, the probabilities can be summarized as shown in the probability tree diagram presented in Figure 1.

If we use the probability information presented in Figure 1, the probability that a randomly selected unit of *blood is infected with the AIDS virus* given that the blood *tests negative* can be found by using Bayes' theorem as follows (be aware that the exact probabilities are not known):

$$P(\textit{Blood is infected} \mid \textit{Test is negative})$$

$$= \frac{P(\textit{Blood is infected} \text{ and } \textit{Test is negative})}{P(\textit{Test is negative})}$$

$$= \frac{P(\textit{Blood is infected})P(\textit{Test is negative} \mid \textit{Blood is infected})}{\begin{array}{c} P(\textit{Blood is infected})P(\textit{Test is negative} \mid \textit{Blood is infected}) + \\ P(\textit{Blood is not infected})P(\textit{Test is negative} \mid \textit{Blood is not infected}) \end{array}}$$

$$= \frac{(.0001)(.00001)}{(.0001)(.00001) + (.9999)(.999)} = \frac{.000000001}{.000000001 + .9989001}$$

$$= \frac{.000000001}{.998900101} = .0000000010011$$

With the screening that began in March of 1985, and from the estimated probabilities given in Figure 1, Bayes' theorem indicates that just 1 in about a billion transfusions will victimize recipients with the AIDS virus. The blood supply is much safer now than it was before tests for the AIDS virus antibodies were started.

Prior to the testing of blood donors for AIDS, it was estimated that about 4 in 10,000 units of blood were infected with the AIDS virus. After the testing began, the probability that a randomly selected unit of blood is infected with the AIDS virus, given that the test for the virus is negative, appears to be just 1 in about 1 billion.

The number of people who will be victimized by AIDS because of virus-infected blood transfusions has been decreased dramatically by testing donors for the virus. Bayes' theorem was invaluable in helping public health managers quantify and understand the chance of the recipient of a blood transfusion becoming infected with the AIDS virus, given that the blood tested negative for the virus.

Case Assignment

a. Use the information given in Figure 1 to find the probability that a blood unit contains the AIDS virus, given that the test for the antibodies is positive.

b. Comment on the loss of blood available for transfusions due to positive tests for the AIDS virus when the blood is actually free of disease.

REFERENCE: Dahl, R. (1987), "Major Epidemics of the Twentieth Century: From Coronary Thrombosis to AIDS," *Journal of the Royal Statistical Society, A* 150: 373–95.

MISUSING PROBABILITY IN A COURT OF LAW

CASE 4.3

In the court case *People v. Collins* a jury found the defendants guilty of second-degree robbery in Los Angeles, California. An appeal of the verdict to the California Supreme Court dealt with whether probability concepts had been properly introduced and used during the prosecution of the criminal case.

On June 18, 1964, a woman walking in an alley was pushed to the ground and her purse was snatched. At trial the victim testified that she had observed a woman who had blonde hair running from the scene of the crime. A witness testified that he had observed a caucasian woman with a blonde ponytail run out of the alley and enter a yellow automobile. The witness also noted as the car drove away that the driver was a black male who had a moustache and a beard.

An expert witness testified at the trial that the probability of the joint occurrence of several independent events is equal to the product of the probabilities for the events. The prosecutor then assumed the probabilities shown in the following table for a set of six characteristics observed by the victim and the witness:

Characteristic	Probability
1. Partly yellow automobile	1/10
2. Man with moustache	1/4
3. Girl with ponytail	1/10
4. Girl with blonde hair	1/3
5. Black male with beard	1/10
6. Interracial couple in car	1/1000

The prosecutor then applied the multiplication rule and arrived at the following:

P(*Partly yellow automobile and Man with
moustache* and *Girl with ponytail* and *Girl
with blonde hair* and *Black male with beard*
and *Interracial couple in car*)

$$= \left(\frac{1}{10}\right)\left(\frac{1}{4}\right)\left(\frac{1}{10}\right)\left(\frac{1}{3}\right)\left(\frac{1}{10}\right)\left(\frac{1}{1000}\right)$$

$$= \frac{1}{12,000,000} = .0000000833$$

The prosecutor indicated that there was but one chance in 12 million that any couple possessed the distinctive characteristics of the defendants. Notice, however, that the sample space and the event under consideration were not clearly specified. The prosecutor also asserted that the chance of someone else having these characteristics is very small.

On appeal to the California Supreme Court the justices commented that the prosecution had provided an inadequate evidentiary foundation for the probabilities presented at trial. No evidence had been provided that indicated that one out of every ten cars was partly yellow, that one of four males had a moustache, and so forth. The prosecutor had simply assumed these probabilities. Furthermore, no proof was presented that the characteristics were mutually independent. The Supreme Court justices concluded that the lower court was in error in admitting the evidence pertaining to probabilities in this case, and the guilty verdict was reversed.

Although evidence dealing with probability is often admitted in courts of law, an adequate evidentiary foundation should be presented and probability concepts must be applied correctly.

Case Assignment

a. Do the probabilities assigned to the characteristics by the prosecutor appear to follow the frequency, classical, or subjective concept of probability?

b. Provide your own subjective probability for the probability that a *man has a moustache* given that the *man is black and has a beard*. Is your subjective probability equal to 1/4, the probability assumed at trial for a man having a moustache?

REFERENCE: Supreme Court of California, *People v. Collins, California Reporter,* Volume 66, pp. 497–507.

REFERENCES

Feller, William. 1968. *An Introduction to Probability Theory and Its Applications,* 3rd ed., vol. I. New York: Wiley.

Mosteller, Frederick, Robert E. K. Rourke, and George B. Thomas, Jr. 1970. *Probability with Statistical Applications,* 2nd ed. Reading, Mass.: Addison-Wesley.

Parzen, Emanuel. 1960. *Modern Probability Theory and Its Applications.* New York: Wiley.

Ross, S. 1988. *A First Course in Probability,* 2nd ed. New York: Macmillan.

Tarasov, Lev. 1988. *The World Is Built on Probability.* Moscow: Mir Publishers.

Optional Topic

Computing a probability by using Equation (4.2) requires computing the number of outcomes favorable to an event and the number of outcomes in the sample space. In the examples thus far these outcomes were typically found by actually counting the simple events in sample spaces. This counting can be done more efficiently at times by using a method of this section that is known as the *multiplication principle* and by using techniques for finding *permutations* and *combinations*.

4.7
COUNTING PRINCIPLES

If we are required to select one object from a set of *a* distinct objects (a first-stage alternative) and one object from another set of *b* distinct objects (a second-stage alternative), then the number of distinct pairs of objects that can be formed can be found by using the following **multiplication principle.**

Multiplication Principle

The number of district pairs of objects that result when we combine one of *a* distinct objects from one set with one of *b* distinct objects from another set is

$$\textit{Number of distinct pairs of objects} = (a)(b) \qquad (4.15)$$

EXAMPLE 4.10   A distributor of small electric appliances sells hand mixers, toasters, and blenders. The distributor can obtain appliances from General Electric, Sunbeam, Westinghouse, and West Bend. How many appliance-manufacturer pairs can a customer purchase from the distributor?

The number of distinct appliances *a* is 3. The number of distinct manufacturers *b* is 4. The number of distinct appliance-manufacturer pairs is

$$(a)(b) = (3)(4) = 12$$

EXAMPLE 4.11   Two ordinary dice, one colored die and one white die, are rolled, and the number of dots on each die is recorded. How many possible outcomes are there?

The number of distinct outcomes for the colored die *a* is 6, and the number of distinct outcomes for the white die *b* is 6. The total number of possible outcomes (see Figure 4.12) is

$$(a)(b) = (6)(6) = 36$$

The multiplication principle extends in the obvious way to the product of a sequence of three or more choices.

To introduce the concept of *permutations,* we begin with a simple example, that of counting the number of three-letter "words" that can be made up from the five letters A, B, C, D, and E. The "word" need not make sense (CDA is counted as well as CAD), a letter may be repeated (BDB is counted), and order matters (AED and ADE are counted as different). Situations like this often arise in computer programming. A programming problem may require several three-letter codes made up from a group of five particular letters.

We must successively choose letters to fill the three blanks __ __ __. Since in each blank we can use any of the five letters, the three-stage multiplication principle with *a*, *b*, and *c* equal to 5 shows that there are 5 × 5 × 5, or 125, possible words. In general, from an "alphabet" of *a* letters one can make up a^n "words" of length *n*. Let us now examine a variation of this situation.

 PRACTICE PROBLEM 4.7

Count the three-letter computer code words that can be made from A, B, C, D, and E when repeats of letters are prohibited (BDB is not counted).

Solution As before, we must successively choose letters to fill the three blanks __ __ __. Since we may fill the first blank with any of the five letters, the number of first-stage alternatives is *a* = 5:

        A __ __
        B __ __
        C __ __
        D __ __
        E __ __

If the first blank was filled with A, the second may be filled with B, C, D, or E; if the first blank was filled with B, the second may be filled with A, C, D, or E; and so on. Thus *b* equals 4 this time; there are always four alternatives at the second stage. Finally, the third blank may be filled with any of the three unused letters; for example, from BC __ we may go on to BCA, BCD, or BCE. So *c* equals 3. And the total number of "words" is thus *a* × *b* × *c* = 5 × 4 × 3, or 60. A systematic listing of the words when letters are not repeated in any given word is given in the following table:

ABC ABD ABE	ACB ACD ACE	ADB ADC ADE	AEB AEC AED
BAC BAD BAE	BCA BCD BCE	BDA BDC BDE	BEA BEC BED
CAB CAD CAE	CBA CBD CBE	CDA CDB CDE	CEA CEB CED
DAB DAC DAE	DBA DBC DBE	DCA DCB DCE	DEA DEB DEC
EAB EAC EAD	EBA EBC EBD	ECA ECB ECD	EDA EDB EDC

Thus the five letters A, B, C, D, and E can be lined up in 5 × 4 × 3, or 60, ways when taken three at a time. A lining up, or ordering, is called a **permutation**.

> **Definition**
>
> A **permutation** of a set of objects is an arrangement of these objects in a definite order.

If we take three people and line them up, we may have Tom, Dick, and Harry *in that order*. That is one permutation. If Tom and Dick change places, we have Dick, Tom, and Harry. This ordering is a different permutation, even though it is the same combination of people. Many of the "words" in the table illustrated in Practice Problem 4.7 have the same letters in them, but they are counted as separate permutations since the ordering of the letters differs from one "word" to the next.

If we denote the number of permutations of five distinct things taken three at a time by $_5P_3$, what we have shown is that $_5P_3 = 5 \times 4 \times 3$.

Answers like this one are most conveniently expressed by using the *factorial notation*. The symbol $n!$, read *n factorial*,* is the product of the integers from 1 to n:

$$n! = n(n - 1)(n - 2) \times \cdots \times 3 \times 2 \times 1 \qquad (4.16)$$

For example,

$$5! = 5 \times 4 \times 3 \times 2 \times 1 = 120$$

One can also consider n factorial to be the number of permutations of n things taken n at a time, $_nP_n = n!$. That is, n factorial is the number of orderings possible for n distinguishable items. Thus five distinguishable items can be arranged into $5! = 120$ different orderings; one could make up 120 different codes with five *different* letters. If the same letter could be used more than once, then $5 \times 5 \times 5 \times 5 \times 5$, or 3125, possible codes could be formed.

From the definition of $n!$ we see that if x is a number between 1 and n, then

$$n! = n(n - 1)! = n(n - 1)(n - 2)!$$
$$= n(n - 1)(n - 2) \cdots (n - x + 1)(n - x)! \qquad (4.17)$$

And if we divide both sides of Equation (4.17) by $(n - x)!$, we see that

$$\frac{n!}{(n - x)!} = n(n - 1)(n - 2) \cdots (n - x + 1) \qquad (4.18)$$

Our case of five things taken three at a time, where $n = 5$ and $x = 3$, helps us test this result:

$$_5P_3 = \frac{5!}{2!} = 5 \times 4 \times 3$$

It is convenient to define $0!$ as 1:

$$0! = 1 \qquad (4.19)$$

---

*The use of the exclamation mark in factorial notation is to be distinguished from the way they are ordinarily used for emphasis.

We can now state the general rule governing permutations, or orderings.

Rule for Permutations

The number of **permutations** of n distinct things taken x at a time $(1 \leq x \leq n)$ is

$$_nP_x = n(n - 1)(n - 2)\cdots(n - x + 1) = \frac{n!}{(n - x)!} \tag{4.20}$$

The argument for Equation (4.20) is the same as for Practice Problem 4.7. We can think of the n objects as distinct symbols or letters with which we are to fill in a succession of x blanks, using a different letter in each blank. The set of alternatives at any stage is the set of letters as yet unused, and the numbers of these for the first stage, second stage, third stage, and so on, are $n, n - 1, n - 2, \ldots$. According to the multiplication principle, the number of permutations is the product of these numbers, of which there are x (one for each blank). This result gives Equation (4.20).

If x equals n, we have a special case described earlier. All n items are to be arranged in order, and we suppress the phrase "taken x at a time."

Rule for Permutations of *n* Distinct Things

The number of permutations of n distinct things is*

$$_nP_n = n! = n(n - 1) \times \cdots \times 3 \times 2 \times 1 \tag{4.21}$$

The number of permutations of the digits 0 through 9 is

$$10! = 3,628,800$$

The number of permutations of the 26 letters of the alphabet is

$$26! = 403,291,461,126,605,635,584,000,000$$

The number of different ways 7 distinct products can be lined up on a display shelf is

$$7! = 5040$$

In these examples the answer certainly must be arrived at *not* by counting all the cases but by the application of general principles.

To arrive at the rule for permutations, we found in Practice Problem 4.7 the number of three-letter computer code words that can be made from A, B, C, D, and E with repetitions disallowed. What was involved was essentially the order in which letters lined up.

---

*This rule is a special case of Equation (4.20). That is,

$$_nP_n = \frac{n!}{(n - n)!} = \frac{n!}{0!} = \frac{n!}{1} = n!$$

This operation shows the necessity for defining 0! as 1.

Since ABC and ACB represented different orderings, they represented different permutations as well. Suppose we change the terms of our problem in such a way that ABC is no longer recognized as different from ACB. This change will lead us to another general principle known as the number of *combinations*.

 PRACTICE PROBLEM 4.8

A smaller grouping of objects chosen from a larger one is called a *subset* of the larger *set*. How many subsets of three people can be formed from a group consisting of office workers named Sue, Tom, Dick, Harry, and Mary? The subset (Tom, Dick, Harry) is distinguishable from (Sue, Tom, Dick), since they contain different people. But (Tom, Dick, Harry) and (Harry, Dick, Tom) differ merely in the order in which the names appear. They are the same subset. In a subset, committee, or group the objects are not considered as being lined up at all; we could write them in a jumble:

Harry		Dick
Tom	or	Harry
Dick		Tom

The usual mathematical notation is {Tom, Dick, Harry}.

Solution Let k be the number of subsets. We can arrive at a solution in two stages. At the first stage we choose any one of the possible subsets k. At the second stage we take the three names chosen and line them up in some order. For example, we may choose the subset {Tom, Dick, Sue} and then line them up in the order Sue, Tom, Dick. The number of alternatives at the first stage is k. The number of alternatives at the second stage is simply the number of permutations of the three names chosen at the first stage, and this number we know to be 3!. So, in the notation of the multiplication principle, $a = k$ and $b = 3!$. But after making these two choices, what we arrive at is some permutation of three of the five names. Each permutation can arise in exactly one way from such a pair of choices. By the two-stage multiplication principle, the product $a \times b$, or $k \times 3!$, is the same as the number $_5P_3$ of permutations of five things taken three at a time. Since we already know that $_5P_3$ is equal to $5!/2!$, we can conclude that

$$k \times 3! = \frac{5!}{2!}$$

Solving for k gives our answer:

$$k = \frac{5!}{3!\,2!} = 10$$

Here is a list of the ten subsets:

{Tom, Dick, Harry}	{Tom, Sue, Mary}
{Tom, Dick, Sue}	{Dick, Harry, Sue}
{Tom, Dick, Mary}	{Dick, Harry, Mary}
{Tom, Harry, Sue}	{Dick, Sue, Mary}
{Tom, Harry, Mary}	{Harry, Sue, Mary}

In the argument in Practice Problem 4.8 we may replace 5 by a general number n and 3 by x, where we suppose $x \leq n$. If from a set of n distinct objects we are to choose a subset of size x, the number of ways this selection can be done is called the number of **combinations** of n things taken x at a time, denoted by $_nC_x$ or, more commonly, by $\binom{n}{x}$.

> **Definition**
>
> A **combination** of a set of objects is a subset of the objects disregarding their order.

We can choose a combination in $_nC_x$ ways and then permute the x objects in the combination in $x!$ ways, arriving at one of the $_nP_x$ permutations. By the two-stage multiplication principle, $_nC_x \times x! = {_nP_x} = n!/(n - x)!$, and we can divide through by $x!$ to get our answer:

> **Rule for Combinations**
>
> The number of **combinations** of n distinct objects taken x at a time (the number of subsets of size x) is
>
> $$\binom{n}{x} = {_nC_x} = \frac{n!}{x!\,(n - x)!} \qquad (4.22)$$

Notice that for the case where x is equal to n, the number of combinations is $n!/(n!\,0!)$, which is 1 by the convention shown in Equation (4.19). This answer is correct: There is but one subset that contains all the n objects. The answer is also 1 for the case where x is equal to zero. This result is itself just a convention: There exists exactly one set with nothing in it.

Sometimes, we cannot tell whether a problem requires combinations or whether it requires permutations. To answer this question, we ask ourselves whether or not the arrangement or order of the objects is relevant. If we need to take order into account, permutations are called for; if order is irrelevant, combinations are called for. As an aid to distinguishing between situations that involve permutations, and those that involve combinations, the reader may wish to determine if the following key words apply to the problem: *arrangements, sequences, orderings,* or *permutations.* If these words apply, then the situation is one involving permutations. The following key words apply to situations involving combinations: *groups, committees, sets, subsets, teams* (unless changing the members' positions on the team gives a new team), and *delegations.*

 PRACTICE PROBLEM 4.9

Assume there are five boxcars that need to be unloaded at a dock. But there is only enough time left in the day to unload three of them. The goods in the car unloaded first will be delivered today. Those unloaded second will be delivered tomorrow morning, and those unloaded last will be delivered tomorrow afternoon. In how many ways can three from among the five cars be unloaded in first, second, and third order?

Solution Here the dock superintendent is definitely interested in order, since goods in each of the cars are needed by customers. Since the order of unloading is important, we use permutations. The answer is

$$_5P_3 = \frac{5!}{2!} = 60$$

⬛ PRACTICE PROBLEM 4.10

How many ways can an executive *committee* of 5 be chosen from a board of directors consisting of 15 members?

Solution Here order is irrelevant; Brown, Jones, Smith, Black, and Williams form the same executive committee regardless of the order in which we list their names. Since order is irrelevant we use combinations. The answer is

$$\binom{15}{5} = \frac{15}{5!\,10!} = 3003$$

Table I in Appendix C at the back of this text is useful in evaluating combinatorial expressions like the one in Practice Problem 4.10. Table I gives the combinations of n things taken x at a time for selected values of n and x. For instance, the answer to Practice Problem 4.10 is found in the $n = 15$ row under the $x = 5$ column of Table I. This table will prove helpful in later chapters too.

Problems: Section 4.7

57. A company sells a product called Make-Your-Doll. It is a toy that consists of three doll bodies, five doll heads, four sets of legs, and two sets of arms. A child assembles the doll he or she wishes to play with. How many complete dolls should the marketing manager advertise that the kit can make if a doll consists of one representative from each of the following groups: body, head, legs, and arms?

58. A company is going to send a delegation of three people to a convention being held in Hawaii. Naturally, many people would like to go. The president of the company has indicated that only one person may go from each of the major departments: production, marketing, and finance. There are four production, six marketing, and three finance people who are eligible to attend the convention. How many different delegations could this company send to the convention?

59. There are six employees who work for one supervisor. The supervisor must rank these people. How many different *rankings* are possible? (Note that order is important for rankings.)

60. In a contest with five finalists, in how many different ways can a winner, a first runner-up, and a second runner-up be selected? (Note that order is important.)

61. If there are nine starters in a race, in how many different ways can first, second, and third prizes be awarded? (Note that order is important.)

62. In how many ways can a committee of three be selected from ten individuals? [*Hint:* You may wish to use Table I in Appendix C.]

63. How many 13-card bridge hands can be dealt from a deck of 52 cards?

64. A display shelf has room to exhibit five items of merchandise. The store manager has eight items she would like to display. How many different displays can she construct under the following conditions?
    a. She counts it as a new display every time she changes the *order* of the five items— even though the same five items may be on the shelf.
    b. She counts it as a new display only if she takes one of the five items off the shelf and puts a new one on (orderings within the five don't get counted separately).

65. A company has 12 geographical areas in which it can market its new product. Determine how many ways the marketing vice president can plan a marketing campaign with the following areas:
    a. One major and one minor target area.
    b. Two minor target areas. (It is *not* necessary to order the minor areas as first and second.)

66. A construction firm builds tract homes for moderate-income families. Its homes have three basic floor plans. Also, each home can be built with a single or double garage or a carport. Each can be outfitted with one of three different types of kitchen cabinets, each can have one of four different fireplaces, and each can have one of seven different carpets laid. How many variations of this firm's tract home can be built?

67. A menu offers a choice of six appetizers, five salads, eight entrees, four kinds of potatoes, seven vegetables, and ten desserts. If a complete meal consists of one of each, in how many ways can one select a dinner?

68. From the six numbers 1, 2, 3, 4, 5, and 6, how many three-digit numbers can be formed if a given integer can only be used once?

69. If a group consists of 10 people, in how many ways can a committee of four be selected?

Discrete Probability Distributions

5

We continue the study of our second major topic, probability, in this chapter by examining the concept of a *random variable*. Random variables, and their associated probability distributions, are important concepts when we collect, analyze, and interpret data.

In this chapter, we discuss two types of random variables: discrete and continuous. Our discussion includes the concepts of expected values, variances, and standard deviations of random variables. We also introduce *probability distributions* for discrete random variables in this chapter, including binomial and Poisson probability distributions.

Probability distributions for continuous random variables, including uniform distributions, normal distributions, which are very important in statistical inference, and exponential distributions are presented in the next chapter.

5.1
RANDOM VARIABLES

Many of the random experiments or observations we have discussed so far have outcomes that can be characterized in a natural way by numbers. In rolling a pair of dice, we can observe the total number of dots; in tossing a coin three times in a row, we can count the number of times heads turns up; in a set of invoices selected for audit, an auditor can count the number of invoices with errors. Outcomes that can be associated with numbers lead us to the following definition of **random variable.**

> **Definition**
>
> A **random variable** is a variable whose numerical value is determined by the outcome of a random trial.*

We denote random variables by X, Y, and so forth.

Before we look further into the idea of random variables consider a physically determined variable that has no element of randomness. To find the area of a square, measure the length l of one of its sides and then multiply that length by itself, computing l^2. Now l is a variable; it may be 1.5 feet or 8.3 cubits, depending on the square. To keep in mind this element of variability, regard l as a name for the length of the side *before the side is measured.* Measuring the side of a specific square converts l into a specific number. The variable l is useful for making general statements such as "The area is l^2."

Now let X be the total on a pair of dice. The dice may show a total of 7 when actually rolled, in which case X is 7; or they may show 5, in which case X is 5; and so on. Now X is a variable whose value depends on the outcome of a random experiment, and the outcome cannot be predicted. To keep in mind this variability and unpredictability, regard the random variable as a name for the number associated with the outcome of the experiment *before the experiment is performed.* Carrying out the experiment converts the random variable into a specific number.

One random experiment is the selection of a random sample from a population. Most random variables of interest to us will arise from sampling. For example, the number of defectives in a sample of microprocessors is a random variable, and in a sample of people their incomes are all random variables. We discuss methods for obtaining random samples in Chapter 7.

There is a classification of random variables that is based on the set of values that the variable can take on. The first type of random variable that we shall consider is a **discrete random variable.**

> **Definition**
>
> A **discrete random variable** is able to take on a countable number of values in an interval.

---

*A random variable can also be defined as a rule that assigns a numerical value to each outcome of a random experiment.

A discrete random variable usually has a definite distance from any possible value of the random variable to the next possible value. For example, the number of defectives in a sample is either 0, 1, 2, ..., or n; consequently, it is a discrete random variable, the distance between successive possible values being 1. The number of customers waiting in line at a checkout counter and the number of calls arriving at a switchboard during a given hour are discrete random variables.

The second type of random variable that we shall consider is a **continuous random variable.**

Definition

A **continuous random variable** is assumed able to take any value in an interval.

A height measured with complete accuracy can, in principle, take on any positive value, and so it is a continuous variable. So can measurements such as weight, temperature, pressure, time, and the like, provided they are measured with complete accuracy. In reality, there is a limit to the precision with which measurements can be made, and continuous random variables are only a useful idealization. Throughout the remainder of this text, we will generally consider physical measurements and monetary values such as lengths, weights, times, costs, and prices to be continuous variables. On the other hand, discrete variables are usually *countable* measures that are referred to as "number of items" that have a particular characteristic.

5.2
GENERAL PROBABILITY DISTRIBUTIONS FOR DISCRETE RANDOM VARIABLES

The behavior of a discrete random variable can be described by giving the probability with which it takes on each of its distinct, discrete values when the experiment is carried out. Thus the behavior of a discrete random variable can be described by using a listing or table of values, a graph of the values, or, sometimes, a formula.

EXAMPLE 5.1   For the roll of a pair of dice, say a white die and a colored die, the sample space (or the set of all 36 possible outcomes) is exhibited in Figure 5.1. By counting outcomes in Figure 5.1, we can determine that the probability of rolling a 7 for the total for the two dice is 1/6.

Letting X be the discrete random variable that represents the total for the two dice (X is a name for the total "before the dice are rolled"), we can express this fact by writing $P(X = 7) = 1/6$. Similarly, the probability of rolling a 4 is 3/36, or $P(X = 4) = 3/36$. Here is a list of all the probabilities (they can be determined by counting outcomes in Figure 5.1):

$$P(X = 2) = 1/36 = .028 \qquad P(X = 3) = 2/36 = .056$$
$$P(X = 4) = 3/36 = .083 \qquad P(X = 5) = 4/36 = .111$$
$$P(X = 6) = 5/36 = .139 \qquad P(X = 7) = 6/36 = .167$$
$$P(X = 8) = 5/36 = .139 \qquad P(X = 9) = 4/36 = .111$$
$$P(X = 10) = 3/36 = .083 \qquad P(X = 11) = 2/36 = .056$$
$$P(X = 12) = 1/36 = .028$$

FIGURE 5.1

Sample Space for Rolling a
Pair of Dice

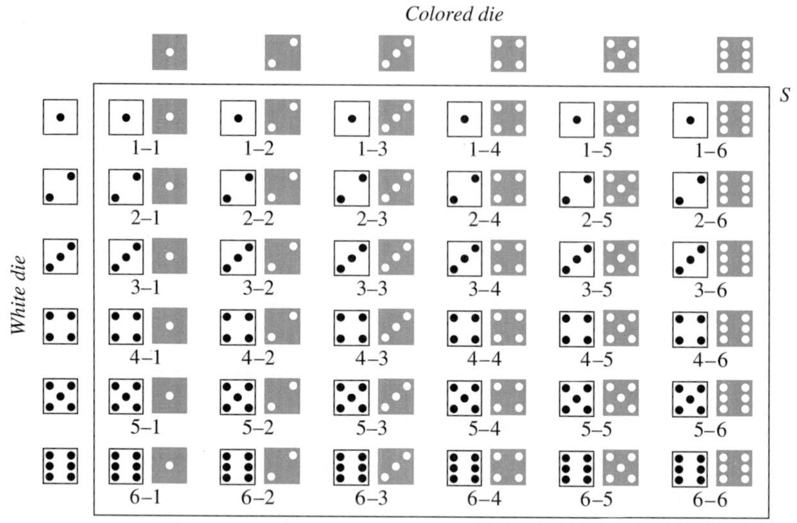

Colored die

White die

This collection of probabilities is called the *distribution* of the discrete random variable X. The distribution can be used to answer any question about probabilities for the values of X. For instance, the chance that "X is 7 or 11" is

$$P(X = 7 \text{ or } X = 11) = P(X = 7) + P(X = 11)$$
$$= 6/36 + 2/36 = 8/36 = 2/9 = .222$$

Because the probabilities come from the events in a sample space, two properties to which the probabilities in a probability distribution must conform are (1) that each probability value must be greater than or equal to zero and (2) that the sum of the probabilities must be equal to one. The values that a discrete random variable can assume, together with their associated probabilities, form a **probability distribution for a discrete random variable.**

> **Definition**
>
> A **probability distribution for a discrete random variable** is the collection of probabilities along with the associated values that the discrete random variable can assume.

A probability distribution can be presented as a listing or a table, a graph, or a formula. Let us continue with our example of rolling a pair of fair dice.

EXAMPLE 5.2 A pair of dice was rolled 100 times, and a record was kept of the number of times each possible value for the sum of the faces occurred. Figure 5.2(a) shows the observed relative frequencies; Figure 5.2(b) shows the corresponding probabilities in a probability distribution. The two figures would agree more closely if the dice had been rolled 1000 times, say. The probabilities in the probability distribution represent a theoretical limit toward which the observed

FIGURE 5.2

Empirical and Theoretical Distributions

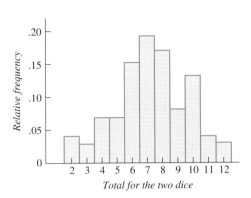

(a) **Empirical relative frequency distribution of the total for a pair of dice rolled 100 times**

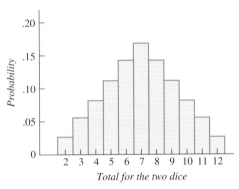

(b) **Theoretical probability distribution of the total for the rolls of a pair of dice**

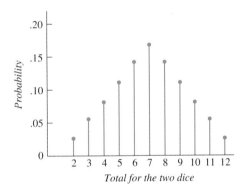

(c) **Theoretical probability distribution of the total for the rolls of a pair of dice, presented as a vertical line chart**

relative frequencies should tend as the number of rolls increases beyond bound.

Figure 5.2(c) is the same as Figure 5.2(b) except that it is presented as a chart of vertical lines rather than a bar chart. The vertical line chart tends to emphasize the discreteness of the random variable, whereas the bar chart tends to emphasize that the area of the bars in a distribution can be used to represent probability.

We can also present the theoretical probability distribution for the dice-rolling example by using the following table:

Probability Distribution for the Total Number of Dots from the Roll of Two Dice

Total	Probability	Total	Probability
2	1/36	8	5/36
3	2/36	9	4/36
4	3/36	10	3/36
5	4/36	11	2/36
6	5/36	12	1/36
7	6/36		1

EXAMPLE 5.3   A warehouse receives goods from Technitrol, Inc. Historical records show that during the past 50 days the warehouse manager indicated 6 trucks arrived on 8 of those days, 7 trucks arrived on 25 of those days, and 8 trucks arrived on 17 of those days. Thus the historical probability distribution for the discrete random variable called "number of truck arrivals" can be shown as in the following table:

Probability Distribution

Number of Truck Arrivals	*Probability*
6	8/50 = .16
7	25/50 = .50
8	17/50 = .34
	1.00

Since the sum of the probabilities is equal to 1, we have accounted for all of the possible outcomes for the number of truck arrivals.

Problems: Sections 5.1–5.2

Answers to odd-numbered problems are given in the back of the text.

1. As determined from long experience, the number of automobiles rented per hour, X, at the Rent-A-Car counter at Honolulu International airport has the probability distribution given in the accompanying table.

   a. What is the sum of the probabilities?
   b. Determine the probability that there will be at most 2 cars rented in an hour.
   c. What is the probability that there will be at least 3 cars rented in an hour?

Automobiles Rented	*Probability*
0	.10
1	.25
2	.40
3	.20
4	.05
	1.00

2. The probability distribution given in the accompanying table shows the number of boxcars unloaded each day by a warehouse crew for Boston Industries over the past year.

   a. Explain the meaning of the .25 associated with $X = 7$.
   b. How often during the past year has the crew unloaded 5 or fewer boxcars? Will the same be true for next year?

Boxcars	*Probability*
6	.10
7	.25
8	.50
9	.10
10	.05
	1.00

3. Suppose that a coin has the number 1 stamped on the head's side and the number 0 stamped on the tail's side. An experiment is conducted by tossing the coin once. The random variable of interest is the number that faces up when the coin is tossed.

   a. Disregarding the stamped numbers, list the outcomes of the experiment.
   b. List the values that the random variable can assume.
   c. What probabilities are associated with the values that the random variable can assume?

For a discrete random variable X the **expected value of X,** denoted $E(X)$, is the weighted mean of the possible values of X, the weights being the probabilities of these values. Here we denote the random variable with an uppercase X and each of the values that the random variable may take with lowercase x_i, where i has k distinct values $(i = 1, 2, 3, \ldots, k)$.*

Definition and Equation for the Expected Value of a Discrete Random Variable X

The **expected value $E(X)$** of a discrete random variable X is a weighted mean μ equal to the sum of the products of each value x of the variable and the associated probability $P(X = x)$, or

$$E(X) = \mu = \sum_{i=1}^{k} x_i P(X = x_i) \qquad (5.1)$$

EXAMPLE 5.4 Suppose that random variable X for the number of cars arriving at a tollbooth during a given minute can take the values 1, 2, and 3 and has the following distribution:

x	1	2	3
$P(X = x)$	.25	.40	.35

Here X takes on one of the $k = 3$ values and

$$E(X) = \mu = \sum_{i=1}^{3} x_i P(X = x_i)$$

$$= 1(.25) + 2(.40) + 3(.35) = 2.10$$

The phrase *expected value* is misleading in that, in this example, 2.10 is not to be expected at all as a value of X, since 1, 2, and 3 are the only possible values. That is, we do not "expect" 2.10 cars to arrive during any single minute. But $E(X)$ is what is to be expected in an *average* sense. The expected value of X is also called the **mean of X,** or $\boldsymbol{\mu}$. This mean is similar to the mean for a set of numerical data, discussed in Chapter 3.

---

*In this chapter we denote random variables with uppercase letters such as X, Y, Z, and so forth, and the values that the random variables may take as the corresponding lowercase letters x, y, or z. The notation helps to clarify the difference between random variables and the values the variables may take. In the following chapters, for simplicity, X, Y, Z, and so forth, are used to denote either the random variable or the values that the variable may take. The context of the discussion will make the meaning clear.

FIGURE 5.3

Expected Value as a
Point of Balance:
Discrete Case

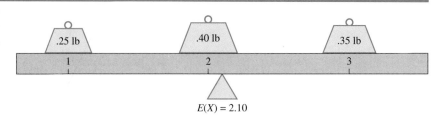

$E(X) = 2.10$

Just as the mean for numerical data can be viewed as a point of balance (see Figure 3.1 and Figure 3.2), so can $E(X)$. For the distribution in Example 5.4, imagine weights positioned on a seesaw at each of the possible values of X, each weight being proportional to the probability of the value it represents, as in Figure 5.3; the expected value 2.10 is where a fulcrum will balance the seesaw.

For the roll of a pair of dice of Example. 5.1 the total number of dots for the two dice takes on one of $k = 11$ different distinct values and the expected value is

$$E(X) = \mu = \sum_{i=1}^{11} x_i P(X = x_i)$$

$$= 2(1/36) + 3(2/36) + 4(3/36) + 5(4/36) + 6(5/36)$$

$$+ 7(6/36) + 8(5/36) + 9(4/36) + 10(3/36)$$

$$+ 11(2/36) + 12(1/36) = 7$$

For the trucks arriving each day at the Technitrol, Inc., warehouse of Example 5.3 the number of trucks takes on one of $k = 3$ different distinct values and the expected value is

$$E(X) = \mu = \sum_{i=1}^{3} x_i P(X = x_i)$$

$$= 6(.16) + 7(.50) + 8(.34) = 7.18$$

The expected value of X, or $E(X)$, is the value that X can be expected to have on the average. The actual value on a trial is typically greater or less than $E(X)$, and it is distributed in such a way as to balance out at $E(X)$.

Suppose a random variable has expected value $E(X)$ equal to μ. The **variance of a discrete random variable** that may take one of the k distinct values x_i $(i = 1, 2, 3, \ldots, k)$ is defined as in Equation (5.2).*

---

*An alternative formula for the variance of a random variable that is useful at times is

$$\text{Variance}(X) = \sigma^2 = E(X - \mu)^2$$

$$= \left[\sum_{i=1}^{k} x_i^2 P(X = x_i)\right] - \mu^2 = E(X^2) - \mu^2$$

Definition and Equation for the Variance of a Discrete Random Variable

The **variance**, denoted Variance(X) or σ^2, of a discrete random variable X is a weighted mean-squared deviation equal to the sum of the products of the squared deviation for each value x of the variable from its mean μ and the associated probability $P(X = x)$, or

$$\text{Variance}(X) = \sigma^2 = E(X - \mu)^2$$

$$= \sum_{i=1}^{k}(x_i - \mu)^2[P(X = x_i)] \qquad (5.2)$$

Now $(X - \mu)^2$, which is itself a random variable, is always positive, and the more distant X is from μ, the greater is $(X - \mu)^2$. The expected value of $(X - \mu)^2$ measures the amount by which the distribution is spread around the mean μ, much as does the variance for a set of numerical data, as defined in Chapter 3.

In all cases, Variance(X) is greater than or equal to zero. In the extreme case in which it is equal to zero, the value of X equals μ all the time, and X is not really random at all.

A standard deviation is always the square root of a variance. Thus we have the following equation for the **standard deviation of a discrete random variable.**

Definition and Equation for the Standard Deviation of a Discrete Random Variable

The **standard deviation**, denoted Standard deviation(X) or σ, of a discrete random variable X is equal to the square root of the variance, or

$$\text{Standard deviation}(X) = \sigma = \sqrt{\text{Variance}(X)} = \sqrt{\sigma^2}$$

$$= \sqrt{E(X - \mu)^2}$$

$$= \sqrt{\sum_{i=1}^{k}(x_i - \mu)^2[P(X = x_i)]} \qquad (5.3)$$

For the probability distribution for the number of cars arriving at a toll booth during a given minute of Example 5.4, the mean μ was 2.10, and so the variance is

$$\text{Variance}(X) = \sigma^2 = \sum_{i=1}^{k}(x_i - \mu)^2[P(X = x_i)]$$

$$= (1 - 2.1)^2(.25) + (2 - 2.1)^2(.40) + (3 - 2.1)^2(.35)$$

$$= .59$$

and the standard deviation is

$$\text{Standard deviation}(X) = \sigma = \sqrt{\sigma^2} = \sqrt{.59} = .768$$

The mean is often denoted by μ, the variance by σ^2 (Greek sigma, squared), and the standard deviation by σ; sometimes μ_X, σ_X^2, and σ_X are used instead, to indicate that the terms refer to computations involving X. Whatever the units of X (feet, trucks, degrees, and so on), its mean and standard deviation have these units, too.

The mean and the standard deviation for a numerical data set were discussed in Chapter 3. The mean and the standard deviation for random variables are related to this earlier discussion in a natural way. Since a random variable is a numerical measure that results from a random experiment or observation, an experimenter or observer making a series of observations will obtain a numerical data set. The mean and the standard deviation for the random variable are the same as the mean and the standard deviation the experimenter or observer would obtain for the numerical data set if the experimenter were theoretically to take infinitely many observations.

Sometimes, the conditions surrounding our probability problems are such that we can apply specific probability formulas to develop our probability distribution. One of the skills of a good probability problem solver is that he or she is adept at recognizing the situations in which a simple probability formula can be applied to develop a probability distribution.

The next two sections of this chapter are devoted to helping the reader develop this skill. Two commonly used probability formulas are presented. The conditions under which these formulas can be used to calculate probability distributions are discussed, along with the formulas themselves and some work-saving probability tables. The probability formulas to be presented are for the following discrete probability distributions: binomial and Poisson.

Problems: Section 5.3

4. The number of bids submitted by a certified public accounting firm prior to winning a competitive government contract from the Small Business Administration is given in the following distribution:

x	1	2	3	4	5
$P(X = x)$	.1	.2	.4	.2	.1

   a. Find the mean. **b.** Find the variance.
   c. Find the standard deviation.

5. The probability distribution for the number of sales calls completed per day by a large number of industrial salespeople with Great Lakes Industries is given in the following table:

x	0	1	2	3	4
$P(X = x)$	1/16	4/16	6/16	4/16	1/16

   a. Find the mean. **b.** Find the variance.
   c. Find the standard deviation.

6. The accompanying table gives the difference between the number of orders X processed on the second and first days at work for a large number of newly hired clerks at The Peoples' Stores, Inc. Negative values indicate that the number of processed orders decreased, and positive numbers indicate that the number of processed orders increased.

x	−1	0	1	2
$P(X = x)$	.1	.2	.3	.4

   a. Find the mean and variance of X.
   b. On the average, how does the number of orders processed on the second day compare with the number processed the first day?

A formula for computing probabilities is often used to work with probability distributions. We will develop a formula that can be used to find probabilities for a family of distributions known as the **binomial probability distributions** in this section. The conditions under which we can apply binomial probability distributions are illustrated in the following example.

5.4
BINOMIAL
PROBABILITY
DISTRIBUTIONS

EXAMPLE 5.5 A coin is to be flipped three times. We would like to know the probability that there will be two heads and one tail in these three flips.

Three conditions must exist if we are to use the binomial probability formula:

1. There is a fixed number of trials n of an experiment with only two possible outcomes for each trial. The two outcomes are labeled "success" and "failure."
2. The probability of a success on any given trial, designated as p, remains constant from trial to trial. Thus the probability of a failure is $1 - p$.
3. The trials are independent of one another in that the probabilities associated with the outcomes in one trial are not influenced by the outcomes in another trial.

Trials of an experiment that meet the three conditions required when we use the binomial probability formula are known as *Bernoulli trials,* generated by a *Bernoulli process,* after the mathematician J. Bernoulli (1654–1705).

Let us examine how the series of coin tosses discussed in Example 5.5 meet the three conditions. First, the three flips of the coin can be considered a series of trials with only two possible outcomes for each trial. That is, each of the three flips results in either "heads" or "tails." In this example, we will call heads a success and tails a failure. The terms *success* and *failure* are arbitrary and have no moral connotations; we use these terms by convention to denote the two possible outcomes. Second, the chance of success remains unchanged from flip to flip. The chance of a head is .5 on the first flip, .5 on the second flip, and .5 on the third flip. Third, the trials are independent of one another in the coin-flipping example.

When the three conditions for the binomial probability formula hold, we desire to find the probability of x successes given a fixed number of trials n. Example 5.5 asked us to find the probability of getting two heads and one tail in the three flips of the coin. Thus we want to know the probability of getting the mixture of two ($x = 2$) successes and one ($n - x = 3 - 2 = 1$) failure in three ($n = 3$) trials.

Consider the following manufacturing problem that illustrates the conditions of Bernoulli trials.

PRACTICE PROBLEM 5.1

Digital Industries uses a manufacturing process to produce memory chips for computers. Memory chips are devices that use thousands of transistors to store information in the form of electrical charges. The production of a chip requires several technical processes, including the slicing of silicon wafers, coating the wafers with oxide film and light-sensitive plastic called photo-resist, masking the photo-resist with a chip design stencil, etching the chip, and doping the chip with chemicals that form conducting zones.

The probability that a chip is defective (prior to quality control testing) is 1/6. One-sixth of the chips are defective due to technical difficulties that can occur during the production process, such as an impurity in the silicon or variability in etching. The problems that cause the defects occur at random and with stability. The defects occur independently from one chip to another.

Digital Industries has selected three ($n = 3$) chips at random from its production process, and the company wants to find the probability that exactly one ($x = 1$) of the three chips is defective. Digital also wants to find the probabilities of getting two defective chips, three defective chips, and no defective chips from the three.

Solution Note for the Digital chip-manufacturing process that each chip is either defective or *not* defective. The probability of any chip being defective is 1/6, and the probability is constant throughout the production process. Defects are also independent from one chip to the next. Finally, the number of trials is fixed at three ($n = 3$), and we desire to know the probability of getting one ($x = 1$) defective chip and two ($n - x = 3 - 1 = 2$) chips that are not defective.

The selection of a memory chip is a Bernoulli trial, and we denote the occurrence of a defective by S (for *success*) and the occurrence of a nondefective by F (for *failure*). The probability tree diagram exhibited in Figure 5.4 lists all the possible sequences of results for the three chips. The sequence SSS indicates that all three chips resulted in success (a defective); SSF indicates that the first two chips resulted in success and the last chip resulted in failure (nondefective); and so forth.

If three chips are selected, what is the probability of obtaining some particular sequence, say SSF? The probability of success on any one trial $P(S)$ is 1/6, and so the probability of failure $P(F)$ is 5/6. We know, as in Example 5.5 of this section, that there is independence from trial to trial, or that the result on one trial can in no way influence the result on another trial. Hence we use the rule of multiplying probabilities for independent events. The chance of getting S for the first chip and then S on the second and then F on the third is $P(SSF) = P(S)P(S)P(F) = (1/6)(1/6)(5/6) = (1/6)^2(5/6)^1$. The other probabilities shown in Figure 5.4 were computed by the same method.

FIGURE 5.4

Probability Tree Diagram of Independent Trials at Digital Industries with $n = 3$, $p = 1/6$

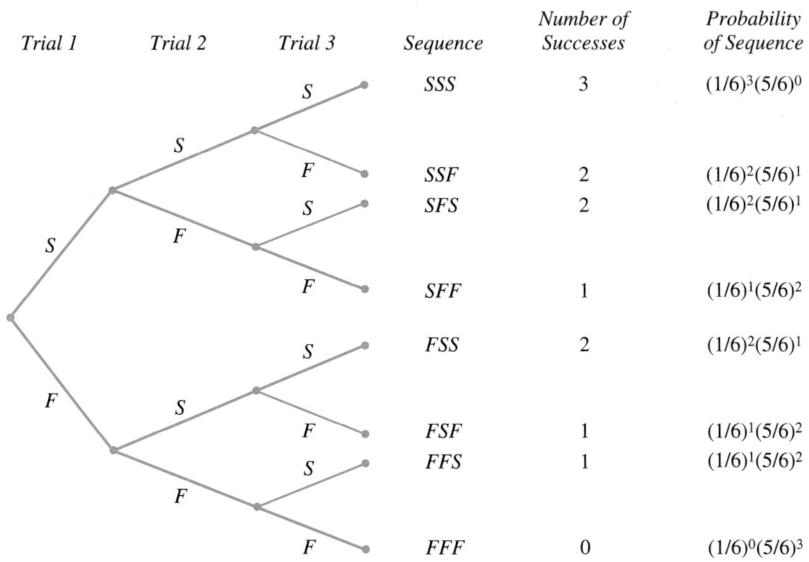

Trial 1	Trial 2	Trial 3	Sequence	Number of Successes	Probability of Sequence
		S	SSS	3	$(1/6)^3(5/6)^0$
	S	F	SSF	2	$(1/6)^2(5/6)^1$
	F	S	SFS	2	$(1/6)^2(5/6)^1$
S		F	SFF	1	$(1/6)^1(5/6)^2$
	S	S	FSS	2	$(1/6)^2(5/6)^1$
F		F	FSF	1	$(1/6)^1(5/6)^2$
	F	S	FFS	1	$(1/6)^1(5/6)^2$
		F	FFF	0	$(1/6)^0(5/6)^3$

Now, what is the probability of getting exactly one ($x = 1$) success [and two ($n - x = 3 - 1 = 2$) failures]? There are three ways this can happen; Figure 5.4 shows them to be *SSF*, *FSF*, and *FFS*. Each of these sequences has probability $(1/6)^1(5/6)^2$, where $(1/6)^1$ is for the one success and $(5/6)^2$ is for the two failures. Thus the probability of getting *exactly one* success is

$$P(X = 1) = 3(1/6)^1(5/6)^2 = 75/216 = .347$$

In addition, there are three sequences containing two ($x = 2$) successes and one ($n - x = 3 - 2 = 1$) failure (they are *SSF*, *SFS*, and *FSS*); and each has probability $(1/6)^2(5/6)^1$. Thus the probability of getting *exactly two* successes is

$$P(X = 2) = 3(1/6)^2(5/6)^1 = 15/216 = .069$$

Finally, the chance of *three* ($x = 3$) successes is the chance of *SSS*, namely.

$$P(X = 3) = 1(1/6)^3(5/6)^0 = 1/216 = .005$$

And the chance of *no* successes is the chance of *FFF*, namely,

$$P(X = 0) = 1(1/6)^0(5/6)^3 = 125/216 = .579$$

In general then, if we want to use the binomial probability formula to compute probabilities for the binomial distribution we repeat an experiment, singling out an event we call success. We let

$p =$ Probability of a success on a single trial

$1 - p =$ Probability of a failure on a single trial

$n =$ Number of trials

$x =$ Observed number of successes

We want to find the probability that the random variable for the number of successes, X, takes on the value x, given n Bernoulli trials and the probability of a success p on each trial. In other words, we want to find the probability that S occurs exactly x times (so that F must occur $n - x$ times) in n trials, where the chance of an S on each trial is p.

The number of different sequences that contain x successes and $n - x$ failures is denoted $\binom{n}{x}$, which is read "the *number of combinations* of n things taken x at a time" and is computed as follows:

$$\binom{n}{x} = \frac{n!}{x!\,(n - x)!} \tag{5.4}$$

The symbol $n!$ is read "*n factorial*" and is the product of the integers from n down to 1:

$$n! = n(n - 1)(n - 2)\cdots(3)(2)(1)$$

The value of *zero factorial* is defined to be one, or $0! = 1$. Values for combinations or $\binom{n}{x}$ are provided in Table I in Appendix C at the back of this text.

With the notation just described, the probability function for **binomial probability distributions** is as follows.

Binomial Probability Function

$$P(X = x \mid n, p) = \binom{n}{x} p^x (1 - p)^{n-x} \qquad x = 0, 1, 2, \ldots, n \qquad (5.5)$$

where $0 \leq p \leq 1$
and n is a positive integer.

The symbol $P(X = x \mid n, p)$ is usually read "the binomial probability of x successes *given n trials, where the chance of success on each trial is p.*" An equivalent notation for $P(X = x \mid n, p)$ is $P(x \mid n, p)$. For this formula to be valid, the three conditions listed at the beginning of this section must hold.

Equation (5.5) may be derived in this way: One sequence that contains x S's and $n - x$ F's is

$$\begin{array}{cc} SS \ldots S & FF \ldots F \\ x & n - x \\ \text{times} & \text{times} \end{array}$$

By independence, the probability of x S's and $n - x$ F's is $p^x(1 - p)^{n-x}$. We do not insist that the x S's occur in the *first x* trials; we insist only that exactly x of the trials produce S and that $n - x$ of them produce F. The probability of each such sequence is $p^x(1 - p)^{n-x}$. Equation (5.5) follows because the number of such sequences is $\binom{n}{x}$. That is, from the n trials there are $\binom{n}{x}$ ways to select a sequence that has x successes and $n - x$ failures.*

EXAMPLE 5.6   To demonstrate that Equation (5.5) gives correct answers that might otherwise have to be derived through detailed study of a problem, perhaps by using a very large probability tree, we will demonstrate how Digital Industries could have used this equation to solve the first part of Practice Problem 5.1. There we sought to find the probability of getting exactly one defective chip out of the three that were selected when the probability of a defective for any chip is 1/6. In that case we called getting a defective a success. Thus $x = 1$ defective, $n = 3$ trials, $p = 1/6$, and Equation (5.5) gives the following:

---

*According to the binomial theorem of algebra,

$$(y + z)^n = \sum_{x=0}^{n} \binom{n}{x} y^x z^{n-x}$$

The quantities $\binom{n}{x}$ are therefore called *binomial coefficients,* and thus the probabilities in Equation (5.5) are called *binomial probabilities.* The values of selected binomial coefficients are given in Table I of Appendix C and can be used to evaluate the number of ways x things can be taken from n things.

$$P(X = 1 \mid n = 3, p = 1/6) = \binom{n}{x} p^x (1 - p)^{n-x}$$

$$= \binom{3}{1} (1/6)^1 (5/6)^2$$

$$= \frac{3!}{1!\,2!} (1/6)^1 (5/6)^2 = .347$$

as we demonstrated previously by using the probability tree in Figure 5.4. ▬▬

▬▬ **EXAMPLE 5.7**   We could also have found the answer to the coin-flipping problem of Example 5.5 by using the binomial probability equation since that example met the conditions for application of the binomial formula. There we sought to find the probability that there will be two heads and one tail in three flips of a coin. Thus if we call a head a success, we have $x = 2$, $n = 3$ again, and $p = 1/2$. Equation (5.5) gives the probability we desire as

$$P(X = 2 \mid n = 3, p = 1/2) = \binom{3}{2} (1/2)^2 (1/2)^1$$

$$= \frac{3!}{2!\,1!} (1/2)^2 (1/2)^1 = 3/8 = .375 \quad \text{═}$$

▬▬ **PRACTICE PROBLEM 5.2**

Travelers Corporation, an insurer, has millions of policyholders. In the past Travelers has made payments to the beneficiaries of 15% of its policyholders on claims relative to accidental death coverage. A random sample of 10 insurance policies is drawn from all of Travelers' policies. Find the probability that the beneficiaries of 2 of the 10 will be paid on accidental death claims.

Solution In this example we have a series of trials ($n = 10$) where there are only two possible outcomes for each trial (beneficiaries will be paid or no payment will be made). The chance of getting a success (a payment due to accidental death) remains at 15% ($p = .15$) from insurance policy to insurance policy selected for the sample, and the trials are independent. The conditions for use of the binomial formula are satisfied, and we seek

$$P(X = 2 \mid n = 10, p = .15) = \binom{10}{2} (.15)^2 (.85)^8$$

$$= \frac{10!}{2!\,8!} (.15)^2 (.85)^8 = .2759 \quad \text{▬▬}$$

▬▬ **EXAMPLE 5.8**   For another example of binomial probabilities, consider the experiment of performing a market survey for General Soaps, Inc. One-fourth of the consumers in a market purchase General's hand soap. Let success S be the random selection of a consumer who purchases the product. In this case $p = 1/4$, and $1 - p = 3/4$. If we randomly select $n = 5$ people to talk to about the soap, the probability that exactly 3 people in the group will be purchasers of the soap is the probability of 3 successes in 5 trials, where the probability of success on each trial is 1/4:

$$P(X = 3 \mid n = 5, p = 1/4) = \binom{5}{3}(1/4)^3(3/4)^2$$

$$= 90/1024 = .0879$$

In Example 5.8 we looked at a situation where a market survey was being performed for General Soaps. We assumed that the probability of a success ($p = 1/4$) remained constant from trial to trial and that the trials were independent of one another. The assumptions would be true only if the market in which surveying was being done were large (say, several thousand people) or if selection were being done *with replacement,* allowing the possibility that once a person was interviewed he or she might be randomly selected and interviewed again later. Thus to meet the condition of independence for binomial distributions, we must sample from large populations, or we must conduct our sampling with replacement.

So far, we have discussed using the binomial formula to calculate the probability that there will be exactly x successes in a fixed number of trials n. In practical problems, however, we are usually interested in the probability that X will exceed a specific value or that X will lie in a given interval. These probabilities can be found more easily if we have a complete table of binomial probabilities that describe all the outcomes of some binomial experiment. We can develop this table by using Equation (5.5). In the next example we fully demonstrate the calculation and use of such a binomial table.

EXAMPLE 5.9   Of the thousands of savings and loan institutions included in a study of liquidity being made by the Federal Home Loan Bank Board, 50% are federally chartered and 50% are state chartered. An auditor randomly selects 5 savings and loans for an audit and is interested in the number of federally chartered institutions among the 5. The probability p of a federally chartered institution (success) on any selection is 1/2, and the probability $1 - p$ of a state chartered institution on any selection is 1/2. Hence the probability of getting $X = x$ federally chartered institutions in the sample of 5 is

$$P(X = x \mid n = 5, p = 1/2) = \binom{5}{x}(1/2)^x(1/2)^{5-x} \qquad x = 0, 1, 2, 3, 4, 5$$

The probability of exactly 3 federally chartered institutions is

$$P(X = 3 \mid n = 5, p = 1/2) = \binom{5}{3}(1/2)^3(1/2)^2$$

$$= 10(1/2)^5 = 10/32$$

and the probability of no federally chartered institutions is

$$P(X = 0 \mid n = 5, p = 1/2) = \binom{5}{0}(1/2)^0(1/2)^5$$

$$= (1)(1)(1/2)^5 = 1/32$$

x	$P(X = x)$
0	1/32
1	5/32
2	10/32
3	10/32
4	5/32
5	1/32
	1

TABLE 5.1

Binomial Probability Distribution
$n = 5, p = 1/2$

Calculating the remaining probabilities in the same way gives the distribution in Table 5.1 Note that these probabilities add to 1. Table 5.1 is an example of a specific *binomial probability distribution,* because the probabilities for this discrete distribution were derived from Equation (5.5).

Each different combination of n and p results in a different binomial distribution. For this reason, binomial distributions are often referred to as a *family* of distributions.

The binomial probability distribution with $n = 5$ and $p = 1/2$ is shown in Figure 5.5.

To find the probabilities that the number of successes X lie in a given interval, we add the terms in the distribution. For example, as illustrated in Figure 5.6, the probability of 2 or more federally chartered institutions in the 5 is

$$P(X \geq 2 \,|\, n = 5, p = 1/2) = \sum_{x=2}^{5} P(x)$$

$$= P(2) + P(3) + P(4) + P(5)$$

$$= 10/32 + 10/32 + 5/32 + 1/32$$

$$= 26/32$$

The probability that the number of federally chartered institutions will be greater than 2 but less than or equal to 4 is

$$P(2 < X \leq 4 \,|\, n = 5, p = 1/2) = \sum_{x=3}^{4} P(x) = P(3) + P(4) = 15/32$$

The sum does not include $P(2)$ since X must be greater than 2.

FIGURE 5.5

Binomial Probability Distribution with $n = 5$, $p = 1/2$

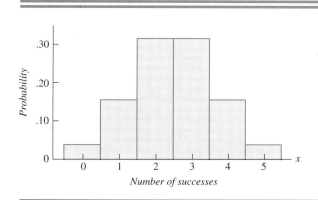

FIGURE 5.6

P(X ≥ 2) for the Binomial
Probability Distribution
(shaded region) with
n = 5, *p* = 1/2

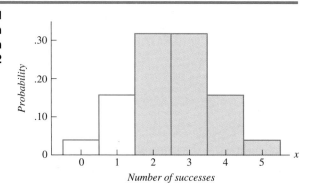

The probability that there will be at least one federally chartered institution is

$$P(X \geq 1 \mid n = 5, p = 1/2) = \sum_{x=1}^{5} P(x) = 31/32$$

Note that we can find this last probability more easily by using the rule for complementary events [Equation (4.8)]; that is, the probability of at least one federally chartered institution is

$$P(X \geq 1) = 1 - P(X = 0) = 1 - 1/32 = 31/32$$

Notice that in writing probabilities associated with *discrete* variables, we must distinguish between *greater than or equal to* ($\geq$) and *greater than* ($>$), and also between *less than or equal to* ($\leq$) and *less than* ($<$).

Evaluating Equation (5.5) every time we want to find a binomial probability or build a binomial probability distribution can be rather tedious. It becomes even more difficult if we desire to know a cumulative probability such as $P(X \leq 10 \mid n = 25,$ $p = .6)$. That is, if we wanted to know the probability of 10 or fewer successes in 25 trials, where the chance of success on each trial is .6, we would have to evaluate Equation (5.5) for the 11 values of *x* from 0 to 10. A set of **binomial tables** has been included in Appendix C of this book to aid in finding some complicated probabilities like the preceding one. These tables were produced by a computer programmed to evaluate Equation (5.5) for various values of *n*, *p*, and *x* and then summing the resulting probabilities from 0 up to *x*. Thus Table III in Appendix C gives cumulative binomial probabilities defined by the following **cumulative binomial probability distribution function:**

$$P(X \leq x \mid n, p) = \sum_{k=0}^{x} \binom{n}{k} p^{k}(1 - p)^{n-k}$$

EXAMPLE 5.10   We want to know the probability of obtaining 3 or fewer defective items (where a defective item is labeled a success) at Micro Devices, Inc., in 5 trials in a situation where the probability of success on each trial is .4. This probability can be found directly from the subtable for *n* = 5 in Table III. It is found in the $p = .40$ column and the $x = 3$ row. The value is .9130. The probability of 3 or fewer successes in $n = 10$ trials can be found in the subtable for $n = 10$. If the probability of a success on each trial is again $p = .4$, then this value is found in the $p = .40$ column and the $x = 3$ row and is .3823. ═══

Sometimes, we would like to know the probability of *exactly* x successes in n trials, not x or fewer. We can find this probability by using Table III also. In order to do so, we must use the following relationship:

$$P(X = x) = P(X \le x) - P(X \le x - 1) \tag{5.6}$$

This expression means that to find the probability of, for example, *exactly* 3, we find the probability of 3 or fewer and subtract the probability of 2 or fewer. Thus the probability of, say, 3 successes in 15 trials, where the probability of success on each trial is .2, can be found in the subtable for $n = 15$ in Table III. Equation (5.6) is then used in the following way:

$$P(X = 3) = P(X \le 3) - P(X \le 2) = .6482 - .3980 = .2502$$

Table III can also be used to find the probability of x *successes or more.* In this case we must use the complement of the probabilities given in Table III. The probability of the random variable X assuming the value of x successes or more is found as follows:

$$P(X \ge x) = 1 - P(X \le x - 1) \tag{5.7}$$

This expression means that to find the probability of, say, 3 or more successes, we find the probability of 2 or fewer and subtract that figure from 1.0. Thus the probability of 3 successes or more in $n = 20$ trials, where the probability of success on each trial is .1, can be found in the subtable for $n = 20$ in Table III. Equation (5.7) is then used in the following way:

$$P(X \ge 3) = 1 - P(X \le 2) = 1 - .6769 = .3231$$

 PRACTICE PROBLEM 5.3

Use Table III to find the following binomial probabilities:

a. $P(X < 8 \mid n = 10, p = .70)$
b. $P(X = 7 \mid n = 15, p = .40)$
c. $P(3 \le X \le 9 \mid n = 20, p = .30)$
d. $P(X \ge 6 \mid n = 10, p = .40)$
e. $P(X = 12 \mid n = 15, p = .85)$
f. $P(X = 25 \mid n = 25, p = .90)$

Solution

a. This value can be read directly from Table III: .8507.
b. Equation (5.6) must be used here: $.7869 - .6098 = .1771$.
c. To find this value, we must take $P(X \le 9) - P(X \le 2) = .9520 - .0355 = .9165$. Note that we subtracted $P(X \le 2)$, which left us with $P(X = 3)$ still in the answer, as we desired, since we wanted the chance that the number of successes ranged from 3 to 9, *inclusive.*

d. We must employ Equation (5.7) here to get $1 - .8338 = .1662$.

e. This answer cannot be found by using Table III since there is no column for $p = .85$. Thus we use Equation (5.5) to obtain the answer:

$$\frac{15!}{12!\,3!}(.85)^{12}(.15)^3 = .2184$$

Note that we did not interpolate between the columns for $p = .80$ and $p = .90$ to obtain this answer. If we did that, we would have obtained an answer of .1894, which is rather different from the correct answer of .2184 given by Equation (5.5). In general, interpolation between columns of the binomial table is not wise, and we must never try to interpolate between sections of the table for different n values.

f. We must use Equation (5.6) to solve this problem, but when we go to Table III to obtain $P(X \le 25 \mid n = 25, p = .90)$, we find that the 25th row in the table is missing. That this row is missing is not a mistake but involves the assumption that the reader realizes that $P(X \le n \mid n, p) = 1$, since this is the probability of n or fewer successes in n trials and takes into account all possibilities. Note that none of Table III's subsections contain a final row. Thus $P(X = 25 \mid n = 25, p = .90) = 1 - .9282 = .0718$. ▬

 PRACTICE PROBLEM 5.4

Managers for the State Department of Transportation know that 70% of the people arriving independently at a toll plaza for a bridge have the correct change. If 25 cars pass independently through the toll plaza in the next 5 minutes, what is the probability that between 10 and 20 cars, inclusive, have the correct change?

Solution This problem is a binomial-sampling situation since we have $n = 25$ trials, with success (correct change) or failure (no change) at each trial. Knowing that the last driver had correct or incorrect change tells nothing about the status of the next driver. Thus $p = .70$ from trial to trial. We desire to know

$$P(10 \le X \le 20 \mid n = 25, p = .70) = P(X \le 20) - P(X \le 9)$$

$$= .9095 - .0005 = .9090 \ ▬$$

We can use Equations (5.1), (5.2), and (5.3) to find the mean, variance, and standard deviation of any randomly distributed variable. However, when the variable follows the binomial probability distribution, we find the mean, variance, and standard deviation of the number of successes X by using the formulas that follow.

Equation for Mean of Binomial-Distributed Random Variable

$$E(X) = np \tag{5.8}$$

Equations for Variance and Standard Deviation of Binomial-Distributed Random Variable

$$\text{Variance}(X) = \sigma_X^2 = np(1 - p)$$

$$\text{Standard deviation}(X) = \sigma_X = \sqrt{\sigma_X^2} = \sqrt{np(1 - p)} \tag{5.9}$$

EXAMPLE 5.11   If we took many samples—say, 50,000 samples of size $n = 5$ with $p = 1/2$—we would get 50,000 numbers, the number of successes in each of the 50,000 samples. We could find the mean of those 50,000 numbers, and it would be approximately $np = 5(1/2) = 2.5$. The variance of those 50,000 numbers would be about $np(1 - p) = 5(1/2)(1/2) = 1.25$. The standard deviation, of course, would be $\sqrt{1.25} = 1.12$.

A binomial probability distribution can take on many different shapes, depending on the values of n and p. Figure 5.7(a) shows three binomial probability distributions; each has $p = .5$, and n begins at 4 and increases to 8 and then to 16. The figure indicates that as n increases, the mean increases, as shown by the shift to the

FIGURE 5.7

Binomial Distributions

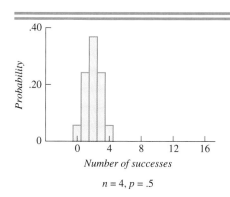

$n = 4, p = .5$

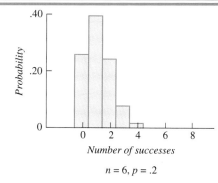

$n = 6, p = .2$

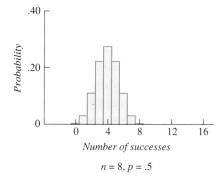

$n = 8, p = .5$

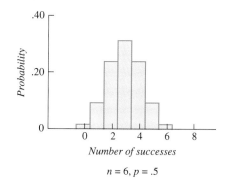

$n = 6, p = .5$

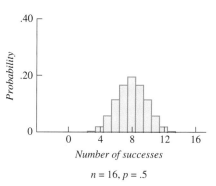

$n = 16, p = .5$

(a) Constant $p = .5$

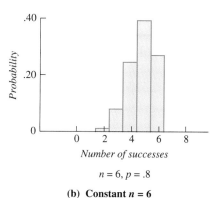

$n = 6, p = .8$

(b) Constant $n = 6$

right of the distributions [see Equation (5.8)]. Also, as n increases, the standard deviation increases, as depicted by the distributions becoming flatter and more widely dispersed [see Equation (5.9)].

Figure 5.7(b) shows three binomial probability distributions; each has the number of trials fixed at 6, and p increases from .2 to .5 to .8. The figure shows that the distribution is skewed right for p less than .5, is symmetric for p equal to .5, and is skewed left for p greater than .5. The binomial probability distribution is more symmetric as p gets closer to .5.

Problems: Section 5.4

7. Use Table III in Appendix C to find the following binomial probabilities.
   a. $P(X \le 5 \mid n = 10, p = .6)$ **b.** $P(X = 12 \mid n = 20, p = .4)$
8. Use Table III in Appendix C to find the following binomial probabilities.
   a. $P(X \ge 16 \mid n = 25, p = .8)$ **b.** $P(3 \le X \le 8 \mid n = 10, p = .7)$
9. Use Equation (5.5) to compute the following binomial probabilities.
   a. $P(X = 1 \mid n = 4, p = .5)$ **b.** $P(X = 3 \mid n = 5, p = .15)$
10. Find the mean and the standard deviation of the number of successes in binomial distributions characterized as follows:
    a. $n = 20, p = .5$ **b.** $n = 100, p = .9$
11. Find the mean and the variance of the number of successes in the binomial distribution, where $n = 5$ and $p = .4$.
    a. Use Equations (5.8) and (5.9).
    b. Use Equations (5.1) and (5.2) and the probabilities for $x = 0, 1, 2, 3, 4, 5$ that can be obtained from Table III in Appendix C.
12. A shop foreman for Northern Manufacturing Corporation wants the supervisor to look at one of the defectives that a new automatic drilling machine is producing. The machine has been producing defects randomly at the rate of one every three parts. What is the probability that the shop foreman will find exactly one defective in a randomly selected sample of $n = 4$ parts? [*Hint:* Use Equation (5.5) with $p = 1/3$.]
13. Large sheets of plate glass made by American Glass Company are inspected for structural defects at each of 5 inspection stations. The probability of successfully identifying a defect at any one station is $1/2$. Give the probability that a defective piece of glass is identified at the following stations [*Hint:* Use Table III in Appendix C.]:
    a. All 5 inspection stations.   **b.** At least 1 inspection station.
    c. Two of the 5 inspection stations.
14. In a triangle taste test conducted at a Lowprice Stores, Inc., grocery store, the taster is presented with three samples, two of which are alike, and is asked to pick the odd one by tasting. If a taster has no well-developed sense and can pick the odd one only by chance, what is the probability that in six trials he or she will make five or more correct decisions? No correct decisions? At least one correct decision? [*Hint:* Equation (5.5) will be useful here.] Discuss the notion of independence for these trials.
15. Three binomial probability distributions are presented in Figures 5.8. All three distributions have the same number of trials (labeled "$n = $?" for each distribution). The probability of a success on a single trial (labeled "$p = $?") is not specified for the distributions.
    a. Determine the value for the number of trials.
    b. Comment on the relative values for the probability of a success on a single trial for each distribution.
    c. Comment on the relative values of the means of the distributions.
    d. Comment on the shape (symmetry or asymmetry) for each distribution.

Binomial Distributions with Different *p*   **FIGURE 5.8**

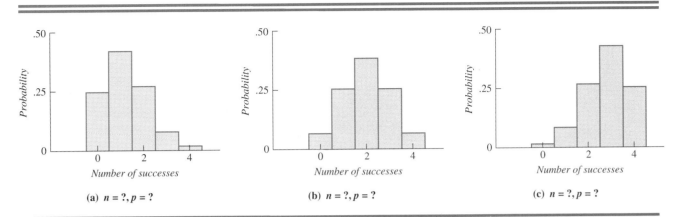

(a) *n* = ?, *p* = ?   (b) *n* = ?, *p* = ?   (c) *n* = ?, *p* = ?

Another family of discrete probability distributions that is useful to managers and administrators is the Poisson distribution family, which was named for the man who developed it, Simeon Poisson (1781–1840). This distribution is useful in quality control situations, waiting line problems, and numerous other applications of probability.

There are few *obvious* conditions under which a Poisson distribution is directly applicable to a probability problem. However, the sampling medium is usually not a discrete trial, as it is for a binomial distribution. The sampling is often done over an interval on some continuous medium such as time, distance, or area.

Four conditions must exist if we are to use a Poisson distribution:

1. Events must occur independently from one interval to another.
2. The probability of the occurrence of an event must be proportional to the size of the interval.
3. The probability of more than one event occurring in a very small interval is negligible.
4. The probability distribution must remain constant from one interval to another.

Events that occur under these conditions are generated by a *Poisson process.*

For instance, a Poisson distribution might be used to answer a question such as "What is the probability of finding seven paint blisters in a paint job that covers 100 square feet of area?" The number of paint blisters *X* is discrete and takes on the values $x = 0, 1, 2, \ldots, 7, 8, \ldots$. But the sampling medium is the painted area, which has no obvious "trials."

Other examples where a Poisson distribution has been applied are as follows:

Number of arrivals at a checkout counter during a 5-minute period

Number of flaws in a pane of glass

Number of telephone calls received at a switchboard during a minute

Number of touchdowns scored in a football game

Number of machine breakdowns during a night shift

**5.5
POISSON
PROBABILITY
DISTRIBUTIONS**

The **Poisson probability function** is as follows.

Poisson Probability Function

$$P(X = x \mid \lambda) = \frac{e^{-\lambda}\lambda^x}{x!} \qquad x = 0, 1, 2, \ldots \tag{5.10}$$

where λ is the mean number of Poisson-distributed events over the sampling medium that is being examined and $\lambda > 0$.

The number e is a mathematical constant, like Greek pi ($\pi = 3.1416$). It occurs frequently in mathematics and statistics. Its value is $e = 2.718\ldots$.

 PRACTICE PROBLEM 5.5

The number of bubbles found in plate glass windows produced by a process at Guardian Industries is Poisson-distributed with a rate of .004 bubble per square foot. A 20-by-5-foot plate glass window is about to be installed. Find the probability that it will have no bubbles in it. Also, what is the probability that it will have two bubbles in it?

Solution The continuous sampling medium in this problem is the area of the plate glass. That area is (20 feet)(5 feet) = 100 square feet. Thus the expected number of bubbles in this piece of plate glass is $\lambda = (100$ square feet)(.004 bubble per square foot) = .4 bubble. Now we can find the two probabilities asked for by using Equation (5.10):

$$P(X = 0 \mid \lambda = .4) = \frac{(2.718)^{-.4}(.4)^0}{0!} = 2.718^{-.4} = .6703$$

and

$$P(X = 2 \mid \lambda = .4) = \frac{(2.718)^{-.4}(.4)^2}{2!} = .0536$$

Evaluating expressions such as those in the preceding problem requires the use of a moderately sophisticated calculator or the use of logarithms. To avoid this problem and obtain the Poisson probabilities of Equation (5.10), the reader can use Table IV in Appendix C. Table IV gives Poisson probabilities that have been calculated from Equation (5.10). The answers to Practice Problem 5.5 can be found under the $\lambda = .4$ column and in the $x = 0$ and $x = 2$ rows.

Table IV has two features that should be noted. First, only selected values of λ are listed. A probability such as $P(X = 5 \mid \lambda = 3.25)$ cannot be read from this table since there is no $\lambda = 3.25$ column. However, interpolation between the columns is valid. For instance, we can find the value midway between $P(X = 5 \mid \lambda = 3.2) = .1140$ and $P(X = 5 \mid \lambda = 3.3) = .1203$; this value is .11715, which can be used as an approximation to $P(X = 5 \mid \lambda = 3.25)$. The exact value of the probability, found by using Equation (5.10), is .11720, which is virtually the same as the interpolated value.

Second, Table IV is a table of individual probabilities. The binomial table, Table III, is cumulative. There is no inherent advantage of one type of table over the other. Tables III and IV were presented as cumulative and individual values, respectively, to demonstrate to the reader the use of the two types of tables.

 PRACTICE PROBLEM 5.6

The number of customers arriving at a teller's window at Commercial Bank is Poisson-distributed with a mean rate of .75 person per minute. What is the probability that 2 or fewer customers will arrive in the next 6 minutes?

Solution In this problem $\lambda = (.75$ customer per minute$)(6$ minutes$) = 4.5$ customers. We desire $P(X \leq 2 | \lambda = 4.5)$, so we must add three probabilities in the $\lambda = 4.5$ column of Table IV corresponding to $x = 0$, 1, and 2; these are $P(X \leq 2 | \lambda = 4.5) = P(X = 0) + P(X = 1) + P(X = 2) = .0111 + .0500 + .1125 = .1736$.

Each column in Table IV represents a different Poisson probability distribution. A unique feature of the Poisson distributions is that $E(X) = \lambda$ and $\text{Variance}(X) = \lambda$. That is, the mean and the variance of a Poisson distribution *both* equal λ. This result could easily be verified by using Equations (5.1) and (5.2) on any column in Table IV.

Three Poisson distributions with λ's of .5, 2, and 4 are shown in Figure 5.9, and these distributions show that Poisson distributions are skewed to the right.

It can be shown mathematically that when there are many trials (n is large) and when the chance of success on each trial is small (p is close to zero), then the binomial probability $P(X = x | n, p)$ is approximately equal to the Poisson probability $P(X = x | \lambda)$, when $\lambda = np$, the mean of the binomial distribution.

EXAMPLE 5.12 Consider the following binomial probability: $P(X = 6 | n = 1000, p = .004)$. We cannot find this probability in Table III of Appendix C since Table III has no section for $n = 1000$ and no columns for $p = .004$. Thus we must find the answer by resorting to Equation (5.5):

$$P(X = 6 | n = 1000, p = .004) = \frac{1000!}{6!\,994!}(.004)^6(.996)^{994} = .1043$$

Poisson Probability Distributions   **FIGURE 5.9**

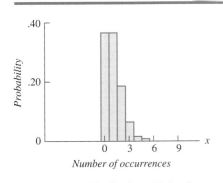

(a) **Poisson distribution with $\lambda = 1$**

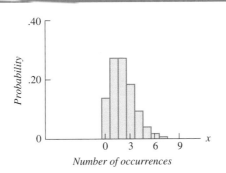

(b) **Poisson distribution with $\lambda = 2$**

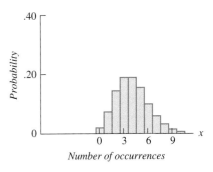

(c) **Poisson distribution with $\lambda = 4$**

MINITAB   **COMPUTER EXHIBIT 5.1A** Binomial and Poisson Probabilities

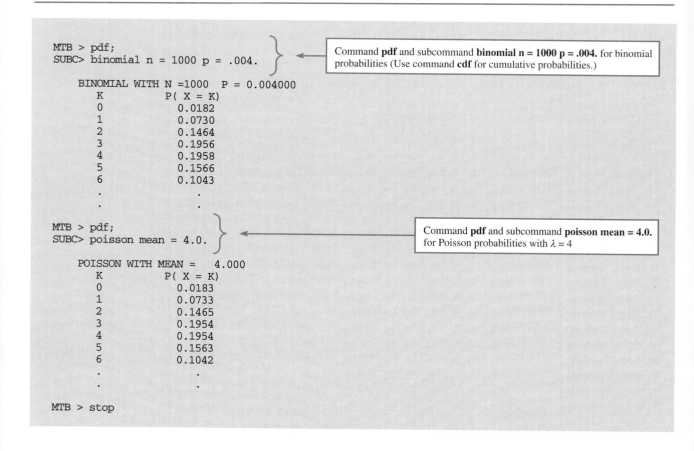

```
MTB > pdf;
SUBC> binomial n = 1000 p = .004.
```
Command **pdf** and subcommand **binomial n = 1000 p = .004.** for binomial probabilities (Use command **cdf** for cumulative probabilities.)

```
     BINOMIAL WITH N =1000   P = 0.004000
         K            P( X = K)
         0              0.0182
         1              0.0730
         2              0.1464
         3              0.1956
         4              0.1958
         5              0.1566
         6              0.1043
         .                .
         .                .
         .                .

MTB > pdf;
SUBC> poisson mean = 4.0.
```
Command **pdf** and subcommand **poisson mean = 4.0.** for Poisson probabilities with $\lambda = 4$

```
     POISSON WITH MEAN =    4.000
         K            P( X = K)
         0              0.0183
         1              0.0733
         2              0.1465
         3              0.1954
         4              0.1954
         5              0.1563
         6              0.1042
         .                .
         .                .
         .                .

MTB > stop
```

Once again, a calculator was required to evaluate this expression. In this example $np = (1000)(.004) = 4.0$. Table IV in Appendix C can be used to find

$$P(X = 6 \mid \lambda = 4.0) = .1042$$

Thus we can see that for this example the Poisson probability approximates the binomial to within .0001.

Computers can be used to find binomial and Poisson probabilities. Computer Exhibits 5.1A and 5.1B present computer outputs showing the binomial and Poisson probabilities for Example 5.12.

Problems: Section 5.5

16. Use Table IV in Appendix C to find the following Poisson probabilities.
    a. $P(X = 3 \mid \lambda = 4.5)$ **b.** $P(X = 8 \mid \lambda = 2.8)$
17. Use Table IV in Appendix C to find the following Poisson probabilities.
    a. $P(X \geq 9 \mid \lambda = 7.7)$ **b.** $P(X = 12 \mid \lambda = 9.2)$

COMPUTER EXHIBIT 5.1B Binomial and Poisson Probabilities **MYSTAT**

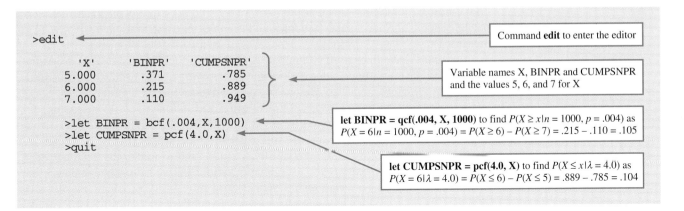

18. Use Table IV in Appendix C to find the following Poisson probabilities.
    a. $P(X \le 4 | \lambda = .8)$ **b.** $P(X = 3 | \lambda = 1.65)$

19. The number of paint blisters produced by an automated painting process at Associated Industries is Poisson-distributed with a mean rate of .06 blister per square foot. The process is about to be used to paint an item that measures 9 by 15 feet.
    a. What is the probability that the finished surface will have no blisters in it?
    b. What is the probability that the finished surface will have between 5 and 8 blisters, inclusive?
    c. What is the probability that the finished surface will have more than 2 blisters?

20. During off-hours, people arrive at a tollbooth on the Will Rogers Turnpike at a mean rate of .5 person per minute. The arrivals are Poisson-distributed.
    a. What is the probability that during the next 10 minutes 7 people will arrive?
    b. What is the probability that during the next 10 minutes none will arrive?
    c. What is the probability that during the next 10 minutes 3 or more people will arrive?
    d. What is the probability that during the next 10 minutes between 6 and 8 people inclusive will arrive?

21. The defects in an automated weaving process at Craft Mills, Inc., are Poisson-distributed at a mean rate of .00025 per square foot. The process is set up to run 1000 square yards of weaving.
    a. What is λ for this problem?
    b. What is the probability that this process will produce 5 defects on this run?
    c. What is the probability that it will produce between 1 and 3 defects inclusive on this run?

22. The defects in an automated weaving process at Allied Mills Company are Poisson-distributed at a mean rate of .00025 per square foot. The process is to be used to weave a piece of material that is 5 by 16 yards.
    a. What is λ for this problem?
    b. What is the probability that this piece will have no defects?
    c. What is the probability that it will have one defect?

23. A production process for an electronic component at an Arizona instruments manufacturing plant results in 4% defectives. One hundred components have been randomly selected from a very large number of components. Find, *approximately,* the probability that 6 of the 100 components are defective.

5.6
Summary

In this chapter we defined two types of random variables—discrete and continuous. Probability distributions for discrete random variables were considered, and two specific discrete probability distributions—binomial and Poisson—were examined. Expected values, variances, and standard deviations of discrete random variables were presented.

Formulas that can be used to compute probabilities for binomial or Poisson distributions and the conditions under which each distribution can be used were discussed. We also provided examples of the use of some work-saving probability tables. Some important equations from Chapter 5 are summarized in Table 5.2.

In Chapters 4 and 5 we have discussed numerous formulas, probability distributions, and relationships. Introductory students cannot be faulted if they are somewhat confused about when to use the probability formulas and relationships that have been presented. A person must answer a series of questions before he or she can determine which formula, table, or relationship applies to a probability problem. Figure 5.10 summarizes relationships among and equations for various probability distributions. Thus, the reader can use the figure to determine the type of distribution, formula, table, or relationship considered in the last two chapters that applies to a given probability problem.

TABLE 5.2   Summary of Equations for Discrete Probability Distributions

Description	Equation	
Expected value of a discrete random variable	$E(X) = \mu = \sum x P(X = x)$	(5.1)
Variance of a discrete random variable	$\text{Variance}(X) = \sigma_X^2 = E(X - \mu)^2 = \sum (x - \mu)^2 [P(X = x)]$	(5.2)
Standard deviation of a discrete random variable	$\text{Standard deviation}(X) = \sigma = \sqrt{\text{Variance}(X)}$ $= \sqrt{E(X - \mu)^2}$ $= \sqrt{\sum (x - \mu)^2 [P(X = x)]}$	(5.3)
Binomial probability distribution (or use Table III)	$P(X = x \mid n, p) = \binom{n}{x} p^x (1 - p)^{n-x}$	(5.5)
Mean of a binomial-distributed random variable	$E(X) = np$	(5.8)
Variance and standard deviation of a binomial-distributed random variable	$\text{Variance}(X) = \sigma_X^2 = np(1 - p)$ $\text{Standard deviation}(X) = \sigma_X = \sqrt{np(1 - p)}$	(5.9)
Poisson probability distribution (or use Table IV)	$P(X = x \mid \lambda) = \dfrac{e^{-\lambda} \lambda^x}{x!} \quad \lambda > 0$	(5.10)

FIGURE 5.10

Probability Summary Chart

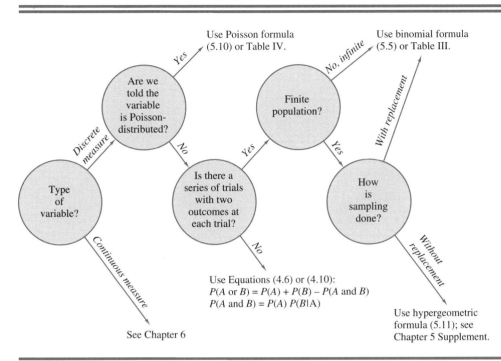

See Chapter 6

Answers to odd-numbered problems are given in the back of the text.

REVIEW PROBLEMS

24. Use Table III in Appendix C to find the following binomial probabilities.
   a. $P(X = 3 \mid n = 5, p = .5)$ **b.** $P(X \leq 1 \mid n = 25, p = .05)$

25. Use Table III in Appendix C to find the following binomial probabilities.
   a. $P(X \geq 7 \mid n = 15, p = .3)$ **b.** $P(18 \leq X \leq 22 \mid n = 25, p = .9)$

26. Use Equation (5.5) to compute the following binomial probabilities.
   a. $P(X \leq 2 \mid n = 6, p = .2)$ **b.** $P(X = 0 \mid n = 7, p = 1.0)$

27. Find the mean and standard deviation of the number of successes in binomial distributions characterized by the following.
   a. $n = 30, p = .7$ **b.** $n = 50, p = .4$

28. The proportion of bids submitted by a certified public accounting firm that win competitive governmental contracts from the Small Business Administration is .30. If the firm submits 10 independently prepared bids, find the following:

   a. The probability of winning exactly 2 contracts.
   b. The probability of winning fewer than 2 contracts.

29. Federally insured financial services institutions are required to provide statements of financial condition to the public. However, some statements are based on regulatory accounting principles and others are based on generally accepted accounting principles. Of the numerous financial institutions that failed in a region, 10% based financial

FIGURE 5.11

Distribution of Number of
Failed Financial Institutions

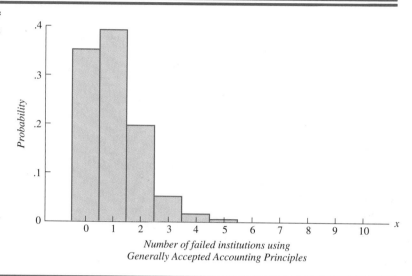

Number of failed institutions using
Generally Accepted Accounting Principles

condition statements on generally accepted accounting procedures. An accounting firm was retained to audit the accounting practices of 10 randomly selected institutions that failed. A binomial distribution with $p = .10$ and $n = 10$ is depicted in Figure 5.11.

a. Use cross-hatching on the figure to depict the probability of the accounting firm finding 2 or fewer failed institutions in the sample that used generally accepted accounting principles in their statements of financial condition that were made available to the public.

b. Find the probability of the accounting firm finding 2 or fewer failed institutions in the sample that used generally accepted accounting principles in their statements of financial condition that were made available to the public.

REFERENCE: Chenok, P. (1989) "Statement of Philip B. Chenok, President, American Institute of Certified Public Accountants, Before the Committee on Banking, Finance and Urban Affairs, United States House of Representatives, February 21, 1989," *Journal of Accountancy,* (April): 143–58.

30. The crisis in the financial services industry escalated to the point that the Financial Institutions Reform, Recovery, and Enforcement Act was passed in 1989. The Garn–St. Germain Act of 1982 had deregulated the industry and allowed the institutions to compete with money market funds in a freer market. About 30% of the numerous savings and loan institutions in one region of the country failed, but the deposits in the institutions were insured by the Federal Deposit Insurance Corporation up to $100,000. An investor had five certificates of deposit with more than $100,000 in different institutions and had essentially selected the institutions randomly since none of the institutions had failed in the region prior to deregulation. Find the probability that the investor suffered losses due to insolvency of any of the financial institutions.

31. At ACME Rivet & Machine Company a process produces 10% defective items. Quality system managers are analyzing how to improve the process. If we take a sample of 20 items, give the probability that we will find the following:

a. No defective items.   **b.** Not more than one defective.

32. A technologically sensitive process with tight tolerance limits is considered to be in control at Chicago Tool Works if it produces no more than 10% defective items. The

process is stopped and checked if a sample of 15 contains more than one defective item. What is the probability that it will be stopped when it is producing 5% defective items?

33. A door-to-door salesperson for Valley Products has a 20% sales success rate. Find the probability that during the next 20 calls (assume these call results are independent of one another) this person will make the following sales:

   a. Five or fewer sales   **b.** Three sales
   c. Four or more sales   **d.** Fail to make 16 or more sales

34. Tree surgeons for Botanical Critical Care Company lose about 5% of their "patients." Tomorrow they will operate on 5 trees.

   a. What is the probability that they will lose 1 patient or fewer from the 5?
   b. What is the probability that they will lose your tree, assuming it is to be operated on and the operation outcomes are independent of one another?

35. In a large Sun City Industries orchard that is experimenting with a biodegradable pesticide, 10% of the apples are wormy. Four apples are selected at random.

   a. What is the probability that exactly one will be wormy?
   b. What is the probability that none will be wormy?
   c. What is the probability that at least one will be wormy?

36. An automatic lathe at a Porker–Hanford Corporation plant is said to be out of control and is checked if a sample of 5 turnings contains any that are defective. If 1% of its output is defective, what is the probability that the lathe will be declared out of control and quality improvements will be instituted after a given sample is checked? Assume the probability of a defective is the same for every item selected.

37. A fair die is rolled 20 times.

   a. What is the probability that an even number will occur between 3 and 5 times, inclusive?
   b. What is the probability that an odd number will occur fewer than 2 times?

38. A process at American Communications, Inc., is being examined by a quality systems management team to determine whether special causes of variation have resulted in the process going out of control. If the process is currently producing 5% defective items, and if we take a sample of ten, what is the probability of fewer than three defective items?

39. Two binomial probability distributions for the number of defective microprocessors in quality control trials at Digital Processors, Inc., are presented in Figure 5.12. The

Binomial Distributions with Different *n*   FIGURE 5.12

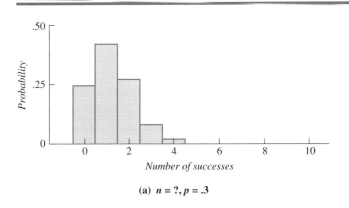

(a) $n = ?, p = .3$

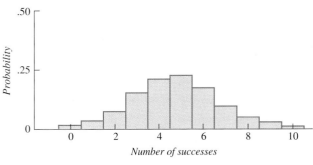

(b) $n = ?, p = .3$

process has gone out of control and is being examined for quality improvement purposes. The probability of a success on a single trial is the same for each distribution ($p = .3$).

a. Comment on the relative value of the number of trials for each distribution.
b. Comment on the relative values of the means of the distributions.
c. Comment on the relative values of the variances of the distributions.
d. Comment on the shapes of the two distributions.

40. According to Internal Revenue Service records at a large service center, 40% of the returns examined at the center contain mathematical errors. An IRS auditor is about to look over 15 returns for mathematical errors. Find the binomial probability that exactly 5 will contain mathematical errors.

41. Consider the list of companies shown in Table 5.3. This table gives changes in market values (as percentages) for companies in the aerospace industry. Some of the companies had increases in market values and some had decreases. An individual investor's portfolio consisted of 10 shares of stock in these companies, and the stocks were purchased at the start of the year. A company was selected at random as each share was purchased. Each company has a large number of shares of outstanding stock.

a. Find the probability that the number of stocks that increased in value is less than or equal to 6 for the investor's portfolio.
b. Find the probability that none of the stocks in the investor's portfolio increased in value.
c. Find the mean or expected number of stocks that increased in value for the investor's portfolio.
d. Find the variance for the number of stocks that increased in value for the investor's portfolio.

TABLE 5.3	*Aerospace*	*Market Value Change (%)*
Aerospace Company Market Value Data	Boeing	−2
	Fairchild Industries	3
	General Dynamics	9
	Grumman	−1
	Lockheed	13
	Martin Marietta	23
	McDonnell Douglas	8
	Northrop	−3
	Rohr Industries	5
	United Technologies	8

SOURCE: *Business Week,* December 29, 1986, p. 141.

42. The Internal Revenue Service uses computers to match randomly selected personal income tax returns against 1099 forms filed by payers of interest, dividends, and miscellaneous nonemployee income. The process is reported to find 60% of 1099 payments unreported on personal income tax returns. Unreported income results in an inquiry from the IRS. A taxpayer has forgotten to report income listed on 1099 forms over serveral years.

a. Find the probability that the taxpayer will not receive an inquiry after two years.
b. Find the probability that the taxpayer will not receive an inquiry after five years.

43. A marketing research polling organization found that 50% of the consumers in a large market would purchase hybrid entertainment systems that include TV, CD, and VCR components. The hybrid entertainment systems are produced by Sony Corporation and other major electronics companies that provide products for the $35 billion consumer electronics business. One retail electronics outlet has 15 consumers during a weekend, and the 15 consumers represent a random selection from the population of consumers in the large market.

   a. Find the expected number of hybrid entertainment systems that would be purchased.
   b. Find the probability that the outlet will have any of its current inventory of 10 hybrid entertainment systems remaining after the weekend.
   c. Should the inventory manager at the retail outlet pay a large shipping premium to rush 10 more hybrid entertainment systems to the outlet before the weekend?

44. Use Table IV in Appendix C to find the following Poisson probabilities.
   a. $P(X = 7 \mid \lambda = 6.2)$ b. $P(X \leq 2 \mid \lambda = 3.4)$

45. Use Table IV in Appendix C to find the following Poisson probabilities.
   a. $P(X = 6 \mid \lambda = 1.3)$ b. $P(5 \leq X \leq 7 \mid \lambda = 3.8)$

46. Use Table IV in Appendix C to find the following Poisson probabilities.
   a. $P(3 \leq X \leq 9 \mid \lambda = 5.5)$ b. $P(X \geq 2 \mid \lambda = 8.8)$

47. A sports utility vehicle has a 7-year, 70,000-mile warranty on all drive train components. The number of breakdowns in drive train components during the third year of the operation of the vehicle is Poisson-distributed with a mean number of breakdowns per year of .40.

   a. Find the probability of no breakdowns in the drive train components during the third year.
   b. Find the probability of one or more breakdowns in the drive train components during the third year.

48. Environmental concerns in Washington County have resulted in a hazardous materials clean-up team organized by the Department of Public Safety. The number of hazardous chemical spills in the county that require the intervention of the clean-up team is Poisson-distributed with a mean of 2 interventions per month. The Poisson distribution with a mean of 2 is depicted in Figure 5.13.

   a. Use cross-hatching on the figure to depict the probability of the team intervening on four spills during a month.
   b. Find the probability of the team intervening on four spills during a month.

FIGURE 5.13

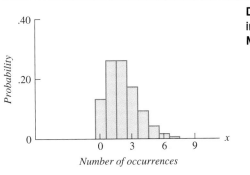

Distribution of Interventions in Spills of Hazardous Materials

49. The number of telephone calls passing through a TCI switchboard has a Poisson distribution with mean λ equal to $3t$, where t is the time in minutes. Find the following probabilities:

   a. Two calls in any one minute.
   b. Five calls in two minutes.
   c. At least one call in one minute.

50. Between Lake Powell and Lee's Ferry on the Colorado River the number of fish caught per angler-hour of fishing effort has a Poisson distribution with λ equal to 1.3 fish per angler-hour. Find the following probabilities:

   a. That 4 fish will be caught by 1 angler fishing 2 hours.
   b. That 4 fish will be caught by 2 anglers fishing 1 hour.
   c. That 6 fish will be caught by 3 anglers fishing 20 minutes.
   d. That 8 fish will be caught by 3 anglers fishing 2 hours.

51. A local area network computer system at Noel, Inc., has several input stations located in various departments. Typically, the central computer can respond to 8 input jobs per minute without causing any noticeable delay at the input stations. That is, when the jobs submitted in 1 minute are 8 or less, each input station behaves as though it had complete control of the network. Internal computer accounting records show that jobs are usually submitted at a Poisson-distributed mean rate of .08333 job per second. What is the probability that during a randomly selected minute of operation the input stations will experience some processing delays?

The following problems are somewhat more challenging than the others in this chapter.

52. The World Series terminates when one team wins its fourth game. If the two teams are evenly matched—that is, if the probability that either team will win any one game is 1/2—what is the probability that the series will terminate at the end of the fourth game? The fifth game? The sixth game? [*Hint:* Use the binomial distribution and formulas from Chapter 4.]

53. Teams A and B are competing in the World Series, and the probability that A wins any one game is .6.

   a. Find the probability that A wins in four games.
   b. Find the probability that B wins in five games.
   c. Find the probability that six games are required to complete the series.

Ratio Data Set   *Refer to the 141 companies listed in the Ratio Data Set in Appendix A.*

Questions   54. Locate the data for net income–total assets (*NI/TA*) ratio as a measure of profitability. If 5 companies are selected randomly with replacement from the data set, find the probability that exactly 3 will have negative *NI/TA* ratios.

55. The ratio data set includes companies that are in the services industry (SIC codes 7000–8999) and others that are in nonservices industries. If 10 companies are selected randomly with replacement from the data set, find the probability that exactly 4 will be from the services industry.

56. The ratio data set includes companies that are in the mining industry (SIC codes 1000–1499) and others that are in nonmining industries. If 8 companies are selected randomly with replacement from the data set, find the probability that one or more will be from the mining industry.

57. The ratio data set includes companies that have a smaller amount of current assets than they have current liabilities; consequently, the data value for the *CA/CL* ratio for these companies is less than one. If 5 companies are selected randomly with replacement from the data set, find the probability that exactly 2 will have *CA/CL* ratio values less than one.

Refer to the 113 applicants for credit listed in the Credit Data Set in Appendix A.

Credit Data Set

Questions

58. A national survey recently showed that 20% of the applicants for credit at department stores are women who apply for credit in their own names. Find the binomial probability that exactly 19 women in a sample of 113 applicants would apply for credit in their own names, as they did in this sample, if the 20% figure is correct.

59. Discuss why the probability found in Question 58 is so small.

60. Find the same probability found in Question 58 by using the Poisson formula with $\lambda = (.20)(113) = 22.6$.

61. Which probability is the correct figure—the binomial or the Poisson probability?

SUBSTANTIATING EMPLOYMENT DISCRIMINATION

CASE 5.1

Title VII of the Civil Rights Act of 1964 prohibits discrimination in the workplace. Numerous lawsuits charging employment discrimination have been filed against employers. The U.S. Supreme Court has ruled that discrimination can be demonstrated by showing statistically that minorities have been impacted disparately by employment practices.

In the case *Hazelwood School District v. United States,* it was alleged that the school district had engaged in a pattern or practice of employment discrimination. The school district was in St. Louis County, Missouri. In the two years 1972–1973 and 1973–1974, only 15 of the 405 teachers hired by the Hazelwood district were black. The proportion of blacks among qualified teachers in the county was .057 (excluding the city of St. Louis). To examine whether a disparate impact had occurred, the statistical question was whether the probability of hiring 15 or fewer black teachers out of 405 hires was small enough to constitute a statistical disparity. A probability that is unreasonably small would be evidence of discrimination; otherwise, the evidence would indicate that the school district was fair in its hiring practices.

Under the assumption of 405 independent hires with the probability of hiring a black of .057 for each trial, what is the probability of hiring 15 or fewer blacks? From the binomial probability equation [see Equation (5.5)] or a data analysis system, the probability of 15 or fewer successes among the 405 trials when $p = .057$ is

$$P(X \leq 15 \mid n - 405, p = .057)$$

$$= \sum_{k=0}^{15} \binom{405}{k}(.057)^k(1 - .057)^{405-k} = .046$$

The *Hazelwood* case is a landmark case that substantiated the use of statistics to demonstrate disparate impact in employment discrimination litigation.

a. A company makes 25 independent hires, with the probability of hiring a minority from the applicant pool on each trial of .10. Find the probability of 5 or fewer minority hires.

b. Why might hiring sequences not be statistically independent?

Case Assignment

REFERENCE: Meier, P., J. Sacks, and S. Zabell (1984), "What Happened in *Hazelwood,*" *American Bar Foundation Journal* (Winter): 139–86.

| **CASE 5.2** | **WINNING THE OIL LEASE LOTTERY** |

Western Petroleum is a small firm located in the southwestern United States. Its only business function is to create and market "oil drilling lease packages." An oil lease package consists of what amounts to "chances" to obtain oil leases on government-owned land in several states in the Southwest.

Each month the federal government conducts a lottery to determine who can obtain oil-drilling leases on various parcels of land the government has opened to oil-drilling activities. The Bureau of Land Management (BLM) releases the numbers and locations of the parcels open for bid on the third Monday of each month. An application card and a $10 filing fee are due at the BLM one week later. Each applicant may submit only one card for each parcel of land. This restriction means that a private individual stands as much chance of winning a lease as does the largest oil company (which can submit only one application for the company, any of its employees, agents, or their immediate families). On the day of the lottery a BLM employee literally draws three application cards out of a barrel. One drawing is made for each land parcel. The applicant who submitted the first card drawn is then given the opportunity to rent the drilling rights on the parcel for $1 per acre. If the winning applicant were disinterested in making the purchase, or unreachable for ten days after the drawing, then the lease would go to the person who submitted the second card drawn. If he or she were not interested, the third person would be contacted.

Almost all leases are rented by the winners, however, since the major oil companies stand ready to purchase any drilling rights at a moderate profit to the holder. The lease holder can seldom realize more than a $2500 profit on a lease since oil company officials realize that the typical winner does not have the resources to exploit the lease and actually conduct drilling operations.

Western Petroleum's president is Malcolm Maxfield. Mr. Maxfield describes his company's services as follows:

> We provide a convenient way for the average citizen to interface with the federal government in this oil-leasing business. We take the hassle and some of the risk out of these transactions for our clients. We screen the properties on which bids are entered and bid on only realistic parcels. Since we are authorized to act as their agents, our clients do not have to endure that rather complicated registration process. The applications we submit are for the individual clients, and the BLM people accept them as though they came straight from the private firms or persons we represent.
>
> We also take some of the risk out of applying for these oil leases. We do this by offering packages that include both high- and low-profit potential lease parcels and low and high probabilities, respectively, of winning the lease. You see, this is a rather speculative business. For a $10 investment someone can obtain a lease they can resell immediately for anywhere from a $200 to $2500 profit. They don't even have to put up the money for the lease purchase. They just sign a form we give them, and their lease rights are turned over to an oil company. The oil company pays the BLM the rental fee and sends us the "profit" for our client. The client interacts only with us. We handle all arrangements and negotiations with the BLM and the oil companies.
>
> We take a 20% fee on everything. Our clients pay $12 for each application they submit through us, and we get 20% of any profit they make on assigning the lease rights they win.

Mr. Maxfield has been concerned lately because some of his clients have been asking questions he can't answer very well. One of these clients is Stewart and Wagstaff Investment Advisors, Inc. Stewart and Wagstaff (S&W), in turn, have clients that entrust them

	Applications	Parcels	**TABLE 1**
	25 for class I parcel leases	Class I: Proven oil or gas reserves found just outside the parcel's immediate area	**Description of Plan B**
	50 for class II parcel leases		
	25 for class III parcel leases	Class II: Geologically promising areas but no proven reserves	
	100 at \$12 each = \$1200 purchase price	Class III: Speculative areas where no geological studies have been conducted and no proven reserves are known	

with investment funds. Some of S&W's clients are currently supplying them with speculation funds. Thus S&W approached Western Petroleum about six weeks ago and expressed some interest in buying several oil lease packages for their clients.

Due to the potential for a great deal of business from S&W, Mr. Maxfield is eager to accommodate them. First, he agreed to cut Western's commission from 20% to 14% on all business with S&W. Second, he agreed to remove from Western's mailing list all S&W clients. Third, he agreed to supply certain information to Mr. Saul Stewart by the first of next month. Meeting this third request is what has been occupying Mr. Maxfield's time for the past several days.

Mr. Stewart has expressed interest in buying several oil lease packages that Western calls its "Plan B—Moderate Return and Moderate Risk."* A table in Western's sales brochure describes plan B in Table 1.

A letter from Mr. Stewart stated his information needs. In part, the letter said:

> Since the selection of the winning application for the drilling rights on any individual parcel of land is a totally random event (a lottery) and since the various lotteries are independent of one another (winning one lease does not change your chance of winning another), it seems to me you could compute some probabilities for winning leases and some expected values for the various plans you offer.

This letter caused Mr. Maxfield to conclude that these were legitimate questions that ought to be answered. He also felt the answers could be used in the marketing of all three of Western's plans. However, Maxfield felt uneasy about how he should proceed. He felt that Saul Stewart was more sophisticated than he in this probability area. Thus he gathered the information he thought was relevant and put it in a letter to Mr. Stewart.

Dear Saul:

Thank you for your excellent letter of last week. You raised some interesting questions we had not considered previously—especially those about the probability of winning leases in plan B. In order to help answer your questions, I have put together this information:

Parcel Type	Profit to Winner (before Commission)	Average No. of Applicants per Parcel	Range in Number of Applicants over Last Two Years
I	\$2500	380	110 to 650
II	1200	80	25 to 125
III	200	15	10 to 40

*Plan A is high return and high risk, and plan C is low return and low risk.

These profit figures do not include the initial application fee. This table shows, as you would expect, that the real action is in the class I parcels. This information would cause me to estimate the chance of winning a lease on any class I parcels to be 1/380, but that chance may go as high as 1/110 or as low as 1/650 (on some of the real "hot" parcels that have high-yield wells nearby). Similarly, the probabilities of winning the leases on class II and class III parcels appear to be as follows:

Class	Probability	Probability Range
II	1/80	1/25 to 1/125
III	1/15	1/10 to 1/40

Since my background in probability is rather weak, I suggest that you and I meet for lunch next Thursday to discuss how this information can be used to answer your questions concerning plan B and other questions that may come to mind. My secretary will be in touch with you to make an appointment.

Sincerely,
Malcolm Maxfield

Case Assignment

a. Find the probability that, on average, plan B will result in the following.
   (1) No winning lease in a type I parcel.
   (2) No winning lease in a type II parcel.
   (3) No winning lease in a type III parcel.

b. Find the mean number of leases that I might expect to win with the purchase of a plan B.

REFERENCES

Cochran, W. G. 1963. *Sampling Techniques,* 2nd ed. New York: Wiley.

Freedman, D., R. Pisani, and R. Purves. 1992. *Statistics,* 2nd ed. New York: Norton.

Hogg, Robert V., and Allen T. Craig. 1978. *Introduction to Mathematical Statistics,* 4th ed. New York: Macmillan.

Mood, A. M., F. A. Graybill, and D. C. Boes. 1974. *Introduction to the Theory of Statistics.* New York: McGraw-Hill.

Mosteller, F., R. Rourke, and G. Thomas. 1970. *Probability with Statistical Applications,* 2nd ed. Reading, Mass.: Addison-Wesley.

Optional Topics

Hypergeometric Probability Distributions

Expected Values and Variances of Linear Functions and Sets of Random Variables

In Example 5.8 we looked at a situation where a market survey was being performed for General Soaps, Inc. We assumed that the probability of a success ($p = 1/4$) remained constant from trial to trial and that the trials were independent of one another. The assumptions would be true only if the market in which surveying was being done were large (say several thousand people) or if selection were being done *with replacement,* allowing the possibility that once a person was interviewed he or she might be randomly selected and interviewed again later. A different computational technique must be used if the surveying takes place in a small, finite population and if the selection is done *without replacement.* That is, we cannot use the binomial probability distribution to compute the probability of various survey outcomes since the second condition required for using the binomial distribution (that of independent trials) is not met. Let us change Example 5.8 to the one that follows and demonstrate how it should be handled.

5.7
Hypergeometric Probability Distributions

EXAMPLE 5.13   As in example 5.8, let us perform a market survey. This time, the survey takes place in an industrial market where one-fourth of the companies use your product. Thus $p = 1/4$. If there are only 40 companies in this market and if once a company is selected it is not selected again, then the outcomes of each trial are not independent, and the binomial formula cannot be used. If there are 40 companies and $p = 1/4$, then there are 10 companies that use your product. But once the first company is a user of your product, then $p = 9/39$ on the next selection. If the first company is not a user of your product, then $p = 10/39$ on the next trial. That is, the outcome of one trial changes the probabilities of the outcomes on the next trial.

Thus a new computational technique is called for. If we desire the probability of getting exactly 3 successes (companies who use your product) in a sample of $n = 5$, the total number of ways in which 5 customers can be selected from 40 is $\binom{40}{5}$, the combination of 40 companies taken 5 at a time. How many favorable cases are there? To obtain exactly three successes, we must make two choices. First, we must choose 3 of the 10 product users, and there are $\binom{10}{3}$ ways to do so. Then we must choose 2 of the $40 - 10 = 30$ nonusers, and there are $\binom{30}{2}$ ways to do so. There are $\binom{10}{3} \times \binom{30}{2}$ favorable cases, so the probability of exactly 3 product users in a survey of 5 companies is

$$\frac{\binom{10}{3}\binom{30}{2}}{\binom{40}{5}} = \frac{725}{9139} = .0793$$

Notice that this probability differs from the binomial probability .0879 in Example 5.8. This difference reflects the effect of selection from a finite population and the resulting variation in the value of *p*. ===

The general rule is as follows: Suppose we have $N = A + B$ distinct objects divided into two classes, say a class of successes and a class of failures. Suppose there are A successes and B failures. And suppose we take at random a sample of size n and ask for the probability that exactly x of the objects in it are successes. Thus we let

A = Number of successes in population of interest

B = Number of failures in population of interest

$N = A + B$ = Total number of objects in population

n = Number of objects drawn in sample

x = Observed number of successes in sample

The probability of the random variable X resulting in exactly x successes (thus there are $n - x$ failures) is known as a hypergeometric probability and is found from the **hypergeometric probability function** as follows

---

Hypergeometric Probability Function

$$P(X = x \,|\, n) = P(x \,|\, n) = \frac{\dbinom{A}{x}\dbinom{N - A}{n - x}}{\dbinom{N}{n}} \tag{5.11}$$

where x is any integer from 0 or $n - (N - A)$ (whichever is larger) up to n or A (whichever is smaller), n and N are positive integers, A is a non-negative integer, and $A \leq N$ and $n \leq N$.

Sometimes, visualizing the three combinations involved in Equation (5.11) is easier if they are presented in word form:

$$P(X = x \,|\, n) = \frac{\begin{pmatrix} \textit{Number of successes} \\ \textit{in population} \\ \textit{taken x at a time} \end{pmatrix}\begin{pmatrix} \textit{Number of failures} \\ \textit{in population} \\ \textit{taken n} - \textit{x at a time} \end{pmatrix}}{\begin{pmatrix} \textit{Total population} \\ \textit{taken n at a time} \end{pmatrix}} \tag{5.12}$$

A notation for the probabilities in Equations (5.11) and (5.12) is $P(X = x \,|\, n)$, which is similar to the binomial probability notation of $P(X = x \,|\, n, p)$ except that the probability of a success changes from trial to trial in the hypergeometric situation because we are sampling from a finite population without replacement. Thus there is no p value that can be indicated in the notation, and $P(X = x \,|\, n)$ is read as "the hypergeometric probability of x successes in n trials."

The mean and the variance for a hypergeometric distribution are $E(X) = n(A/N)$ and $\text{Variance}(X) = [(N - n)/(N - 1)][(n)(A/N)(1 - A/N)]$.

EXAMPLE 5.14 In Example 5.13 there were 10 successes (product users) in the population and 30 failures (nonusers). We wanted the probability of getting 3 successes in a sample of 5 from the total population of 40 people. Thus by Equation (5.11) we arrive at the same answer we reasoned to in Example 5.13:

$$P(X = 3 \,|\, n = 5) = \frac{\dbinom{10}{3}\dbinom{40 - 10}{5 - 3}}{\dbinom{40}{5}} = .0793$$

EXAMPLE 5.15 Suppose 9 families live on a street and 6 of them own their homes while 3 of them rent. If we interview 4 families at random without replacement (that is, if we take a random sample of size 4), and if we let X be the number of homeowners in the sample, then X is a random variable that can take the values $x = 1, 2, 3$, and 4 (there must be at least one homeowner since we select 4 families and there are only 3 renters on the street). The probability of getting $X = x$ homeowners in a sample of 4 is the same as the probability of getting x successes and $4 - x$ failures. This probability is given by Equation (5.11) or (5.12):

$$P(X = x \,|\, n = 4) = \frac{\dbinom{6}{x}\dbinom{9 - 6}{4 - x}}{\dbinom{9}{4}} \qquad x = 1, 2, 3, 4$$

Table I in Appendix C can be used to assist us in evaluating these combinations. The individual probabilities are as follows:

$$P(X = 1) = 6/126 = .048 \qquad P(X = 2) = 45/126 = .357$$
$$P(X = 3) = 60/126 = .476 \qquad P(X = 4) = 15/126 = .119$$

A plot of the hypergeometric distribution for the number of homeowners in the sample is shown in Figure 5.14.

The set of probabilities for Example 5.15 is the distribution of X. The probabilities sum to 1, and they suffice to answer any question about the probability that X will have a given property. For instance, the probability that X *is even* in this situation is

$$P(X \text{ is even}) = P(X = 2) + P(X = 4) = 60/126$$

FIGURE 5.14

Distribution of Number of Homeowners

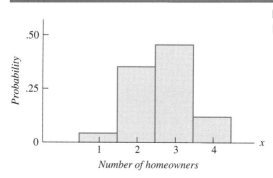

The probability that *X is 3 or more* is

$$P(X \geq 3) = P(X = 3) + P(X = 4) = 75/126$$

 PRACTICE PROBLEM 5.7

A group consists of 8 management and 4 union people. A committee of 3 is to be selected. Find the probability that a randomly selected committee will consist of 1 management person and 2 union people.

Solution In this problem we have a series of $n = 3$ trials where there are two possible outcomes (management or union person) on each trial. But the trials are dependent because we are sampling from a population of 12 people without replacement. Thus Equations (5.11) and (5.12) hold, and, if we call selection of a management person a "success," we have

$$P(X = 1 \mid n = 3) = \frac{\binom{8}{1}\binom{12-8}{3-1}}{\binom{12}{3}}$$

This expression can be evaluated by consulting Table I in Appendix C to find the three combination values. So,

$$P(X = 1 \mid n = 3) = \frac{(8)(6)}{220} = .2182$$

There are many sampling situations in which we sample without replacement from a finite population and thus have a hypergeometric probability situation. However, when the population is large, say several thousand, and the sample is small, the probability of a success does not change very much from trial to trial. Thus, in this situation the binomial probability Equation (5.5) will give a good approximation to the hypergeometric Equation (5.11).

Problems: Section 5.7

62. A population of 20 items contains 12 successes and 8 failures. The hypergeometric probability that a sample of 7 items from this population will have 5 successes in it is given by the expression

$$P(X = 5 \mid n = 7) = \frac{\binom{12}{5}\binom{20-12}{7-5}}{\binom{20}{7}}$$

Use Table I in Appendix C to help evaluate this probability expression.

63. There are 7 people who work in an office. Of the 7, 4 would like to be transferred. If 3 people from this office are randomly selected for transfer, give the following probabilities:

    a. That all 3 will be people who wanted the transfer.
    b. That 2 of the 3 will be those who wanted a transfer.

64. A dishonest accountant has "adjusted" 6 of the entries in a particular account. That account has 18 total entries, and an auditor is about to randomly select 4 of the entries to be examined in detail.

    a. Find the probability that 2 of the "adjusted" entries will be examined.
    b. Find the probability that 1 or more of the "adjusted" entries will be examined.

65. A salesperson has 15 major accounts. Eight of those accounts were sent the wrong billing last month (they were billed too much by the accountant). The salesperson wants to call her accounts to explain what happened, but she only has time to call 10 of them before quitting time (assume a call to a wrongly billed account takes just as long as a call to one with correct billing). If she selects accounts and calls them randomly, what are the following probabilities?

    a. That she will reach 6 of the 8 she wants to contact.
    b. That she will reach only 5.
    c. That she will reach none of these incorrectly billed.

66. To test the skills of its auditors, an accounting firm asks its auditors to examine 25 accounting transactions, 5 of which are in error. Assume the auditors are told in advance that there are 5 incorrect transactions. If one of the auditors is simply guessing, what is the probability that he will correctly select the 5 incorrect transactions? What is the probability that he will select 2 of the incorrect transactions?

67. Production inspection records at International Pharmaceuticals, Inc., show that 8 of the 35 items in a production lot are defective, but the defective items were not marked. A rush order for 5 of these items has just arrived. If 6 items are selected from the lot and sent, what is the probability that 5 or more of them will be good?

68. In Problem 67, what is the probability that 2 or fewer of the items are good?

69. A company has 10 customers, 6 of whom are located out of state. Five orders were randomly selected recently. Find the hypergeometric probability that 2 of the 5 will be from out-of-state customers.

70. In a batch of 20 bottles of their product the Restin Chemical Company knows that 10 bottles were contaminated. Assume that these 10 bottles were spread randomly through the batch. Restin's best customer was recently sent 6 bottles from this batch and claims that 4 of the bottles were contaminated. Your boss would like to know the probability that this could happen.

5.8
Expected Values and Variances of Linear Functions and Sets of Random Variables

We often use linear functions of random variables in statistical applications. If X is a random variable, discrete or continuous, and if A and B are fixed numbers, then U, as determined by the linear function

$$U = AX + B$$

is another random variable. The expected values of these two random variables are related by Equation (5.13).

Equation for Expected Value of Linear Function $U = AX + B$

$$E(U) = AE(X) + B \qquad (5.13)$$

EXAMPLE 5.16 If X is a random temperature in Celsius degrees, the same temperature in Fahrenheit degrees, U, is $(9/5)X + 32$. Thus the expected values of U and X are related by the formula $E(U) = (9/5)E(X) + 32$.

The variance and the standard deviation of U are given by the following rules.

Equation for Variance of Linear Function $U = AX + B$

$$\text{Variance}(U) = A^2 \times \text{Variance}(X) \tag{5.14}$$

Equation for Standard Deviation of Linear Function $U = AX + B$

$$\text{Standard deviation}(U) = A \times \text{Standard deviation}(X) \tag{5.15}$$

for positive A.

Suppose X has mean μ and variance σ^2 (standard deviation σ), and consider the related variable

$$Z = \frac{X - \mu}{\sigma} = \frac{1}{\sigma}X - \frac{\mu}{\sigma}$$

According to Equation (5.13),

$$E(Z) = \frac{1}{\sigma}E(X) - \frac{\mu}{\sigma} = \frac{1}{\sigma}\mu - \frac{\mu}{\sigma} = 0$$

And according to Equation (5.14),

$$\text{Variance}(Z) = \left(\frac{1}{\sigma}\right)^2 \times \text{Variance}(X)$$

$$= \left(\frac{1}{\sigma}\right)^2 \times \sigma^2 = 1$$

Thus,

$$E(Z) = E\left(\frac{X - \mu}{\sigma}\right) = 0 \tag{5.16}$$

and

$$\text{Variance}(Z) = \text{Variance}\left(\frac{X - \mu}{\sigma}\right) = 1 \tag{5.17}$$

if X has mean μ and standard deviation σ. Subtracting μ from X centers Z at 0, and dividing by σ standardizes the variability; Z, or the term $(X - \mu)/\sigma$, is the random variable X **standardized** to have mean 0 and standard deviation 1.

Random variables often come in pairs. For instance, a pair of numbers may be associated with each of the various outcomes of an experiment. Thus if a man is drawn at random from some population, his height X in inches is one random variable and his weight Y in pounds is another. These random variables are associated with one another because they are associated with one experiment (the drawing of the person from the population); X and Y attach to the same man.

Expected values of sums $X + Y$ and differences $X - Y$ satisfy the next formulas.

Equations for Expected Values of $X + Y$ and $X - Y$

$$E(X + Y) = E(X) + E(Y) \qquad (5.18)$$

$$E(X - Y) = E(X) - E(Y) \qquad (5.18a)$$

For example, if X is the income of the husband in a family and Y is the income of the wife, $X + Y$ is the family income (other sources excluded). If in a population of families the average income for husbands is $9000 and the average income for wives is $5000, certainly the average income for families must be $14,000. Equation (5.18) expresses the general form of this fact.

In Chapter 4 we gave a definition of independence of events [Equation (4.11)], a definition embodying the idea that knowing whether or not one of the two events occurred does not in any way help us to guess whether or not the other occurred. There is a similar notion of independence of random variables X and Y; it embodies the idea that knowing the value of X does not in any way help us to guess the value of Y, and vice versa.*

For instance, suppose two dice are rolled, one white die and one colored die to keep them straight; let X be the number showing on the white die and let Y be the number showing on the colored die. Since there is no interaction between the dice, the value that X takes on when the dice are rolled has no influence on the value Y takes on; X and Y are independent. In contrast, if X and Y are the height and the weight of a man, then X and Y are not independent: If you know that X is very large, then you know that Y is likely to be large also.

The fact of use to us is that *if X and Y are independent,* then the following equations give the variance of their sum or difference.

Equations for Variances of Sums or Differences Between Two Independent Random Variables

If X and Y are *independent* random variables, then the variance of the sum of the two variables is equal to the sum of the individual variances, or

$$\text{Variance}(X + Y) = \text{Variance}(X) + \text{Variance}(Y) \qquad (5.19)$$

and the variance of the difference between the two variables is also equal to the sum of the individual variances, or

$$\text{Variance}(X - Y) = \text{Variance}(X) + \text{Variance}(Y) \qquad (5.19a)$$

A standard deviation is the square root of a variance. Consequently, we have the following equations for the standard deviations of the sums or differences between two independent random variables.

---

*Mathematically, two discrete random variables X and Y are statistically independent if and only if $P(X = x_i \text{ and } Y = y_j) = P(X = x_i)P(Y = y_j)$ for all x_i and y_j.

Equations for Standard Deviations of Sums or Differences Between Two Independent Random Variables

If X and Y are *independent* random variables, then the standard deviation of the sum of the two variables is equal to the square root of the sum of the individual variances, or

$$\text{Standard deviation}(X + Y) = \sqrt{\text{Variance}(X + Y)}$$

$$= \sqrt{\text{Variance}(X) + \text{Variance}(Y)} \qquad (5.20)$$

and the standard deviation of the difference between the two variables is also equal to the square root of the sum of the individual variances, or

$$\text{Standard deviation}(X - Y) = \sqrt{\text{Variance}(X - Y)}$$

$$= \sqrt{\text{Variance}(X) + \text{Variance}(Y)} \qquad (5.20a)$$

Thus the standard deviation of a sum of two variables is *not* equal to the sum of the individual standard deviations; it is equal to the square root of the sum of the individual variances.

We will often be concerned with sets of random variables. Suppose we have at hand a random sample of size n from some population (say a sample of corporate stocks), and suppose we have measurements $X_1, X_2, \ldots, X_n$ (say the prices of the stocks), one measurement for each element of the sample. Choosing a random sample is a random experiment, and each X_i is a number associated with the outcome; thus $X_1, X_2, \ldots, X_n$ are random variables. The procedures of Chapter 3—the computations of means and variances for sets of data—are usually performed on samples.

Equation (5.18) extends to sets of random variables:

$$E\left(\sum_{i=1}^{n} X_i\right) = \sum_{i=1}^{n} E(X_i) \qquad (5.21)$$

Equation (5.19) can be extended to more than two random variables *if they are independent* of one another:

$$\text{Variance}\left(\sum_{i=1}^{n} X_i\right) = \sum_{i=1}^{n} \text{Variance}(X_i) \qquad (5.22)$$

Problems: Section 5.8

---

71. A dime and a quarter each have the number 1 stamped on the head's side and the number 0 stamped on the tail's side. Let X be a random variable representing the number facing up on a single toss of the dime. Let Y be a random variable representing the number facing up on a single toss of the quarter. Let W be a random variable representing the sum of X and Y when the coins are each tossed once: that is, $W = X + Y$.

    a. Determine the expected values and variances of X and Y, individually.
    b. Determine the expected value of W.
    c. Determine the variance of W.
    d. Determine the standard deviation of W.

72. An individual investor has a portfolio that is composed of one stock and one bond. The expected value for the amount of profit for the stock (for the investor's time frame) is $E(X) = \$15$, and the expected value for the bond's profit is $E(Y) = \$10$. The variance for the amount of profit for the stock is 6 (dollars)^2 and for the bond is 3 (dollars)^2. The amount of profit for the portfolio, W, is the sum of the profits for the stock and the bond; that is, $W = X + Y$. For this simplified problem the profit amounts for the stock and the bond are independent.

   a. Determine the expected value of the profit for the portfolio.
   b. Determine the variance for the amount of the portfolio profit.
   c. Determine the standard deviation for the amount of the portfolio profit.

73. An accounting firm has offices in New York and Los Angeles. The billable hours per day at New York, X, show $E(X) = 100$ hours and Standard deviation$(X) = 30$ hours; and the billable hours per day at Los Angeles, Y, show $E(Y) = 150$ hours and Standard deviation$(Y) = 40$ hours. Assume that the billable hours at the two offices are independent.

   a. What is the expected value of the total billable hours per day for the two offices?
   b. What is the variance of the total billable hours per day for the two offices?
   c. What is the standard deviation of the total billable hours per day for the two offices?

Continuous Probability Distributions

6

The study of probability is continued in this chapter with the examination of probability distributions for continuous random variables. Random variables, including continuous random variables, were defined in the previous chapter. We make the transition from the discrete probability distributions of the previous chapter to general probability distributions for continuous random variables in this chapter. We also discuss two specific types of probability distributions for continuous random variables, *uniform* distributions and *normal* distributions.

Normal distributions can be used to find the probability that many continuous measures, like weights, lengths, and times, will be within a particular range of values. We will find probabilities for normally distributed variables by converting to the *standard normal distribution*. Probabilities (areas) for the standard normal distribution are given in Table V in Appendix C. Sketches or plots of the distributions are important aids in finding probabilities (areas) when we use normal distributions.

In addition, we will see how normal distributions can be used to *approximate binomial probabilities* when the number of trials n is large. A continuity correction that results in even better approximations will be presented. The number of successes in n trials for the binomial distribution will be transformed into the *fraction*, or *proportion, of successes* $\hat{p}$ (read as p-hat) by dividing the number of successes X by the number of trials n; thus $\hat{p} = X/n$. Normal distributions will then be used to approximate the distributions of the fractions of successes.

Normal distributions are the most commonly used of all probability distributions, so you should be sure to develop proficiency with the techniques of this chapter. Normal probability distributions will be applicable as the probability distributions for several sample statistics in the next chapter. Thus normal distributions will be very important in the chapters on statistical inference.

6.1

GENERAL PROBABILITY DISTRIBUTIONS FOR CONTINUOUS RANDOM VARIABLES

A *discrete random variable* is described by its distribution—that is, by the list of probabilities for its various possible values. This sort of distribution does not work for a *continuous random variable,* because such a variable takes on any given value with probability *zero:* If X is continuous, then $P(X = 3.1)$ is 0, $P(X = 2.854)$ is 0, and so on.

The fact that the probability of any specific value of a continuous random variable is zero seems paradoxical at first, since X must assume *some* value when the experiment is carried out. But consider an example involving weights. Weight is a continuous measure. What is the probability of finding a bag of sugar in your grocery store that weighs exactly 10 pounds? This chance is zero, since if you found a bag that weighed 10 pounds on your scale, then someone else could theoretically find a more accurate scale that would show your 10-pound bag to be slightly different from 10 pounds. Then if you found a bag that weighed 10 pounds on that finer scale, someone else could theoretically find an even more accurate scale and show you that your bag did not weigh exactly 10 pounds out to, say, the 51st decimal place. In the extreme we have to concede that it would be virtually impossible to find a bag of sugar that weighed 10 pounds exactly (say, out to a billion decimal places). Thus $P(X = 10.0) = 0$.

Since there are infinitely many possible values for a continuous variable, we cannot possibly list those values in a table, as we did with probability distributions for discrete random variables. In consequence, the distribution of a continuous X must be given not by a list or table of probabilities but by a *continuous curve* such as the one to be considered in Figure 6.1. The curve is called a *frequency curve,* or a *probability density curve,* and the **area** under the curve between two limits on the horizontal scale is the *probability* that the random variable will take on a value lying between those two limits. The height of the curve over a point on the horizontal scale is called a **probability density.** Unlike the areas under the frequency curve, the height of the curve has no direct probability meaning. Since the area under a probability density curve corresponds to probability, the total area under the curve must be equal to one. The height of a probability density curve must be greater than or equal to zero. Frequency curves are also termed **probability distributions for continuous random variables.**

FIGURE 6.1

Theoretical Distribution of Blast Yields

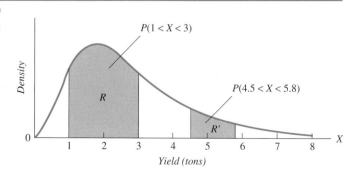

> **Definition**
>
> **A probability distribution for a continuous random variable** is specified by a probability density curve (or frequency curve). **Areas** under the probability density curve are probabilities.

EXAMPLE 6.1 Great Basin construction company is involved in blasting a path for a new highway through a mountain pass. The weight X in tons of rock that is hauled off after each small blast is measured. The behavior of X is described by a curve such as that in Figure 6.1. The curve lies entirely above the horizontal axis, and the area under the curve is 1; that is, the area between the curve and the horizontal axis is 1. The probability that when the experiment is carried out, X will assume a value in a given interval equals the area under the curve and over that interval. The chance that X lies between 1 and 3 is the area of the shaded region R in Figure 6.1; the chance that it lies between 4.5 and 5.8 is the area of the shaded region R'. Notice that since an individual point has probability 0, $P(4.5 < X < 5.8)$ is the same as $P(4.5 \leq X \leq 5.8)$; the endpoints do not make any difference. This feature does not hold for discrete probability distributions.

Since we started studying probability, we have interpreted probabilities in terms of limiting relative frequencies, and we can give a similar interpretation to the curve in Figure 6.1 by using the histograms of Chapter 2. If we were to set off 100 blasts and measure the rock yield to the nearest ton, using a class interval of 1 ton, then the histogram would resemble that in Figure 6.2(a). If we were to set off 10,000 blasts instead of 100, we could use shorter class intervals (say quarter-ton intervals) and more of them, and the histogram would resemble that in Figure 6.2(b).

If we continue to set off blasts, using more and more intervals of smaller and smaller size, the histogram will approach the curve in Figure 6.1. In each histogram the area of a bar is proportional to the observed relative frequency in that class, and so an *area* under the histogram (such as the areas of the shaded regions in Fig-

FIGURE 6.2

Empirical Relative Frequency Distribution of Blast Yields

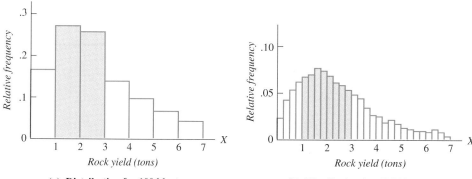

(a) **Distribution for 100 blasts**   (b) **Distribution for 10,000 blasts**

ure 6.2) represents an *observed relative frequency,* which converges to a *probability* such as that represented by the *area* of the shaded region in R in Figure 6.1.

If the distribution of a continuous variable is specified by a frequency curve, the question arises as to how the frequency curve is itself specified. This specification is only rarely done by an actual curve carefully drawn on graph paper. Sometimes, as in the following example, the distribution is specified by a geometric description.

=== **EXAMPLE 6.2**   A continuous random variable X always has a value between 0 and 2, and its distribution is given by the straight line in Figure 6.3. The equation of the line is *density* $= (1/2)(X)$. By the rule for the area of a triangle, the area under the line is 1. To get the probability that X lies between 1/2 and 3/2, we compute the *area* of the shaded trapezoid in the figure. The base of the trapezoid has length 1, and the sides have lengths 1/4 and 3/4 (found by substituting 1/2 and 3/2 in the density equation). Then the area rule for trapezoids gives

$$Area = 1 \times \frac{1/4 + 3/4}{2} = \frac{1}{2}$$

The probability sought is

$$P\left(\frac{1}{2} \le X \le \frac{3}{2}\right) = \frac{1}{2}$$

=

Usually, distribution curves for continuous random variables are specified by mathematical formulas called **probability density functions,** and the relevant *probabilities*—that is, the relevant *areas*—are determined by the methods of integral calculus. Such determinations lie outside the scope of this book. For various important frequency curves often encountered in statistical practice, tables of these areas have been constructed. Several such tables are given in Appendix C; they represent the frequency curves we will discuss and use in succeeding chapters.

A continuous random variable X has an expected value or mean noted as $E(X) = \mu$. For a continuous X, as for a discrete one, the mean $E(X)$ represents an average. It is the value that X can be expected to have on the average, the actual value on a trial being greater or less than $E(X)$ and distributed in such a way as to balance out at $E(X)$. Just as in the case of the discrete variable, $E(X)$ can be viewed as a point of balance in the continuous case as well. If we were to draw the frequency curve on some stiff material of uniform density, such as plywood or metal, cut it out,

FIGURE 6.3

Triangular Distribution for
Example 6.3

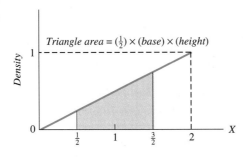

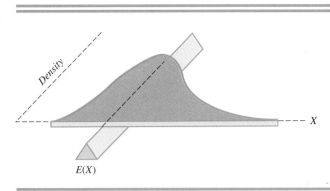

FIGURE 6.4

Expected Value as a Point of Balance: Continuous Case

and balance it on a knife-edge arranged perpendicular to the horizontal scale, the value corresponding to the point of balance would be the expected value. Figure 6.4 illustrates this for the curve of Figure 6.1; compare it with the analogous Figure 3.2, which relates to the mean for empirical frequency distributions.

For the frequency curve in Example 6.2 the expected value works out to 4/3. The computation of expected values in the continuous case, requiring as it does the methods of integral calculus, lies outside the scope of this book. Nonetheless, we can use the results of such computations for statistical purposes.

A continuous random variable has variance $\sigma^2 = E(X - \mu)^2$ and standard deviation $\sigma = \sqrt{\sigma^2}$. We will study one specific type of continuous distribution, the continuous uniform distribution, in the next section.

Problems: Section 6.1

Answers to odd-numbered problems are given in the back of the text.

1. The random variable X has the probability distribution shown in Figure 6.5.
   a. Find $P(-5 \le X \le 5)$.   **b.** Find $P(X \le 0)$.   **c.** Find $P(X \ge 0)$.

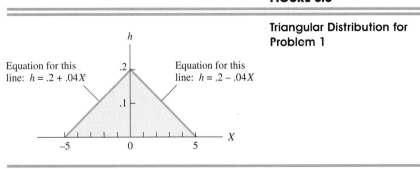

FIGURE 6.5

Triangular Distribution for Problem 1

2. Suppose X has the triangular distribution shown in Figure 6.5.
   a. Find $P(X \ge 2)$.   **b.** Find $P(0 \le X \le 3)$.   **c.** Find $P(X \le -1)$.
   [*Remember: Triangle area* $= (1/2)(base \times height)$.]
3. Refer to the continuous probability distribution shown in Figure 6.5.
   a. Find $P(-2 \le X \le 2)$.   **b.** Find $P(X \ge -2)$.

6.2
UNIFORM
PROBABILITY
DISTRIBUTIONS

A *continuous uniform probability distribution* is a simple distribution with a rectangular shape, and it is useful in a diverse number of applications. For example,

1. The time X that a commuter waits to board a train from Connecticut to Manhattan has a uniform distribution.
2. A lathe operator at Mega Torque Milling observes that it is equally likely that any point on the spindle stops directly opposite of the tool when the power switch is turned off.

Random numbers such as those given in Table II in Appendix C are generated from uniform distributions. For each random number of a given number of digits in the table, any number has an equal probability of being in that position, and the number in any position is independent of the numbers in other positions. Most computer systems can be used to generate random numbers (or essentially random numbers) from a uniform continuous probability distribution.

A *continuous uniform random variable* has uniform or constant probability density over an interval of values taken by the variable. The **uniform probability density function** follows.

Uniform Probability Density Function

$$f(X) = \frac{1}{b - a} \tag{6.1}$$

where: $a \le X \le b$

The values of a and b are parameters or constants, and their particular values depend on the probability problem. The smallest value the random variable can take is a and the largest is b. For any value of X substituted into Equation (6.1), where $a \le X \le b$, $f(X)$ is the constant height of the uniform probability density function at that value of X. A uniform distribution is shown in Figure 6.6.

There is a different uniform distribution from the family of uniform distributions for each pair of values for a and b. The mean and the standard deviation of a continuous uniform probability distribution are $\mu_X = (a + b)/2$ and $\sigma_X = \sqrt{(b - a)^2/12}$.

The probability that a continuous random variable X takes on a value in a specified interval is found by determining the corresponding area of a rectangle under the probability density function.

FIGURE 6.6

Uniform Distribution

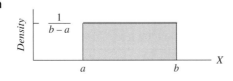

FIGURE 6.7

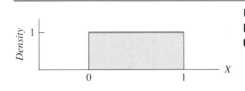

Uniform Distribution for
Random Numbers Between
0 and 1

EXAMPLE 6.3   A data analysis system generates random numbers between 0 and 1 from the uniform distribution shown in Figure 6.7. We want to find the probability that a random number generated from this distribution will be less than or equal to .25, or we want to find $P(X \leq .25)$.

The probability that a random number generated from the uniform distribution with $a = 0$ and $b = 1$ will be less than or equal to .25 is found by determining the area of the rectangle under the uniform density line and bounded on the right by .25. The rectangle has a width of .25 and a height of 1, so the area or probability is $(.25)(1) = .25$, or $P(X \leq .25) = .25$.

Find the probability that a random number generated from the distribution shown in Figure 6.7 will be greater than or equal to .25 but less than or equal to .75, or find $P(.25 \leq X \leq .75)$. The corresponding rectangle has a width of .50, since $.75 - .25$ is .50, and a height of 1; thus, the area or probability is $(.5)(1) = .5$, or $P(.25 \leq X \leq .75) = .5$.

Find the probability that a random number generated from the distribution will be less than or equal to 1.0, or find $P(X \leq 1.0)$. The area is the total area under the uniform line and is 1.0, or $P(X \leq 1.00) = 1.0$.

Find the mean and standard deviation for the distribution that is shown in Figure 6.7. The mean is

$$\mu_X = \frac{a + b}{2} = \frac{0 + 1}{2} = .5$$

And the standard deviation is

$$\sigma_X = \sqrt{(b - a)^2/12} = \sqrt{(1 - 0)^2/12} = .289$$

Problems: Section 6.2

4. A data analysis system generates random numbers according to the uniform distribution shown in Figure 6.8. We want to find the following probabilities for a random number generated from this distribution:
   a. $P(.5 \leq X \leq 1)$ **b.** $P(X \geq .75)$

FIGURE 6.8

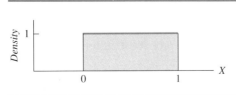

Uniform Distribution for
Problem 4

FIGURE 6.9

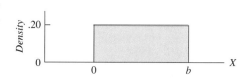

Uniform Distribution for
Time to Board Trolley

5. The time X that a tourist waits to board a trolley car on Polk Street in San Francisco has the rectangular or uniform distribution shown in Figure 6.9.
   a. Find b.   **b.** Find $P(X \le 2)$.
6. Suppose X has the rectangular or uniform distribution shown in Figure 6.9. Find C such that $P(X \le C) = .6$.

6.3
NORMAL PROBABILITY DISTRIBUTIONS

To find the probability that many continuous measures, such as weights, lengths, times, measurement errors, costs, and prices will be within a particular range of values, we can use the **normal distribution.** We will find probabilities for normally distributed variables by converting to the *standard normal distribution.* Probabilities (areas) for the standard normal distribution are given in Table V in Appendix C. Sketches or plots of the normal distributions are important aids when we use normal distributions. Normal probability distributions are the most commonly used of all probability distributions, so you should be sure to develop proficiency with the techniques of this section.

The **probability density function** for **normal distributions** follows.

Normal Probability Density Function

$$f(X) = \frac{1}{\sqrt{2\pi}\,\sigma} e^{(-1/2)\,[(X-\mu)/\sigma]^2} \qquad -\infty < X < +\infty \qquad (6.2)$$

Here π is the mathematical constant $3.1416\ldots$, and e is another constant that often occurs in mathematics; its value is $2.718\ldots$.

The values of μ and σ are parameters or constants (μ can be any real value and σ is positive), but their particular values depend on the probability problem. For any value of X substituted into Equation (6.2), $f(X)$ is the height of the normal frequency curve at that value of X.

Working with Equation (6.2) mathematically is not important to us. What is important is the fact that the normal frequency curve it describes is a special continuous probability distribution. The curve is a symmetrical, bell-shaped distribution, as shown in Figure 6.10. Any random variable whose values follow a normal distribution occurs with probabilities that can be developed from Equation (6.2). In our computations we will sidestep the formula and mathematics associated with it and instead find the probabilities for normally distributed random variables using a table in the appendix.

There is a different normal distribution from the family of normal distributions for each pair μ and σ, where μ is the mean and σ is the standard deviation of the

FIGURE 6.10

Normal Distribution with Mean μ and Standard Deviation σ

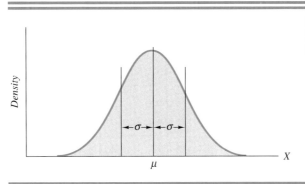

distribution. The mean μ determines the location of the distribution, and the standard deviation σ determines the spread or dispersion of the distribution. Figure 6.11 shows three different normal distributions with means and standard deviations of $\mu = 3$ and $\sigma = 1$, $\mu = 6$ and $\sigma = 1$, and $\mu = 6$ and $\sigma = 2$. The normal distributions in Figures 6.11(a) and 6.11(b) have the same standard deviation, but the distribution with the larger mean $\mu = 6$ is located or shifted to the right of the distribution with $\mu = 3$. The distributions in Figures 6.11(b) and 6.11(c) have the same mean, but the distribution with the larger standard deviation $\sigma = 2$ is lower and more spread out than the distribution with $\sigma = 1$. The normal distribution covers the entire axis for the variable of interest; however, it gets closer and closer to the axis as it gets further from the mean.

FIGURE 6.11

Normal Probability Distributions with Different Means or Different Standard Deviations

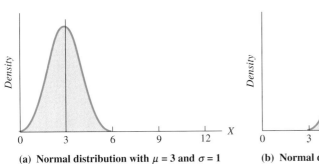

(a) Normal distribution with $\mu = 3$ and $\sigma = 1$

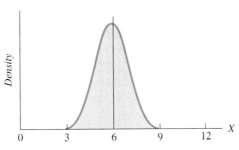

(b) Normal distribution with $\mu = 6$ and $\sigma = 1$

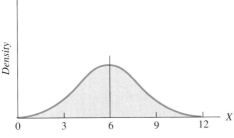

(c) Normal distribution with $\mu = 6$ and $\sigma = 2$

Experience has shown that many continuous random variables in diverse fields of application have distributions for which a normal distribution may serve as a mathematical or theoretical model or a good approximation. The normal probability distribution also approximates the distribution of several sample statistics, as we will see in the next chapter. For instance, boxcars arriving at a Capstar Corporation warehouse are filled with insulation material. The weights of material shipped in these cars are approximately normally distributed with a mean weight of 25 tons per car and a standard deviation of .5 ton per car. The warehouse operator may be interested in knowing what proportion of the boxcars arriving at her warehouse have a load of less than 24 tons. That is, she would like to know the probability of randomly selecting a boxcar of insulation material and finding that it contains less than 24 tons. We can use Equation (6.2) to find the height of the normal curve at any weight X by substituting into the formula the value for X, $\mu = 25$, and $\sigma = .5$. But it is the areas under the normal curve that give probabilities. In particular, the warehouse operator will be interested in the area under the normal curve to the left of 24 tons. Before she could use the table in the appendix to find that area, however, she would have to know about the **standard normal distribution.**

As we have seen, there is a normal distribution for each pair μ and σ. Although constructing a table for each of them is impossible, we can select one, tabulate its areas, and use this table with appropriate conversion formulas to find probabilities for any normally distributed variable. Thus we have the following definition.

> **Definition**
>
> The **standard normal distribution** is that normal distribution that has a mean of 0 and a standard deviation of 1.

The standard normal distribution is tabulated in Table V of Appendix C. A plot of the standard normal distribution is presented in Figure 6.12.

Let Z be a **standard normal variable.** That is, Z has a mean value $\mu_Z = 0$ and a standard deviation $\sigma_Z = 1$. We can find the height of the normal curve that describes the variation of Z for any value of Z by using Equation (6.2), with $\mu = 0$, $\sigma = 1$, and $X = Z$. But for continuous frequency curves we are interested not in the height of the curve but in the area under the curve. In the table of areas for the stan-

FIGURE 6.12

Standard Normal
Distribution with Mean 0
and Standard Deviation 1

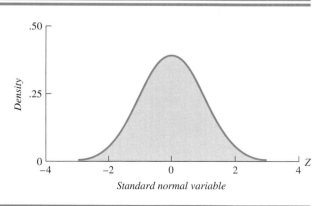

Standard normal variable

FIGURE 6.13

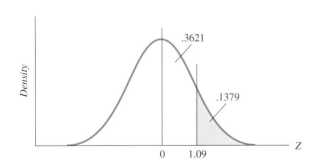

Area under the Standard Normal Distribution for Example 6.4

dard normal curve, Table V, values of Z from .00 to 3.89 are given in the body of the table. The table gives the area under the curve *between* 0 *and the given value of Z.*

EXAMPLE 6.4   If we want the area between $Z = 0$ and $Z = 1.96$, we find the $Z = 1.9$ row and the .06 column in Table V (for $Z = 1.9 + .06 = 1.96$) and read .4750 at the intersection. The shaded area in Figure 6.13 shows the probability. Since the standard normal curve is symmetric about 0, the area between $-Z$ and 0 is equal to the area between 0 and Z. Therefore only positive values of Z are given in the table. Since area is proportional to probability, the total area under the curve is equal to 1.

We can illustrate the use of the table of standard normal areas, Table V, by several problems. A rough sketch of the curve together with the area desired is a great help in finding probabilities by means of the table.

PRACTICE PROBLEM 6.1

What is the probability that Z will be greater than or equal to 1.09?

Solution The shaded area to the right of $Z = 1.09$ in Figure 6.14(a) is the probability that Z will be greater than 1.09. In the table the area between 0 and 1.09 is .3621.

Areas under the Standard Normal Distribution   **FIGURE 6.14**

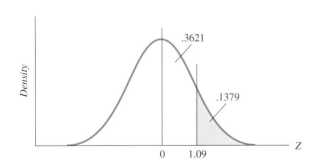
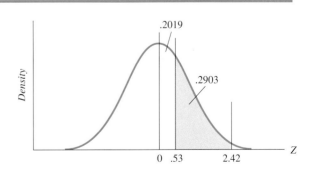

(a) **Area under the standard normal distribution for Practice Problem 6.1**

(b) **Area under the standard normal distribution for Practice Problem 6.2**

FIGURE 6.15

**Areas under the Standard
Normal Distribution for
Example 6.5**

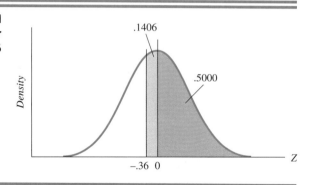

Watson, Statistics 5/e

The total area under the curve is equal to 1, so the area to the right of 0 must be
.5000. Therefore

$$P(Z \geq 1.09) = .5000 - .3621 = .1379$$

 PRACTICE PROBLEM 6.2

Find the probability that a single random value of Z will be between .53 and 2.42.

Solution: The shaded area in Figure 6.14(b) is equal to the desired probability. The
area given in the table for $Z = 2.42$ is .4922, but this area is from 0 to 2.42. The
tabled area for .53 is .2019. Referring to the sketch, we see that the area between .53
and 2.42 is the difference between their tabular values:

$$P(.53 \leq Z \leq 2.42) = .4922 - .2019 = .2903$$

EXAMPLE 6.5   As illustrated in Figure 6.15, the probability that Z will be
greater than or equal to $-.36$ is the total area to the right of $Z = -.36$ and consists of the area between $-.36$ and 0, which is .1406, plus the area under the
right-hand half of the curve:

$$P(Z \geq -.36) = .1406 + .5000 = .6406$$

EXAMPLE 6.6   We find the probability that a random value of Z is between
-1.00 and 1.96 by adding the corresponding areas indicated in Figure 6.16:

$$P(-1.00 \leq Z \leq 1.96) = .3413 + .4750 = .8163$$

FIGURE 6.16

**Areas under the Standard
Normal Distribution for
Example 6.6**

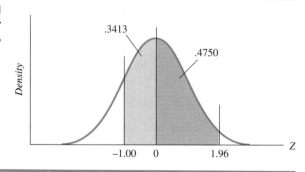

FIGURE 6.17

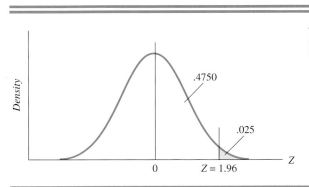

Area under the Standard
Normal Distribution for
Practice Problem 6.3

As a general rule, if we desire to know the area under the normal curve between two values *a* and *b*, and if *a* and *b* are on the *same side of the origin,* we *subtract* the corresponding areas. If they are on *opposite sides,* we *add.*

A slightly different use of the table is required in the following problems.

PRACTICE PROBLEM 6.3

Find a value of *Z* such that the probability of a larger value is equal to .0250.

Solution The probability of a larger value is the area to the right of *Z*. The area between 0 and *Z* must be .5000 − .0250, or .4750. We look in the body of the table until we find .4750, and we take the desired value, *Z* = 1.96, since .4750 appears in the 1.9 row and the .06 column. (See Figure 6.17.)

PRACTICE PROBLEM 6.4

Find the value of *Z* such that the probability of a larger value is .7881.

Solution Since the area to the right of *Z* is greater than .5000, we know that *Z* must be negative and that the area between *Z* and 0 must be .7881 − .5000, or .2881. Reading in the body of the table, we locate .2881 and find that the corresponding *Z* is −.80. (See Figure 6.18.)

FIGURE 6.18

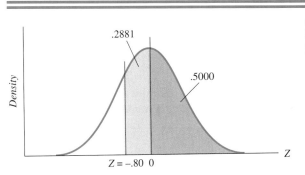

Areas under the Standard
Normal Distribution for
Practice Problem 6.4

FIGURE 6.19

Area under the Standard
Normal Distribution for
Practice Problem 6.5

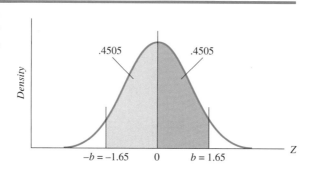

PRACTICE PROBLEM 6.5

Find the value for b such that

$$P(-b \leq Z \leq b) = .9010$$

Solution The endpoints $-b$ and b are symmetrical about the mean, therefore half the area between them must be between 0 and b. In the table we find the area value $.9010/2$ or $.4505$ in the 1.6 row and .05 column, so $Z = 1.65$. (See Figure 6.19.)

Now that we have seen how the table of normal areas is used to find probabilities for the standard normal variable, we turn our attention to the general case of the normal variable with mean μ, usually some value other than 0, and standard deviation σ, usually some value other than 1. The key to using the standard normal areas for finding probabilities for the general normal variable is to employ the **standardized variable.**

Definition

If X is a random variable with mean μ and standard deviation σ, then the **standardized variable**

$$Z = \frac{X - \mu}{\sigma} \tag{6.3}$$

has mean 0 and standard deviation 1. If X is normally distributed, Z is a **standard normal variable.***

Let X be normal with mean μ and standard deviation σ, and suppose we want to find the probability that a randomly selected value for X will be between a and b (where a is less than or to the left of b). We use the standardization Equation (6.3)

*Any linear function of a normal random variable is also a normal random variable.

and find that when $X = a$,

$$Z = \frac{a - \mu}{\sigma}$$

and that when $X = b$,

$$Z = \frac{b - \mu}{\sigma}$$

Therefore, whenever X is between a and b, the standard variable Z will be between $Z = (a - \mu)/\sigma$ and $Z = (b - \mu)/\sigma$. Then we have

$$P(a \leq X \leq b) = P\left(\frac{a - \mu}{\sigma} \leq Z \leq \frac{b - \mu}{\sigma}\right)$$

Thus we can turn to the tables and find the probability that Z is between $Z = (a - \mu)/\sigma$ and $Z = (b - \mu)/\sigma$. These relationships between the normal distribution for X and the standard normal distribution for Z are depicted in Figure 6.20.

Commonly, in working problems, we draw a single plot to represent both the X and the Z distributions, and we label the horizontal axis of the plot as both X and Z. We will use this method in our problems, but you should realize that two distributions are involved, one for X and one for Z.

PRACTICE PROBLEM 6.6

We are now ready to return to the example of the Capstar Corporation warehouse operator who receives boxcars loaded with insulation materials. (This example was introduced just prior to the discussion of the standard normal distribution.) The weights of the materials are normally distributed with a mean of 25 tons and a stan-

FIGURE 6.20

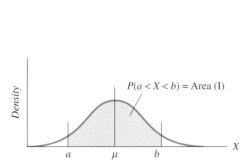

(a) Normal distribution for X with mean μ and standard deviation σ

(b) Standard normal distribution for $Z = (X - \mu)/\sigma$ with mean 0 and standard deviation 1

Relationship Between the Normal Distribution for X and the Standard Normal Distribution for Z

FIGURE 6.21

Determining the Proportion
of Boxcars

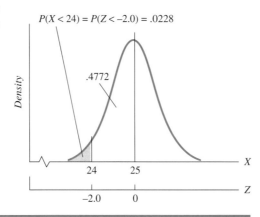

$P(X < 24) = P(Z < -2.0) = .0228$

.4772

dard deviation of .5 ton. The operator desires to know the proportion of the cars that have less than or equal to 24 tons in them.

Solution The standardized value of $X = 24$ when $\mu = 25$ and $\sigma = .5$ is

$$Z = \frac{X - \mu}{\sigma} = \frac{24 - 25}{.5} = -2.0$$

The probability that a boxcar load X is less than or equal to 24 tons is equal to the probability that Z is less than or equal to -2.0. The standard normal table, Table V, gives us an area of .4772 between $Z = 0$ and $Z = 2.0$ (or $Z = -2.0$ due to symmetry), and as can be seen in Figure 6.21, we find

$$P(X \le 24) = P(Z \le -2.0) = .5000 - .4772 = .0228$$

The warehouse operator may conclude that 2.28% of the cars arrive loaded with less than or equal to 24 tons.

In the previous practice problem, we found the proportion of loaded boxcars that weigh less than or are equal to 24 tons when the weights are normally distributed with $\mu = 25$ and $\sigma = .5$ to be $P(X \le 24) = .0228$. A probability such as $P(X \le 24) = .0228$ is known as a **cumulative probability.** Cumulative probabilities for normal distributions are sometimes found by using a computer, as illustrated by examples in Computer Exhibits 6.1A and 6.1B.

MINITAB

COMPUTER EXHIBIT 6.1A FINDING THE CUMULATIVE PROBABILITY $P(X \le 24) = .0228$
FOR A NORMAL DISTRIBUTION WITH $\mu = 25$ AND $\sigma = .5$

```
MTB > cdf 24;
SUBC> normal mu = 25, sigma = .5.
   24.0000      0.0228
```

Command **cdf** and subcommand **normal** to find the cumulative probability
$P(X \le 24) = .0228$ for a normal distribution with $\mu = 25$ and $\sigma = .5$

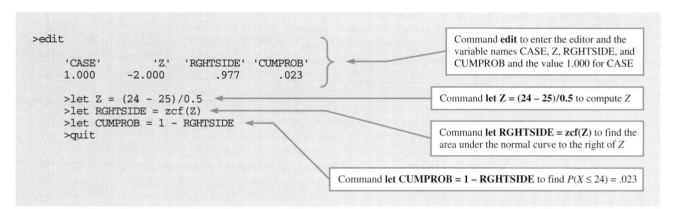

COMPUTER EXHIBIT 6.1B FINDING THE CUMULATIVE PROBABILITY $P(X \leq 24) = .0228$ FOR A NORMAL DISTRIBUTION WITH $\mu = 25$ AND $\sigma = .5$

MYSTAT

```
>edit

    'CASE'        'Z'   'RGHTSIDE' 'CUMPROB'
    1.000   -2.000       .977       .023

    >let Z = (24 - 25)/0.5
    >let RGHTSIDE = zcf(Z)
    >let CUMPROB = 1 - RGHTSIDE
    >quit
```

Command **edit** to enter the editor and the variable names CASE, Z, RGHTSIDE, and CUMPROB and the value 1.000 for CASE

Command **let Z = (24 − 25)/0.5** to compute Z

Command **let RGHTSIDE = zcf(Z)** to find the area under the normal curve to the right of Z

Command **let CUMPROB = 1 − RGHTSIDE** to find $P(X \leq 24) = .023$

 PRACTICE PROBLEM 6.7

A wooden beam produced by Carolina Atlantic Corporation has a mean breaking strength of 1500 pounds, has a standard deviation of 100 pounds, and is normally distributed. We want to know the relative frequency of all such beams whose breaking strength is between 1450 and 1600 pounds.

Solution As shown in Figure 6.22, the standardized value of $X = 1450$ when $\mu = 1500$ and $\sigma = 100$ is

$$Z = \frac{X - \mu}{\sigma} = \frac{1450 - 1500}{100} = -.5$$

The standardized value of $X = 1600$ is

$$Z = \frac{X - \mu}{\sigma} = \frac{1600 - 1500}{100} = 1.0$$

FIGURE 6.22

Determining the Proportion of Beams

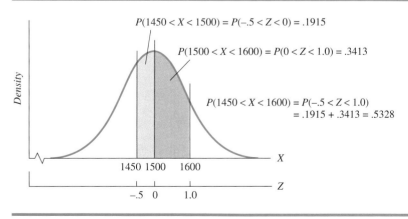

$P(1450 < X < 1500) = P(-.5 < Z < 0) = .1915$

$P(1500 < X < 1600) = P(0 < Z < 1.0) = .3413$

$P(1450 < X < 1600) = P(-.5 < Z < 1.0)$
$= .1915 + .3413 = .5328$

The probability is

$$P(1450 \le X \le 1600) = P(-.5 \le Z \le 1.0) = .1915 + .3413$$
$$= .5328$$

We may conclude that about 53% of the beams have breaking strengths between 1450 and 1600 pounds.

Problems: Section 6.3

7. Given that Z is the standard normal variable, use the table of normal areas to find the following:
   a. $P(0 \le Z \le 1)$
   b. $P(Z \ge 1)$

8. If Z is the standard normal variable, find the following:
   a. $P(Z \le .75)$
   b. $P(.25 \le Z \le .75)$

9. Given that Z is the standard normal variable, find the following:
   a. $P(-1.96 \le Z)$
   b. $P(-1.96 \le Z \le -1.5)$

10. Given that Z is the standard normal variable, find C in the following.
    a. $P(Z \ge C) = .025$
    b. $P(-C \le Z \le C) = .95$

11. Given that Z is the standard normal variable, find C in the following:
    a. $P(Z \le C) = .0287$
    b. Finding C such that $P(Z \le C) = .0287$ is known as finding the inverse of a cumulative probability. MINITAB and MYSTAT were used to find $C = -1.9003$ as follows:

    MINITAB:

    ```
    MTB > invcdf .0287;
    SUBC> normal mu = 0, sigma = 1.
        0.0287    -1.9003
    ```

    MYSTAT:

    ```
    >edit
          'PROB'        'C'
           .0287       -1.9003
        >let C = zif(PROB)
        >quit
    ```

    Is the answer from part a the same as the computer solution?

12. Given that X is a normal variable with $\mu = 50$ and $\sigma = 10$, find the following:
    a. $P(X \le 65)$ b. $P(42 \le X \le 62)$

13. Given that X is normal with $\mu = 15$ and $\sigma^2 = 25$, find the following:
    a. $P(X \le 20)$ b. $P(10 \le X \le 18)$

14. Given that X is normal with $\mu = .05$ and $\sigma = .012$, find $P(X \ge .074)$.

15. A new brand of disposable flashlight is guaranteed to last for at least one year of normal use. Tests indicate that the lifetime of these lights under normal use is approximately normally distributed with a mean of 1.5 years and a standard deviation of .4 year.

    a. What proportion of the flashlights will fail to meet the guarantee?
    b. What proportion of the flashlights will last longer than one year and nine months?
    c. Would you expect that the proportion of flashlights returned under the guarantee would be higher or lower than the proportion determined in part a?

16. Historical records of the Boonville Redistribution Center show that weights of the trucks arriving at its docks are normally distributed with a mean of 7 tons and a standard deviation of 2 tons.

    a. What proportion of the trucks weigh more than 9 tons?
    b. What proportion of the trucks weigh less than 5.24 tons?
    c. What proportion of the trucks weigh between 4.6 and 10.6 tons?

17. The mean monthly carbon monoxide count for June in a western city is approximately normally distributed with a mean of 7.5 parts per million and a standard deviation of .8 part per million. The local air pollution control agency has asked the major employers in the area to allow their employees to stay at home when the carbon monoxide count reaches or exceeds 9.5 parts per million. If the companies agree to this proposal, on what proportion of June days will their employees receive air pollution emergency vacations?

18. Huntson Chemical Company sells its major product, HBC–50K, in gallon containers. The company's records show that the amount of caustic materials in the product averages 4.8 ounces with a standard deviation of .5 ounce and is normally distributed. Find the proportion of the gallon containers that will have the following amounts of caustic material:

    a. More than 5 ounces of caustic material.
    b. Less than 2 ounces of caustic material.
    c. Between 3.2 and 5.2 ounces of caustic material.

19. Excess returns for stocks are determined by finding the difference between the return for a stock and the returns of firms in the market that have similar levels of risk. Suppose stocks listed on an exchange have a mean monthly excess return of .005 and a standard deviation of .004. Find the probability of a randomly selected stock resulting in a monthly excess return less than zero. Assume monthly excess returns are normally distributed.

    REFERENCE: Lubatkin, M., K. Chung, R. Rogers, and J. Owers (1989), "Stockholder Reactions to CEO Changes in Large Corporations," *Academy of Management Journal*, 32:47–68.

20. The diameters of shafts manufactured at Rice Machining is normally distributed with a mean of 10 centimeters and a standard deviation of .02 centimeter. A manufacturer of an electric motor requires shafts with diameters from 9.98 to 10.02 centimeters for the motors to operate satisfactorily. The electric motor manufacturer inspects incoming shafts and wishes to use Rice as a supplier. What proportion of shafts supplied by Rice will fail the inspection?

We used Equation (5.5) to compute the binomial probability of x successes in n trials, where p is the probability of a success on a single trial. However, if n is large, using Equation (5.5) to compute $P(X = x \mid n, p)$ is cumbersome. Also, Table III does not give binomial probabilities beyond $n = 25$. In this section we consider how the normal distribution can be used to find *approximate* binomial probabilities when n is large, thereby saving us from making some tedious calculations.

6.4
NORMAL APPROXIMATION OF BINOMIAL DISTRIBUTIONS

In the next three subsections we discuss the *normal approximation* of the binomial *number of successes X*, the normal approximation of the binomial *proportion* or *fraction of success p̂*, where the proportion of successes is the number of successes divided by the number of trials or $\hat{p} = X/n$, and a *continuity correction* to improve the normal approximation of the binomial number of successes.*

6.4.1 Normal Approximation of the Binomial Number of Successes X

When the sample size (or number of trials) n is large, and p is close to .5, the binomial distribution becomes very close to a normal distribution. In fact, if n is very large and p is *not* very close to 0 or 1, then the binomial distribution still becomes very close to a normal distribution. The normal distribution that can be used to approximate the binomial must have the same mean and standard deviation as the binomial distribution that is to be approximated. In Section 5.4 we found that a binomial distribution has mean np and standard deviation $\sqrt{np(1-p)}$.

Figure 6.23 shows why the normal curve can be used to give binomial probability approximations. The bars in this figure represent the binomial probabilities for the distribution of X successes in $n = 100$ trials, where $p = .5$ is the chance of a success on each trial. Earlier, we mentioned that it is the *area* in the bars of the histogram that represents probabilities, and it is the *area* under the normal curve that represents normal distribution probabilities. The normal curve that is superimposed over the binomial distribution has a mean $\mu = np = (100)(.5) = 50$ and a standard deviation $\sigma = \sqrt{np(1-p)} = \sqrt{100(.5)(.5)} = 5$. Thus we see that the area in any set of bars (the binomial probabilities) is approximately equal to the corresponding area under the curve (the normal probabilities).

A convincing way to show how the binomial distribution can be approximated by the normal distribution is to use graphics. In Figure 6.24(a) we have plotted four binomial probability distributions with constant $p = .5$ and n increasing from 4, to 8, to 18, and then to 30. Normal distributions with means of np and variances of $np(1-p)$ have been superimposed on the binomial distributions. As the fig-

*Analysts use a variety of notations for the proportion or fraction of successes in n Bernoulli trials such as $\hat{p} = X/n, f = X/n$, or $\bar{p} = X/n$ ($\hat{p}$ is read p-hat).

FIGURE 6.23

Normal Approximation of a
Binomial Distribution for
$n = 100$ and $p = .5$

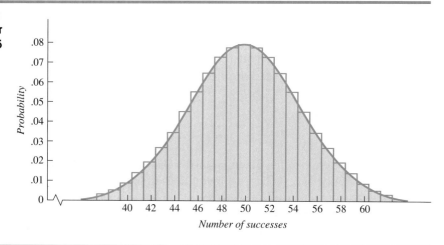

FIGURE 6.24

Normal Approximation of Binomial Distributions

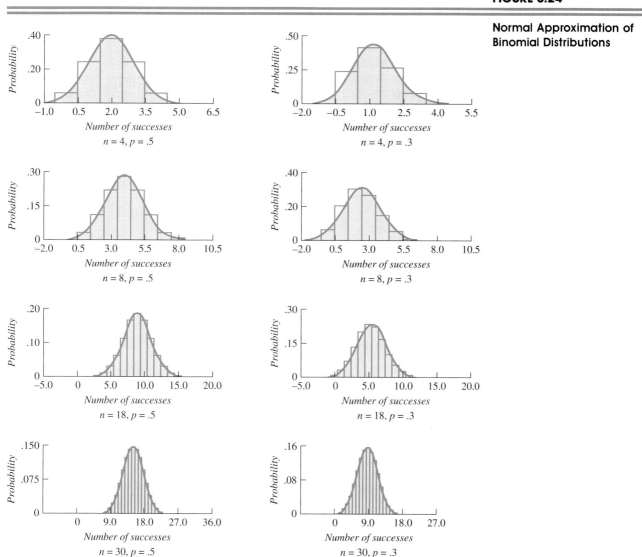

(a) Constant $p = .5$

(b) Constant $p = .3$

ure indicates, the normal distribution more closely approximates the binomial as n increases.

In Figure 6.24(b) we have plotted the same distributions, except that the probability of a success on a single trial is a constant $p = .3$. In this case the binomial distributions are skewed; but as n increases, the skewness decreases, and again the normal distributions more closely approximate the binomial distributions. (Be careful if you attempt to compare means and standard deviations for the distributions in Figure 6.24; the scales are not the same from one distribution to the next. The scales have been allowed to change because the purpose of the figure is to provide a visual assessment of the normal approximation of the binomial distribution, not the means and standard deviations. Also, the vertical axis shows probabilities and not densities; the density scale was omitted to avoid clutter.)

Figure 6.24 uses somewhat small n's in order to show that the normal approximation of the binomial distribution gets better as n increases. In many applications of the normal approximation of the binomial, the number of trials will be much larger than those in the figure, making the approximation even better.

Thus when n is large and p is close to .5, the binomial distribution of X, the number of successes in n trials, is nearly normal around the mean of np with a standard deviation of $\sqrt{np(1-p)}$. For use of the normal approximation to the binomial, the number of successes X must be standardized. The **standardized binomial variable** is given by the following equation.

Equation for Standardized Binomial Variable

$$Z = \frac{X - np}{\sqrt{np(1-p)}} \tag{6.4}$$

The standardized binomial variable is approximately a standard normal variable when n is large. Therefore the probability that X will lie between a and b will be approximately the probability that a standard normal variable Z will lie between

$$Z = \frac{a - np}{\sqrt{np(1-p)}} \quad \text{and} \quad Z = \frac{b - np}{\sqrt{np(1-p)}}$$

Thus,

$$P(a \le X \le b) \approx P\left(\frac{a - np}{\sqrt{np(1-p)}} \le Z \le \frac{b - np}{\sqrt{np(1-p)}}\right)$$

where the symbol $\approx$ means "is approximately equal to."

Readers sometimes wonder how large n must be and how close p must be to .5 before the normal distribution can be used to approximate binomial probabilities. A conservative general *rule of thumb* is that the value of $np(1-p)$ should equal or exceed 5 if the approximation is to be used. This value is the variance of the binomial distribution being approximated. Note that in all the problems in this section, $np(1-p) \ge 5$.

PRACTICE PROBLEM 6.8

Mountain Manufacturing Company is located in a right-to-work state, and 36% of its 7000 employees belong to unions. The personnel manager of the company takes a random sample of 100 employees in order to find out union members' attitudes toward management. He considers the selection of a union member in his sample a "success." Find the probability that the number X of successes will be between 24 and 42, inclusive.

Solution Here the mean and the standard deviation of X are

$$\mu = np = (100)(.36) = 36$$
$$\sigma = \sqrt{np(1-p)} = \sqrt{(100)(.36)(.64)} = \sqrt{23.04} = 4.8$$

FIGURE 6.25

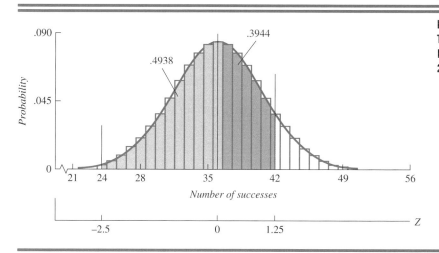

Determining the Probability That the Number of Union Members Will Be Between 24 and 42 When $p = .36$

The standardized value of $X = 24$ when $\mu = 36$ and $\sigma = 4.8$ is

$$Z = \frac{X - \mu}{\sigma} = \frac{24 - 36}{4.8} = \frac{-12}{4.8} = -2.5$$

The standardized value of $X = 42$ is

$$Z = \frac{X - \mu}{\sigma} = \frac{42 - 36}{4.8} = \frac{6}{4.8} = 1.25$$

As illustrated in Figure 6.25, the probability is found by using the table of standard normal areas as follows:

$$P(24 \le X \le 42) \approx P(-2.5 \le Z \le 1.25) = .4938 + .3944 = .8882$$

The exact probability (taken from binomial tables) is .9074. Our approximation, .8882, is accurate to about 2%. Rounding .8882 to .89, we see that the normal approximation gives

$$P(24 \le X \le 42) \approx .89$$

PRACTICE PROBLEM 6.9

Mountain Manufacturing Company's personnel manager takes a sample of 100 employees. Again, find the probability that X, the number of union members that appear in the sample, or the number of successes, will be between 24 and 42. This time, however, suppose that the proportion of union members working for the firm is .40.

Solution Here,

$$\mu = np = (100)(.40) = 40$$
$$\sigma = \sqrt{np(1 - p)} = \sqrt{(100)(.40)(.60)} = \sqrt{24} = 4.9$$

FIGURE 6.26

Determining the Probability
That the Number of Union
Members Will Be Between
24 and 42 When $p = .40$

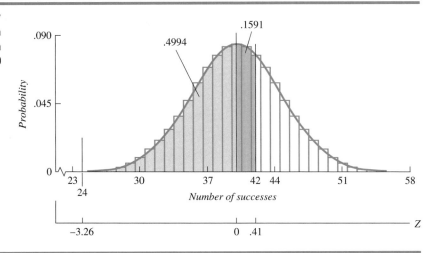

The standardized value of $X = 24$ when $\mu = 40$ and $\sigma = 4.9$ is

$$Z = \frac{X - \mu}{\sigma} = \frac{24 - 40}{4.9} = \frac{-16}{4.9} = -3.26$$

and the standardized value of $X = 42$ is

$$Z = \frac{X - \mu}{\sigma} = \frac{42 - 40}{4.9} = \frac{2}{4.9} = .41$$

As illustrated in Figure 6.26, the probability is found by using the table of standard
normal areas as follows:

$$P(24 \le X \le 42) \approx P(-3.26 \le Z \le .41) = .4994 + .1591 = .6585$$

Rounding gives

$$P(24 \le X \le 42) \approx .66$$

We can also approximate probabilities defined by a single inequality, as the following problem illustrates.

PRACTICE PROBLEM 6.10

What is the probability that the number X of successes (union members) in the personnel manager's sample of 100 at Mountain Manufacturing is 45 or more if
$p = .36$? If $p = .40$?

Solution If p, the proportion of union members working for the company, is .36, we
standardize by the mean and the standard deviation:

$$\mu = np = (100)(.36) = 36$$
$$\sigma = \sqrt{np(1 - p)} = \sqrt{(100)(.36)(.64)} = \sqrt{23.04} = 4.8$$

FIGURE 6.27

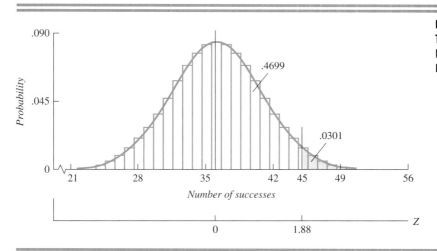

As illustrated in Figure 6.27, the probability is

$$P(X \geq 45) = P\left(\frac{X - 36}{4.8} \geq \frac{45 - 36}{4.8}\right)$$

$$\approx P\left(Z \geq \frac{45 - 36}{4.8}\right) = P(Z \geq 1.88)$$

$$= .5000 - .4699 = .0301$$

If p is .40, we use the following mean and standard deviation

$$\mu = np = (100)(.40) = 40$$
$$\sigma = \sqrt{np(1 - p)} = \sqrt{(100)(.40)(.60)} = \sqrt{24} = 4.9$$

Thus, as shown in Figure 6.28, the probability is

$$P(X \geq 45) = P\left(\frac{X - 40}{4.9} \geq \frac{45 - 40}{4.9}\right)$$

$$\approx P\left(Z \geq \frac{45 - 40}{4.9}\right) = P(Z \geq 1.02)$$

$$= .5000 - .3461 = .1539$$

Rounding to two places in each result, we have

$$P(X \geq 45) \approx .03 \qquad \text{if} \quad p = .36$$

and

$$P(X \geq 45) \approx .15 \qquad \text{if} \quad p = .40$$

Of course, the larger p is, the more likely we are to get 45 or more successes.

FIGURE 6.28

Determining the Probability
That the Number of Union
Members Will Be 45 or
More When p = .40

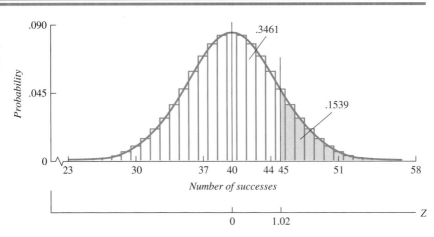

6.4.2 Normal Approximation of a Binomial Proportion of Successes $\hat{p}$

Sometimes, an appropriate procedure is to work not with the number X of successes in n trials but, instead, with the **proportion** or **fraction of successes,** $\hat{p} = X/n$. If we convert the number of successes X in Equations (5.8) and (5.9) to the fraction or proportion of successes $\hat{p} = X/n$, then we find the mean, variance, and standard deviation for the binomial fraction or proportion of successes $\hat{p}$ as follows.

Equation for Mean of Binomial Fraction of Successes

$$E(\hat{p}) = p \qquad (6.5)$$

Equations for Variance and Standard Deviation of Binomial Fraction of Successes

$$\text{Variance}(\hat{p}) = \sigma_{\hat{p}}^2 = \frac{p(1 - p)}{n}$$

$$\text{Standard deviation}(\hat{p}) = \sigma_{\hat{p}} = \sqrt{\frac{p(1 - p)}{n}} \qquad (6.6)$$

Thus the equation for the **standardized binomial proportion** or **fraction of successes** is as follows.

Equation for the Standardized Binomial Proportion or Fraction of Successes

$$Z = \frac{\hat{p} - p}{\sqrt{p(1 - p)/n}} \qquad (6.7)$$

The standardized binomial fraction of successes is approximately the standard normal variable when n is large. Thus this ratio has approximately a standard normal distribution for large n; the Z of Equation (6.7) is, in fact, $\hat{p}$ standardized, because $\hat{p}$ has mean $E(\hat{p}) = p$ and standard deviation $\sigma_{\hat{p}} = \sqrt{p(1-p)/n}$.

 PRACTICE PROBLEM 6.11

The proportion of Mountain Manufacturing Company employees that belong to the union (a success) is $p = .36$. A sample of $n = 100$ employees is selected randomly from Mountain's 7000 employees to participate in a survey about attitudes. What is the probability that the fraction of union members (fraction of successes) $\hat{p}$ in the sample will be between .24 and .42?

Solution Here the mean and standard deviation of $\hat{p}$ are

$$E(\hat{p}) = p = .36$$

$$\text{Standard deviation}(\hat{p}) = \sqrt{\frac{p(1-p)}{n}} = \sqrt{\frac{(.36)(.64)}{100}}$$

$$= \sqrt{.002304} = .048$$

Hence by Equation (6.7), the standardized $\hat{p}$ is

$$Z = \frac{\hat{p} - .36}{.048}$$

So, as shown in Figure 6.29,

$$P(.24 \le \hat{p} \le .42) = P\left(\frac{.24 - .36}{.048} \le \frac{\hat{p} - .36}{.048} \le \frac{.42 - .36}{.048}\right)$$

$$\approx P\left(\frac{.24 - .36}{.048} \le Z \le \frac{.42 - .36}{.048}\right)$$

$$= P(-2.5 \le Z \le 1.25) = .4938 + .3944$$

$$= .8881 \approx .89$$

FIGURE 6.29

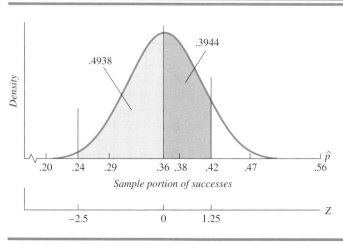

Determining the Probability That the Sample Fraction of Union Members Will Be Between .24 and .42

Note that the answer arrived at in Practice Problem 6.11 is the same as the answer to Practice Problem 6.8, because n and p are the same in the two problems and because $.24 \leq \hat{p} \leq .42$ is the same thing as $24 \leq X \leq 42$. Any given problem can be solved in terms of X or in terms of $\hat{p}$; the answer will be the same in either case (if the continuity correction discussed in the next subsection is ignored for both).

6.4.3 Continuity Correction

We noted in Practice Problem 6.8 for the Mountain Manufacturing Company that the normal approximation gave an answer about 2% off from the true answer. This accuracy suffices for many practical purposes. Further accuracy can be achieved by using **continuity correction.**

Definition

A **continuity correction** is an adjustment that we make by adding or subtracting 1/2 to a discrete value when we use a continuous distribution to approximate a discrete distribution.

The purpose of a continuity correction is to improve a probability approximation. The exact value of the probability sought in Practice Problem 6.8 is

$$P(24 \leq X \leq 42) = \sum_{k=24}^{42} \binom{100}{k}(.36)^k(.64)^{100-k}$$

It is represented by the combined areas of the bars in Figure 6.30. In Practice Problem 6.8 we approximated this probability by the area under the normal curve (the one for a μ of 36 and a σ of 4.8) for the interval between 24 and 42. Examination of the figure shows that we should obtain a better approximation if we subtract 1/2 from

FIGURE 6.30

Normal Approximation of a Binomial Distribution with Continuity Correction

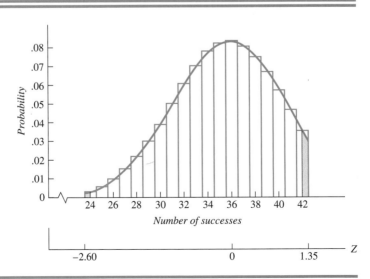

the smallest value in the interval, giving us $24 - .5$ or 23.5, and if we add $1/2$ to the largest value included in the interval, giving us $42 + .5 = 42.5$, and then we find the area under the normal curve for the adjusted interval with continuity correction between 23.5 and 42.5. This area will include the areas of the two shaded regions, omitted in the previous approximation. With this procedure, using the continuity correction, the standardized value of $X = 23.5$ when $\mu = 36$ and $\sigma = 4.8$ is

$$Z = \frac{X - \mu}{\sigma} = \frac{23.5 - 36}{4.8} = \frac{-12.5}{4.8} = -2.60$$

and the standardized value of $X = 42.5$ is

$$Z = \frac{X - \mu}{\sigma} = \frac{42.5 - 36}{4.8} = \frac{6.5}{4.8} = 1.35$$

As illustrated in Figure 6.30, the probability of X taking a value between 24 and 42, $P(24 \leq X \leq 42)$ is approximated by using the table of standard normal areas for the adjusted interval with continuity correction as follows:

$$P(23.5 \leq X \leq 42.5) \approx P(-2.60 \leq Z \leq 1.35)$$
$$= .4953 + .4115 = .9068$$

The error in this second approximation is less than .1%. The reader should work all of the problems at the end of this section and at the end of the chapter that use the normal approximation to the binomial by using the continuity correction, in order to obtain the same answers as those listed in the back of the text.

To apply a **continuity correction,** we use the following rule.

> ### Rule for Applying Continuity Correction
>
> To apply a **continuity correction** to the discrete values included in an interval, we subtract $1/2$ from the smallest value included in the interval, add $1/2$ to the largest value included in the interval, and then proceed as before.

The rule for applying continuity correction to the discrete values included in an interval enables us also to approximate the probability that X will take on a single discrete value, as shown in the following problem.

 PRACTICE PROBLEM 6.12

A batch of $n = 80$ items is taken from a manufacturing process at Pratt-Zungia Limited. The process produces a fraction $p = .16$ of defectives. What is the probability that the batch of 80 items will contain exactly 20 defectives (call finding a defective a success)?

Solution Here we have

$$\mu = np = (80)(.16) = 12.8$$
$$\sigma = \sqrt{np(1 - p)} = \sqrt{(80)(.16)(.84)} = 3.279$$

FIGURE 6.31

Determining the Probability
for Exactly 20 Defective
Items

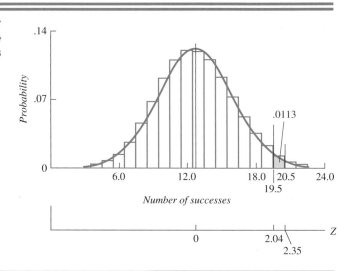

Using the continuity correction to approximate $P(X = 20)$, we subtract $1/2$ from 20, giving us $20 - .5$ or 19.5, and we add $1/2$ to 20, giving us $20 + .5$ or 20.5. The standardized value of $X = 19.5$ when $\mu = 12.8$ and $\sigma = 3.279$ is

$$Z = \frac{X - \mu}{\sigma} = \frac{19.5 - 12.8}{3.279} = \frac{6.7}{3.279} = 2.04$$

and the standardized value of $X = 20.5$ is

$$Z = \frac{X - \mu}{\sigma} = \frac{20.5 - 12.8}{3.279} = \frac{7.7}{3.279} = 2.35$$

As illustrated in Figure 6.31, the probability of X taking the value 20, $P(X = 20)$, is approximated by using the table of standard normal areas for the interval with continuity correction $19.5 \leq X \leq 20.5$ as follows:

$$P(19.5 \leq X \leq 20.5) \approx P(2.04 \leq Z \leq 2.35)$$

$$= .4906 - .4793 = .0113$$

The exact binomial probability is .0122.

Problems: Section 6.4

21. An operator for a telephone-answering service answers a large number of calls for two insurance agents. The two agents are charged the same monthly fee for this service under the assumption that they will each receive approximately half the calls coming in. (The operator receives calls only for these two agents.)

   a. If this assumption is correct, find the probability that in the next 225 calls 125 or more will be for agent A.

   b. Find the probability that one of the agents will get 130 or more of the next 225 calls.

c. During the past week the operator has answered 400 calls. Of these, 300 were for agent A. Is this sufficient evidence to indicate that agent A gets more calls than agent B and thus should be charged more? [*Hint:* Find the probability of 300 or more calls for A in 400 trials, where the probability of a call for A is $p = .5$.]

22. Electronics Corporation of America, a company with a very large number of employees, has 44.4% women employees. If a group of 50 employees is randomly selected to serve on a grievance review board, what is the probability that 30 or more women will be on the board?

23. In a very large company only 6% of the people have had the training required by a special unit that is being formed. If 100 people are randomly selected for this unit, find the following probabilities:

   a. That 2 or more will have the special training.
   b. That between 4 and 8, inclusive, will have the special training.

24. Two binomial probability distributions with normal curves superimposed are presented in Figure 6.32.

   a. Comment on the degree of difficulty of determining the probability of 11 or more successes for each distribution by using the binomial formula (5.5).
   b. Would you be willing to compute approximate binomial probabilities by using the normal approximation for both distributions?

Normal Approximations of Binomial Distributions for Problem 24   FIGURE 6.32

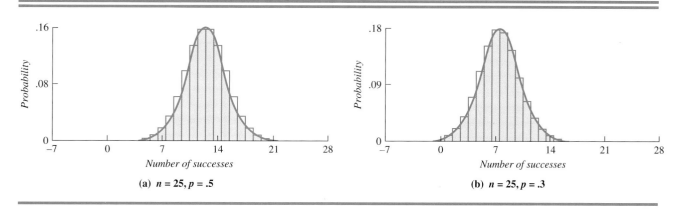

(a) $n = 25, p = .5$ (b) $n = 25, p = .3$

25. The mechanism of heredity is due to genes (fragments of DNA on chromosomes). Gregor Mendel (1866) used statistics to make the important discovery that genetic heredity is based on genes by growing and studying garden peas. Mendel's discovery has enabled modern geneticists to map segments of DNA, the building block of life itself. If a Biotech, Inc., employee crosses hybrid peas that have the gene pair [black, white], then the pollen and ovule of the cross combine to produce gene pairs of the progeny that are [black, black], [black, white], [white, black], or [white, white]. Pollen and ovules are numerous, and the black or white gene outcomes are essentially equally likely and independent. Since black is dominant, just 25% of the peas grown from the crossed seeds are expected to be white (owing to the gene pair [white, white]). In 100 peas from crosses of hybrid peas, what is the probability that the fraction of white peas in the progeny will be less than or equal to .15?

REFERENCE: Mendel, G. (1866), "Experiments in Plant Hybridization," *Transactions of the Natural History Society of Brunn,* Bohemia, Czechoslovakia.

26. In the case *Hazelwood School District v. United States,* it was alleged that the school district had engaged in a pattern or practice of employment discrimination. Only 15 of the 405 teachers hired by the Hazelwood district were members of a minority group (for a fraction of 15/405, or .037). The proportion of the minority group members among qualified teachers in the county was .057 according to Hazelwood and was .154 according to the government. Under the assumption of 405 independent hires, find the probability of hiring a fraction of .037 minority group members or less in the following cases:

   a. Assuming the proportion of minority group members among qualified teachers was .057, as asserted by Hazelwood.
   b. Assuming the proportion of minority group members among qualified teachers was .154, as asserted by the government.

6.5 SUMMARY

In this chapter we considered probability distributions for continuous random variables in general, and then we discussed uniform and normal probability distributions. We can use normal distributions to find the probability that many continuous measures, such as weights, lengths, and times, will be within a particular range of

TABLE 6.1 Summary of Equations for Continuous Probability Distributions

Description	Equation
If X is uniformly distributed, then use a sketch to find probability as a rectangular area under the uniform line.	$Rectangular\ area = (Height)(Width)$
If X is normally distributed, then Z is the standard normal variable. Find the value for Z; then use a sketch and Table V to find the probability (or area).	$Z = \dfrac{X - \mu}{\sigma}$ (6.3)
If X is the number of successes for the binomial distribution, and if n is large, then Z is approximately the standard normal variable. Find the value for Z; then use a sketch and Table V to determine the approximate probability (or area). Use the continuity correction for a better approximation.	$Z = \dfrac{X - np}{\sqrt{np(1 - p)}}$ (6.4)
The mean of the binomial proportion of successes ($\hat{p} = X/n$) can be found with this equation.	$E(\hat{p}) = p$ (6.5)
The variance and standard deviation of the binomial proportion of successes can be found with these equations.	$Variance(\hat{p}) = \sigma_{\hat{p}}^2 = \dfrac{p(1 - p)}{n}$ $Standard\ deviation(\hat{p}) = \sigma_{\hat{p}} = \sqrt{\dfrac{p(1 - p)}{n}}$ (6.6)
If $\hat{p}$ is the fraction of successes for the binomial distribution ($\hat{p} = X/n$) and if n is large, then Z is approximately the standard normal variable. Find the value for Z; then use a sketch and Table V in Appendix C to determine the approximate probability (or area).	$Z = \dfrac{\hat{p} - p}{\sqrt{p(1 - p)/n}}$ (6.7)

FIGURE 6.33

Probability Summary Chart

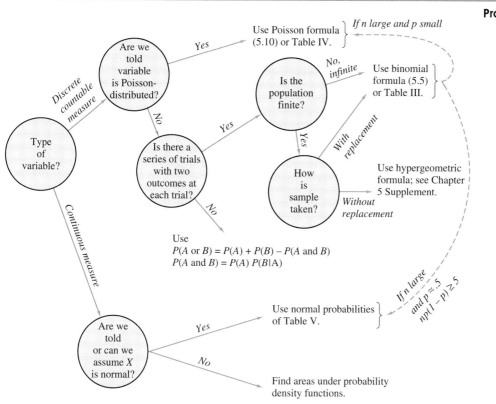

values. We found probabilities for normally distributed variables by converting to the standard normal distribution. Probabilities (areas) for the standard normal distribution are given in Table V of Appendix C. Sketches or plots are important aids in finding probabilities (areas) when we use normal distributions.

We also saw how normal distributions can be used to approximate binomial probabilities when the number of trials n is large. A continuity correction that results in even better approximations was presented. The number of successes for the binomial distribution was transformed into the fraction of successes $\hat{p} = X/n$. We then used normal distributions to approximate the distributions of the fractions of successes. Some important equations from this chapter are summarized in Table 6.1.

In Chapters 4, 5, and 6 we discussed numerous formulas, probability distributions, and relationships. The introductory student cannot be faulted if he or she is somewhat confused about when to use the probability formulas and relationships that have been presented. A person must answer a series of questions before he or she can determine which formula, table, or relationship applies to a probability problem. Figure 6.33 presents a diagram that shows (with dashed lines) the relationship between the various probability distributions and formulas. Thus the reader can use the figure to determine the type of distribution, formula, table, or relationship considered in Chapters 4, 5, and 6 to apply to a given probability problem.

REVIEW PROBLEMS

Answers to odd-numbered problems are given in the back of the text.

27. A computer system generates random numbers according to the uniform distribution shown in Figure 6.8. We want to find the following probabilities for a random number generated from this distribution:

    a. $P(.20 \leq X \leq .80)$ **b.** $P(X \leq .15)$

28. The time X that a tourist waits to board a trolley car on Polk Street in San Francisco has the uniform distribution shown in Figure 6.9.

    a. Find $P(X \geq 3)$.   **b.** Find $P(1 \leq X \leq 4)$.

29. For X with the uniform distribution shown in Figure 6.34, find the following:

    a. d **b.** $P(-1 \leq X \leq .5)$ **c.** $P(X \geq .6)$

FIGURE 6.34

Rectangular or Uniform Distribution for Problems 29 and 30

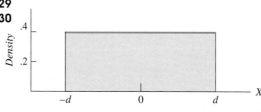

30. For X with the uniform distribution shown in Figure 6.34, find the following:

    a. C such that $P(-C \leq X \leq C) = .6$ **b.** C such that $P(X \geq C) = .8$

31. The following probability distribution represents the distribution of times until failure of those microprocessors that do not pass inspection during testing of their electric conductivity at Intertech Industries. The distribution applies only to those that fail during the 2-second test. The density function for the time (in seconds) to failure during the test (for those microprocessors that fail) is

    $$f(X) = .5 \quad \text{for} \quad 0 \leq X \leq 2 \text{ seconds}$$

    a. Find the probability that a microprocessor that fails will do so in less than .5 second after the test starts.
    b. Find the probability that a microprocessor that fails will do so between .2 and 1.0 second from the start of the test.

32. A center-pivot sprinkler system operated by Colorado Land & Agriculture, Inc., is used to water the alfalfa in a circular field. The circumference of (distance around) the field is 2 miles. As the sprinkler travels in an arc around the field, sprinkler nozzles get clogged and must be cleaned. The distance around the circumference of the field from the starting point to the point located opposite the outermost nozzle when a nozzle becomes clogged is the value of a random variable X that is uniformly distributed over the 2-mile circumference of the field. For X a random variable for the distance covered before a nozzle clogs with a uniform distribution, where $0 \leq X \leq 2$, find the following probabilities:

    a. $P(X \leq 1.0)$ **b.** $P(.5 \leq X \leq 1.0)$ **c.** $P(X \geq 1.5)$ **d.** $P(0 \leq X \leq 2.0)$

33. Given that Z is the standard normal variable, use the table of normal areas to find the following:
    a. $P(-1 \leq Z \leq 1)$ b. $P(Z \geq -1)$
34. If Z is the standard normal variable, find the following:
    a. $P(-.25 \leq Z \leq .75)$ b. $P(Z \leq -.25)$
35. Given that Z is the standard normal variable, find the following:
    a. $P(.38 \leq Z \leq 1.42)$ b. $P(-.49 \leq Z \leq 1.05)$
36. Given that Z is the standard normal variable, find the following:
    a. $P(Z \leq 1.23)$ b. $P(Z \leq -2.12)$
37. Given that Z is the standard normal variable, find the following:
    a. $P(Z \geq -.62)$ b. $P(-1.56 \leq Z \leq -.64)$
38. Given that Z is the standard normal variable, find the following:
    a. $P(Z \geq 1.17)$ b. $P(-.72 \leq Z \leq 1.89)$
39. Given that Z is the standard normal variable, find C.
    a. $P(Z \leq C) = .9554$ b. $P(Z \leq C) = .3085$
40. Given that Z is the standard normal variable, find C.
    a. $P(Z \geq C) = .0322$ b. $P(-C \leq Z \leq C) = .4515$
41. Given that Z is the standard normal variable, find C.
    a. $P(Z \geq C) = .9998$ b. $P(1 \leq X \leq C) = .1219$
42. Given that X is normal with $\mu = .130$ and $\sigma^2 = .000625$, find the following:
    a. $P(X \geq .126)$ b. $P(.110 \leq X \leq .165)$
43. Given that X is a normal variable with $\mu = 50$ and $\sigma = 10$, find the following:
    a. $P(X \leq 25)$ b. $P(38 \leq X \leq 47)$
44. Given that X is normal with $\mu = 15$ and $\sigma^2 = 25$, find the following:
    a. $P(X \leq 13)$ b. $P(19 \leq X \leq 40)$
45. Given that X is normal with $\mu = .05$ and $\sigma = .012$, find $P(.071 \leq X \leq .077)$.
46. The resistances of carbon resistors of 1300 ohms nominal value are normally distributed with $\mu = 1300$ ohms and $\sigma = 150$ ohms.
    a. What proportion of these resistors will have resistances greater than 1000 ohms?
    b. What proportion will have resistances that do not differ from the mean by more than 1% of the mean?
    c. What proportion do not differ from the mean by more than 5% of the mean?
47. The service lives of electron tubes produced by Tensor Corporation are normally distributed; 92.5% of the tubes have lives greater than 2160 hours, and 3.92% have lives greater than 17,040 hours. What are the mean and the standard deviation of the service lives?
48. The life of a nickel-cadmium battery produced by Batterypac is normally distributed with a mean of 20 hours and a standard deviation of 10 hours. Batterypac is considering a warranty for the battery.
    a. What proportion of the batteries have a life of more than 30 hours?
    b. What proportion of batteries last between 15 and 30 hours?
    c. If the managers of the firm want to replace less than 5% of the batteries under a warranty, how many hours of use should the warranty cover?
49. A soft-drink-dispensing machine uses paper cups that hold a maximum of 12 ounces. The machine is set to dispense a mean of 10 ounces of the drink. Because of machine tolerances, the amount that is actually dispensed varies; it is normally distributed and has a standard deviation of 1 ounce.
    a. What proportion of cups are filled with between 11 and 12 ounces?
    b. What proportion of the cups have an overflow?

50. The scores on a national high school aptitude test are normally distributed with a mean of 500 and a standard deviation of 80. The scores may have fractional parts.

    a. What proportion of the scores exceed 600?

    b. What proportion of the scores are between 400 and 600?

    c. What value do the upper 10% of the scores exceed?

51. A Bell's Root Beer, Inc., soft-drink-dispensing machine has been reset so that it dispenses an amount that results in a cup overflow on only 1% of the 12-ounce cups that are used. The amount dispensed has a standard deviation of 1 ounce and is normally distributed. What is the new setting for the mean amount of soft drink that is dispensed?

52. International Waste Deposits, Inc., has applied to the county government for a permit to construct a toxic waste disposal facility. Trucks carrying toxic wastes to the proposed site must go under an overpass on Interstate 80, or they must exit the freeway and detour through the center of a business district. Managers are worried about possible health effects in case of an accident. The overpass has a clearance of 23 feet. The height of the trucks that carry toxic wastes is normally distributed with a mean of 18 feet and a standard deviation of 3 feet.

    a. Find the proportion of trucks carrying toxic wastes that must take the detour through the center of the business district.

    b. Find the height of a new overpass structure that would be required to accommodate 99.9% of the trucks.

53. The project of planning and scheduling the budget requires a sequence of tasks including forecasting demand, preparing the production schedule, estimating costs, and preparing the budget. Project schedulers have found that the time required to complete the budget is normally distributed with a mean of 8 days and a standard deviation of 2 days. To arrange financing at the current rate of interest, the treasurer must know the budget within 5 days. Otherwise, a penalty will be imposed by the bank. What is the probability that the penalty will be incurred if the job is not expedited?

REFERENCE: Wiest, J., and F. Levy (1969), *A Management Guide to PERT/CPM* (New York: Prentice-Hall).

54. The Securities Act of 1933 and the Securities Exchange Act of 1934 require that companies that have their securities traded publicly must register and file periodic financial reports with the Securities and Exchange Commission. During an audit a certified public accountant examines the accounting records, supporting documents, physical properties, and financial statements and issues an opinion as to whether the statements are in accord with generally accepted accounting principles. An auditor may issue a modified opinion owing to a material uncertainty, such as a lawsuit. Media disclosures of an auditor's modified opinion tend to result in negative stock price reactions. The abnormal return because of media disclosures of a modified opinion is normally distributed with a mean of -5% and a standard deviation of 3%. A shareholder of a company's stock has just learned from a media disclosure that the company has received a modified opinion from an auditor.

    a. Find the probability that the shareholder's abnormal return was less than or equal to -6%.

    b. Find the probability that the shareholder's wealth decreased.

REFERENCE: Dopouch, N., R. Holthausen, and R. Leftwich (1986), "Abnormal Stock Returns Associated with Media Disclosures of Qualified Audit Opinions," *Journal of Accounting and Economics,* 8: 93–117.

55. National Brick and Refractory manufactures bricks for the building industry. The process includes mixing the clay, extruding and cutting the brick, firing the brick in a gas-fired kiln, glazing the brick, and firing the glazed brick in the kiln. The kiln is heated by natural gas burners placed throughout the kiln. The heat from the burners is

not distributed evenly throughout the kiln; it decreases as the distance from the burner increases. The temperature at any random location in the kiln is normally distributed with a mean of 1000 and a standard deviation of 50 degrees Fahrenheit.

   a. If bricks are fired at a temperature above 1125°F, then they crack and must be disposed of or reprocessed. Find the proportion of bricks that crack during the firing process.

   b. If glazed bricks are fired below 900°F, then they will be miscolored. Find the proportion of glazed bricks that are miscolored during the firing process.

56. The heights of women who purchase dresses from Tall Sizes Clothing, a direct mail marketing clothier, are normally distributed with a mean of 70 inches and a standard deviation of 3 inches. What proportion of dresses stocked by Tall Sizes should be for women over six feet tall?

57. The weight of one-ton nominal cold-rolled slabs of steel produced by Ganiving Steel, Inc., is normally distributed with a mean of 1950 pounds and a standard deviation of 50. Find the proportion of one-ton nominal cold-rolled steel slabs that weigh less than 2000 pounds.

58. Atlantic Bank receives 42% of its bank card applications from unmarried people. What is the probability that in the next 150 applications 66 or fewer applicants will be unmarried? [*Hint:* Remember the continuity correction.]

59. Allied Auto Parts has found that 4% of the spark plugs it receives from a supplier are misgapped. A shipment of 1000 spark plugs just arrived from this supplier.

   a. What is the expected number of misgapped plugs?

   b. What is the probability that there will be 35 or fewer misgapped plugs in the shipment? Use the normal approximation.

   c. What is the probability of 50 or fewer? Use the normal approximation again. [*Hint:* Remember the continuity correction.]

60. There is a worldwide market for your product that encompasses a very large number of customers. Your company has a 40% share of that market. You randomly select the names of 25 purchasers of this type of product.

   a. What is the probability that between 8 and 12 of these purchasers, inclusive, will be your customers if the normal approximation for the binomial is used with the continuity correction?

   b. Find the same probability by using the binomial table.

61. An employment agency claims that it finds jobs for 85% of the people who use its services. A newspaper reporter investigating the claims of the agency did a survey of people who had used the agency. The results showed that of the 90 people polled only 70 had found jobs through the agency. Find the probability that 70 or fewer have found jobs through the agency if the agency's 85% placement success claim is true.

62. The administrators at a state social services agency have found that 65% of the applicants they process need help in filling out the agency's forms. There will be 120 people making applications this week.

   a. What is the probability that the proportion or fraction of people needing help with the forms will be less than .70?

   b. What is the probability that the proportion or fraction of people needing help with the forms will be between .60 and .75?

   c. What is the probability that the proportion or fraction of people needing help with the forms will be less than .50? (Do *not* use a continuity correction.)

63. A manufacturing process for a mass-produced glassware item at Stemware Corporation results in 5% defective. A sample of 100 of the items is taken.

   a. What is the probability that the proportion or fraction of defective items in the sample is less than .04? (Do *not* use a continuity correction.)

   b. If the normal approximation of the binomial was used in part a, then is your answer very exact, given the rule of thumb stated in this section?

64. Of the customers who rent automobiles at International Rentals in Kona, Hawaii, 50% prefer an intermediate-sized automobile. If International has 40 intermediate automobiles available for rental, what is the probability that the demand for intermediate cars will not be met in a group of 100 randomly selected customers?

Ratio Data Set   *Refer to the 141 companies listed in the Ratio Data Set in Appendix A.*

Questions   65. Locate the data for the quick asset–sales ratio (QA/S). Standardize the sample of QA/S data values by subtracting the sample mean from each data value and then dividing the result by the sample standard deviation. How many of the standardized values exceed 3 or are less than -3?

66. Omit the QA/S values that have standardized values found in Problem 65 that exceed 3 or are less than -3, and find the sample mean and sample standard deviation of the reduced data set.

67. If the QA/S ratio, reduced so that none of the standardized values is farther than three standard deviations from the mean, is assumed to be approximately normally distributed with mean and standard deviation found in Problem 66 for the reduced data set, then find the proportion of companies in this situation that have values of QA/S less than .50.

68. Locate the data for the current-asset–current liabilities ratio (CA/CL) as a measure of liquidity. Standardize the sample of CA/CL data values by subtracting the sample mean from each data value and then dividing the result by the sample standard deviation. How many of the standardized values exceed 3 or are less than -3?

69. Omit the CA/CL values that have standardized values found in Problem 68 that exceed 3 or are less than -3, and find the sample mean and sample standard deviation of the reduced data set.

70. If the CA/CL ratio, reduced so that none of the standardized values is farther than three standard deviations from the mean, is assumed to be approximately normally distributed with mean and standard deviation found in Problem 69 for the reduced data set, then find the proportion of companies in this situation that have CA/CL ratio values less than 2.5.

Credit Data Set   *Refer to the 113 applicants for credit listed in the Credit Data Set in Appendix A.*

Questions   71. A protest committee of 5 people is to be selected from the people who were denied credit and who are listed in the Credit Data Set. Assume that sampling with replacement is employed. Use the binomial formula to calculate the probability that this committee will contain 4 or more women.

72. Use the normal approximation to the binomial distribution to find the probability of the event described in Question 71.

CASE 6.1	**INVESTING IN COMMON STOCKS**

The rate of return realized by an investment in a common stock measures the profit or loss as a percentage of the initial investment. Many investors try to invest in stocks to achieve high returns. A rate of return is determined over a holding period (one day, one week, one month, and so forth) by subtracting the initial outlay from the selling price, adding any dividends or cash payouts, and dividing the remaining quantity by the initial outlay.

Distributions of daily and weekly returns for common stocks tend to be heavier in the tails than the normal distribution. However, monthly returns are generally considered to be approximately normally distributed.

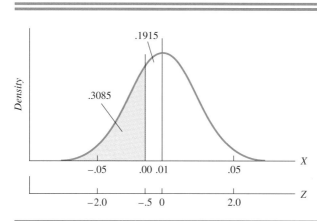

Stocks listed on the National Exchange have a mean monthly return of .01 and a standard deviation of .02. Find the probability of a randomly selected stock resulting in a loss or monthly return less than zero.

The standardized value for a return of .00 is

$$Z = \frac{X - \mu}{\sigma} = \frac{.00 - .01}{.02} = -.5$$

From the standard normal table the area under the normal density between $Z = 0$ and $Z = -.5$ is .1915. As Figure 1 shows, the probability of a loss is

$$P(X \leq 0) = P(Z \leq -.5) = .5000 - .1915 = .3085$$

Find the proportion of stocks with monthly returns greater than or equal to .03.

Case Assignment

ALLOCATING NATURAL GAS

CASE 6.2

Northeastern Gas Company is a natural gas utility that serves customers in a small area covering parts of two states on the East Coast. The company has limited natural gas sources and has had to put its business and industrial customers on an allocation during the past two winters. The company has worked for the most part on a voluntary basis with these customers and has had good success in reducing use during times when the demand for gas would exceed the company's supply. The allocation periods have been limited to January and February, the coldest months of the year in Northeastern's service area.

In order to implement its allocation policies, the company has, in the past, established target "sendout" figures for its business and industrial customers. The sendout is the volume of natural gas sent out from the company in a single day and is measured in units called MCF (thousands of cubic feet).

The company has a sendout capacity of 310,000 MCF. On severely cold winter days it has been the official policy of the company to reserve 180,000 MCF for its nonbusiness customers (homes, schools, hospitals, etc.). This reserve has proven adequate over the past few years but is not expected to prove adequate in the coming winter. That is, in the past a target sendout of 130,000 MCF for business and industrial customers has been adequate.

TABLE 1	Predicted Low Temperature	Suggested Nonbusiness Reserves	Predicted Low Temperature	Suggested Nonbusiness Reserves
	25°F or higher	180,000 MCF	−5°F up to 0°F	225,000 MCF
	20°F up to 25°F	185,000 MCF	−10°F up to −5°F	238,000 MCF
	15°F up to 20°F	192,000 MCF	−15°F up to −10°F	253,000 MCF
	10°F up to 15°F	199,000 MCF	−20°F up to −15°F	271,000 MCF
	5°F up to 10°F	207,000 MCF	−25°F up to −20°F	290,000 MCF
	0°F up to 5°F	216,000 MCF		

Recommended Reserve Levels

However, experience shows that if the target sendout is set at 130,000 MCF for business and industrial customers, then actual sendout might be slightly above or below that figure. In fact, on those days during the current winter when the target sendout was 130,000 MCF, the actual sendout averaged 130,500 MCF, with a standard deviation of 1230 MCF.

Mr. Ralph Maybe, director of operations for Northeastern, noted these figures and stated, "It looks as if we've got to set our business target sendout considerably lower than 130,000 MCF if we don't want to cut into the 180,000 MCF reserves for nonbusiness customers." In addition, he suggested that the 180,000 MCF figure was no longer a realistic reserve figure. "The growth of home building in our service area and the last few cold winters have shown us that 180,000 MCF is an adequate reserve under normal conditions, but we should be more realistic and establish variable reserve figures. Reserves for residential users should be higher on days that are predicted to be severely cold than on days that are only mildly cold."

On the basis of this reasoning, Mr. Maybe had one of his staff assistants prepare a list of recommended reserve levels (see Table 1). The assistant's report also noted that during the past few winters the Operations Department had been experimenting with the demand for gas when different target sendout figures were established (see Table 2).

The report concluded that "the coefficient of variation for the sendout figures seems to be rather stable over a fairly wide range of target sendout levels. It appears that this information could be used in conjunction with our knowledge that the sendout figures tend to follow a normal distribution about their mean to help us set daily target sendout levels for business and industrial users based on the forecasted low temperature for the following day."

Mr. Maybe was pleased with these results and gave his assistant permission to begin development of target sendout figures to correspond to the ranges of forecasted low temperatures identified in Table 1. The only restriction he established was to "make sure your

TABLE 2	Experimental Target Sendout	Number of Days at This Level	Mean Sendout Achieved	Standard Deviation	Coefficient of Variation
	100,000 MCF	7	99,600 MCF	1050 MCF	1.05%
	110,000 MCF	12	111,200 MCF	1100 MCF	0.99%
	115,000 MCF	10	114,800 MCF	1090 MCF	0.95%
	120,000 MCF	8	118,900 MCF	1280 MCF	1.08%
	125,000 MCF	20	125,400 MCF	1225 MCF	0.98%
	130,000 MCF	35	130,500 MCF	1230 MCF	0.94%

Experimental Demand Data

target levels are set low enough so that we run only a 10% chance of having the business and industrial customers eating into the suggested reserves."

a. Find the maximum business sendout given the nonbusiness reserves for each range of temperatures.

b. Find the weighted average of the coefficients of variation.

c. Use the weighted-average coefficient of variation with days as weights to estimate the standard deviation of the business sendout as a function of the mean business sendout.

d. Set the target or mean business sendout levels for each temperature range so that the chance of exceeding the maximum business sendout is 10%.

Case Assignment

REFERENCES

Cochran, W. G. 1963. *Sampling Techniques,* 2nd ed. New York: Wiley.

Freedman, D., R. Pisani, and R. Purves. 1992. *Statistics,* 2nd ed. New York: Norton.

Hogg, Robert V., and Allen T. Craig. 1978. *Introduction to Mathematical Statistics,* 4th ed. New York: Macmillan.

Mood, A. M., F. A. Graybill, and D. C. Boes. 1974. *Introduction to the Theory of Statistics.* New York: McGraw-Hill.

Mosteller, F., R. Rourke, and G. Thomas. 1970. *Probability with Statistical Applications,* 2nd ed. Reading, Mass.: Addison-Wesley.

Optional Topic

Exponential probability distributions are a family of continuous probability distributions useful in describing the interval prior to the occurrence of an event. The interval X is often the time that it takes to complete a task, the time between arrivals at a waiting line, or, in a product reliability situation, the time or distance before the failure of a product. For example, exponential probability distributions may be used to find the following probabilities:

6.6
EXPONENTIAL
PROBABILITY
DISTRIBUTIONS

1. What proportion of customers arriving at a teller's window at Commercial Bank will approach the window within three minutes of the previous customer. If most of the customers at the teller's window arrive within three minutes of the previous customer, then customer service is exemplary.
2. What proportion of long-distance telephone calls received at a switchboard will be completed within three minutes. If a high proportion of long-distance calls are completed within three minutes, then the telephone company may have excess capacity.
3. What proportion of machines will break down during an eight-hour night shift. If a high proportion break down, then the reliability of the production system is problematical.
4. What proportion of new automobiles will have transmission failures in 50,000 or fewer miles. If a high proportion of automobiles have transmission failures within 50,000 miles, then the manufacturer will suffer losses due to its warranty.

The **exponential probability density function** follows.

Exponential Probability Density Function

$$f(X) = \lambda e^{-\lambda X}$$

where: $X > 0$
$\lambda > 0$

The value for e is the mathematical constant $e = 2.718\ldots$. The parameter of an exponential probability distribution is λ, and it must be positive. Different values of λ give us different exponential probability distributions. Figure 6.35 shows the exponential distributions with $\lambda = .25$, $\lambda = .5$, and $\lambda = 1$ and enables us to see that exponential distributions are skewed appreciably to the right.

FIGURE 6.35

Exponential Probability Distributions with $\lambda = 1$,
$\lambda = .5$, and $\lambda = .25$

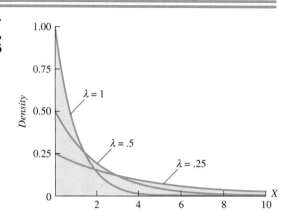

Exponential distributions have a distinctive association with Poisson distributions. If the mean number of occurrences from a Poisson process is λ, then the interval between occurrences from the process will have an exponential distribution with the same parameter λ. The mean of an exponential distribution is $E(X) = 1/\lambda$ and the standard deviation is also $\sigma_X = 1/\lambda$.

For an example of the association between Poisson and exponential distributions, if the number of customers arriving at a teller's window at Commercial Bank is a Poisson random variable with a mean of $\lambda = .75$ customer per minute, then the time between customer arrivals at the window will be exponentially distributed with a mean of $E(X) = 1/\lambda = 1/(.75$ customer arrivals/minute$) = 1.33$ minutes per customer arrival, and the standard deviation has the same value as the mean.

A **cumulative probability for an exponential distribution** can be computed quite simply as follows.

Cumulative Exponential Distribution Function

The probability of obtaining a value for the exponentially distributed random variable X that is less than or equal to a specific value a is

$$P(X \le a) = 1 - e^{-\lambda a} \qquad \begin{array}{l} a > 0 \\ \lambda > 0 \end{array}$$

 PRACTICE PROBLEM 6.13

The number of customers arriving at a teller's window at Commercial Bank follows the Poisson distribution with a mean rate of $\lambda = .75$ customer per minute. If the time between arrivals is less than or equal to three minutes, then the teller can provide banking services without irritating customers with annoying waiting times.

a. Find the mean and standard deviation of the time between customer arrivals at the teller's window, X.
b. Find the proportion of customers for whom the teller provides service without an annoying delay.

FIGURE 6.36

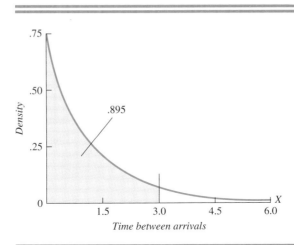

Exponential Probability Distribution for the Time Between Arrivals

Solution Since the number of arrivals in one minute has a Poisson distribution, the time between arrivals is exponentially distributed.

a. The mean of the exponential distribution of X is

$$E(X) = \frac{1}{\lambda} = \frac{1}{(.75 \text{ customer arrival/minute})} = 1.33 \text{ minutes per customer}$$

and the standard deviation is

$$\sigma_X = \frac{1}{\lambda} = \frac{1}{.75} = 1.33$$

b. The teller can provide services without an annoying delay if $X \leq 3$ minutes. Thus, as illustrated in Figure 6.36, with the aid of a calculator to compute $e^{-\lambda a}$, we find

$$P(X \leq 3) = 1 - e^{-(.75)(3)} = 1 - e^{-2.25} = 1 - .105 = .895$$

Consequently, about 90% of the customers will not be irritated by an annoying delay.

Problems: Section 6.6

73. During off-hours, people arrive at a toll booth on the Will Rogers Turnpike at a mean rate of .5 person per minute. The arrivals are Poisson-distributed.
    a. What is the probability that the next person will arrive within 2 minutes of the previous arrival?
    b. What is the probability that the next arrival will not approach the booth while the operator is on a personal break if the break lasts for 5 minutes?

74. The defects in an automated weaving process at Craft Mills, Inc., are Poisson-distributed at a mean rate of .025 defect per linear foot.
    a. What is the mean distance between defects for this problem?
    b. What is the probability that the next defect will be within 10 feet of the previous defect?
    c. What is the probability that the next defect will be between 10 and 20 feet from the previous defect?

75. The number of telephone calls passing through a TCI switchboard has a Poisson distribution with mean λ equal to 3 calls per minute. Find the following probabilities.

    a. What is the probability distribution of the time between calls at the switch board?
    b. What is the mean time between calls at the switchboard?
    c. What is the probability that the next call will pass through the switchboard within one minute?

76. Between Lake Powell and Lee's Ferry on the Colorado River the number of fish caught per angler-hour of fishing effort has a Poisson distribution with λ equal to 1.3 fish per angler-hour. Find the following probabilities:

    a. That the time between fish landings by 1 angler will be less than or equal to 30 minutes.
    b. That the time between fish landings by 2 anglers will be less than or equal to 30 minutes.

77. A local area network computer system at Noel, Inc., has several input stations located in various departments. Typically, the central computer can respond to 8 input jobs per minute without causing any noticeable delay at the input stations. That is, when the jobs submitted in one minute are 8 or less, each input station behaves as though it had complete control of the network. Internal computer accounting records show that jobs are usually submitted at a Poisson-distributed mean rate of .08333 per second. What proportion of input jobs are submitted within one second of one another?

78. A sports utility vehicle has a 7-year, 70,000-mile warranty on all drive train components. The number of breakdowns in drive train components during the third year of the operation of the vehicle is Poisson-distributed with a mean number of breakdowns per year of .40. Find the proportion of vehicles that will be expected to break down in the next six months.

79. Environmental concerns in Washington County have resulted in a hazardous materials clean-up team organized by the Department of Public Safety. The number of hazardous chemical spills in the county that require the intervention of the clean-up team is Poisson-distributed with a mean of 2 interventions per month.

    a. Find the mean time between interventions for hazardous chemical spills.
    b. Find the proportion of interventions that occur within two weeks of a previous intervention where a month has four weeks.

Sampling Distributions

7

To complete our study of probability, we consider the probability distributions for several sample statistics in this chapter. Probability distributions for sample statistics are called *sampling distributions.*

In this chapter, we discuss sampling distributions for the sample mean $\overline{X}$ when the sample is taken from discrete or continuous populations, the number of successes X when the sample is taken from a binomial population, and the sample proportion (or fraction) of successes $\hat{p}$ when the sample is taken from a binomial population. Also, *central limit theorems,* the most important theorems in all of statistics, are presented in this chapter.

The sampling distributions that are discussed in this chapter and the central limit theorem are very important for making statistical inferences. Thus this chapter is essential for the chapters on estimation and hypothesis testing that follow.

7.1
Sampling and Inference

A statistical **population**, or *universe,* is simply a set or collection; the things of which the population is composed are called its *items* or its *elements.* Sometimes, the population is specified by a complete list of its items, as when the voting population of a town is explicitly listed on the voting rolls. More commonly, a population is specified by a definition of some kind or by the singling out of a characteristic common to its elements, as when we speak of the population of microprocessors produced by a given manufacturer during a given year.

The populations in the preceding examples are all *finite.* Statistics also involves hypothetical *infinite populations.* The analysis of a manufacturing **process** often involves an infinite population, the hypothetical population of all the items—say all the microprocessors—that the process would produce if it were to run indefinitely under constant conditions. There is an infinite population of the random experiment of tossing a coin, and to toss the coin is to observe an element of the population.

Often a finite population is so large as to be effectively infinite for the purposes of statistical analysis. An effective way to acquire an initial understanding of what it means for a population to be infinite is to consider the population to be finite but so very large that removing a number of elements from it has no discernible effect on the composition of the population.

For inquiring into the nature of a population it would be ideal if we could examine each one of its elements, but this is often out of the question. Sometimes, it is impossible because some of the elements are physically inaccessible. In other cases it is too costly. In addition, we will not test every item produced if the test destroys the item. For many practical purposes we can obtain sufficiently accurate results more quickly and inexpensively by examining only a small part of the population.

In many situations, then, we are content with investigating only a part of the whole population. The part investigated is a **sample.** Samples may be collected or selected in a variety of ways, but most statistical techniques presuppose an element of randomness in the sampling. We shall be concerned almost exclusively with *random samples.*

For example, consider first a finite population of size 50, from which we take a sample of size 5. There are $\binom{50}{5} = 50!/[5!(50-5)!]$, or 2,118,760, such samples, a sample being a subset or combination of size 5. As mentioned previously, the sample is random if it is selected in such a way that all samples of size 5 have the same probability of being chosen. Randomness is a property of the selection procedure. In general, there are $\binom{N}{n} = N!/[n!(N-n)!]$ samples of size n in a finite population of size N, and **random sampling** means choosing one in such a way that all are equally likely to be chosen.

> ### Definition
>
> A sample that is selected from a finite population in such a way that all samples of the same size have the same probabiltiy of being chosen is called a **random sample.** This type of random sample is also known as a **simple random sample.**

Although the concept of a random sample is fairly easy to grasp, there are situations in which it is not clear how to obtain one. If we can list all the elements of a population and number them, we can get a random sample by drawing thoroughly mixed numbered tags or tickets from a bowl, or we can use a table of random numbers to decide which elements to include in the sample. Table II in Appendix C is a table of random numbers. As its name implies, this table consists of numbers that have no pattern or scheme. The numbers are grouped into columns that are five digits wide. Each column of digits can be thought of as a column of one-digit random numbers.

EXAMPLE 7.1   The marketing manager of a firm with 1000 customers desires to take a random sample of 5 customers to determine their attitudes concerning a new product. The marketing manager could obtain a list of the customers and number them from 1 to 1000. Then she could satisfy herself that she obtained a truly random sample by going to the random numbers of Table II in Appendix C and picking any set of 5 three-digit random numbers. These numbers would then represent the numbers of the customers whose opinions would be sought. If, for simplicity, we assume that the marketing manager began at the upper left-hand corner of the random number table and selected the first three digits in the left column for her first number, she would obtain the number 395. She might then move down to the second row and select the number 463. Continuing in this manner, she would obtain a sample consisting of customers numbered 395, 463, 995, 67, and 695. It might at first appear that in using three-digit random numbers, the marketing manager would never select customer number 1000. However, the random number 000 could be assigned to this customer.

The process of selecting random numbers from a random number table can become tedious, especially if large samples are required, if the random numbers need to be sorted into an ascending sequence, or if the range from the smallest to the largest random numbers that are needed does not fit the random number table well. In such cases, a computer can be used to generate and sort random numbers, as shown in Computer Exhibits 7.1A and 7.1B.

If we cannot enumerate the population, we cannot use these techniques. Obtaining a random sample of oranges from a tree, fish from a river, trees from a forest, or people in a city is more difficult (see chapter references). One method that is used often is **systematic sampling,** wherein a sample is selected according to some fixed system—taking every hundredth name from the phone book, for example, or selecting

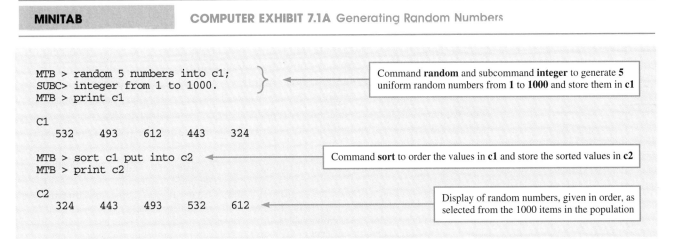

MINITAB	COMPUTER EXHIBIT 7.1A Generating Random Numbers

```
MTB > random 5 numbers into c1;
SUBC> integer from 1 to 1000.
MTB > print c1
```
Command **random** and subcommand **integer** to generate **5** uniform random numbers from **1** to **1000** and store them in **c1**

```
C1
    532    493    612    443    324
```

```
MTB > sort c1 put into c2
MTB > print c2
```
Command **sort** to order the values in **c1** and store the sorted values in **c2**

```
C2
    324    443    493    532    612
```
Display of random numbers, given in order, as selected from the 1000 items in the population

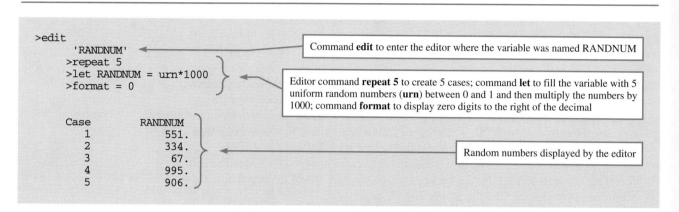

```
>edit
    'RANDNUM'
    >repeat 5
    >let RANDNUM = urn*1000
    >format = 0
```
Command **edit** to enter the editor where the variable was named RANDNUM

Editor command **repeat 5** to create 5 cases; command **let** to fill the variable with 5 uniform random numbers (**urn**) between 0 and 1 and then multiply the numbers by 1000; command **format** to display zero digits to the right of the decimal

```
Case       RANDNUM
 1           551.
 2           334.
 3            67.
 4           995.
 5           906.
```
Random numbers displayed by the editor

machined parts from a tray according to some definite pattern. **Stratified random sampling** and **cluster sampling** are two additional methods that are used to obtain samples (see Cochran, 1977).

To obtain a random sample from a hypothetical infinite population, or from a **process,** the sample of n random variables $X_1, X_2, \ldots, X_n$ must be generated from the same probability distribution or be identically distributed and they must be statistically independent of one another.* Any set of independent and identically distributed random variables generated by a process is a random sample. If we have successfully obtained a random sample from the population in which we are interested, then we can investigate the inferences that can be based on such a sample.

Random sampling as illustrated previously is sometimes called **sampling without replacement,** because an element sampled is not returned to the population before the next element of the sample is drawn. Sampling that is done **with replacement** involves putting each element back into the population before the next one is chosen. In practice, sampling from finite populations is generally done without replacement.

We take random samples from populations to study or estimate facts about the populations. Facts about populations are usually expressed in terms of numbers called *parameters.* In general, a **parameter** is a number describing some aspect of a population. Population parameters are often unknown. In making inferences about a parameter on the basis of a sample, we usually deal with a **statistic,** which is often one of the descriptive measures of Chapter 3. Values for statistics can be computed from samples. Generally, we are concerned with one or more parameters that help describe a population. We do not know the values of the parameters and usually never will. We draw a random sample and compute a value or values for one or more statistics on the basis of the sample. We have actual numerical values for the statistics. And we use the statistics to make **inferences** in the form of estimates or tests of hypotheses about the unknown parameters. To know how to make these inferences, we

---

*Mathematically, two discrete random variables, X and Y, are statistically **independent** if $P(X = x_i$ and $Y = y_j) = P(X = x_i)P(Y = y_j)$ for all x_i and y_j; two continuous random variables, X and Y, are statistically independent if $f(X$ and $Y) = f(X)f(Y)$. $P(X = x_i$ and $Y = y_j)$ is the **joint distribution** of the discrete random variables and $f(X$ and $Y)$ is the **joint probability density** of the continuous random variables. For a set of n random variables, the variables are independent if the joint probability or density function is equal to the product of the individual or marginal probability or density functions.

must know how the values of the statistics vary from sample to sample, or we must know the **sampling distributions** of the statistics.

> **Definition**
>
> The probability distribution of a statistic is known as the statistic's **sampling distribution.**

EXAMPLE 7.2   Consider the average weight μ (in pounds) of fertilizer in bags coming off an automatic packaging line at Grace & Company. We do not know the value of the parameter μ and never will. To get an idea of what μ is, we take a sample of bags and compute $\overline{X}$, the mean of their weights (as in Chapter 3); the statistic $\overline{X}$ is something we find the actual numerical value for. On the basis of $\overline{X}$, we make inferences about μ. Without statistical theory, we know that $\overline{X}$ somehow estimates μ; but to know just how to make the inferences and how exact they will be, we need to know the distribution of the random variable $\overline{X}$. That is, we need to know its sampling distribution.

7.2
EXPECTED VALUES AND VARIANCES OF STATISTICS

In this section, we discuss the expected value and variance of the sample mean $\overline{X}$ and the expected value of the sample variance s^2. The expected value and variance for the sample mean will help in defining the sampling distribution of $\overline{X}$, which in turn will be used extensively for purposes of statistical inference.

The mean and the variance of a population of numbers are defined as the mean and the variance of all possible single random observations from that population, that is, $E(X_1)$ and Variance(X_1) for all random samples X_1 of size 1. The population is said to be *discrete or continuous* according to whether a single observation X_1 has a discrete or a continuous distribution. A discrete population may be *finite* or *infinite;* a continuous population must be *infinite,* although it may be approximated by a large but finite population.

EXAMPLE 7.3   Suppose the population consists of five small businesses that have been assisted by the Small Business Development Corporation whose lives, or years of existence, are 2, 4, 6, 8, and 10. In this case the population mean μ is

$$\mu = (1/5)(2 + 4 + 6 + 8 + 10) = 6$$

Now suppose we take a random sample of size 2 without replacement. There are $\binom{5}{2} = 5!/[2!\,(5 - 2)!]$, or 10, different outcomes—different samples of size 2—and each of them has a sample mean $\overline{X}$ as follows:

Sample	2, 4	2, 6	2, 8	2, 10	4, 6	4, 8	4, 10	6, 8	6, 10	8, 10
$\overline{X}$	3	4	5	6	5	6	7	7	8	9

Here the notation 2, 4, for instance, indicates the sample that includes the businesses of ages 2 and 4.

Since the sample is random, each outcome has probability $1/10$. The probability that $\overline{X}$ has the value 5 is the probability of getting the sample 2, 8 or the sample 4, 6, so $P(\overline{X} = 5) = 2/10$. This and analogous computations give the sampling distribution of $\overline{X}$:

$\overline{X}$	3	4	5	6	7	8	9
$P(\overline{X})$	.1	.1	.2	.2	.2	.1	.1

By Equation (5.1), for the expected value for a discrete random variable,

$$E(\overline{X}) = (3)(.1) + (4)(.1) + (5)(.2) + (6)(.2)$$
$$+ (7)(.2) + (8)(.1) + (9)(.1) = 6$$

The point is that $E(\overline{X})$ coincides with the population mean μ. ▬▬

The expected value $E(\overline{X})$ always coincides with the population mean μ. Let $X_1, X_2, \ldots, X_n$ be a random sample of size n from a population, finite or infinite, discrete or continuous. The sample mean is, as in Chapter 3,

$$\overline{X} = \frac{1}{n} \sum_{i=1}^{n} X_i \tag{7.1}$$

Its expected value is as shown in Equation (7.2).*

Equation for Expected Value of Sample Mean

$$E(\overline{X}) = \mu \tag{7.2}$$

Equation (7.2) is sometimes expressed as follows: The expected value of the sample mean $\overline{X}$ is the population mean μ. The meaning of Equation (7.2) can be demonstrated with an example.

▬▬ **EXAMPLE 7.4** We are interested in the weights of packages coming off a production line at an Amstar Corporation sugar-packaging plant. The production line fills the packages to a mean weight of 10 pounds. Thus the population mean is $\mu = 10$. Each day at noon a quality control engineer takes a random sample of nine packages that have come off the line during the previous 24 hours, and he computes the sample mean weight $\overline{X}$ of the nine packages. If this procedure were followed every day for 400 working days, the engineer would have collected 400 sample means. These 400 sample means themselves could be added and divided

---

*This equation and the other numbered formulas in this section will be used in the rest of the book. Although we give derivations of them in footnotes, the derivations are not needed; an understanding of the *meaning* of the formulas will suffice. The derivation of Equation (7.2) can be shown by using the rules for manipulating expected values. We can apply two rules to Equation (7.1): (1) that the expected value of a constant multiplied by a random variable is equal to the constant multiplied by the expected value of the random variable and (2) that the expected value of a sum of random variables is the sum of the expected values of the random variables. Using these rules and the facts that each individual X_i is a single observation from the population and $E(X_i)$ has the same value as μ, we have

$$E(\overline{X}) = E\left(\frac{1}{n} \sum_{i=1}^{n} X_i\right) = \frac{1}{n} E\left(\sum_{i=1}^{n} X_i\right) = \frac{1}{n} \sum_{i=1}^{n} E(X_i) = \frac{1}{n} \sum_{i=1}^{n} \mu = \frac{1}{n}(n\mu) = \mu$$

by 400 to obtain their mean. Equation (7.2) indicates that the 400 sample means would have a mean close to 10 pounds. In the long run, after infinitely many working days, the mean of the infinitely many sample means would be *exactly* 10 pounds. ▬▬

If the population is **infinite** and the elements $X_1, X_2, \ldots, X_n$ in the random sample are *independent*, then we have, for the **variance** and **standard deviation** of $\overline{X}$, the following equations.*

Equation for Variance of Sample Mean (Infinite Population)

$$\text{Variance}(\overline{X}) = \sigma_{\overline{X}}^2 = \frac{\sigma^2}{n} \tag{7.3}$$

where σ^2 is the population variance.

Equation for Standard Deviation of Sample Mean (Infinite Population)

$$\text{Standard deviation}(\overline{X}) = \sigma_{\overline{X}} = \frac{\sigma}{\sqrt{n}} \tag{7.4}$$

The standard deviation of the sample mean $\sigma_{\overline{X}}$ is often referred to as the **standard error of the mean.**

▬▬ **EXAMPLE 7.5** Equations (7.3) and (7.4) give us a method of computing the variance and standard deviation of the 400 sample mean values collected by the quality control engineer. If in addition to having a mean package weight of

---

*Equations (7.3) and (7.4) apply when the population is *infinite*, or the population is *finite* and sampling is done with replacement. If the population is finite and sampling is done without replacement, Equation (7.3) must be replaced by the formula

$$\text{Variance}(\overline{X}) = \frac{N - n}{N - 1} \times \frac{\sigma^2}{n}$$

where N is the size of the population. The factor $(N - n)/(N - 1)$ is a *finite population correction factor*. In Example 7.3, $E(\overline{X}) = \mu = 6$, so the population variance is

$$\sigma^2 = E(X_i - 6)^2 = (1/5)(16 + 4 + 0 + 4 + 16) = 8$$

The variance of $\overline{X}$ is

$$E(\overline{X} - 6)^2 = 9 \times .1 + 4 \times .1 + 1 \times .2 + 1 \times .2 + 4 \times .1 + 9 \times .1 = 3$$

which agrees with the preceding formula for $N = 5$ and $n = 2$ as follows:

$$\text{Variance}(\overline{X}) = \frac{5 - 2}{5 - 1} \times \frac{8}{2} = 3$$

If the population size N is large in comparison with the sample size n, then the correction factor is nearly 1; so the formula for Variance($\overline{X}$) for a finite population is practically the same thing as Equation (7.3) in this case. We use infinite populations or large populations in comparison to the sample size throughout essentially all of the book.

10 pounds, the package-filling process fills individual packages with a standard deviation of $\sigma = .6$ pound, then the standard deviation of the 400 mean values around 10 will be, according to Equation (7.4), $\sigma/\sqrt{n} = .6/\sqrt{9} = .2$. This result says that the *means* of many samples of size $n = 9$ will be three times more tightly clustered around the value of 10 than are the *individual package weights,* where we measure the degree of clustering by the standard deviation. ▬

Another example will help to clarify these points.

▬ **EXAMPLE 7.6** We took 50 samples of size $n = 36$ from a population consisting of savings accounts and examined the account balances at First International Bankshares. These samples are made up of 36 randomly selected account balances taken at the close of business each working day for ten weeks. The mean balance in the population of accounts (which will be considered to be very large) is $3900, and the standard deviation of the population is $1200. That is,

$$\mu = \$3900 \quad \text{and} \quad \sigma = \$1200$$

After the 50 samples have been taken, there will be 50×36, or 1800, observations. But there will be *only 50 means,* one for each sample. According to Equation (7.2), the mean of these 50 means will be approximately $3900, the mean of the population. [We say *approximately* $3900 since Equation (7.2) assumes that infinitely many samples, not just 50, have been taken.] Also, according to Equation (7.4), the standard deviation of the 50 mean values will be approximately

$$\text{Standard deviation}(\overline{X}) = \sigma_{\overline{X}} = \frac{\sigma}{\sqrt{n}} = \frac{\$1200}{\sqrt{36}} = \frac{\$1200}{6} = \$200$$

That is, the sampling distribution of the means of samples of size $n = 36$ will have a mean value of $3900 and a standard deviation of $200. ▬

▬ **EXAMPLE 7.7** A bimodal distribution of the amount of time required for people to pass through a border inspection station is shown in Figure 7.1. The distribution shows that many people pass with a very short delay or no delay at all, whereas many others are detained a rather long period for inspections of baggage and personal belongings. Very few people are delayed a moderate amount of time.

The mean and the standard deviation of this distribution of times are $\mu = 8$ minutes and $\sigma = 6$ minutes. If 150 random samples of size $n = 64$ delay times at this inspection station were taken, then a total of $(150)(64) = 9600$ individual times would be available, but only 150 sample means would be calculated. According to Equation (7.2), these 150 sample means would have a mean value very close to the population mean $\mu = 8$ minutes. (The many sample means would have a mean of exactly 8 minutes in the long run; that is, after infinitely many samples of $n = 64$ were taken.) According to Equation (7.4), these 150 sample means would have a standard deviation of approximately $\sigma/\sqrt{n} = (6 \text{ minutes})/\sqrt{64} = .75$ minute.

If we use the rule about the standard deviation mentioned in Section 3.12, we can say that about all of the 150 sample means will fall within ± 3 standard deviations of their central value, or 8 minutes. That is, it would be most unlikely for

FIGURE 7.1

Distribution of the Time
Needed to Pass the Border
Inspection Station

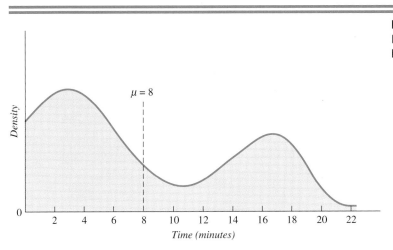

us to observe the mean time for any sample of $n = 64$ outside the interval defined by $8 \pm 3(.75) = 8 \pm 2.25$, or 5.75 to 10.25 minutes. If someone were to tell us that he observed the amount of time that it took a randomly chosen sample of $n = 64$ people to pass through the inspection station and found that the time averaged $\overline{X} = 11$ minutes per person, then since 11 exceeds 10.25 minutes, which is about the highest mean we would expect to see, we could reach one of three conclusions:

1. The person has observed a *very* unusual sample.
2. The person is not telling us the truth.
3. For some reason the population mean time has shifted upward from $\mu = 8$ minutes.

If we believe that a very unusual sample does not happen very often and that the person is telling the truth, then we would draw the third conclusion.

In addition to finding the mean and the variance of $\overline{X}$, we can find the mean of the sample variance s^2. Recall from Chapter 3 that

$$s^2 = \frac{1}{n-1} \sum_{i=1}^{n} (X_i - \overline{X})^2 \tag{7.5}$$

For a sample from an *infinite* population, we have the following formula.*

Equation for Expected Value of Sample Variance (Infinite Population)

$$E(s^2) = \sigma^2 \tag{7.6}$$

*For sampling without replacement from a finite population, we use $E(s^2) = [N/(N-1)]\sigma^2$. As N increases, $N/(N-1)$ gets closer to 1.

Equation (7.6) concerns the mean of the variance, whereas Equation (7.3) concerns the variance of the mean. To keep these equations straight, we must expand the terms: Equation (7.6) concerns the expected value of the sample variance, and Equation (7.3) concerns the variance of the sample mean.

=== **EXAMPLE 7.8**   Equation (7.6) can be explained in terms of the 400 samples taken by the quality control engineer in Example 7.4. Each of the 400 samples would have a variance, which would be computed by dividing the sum of squared deviations from the $\overline{X}$ value by 8. These 400 sample variances would average out to about the population variance. That is, if all the 400 sample variances were added and divided by 400, their value would be close to $\sigma^2 = (.6)^2 = .36$.

═══

Problems: Sections 7.1–7.2

Answers to odd-numbered problems are given in the back of the text.

1. A sample of size $n = 100$ is drawn from an infinite population with mean $\mu = 30$ and standard deviation $\sigma = 20$.
   a. Find $E(\overline{X})$, the mean of the sampling distribution of the sample mean $\overline{X}$.
   b. Find Standard deviation($\overline{X}$), the standard deviation of the sampling distribution of the sample mean $\overline{X}$.

2. A sample of size $n = 144$ is drawn from an infinite population with mean $\mu = 30$ and standard deviation $\sigma = 20$.
   a. Find $E(\overline{X})$, the mean of the sampling distribution of the sample mean $\overline{X}$.
   b. Find Standard deviation($\overline{X}$), the standard deviation of the sampling distribution of the sample mean $\overline{X}$.

3. The number of personal computers sold by a Business Computers retail store each day for the past five days is presented in the accompanying table. Ten possible random samples of size two days each can be selected, if the days are selected without replacement.

Day	Mon.	Tues.	Wed.	Thurs.	Fri.
Units Sold	1	3	7	5	9

   a. List the ten possible random samples with respect to days.
   b. Compute the ten sample means for the number of units sold per day, and list the possible values along with their corresponding probabilities.
   c. Determine the expected value of the sample mean.

4. The price-earnings ratios of a very large (essentially infinite) set of common stocks are approximately normally distributed with a mean of 15 and a standard deviation of 5. If a large number of samples, each of size 25, are selected from the population of stocks and a sample mean is computed for each sample, what will be the expected value and the variance of the large number of sample means for the price-earnings ratios?

7.3
SAMPLING FROM NORMAL POPULATIONS

Let $\overline{X}$ be the sample mean for a sample of independent observations of size n. As we saw in the previous section, if the population mean and standard deviation are μ and σ, then $\overline{X}$ has mean μ and standard deviation $\sigma_{\overline{X}} = \sigma/\sqrt{n}$, which is true whatever the form of the parent population may be. Suppose now that the population is *normal*. In this case $\overline{X}$ is normally distributed with mean μ and standard deviation $\sigma_{\overline{X}} = \sigma/\sqrt{n}$. This fundamental fact about **sample means from normal populations** is summarized as the following rule.

> **Rule for Distribution of Sample Means for Normal Population**
>
> If the population for X is normally distributed with mean μ and standard deviation σ, the sample mean $\overline{X}$ is normally distributed with mean μ and standard deviation $\sigma_{\overline{x}} = \sigma/\sqrt{n}$.

The fact that the sampling distribution of $\overline{X}$ is normal when the population is normal is clearly shown in the results of a sampling exercise. Statistics students took 1000 different random samples of 5 values ($n = 5$) each, and another 1000 different random samples of 10 values ($n = 10$) each from a normal population with mean 40 and standard deviation 10. Figure 7.2(a) shows the normal probability (or population) distribution.

The students then computed the 1000 sample means when $n = 5$ and the 1000 sample means when $n = 10$. Figures 7.2(b) and 7.2(c) show the resulting empirical percentage histograms for the sample means when $n = 5$ and $n = 10$, respectively. Normal sampling distributions for the sample means have been superimposed on the empirical histograms. The very close fit of the empirical histograms to the normal

Normal Probability Distribution, Empirical Histograms, and Normal Sampling Distributions FIGURE 7.2

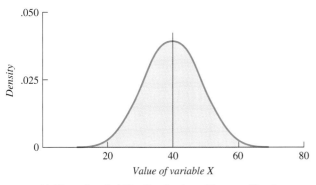
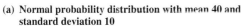

(a) **Normal probability distribution with mean 40 and standard deviation 10**

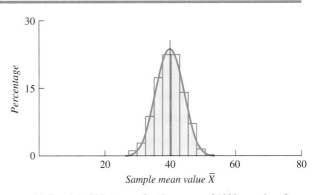

(b) **Empirical histogram for the means of 1000 samples of size 5 and a normal sampling distribution**

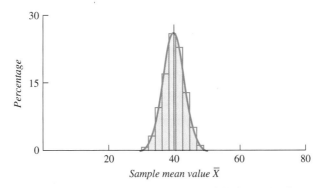

(c) **Empirical histogram for the means of 1000 samples of size 10 and a normal sampling distribution**

sampling distribution curves in Figures 7.2(b) and 7.2(c) demonstrate that the sampling distribution of the sample mean is normally distributed if the population is normally distributed, a fact that we asserted earlier.

Figure 7.2 shows clearly that the sampling distribution of the sample mean is normally distributed if the population distribution is normal. It also shows that the mean of the sampling distribution of the sample mean is equal to the mean of the population distribution. In addition, the figure shows that the standard deviations of the sampling distributions are smaller than the standard deviation of the population distribution ($\sigma_{\bar{x}} = \sigma/\sqrt{n}$) and that the standard deviations of the sampling distributions decrease as the sample size n increases.

7.4
STANDARDIZED SAMPLE MEAN

We used Equation (6.3) to standardize the random variable X for a normal distribution in Section 6.3. To **standardize a sample mean** $\bar{X}$ for a sampling distribution, we use the following equation.

Equation for Standardized Sample Mean

Since $\bar{X}$ has mean μ and standard deviation $\sigma_{\bar{x}} = \sigma/\sqrt{n}$, the **standardized sample mean** is

$$Z = \frac{\bar{X} - \mu}{\sigma/\sqrt{n}} \qquad\qquad (7.7)$$

The distribution of the standardized sample mean has a mean of 0 and a variance of 1. If the population is normal, then Z, defined by Equation (7.7), is a standard normal variable; and we may use the table of normal areas to calculate probabilities for the sample mean.

▬ PRACTICE PROBLEM 7.1

The lengths of individual machined parts coming off a production line at Morton Metalworks are normally distributed around their mean of $\mu = 30$ centimeters. Their standard deviation around the mean is $\sigma = .1$ centimeter. An inspector just took a sample of $n = 4$ of these parts and found that $\bar{X}$ for this sample is 29.875 centimeters. What is the probability of getting a sample mean this low or lower if the process is still producing parts at a mean of $\mu = 30$?

Solution We know that sample means from normal populations are normally distributed around μ with a standard deviation of $\sigma_{\bar{x}} = \sigma/\sqrt{n}$. Knowing this fact, we can, as illustrated in Figure 7.3, standardize our sample mean of $\bar{X} = 29.875$ and use the normal table to find the answer we seek as follows:

$$P(\bar{X} \leq 29.875) = P\left(Z \leq \frac{\bar{X} - \mu}{\sigma/\sqrt{n}}\right) = P\left(Z \leq \frac{29.875 - 30.0}{.1/\sqrt{4}}\right)$$

$$= P\left(Z \leq \frac{-.125}{.05}\right) = P(Z \leq -2.50)$$

FIGURE 7.3

Determining the Probability
for Morton Metalworks

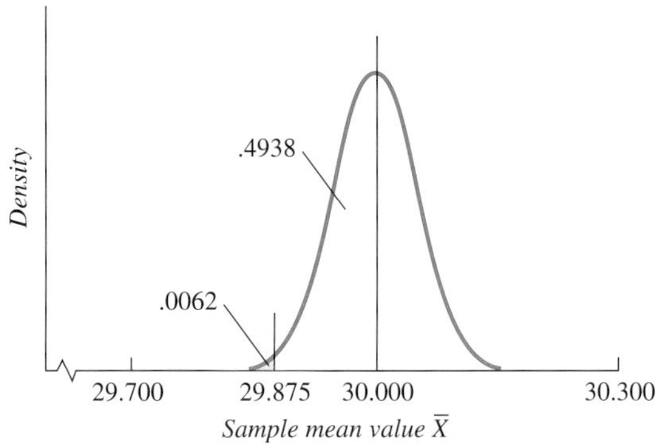

In Table V in Appendix C, we find that the area under the standard normal density function between $Z = 0$ and $Z = -2.50$ is .4938, so

$$P(\overline{X} \le 29.875) = P(Z \le -2.50) = .5000 - .4938 = .0062$$

The inspector could conclude from this result that either (1) a sample has been taken that occurs with very small probability or (2) the assumption that $\mu = 30$ is not valid.

PRACTICE PROBLEM 7.2

The increase in the yield of wheat (in bushels) when a pesticide is used on a plot of ground at California Agronomics Land & Farming, Inc., is normally distributed with mean $\mu = 50$ and variance $\sigma^2 = 100$. Find the probability that the mean increase in yield of a sample of 25 plots will differ from the population mean by less than 4 bushels.

Solution We want $P(-4 < \overline{X} - \mu < 4)$. As illustrated in Figure 7.4, if we divide each member of the inequality by $\sigma/\sqrt{n}$, we have*

$$P(-4 < \overline{X} - \mu < 4) = P\left(\frac{-4}{\sigma/\sqrt{n}} < \frac{\overline{X} - \mu}{\sigma/\sqrt{n}} < \frac{4}{\sigma/\sqrt{n}}\right)$$

$$= P\left(\frac{-4}{10/\sqrt{25}} < Z < \frac{4}{10/\sqrt{25}}\right)$$

$$= P(-2 < Z < 2) = .9544$$

---

*This step can be taken since the same positive amount has been divided into every term in the inequality string. In general, if $a \le b \le c$, then $a/d \le b/d \le c/d$ as long as $d > 0$. Also, the same amount can be added to or subtracted from each term in an inequality string without changing the relationship. Thus if $a \le b \le c$, then $a + d \le b + d \le c + d$. This fact is used in the solution here. Finally, the same positive amount can be multiplied by each term in an inequality string without changing the relationship. If $a \le b \le c$, then $ad \le bd \le cd$ as long as $d \ge 0$.

FIGURE 7.4

Determining the Probability
for the Yield of Wheat

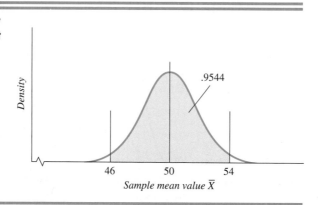

Sample mean value $\overline{X}$

PRACTICE PROBLEM 7.3

For the normal distribution of California Agronomics Land & Farming, Inc., for
Practice Problem 7.2, find two values equidistant from the mean such that 90% of the
means of all samples of size $n = 64$ will be contained between them.

Solution See Figure 7.5. From the table of normal areas, we have

$$.90 = P(-1.64 \le Z \le 1.64)$$

From Equation (7.7),

$$.90 = P\left(-1.64 \le \frac{\overline{X} - \mu}{\sigma/\sqrt{n}} \le 1.64\right)$$

Therefore, since $\sigma/\sqrt{n} = 10/8 = 1.25$,

$$.90 = P\left(-1.64 \le \frac{\overline{X} - 50}{1.25} \le 1.64\right) = P(-2.05 \le \overline{X} - 50 \le 2.05)$$

$$= P(50 - 2.05 \le \overline{X} \le 50 + 2.05) = P(47.95 \le \overline{X} \le 52.05)$$

The desired values are 47.95 and 52.05.

FIGURE 7.5

Determining the Values
Equidistant from the Mean

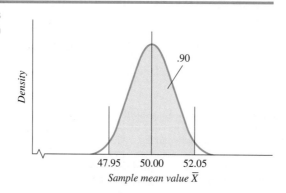

Sample mean value $\overline{X}$

We will use a procedure similar to the one illustrated in Practice Problem 7.3 to construct confidence intervals in the following chapter.

Problems: Sections 7.3–7.4

5. A sample of size $n = 100$ is drawn from a normally distributed population with mean $\mu = 30$ and standard deviation $\sigma = 20$.

   a. Find $P(\overline{X} \geq 34)$. **b.** Find $P(26 \leq \overline{X} \leq 34)$.

6. A sample of size $n = 49$ is drawn from a normally distributed population with mean $\mu = 10$ and standard deviation $\sigma = 21$.

   a. Find $P(\overline{X} \geq 13)$. **b.** Find $P(4 \leq \overline{X} \leq 13)$.

7. A food-processing company packages a product that is periodically inspected by the Food and Drug Administration (FDA). The FDA has ruled that the company's product may have no more than 2.0 grams of a certain toxic substance in it. Past records of the company show that packages of this product have a mean weight of toxic substance equal to 1.25 grams per package and that the weights are normally distributed around 1.25 grams with a standard deviation of .50 gram.

   a. What proportion of the individual packages exceed the FDA limit?
   b. What is the probability of selecting an individual package that has between 1.75 and 2.00 grams of toxic substance in it?
   c. A team of FDA inspectors is on its way to inspect a random sample of the company's output. The inspectors plan to take a preliminary sample of 25 packages. If they find that the mean weight of toxic substance in this sample exceeds 1.6 grams, they will close down the plant and have an extensive inspection of the company's entire inventory. What is the probability that they will close the plant?

8. The Bills Processing Company makes an item that it sells to the U.S. Navy. The item is called the KM–2. During the manufacturing process the individual KM–2s are placed in a baking oven. The time it takes to bake the items is normally distributed around a mean of 64 minutes with a standard deviation of 5 minutes. Thus, the population of baking times is normal in shape.

   a. If 100 KM–2s are baked, what is the probability that the mean baking time will be 64 minutes and 45 seconds or greater?
   b. What proportion of the individual KM–2s bake in 57 minutes or less?

9. Hunter Chemical Company claims that its major product contains on the average 4.0 fluid ounces of caustic materials per gallon. It further states that the distribution of caustic materials per gallon is normal and has a standard deviation of 1.3 fluid ounces.

   a. What proportion of the individual gallon containers for this product will contain more than 5.0 fluid ounces of caustic materials?
   b. A government inspector randomly selects 100 gallon-size containers of the product and finds the mean weight of caustic material to be 4.5 fluid ounces per gallon. What is the probability of finding the mean of a sample of 100 that is 4.5 or greater? Do you think the production process was producing its usual level of caustic materials when this sample was taken?

7.5
CENTRAL LIMIT THEOREM

Central limit theorems concern the approximate normality of means of random samples or of sums of random variables; we accordingly state central limit theorems in two forms.

Suppose $X_1, X_2, \ldots, X_n$ is a random sample from an infinite population of any shape with mean μ and variance σ^2; the X_i are independent random variables. The

first form of the **central limit theorem** is as follows.

Central Limit Theorem: First Form

If n is large, then

$$Z = \frac{\overline{X} - \mu}{\sigma/\sqrt{n}} \tag{7.8}$$

has approximately a standard normal distribution, or (in terms of what is the same) $\overline{X}$ has approximately a normal distribution with mean μ and standard deviation $\sigma_{\overline{X}} = \sigma/\sqrt{n}$.

We know that whatever the parent population may be, the standardized variable of Equation (7.8) has mean 0 and standard deviation 1; and we know that if the parent population is normal, then the variable of Equation (7.8) has exactly a standard normal distribution. The remarkable fact is that even if the parent population is *not* normal, the standardized mean is approximately normal if n is large. The importance of the theorem is that it permits us to use normal theory for inferences about the population mean regardless of the form of the population, provided only that the sample size is large enough.

The central limit theorem opens up an entirely new class of problems that we can solve by using the normal probability distribution and Table V of Appendix C. In Chapter 6, we found that the normal distribution could be used to find the probability that an *individual* measure would lie in a particular interval by using Table V, assuming that the measure was known to be normally distributed. However, the central limit theorem tells us that we can use the normal distribution to find the probability that a *sample mean* $\overline{X}$ will lie in a particular interval—regardless of the form of the population distribution, as long as the sample size we use is large enough (often in excess of $n = 30$). The use of the normal distribution to find probabilities for sample means is demonstrated in the next problem.

 PRACTICE PROBLEM 7.4

The probability density function shown in Figure 7.6 is a triangular distribution showing the probability that a delicate new medical device produced by KJ Industries will fail between 0 and 20 years after it is implanted in the human body. The mean time to failure is $\mu = 6.7$ years, and the standard deviation is $\sigma = 3$ years.

   a. Verify that this density function has a total area under it of 1.0.
   b. Find the probability that an individual device will fail 15 years or more after implantation.
   c. Find the probability that in a sample of $n = 36$ of these devices the sample mean time to failure $\overline{X}$ will be 8 years or less.

Solution
   a. To verify that the total area under the density function is 1.0, we must use the relationship for the area of a triangle: *Area* = $(1/2)(base)(height)$. In this case *Area* = $(1/2)(20)(.1) = (10)(.1) = 1.0$.

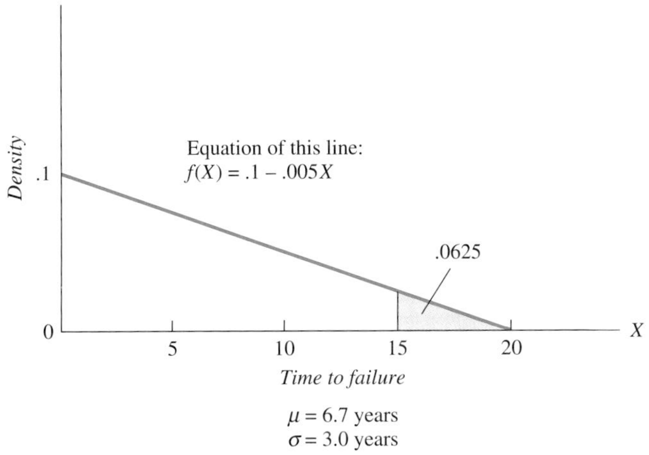

FIGURE 7.6

Distribution of Times to Failure for the Medical Device

b. Since this question concerns the probability that an *individual* device will fail after 15 years or more, we must use the probability distribution for individual times to failure given in Figure 7.6. The probability we seek is given by the shaded area under the density function from 15 to 20 years. This area is the area in a triangle of base$(20 - 15) = 5$. To find the height of the function at $X = 15$, we must find

$$f(15) = .1 - .005(15) = .025$$

And the area of the triangle from $X = 15$ out to $X = 20$ is

$$Area = (1/2)(5)(.025) = .0625$$

Thus the chance that one of these devices will last 15 years or longer is .0625, or 1 out of 16.

c. The question in part c asks about the probability that a sample mean will be 8 years or less. To answer this question, we must use the probability distribution shown in Figure 7.7. From Equation (7.2) we know that the mean of the $\bar{X}$'s for samples of size $n = 36$ is the population mean, $\mu = 6.7$ years. From Equation (7.4) we know that the distribution of the sample means has a standard deviation of

$$\sigma_{\bar{X}} = \frac{\sigma}{\sqrt{n}} = \frac{3}{\sqrt{36}} = .5$$

The central limit theorem tells us that even though the population of times from which the sample was drawn was triangular in shape (Figure 7.6), the shape of the distribution of sample means is almost normal since the sample size, $n = 36$, is large. Thus we seek the shaded area under the normal

FIGURE 7.7

**Distribution of Means of
Samples of *n* = 36 Devices**

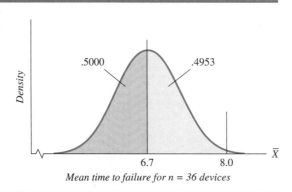

Mean time to failure for n = 36 devices

curve in Figure 7.7 to the left of $\overline{X} = 8$:

$$P(\overline{X} \le 8) = P\left(Z \le \frac{8 - 6.7}{3/\sqrt{36}}\right) = P\left(Z \le \frac{1.3}{.5}\right) = P(Z \le 2.6)$$

$$= .5000 + .4953 = .9953$$

Thus the chances are nearly certain that the mean of a sample of size $n = 36$ will be 8 years or less.

To *prove* the central limit theorem would require a full use of mathematical probability that is beyond the scope of this text. But we can illustrate the central limit theorem further by using the results of a sampling exercise. We start with two specific nonnormal distributions: a uniform distribution [Figure 7.8(a)] and a skewed distribution [Figure 7.8(b)].

Statistics students then took 1000 different random samples each for sample sizes of 4, 8, and 25 in succession from the uniform and skewed distributions, computed the sample means, and produced the empirical percentage histograms that are presented in Figure 7.8. Normal sampling distributions for the sample means have been superimposed on the empirical histograms. The plots indicate that the empirical histograms for the sample mean values approximate the normal distribution more closely as the sample size *n* increases. This sampling exercise demonstrates that the sampling distribution of the sample mean has an approximate normal distribution for large samples even when the population distribution is nonnormal, just as the central limit theorem indicates that it will.

Figure 7.8 also shows that the normal approximation is better for a symmetric population distribution (such as the uniform distribution) than it is for a skewed population distribution, especially for smaller sample sizes. As indicated by this sampling exercise, however, the sample need not be excessively large before we can feel reasonably safe in using the central limit theorem.

Figure 7.8 shows once again that the mean of a sampling distribution of the sample mean $\overline{X}$ is equal to the mean of the corresponding population distribution. It also shows that the standard deviations of the sampling distributions of the sample mean get smaller as the sample size increases, because $\sigma_{\overline{x}} = \sigma/\sqrt{n}$.

The material in this section deals with the probability distribution of sample means. The reader might ask, "Who cares about the distribution of sample means?"

FIGURE 7.8

Central Limit Theorem

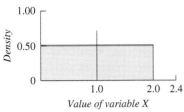

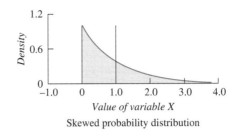

Uniform probability distribution

Skewed probability distribution

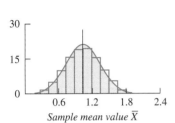

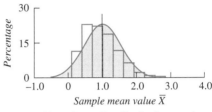

Normal sampling distribution, $n = 4$

Normal sampling distribution, $n = 4$

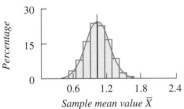

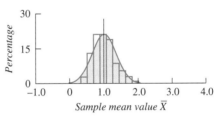

Normal sampling distribution, $n = 8$

Normal sampling distribution, $n = 8$

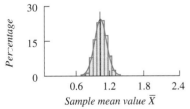

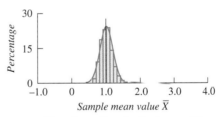

Normal sampling distribution, $n = 25$

Normal sampling distribution, $n = 25$

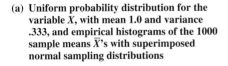

(a) **Uniform probability distribution for the variable X, with mean 1.0 and variance .333, and empirical histograms of the 1000 sample means $\overline{X}$'s with superimposed normal sampling distributions**

(b) **Skewed probability distribution for the variable X, with mean 1.0 and variance 1.0, and empirical histograms of 1000 sample means $\overline{X}$'s with superimposed normal sampling distributions**

The answer is that many people do. And the reason that they do is that in many probability applications we want to determine the probability of getting a sample result like the one we have just observed, and the sample result we have observed is very often a sample mean, an $\overline{X}$.

EXAMPLE 7.9 In Example 7.7 we considered a situation where the delay time for inspection of baggage at a border station had a mean of $\mu = 8$ minutes and $\sigma = 6$ minutes. The distribution of those times was very nonnormal and was shown in Figure 7.1. Assume that a representative of a particular minority group took a sample of $n = 64$ people from her group and found that their mean time to get through the inspection station was $\overline{X} = 10$ minutes. Would the representative be justified in suggesting that the minority group people were being delayed longer than one would expect if the minority group people are being processed the same as people in general? That is, we desire the probability that the sample mean, $\overline{X} \geq 10$, could occur by random chance alone.

In Example 7.7 we suggested that a sample mean of 10 minutes or more would be a rather unusual result. But now we have the tools to specify exactly how unusual this event would be. Figure 7.9 shows the distribution of sample means for samples of size $n = 64$ from a population with $\mu = 8$ and $\sigma = 6$. Our knowledge of the central limit theorem tells us that even though the population (shown in Figure 7.1) is not normal in shape, the distribution of sample means in Figure 7.9 is very close to normal in shape since the sample size $n = 64$ is large. Thus,

$$P(\overline{X} \geq 10) = P\left(Z \geq \frac{10 - 8}{6/\sqrt{64}}\right) = P\left(Z \geq \frac{2}{.75}\right) = P(Z \geq 2.67)$$

$$= .5000 - .4962 = .0038$$

Our knowledge of the central limit theorem has allowed us to calculate exactly how unusual a sample mean the minority group representative has observed, if the mean processing time for members of her group is the same as for people in general. This value of .0038 is so unusual that we would be likely to conclude that people from the minority group are being processed in a significantly longer time than the population mean of 8 minutes, and they might have some justification to complain about discrimination against them.

FIGURE 7.9

Distribution of Sample Mean Delay Times for Samples of $n = 64$

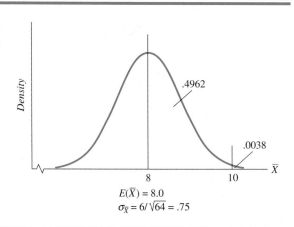

A second version of the central limit theorem concerns the sum ΣX_i of a set $X_1, X_2, \ldots, X_n$ of independent random variables all having the same distribution with mean μ and variance σ^2. The **second form** of the **central limit theorem** follows.

Central Limit Theorem: Second Form

If n is large, then

$$Z = \frac{\left(\sum_{i=1}^{n} X_i\right) - n\mu}{\sigma\sqrt{n}} \qquad (7.9)$$

has approximately a standard normal distribution, or (in terms of what is the same) ΣX_i has approximately a normal distribution with mean $n\mu$ and standard deviation $\sigma\sqrt{n}$.*

This theorem (together with more general versions of it) is one of the reasons why the normal distribution often arises in nature. If one performs a complicated physical measurement, the measurement error is the sum ΣX_i of many small independent random errors X_i, and the height of a plant is the sum ΣX_i of many small independent increments X_i. In such cases the normal distribution at least roughly approximates the distribution of the sum.

PRACTICE PROBLEM 7.5

The daily catch of a small tuna-fishing fleet from Ocean Foods Company averages 130 tons. The fleet's logbook shows that the weight of the catch varies from day to day, and this variation is measured by the standard deviation of the daily catch, $\sigma = 42$ tons. What is the probability that during a sample of $n = 36$ fishing days the total weight of the catch will be 4320 tons or more?

Solution We don't know the shape of the distribution of weight for individual daily catches, but we don't need to know it. The question concerns the probability that the *total* weight in 36 days is 4320 or more. Thus we can use Equation (7.9) and the second form of the central limit theorem, which says the sum of a large number of measurements is approximately normal. Figure 7.10 shows the distribution of the sum of $n = 36$ measurements from a population that has mean $\mu = 130$ and $\sigma = 42$. The probability we seek is in the shaded area:

$$P\left(\sum X \geq 4320\right) = P\left(Z \geq \frac{4320 - (36)(130)}{42\sqrt{36}}\right)$$

$$= P\left(Z \geq \frac{4320 - 4680}{252}\right)$$

$$= P(Z \geq -1.43) = .5000 + .4236 = .9236$$

---

*The second form of the central limit theorem is the same as the first, because ΣX_i is $n\overline{X}$, so the variable in Equation (7.9) is just $(n\overline{X} - n\mu)/(\sigma\sqrt{n})$, and algebra reduces this expression to the variable in Equation (7.8).

FIGURE 7.10

Distribution of Total Catch in $n = 36$ Days

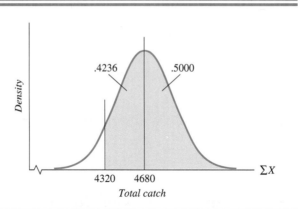

Thus the chances are quite good that the total catch will be 4320 tons or more in a sample of $n = 36$ days. The owner of this fleet might thus be willing to sign a contract to deliver that much tuna to his buyers over the next 36 days of fishing if he is willing to assume that the next 36 days will be a random sample and will have no unusual deviations from past fishing results.

As a final note in this section, we should point out that the normal approximation to the binomial, as treated in Section 6.4, is due to the central limit theorem. In an infinite (or a finite but large) population split into two categories, success and failure, label each element in the success category with a 1 and each element in the failure category with a 0. Now a random sample of size n from the population becomes converted into a set $X_1, X_2, \ldots, X_n$ of independent random variables, each having a value of either 0 or 1. The number of successes X is the number of 1s, which is $X = \Sigma X_i$, so the central limit theorem applies. In this case [see Equations (5.8) and (5.9)], the mean of the number of successes X is $E(X) = np$, and the standard deviation of the number of successes is

$$\text{Standard deviation}(X) = \sqrt{np(1 - p)}$$

The standardized number of successes [see Equation (6.4)] is

$$Z = \frac{X - np}{\sqrt{np(1 - p)}}$$

and is the same as the standardized variable in Equation (7.9). [The corresponding terms are $X = \Sigma X_i$, $np = n\mu$, and $\sqrt{np(1 - p)} = \sigma \sqrt{n}$.] The standardized variable Z has approximately a standard normal distribution if n is reasonably large. Also recall that the fraction of successes $\hat{p}$, where $\hat{p} = X/n$, has mean $E(\hat{p}) = p$ and standard deviation $\sigma_{\hat{p}} = \sqrt{p(1 - p)/n}$ [see Equations (6.5) and (6.6)] and that the standardized fraction of successes [see Equation (6.7)] is

$$Z = \frac{\hat{p} - p}{\sqrt{p(1 - p)/n}}$$

Due to Equation (7.9), the standardized fraction of successes has approximately a standard normal distribution when n is reasonably large.

Problems: Section 7.5

10. A sample of size $n = 169$ is drawn from a skewed infinite population with mean $\mu = 30$ and standard deviation $\sigma = 26$.

    a. Find $P(\overline{X} \geq 34)$.
    b. Why is the sampling distribution of the sample mean $\overline{X}$ approximately normally distributed for this problem?

11. A sample of size $n = 81$ is drawn from a skewed infinite population with mean $\mu = 10$ and standard deviation $\sigma = 18$.

    a. Find $P(\overline{X} \geq 12)$.
    b. Why is the sampling distribution of the sample mean $\overline{X}$ approximately normally distributed for this problem?

12. The probability distribution of six-month incomes of account executives has a mean of $20,000 and a standard deviation of $5000.

    a. A single account executive is selected at random, and the executive's income is $20,000. Can it be said that this executive's income exceeds 50% of all account executive incomes?
    b. A random sample of 64 account executives is taken. What is the probability that the sample mean value exceeds $20,500?
    c. What part of this problem requires the use of the central limit theorem?

13. Herculean, Inc., a producer of composite (graphite) material, uses a process that in the past has resulted in material that breaks, on the average (mean), when submitted to a 100-pound load. The standard deviation is 10 pounds. A prospective user of the material plans to use the producer as a supplier if the mean breaking strength of the material exceeds 95 pounds. The prospective user wants to check the breaking strength of the material independently, so she sends an employee of her quality control department to the production plant to randomly select 36 units of the material and check the mean breaking strength. What is the probability that the sample mean will exceed the prospective graphite material user's requirement, given the past history of the material?

14. Time lost at work from employee absenteeism is an important problem for many companies. The human resources department of Western Electronics has studied the distribution of time lost from absenteeism by individual employees. During a one-year period the department found a mean of 21 days and a standard deviation of 10 days. A group of 49 employees is selected at random to participate in a program that allows a flexible work schedule, which the human resources department hopes will decrease the amount of absenteeism (in the future).

    a. What is the probability that the sample mean value for time lost from absenteeism for this group of employees exceeded 21 days?
    b. What is the probability that the sample mean value was between 19 and 23 days?
    c. What values that are an equal distance from the mean amount of time lost from absenteeism encompass 95% of the possible sample mean values?

    REFERENCE: Driver, Russell, and Collin J. Watson (1989), "Construct Validity of Voluntary and Involuntary Absenteeism," *Journal of Business and Psychology,* 4: 109–18.

15. A tourist board reported that the distribution of daily expenditures for vacationing skiers at Greatest Snow Ski Resort has a mean of $100 and a standard deviation of $18.

    a. Do you think that the distribution of vacationing skier daily expenditures is skewed? Explain.

b. Assuming that the board's claim is true, describe the sampling distribution of the sample mean daily expenditure per skier for 36 randomly selected vacationing skiers.

c. Assuming that the board's claim is true, determine the probability that the sample mean daily expenditure per skier for the sample of 36 vacationing skiers will exceed $97.

d. Assuming that the board's claim is true, determine the probability that the sample mean daily expenditure for the sample of 36 vacationing skiers will exceed $103.

e. Do you think that the board's claim is true, given that the sample mean daily expenditure for the 36 sampled skiers turned out to be $106? Explain.

16. The distribution of annual returns for common stocks listed on the New York and American Stock Exchanges historically has been approximately symmetrically distributed with a mean of 9.6% and a standard deviation of 21.4%. However, the distribution has heavier tails than the tails for a normal distribution. A publisher of an investing newsletter takes a random sample of 36 individual investors who subscribed to the newsletter and found that the sample mean annual return for common stocks held in the portfolios of the investors was 7%.

a. Find the probability of a sample mean return of 7% or less if the mean returns for the newsletter subscribers have a true mean return equal to that of the stocks listed on the exchanges.

b. Is the sampling distribution for the sample mean exactly normally distributed or approximately normally distributed? Why?

c. Would you conclude that the true mean return for the newsletter subscribers is less than that of the stocks listed on the exchanges?

REFERENCE: Malkiel, B. (1985), *A Random Walk Down Wall Street* (New York: Norton).

17. Seventy percent of newly organized small businesses experience cash flow problems during their first year of operation. A consultant for the Small Business Administration takes a random sample of 50 small businesses that have been in operation for one year.

a. Determine the mean and the variance of the fraction of new small businesses that experience cash flow problems during the first year of operation.

b. Determine the probability that more than 80% of the businesses sampled have experienced cash flow problems.

18. Seventy-five percent of a school's law class passes the state bar examination on the first attempt. If a randomly selected group of 60 of this school's law graduates take the examination, what is the probability that 80 % or more of them will pass on the first attempt?

19. A large mining company has been forced to terminate 10% of its work force because of low market prices for precious metals (employees are therefore either terminated or not terminated). A group of mature (age 55 or older) terminated employees has filed a lawsuit claiming that age discrimination was practiced by the company during the termination process. Analysts have determined that 12 of the 64 (12/64 = .1875) employees age 55 or older in the total work force were terminated.

a. Determine the probability that the sample fraction of age 55 or older employees who were terminated would be .1875 or greater if employees were selected at random for termination.

b. Does it appear that age discrimination may have occurred during the selection of employees for termination?

c. Determine the probability that the sample fraction of age 55 or older employees who were terminated would be between .0625 and .125 if employees were selected at random for termination.

Chapters 4 through 7 of the text have dealt with the subject of probability. Subsequent chapters will cover the area of statistics known as statistical inference. Many of the realistic problems of statistics lie in this latter area.

To work a probability problem, we need to know the population parameters μ, σ, or p. Then we can ask questions about the probability of obtaining certain sample results.

In most realistic situations, however, we do not know the population parameters μ, σ, or p. We usually have sample results like $\overline{X}$, s, or $\hat{p}$. From these sample results, we often want to estimate the population characteristics. However, to do so, we must use the knowledge of probability we have gained in these chapters. Thus before you proceed to the subject of statistical inference, a good grasp of the material covered in this chapter is essential.

FIGURE 7.11

Probability Summary Chart

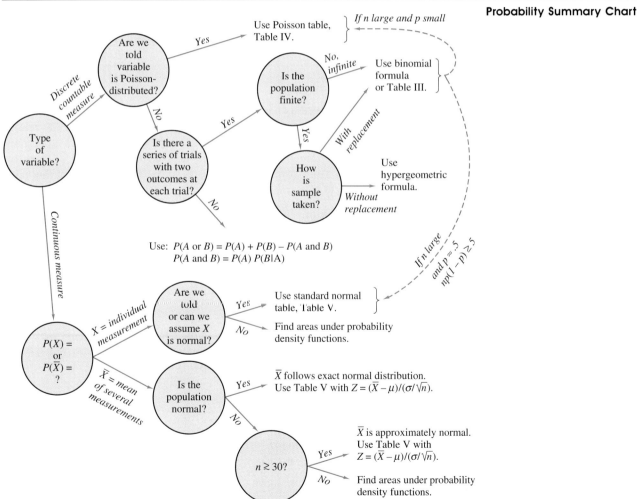

TABLE 7.1 Summary of Important Sampling Distributions

Sampling Situation	Expected Value of Sample Statistic	Standard Deviation of Sampling Distribution	Shape of the Sampling Distribution
One sample of n items from a continuous population with mean μ and standard deviation σ	$E(\overline{X}) = \mu$, where the sample statistic is $\overline{X}$, sample mean	$\sigma_{\overline{x}} = \dfrac{\sigma}{\sqrt{n}}$	—Exactly normal if the population is normal in shape —Approximately normal if the population is not normal but n is large, due to central limit theorem (C.L.T.)
One sample of n items from a binomial population with a proportion of successes p	$E(\hat{p}) = p$, where the sample statistic is $\hat{p} = X/n$, sample proportion of successes	$\sigma_{\hat{p}} = \sqrt{\dfrac{p(1-p)}{n}}$	—Approximately normal if n is large, due to C.L.T. *Rule of thumb:* $np(1-p) \geq 5$

In Chapter 6, we reviewed a chart that showed the logic of how to solve a probability problem. That chart was presented as Figure 6.33. Now we can use the knowledge we have gained in this chapter concerning the central limit theorem to expand the chart. The revised Figure 6.33 is presented here as Figure 7.11. The bottom branches of the chart summarize the material covered by the central limit theorem.

Table 7.1 summarizes two important probability situations that will be used often in the next two chapters. These concepts were covered in this chapter and in Section 6.4. These concepts are basically outgrowths of the central limit theorem, the most important theorem in all of statistics.

REVIEW PROBLEMS

Answers to odd-numbered problems are given at the end of the text.

20. If many samples of size 100 (that is, each sample consists of 100 items) were taken from an infinite normal population with mean 10 and variance 16, what would be the mean and the variance of these samples' means?

21. If many samples of size 64 were taken from a continuous nonnormal population with mean of 50 and standard deviation of 20, what would be the mean and the standard deviation of these samples' means?

22. A sample of size $n = 16$ is drawn from a normally distributed population with mean $\mu = 20$ and standard deviation $\sigma = 8$.
    a. Find $P(\overline{X} \geq 24)$.   **b.** Find $P(16 \leq \overline{X} \leq 24)$.

23. A sample of size $n = 144$ is drawn from a skewed infinite population with mean $\mu = 30$ and standard deviation $\sigma = 20$.
    a. Find $P(\overline{X} \geq 34)$.
    b. Why is the sampling distribution of the sample mean $\overline{X}$ approximately normally distributed for this problem?

24. A sample of size $n = 64$ is drawn from skewed infinite population with mean $\mu = 40$ and standard deviation $\sigma = 16$.
    a. Find $P(\overline{X} \geq 44)$.
    b. Why is the sampling distribution of the sample mean $\overline{X}$ approximately normally distributed for this problem.

25. What differences, if any, will there be between the distributions of the sample means in Problems 20 and 21?

26. Corporate performance plays an important role in productivity, quality, and living standards. Motorola, Inc., was awarded the first annual Malcolm Baldrige National Quality Award (the United States counterpart to Japan's Deming Prize) for attaining preeminent leadership in quality. Motorola has automated its factories, eliminated barriers in the work place and trained workers to improve company performance. Computer-integrated manufacturing was instituted to reduce the time it takes to get new products from design, through production, and to the market. Historically, Motorola took 3 to 5 years to bring new products to market. Suppose that the time was normally distributed with a mean of 4 years and a standard deviation of .3 year. With computer-integrated manufacturing, the times in years that it took the company to bring a random sample of four new products to market were 1, 3, 1, and 3 years.
    a. Find the probability of a sample mean time in years for bringing a product to market being less than or equal to that found with computer-integrated manufacturing if the actual mean time continues to be 4 years.
    b. Does it appear that the mean time for bringing a product to market has decreased with computer-integrated manufacturing?

    REFERENCE: Henkoff, R. (1989), "What Motorola Learns from Japan," *Fortune* (April 24).

27. The increase in the yield of corn (in bushels) when a fungicide is used on a plot of ground at Nebraska Land & Farming, Inc., is normally distributed with mean 50 and standard deviation 12. Find the probability that the sample mean increase in yield of a sample of 36 plots will differ from the population mean by less than 4 bushels.

28. Digital industries uses statistical quality control to monitor the production process during the manufacture of memory chips. The electrical charge across a conducting zone for a memory chip is normally distributed with a mean of 10.05 microvolts and the standard deviation is .92 microvolt. Digital takes samples of 4 chips and if the sample mean electrical charge is between 8.67 and 11.43 microvolts then the process is considered to be in control. Find the probability that a random sample of 4 chips will result in the process being declared in control.

29. In a manufacturing process at Newco, Inc., the standard length of a machined part is 100 centimeters. From past measurements we know that when the process is in adjustment, this measurement is a random variable with a mean of 100 centimeters and a standard deviation of .5 centimeter. The individual measurements are normally distributed around their mean.
    a. What is the probability that a randomly selected part will have a length longer than 100.2 centimeters?
    b. If many samples of size 25 were taken from this process, what would be the mean and the standard deviation of these samples' means?
    c. What probability distribution would the distribution of these sample means follow?
    d. If a single sample of size 25 is taken, what is the probability that the mean length of the parts in this sample will exceed 100.2 centimeters?
    e. What is the difference between the problems in parts a and d?

30. The life of a nickel-cadmium battery produced by Batterypac is normally distributed with a mean of 20 hours and a standard deviation of 10 hours. Batterypac guarantees that the average life of the batteries in a case of 24 batteries exceeds 16 hours. Find the probability that a randomly selected case of batteries meets the guarantee.

31. The weight of one-ton nominal cold-rolled slabs of steel produced by Ganiving Steel, Inc., is normally distributed with a mean of 1950 pounds and a standard deviation of 50. Fisher Bodies recently purchased 4 one-ton slabs from Ganiving. Find the probability that the average weight of the 4 slabs is less than 2000 pounds.

32. A type of cathode ray tube produced at Amp, Inc., has a mean life of 10,000 hours and a variance of 3600. A few tubes have extraordinarily long lives, so the distribution is slightly skewed. If we take samples of 36 tubes each and for each sample we find the mean life, between what limits (symmetric with respect to the mean) would 50% of the sample means be expected to lie?

33. The population of times that it takes employees to perform an assembly task at Custom Motors, Inc., is skewed, with μ equal to 3 minutes and σ equal to .2 minute; and we test samples of 36 employees. Find the time that would be exceeded by 95% of the sample means.

34. The probability density function $f(X) = .1 - .005X$ is for a triangular distribution that defines the probability that a delicate new medical device produced by KJ Industries will fail between 0 and 20 years after it is implanted in the human body. The mean time to failure is $\mu = 6.7$ years, and the standard deviation is $\sigma = 3$ years. Find the probability that in a sample of $n = 81$ devices the mean time to failure $\overline{X}$ will be six years or less.

35. The accounting vice president for Ingersoll Sand Company indicates that from her experience the mean amount of error per invoice is 50 cents with a standard deviation of 12 cents and that the distribution is skewed to the right. An internal auditor randomly samples 36 invoices and computes the sample mean amount of error.

    a. Determine the probability that the sample mean amount of error is less than 52 cents.

    b. What two values that are an equal distance on either side of the mean will encompass 95% of all possible sample means?

    c. The sample mean computed by the auditor is 56 cents. Determine the probability of finding a sample mean this high or higher given the values specified by the vice president.

36. Rework Problem 35 assuming that the internal auditor has used a sample size of 144 (rather than 36).

37. Capital structure decisions are made by the managers of corporations. One such decision involves the relative amounts of debt and equity financing. Consider the list of companies from the surface transportation industry shown in Table 7.2. This table gives each company's debt as a percentage of equity. Assume that the companies on the list are a random sample of all of the numerous surface transportation companies and that the standard deviation is equal to the sample standard deviation.

    a. Find the sample mean and the standard error of the sample mean for debt as a percentage of equity. Assume that the sample standard deviation is equal to the population standard deviation.

    b. A financial analyst has told you that surface transportation companies on average use equal amounts of debt financing and equity financing, or 100% debt as a percentage of equity. Find the probability of finding a sample mean as large as or larger than the sample mean for the data, assuming that the analyst is correct. Is the analyst's statement questionable in light of this probability?

    c. Did your solution for this problem require the central limit theorem? Provide an explanation for your answer.

REFERENCE: Stowe, J. D., C. J. Watson, and T. Robertson (1980), "Relationships Between the Two Sides of the Balance Sheet," *Journal of Finance,* 35 (September): 973–80.

38. An auditor has been retained to determine whether errors may have occurred during

Company	Debt as % of Equity	Company	Debt as % of Equity	**TABLE 7.2**
Trucks and Leasing		**Railroads**		**Debt as a Percentage of Equity for Surface Transportation Companies**
Minstar	175.9	Kansas City Southern	79.3	
Carolina Freight	70.6	Burlington Northern	81.2	
Mayflower Group	106.7	Norfolk Southern	16.6	
Roadway Services	0.0	Union Pacific	52.9	
Consol Freightways	9.3	Santa Fe Southern	27.9	
Ryder System	168.3	CNW	137.4	
Leaseway Transport	86.8	American Standard	18.3	
Yellow Freight System	17.7	IC Industries	100.7	
Household International	368.3	CSX	69.5	
Arkansas Best	126.0	Soo Line	124.7	
Gelco	548.9	GATX	591.5	
Dorsey	30.6	**Other Surface**		
RLC	281.9	McLean Industries	892.2	
IU International	90.9	Alexander & Baldwin	32.7	
		Greyhound	58.3	
		American President	62.8	

SOURCE: *Forbes*, January 12, 1987, pp. 210–11.

the payment of invoices at CO-OP Energy, Inc., due to data processing difficulties for accounts payable. The distribution of payment errors due to data processing difficulties historically has been approximately normally distributed with a standard deviation of $2 per invoice. An audit of a random sample of payments was conducted and the payment errors for the sample are given in the following listing:

5   1   3   −1   3   1   5   0   1

a. Find the probability of a sample mean overpayment being greater than that found in the audit if the mean payment error owing to data processing difficulties during the billing process is zero.

b. Does it appear that the mean payment error due to data processing difficulties is not zero?

39. Annual percentage yields for money market funds during a period of average economic growth had a mean of 8% and a standard deviation of 2%. Annual yields were approximately normally distributed. A random sample of money market funds resulted in the percentage yields listed below. Find the probability of a randomly selected portfolio of 7 money market funds resulting in a sample mean annual yield less than that for the sample.

7.77   8.23   7.82   7.74   8.06   8.10   8.17

REFERENCE: National Association of Security Dealers, Inc. (1992), *Money Market Summary* (New York).

40. Excess Engine Wear, Inc., uses an automatic filling machine to fill plastic one-quart containers with 10-40 weight motor oil. The volume of oil in the containers has a mean of .9 quart and a standard deviation of .05. Due to imperfections in the filling machine, the containers overflow on occasion.

a. Is it likely that 50% of the containers have more than .9 quart of oil?

b. What is the probability that the mean volume of oil in a case of 24 containers selected at random is less than .92 quart.

41. Excess returns for stocks are determined by finding the difference between the return for a stock and the returns of firms in the market that have similar levels of risk. Stocks listed on the Upper Isthmus Exchange have a mean monthly excess return of .002 and a standard deviation of .004. Monthly excess returns on the Upper Isthmus Exchange are approximately normally distributed. Find the probability of a randomly selected portfolio of 16 stocks resulting in a sample mean monthly excess return less than zero.

REFERENCE: Lubatkin, M., K. Chung, R. Rogers, and J. Owers (1989), "Stockholder Reactions to CEO Changes in Large Corporations," *Academy of Management Journal*, 32: 47–68.

42. Fifty-five percent of Pacific Labs, Inc., customers are women. Many samples of $n = 200$ were taken from this company's customer list.
   a. What is the expected proportion of women customers in these samples over the long run?
   b. What is the standard deviation of the proportion of women customers found in these samples?

43. Digital industries uses statistical quality control to monitor the production process during the manufacture of memory chips. Historically, Digital's manufacturing process has produced 5.5% defective chips. Digital takes random samples of 200 chips and if the fraction of chips that are defective in the sample is between .007 and .103, then the process is considered to be in the control limits. Find the probability that a random sample of chips will result in the process being declared in the control limits if it is presently producing 5.5% defective chips.

44. A company that assembles electronic components has been forced to terminate 40% of its employees because international competitors have taken over a great deal of the assembly business due to lower labor costs. Consequently, each employee is either terminated or not terminated. A group of minority employees that was terminated has filled a lawsuit claiming that discrimination was practiced by the company during the termination process. Analysts have determined that 100 of the 200 ($100/200 = .50$) minority employees were terminated.
   a. Determine the probability that the sample fraction of minority employees who were terminated would be .50 or greater if employees were selected at random for termination.
   b. Does it appear that discrimination was practiced during the termination process?

45. Historically, Upton, Inc., has had a 25% market share in the Pharmaceutical Preparations Industry. A random sample taken recently of 200 pharmaceutical preparation users resulted in 30 customers who used Upton products.
   a. Determine the probability that the sample fraction of pharmaceutical preparation consumers would be less than or equal to that found during sampling if Upton continues to have a 25% market share.
   b. Would it be reasonable to conclude that Upton's market share has decreased?

46. Eighty percent of a school's accounting class passes the state CPA examination on the first attempt. If a randomly selected group of 100 of this school's graduates take the examination, what is the probability that 85% or more of them will pass on the first attempt?

47. A production line at Spectrum-Physics produces 10% defective items. If a sample of $n = 64$ items is taken, what is the probability that 5 or fewer items in this sample will be defective? [*Hint:* This problem uses the normal approximation of a binomial distribution, which is a special case of the central limit theorem. Use the continuity correction.]

48. A light bulb manufacturer claims that 90% of the bulbs it produces meet tough, new standards imposed by the Consumer Protection Agency. You just received a shipment containing 400 bulbs from this manufacturer. What is the probability that 375 or more of the bulbs in your shipment meet the new standards? [See the hint in Problem 47.]

Refer to the 141 companies listed in the Ratio Data Set in Appendix A.

49. Locate the data for the current assets–sales ratio (CA/S). Assume that the companies on the list are a random sample of all of the numerous possible companies.

   a. Find the sample mean and the standard error of the sample mean for the ratio assuming that the sample standard deviation is equal to the population standard deviation.
   b. A financial analyst has told you that companies on average have CA/S equal to 1.40. Find the probability of finding a sample mean smaller than the sample mean for the data assuming that the analyst is correct.
   c. Is the analyst's statement questionable in light of this probability?

50. Locate the data for the quick assets–sales ratio (QA/S). Assume that the companies on the list are a random sample of all of the numerous possible companies.

   a. Find the sample mean and the standard error of the sample mean for the ratio assuming that the sample standard deviation is equal to the population standard deviation.
   b. A financial analyst has told you that companies on average have QA/S equal to 1.40. Find the probability of finding a sample mean smaller than the sample mean for the data assuming that the analyst is correct.
   c. Is the analyst's statement questionable in light of this probability?

51. Locate the data for the net income–total assets ratio (NI/TA). Assume that the companies on the list are a random sample of all of the numerous possible companies.

   a. Find the sample mean and the standard error of the sample mean for the ratio assuming that the sample standard deviation is equal to the population standard deviation.
   b. A stockbroker trying to sell stock for the companies has told you that companies on average have NI/TA equal to .04. Find the probability of finding a sample mean smaller than the sample mean for the data assuming that the broker is correct.
   c. Is the broker's statement questionable in light of this probability?

52. Refer again to the data for NI/TA. Some of the companies suffered financial losses as reflected by negative values for the ratio. Assume that the companies on the list are a random sample of all of the numerous possible companies.

   a. Find the sample proportion of companies that suffered losses.
   b. A stockbroker trying to sell stock for the companies has told you that 85% of the companies in the population are profitable or at least have not suffered losses. Find the probability of finding a sample proportion of companies that have suffered losses smaller than the sample proportion for the data assuming that the broker is correct.
   c. Is the broker's statement questionable in light of this probability?

Refer to the 113 applicants for credit listed in the Credit Data Set in Appendix A.

53. Find the sample mean and standard deviation of age for the applicants who were granted credit. Find the probability that a sample mean age for a random sample will be greater than or equal to the sample mean age of the applicants who were granted credit if the actual mean age for the population of people granted credit from which the data were drawn in 35 years. (Assume that the variance computed from the sample is equal to the population variance.)

54. Assume that the proportion of women in the population of people who were denied credit is .30. Find the sample fraction of women who were denied credit from the people in the data set who were denied credit. Find the probability that the sample

fraction will be greater than or equal to the fraction found in the data set, assuming that the people in the data set who were denied credit is a random sample from the population of all people denied credit.

CASE 7.1   **SAMPLING INVENTORY FOR LAST-IN–FIRST-OUT VALUATION**

Many companies use the last-in–first-out (LIFO) inventory method because it can be advantageous for income tax purposes. However, to determine end-of-year inventories is a difficult task. American Stores, Inc., has warehouses and retail stores throughout the nation that contain hundreds of thousands of inventory items. Not only are the items physically inaccessible, but the thousands of items are purchased at different costs.

In many studies we take random samples from populations to make inferences or estimate facts about populations. Samples are often used when a data set is very large or when the items in the population are physically inaccessible. If a test destroys an item, then a

---

MINITAB   **COMPUTER EXHIBIT 1A** Random Samples of Companies and Current Assets–Sales Ratios from the Ratio Data Set in Appendix A

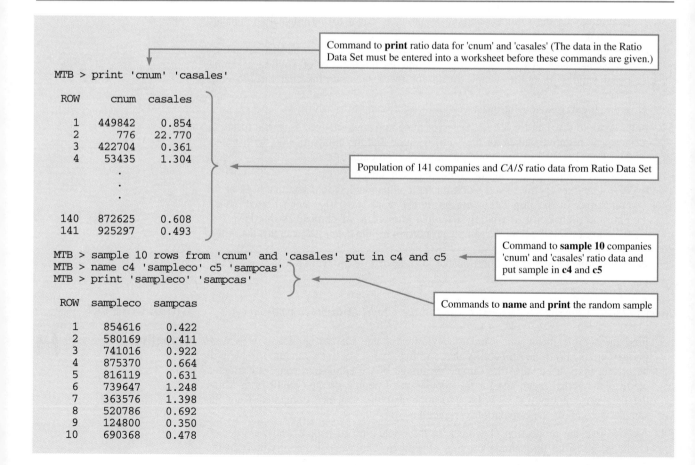

Command to **print** ratio data for 'cnum' and 'casales' (The data in the Ratio Data Set must be entered into a worksheet before these commands are given.)

```
MTB > print 'cnum' 'casales'

ROW      cnum   casales

  1    449842     0.854
  2       776    22.770
  3    422704     0.361
  4     53435     1.304
          .
          .
          .
140    872625     0.608
141    925297     0.493
```

Population of 141 companies and *CA/S* ratio data from Ratio Data Set

```
MTB > sample 10 rows from 'cnum' and 'casales' put in c4 and c5
MTB > name c4 'sampleco' c5 'sampcas'
MTB > print 'sampleco' 'sampcas'
```

Command to **sample 10** companies 'cnum' and 'casales' ratio data and put sample in **c4** and **c5**

Commands to **name** and **print** the random sample

```
ROW  sampleco  sampcas

  1    854616     0.422
  2    580169     0.411
  3    741016     0.922
  4    875370     0.664
  5    816119     0.631
  6    739647     1.248
  7    363576     1.398
  8    520786     0.692
  9    124800     0.350
 10    690368     0.478
```

sample of items must be used. If data collection is costly, then a sample can be used to obtain information that is sufficiently accurate for most practical purposes. The data in a sample may be even more accurate than population data if greater control over errors can be achieved during sampling than during a complete census.

The Internal Revenue Service allows American Stores to use sampling to estimate its cost of goods sold by the LIFO method. Accountants sample inventory items from the inventory data base by using the Statistical Analysis System (SAS). The LIFO method reduces income tax liabilities by hundreds of thousands of dollars when inventory valuations are increasing.

Computers are often used to select a random sample from a large data file. The random sample may then be analyzed or additional information may be collected for just the sampled items. A data set containing information on financial ratios for 141 companies is given in Appendix A. To illustrate random sampling using data analysis software, we obtained random samples of companies and their current assets–sales ratio data from the Ratio Data Set in Appendix A, and the results are presented in Computer Exhibits 1A, 1B, and 1C.

COMPUTER EXHIBIT 1B Random Samples of Companies and Current Assets–Sales Ratios from the Ratio Data Set in Appendix A

SAS

SAS program listing to obtain a random sample

```
TITLE 'SAMPLE FINANCIAL RATIO DATA';
DATA SAMPLE1;
INPUT CNUM CAS;
     IF RANUNI(0) <= .07;
CARDS;
449842    0.854
000776   22.770
422704    0.361
053435    1.304
   .        .
   .        .
   .        .
872625    0.608
925297    0.493
;
PROC PRINT;
    TITLE 'SAMPLE1 DATA';
    VAR CNUM CAS;
RUN;
```

IF statement and RANUNI function to include about 7% of the companies in the sample

```
        SAMPLE1 DATA

   OBS    CNUM     CAS

    1    449842   0.854
    2    688222   0.369
    3    949391   0.722
    4    317312   0.535
    5    544118   0.414
    6    981904   0.499
    7    718009   0.227
    8    220291   1.592
    9    934436   0.486
```

Companies sampled from Ratio Data Set

SPSS/PC+

COMPUTER EXHIBIT 1C Random Samples of Companies and Current Assets–Sales Ratios from the Ratio Data Set in Appendix A

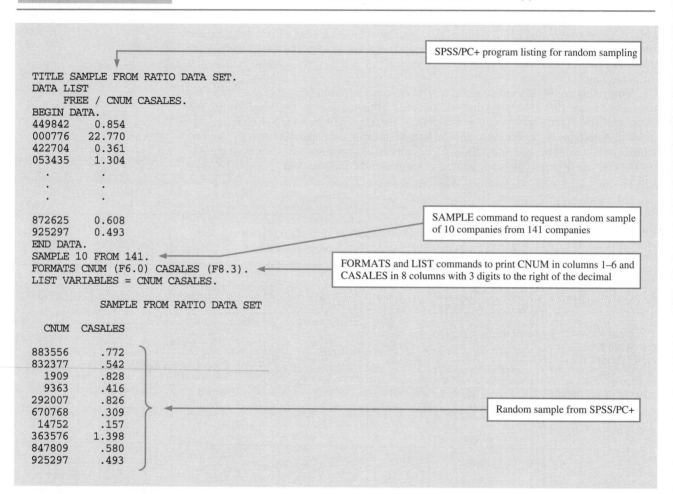

```
                                                    ┌─────────────────────────────────────┐
                                                    │ SPSS/PC+ program listing for random sampling │
                                                    └─────────────────────────────────────┘
TITLE SAMPLE FROM RATIO DATA SET.
DATA LIST
       FREE / CNUM CASALES.
BEGIN DATA.
449842    0.854
000776   22.770
422704    0.361
053435    1.304
  .         .
  .         .
  .         .

872625    0.608                    ┌─────────────────────────────────────┐
925297    0.493                    │ SAMPLE command to request a random sample │
END DATA.                          │ of 10 companies from 141 companies   │
SAMPLE 10 FROM 141.                └─────────────────────────────────────┘
FORMATS CNUM (F6.0) CASALES (F8.3).   ┌─────────────────────────────────────────┐
LIST VARIABLES = CNUM CASALES.        │ FORMATS and LIST commands to print CNUM in columns 1–6 and │
                                      │ CASALES in 8 columns with 3 digits to the right of the decimal │
                                      └─────────────────────────────────────────┘

          SAMPLE FROM RATIO DATA SET

   CNUM   CASALES

  883556    .772
  832377    .542
    1909    .828
    9363    .416
  292007    .826                    ┌─────────────────────────┐
  670768    .309                    │ Random sample from SPSS/PC+ │
   14752    .157                    └─────────────────────────┘
  363576   1.398
  847809    .580
  925297    .493
```

Case Assignment

Use a computer to obtain a random sample of about ten companies from the Ratio Data Set in Appendix A.

REFERENCES: Minitab, Inc. (1988), *Minitab Reference Manual* (State College, Penn.: Minitab, Inc); SAS Institute, Inc. (1987), *SAS Applications Guide* (Cary, N.C.: SAS Institute, Inc.); SPSS, Inc. (1991), *SPSS/PC+ Users Guide* (Chicago, Ill.: SPSS, Inc.); and Internal Revenue Service (1984), "Position Paper on LIFO Inventory Valuation Methods" (Washington, D.C.: Statistical Methods Group, Research Division).

CASE 7.2

DETERMINING THE PROBABILITY OF OVERPAYMENTS TO HEALTH CARE PROVIDERS

Health care costs have increased dramatically in the past decade. The percentage of the nation's gross national product that is consumed for health care is approaching 20%, the highest of any country. Health services providers are reimbursed by a national health in-

FIGURE 1

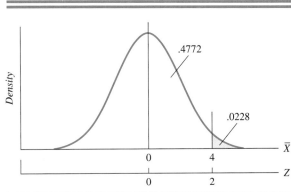

Sampling Distribution of the Mean Overpayment to a Health Care Provider

surance organization for an inordinately large number of services. It appears that overpayments may have occurred on many of the reimbursements due to errors made during the billing process or due to upcoding. Upcoding occurs when providers bill for a more complicated service than is actually provided. It is simple and lucrative for health care services providers to upcode because they only need to check a box on the insurance forms for a more complicated service than the service that was actually provided.

In the case *Illinois Physicians Union v. Illinois Department of Public Aid,* the U.S. Court of Appeals, Seventh Circuit, held that the use of a sampling and extrapolation auditing procedure is not arbitrary, capricious, or invidiously discriminatory, if the provider has the opportunity to rebut the determination of an overpayment. The court ordered the provider to pay the department the full amount of recoupment for the overpayments that had been received.

In one state, overpayments to health care providers in the past have been skewed with a standard deviation of $12 per service provided. A health services provider undergoing an audit contended that no overpayments had occurred on average or, in other words, that the actual mean overpayment was $0. A random sample of 36 reimbursements was conducted and the sample mean overpayment was $4.

To assess the validity of the provider's contention, the state wanted to find the probability of a sample mean overpayment being greater than or equal to that found in the audit if the provider's claim is true. If the provider's contention that the average overpayment is $0, then the mean and standard deviation of the sampling distribution of the sample mean are $E(\overline{X}) = \$0$ and Standard deviation$(\overline{X}) - 12/\sqrt{36} = 2$. The standardized value of the sample mean $\overline{X} = \$4$ is

$$Z = \frac{4 - 0}{2} = 2$$

As depicted in Figure 1, the probability is

$$P(\overline{X} \geq 4) = P(Z \geq 2) = .5000 - .4772 = .0228$$

Does it appear that the provider either makes an error in favor of the provider or upcodes on the average on reimbursement claims? Yes, since the probability, .0228, of getting a sample mean of $4 or more under the assumption that the true mean overpayment is $0 is so small.

a. Find the probability that the sample mean overpayment for the audited reimbursements would be between $2 and $4 if the provider's claim is correct.

Case Assignment

b. Find the probability that the total overpayment to the provider for the sampled reimbursements would be greater than or equal to $144 if the provider's claim is correct.

c. Does this problem require the central limit theorem? Explain.

REFERENCE: *Illinois Physicians Union v. Illinois Department of Public Aid* (1982), No. 81–1048, 675 *Federal Reporter,* (2nd Series): 151–58.

REFERENCES

Cochran, W. G. 1977, *Sampling Techniques,* 3rd ed. New York: Wiley.

Hoel, Paul G. 1971. *Elementary Statistics,* 3rd ed. New York: Wiley, Chap. 6.

Hogg, Robert V., and Allen T. Craig. 1978. *Introduction to Mathematical Statistics,* 4th ed. New York: Macmillan, Chap. 4.

Freedman, D., R. Pisani, and R. Purves. 1992. *Statistics,* 2nd ed. New York: Norton.

Kish, L. 1965. *Survey Sampling.* New York: Wiley.

Optional Topics

Sampling Distributions of $\overline{X}_1 - \overline{X}_2$
Sampling Distributions of $\hat{p}_1 - \hat{p}_2$

Many times, we are interested in comparing the results of samples that come from two populations. For instance, if a company has two different sources that supply it with raw materials, it might be interested in which source, on the average, provides higher-quality material. If a personnel manager has two different training programs, he or she might be interested in which program, on the average, produces more efficient workers. The manager of a factory may have two production lines that produce the same goods. The manager may be interested in knowing which line, on the average, gives higher output.

To make comparisons like those just suggested, we often take two independent samples—one sample from each of the two populations we are interested in comparing selected such that the items included in one sample have no influence on the items included in the other sample. To compare the differences between the two samples' mean values is often useful. That is, we may have two populations that have means μ_1 and μ_2. If we take samples of n_1 items from the first population and n_2 items from the second population, we can calculate two sample means, $\overline{X}_1$ and $\overline{X}_2$. These two sample means can then be compared.

If we wish to make probability statements about the differences that can arise between two sample means, then we must study the distribution of the difference $\overline{X}_1 - \overline{X}_2$. The distribution of the difference between sample means has many of the same characteristic features as the distribution of single sample means.

The equation for the mean or **expected value of the sampling distribution of the difference between two sample means** is given next.

Equation for Expected Value of Difference Between Two Sample Means

$$E(\overline{X}_1 - \overline{X}_2) = \mu_1 - \mu_2 \tag{7.10}$$

EXAMPLE 7.10 If we took a set of many, many samples from two populations that have means μ_1 and μ_2, then the difference between the sample means of the samples from the two populations would average out to $\mu_1 - \mu_2$ in the long run. Table 7.1 gives an indication of what this statement means. This table shows ten samples from each of two populations, where the sample sizes are $n_1 = 5$ and $n_2 = 5$. The fourth and fifth columns of the table give the ten sample means for each population. The sixth column shows the difference between the two sample means. The populations

TABLE 7.1	Sample Number	Sample from First Population	Sample from Second Population	$\overline{X}_1$	$\overline{X}_2$	$\overline{X}_1 - \overline{X}_2$
Distribution of Differences in Sample Means for Samples from Populations	1	3, 9, 5, 9, 1	4, 6, 3, 0, 4	5.4	3.4	2.0
	2	9, 9, 5, 4, 7	0, 6, 7, 4, 3	6.8	4.0	2.8
	3	6, 9, 5, 6, 8	9, 1, 2, 4, 0	6.8	3.2	3.6
	4	3, 5, 2, 4, 9	9, 7, 4, 5, 8	4.6	6.6	−2.0
	5	3, 8, 9, 8, 0	1, 0, 7, 5, 0	5.6	2.6	3.0
	6	3, 6, 2, 4, 7	7, 0, 9, 9, 4	4.4	5.8	−1.4
	7	9, 9, 6, 3, 8	7, 2, 0, 5, 5	7.0	3.8	3.2
	8	2, 4, 0, 3, 8	7, 4, 9, 7, 6	3.4	6.6	−3.2
	9	3, 5, 5, 5, 3	3, 5, 6, 7, 6	4.2	5.4	−1.2
	10	7, 4, 8, 1, 5	4, 5, 2, 4, 6	5.0	4.2	0.8

The sample mean of $(\overline{X}_1 - \overline{X}_2)$ is .76.
The sample variance of $(\overline{X}_1 - \overline{X}_2)$ is 6.28.
The sample standard deviation of $(\overline{X}_1 - \overline{X}_2)$ is 2.51.

from which these samples were drawn by sampling with replacement are both composed of the set of integers from 0 through 9. Thus, both populations have the same mean. That is, $\mu_1 = \mu_2 = (0 + 1 + 2 + \cdots + 8 + 9)/10 = 4.5$. To obtain the samples, we merely took the first 20 sets of five random digits from Table II in Appendix C.

Since $\mu_1 = \mu_2$ for this example, Equation (7.10) tells us that in the long run (after infinitely many pairs of $\overline{X}_1$ and $\overline{X}_2$ have been calculated) the mean value of the sixth column in Table 7.1 should be $\mu_1 - \mu_2 = 4.5 - 4.5 = 0$. After taking only ten pairs of samples, this difference averages to .76.

Consider another example.

EXAMPLE 7.11   Two chemical processes are being run at the same Hunter Chemical plant, side by side. The first process produces $\mu_1 = 2.00$ ounces of impurities per gallon of product. The second process produces $\mu_2 = 1.25$ ounces of impurities per gallon of output. Also, on each shift there is one quality control engineer assigned to each process. Each day, the engineers take random samples of 24 gallons (3 each hour of their 8-hour day) from the line to which they are assigned, and they measure the weight of impurities in their sample gallons. At the end of a day each engineer will have 24 measurements, which are then averaged to get $\overline{X}_1$ for the mean impurities per gallon coming off the first line that day, and $\overline{X}_2$ for the mean impurities per gallon coming off the second line. After many, many working days, Equation (7.10) tells us that the difference between the mean daily impurities measures will average out to $\mu_1 - \mu_2 = 2.00 - 1.25 = .75$ ounce.

Theoretically, we could take infinitely many independent samples of, say sizes $n_1 = 5$ and $n_2 = 5$ for the two normal populations with $\mu_1 = 30$ and $\sigma_1 = 2$, and $\mu_2 = 10$ and $\sigma_2 = 3$ [see Figure 7.12(a)]. Again theoretically, we could plot the difference between the infinitely many pairs of sample means. (Imagine a set of values like column 6 of Table 7.1, except that the column of numbers would be infinitely long.) An example of what the distribution of these values might look like is given in Figure 7.12(b). Equation (7.10) tells us that the mean of the sampling distribution of $\overline{X}_1 - \overline{X}_2$ in Figure 7.12(b) is $\mu_1 - \mu_2 = 30 - 10 = 20$. But if we want to make probability statements when we compare two sample means, we need to know the standard deviation and shape of the distribution in Figure 7.12(b).

FIGURE 7.12

Two Population Distributions and the Sampling Distribution of $\overline{X}_1 - \overline{X}_2$

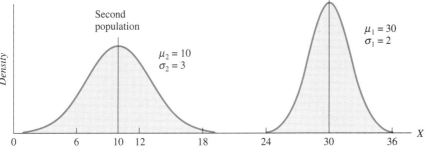

(a) **Two population distributions**

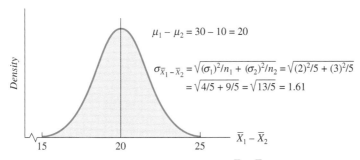

(b) **Sampling distribution of $\overline{X}_1 - \overline{X}_2$**

If X and Y are independent random variables, then

$$\text{Variance}(X - Y) = \text{Variance}(X) + \text{Variance}(Y)$$

(One might at first expect a minus on the right, but a minus would be incorrect because it could even give a negative variance.)

Since $\overline{X}_1$ and $\overline{X}_2$ are the means of independent random samples, they are independent of one another. They have variances σ_1^2/n_1 and σ_2^2/n_2, respectively. Therefore for the difference in sample means $\overline{X}_1 - \overline{X}_2$, the variance of the difference $\sigma^2_{\overline{X}_1 - \overline{X}_2}$ is as follows:

$$\sigma^2_{\overline{X}_1 - \overline{X}_2} = \frac{\sigma_1^2}{n_1} + \frac{\sigma_2^2}{n_2} \qquad (7.11)$$

Thus to find the **standard deviation of the sampling distribution of the difference between two sample means $\overline{X}_1 - \overline{X}_2$,** we use the following equation.

Equation for Standard Deviation of Difference Between Two Sample Means of Independent Random Samples

$$\sigma_{\overline{X}_1 - \overline{X}_2} = \sqrt{\frac{\sigma_1^2}{n_1} + \frac{\sigma_2^2}{n_2}} \qquad (7.12)$$

EXAMPLE 7.12   Let us return to Table 7.1 to demonstrate the meaning of Equations (7.11) and (7.12). These formulas tell us what the variance and the standard deviation of the sixth column in Table 7.1 would be after infinitely many pairs of samples had been taken. Since the population from which the samples are being drawn with replacement is the uniform distribution of digits between 0 and 9, we could calculate that $\sigma_1^2 = \sigma_2^2 = 8.25$. The long-run variance of column 6 would then be given by Equation (7.11) as

$$\sigma_{\overline{X}_1 - \overline{X}_2}^2 = \frac{8.25}{5} + \frac{8.25}{5} = 3.3$$

and $\sigma_{\overline{X}_1 - \overline{X}_2} = \sqrt{3.3} = 1.82$. The variance and the standard deviation of column 6 after only ten pairs of samples are 6.28 and 2.51. These differ from the long-run values, and the error is due to taking only ten pairs of samples instead of infinitely many. If more and more pairs of samples were taken, column 6 of the table would get longer and longer. Its mean would get closer to 0, and its variance would get closer to 3.3.

Since the standard deviations of the two population distributions in Figure 7.12(a) are $\sigma_1 = 2$ and $\sigma_2 = 3$, and the sample sizes are $n_1 = 5$ and $n_2 = 5$, the standard deviation of the sampling distribution of $\overline{X}_1 - \overline{X}_2$ shown in Figure 7.12(b) is

$$\sigma_{\overline{X}_1 - \overline{X}_2} = \sqrt{\frac{\sigma_1^2}{n_1} + \frac{\sigma_2^2}{n_2}} = \sqrt{\frac{(2)^2}{5} + \frac{(3)^2}{5}} = \sqrt{\frac{4}{5} + \frac{9}{5}} = \sqrt{\frac{13}{5}} = 1.61$$

Finally, we ask, "What is the functional form or the **shape** of the distribution of differences in sample means shown in Figure 7.12(b)?" The answer is given by extension of the concepts discussed in Section 7.3 and by extension of the central limit effect.

Rules for Shapes of Distributions of Differences in Sample Means

1. If the populations from which the samples are drawn are normal in shape, then the distribution of $\overline{X}_1 - \overline{X}_2$ will be normal in shape.*
2. If the populations from which the samples are drawn are not normal in shape, then the distribution of $\overline{X}_1 - \overline{X}_2$ will be approximately normal, owing to the central limit effect, if the sample sizes n_1 and n_2 are both large.

Since the populations in Figure 7.12(a) are normal in shape, the first rule applies, and the distribution of $\overline{X}_1 - \overline{X}_2$ values in Figure 7.12(b) is normal.

To **standardize a difference between two sample means** $\overline{X}_1 - \overline{X}_2$ for a sampling distribution, we use the following equation.

Equation for Standardized Difference Between Two Sample Means

$$Z = \frac{(\overline{X}_1 - \overline{X}_2) - (\mu_1 - \mu_2)}{\sigma_{\overline{X}_1 - \overline{X}_2}} \tag{7.13}$$

---

*The sum or difference of independent normal variables is normally distributed.

The distribution of the standardized difference between two sample means has a mean of 0 and a standard deviation of 1. If the sampling distribution of $\overline{X}_1 - \overline{X}_2$ is normal, as indicated by the above rules for the shape of the distribution, then Z, defined by Equation (7.13), is a standard normal variable; and we may use the table of normal areas to find probabilities for the difference between two means. The next problem demonstrates the procedure.

 PRACTICE PROBLEM 7.6

DP Minerals, Ltd., operates two titanium mines. The daily production of ore from mine 1 averages $\mu_1 = 150$ tons and is normally distributed with a standard deviation $\sigma_1 = 20$ tons. Mine 2's daily production is also normally distributed, but the mean is $\mu_2 = 125$ tons and the second standard deviation is $\sigma_2 = 25$ tons. A sample of five randomly selected daily production figures is taken from each of the mines. What is the probability that the sample mean production for mine 1 will be less than or equal to the sample mean production for mine 2? That is, what is $P(\overline{X}_1 - \overline{X}_2 \leq 0)$?

Solution From the discussion in this section we know that the random variable $\overline{X}_1 - \overline{X}_2$ for samples taken from these two mines should average to $E(\overline{X}_1 - \overline{X}_2) = \mu_1 - \mu_2 = 150 - 125 = 25$ tons. Also, the standard deviation of the distribution of $\overline{X}_1 - \overline{X}_2$ is

$$\sigma_{\overline{X}_1 - \overline{X}_2} = \sqrt{\frac{\sigma_1^2}{n_1} + \frac{\sigma_2^2}{n_2}} = \sqrt{\frac{(20)^2}{5} + \frac{(25)^2}{5}} = \sqrt{\frac{400}{5} + \frac{625}{5}} = \sqrt{\frac{1025}{5}}$$

$$= 14.32$$

The distribution of the differences of sample means taken from the two mines is shown in Figure 7.13. The shape of this distribution is normal since the two populations are normal, and the shaded area shows the probability we seek. The Z value associated with this area is

$$Z = \frac{0 - (150 - 125)}{14.32} = \frac{-25}{14.32} = -1.75$$

The area under the standard normal density function and between $Z = 0$ and $Z = -1.75$ is .4599. The area for $P(\overline{X}_1 - \overline{X}_2 \leq 0)$ is the area to the left of $Z = -1.75$, so

$$P(\overline{X}_1 - \overline{X}_2 \leq 0) = P(Z \leq -1.75) = .5000 - .4599 = .0401$$

Thus there is only a 4% chance that the first sample's mean will be smaller than the second sample's mean. So if the owner of the two mines found a smaller first sample mean, say $\overline{X}_1 = 130$ tons, and a larger second sample mean, say $\overline{X}_2 = 135$ tons, in independent

FIGURE 7.13

Distribution of Differences of Sample Means from Two Normal Populations

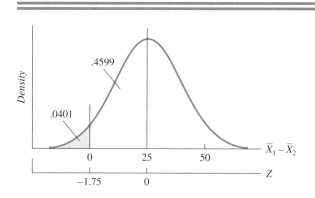

random samples of five randomly selected days from each mine, he would suspect that either the sampling was faulty or that the difference in the mines' mean daily outputs had changed.

 PRACTICE PROBLEM 7.7

Two production processes at Eastern Refractories, Inc., are, on the average, identical. Both use an average of $\mu_1 = \mu_2 = 500$ kilograms of raw material per day. Both have the same standard deviation of daily use, $\sigma_1 = \sigma_2 = 9$ kilograms per day. Thus the daily use of material may vary for the two processes, but on the average they are the same. What is the probability that two samples of $n_1 = 100$ and $n_2 = 100$ randomly selected daily use figures would show that the sample means $\overline{X}_1$ and $\overline{X}_2$ differ by less than or equal to .5 kilogram? That is, find

$$P(-.5 \le \overline{X}_1 - \overline{X}_2 \le .5)$$

Solution Since the two populations have the same mean, Equation (7.10) tells us that the distribution of differences between the sample means is centered at $E(\overline{X}_1 - \overline{X}_2) = 500 - 500 = 0$. The standard deviation of the distribution is given by Equation (7.12):

$$\sigma_{\overline{X}_1 - \overline{X}_2} = \sqrt{\frac{9^2}{100} + \frac{9^2}{100}} = \sqrt{1.62} = 1.27$$

The distribution of the differences $\overline{X}_1 - \overline{X}_2$ for this problem is shown in Figure 7.14. The shape of this distribution is approximately normal. It is approximately normal because we don't know the shape of the distributions of material use in the two populations, but we do know that the samples are large ($n_1 = 100$ and $n_2 = 100$). Thus the differences in the sample means are approximately normal at a minimum, and we can use the standard normal distribution and Table V to answer the probability question posed in this problem.

The probability we seek is shown by the shaded area in Figure 7.14. Note that the difference between the sample means may be positive or negative, as long as the two sample means differ by no more than .5 kilogram. Because of the symmetry of the normal curve, we need only find the area between 0 and .5 and double it to get the answer we seek:

$$P(-.5 \le \overline{X}_1 - \overline{X}_2 \le .5) = P\left(\frac{-.5 - 0}{1.27} \le Z = \frac{\overline{X}_1 - \overline{X}_2}{\sigma_{\overline{X}_1 - \overline{X}_2}} \le \frac{.5 - 0}{1.27}\right)$$

$$= P(-.39 \le Z \le .39) = (2)[P(0 \le Z \le .39)]$$

$$= (2)(.1517) = .3034$$

FIGURE 7.14

Distribution of Differences of Sample Means for Material Use

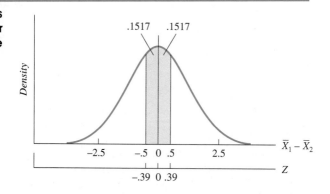

One might wonder what happens if we take pairs of *small* samples from two *nonnormal* populations. What is the shape of the $\overline{X}_1 - \overline{X}_2$ distribution? The answer is that the distribution may be complicated and we thus cannot make probability statements about the $\overline{X}_1 - \overline{X}_2$ differences. The only choice we have in a situation like that is to continue sampling until we obtain large samples (which might be expensive) or to use a completely different approach to the problem.

Problems: Section 7.7

55. Two populations of measurements are normally distributed with $\mu_1 = 57$ and $\mu_2 = 25$. The two populations' standard deviations are $\sigma_1 = 12$ and $\sigma_2 = 6$. Two independent samples of $n_1 = n_2 = 36$ are taken from the populations.

   a. What is the expected value of the difference in the sample means $\overline{X}_1 - \overline{X}_2$?
   b. What is the standard deviation of the distribution of $\overline{X}_1 - \overline{X}_2$?
   c. What is the shape of the distribution of $\overline{X}_1 - \overline{X}_2$? How do you know?

56. What proportion of the time will the means of the samples described in Problem 55 differ by +35 or more? That is, find $P(\overline{X}_1 - \overline{X}_2 \geq 35)$.

57. Two production processes at Eastern Refractories, Inc., are, on the average, identical. Both use an average of $\mu_1 = \mu_2 = 500$ kilograms of raw material per day. Both have the same standard deviation of daily use, $\sigma_1 = \sigma_2 = 9$ kilograms per day. Thus the daily use of material may vary for the two processes, but on the average they are the same. Find the probability that two samples of $n_1 = 81$ and $n_2 = 36$ will produce sample means $\overline{X}_1$ and $\overline{X}_2$ that differ by no more than 1.0 kilogram. [*Hint:* "Differ by no more..." implies that you must consider a negative difference or a positive difference.]

58. Yields for municipal bonds (X_1) over a period have had a mean of .10 and a standard deviation of .02, whereas yields for industrial bonds (X_2) have had a mean of .13 and a standard deviation of .03. Independent random samples, each of size 49, are obtained from the two very large populations of bonds.

   a. What is the expected value of the difference in sample mean yields $(\overline{X}_1 - \overline{X}_2)$?
   b. What is the standard deviation of the difference in sample mean yields?
   c. What is the shape of the distribution of the difference in sample mean yields?
   d. Find the probability that the differences between the sample mean municipal bond yield and the sample mean industrial bond yield will be less than or equal to $-.04$. That is, find $P(\overline{X}_1 - \overline{X}_2 \leq -.04)$.

59. There are two classes of land investment in a company's investment plans. Both have five-year mean yields of 220% return on investment. That is, $\mu_1 = \mu_2 = 220$. They also have the same variances of returns for the investment purchases within the two classes: $\sigma_1^2 = \sigma_2^2 = 30$. If the company purchases $n_1 = 50$ pieces of land in the first class and $n_2 = 75$ pieces of land in the second class, what is the probability that the mean yields of these two sample portfolios will differ by less than 2.3%? That is, find $P(-2.3 \leq \overline{X}_1 - \overline{X}_2 \leq 2.3)$.

7.8
SAMPLING DISTRIBUTIONS OF $\hat{p}_1 - \hat{p}_2$

Many times, we want to know about the differences in the proportion of successes in two binomial populations. We may have two production processes. If the first one produces defectives at a rate p_1 and the second produces defectives at a rate p_2, we may wish to know about the distribution of $\hat{p}_1 - \hat{p}_2$, the difference between the rates or fractions of defectives found in two independent samples taken from the two processes. A company may have two products that are bought by both men and women. If p_1 is the proportion of male buyers for the first product and p_2 is the proportion of male buyers for the second product, we may wish to know about the chance of getting a value of $\hat{p}_1 - \hat{p}_2$, the difference in proportion of male buyers in two independent samples of these products' customers.

The formula for the mean or **expected value of the sampling distribution of $\hat{p}_1 - \hat{p}_2$** is as follows.

Equation for Expected Value of Difference Between Sample Proportions from Binomial Populations

$$E(\hat{p}_1 - \hat{p}_2) = p_1 - p_2 \qquad (7.14)$$

That is, if one population has a success rate of $p_1 = .50$ and another has a success rate of $p_2 = .35$, then on the average two independent samples will yield a difference in sample success rates of $.50 - .35 = .15$.

Since we are assuming that two *independent* random samples are drawn from the two binomial populations, the **standard deviation** of the **differences $\hat{p}_1 - \hat{p}_2$, $\sigma_{\hat{p}_1 - \hat{p}_2}$,** is the square root of the sum of their individual variances.

Equation for Standard Deviation of Difference Between Sample Proportions from Binomial Populations

$$\sigma_{\hat{p}_1 - \hat{p}_2} = \sqrt{\frac{p_1(1 - p_1)}{n_1} + \frac{p_2(1 - p_2)}{n_2}} \qquad (7.15)$$

By extending the normal approximation to the binomial from Section 6.4, we know that when n_1 and n_2 are *large,* or (by the rule of thumb given in Section 6.4) when $n_1 p_1(1 - p_1) \geq 5$ *and* $n_2 p_2(1 - p_2) \geq 5$, the shape of the $\hat{p}_1 - \hat{p}_2$ distribution is approximately normal. Thus the equation for the **standardized difference between sample proportions from binomial populations** is as follows.

Equation for Standardized Difference Between Sample Proportions from Binomial Populations

$$Z = \frac{(\hat{p}_1 - \hat{p}_2) - (p_1 - p_2)}{\sqrt{[p_1(1 - p_1)/n_1] + [p_2(1 - p_2)/n_2]}} \qquad (7.16)$$

The standardized Z in Equation (7.16) has approximately a standard normal distribution. So we can use the standard normal distribution and Table V in Appendix C to answer probability questions about the differences $\hat{p}_1 - \hat{p}_2$. The next problem demonstrates the procedure.

 PRACTICE PROBLEM 7.8

Southern Industries, Inc., has two sales outlets. At both outlets 40% of the customers charge their purchases. That is, $p_1 = p_2 = .40$. In doing an audit, the company accountant took random samples of $n_1 = 100$ and $n_2 = 100$ sales slips from the two outlets. Forty-one and 36 charge customers were found in the samples from the first and second

FIGURE 7.15

Distribution of Differences in Proportions of Charge Customers

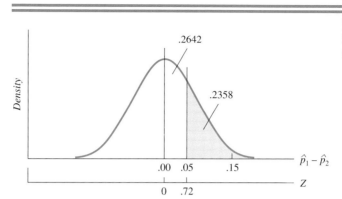

outlets, respectively. What is the probability that a result would be achieved whereby the first outlet's proportion of charge customers exceeded the second outlet's proportion of charge customers by this much or more?

Solution In this problem $\hat{p}_1 = X_1/n_1 = 41/100 = .41$, and $\hat{p}_2 = X_2/n_2 = 36/100 = .36$. Since $p_1 = p_2$, on the average we would expect $\hat{p}_1 - \hat{p}_2$ to be 0, according to Equation (7.14). The probability we seek can be written as $P(\hat{p}_1 - \hat{p}_2 \geq .41 - .36)$, or $P(\hat{p}_1 - \hat{p}_2 \geq .05)$. Using Equation (7.16), we can find this probability as follows (see Figure 7.15):

$$P(\hat{p}_1 - \hat{p}_2 \geq .05) = P\left(Z \geq \frac{.05 - 0}{\sqrt{[(.4)(.6)/100] + [(.4)(.6)/100]}} \right)$$

$$= P\left(Z \geq \frac{.05}{.069} \right) = P(Z \geq .72)$$

Thus we wish to find the area under the normal curve to the right of $Z = .72$. This area is $.5000 - .2642 = .2358$. Hence there is about a 24% chance of getting two independent random samples that differ by .05 or more in the rate of charge customers from the first outlet over the rate of charge customers from the second outlet. Thus the accountant did not obtain an unusual difference. In fact, because of the symmetry of the problem, we can say that in approximately 48% of the samples taken from these two outlets' sales slips, one of the outlet's charge rate will exceed the other by .05 or more.

 PRACTICE PROBLEM 7.9

Commercial Printers, Ltd., receives parts from two suppliers. Historical records show that the first supplier's goods are rejected at a rate of $p_1 = .08$ and the second supplier's goods have a rejection rate of $p_2 = .05$. Commercial's production operation uses $n_1 = 150$ items from the first supplier and $n_2 = 300$ items from the second supplier each day. On what proportion of the days will the difference in the rejection rates of the two suppliers, $\hat{p}_1 - \hat{p}_2$, be 1% or less?

Solution The probability we want, $P(\hat{p}_1 - \hat{p}_2 \leq .01)$, is represented by the shaded area of Figure 7.16. Since the daily uses of the parts constitute *large* random samples of output from the two suppliers, the differences $\hat{p}_1 - \hat{p}_2$ each day will be approximately normally distributed around $p_1 - p_2 = .08 - .05 = .03$. By Equation (7.16), we want the area to the left of Z:

$$Z = \frac{.01 - .03}{\sqrt{[(.08)(.92)/150] + [(.05)(.95)/300]}} = \frac{-.02}{\sqrt{.0006}} = -.82$$

FIGURE 7.16

Distribution of Differences
in Sample Rejection Rates

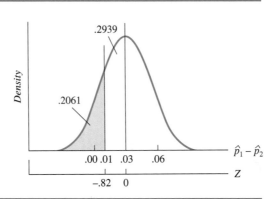

This value is $.5000 - .2939 = .2061$, or $P(\hat{p}_1 - \hat{p}_2 \le .01) = .2061$. Thus on about 21% of the days the difference in the rejection rates of the two suppliers, $\hat{p}_1 - \hat{p}_2$, will be 1% or less.

We should note that in order to solve problems like the preceding two, we must have large samples. When n_1 and n_2 are both large, then the normal approximation to the binomial holds, and the differences of sample proportions of success, $\hat{p}_1 - \hat{p}_2$, are approximately normally distributed. We noted previously that a good rule of thumb for determining when this approximation can be used is that both $n_1 p_1(1 - p_1) \ge 5$ and $n_2 p_2(1 - p_2) \ge 5$.

Problems: Section 7.8

60. A weight reduction clinic has offices in New York and Chicago. It has found that the proportion of people signed up for its weight-reducing classes who actually complete their entire program is $p_1 = .80$ in New York and $p_2 = .72$ in Chicago. A class of $n_1 = 80$ participants just started the program in New York and a class of $n_2 = 60$ just started the program in Chicago. Consider these two groups to be independent random samples.

    a. What is the expected value of the difference in completion rates, $\hat{p}_1 - \hat{p}_2$, between these New York and Chicago groups?
    b. What is the standard deviation of this difference over many pairs of classes of this size?

61. What is the probability that the difference $\hat{p}_1 - \hat{p}_2$ will be 10% or larger in Problem 60?

62. Two different salespeople have the same sales success rate when they call on customers: $p_1 = p_2 = .30$. During the past month each salesperson has called on the same number of randomly selected customers: $n_1 = n_2 = 210$.

    a. What is the expected difference in their sales success rates?
    b. What is the standard deviation of the difference in success rates over many months where each salesperson visits 210 customers?
    c. What is the probability that the sales success rates will differ by less than 2% this month? That is, what is $P(|\hat{p}_1 - \hat{p}_2| \le .02)$?

63. Two large national companies in the same industry differ in the proportions of women in their production labor forces. The first company has a proportion of women

that is $p_1 = .30$. The second company's proportion of women is $p_2 = .18$. Randomly selected groups of $n_1 = 80$ and $n_2 = 70$ production workers are being sent by their companies to an industry-sponsored training program in Miami. What is the probability that the proportion of women in the first company's group will exceed the proportion of women in the second company's group by 20% or more? That is, what is $P(\hat{p}_1 - \hat{p}_2 \geq .20)$?

64. Airlines often overbook flights in order to minimize the effects of people who have reservations but fail to show up for the flights. The fraction of no-shows is not the same for two airlines. Airline 1 has a no-show rate of .10, and airline 2 has a no-show rate of .12. If each airline has a randomly selected flight with 120 reservations for each flight, what is the probability that the fraction of no-shows for the two airlines will be different by more than $-.02$ and less than $.02$?

Estimation

8

In Chapters 4 through 7 of this text, we dealt with probability topics. In this chapter, we begin discussing the third major area of statistics—the area called *statistical inference*. The problems in this area of statistics have some features in common. An analyst collects data by experiment, sample, or sample survey with the hope of drawing conclusions about the phenomenon under investigation. From the experimental results or sample data values the analyst wants to make *inferences* about the underlying population. The analyst may use the sample data for the *estimation* of the values of unknown parameters or for *tests of hypotheses* concerning these parameters.

Consider a personnel manager who takes a sample from a population of employees divided into those who belong to a union and those who don't. The unknown parameter is the proportion of employees p who belong to a union. The personnel manager may *estimate p* (try to assess its value) or may *test a hypothesis* about p, for example, test the hypothesis, say, that $p \geq 1/2$. In either case, the manager makes an inference about p.

In this chapter, we discuss principles and methods of estimation. We will take up hypothesis testing in Chapter 9.

8.1
ESTIMATORS

We have a sample from a population involving an unknown parameter. The problem is to construct a sample quantity that will serve to estimate the unknown parameter. Such a sample quantity is called an **estimator**, and the actual numerical value obtained by evaluating an estimator in a given instance is the **estimate** or **point estimate.**

> **Definitions**
>
> An **estimator** is a sample statistic that is used to estimate an unknown population parameter.
>
> An **estimate** or **point estimate** is an actual numerical value obtained for an estimator.

For example, the sample mean $\overline{X}$ is an estimator for the population mean μ. If for a specific sample the sample mean is 10.31, we say 10.31 is our estimate, or *point estimate,* for μ. An estimator must be a statistic; it must depend only on the sample and not on the parameter to be estimated.

Estimators have many characteristics that are important for making inferences about population parameters. One desirable characteristic for estimators is that their sampling distributions be balanced at the population parameter we wish to estimate. In other words, it is desirable for the sampling distribution of an estimator to be centered on the population parameter. If the expected value of an estimator is equal to the population parameter, then the estimator is an **unbiased estimator.**

> **Definition**
>
> The expected value of an **unbiased estimator** is equal to the population parameter; that is, an estimator is unbiased if
>
> $$E(Sample\ estimator) = Population\ parameter \qquad (8.1)$$

EXAMPLE 8.1   An auditor for the accounting firm of Princeton & Princeton wishes to use sampling to estimate the value of the end-of-year inventory for an audit client. What estimator should the auditor use?

In Chapter 7 we found that the sampling distribution of the sample mean $\overline{X}$ balances at the population mean μ in the sense that the expected value of $\overline{X}$ is equal to μ, or in the sense that

$$E(\overline{X}) = \mu \qquad (8.2)$$

Thus the sample mean $\overline{X}$ is an unbiased estimator for the population mean μ. But what if we wanted to consider using $\overline{X} + 1000$ as an estimator for the population mean μ? Since $E(\overline{X} + 1000) = \mu + 1000 \neq \mu$, the estimator $\overline{X} + 1000$ is a biased estimator of the population mean μ. From the property of unbiasedness we can say that $\overline{X}$ is a more desirable estimator of μ than is $\overline{X} + 1000$.

For a second example, what if we know that a population distribution is skewed, and we want to consider using the sample median Md as an estimator for the population mean μ? In this case $E(Md) \neq \mu$. When the population distribution is skewed, the sample median Md is a biased estimator of the population mean μ. If a population distribution is skewed, then we can say, from the property of unbiasedness, that $\overline{X}$ is a more desirable estimator of μ than is Md.

The sample variance

$$s^2 = \frac{1}{n-1} \sum_{i=1}^{n} (X_i - \overline{X})^2 \tag{8.3}$$

is an unbiased estimator of σ^2 for an infinite population since, as we saw in Chapter 7,

$$E(s^2) = \sigma^2 \tag{8.4}$$

Also, $s_{\overline{X}}^2 = s^2/n$ is an unbiased estimator of $\sigma_{\overline{X}}^2 = \sigma^2/n$ for an infinite population. Notice that s^2 is a statistic; it does not depend on the unknown μ and σ^2. The point of Equation (8.1), the condition for unbiasedness, is not that we can somehow check it for the true values of μ and σ^2; we do not know what these true values are. The point is that Equation (8.1) holds *whatever* values μ and σ^2 may happen to have. The same remark applies to any unbiased estimator.*

To make the sample variance unbiased in the original definition in Chapter 3, we divided by $n - 1$; furthermore, if we had divided by n, we would have introduced bias. As an estimator of σ, we ordinarily use the sample standard deviation

$$s = \sqrt{\frac{1}{n-1} \sum_{i=1}^{n} (X_i - \overline{X})^2} \tag{8.5}$$

and as an estimator of the standard error $\sigma_{\overline{X}} = \sigma/\sqrt{n}$, we ordinarily use $s_{\overline{X}} = s/\sqrt{n}$. Nevertheless, s is a somewhat biased estimator, in that on the average s is not exactly equal to σ. Fortunately, the bias also gets even smaller when we use larger samples.

Figure 8.1 shows the sampling distribution of an unbiased estimator centered above the population parameter; whereas the sampling distribution for a biased estimator is not centered above the parameter.

*The sample mean $\overline{X}$ is known as the first moment of the sample data and the sample variance s^2 is essentially the second moment about the mean. Consequently, $\overline{X}$ and s^2 are estimators obtained by the method of *moments*. Other methods, such as the method of *maximum likelihood*, are also used to obtain estimators.

FIGURE 8.1

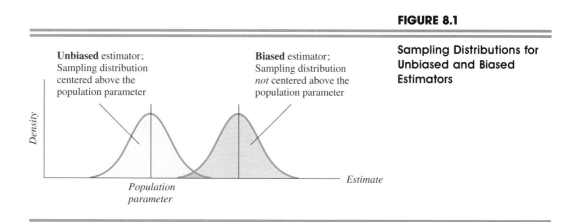

Sampling Distributions for Unbiased and Biased Estimators

Another important characteristic of an estimator is its variability. The variability of an estimator can be measured by the variance of its sampling distribution. Since variance measures spread, if we have two unbiased estimators, we naturally prefer the one with smaller variance. Estimators that are unbiased and have minimum variance in comparison to other estimators are called **minimum-variance unbiased estimators.**

Definition

A **minimum-variance unbiased estimator** is an estimator that not only is unbiased but also has smaller variance than *any other* unbiased estimator.

A minimum-variance unbiased estimator is also termed a **best estimator.**

EXAMPLE 8.2   The auditor for the accounting firm of Princeton & Princeton in Example 8.1 who wishes to use sampling to estimate the value of the end-of-year inventory for an audit client can decide which estimator to use: If the end-of-year inventory value is normally distributed, then both the sample mean $\overline{X}$ and the sample median Md are unbiased estimators of the population mean μ. But we know that Variance$(\overline{X}) = \sigma^2/n$; whereas in the case of a normal distribution and large n, Variance$(Md) \approx \pi\sigma^2/(2n) = 1.57\sigma^2/n$. In this case Variance$(\overline{X}) = \sigma^2/n$ is less than Variance$(Md) \approx 1.57\sigma^2/n$.

When two estimators for a population parameter are unbiased, we prefer the estimator with the smaller variance. Consequently, in this situation we can say that $\overline{X}$ is a more desirable estimator of μ than is Md.

Figure 8.2 shows the sampling distributions of two unbiased estimators centered above the population parameter. However, the sampling distribution for one of the estimators shows less variability than does the sampling distribution of the other. In this case, the estimator with the smaller variability is the preferred estimator.

We have seen that the sample mean $\overline{X}$ is an unbiased estimator of the population mean μ. Now if the parent population is *normal,* $\overline{X}$ is a minimum-variance unbiased

FIGURE 8.2

Sampling Distributions for Two Unbiased Estimators with Different Variances

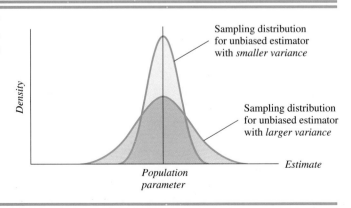

estimator of μ. In this sense, we may say that for a given sample size $\overline{X}$ is the *best* estimator of μ.

In this section, we have discussed *point estimators,* sample quantities that estimate population parameters with specific numerical values. More often, however, we estimate a parameter of a population by constructing an interval, within which we think the parameter might lie. The following sections of this chapter show how these *confidence intervals* are constructed.

Problems: Section 8.1

Answers to odd-numbered problems are given in the back of the text.

1. The results from a sample taken from a population are

$$n = 9 \qquad \sum X_i = 36 \qquad \sum (X_i - \overline{X})^2 = 288$$

   Find the point estimates for the following:
   a. The mean μ
   b. The variance σ^2
   c. The standard deviation σ
   d. The standard deviation of the sample mean $\sigma_{\overline{X}}$

2. The results from a sample taken from a population are

$$n = 16 \qquad \sum X_i = 64 \qquad \sum (X_i - \overline{X})^2 = 180$$

   Find the point estimates for the following:
   a. The mean μ
   b. The variance σ^2
   c. The standard deviation σ
   d. The standard deviation of the sample mean $\sigma_{\overline{X}}$

3. The results from a sample taken from a population are

$$n = 25 \qquad \sum X_i = 500 \qquad \sum X_i^2 = 12{,}400$$

   Find the point estimates for the following:
   a. The mean μ
   b. The variance σ^2
   c. The standard deviation σ
   d. The standard deviation of the sample mean $\sigma_{\overline{X}}$

4. A sample of data for the lives of Collatarized Mortgage Obligations taken from a normal population is

   6   15   3   12   6   21   15   18   12

   Find the point estimates for the following:
   a. The mean μ
   b. The variance σ^2
   c. The standard deviation σ
   d. The standard deviation of the sample mean $\sigma_{\overline{X}}$

5. A sample of data for the price changes in ten-year Treasury bonds taken from a normal population is

   4   10   2   8   4   14   10   12   8

   Find the point estimates for the following:
   a. The mean μ
   b. The variance σ^2
   c. The standard deviation σ
   d. The standard deviation of the sample mean $\sigma_{\overline{X}}$

6. A sample of data for the monthly changes in the price of one troy ounce of gold on a commodity exchange taken from a normal population is

   2   5   9   11   13

   Find the point estimates for the following:
   a. The mean μ
   b. The variance σ^2
   c. The standard deviation σ
   d. The standard deviation of the sample mean $\sigma_{\overline{X}}$

7. A real estate broker for Geselman and Associates randomly selected four properties from a subdivision, and an appraiser gave an opinion on the value of each property as $17, $21, $21, and $21 thousand.

   a. Find the best point estimate for the mean value of the properties in the subdivision.

   b. Find the best point estimate for the variance of the property values in the subdivision.

   c. Find the point estimate for the standard deviation of the property values in the subdivision.

   d. Find the point estimate for the standard deviation of the sample mean value for the properties in the subdivision.

 8. A consumer advocate randomly selected nine prescription drug items at an American Drug pharmacy and found the values of the suggested retail prices minus the actual retail prices as $0, $8, $0, $0, $3, $0, $3, $5, and −$1.

   a. Find the best point estimate for the mean difference in suggested and actual prices of the prescription drug items in the pharmacy.

   b. Find the best point estimate for the variance of the difference in suggested and actual retail prices of the prescription drug items in the pharmacy.

   c. Find the point estimate for the standard deviation of the difference in suggested and actual and retail prices of the drug items in the pharmacy.

   d. Find the point estimate for the standard deviation of the sample mean difference in suggested and actual retail prices of the drug items in the pharmacy.

8.2
CONFIDENCE INTERVALS FOR NORMAL MEANS: KNOWN STANDARD DEVIATION

This section deals with the problem of estimating the mean μ of a normal population when the standard deviation σ is known. This circumstance is not typical: If one does not know the mean, one usually does not know the standard deviation as well. Assuming that σ is known, however, serves to simplify the reasoning and make clear the principles underlying estimation; it also serves to introduce the more complicated and more realistic case, treated in the next two sections, where σ is unknown.

EXAMPLE 8.3   Gosset Pharmaceuticals, Inc., has a sleep-enhancing drug on the market. The company needs to estimate the effectiveness of the drug in terms of the mean amount of increase in sleep that patients might expect. If the drug is an effective sleep enhancer, then the profits from the sales of the drug are likely to increase Gosset's stock price, thereby increasing the wealth of their shareholders. Ten patients were given the sleep-enhancing drug. In almost every case, the patient slept longer under the effect of the drug than usual. Table 8.1 shows the amount of the increase in sleep (in hours) in each case. Because the amount of increase in sleep varies, we want to estimate the increase in such a

TABLE 8.1	*Patient*	*Increase*	*Patient*	*Increase*
Additional Hours of Sleep	1	1.2	6	1.0
Gained by Using Drug	2	2.4	7	1.8
	3	1.3	8	0.8
	4	1.3	9	4.6
	5	0.0	10	1.4
			Total	15.8

$$\overline{X} = \frac{15.8}{10} = 1.58 \text{ hours}$$

way that the reliability of our conclusions concerning the effects of the drug may be evaluated objectively by means of probability or confidence statements. Accordingly, we want to find a method for obtaining an *interval estimate* for the amount of increase in sleep. We will find a method that will allow us to construct the interval in such a way that a probability or confidence statement can be made about its reliability.

An interval estimate is known as a **confidence interval.**

Definition

A **confidence interval** is an interval, bounded on the left by L and on the right by R, that is used to estimate an unknown population parameter. The interval is constructed in such a way that the reliability of the estimate may be evaluated objectively by means of a confidence statement.

The L and R bounds are known as **confidence limits.**

To construct confidence intervals, we need some auxiliary concepts. If Z is a standard normal variable, the quantity Z_α, in the relationship

$$P(Z > Z_\alpha) = \alpha$$

is the **upper percentage cutoff point** of the standard normal distribution corresponding to a probability of α (Greek alpha). It is the point on the Z scale such that the probability of a *larger* value is equal to α (see Figure 8.3). The upper 2.5 percentage point for the standard normal distribution is denoted by $Z_{.025}$ and is the value on the Z scale such that the area to the right of it under the standard normal density curve is .025. By the table of standard normal areas, Table V in Appendix C, we find that $Z_{.025}$ is equal to 1.96 (since the area .5000 − .0250, or .4750, falls between $Z = 0$ and $Z = 1.96$). We may say that at $Z = 1.96$, 2.5% of the area under the normal curve is *cut off* in the curve's tail (see Figure 8.4).

Because of the symmetry of the standard normal distribution, its lower percentage cutoff points are equal in absolute value to the corresponding upper ones, but they are negative. Thus $-Z_\alpha$ is the point on the Z scale such that the probability of a smaller value is α; the area to the left of $-Z_\alpha$ is α. There is probability .025 that a standard normal variable is less than $-Z_{.025}$, or -1.96 (see Figure 8.4).

FIGURE 8.3

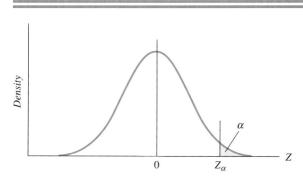

Standard Normal Curve Showing Area α for the Percentage Cutoff Point Z_α

FIGURE 8.4

Standard Normal Curve

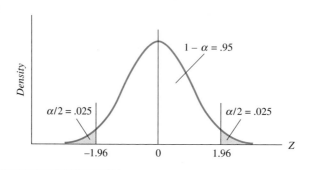

In terms of cutoff points, we have, by definition,

$$P(-Z_{\alpha/2} \leq Z \leq Z_{\alpha/2}) = 1 - \alpha \tag{8.6}$$

Note that we use the $(1/2)\alpha$ cutoff points in order to have $100(1 - \alpha)\%$ of the area between them; each *tail* has area $(1/2)\alpha$, so the two together have area α.

Now consider a normal population with unknown mean μ and known standard deviation σ. If we take a random sample of size n from the normal population and compute $\overline{X}$, then the sample mean $\overline{X}$ is normally distributed, and the standardized variable $(\overline{X} - \mu)/(\sigma/\sqrt{n})$ has the standard normal distribution. Then

$$P\left(-Z_{\alpha/2} \leq \frac{\overline{X} - \mu}{\sigma/\sqrt{n}} \leq Z_{\alpha/2}\right) = 1 - \alpha \tag{8.7}$$

Using the algebraic rules of inequalities,* we can show that Equation (8.7) is the same as

$$P\left(\overline{X} - Z_{\alpha/2}\frac{\sigma}{\sqrt{n}} \leq \mu \leq \overline{X} + Z_{\alpha/2}\frac{\sigma}{\sqrt{n}}\right) = 1 - \alpha \tag{8.8}$$

Equation (8.9) is the basis for the definition of a **confidence interval for the population mean μ** when the population standard deviation is known.

Equations for Limits of a Confidence Interval for μ: σ Known

The $100(1 - \alpha)\%$ **confidence interval for the population mean μ** when the population is normally distributed and the standard deviation σ is known is the interval bounded by the confidence limits

$$L = \overline{X} - Z_{\alpha/2}\frac{\sigma}{\sqrt{n}} \quad \text{and} \quad R = \overline{X} + Z_{\alpha/2}\frac{\sigma}{\sqrt{n}} \tag{8.9}$$

---

*These rules are given in a footnote at the beginning of Section 7.4.

Equation (8.9) shows that the expression for finding the limits of these confidence interval estimates is

$$(\textit{Point estimate}) \pm Z_{\alpha/2}(\textit{Standard error of point estimate})$$

By Equation (8.8), there is probability $1 - \alpha$ that the random variables L and R will contain the unknown μ, whatever μ may be.

Definitions

The value $1 - \alpha$ is termed the **confidence coefficient.**

The value $100(1 - \alpha)\%$ is known as the **confidence level.**

EXAMPLE 8.4 Gosset Pharmaceuticals, Inc., of Example 8.3 gave ten patients a sleep-enhancing drug. Table 8.1 shows the amount of the increase in sleep (in hours) in each case. Assume we *know from past experience* that the increase in sleep is normally distributed with some mean μ and that the population standard deviation is $\sigma = 1.29$. The estimator $\overline{X}$ then has the following standard deviation:

$$\sigma_{\overline{X}} = \frac{\sigma}{\sqrt{n}} = \frac{1.29}{\sqrt{10}} = .408$$

We want to use a confidence level of 95% and the data of Table 8.1 to find an *interval estimate* for the mean amount of increase in sleep μ. In other words, we want to find the 95% confidence interval estimate for the mean increase in sleep.

Since the confidence level is $100(1 - \alpha) = 95\%$, we have

$$1 - \alpha = .95 \qquad \alpha = .05$$
$$\alpha/2 = .025 \qquad Z_{.025} = 1.96$$

Now the data of Table 8.1 for the ten patients give $\overline{X} = 1.58$ hours, and we know that $\sigma_{\overline{X}} = \sigma/\sqrt{n} = 1.29/\sqrt{10} = .408$. Thus Equation (8.9) gives confidence limits of

$$L = \overline{X} - Z_{\alpha/2}\frac{\sigma}{\sqrt{n}} = \overline{X} - (1.96)(.408) = 1.58 - .80 = .78$$

and

$$R = \overline{X} + Z_{\alpha/2}\frac{\sigma}{\sqrt{n}} = \overline{X} + (1.96)(.408) = 1.58 + .80 = 2.38$$

Thus we can feel 95% confident that the population mean lies between .78 and 2.38 hours of increased sleep.

The confidence we have in the limits .78 and 2.38 in Example 8.4 derives from our confidence in the *statistical procedure* that gave rise to them. The procedure gives random variables L and R that have a 95% chance of enclosing the true but unknown mean μ; whether their specific values .78 and 2.38 enclose μ we have no way of knowing.

The reason that "we can feel 95% confident" is as follows: If we were to take 100 different samples from the same population and calculate the confidence limits for each sample, then we would expect that about 95 of these 100 intervals would contain the true value of μ, and 5 would not contain the true value of μ. Since we usually only have one sample and hence one confidence interval, we don't know whether our interval is one of the 95 or one of the 5. In this sense, then, we are 95% confident.

The meaning of our being 95% confident is shown in the results of a sampling exercise. Statistics students took 100 different random samples of size $n = 10$ each from a normal population with a mean μ and a standard deviation $\sigma = 1.29$. One hundred sample means, $\overline{X}$'s, and corresponding confidence intervals were then computed, and the results are presented in Figure 8.5. The mean μ was contained in 94 of the 100 intervals, as shown in the figure. This result conforms to our expectation that about 95 of our 100 intervals should encompass the mean μ.

 PRACTICE PROBLEM 8.1

A package-filling process at Portland Masonry fills bags of cement to an average weight of μ, but μ changes from time to time as the process adjustment changes through wear. However, even though the mean changes over time, the standard deviation of the weights is a constant $\sigma = 3$ pounds. A sample of 25 bags has just been taken, and their mean weight $\overline{X}$ was found to be 150 pounds. Assume that the weights of the individual bags at Portland Masonry are normally distributed, and find the 90% confidence limits for μ.

Solution Here

$$1 - \alpha = .90 \qquad \alpha = .10$$
$$\alpha/2 = .05 \qquad Z_{.05} = 1.645$$

By computation,

$$L = 150 - 1.645\frac{3}{\sqrt{25}} = 150 - .987 = 149.013$$

$$R = 150 + 1.645\frac{3}{\sqrt{25}} = 150 + .987 = 150.987$$

We are 90% confident that μ lies between 149.013 and 150.987 pounds, in that the procedure just followed will give limits enclosing the population mean 90% of the time.

 PRACTICE PROBLEM 8.2

For the Portland Masonry package-filling process, where a sample of 25 packages had a mean $\overline{X} = 150$ and process standard deviation $\sigma = 3$, calculate 95% confidence limits.

FIGURE 8.5

100 Confidence Interval
Estimates: Known
Standard Deviation

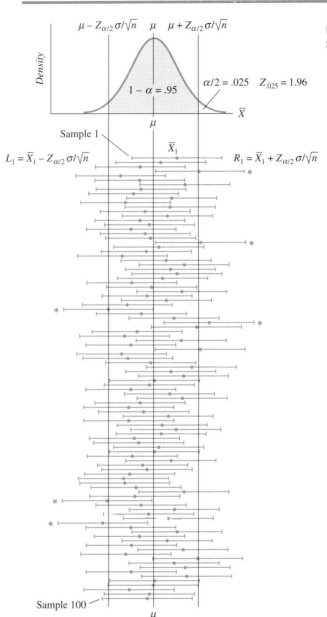

* These six intervals do not contain the mean μ.
  Of the 100 confidence intervals, 94 contain the mean μ.

Solution   Here, $Z_{\alpha/2}$ is $Z_{.05/2} = Z_{.025} = 1.96$, and so

$$L = 150 - 1.96\frac{3}{\sqrt{25}} = 150 - 1.176 = 148.824$$

$$R = 150 + 1.96\frac{3}{\sqrt{25}} = 150 + 1.176 = 151.176$$

The width of the 95% confidence interval in this problem is 2×1.176, or 2.352; the width of the 90% confidence interval computed in Practice Problem 8.1 is $2 \times .987$, or 1.974. The 90% confidence interval is better than the 95% confidence interval in that it is narrower; it apparently gives greater *precision*. But of course, we have less confidence in it (90% as opposed to 95%). Confidence has been traded for precision, and neither interval can be said to be better than the other. ▬▬

Thus, to find a confidence interval estimate for an unknown population mean μ of a normal distribution with known standard deviation σ, we take a random sample and find the point estimate $\overline{X}$, and then we subtract and add the amount $Z_{\alpha/2}\sigma/\sqrt{n}$ to find the confidence limits L and R as follows:

$$\overline{X} \pm Z_{\alpha/2}\frac{\sigma}{\sqrt{n}}$$

Problems: Section 8.2

9. For a random sample from a normal population with known standard deviation, where

$$n = 9 \qquad \overline{X} = 20 \qquad \sigma = 3 \qquad confidence\ level = 90\%$$

find the confidence interval for the unknown population mean for the specified level of confidence. The data values (which are only needed if you wish to use a computer) are as follows:

18.6518 22.8106 16.7165 24.6776 18.1082
23.1856 17.0271 21.1952 17.6274

10. For a random sample from a normal population with known standard deviation, where

$$n = 16 \qquad \overline{X} = 52 \qquad \sigma = 8 \qquad confidence\ level = 98\%$$

find the confidence interval for the unknown population mean for the specified level of confidence. The data values (which are only needed if you wish to use a computer) are as follows:

49.5863 45.5248 60.5368 60.8315 58.5781 59.2677
54.6502 49.0426 55.9083 49.5994 62.3354 45.5224
59.5915 43.3454 39.4638 38.2158

11. A Florida State Highway Department inspector is interested in knowing the mean weight of commercial vehicles traveling on the roads in his state. He takes a sample of 100 randomly selected trucks passing through state weigh stations and finds the mean gross weight $\overline{X} = 15.8$ tons. The standard deviation is known and equals 4.2 tons. Construct a 95% confidence interval for the mean gross weight of commercial vehicles traveling the highways of this state. Assume that the weights are normally distributed.

12. Assuming you have a sample from a normal population with known standard deviation, find the level of confidence used if $n = 16$, $\sigma = 8$, and the total width of a confidence interval for the mean is 3.29 units.

13. Assuming you have a sample from a normal population with known standard deviation, find the sample size when $\sigma = 10$ and the 95% confidence interval for the mean is from 17.2 to 22.8.

The interval estimates of the previous section, where σ was assumed *known,* were based on the fact that in normal sampling the standardized mean

$$Z = \frac{\overline{X} - \mu}{\sigma/\sqrt{n}} \tag{8.10}$$

8.3

t **DISTRIBUTIONS**

has a distribution that does not depend on μ and σ, namely, the standard normal distribution. In trying to construct a confidence interval for the case of *unknown* variance, we need to know what happens if in Equation (8.10) we merely replace σ by its estimator s, the sample standard deviation. We then have the following equation for a statistic known as a *t* **statistic.**

Equation for *t* Statistic

$$t = \frac{\overline{X} - \mu}{s/\sqrt{n}} \tag{8.11}$$

The value of Z from Equation (8.10) varies from one sample to the next because each sample has a different $\overline{X}$. Thus the Z of Equation (8.10) has one source of variation. The t of Equation (8.11), however, differs in that it varies from one sample to the next because of *two* sources of variation. The sample mean $\overline{X}$ and the sample standard deviation s change from sample to sample. Thus the term on the right side of Equation (8.11) follows a sampling distribution different from the normal distribution. For use of Equation (8.11), however, the *population* from which the n sample items was drawn must be normally distributed. The denominator of Equation (8.11), $s_{\overline{X}} = s/\sqrt{n}$, is the **estimated** or **sample standard error of the mean.**

Equation (8.11) is the first of a number of *t statistics* we will encounter. The *t* **distributions** form a family of distributions dependent on a parameter, the degrees of freedom. For the *t* variable in Equation (8.11), the value of the degrees of freedom ν (Greek nu) is $n - 1$, where n is the sample size. In general, the degrees of freedom for a *t* statistic are the degrees of freedom associated with the sum of squares used to obtain an estimate of a variance.

A *t* distribution is a symmetric distribution with mean 0 and standard deviation $\sqrt{\nu/(\nu - 2)}$, $\nu > 2$. Its graph is similar to that of the standard normal distribution, as Figure 8.6 shows. There is more area in the tails of the *t* distribution, and the standard normal distribution is higher in the middle. The larger the number of degrees of freedom, the more closely a *t* distribution resembles the standard normal. As the number of degrees of freedom increases without limit, a *t* distribution approaches the standard normal distribution, and it is convenient to regard the standard normal distribution as the *t* distribution with an infinite number of degrees of freedom.

Table VI of Appendix C gives percentage cutoff points of *t* distributions, those points on the *t* scale such that the probability of a larger *t* is equal to α for various values for the degrees of freedom ν. The percentage cutoff point t_α is defined as that point at which

$$P(t > t_\alpha) = \alpha$$

FIGURE 8.6

t Distributions with 1 and 2
Degrees of Freedom and
the Standard Normal
Distribution

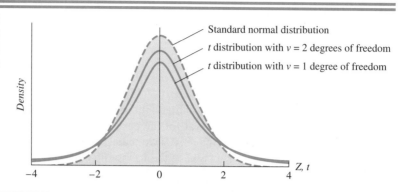

Since the distribution is symmetric about 0, only positive *t* values are tabulated. The lower α percentage cutoff point is $-t_\alpha$, because

$$P(t < -t_\alpha) = P(t > t_\alpha) = \alpha$$

In general, we denote a cutoff point for *t* by $t_{\alpha, \nu}$, where α is the probability level and ν is the degrees of freedom.

EXAMPLE 8.5 To find a cutoff point in the *t* table, Table VI, we locate the row of the table that corresponds to the given degrees of freedom ν and then take the value in that row that is also in the column headed by the given probability level α. For example, $t_{.05, 15}$ is the .05 percentage point of *t* with 15 degrees of freedom. Using Table VI, we find that it is equal to 1.753 (see Figure 8.7). If 15 is the degrees of freedom for *t*, then $P(t > 1.753) = .05$. The bottom line of the table corresponds to an infinite number of degrees of freedom, that is, to the standard normal distribution. Thus the .05 percentage cutoff point for the standard normal, $Z_{.05}$, is $t_{.05, \infty} = 1.645$, and the .025 percentage cutoff point, $Z_{.025}$, is $t_{.025, \infty} = 1.96$.

FIGURE 8.7

t Distribution with 15
Degrees of Freedom and
Percentage Cutoff Point
$t_{\alpha, \nu} = t_{.05, 15} = 1.753$

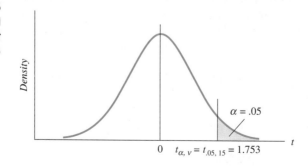

14. Find the value of t when the probability of a larger value is .005 and ν (degrees of freedom) is equal to 28.

15. Find the value of t when the probability of a smaller value is .975 and $\nu = 24$.

16. Find the value of t when the probability of a more extreme value (a larger value, sign ignored) is .10 when $\nu = 20$.

17. Find the value of t such that the probability of a larger value is .005 and the value for the degrees of freedom is very large.

18. Find the value of t such that the probability of a smaller value is .975 and the value for the degrees of freedom is very large.

19. Are the percentage cutoff points or t values essentially the same as the corresponding Z values when the value of the degrees of freedom is very large.

8.4
CONFIDENCE INTERVALS FOR NORMAL MEANS: UNKNOWN STANDARD DEVIATION

By the definition of percentage cutoff points, the following equality holds if t has a t distribution with ν degrees of freedom and each tail of the distribution has area $\alpha/2$:

$$P(-t_{\alpha/2, \nu} \le t \le t_{\alpha/2, \nu}) = 1 - \alpha$$

The fact underlying the construction of confidence intervals in the case where σ is estimated by s is that the statistic in Equation (8.11), $t = (\overline{X} - \mu)/(s/\sqrt{n})$, has a t distribution with $n - 1$ degrees of freedom, where n is the sample size and the population is assumed normal. Therefore,

$$P\left(-t_{\alpha/2, n-1} \le \frac{\overline{X} - \mu}{s/\sqrt{n}} \le t_{\alpha/2, n-1}\right) = 1 - \alpha \tag{8.12}$$

If we multiply each term of the expression within the parentheses by $s/\sqrt{n}$, subtract $\overline{X}$ from each, and then multiply through by -1, we arrive at

$$P\left(\overline{X} - t_{\alpha/2, n-1}\frac{s}{\sqrt{n}} \le \mu \le \overline{X} + t_{\alpha/2, n-1}\frac{s}{\sqrt{n}}\right) = 1 - \alpha \tag{8.13}$$

Equation (8.13) is the basis for the definition of a **confidence interval for the population mean μ** when the population variance is estimated by s.

Equations for Limits of a Confidence Interval for μ: σ Unknown

The $100(1 - \alpha)\%$ **confidence interval for the population mean μ** when the population is normally distributed and the population standard deviation σ is estimated by s is the interval bounded by the limits

$$L = \overline{X} - t_{\alpha/2, n-1}\frac{s}{\sqrt{n}} \quad \text{and} \quad R = \overline{X} + t_{\alpha/2, n-1}\frac{s}{\sqrt{n}} \tag{8.14}$$

The equations show that the expression for finding the limits of these confidence interval estimates is

$$(Point\ estimate) \pm t_{\alpha/2,\,n-1}(Sample\ standard\ error\ of\ point\ estimate)$$

Note that the confidence limits of Equation (8.9) involve the population standard deviation σ, assumed known in Section 8.2, but not here. The confidence limits of Equation (8.14) involve the sample standard deviation s, instead of σ. Thus these limits can be computed from the sample alone under the following conditions:

1. The population standard deviation is estimated by s.
2. The sample is a random sample taken from a normally distributed population.

By Equation (8.13), the probability that the confidence interval includes μ, the probability that the confidence limits L and R surround μ, is just $1 - \alpha$. The interpretation of these limits is the same as for those in Section 8.2.

The meaning of our being 95% confident is shown in the results of a sampling exercise. Students took 100 different random samples of size $n = 10$ each from a normal population with mean μ and standard deviation $\sigma = 1.29$ (the standard deviation σ was not known by the students). One hundred sample means, $\overline{X}$'s, sample standard deviations, s's, and corresponding confidence intervals were then computed, and the results are presented in Figure 8.8. The mean μ was contained in 95 of the 100 intervals, as shown in the figure. This result conforms to our expectation that about 95 of our 100 intervals should encompass the mean μ. Notice that the lengths of the intervals in Figure 8.8 are different, in contrast to the intervals depicted previously in Figure 8.5. The lengths of the 100 intervals are different because of the different standard deviations s's for the samples. Both the sample means $\overline{X}$'s and the sample standard deviations s's are contributing to the variability among the intervals. Consequently, it is appropriate to use the t distribution to construct the intervals.

The condition of randomness is necessary if we want to make probability statements about the results we obtain; therefore it cannot be relaxed. The condition of normality is not too strong and may be relaxed. Because of the central limit theorem, we may say that even though the population has a distribution that is not normal, a t distribution may still be used and the probabilities will not be greatly affected, provided we have a sufficiently large sample. What size sample may be considered sufficiently large depends on how much the population under consideration differs from normality. As a rough rule of thumb, many statisticians use t distributions if the sample size is greater than or equal to 30, even if the population is not normal. However, if the population distribution is unimodal and not too skewed, then adequate approximations to normality are often achieved with smaller samples. Also, if the sample is sufficiently large and Table VI in Appendix C does not contain the t value for the appropriate degrees of freedom, then $Z_{\alpha/2}$ can be used to approximate $t_{\alpha/2,\,\nu}$ in finding the confidence limits of Equation (8.14).

PRACTICE PROBLEM 8.3

A random sample of 20 homeowners in Akron, Ohio, showed that the mean mortgage payment being made by the people in the sample was $800 per month. The sample standard deviation was $60 per month. Find an interval within which you can be 98% confident that the mean monthly mortgage payment of all the people in the population will lie.

FIGURE 8.8

100 Confidence Interval Estimates: Unknown Standard Deviation

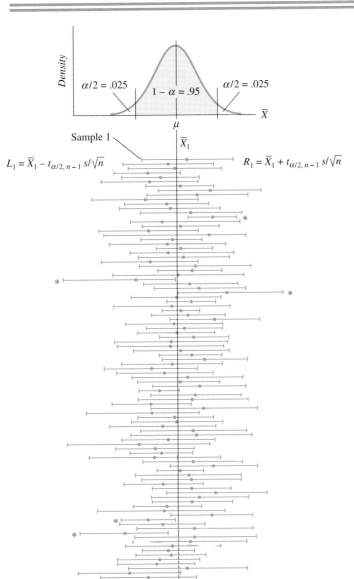

$$L_1 = \overline{X}_1 - t_{\alpha/2,\, n-1}\, s/\sqrt{n} \qquad\qquad R_1 = \overline{X}_1 + t_{\alpha/2,\, n-1}\, s/\sqrt{n}$$

* These five intervals do not contain the mean μ.
  Of the 100 confidence intervals, 95 contain the mean μ.

Solution From the problem statement we have

$$\overline{X} = \$800/\text{month} \qquad s = \$60/\text{month} \qquad n = 20$$

The 98% confidence limits can be found if we note two things. First, we must assume the population of payments from which we are sampling is normally distributed. Second, since we don't know the population standard deviation, we must use the confidence limits involving t and Equation (8.14), which incorporates the sample standard deviation s. Thus the limits are

$$L = \overline{X} - t_{.01, 19} \frac{s}{\sqrt{n}} = 800 - 2.539 \frac{60}{\sqrt{20}} = 800 - 34 = 766$$

and

$$R = \overline{X} + t_{.01, 19} \frac{s}{\sqrt{n}} = 800 + 2.539 \frac{60}{\sqrt{20}} = 800 + 34 = 834$$

Thus we can be 98% confident that the mean monthly mortgage payment of the people in the population from which we sampled is between \$766 and \$834. ▬

EXAMPLE 8.6 Let us return to Example 8.3, where ten patients took a sleep-enhancing drug, produced by Gosset Pharmaceuticals, Inc. This time we will not use the unrealistic assumption that we know the population variance. The sample variance s^2 for the data in Table 8.1 is 1.513, so the sample standard deviation s is $\sqrt{1.513}$, or 1.23, and $s/\sqrt{n}$ is $1.23/\sqrt{10}$, or .389. The value for the degrees of freedom $n - 1$ is 9. If we are to compute 95% confidence limits ($\alpha = .05$, $\alpha/2 = .025$), the appropriate percentage cutoff point is $t_{.025, 9}$, or 2.262. Since $\overline{X}$ is equal to 1.58, the 95% confidence limits are

$$L = \overline{X} - 2.262 \frac{1.23}{\sqrt{10}} = 1.58 - (2.262)(.389) = 1.58 - .88 = .70$$

and

$$R = \overline{X} + 2.262 \frac{1.23}{\sqrt{10}} = 1.58 + (2.262)(.389) = 1.58 + .88 = 2.46$$

We are 95% confident that μ lies between .70 and 2.46, in the sense that the random variables L and R (which for our data take values .70 and 2.46) have probability .95 of enclosing μ between them, whatever μ may be. ▬

We can use a computer to find confidence intervals for means of normal distributions with unknown variances. Confidence intervals are shown in Computer Exhibits 8.1A and 8.1B for the Gosset data of Example 8.3 concerning the additional

---

MINITAB

COMPUTER EXHIBIT 8.1A Confidence Interval for a Mean with Unknown Variance for the Gosset Data

```
MTB > set c1
DATA > 1.2  2.4  1.3  1.3  0.0  1.0  1.8  0.8  4.6  1.4
DATA > end

MTB > tinterval 95 c1  ◄─────────────

              N      MEAN    STDEV   SE MEAN     95.0 PERCENT C.I.
C1           10      1.58     1.23      0.39   (  0.70,     2.46)

MTB > stop
```

Command for a **tinterval** with a **95%** confidence level for **c1** (a 95% confidence interval using the t distribution for c1)

COMPUTER EXHIBIT 8.1B Confidence Interval for a Mean with Unknown Variance for the Gosset Data

MYSTAT

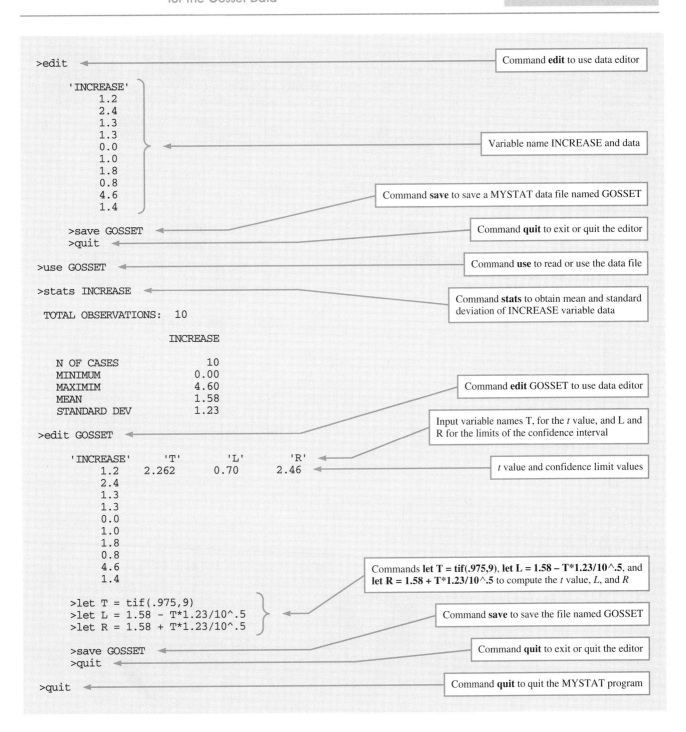

```
>edit                                          Command edit to use data editor

    'INCREASE'
       1.2
       2.4
       1.3
       1.3
       0.0                                      Variable name INCREASE and data
       1.0
       1.8
       0.8
       4.6                                      Command save to save a MYSTAT data file named GOSSET
       1.4

    >save GOSSET                                Command quit to exit or quit the editor
    >quit

>use GOSSET                                     Command use to read or use the data file

>stats INCREASE                                 Command stats to obtain mean and standard
                                                deviation of INCREASE variable data
  TOTAL OBSERVATIONS:   10

                      INCREASE

    N OF CASES              10
    MINIMUM              0.00
    MAXIMIM              4.60                    Command edit GOSSET to use data editor
    MEAN                 1.58
    STANDARD DEV         1.23                    Input variable names T, for the t value, and L and
                                                R for the limits of the confidence interval
>edit GOSSET

    'INCREASE'     'T'       'L'       'R'
       1.2        2.262      0.70      2.46      t value and confidence limit values
       2.4
       1.3
       1.3
       0.0
       1.0
       1.8
       0.8
       4.6                                       Commands let T = tif(.975,9), let L = 1.58 – T*1.23/10^.5, and
       1.4                                       let R = 1.58 + T*1.23/10^.5 to compute the t value, L, and R

    >let T = tif(.975,9)
    >let L = 1.58 - T*1.23/10^.5                 Command save to save the file named GOSSET
    >let R = 1.58 + T*1.23/10^.5

    >save GOSSET                                 Command quit to exit or quit the editor
    >quit

>quit                                           Command quit to quit the MYSTAT program
```

hours of sleep gained by taking a drug. Notice that the results for the confidence limits given in Computer Exhibit 8.1. match our results for this example.*

 PRACTICE PROBLEM 8.4

Electronic Systems claims that it has developed a new, cheap process for manufacturing high definition color television picture tubes. Assume they have approached you about investing in their company so that they can begin mass production of the picture tubes. One of the things you might want to know is how durable the tubes manufactured by the cheaper process may be. To answer this question, the company tested 16 picture tubes and found the following:

Mean length of life for 16 tubes	3220 hours
Standard deviation of life length in sample	120 hours

Construct a 90% confidence interval for the mean lifetime of the new picture tubes.

Solution Since $n = 16$, there are 15 degrees of freedom, and the t value associated with 5% in each tail of the t distribution is 1.753. Thus

$$L = \overline{X} - 1.753\frac{120}{\sqrt{16}} = 3220 - 52.6 = 3167.4$$

$$R = \overline{X} + 1.753\frac{120}{\sqrt{16}} = 3220 + 52.6 = 3272.6$$

Thus, we can be 90% confident that the mean lifetime of new picture tubes manufactured under the cheaper process is between 3167.4 and 3272.6 hours.

Problems: Section 8.4

20. For the following results from a sample taken from a normal distribution with unknown standard deviation,

$$n = 9 \qquad \sum X_i = 36 \qquad \sum (X_i - \overline{X})^2 = 288$$

find the 95% confidence interval estimate for the unknown mean μ.

21. For the following results from a sample taken from a normal distribution with unknown standard deviation,

$$n = 16 \qquad \sum X_i = 64 \qquad \sum (X_i - \overline{X})^2 = 180$$

find the 95% confidence interval estimate for the unknown mean μ. The data values (which are only needed if you wish to use a computer) are as follows:

−0.35091	7.95390	7.19857	5.08392	3.50356	4.49021
7.03274	3.75153	0.02868	−1.09935	9.70505	3.49417
7.73377	3.03312	−1.89230	4.33335		

*The t distribution was discovered in 1908 by W. S. Gosset, who wrote under the name "Student." The data in the example are his; see Fisher (1963).

22. For the following results from a sample taken from a normal distribution with un-known standard deviation,

$$n = 25 \qquad \sum X_i = 500 \qquad \sum X_i^2 = 12{,}400$$

find the 95% confidence interval estimate for the unknown mean μ.

23. For the following sample of data for the lives of collateralized mortgage obligations taken from a normal population,

    6   15   3   12   6   21   15   18   12

find the 95% confidence interval estimate for the unknown mean μ.

24. For the following sample of data for the price changes in ten-year Treasury bonds taken from a normal population,

    4   10   2   8   4   14   10   12   8

find the 95% confidence interval estimate for the unknown mean μ.

25. For the following sample of data for the monthly changes in the price of one troy ounce of gold on a commodity exchange taken from a normal population,

    2   5   9   11   13

find the 99% confidence interval estimate for the unknown mean μ.

26. A health services provider has been reimbursed by Gray Cross and Gray Shield for a large number of services. It appears that overpayments may have occurred on many of the reimbursements due to errors made during the billing process or owing to upcoding (billing for a more complicated service than was actually provided). Find the 95% confidence interval estimate of the mean overpayment per service given the following random sample of overpayments. (Assume normality.)

    $7 $3 $3 $3

27. A sample of 25 tanks holding compressed air at Eastco Corporation had a mean loss of pressure over a year of 15 pounds per square inch. The standard deviation of these losses was 4 pounds per square inch. Find a 95% confidence interval for the mean yearly pressure loss for tanks of this type. Assume normality. The data values (which are only needed if you wish to use a computer) are as follows:

8.1920	12.0691	13.6972	15.3893	16.7046	15.8237	13.8855
13.7947	8.3201	12.5442	17.9353	16.1431	17.1104	18.8413
21.2489	11.2440	13.0408	22.0610	8.7755	18.9428	16.5779
16.7327	14.1595	9.8727	21.8935			

28. A consumer protection agency took a random sample of a product whose package was marked as weighing 1 pound. The 16 packages in the sample had a mean weight of 1.10 pounds and a standard deviation of .36 pound. Assume normality.

    a. Find a 95% confidence interval for the mean weight of the 1-pound packages.
    b. On the basis of this interval, would the agency be justified in challenging the company's contention that the mean weight of packages is 1 pound?
    c. On what ground might the agency challenge the company's designation of a 1-pound weight on its packages?

29. During a recent maintenance check at Thio Corporation, the maximum thrust of 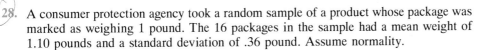 40 solid-fuel rocket motors was measured. The accompanying data give the differences

between the thrusts of the port and starboard motors, measured in pounds.

a. Find the 90% confidence interval estimate for the difference between motor thrusts.

b. If there is a consistent difference in thrusts of the port and starboard motors, then the rocket's trajectory may be affected. Would you conclude that the rocket's trajectory is consistently being affected by the thrust differences?

70	−40	−330	−20	−40	220	90	−80
0	−20	360	−50	200	30	130	20
310	−90	−270	−150	80	210	10	−450
−340	480	60	440	60	−30	−240	110
50	190	−400	−180	−100	−230	90	−200

30. A random sample of 28 families in New York City revealed that the children in these families average 130 minutes of television viewing each day. The standard deviation of the sample was 50 minutes. Find a 98% confidence interval for the mean daily viewing time of children in this city. Assume normality.

31. In an investigation of electronic equipment reliability at United Systems, Inc., a record of the time between equipment malfunctions was kept for nearly two years. Forty pieces of equipment were involved, giving rise to a sample of 850 observations on times between malfunctions. For the 850 observations the sample mean was 550.35 hours, and the sample standard deviation was 207 hours.

a. Find the 99% confidence interval estimate for the mean time between malfunctions for all equipment of this type.

b. United supplies the electronic equipment to National Electronics assemblers and guarantees an average time between malfunction of 600 or more hours. Does it appear that United has a problem with its guarantee?

32. A financial analyst with the Goldman Sachs group is examining the stock prices of companies in the paper and forest products industry. Consider the list of companies shown in Table 8.2, which gives each company's price–earnings (*PE*) ratio. Assume that the companies on the list are a random sample of all of the numerous companies in the paper and forest products industry.

a. Find the best point estimate for the mean *PE* ratio.

b. Find the 95% confidence interval estimate for the mean *PE* ratio.

TABLE 8.2	*Paper and Forest Products*	*PE Ratio (Estimate)*	*Paper and Forest Products*	*PE Ratio (Estimate)*
PE Ratio Data	Boise Cascade	19	Longview Fibre	12
	Bowater	20	Louisiana-Pacific	19
	Champion International	17	Mead	16
	Chesapeake	28	Pentair	15
	Consolidated Papers	13	Potlatch	12
	Federal Paper Board	19	Scott Paper	13
	Fort Howard Paper	18	Sonoco Products	15
	Georgia-Pacific	15	Stone Container	47
	Great Northern Nekoosa	19	Temple-Inland	18
	International Paper	17	Union Camp	21
	James River Corp. of Virginia	15	Westvaco	16
	Jefferson Smurfit	23	Weyerhaeuser	20
	Kimberly-Clark	15	Willamette Industries	16

SOURCE: *Business Week*, December 29, 1986, p. 155.

In Section 5.4 we defined a binomial population as a population whose elements are classified as belonging to one of two classes conventionally labeled success and failure. We represent the proportion of the population that belongs to the first class by p and the proportion that belongs to the second class by $1 - p$. In general, p is unknown, and the problem is to estimate it. That is, someone may ship us a large order of goods, and we wish to estimate p, the proportion of the order that meets our standards.

Suppose we take an independent sample of size n; let X be the number of successes in the sample (those that meet standards), and let $\hat{p}$ (equal to X/n) be the sample's proportion or fraction of successes. As we stated in Section 5.4, X has mean np and variance $np(1 - p)$, and $\hat{p}$ has the following mean, variance, and standard deviation:

$$E(\hat{p}) = p \tag{8.15}$$

$$\text{Variance}(\hat{p}) = \sigma_{\hat{p}}^2 = \frac{p(1 - p)}{n} \tag{8.16}$$

$$\text{Standard deviation}(\hat{p}) = \sigma_{\hat{p}} = \sqrt{\frac{p(1 - p)}{n}} \tag{8.17}$$

Equation (8.15) shows that $\hat{p}$ is an unbiased estimator of p. Equation (8.17) gives the standard deviation of this estimator. Its value, of course, cannot be found without knowledge of the value of the parameter p to be estimated. But we can get an idea of the standard deviation by replacing p with its estimator $\hat{p}$. When we do so, the standard deviation $\sigma_{\hat{p}}$ is estimated by

$$s_{\hat{p}} = \sqrt{\frac{\hat{p}(1 - \hat{p})}{n}} \tag{8.18}$$

The normal approximation to the binomial provides a method of finding approximate confidence limits for the binomial proportion of successes p for large sample sizes.

As we stated in Section 6.4, the standardized fraction of success

$$Z = \frac{\hat{p} - p}{\sqrt{p(1 - p)/n}} \tag{8.19}$$

has, for large n, approximately a standard normal distribution. If in Equation (8.19) we replace each p in the denominator by its estimator $\hat{p}$, we get a ratio

$$Z \approx \frac{\hat{p} - p}{\sqrt{\hat{p}(1 - \hat{p})/n}} \tag{8.20}$$

which also approximates a normal distribution for large n. A rule of thumb is that $n\hat{p}(1 - \hat{p}) \geq 5$ if we desire a good approximation.

By the definition of the percentage cutoff point $Z_{\alpha/2}$ for the standard normal distribution, we therefore have the approximate equation

$$P\left[-Z_{\alpha/2} \leq \frac{\hat{p} - p}{\sqrt{\hat{p}(1 - \hat{p})/n}} \leq Z_{\alpha/2}\right] \approx 1 - \alpha$$

If we operate on the inequalities here just as we did in constructing our previous confidence intervals, we arrive at

$$P\left[\hat{p} - Z_{\alpha/2}\sqrt{\frac{\hat{p}(1 - \hat{p})}{n}} \leq p \leq \hat{p} + Z_{\alpha/2}\sqrt{\frac{\hat{p}(1 - \hat{p})}{n}}\right] \approx 1 - \alpha$$

The random variables L and R defined by Equation (8.21) that follows have approximate probability $1 - \alpha$ of containing between them the true value of p, whatever it may be. They therefore can be used as approximate confidence limits for a $100(1 - \alpha)\%$ **confidence interval for the binomial proportion of successes p** if n is large.

Equations for Limits of a Confidence Interval for p

The $100(1 - \alpha)\%$ **confidence interval for the binomial proportion of successes p** if n is large is the interval bounded by the following confidence limits:

$$L = \hat{p} - Z_{\alpha/2}\sqrt{\frac{\hat{p}(1 - \hat{p})}{n}}$$

$$R = \hat{p} + Z_{\alpha/2}\sqrt{\frac{\hat{p}(1 - \hat{p})}{n}}$$

(8.21)

As a rough rule of thumb, most analysts feel that $n\hat{p}(1 - \hat{p})$ should be at least 5 before this approximation is adequate.

 PRACTICE PROBLEM 8.5

A sample of 100 voters in Laguna Beach contained 64 persons who favored a bond issue. Between what limits can we be 95% confident that the proportion of voters in the community who favor the issue is contained?

Solution Here,

$$n = 100 \qquad \textit{Number of successes} = 64$$
$$\hat{p} = 64/100 = .64 \qquad Z_{.025} = 1.96$$

Hence,

$$s_{\hat{p}} = \sqrt{\frac{\hat{p}(1 - \hat{p})}{n}} = \sqrt{\frac{(.64)(.36)}{100}} = .048$$

is our approximation for the population standard deviation of $\hat{p}$. And

$$L = .64 - (1.96)(.048) = .64 - .094 = .546$$
$$R = .64 + (1.96)(.048) = .64 + .094 = .734$$

We are 95% confident that the true proportion is covered by the interval, rounding to two decimal places, from .55 to .73.

PRACTICE PROBLEM 8.6

Auto Motor Company has received a shipment of several thousand parts. A random sample of 81 parts is selected, and 8 are found to be defective. Find the 90% confidence limits for the proportion of defects in the entire shipment.

Solution Here,

$$n = 81 \qquad\qquad \textit{Number of successes} = 8$$
$$\hat{p} = 8/81 = .099 \qquad Z_{.05} = 1.645$$

Hence,

$$s_{\hat{p}} = \sqrt{\frac{\hat{p}(1 - \hat{p})}{n}} = \sqrt{\frac{(.099)(.901)}{81}} = .033$$

approximates the population standard deviation of $\hat{p}$, $\sigma_{\hat{p}}$. And

$$L = .099 - (1.645)(.033) = .099 - .054 = .045$$
$$R = .099 + (1.645)(.033) = .099 + .054 = .153$$

We are 90% confident that the proportion of defects in the shipment is between .045 and .153.

Problems: Section 8.5

33. If $n = 25$ and the number of successes is 5 for a random sample from a binomial population, find the estimates for the following:

   a. The binomial proportion of successes p
   b. The variance of the binomial number of successes σ_X^2
   c. The standard deviation of the binomial number of successes σ_X
   d. The variance of the binomial proportion of successes $\sigma_{\hat{p}}^2$
   e. The standard deviation of the binomial proportion of successes $\sigma_{\hat{p}}$

34. If $n = 64$ and the number of successes is 32 for a random sample from a binomial population, find the estimates for the following:

   a. The binomial proportion of successes p
   b. The variance of the binomial number of successes σ_X^2
   c. The standard deviation of the binomial number of successes σ_X
   d. The variance of the binomial proportion of successes $\sigma_{\hat{p}}^2$
   e. The standard deviation of the binomial proportion of successes $\sigma_{\hat{p}}$

35. In a sample of 100 small castings taken from a very large production lot at Kisor Aluminum & Chemical Corporation, 98 were *not* defective.

   a. Find the best estimate for the proportion of nondefectives in the lot from which the sample was taken.
   b. Find the sample standard deviation of the proportion of nondefectives, $s_{\hat{p}}$.
   c. Find the 95% confidence interval estimate for the proportion of nondefectives, p.

36. In a public opinion poll of $n = 100$ subscribers of periodicals published by the giant publishing house TW that was conducted by Macrostat, Inc., 23 respondents indicated that the publisher should use recycled paper even if the subscription rates would need to be increased by doing so. Find the 95% confidence interval for p, the proportion of subscribers in favor of using recycled paper even if the subscription rate was increased.

37. Find the 95% confidence interval for p, the proportion of employees enrolled in a health maintenance organization's medical plan, if a random sample of size $n = 100$ employees found the number enrolled in the plan was 64.

38. Find the 95% confidence interval for p, the proportion of employees in the health care industry who are concerned about acquiring an immune deficiency virus on the job, if a random sample of size $n = 250$ found that 150 expressed concern about acquiring the virus on the job.

39. Find the 95% confidence interval for p, the proportion of small businesses in favor of increased tax rates to decrease the national debt, if a random sample of size $n = 1000$ found the number of businesses in favor of increased taxes was 50.

40. In a random sample of 100 physicians 30 specified their primary specialty as surgery. Find the 95% confidence interval estimate of the proportion of physicians who specify their primary specialty as surgery.

41. A sample of 100 stocks from the Oil and Gas Field Services industry revealed that 40 of the stocks had a negative annual return on investment the previous year. Under economic conditions similar to those of the previous year, find the 95% confidence interval estimate for the fraction of stocks that will have negative returns for this industry.

42. In a sample of 100 married purchasers of trucks made by Pickup Motors, the major decision maker was the husband for 70 of the purchases. Find the 95% confidence interval estimate for the proportion of truck purchases by married couples in which the major decision was made by the husband.

43. A local government agency has just purchased a new computer. The agency chief is interested in determining how many of the local government's 17,000 employees have computer programming experience. She takes a sample of 50 people and finds 6 who have had this experience. Construct a 95% confidence interval for the proportion of people who have had this experience. Use the normal approximation to the binomial.

44. If 10% of a sample of 400 computer memory chips made at Digital Devices, Inc., were found to be defective, what is the 99% confidence interval for the proportion defective in the lot from which the sample was taken? Use the normal approximation to the binomial.

8.6
SAMPLE SIZE

In all of the examples in the previous two sections, we assumed that a sample *had been taken*. That is, we looked at the data *after* the sample results had been obtained. There are many situations, however, when we know *before* we take the sample that we will be constructing a confidence interval with the results we obtain. In these situations we often ask the question, "How big a sample should we take?" Three formulas will be given in this section for determining the sample size needed to obtain confidence intervals for the population mean μ and the population proportion of successes p.

To determine the sample size needed to construct a confidence interval before the sample is taken, we need three pieces of information:

1. The confidence level (usually 90%, 95%, or 98%).
2. The desired maximum width of the confidence interval that is to be constructed, that is, the value $w = R - L$. The amount of error we can tolerate in our estimate of the population mean is thus $w/2$.
3. The standard deviation associated with the population from which we are about to sample. That is, we need to know σ if our interval is estimating μ, or $\sigma_{\hat{p}}$ if our interval is estimating p.

If we desire a confidence interval to estimate the population mean μ, then we form the interval by finding a sample mean $\overline{X}$ and adding and subtracting the following term:

$$Z_{\alpha/2} \frac{\sigma}{\sqrt{n}}$$

Since this amount is both added and subtracted, the width of the confidence interval will be twice this amount:

$$w = 2Z_{\alpha/2} \frac{\sigma}{\sqrt{n}} \tag{8.22}$$

If we know the three items of information listed previously, we can use Equation (8.22) to solve for n, the sample size we need to obtain the confidence interval specified by the three features. That is, we must specify w, the width of the confidence interval we want to construct; Z, which indicates the confidence level; and σ, the population standard deviation. Once these items have been specified, we can solve Equation (8.22) for the **sample size n** and obtain Equation (8.23).

Equation for Sample Size for Estimating Mean, σ Known

$$n = \frac{4(Z_{\alpha/2})^2 \sigma^2}{w^2} \tag{8.23}$$

 PRACTICE PROBLEM 8.7

We know that the standard deviation of daily output on a production line for steel pipe at Republic Supply is $\sigma = 10$ tons. We want to estimate the mean daily output of the production line, μ, to within ± 2.5 tons with 95% confidence. What sample size is required?

Solution In this problem $w = 5$ tons (since ± 2.5 tons implies a total width of $2 \times 2.5 = 5$). We also know from the problem statement that $\sigma = 10$ and $Z_{.025} = 1.96$, since the confidence level is 95%. Using Equation (8.23) yields the following sample size:

$$n = \frac{4(1.96)^2(10)^2}{(5)^2} = 61.5$$

We will be conservative and round up, so a sample size of $n = 62$ will give us an interval that allows us to estimate the population mean daily output to within ± 2.5 tons with 95% confidence (or one slightly narrower).

Practice Problem 8.7 is somewhat artificial since there will be few actual applications where one wishes to estimate the population mean μ and can use the population standard deviation σ in Equation (8.23). That is, if we don't know μ, we probably don't know σ either. Thus we are often forced to replace σ^2 in Equation (8.23) with some reasonable estimate s^2, which has been obtained from a previous sample of the population or from a preliminary sample. But we learned in Section 8.4 that confidence intervals constructed with sample standard deviations (since σ is not known) require the use of $t_{\alpha/2, n-1}$ instead of $Z_{\alpha/2}$.

These intervals were formed in Equation (8.14) by adding and subtracting the amount

$$t_{\alpha/2, n-1} \frac{s}{\sqrt{n}}$$

around the sample mean. Thus this expression is half the width of the confidence interval, and the full width of the interval is

$$w = 2t_{\alpha/2, n-1} \frac{s}{\sqrt{n}} \tag{8.24}$$

When Equation (8.24) is solved for n, we obtain the following equation, which is exactly parallel to (8.23), the **sample size** formula used when the population standard deviation σ is known.

Equation for Sample Size for Estimating Mean, σ Unknown

$$n = \frac{4(t_{\alpha/2, n-1})^2 s^2}{w^2} \tag{8.25}$$

The use of Equation (8.25) is demonstrated in the following problem.

 PRACTICE PROBLEM 8.8

A manager for Seers Company wishes to estimate the mean length of time μ it takes company crews to do an installation job. She wishes to estimate μ to within ± 5 minutes with 90% confidence. Since she has no idea about the value of σ, the population standard deviation, she took a preliminary sample of $n = 15$ jobs and found the 15 job completion times had a standard deviation of $s = 20$ minutes. How much larger should she make her sample to obtain the confidence interval she desires?

Solution From the problem statement we know that

$$w = 10 \qquad s = 20 \qquad \alpha = .10$$

Since the population standard deviation is unknown, we must use Equation (8.25) to determine the sample size. We will use a t value of $t_{\alpha/2, n-1} = t_{.05, 14} = 1.761$. Then

$$n = \frac{4(1.761)^2(20)^2}{(10)^2} = 49.6$$

Thus the manager must round up and take $50 - 15 = 35$ additional observations to complete her total sample.

Two important points should be observed about this problem. First, it demonstrates a process known as *two-stage sampling*. The first stage of the sampling was the preliminary step involving 15 observations. This step allowed the manager to obtain a sample standard deviation s to use in Equation (8.25). The second stage of the sampling involved 35 observations to fill out the total sample to the full 50 items.

The second point about the problem concerns the use of the t value. We chose to use $t_{.05, 14}$. Technically, however, the final confidence interval following the second stage of sampling will be constructed with the *total* sample's standard deviation, based on 50 observations and a t value with $n - 1 = 49$ degrees of freedom. Since $t_{.05, 49}$ is about 1.678, the actual confidence interval obtained will likely be narrower than the ±5 minutes specified in the problem. Thus Equation (8.25) and the method demonstrated in Practice Problem 8.8 usually give a conservatively large sample size—one that is as large as, or larger than, what will be needed to obtain the desired confidence interval.*

 PRACTICE PROBLEM 8.9

A claims manager for the Chubbs Group would like to know the mean size of the automobile insurance repair claims paid by his company. Thus he took a sample of $n = 20$ claims and found $\overline{X} = \$900$ and $s = \$300$. How much larger should his sample be if he wants to estimate the mean payment to within ±$100 with 95% confidence?

Solution From the problem statement we have $w = 200$, $s = 300$, and $t_{.025, 19} = 2.093$, to be used in Equation (8.25) (since we only know the sample standard deviation):

$$n = \frac{4(t_{\alpha/2, n-1})^2 s^2}{w^2} = \frac{4(2.093)^2(300)^2}{(200)^2} = 39.4$$

Thus if he examines $40 - 20 = 20$ more payments, the claims manager will be able to obtain the interval he desires (or one slightly narrower). ▬▬▬

The preceding three problems have concerned methods of determining the sample size needed to estimate μ, the population mean. Comparable methods can be used to estimate p, the proportion of successes in a binomial population. According to Equation (8.21), a confidence interval used to estimate p is formed by taking a sample, obtaining the sample proportion of successes $\hat{p}$, then adding to $\hat{p}$ and subtracting from $\hat{p}$ the following term:

$$Z_{\alpha/2} \sqrt{\frac{\hat{p}(1 - \hat{p})}{n}}$$

Thus any confidence interval is twice this wide, and

$$w = 2Z_{\alpha/2} \sqrt{\frac{\hat{p}(1 - \hat{p})}{n}} \tag{8.26}$$

*There is an iterative method for determining the sample size more exactly than Equation (8.25) does. However, such a method is beyond the scope of this material. Our method does not *always* give an interval that is too large, however, since when the sample is filled out, the sample standard deviation s may become large and offset the increased sample size.

We then solve Equation (8.26) for the **sample size n** to estimate a proportion according to the following equation.

> **Equation for Sample Size for Estimating Proportion**
>
> $$n = \frac{4Z_{\alpha/2}^2\,\hat{p}(1 - \hat{p})}{w^2}$$
>
> (8.27)

We can use this formula to find the sample size necessary to obtain a confidence interval as long as we can specify the level of confidence desired (and thus the $Z_{\alpha/2}$ value), the width of the desired interval (and thus w), and some estimate of the proportion of successes that will be obtained from the sample (and thus $\hat{p}$).

The value of $\hat{p}$ inserted in Equation (8.27) can be obtained in several ways. First, it might be estimated by using a preliminary sample in a two-stage sampling procedure like that described earlier. Also, a person experienced with the situation where the sampling is being done might be able to supply a good estimate of what $\hat{p}$ will be. Finally, where there is no suggestion as to the possible value for $\hat{p}$ and when two-stage sampling is not practical, we should let $\hat{p} = .5$. When $\hat{p}$ is assigned the value of .5, the product $\hat{p}(1 - \hat{p})$ has a higher value than when any other value is used for $\hat{p}$. Table 8.3 shows how the value of $\hat{p}(1 - \hat{p})$ changes with various assignments for $\hat{p}$. When $\hat{p} = .5$, the product term reaches the maximum value of .25. Thus when this product is inserted in Equation (8.27), the largest numerator of the fraction is obtained, and the largest possible value of n results. In other words, assigning $\hat{p} = .5$ is the most conservative choice we can make since it will yield a sample size as large as, or possibly a little larger than, we need for the confidence interval we seek.

 PRACTICE PROBLEM 8.10

A quality control engineer with Systems Planning Corporation would like to estimate the proportion of defects being produced on a production line that needs quality improvements to within $\pm.04$ with 95% confidence. In a preliminary sample of 25 items the engineer found eight defectives. How much larger a sample needs to be taken?

Solution From the problem statement, we know

$$w = (2)(.04) = .08 \qquad Z_{\alpha/2} = 1.96 \qquad \text{Preliminary } \hat{p} = 8/25 = .32$$

TABLE 8.3	$\hat{p}$	$\hat{p}(1 - \hat{p})$	$\hat{p}(1 - \hat{p})$	$\hat{p}$	$\hat{p}(1 - \hat{p})$	$\hat{p}(1 - \hat{p})$
Product $\hat{p}(1 - \hat{p})$ for	.1	.9	.09	.6	.4	.24
Selected Values of $\hat{p}$	.2	.8	.16	.7	.3	.21
	.3	.7	.21	.8	.2	.16
	.4	.6	.24	.9	.1	.09
	.5	.5	.25	1.0	.0	.00

From Equation (8.27), we find

$$n = \frac{4(1.96)^2(.32)(.68)}{(.08)^2} = 522.5$$

Thus, the quality control engineer needs to examine $523 - 25 = 498$ more items.

We can use this problem to point out the value of a preliminary sample. If the preliminary sample had not been taken, the engineer would have had to assume, conservatively, that the defect rate, p, was close to .5. In that case the engineer would propose a sample of

$$n = \frac{4(1.96)^2(.5)(.5)}{(.08)^2} = 600.3$$

Thus the preliminary sample resulted in a total sample that is $601 - 523 = 78$ smaller than the sample that would have been taken without it.

 PRACTICE PROBLEM 8.11

The credit manager of Northwest Department Store would like to know what proportion of the charge customers take advantage of the store's deferred-payment plan each year. The manager would like to estimate this figure to within 10% at a 90% confidence level, but she has no idea right now about what this proportion might be.

Solution From the problem statement we can find

$$w = (2)(.10) = .20 \qquad Z_{\alpha/2} = Z_{.05} = 1.645$$

Since there is no preliminary sample and the credit manager has no idea about the possible $\hat{p}$ value, we should assign $\hat{p} = .5$. Then from Equation (8.27) we obtain

$$n = \frac{4Z_{\alpha/2}^2 \hat{p}(1 - \hat{p})}{w^2} = \frac{4(1.645)^2(.5)(.5)}{(.20)^2} = 67.7$$

This number seems like a reasonable sample to take, but the credit manager could probably save a little work by taking a preliminary sample first to get an estimate of $\hat{p}$. If this estimate is very much different from .5, it should be used in Equation (8.27) again to reestimate the sample size, which would necessarily be less than 68. ▬▬

Equations (8.23), (8.25), and (8.27) can be examined to yield a few generalizations about the size of samples needed to obtain confidence intervals. First, the more confidence we desire, all other things being constant, the larger the sample size will be (since the Z and t values in these formulas are larger). Second, the larger the variation in the population (as measured by σ or s or indirectly by $\hat{p}$ in these formulas), the larger is the sample. And finally, the most influential element of these formulas is the confidence interval width w. The narrower the confidence interval, the larger the sample size will be. Since w^2 appears in the denominator of all these formulas, a general rule is that, to be twice as precise, we need to increase our sample size by a factor of 4.

In many sampling situations such as auditing, the population from which the sample is taken is finite. That is, the population of invoices being audited may have some finite size we will call N. In that case the standard deviation of the sample result is smaller by a factor that is called the *sample finite population correction factor*.

When sampling is done from a finite population, then the confidence interval widths in Equations (8.22), (8.24), and (8.26) are all reduced by multiplying sample finite population correction factors by the terms on the right sides of the equations. The factors then have an impact on the sample sizes given by Equations (8.23), (8.25), and (8.29). For instance, when the finite population correction factor is included, Equation (8.27) becomes

$$n = \frac{4Z_{\alpha/2}^2 \hat{p}(1 - \hat{p})N}{(N - 1)w^2 + 4Z_{\alpha/2}^2 \hat{p}(1 - \hat{p})} \qquad (8.28)$$

This revised equation seems quite complicated, and we will not use it in the problems at the end of this section and the end of the chapter. The impacts of the finite population correction factors are quite minimal in terms of reducing the required sample sizes unless the samples become more than about 10% of the populations. Since there are few realistic situations in which the sample size turns out to be more than 10% of the population, we will ignore the revised sample size equations and stay with the simpler versions of (8.23), (8.25), and (8.27). However, auditors sometimes use standard tables for determining sample sizes, and the standard tables were developed by using the more complicated formulas.

Problems: Section 8.6

45. Hercuan, Inc., manufactures fly rods made of composite material that is baked for a given amount of time and at a given temperature. A normal population is taken of the strengths of the fly rods, and it has a known population standard deviation of .75 pound. How large a sample must be taken in order that the total width w of the 95% confidence interval for the population mean will not be greater than .10?

46. If a normal population of overpayments by National Insurance to health service provider is known to have a population standard deviation equal to $5, how large a sample would we take in order to be 95% confident that the sample mean overpayment will not differ from the population mean overpayment by more than $\pm\$.80$?

47. Southwest Oil and Gas, Inc., wishes to estimate the mean amount of water that has seeped into the oil storage tanks at its refineries in the southern United States. A preliminary sample of $n = 21$ tanks showed that $s = 45$ gallons. How much larger should the sample be in order to estimate the mean water content of the tanks to within ±10 gallons with 95% confidence?

48. The managers at STAT-SOFTWARE wish to determine what proportion of personal computer users who have statistics packages use the company's software. Find the sample size needed to be about 95% confident that their sample results $\hat{p}$ will not differ from p by more than $\pm.05$. Assume that the managers believe they currently have about 30% of the market.

49. How large a sample should we take from a large population of microprocessors produced by International Rectifier Corporation in order to be about 95% sure that the sample proportion of defectives $\hat{p}$ will be within $\pm.04$ of p? Assume that we suspect that p is about .20.

50. A sales manager with Ideal Industrial Corporation wants to know what proportion of her accounts are inactive. A preliminary sample of $n = 25$ accounts showed that 4 were inactive. How many more accounts should she examine if she wants her confidence interval to be no more than $w = .08$ wide with 90% confidence?

In this chapter, we discussed statistical estimation. We used procedures described in the previous sections in situations where we were interested in taking a sample in order to estimate some characteristic of the population.

When one deals with a population of continuous measurements, that population is characterized by a mean μ and a standard deviation σ. These characteristics are estimated by the sample mean $\overline{X}$ and the sample standard deviation s. When one deals with a population composed only of successes and failures, a binomial population, then the population is characterized by the proportion of successes p in the population. The sample proportion of successes $\hat{p}$ is the best single estimator of p.

However, a single point estimate of a population characteristic is almost surely off the mark to some extent. Thus in estimating population characteristics, we often construct confidence intervals that show an interval of values within which the population characteristic might lie. All of the $100(1 - \alpha)\%$ confidence limits we discussed have the same general form:

$$(\textit{Point estimate}) \pm \begin{pmatrix} Z_{\alpha/2} \\ \text{or} \\ t_{\alpha/2,\nu} \end{pmatrix} \begin{pmatrix} \textit{Standard error of point estimate} \\ \text{or} \\ \textit{Sample standard error of point estimate} \end{pmatrix}$$

That is, we find the left confidence limit by calculating the point estimate and then subtracting a Z or a t value times the standard error of the point estimate. We find the right confidence limit by adding the same quantity to the point estimate.

The Z value is used when we have a sample from a normal population with a known variance (which is seldom the case in actual practice). The t value is used when we have a sample from a normal population whose variance is estimated by s. The number of degrees of freedom associated with the t value is $\nu = n - 1$, and its value is determined by the number of degrees of freedom in the standard error of the point estimate.

If n is large, say greater than or equal to 30, then the population does not need to be exactly normally distributed due to the central limit effect. If the sample size is so large that the t table does not contain the t value for the appropriate degrees of freedom, then $Z_{\alpha/2}$ can be used to approximate $t_{\alpha/2,\nu}$. However, computers use the t distribution even if the sample size is large.

When the population characteristic we are estimating is p, then the populations we deal with are binomial (not normal), and the $Z_{\alpha/2}$ value is used if n is large.

In addition, in this chapter we considered the question of how large a sample size is needed to construct confidence intervals for a population mean μ and for a population proportion of successes p. To determine the same size, one must specify three pieces of information: the confidence level desired, the maximum width of the confidence interval to be constructed, and an estimate of the population dispersion. The third item is often estimated by using a preliminary sample, which is the first stage in two-stage sampling. Equations (8.23), (8.25), and (8.27) can be used to determine sample sizes once the appropriate parameters have been identified.

Table 8.4 summarizes the types of confidence intervals discussed in this chapter. Note that the formulas given in the second column of the table all have the general form for confidence intervals indicated previously. When the plus is used in the plus-or-minus operation, the right confidence limit R is obtained. Limit L is obtained when the minus is used.

TABLE 8.4   Summary of Confidence Interval Formulas

Population Characteristic Being Estimated	*Expression for Limits of a 100 (1 − α)% Confidence Interval*	
μ, the mean of a normal population of continuous measurements	$\overline{X} \pm Z_{\alpha/2} \dfrac{\sigma}{\sqrt{n}}$ or $\overline{X} \pm t_{\alpha/2,\,n-1} \dfrac{s}{\sqrt{n}}$	(8.9) (8.14)

Comment: If σ is unknown, use Equation (8.14), s, and the appropriate t value. If n is large, say, greater than or equal to 30, then $Z_{\alpha/2}$ can be used in place of $t_{\alpha/2,\,n-1}$ in Equation (8.14).

p, the proportion of successes in a binomial population	$\hat{p} \pm Z_{\alpha/2} \sqrt{\dfrac{\hat{p}(1 - \hat{p})}{n}}$	(8.21)

Comment: The formula applies when $n\hat{p}(1 - \hat{p}) \geq 5$.

REVIEW PROBLEMS

Answers to odd-numbered problems are given in the back of the text.

51. The results from a sample taken from a population are

$$n = 9 \qquad \sum X_i = 450 \qquad \sum(X_i - \overline{X})^2 = 32$$

Find the point estimates for the following:

a. The mean μ **b.** The variance σ^2
c. The standard deviation σ **d.** The standard deviation of the sample mean $\sigma_{\overline{X}}$

52. The results from a sample taken from a population are

$$n = 16 \qquad \sum X_i = 320 \qquad \sum X_i^2 = 6640$$

Find the point estimates for the following:

a. The mean μ **b.** The variance σ^2
c. The standard deviation σ **d.** The standard deviation of the sample mean $\sigma_{\overline{X}}$

53. The sample of data for the monthly increase in the benchmark price of a barrel of West Texas crude oil taken from a normal population is

$$1 \quad 3 \quad 3 \quad 5 \quad 6 \quad 6$$

Find the point estimates for the following:

a. The mean μ **b.** The variance σ^2
c. The standard deviation σ **d.** The standard deviation of the sample mean $\sigma_{\overline{X}}$

54. The sample of data for the semi-annual rate of return for stocks taken from the financial services industry from a normal population is

$$-4 \quad 2 \quad -6 \quad 0 \quad -4 \quad 6 \quad 2 \quad 4 \quad 0$$

Find the point estimates for the following:

a. The mean μ **b.** The variance σ^2
c. The standard deviation σ **d.** The standard deviation of the sample mean $\sigma_{\overline{X}}$

55. The sample of data for the weights of three and one-third ton billets of hot-rolled steel taken from a normal population is

    6676   6678   6681   6680   6681   6678

    Find the point estimates for the following:
    a. The mean μ **b.** The variance σ^2
    c. The standard deviation σ **d.** The standard deviation of the sample mean $\sigma_{\overline{X}}$

56. For a random sample from a normal population with known standard deviation, where

    $$n = 25 \qquad \overline{X} = 120 \qquad \sigma = 20 \qquad confidence\ level = 95\%$$

    find the confidence interval for the unknown population mean for the specified level of confidence. The data values (which are only needed if you wish to use a computer) are as follows:

146.761	97.811	73.796	112.885	119.933	119.510	110.584
150.504	115.196	133.734	129.046	119.677	105.539	93.023
115.591	110.371	169.172	148.292	135.707	107.916	118.502
120.220	106.037	115.874	124.319			

57. Assuming you have a sample from a normal population with known standard deviation, find the known standard deviation when $n = 100$ and the 98% confidence interval for the mean is 23.26 units in width.

58. The results from a sample taken from a normal distribution with unknown standard deviation are

    $$n = 9 \qquad \sum X_i = 450 \qquad \sum(X_i - \overline{X})^2 = 32$$

    Find the 95% confidence interval estimate for the unknown mean μ. The data values (which are only needed if you wish to use a computer) are as follows:

50.8241	48.8784	48.1478	50.6697	51.4943
51.1064	46.9768	48.5247	53.3778	

59. The results from a sample taken from a normal distribution with unknown standard deviation are

    $$n = 16 \qquad \sum X_i = 320 \qquad \sum X_i^2 = 6640$$

    Find the 95% confidence interval estimate for the unknown mean μ.

60. For the following sample of data for the monthly increase in the benchmark price of a barrel of West Texas crude oil taken from a normal population,

    1   3   3   5   6   6

    find the 99% confidence interval estimate for the unknown mean μ.

61. For the following sample of data for the semi-annual percentage rate of return for stocks taken from the financial services industry from a normal population,

    -4 2 -6 0 -4 6   2   4   0

    find the 99% confidence interval estimate for the unknown mean μ.

62. For the following sample of data for the weights of three and one-third ton billets of hot-rolled steel taken from a normal population,

    6676   6678   6681   6680   6681   6678

    find the 98% confidence interval estimate for the unknown mean μ.

63. The mean height of 16 children whose parents bought a brand XBM bicycle was 56 inches. The standard deviation of heights for the population of children was 8 inches, and heights were normally distributed. Find a 90% confidence interval for the true mean height of children whose parents buy this bicycle.

64. A real estate broker for Geselman and Associates randomly selected four properties from a subdivision, and an appraiser gave an opinion on the value of each property as $17, $21, $21, and $21 thousand. Find the 95% confidence interval estimate for the mean value of a property located in the subdivision. Assume normality.

65. A consumer advocate randomly selected nine prescription drug items at an American Drug pharmacy and found the values of the suggested retail prices minus the actual retail prices as $0, $8, $0, $0, $3, $0, $3, $5, and −$1. Find the 95% confidence interval estimate for the mean difference in retail prices. Assume normality.

66. Sensitivity tests were made on 25 randomly selected cathode ray tubes produced by West House Electonics Corporation. The mean was 3.2 microvolts, and the estimated variance was .20. Find a 95% confidence interval for the mean sensitivity of this type of tube. Assume normality. The data values (which are only needed if you wish to use a computer) are as follows:

2.64574	2.59041	3.43048	2.60505	3.29543	3.94281	3.21996
3.11873	3.07645	3.59652	3.46971	2.72210	3.37996	2.71890
4.00157	2.83258	4.17699	3.00077	3.39857	2.93425	3.27506
2.80532	3.27043	3.66823	2.82398			

67. A soft-drink bottler took a sample of nine families and found a mean consumption of his drink of 100 ounces per week with a standard deviation of 18 ounces. Find a 98% confidence interval for the mean weekly consumption of this drink in the entire population. Assume normality.

68. Acme Milling, Inc., has just installed a new automatic milling machine. The time it takes the machine to mill a particular part is recorded for a sample of nine observations. The mean time is found to be $\overline{X} = 8.50$ seconds, and $s^2 = .0064$. Find a 90% confidence interval for the true unknown mean time for milling this part. Assume normality.

69. Twenty secretaries were given a spelling test. They were then given a special short course designed to improve spelling ability and were tested again at the end of the course. The differences between the first and the second scores had a mean equal to 4 (that is, there was a 4-point improvement). The variance of the improvement was 16. Find a 90% confidence interval for the mean of the score improvement if this course were given to all the secretaries in the company. Assume normality.

70. A stimulus was tested at a Humanon hospital for its effect on diastolic blood pressure. Twenty men had their blood pressure measured before and after the stimulus. The results for blood pressure changes are as follows:

8	7	1	9	−8	−3	1	2	−8	2
3	8	1	7	−5	−4	0	7	−1	5

Find a 95% confidence interval for the mean change in blood pressure. Assume normality.

71. Cattle are sold by weight. A sample of six animals from one herd at American Livestock Corporation yielded the following weights (in pounds):

692   800   685   790   695   793

Find a 95% confidence interval for the mean weight per animal in the herd. Assume normality.

72. The accompanying values are for the weights of a random sample of 30 cattle (in pounds) after they have been in a Mountain States Feed Exchange feedlot for 60 days. Find the 95% confidence interval estimate for the mean weight of the feedlot cattle.

982	1205	258	927	620	1023
395	1406	1012	762	840	960
1056	793	713	736	1582	895
1384	862	1152	1230	1261	624
862	1650	368	358	956	1425

73. An auditor with Ernie & Whinst found that the billing errors that occurred on a random sample of invoices were $32, $-\$45$, $66, $2, $-\$8$, $-\$51$, $12, and $18. Positive errors indicate that the customer was billed too much, and negative errors indicate that the customer was billed too little. Find the 99% confidence interval estimate for the mean billing error. Assume normality.

74. If $n = 400$ and the number of successes is 144 for a random sample from a binomial population, find the estimates for the following:
   a. The binomial proportion of successes p
   b. The variance of the binomial number of successes σ_X^2
   c. The standard deviation of the binomial number of successes σ_X
   d. The variance of the binomial proportion of successes $\sigma_{\hat{p}}^2$
   e. The standard deviation of the binomial proportion of successes $\sigma_{\hat{p}}$

75. If $n = 100$ and the number of successes is 64 for a random sample from a binomial population, find the estimates for the following:
   a. The binomial proportion of successes p
   b. The variance of the binomial number of successes σ_X^2
   c. The standard deviation of the binomial number of successes σ_X
   d. The variance of the binomial proportion of successes $\sigma_{\hat{p}}^2$
   e. The standard deviation of the binomial proportion of successes $\sigma_{\hat{p}}$

76. Find the 95% confidence interval estimate for the binomial proportion of successes p if a random sample of $n = 100$ has been selected and the number of successes is 87.

77. Find the 95% confidence interval estimate for the binomial proportion of successes p if a random sample of $n = 250$ has been selected and the number of successes is 125.

78. Find the 95% confidence interval estimate for the binomial proportion of successes p if a random sample of $n = 1000$ has been selected and the number of successes is 120.

79. Find the 95% confidence interval estimate for the binomial proportion of successes p if a random sample of $n = 250$ has been selected and the number of successes is 200.

80. From 1000 heads of families questioned by Dune & Broadstreet Corporation, it was found that 930 owned one or more television sets. Find the 95% confidence interval estimate for the proportion of families that own one or more television sets.

81. A sample of 30 certified public accountants in one geographical region was asked if the Financial Accounting Standards Board (FASB) is too slow in issuing new accounting rules, and 20 of those surveyed thought that FASB was too slow. Find the 95% confidence interval estimate for the proportion of certified public accountants in this region who think that FASB is too slow in issuing new accounting rules.

82. In a city a sample of 1000 people was interviewed. In the sample, 290 people drove Fords. Give the 95% confidence interval for the percentage of people in this city who drive Fords.

83. In a sample of 100 urban families, 84 had checking accounts with banks insured by the Federal Deposit Insurance Corporation (FDIC). Find a 95% confidence interval for the proportion of families who have FDIC-insured checking accounts.

84. A sample of 200 people who had filed tax returns with the Internal Revenue Service (IRS) during each year of a three-year period was taken, and 8 of the people had been selected for at least one tax audit for the period. Find the 90% confidence interval estimate for the proportion of tax filers who undergo at least one IRS audit during a three-year period.

85. In a sample of 100 high-pressure castings at General Corporation, 20 were found to have chill folds resulting from improper heating of the mold. Find the 90% confidence interval for the true percentage of all castings that have chill folds.

86. One-fourth of 300 persons interviewed were found to be opposed to the policies of a county school superintendent. Calculate a 95% confidence interval for the proportion of the population that is opposed to these policies. Use the normal approximation.

87. If X is the amount of the markup in the price of new automobiles at a local dealership for automobiles manufactured in Japan and X is a normal variable with known variance equal to 625 (squared dollars), how large a sample must we take to be 95% confident that the sample mean will not differ from the true mean by more than ±$6?

88. A preliminary sample of 10 ball bearings made by a Low-Shear Industries process yielded a standard deviation of the diameter that was $s = .3$ millimeter. How much larger a sample do we need in order to estimate the mean diameter of all ball bearings made by this process to within ±.01 millimeter and with 95% confidence?

89. Getup polling organization wishes to determine which of two presidential candidates will win an election. The researchers want to find out what proportion of the voters favor candidate A, and they desire to be accurate to within ±.01. Find the sample size they need, assuming that the candidates seem to be running about even in the race at this point; that is, p is about .5, and confidence is 95%.

Ratio Data Set   *Refer to the 141 companies listed in the Ratio Data Set in Appendix A.*

Questions   90. Locate the data for the current assets–sales ratio (CA/S). The companies in the file are a random sample from the numerous companies in the COMPUSTAT files.

    a. Find the 95% confidence interval estimate for the mean CA/S ratio.

    b. A financial analyst has told you that companies on average have CA/S equal to 1.40. Is the analyst's statement questionable in light of this interval?

91. Locate the data for the quick assets–sales ratio (QA/S). The companies in the file are a random sample from the numerous companies in the COMPUSTAT files.

    a. Find the 95% confidence interval estimate for the mean QA/S ratio.

    b. A financial analyst has told you that companies on average have QA/S equal to 1.40. Is the analyst's statement questionable in light of this interval?

92. Locate the data for the current assets–current liabilities ratio (CA/CL). The companies in the file are a random sample of the numerous companies in the COMPUSTAT files. Find the 95% confidence interval estimate for the mean CA/CL ratio.

93. Locate the data for the net income–total assets ratio (NI/TA). The companies on the list are a random sample of all of the numerous companies included in the COMPUSTAT data base.

    a. Find the 95% confidence interval estimate for the mean NI/TA ratio.

    b. A stock broker trying to sell stock for the companies has told you that companies on average have NI/TA equal to .04. Is the broker's statement questionable given the confidence interval estimate?

94. Refer again to the data for the net income–total assets ratio (NI/TA). Some of the companies suffered financial losses as reflected by negative values for the ratio. The companies on the list are a random sample of the numerous companies included in the COMPUSTAT files.

a. Find the 95% confidence interval estimate of the proportion of companies that did not suffer losses.

b. A stock broker trying to sell stock for the companies has told you that 85% of the companies in the population are profitable or at least have not suffered losses. Is the broker's statement questionable given the confidence interval estimate?

Refer to the 113 *applicants for credit listed in the Credit Data Set in Appendix A.*

Credit Data Set

Questions

95. Find an interval within which you can be 95% confident that the population mean value of debt owed by successful credit applicants lies.

96. Discuss whether you must assume the variable in Question 95 has a normal distribution in order for the confidence interval you constructed to be a valid one. Do the data look normally distributed?

97. Assume you wanted to know the population mean value of TOTBAL for successful applicants to within ±$100 with 90% confidence. How much larger would your sample have to be?

98. Find the 95% confidence interval for the proportion of all applicants in the population who are married.

ESTIMATING LIFETIMES OF CONSUMER PRODUCTS

CASE 8.1

There are only a few laboratories across the country that test consumer products. Underwriters Laboratories of Chicago is the best known. Gold Seal Laboratories is another of these testing organizations, and it is located in Dayton, Ohio.

Paula Sear is the chief testing engineer for the company and supervises the work of seven other engineers. The engineers have between four and six testing technicians reporting to each of them. If a company wishes to obtain the Gold Seal of approval from the laboratory to display on its products, it must allow a laboratory technician to come to its plant, randomly select items from the production line, and bring them back to Dayton for testing. The tests involve examining the items to see that they meet their design specifications and federal or local safety standards.

For instance, a clock radio might be tested to make sure that the clock keeps reasonably accurate time, the alarm goes off when it is set, the case is shockproof, and the circuitry has been connected just as the electrical schematic diagram shows it should be. Many items are tested for parts reliability. That is, an electrical appliance might be operated continuously for several hundred hours or until it breaks down in order to determine the operating life of the item.

Ms. Sear has just received some data on United Lighting, Inc.'s new ornamental outdoor arc lamp. The lamp is used on a post in the front yards of homeowners living on poorly lighted streets. United's lamp sells for a substantially higher price than the conventional lamps used for this purpose. However, it lights a much wider area, and United believes that it lasts much longer. Unfortunately, United has no hard data to prove this second claim. Sales of the lamp have been slow since the competition has been successful in making dealers believe that United's lamp is too expensive and has a rather short life.

Nearly two years ago, Gold Seal was asked to bid on the cost of testing some of United's arc lamps. Since United wanted to advertise the length of life of their lamp to dealers, they felt it necessary to use an independent testing agent.

Gold Seal won the contract to perform the tests. A Gold Seal technician visited the St. Louis plant of United Lighting and randomly selected 20 lamps for testing. The lamps were wired to a special control board that regulated the amount of time the lamps were lighted. This automatic board turned on all the lamps simultaneously and left them on for a randomly determined period of time, but the time was controlled so that it never exceeded 12 hours of continuous operation. When the lights were turned off, the control board monitored their temperature. Once the lights cooled to room temperature, they were turned on again for another randomly determined period.

When a lamp burned out, a small alarm sounded, and a technician recorded, from a timer mounted on the control board, the total amount of time the lamp had been operated. The test data showing the times until failure for all 20 lamps are shown in the accompanying table. All that remains is for Ms. Sear to write her report to United Lighting concerning the length of life that can be expected from all their arc lamps. In her report Ms. Sear intends to note that 8 of 40 lamps in a separate study had popped, cracked, or exploded when they failed. Since such events might involve a hazard to lamp users, she intends to make an estimate of the extent to which this characteristic can be expected among all United's new arc lamps.

Item Number	Hours to Failure	Item Number	Hours to Failure
1	13,140	11	14,990
2	17,555	12	7,901
3	9,490	13	13,893
4	8,090	14	15,737
5	12,234	15	12,459
6	13,876	16	13,952
7	14,570	17	9,333
8	18,442	18	10,873
9	11,109	19	11,537
10	14,007	20	10,755

Case Assignment

a. Find a 95% confidence interval for the mean time to failure.
b. Find a 95% confidence interval for the proportion of lamps that exhibit cracking, popping, or exploding on failure.

CASE 8.2

ESTIMATING OVERPAYMENTS ON LOSSES INCURRED WITH FEDERALLY INSURED MORTGAGES

The Department of Housing and Urban Development (HUD) guarantees the timely payment of principal and interest on mortgage-backed securities (GNMA securities), representing interest in mortgages insured or guaranteed by the Federal Home Administration, Veterans Administration, and the Farmers Home Administration.

When a lender, such as a bank or savings and loan association, experiences a loss on an insured mortgage, the lender submits a claim to HUD. HUD honors each of these claims. HUD pays most of the claims, then HUD takes a sample of the claims from the lender's portfolio of losses and audits the losses to determine whether the loans on which losses have occurred are correctly insured, appropriately serviced, and correctly forwarded for payment by HUD. If HUD finds that the lender has overcharged for foreclosure procedures, such as window boarding, or took too long to foreclose and therefore charged HUD for too much interest, then HUD requests repayment from the lender.

In its past procedures HUD used the sample mean "incorrect payment request" per loan sampled as an estimate of the population mean for the entire set of claims submitted by the lender. HUD then asked the lender to repay HUD an amount based on the estimate of the population mean multiplied by the number of claims sent to HUD.

A dispute over HUD payments centered around the fact that mortgage lenders, through the Mortgage Bankers Association of America, complained that the sample mean may be "an incorrect estimate" of the population mean. Mortgage lenders argued in their suit that it would be fairer for HUD to construct a confidence interval for the sample mean and then base their requested repayment on "the low point" of that estimate. It turns out that the "low point" is the lower limit of the 90% confidence interval for the population mean overpayment by HUD to the lender.

HUD made major concessions to the mortgage lenders and adopted the procedure recommended by the lenders. While HUD continues to rely on statistical sampling, charges for overpayments are now based on the low point, rather than the midpoint of the interval of possible recovery amounts. The new procedure reduced reimbursements to HUD by over $15 million.

To see how the new procedure reduced reimbursements to HUD, consider the Rock Solid Thrift that submitted 1000 insured claims to HUD. An audit of a sample of $n = 9$ claims uncovered the following overpayments to Rock Solid:

1000   200   0   250   150   500   100   210   190

The sample mean and standard deviation of the overpayments are $\overline{X} = 288.9$ and $s = 298.8$. The t table gives $t_{.05, 8} = 1.860$ and the limits of the 90% confidence interval estimate for the mean overpayment are

$$L = 288.9 - 1.860 \frac{298.8}{\sqrt{9}} = 288.9 - 185.3 = 103.6$$

and

$$R = 288.9 + 1.860 \frac{298.8}{\sqrt{9}} = 288.9 + 185.3 = 474.2$$

Under the old procedure, Rock Solid would have reimbursed HUD by the following amount: (1000 claims) × ($288.9 per claim), or $288,900. Under the new procedure, Rock Solid would reimburse HUD by the following amount: (1000) × ($103.6), or $103,600. The savings to Rock Solid and cost to HUD of adopting the new procedure are substantial.

a. Is the lower limit of the 90% confidence interval an unbiased estimate of the population mean overpayment?
b. Do you think that HUD should have adopted the new procedure?
c. How might the new procedure be exploited to take advantage of HUD?

Case Assignment

REFERENCE: Newsletter Division of American Banker-Bond Buyer (1989), "Industry Could Save Millions in HUD Audit Claim Settlements," *The Mortgage Marketplace*, 89–23: 1.

REFERENCES

Fisher, R. A. 1963. *Statistical Methods for Research Workers,* 13th ed. New York: Hafner Press.

Hogg, Robert V., and Allen T. Craig. 1978. *Introduction to Mathematical Statistics,* 4th ed. New York: Macmillan.

Lehmann, E. L. 1983. *Theory of Point Estimation.* New York: Wiley.

Mosteller, Frederick, Robert E. K. Rourke, and George B. Thomas, Jr. 1970. *Probability with Statistical Applications,* 2nd ed. Reading, Mass.: Addison-Wesley.

Snedecor, George W., and William G. Cochran. 1989. *Statistical Methods,* 8th ed. Ames: Iowa State University Press.

Student. 1908. "The Probable Error of a Mean." *Biometrika,* 6: 1–25.

In this chapter, we looked at some of the inferences that are possible when we have a single random sample from a normal population and also when our random sample is assumed to have been taken from a binomial population. Here we make inferences from two independent random samples taken from two populations.

When we have two samples, an observation X_{ij} has two subscripts; the first denotes the particular element within the sample, and the second designates the sample. Thus X_{42} is the fourth value in the second sample. Let $X_{11}, X_{21}, \ldots, X_{n_1 1}$ represent observed values in a sample of size n_1 taken from a normal population with mean μ_1 and variance σ^2, and let $X_{12}, X_{22}, \ldots, X_{n_2 2}$ be the observations in a sample of size n_2 taken from a second normal population with mean μ_2 and variance σ^2 (the same σ^2 as for the first population). Such two-sample situations are characterized by three basic assumptions:

1. The samples are independent random samples.
2. The populations are normal.
3. The populations have a common variance.

The methods to be presented in this section are valid only if these assumptions are satisfied.

Data that satisfy our three assumptions and that fit exactly into the framework of the two-sample situation may be obtained as the result of an experiment for comparing the effects of two **treatments**.

Definition

In statistical usage, **treatments** are any procedures, methods, or stimuli whose effects we want to estimate and compare.

Treatments in one situation might represent different machines, but in another they could be different operators; they could be different chemicals or different rates of application of one chemical; they could be different advertisements; and so on.

After the data have been obtained, either by selecting the samples or by performing an experiment, the information they contain may be summarized in tabular form, as shown in Table 8.5: The *pooled sum of squares, Pooled SS,* can be obtained by adding the sums of squares for the individual samples in the last column of the table.

TABLE 8.5	Sample	Size	Degrees of Freedom	Mean	Sum of Squares
Summary Table for Two Samples from Continuous Populations	1	n_1	$n_1 - 1$	$\overline{X}_1$	$\Sigma(X_{i1} - \overline{X}_1)^2 = \Sigma X_{i1}^2 - \dfrac{(\Sigma X_{i1})^2}{n_1}$
	2	n_2	$n_2 - 1$	$\overline{X}_2$	$\Sigma(X_{i2} - \overline{X}_2)^2 = \Sigma X_{i2}^2 - \dfrac{(\Sigma X_{i2})^2}{n_2}$
	Total	$n_1 + n_2$	$n_1 + n_2 - 2$		Pooled SS

After the information contained in the data has been summarized in the table, we turn our attention to the estimation of the parameters of interest. The best unbiased estimators for the population means μ_1 and μ_2 are the respective sample means $\overline{X}_1$ and $\overline{X}_2$. The best unbiased estimator for the difference between the means, $\mu_1 - \mu_2$, is the difference between the sample means, $\overline{X}_1 - \overline{X}_2$. The latter statistic is of particular interest because two populations are involved, and our primary concern is making inferences about the difference between the means. In experimental situations the difference between the means is also the difference between the effects of the two treatments.

To form a **confidence interval for $\mu_1 - \mu_2$** when σ_1 and σ_2 are known, we find $\overline{X}_1 - \overline{X}_2$ and then add and subtract a Z value times $\sigma_{\overline{X}_1 - \overline{X}_2}$ to obtain the left and right confidence limits, according to the following equations.

Equations for Limits of a Confidence Interval for $\mu_1 - \mu_2$: Known Standard Deviations

$$L = (\overline{X}_1 - \overline{X}_2) - Z_{\alpha/2}\sigma_{\overline{X}_1 - \overline{X}_2}$$
$$R = (\overline{X}_1 - \overline{X}_2) + Z_{\alpha/2}\sigma_{\overline{X}_1 - \overline{X}_2}$$

(8.29)

Equation (8.29) needs a little explanation. Since the standard deviation of the difference between two independent random variables is the square root of the sum of the variances, we know that $\sigma_{\overline{X}_1 - \overline{X}_2}$ has the following value:

$$\sigma_{\overline{X}_1 - \overline{X}_2} = \sqrt{\frac{\sigma_1^2}{n_1} + \frac{\sigma_2^2}{n_2}}$$

(8.30)

However, in this section we assume that the two populations have the same variance: $\sigma^2 = \sigma_1^2 = \sigma_2^2$. When we know this value of the common variance, we can obtain $\sigma_{\overline{X}_1 - \overline{X}_2}$ by using Equation (8.30).

In the general case where the common standard deviations $\sigma_1 = \sigma_2 = \sigma$ are *not* known, σ^2 can be factored out of Equation (8.30). Since the variance is assumed to be the same for both populations, we can obtain an unbiased estimate for σ^2 from each sample. But the best unbiased estimate is obtained by pooling the information contained in both samples to get what we call the *pooled estimate of variance* s_{pooled}^2. This pooled estimate of variance is an estimate of the common population variance σ^2 and can be found in two equivalent ways. The first method involves pooling (or adding) the sum of squares from each of the samples and then dividing by the **pooled degrees of freedom** $(n_1 + n_2 - 2)$. The second method is to find the pooled estimate of variance as the weighted mean of the individual sample estimates, where the weights are the degrees of freedom.

The equations for the two equivalent methods of finding the **pooled estimate of variance** are given next.

Equation for Pooled Estimate of Variance: First Method

$$s^2_{pooled} = \frac{Pooled\ SS}{Pooled\ df} = \frac{\sum(X_{i1} - \overline{X}_1)^2 + \sum(X_{i2} - \overline{X}_2)^2}{n_1 + n_2 - 2} \qquad (8.31)$$

Equation for Pooled Estimate of Variance: Second Method

$$s^2_{pooled} = \frac{(n_1 - 1)s_1^2 + (n_2 - 1)s_2^2}{n_1 + n_2 - 2} \qquad (8.32)$$

where:
$$s_1^2 = \frac{\sum(X_{i1} - \overline{X}_1)^2}{n_1 - 1} = \frac{\sum X_{i1}^2 - (\sum X_{i1})^2/n_1}{n_1 - 1}$$

$$s_2^2 = \frac{\sum(X_{i2} - \overline{X}_2)^2}{n_2 - 1} = \frac{\sum X_{i2}^2 - (\sum X_{i2})^2/n_2}{n_2 - 1}$$

The first method for finding the pooled estimate of variance is ordinarily used with raw data, and the second method is ordinarily used if the sample variances are available.

The standard error of the difference between two means $\sigma_{\overline{X}_1 - \overline{X}_2}$ can then be estimated by the **sample standard error of the difference between two sample means** $s_{\overline{X}_1 - \overline{X}_2}$ according to the following equations.

Equation for Estimating Standard Deviation $\sigma_{\overline{X}_1 - \overline{X}_2}$ of Difference Between Two Sample Means

$$s_{\overline{X}_1 - \overline{X}_2} = \sqrt{s^2_{pooled}\left(\frac{1}{n_1} + \frac{1}{n_2}\right)} \qquad (8.33)$$

To form a **confidence interval for $\mu_1 - \mu_2$** when the common standard deviations $\sigma_1 = \sigma_2 = \sigma$ are *not* known, we use the following equations.

Equations for Limits of a Confidence Interval for $\mu_1 - \mu_2$: Unknown Standard Deviations

$$L = (\overline{X}_1 - \overline{X}_2) - t_{\alpha/2, n_1 + n_2 - 2}(s_{\overline{X}_1 - \overline{X}_2})$$
$$R = (\overline{X}_1 - \overline{X}_2) + t_{\alpha/2, n_1 + n_2 - 2}(s_{\overline{X}_1 - \overline{X}_2}) \qquad (8.34)$$

Note that the t value in Equation (8.34) has $n_1 + n_2 - 2$ degrees of freedom. As always, the degrees of freedom for t are those associated with the sum of squares used in finding our unbiased estimate of σ^2.

TABLE 8.6	Paint 1			Paint 2		
Reflectivity Data for	12.5	10.3	8.7	9.4	6.9	7.0
Two Paints	11.7	9.6	11.5	11.6	7.3	8.2
	9.9	9.4	10.6	9.7	8.4	12.7
	9.6	11.3	9.7	10.4	7.2	9.2

PRACTICE PROBLEM 8.12

To compare the durabilities of two paints suggested by a Sherman-Wilson Company representative for highway use, twelve 4-inch-wide lines of each paint were laid down across a heavily traveled road. The order was decided at random. After a period of time reflectometer readings were obtained for each line. The higher the readings, the greater is the reflectivity and the better is the durability of the paint. The data are given in Table 8.6. Find a 95% confidence interval for the difference between the two paints.

Solution For these sets of data, we find the results presented in Table 8.7. The summary table is presented in Table 8.8. Hence,

$$\overline{X}_1 - \overline{X}_2 = 1.40$$

The value for the pooled sums of squares is *Pooled SS* = 14.08 + 38.64 = 52.72 and for the pooled degrees of freedom is *Pooled df* = $n_1 + n_2 - 2 = 12 + 12 - 2 = 22$. Equation (8.31) gives the pooled estimate of variance:

$$s_{pooled}^2 = \frac{Pooled\ SS}{Pooled\ df} = \frac{52.72}{22} = 2.40$$

The estimated standard error of the difference between means by Equation (8.33) is

$$s_{\overline{X}_1 - \overline{X}_2} = \sqrt{s_{pooled}^2\left(\frac{1}{n_1} + \frac{1}{n_2}\right)} = \sqrt{2.40\left(\frac{1}{12} + \frac{1}{12}\right)}$$

$$= \sqrt{2.40\left(\frac{2}{12}\right)} = \sqrt{.40} = .63$$

TABLE 8.7	Paint 1	Paint 2
Calculation Results	$\sum X_{i1} = 124.8$	$\sum X_{i2} = 108.0$
	$\overline{X}_1 = 10.4$	$\overline{X}_2 = 9.0$
	$\sum X_{i1}^2 = 1312.00$	$\sum X_{i2}^2 = 1010.64$
	$(1/n_1)(\sum X_{i1})^2 = 1297.92$	$(1/n_2)(\sum X_{i2})^2 = 972.00$
	$\sum(X_{i1} - \overline{X}_1)^2 = 14.08$	$\sum(X_{i2} - \overline{X}_2)^2 = 38.64$

TABLE 8.8	Paint	n	df	$\overline{X}_j$	Sum of Squares
Summary Table	1	12	11	10.4	14.08
	2	12	11	9.0	38.64
	Total	24	22		52.72

COMPUTER EXHIBIT 8.2 Confidence Interval for the Difference Between Mean Reflectivity of Sherman-Wilson Paints

MINITAB

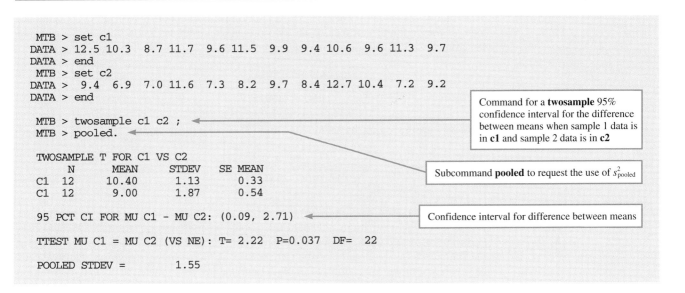

```
MTB > set c1
DATA > 12.5 10.3  8.7 11.7  9.6 11.5  9.9  9.4 10.6  9.6 11.3  9.7
DATA > end
MTB > set c2
DATA >  9.4  6.9  7.0 11.6  7.3  8.2  9.7  8.4 12.7 10.4  7.2  9.2
DATA > end

MTB > twosample c1 c2 ;
MTB > pooled.

TWOSAMPLE T FOR C1 VS C2
      N       MEAN      STDEV    SE MEAN
C1   12      10.40       1.13       0.33
C1   12       9.00       1.87       0.54

95 PCT CI FOR MU C1 - MU C2: (0.09, 2.71)

TTEST MU C1 = MU C2 (VS NE): T= 2.22  P=0.037  DF= 22

POOLED STDEV =          1.55
```

Command for a **twosample** 95% confidence interval for the difference between means when sample 1 data is in **c1** and sample 2 data is in **c2**

Subcommand **pooled** to request the use of s^2_{pooled}

Confidence interval for difference between means

If we desire a 95% confidence interval, we must obtain $t_{.025, 22} = 2.074$ from Table VI of Appendix C. The 95% confidence interval can now be formed by using Equation (8.34):

$$L = 1.40 - (2.074)(.63) = .09$$

$$R = 1.40 + (2.074)(.63) = 2.71$$

We can be 95% confident that the difference between the actual means is between .09 and 2.71.

Note that this confidence interval is for $\mu_1 - \mu_2$. If we wanted the interval for $\mu_2 - \mu_1$, the estimator would be $\overline{X}_2 - \overline{X}_1$, or -1.40, and the confidence limits would be -2.71 and $-.09$ (the variance of $\overline{X}_2 - \overline{X}_1$ is the same as the variance of $\overline{X}_1 - \overline{X}_2$). Thus there is no difference in these two confidence intervals except for the algebraic signs on the confidence limits.

Computers can be used to find confidence intervals for the differences between two population means. A computer was used to analyze the Sherman-Wilson paint reflectivity problem, and the results are presented in Computer Exhibit 8.2.

 PRACTICE PROBLEM 8.13

Random samples of weights were taken from two package-filling processes at Oregon Cement, Inc. The first sample consisted of $n_1 = 160$ bags, and the second consisted of $n_2 = 200$ bags. The two sample means and standard deviations are as follows (in kilograms):

$$\overline{X}_1 = 10.2 \text{ kg} \qquad \overline{X}_2 = 9.9 \text{ kg}$$

$$\overline{X}_1 - \overline{X}_2 = 10.2 - 9.9 = .3 \text{ kg}$$

$$s_1 = .3 \text{ kg} \qquad s_2 = .4 \text{ kg}$$

Find an interval within which we can be 90% sure that $\mu_1 - \mu_2$ lies.

Solution Since the samples are large, even if the populations of weights in the two processes are not normally distributed, each sample mean will be approximately normally distributed owing to the central limit theorem; and the difference between the means will be normally distributed. The first step in solving this problem is to find s^2_{pooled}, the estimate of the common variance of the two processes. Since we know the sample standard deviations, we can use Equation (8.32) to obtain

$$s^2_{pooled} = \frac{(160 - 1)(.3)^2 + (200 - 1)(.4)^2}{160 + 200 - 2} = .13$$

Now we can find $s_{\overline{X}_1 - \overline{X}_2}$ by using Equation (8.33):

$$s_{\overline{X}_1 - \overline{X}_2} = \sqrt{.13\left(\frac{1}{160} + \frac{1}{200}\right)} = .038$$

Finally, we use Equation (8.34) to form the confidence interval. Technically, since we do not know the populations' standard deviations, we must use Equation (8.34). However, $t_{.05, 358}$ is the t value with 358 degrees of freedom. We noted earlier that for large values for the degrees of freedom the t distribution approaches the normal distribution. Thus we seek the t value from the .05 column of Table VI and the last row, where the normal distribution values are found. So $t_{.05, 358} \approx 1.645$, and the confidence interval is given by

$$L = .3 - (1.645)(.038) = .237$$

$$R = .3 + (1.645)(.038) = .363$$

Thus we can be 90% sure that the first process fills packages to a mean weight that is between .237 and .363 kilogram heavier than the mean of the second process. ▬▬

Problems: Section 8.8

99. Two samples, one from each of two normal populations having the same variance, gave the results shown in the accompanying table.
   a. What are the estimates for $\mu_1 - \mu_2$ and $\sigma_{\overline{X}_1 - \overline{X}_2}$?
   b. Find the 95% confidence interval for $\mu_1 - \mu_2$.

Sample	n	$\overline{X}_j$	$\Sigma(X_{ij} - \overline{X}_j)^2$
1	11	150	5600
2	11	120	5400

The data values (which are only needed if you wish to use a computer) are as follows:

Sample 1	188.669	174.227	160.138	140.992	168.326	117.260
	151.146	153.313	125.689	156.094	114.145	

Sample 2	157.972	143.791	129.955	111.154	137.996	87.850
	121.126	123.254	96.127	125.984	84.791	

100. Use the accompanying summary table, and assume two normal populations and common variance.

a. Find the estimates for $\mu_1 - \mu_2$ and $\sigma_{\overline{X}_1 - \overline{X}_2}$.
b. Find the 90% confidence interval for $\mu_1 - \mu_2$.

Sample	n	$\overline{X}_j$	$\Sigma(X_{ij} - \overline{X}_j)^2$
1	6	30	300
2	4	20	180

101. In an experiment designed to compare the means of normally distributed service lives of two types of tires at Eagle, Inc., the difference between the means was 2250 miles, and the pooled estimate of variance was equal to 625,000. There were 10 tires of each type. Find a 95% confidence interval for the difference between the mean lives of these types of tires.

102. Fifteen of each of two types of fabricated wood beams were tested for breaking load at Georgia Atlantic, Inc., with the results as shown in the accompanying table. Find a 95% confidence interval for the difference between the mean breaking loads, $\mu_1 - \mu_2$.

Type	n	$\overline{X}_j$	s_j^2
1	15	1560	3500
2	15	1600	2500

103. An employee of American Manufacturing, Inc., wishes to know if normally distributed managerial salaries are higher for those who have worked their way up from nonmanagerial positions in the company or for those who were hired from outside the company. The employee randomly selects 10 of each type of manager. The mean annual income for those hired from outside he computes to be $31,750, and the mean figure for the insiders is $27,000. The corresponding standard deviations are $6350 and $4050. Find a 90% confidence interval for the difference in annual income of the two groups. Should this interval lead the employee to conclude that there is a real difference?

104. Two assembly lines produce the same item. The amount of time to produce one finished piece varies because of the speed of the operators and the number of times the lines must be stopped for malfunctions or repairs. The amount of time (in minutes) it took to finish each piece in a sample of 9 items on the first line and 16 items on the second line is presented in the acompanying summary table.

a. Find 95% confidence intervals for $\mu_1 - \mu_2$.
b. Would the confidence interval for $\mu_1 - \mu_2$ lead you to conclude that the mean times per piece for the two lines are different? Why?

Sample	n	$\overline{X}_j$	$\Sigma(X_{ij} - \overline{X}_j)^2$
1	9	52	68
2	16	55	54

105. A leveraged buy-out (LBO) is one form of acquiring a company. In a leveraged buy-out a group of investors buys the stock of a company by using debt financing. The assets of the company are typically used as collateral for the debt. An increase in debt tends to drive a company's bond price down because of the increased chance of a default. Samples of the decreases in bond prices for 9 companies that were recently acquired by LBOs (sample 1) and for 16 companies that had not been taken over (sample 2) are summarized in the accompanying table. Assume normality and equal variances.

a. Find a 95% confidence interval for the difference in mean bond price decreases $\mu_1 - \mu_2$.

b. Would you conclude that the bond price decrease after an LBO is greater than for non-acquired companies?

Sample	n	$\overline{X}_j$	s^2
1	9	96	24
2	16	80	34

 106. Greenmail is a term used for the situation in which investors are paid a premium for their shares by managers who want to avoid a takeover of their company. Speculators in the stock market sometimes accumulate a block of stock in a company to attempt to gain a seat on the board of directors or to acquire the company through a stock tender offer. In an attempt to avoid a takeover, some corporations have then purchased the speculators' stock at a premium price. Since takeovers tend to increase stock prices and thereby shareholder wealth, payment of greenmail tends to decrease shareholder wealth. Samples of the decreases in stock prices for 15 companies that recently paid greenmail (sample 1) and for 12 companies that had refused to pay greenmail (sample 2) are summarized in the accompanying table. Assume normality and equal variances. Find a 98% confidence interval for the difference in mean stock price decreases $\mu_1 - \mu_2$.

Sample	n	$\overline{X}_j$	$\Sigma(X_{ij} - \overline{X}_j)^2$
1	15	25	55
2	12	20	45

The data values (which are only needed if you wish to use a computer) are as follows:

1. Paid Greenmail	24.8308	24.1090	26.6375	26.4940
	23.2384	29.4480	24.1787	26.6405
	24.8536	23.9649	27.0692	22.1427
	22.2934	23.6175	25.4815	

2. Refused Greenmail	20.6986	18.6262	21.0268	22.8245
	17.7578	19.8537	20.6198	17.9892
	21.3134	22.9642	16.3115	20.0140

 107. Independent random samples from normal distributions with equal standard deviations of sales growth percentages by (1) firms that have been takeover targets and (2) firms that have not been takeover targets gave the results listed in the accompanying table. Find the 95% confidence interval estimate for the difference between means $\mu_1 - \mu_2$.

1. Target Firm	12	0	10	2	4	8	6	6
2. Nontarget Firm	8	8	−2	−2	0	0	2	2

108. Independent random samples from normal distributions with equal standard deviations of the amount of insider trading during mergers by employees of (1) investment banking firms and (2) brokerage houses gave the results listed in the accompanying table. Find the 95% confidence interval estimate for the difference between means $\mu_1 - \mu_2$. (You may want to compute sample means and standard deviations for this problem by using hand calculations.)

1. Investment Banking Firm	10	5	5	5	8	8	8
2. Brokerage House	6	0	5	2	2	3	3

We now examine confidence intervals when we have two independent random samples, one of size n_1 from a binomial distribution with a proportion of successes equal to p_1, the other of size n_2 from a binomial population with proportion of successes equal to p_2. We let X_1 and X_2 be the number of successes in the first and second samples. Our primary objective will be to estimate the difference $p_1 - p_2$. To achieve this objective, we make use of the normal approximation to the binomial.

The proportion p_j is estimated unbiasedly by the observed value of the corresponding sample fraction $\hat{p}_j$ (that is, X_j/n_j), and the difference $p_1 - p_2$ is estimated unbiasedly by $\hat{p}_1 - \hat{p}_2$. The variance of $\hat{p}_j$ is estimated by

$$s_{\hat{p}_j}^2 = \frac{\hat{p}_j(1 - \hat{p}_j)}{n_j} \tag{8.35}$$

which is the sample estimate of the variance of the binomial fraction of successes [see Equation (6.6)]. Since $\hat{p}_1$ and $\hat{p}_2$ are independent, the variance of their difference, $\hat{p}_1 - \hat{p}_2$, is the sum of their variances and is estimated by

$$s_{\hat{p}_1-\hat{p}_2}^2 = \frac{\hat{p}_1(1 - \hat{p}_1)}{n_1} + \frac{\hat{p}_2(1 - \hat{p}_2)}{n_2} \tag{8.36}$$

So the standard deviation of $\hat{p}_1 - \hat{p}_2$ is estimated by

$$s_{\hat{p}_1-\hat{p}_2} = \sqrt{\frac{\hat{p}_1(1 - \hat{p}_1)}{n_1} + \frac{\hat{p}_2(1 - \hat{p}_2)}{n_2}} \tag{8.37}$$

For large n_1 and n_2 the quantity $\hat{p}_1 - \hat{p}_2$ is approximately normally distributed, as we learned in Section 7.8. So

$$Z = \frac{(\hat{p}_1 - \hat{p}_2) - (p_1 - p_2)}{\sqrt{[p_1(1 - p_1)/n_1] + [p_2(1 - p_2)/n_2]}} \tag{8.38}$$

has approximately a standard normal distribution. If we replace the p_j in the denominator by their estimates $\hat{p}_j$, we obtain

$$Z = \frac{(\hat{p}_1 - \hat{p}_2) - (p_1 - p_2)}{\sqrt{[\hat{p}_1(1 - \hat{p}_1)/n_1] + [\hat{p}_2(1 - \hat{p}_2)/n_2]}} \tag{8.39}$$

and this variable too has approximately a standard normal distribution. This result leads to the approximate $100(1 - \alpha)\%$ **confidence interval for $p_1 - p_2$.**

Equations for Limits of a Confidence Interval for $p_1 - p_2$

$$L = (\hat{p}_1 - \hat{p}_2) - Z_{\alpha/2} \sqrt{\frac{\hat{p}_1(1 - \hat{p}_1)}{n_1} + \frac{\hat{p}_2(1 - \hat{p}_2)}{n_2}}$$

$$R = (\hat{p}_1 - \hat{p}_2) + Z_{\alpha/2} \sqrt{\frac{\hat{p}_1(1 - \hat{p}_1)}{n_1} + \frac{\hat{p}_2(1 - \hat{p}_2)}{n_2}} \tag{8.40}$$

 PRACTICE PROBLEM 8.14

Two different methods of manufacture, casting and die forging, were used to make parts at Northeastern Foundaries, Inc., for an appliance. In service tests of 100 of each type, 10 castings failed but only 3 forged parts were found to be defective. Find 95% confidence limits for the difference between the proportions of the cast and forged parts that would fail under similar conditions.

Solution Here,

$$\hat{p}_1 = \frac{10}{100} = .10 \qquad \hat{p}_2 = \frac{3}{100} = .03$$

$$s_{\hat{p}_1 - \hat{p}_2} = \sqrt{\frac{.10(90)}{100} + \frac{.03(.97)}{100}}$$

$$= \sqrt{.001191} = .0345$$

From the last row of Table VI, $Z_{.025}$ is found to be 1.96, and approximate confidence limits for $p_1 - p_2$ are given by

$$L = (\hat{p}_1 - \hat{p}_2) - Z_{.025}s_{\hat{p}_1 - \hat{p}_2} = .07 - 1.96(.0345) = .002$$

and

$$R = (\hat{p}_1 - \hat{p}_2) + Z_{.025}s_{\hat{p}_1 - \hat{p}_2} = .07 + 1.96(.0345) = .138$$

We are approximately 95% confident that the difference between the proportions of failures is between .00 and .14. Thus the two different methods might differ in failure rates by as little as 0 ($p_1 - p_2 = .00$), or the first method might produce failures at a rate as high as 14% greater than the second method ($p_1 - p_2 = .14$), with 95% confidence.

Problems: Section 8.9

109. In a sample of 100 from one binomial population, there were 28 successes. A sample of 200 from a second binomial population contained 92 successes.
   a. Find estimates for p_1, p_2, and $p_2 - p_1$.
   b. Estimate the standard deviation of $\hat{p}_2 - \hat{p}_1$.
   c. Find a 98% confidence interval for $p_2 - p_1$.

110. Samples of 200 were taken from each of two binomial populations. They contained 104 and 96 successes, respectively.
   a. Find the estimated standard deviation of $\hat{p}_1 - \hat{p}_2$.
   b. Find a 90% confidence interval for $p_1 - p_2$.

111. A sample of 400 voters was classified by designating each voter as a "minority" or a "nonminority." Among the 250 classified as minority, 150 planned to vote for candidate Smith. Of the 150 nonminority voters, 60 planned to vote for Smith. Find a 95% confidence interval for the difference in proportions who plan to vote for Smith ($p_{min} - p_{non}$).

112. A sample of 500 people were classified as being "athletic" or "nonathletic." Among 300 classified as athletic, 60 regularly eat a certain breakfast food. Among the 200 nonathletic persons, 50 regularly use the product. Find a 95% confidence interval for the difference in the proportions that use the product ($p_{ath} - p_{non}$).

113. In a sample of 50 turnings from an automatic lathe 8 were found to be outside specifications. The cutting bits were then changed and the machine restarted. A new sample of 50 contained 3 defective turnings. Find a 90% confidence interval for the decrease in the proportion defective after changing bits.

Tests of Hypotheses

9

We discussed methods of estimation in Chapter 8. In this chapter, we introduce the theory and practice of hypothesis testing. We use estimation and hypothesis tests to make statistical inferences about populations, including population parameters and population distributions.

We examine how to establish hypotheses in this chapter. In addition, we introduce a procedure that employs six steps for testing hypotheses. Since the results of a hypothesis test may be incorrect, we also discuss two types of errors that may result from a hypothesis test and the probabilities for these errors. We cover several forms of hypothesis tests about *means, proportions,* and *variances* of *one* or *two populations* in the chapter. In addition, we consider *one-sided* and *two-sided* forms of hypothesis tests.

Hypothesis-testing procedures will be used in many of the chapters that follow. Thus, it is important to become proficient with the methods of hypothesis testing.

9.1
ESTABLISHING HYPOTHESES

A *triangle test* is used in marketing research to check whether a candidate for a consumer taste panel can detect subtle differences in food tastes. In a single trial of a triangle test, a subject is presented with three food portions; two of the portions are the same and one is different. After tasting the three food portions, the subject is asked to select the food that is different. Except for the taste difference, the portions are as similar as possible, and the order of presentation is random. In the absence of any ability to distinguish tastes, the subject has a one-third chance of correctly distinguishing the food that is different, and the question is whether the candidate has the ability to detect differences between food tastes. Let p be the probability that a subject correctly identifies the food that tastes different in a single trial. Then $p \leq 1/3$ if the subject does *not* have the ability to distinguish between tastes, and $p > 1/3$ if the subject has the ability to distinguish between tastes.

To place the consumer triangle taste test problem in a standard hypothesis testing framework, we establish two hypotheses known as the **null hypothesis H_0** and the **alternative hypothesis H_a.** For a triangle test, the hypotheses are

$$\text{Null hypothesis:} \quad H_0\colon \ p \leq 1/3$$

$$\text{Alternative Hypothesis:} \quad H_a\colon \ p > 1/3$$

In this case, the null hypothesis is that the subject does *not* have the ability to distinguish between tastes. The *null hypothesis* is generally established in such a way that it states *nothing is different* from what it is supposed to be or what it has been in the past. We shall augment this guideline for establishing null hypotheses in subsequent discussion; however, in this chapter, a null hypothesis always includes an equality.

We normally assume that the null hypothesis is true unless we see strong evidence to the contrary. When we have strong evidence that is contrary to the null hypothesis H_0, we *reject H_0*. That is, we no longer consider the null hypothesis to be true. Whereas, when we do *not* have strong evidence contrary to the null hypothesis H_0, we *do not reject H_0*. If we reject the null hypothesis, the evidence that is contrary to H_0 may very well better support something else. The "something else" that is generally better supported by the evidence is the *alternative hypothesis H_a*.

In a triangle test, we reject the null hypothesis of no tasting ability if the subject detects taste differences correctly on a reasonably high proportion of trials. In this case, the taste-testing results better support the alternative hypothesis, that the subject has some tasting ability or that $p > 1/3$.

In some situations, as in a triangle test, hypotheses involve a proportion p. In other situations, we test hypotheses about a population mean μ or about population variances. Several hypothesis-testing situations are presented in the following examples. We establish the null and alternative hypotheses for each example.

EXAMPLE 9.1 The International Rubber Band Company desires to fill its number 6 box of rubber bands with an average of "at most 6 ounces of little office helpers" (rubber bands, that is). If a quality control inspector is concerned that the process for filling the boxes is filling the boxes with more than 6 ounces of rubber bands, the inspector will test the following hypotheses:

$$H_0\colon \ \mu \leq 6$$

$$H_a\colon \ \mu > 6$$

If the inspector took a sample of boxes and found that the sample mean weight of boxes of little office helpers was much greater than 6 ounces, the inspector would reject the null hypothesis. The sample evidence would suggest that the quality of the filling process had been stretched too far. ≡

EXAMPLE 9.2 As required by law, the Deluxe Pizza Sauce Company has produced cans of sauce that have had less than or equal to .01 milligram or more of a food preservative in them. The Food and Drug Administration (FDA) may examine several of the company's cans to test the null hypothesis against the alternative hypothesis:

$$H_0: \mu \leq .01$$
$$H_a: \mu > .01$$

If the sample mean for the cans is too far above .01 milligram of perservative, then the FDA will reject the null hypothesis. The sample evidence better supports the alternative hypothesis, and the FDA may recall some of the company's product from the market. ≡

EXAMPLE 9.3 Production workers at the American Standard Products Company have been trained in their jobs by using two different training programs. The company training director would like to know whether there is a difference in mean productivity for workers trained in the two programs. The director might take a sample of workers and compare the productivity of those trained under program 1 with the productivity of those trained under program 2. The director would test the hypotheses

$$H_0: \mu_1 = \mu_2 \quad \text{or} \quad \mu_1 - \mu_2 = 0$$
$$H_a: \mu_1 \neq \mu_2 \quad \text{or} \quad \mu_1 - \mu_2 \neq 0$$

If the director found that the mean productivity of one of the sample groups was quite different from the mean productivity of the other, then the director would be justified in rejecting the null hypothesis. ≡

EXAMPLE 9.4 The Skills Unlimited Company specializes in training people for jobs in the building trades. The experience of the company has been that 90% or more of their graduates have been placed in jobs immediately. A person considering training with Skills Unlimited who wishes to test whether demand for people who have undertaken the training has decreased might take a sample of recent graduates and find out what proportion of them were placed in jobs. The hypotheses are

$$H_0: p \geq .90$$
$$H_a: p < .90$$

If the sample showed that a small proportion had been placed in jobs, then we would reject the null hypothesis H_0. The sample evidence would better support the alternative hypothesis that the demand for people who have had the training has decreased. ≡

EXAMPLE 9.5 Hammond Manufacturing, Inc., has two manufacturing plants that produce special photographic lenses. The proportion of lens that fail to pass final inspection has historically been the same at the two plants. If a quality control engineer desired to check whether these proportions are still equal, she could sample the inspection records at each plant and compare the proportion of rejects at each. If she calls the proportions at the two plants p_1 and p_2, then she will look at the hypotheses

$$H_0: \ p_1 = p_2 \quad \text{or} \quad p_1 - p_2 = 0$$

$$H_a: \ p_1 \neq p_2 \quad \text{or} \quad p_1 - p_2 \neq 0$$

If the engineer found the proportions of rejects in her samples to be quite different, then she would reject the null hypothesis.

Problems: Section 9.1

Answers to odd-numbered problems are given in the back of the text.

1. The manager at Costello Drug Store assumes that the company's employees are honest. However, there have been many shortages from the cash register lately. There is only one employee who could have taken any money from the register during these periods. Realizing that the shortages might have resulted from the employee's inadvertently giving incorrect change to customers, the employer does not know whether to forget the situation or accuse the employee of theft. In words, what is the null hypothesis?

2. A salesman for Hercules, a producer of composite (graphite) material, tells a potential customer that the mean material breaking strength is at least 10 pounds per unit.

   a. Specify the null and alternative hypotheses the potential customer should investigate if she wants to test to see whether the strength of the material might have decreased.

   b. If the potential customer's specifications require a mean breaking strength of at least 10 pounds, but a sample results in a sample mean that is quite far below 10 pounds, what hypothesis should probably be rejected? What is the potential customer's business decision that corresponds to the decision regarding the hypotheses?

3. A training program is designed to assist accountants to achieve higher scores on the Certified Public Accounting (CPA) exam. The program director maintains from past experience and has advertised that 10% or less of those accountants who complete the training program do not pass the exam. Specify the null and alternative hypotheses, assuming that a candidate for the exam wanted to test whether the advertising is false.

9.2
TESTING HYPOTHESES

Let us return to Example 9.1, where the International Rubber Band Company has been filling their number 6 box of rubber bands on average with 6 or fewer ounces of rubber bands. We confront the problem of testing the null hypothesis $H_0: \ \mu \leq 6$ against the alternative $H_a: \ \mu > 6$. The weights for the boxes are normally distributed.

If we take a random sample of $n = 9$ of the number 6 boxes of rubber bands, then the sample mean $\overline{X}$ is normally distributed; and t computed from

$$t = \frac{\overline{X} - \mu_0}{s/\sqrt{n}} \tag{9.1}$$

follows the t distribution with $n - 1$ degrees of freedom, where μ_0 is the hypothesized value for μ.

We want to use the sample mean $\overline{X}$ to compute a value for the t statistic to decide whether or not the alternative hypothesis H_a is more reasonable or plausible than the null hypothesis H_0—that is, whether or not the average weight of rubber bands in International's filling process is greater than six ounces. And we want to set up, in advance of sampling, a rule for making the decision. We want to set up a statistical procedure in the form of a **hypothesis test.** This process parallels what we do in constructing confidence intervals: for a given problem settling on confidence limits $\overline{X} \pm t_{\alpha/2, n-1}(s/\sqrt{n})$, in advance of sampling, and letting the sample give to these limits its actual numerical values.

In the present problem, we want to use a *sample statistic* or a *test statistic* to set up a *rejection region,* or a *critical region*—a set R of values from the t distribution with $n - 1$ degrees of freedom that will lead us to reject H_0 and prefer H_a. Associated with a rejection region is a *rejection rule.* If the sample results give to t of Equation (9.1) a value lying in the rejection region, the rejection rule will tell us to reject H_0; if the sample result for t is not a value lying in the rejection region, then we will fail to reject H_0.

A **rejection rule** is defined as follows.

Definition

A **rejection rule** indicates the values of a sample statistic, or test statistic, that will lead us to reject a null hypothesis.

EXAMPLE 9.6 To illustrate, suppose that we take a random sample of $n = 9$ of International's number 6 boxes of rubber bands and that the standard deviation for the sample is $s = 3$. We will determine the sample mean $\overline{X}$ and then use Equation (9.1) to compute a value for t. Notice that the degrees of freedom will be $n - 1 = 9 - 1 = 8$. Consider a rejection region—say for example, $t > 1.860$—and the corresponding rejection rule:

Rejection rule: If $t > 1.860$, then reject H_0.

The value 1.860 is known as a **critical value** and in this case is the value for $t_{.05, 8}$. Subsequent discussion will make clear why we have selected 1.860 as the critical value. If we use this rejection rule, the test is as follows: If the sample results give a value for t in Equation (9.1) that is greater than 1.860, then we reject H_0. In this situation, the sample results are contrary to H_0 to such an extent that we reject the hypothesis that International's filling process is providing number 6 boxes with 6 or fewer ounces of rubber bands. If the sample results give a value to t in Equation (9.1) that is less than or equal to 1.860, then we do not reject H_0.

The rule for the rejection region is logical and reasonable. The sample results support the alternative hypothesis H_a that $\mu > 6$ when t takes on a value that is large because of a large sample mean $\overline{X}$. It is contrary to common sense to use a rule that says the sample results support H_a: $\mu > 6$, if the sample mean $\overline{X}$ is less than or equal to 6, for example. But even though the rejection region in this case makes sense, we are left with the question: Why are we under the impression that this rejection region is a "good" rejection region? The answer to this question depends on the kinds of errors we might make by using the rejection rule, as we discuss in the next section.

9.3
TYPE I AND TYPE II ERRORS

Applying a hypothesis test may lead to the wrong conclusion. There are, in fact, two kinds of error possible, called **type I** and **type II errors.**

> **Definitions**
>
> A **type I error** occurs if we reject H_0 when H_0 is true.
>
> A **type II error** occurs if we do not reject H_0 when H_0 is false.

Notice that it is possible to make a type I error only when we reject H_0, and it is possible to make a type II error only when we do not reject H_0.

We can assess a test's cabability to lead to correct conclusions by knowing about the probabilities of type I and type II errors. The probabilities of type I and type II errors are known as **alpha (α)** and **beta (β)**, which are defined as follows.

> **Definitions**
>
> $$\alpha = P(\textit{Rejecting } H_0 \textit{ when } H_0 \textit{ is true}) = P(\textit{Type I error})$$
>
> $$\beta = P(\textit{Not rejecting } H_0 \textit{ when } H_0 \textit{ is false}) = P(\textit{Type II error})$$

Naturally, we want probabilities for errors, α and the possible β values, to be small. The probability α of a type I error is also called the **level of significance** or the **size** of the test, and $1 - \beta$ is known as a test's **power.** Table 9.1 presents the pos-

TABLE 9.1

Conclusions and Outcomes for a Hypothesis Test	*(a) If H_0 Is True*	*Conclusion*	
		Reject H_0	*Do Not Reject H_0*
	Outcome	Type I error	Correct conclusion
	Probability	*Probability = Significance level = α*	*Probability = $1 - \alpha$*
	(b) If H_0 Is False	*Conclusion*	
		Reject H_0	*Do Not Reject H_0*
	Outcome	Correct conclusion	Type II error
	Probability	*Probability = Power = $1 - \beta$*	*Probability = β*

sible conclusions and errors in a test. The type I error probability α is computed when we are testing a null hypothesis for a mean, where $\mu = \mu_0$.

EXAMPLE 9.7 Consider again our International Rubber Band Company example and our rejection rule:

Rejection rule: If $t > 1.860$, then reject H_0.

Here a type I error occurs when the population mean weight of the number 6 boxes of rubber bands is in truth less than or equal to 6 ounces, but the sample mean $\overline{X}$ is large enough that the value for $t = (\overline{X} - \mu_0)/(s/\sqrt{n})$ [see Equation (9.1)] is greater than 1.860. We want to find α, the probability of a type I error if $\mu = 6$ so that H_0 is true. The value for the degrees of freedom in this example is $n - 1 = 9 - 1 = 8$. The probability of a type I error is as follows:

$$\alpha = P\left(t > \frac{\overline{X} - \mu_0}{s/\sqrt{n}} \quad \text{when } \mu = 6\right)$$
$$= P(t > 1.860) = .05$$

The probability value of .05 for α, when t has 8 degrees of freedom, can be found in Table VI of Appendix C, since $t_{.05,8} = 1.860$. (See Figure 9.1.)

We are now in a position to answer the following question that was asked in Example 9.6: Why are we under the impression that the rejection rule for the International Rubber Band Company example is appropriate? We found in Example 9.7 that the probability of committing a type I error is only .05 using the rejection rule:

Rejection rule: If $t > 1.860$, then reject H_0.

Since we naturally want the probability of making an error to be small, like .05, this rejection rule is reasonable and sensible for testing whether International's filling process has undergone a shift in the average weight of rubber bands in number 6 boxes.

FIGURE 9.1

Probability of a Type I Error, α, for Rejection Region for International Rubber Band Company

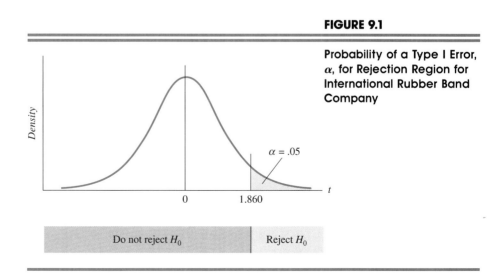

For many hypothesis tests, the null hypothesis is chosen so that the type I error is a more serious error than the type II error. Thus, the most important probability in hypothesis tests is usually α, the probability that the hypothesis test will lead to a type I error—that is, rejection of the null hypothesis when it is true. Thus, α is set explicitly to a small value such as .01 or .05. Seldom would α be allowed to exceed .10.

To deal with β is somewhat more difficult, since we can calculate a β probability for each possible true value of the parameter, say, μ or p. One can be safe, however, with using large samples whenever possible. Under these circumstances, both the α probability and the important β probabilities are generally small. We also note that α and β almost never sum to unity. Under large samples both probabilities are generally small.

Problems: Sections 9.2–9.3

4. El Pez Foods Company suspects that several thousand cases of recently canned tuna may be contaminated. The company decides to examine several randomly selected cases. Its null hypothesis is that the cases are contaminated.

   a. What are the type I and type II errors?
   b. Would the company want a small α or a small β if it had to choose one or the other to be small? Explain.

5. Northway Insurance Company has maintained a record of overpaying invoices by a mean amount less than or equal to $5. An auditor of Northway Insurance Company suspects that the mean overpayment for a large, essentially normal population of invoices may be greater than $5. For a test, the auditor takes a random sample of 9 invoices and finds a sample standard deviation of $3. The auditor rejects the null hypothesis if

$$t = \frac{\overline{X} - \mu_0}{s/\sqrt{n}}$$

is greater than 2.306.

   a. Give the null and alternative hypotheses.
   b. Find the probability of a type I error if the population mean overpayment is equal to $5.

6. A box contains four circuit boards that were produced by Digital Devices, Inc. Some of the boards are defective. The others are not defective. To test the hypothesis that there are two defective circuit boards and two nondefective circuit boards, we select two without replacement and conclude that there are not two of each if both circuit boards are defective or if both circuit boards are nondefective. What are the type I and type II errors?

9.4
STEPS OF HYPOTHESIS TESTING

In this text, we will encounter many hypothesis-testing problems. They have features in common with those of our International Rubber Company example, and they also can be conveniently analyzed according to the following *standard sequence of steps.*

> **Hypothesis-Testing Steps**
>
> **Step One**   Formulate the null hypothesis H_0.
>
> **Step Two**   Formulate the alternative hypothesis H_a in statistical terms.
>
> **Step Three**   Set the level of significance α and the sample size n.
>
> **Step Four**   Select the appropriate test statistic and the rejection rule.
>
> **Step Five**   Collect the data and calculate the test statistic.
>
> **Step Six**   If the calculated value of the test statistic falls in the rejection region, then reject H_0. If the calculated value of the test statistic does not fall in the rejection region, then do not reject H_0.

Each of these steps calls for comment.

Step One   The null hypothesis must be stated in statistical terms. It would not suffice in our example merely to hypothesize that the International filling process has not changed; we need the specific t distribution, together with the identification of $\mu \le 6$ representing International's filling process.

Step Two   The alternative hypothesis is essential to the problem. Recall that International's filling process gave the average weight of their number 6 boxes of rubber bands as less than or equal to 6 ounces, so the alternative hypothesis was $H_a: \mu > 6$. Suppose that International's filling process produced number 6 boxes of rubber bands with an average weight greater than or equal to 6 ounces, that is $H_0: \mu \ge 6$. Should the alternative hypothesis again be $H_a: \mu > 6$ ounces? Obviously not, since on average the boxes must hold greater than or equal to 6 ounces or *less* than 6 ounces of rubber bands. Thus in this case the alternative hypothesis would be $H_a: \mu < 6$ ounces, and we would need a small sample mean $\overline{X}$, which in turn would lead to a small value for t in Equation (9.1) for us to reject the null hypothesis. Finally, International's filling process could be producing number 6 boxes with an average of exactly 6 ounces of rubber bands, or $H_0: \mu = 6$. Here the alternative hypothesis would be that their number 6 box has a mean weight that is *not* equal to 6 ounces, or $H_a: \mu \ne 6$ ounces; and either a large sample mean $\overline{X}$ would lead to a large value for t to reject the null hypothesis or a small sample mean $\overline{X}$ would lead to a small value of t to reject the null hypothesis. Since the form of the alternative hypothesis determines the rejection region and rejection rule we will choose in Step 4, the alternative hypothesis must be carefully stated to represent the problem.

Step Three   Since the null hypothesis is usually set up in such a way that we want strong evidence against it before we reject it, the level of α is usually rather small. Values often used for α are .01, .02, .05, and .10. Researchers and decision makers seldom allow more than a 10% chance of a type I error in hypothesis tests. Larger values of α are sometimes tolerated when a type I error is not as serious as a type II error.

Step Four   In the International Rubber Band Company example the sample results are used to compute a value for the t statistic defined in Equation (9.1); and since $t_{.05,8} = 1.860$, the rejection region is $R: t > 1.860$, and the corresponding rejection rule is as follows:

Rejection rule: If $t > 1.860$, then reject H_0.

The rejection rule accords with common sense, since a large value for $\overline{X}$ gives a large value for t in Equation (9.1). In all our testing problems, logic and reasoning dictate the proper form of the rejection rule; choosing the exact rejection region (choosing the critical value of t in our example) requires knowing the distribution of the statistic. We used the t distribution with $n - 1 = 9 - 1 = 8$ in the International Rubber Band Company example, but other distributions may be used in other situations. For example, the Z statistic is generally used to test hypotheses about a population proportion p. In general, when the alternative hypothesis is a strict inequality, as in the International example where $H_a: \mu > 6$, then the rejection region consists of one set of values, such as reject the null hypothesis if $t > 1.860$. But when the alternative hypothesis is of the "not equal to" type, as in the case where $H_a: \mu \neq 6$, then the rejection region consists of two sets of values, either very high or very low values for the sample mean $\overline{X}$, which lead to very high or very low values for t.

Step Five   The sampling procedure must accord with the model—that is, the t distribution model in our International Rubber Band Company example. Random samples must be used to test hypotheses.

Step Six   The hypotheses H_0 and H_a are not treated in a symmetric fashion. That is, if we reject H_0, then the sample results may very well better support H_a. However, if we do not reject H_0, that does not mean that the data support H_0. Just because we do not have sufficient evidence to reject a hypothesis does not mean that we have strong evidence that it is true.

EXAMPLE 9.8   To summarize the steps of hypothesis testing, we list the steps that we have discussed previously for the International Rubber Band Company example, and we take a random sample of $n = 9$ number 6 boxes of rubber bands from the normal population, summarize the sample results, compute the t statistic, and complete the hypothesis test.

Step One $H_0: \mu \leq 6$.

Step Two $H_a: \mu > 6$.

Step Three $\alpha = .05, n = 9$.

Step Four $t_{.05,9-1} = t_{.05,8} = 1.860$. So the rejection rule is as follows:

Rejection rule: If $t > 1.860$, then reject H_0.

Step Five   The weights of the 9 boxes of number 6 rubber bands in the sample were

14   6   7   6   6   10   7   11   5

The sample mean and standard deviation are

$$\overline{X} = (14 + 6 + 7 + \cdots + 5)/9 = 72/9 = 8$$

and

$$s = \sqrt{\Sigma(X - \overline{X})^2/(n - 1)}$$
$$= \sqrt{[(14 - 8)^2 + (6 - 8)^2 + \cdots + (5 - 8)^2]/(9 - 1)}$$
$$= \sqrt{72/8} = \sqrt{9} = 3$$

Thus, we have

$$t = \frac{\overline{X} - \mu_0}{s/\sqrt{n}} = \frac{8 - 6}{3/\sqrt{9}} = \frac{2}{1} = 2$$

Step Six The t value for the sample, 2, is greater than 1.860 (see Figure 9.2), so we reject H_0. The sample results differ by too much from what we would expect if the null hypothesis H_0: $\mu \leq 6$ were true to maintain that hypothesis. The sample results better support the alternative hypothesis, H_a: $\mu > 6$, that the mean weight of the number 6 boxes of rubber bands is greater than 6 ounces. As the quality control inspector suspected, the quality of the filling process has apparently been stretched too far. Based on the test results, International should adjust or improve the filling process.

Sometimes, tests of hypothesis are called **tests of significance.** If the α we use in such a test is .05 and if we reject the null hypothesis, we sometimes say, "The hypothesis was rejected at the 5% **significance level.**" This statement means that the sample results we obtained were *significantly* different from those we would expect to see if the hypothesis were true. In fact, the sample result will have only a 5% chance or less of occurring if the hypothesis is true, and thus the result is significantly different from the hypothesized or expected result.

In a managerial or decision-making context, the results of a hypothesis test are often used to *aid* decision making. For instance, if we conclude that the filling process at International Rubber Band has been stretched too far, then we may decide to adjust the process. However, *statistical significance* does not always imply *practical significance,* and decisions are often influenced by additional factors such as economic and practical implications of the decision. Consequently, both statistical significance and practical significance should be considered during decision making.

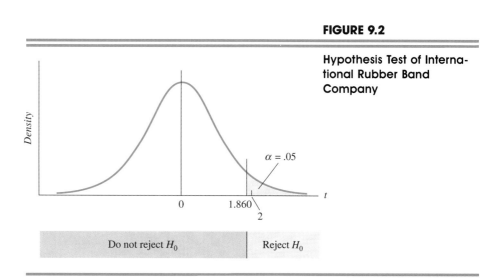

FIGURE 9.2

Hypothesis Test of International Rubber Band Company

In the following sections, we construct a number of tests of hypotheses. In each case, we must actually specify the rejection rule, and these constructions will clarify the general principles set out previously. Tests of hypotheses are used extensively in research and decision making. The theoretical framework allows us to determine what test is appropriate for any given problem and to assess the test's capability to lead us to a correct conclusion.

Hypothesis testing steps are loosely analogous to the rules that are used in a criminal trial. Case 9.1 at the end of this chapter discusses how the procedures in a court of law parallel the steps in a hypothesis test.

9.5
HYPOTHESES ON A NORMAL MEAN

Consider a normal population with the standard deviation σ unknown and estimated by s and three forms of a null hypothesis on a specified value of the mean μ_0:

$$H_0: \mu \leq \mu_0$$

$$H_0: \mu \geq \mu_0$$

$$H_0: \mu = \mu_0$$

Under these hypotheses, the statistic $t = (\overline{X} - \mu_0)/(s/\sqrt{n})$ [see Equation (9.1)] has a t distribution with $n - 1$ degrees of freedom (see Section 8.3), which can be made the basis of three tests.

If the null hypothesis is $H_0: \mu \leq \mu_0$ and the alternative hypothesis is $H_a: \mu > \mu_0$, then it is reasonable to take a sample and to reject H_0 in favor of H_a when $\overline{X}$ is too large—that is, when $\overline{X}$ is greater than μ_0 by too much. Since $t = (\overline{X} - \mu_0)/(s/\sqrt{n})$ increases when $\overline{X}$ increases, we can just as well reject H_0 when t is large. How large? Recall that the upper α percentage cutoff point $t_{\alpha, n-1}$ is defined so that $P(t > t_{\alpha, n-1})$ is equal to α. Therefore, if we adopt the *rejection rule* of rejecting H_0 when the statistic $t = (\overline{X} - \mu_0)/(s/\sqrt{n})$ is greater than $t_{\alpha, n-1}$, then the chance of rejecting H_0 when the true mean μ is exactly equal to μ_0 so that H_0 is true is just α.

FIGURE 9.3

Determining the One-Sided Rejection Region for the Alternative Hypothesis H_a: $\mu > \mu_0$

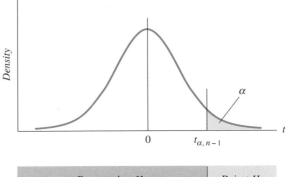

The rejection region R is specified by the inequality $t > t_{\alpha, n-1}$. The shaded area in Figure 9.3 represents α, and all t values to the right of $t_{\alpha, n-1}$ form the rejection region. The alternative hypothesis in this case, H_a: $\mu > \mu_0$, is *one-sided* (or one-tailed), and so is the corresponding rejection region.

The hypothesis-testing procedure for this situation is summarized as follows.*

**Hypotheses on Normal Mean and Rejection Rule
for One-Sided Test Where H_a: $\mu > \mu_0$**

H_0: $\mu \leq \mu_0$

H_a: $\mu > \mu_0$ (9.2)

Rejection rule: If $t > t_{\alpha, n-1}$, then reject H_0;

where: $t = \dfrac{\overline{X} - \mu_0}{s/\sqrt{n}}$

Similarly, if the null hypothesis is H_0: $\mu \geq \mu_0$, and the alternative hypothesis is H_a: $\mu < \mu_0$, it is sensible to reject the null hypothesis when t is too small. Figure 9.4 shows the rejection values of t for this hypothesis test, and the shaded area represents the probability of a type I error α. Notice again that $t < -t_{\alpha, n-1}$ has probability α (due to symmetry) if μ is exactly equal to μ_0.

---

*The rejection rule may also be stated in terms of the sample mean $\overline{X}$. If we substitute the ratio for t for the t of the rejection rule in Summary (9.2), multiply by $s/\sqrt{n}$, and then add μ_0, the inequality becomes $\overline{X} > \mu_0 + t_{\alpha, n-1}(s/\sqrt{n})$. Thus we can also reject the hypothesis that $\mu \leq \mu_0$ when $\overline{X}$ is too large, or when $\overline{X}$ is greater than $\mu_0 + t_{\alpha, n-1}(s/\sqrt{n})$. Nevertheless, we give rejection rules in terms of test statistics in the hypothesis tests that follow. By convention, computers also generally give test results in the form of the same test statistics that we use.

FIGURE 9.4

Determining the One-Sided
Rejection Region for the
Alternative Hypothesis H_a:
$\mu < \mu_0$

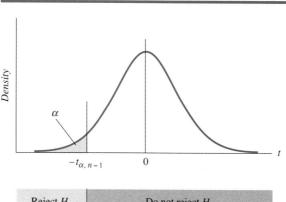

The hypothesis-testing procedure for this situation is summarized as follows.

**Hypotheses on Normal Mean and Rejection Rule
for One-Sided Test Where H_a: $\mu < \mu_0$**

$$H_0: \mu \geq \mu_0$$

$$H_a: \mu < \mu_0$$

Rejection rule: If $t < -t_{\alpha,n-1}$, then reject H_0;

where: $t = \dfrac{\overline{X} - \mu_0}{s/\sqrt{n}}$

(9.3)

The alternative hypotheses in Summaries (9.2) and (9.3) are one-sided (or one-tailed), and so are the corresponding rejection regions. If the null hypothesis is H_0: $\mu = \mu_0$, then the alternative hypothesis H_a: $\mu \neq \mu_0$ is *two-sided* because it includes both possibilities $\mu < \mu_0$ and $\mu > \mu_0$. The rejection region is also two-sided (or two-tailed). The cutoff points are $\alpha/2$ percentage points, since the test is two-tailed. As Figure 9.5 indicates, the total area of the two tails taken together is α. The rejection region is specified by the two sets of values $t < -t_{\alpha/2,n-1}$ and $t > t_{\alpha/2,n-1}$. We reject H_0 if $\overline{X}$ is excessively far from μ_0 in either direction, that is, if $\overline{X}$ is many standard deviations above or below the value μ_0.

The hypothesis-testing procedure for the two-sided situation is summarized as follows.

**Hypotheses on Normal Mean and Rejection Rule
for Two-Sided Test**

$$H_0: \mu = \mu_0$$

$$H_a: \mu \neq \mu_0$$

Rejection rule: If $t < -t_{\alpha/2,n-1}$ or $t > t_{\alpha/2,n-1}$,
                            then reject H_0;

where: $t = \dfrac{\overline{X} - \mu_0}{s/\sqrt{n}}$

(9.4)

 PRACTICE PROBLEM 9.1

Telsyme, Inc., a manufacturer of small electric motors, maintains that on average their motors have always drawn .8 ampere or more under normal load conditions. A sample of 16 of the motors was tested, and the results (in amperes) are presented below. Are we justified in rejecting the null hypothesis that the motors draw .8 ampere or more?

.5   .9   .7   1.1   1.1   .8   .3   .7
.6   .8   .6   .4   .8   .2   .5   .6

FIGURE 9.5

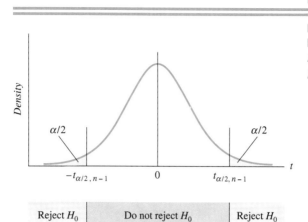

Determining the Two-Sided
Rejection Region for the
Alternative Hypothesis H_a:
$\mu \neq \mu_0$

Solution The null hypothesis is that they draw .8 ampere or more on average, or H_0:
$\mu \geq .8$. Since we want to reject H_0 only if the evidence indicates that the mean current consumption (the population value) is less than .8, we have a one-sided alternative, or $H_a: \mu < .8$. If we are willing to take a risk of 1 in 20 of rejecting the assertion
if it is true, we set α at .05. Under the assumption of normality we use a t statistic with
$n - 1 = 16 - 1 = 15$ degrees of freedom and a one-tailed test. The cutoff point is
$-t_{.05,15} = -1.753$. (See Figure 9.6.)

Step One $H_0: \mu \geq 8$.
Step Two $H_a: \mu < .8$.
Step Three $\alpha = .05, n = 16$.
Step Four   Rejection rule: If $t < -1.753$, then reject H_0.
Step Five   For our data,

$$\overline{X} = .663 \qquad s = .255$$

FIGURE 9.6

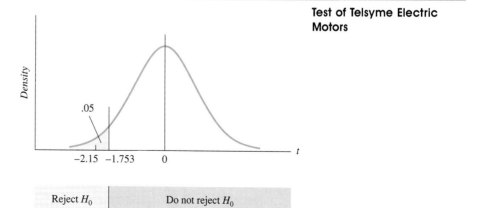

Test of Telsyme Electric
Motors

and the calculated t is

$$t = \frac{\overline{X} - \mu_0}{s/\sqrt{n}} = \frac{.663 - .8}{.255/\sqrt{16}} = -2.15$$

Step Six   Since $t = -2.15$ is less than -1.753, we reject the null hypothesis and conclude that the mean current is less than .8 ampere. The chance that our conclusion is incorrect is .05 or less.*

Two statistical software packages were also used to solve this problem, and the results are presented in Computer Exhibits 9.1A and 9.1B.

PRACTICE PROBLEM 9.2

In an attempt to determine whether or not special training increases the speed with which assembly line workers can do an assembly job at AMTEL, Inc., 25 workers are timed doing this job. Then they are given a special course designed to increase their assembly efficiency. At the end of the course they are timed doing the same job. The differences between the first and the second times are recorded for each assembly line worker: the mean reduction for the 25 workers is found to be 2.5 minutes, and the sample standard deviation is 9 minutes. Has the training reduced the time required on this job?

Solution   The null hypothesis that the training has no effect or a detrimental effect is that the population mean reduction is $\mu \leq 0$. We will reject H_0 only if we think the mean reduction is positive, so the alternative is H_a: $\mu > 0$. We need a one-tailed test, and if we set α at .05, the cutoff point is $t_{.05,24}$, or 1.711. (See Figure 9.7.)

Step One H_0: $\mu \leq 0$.
Step Two H_a: $\mu > 0$.

*In reality, the chance that we are in error is either 0 or 1, since our conclusion must be either right or wrong. However, it is useful to think of α as the level of risk we expose ourselves to in rejecting the null hypothesis, and little is lost if we think of α as the probability of being in error.

MINITAB   **COMPUTER EXHIBIT 9.1A** Hypothesis Test for Telsyme, Inc., Data

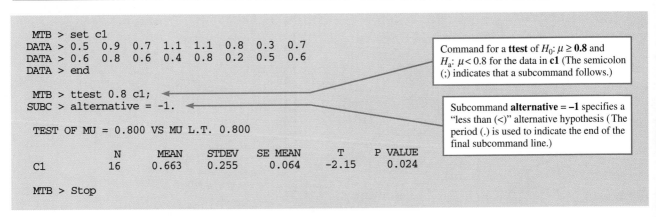

```
MTB > set c1
DATA > 0.5  0.9  0.7  1.1  1.1  0.8  0.3  0.7
DATA > 0.6  0.8  0.6  0.4  0.8  0.2  0.5  0.6
DATA > end

MTB > ttest 0.8 c1;
SUBC > alternative = -1.

 TEST OF MU = 0.800 VS MU L.T. 0.800

            N      MEAN    STDEV   SE MEAN      T    P VALUE
C1         16     0.663    0.255    0.064    -2.15    0.024

MTB > Stop
```

Command for a **ttest** of H_0: $\mu \geq$ **0.8** and H_a: $\mu < 0.8$ for the data in **c1** (The semicolon (;) indicates that a subcommand follows.)

Subcommand **alternative = −1** specifies a "less than (<)" alternative hypothesis (The period (.) is used to indicate the end of the final subcommand line.)

Note: Subcommand **alternative = +1** is used for a "greater than (>)" alternative hypothesis. The semicolon and alternative subcommand are omitted for a "not equal to (≠)" alternative hypothesis.

COMPUTER EXHIBIT 9.1B Hypothesis Test for Telsyme, Inc., Data

MYSTAT

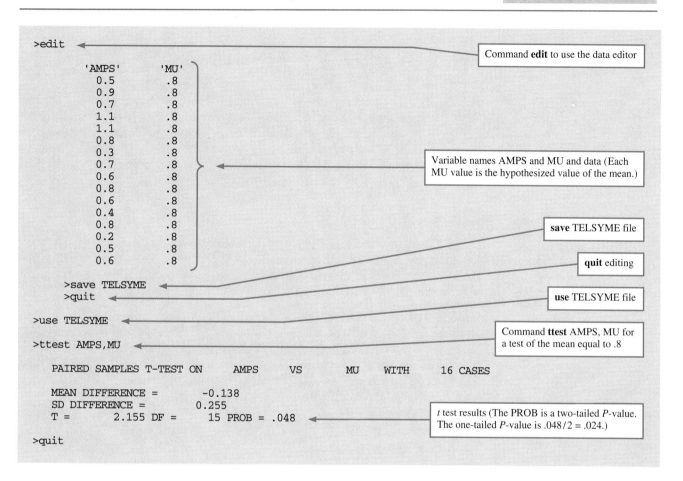

```
>edit                                          Command edit to use the data editor
        'AMPS'     'MU'
         0.5        .8
         0.9        .8
         0.7        .8
         1.1        .8
         1.1        .8
         0.8        .8
         0.3        .8          Variable names AMPS and MU and data (Each
         0.7        .8          MU value is the hypothesized value of the mean.)
         0.6        .8
         0.8        .8
         0.6        .8
         0.4        .8
         0.8        .8                          save TELSYME file
         0.2        .8
         0.5        .8                          quit editing
         0.6        .8
    >save TELSYME
    >quit                                       use TELSYME file
>use TELSYME
                                      Command ttest AMPS, MU for
>ttest AMPS,MU                        a test of the mean equal to .8

   PAIRED SAMPLES T-TEST ON    AMPS    VS    MU    WITH    16 CASES

   MEAN DIFFERENCE =        -0.138
   SD DIFFERENCE =         0.255
   T =        2.155 DF =      15 PROB = .048        t test results (The PROB is a two-tailed P-value.
                                                   The one-tailed P-value is .048/2 = .024.)
>quit
```

FIGURE 9.7

Test of Assembly Time at AMTEL

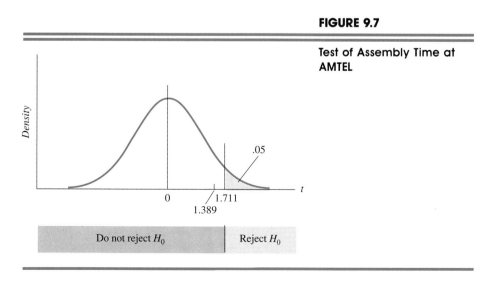

Step Three $\alpha = .05, n = 25.$
Step Four   Rejection rule: If $t > 1.711$, then reject H_0.
Step Five   The calculated t is

$$t = \frac{\overline{X} - 0}{s/\sqrt{25}} = \frac{2.5 - 0}{9/5} = 1.389$$

Step Six   The t value 1.389 is less than 1.711, so we do not reject H_0. Thus the sample evidence and the null hypothesis are consistent with one another since the reduction of 2.5 minutes is not large enough to cause rejection of H_0. However, note that not rejecting H_0 does not mean we have accepted H_0 and concluded that $\mu \leq 0$. We simply do not have enough evidence to reject the hypothesis that $\mu \leq 0$.

PRACTICE PROBLEM 9.3

Crager Company's last year's retail sales records show that the average monthly expenditure per person for a food product was $5.50. We wish to know whether there has been any significant increase or decrease in this average during the first quarter of this year. Thus we are looking for both positive and negative changes.

Solution To make a comparison, we sample 30 families and find that the mean expenditure is $5.10 with a standard deviation of $.90. Here the null hypothesis of no increase or decrease is $H_0: \mu = 5.50$. The two-sided alternative, $H_a: \mu \neq 5.50$, represents an increase or a decrease. Therefore we want a two-tailed test; if α is set at .01, the cutoff is the $\alpha/2 = .01/2 = .005$, percentage point for 29 degrees of freedom: $t_{.005,29}$ is 2.756. (See Figure 9.8.)

Step One $H_0: \mu = 5.50.$
Step Two $H_a: \mu \neq 5.50.$
Step Three $\alpha = .01, n = 30.$
Step Four   Rejection rule: If $t < -2.756$ or $t > 2.756$, then reject H_0.

FIGURE 9.8

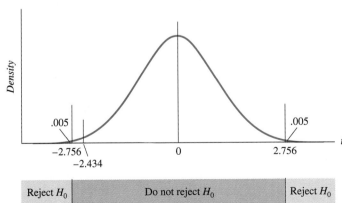

Test of Expenditures at Crager

Step Five   The calculated value of t is

$$t = \frac{\overline{X} - 5.50}{s/\sqrt{30}} = \frac{5.1 - 5.50}{.9/\sqrt{30}} = -2.434$$

Step Six   Since the t value -2.434 is not less than -2.756, we do not reject the null hypothesis. ▬▬

In each of our three problems we observe the following features:

1. The null hypothesis always contains in part the equality statement, and the alternative hypothesis is determined by the question implicit in the statement of the problem.
2. The calculated statistic is based on the difference between the observed mean $\overline{X}$ and the mean μ_0 under the null hypothesis. Since the difference between $\overline{X}$ and μ_0 is meaningful only in relation to the amount of variation in the data, the statistic is the ratio of the difference to its estimated standard deviation.
3. When the difference between $\overline{X}$ and μ_0 is large enough that the calculated t value falls into the rejection region (the appropriate rejection region is determined by the alternative hypothesis), we reject the null hypothesis, because the probability of such a t value, given that the null hypothesis is true, is too small to be attributed to ordinary sampling variation.

Whenever we use the t distribution for calculating confidence intervals or for testing hypotheses, we are assuming that the data are a random sample from a normal population. The first assumption, randomness, is necessary if we want to make probability statements about the results we obtain; therefore, it cannot be relaxed.

The second assumption, normality, is not too strong and may be relaxed. Because of the central limit theorem, we may say that even though the population has a distribution that is not normal, the t statistic may be used and the probabilities will not be greatly affected, provided we have a sufficiently large sample. What size sample may be considered sufficiently large depends on how much the population concerned departs from normality. As a rough rule of thumb, many statisticians use t distributions when the population standard deviation is not known if the sample size is greater than or equal to 30, even if the population is not normal. However, when the population distribution is unimodal and not too skewed, adequate approximations to normality can be achieved with sample sizes that are smaller than 30. Also, if the sample is sufficiently large and Table VI in Appendix C does not contain the t value for the appropriate value for the degrees of freedom, then Z_α or $Z_{\alpha/2}$ can be used to approximate $t_{\alpha,\nu}$ or $t_{\alpha/2,\nu}$ in testing the hypotheses of Summaries (9.2), (9.3), and (9.4).

Finally, if for some reason we are fortunate enough to know the standard deviation σ of the normal population for which the hypothesis test is being conducted, then σ can be substituted for s in Equation (9.1). With this change the ratio in (9.1) is exactly normally distributed as the standardized Z variable:

$$Z = \frac{\overline{X} - \mu_0}{\sigma/\sqrt{n}} \qquad (9.5)$$

Hypothesis tests using Equation (9.5) are conducted just as those using Equation (9.1) are, although it is rather rare that such tests are conducted since we seldom know the population standard deviation. In this case the standard normal distribution would be used to determine the rejection region. The Z of Equation (9.5) is approximately normally distributed for large samples even if the population is not normal due to the central limit effect.

 PRACTICE PROBLEM 9.4

Compudyne Corporation has a cutting machine that is designed so that the lengths of cut parts turned out by the machine may be adjusted to different settings. However, the variance of the measurements, once a setting is chosen, is always $\sigma^2 = .5$. The machine was set to cut at a mean length of 30. Use the following sample data to test the null hypothesis that the mean length of the normally distributed cuts is 30 against the alternative hypothesis that μ is not 30. Use $\alpha = .05$.

$$n = 16 \qquad \overline{X} = 30.25 \qquad s = .6$$

Solution Since we know that the population standard deviation is $\sigma = \sqrt{.5} = .707$, we will use the ratio in Equation (9.5) and ignore the $s = .6$ value. If we use $\alpha = .05$, then the Z value that is used to define our rejection region is $Z_{.025}$. This value is found in the normal table or in the last row of the t table, and it is 1.96.

Step One H_0: $\mu = 30$.
Step Two H_a: $\mu \neq 30$.
Step Three $\alpha = .05, n = 16$.
Step Four   Rejection rule: If $Z < -1.96$ or $Z > 1.96$, then reject H_0.
Step Five   The calculated Z is

$$Z = \frac{30.25 - 30}{.707/\sqrt{16}} = 1.41$$

Step Six   As illustrated in Figure 9.9, the value of Z, 1.41, does not fall in the rejection region, so we cannot reject the hypothesis that the mean length of cuts is $\mu = 30$.

FIGURE 9.9

Test of Lengths of Parts at
Compudyme

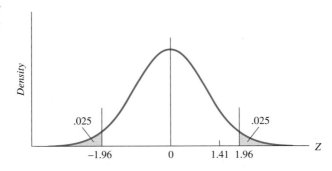

Problems: Sections 9.4-9.5

7. For the following information from a normal distribution,

$$n = 25 \qquad \overline{X} = 28 \qquad s = 3$$

$$H_0: \mu \leq 20 \qquad H_a: \mu > 20 \qquad \alpha = .05$$

test the hypothesis at the given level of significance. The data (which are only needed if you want to use them with a computer) are as follows:

29.7029	26.2284	25.9161	35.0564	26.9919	33.2668	29.4872
29.6590	32.5816	28.6065	28.7908	28.1814	22.4731	26.1621
26.3596	24.4611	26.9709	21.7426	27.5778	27.3449	30.7388
27.8609	28.8086	26.6535	28.3772			

8. For the following information from a normal distribution,

$$n = 9 \qquad \overline{X} = 324.7 \qquad s^2 = 9$$

$$H_0: \mu \geq 327 \qquad H_a: \mu < 327 \qquad \alpha = .10$$

test the hypothesis at the given level of significance. The data (which are only needed if you want to use them with a computer) are as follows:

327.319	326.107	323.427	324.688	320.369
325.423	328.895	326.128	319.944	

9. For the following information from a normal distribution,

$$n = 25 \qquad \overline{X} = 50 \qquad s^2 = 100$$

$$H_0: \mu = 55 \qquad H_a: \mu \neq 55 \qquad \alpha = .01$$

test the hypothesis at the given level of significance. The data (which are only needed if you want to use them with a computer) are as follows:

47.0697	50.5307	52.2970	54.5404	46.6408	37.6321	56.7735
45.1039	69.8864	46.7952	64.2181	39.7026	52.6325	54.8496
48.2739	30.1089	44.0208	40.8290	62.5978	61.5951	48.4856
38.2399	38.7552	50.8920	67.5293			

10. For the following information from a normal distribution,

$$n = 9 \qquad \overline{X} = 76 \qquad \sum(X_i - \overline{X})^2 = 32$$

$$H_0: \mu = 75 \qquad H_a: \mu \neq 75 \qquad \alpha = .10$$

test the hypothesis at the given level of significance. The data (which are only needed if you want to use them with a computer) are as follows:

74.1058	75.4106	76.2693	73.0437	76.5919
75.8299	77.7728	75.1109	79.8652	

11. The mean time between malfunctions for an electronic component produced by International Semiconductor Corporation has been 600 hours. A model using a new microprocessor was designed, on the average, to operate more than three times as long without malfunctioning. These supposedly superior pieces of equipment are put into service and give rise to $n = 61$ malfunctions, with a mean time between malfunctions equal to 2250 hours. The sample standard deviation is 980 hours. At the 5% level, test the null hypothesis that the new model does not operate more than three times as long without malfunctioning. Use $H_a: \mu > 1800$ hours. Note that the sample

statistics summarized in this problem were computed from the accompanying data; however, the data may be useful if you wish to use a computer.

724.39	872.73	1721.40	1108.54	2494.94	2493.62	3265.96
426.18	1612.38	3469.40	1863.38	2760.38	1384.17	2596.98
2965.30	2284.19	2162.03	1721.09	2289.63	1087.40	696.57
1714.15	2248.61	2831.35	2632.19	3226.99	4653.22	1802.96
1626.53	1068.61	2713.42	3925.77	2147.13	1296.11	3584.61
2174.88	1857.92	1230.92	2974.12	3826.72	3606.87	2094.77
1925.60	2359.83	1441.81	1504.81	3290.71	1813.65	957.01
2219.35	3767.31	2571.02	4309.75	2104.75	2319.71	3249.72
4214.07	1546.41	1366.46	1220.40	1829.12		

12. Five years ago a state's Department of Transportation did a study of automobile pollution at a particular location. During the month of November it found that the mean pollution measurement was 132. During November of this year the department took a random sample of $n = 8$ measurements and found that the sample mean pollution measurement was 120 and the sample standard deviation was $s = 10$. Test the hypothesis that the population mean measurement is still 132 or more against the alternative that the mean is now less than 132. Use $\alpha = .025$. Pollution measurements are normally distributed. The data (which are only needed if you want to use them with a computer) are as follows:

120.064	128.824	131.840	129.326
105.479	121.341	107.093	116.033

13. The accompanying data are the times (in seconds) that it took a sample of employees to assemble a toy truck at a Cole Industries assembly plant. Assembly times are normally distributed. At the 10% level, can we conclude that the mean assembly time for this toy truck is not equal to 3 minutes? Use H_a: $\mu \neq 180$ seconds as the alternative hypothesis.

190	198	180	181	208	198
199	176	174	183	188	165

14. A financial analyst for James, Stephen & Associates maintains the historical view that takeover target firms have always experienced percentage sales growths with a mean of 23 percent or less. A sample of $n = 9$ firms was taken, and the sales growth percentages are given in the accompanying listing. Test the hypothesis that firms that are now takeover targets have increased their average percentage sales growth. Assume normality and use a .05 level of significance.

31   21   21   26   26   25   25   25   25

15. The manufacturer of a four-wheel-drive vehicle maintains that a driver has always been able to travel an average of 150 miles or more over rough terrain in an 8-hour day. An independent testing agency was hired by the manufacturer's competitor to test whether the average traveling distance has declined. The agency drove 18 of the vehicles an average of 145 miles over rough terrain during an 8-hour period. The standard deviation of the sample was 8.3 miles. Assume normality.

a. What is the null hypothesis in this problem? [*Hint:* Give the manufacturer the benefit of the doubt.]

b. Test the hypothesis, using 5% as the risk you are willing to accept for a type I error.

16. In Example 7.7 there is a discussion of a minority group whose members were being delayed in crossing a border. A sample of $n = 64$ showed that the sample mean delay time for members of the minority group was $\overline{X} = 10$ minutes. If the minority

group representative mentioned in that discussion were to test the null hypothesis H_0: $\mu \leq 8.0$ against the alternative hypothesis H_a: $\mu > 8.0$ with $\sigma = 6$ and were to reject the null, what would be the chance of a type I error, α?

17. Vitzer Pharmaceutical, Inc., has developed 6-voricil, a new cauterizing solution for local application on the benign tumor condylomata acuminata. Ten milliliters of the currently used resin solution has provided curative therapy for an average of 5 square centimeters of tumor. The areas of tumors that receive adequate therapy are approximately normally distributed. A sample of clinical trials of 10 ml of 6-voricil resulted in adequate therapy for the tumor areas given in the accompanying listing. Test to determine whether 6-voricil is more effective in cauterizing the benign tumors than is the current resin solution. Use a significance level of .01.

    14   6   10   6   7   6   7   11   5

REFERENCE: Ferenczy, A. (1987), "Laser Treatment of Patients with Condylomata...," *CA*, 37: 334–47.

18. An article in the *Journal of Finance* reported that large nonfinancial firms have a mean debt as a percentage of equity of 119%. Debt as a percentage of equity is a measure of financial leverage. Consider the list of brokerage houses shown in the accompanying listing, which gives debt as a percentage of equity. Assume normality and that the list is a random sample of all brokerage houses. Should an investor conclude that brokerage houses use less financial leverage than nonfinancial firms based on these data? Use .05 as the level of significance.

Brokerage Houses	Debt as % of Equity
First Boston	43.3
Salomon	151.6
Paine Webber Group	130.9
EF Hutton Group	36.7
Merrill Lynch	162.2
Bear Stearns Cos	45.0
Morgan Stanley	57.9

SOURCE: *Forbes,* January 12, 1987, p. 126.

REFERENCE: Stowe, J. D., C. J. Watson, and T. Robertson (1980), "Relationships Between the Two Sides of the Balance Sheet," *Journal of Finance* 35 (September): 973–80.

19. The data entry department at Keydata, Inc., has recently installed some new data entry equipment. Unfortunately, the equipment has required a great deal of service. With the previous equipment, Keydata had been able to give its customers a turnaround time of 6 hours or less. The standard deviation of the turnaround time over many years has been $\sigma = 1.5$ hours. To examine whether the turnaround time has increased, one of the customers took a sample of 36 jobs and found that the sample mean turnaround time was $\overline{X} = 6.5$ hours.

  a. Let the significance level be .10. Test whether the turnaround time has increased.

  b. Did you use the standard normal distribution or a t distribution for the test of part a? Why?

  c. The problem does not specify that turnaround times are normally distributed. Comment on the test in part a given that turnaround times may be skewed.

Note that the sample statistics summarized in this problem were computed from the accompanying data; however, the data may be useful if you wish to use a calculator or computer.

5.2435	5.5074	6.3609	10.8858	5.6274	6.5553
5.4539	6.9740	8.2565	5.5829	5.3427	7.9773
5.2514	5.3555	5.5051	5.8743	5.4756	6.9756
9.2847	11.5664	6.3081	7.0333	5.5612	5.6098
5.6064	5.2454	5.8691	7.5146	6.7131	5.4091
6.0849	5.6156	6.8077	7.0023	6.1719	6.3914

9.6
P-VALUES AND HYPOTHESIS TESTING

Sometimes when hypothesis tests are conducted, especially when a computer is used, a *P*-value, defined in the following box, is employed to decide whether or not to reject a null hypothesis or to determine the plausibility of a null hypothesis.

Definition

A *P*-value is the probability that the test statistic employed in the hypothesis test would assume a value as extreme as or more extreme than the observed value of the test statistic if the null hypothesis is true.

We compute a *P*-value under the null hypothesis—that is, under the assumption that the null hypothesis is true at the value indicated for the test.

=== **EXAMPLE 9.9**   Consider the International Rubber Band Company problem summarized in Example 9.8, where the quality control inspector was concerned that the mean weight of number 6 boxes of rubber bands was greater than 6 ounces. The null and alternative hypotheses were H_0: $\mu \leq 6$ and H_a: $\mu > 6$, and the significance level was .05.

In Example 9.8 the value of $t_{.05, 8}$ was 1.860, so the rejection rule was as follows:

Rejection rule: If $t > 1.860$, then reject H_0.

The t statistic computed from the sample was

$$t = \frac{\overline{X} - \mu_0}{s/\sqrt{n}} = \frac{8 - 6}{3/\sqrt{9}} = 2$$

so we rejected the null hypothesis.

The probability that t would assume a value as extreme as 2 (as large as or larger than 2) is .040, assuming that μ is 6 [see Figure 9.10(b)]. Thus the *P*-value for the International Rubber Band hypothesis test is .040. In other words, for the t distribution with degrees of freedom of 8, $P(t \geq t_{.040, 8}) = P(t \geq 2) = .040$, assuming that the null hypothesis is true. The value .040 can be found by interpolation from Table VI or by using a computer. ══

The *P*-value is less than α when the sample statistic is more extreme than the α level percentage point for the statistic. For instance, in the International Rubber Band

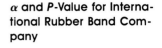

FIGURE 9.10

α and *P*-Value for International Rubber Band Company

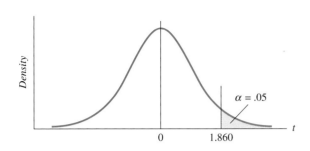

(a) α

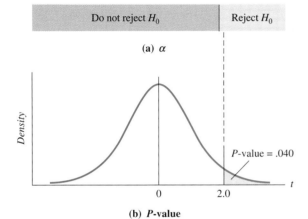

(b) *P*-value

Note: Reject H_0 when *P*-value is less than α.

Company hypothesis test of Example 9.8, the *P*-value of .040 is less than the α value of .05 and $t = 2$ is greater than $t_{.05, 8} = 1.860$ [see Figure 9.10(a)]. Also, recall that $t = 2$ falls in the rejection region for the International hypothesis test. Therefore, since we reject H_0 when the value of the sample statistic falls in the rejection region, to use *P*-values to decide whether or not to reject a null hypothesis, we can use the following criteria.

Criteria for Rejecting or Not Rejecting a Null Hypothesis by Using *P*-Values

If *P*-value $< \alpha$, then reject H_0.

If *P*-value $\geq \alpha$, then do not reject H_0.

EXAMPLE 9.10 For the International Rubber Band Company hypothesis test of Example 9.8 we found that the *P*-value of .040 is less than the α value of .05, so we reject the null hypothesis. We reject the null hypothesis by using the *P*-value in this example just as we rejected the null hypothesis by using the *t* statistic in Example 9.8.

COMPUTER EXHIBIT 9.2 Hypothesis Test for International Rubber Band Company   **MINITAB**

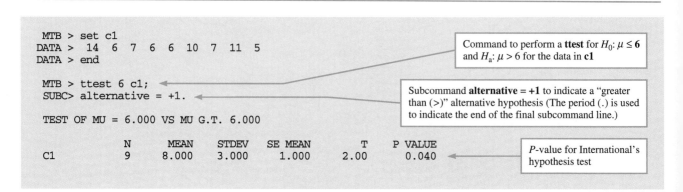

```
MTB > set c1
DATA >  14  6  7  6  6  10  7  11  5
DATA > end

MTB > ttest 6 c1;
SUBC> alternative = +1.

TEST OF MU = 6.000 VS MU G.T.  6.000

              N      MEAN    STDEV   SE MEAN      T    P VALUE
C1            9     8.000    3.000     1.000   2.00      0.040
```

Command to perform a **ttest** for $H_0: \mu \leq 6$ and $H_a: \mu > 6$ for the data in **c1**

Subcommand **alternative = +1** to indicate a "greater than (>)" alternative hypothesis (The period (.) is used to indicate the end of the final subcommand line.)

P-value for International's hypothesis test

Most data analysis software packages routinely provide P-values for statistics used in hypothesis tests. Such software was used to test International's hypothesis, and Computer Exhibit 9.2 shows that the P-value is equal to .040.

The P-value that we found in Example 9.9 is for a one-tailed test. A P-value for a two-tailed test generally includes the area in both tails of the distribution since both tails include values that are more *extreme* (larger in absolute value) than the sample statistic. A P-value computed in this manner is sometimes referred to as a two-tailed P-value. Most data analysis packages provide two-tailed P-values for two-tailed tests. The criteria given above for deciding whether or not to reject a null hypothesis hold for the two-tailed P-values of two-tailed tests as well as for the one-tailed P-values of one-tailed tests.

The P-value we found in Example 9.9 was for a t statistic. P-values can be found for Z statistics or for statistics for other distributions by following similar procedures to those illustrated for the t distribution.

The criteria summarized previously are used to reject H_0 or not reject H_0 according to the classical approach to hypothesis testing that was discussed in Section 9.4. Under that approach, a value for α is specified and a rejection rule is formulated before the sample results are observed. However, sometimes when we test hypotheses, particularly when we use data analysis software, we do *not* specify a value for α. In this case, we simply report or use P-values from computer output. When we use P-values from computer output, individuals can then select their own α values and make conclusions based on their own dispositions regarding α.

Since it is burdensome at times to compute P-values without the aid of a computer, we follow the classical approach of specifying α and a rejection rule when the P-value is not available from computer output. If a P-value is available on computer output, then we may use the criteria that we summarized previously for rejecting or not rejecting a null hypothesis by using P-values.

9.7
HYPOTHESES ON A BINOMIAL p

If we take a sample from a binomial population with the intention of using the information it contains to test the hypothesis that the *proportion* of successes in the population p has a specified value p_0, we can use an approximate test based on the central limit theorem.

Under the null hypothesis $H_0: p = p_0$, the ratio

$$Z = \frac{X - np_0}{\sqrt{np_0(1 - p_0)}} \tag{9.6}$$

has approximately a standard normal distribution if n is large, where X is the number of successes in the sample (see Section 7.5). The denominator in the ratio is the standard deviation of X; we know its value under the null hypothesis, and there is no need to estimate it from the sample.

Testing H_0 is like testing for a normal mean when there are infinitely many degrees of freedom—that is, with the normal distribution. The rejection region will be one-tailed or two-tailed depending on whether the alternative hypothesis is one-sided or two-sided.

Dividing numerator and denominator in Equation (9.6) by sample size n gives

$$Z = \frac{\hat{p} - p_0}{\sqrt{p_0(1 - p_0)/n}} \tag{9.7}$$

where $\hat{p}$ is X/n, the fraction of successes in the sample. Clearly, the test may be performed either with X or with $\hat{p}$, as convenience dictates. We must, however, be sure to use the appropriate standard deviation in the denominator.

The hypothesis test of a binomial p for the one-sided alternative $H_a: p > p_0$ can be summarized as follows.

Hypotheses on Binomial p and Rejection Rule for One-Sided Test Where $H_a: p > p_0$

$H_0: p \leq p_0$

$H_a: p > p_0$

Rejection rule: If $Z > Z_\alpha$, then reject H_0;

where: $Z = \dfrac{X - np_0}{\sqrt{np_0(1 - p_0)}} = \dfrac{\hat{p} - p_0}{\sqrt{p_0(1 - p_0)/n}}$

and $\hat{p}$ is X/n, the proportion or fraction of successes in the sample.

If the alternative hypothesis is $H_a: p < p_0$, the rejection rule is

Rejection rule: If $Z < -Z_\alpha$, then reject H_0.

If the alternative hypothesis is two-sided, $H_a: p \neq p_0$, the rejection rule is

Rejection rule: If $Z < -Z_{\alpha/2}$ or $Z > Z_{\alpha/2}$, then reject H_0.

 PRACTICE PROBLEM 9.5

In a sample of 400 bushings manufactured by Algonquin Corporation, there were 12 whose internal diameters were not within the specifications. Is this sufficient evidence

FIGURE 9.11

Test of the Proportion of
Defective Bushings at
Algonquin

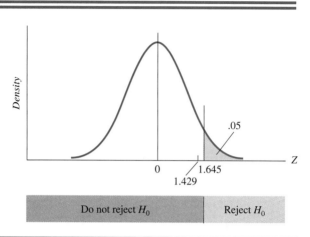

in a quality improvement study for rejecting that the manufacturing process is turning out 2% or fewer defective bushings? Let α be .05.

Solution The null hypothesis is H_0: $p \leq .02$. The alternative hypothesis is H_a: $p > .02$. The upper .05 percentage cutoff point $Z_{.05}$ for the standard normal distribution is 1.645. We should reject H_0 if we calculate larger Z values, since a one-sided test calls for a one-sided rejection region.

Step One H_0: $p \leq .02$.

Step Two H_a: $p > .02$.

Step Three $\alpha = .05$, $n = 400$.

Step Four   Rejection rule: If $Z > 1.645$, then reject H_0.

Step Five   The test statistic for our data is

$$Z = \frac{X - np_0}{\sqrt{np_0(1 - p_0)}} = \frac{X - (400)(.02)}{\sqrt{(400)(.02)(.98)}} = \frac{X - 8}{2.8}$$

$$= \frac{12 - 8}{2.8} = 1.429$$

Step Six   Since the value of Z, 1.429, is less than 1.645 (see Figure 9.11), the data favor the null hypothesis; 12 defective bushings out of 400 is not excessively large, and we do not reject the null hypothesis.   ▬

 PRACTICE PROBLEM 9.6

A package designer for American Cookies, Inc., wishes to determine whether there is any difference in the way customers react to two new package designs. She places equal numbers of packages on the shelf in a supermarket and observes how many of each package are purchased each day. She rotates the position of the two package types from one day to the next to eliminate the effect of shelf position, and she makes sure that there are always equal numbers of the two package types on the shelf. After 500 purchases have been made, 200 of the first package design and 300 of the second design have been purchased. Use $\alpha = .05$.

FIGURE 9.12

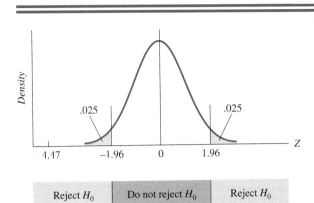

Test of Package Design Preference Proportion at American Cookies

Solution Let p equal the proportion of purchasers of the first design in the population (we could as easily have chosen p to be those preferring the second design). We have observed a sample value $\hat{p} = 200/500$, or .40. The null hypothesis here is that there is no difference in the two package designs, and $H_0: p = .5$. If we test this hypothesis with α of .05, then the percentage cutoff points are $\pm Z_{.025}$, or ± 1.96, on a two-tailed test. This test is a two-tailed test since very large values of $\hat{p}$ indicate preference for the first design, and very small values of $\hat{p}$ indicate preference for the second.

Step One $H_0: p = .5$.

Step Two $H_a: p \neq .5$.

Step Three $\alpha = .05$, $n = 500$.

Step Four Rejection rule: If $Z < -1.96$ or $Z > 1.96$, then reject H_0.

Step Five Since $\hat{p} = .4$, we have

$$Z = \frac{\hat{p} - p_0}{\sqrt{p_0(1 - p_0)/n}} = \frac{.4 - .5}{\sqrt{(.5)(.5)/500}} = \frac{-.1}{.02236} = -4.47$$

Step Six The value for Z, -4.47, is less than -1.96, as shown in Figure 9.12, so we reject the null hypothesis and conclude that the first package design is inferior to the second. Our chance of error in this conclusion is 5% or less.

Problems: Section 9.7

20. In a sample of 400 seeds purchased at American Agronomics, 326 germinated. At the 2.5% level, would we reject the null hypothesis $H_0: p \geq .90$?

21. In a sample of 70 die castings taken at Eastern Enterprises, we found 8 with defects. Does the evidence support the supposition that more than 8% of all such castings would show defects? Use a 5% level of significance to test $H_0: p \leq .08$.

22. The rate of defectives produced by a small gasoline engine production line at a Brick & Stratford production facility is usually 15%. The quality control engineer took a sample of $n = 80$ items from this line and found 18 defectives. Test $H_0: p \leq .15$ against $H_a: p > .15$, using $\alpha = .10$.

23. The Department of Transportation in California did a study five years ago that showed that the proportion of cars tested failing to meet state pollution control standards was .37. In a sample of $n = 100$ cars this year, the proportion not meeting the standards was .28. Test the hypothesis $H_0: p \geq .37$ this year against the alternative $H_a: p < .37$, using a significance level of .025.

24. In a sample of 144 voters 84 were in favor of a bond issue. Test the hypothesis that opinion is equally divided on this issue. Let $\alpha = .10$ and $H_a: p \neq .50$.

9.8

TESTING THE HYPOTHESIS $\mu_1 = \mu_2$

Consider the null hypothesis that the *difference between the means* of two normal populations is equal to a given value δ_0:

$$H_0: \mu_1 - \mu_2 = \delta_0$$

To test this null hypothesis, we rely on the results from two samples.

When we have two samples, an observation X_{ij} has two subscripts; the first denotes the particular element within the sample, and the second designates the sample. Thus X_{42} is the fourth value in the second sample. Let $X_{11}, X_{21}, \ldots, X_{n_1 1}$ represent observed values in a sample of size n_1 taken from a normal population with mean μ_1 and standard variance σ_1, and let $X_{12}, X_{22}, \ldots, X_{n_2 2}$ be the observations in a sample of size n_2 taken from a second normal population with mean μ_2 and standard deviation σ_2. Such a two-sample situation is characterized by two basic assumptions:

1. The samples are independent random samples.
2. The populations are normally distributed.

The methods to be presented in this section are valid only if these assumptions are satisfied.

The difference between the two sample means $\overline{X}_1 - \overline{X}_2$ will be used to test the null hypothesis $H_0: \mu_1 - \mu_2 = \delta_0$.

To test the difference between two means, we differentiate between two cases: (1) the case when the populations have a common standard deviation, so $\sigma_1 = \sigma_2 = \sigma$, and (2) the case when the populations have different standard deviations, so $\sigma_1 \neq \sigma_2$.

9.8.1 Testing the Hypothesis $H_0: \mu_1 = \mu_2$ When $\sigma_1 = \sigma_2 = \sigma$

For the first case we wish to test $H_0: \mu_1 - \mu_2 = \delta_0$ when we assume that the populations have a common standard deviation, so $\sigma_1 = \sigma_2 = \sigma$. Under the assumption of independent random samples the standard deviation of the sampling distribution of the difference between the sample means $\sigma_{\overline{X}_1 - \overline{X}_2}$ is as follows:

$$\sigma_{\overline{X}_1 - \overline{X}_2} = \sqrt{\frac{\sigma_1^2}{n_1} + \frac{\sigma_2^2}{n_2}} \tag{9.8}$$

Since the variance is assumed to be the same for both populations, we can obtain an unbiased estimate for the common variance $\sigma^2 = \sigma_1^2 = \sigma_2^2$ from each sample. But we obtain the best unbiased estimate by pooling the information contained in both samples to get what we call the **pooled estimate of variance,** s_{pooled}^2. This pooled estimate of variance is an estimate of the common population variance σ^2 and can be found in two equivalent ways. The *first* method involves pooling (or adding) the sum of squares *SS* from each of the samples and then dividing by the pooled degrees of freedom (*Pooled df*, which is equal to $n_1 + n_2 - 2$):

$$s_{pooled}^2 = \frac{Pooled\ SS}{Pooled\ df} = \frac{\sum_{i=1}^{n_1}(X_{i1} - \overline{X}_1)^2 + \sum_{i=1}^{n_2}(X_{i2} - \overline{X}_2)^2}{n_1 + n_2 - 2} \tag{9.9}$$

The *second* method for finding the pooled estimate of variance involves a weighted mean of the individual sample estimates of variance, where the weights are the individual degrees of freedom as follows:

$$s_{pooled}^2 = \frac{(n_1 - 1)s_1^2 + (n_2 - 1)s_2^2}{n_1 + n_2 - 2} \tag{9.10}$$

where: $\quad s_1^2 = \dfrac{\sum(X_{i1} - \overline{X}_1)^2}{n_1 - 1}$

$\quad\quad\quad s_2^2 = \dfrac{\sum(X_{i2} - \overline{X}_2)^2}{n_2 - 1}$

The two methods for finding s_{pooled}^2 are equivalent, so we can use whichever method is more convenient.

Now we can use s_{pooled}^2 to compute the sample estimate $s_{\overline{X}_1 - \overline{X}_2}$ of the population standard deviation $\sigma_{\overline{X}_1 - \overline{X}_2}$ as follows:

$$s_{\overline{X}_1 - \overline{X}_2} = \sqrt{s_{pooled}^2\left(\frac{1}{n_1} + \frac{1}{n_2}\right)} \tag{9.11}$$

To test the null hypothesis H_0: $\mu_1 - \mu_2 = \delta_0$, we rely on the value computed from the following equation:

$$t = \frac{\overline{X}_1 - \overline{X}_2 - \delta_0}{s_{\overline{X}_1\ \overline{X}_2}} \tag{9.12}$$

When the null hypothesis holds, this statistic has a t distribution with $n_1 + n_2 - 2$ degrees of freedom. The test is a simple t test, except that we must assume the populations have the same variance.

A hypothesis of particular interest in the two-sample case is the one that states that there is no difference between the means: H_0: $\mu_1 = \mu_2$. Under this hypothesis δ_0 is 0, and the test statistic as given in Equation (9.12) reduces to

$$t = \frac{\overline{X}_1 - \overline{X}_2}{s_{\overline{X}_1 - \overline{X}_2}} \tag{9.13}$$

Here, as with other t tests, the rejection region depends on the alternative hypothesis. If it states simply that the means are not equal (H_a: $\mu_1 \neq \mu_2$), we use the two-tailed critical region $|t| \geq t_{\alpha/2, n_1+n_2-2}$. If the alternative is one-sided (that is, H_a: $\mu_1 > \mu_2$ or H_a: $\mu_1 < \mu_2$), we use one-tailed critical regions (that is, $t > t_{\alpha, n_1+n_2-2}$ or $t < -t_{\alpha, n_1+n_2-2}$).

For a two-sided test, the testing procedure for H_0: $\mu_1 = \mu_2$ is summarized as follows.

Hypotheses on Difference Between Two Means and Rejection Rule for Two-Sided Tests When $\sigma_1 = \sigma_2 = \sigma$

H_0: $\mu_1 = \mu_2$

H_a: $\mu_1 \neq \mu_2$

Rejection rule: If $t < -t_{\alpha/2, n_1+n_2-2}$ or $t > t_{\alpha/2, n_1+n_2-2}$, then reject H_0;

where: $t = \dfrac{\overline{X}_1 - \overline{X}_2}{s_{\overline{X}_1 - \overline{X}_2}}$

For one-tailed test there is often some doubt as to which tail should be used. This difficulty is resolved if we always write our t, Equation (9.13), and the alternative hypothesis in such a way that the subscripts are in the same order. If the alternative is H_a: $\mu_1 > \mu_2$, and if we write the numerator of our t statistic as $\overline{X}_1 - \overline{X}_2$, then the inequality in H_a is an arrowhead pointing to the right, and we use the upper tail.

The determination of the proper tail to use when H_a: $\mu_1 > \mu_2$ is summarized as follows.

Hypotheses on Difference Between Two Means and Rejection Rule for One-Sided Test Where H_a: $\mu_1 > \mu_2$ and $\sigma_1 = \sigma_2 = \sigma$

H_0: $\mu_1 \leq \mu_2$

H_a: $\mu_1 > \mu_2$

Rejection rule: If $t > t_{\alpha, n_1+n_2-2}$, then reject H_0;

where: $t = \dfrac{\overline{X}_1 - \overline{X}_2}{s_{\overline{X}_1 - \overline{X}_2}}$

We ought to reject H_0 in favor of H_a if $\overline{X}_1$ (which estimates μ_1) is greater than $\overline{X}_2$ (which estimates μ_2) by an excessive amount. Since $\overline{X}_1$ is greater than $\overline{X}_2$ by an excessive amount exactly when $\overline{X}_1 - \overline{X}_2$ is greater than 0 by an excessive amount, we reject H_0 if the value of the statistic in Equation (9.13) is greater than t_{α, n_1+n_2-2}.

If the alternative hypothesis is written as H_a: $\mu_1 < \mu_2$, we use $\overline{X}_1 - \overline{X}_2$. The inequality sign then points to the left, showing that we use the lower tail for the critical region, as summarized next.

> ### Hypotheses on Difference Between Two Means and Rejection Rule for One-Sided Test Where H_a: $\mu_1 < \mu_2$ and $\sigma_1 = \sigma_2 = \sigma$
>
> H_0: $\mu_1 \geq \mu_2$
>
> H_a: $\mu_1 < \mu_2$
>
> Rejection rule: If $t < -t_{\alpha, n_1+n_2-2}$, then reject H_0;
>
> where: $t = \dfrac{\overline{X}_1 - \overline{X}_2}{s_{\overline{X}_1-\overline{X}_2}}$

We should prefer H_a to H_0 if $\overline{X}_1$ is greatly less than $\overline{X}_2$, or if $\overline{X}_1 - \overline{X}_2$ is greatly less than 0; so we reject H_0 if the value of Equation (9.13) is less than $-t_{\alpha, n_1+n_2-2}$.

 PRACTICE PROBLEM 9.7

A large mining company has experienced a decline in business and has terminated many of its employees. Terminated employees are referred to as population 2. A terminated employee feels that age discrimination was practiced by the company during the termination process. A sample of 15 employees who were not terminated and a sample of 9 employees who were terminated resulted in the accompanying summary table for a two-sample situation. Is there any reason to believe that the mean age of the first population is less than that of the second? Let $\alpha = .05$ in testing the hypothesis that the population means are equal against the alternative that the first population mean is smaller.

Sample	n	df	$\overline{X}_j$	$\Sigma(X_{ij} - \overline{X}_j)^2$
1	15	14	51	1224
2	9	8	59	756
Total	24	22		1980

Calculating from the information in the summary table gives

$$s^2_{\text{pooled}} = \frac{1980}{22} = 90 \quad \text{and} \quad s_{\overline{X}_1-X_2} = \sqrt{90\left(\frac{1}{15} + \frac{1}{9}\right)} = \sqrt{16} = 4$$

Solution

Step One H_0: $\mu_1 \geq \mu_2$.

Step Two H_a: $\mu_1 < \mu_2$.

Step Three $\alpha = .05$, $n_1 = 15$, $n_2 = 9$.

Step Four   Since $-t_{.05, 22} = -1.717$, the rejection rule is as follows:

   Rejection rule: If $t < -1.717$, then reject H_0.

Step Five   We calculate

$$t = \frac{\overline{X}_1 - \overline{X}_2}{s_{\overline{X}_1-\overline{X}_2}} = \frac{51 - 59}{4} = -2$$

FIGURE 9.13

Test of Mean Ages of Terminated and Nonterminated Employees

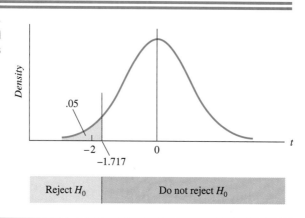

Step Six   The calculated t, -2, is less than the tabular value, -1.717 (see Figure 9.13); therefore, we reject the hypothesis and conclude that $\mu_1 < \mu_2$, or that discrimination was practiced against older employees.

   PRACTICE PROBLEM 9.8

In a comparison of the durabilities of two paints suggested by a Sherman–Wilson Company representative for highway use, 12 four-inch-wide lines of each paint were laid down across a heavily traveled road. The order was decided at random. After a period of time reflectometer readings were obtained for each line. The higher the readings, the greater is the reflectivity and the better is the durability of the paint. The data are given in Table 9.2. Test the null hypothesis that the two paints are equal in reflectivity. Let the significance level be .05.

Solution   For these sets of data, we obtain the results that are presented in Table 9.3,

TABLE 9.2

Data for Sherman–Wilson

	Paint 1			*Paint 2*	
12.5	10.3	8.7	9.4	6.9	7.0
11.7	9.6	11.5	11.6	7.3	8.2
9.9	9.4	10.6	9.7	8.4	12.7
9.6	11.3	9.7	10.4	7.2	9.2

TABLE 9.3

Calculation Results

Paint 1	*Paint 2*
$\sum X_{i1} = 124.8$	$\sum X_{i2} = 108.0$
$\overline{X}_1 = 10.4$	$\overline{X}_2 = 9.0$
$\sum X_{i1}^2 = 1312.00$	$\sum X_{i2}^2 = 1010.64$
$(1/n_1)(\sum X_{i1})^2 = 1297.92$	$(1/n_2)(\sum X_{i2})^2 = 972.00$
$\sum(X_{i1} - \overline{X}_1)^2 = 1312.00 - 1297.92$	$\sum(X_{i2} - \overline{X}_2)^2 = 1010.64 - 972.00$
$= 14.08$	$= 38.64$

Paint		n	df	$\overline{X}_j$	Sum of Squares	**TABLE 9.4**
1		12	11	10.4	14.08	**Summary Results**
2		12	11	9.0	38.64	
	Total	24	22		52.72	

and a summary table is presented in Table 9.4.

Hence,

$$\overline{X}_1 - \overline{X}_2 = 10.4 - 9.0 = 1.4$$

The pooled estimate of variance is

$$s^2_{\text{pooled}} = \frac{52.72}{22} = 2.40$$

The estimated standard deviation of the difference between the means is

$$s_{\overline{X}_1 - \overline{X}_2} = \sqrt{s^2_{\text{pooled}}\left(\frac{1}{n_1} + \frac{1}{n_2}\right)} = \sqrt{2.40\left(\frac{1}{12} + \frac{1}{12}\right)} = \sqrt{.40} = .63$$

To test the hypothesis that the two paints are equal in reflectivity, we use the following steps.

Step One $H_0: \mu_1 = \mu_2$.

Step Two $H_a: \mu_1 \neq \mu_2$.

Step Three $\alpha = .05$, $n_1 = 12$, $n_2 = 12$.

Step Four Using $t_{.025, 22} = 2.074$ from Table VI gives the following rejection rule:

Rejection rule: If $t < -2.074$ or $t > 2.074$, then reject H_0.

Step Five The t statistic is

$$t = \frac{\overline{X}_1 - \overline{X}_2}{s_{\overline{X}_1 - \overline{X}_2}} = \frac{1.4}{.63} = 2.22$$

Step Six Our calculated t, 2.22, as shown in Figure 9.14, is greater than 2.074; therefore, we reject the null hypothesis and conclude that the means are not equal.

A computer was also used to analyze the Sherman–Wilson paint reflectivity problem, and the results are presented in Computer Exhibits 9.3A, 9.3B, and 9.3C.

When we first introduced statistical analysis involving two populations, we stated two assumptions that must be met for the analyses to be valid. Let us now reconsider

FIGURE 9.14

Test of Two Sherman–Wilson
Paints

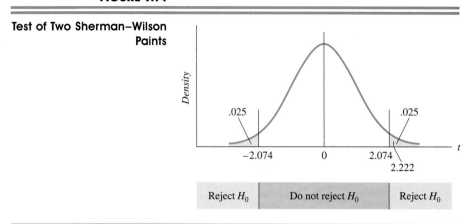

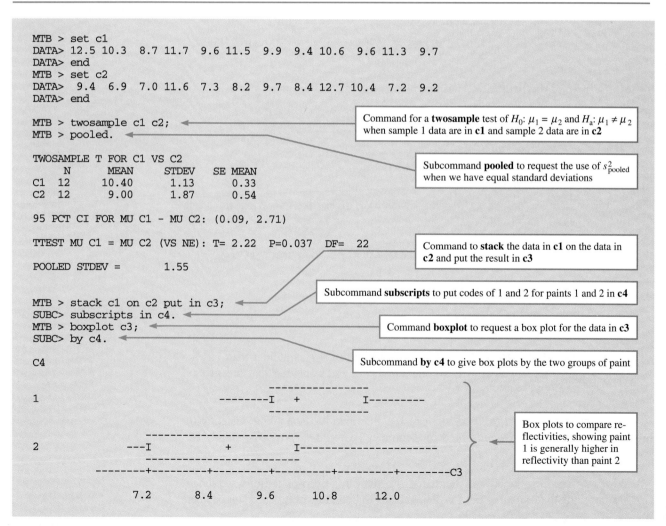

| MINITAB | **COMPUTER EXHIBIT 9.3A** Hypothesis Test of a Difference Between Means for Sherman–Wilson Paints |

```
MTB > set c1
DATA> 12.5 10.3  8.7 11.7  9.6 11.5  9.9  9.4 10.6  9.6 11.3  9.7
DATA> end
MTB > set c2
DATA>  9.4  6.9  7.0 11.6  7.3  8.2  9.7  8.4 12.7 10.4  7.2  9.2
DATA> end

MTB > twosample c1 c2;
MTB > pooled.

TWOSAMPLE T FOR C1 VS C2
      N      MEAN     STDEV    SE MEAN
C1   12     10.40     1.13     0.33
C2   12      9.00     1.87     0.54

95 PCT CI FOR MU C1 - MU C2: (0.09, 2.71)

TTEST MU C1 = MU C2 (VS NE): T= 2.22  P=0.037  DF=  22

POOLED STDEV =        1.55


MTB > stack c1 on c2 put in c3;
SUBC> subscripts in c4.
MTB > boxplot c3;
SUBC> by c4.

C4


1                              -----------------
                     --------I    +       I---------
                              -----------------


2          -------------------------
           ---I         +          I----------------------
           -------------------------
   --------+---------+---------+---------+---------+-------C3
         7.2       8.4       9.6      10.8      12.0
```

Command for a **twosample** test of H_0: $\mu_1 = \mu_2$ and H_a: $\mu_1 \neq \mu_2$ when sample 1 data are in **c1** and sample 2 data are in **c2**

Subcommand **pooled** to request the use of s^2_{pooled} when we have equal standard deviations

Command to **stack** the data in **c1** on the data in **c2** and put the result in **c3**

Subcommand **subscripts** to put codes of 1 and 2 for paints 1 and 2 in **c4**

Command **boxplot** to request a box plot for the data in **c3**

Subcommand **by c4** to give box plots by the two groups of paint

Box plots to compare reflectivities, showing paint 1 is generally higher in reflectivity than paint 2

Note: Use subcommand **alternative = –1** for H_a: $\mu_1 < \mu_2$. Use subcommand **alternative = +1** for H_a: $\mu_1 > \mu_2$.

COMPUTER EXHIBIT 9.3B Hypothesis Test of a Difference Between Means for Sherman–Wilson Paints

SAS

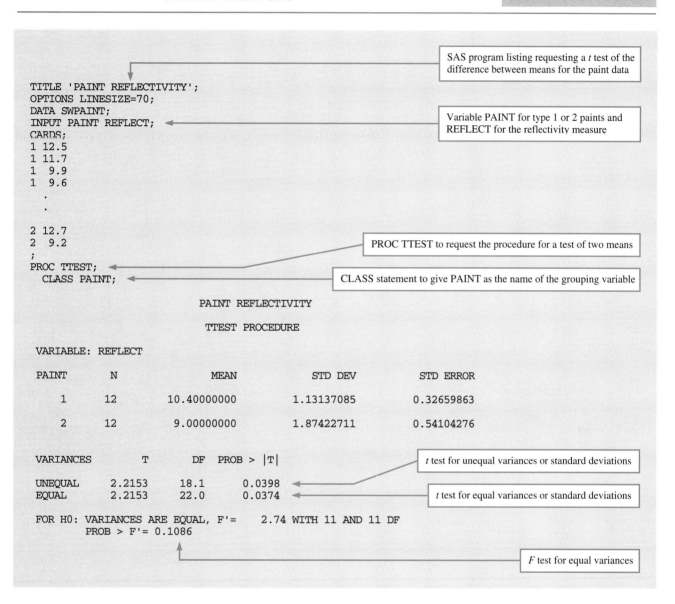

```
                                                    ┌─────────────────────────────────────┐
                                                    │ SAS program listing requesting a t test of the │
                                                    │ difference between means for the paint data    │
                                                    └─────────────────────────────────────┘
TITLE 'PAINT REFLECTIVITY';
OPTIONS LINESIZE=70;
DATA SWPAINT;
INPUT PAINT REFLECT;                                ┌─────────────────────────────────────┐
CARDS;                                              │ Variable PAINT for type 1 or 2 paints and │
 1 12.5                                             │ REFLECT for the reflectivity measure      │
 1 11.7                                             └─────────────────────────────────────┘
 1  9.9
 1  9.6
   .
   .
 2 12.7                                             ┌─────────────────────────────────────┐
 2  9.2                                             │ PROC TTEST to request the procedure for a test of two means │
;                                                   └─────────────────────────────────────┘
PROC TTEST;
   CLASS PAINT;                                     ┌─────────────────────────────────────┐
                                                    │ CLASS statement to give PAINT as the name of the grouping variable │
                                                    └─────────────────────────────────────┘
                       PAINT REFLECTIVITY

                       TTEST PROCEDURE

  VARIABLE: REFLECT

  PAINT       N          MEAN          STD DEV         STD ERROR

    1        12      10.40000000     1.13137085       0.32659863

    2        12       9.00000000     1.87422711       0.54104276


  VARIANCES        T      DF   PROB > |T|            ┌─────────────────────────────────┐
                                                     │ t test for unequal variances or standard deviations │
  UNEQUAL       2.2153   18.1     0.0398             └─────────────────────────────────┘
  EQUAL         2.2153   22.0     0.0374             ┌─────────────────────────────────┐
                                                     │ t test for equal variances or standard deviations │
  FOR H0: VARIANCES ARE EQUAL, F'=  2.74 WITH 11 AND 11 DF
           PROB > F'= 0.1086

                                                     ┌─────────────────────────────────┐
                                                     │ F test for equal variances │
                                                     └─────────────────────────────────┘
```

those assumptions and see which, if any, can be relaxed. The first assumption, *independent random samples,* must be satisfied if we are to make probability statements in connection with our inferences.

As is frequently the case, the second assumption, *normality,* is not strictly required. Because of the central limit effect, if the samples are sufficiently large, even large departures from normality will not affect the probabilities to any great extent.

A third assumption adopted in this subsection, *common* or *homogeneous variance,* cannot be relaxed. If the population variances are not the same, a significant result for the t test may be due to the different variances and not to different means. If this assumption is not satisfied, there is no exact test of the hypothesis

MYSTAT

COMPUTER EXHIBIT 9.3C Hypothesis Test of a Difference Between Means for Sherman–Wilson Paints

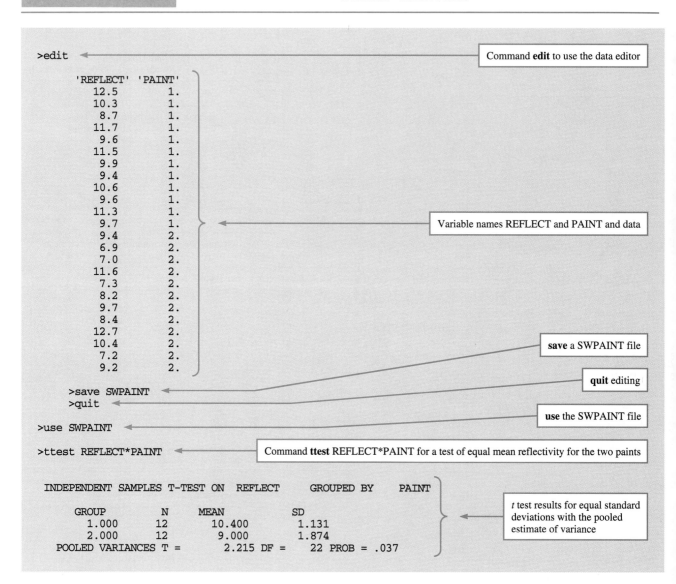

```
>edit                                                    Command edit to use the data editor

       'REFLECT'  'PAINT'
          12.5      1.
          10.3      1.
           8.7      1.
          11.7      1.
           9.6      1.
          11.5      1.
           9.9      1.
           9.4      1.
          10.6      1.
           9.6      1.
          11.3      1.
           9.7      1.                                   Variable names REFLECT and PAINT and data
           9.4      2.
           6.9      2.
           7.0      2.
          11.6      2.
           7.3      2.
           8.2      2.
           9.7      2.
           8.4      2.
          12.7      2.
          10.4      2.                                   save a SWPAINT file
           7.2      2.
           9.2      2.
                                                         quit editing
      >save SWPAINT
      >quit
                                                         use the SWPAINT file
>use SWPAINT

>ttest REFLECT*PAINT       Command ttest REFLECT*PAINT for a test of equal mean reflectivity for the two paints


 INDEPENDENT SAMPLES T-TEST ON  REFLECT     GROUPED BY     PAINT
                                                                    t test results for equal standard
     GROUP       N       MEAN            SD                         deviations with the pooled
     1.000      12      10.400         1.131                        estimate of variance
     2.000      12       9.000         1.874
  POOLED VARIANCES T =    2.215 DF =    22 PROB = .037
```

H_0: $\mu_1 - \mu_2 = \delta_0$. However, an approximate test is described in the following subsection.

9.8.2 Testing the Hypothesis H_0: $\mu_1 = \mu_2$ When $\sigma_1 \neq \sigma_2$

For the second case we wish to test H_0: $\mu_1 - \mu_2 = \delta_0$ when we assume that the populations have different standard deviations, so $\sigma_1 \neq \sigma_2$. Since $\overline{X}_j$ has mean μ_j and standard deviation σ_j/n_j, the difference $\overline{X}_1 - \overline{X}_2$ has mean $\mu_1 - \mu_2$ and standard deviation

$$\sigma_{\overline{X}_1 - \overline{X}_2} = \sqrt{\frac{\sigma_1^2}{n_1} + \frac{\sigma_2^2}{n_2}} \tag{9.14}$$

The sample estimate $s_{\overline{X}_1 - \overline{X}_2}$ of the population standard deviation $\sigma_{\overline{X}_1 - \overline{X}_2}$ is

$$s_{\overline{X}_1 - \overline{X}_2} = \sqrt{\frac{s_1^2}{n_1} + \frac{s_2^2}{n_2}} \tag{9.15}$$

where s_1^2 and s_2^2 are the sample variances. Since we are *not* assuming $\sigma_1^2 = \sigma_2^2 = \sigma^2$, we use the estimates s_1^2 and s_2^2 of σ_1^2 and σ_2^2 and *not* the pooled estimate of variance that we used in the case where we assumed that $\sigma_1^2 = \sigma_2^2$.

To test the null hypothesis H_0: $\mu_1 - \mu_2 = \delta_0$, we use the following statistic, which is approximately *t*-distributed:

$$t = \frac{(\overline{X}_1 - \overline{X}_2) - (\mu_1 - \mu_2)}{s_{\overline{X}_1 - \overline{X}_2}} \tag{9.16}$$

where $s_{\overline{X}_1 - \overline{X}_2}$ is found by using Equation (9.15). The degrees of freedom ν for this *t* statistic are computed by the formula

$$\nu = \frac{(s_1^2/n_1 + s_2^2/n_2)^2}{\dfrac{(s_1^2/n_1)^2}{n_1 - 1} + \dfrac{(s_2^2/n_2)^2}{n_2 - 1}} \tag{9.17}$$

To use the *t* table with degrees of freedom as integers, round ν *down* to the nearest integer.

The *t* statistic of Equation (9.16) can be used to **test the null hypothesis H_0: $\mu_1 - \mu_2 = 0$ or, equivalently, H_0: $\mu_1 = \mu_2$, without the assumption of equal population variances,** as follows. This test is sometimes referred to as a two-sample *t* test with unequal or separate variances.*

Hypotheses on Difference Between Two Means and Rejection Rule for Two-Sided Tests When $\sigma_1 \neq \sigma_2$

H_0: $\mu_1 = \mu_2$

H_a: $\mu_1 \neq \mu_2$

Rejection rule: If $t < -t_{\alpha/2, \nu}$ or $t > t_{\alpha/2, \nu}$, then reject H_0;

where: $t = \dfrac{(\overline{X}_1 - \overline{X}_2)}{s_{\overline{X}_1 \ \overline{X}_2}}$

$s_{\overline{X}_1 - \overline{X}_2} = \sqrt{\dfrac{s_1^2}{n_1} + \dfrac{s_2^2}{n_2}}$

and the value for the degrees of freedom ν is computed by the formula

$$\nu = \frac{(s_1^2/n_1 + s_2^2/n_2)^2}{\dfrac{(s_1^2/n_1)^2}{n_1 - 1} + \dfrac{(s_2^2/n_2)^2}{n_2 - 1}}$$

To use the *t* table with degrees of freedom as integers, round ν down to the nearest integer.

---

*The test is also referred to as the Aspin, Satterthwaite, or Welch test, or combinations thereof, in recognition of its developers.

For a one-sided test where the alternative hypothesis is $H_a: \mu_1 > \mu_2$, the rejection rule is as follows: If $t > t_{\alpha, \nu}$, then reject H_0. For a one-sided test where the alternative hypothesis is $H_a: \mu_1 < \mu_2$, the rejection rule is as follows: If $t < -t_{\alpha, \nu}$, then reject H_0.

PRACTICE PROBLEM 9.9

Excess returns can sometimes accrue to investors when important financial events occur. In addition, variances of returns can be influenced by financial events. An important event might be the announcement that a firm is "in play" as a takeover target. Independent random samples from normal distributions of the percentage yields for firms when (1) a takeover announcement has been made and when (2) no takeover announcement has been made under similar economic conditions gave the yields and descriptive measures shown in the accompanying listing and table. Can we conclude that the mean return with a takeover announcement is equal to the mean return with no takeover announcement? That is, test $H_0: \mu_1 - \mu_2 = 0$ against $H_a: \mu_1 - \mu_2 \neq 0$. Use a .05 level of significance.

| 1. Takeover | 16 | 10 | 4 | 10 | 14 | 12 | 6 | 8 | |
| 2. No Takeover | 10 | 6 | 4 | 5 | 4 | 7 | 5 | 8 | 5 |

Sample	n_j	$\overline{X}_j$	s_j
1	8	10	4
2	9	6	2

Solution Since variances of returns can be different due to financial events, we may not assume $\sigma_1 = \sigma_2$. Our hypothesis test is as follows.

Step One $H_0: \mu_1 - \mu_2 = 0.$

Step Two $H_a: \mu_1 - \mu_2 \neq 0.$

Step Three $\alpha = .05, n_1 = 8, n_2 = 9.$

Step Four The value for the degrees of freedom ν for the t statistic is computed by Equation (9.17):

$$\nu = \frac{[(4)^2/8 + (2)^2/9]^2}{\dfrac{[(4)^2/8]^2}{8-1} + \dfrac{[(2)^2/9]^2}{9-1}} = \frac{(2 + .444)^2}{\dfrac{(2)^2}{7} + \dfrac{(.444)^2}{8}} = \frac{5.97}{.596} = 10.02$$

So we round down for the degrees of freedom to get

$$\nu = 10$$

Since $t_{.025, 10} = 2.228$, our rejection rule is as follows:

Rejection rule: If $t < -2.228$ or $t > 2.228$, then reject H_0.

Step Five Since we are *not* assuming $\sigma_1^2 = \sigma_2^2 = \sigma^2$, we use the estimates s_1^2 and s_2^2 of σ_1^2 and σ_2^2 and *not* the pooled estimate of variance, and by

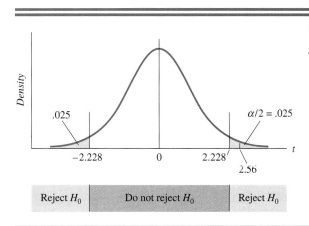

FIGURE 9.15

Hypothesis Test of Equal Mean Yields for Takeover Target Companies and Nontarget Companies

Equation (9.15) our sample estimate $s_{\overline{X}_1 - \overline{X}_2}$ of the population standard deviation $\sigma_{\overline{X}_1 - \overline{X}_2}$ is

$$s_{\overline{X}_1 - \overline{X}_2} = \sqrt{\frac{s_1^2}{n_1} + \frac{s_2^2}{n_2}} = \sqrt{\frac{(4)^2}{8} + \frac{(2)^2}{9}} = 1.56$$

And the t statistic is

$$t = \frac{\overline{X}_1 - \overline{X}_2}{s_{\overline{X}_1 - \overline{X}_2}} = \frac{10 - 6}{1.56} = 2.56$$

Step Six   Since $t = 2.56$ is greater than 2.228 (see Figure 9.15), we reject H_0 and conclude that the mean yield for companies in play as takeover targets is not equal to the mean yield of companies that are not takeover targets.

A computer was also used to test the difference between means for the takeover data, and the results are presented in Computer Exhibits 9.4A and 9.4B.

For the test of H_0: $\mu_1 = \mu_2$ when $\sigma_1 \neq \sigma_2$, the assumption of *independent random samples* must be satisfied if we are to make probability statements in connection with our conclusions. The assumption of *normality* is not a strong assumption because of the central limit effect. If the samples are sufficiently large, then departures from normality for the populations will not hinder the conclusions, for practical purposes. If the standard deviations are equal, $\sigma_1 = \sigma_2$, and we use the methods of this section assuming they are unequal, $\sigma_1 \neq \sigma_2$, then the true significance level for the test may be slightly less than α—or, in other words, the test is slightly conservative. Larger samples alleviate the conservativeness of the test in this situation. If the sample sizes are different—say n_1 is greater than n_2—and $\sigma_1 \neq \sigma_2$, then the test of H_0: $\mu_1 = \mu_2$ assuming $\sigma_1 \neq \sigma_2$ is the preferred test. It is preferred because incorrectly assuming that $\sigma_1 = \sigma_2$ in this case can result in a true significance level that differs substantially from that designed for the test. For this reason, if it is not known whether the two standard deviations are equal, then some analysts recommend always testing H_0: $\mu_1 = \mu_2$ by assuming $\sigma_1 \neq \sigma_2$.

| MINITAB | COMPUTER EXHIBIT 9.4A Hypothesis Test of Equal Mean Yields for Takeover Target Companies and Nontarget Companies |

```
MTB > set c1
DATA > 16 10   4 10 14 12   6   8
DATA > end
MTB > set c2
DATA > 10   6   4   5   4   7   5   8   5
MTB > end

MTB > twosample c1 c2

TWOSAMPLE T FOR C1 VS C2
     N       MEAN      STDEV    SE MEAN
C1   8      10.00       4.00        1.4
C2   9       6.00       2.00       0.67

95 PCT CI FOR MU C1 - MU C2: (0.5, 7.48)

TTEST MU C1 = MU C2 (VS NE): T= 2.56   P=0.028   DF=  10
```

Command for a **twosample** test of H_0: $\mu_1 = \mu_2$ and H_a: $\mu_1 \neq \mu_2$ when sample 1 data is in **c1** and sample 2 data is in **c2** and when we have unequal standard deviations

Note: Use subcommand **alternative = –1** for H_a: $\mu_1 < \mu_2$. Use subcommand **alternative = +1** for H_a: $\mu_1 > \mu_2$.

Problems: Section 9.8

 25. Independent random samples were taken from normal distributions with equal standard deviations of start-up owners' equity (in hundreds of thousands of dollars) for small business firms that within the first year of operation have resulted in (1) non-bankruptcy or going concerns and (2) bankruptcy. The results are given in the accompanying table. Test the null hypothesis that the mean start-up owners' equity for small business firms that are not bankrupt within the first year is less than or equal to that for bankrupt firms. Use .05 for the level of significance.

| *1. Nonbankrupt Firm* | 14 | 2 | 12 | 4 | 6 | 10 | 8 | 8 |
| *2. Bankrupt Firm* | 10 | 10 | 0 | 0 | 2 | 2 | 4 | 4 |

26. Independent random samples were taken from normal distributions with equal standard deviations of the weekly downtime during production of hay bailers at a Houston Corporation assembly plant under (1) a machine-paced assembly line and (2) a worker-paced assembly line. The results are given in the accompanying table. Can we conclude that H_0: $\mu_1 \geq \mu_2$ should be rejected in favor of H_a: $\mu_1 < \mu_2$ when α is .025?

Sample	n	Mean	Sum of Squares
1	10	17	106
2	6	21	124

The sample statistics summarized in the table were computed from the accompanying data; however, these data may be useful if you wish to use a computer.

1. Machine Paced	14.2833	14.6227	18.2881	14.6686	24.0676	15.8960
	13.8720	19.9058	14.3347	20.0612		
2. Worker Paced	24.8831	20.0909	14.0523	17.4255	21.7277	27.8204

COMPUTER EXHIBIT 9.4B Hypothesis Test of Equal Mean Yields for Takeover Target Companies and Nontarget Companies

MYSTAT

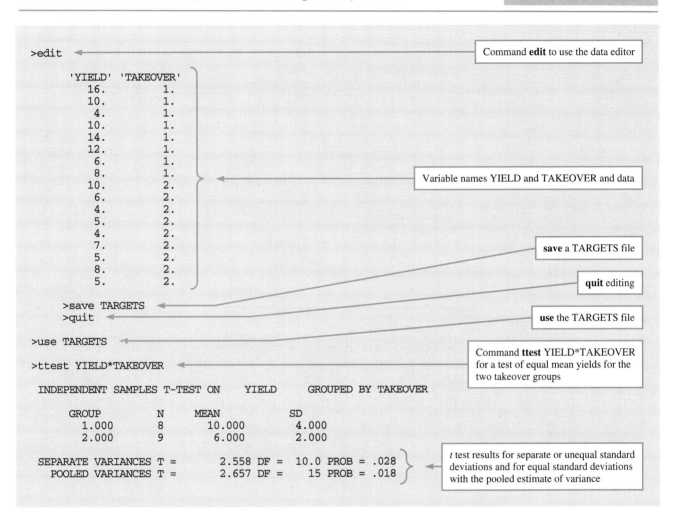

```
>edit  ◄───────────────────────────────────────────────   Command edit to use the data editor

     'YIELD'  'TAKEOVER'
       16.       1.
       10.       1.
        4.       1.
       10.       1.
       14.       1.
       12.       1.
        6.       1.
        8.       1.
       10.       2.  ◄──────────────────────   Variable names YIELD and TAKEOVER and data
        6.       2.
        4.       2.
        5.       2.
        4.       2.
        7.       2.                                              save a TARGETS file
        5.       2.
        8.       2.                                              quit editing
        5.       2.
     >save TARGETS  ◄────────────
     >quit  ◄──────────────────────────────────────        use the TARGETS file
  >use TARGETS  ◄──────────────

  >ttest YIELD*TAKEOVER  ◄────────────────────────          Command ttest YIELD*TAKEOVER
                                                            for a test of equal mean yields for the
   INDEPENDENT SAMPLES T-TEST ON    YIELD    GROUPED BY TAKEOVER    two takeover groups

        GROUP       N      MEAN        SD
        1.000       8     10.000      4.000
        2.000       9      6.000      2.000

   SEPARATE VARIANCES T =      2.558 DF =  10.0 PROB = .028
     POOLED VARIANCES T =      2.657 DF =    15 PROB = .018
```

t test results for separate or unequal standard deviations and for equal standard deviations with the pooled estimate of variance

27. The accompanying data are the amounts of fat (in ounces of fat per 100 ounces of meat) found in samples of two types of meat products from Meat Products Company. The fat contents are normally distributed and have equal standard deviations for the two meat types. Do the meats have different fat contents? That is, test $H_0: \mu_1 = \mu_2$ against $H_a: \mu_1 \neq \mu_2$, using $\alpha = .10$.

| Meat 1 | 30 | 26 | 30 | 19 | 25 | 37 | 27 | 38 | 26 | 31 |
| Meat 2 | 40 | 34 | 28 | 29 | 26 | 36 | 28 | 37 | 35 | 42 |

28. Percentage return on equity, defined loosely as earnings per share divided by common shareholders' equity per share, is a measure of profitability. The accompanying listing gives (1) medical supply companies and (2) health care services companies from the health industry and their percentage returns on equity for a one-year period. Assume normality and equal standard deviations and that the lists are independent random samples. Should an investor reject the idea that medical supply companies and

health care services had equal profitability for the period based on these data? Use .05 as the level of significance.

Medical Supplies	Return on Equity (%)	Health Care Services	Return on Equity (%)
SmithKline Beckman	22.2	Charter Medical	20.1
Abbott Laboratories	28.3	Humana	6.0
Johnson & Johnson	10.9	Maxicare Health	15.1
Hillenbrand Inds	17.3	Manor Care	24.9
Baxter Travenol Labs	15.7	National Medical	9.5
Becton Dickinson	15.6	Hospital Corp	11.0
Corning Glass Works	13.6	Beverly Enterprises	11.8

SOURCE: *Forbes,* January 12, 1987, p. 142.

29. Recent studies in financial economics suggest that excess returns can accrue to shareholders when events that are financially important to companies occur. Independent random samples were taken from normal distributions with equal standard deviations of excess percentage returns during similar economic conditions for stocks (1) just after and (2) just before stock split announcements. The results are given in the accompanying table. Can we conclude that the difference between the means $\mu_1 - \mu_2$ is greater than 5? That is, test $H_0: \mu_1 - \mu_2 \leq 5$ with $\alpha = .05$.

Sample	n	Mean	Standard Deviation
1	127	9	9
2	73	2	7

The sample statistics summarized in the above table were computed from the data in Table 9.5; however, these data may be useful if you wish to use a computer.

30. The "triple witching hour" for the stock market occurs when futures and options for stocks expire around the same time. A future is a contract whereby a security is bought or sold with a specified price for delivery in the future. A stock option is a contract that gives the owner the right to buy or sell securities at a given price during a given period. Extreme changes in stock prices may accompany the triple witching hour. Independent random samples from normal distributions with *unequal* standard deviations of stock returns for a stock exchange (1) at the triple witching hour and (2) not at the triple witching hour gave the accompanying results. Test whether the mean returns are equal or are not equal at the triple witching hour and at other times. Use .05 for the level of significance. (You may want to compute sample means and standard deviations for this problem by using hand calculations.)

1. Triple Witching	10	4	2	0	10	0	2	4
2. Not Triple Witching	11	8	7	8	5	10	6	9

31. Independent random samples from normal distributions with *unequal* standard deviations of auditor fees when the audit resulted in (1) a lawsuit or (2) no lawsuit gave the following results. Test to see whether the mean fee when a lawsuit occurs is equal or unequal to the mean fee when no lawsuit occurs. Use .05 for the level of significance.

1. Lawsuit	17	11	5	11	15	13	7	9	
2. No Lawsuit	9	5	3	4	3	6	4	7	4

Sample	n_j	$\overline{X}_j$	s_j
1	8	11	4
2	9	5	2

1. After Stock Split Announcement							TABLE 9.5
−2.7729	3.4013	−4.9097	−3.3245	4.7984	11.5385	8.6349	Excess Stock Returns
14.6106	19.1836	17.4244	13.2896	15.2881	4.1479	7.0051	
9.9159	3.6231	−3.0841	27.9099	3.7380	11.1324	10.1939	
4.8462	−9.8542	−0.4928	11.2589	10.9606	5.3959	13.5501	
21.4540	17.7074	12.0173	9.8564	15.5856	2.6871	9.0254	
10.2090	−1.3478	13.6646	14.0460	0.9031	0.9963	−3.5950	
−1.2306	4.6856	5.5477	10.6969	12.1177	0.9794	−4.7787	
0.8284	−0.9290	17.8896	17.8679	−2.1871	18.8534	20.4224	
19.2577	6.6250	7.7375	26.9461	7.0852	10.9237	18.0942	
13.8071	2.8859	21.3429	−6.0747	−2.6437	21.4472	7.5019	
−2.0257	−2.3400	21.0896	11.7094	27.2429	19.1432	−5.6641	
11.6648	12.7529	25.8458	10.2331	6.2171	10.5074	−5.6964	
16.7806	20.1195	10.1684	13.4210	14.1390	15.5163	8.9811	
−14.2967	20.2591	18.2312	15.1361	−3.2610	11.4302	12.4925	
14.6985	−5.6841	−2.8302	12.3770	−9.3794	1.5595	14.2293	
6.9247	11.7632	−9.3623	4.0250	15.0896	4.5979	16.0885	
0.0554	22.7592	20.3617	23.6499	9.0528	18.2330	12.5025	
20.4995	15.8919	8.7434	2.1146	10.6053	2.6977	0.9503	
14.6705							

2. Before Stock Split Announcement						
2.6133	−5.7178	4.1453	18.9659	4.1570	5.7270	6.8963
2.4416	−0.0312	−2.9543	−4.1581	2.6738	1.2447	11.8593
5.7487	6.4257	−6.1773	−8.5555	−12.0808	−3.5536	1.1084
9.4063	5.3647	8.2805	2.9315	−4.3973	2.1159	−9.3659
5.9276	3.3862	−2.0730	−5.7798	7.2575	6.5643	5.6289
3.2487	6.8507	8.9953	9.4351	−3.2274	5.9324	−5.3959
5.5116	3.0195	14.3507	13.9319	−7.3500	4.9853	2.1925
1.2930	0.9581	−1.0863	−1.0387	8.5999	9.7485	6.7364
−3.3670	8.1792	−2.0570	−15.0672	9.3113	−14.3790	7.2623
−8.0087	−9.0918	−0.7146	6.3880	−8.8612	1.4831	1.1343
16.0750	−1.5121	5.5082				

32. The Equal Credit Opportunity Act prohibits banks from discriminating against borrowers on the basis of marital status, sex, age, and race or because they receive alimony or public assistance. Independent random samples from normal distributions with *unequal* standard deviations of lines of credit (in thousands of dollars) for (1) married and (2) not married individuals of the same economic status gave the accompanying results. Test whether the mean lines of credit are greater for married people than they are for those who are not married. Use a significance level of .05.

1. Married	12	15	12	9	13	14	10	11
2. Not married	18	1	3	10	9	7		

When we have two binomial populations, we often must test the hypothesis that the *proportions of successes are equal* in both. For instance, we may wish to test the hypothesis that the rate of defectives produced by two production lines is the same. To test the hypothesis that $p_1 = p_2$, we obtain independent random samples of n_1 items from the first production line and n_2 items from the second. For the first sample

9.9

TESTING THE HYPOTHESIS $p_1 = p_2$

$\hat{p}_1 = X_1/n_1$, which is the proportion of defects in that sample. For the second sample $\hat{p}_2 = X_2/n_2$. Then we use a test based on the standard normal distribution. The hypothesis states that the proportions are equal to each other, but it does not specify the common value. Therefore we must estimate the variance of the difference under the assumption that the proportions are the same.

If $p_1 = p_2 = p$, both $\hat{p}_1$ and $\hat{p}_2$ are unbiased estimates for p, but the best estimate will be obtained by pooling the two samples into one sample of size $n_1 + n_2$ with the pooled number of successes $X_1 + X_2$ or $n_1\hat{p}_1 + n_2\hat{p}_2$. The fraction of successes for the pooled samples,

$$\hat{p}_{\text{pooled}} = \frac{X_1 + X_2}{n_1 + n_2} \tag{9.18}$$

is used to find our best estimate for the common value of p.

Now the standard deviation of $\hat{p}_1 - \hat{p}_2$, $\sigma_{\hat{p}_1 - \hat{p}_2}$, is estimated by

$$s_{\hat{p}_1 - \hat{p}_2} = \sqrt{\frac{\hat{p}_1(1 - \hat{p}_1)}{n_1} + \frac{\hat{p}_2(1 - \hat{p}_2)}{n_2}} \tag{9.19}$$

If $p_1 = p_2$, then this expression can be reduced to

$$s_{\hat{p}_1 - \hat{p}_2} = \sqrt{\hat{p}_{\text{pooled}}(1 - \hat{p}_{\text{pooled}})\left(\frac{1}{n_1} + \frac{1}{n_2}\right)} \tag{9.20}$$

Under the null hypothesis $H_0: p_1 = p_2$ or $H_0: p_1 - p_2 = 0$, the quantity

$$Z = \frac{\hat{p}_1 - \hat{p}_2}{\sqrt{\hat{p}_{\text{pooled}}(1 - \hat{p}_{\text{pooled}})(1/n_1 + 1/n_2)}} \tag{9.21}$$

has approximately a standard normal distribution if n_1 and n_2 are large. This quantity, then, provides a basis for an approximate test procedure. The hypothesis test consists of computing a Z value from the sample results and comparing the computed Z with the standard normal $Z_{\alpha/2}$ and $-Z_{\alpha/2}$ if the test is two-tailed. The procedure of a two-sided test is summarized as follows.

Hypotheses on Difference Between Two Proportions and Rejection Rule for Two-Sided Test

$H_0: p_1 = p_2$

$H_0: p_1 \neq p_2$

Rejection rule: If $Z < -Z_{\alpha/2}$ or $Z > Z_{\alpha/2}$, then reject H_0;

where: $Z = \dfrac{\hat{p}_1 - \hat{p}_2}{\sqrt{\hat{p}_{\text{pooled}}(1 - \hat{p}_{\text{pooled}})(1/n_1 + 1/n_2)}}$

$\hat{p}_1 = \dfrac{X_1}{n_1} \qquad \hat{p}_2 = \dfrac{X_2}{n_2} \qquad \hat{p}_{\text{pooled}} = \dfrac{X_1 + X_2}{n_1 + n_2}$

If the test is one-sided, then we compare the computed Z value from the sample to Z_α for $H_a: p_1 > p_2$ or to $-Z_\alpha$ for $H_a: p_1 < p_2$.

PRACTICE PROBLEM 9.10

General Refractories Company makes bricks by two different processes. In samples of 200 bricks from the first process and 300 bricks from the second process, 20 of the first type broke during baking in the kiln, and 45 of the second type broke in the kiln. Test the hypothesis that the kiln breakage rates are the same against the alternative hypothesis that they are different. Let α be .02.

Solution For this problem we have the following test.

Step One $H_0: p_1 = p_2$.

Step Two $H_a: p_1 \neq p_2$.

Step Three $\alpha = .02$, $n_1 = 200$, $n_2 = 300$.

Step Four Since $Z_{.01} = 2.326$, we have the following rejection rule:

Rejection rule: If $Z < -2.326$ or $Z > 2.326$, then reject H_0.

Step Five $\hat{p}_1 = 20/200 = .10$ and $\hat{p}_2 = 45/300 = .15$. The pooled value of the breakage rate in the two populations, assuming these rates are equal, is

$$\hat{p}_{\text{pooled}} = \frac{20 + 45}{200 + 300} = \frac{65}{500} = .13$$

Then by Equation (9.20)

$$s_{\hat{p}_1 - \hat{p}_2} = \sqrt{(.13)(.87)\left(\frac{1}{200} + \frac{1}{300}\right)} = \sqrt{.000943} = .031$$

Our calculated Z is

$$Z = \frac{\hat{p}_1 - \hat{p}_2}{s_{\hat{p}_1 - \hat{p}_2}} = \frac{.10 - .15}{.031} = -1.6$$

Step Six Since the value -1.6 is not within the rejection region, as shown in Figure 9.16, we cannot reject the null hypothesis that the breakage rates of the two processes are the same.

FIGURE 9.16

Test of Kiln Breakage Rates at General Refractories

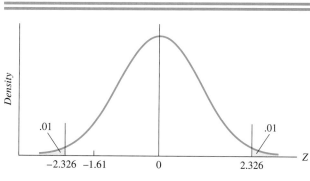

PRACTICE PROBLEM 9.11

To test the effectiveness of the approach and layout of two direct-mail brochures, a marketing manager for J. J. Hart, Inc., mailed out 150 copies of each brochure and recorded the number of responses generated by each. There were 30 responses generated by the first brochure and 10 generated by the second. Can the marketing manager conclude that the first brochure is more effective? Let $\alpha = .05$.

Solution Here our test is as follows.

Step One $H_0: p_1 \le p_2$.

Step Two $H_a: p_1 > p_2$, where p_1 is the proportion of responses that will be generated from a nationwide mailing of the first brochure and p_2 is the proportion of responses generated by the second brochure.

Step Three $\alpha = .05$, $n_1 = 150$, $n_2 = 150$.

Step Four   Since $Z_{.05} = 1.645$, the rejection rule is as follows:

Rejection rule: If $Z > 1.645$, then reject H_0.

Step Five   We have

$$\hat{p}_1 = \frac{30}{150} = .2 \qquad \hat{p}_2 = \frac{10}{150} = .0667$$

for which the data are

$$\hat{p}_{\text{pooled}} = \frac{X_1 + X_2}{n_1 + n_2} = \frac{30 + 10}{150 + 150} = \frac{40}{300}$$

Also, by Equation (9.20)

$$s_{\hat{p}_1 - \hat{p}_2} = \sqrt{\frac{40}{300}\left(\frac{260}{300}\right)\left(\frac{1}{150} + \frac{1}{150}\right)} = .03925$$

FIGURE 9.17

Test of the Effectiveness of
Two Direct-Mail Brochures
at J. J. Hart

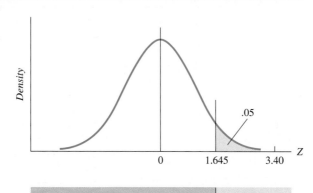

Our calculated Z is

$$Z = \frac{\hat{p}_1 - \hat{p}_2}{s_{\hat{p}_1 - \hat{p}_2}} = \frac{.2 - .0667}{.03925} = \frac{.1333}{.03925} = 3.40$$

Step Six The calculated value $Z = 3.40$ is greater than 1.645 (see Figure 9.17), so we reject the null hypothesis and conclude that the first brochure is more effective in generating responses. ▬▬▬

Problems: Section 9.9

33. In a sample of 200 men with MBA degrees 120 had received promotions within two years on the job. However, a sample of 90 women with MBA degrees showed only 45 had received promotions within two years on the job.

   a. What are the appropriate null and alternative hypotheses, assuming a one-tailed test is desired?
   b. Perform the test, using $\alpha = .05$.

34. Random samples of VCRs assembled in a Unitech assembly plant in South Korea included 50 assembled during the first shift and 50 assembled during the second shift. Of the VCRs assembled during the first shift 10 were defective; 20 were defective from the second shift. From these data, would a production foreman reject the hypothesis that the proportion of defectives assembled by the first shift is greater than or equal to that for the second shift? Use a .05 level of significance.

35. Financial intermediaries, such as insurance companies, must decide on the mixture of assets to buy and liabilities to sell. The asset-liability mixture is a portfolio theory decision, and different mixtures reflect different risk-return trade-offs. (1) Stock insurers are owned by stockholders and (2) mutual insurers are owned by policyholders. A random sample of 50 stock insurers and 50 mutual insurers included 30 stock insur- ers that had more than half of their assets in bonds and 40 mutual insurers that had more than half of their assets in bonds. From these data, would you reject the hypothesis that the proportion of stock insurers that have more than half of their assets in bonds is the same as the proportion for mutual insurers? Use a .05 level of significance.

36. A random sample of 150 physicians included 50 general practitioners and 100 specialists. Of the (1) generalists 10 were married and had children when they were admitted to medical school. Of the (2) specialists 10 were married and had children at admission to medical school. From these data, could the AMA reject the hypothesis that the proportion of generalists who are married and have children at admission to medical school is less than or equal to the proportion for specialists? Let the level of significance be .05.

37. The only randomized study of the effectiveness of cancer screening by mammography and clinical examination in the United States was undertaken by the Health Insurance Plan (HIP) of Greater New York. It was designed to test whether annual mammography and clinical examination decrease the mortality rate for breast cancer. Of the patients 40 to 49 years of age at the start of the study, one group of randomly selected patients were (1) offered annual screening by mammography and clinical examination for cancer, whereas a second group of randomly selected patients were (2) not offered screening. The accompanying table shows the number of subjects and the number that had died from the cancer after nine years for each group. Do the sample results support the hypothesis that a program offering annual cancer screening by mammography and clinical examination decreases the mortality rate for breast cancer for those 40 to 49 years of age? Use a .05 level of significance.

	1. Screened Annually	2. Not Screened
Number of Subjects	14,849	14,911
Number of Deaths	39	48

REFERENCE: Shapiro, S. (1977), "Evidence on Screening for Breast Cancer from a Randomized Trial," *Cancer* 39: 2272–82.

38. A random sample of retired CPAs whose first jobs were in public accounting included 100 accountants from firm 1 and 80 accountants from firm 2. Of the firm 1 accountants 5 had been promoted to partner, while of the firm 2 accountants 8 had been promoted to partner. From these data, could you reject the hypothesis that the proportion of accountants who make partner at the two firms is the same? Let the significance level be .05.

9.10

TESTING THE HYPOTHESIS $\sigma_1^2 = \sigma_2^2$ AND F DISTRIBUTIONS

In Section 9.8, we noted that in order to test the hypothesis $H_0: \mu_1 = \mu_2$, we had to assume that two conditions existed:

1. The samples taken for the hypothesis test were random and independent.
2. The populations from which the samples were drawn were normally distributed.

Then in Section 9.7.1 we assumed that the two populations had the same standard deviation, $\sigma_1 = \sigma_2 = \sigma$; whereas in Section 9.7.2 we assumed that the populations had different standard deviations, $\sigma_1 \neq \sigma_2$.

We can ensure the independence and randomness of samples by using a careful sampling design and random number tables to obtain the samples. How, one might ask, can we test whether the two populations have the same standard deviation—or, equivalently, whether the two populations have the same variance? Also, there are numerous situations in which we might want to test $H_0: \sigma_1^2 = \sigma_2^2$ completely outside the context of the problems in Section 9.7. This section introduces *F* probability distributions and demonstrates how they can be used to test the hypothesis that *two populations have equal variances*.

If two populations have the same variance, then $\sigma_1^2 = \sigma_2^2$. If we take a sample from each of these populations, the best estimates we will obtain for the population variances are s_1^2 and s_2^2. If these two values differ from one another a great deal, the two population variances are probably not equal. We can compare the sample variances by computing their ratio s_1^2/s_2^2. When repeated independent random samples are drawn from two normal populations with the same variance, the ratio

$$F = \frac{s_1^2}{s_2^2} \tag{9.22}$$

follows an *F* probability distribution. If this ratio is very large or very small, then the hypothesis that the two populations have the same variance is probably false.

Equation (9.22) is the first of several *F* statistics we will encounter. **F distributions** form a family of distributions dependent on two parameters or two degrees of freedom, ν_1 and ν_2. The first of the two degrees of freedom, ν_1, is associated with the numerator of the *F* ratio, and the second, ν_2, is associated with the denominator.

FIGURE 9.18

F Distributions with Different Degrees of Freedom

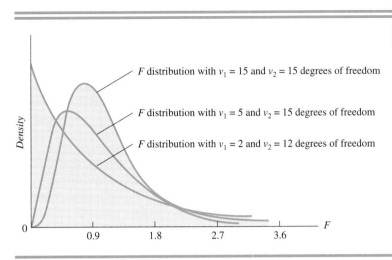

Three F distributions with different degrees of freedom are shown in Figure 9.18. Notice that these F distributions are truncated at zero and skewed to the right.*

To perform a test of the hypothesis H_0: $\sigma_1^2 = \sigma_2^2$, we proceed as follows. First, we set up the null hypothesis H_0: $\sigma_1^2 = \sigma_2^2$ and the alternative hypothesis.

Second, we calculate the F ratio as $F = s_1^2/s_2^2$. However, since the designation of which population is the first and which is the second is somewhat arbitrary, it is convenient if we always place the *larger* sample variance in the numerator of this ratio. If the null hypothesis is true, then the ratio should not be too much larger than 1.0. Calculated F values a great deal larger than 1.0 cause us to reject the null hypothesis of equal variances.

Thus, the following equation is used to obtain the F statistic for testing H_0: $\sigma_1^2 = \sigma_2^2$.

Equation for F Statistic for Testing Hypothesis $\sigma_1^2 = \sigma_2^2$

$$F = \frac{s_1^2}{s_2^2} \qquad (9.23)$$

The *larger* sample variance is always placed in the numerator of the ratio.

Of course, the big question is, How large must the F ratio of s_1^2/s_2^2 become before we can safely reject the hypothesis that the two populations have the same variance? To answer this question, we must compare the F ratio calculated from our data with the F ratio percentage point located in Table VII in Appendix C.

As we noted above, associated with each F value are two values for the degrees of freedom. For the F statistic of Equation (9.23), we call ν_1 the number of degrees

---

*The mean for an F distribution is $\nu_2/(\nu_2 - 2)$, $\nu_2 > 2$; the standard deviation is
$$\sqrt{2\nu_2^2(\nu_1 + \nu_2 - 2)/[\nu_1(\nu_2 - 2)^2(\nu_2 - 4)]}, \ \nu_2 > 4$$

of freedom in the larger sample variance, s_1^2, and ν_2 the number of degrees of freedom in the smaller sample variance, s_2^2. Thus

$$\nu_1 = n_1 - 1 \quad \text{and} \quad \nu_2 = n_2 - 1$$

To determine whether a calculated $F = s_1^2/s_2^2$ is so large that we should reject the hypothesis that the two population variances are equal, we must consult the F table, Table VII of Appendix C. In that table we find the F values that are exceeded by pure chance with only 5% probability for various combinations of degrees of freedom in the numerator and the denominator of the ratio. For example, if $\alpha = .05$, $\nu_1 = n_1 - 1 = 4$, and $\nu_2 = n_2 - 1 = 7$, from Table VII we find that $F_{\alpha, \nu_1, \nu_2} = F_{.05, 4, 7} = 4.1203$ or, in other words, that an F ratio of 4.1203 is exceeded with only a 5% probability when H_0 is true and we have 4 and 7 degrees of freedom (see Figure 9.19).

Thus, if $n_1 = 5$ and $n_2 = 8$, and if $\alpha/2$ (or one-half of the probability of a type I error that we are willing to tolerate, since this is a two-tailed test) for testing the hypothesis H_0: $\sigma_1^2 = \sigma_2^2$ were .05, then we would reject the hypothesis if s_1^2/s_2^2 exceeded 4.1203.

Finally, to determine if the calculated ratio is so large as to cause us to reject the hypothesis that $\sigma_1^2 = \sigma_2^2$, we compare the calculated F value with the $F_{\alpha/2, \nu_1, \nu_2}$ value in Table VII, using $n_1 - 1$ for ν_1 and $n_2 - 1$ for ν_2 and remembering that the sample with the larger variance is designated by the subscript 1. Reject the hypothesis H_0: $\sigma_1^2 = \sigma_2^2$ if the calculated F ratio exceeds the table value of $F_{\alpha/2, \nu_1, \nu_2}$. Table VII contains F values for several percentage points.

The test of the hypothesis of the equality of two variances is summarized as follows.

Hypotheses on Equality of Variances and Rejection Rule for Two-Sided Test

$$H_0: \sigma_1^2 = \sigma_2^2$$

$$H_a: \sigma_1^2 \neq \sigma_2^2 \qquad\qquad (9.24)$$

Rejection rule: If $F > F_{\alpha/2, \nu_1, \nu_2}$, then reject H_0;

where: $F = s_1^2/s_2^2$

and the *larger* sample variance is always placed in the numerator of the ratio.

If the test is one-sided, the rejection rule is as follows:

Rejection rule: If $F > F_{\alpha, \nu_1, \nu_2}$, then reject H_0.

 PRACTICE PROBLEM 9.12

Overnite Express offers a service in which they collect packages to be sent by air freight from businesses in downtown Washington, D.C., to New York City. The company likes to send out its last delivery to the air freight office at National Airport

FIGURE 9.19

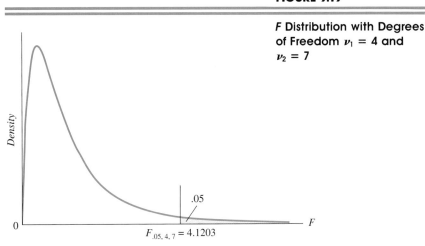

F Distribution with Degrees of Freedom $\nu_1 = 4$ and $\nu_2 = 7$

in Washington in time to make the 6:00 P.M. flight to New York. Since they must leave for the airport during the rush hour, the drivers are interested in which of two routes is faster. Thus they took route 1 for five days one week and route 2 for five days the next week. Let $\alpha = .10$. The means and standard deviations of the samples' travel times follow:

$$\overline{X}_1 = 34.2 \text{ min} \qquad \overline{X}_2 = 33.8 \text{ min}$$
$$s_1 = 6.1 \text{ min} \qquad s_2 = 16.4 \text{ min}$$

Solution To proceed with this problem, we must assume that the route travel times are normally distributed, the samples are random weeks, and the weeks' times are independent of one another. Route 2 looks slightly better than route 1 since the sample mean time is smaller (but not by much). To test whether the difference in population mean times for the two routes are the same, $H_0: \mu_1 = \mu_2$, we would have to be able to conclude whether the two population variances are the same, $\sigma_1^2 = \sigma_2^2$; and the sample data suggest that they are not. Thus we might begin with testing the hypothesis that the two routes have the same travel time variance:

$$H_0: \sigma_1^2 = \sigma_2^2$$
$$H_a: \sigma_1^2 \neq \sigma_2^2$$

We find the calculated F value by taking the ratio of the larger to the smaller sample variance:

$$F = \frac{(16.4)^2}{(6.1)^2} = 7.23$$

We now compare this value with the Table VII F value for an $\alpha/2$ value of .05, $\nu_1 = 4$, and $\nu_2 = 4$. This value is $F_{.05, 4, 4} = 6.3883$. We find that (see Figure 9.20) 7.23 is greater than 6.3883. Thus we reject the hypothesis that the two routes have the same travel time variances.

FIGURE 9.20

Hypothesis Test of Delivery
Time Variances of Routes
for Overnite Express

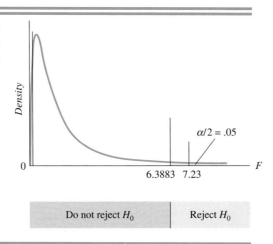

This conclusion tells us to use the *t* test of H_0: $\mu_1 = \mu_2$ with $\sigma_1 \neq \sigma_2$ to test the hypothesis that the population mean travel times are the same. Note that we would need to take a separate sample to test H_0: $\mu_1 = \mu_2$. That is, the sample to check an assumption underlying a test and the sample to actually perform the test should be independent—although in practice some statisticians ignore this restriction. In this problem we would use common sense and probably advise the company to use route 1. The route *appears* to have almost the same travel time, on the average, and it is more reliable. That is, the smaller sample standard deviation for this route indicates that each trip on this route takes about the same time. The large sample variance for route 2 indicates that on this route some of the times are very fast and some are very slow (perhaps because of having to wait at a railroad crossing). Since route 1 is more predictable, it would likely be the preferred route.

 PRACTICE PROBLEM 9.13

Mesta Machine Tool Company has been using a product line that mills parts to a predetermined length. A second production line has recently been installed. Quality assurance engineers are concerned that the older line may be resulting in parts that have lengths with a variance larger than that of the newer line. Thus they took a sample of 16 parts from the older line and 20 parts from the newer line. They found sample standard deviations of

$$s_{\text{older}} = 12 \quad \text{and} \quad s_{\text{newer}} = 9$$

Test the hypothesis that the old line produces parts that have lengths with a variance less than or equal to that of the new line. Let $\alpha = .05$.

Solution The test follows.

  Step One H_0: $\sigma^2_{\text{older}} \leq \sigma^2_{\text{newer}}$.
  Step Two H_a: $\sigma^2_{\text{older}} > \sigma^2_{\text{newer}}$.
 Step Three $\alpha = .05$, $n_1 = 16$, $n_2 = 20$.

FIGURE 9.21

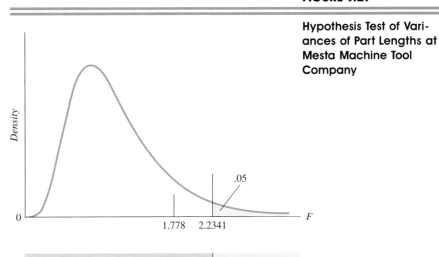

Step Four   We must find the value of F that is exceeded by chance with $\alpha = .05$ probability, F_{α, ν_1, ν_2} (since this test is one-tailed). To find this value, we enter the second section of Table VII where $\nu_1 = 16 - 1 = 15$ and $\nu_2 = 20 - 1 = 19$, and we find $F_{.05, 15, 19} = 2.2341$. Our rejection rule is as follows:

Rejection rule: If $F > 2.2341$, then reject H_0.

Step Five   We can now calculate the sample F ratio, remembering to put the larger sample variance in the numerator of the ratio. Thus

$$F = \frac{s^2_{\text{older}}}{s^2_{\text{newer}}} = \frac{(12)^2}{(9)^2} = \frac{144}{81} = 1.778$$

Step Six   Since our calculated value $F = 1.778$ is less than $F_{.05, 15, 19} = 2.2341$ (see Figure 9.21), we cannot reject the hypothesis that the older line has a variance from the predetermined length less than or equal to that for the newer line. In other words, the sample variances do not differ enough for us to conclude that the older line has a variance from the predetermined length that is greater than the variance for the newer line.

Problems: Section 9.10

39. Two different package designs were tested for a drug produced by Fitzer, Inc. The shelf lives of the contents were measured for a sample of $n_1 = 25$ packages of the first design and $n_2 = 25$ packages of the second design. Use the following sample variances to test the hypothesis that the two package designs have the same variance

of shelf life. Use $\alpha = .05$. What do you have to assume about the way the population values are distributed?

$$s_1^2 = 720 \text{ hours}^2 \qquad s_2^2 = 1455 \text{ hours}^2$$

40. Independent random samples were taken from normal distributions of the yearly production of ships built by the International Ship Building Company under (1) a fixed-position layout and (2) a project layout. The results are given in the accompanying table. Test the null hypothesis that the variance for the fixed-position layout is equal to that for the project layout. Use .05 for the level of significance.

1. Fixed-Position Layout	8	0	1	5	1	5	1
2. Project Layout	6	1	1	1	4	4	4

41. Studies in financial economics suggest that excess returns can sometimes accrue to shareholders when important financial events occur. In addition, variances of returns can be influenced by financial events. An important event might be the announcement that a firm is "in play" as a takeover target. Independent random samples were taken from normal distributions of the percentage daily yields for firms (1) just after a takeover announcement and (2) immediately before a takeover announcement under similar economic conditions. The results are given in the accompanying table. Can we conclude that the variance in returns after a takeover announcement is greater than the variance in returns before a takeover announcement? Use a .05 level of significance.

Sample	n	Standard Deviation
1	15	10
2	9	4

42. The Equal Credit Opportunity Act prohibits banks from discriminating against borrowers on the basis of marital status, sex, age, and race or because they receive alimony or public assistance. Independent random samples from normal distributions of lines of credit (in thousands of dollars) for (1) married and (2) not married individuals of the same economic status gave the accompanying results. Test to see whether the variances of lines of credit are different for married people than they are for those who are not married. Use a significance level of .05.

1. Married	15	12	9	13	14	10	11
2. Not married	1	3	10	9	7		

43. The data given in the accompanying table are yields in bushels per acre for two oat varieties planted by Kansas Land & Farming. Yields are normally distributed for the two varieties. Should we reject the hypothesis that the variance of the yields for variety 1 is less than or equal to the variance for variety 2? Let the level of significance be .01.

Variety 1	81.2	72.6	56.8	76.9	42.5	49.6	62.8	48.2
Variety 2	56.6	58.6	45.4	39.1	42.8	65.2	40.7	49.9

9.11
SUMMARY

In this chapter we have discussed several types of hypothesis testing. Two tests involved single populations and required our comparing the sample evidence from one sample to the hypothesized value of the population characteristic. For a single population with a continuous measure we computed the sample $\overline{X}$ and compared it with the hypothesized value for μ. For a single binomial population we computed the sample proportion of successes $\hat{p}$ and compared it with the hypothesized value of p. Table 9.6 presents the statistics and assumptions used in these situations as the first

TABLE 9.6 Summary of Hypothesis Tests

Population Type	Hypothesis to Be Tested	Test Statistic

Comments:
—All the tests described assume that random samples have been taken. When two samples are taken for the same test, the two samples must be independent.

One population, continuous measure $H_0: \mu = \mu_0$

$$t = \frac{\overline{X} - \mu_0}{s/\sqrt{n}}$$

Comments and assumptions:
—If $n < 30$, then the population should be normally distributed or approximately normally distributed.
—If σ is known and used rather than s in this calculation and if the population is normal, then the ratio is a Z value. If the population is not normal, then the ratio is an approximate Z value according to the central limit theorem.
—When $n > 30$, then $t \approx Z$ even if the population is not normally distributed.
—The t value has $n - 1$ degrees of freedom.

One binomial population $H_0: p = p_0$

$$Z = \frac{X - np_0}{\sqrt{np_0(1 - p_0)}}$$

or

$$Z = \frac{\hat{p} - p_0}{\sqrt{p_0(1 - p_0)/n}}$$

Comments and assumptions:
—Use the first form of the test statistic when the sample results are presented in terms of *number* of successes. Use the second form when the sample results are presented as *proportion* or *fraction* of successes.
—This test requires large samples, say, $np_0(1 - p_0) \geq 5$.

Two populations with continuous measures X_1 and X_2 and with $\sigma_1 = \sigma_2$ $\quad H_0: \mu_1 = \mu_2$

$$t = \frac{\overline{X}_1 - \overline{X}_2}{s_{\overline{X}_1 - \overline{X}_2}}$$

where:

$$s_{\overline{X}_1 - \overline{X}_2} = \sqrt{s_{pooled}^2 \left(\frac{1}{n_1} + \frac{1}{n_2} \right)}$$

Comments and assumptions:
—If either sample is small ($n_j < 30$), then we must assume that the population is normally distributed.
—Both populations must have the same variance, which is estimated by s_{pooled}^2, the weighted average of s_1^2 and s_2^2:

$$s_{pooled}^2 = \frac{(n_1 - 1)s_1^2 + (n_2 - 1)s_2^2}{n_1 + n_2 - 2}$$

—The t value has $n_1 + n_2 - 2$ degrees of freedom.

Two populations with continuous measures X_1 and X_2 with $\sigma_1 \neq \sigma_2$ $\quad H_0: \mu_1 = \mu_2$

$$t = \frac{\overline{X}_1 - \overline{X}_2}{s_{\overline{X}_1 - \overline{X}_2}}$$

where:

$$s_{\overline{X}_1 - \overline{X}_2} = \sqrt{\frac{s_1^2}{n_1} + \frac{s_2^2}{n_2}}$$

(continues)

TABLE 9.6   Summary of Hypothesis Tests Continued

Population Type	Hypothesis to Be Tested	Test Statistic

Comments and assumptions:
—If either sample is small ($n_j < 30$), then we must assume that the population is normally distributed.
—The t value has ν degrees of freedom where:

$$\nu = \frac{(s_1^2/n_1 + s_2^2/n_2)^2}{\dfrac{(s_1^2/n_1)^2}{n_1 - 1} + \dfrac{(s_2^2/n_2)^2}{n_2 - 1}}$$

—To use the t table with degrees of freedom as integers, round ν down to the nearest integer.

| Two binomial populations | $H_0: p_1 = p_2$ | $Z = \dfrac{\hat{p}_1 - \hat{p}_2}{\sqrt{\hat{p}_{\text{pooled}}(1 - \hat{p}_{\text{pooled}})(1/n_1 + 1/n_2)}}$ |

Comments and assumptions:
—Both sample sizes must be large, say $n_1 \hat{p}_{\text{pooled}}(1 - \hat{p}_{\text{pooled}}) \geq 5$ and $n_2 \hat{p}_{\text{pooled}}(1 - \hat{p}_{\text{pooled}}) \geq 5$.
—$\hat{p}_{\text{pooled}}$ is the pooled estimate of the common value for p_1 and p_2. It is found as follows:

$$\hat{p}_{\text{pooled}} = \frac{Total\ number\ of\ successes}{n_1 + n_2} = \frac{X_1 + X_2}{n_1 + n_2}$$

| Two populations with continuous measures X_1 and X_2 | $H_0: \sigma_1^2 = \sigma_2^2$ | $F = \dfrac{s_1^2}{s_2^2}$ |

Comments and assumptions:
—The populations must be normally distributed.
—The larger sample variance is always placed in the numerator.
—The F value has two degrees-of-freedom values:

$$\nu_1 = n_1 - 1 \qquad \nu_2 = n_2 - 1$$

two rows of the table. If the assumptions are not met, then the nonparametric tests in Chapter 18 can be used.

We have discussed several hypothesis tests when two populations are being considered. Tests of $H_0: \mu_1 = \mu_2$ were discussed for the case when $\sigma_1 = \sigma_2$ and for the case when $\sigma_1 \neq \sigma_2$. Tests of $H_0: \mu_1 = \mu_2$ require two samples, one from each population, and the sample means and standard deviations are used according to the methods summarized in the third and fourth rows of Table 9.6.

The fifth row of Table 9.6 summarizes the methods for testing $H_0: p_1 = p_2$ for two binomial populations. Again, two samples are taken from the populations, and the proportions of successes $\hat{p}_1$ and $\hat{p}_2$ are compared.

The sixth row of Table 9.6 summarizes the methods for testing the equality of variances, $H_0: \sigma_1^2 = \sigma_2^2$. The test of the equality of variances is useful for checking the assumption of equal variances in the test of equal means in addition to practical problems where examining the equality of variances is an important conclusion for solving the problem or making decisions.

Answers to odd-numbered problems are given in the back of the text.

44. For the following information from a normal distribution,

$$n = 16 \qquad \overline{X} = 1550 \qquad s^2 = 12$$
$$H_0: \mu \le 1500 \qquad H_a: \mu > 1500 \qquad \alpha = .01$$

test the hypothesis at the given level of significance. The data (which are only needed if you want to use them with a computer) are as follows:

1552.47	1551.60	1550.07	1553.35	1546.00	1542.53
1546.61	1546.74	1550.38	1551.69	1555.00	1549.81
1550.82	1550.15	1555.49	1547.29		

45. For the following information from a normal distribution,

$$n = 49 \qquad \overline{X} = 17 \qquad s = 1$$
$$H_0: \mu \ge 18 \qquad H_a: \mu < 18 \qquad \alpha = .05$$

test the hypothesis at the given level of significance. The data (which are only needed if you want to use them with a computer) are as follows:

17.6666	17.6702	16.7602	16.3096	18.0860	17.1423	18.7986
16.3852	17.7342	17.4272	17.4937	16.3416	15.9673	16.5035
15.8226	17.3253	17.0454	16.6283	15.7324	16.9567	16.3561
15.7844	15.5585	18.3750	16.6879	16.1551	17.7948	18.2603
16.4459	18.8610	17.6091	17.5592	17.1816	17.6723	17.0552
15.7854	16.5432	14.5121	18.4449	16.3944	17.2406	17.8385
16.9674	18.6206	15.7646	16.1308	17.0974	18.9617	15.5449

46. For the following information from a normal distribution

$$n = 9 \qquad \overline{X} = 10.1 \qquad s^2 = .81$$
$$H_0: \mu = 12 \qquad H_a: \mu \ne 12 \qquad \alpha = .05$$

test the hypothesis at the given level of significance. The data (which are only needed if you want to use them with a computer) are as follows:

11.3357	9.9665	9.9126	9.4319	8.7551
9.7832	11.2546	10.9976	9.4628	

47. For the following information from a normal distribution

$$n = 25 \qquad \sum X_i = 500 \qquad \sum X_i^2 = 12,400$$
$$H_0: \mu \le 17 \qquad H_a: \mu > 17 \qquad \alpha = .10$$

test the hypothesis at the given level of significance. The data (which are only needed if you want to use them with a computer) are as follows:

28.2579	17.7103	−0.7540	8.2166	26.9483	27.8785	12.3270
7.5619	14.6526	14.4479	24.0254	26.2116	21.7087	21.9608
27.6207	21.2185	29.9012	34.4570	34.8762	6.7107	21.9753
11.6338	15.3323	37.3722	7.7484			

48. Sensitivity tests were made on 18 randomly selected p-doped transistors produced by Dataram Devices Corporation. The mean was 2.5 microvolts and the sample variance s^2 was .48. Would the sample results support the hypothesis that the mean sensitivity of transistors of this type in the given circuit is greater than 2.0 microvolts? That is, use $H_a: \mu > 2.0$. Let $\alpha = .05$, and assume normality.

49. A sample of $n = 25$ recently promoted managers for a regional certified public accounting firm had a mean salary equal to \$33,000 and a sample standard deviation of \$10,000. A partner for the firm has indicated to a recruit that the mean salary for managers has been \$37,000 or greater. Would we have reason to reject the partner's recollection of what the salary had been? Let the level of significance be .025, and assume normality. Note that sample statistics summarized in this problem were computed from the accompanying data; however, the data may be useful if you wish to use a computer.

27451.3	26063.0	45195.2	34601.5	28392.0	20548.3	27264.4
46210.4	31908.9	20931.1	21564.5	30305.3	31965.7	16204.7
53806.6	35650.7	34901.7	36002.2	20443.9	35765.2	36039.2
31851.7	55438.1	34302.6	42191.8			

50. A sample of 15 rulers from Kiffel and Essex has an average length of 12.04 inches and the sample standard deviation is .015 inch. If $\alpha = .02$, test the hypothesis that the average length of rulers produced by this supplier is 12 inches. Use $H_a: \mu \neq 12.00$? Ruler lengths are approximately normally distributed.

51. Varner Corporation produces motor vehicle parts and accessories, including hold-down springs for disk brakes. The estimated variance based on four measurements of a normally distributed spring tension was .25 gram. The mean was 37 grams. Test the hypothesis that the true value is 35 grams or less. Use $\alpha = .10$ and $H_a: \mu > 35$.

52. Nine rafters were tested by Great Basin Construction Company for breaking load, giving a mean breaking strength of 1500 pounds and a sample standard deviation s of 110 pounds. At the 5% level of significance, would we reject the null hypothesis that the mean breaking load is 1600 pounds or more? The breaking strengths of these rafters are normally distributed. Note that sample statistics summarized in this problem were computed from the accompanying data; however, the data may be useful if you wish to use a calculator or computer.

1411.71	1540.12	1428.14	1381.77	1578.60
1597.68	1702.36	1400.40	1459.22	

53. A financial analyst at Transpacific Corporation maintains the notion that historically, the mean equity–debt ratio for firms that are going concerns has been greater than or equal to 3. A sample of $n = 9$ going concerns is taken and the resulting equity to debt ratios are presented in the accompanying list. Do the sample results contradict the historical notion maintained by the analyst to such a degree that an investor should reject the notion? Assume normality and use a significance level of .05.

     8   0   0   0   4   1   1   2   2

54. A manager of public health services in an area downwind of a nuclear test site wants to test the hypothesis that the mean amount of radiation in the form of strontium 90 in the bone marrow (measured in picocuries) for citizens who live downwind of the site does not exceed that of citizens who live upwind from the site. It is known that upwinders have a mean level of strontium 90 of 1 picocurie. Measurements of strontium 90 radiation for a sample of $n = 9$ citizens who live downwind of the site was taken, with the results given in the accompanying list. Assume normality and use a significance level of .05.

     10   2   2   2   2   5   5   7   1

55. The drained weights in ounces for a sample of 15 cans of fruit from Pineapple Corporation are given in the accompanying list. At the 5% level of significance, test the hypothesis that on the average a 12-ounce drained-weight standard is being main-

tained. Use H_a: $\mu \neq 12.0$ as the alternative hypothesis. The drained weights of these cans are approximately normally distributed.

12.1	12.1	12.3	12.0	12.1
12.4	12.2	12.4	12.1	11.9
11.9	11.8	11.9	12.3	11.8

56. A random sample of $n = 8$ chief executive officers (CEOs) of large corporations were asked about the number of hours they worked per week, and the sample mean was 62 hours. Assuming a normal population and a known population variance of 2 hours, can we reject the null hypothesis H_0: $\mu \leq 60$? Let the significance level be .10.

57. The ages (in years) of all summer employees of Resorts America over many years have been normally distributed with a variance of 100. A sample of 25 employees from the summer work force had a mean age equal to 17 years. Can we reject the null hypothesis that the mean age of summer employees at the resort is equal to 21 years, H_0: $\mu = 21$, in favor of H_a: $\mu \neq 21$? Use a significance level of .05.

58. On 384 out of 600 randomly selected farms the farm operator was also the owner. Is there sufficient evidence for us to reject H_0: $p \leq .60$ if we use $\alpha = .10$?

59. A United States automobile manufacturer asked each of 50 randomly selected drivers to compare the ride in his car with the smoothness of ride in a \$40,000 European touring car. The number of drivers who preferred the ride in the American car was 38. Would the manufacturer be justified in advertising that his car has the smoother ride? That is, can he reject H_0: $p \leq .50$ in favor of H_a: $p > .50$, where p is the proportion of people preferring the American car's ride? Let $\alpha = .05$.

60. A large mining company has been forced to terminate 5% of its work force because of low market prices for precious metals (employees are therefore either terminated or not terminated). A group of older (age 55 or older) terminated employees has filed a lawsuit claiming that age discrimination was practiced by the company during the termination process. Analysts have determined that 10 of the 100 employees age 55 or older in the total work force were terminated. As an expert witness for this lawsuit, would you testify that age discrimination has taken place during the termination of these employees? That is, for the older employees, test H_0: $p \leq .05$ against H_a: $p > .05$, using .05 for the level of significance.

61. A health services provider has been reimbursed by National Health Insurance, Inc., for a large number of services. It appears that overpayments may have occurred on many of the reimbursements due to errors made during the billing process or due to upcoding (billing for a more complicated service than was actually provided). The health services provider maintains that the proportion of reimbursements for which overpayments have occurred has not significantly exceeded 1%. An audit of a random sample of 500 reimbursements was conducted for National, and 100 of these reimbursements involved overpayments. Would you reject the hypothesis that the proportion of reimbursements for which overpayments have occurred has not exceeded 1% and, as a consequence, require that the provider refund the overpayments? Use a significance level of .05.

62. Astro Company, an aircraft manufacturer, receives large lots of wing struts from a subcontractor who guarantees that no more than 5% of each lot will fail to meet design specifications. One hundred struts are inspected in each lot.

   a. What are the null and alternative hypotheses?
   b. How many struts in the sample should fail to meet specifications before the hypothesis is rejected, if we desire an α of about .05? (*Note:* In this special problem you are asked to find the rejection region.)

63. Sunshine Minerals Company conducted a study of productivity at two of its mines. The number of tons of ore taken from each of the mines each day is normally distrib-

uted with a common standard deviation. The amount of ore extracted was recorded over 10 randomly selected days at mine 1 and 15 days at mine 2. Given the accompanying summary of data, will the company be justified in rejecting the supposition that the mean daily yield of the first mine is equal to that of the second? That is, test H_0: $\mu_1 = \mu_2$ against H_a: $\mu_1 \neq \mu_2$, using $\alpha = .10$.

Sample	n	$\overline{X}_j$	$\Sigma(X_{ij} - \overline{X}_j)^2$
1	10	25	250
2	15	30	302

The sample statistics summarized in the problem were computed from the accompanying data; however, the data may be useful if you wish to use a computer.

Mine 1	24.0496	18.6801	30.2698	22.7758	31.7565	32.5576
	22.7561	27.2336	22.4955	17.4254		

Mine 2	24.2967	27.7678	28.5166	24.4041	26.3746	34.7172
	32.4398	37.6339	29.3964	30.0468	24.1086	32.3867
	23.7086	26.2704	37.9319			

 64. Independent random samples of the number of hours spent on decision making per month by chief executive officers of (1) mining and (2) computer companies, which we assume have normal distributions and equal standard deviations, gave the results listed in the accompanying table. Can we reject the null hypothesis H_0: $\mu_1 \geq \mu_2$ with $\alpha = .025$?

Sample	n	$\overline{X}_j$	$\Sigma(X_{ij} - \overline{X}_j)^2$
1	10	62	340
2	20	70	302

The sample statistics summarized in the table were computed from the accompanying data; however, the data may be useful if you wish to use a computer.

1. Mining	55.9700	57.3885	65.4015	63.9793	52.9981
	67.9166	65.4484	72.9557	58.4341	59.5078

2. Computer	63.0698	73.8480	69.6745	65.8673	71.8822
	68.7609	74.4337	75.3603	74.4338	66.9954
	74.6010	68.9460	69.8818	73.2719	69.2337
	71.8457	72.9083	63.7395	68.5273	62.7191

 65. Independent random samples from normal distributions with equal standard deviations of sales growth percentages by (1) firms that have been takeover targets and (2) firms that have not been takeover targets gave the results listed in the accompanying table. Test the null hypothesis that the mean growth in sales of takeover target firms does not exceed that for nontarget firms. Use .05 for the level of significance.

1. Target Firm	12	0	10	2	4	8	6	6
2. Nontarget Firm	8	8	-2	-2	0	0	2	2

66. The daily catch of two fishing boats in the United Brands fleet was recorded on a random basis. The results of two independent random samples from normal populations with equal standard deviations are given in the accompanying table. Test H_0: $\mu_1 = \mu_2$ against the alternative H_a: $\mu_1 \neq \mu_2$ with $\alpha = .02$.

Boat 1	108	110	103	100	107	107	101
Boat 2	113	110	108	98	111	112	110

67. Independent random samples from normal distributions with equal standard deviations of the amount of insider trading during mergers by employees of (1) investment

banking firms and (2) brokerage houses gave the results listed in the accompanying table. Test the null hypothesis that the mean amount of insider trading at investment banking firms is less than or equal to that at brokerage houses. Let the significance level be .05.

1. Investment Banking Firm	10	5	5	5	8	8	8
2. Brokerage House	6	0	5	2	2	3	3

68. Independent random samples from normal distributions with equal standard deviations of the hourly production of a radio transmitter at Electronics, Inc., under (1) a project layout and under (2) a fixed-position layout gave the results listed in the accompanying table. Test the null hypothesis that the mean hourly production under the project layout is less than or equal to that under the fixed-position layout. Use .05 for the level of significance.

1. Project Layout	4	8	4	4
2. Fixed-Position Layout	4	4	0	4

69. The accompanying table lists yields in bushels per acre for two oat varieties planted by California Land and Farming Company. Yields are normally distributed and have equal standard deviations for the two varieties. Test the hypothesis that the yields are the same for variety 1 and variety 2 against the alternative hypothesis H_a: $\mu_1 \neq \mu_2$. Use .10 as the level of significance.

Variety 1	81.2	72.6	56.8	76.9	42.5	49.6	62.8	48.2
Variety 2	56.6	58.6	45.4	39.1	42.8	65.2	40.7	49.9

70. Independent random samples from normal distributions with *unequal* standard deviations of start-up owners' equity (in hundreds of thousands of dollars) for (1) small business firms that within the first year of operation have resulted in bankruptcy and (2) firms that are not bankrupt gave the results in the accompanying table. Test whether the mean of start-up owners' equity for small business firms that are bankrupt within the first year is equal or unequal to that for nonbankrupt firms. Use .05 for the level of significance. (You may want to compute sample means and standard deviations for this problem by using hand calculations.)

1. Bankrupt Firm	11	5	3	1	11	1	3	5
2. Nonbankrupt Firm	12	9	8	9	6	11	7	10

71. Independent random samples were taken from normal distributions with *unequal* standard deviations of sales (in thousands of dollars) of a detergent by using (1) a talking display and (2) a nontalking display in similar retail stores. The results are listed in the accompanying table. Test the hypothesis that $\mu_1 - \mu_2 = 10$ at the 2% level of significance. Use H_a: $\mu_1 - \mu_2 \neq 10$.

Sample	n	Mean	Standard Deviation
1	20	77.1	21
2	12	62.3	10

The sample statistics summarized in the table were computed from the accompanying data; however, the data may be useful if you wish to use a calculator or computer.

1. Talking Display	64.313	98.019	86.876	88.962	32.215	
	101.568	111.208	76.391	56.948	75.578	
	77.004	55.879	83.427	58.859	91.279	
	49.941	87.236	65.029	112.763	68.505	
2. Nontalking Display	60.0188	61.3116	57.6571	45.6422	65.1234	60.0603
	67.2181	74.8671	49.9339	71.2527	54.6589	79.8560

72. A random sample of 900 persons included 467 smokers and 433 nonsmokers. Of the smokers 18 had required hospitalization in the past year, but only 2 nonsmokers had required hospitalization. From these data, could a health insurance company reject the hypothesis $H_0: p_s \leq p_{ns}$ in favor of $H_a: p_s > p_{ns}$, where p_s is the proportion of smokers requiring hospitalization and p_{ns} is the proportion of nonsmokers requiring hospitalization? Use $\alpha = .01$.

73. In a test of the applicability of alloy metals, parts for a washing machine produced by Whirlmor, Inc., were manufactured first from one alloy and then from another. One hundred samples of parts made from each type of alloy were subjected to shock testing. Defects developed in 18 of those made from alloy I and in 26 of those made from alloy II. Can we reject $H_0: p_I \geq p_{II}$ in favor of $H_a: p_I < p_{II}$, using $\alpha = .025$? Note that p represents the defective rate.

74. Managers at the Crager Company supermarket chain were interested in a report that consumers in New England tend to prefer brown eggs to a greater extent than do consumers in western states. A random sample of 50 consumers included (1) 25 persons from New England and (2) 25 persons from western states. Of the New England consumers 20 preferred brown eggs to white eggs, and 10 consumers in the West preferred brown eggs. From these data, could the Crager supermarket chain reject the hypothesis that the proportion of consumers who prefer brown eggs does not differ between consumers in New England and consumers in western states? Use a .05 level of significance.

75. An official of the U.S. Savings and Loan League feels that savings and loan associations should be more active in soliciting the savings of minority group members. To support this position, the official took a random sample of 1000 people in various midwestern cities. Of the 100 minority group members in this sample, only 20 had accounts with a savings and loan association. But 480 of the 900 other people in the sample had S&L accounts. Test the hypothesis that the proportion of account holders among nonminority group members is less than or equal to that for minority group members against the alternative hypothesis that the proportion of account holders is less among minority group members. Let $\alpha = .10$.

76. A random sample of loan applicants at a financial institution included 50 women and 50 men. Of the (1) women applicants 20 were granted credit, and of the (2) men applicants 30 were granted credit. From these data, should an executive reject the hypothesis that the institution discriminates against women in granting credit if discrimination is interpreted to mean that a smaller proportion of women than men are granted credit? Use a .05 level of significance.

77. A study was done by a hospital's quality control administrators to determine what factors influence the infection rates for surgical incisions. Hospital administrators wanted to decrease hospitalization costs and morbidity for health care consumers and increase profitability at the same time. One factor that was studied was the traditional procedure of shaving the operation site, which is explained at times by health care providers as a procedure to decrease infection rates. However, shaving can result in skin nicks that are favorable for or increase infection. For (1) patients where the operation site was shaved and for (2) patients where the operation site was not shaved, the accompanying table shows the number of patients in each group and the number having infected incisions. Do the sample results support the hypothesis that shaving increases the infection rate of surgical incisions? Use a .05 level of significance.

	1. Shaved	2. Not Shaved
Number of Patients	1000	1000
Number of Infected incisions	23	9

REFERENCE: Cruse, P. and R. Foord (1973), "A Five-Year Prospective Study of 23,649 Surgical Wounds," *Archives of Surgery,* 107: 206–10.

78. Independent random samples were taken from normal distributions for sales of a detergent using (1) a talking display and (2) a nontalking display in similar retail stores. The results are given in the accompanying table.

    a. Test the hypothesis that the two populations have the same variance. Let the significance level be .10.

    b. Compare your result from part a with the assumption of unequal standard deviations in Problem 71. Comment on the validity of the test in Problem 71.

Sample	n	Mean	Standard Deviation
1	20	77.1	21
2	12	62.3	10

79. Two economic forecasting services have the same mean error in predicting the gross national product (GNP) over the past ten years. However, the standard deviation of the first service's error is $10 billion, and the other service's standard deviation is $25 billion.

    a. Test the hypothesis that the two services have the same variance of error, using $\alpha = .05$.

    b. Since the mean error figures are the same, would you be satisfied with the prediction of either service as to what next year's GNP will be? Why?

80. Independent random samples were taken from normal distributions of the ratio of municipal debt to the value of real estate (as a percentage) for municipalities with (1) A or lower bond ratings and (2) AA bond ratings. The results are given in the accompanying table. Test the null hypothesis that the variances of the ratios in the two groups are the same. Use .05 for the level of significance.

1. A or Lower Bond Rating	1	9	1	1
2. AA Bond Rating	1	1	5	1

81. Independent random samples were taken from normal distributions of start-up owners' equity (in hundreds of thousands of dollars) for small business firms that within the first year of operation have resulted in (1) bankruptcy and (2) no bankruptcy. The results are given in the accompanying table. Test the null hypothesis that the variance of start-up owners' equity for small business firms that are bankrupt within the first year is less than or equal to that for nonbankrupt firms. Use .05 for the level of significance.

1. Bankrupt Firm	11	11	1	1	3	3	5	5
2. Nonbankrupt Firm	7	1	6	2	5	3	4	4

Refer to the 141 companies listed in the Ratio Data Set in Appendix A.

Ratio Data Set

Questions

82. Locate the data for the current assets–sales ratio (*CA/S*). The companies in the file are a random sample from the numerous companies in the COMPUSTAT files. A financial analyst maintains that companies in the population historically on average have had *CA/S* equal to 1.40 or more. Has the ratio decreased on average? Use a significance level of .05.

83. Locate the data for the quick assets–sales ratio (*QA/S*). The companies in the file are a random sample from the numerous companies in the COMPUSTAT file. A financial analyst maintains that the population of companies historically average have *QA/S* equal to 1.40 or more. Has the ratio decreased on average? Use a significance level of .05.

84. Locate the data for the net income–total assets ratio (*NI/TA*). Some of the companies (1) suffered financial losses as reflected by negative values for *NI/TA*; whereas, other

companies (2) did not suffer losses. The data set also includes the current assets to current liabilities (*CA/CL*) ratio as a measure of liquidity. The companies in the file are a random sample of the numerous companies in the COMPUSTAT files. Test the hypothesis that mean liquidity is higher for companies that suffer losses. Use a significance level of .05 and assume *unequal* standard deviations.

85. Locate the data for the net income–total assets ratio (*NI/TA*). The companies on the list are a random sample of all of the numerous companies included in the COMPUSTAT data base. A stockbroker trying to sell stock for the companies maintains that companies on average have historically had *NI/TA* equal to .04 or more. Has the ratio decreased on average? Use a significance level of .05.

86. Locate the data set for the net income–total assets ratio (*NI/TA*). Some of the companies suffered financial losses as reflected by negative values for the ratio. The companies on the list are a random sample of the numerous companies included in the COMPUSTAT files. A stockbroker trying to sell stock for the companies maintains that historically 85% or more of the companies in the population are profitable or at least have not suffered losses. Has the population proportion of companies that are profitable or have not suffered losses decreased? Use a significance level of .05.

Credit Data Set   *Refer to the 113 applicants for credit listed in the Credit Data Set in Appendix A.*

Questions   87. Test the hypothesis that the population means for JOBINC are the same for those who were granted credit and those who were denied credit. When you make this comparison, leave out those applicants who did not list a JOBINC value. As your alternative hypothesis, use "Applicants in the two groups do not have equal incomes." Use an α value of .02.

88. Test the hypothesis that the proportion of women is the same for the populations of people who were granted and denied credit. As an alternative hypothesis, use "The proportions of women in the two groups are not the same." Use an α value of .05.

89. Test the hypothesis that the variance of AGE is the same in the groups where credit was granted and where it was denied. Let $\alpha = .05$.

CASE 9.1   **COMPARING HYPOTHESIS TESTING AND COURT PROCEDURES**

Procedures in a court of law afford a parallel to tests of hypotheses. Let the null hypothesis be

$$H_0: \text{The defendant is innocent.}$$

And let the alternative hypothesis be

$$H_a: \text{The defendant is guilty.}$$

To condemn an innocent person is to commit a type I error; to acquit a guilty person is to commit a type II error.

   The rules that govern a trial are loosely analogous to a statistical test or a rejection rule. Any rule that decreases the chance α of a type I error necessarily increases the chance β of a type II error. For example, a rule that defendants need not testify against themselves decreases the chance of condemning an innocent person, but it also increases the chance of acquitting a guilty person.

   Also, a rule that decreases the chance β of a type II error necessarily increases the chance α of a type I error. For example, a rule allowing a split jury to find a defendant guilty would decrease the chance of acquitting a guilty person, but it would also increase the chance of condemning an innocent person. To arrange that both α and β are equal to 0 is impossible. But to arrange that both α and β are 1 is equally impossible.

Random samples are used in hypothesis tests, and jurors must be selected so that they are representative of the citizenry.

Conclusions reached by juries in court proceedings are parallel to conclusions reached in hypothesis tests. Jurors in a court case may feel a reasonable doubt about the guilt of a defendant because there is not enough evidence to convict. Thus they may fail to reject the null hypothesis of innocence. But that does not mean that the juror must accept the notion that the defendant is innocent. The juror will vote "not guilty," which is a middle ground between "guilty" and "innocent." In the same way, when we do not have enough evidence to reject the null hypothesis in testing a hypothesis, we typically simply say we "do not reject H_0" rather than saying we "accept" H_0. This statement too is a middle-ground position.

In most hypothesis testing the null hypothesis is chosen so that the type I error is a more serious error than the type II error. In parallel to the procedures in a court of law, convicting someone who is innocent is generally viewed as more serious than letting a guilty person go. Thus the most important probability in hypothesis tests is usually α, the probability that the hypothesis test will lead to a type I error—rejection of the null hypothesis when it is true. The rules in a court of law are generally structured so that there will be but a modest probability of convicting an innocent person. For example, unanimous agreement of guilt beyond a reasonable doubt is required for jurors to condemn a defendant.

In a brutal crime in Iron County in the state of Utah, a student at Southern Utah University was kidnapped and murdered. An informant, later identified as an itinerant laborer named R. Bott, implicated another man named Hamilton in the murder. Bott was granted immunity from prosecution for providing evidence against Hamilton. However, after being granted immunity, Bott was implicated as a participant in the crime and the prosecution sought to rescind the immunity. At trial, Hamilton was convicted of murder. The court ruled that immunity could not be rescinded for Bott, even though he had been implicated subsequently as a participant in the crime. By order of the court, Bott was freed and paid witness fees at the conclusion of the trial.

Case Assignment

a. Assuming that Bott was guilty, what type of error occurred in this case?
b. Comment on the philosophy of the court when the grant of immunity to prosecution was not rescinded and on how the same philosophy pertains to tests of hypotheses.

TESTING THE EFFECTIVENESS OF A CATTLE FEEDING PROGRAM

CASE 9.2

The Mountain States Feed Exchange (MSFE) has developed a new feed that it thinks will result in high weight gains for cattle in feed lots. At the same time, MSFE believes the new feed will help the animals resist certain diseases. To test the new feed, MSFE obtained the cooperation of Wilma Maxwell, who operates a feed lot in southern Colorado. Wilma fed the new feed to a group of animals in her feed lot between February 1 and May 12, inclusive. The information on these animals is presented in Table 1.

MSFE also obtained information concerning a number of animals that Wilma had in her southern Utah feed lot between January 15 and May 5, inclusive, of the previous year. This group of animals was fed the standard MSFE feed. The record for these animals follows the table in summary form.

MSFE would like you to examine these data and give them a report discussing what the data show. Mr. Jeremiah Kroft, owner of MSFE, will review your report. He has a fair knowledge of statistics.

TABLE 1	Beginning Weight	Ending Weight	Free of Disease?	Beginning Weight	Ending Weight	Free of Disease?
Data for Cattle that Received the New Feed	375	552	Yes	450	738	No
	312	498	Yes	349	654	Yes
	433	720	Yes	387	729	Yes
	343	451	No	401	613	No
	329	487	Yes	378	492	No
	396	556	Yes	444	721	Yes
	430	701	Yes	316	617	Yes
	438	578	No	359	596	Yes
	450	739	Yes	369	637	Yes
	303	430	No	355	640	Yes
	412	645	Yes	323	589	Yes
	440	689	Yes	341	612	Yes
	357	599	Yes	377	712	Yes
	437	687	Yes	301	Died	Died

The record of animals receiving standard feed is as follows:

1. Mean beginning weight of 43 animals = 392.49 pounds. Standard deviation of beginning weights = 54.80 pounds.
2. Mean ending weight of 41 animals = 552.65 pounds. Standard deviation of ending weights = 133.77 pounds.
3. Mean daily weight gain (41 animals only) = 2.01 pounds. Standard deviation of daily weight gains = .45 pounds.
4. Number of sick animals (including two that died) = 12.

Case Assignment

a. Test the hypothesis that the mean beginning weights are the same for the populations of animals fed the new feed and the standard feed.
b. Test the hypothesis that the mean daily weight gains are the same for the two populations.
c. Test the hypothesis that the rate of sickness is the same for the two populations of animals.

REFERENCES

Hoel, Paul G. 1971. *Elementary Statistics,* 3rd ed. New York: Wiley.
Hogg, Robert V., and Allen T. Craig. 1978. *Introduction to Mathematical Statistics,* 4th ed. New York: Macmillan.
Lehmann, E. 1986. *Testing Statistical Hypotheses,* 2nd ed. New York: Wiley.
Mosteller, Frederick, Robert E. K. Rourke, and George B. Thomas, Jr. 1970. *Probability with Statistical Applications,* 2nd ed. Reading, Mass.: Addison-Wesley.
Snedecor, George W., and William G. Cochran. 1989. *Statistical Methods,* 8th ed. Ames: Iowa State University Press.

Optional Topics

9.12 Type II Error Probabilities and Power of a Hypothesis Test
9.13 Testing the Hypothesis $\mu_1 = \mu_2$ with Paired Data
9.14 Groups Versus Pairs

In this section, we show how to calculate *probabilities for type II errors,* and we discuss the *power* of a hypothesis test for a mean μ. Throughout this section, we assume that the population is normally distributed and that the population standard deviation σ is known. In our discussion of hypothesis testing, we defined the **probability of a type II error** as

$$\beta = P(type\ II\ error) = P(Not\ rejecting\ H_0\ when\ H_0\ is\ false)$$

Naturally, since a β value is the probability of an error, we want β values to be reasonably small.

We can calculate a value of β, for each different value of μ that is included under the alternative hypothesis. We use the following procedure to find β values:

1. Formulate H_0 and H_a; set the values of α and n; determine the rejection rule and find the values of the test statistic Z that do *not* fall in the rejection region, where

$$Z = \frac{\overline{X} - \mu_0}{\sigma/\sqrt{n}}$$

2. Find the values of the sample mean that correspond to the Z values that do *not* fall in the rejection region.
3. Select a value of μ that falls under the alternative hypothesis and find the probability of $\overline{X}$ *not* falling in the rejection region under the assumption that the value of μ that was selected is the true value of the population mean. This probability is β, the probability of a type II error.

The following example will clarify the procedure for finding β.

EXAMPLE 9.11 Recall that the International Rubber Band Company had filled its number 6 box of rubber bands with an average of "at most 6 ounces of little office helpers" (rubber bands, that is). Thus the hypotheses are

$$H_0: \mu \le 6$$

$$H_a: \mu > 6$$

For the present example, the weight of the rubber bands is normally distributed; the population standard deviation is $\sigma = 3$; the significance level is $\alpha = .05$; and the sample size is $n = 36$. In this situation, the critical value for the test statistic from the standard normal table is

$$Z_{.05} = 1.645$$

and the rejection rule is

Rejection rule: If $Z > 1.645$, then reject H_0.

Thus, values of Z that are less than or equal to 1.645 are *not* in the rejection region. The value of the sample mean that corresponds to $Z = 1.645$, or the critical value of the sample mean, can be found by solving the following equation for $\overline{X}$:

$$Z = \frac{\overline{X} - \mu_0}{\sigma/\sqrt{n}} = \frac{\overline{X} - 6}{3/\sqrt{36}} = 1.645$$

Solving for the value of the sample mean $\overline{X}$ gives us

$$\overline{X} = 6 + (1.645)(3/\sqrt{36}) = 6 + .8225 = 6.8225$$

Thus, an equivalent rejection rule in terms of $\overline{X}$ is

Rejection rule: If $\overline{X} > 6.8225$, then reject H_0.

Consequently, any value of $\overline{X}$ that is less than or equal to 6.8225 is *not* in the rejection region. The rejection region and the region where we do not reject H_0 are shown on Figure 9.22 (a) for both Z values and $\overline{X}$ values.

We want to find the value of β assuming that the true value of the population mean μ is 8. Notice that the value of μ that we have selected, $\mu = 8$, is included under H_a: $\mu > 6$. The probability of a type II error in this situation is the probability of an $\overline{X}$ not falling in the rejection region, assuming that the true value of the population mean is $\mu = 8$. This probability is

$$\beta = P(\overline{X} \le 6.8225 \text{ assuming that } \mu = 8)$$

The sampling distribution of $\overline{X}$, assuming $\mu = 8$, and β are shown in Figure 9.22(b). The standardized value of $\overline{X} = 6.8225$ for this distribution is

$$Z = \frac{6.8225 - 8}{3/\sqrt{36}} = -2.35$$

The standard normal table gives the area between $Z = 0$ and $Z = -2.35$ as .4906. Thus the probability of a type II error is merely

$$\beta = P(\overline{X} \le 6.8225 \text{ assuming that } \mu = 8)$$

$$= P(Z \le -2.35) = .5000 - .4906 = .0094$$

The type II error probability is small, just as we wanted.

We can use the same procedure repeatedly to find a series of β values. Let us now find β, assuming that the true value of μ is 7, which is also included under H_a: $\mu > 6$. This probability is

$$\beta = P(\overline{X} \le 6.8225 \text{ assuming that } \mu = 7)$$

FIGURE 9.22

Calculation of β Assuming $\mu = 8$ and $\mu = 7$ for International Rubber Band Company Hypothesis Test (H_0: $\mu \leq 6$, H_a: $\mu > 6$, $\sigma = 3$, and $n = 36$)

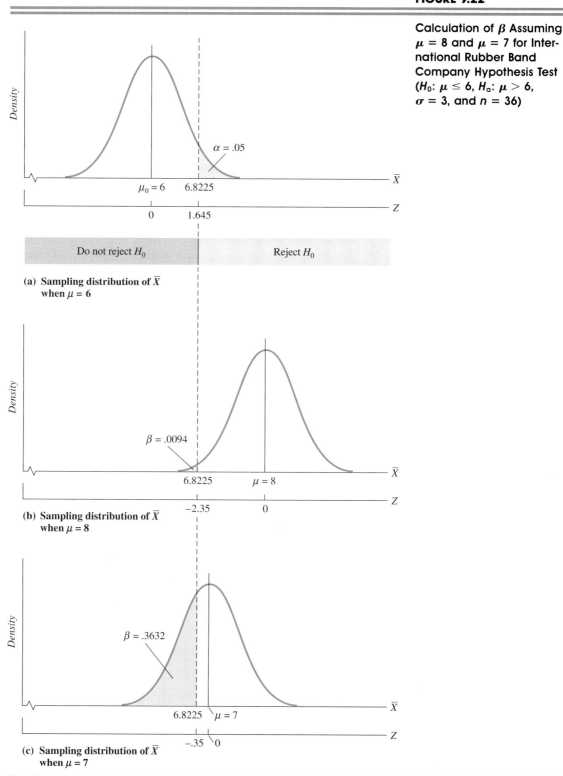

$\alpha = .05$

$\mu_0 = 6$ 6.8225

$\overline{X}$

Z

0   1.645

Do not reject H_0 Reject H_0

(a) Sampling distribution of $\overline{X}$ when $\mu = 6$

$\beta = .0094$

6.8225 $\mu = 8$

$\overline{X}$

Z

-2.35 0

(b) Sampling distribution of $\overline{X}$ when $\mu = 8$

$\beta = .3632$

6.8225 $\mu = 7$

$\overline{X}$

Z

$-.35$ 0

(c) Sampling distribution of $\overline{X}$ when $\mu = 7$

The sampling distribution of $\overline{X}$, assuming $\mu = 7$, and β are shown in Figure 9.22(c). The standardized value of $\overline{X} = 6.8225$ for this distribution is

$$Z = \frac{6.8225 - 7}{3/\sqrt{36}} = -.35$$

The standard normal table gives the area between $Z = 0$ and $Z = -.35$ as .1368. Thus the probability of a type II error is

$$\beta = P(\overline{X} \leq 6.8225 \text{ assuming that } \mu = 7)$$

$$= P(Z \leq -.35) = .5000 - .1368 = .3632$$

The error probability has increased somewhat.

In Table 9.7 we present β values for a series of μ values. Notice that the probability of a type II error decreases as the process fills boxes way above the intended weight of 6 ounces; this result is important to International because they do not want to overfill the boxes way above the intended weight. The β values increase as the process fills boxes closer to the intended weight of 6 ounces; but International is not too concerned about a small overfill as long as the boxes are filled close to the intended weight. ==

We also noted in discussing hypothesis tests that $1 - \beta$ is the **power** of a test. Since a value of β is the probability of a type II error, $1 - \beta$ is the probability of making a correct decision. It is the probability of rejecting H_0 when H_0 is false. Naturally, we want power to be high. Observe that if the error probability β is small then $1 - \beta$ or power will be large. If we let μ take different values under H_a, then we can find the corresponding $1 - \beta$ values. The different values of μ and the corresponding power values make up the **power function** of the test. A complete understanding of the power of a test can be obtained by graphing the power function to get a **power curve**. The height of the power curve gives us the probability of correctly rejecting H_0 when H_0 is false for any value of μ under H_a.*

== **EXAMPLE 9.12**   Values of power for the International Rubber Band Company's hypothesis test are also given in Table 9.7, and the graph of the paired values

*If H_0 is true, some analysts include the probability of rejecting H_0 as part of the power curve. An **operating characteristic curve** is sometimes used to display values of β (rather than $1 - \beta$ shown by the power curve) against corresponding values of μ. The height of the operating characteristic curve gives us the probability of not rejecting H_0 when H_0 is false. The operating characteristic curve and the power curve are merely different representations of the same information.

TABLE 9.7	Value of μ	$Z = \dfrac{6.8225 - \mu}{3/\sqrt{36}}$	Probability of a Type II Error, β	Power, $1 - \beta$
Type II Error Probabilities	6.001	1.645	.9500	.0500
for the International	6.2	1.25	.8944	.1056
Rubber Band Company.	6.5	.65	.7422	.2578
Hypothesis Test (H_0: $\mu \leq 6$,	6.8225	0	.5000	.5000
H_a: $\mu > 6$, $\alpha = .05$, and	7	−.35	.3632	.6358
$n = 36$)	7.5	−1.35	.0885	.9115
	7.8	−1.95	.0256	.9744
	8	−2.35	.0094	.9906

FIGURE 9.23

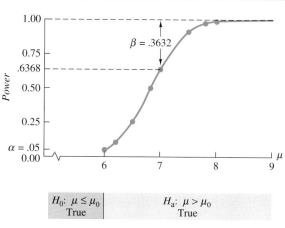

Power Curves for International Rubber Band Company Hypothesis Test (H_0: $\mu \leq 6$, H_a: $\mu > 6$, $\sigma = 3$, and $\alpha = .05$)

(a) **Power curve with $n = 36$**

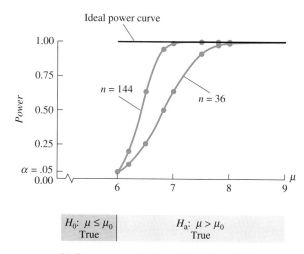

(b) **Power curves (with $n = 36$ and $n = 144$) and ideal power curve**

(μ, $1 - \beta$) in Figure 9.23(a) is the *power curve* for International's hypothesis test. The curve shows that the test gets more powerful as the process fills boxes way above the intended weight of 6 ounces. That is, power increases as the distance increases between the true value of μ under H_a and the value of μ specified under H_0. The power decreases as the process fills boxes closer to the intended weight. Figure 9.23(a) shows that power approaches α as μ approaches the value specified by H_0, implying that we can increase the power of a test if we are willing to increase α (or β decreases as α increases). The results for the power curve are important to International because they do not want to overfill the boxes way above the intended weight and the power is near 1 in this situation. International is not too concerned about a small overfill where the test is not as powerful.

The ideal power curve would have a height of 1 for any value of μ that falls under H_a. To get a power curve closer to the ideal, we can increase our sample size n. But, to increase the power or accuracy of our test by increasing n, we must increase the effort, time or expense required to obtain the larger sample. ≡

≡ **EXAMPLE 9.13**   To increase the power of the International Rubber Band Company hypothesis test of H_0: $\mu \le 6$ against H_a: $\mu > 6$ with $\sigma = 3$ and $\alpha = .05$, we increased the sample size to $n = 144$. The values of μ and power when $n = 36$ and $n = 144$ are shown in Figure 9.23(b), along with the ideal power curve. The figure shows that an increase in the sample size n results in an increase in the power of the test. ≡

We developed power curves for a one-sided test in our examples in this section. Two-sided tests have power curves that can be found by the three-step procedure we described. The power curve for a two-tailed test simply increases on both sides of the hypothesized value of μ. If the sample size is large, the assumption of a normal distribution can be relaxed in calculating β, due to the central limit effect, and s can be used for an unknown σ. It is also possible to find values of β for a t test, but an understanding of distributions that are beyond the scope of our discussion is required.

The properties of the power of a hypothesis test that we have shown in our examples in this section, assuming other conditions are held constant, are as follows:

1. The power of a test increases if the sample size n is increased.
2. The power of a test increases as the distance between the true value of μ under H_a and the value of μ specified under H_0 increases.
3. The power of a test increases if α is increased.

Problems: Section 9.12

90. The mean time between malfunctions for an electronic component produced by International Semiconductor Corporation has been 600 hours. A model using a new microprocessor was designed, on the average, to operate more than three times as long without malfunctioning. These supposedly superior pieces of equipment are put into service and give rise to $n = 61$ malfunctions. The population standard deviation is $\sigma = 980$ hours. The null hypothesis is that the new model does not operate more than three times as long without malfunctioning, and $\alpha = .05$. Use H_a: $\mu > 1800$ hours.

   a. What would it mean to International to make a type II error in this situation?
   b. What is the probability of making a type II error if the actual population mean operating time without malfunctioning is 2200 hours?
   c. Sketch the power curve for this test.

91. Five years ago a state's Department of Transportation did a study of automobile pollution at a particular location. During the month of November, it found that the mean pollution measurement was 132. During November of this year the department took a random sample of $n = 8$ measurements and examined the pollution measurement. The population standard deviation is $\sigma = 10$. The null hypothesis is that the population mean measurement is still 132 or more, and the alternative is that the mean is now less than 132. Use $\alpha = .05$. The pollution measurements are normally distributed.

   a. What would it mean to the department to make a type II error in this situation?
   b. What is the probability of making a type II error if the actual population mean is 126.184?
   c. Sketch the power curve for this test.

92. The times (in seconds) that it takes employees to assemble a toy truck at a Cole Industries assembly plant is normally distributed. At the 10% level, we want to test whether the mean assembly time for this toy truck is equal to 3 minutes. Use H_a:

$\mu \neq 180$ seconds as the alternative hypotheses. The population standard deviation is $\sigma = 12.471$ seconds and $n = 12$.

a. What would it mean to Cole to make a type II error in this situation?

b. What is the probability of making a type II error if the actual population mean is 185.92?

c. What is the probability of making a type II error if the actual population mean is 174.08?

d. Sketch the power curve for this test.

93. A financial analyst for James, Stephen & Associates maintains the historical view that takeover target firms have always experienced percentage sales growth with a mean of 23% or less. A sample of $n = 9$ firms was taken to the test the hypothesis that firms that are now takeover targets have increased their average percentage sales growth. Assume normality, and use a .05 level of significance. The population standard deviation is $\sigma = 3$.

a. What is the probability of making a type II error if the actual population mean is 24.645?

b. What is the probability of making a type II error if the sample size is $n = 36$ and the actual population mean is 24.645?

c. Sketch the power curves for $n = 9$ and $n = 36$ for this test.

9.13
TESTING THE HYPOTHESIS $\mu_1 = \mu_2$ WITH PAIRED DATA

In the *completely randomized experiment* for comparing two treatments, the treatments are randomized over, or randomly assigned to, the whole set of experimental units. That is, a random device is used to determine which of the two treatments is applied to any given experimental unit.

Usually, we restrict the randomization to the extent that each treatment is applied to the same number of units. As we have seen, the data from this type of experiment can be analyzed by using the two-sample techniques discussed in Section 9.8.

For a completely randomized experiment with equal population standard deviations the efficiency and the precision of estimation are inversely proportional to the *pooled estimate* of variance, s^2_{pooled}; the smaller the variance, the greater is the precision. To increase the precision of an experiment with a fixed amount of experimental material, we need to alter the design of the experiment in order to reduce the unexplained variation in the data as measured by s^2_{pooled}. Since s^2_{pooled} is based on the variation among experimental units treated alike, either we must reduce this variation by using more homogeneous experimental material, or we must eliminate some of this variation by using prior information about the responses of the experimental units. It is the latter approach we now consider.

If the experimental units occur in pairs or can be grouped into pairs in such a way that the variation in the responses between the members of any pair is less than the variation between members of different pairs, we can improve the efficiency of our experiment by randomizing the two treatments over the two members of each pair. We restrict our randomization so that each treatment is applied to one member of each pair; hence we obtain a separate estimate of the difference between the treatment effects from each pair, and the variation between or among the pairs is not included in our estimate of the variance. If the variation among the pairs of units is large relative to the variation within the pairs, the variance will be smaller than if a completely randomized design were used. In addition, pairing can reduce the number of units or subjects and result in a more economical study.

EXAMPLE 9.14 International Shipping, Inc., wants to compare the efficiencies of two different barnacle-resistant paints, and ten ships are available for participation in the test. We could randomly assign the two paints to five ships each and then, after a suitable length of time, obtain a measure of the weight of barnacles clinging to each ship. Since the variance would be based on the variation among ships painted alike, we could expect it to be large, because the ships would very likely be sailing in different waters for different periods of time.

However, where the port side of a hull goes, the starboard goes also. Therefore we would probably have a much more precise experiment if we were to paint one side of each hull with one paint and the other side with the second paint, tossing a coin to decide which side gets which paint. Each ship would then provide a measure of the difference in effectiveness of the two paints, and the variation from other factors, such as time at sea and the parts of the world where the ships sailed, would be eliminated from our estimate of variance.

These same considerations lead to the use of identical twins or of littermates in animal experiments and to grouping experimental units into pairs according to some factor such as age, weight, education, or environment. If the pairing is successful, much of the variation due to the factor on which it is based is eliminated from the estimate of variance, and the efficiency of the experiment is increased accordingly. The **paired experiment** is the simplest example of a class of experimental designs known as the **randomized complete block** designs. In this experimental design each pair is called a **block**.

The analysis of paired comparisons reduces a problem to the single-sample techniques discussed in the early sections of Chapter 8 and this chapter. The data for a paired comparison may be represented as shown in Table 9.8. In this table a single observation is denoted by X_{ij}, where the first subscript refers to the pair name and the second to the treatment name. Thus X_{32} refers to the observation produced by assigning a member of the third pair to the second treatment.

To estimate the mean difference and to test hypotheses about the difference between the treatment effects, we use the differences between the data values of each pair D_i, which are computed as follows.

Equation for Difference Between Data Values of ith Pair

$$D_i = X_{i1} - X_{i2} \tag{9.25}$$

We treat these as a random sample from the population of differences. We then have the usual equations for the **sample mean** and **sample standard deviation of the difference** between pairs.

Equations for Sample Mean and Sample Standard Deviation of Differences Between Pairs

$$\overline{D} = \frac{1}{n} \sum D_i \tag{9.26}$$

and

$$s_D = \sqrt{\frac{1}{n-1} \sum (D_i - \overline{D})^2} \tag{9.27}$$

Pair	Treatment 1	Treatment 2	Difference	**TABLE 9.8**
1	X_{11}	X_{12}	$D_1 = X_{11} - X_{12}$	**Paired Data and D_i**
2	X_{21}	X_{22}	$D_2 = X_{21} - X_{22}$	
3	X_{31}	X_{32}	$D_3 = X_{31} - X_{32}$	
$\vdots$	$\vdots$	$\vdots$	$\vdots$	
n	X_{n1}	X_{n2}	$D_n = X_{n1} - X_{n2}$	

The value of $\overline{D}$ is an unbiased estimate for μ_D, the mean difference, and the value of s_D^2 is an unbiased estimate for σ_D^2, the variance of the differences. The **standard deviation of the mean difference** $\overline{D}$ is estimated by $s_{\overline{D}}$ as follows.

> **Equation for Sample Standard Deviation of Sample Mean Paired Difference**
>
> $$s_{\overline{D}} = \frac{s_D}{\sqrt{n}} \qquad (9.28)$$

When the differences D_i constitute a single random sample from a normal population, the appropriate methods for finding confidence intervals and for testing hypotheses are the methods for a single normal population as considered in Chapter 8 and this chapter. Thus $100(1 - \alpha)\%$ confidence limits for μ_D are given by

$$L = \overline{D} - t_{\alpha/2, n-1} s_{\overline{D}} \qquad R = \overline{D} + t_{\alpha/2, n-1} s_{\overline{D}} \qquad (9.29)$$

And the test of the hypothesis $H_0: \mu_D = \mu_{D_0}$ is the single-sample t test with $n - 1$ degrees of freedom based on the statistic

$$t = \frac{\overline{D} - \mu_{D_0}}{s_{\overline{D}}} \qquad (9.30)$$

The test of the hypothesis of no difference in the effects of the two treatments is the special case $H_0: \mu_D = 0$ as summarized below.

> **Hypotheses on Mean Difference for Paired Comparisons and Rejection Rule for Two-Sided Test**
>
> $H_0: \mu_D = 0$
>
> $H_a: \mu_D \neq 0$
>
> Rejection rule: If $t < -t_{\alpha/2, n-1}$ or $t > t_{\alpha/2, n-1}$, then reject H_0;
>
> where: $t = \dfrac{\overline{D}}{s_{\overline{D}}}$

		log(DMC)		
TABLE 9.9	Sample, i	Before Treatment, X_{i1}	After Treatment, X_{i2}	Difference, $D_i = X_{i1} - X_{i2}$
Data for Kroft Foods, Inc.	1	6.98	6.95	0.03
	2	7.08	6.94	0.14
	3	8.34	7.17	1.17
	4	5.30	5.15	0.15
	5	6.26	6.28	−0.02
	6	6.77	6.81	−0.04
	7	7.03	6.59	0.44
	8	5.56	5.34	0.22
	9	5.97	5.98	−0.01
	10	6.64	6.51	0.13
	11	7.03	6.84	0.19
	12	7.69	6.99	0.70
			Total	3.10

 PRACTICE PROBLEM 9.14

For determination of whether or not a heat treatment is effective in reducing the number of bacteria in skim milk at Kroft Foods, Inc., counts were made before and after treatment on 12 samples of skim milk, with the results as shown in Table 9.9. The data are in the form of log(DMC), the logarithms of direct microscopic counts. The logarithmic function was used to transform the data because the logarithm of bacteria counts more closely approximates the normal distribution. The heat treatment is effective if the bacteria counts decrease after treatment, or if the difference after subtracting the after treatment counts from the before treatment counts is positive for the population. We want to test the null hypothesis that the heat treatment is not effective, H_0: $\mu_D \leq 0$, against the alternative hypothesis that the heat treatment effectively reduces the number of bacteria, H_a: $\mu_D > 0$. Let the significance level be .05.

Solution For these data,

$$\sum D_i = 3.10 \qquad \overline{D} = 3.10/12 = .258$$

$$\sum D_i^2 = 2.1990 \qquad \frac{(\sum D_i)^2}{n} = .8008$$

$$\sum (D_i - \overline{D})^2 = \sum D_i^2 - \frac{(\sum D_i)^2}{n} = 2.1990 - .8008 = 1.3982$$

$$s_D = \sqrt{\frac{1}{n-1}\sum (D_i - \overline{D})^2} = \sqrt{\frac{1}{11}(1.3982)} = \sqrt{.12711} = .357$$

$$s_{\overline{D}} = \frac{s_D}{\sqrt{n}} = \frac{.357}{\sqrt{12}} = .103$$

Step One H_0: $\mu_D \leq 0$.

Step Two H_a: $\mu_D > 0$.

FIGURE 9.24

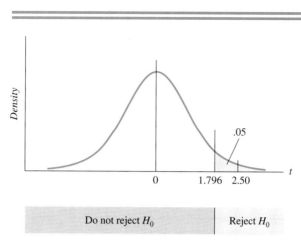

Hypothesis Test for Kroft Foods, Inc.

Step Three $\alpha = .05$, $n = 12$.

Step Four From the t table with $n - 1 = 12 - 1 = 11$ degrees of freedom, we find $t_{.05, 11}$ equals 1.796; the rejection rule is as follows:

Rejection rule: If $t > 1.796$, then reject H_0.

Step Five From the summary measures for the difference scores we have

$$t = \frac{\overline{D}}{s_{\overline{D}}} = \frac{.258}{.103} = 2.50$$

Step Six Since the sample t, 2.50, exceeds 1.796 (see Figure 9.24), we reject the null hypothesis and conclude that the heat treatment has reduced the number of bacteria.

A computer was also used to test the hypothesis of Kroft Food, Inc. Computer Exhibits 9.5A and 9.5B give the results for the test with paired data.

 PRACTICE PROBLEM 9.15

The training director of National Retail department stores wants to know whether a special one-day training program will help the chain's sales personnel improve customer service and reduce the number of complaints turned in by customers.

Solution To test the program's effectiveness, the training director could present the program to 100 salespeople. She could then compare the number of complaints these people get in the following month with the number of complaints received on another 100 salespeople who have not had the program presented to them. However, she can receive more information about the effectiveness of the program by making a paired comparison of the first 100 salespeople's complaint records *before* and *after* the training program.

Suppose that this procedure was followed, and the D_i values were recorded for the 100 salespeople as the difference between the number of complaints received in the month before and the month after the program. This pairing eliminates the variations in complaint

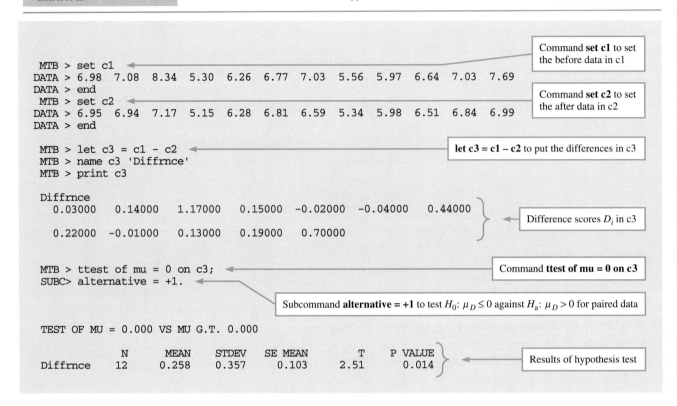

MINITAB   **COMPUTER EXHIBIT 9.5A** Hypothesis Test for Kroft Foods, Inc., with Paired Data

records between salespeople. The data showed $\overline{D} = 1.2$—that is, the group averaged 1.2 more complaints in the month prior to the training program. Also, $s_D^2 = 144$. Thus

$$s_{\overline{D}}^2 = 144/100 = 1.44 \qquad \text{and} \qquad s_{\overline{D}} = \sqrt{1.44} = 1.2$$

Step One $H_0: \mu_D \le 0.$

Step Two $H_a: \mu_D > 0.$

Step Three $\alpha = .05, n = 100.$

Step Four   Table VI in Appendix C does not give $t_{.05, 99}$, but interpolation of the .05 column between 60 and 120 degrees of freedom gives a value for t of 1.660. Thus our rejection rule is as follows:

Rejection rule: If $t > 1.660$, then reject H_0.

Step Five   From the D_i summary measures we have

$$t = \frac{1.2}{1.2} = 1.0$$

Step Six   Since our t value, 1.0, is less than 1.660, we do not reject H_0 and we cannot conclude that the training program reduces the number of complaints on salespeople.

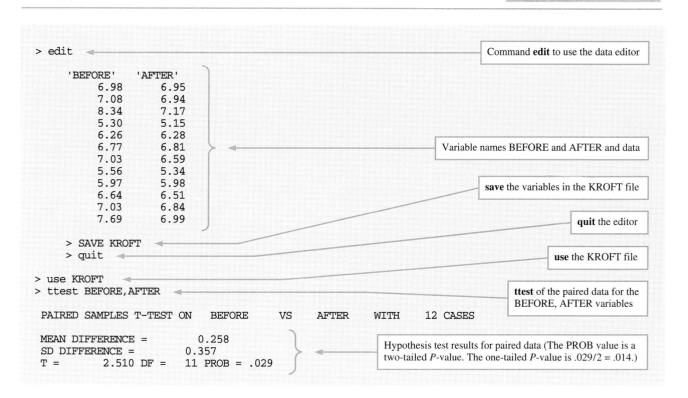

```
> edit                                          Command edit to use the data editor

    'BEFORE'   'AFTER'
       6.98      6.95
       7.08      6.94
       8.34      7.17
       5.30      5.15
       6.26      6.28
       6.77      6.81                            Variable names BEFORE and AFTER and data
       7.03      6.59
       5.56      5.34
       5.97      5.98                            save the variables in the KROFT file
       6.64      6.51
       7.03      6.84                            quit the editor
       7.69      6.99

    > SAVE KROFT
    > quit                                       use the KROFT file

> use KROFT
> ttest BEFORE,AFTER                             ttest of the paired data for the
                                                 BEFORE, AFTER variables

PAIRED SAMPLES T-TEST ON   BEFORE   VS   AFTER   WITH   12 CASES

MEAN DIFFERENCE =      0.258
SD DIFFERENCE =        0.357                      Hypothesis test results for paired data (The PROB value is a
T =      2.510 DF =   11 PROB = .029             two-tailed P-value. The one-tailed P-value is .029/2 = .014.)
```

Note that in Practice Problem 9.15 there is an unstated assumption that the salespeople served approximately the same number of customers in the two months for which the data were gathered. If this assumption were not true and if, say, the second month were much busier than the first, then the smaller number of complaints was spread over a larger number of customers, and the reduction in complaints might be significant after all. Under these circumstances the training director should do one of two things. First, the number of complaints per customer might be compared before and after the training program. Second, if records on the numbers of customers served are not available, then the complaint records of the trained 100 should be compared with those for another 100 untrained personnel over the same sales period.

Problems: Section 9.13

94. In a test of the effectiveness of two different sales approaches at Upjames, Inc., each of 16 commercial cleaning compound salespeople used alternately both of the two approaches for the same period of time and the same number of sales contacts. The sales of the compound (in pounds) for each salesperson (1–16) are given in the accompanying table.

   a. Why would a test with paired data be expected to provide a better test in this problem?

   b. Test the hypothesis that the mean improvement in sales is less than or equal to zero against the alternative that $\mu_1 - \mu_2 > 0$, using $\alpha = .05$.

Approach 1			Approach 2		
1	130	9	73	1	44	9	110
2	120	10	56	2	62	10	38
3	61	11	65	3	77	11	66
4	111	12	71	4	58	12	120
5	93	13	109	5	88	13	81
6	56	14	122	6	101	14	54
7	25	15	85	7	42	15	31
8	123	16	131	8	57	16	11

95. In a comparison of the average weight gains of pigs fed two different rations at Colorado Agronomics, Inc., nine pairs of pigs were used. The pigs within each pair were littermates, the rations were assigned at random to the two animals within each pair, and they were individually housed and fed. The gains (in pounds) after 30 days are given in the accompanying table. Test the hypothesis that the difference in mean gains is less than or equal to zero against the alternative that feed A produces a larger gain. Let $\alpha = .01$.

	Litter									
Ration	*1*	*2*	*3*	*4*	*5*	*6*	*7*	*8*	*9*	*Sum*
A	60	38	39	49	49	62	53	42	58	450
B	53	39	29	41	47	50	56	47	52	414

96. At National Appliances, Inc., ten kitchen timers were timed in two positions: vertical and at 20 degrees from vertical. The values (in seconds) are given in the accompanying table. Use paired data methods to find the value of t for testing the hypothesis that the mean time is the same in both positions. Use $H_a: \mu_1 \neq \mu_2$ and $\alpha = .10$.

	Timer									
Position	*1*	*2*	*3*	*4*	*5*	*6*	*7*	*8*	*9*	*10*
Vertical	170	191	205	181	210	192	183	205	185	216
Tipped	160	197	175	181	163	172	177	185	183	177

97. The management in a large assembly plant wishes to make changes in assembly techniques, but union officials are afraid that the changes will result in lower wages for their workers, who are paid on the basis of output. Thus management set up a test, using a randomly selected group of 14 employees who had varying levels of experience in the assembly process. The assembly process was run by using the old assembly techniques; and then, after a suitable training period, it was run by using the new techniques. The hourly wages earned by the 14 employees using the two techniques are given in the accompanying table. Is there sufficient evidence to indicate that the labor force will be able to make more money by using the new techniques? Test the hypothesis that the mean wage difference $\mu_1 - \mu_2$ is zero or less against the alternative hypothesis that the difference is positive. Use paired data, and let the probability of a type I error be .10.

| | Technique | | | | Technique | |
Employee	New	Old	Employee	New	Old
1	4.61	3.96	8	8.12	7.35
2	6.33	6.13	9	4.75	4.13
3	7.38	8.21	10	6.43	6.37
4	6.87	6.05	11	7.38	7.08
5	6.62	5.21	12	6.45	7.56
6	6.82	5.25	13	6.16	6.24
7	6.87	6.32	14	8.14	8.05

9.14 GROUPS VERSUS PAIRS

Paired comparisons can give greater precision of estimation than group comparisons, but only if the pairing is effective. To be effective, the pairs must be such that the variation among the pairs is greater than the variation between the units within the pairs. The degrees of freedom for the t test based on paired comparisons are equal to $n - 1$, compared with $2(n - 1)$ for the group comparison based on the same number of experimental units. Since degrees of freedom are like money in the bank, they should not be invested unless a suitable return can be expected.

With pairing, then, as compared with grouping, we lose $n - 1$ degrees of freedom. But if the variance is reduced enough to more than compensate for their loss, then a gain in efficiency is achieved. If, however, the experimental material is nearly homogeneous, we have no justification for pairing, because the variation among pairs will be but little greater (if at all) than that within the pairs. We would squander degrees of freedom without getting a sufficient reduction in the variance.

For data from experiments involving two treatments, the proper method of analysis must be employed. If the data are from a paired experiment, and if we ignore the pairing and use the group analysis of Section 9.8.1, the estimate of variance is inflated by the variation among pairs and will be too large. A difference between the treatment effects that is actually significant might not be detected because of the inflated variance estimate.

EXAMPLE 9.16   To illustrate what can happen when the two-sample techniques are used to analyze paired data, we use the data of Kroft Foods, Inc., of Practice Problem 9.14. The summary table is as follows:

Sample	n	df	$\overline{X}_j$	Sum of Squares
Before	12	11	6.721	8.0997
After	12	11	6.463	4.7750
Total	24	22		12.8747

From the information in the summary table, we find

$$s^2_{pooled} = \frac{12.8747}{22} = .5852$$

and

$$s_{\overline{X}_1 - \overline{X}_2} = \sqrt{s^2_{pooled}\left(\frac{1}{n_1} + \frac{1}{n_2}\right)} = \sqrt{.5852\left(\frac{1}{12} + \frac{1}{12}\right)} = \sqrt{.0975} = .312$$

For the hypothesis test, we have the following steps.

Step One $H_0: \mu_1 \le \mu_2$.

Step Two $H_a: \mu_1 > \mu_2$.

Step Three $\alpha = .05$, $n_1 = 12$, $n_2 = 12$.

Step Four   Since $n_1 + n_2 - 2 = 12 + 12 - 2 = 22$ and $t_{.05, 22} = 1.717$, we have the following rejection rule:

Rejection rule: If $t > 1.717$, then reject H_0.

Step Five   The t value is

$$t = \frac{\overline{X}_1 - \overline{X}_2}{s_{\overline{X}_1 - \overline{X}_2}} = \frac{.258}{.312} = .83$$

Step Six   Since .83 is less than 1.717, we do not reject H_0. Using this analysis from Section 9.8.1 without pairing, we would conclude that the treatment had no effect on the bacteria in the skim milk; whereas in Section 9.13 with pairing, we concluded that the treatment did have an effect in lowering the bacteria.

Notice that the standard deviation of the difference between the means here is over three times as large as when these paired data were analyzed by the proper pairing techniques of Section 9.13 (.312 here and .103 in Practice Problem 9.14).   ═══

Just as we have run into trouble by using group techniques to analyze paired data, we are also in trouble if the data from a group experiment are paired. Random pairing would result in a loss of degrees of freedom, and at the same time, the variance estimate could be either larger or smaller than it should be and could lead to the wrong conclusion.

Suppose we were to systematically pair the results of a completely randomized experiment by ranking both sets of responses from high to low and then pairing the largest of each group, the next largest, and so on. Such a procedure would generally result in a variance estimate that is much smaller than it should be and, therefore, in t values that are too large. We might well detect a difference that does not exist.

The design of the study dictates the method of analysis that should be used. The reader will have an opportunity to examine situations where either group or paired techniques are appropriate in the problems at the end of this section.

Problems: Section 9.14

98. Assume that the data on the effectiveness of two different sales approaches at Up-james, Inc., presented in Problem 94 came from 32 different salespeople, 16 of whom were randomly assigned to sales approach 1 and 16 of whom were randomly assigned to sales approach 2.

   a. Test the hypothesis that the mean difference in sales is less than or equal to 0 against the alternative that $\mu_1 - \mu_2 > 0$, using $\alpha = .05$, with $\sigma_1 = \sigma_2$.

   b. What conclusion is reached in this analysis compared with that for Problem 94?

Approach 1		Approach 2	
130	73	44	110
120	56	62	38
61	65	77	66
111	71	58	120
93	109	88	81
56	122	101	54
25	85	42	31
123	131	57	11

99. Assume that the pigs at Colorado Agronomics described in Problem 95 are not litter-mates; rather, they constitute two groups that were randomly assigned to feed A and feed B and that attained the weight gains (after 30 days) given in the accompanying table.

    a. Test the hypothesis that the difference between the mean gains is less than or equal to 0 against the alternative that feed A produces a larger gain. Let $\alpha = .01$ with $\sigma_1 = \sigma_2$.
    b. What conclusion is reached in this analysis compared with that for Problem 95?

Ration A	60	38	39	49	49	62	53	42	58
Ration B	53	39	29	41	47	50	56	47	52

100. Assume that 20 National Appliances kitchen timers were used in Problem 96. Ten were timed at a vertical position and ten were timed at 20 degrees from vertical. The position assignments were made randomly, and the results (in seconds) are given in the accompanying table. Use H_a: $\mu_1 \neq \mu_2$ and $\alpha = .10$ with $\sigma_1 = \sigma_2$.

    a. Treat the data as two groups, and calculate t for testing the hypothesis that the mean time is the same in both positions.
    b. What conclusion is reached in this analysis compared with that for Problem 96?

Vertical	170	191	205	181	210	192	183	205	185	216
Tipped	160	197	175	181	163	172	177	185	183	177

Analysis of Variance

In Section 9.8, we discussed testing the hypothesis that two populations have the same mean value. The hypothesis was stated as H_0: $\mu_1 = \mu_2$ and was tested by calculating a t value from two samples.

In this chapter, we discuss *analysis of variance* (ANOVA), a method that allows us to extend the two-sample test of hypothesis for continuous populations with common standard deviations covered in Chapter 9 to several, say k, populations. The null hypothesis for one-way, or one-factor, analysis of variance is H_0: $\mu_1 = \mu_2 = \mu_3 = \cdots = \mu_k$, and the alternative hypothesis is H_a: One or more of the population means is not equal to the others. If we reject the hypothesis of several equal means in an analysis of variance, then we can use multiple pairwise comparisons of means to examine which means are different.

In Section 10.4, we discuss partitioning the sum of squares. We also use partitioning of a sum of squares in Chapters 12 and 13.

10.1
ANALYSIS OF VARIANCE CONCEPTS AND COMPUTATIONS

A test of the equality of several means is required in many applied problems. For example, Sierra Pacific Electric Company uses water-cooled turbines at one of its power-generating plants. If the water used in the cooling system is too polluted, then the system will become corroded. For this reason filters are used to reduce the pollution before the water enters the system. Managers at Sierra Pacific want to test the effectiveness of four different types of filters in reducing water pollution.

To test the equality of several means, we use a procedure known as the **analysis of variance.**

> **Definition**
>
> **Analysis of variance (ANOVA),** or **one-factor analysis of variance,** is a procedure to test the hypothesis that several populations have the same mean.

The name *analysis of variance* stems from the somewhat surprising fact that a set of computations on several variances is used to test the equality of several means.

To conduct an analysis of variance and test the null hypothesis $H_0: \mu_1 = \mu_2 = \mu_3 = \cdots = \mu_k$, we must take independent random samples from each of the populations and obtain sample means and sample variances (or standard deviations). We will illustrate the method with the Sierra Pacific Electric Company example.

To test the effectiveness of the four different types of filters in reducing water pollution, managers at Sierra arranged for independent random samples of size $n = 3$ containers of filtered water to be taken from the cooling system when it was fitted with the four types of filters. The sample water pollution count values are shown in Table 10.1, and they are arranged so that the three containers of water (items) are in rows and the four filter types (groups) are in columns. We want to test the hypothesis that the mean pollution count is the same for all four filters. That is, we wish to test $H_0: \mu_1 = \mu_2 = \mu_3 = \mu_4$ against the alternative hypothesis H_a: One or more of the filters produce different pollution counts.

We can summarize the sample results for the pollution counts as shown in Table 10.2. The sample mean for the jth group (column) is $\overline{X}_j$, so the four sample means are denoted $\overline{X}_1, \overline{X}_2, \overline{X}_3$, and $\overline{X}_4$. To compute the sample mean for the jth group (column), we sum the values in the column to find the total, and divide the column total by the number of items in that column:

$$\overline{X}_j = \frac{\sum\limits_{i=1}^{n_j} X_{ij}}{n_j} \tag{10.1}$$

	Group			
TABLE 10.1	*Filter 1*	*Filter 2*	*Filter 3*	*Filter 4*
Pollution Counts at Sierra	10	11	13	18
Pacific, Four Groups of	9	16	8	23
Size 3	5	9	9	25

	Group			
	Filter 1	*Filter 2*	*Filter 3*	*Filter 4*
	10	11	13	18
	9	16	8	23
	5	9	9	25
Total	24	36	30	66
	$n_1 = 3$	$n_2 = 3$	$n_3 = 3$	$n_4 = 3$
	$\overline{X}_1 = 8$	$\overline{X}_2 = 12$	$\overline{X}_3 = 10$	$\overline{X}_4 = 22$
	$s_1^2 = 7$	$s_2^2 = 13$	$s_3^2 = 7$	$s_4^2 = 13$

$\overline{\overline{X}} = 13$

TABLE 10.2

Summary of Results for Pollution Counts at Sierra Pacific

where: X_{ij} = the value of the dependent variable for the ith item (row) in the jth group (column)

n_j = the number of items (rows) in the jth group (column)

For instance, $j = 1$ for filter 1 (or column 1 in Table 10.1) and $n_1 = 3$. Thus,

$$\overline{X}_1 = \frac{10 + 9 + 5}{3} = \frac{24}{3} = 8$$

The other sample means were determined by the same method. The $\overline{\overline{X}}$ value is the **grand mean** of all 12 sample values in the table.

> **Definition**
>
> The **grand mean** $\overline{\overline{X}}$ is the overall sample mean for all of the observations included in the study. To find the value of the grand mean, add all of the sample values from each of the populations, and divide the resulting sum by the total number of observations in the study, or
>
> $$\overline{\overline{X}} = \textit{the grand mean} = \frac{\sum\limits_{i=1}^{n_j} \sum\limits_{j=1}^{k} X_{ij}}{n_T} \tag{10.2}$$
>
> where: k = the number of populations being tested or the number of groups (columns)
>
> n_T = the total number of items sampled
>
> $\quad\;\; = n_1 + n_2 + \cdots + n_k \tag{10.2a}$

For the pollution count data of Table 10.1 the total number of observations is

$$n_T = n_1 + n_2 + n_3 + n_4 = 3 + 3 + 3 + 3 = 12$$

and the sum of all 12 pollution counts is

$$\sum_{i=1}^{n_j} \sum_{j=1}^{k} X_{ij} = 10 + 11 + 13 + \cdots + 25 = 156$$

So the overall mean or grand mean is

$$\overline{\overline{X}} = \frac{\sum\limits_{i=1}^{n_j} \sum\limits_{j=1}^{k} X_{ij}}{n_T} = \frac{156}{12} = 13$$

The values in the bottom row of Table 10.2 are the sample variances for each group. The sample variance for the jth group (column) can be computed as follows:

$$s_j^2 = \frac{\sum\limits_{i=1}^{n_j} (X_{ij} - \overline{X}_j)^2}{n_j - 1} \tag{10.3}$$

For instance, the column for filter 3 is $j = 3$, and we find s_3^2 by determining the sample variance of the three values in column 3 around their sample mean of $\overline{X}_3 = 10$. That is,

$$\begin{aligned} s_3^2 &= \frac{\sum\limits_{i=1}^{n_3} (X_{i3} - \overline{X}_3)^2}{n_3 - 1} \\ &= \frac{(13 - 10)^2 + (8 - 10)^2 + (9 - 10)^2}{3 - 1} = \frac{14}{2} = 7 \end{aligned}$$

The other sample variances were determined by the same method.

The hypothesis that all four of these samples came from populations with the same mean can be tested under the following conditions:

1. All the samples were **randomly** selected and are **independent** of one another.
2. The populations from which the samples were drawn are **normally distributed.**
3. All the populations have the **same variance** σ^2.

For the Sierra Pacific filter example these conditions mean that the containers of water were selected randomly and that the pollution count for any one container is not influenced by the pollution count for other containers. In addition, pollution counts are normally distributed and have equal variances from one type of filter to the next.

We test the hypothesis that all the filters produce the same mean by examining the third condition more closely. The following paragraphs outline a general procedure for testing $H_0: \mu_1 = \mu_2 = \mu_3 = \mu_4$.

If the third condition can be assumed true, then we might ask, What is the value of σ^2? What is the value of the variance that is common to all k populations? There are *two* ways of estimating the population variance σ^2.

To find the *first* estimate of the common population variance, use the now familiar relationship

$$\text{Standard deviation}(\overline{X}) = \sigma_{\overline{X}} = \sigma/\sqrt{n}$$

That is, if many samples of size n were taken from a population with a variance of σ^2 and if many sample means $\overline{X}_j$'s were examined, we would find that the sample standard deviation of the many sample means would be equal to the standard deviation of the population divided by the square root of the sample size.

In our example we did not take *many* samples; we took only four samples. But from the four samples we got four means: $\overline{X}_1$, $\overline{X}_2$, $\overline{X}_3$, and $\overline{X}_4$. The four sample means can be used to estimate the value $\sigma_{\overline{X}}$. If we have this value, then we can estimate the population variance σ^2, because

$$\sigma_{\overline{X}} = \sigma/\sqrt{n}$$

Or squaring both sides of the equation, we get

$$\sigma_{\overline{X}}^2 = \sigma^2/n \quad \text{and} \quad n\sigma_{\overline{X}}^2 = \sigma^2$$

Thus n times the variance of the sample means around their grand mean is the **between-groups** estimate s_B^2 of the population variance of pollution counts, as follows.

Equation for Between-Groups Variance Estimate When $n_1 = n_2 = \cdots = n_k = n$

$$s_B^2 = \frac{n[(\overline{X}_1 - \overline{\overline{X}})^2 + (\overline{X}_2 - \overline{\overline{X}})^2 + \cdots + (\overline{X}_k - \overline{\overline{X}})^2]}{k - 1} \tag{10.4}$$

$$= \frac{n\sum\limits_{j=1}^{k}(\overline{X}_j - \overline{\overline{X}})^2}{k - 1}$$

This s_B^2 variance estimate is called the *between-groups variance estimate* because it is based on the variation from one sample mean to the next, or the variation *between* the samples. The s_B^2 figure has $k - 1$ degrees of freedom.

EXAMPLE 10.1   In the Sierra Pacfic pollution count example we calculate the between-groups variance estimate as follows:

$$s_B^2 = \frac{n\sum\limits_{j=1}^{k}(\overline{X}_j - \overline{\overline{X}})^2}{k - 1}$$

$$= \frac{3[(8 - 13)^2 + (12 - 13)^2 + (10 - 13)^2 + (22 - 13)^2]}{4 - 1}$$

$$= \frac{348}{3} = 116$$

We will return to this value when we complete the test of $H_0: \mu_1 = \mu_2 = \mu_3 = \mu_4$.

The *second* method of estimating the population variance is straightforward. Since all the samples have sample variances, s_j^2, these variances should be good estimates of the common population variance. Thus, we can take the weighted average of all the sample variances. Note that when we pooled the sample variances in Chapter 9, we were just taking the weighted average of the two-sample variances. Here we will proceed in the same fashion and estimate the common population variance as the **within-groups** estimate of variance s_W^2 as given by the weighted average of the sample variances, using the number of degrees of freedom in each sample, $n_j - 1$, as the weights.

Equation for Within-Groups Variance Estimate

$$s_W^2 = \frac{\sum_{j=1}^{k} (n_j - 1)s_j^2}{n_T - k} \qquad (10.5)$$

The value s_W^2 is the second estimate of the population variance. It is called the *within-groups variance estimate* (that is the reason for the subscript W) because the sample variances measure the variation *within* each sample, and this estimate is the weighted average of those within-groups estimates. The s_W^2 figure has $n_T - k$ degrees of freedom.

EXAMPLE 10.2   In our Sierra Pacific example the within-groups variance estimate is

$$
\begin{aligned}
s_W^2 &= \frac{\sum_{j=1}^{k} (n_j - 1)s_j^2}{n_T - k} \\[2mm]
&= \frac{(3-1)(7) + (3-1)(13) + (3-1)(7) + (3-1)(13)}{12 - 4} \\[2mm]
&= \frac{14 + 26 + 14 + 26}{8} = \frac{80}{8} = 10
\end{aligned}
$$

10.2
TESTING THE HYPOTHESIS OF SEVERAL EQUAL MEANS

The key to testing the hypothesis that all the k populations have the same mean lies in the relationship between the two estimates of variance, s_B^2 and s_W^2. We can think of the second estimate, s_W^2, as being a more reasonable estimate of the variance of all the populations because we assumed all the populations had the same variance, and the weighted average of the sample variances is a good estimate of that common variance's value.

However, the first estimate of the population's variance, s_B^2, is somewhat suspect since it is based on the notion that all the populations have the same mean. That is, the estimate s_B^2 is a good estimate of the population variance *only if* the hypothesis is true and all the populations' means are equal: $\mu_1 = \mu_2 = \cdots = \mu_k$.

Let's assume for a moment that, unknown to us, all the population means are not equal. In fact, let's assume that the populations' means, the μ_j values, are *radically* different from one another. Then the sample means, the $\overline{X}_j$ values, will most likely be radically different from each other too. This difference will have a marked effect on the first variance estimate, s_B^2, of Equation (10.4). That is to say, the $\overline{X}_j$ values will vary a great deal and the $(\overline{X}_j - \overline{\overline{X}})^2$ terms will be large. Thus, if the population means are *not* all equal, then the s_B^2 variance estimate will be large relative to the s_W^2 estimate. That is, if the between-groups variance estimate is large relative to the within-groups variance estimate, the hypothesis that all the populations have the same mean is not likely to be true.

The important question is, of course, How large is "large?" Also, how do we measure the relative sizes of the two variance estimates? The answer to these questions is given to us by the F distribution introduced in Section 9.9. If k samples of n_j $(j = 1, \ldots, k)$ items each are taken from k normal populations that have equal variances and for which the hypothesis $H_0: \mu_1 = \mu_2 = \cdots = \mu_k$ is true, then the ratio of the between-groups estimate of variance to the within-groups estimate of variance is an F value that follows an F probability distribution.

Equation for ANOVA F Statistic

$$F = \frac{s_B^2}{s_W^2}$$

(10.6)

Each F value has two different degrees-of-freedom figures associated with it: ν_1 is the number of degrees of freedom in the numerator's variance estimate, and ν_2 is the number of degrees of freedom in the denominator's variance estimate. Thus for the F ratio in Equation (10.6),

$$\nu_1 = k - 1$$

(10.7)

and

$$\nu_2 = n_T - k$$

(10.8)

A set of samples gives a large F value only if s_B^2 is large relative to s_W^2. But this condition exists only if the sample means, the $\overline{X}_j$ values, differ greatly from one another. And this condition usually exists only if the population means, the μ_j values, differ greatly from one another. This point is illustrated in the results of a sampling exercise as presented in Figures 10.1(a) and 10.1(b). Figure 10.1(a) shows three normal populations with equal variance σ^2 and *unequal* means $\mu_1 < \mu_2 < \mu_3$. Three samples of size $n = 50$ each were taken, one from each of the three populations, and the sample results are presented as histograms. The value of s_W^2 is the weighted average spread or dispersion of the histograms. The dispersions of the histograms in Figure 10.1(a) [and Figure 10.1(b)] are all very similar since all of the populations have the same variance. Notice that the three sample means $\overline{X}_1$, $\overline{X}_2$, and $\overline{X}_3$ of Figure 10.1(a) are substantially different and that s_B^2 increases as the sample means get further apart. However, the dispersions of the histograms are all very similar, so s_W^2 does *not* increase as the sample means get further apart. Thus in this case, where $\mu_1 < \mu_2 < \mu_3$, the value of s_B^2 will be large relative to s_W^2.

Figure 10.1(b) shows three normal populations with equal variance σ^2 and *equal* means $\mu_1 = \mu_2 = \mu_3$. Three samples of size $n = 50$ each were taken, one sample from each of these three populations, and the results are presented as histograms.

FIGURE 10.1

Analysis of Variance When
Population Means Are *Not*
Equal and When Population
Means Are Equal

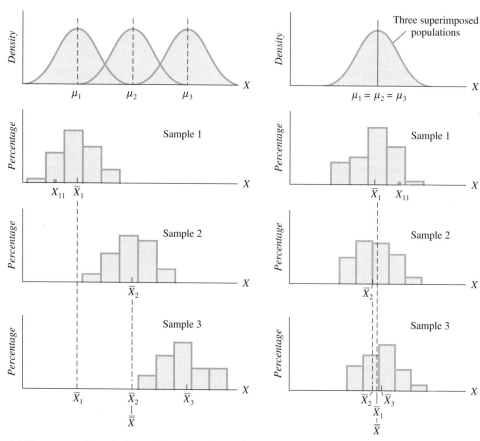

(a) **Three normal populations with equal variance σ^2
and unequal means $\mu_1 < \mu_2 < \mu_3$, and sample
results**

(b) **Three normal populations with equal
variance σ^2 and equal means $\mu_1 = \mu_2 = \mu_3$,
and sample results**

Notice that the three sample means $\overline{X}_1, \overline{X}_2$, and $\overline{X}_3$ of Figure 10.1(b) are very similar. Thus, in this case, where $\mu_1 = \mu_2 = \mu_3$, the value of s_B^2 will *not* be very large relative to s_W^2.

In sum, then, a large F value, where $F = s_B^2/s_W^2$, usually indicates that the null hypothesis $H_0: \mu_1 = \mu_2 = \cdots = \mu_k$ is false. Hence this test is always a one-sided test, with the rejection region always in the upper tail of the F distribution.

The analysis will result in either of two conclusions: If the calculated F ratio is not larger than the table F_{α, ν_1, ν_2} value, then the conclusion is that there is not sufficient evidence to indicate that one or more of the population means is not equal to the others. If the calculated F ratio is larger than the table F_{α, ν_1, ν_2} value, then the conclusion is that one or more of the population means is not equal to the others. The procedure for **testing the equality of several means** is summarized as follows.*

*An equivalent way to specify the alternative hypothesis is

H_a: Not all μ_j are equal.

> **Hypotheses on Equality of Several Population Means and Rejection Rule for ANOVA**
>
> $H_0: \mu_1 = \mu_2 = \mu_3 = \cdots = \mu_k.$
>
> H_a: One or more of the population means is not equal to the others.
>
> Rejection rule: If $F > F_{\alpha, \nu_1, \nu_2}$, then reject H_0;
>
> where: $F = \dfrac{s_B^2}{s_W^2}$
>
> $\nu_1 = k - 1$
>
> $\nu_2 = n_T - k$

EXAMPLE 10.3   For the Sierra Pacific Power filter problem, the pollution count data, and with a level of significance of .05, we now want to perform the following test.

Step One $H_0: \mu_1 = \mu_2 = \mu_3 = \mu_4.$

Step Two H_a: One or more of the population means is not equal to the others.

Step Three $\alpha = .05, n_1 = 3, n_2 = 3, n_3 = 3, n_4 = 3.$

Step Four   To determine the rejection rule, we must consult the F probability distribution table, Table VII of Appendix C. In that table we find the F values that are exceeded by chance with only .05 probability for various combinations of degrees of freedom in the numerator and the denominator. In our example $\nu_1 = k - 1 = 4 - 1 = 3$, $\nu_2 = n_T - k = 12 - 4 = 8$, and $\alpha = .05$. From Table VII we find that $F_{\alpha, \nu_1, \nu_2} = F_{.05, 3, 8} = 4.0662$, and we have the following rejection rule.

Rejection rule: If F > 4.0662, then reject H_0.

Step Five   For the Sierra Pacific Power pollution count data of Table 10.1 and the variance estimates $s_B^2 = 116$ and $s_W^2 = 10$ that we found in the previous section, the value of the F ratio is

$$F = \frac{s_B^2}{s_W^2} = \frac{116}{10} = 11.6$$

Step Six   The sample F of 11.6 exceeds 4.0662 (see Figure 10.2), so we reject H_0. We conclude that one or more of the filters produce mean pollution counts different from the others.

We can see why this procedure for testing the hypothesis that several populations have the same mean is called *analysis of variance*. We reach our conclusion about the hypothesis by *analyzing* the *variances* within the samples and between the samples.

The results of an analysis of variance are usually given in a table known as an **analysis of variance** or **ANOVA table** such as Table 10.3. The purpose of such a table is to provide a *summary* of the results of an analysis of variance. An ANOVA

FIGURE 10.2

The *F* Distribution for Testing
the Equality of the Mean
Water Pollution Counts

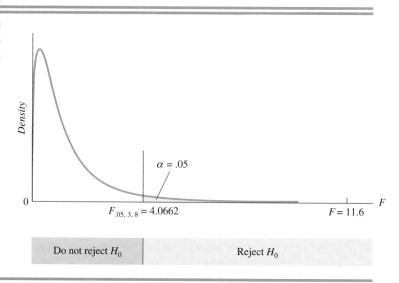

table shows the sources of variation that are between the groups and that are within the groups. The number of degrees of freedom for each of the variance estimates is shown in the second column. Variance estimates are obtained by dividing a sum of squares by its degrees of freedom. Thus, the *SS* column in Table 10.3 shows the sum of squares associated with each of the variance estimates. In other words, the *SS* values are the numerators of Equations (10.4) and (10.5). The fourth column shows the two variance estimates (which are also known as *mean squares*). These figures are simply s_B^2 and s_W^2, respectively, and can be found by dividing the *SS* column values by the *df* column values. The sample *F* is the ratio of the two variance estimates s_B^2/s_W^2, and its value is usually presented in the table. The table also conveniently displays the degrees of freedom for each of the variance estimates, which facilitates looking up the F_{α, ν_1, ν_2} value in Table VII.

Problems: Sections 10.1–10.2

Answers to odd-numbered problems are given in the back of the text.

 1. An administrator at Mood's Investors Service, Inc., is interested in the ratio of municipal debt to real estate value for cities that have been assigned bond ratings of *AA*, *A*, and *B*. Random samples of debt to real estate value (as percentages) for three cities were taken from each bond rating. The results are presented in the accompanying table.

   a. Find the grand mean $\overline{\overline{X}}$ for all of the sample data, the sample group means $\overline{X}_1$, $\overline{X}_2$, and $\overline{X}_3$, and the sample variances for each group, s_1^2, s_2^2, and s_3^2.

TABLE 10.3

Analysis of Variance Table

Source	df	SS	Variance or Mean Square	F
Between groups	3	348	116	11.6
Within groups	8	80	10	
Total	11	428		

b. Find the between-groups estimate of variance.
c. Find the within-groups estimate of variance.
d. Test the hypothesis that the mean debt-to-real-estate-value percentages are the same for all three bond ratings. Use .05 for the level of significance.

AA	A	B
2	1	9
4	6	10
6	5	11

2. Mutual funds invest in stocks and other investments in an attempt to meet investor goals. An investor studied the percentage rates of return of a growth fund, an income fund, a stability fund, and a liquidity fund. Random samples of percentage rates of return for three periods were taken from each fund. The results are presented in the accompanying table.

   a. Find the grand mean, $\overline{\overline{X}}$ for all of the sample data, the sample group means $\overline{X}_1$, $\overline{X}_2$, $\overline{X}_3$, and $\overline{X}_4$, and the sample variances for each group, s_1^2, s_2^2, s_3^2, and s_4^2.
   b. Find the between-groups estimate of variance.
   c. Find the within-groups estimate of variance.
   d. Test the hypothesis that the mean percentage rates of return are the same for all four funds. Use .05 for the level of significance.

Growth Fund	Income Fund	Stability Fund	Liquidity Fund
13	9	4	8
14	4	6	6
15	5	8	4

3. The accompanying table shows the sales figures (in hundreds of dollars) for three randomly selected days for four salespeople.

   a. What are the n_j values for this problem?
   b. What is the k value in this problem?
   c. Use analysis of variance to test the hypothesis that the mean daily sales figures are the same for all four salespeople. That is, test H_0; $\mu_1 = \mu_2 = \mu_3 = \mu_4$, and use $\alpha = .01$.

	Salesperson			
	1	*2*	*3*	*4*
	15	9	17	13
	17	12	20	12
	22	15	23	17
$\overline{X}_j$	18	12	20	14
s_j^2	13	9	9	7

4. Kroft Foods Company tested three packaging materials for moisture retention by storing the same food product in each of them for a fixed period of time and then determining the moisture loss. Each material was used to wrap 10 food items. The results are given in the accompanying table.

a. Construct the analysis of variance table.
b. Can we reject the hypothesis that the materials are equally effective? Let $\alpha = .05$.

	Material 1	Material 2	Material 3
Number of Packages	10	10	10
Mean Loss	224	232	228
Sample Variance	30	40	38

10.3
ANALYSIS OF VARIANCE COMPUTATIONS FOR UNEQUAL SAMPLE SIZES

In this section, we consider a situation where there are three populations but the samples taken from each are different in size. The following Practice Problem uses the analysis of variance method of the previous section with samples of unequal size.

 PRACTICE PROBLEM 10.1

A regional fabric store chain owned by Pierpoint Industries consisted of 6 stores. Three years ago the chain acquired two other groups of stores. One of these, constituting group 2 of the enlarged chain, contained 4 stores, and the other (group 3) contained 5 stores. Turnover of salespeople is a problem at all 15 stores. The number of salespeople who quit at each of the stores last year is shown in the accompanying table. Assume, for simplicity, that all the stores have the same-size sales force to start with. The employee compensation and benefit policies of the three groups of stores have not been standardized, and each group of stores has retained its old policies. The personnel manager of the entire 15-store chain is interested in knowing whether there is a real difference in the turnover of personnel in the three groups of stores. If there is, he will consider standardizing compensation and benefits soon.

Group 1	Group 2	Group 3
10	6	14
8	9	13
5	8	10
12	13	17
14		16
11		

Solution We can examine this problem by using the analysis of variance method for unequal-sized groups. The hypothesis we will test is $H_0: \mu_1 = \mu_2 = \mu_3$. That is, the mean annual turnover is the same for all three groups of stores. Before we proceed, we must make the following assumptions:

1. The samples are random and independent (that is, last year was a typical, random year in terms of turnover, and turnover experiences in the three groups of stores were in no way related).
2. The turnover figures are normally distributed.
3. The three groups have equal population variances.

   The first step in performing our analysis of variance is to find the grand mean turnover for the 15 stores and the three samples' means and variances. These figures are

$$\overline{\overline{X}} = 11.07$$

$$\overline{X}_1 = 10.0 \qquad \overline{X}_2 = 9.0 \qquad \overline{X}_3 = 14.0$$

$$s_1^2 = 10.0 \qquad s_2^2 = 8.67 \qquad s_3^2 = 7.50$$

We cannot find the between-groups estimate of the common population variance by using Equation (10.4). That formula was developed for a situation where many samples, all of the same n, were taken. In this problem, however, each sample has a different size. Thus, we must modify Equation (10.4) to allow for sample sizes that are different, n_j. Then in Equation (10.4) we can bring the n_j figure inside the summation sign to yield the following formula:

$$s_B^2 = \frac{\sum\limits_{j=1}^{k} n_j(\overline{X}_j - \overline{\overline{X}})^2}{k - 1} \qquad (10.9)$$

In this formula, each sample mean's squared deviation from the grand mean is weighted by that sample's size. When we apply this formula to the data in our problem, we obtain the following between-groups estimate of the common population variance:

$$s_B^2 = \frac{\sum\limits_{j=1}^{k} n_j(\overline{X}_j - \overline{\overline{X}})^2}{k - 1}$$

$$= \frac{6(10.0 - 11.07)^2 + 4(9.0 - 11.07)^2 + 5(14.0 - 11.07)^2}{3 - 1}$$

$$= \frac{66.93}{2} = 33.47$$

This estimate has 2 degrees of freedom.

We find the within-groups (store groups in this case) variance estimate by using Equation (10.5) as follows:

$$s_W^2 = \frac{\sum\limits_{j=1}^{k} (n_j - 1)s_j^2}{n_T - k}$$

$$= \frac{(6 - 1)(10.0) + (4 - 1)(8.67) + (5 - 1)(7.5)}{15 - 3} = \frac{106}{12} = 8.83$$

This estimate has 12 degrees of freedom, and this value is equal to the sum of the degrees of freedom within each of the three groups.

The ratio of the two variance estimates given by Equations (10.9) and (10.5) is our calculated F figure. A large ratio will cause us to reject the hypothesis that the three groups of stores have equal mean annual turnover. From Table VII we find that $F_{.05,2,12} = 3.8853$. Thus, if we are willing to accept a 5% chance of rejecting the hypothesis of equal means when, in fact, it is true, then we will reject if our calculated F value exceeds 3.8853. Our F value is

$$F = \frac{s_B^2}{s_W^2} = \frac{33.47}{8.83} = 3.79$$

which is less than 3.8853. Thus we do not reject the hypothesis. The personnel manager may not want to consider standardizing compensation and benefits solely on the basis of these data. (Note that if he were willing to accept a slightly higher chance of a type I error, he would probably reject the hypothesis of equal mean annual turnover since the calculated and table F values are quite close.)

The analysis of variance table that summarizes the calculations in this problem is as follows:

Source	df	SS	Variance	F
Between groups	2	66.93	33.47	3.79
Within groups	12	106.00	8.83	
Total	14	172.93		

Problems: Section 10.3

5. A financial analyst with Donaldson, Lufkin & Jenrette is interested in the ratio of debt to total assets (as a percentage) for corporations that have been assigned bond ratings of (1) *Baa*, (2) *AA*, and (3) *AAA* by a ratings service. Random samples of debt to total assets (as percentages) for five, four, and three corporations were taken from the *Baa*, *AA*, and *AAA* ratings groups, respectively. The results are presented in the accompanying table.
   a. Find the grand mean $\overline{\overline{X}}$ for all of the sample data.
   b. Find the between-groups estimate of variance.
   c. Find the within-groups estimate of variance.
   d. Test the hypothesis that the mean debt-to-total-assets-value percentages are the same for all three bond ratings. Use .05 for the level of significance.

	Baa	*AA*	*AAA*
	41	25	15
	31	20	18
	38	17	20
	31	20	
	26		
n_j	5	4	3
$\overline{X}_j$	33.40	20.50	17.67
s_j	6.02	3.32	2.52

6. A career consultant wants to know whether individuals with degrees in business administration who work in the functional areas of (1) management, (2) accounting, and (3) finance have the same interest in business details. Interest in business detail is measured by an index. Random samples of interest in business detail were taken from individuals who work in the three functional areas, and the results are presented in the accompanying table. The samples included four individuals in management, four in accounting, and three in finance.
   a. Find the grand mean $\overline{\overline{X}}$ for all of the sample data.
   b. Find the between-groups estimate of variance.
   c. Find the within-groups estimate of variance.
   d. Test the hypothesis that the means are the same for all three groups. Use .05 for the level of significance.

	Management	Accounting	Finance
	6	10	10
	5	13	7
	6	9	8
	5	8	
n_j	4	4	3
$\overline{X}_j$	5.50	10.00	8.33
s_j	.577	2.16	1.53

7. A public administrator for the Federal Aviation Administration, in light of terrorist bombings and hijackings of airplanes, was concerned about how four different methods of screening passengers affect air traveler satisfaction. The four methods were (1) no screen, (2) walk-through metal detector, (3) frisk search, and (4) handheld metal detector. Random samples of traveler satisfaction were taken when the travelers were screened by one of the four methods. The sample sizes were $n_1 = 2$, $n_2 = 3$, $n_3 = 3$, and $n_4 = 2$ for the four methods. The results are presented in the accompanying table.

   a. Find the grand mean $\overline{\overline{X}}$ for all of the sample data, the sample group means $\overline{X}_1, \overline{X}_2,$ $\overline{X}_3,$ and $\overline{X}_4$, and the sample variance for each group, $s_1^2, s_2^2, s_3^2,$ and s_4^2.
   b. Find the between-groups estimate of variance.
   c. Find the within-groups estimate of variance.
   d. Test the hypothesis that the mean satisfaction values are the same for all four methods. Use .05 for the level of significance, and assume normal distributions and equal variance for each method.

No Screen	Walk-Through	Frisk	Hand-Held
8	7	11	3
14	3	9	7
	5	13	

8. A market study was done by Rulon, Inc., to examine the differences in hourly sales (in dollars) for a product that was packaged in four different package designs. Random samples of sales were taken under the four designs. The sample sizes were $n_1 = 2$, $n_2 = 4$, $n_3 = 3$, and $n_4 = 3$ for the four designs; other factors that influence sales were essentially held constant. The results are presented in the accompanying table.

   a. Find the grand mean $\overline{\overline{X}}$ for all of the sample data, the sample group means $\overline{X}_1, \overline{X}_2,$ $\overline{X}_3,$ and $\overline{X}_4$, and the sample variances for each group, $s_1^2, s_2^2, s_3^2,$ and s_4^2.
   b. Find the between-groups estimate of variance.
   c. Find the within-groups estimate of variance.
   d. Test the hypothesis that the means are the same for all four designs. Use .05 for the level of significance, and assume normal distributions and equal variances.

Design 1	Design 2	Design 3	Design 4
8	7	7	6
6	3	3	4
	3	5	5
	3		

10.4
ANALYSIS OF VARIANCE BY PARTITIONING THE SUM OF SQUARES

Some analysts like to approach the topic of analysis of variance from a different point of view. In this approach, they examine the sum of squares SS of the individual items in a set of groups from their grand mean; this sum of squares is called the **total sum of squares,** or SST.

We will use the data from the Sierra Pacific filter study of Section 10.1 to demonstrate this approach. A portion of Table 10.2 is reproduced here in Table 10.4 for convenient reference.

Figure 10.3 shows graphically how the total sum of squares around the grand mean is obtained. Each of the twelve items in Table 10.4 is scattered around the grand mean of $\overline{\overline{X}} = 13$. The deviations of the individual items from this mean value are represented by the vertical deviations drawn in the graph. Deviations above the grand mean are positive, and those below the grand mean are negative. To get the **total sum of squares SST,** we square each of these deviations from the grand mean $\overline{\overline{X}}$ and add these squared values.

Equation for Computing Total Sum of Squares SST

$$Total\ SS = SST = \sum_{j=1}^{k} \sum_{i=1}^{n_j} (X_{ij} - \overline{\overline{X}})^2 \qquad (10.10)$$

FIGURE 10.3

Deviations Around the Grand Mean

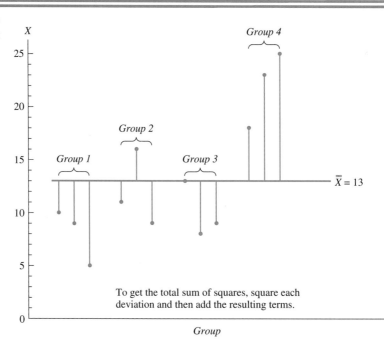

To get the total sum of squares, square each deviation and then add the resulting terms.

Filter 1	Filter 2	Filter 3	Filter 4		TABLE 10.4
10	11	13	18		**Pollution Counts at Sierra**
9	16	8	23	$\overline{\overline{X}} = 13$	**Pacific**
5	9	9	25		
$\overline{X}_1 = 8$	$\overline{X}_2 = 12$	$\overline{X}_3 = 10$	$\overline{X}_4 = 22$		

For the Sierra Pacific pollution count data presented in Table 10.4 and Figure 10.3, with a grand mean $\overline{\overline{X}} = 13$, this total sum of squares is

$$Total\ SS = SST = \sum_{j=1}^{k} \sum_{i=1}^{n_j} (X_{ij} - \overline{\overline{X}})^2$$

$$= \begin{bmatrix} (10-13)^2 \\ +(9-13)^2 \\ +(5-13)^2 \end{bmatrix} + \begin{bmatrix} (11-13)^2 \\ +(16-13)^2 \\ +(9-13)^2 \end{bmatrix} + \begin{bmatrix} (13-13)^2 \\ +(8-13)^2 \\ +(9-13)^2 \end{bmatrix} + \begin{bmatrix} (18-13)^2 \\ +(23-13)^2 \\ +(25-13)^2 \end{bmatrix}$$

$$= 428$$

Note that *all* 12 of the items in the four samples had their squared deviation from the grand mean $\overline{\overline{X}}$ added into this sum.

The total sum of squares *SST* can be partitioned into two separate terms known as the **between-groups sum of squares** *SSB* and the **within-groups sum of squares** *SSW* as follows.

Definition Formula for Partitioning Sum of Squares

$$\sum_{j=1}^{k} \sum_{i=1}^{n_j} (X_{ij} - \overline{\overline{X}})^2 = \sum_{j=1}^{k} n_j (\overline{X}_j - \overline{\overline{X}})^2 + \sum_{j=1}^{k} \sum_{i=1}^{n_j} (X_{ij} - \overline{X}_j)^2$$

or (10.11)

$$SST = SSB + SSW$$

The total sum of squares *SST* on the left of Equation (10.11) has been partitioned on the right into two sums of squares. The first term on the right of the equality is the between-groups sum of squares *SSB*.

Figure 10.4 gives a graphical presentation of what the between-groups sum of squares represents. The deviation of the four groups' means from the grand mean of $\overline{\overline{X}} = 13$ are shown. However, the three boxes at the ends of the deviations show that each deviation is to be squared and then counted $n = 3$ times. If one of the groups had contained four items, then four boxes would have been placed at the end of the deviation for that group.

FIGURE 10.4

Deviations Used to Obtain
Between-Groups Sum of
Squares

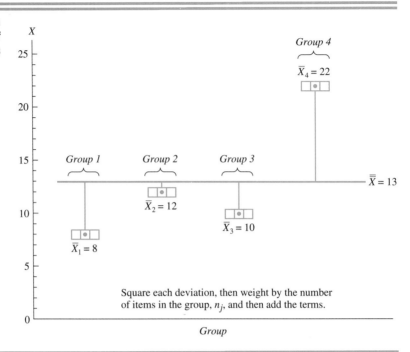

Square each deviation, then weight by the number
of items in the group, n_j, and then add the terms.

EXAMPLE 10.4   For the Sierra Pacific pollution count data in Table 10.4 the mathematical calculation of the between-groups sum of squares *SSB* is

$$\textit{Between-groups } SS = SSB = \sum_{j=1}^{k} n_j (\overline{X}_j - \overline{\overline{X}})^2$$

$$= 3(8 - 13)^2 + 3(12 - 13)^2 + 3(10 - 13)^2$$
$$+ 3(22 - 13)^2 = 348$$

The first term on the right side of Equation (10.11) is a weighted sum. Each term is weighted by the sample size $n = 3$. However, in the general case each sample has a different sample size n_j. This first term is often called the *between-groups sum of squares*. The name derives from the fact that there is a different $(\overline{X}_j - \overline{\overline{X}})^2$ term for each group, and thus the squared deviations differ between the groups. The between-groups sum of squares *SSB* increases substantially if the sample means, the $\overline{X}_j$ values, differ greatly from one another.

The second term on the right of Equation (10.11) is the within-group sum of squares *SSW*. Notice that it includes a double summation. We must remember that a double summation is evaluated from the inside out. Thus it begins with the sum of squared deviations of the individual items in each sample from that sample's mean. This summation goes from $i = 1$ to $i = n_j$. Then these sums for each sample are added for all the samples, and this summation goes from $j = 1$ to $j = k$.

The result of this double-summation operation is called the *within-groups sum of squares*. The name comes from the fact that the sum of squared deviations *within* each group (the first summation operation of the double sum) is taken before the terms are added for each group.

FIGURE 10.5

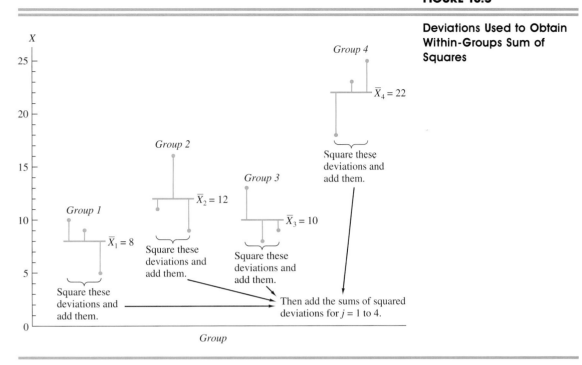

Figure 10.5 shows graphically what the within-groups sum of squares represents. The figure shows that each sample in Table 10.4 has a different mean. The individual items in each sample are scattered around their mean. The deviations of the individual items from the mean are represented by the vertical deviations drawn in the graph. Deviations above the mean are positive, and those below the mean are negative. If we were to add these deviations for any one sample, the positive and negative deviations would cancel. That is, the sum of the deviations would be zero for all four samples.

If, instead, we square each of the deviations before we add them, we will get four different sums of squares. These values are the four sums of squares we get by doing the first summation in Equation (10.11), the summation from $i = 1$ to $i = n_j$. Then we can add the four sums of squares to obtain the within-groups sum of squares. This second summation goes from $j = 1$ to $j = k$.

EXAMPLE 10.5   For the Sierra Pacific pollution count data in Table 10.4 the value for the within-groups sum of squares SSW from the second term on the right side of Equation (10.11) is found as follows:

$$\text{Within-groups } SS = SSW = \sum_{j=1}^{k} \sum_{i=1}^{n_j} (X_{ij} - \overline{X}_j)^2$$

$$= \begin{bmatrix} (10 - 8)^2 \\ +(9 - 8)^2 \\ +(5 - 8)^2 \end{bmatrix} + \begin{bmatrix} (11 - 12)^2 \\ +(16 - 12)^2 \\ +(9 - 12)^2 \end{bmatrix} + \begin{bmatrix} (13 - 10)^2 \\ +(8 - 10)^2 \\ +(9 - 10)^2 \end{bmatrix} + \begin{bmatrix} (18 - 22)^2 \\ +(23 - 22)^2 \\ +(25 - 22)^2 \end{bmatrix}$$

$$= (14) + (26) + (14) + (26) = 80$$

Note that

$$Total\ SS = Between\text{-}groups\ SS + Within\text{-}groups\ SS \tag{10.12}$$

in that

$$428 = 348 + 80$$

Thus if we first determine the values of *Total SS* and *Between-groups SS,* we can use the following formula to compute the *Within-groups SS*:

$$Within\text{-}groups\ SS = Total\ SS - Between\text{-}groups\ SS \tag{10.13}$$

We mentioned in Chapter 3 that each sum of squares has a degree-of-freedom figure associated with it. The total sum of squares has just one less than the combined sample size, $n_T - 1$. The within-groups sum of squares has $[(n_1 - 1) + (n_2 - 1) + \cdots + (n_k - 1)]$ degrees of freedom. This figure is simply $(n_1 + n_2 + \cdots + n_k - k)$, which is equal to $n_T - k$, or the sum of the samples' sizes minus the number of populations from which the samples were taken. The between-groups sum of squares has $k - 1$ degrees of freedom. The degrees of freedom are additive, meaning that the value of the degrees of freedom for *Total SS* equals the value of the degrees of freedom for the *Between-groups SS* plus the value of the degrees of freedom for the *Within-groups SS,* as shown in the following equation:

$$Total\ df = Between\text{-}groups\ df + Within\text{-}groups\ df \tag{10.14}$$

or

$$(n_T - 1) = (k - 1) + (n_T - k)$$

We noted in the previous section that the results of an analysis of variance are usually summarized in a table such as Table 10.3, which is presented again here as Table 10.5 for convenient reference. Now the column labels on the table may be more clear. The *SS* column shows the sum of squares as they were developed in the previous paragraphs of this section. The *MS* column stands for **mean square** and shows the "mean of the squares," if we consider division by the number of degrees of freedom as a sort of averaging process as defined in the following box.

> **Definition**
>
> A **mean square** is a sum of squares divided by its associated degrees of freedom.

Thus, we have

$$Between\text{-}groups\ MS = MSB = \frac{Between\text{-}groups\ SS}{Between\text{-}groups\ df} = \frac{SSB}{k - 1} \tag{10.15}$$

	Source	df	SS	MS	F
TABLE 10.5					
Analysis of Variance Table	Between groups	3	348	116	11.6
for the Sierra Pacific	Within groups	8	80	10	
Pollution Count Data	Total	11	428		

$$Within\text{-}groups\ MS = MSW = \frac{Within\text{-}groups\ SS}{Within\text{-}groups\ df} = \frac{SSW}{n_T - k} \qquad (10.16)$$

EXAMPLE 10.6   For the Sierra Pacific pollution count data of Table 10.4 we have the following values for the between-groups mean squares *MSB* and the within-groups mean squares *MSW*:

$$Between\text{-}groups\ MS = MSB = \frac{SSB}{k-1} = \frac{348}{4-1} = \frac{348}{3} = 116$$

$$Within\text{-}groups\ MS = MSW = \frac{SSW}{n_T - k} = \frac{80}{12-4} = \frac{80}{8} = 10$$

Note that these values are exactly the same as those obtained in Examples 10.1 and 10.2, where we estimated the common population variance for the four groups. That is, $s_B^2 = 116$ and $s_W^2 = 10$. A careful examination of formulas for these values will reveal that the between-groups variance estimate is the same as the between-groups mean square. Also, the within-groups variance estimate is the same as the within-groups mean square. That is,

$$s_B^2 = Between\text{-}groups\ MS = MSB$$
$$s_W^2 = Within\text{-}groups\ MS = MSW \qquad (10.17)$$

Since this relationship holds, to test the hypothesis that several groups have the same mean, $H_0: \mu_1 = \mu_2 = \mu_3 = \cdots = \mu_k$, we can compare the value of the *F* ratio calculated as

$$F = \frac{Between\text{-}groups\ MS}{Within\text{-}groups\ MS} = \frac{MSB}{MSW} \qquad (10.18)$$

with the critical *F* value in Table VII for the appropriate level of α and number of degrees of freedom.

Now a set of samples gives a large *F* value only if *MSB* is large relative to *MSW*. But this condition exists only if the sample means, the $\overline{X}_j$ values, differ greatly from one another. And this condition usually exists when the population means, the μ_j values, differ greatly from one another. In sum, then, a large *F* value indicates that the null hypothesis $H_0: \mu_1 = \mu_2 = \cdots = \mu_k$ is false, and we have the following procedure.

Hypotheses on Equality of Several Population Means and Rejection Rule for ANOVA

$H_0: \mu_1 = \mu_2 = \cdots = \mu_k$.

H_a: One or more of the population means is not equal to the others.

Rejection rule: If $F > F_{\alpha, \nu_1, \nu_2}$, then reject H_0;

where: $F = \dfrac{MSB}{MSW}$

$\nu_1 = k - 1$

$\nu_2 = n_T - k$

━━ **EXAMPLE 10.7** For the Sierra Pacific filter problem and the pollution count data of Table 10.4, and with a significance level of .05, we want to perform the following test.

Step One $H_0: \mu_1 = \mu_2 = \mu_3 = \mu_4$.

Step Two H_a: One or more of the population means is not equal to the others.

Step Three $\alpha = .05$, $n_1 = 3$, $n_2 = 3$, $n_3 = 3$, $n_4 = 3$.

Step Four In our example, $v_1 = k - 1 = 4 - 1 = 3$ and $v_2 = n_T - k = 12 - 4 = 8$. We find that $F_{\alpha,v_1,v_2} = F_{.05,3,8} = 4.0662$, so we have the following rejection rule.

Rejection rule: If $F > 4.0662$, then reject H_0.

Step Five For the Sierra Pacific pollution count data of Table 10.1 and the mean squares $MSB = 116$ and $MSW = 10$ that we found in Example 10.6, we have

$$F = \frac{MSB}{MSW} = \frac{116}{10} = 11.6$$

Step Six Since $F = 11.6$ exceeds 4.0662, we reject the null hypothesis and conclude that one or more of the filters produce mean pollution counts that are different from the others. ━━

Since the calculated F value and the conclusions reached will be the same, regardless of the method used, the results of the hypothesis test of the equality of several means will be the same regardless of whether one approaches it by estimating the common population variance or by partitioning the total sum of squares into two parts.

Even though an analysis of variance hypothesis-testing problem can be solved either way, we prefer calculating the F ratio by the second method, which involves dividing the total sum of squares into two parts because it is useful in additional applications. One of these applications will be discussed in Chapter 12.

The procedures for determining the values presented in a table for showing the results of a one-way analysis of variance (ANOVA) are summarized in Table 10.6.

We have used the terms of Equation (10.11), which define the sums of squares, to find these values in our examples. We have followed this practice so that we can emphasize the meaning of the various sums of squares. However, data analysis systems are often used to find ANOVA tables, thereby easing the computational burden.

TABLE 10.6

ANOVA Table

Source	df	SS	MS	F
Between groups	$k - 1$	$SSB = \sum_{j=1}^{k} n_j(\overline{X}_j - \overline{\overline{X}})^2$	$MSB = \dfrac{SSB}{k-1}$	$F = \dfrac{MSB}{MSW}$
Within groups	$n_T - k$	$SSW = \sum_{j=1}^{k} \sum_{i=1}^{n_j}(X_{ij} - \overline{X}_j)^2$	$MSW = \dfrac{SSW}{n_T - k}$	
Total	$n_T - 1$	$SST = \sum_{j=1}^{k} \sum_{i=1}^{n_j}(X_{ij} - \overline{\overline{X}})^2$		

You may want to use the definitional terms of Equation (10.11) to obtain the sums of squares for a few simple problems as you become acquainted with the methods of analysis of variance and then, when you are familiar with the meaning of the various sums of squares, use a data analysis system to obtain the analysis of variance table. We used a computer to analyze the pollution count data of Table 10.4, and the results are presented in Computer Exhibits 10.1A, 10.1B, and 10.1C. The analysis of variance tables given in the printouts are comparable to the results we gave in Table 10.5. Problems for this section and supplementary problems at the end of the chapter include printouts from other data analysis software packages. It is usually quite simple to adapt from one program to another.

If data sets are large, or if the computations are burdensome, and a computer is not available, then the following **computational formulas** can be used to determine the values of the various sums of squares, although the general availability of data analysis software tends to prevail over the use of computational formulas.

COMPUTER EXHIBIT 10.1A Analysis of Variance for the Sierra Pacific Pollution Count Data   **MINITAB**

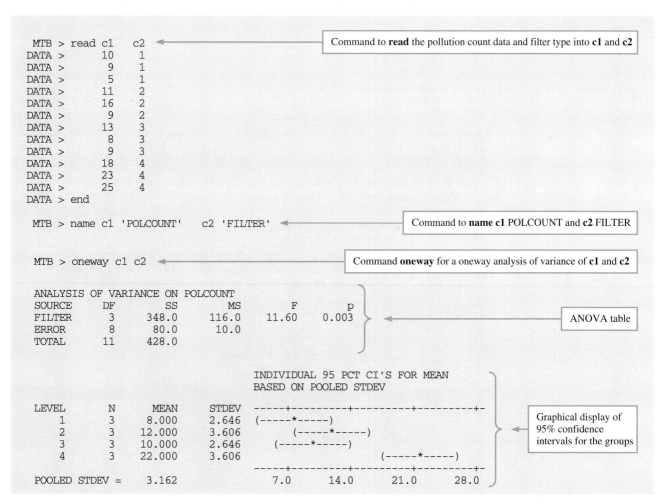

```
MTB > read c1   c2            ◄─────    Command to read the pollution count data and filter type into c1 and c2
DATA >     10   1
DATA >      9   1
DATA >      5   1
DATA >     11   2
DATA >     16   2
DATA >      9   2
DATA >     13   3
DATA >      8   3
DATA >      9   3
DATA >     18   4
DATA >     23   4
DATA >     25   4
DATA > end

MTB > name c1 'POLCOUNT'   c2 'FILTER'   ◄──────   Command to name c1 POLCOUNT and c2 FILTER

MTB > oneway c1 c2    ◄────   Command oneway for a oneway analysis of variance of c1 and c2

ANALYSIS OF VARIANCE ON POLCOUNT
SOURCE       DF        SS        MS        F       p
FILTER        3     348.0     116.0    11.60   0.003   ◄────   ANOVA table
ERROR         8      80.0      10.0
TOTAL        11     428.0

                              INDIVIDUAL 95 PCT CI'S FOR MEAN
                              BASED ON POOLED STDEV

LEVEL        N      MEAN     STDEV    -----+---------+---------+---------+-
    1        3     8.000     2.646    (-----*-----)
    2        3    12.000     3.606          (-----*-----)
    3        3    10.000     2.646      (-----*-----)                          Graphical display of
    4        3    22.000     3.606                        (-----*-----)        95% confidence
                                       -----+---------+---------+---------+-   intervals for the groups
POOLED STDEV =     3.162                7.0       14.0      21.0      28.0
```

(continues)

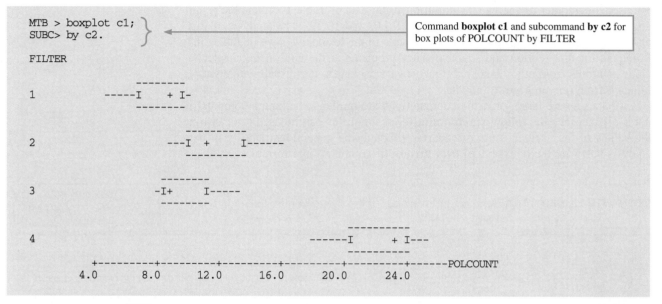

MINITAB

COMPUTER EXHIBIT 10.1A Analysis of Variance for the Sierra Pacific Pollution Count Data (Continued)

```
MTB > boxplot c1;
SUBC> by c2.
```
Command **boxplot c1** and subcommand **by c2** for box plots of POLCOUNT by FILTER

```
FILTER

                       ---------
1             -----I    + I-
                       ---------

                          ----------
2                    ---I   +      I------
                          ----------

                    ---------
3              -I+       I-----
                    ---------

                                       ----------
4                                ------I    + I---
                                       ----------
         +---------+---------+---------+---------+---------+------POLCOUNT
        4.0       8.0      12.0      16.0      20.0      24.0
```

Note: POOLED STDEV = $\sqrt{MSW}$. The following labels are used by various computer programs, and the labels can be regarded as equivalent for our purposes: (1) Between-groups = FACTOR = MODEL = TREATMENTS; (2) Within-groups = ERROR = RESIDUAL; (3) Total = CORRECTED TOTAL; (4) Group = LEVEL.

Computational Formulas for One-Factor Analysis of Variance

For T = the sum of the values of the dependent variable X_{ij} for all experimental items, or

$$T = \sum_i \sum_j X_{ij}$$

and T_j = the sum of the values of the dependent variable for the experimental items in treatment j (column j), or

$$T_j = \sum_i X_{ij}$$

the *computational formulas* are as follows:

$$\text{Total SS} = \sum_i \sum_j X_{ij}^2 - \frac{T^2}{n_T} \tag{10.19}$$

$$\text{Between-groups SS} = \sum_j \frac{T_j^2}{n_j} - \frac{T^2}{n_T} \tag{10.20}$$

$$\text{Within-groups SS} = \sum_i \sum_j X_{ij}^2 - \sum_j \frac{T_j^2}{n_j} \tag{10.21}$$

COMPUTER EXHIBIT 10.1B Analysis of Variance for the Sierra Pacific Pollution Count Data

SAS

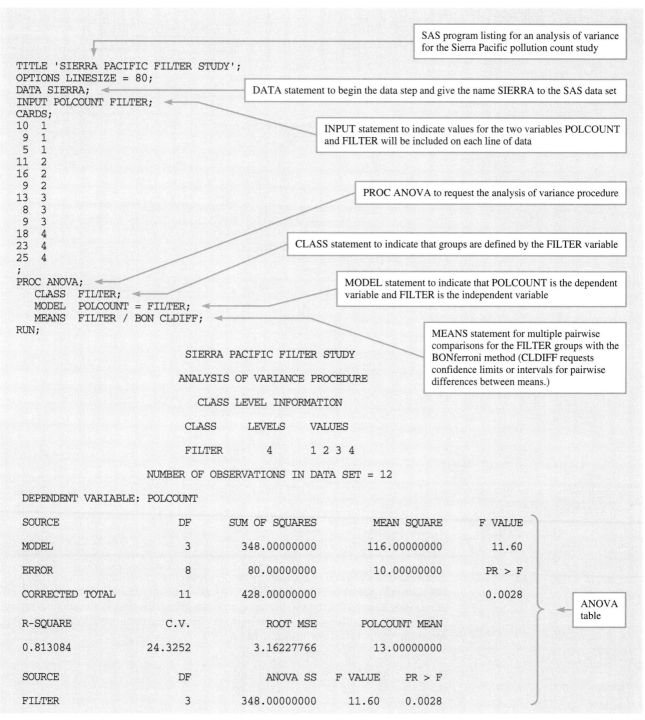

SAS program listing for an analysis of variance for the Sierra Pacific pollution count study

```
TITLE 'SIERRA PACIFIC FILTER STUDY';
OPTIONS LINESIZE = 80;
DATA SIERRA;
INPUT POLCOUNT FILTER;
CARDS;
10  1
 9  1
 5  1
11  2
16  2
 9  2
13  3
 8  3
 9  3
18  4
23  4
25  4
;
PROC ANOVA;
    CLASS  FILTER;
    MODEL  POLCOUNT = FILTER;
    MEANS  FILTER / BON CLDIFF;
RUN;
```

DATA statement to begin the data step and give the name SIERRA to the SAS data set

INPUT statement to indicate values for the two variables POLCOUNT and FILTER will be included on each line of data

PROC ANOVA to request the analysis of variance procedure

CLASS statement to indicate that groups are defined by the FILTER variable

MODEL statement to indicate that POLCOUNT is the dependent variable and FILTER is the independent variable

MEANS statement for multiple pairwise comparisons for the FILTER groups with the BONferroni method (CLDIFF requests confidence limits or intervals for pairwise differences between means.)

```
                    SIERRA PACIFIC FILTER STUDY

                  ANALYSIS OF VARIANCE PROCEDURE

                     CLASS LEVEL INFORMATION

              CLASS      LEVELS      VALUES

              FILTER        4        1 2 3 4

          NUMBER OF OBSERVATIONS IN DATA SET = 12
```

DEPENDENT VARIABLE: POLCOUNT

SOURCE	DF	SUM OF SQUARES	MEAN SQUARE	F VALUE
MODEL	3	348.00000000	116.00000000	11.60
ERROR	8	80.00000000	10.00000000	PR > F
CORRECTED TOTAL	11	428.00000000		0.0028

R-SQUARE	C.V.	ROOT MSE	POLCOUNT MEAN
0.813084	24.3252	3.16227766	13.00000000

SOURCE	DF	ANOVA SS	F VALUE	PR > F
FILTER	3	348.00000000	11.60	0.0028

ANOVA table

(continues)

COMPUTER EXHIBIT 10.1B Analysis of Variance for the Sierra Pacific Pollution Count Data (Continued)

To test H_0: $\mu_j = \mu_{j'}$ with SAS, use the following rule.

Rejection rule: If $|\bar{X}_j - \bar{X}_{j'}| > t_{\alpha/(2m),\, n_T - k} \sqrt{MSW(1/n_j + 1/n_{j'})}$, then reject H_0.

For the Sierra Pacific data,

Rejection rule: If $|\bar{X}_j - \bar{X}_{j'}| > 8.9824$, then reject H_0.

```
BONFERRONI (DUNN) T TESTS FOR VARIABLE: POLCOUNT
NOTE: THIS TEST CONTROLS THE TYPE I EXPERIMENTWISE ERROR RATE
      BUT GENERALLY HAS A HIGHER TYPE II ERROR RATE THAN TUKEY'S

    ALPHA=0.05  CONFIDENCE=0.95  DF=8  MSE=10
    CRITICAL VALUE OF T=3.47888
    MINIMUM SIGNIFICANT DIFFERENCE=8.9824

COMPARISONS SIGNIFICANT AT THE 0.05 LEVEL ARE INDICATED BY '***'
```

Limits of confidence intervals for multiple pairwise comparisons of means using the Bonferroni method

FILTER COMPARISON		SIMULTANEOUS LOWER CONFIDENCE LIMIT	DIFFERENCE BETWEEN MEANS	SIMULTANEOUS UPPER CONFIDENCE LIMIT	
4	- 2	1.018	10.000	18.982	***
4	- 3	3.018	12.000	20.982	***
4	- 1	5.018	14.000	22.982	***
2	- 4	-18.982	-10.000	-1.018	***
2	- 3	-6.982	2.000	10.982	
2	- 1	-4.982	4.000	12.982	
3	- 4	-20.982	-12.000	-3.018	***
3	- 2	-10.982	-2.000	6.982	
3	- 1	-6.982	2.000	10.982	
1	- 4	-22.982	-14.000	-5.018	***
1	- 2	-12.982	-4.000	4.982	
1	- 3	-10.982	-2.000	6.982	

Note: ROOT MSE $= \sqrt{MSE} = \sqrt{MSW}$

Problems: Section 10.4

9. A sample of 15 beginning typists was randomly separated into three groups. The typists then took speed tests, with groups 1, 2, and 3 using electronic typewriters, word processors, and electric typewriters, respectively. Given the data in the accompanying table, test the hypothesis that the three groups have the same population mean typing speed (in words typed per minute). Let $\alpha = .05$.

Group 1	50	48	53	48	51
Group 2	66	72	72	74	66
Group 3	50	44	43	45	43

10. A manager with Federated Insurance Corporation recently heard rumors that different types of hospitals charge different fees for the same service. The insurance company, whose clients use private, government, and university hospitals, hired a statistical

COMPUTER EXHIBIT 10.1C Analysis of Variance for the Sierra Pacific Pollution Count
Data

MYSTAT

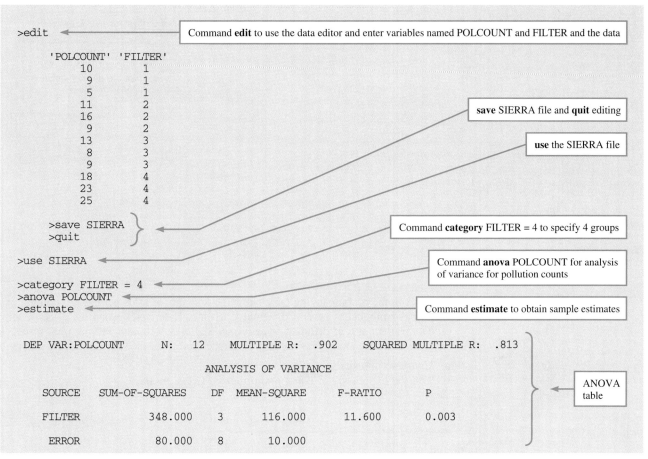

```
>edit ←———— Command edit to use the data editor and enter variables named POLCOUNT and FILTER and the data

      'POLCOUNT' 'FILTER'
          10        1
           9        1
           5        1
          11        2             save SIERRA file and quit editing
          16        2
           9        2
          13        3             use the SIERRA file
           8        3
           9        3
          18        4
          23        4
          25        4

      >save SIERRA  }←———— Command category FILTER = 4 to specify 4 groups
      >quit

>use SIERRA ←————               Command anova POLCOUNT for analysis
                                 of variance for pollution counts
>category FILTER = 4 ←————
>anova POLCOUNT ←————
>estimate ←————                  Command estimate to obtain sample estimates


DEP VAR:POLCOUNT    N:  12   MULTIPLE R:  .902   SQUARED MULTIPLE R:  .813

                        ANALYSIS OF VARIANCE
                                                                          ANOVA
   SOURCE   SUM-OF-SQUARES   DF  MEAN-SQUARE   F-RATIO     P             ←  table

   FILTER        348.000      3     116.000    11.600    0.003

   ERROR          80.000      8      10.000
```

(continues)

consultant to conduct a study of the fees charged by the different types of hospitals
for penicillin injections. The accompanying data shows the results of a random sampling of the fees (in dollars) at the different types of hospitals. The manager wants to know if the rumor is consistent with the data.

a. Find the grand mean $\overline{\overline{X}}$ for all of the sample data and the sample group means $\overline{X}_1$, $\overline{X}_2$, and $\overline{X}_3$.
b. Find the between-groups sum of squares.
c. Find the between-groups degrees of freedom.
d. Find the mean square between groups.
e. Find the within-groups sum of squares.
f. Find the within-groups degrees of freedom.
g. Find the mean square within groups.
h. Test the hypothesis that the mean fees are the same for all three hospital types. Use .05 for the level of significance.

Private	11	12	13
Government	3	8	7
University	4	6	8

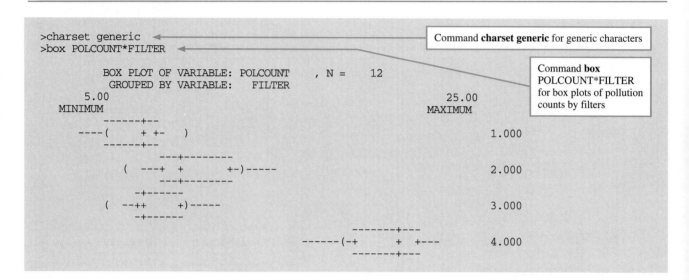

| MYSTAT | **COMPUTER EXHIBIT 10.1C** Analysis of Variance for the Sierra Pacific Pollution Count Data (Continued) |

```
>charset generic
>box POLCOUNT*FILTER

        BOX PLOT OF VARIABLE: POLCOUNT    , N =    12
          GROUPED BY VARIABLE:    FILTER
      5.00                                                      25.00
    MINIMUM                                                    MAXIMUM
        ------+--
    ----(     + +-   )                                           1.000
        ------+--
              ---+---------
        (   ---+  +       +-)-----                               2.000
              ---+---------
          -+-------
        (   --++     +)-----                                     3.000
          -+-------
                            -------+---
                    ------(-+     +  +---                        4.000
                            -------+---
```

Command **charset generic** for generic characters

Command **box** POLCOUNT*FILTER for box plots of pollution counts by filters

11. The data in the accompanying table show the amounts of sales per day (in dollars) of a particular product at several different Carpenter Hale stores when the price of the product was set at four different price levels: $20, $15, $10, and $12 per unit, respectively. Each of the stores was randomly assigned to one of the four price-level groups.

    a. Construct the analysis of variance table.
    b. Test the hypothesis of equal group means. Let $\alpha = .05$.

Group 1	Group 2	Group 3	Group 4
9	17	22	13
6	10	17	14
8	12	21	22
13			16
			10

12. A money manager with Haddon Group studied the difference between the closing daily federal funds rate (a short-term interest rate) and the weekly mean federal funds rate for three different days: Monday, Wednesday, and Friday. Random samples were taken for the three different days, and the results are given in the accompanying table (as percentages). Use ANOVA to test the hypothesis of equal means for the three days. Use a significance level of .05. An edited SAS printout for the analysis of variance is given in Computer Exhibit 10.2.

Monday	Wednesday	Friday
−1.0	2.0	−1.2
−1.6	1.0	−1.1
1.2	1.2	−1.4
−0.8	0.4	−0.1

COMPUTER EXHIBIT 10.2 ANOVA Table for Haddon Group

```
ANALYSIS OF VARIANCE PROCEDURE

DEPENDENT VARIABLE:   RATE

SOURCE              DF       SUM OF SQUARES     MEAN SQUARE    F VALUE

MODEL                2          9.94666667      4.97333333       6.63

ERROR                9          6.75000000      0.75000000     PR > F

CORRECTED TOTAL     11         16.69666667                      0.0170
```

When the null hypothesis of equal means, $H_0: \mu_1 = \mu_2 = \cdots = \mu_k$, has been rejected in an analysis of variance problem, we may then desire to know which of the means are not equal.

To examine which of the means are not equal when we have rejected the null hypothesis of equal means in an analysis of variance, we discuss two procedures: (1) informal multiple comparisons of means and (2) multiple pairwise comparisons of means.

10.5
MULTIPLE COMPARISONS OF MEANS

10.5.1 Informal Multiple Comparisons of Means

If the null hypothesis of equal means has been rejected in an analysis of variance, then we can obtain information about which groups are different and make an informal decision about which means are not equal or about the highest or lowest means (1) by using graphs of confidence intervals for the group means μ_j or (2) by comparing the distributions for the groups through the use of box plots.

To find confidence intervals for the group means, we use the procedures from Chapter 8. Since we assume equal population variances in an analysis of variance, our estimate for the common population standard deviation is $\sqrt{MSW}$. The estimated standard deviation $\sqrt{MSW}$ is parallel to s_{pooled} from Chapter 9, and the value of the degrees of freedom for this estimate is $n_T - k$. The following expression gives us the limits of a $100(1 - \alpha)\%$ confidence interval for the group mean μ_j:

$$\overline{X}_j \pm (t_{\alpha/2, n_T - k}) (\sqrt{MSW}/\sqrt{n_j}) \tag{10.22}$$

Suppose we find confidence intervals for all k group means and graph them. If the confidence intervals for any pair of groups do not overlap, then, loosely speaking, we have information that suggests that the two groups have different means.

Another method for obtaining information about which groups are different when the equality of means has been rejected in an analysis of variance is to compare the distributions for the groups by using box plots. If the boxes for pairs of groups do not overlap markedly, then we have information suggesting that the two groups have different means.

EXAMPLE 10.8   We rejected the hypothesis of equal mean pollution counts for the four filters used by Sierra Pacific Electric Company in the previous section.

However, we did not decide which filter (or filters) gives the lowest mean pollution count. We know from previous calculations that

$$\overline{X}_1 = 8 \qquad \sqrt{MSW} = \sqrt{10} \qquad n_T - k = 12 - 4 = 8$$

To find the 95% confidence interval for filter 1, we have $\alpha/2 = .05/2 = .025$ and $t_{.025,8} = 2.306$; and Expression (10.22) gives

$$8 \pm (2.306)(\sqrt{10}/\sqrt{3})$$

or the interval limits for the mean pollution count for the first type of filter are

$$L_1 = 8 - 4.21 = 3.79 \quad \text{and} \quad R_1 = 8 + 4.21 = 12.21$$

The 95% confidence intervals for filters 2, 3, and 4 were also determined, and a plot of the four intervals is given in Figure 10.6(a).

FIGURE 10.6

Informal Comparisons of Mean Pollution Counts for Four Filters at Sierra Pacific

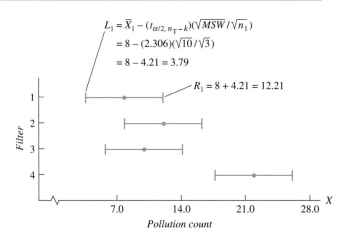

(a) **95% confidence intervals for the mean pollution counts for four filters**

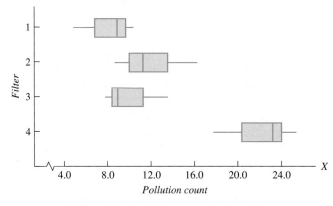

(b) **Box plots of pollution counts for four filters**

To compare the pollution count distributions for the filter groups, we can also use the four box plots given in Figure 10.6(b). Box plots from data analysis systems and a plot of the confidence intervals were also given in Computer Exhibit 10.1.

Figure 10.6(a) shows that the confidence interval for filter 4 does not overlap any of the other intervals and it has higher pollution counts than do the other three. Figure 10.6(b) shows that the box plot for filter 4 does not overlap any of the other box plots and it has higher pollution counts than do the other three.

Thus we decide, by using the informal procedures, *not* to use type 4 filters since they apparently result in higher pollution counts. From informal comparisons of means we would use any of the first three types of filters. ═══

10.5.2 Multiple Pairwise Comparisons of Means

If the hypothesis of equal means for the k groups in an analysis of variance has been rejected, then we may wish to conduct a formal test of the equality of multiple pairs of means to examine which means are different. To test which μ_j's are not equal, we can use **multiple pairwise comparisons of means.**

Definition

Multiple pairwise comparisons of means are made by testing multiple hypotheses of the equality of pairs of means. The null hypotheses for the pairwise comparisons can be represented as H_0: $\mu_j = \mu_{j'}$, where j and j' denote the two groups that are being compared. The alternative hypotheses are that the pairs of means are not equal.

We can test all of the combinations of pairs of means, or we can select just the pairs that are of interest before we see the sample results. For the Sierra Pacific Electric Company example concerning the water pollution counts for the four types of filters, a comparison of all pairs of means would require us to test six null hypotheses:

$$H_{0_1}: \mu_1 = \mu_2 \qquad H_{0_2}: \mu_1 = \mu_3 \qquad H_{0_3}: \mu_1 = \mu_4$$
$$H_{0_4}: \mu_2 = \mu_3 \qquad H_{0_5}: \mu_2 = \mu_4 \qquad H_{0_6}: \mu_3 = \mu_4$$

The results of the six hypothesis tests enable us to make a conclusion about which type of filter (or filters) results in the lowest mean pollution count.

To test the multiple pairwise comparison hypotheses, we will use the approach for testing the equality of two means that was presented in Section 9.8 with a few adjustments. First, we will use multiple confidence intervals to test the hypotheses. If a confidence interval does *not* include the value 0, then we reject the pairwise equality of means. If a confidence interval includes 0, then we do not reject the pairwise equality of means. Second, we estimate the common variance by using the within-groups estimate of variance MSW (or s_W^2). Third, since MSW has $n_T - k$ degrees of freedom, we will use $n_T - k$ degrees of freedom to determine the t value that will be used to compute the confidence intervals. And finally, we will make an adjustment to α, as explained in the next paragraph.

If the number of pairwise comparisons of means we wish to make is m, and if we want to be conservative and have the confidence level be at least $100(1 - \alpha)\%$

for all m confidence intervals considered simultaneously, then we must make an adjustment to α. A confidence level for multiple confidence intervals considered simultaneously—in other words, a family of confidence intervals—is called a *joint,* or *family, confidence level.* One way to make a family confidence level for m confidence intervals be at least $100(1 - \alpha)\%$ is to construct each confidence interval separately, with each interval having a confidence level of $100(1 - \alpha/m)\%$.

The reason that α is adjusted to α/m can be explained as follows: If a hypothesis test is made with $\alpha = .05$ and the null hypothesis is true, then the probability of rejecting the null hypothesis when it is true (a type I error) is .05. The probability of a correct conclusion, then, is $1 - .05 = .95$. As a simple example, suppose that we want to test a family of two null hypotheses and we desire the family confidence level to be 95%. If *two* null hypotheses that are true are tested *independently,* and $\alpha = .05$ is used for each test, then the probability that the conclusions on both tests are correct is only $(.95)(.95) = .90$. The probability that at least one of the tests results in the rejection of a true null hypothesis is $1 - .90 = .10$. Thus in this situation the joint, or family, confidence level is only 90%, not 95%, as was desired. Furthermore, multiple pairwise comparisons in analysis of variance are not made independently, because MSW is used in each comparison. Thus the family confidence level, without any adjustment to α, is likely to be even lower than the 90% of our simple example, which included just two tests and assumed independence.

The implication of this discussion is quite straightforward. Let us say that we want to make m pairwise comparisons while maintaining a family confidence level of at least $100(1 - \alpha)\%$. Then one way to accomplish this set of comparisons is to adjust α to $\alpha/(m)$. Now since we usually put an area of $\alpha/2$ in each tail for a two-sided test, then with this adjustment we will put an area of $\alpha/(2m)$ *in each tail* when we find the t value for computing confidence intervals for conducting multiple pairwise comparisons.

We now want to find m confidence intervals with a family confidence level of $100(1 - \alpha)\%$. The formulas for finding the confidence limits L and R for the intervals that can be used to test the m null hypotheses, H_0: $\mu_j = \mu_{j'}$, where j and j' denote the groups being compared, and the difference between the two sample means is $\overline{X}_j - \overline{X}_{j'}$, are as follows.

Equations for Confidence Limits for Multiple Pairwise Comparisons of Means

$$L = (\overline{X}_j - \overline{X}_{j'}) - t_{\alpha/(2m), n_T - k} \sqrt{MSW\left(\frac{1}{n_j} + \frac{1}{n_{j'}}\right)}$$

$$R = (\overline{X}_j - \overline{X}_{j'}) + t_{\alpha/(2m), n_T - k} \sqrt{MSW\left(\frac{1}{n_j} + \frac{1}{n_{j'}}\right)}$$

(10.23)

If the confidence interval formed by the values of L and R, as determined from Equation (10.23), for the pairwise comparison of groups denoted j and j' does *not* include the value 0, then we reject the corresponding null hypothesis, H_0: $\mu_j = \mu_{j'}$. If the confidence interval includes the value 0, then we do not reject the null hypothesis.

PRACTICE PROBLEM 10.2

Let us reconsider the Sierra Pacific Electric Company example, where we examined the water pollution counts that result when four different filters are used. Recall that we rejected the null hypothesis that the mean water pollution counts were equal for all four filters. We now want to decide which filter (or filters) results in lower mean pollution counts. In other words, we want to determine which pairs of means are significantly different by testing the null hypotheses $H_0: \mu_j = \mu_{j'}$, where j and j' denote the groups. Let the family confidence level be at least 95%.

Solution To determine which filter (or filters) results in the lower mean water pollution counts, we can make $m = 6$ multiple comparisons of means by testing the following six null hypotheses:

$$H_{0_1}: \mu_1 = \mu_2 \qquad H_{0_2}: \mu_1 = \mu_3 \qquad H_{0_3}: \mu_1 = \mu_4$$
$$H_{0_4}: \mu_2 = \mu_3 \qquad H_{0_5}: \mu_2 = \mu_4 \qquad H_{0_6}: \mu_3 = \mu_4$$

The alternative hypothesis in each case is that the two means are not equal.

The sample mean water pollution counts taken by Sierra Pacific Electric Company and the sample sizes for the four different types of filters were presented in Section 10.1, and they are as follows:

$$\overline{X}_1 = 8 \qquad \overline{X}_2 = 12 \qquad \overline{X}_3 = 10 \qquad \overline{X}_4 = 22$$
$$n_1 = 3 \qquad n_2 = 3 \qquad n_3 = 3 \qquad n_4 = 3$$

The value for MSW (or s_W^2) from Table 10.5 is 10, and the value for the degrees of freedom associated with MSW is $n_T - k = 12 - 4 = 8$. Because the family confidence level must be at least 95%, we will let $\alpha = .05$. Each of the $m = 6$ individual confidence intervals will be found by letting $\alpha/m = .05/6 = .008$, so the area we place in a single tail for determining the value of t in computing a confidence interval is $\alpha/(2m) = .05/[(2)(6)] = .004167$. Thus the value of t we will use in Equation (10.23) to determine the confidence limits is

$$t_{\alpha/(2m), n_T - k} = t_{.004167, 8} = 3.4783$$

Now to compare the pair of means for filter 1 and filter 2, we test the null hypothesis $H_{0_1}: \mu_1 = \mu_2$. For this hypothesis $\overline{X}_1 - \overline{X}_2 = 8 - 12 = -4$. The confidence limits that we find by using Equation (10.23) are as follows:

$$L = (\overline{X}_j - \overline{X}_{j'}) - t_{\alpha/(2m), n_T - k} \sqrt{MSW\left(\frac{1}{n_j} + \frac{1}{n_{j'}}\right)}$$

$$= -4 - (3.4783)\sqrt{10\left(\frac{1}{3} + \frac{1}{3}\right)} = -4 - 8.98 = -12.98$$

and

$$R = (\overline{X}_j - \overline{X}_{j'}) + t_{\alpha/(2m), n_T - k} \sqrt{MSW\left(\frac{1}{n_j} + \frac{1}{n_{j'}}\right)}$$

$$= -4 + (3.4783)\sqrt{10\left(\frac{1}{3} + \frac{1}{3}\right)} = -4 + 8.98 = 4.98$$

We will use the format of the following expression to present the confidence intervals:

$$L \le \mu_j - \mu_{j'} \le R \tag{10.24}$$

The interval for comparing filter 1 and filter 2 is thus

$$-12.98 \le \mu_1 - \mu_2 \le 4.98$$

And because the interval includes 0, we do *not* reject the null hypothesis H_{0_1}: $\mu_1 = \mu_2$. Our conclusion is that the mean water pollution counts that result from the use of filter 1 and filter 2 are not significantly different.

Confidence intervals for testing the null hypotheses H_{0_2} through H_{0_6} were also computed, and they are presented next by using the format of Equation (10.24).

$-10.98 \le \mu_1 - \mu_3 \le 6.98$	(Includes 0; do not reject H_{0_2}.)
$-22.98 \le \mu_1 - \mu_4 \le -5.02$	(Does not include 0; reject H_{0_3}.)
$-6.98 \le \mu_2 - \mu_3 \le 10.98$	(Includes 0; do not reject H_{0_4}.)
$-18.98 \le \mu_2 - \mu_4 \le -1.02$	(Does not include 0; reject H_{0_5}.)
$-20.98 \le \mu_3 - \mu_4 \le -3.02$	(Does not include 0; reject H_{0_6}.)

From the family of confidence intervals we conclude that the mean pollution count for filter 1 is lower than the mean pollution count for filter 4 by an amount somewhere between 5.02 and 22.98. That these amounts are negative in the confidence interval indicates that the mean for filter 1 is lower than the mean for filter 4. We also conclude that the mean pollution count for filter 2 is lower than the mean pollution count for filter 4 by an amount somewhere between 1.02 and 18.98, and that the mean pollution count for filter 3 is lower than the mean pollution count for filter 4 by an amount somewhere between 3.02 and 20.98.

The results for the six pairwise comparisons enable us to conclude that μ_1, μ_2, and μ_3 are not different from each other on a pairwise basis, but each of these means is different from (lower than) μ_4. Thus our decision is *not* to use filter 4. We would use any of the first three types of filters.

The method that we have presented for making multiple (m) pairwise comparisons by using $100(1 - \alpha/m)\%$ as the confidence level for each pairwise comparison when we want the family confidence level to be at least $100(1 - \alpha)\%$ is known as *Bonferroni's method*. Bonferroni's method works quite well if m is not very large. Other multiple comparison methods are presented in the chapter references. Two of these methods are referred to as Tukey's and Sheffé's methods. Methods for making multiple comparisons among groups in an analysis of variance should be considered when the hypothesis of the equality of several means has been rejected and one wishes to study the differences among pairs of groups.

Problems: Section 10.5

13. Refer to Problem 1.
    a. If it is appropriate to do so, use informal comparisons of means to conclude which bond-rating group has the highest debt-to-real-estate-value percentage.
    b. If it is appropriate to do so, use multiple pairwise comparisons of means and a family confidence level of at least 95% to decide which type of bond-rating groups have the higher debt-to-real-estate-value percentages. (*Note:* $t_{.008333,6} = 3.2875$.)

14. Refer to Problem 2.
    a. If it is appropriate, determine which type of mutual fund has the highest rate of return, using informal comparisons of means.
    b. If it is appropriate, determine which type of mutual fund has the highest rate of return, using multiple pairwise comparisons of means. Use at least 95% for the family confidence level. (*Note:* $t_{.004167,8} = 3.4783$.)
15. Refer to Problem 3. If we want to select a salesperson to receive an achievement award based on the sample data of Problem 3, should we use multiple pairwise comparisons of means?
16. Refer to Problem 4. Use multiple pairwise comparisons with a family confidence level of at least 95% to decide which kind of packaging materials result in lower moisture loss. (*Note:* $t_{.008333,27} = 2.5526$.)

10.6 SUMMARY

In this chapter we have considered two different approaches to viewing analysis of variance as it is used to test $H_0: \mu_1 = \mu_2 = \cdots = \mu_k$. The two approaches produce identical results and may be used interchangeably. In addition, we examined a proce-

TABLE 10.7 Summary of Important Points in Chapter 10

Hypothesis To Be Tested	Test Statistic

$$H_0: \mu_1 = \mu_2 = \cdots = \mu_k$$

$$F = s_B^2 / s_W^2$$

where:

$$s_B^2 = \frac{\sum_{j=1}^{k} n_j (\overline{X}_j - \overline{\overline{X}})^2}{k - 1}$$

$$s_W^2 = \frac{\sum_{j=1}^{k} (n_j - 1) s_j^2}{n_T - k}$$

Comments and assumptions:
    —The F value has two degrees of freedom values associated with it: $\nu_1 = k - 1$ and $\nu_2 = n_T - k$.
    —The populations are all assumed to be normally distributed.
    —All the populations must have the same variance.

Alternative view of test for

$$H_0: \mu_1 = \mu_2 = \cdots = \mu_k$$

$$F = \frac{Between\text{-}groups\ MS}{Within\text{-}groups\ MS}$$

where *Between-groups MS* has same formula as s_B^2 above and *Within-groups MS* is

$$MSW = \frac{\sum_{j=1}^{k} \sum_{i=1}^{n_j} (X_{ij} - \overline{X}_j)^2}{n_T - k}$$

Comments and assumptions:
    —This method gives results identical to those obtained by using the preceding method.
    —All comments for the preceding method apply here.

For multiple pairwise comparisons of means, use Equation (10.23).

dure for making multiple pairwise comparisons of means that is appropriate when the null hypothesis of the equality of several means has been rejected.

We have only considered analysis of variance problems where our conclusions are limited to the specific levels of the factors that have been included; that is, the effects of the treatments are *fixed*. Discussions of extensions of these designs are found in the references.

Because of the computational effort involved in an analysis of variance, data analysis systems are often employed to obtain ANOVA tables. For that reason several edited computer printouts were presented with the problems. Some important points of the chapter are summarized in Table 10.7

REVIEW PROBLEMS   *Answers to odd-numbered problems are given in the back of the text.*

 17. An executive with the Hand & Welding Communications advertising agency is interested in the effectiveness of three types of advertisements: movies, television, and free samples of a product. Random samples of the effectiveness (in thousands of exposures per dollar) for three advertisement campaigns for each type of advertisement were taken, and the results are presented in the accompanying table.

    a. Find the grand mean $\overline{\overline{X}}$ for all of the sample data, the sample group means $\overline{X}_1, \overline{X}_2$, and $\overline{X}_3$, and the sample variances for each group, s_1^2, s_2^2, and s_3^2.

    b. Find the between-groups estimate of variance.

    c. Find the within-groups estimate of variance.

    d. Test the hypothesis that the mean effectiveness values are the same for all three types of advertising. Use .05 for the level of significance.

Movies	Television	Free Samples
6	4	1
7	5	4
8	9	1

 18. Companies sometimes experience financial difficulty that eventually results in bankruptcy. An analyst for Delta Industries studied cumulative profitability, as measured by the firm's retained-earnings-to-total-assets ratio as a percentage, for firms that are going concerns, in reorganization, and bankrupt. Random samples of three firms for each category of financial condition were taken, and the results are presented in the accompanying table.

    a. Find the grand mean $\overline{\overline{X}}$ for all of the sample data, the sample group means $\overline{X}_1, \overline{X}_2$, and $\overline{X}_3$, and the sample variances for each group, s_1^2, s_2^2, and s_3^2.

    b. Find the between-groups estimate of variance.

    c. Find the within-groups estimate of variance.

    d. Test the hypothesis that the mean retained-earnings-to-total-assets ratios as percentages are the same for all three groups. Use .05 for the level of significance.

Going Concern	Reorganizing	Bankrupt
11	1	6
12	6	8
13	5	10

19. A director of In-Vest, Inc., believes that the percentage return on assets should be different for companies pursuing single-business, related-business, and conglomerate strategies, because these types of companies undertake different levels of risk. Random samples of percentage return on assets for three, three, and four companies were taken from the single-business, related-business, and conglomerate-business strategy groups, respectively. The results are presented in the accompanying table.

   a. Find the grand mean $\overline{\overline{X}}$ for all of the sample data.
   b. Find the between-groups estimate of variance.
   c. Find the within-groups estimate of variance.
   d. Test the hypothesis that the means are the same for all three groups. Use .05 for the level of significance.

	Single	Related	Conglomerate
	22	13	8
	22	11	4
	20	14	5
			7
n_j	3	3	4
$\overline{X}_j$	21.33	12.67	6.00
s_j	1.15	1.53	1.83

20. An information systems manager with ICX Corporation conducted a study of the operating time prior to failure of the flexible-disk drives for brand A, brand B, and brand C microcomputers. Random samples of computers were selected, and a software package designed to test disk drives was used on the different brands until the disk drives failed. The results for the time to failure (in thousands of hours) are presented in the accompanying table. The samples were of different sizes.

   a. Find the grand mean $\overline{\overline{X}}$ for all of the sample data.
   b. Find the between-groups estimate of variance.
   c. Find the within-groups estimate of variance.
   d. Test the hypothesis that the means are the same for all three brands. Use .05 for the level of significance.

	Brand A	Brand B	Brand C
	3	7	7
	3	7	8
	4	7	8
	6	7	7
	3	6	
	4		
n_j	6	5	4
$\overline{X}_j$	3.83	6.80	7.50
s_j	1.17	.447	.577

21. Many companies have recently attempted to enhance the quality of work life and to increase employee job satisfaction by changing work schedules or by making work schedules more flexible. Three types of work schedules that are now popular are four-day work weeks, five-day work weeks, and flextime. Random samples of employee job satisfaction, as measured by a job satisfaction index, were taken for four employees from three different work schedules. The results are presented in the accompanying table.

a. Find the grand mean $\overline{\overline{X}}$ for all of the sample data and the sample group means $\overline{X}_1$, $\overline{X}_2$, and $\overline{X}_3$.
b. Find the between-groups sum of squares.
c. Find the between-groups degrees of freedom.
d. Find the mean square between groups.
e. Find the within-groups sum of squares.
f. Find the within-groups degrees of freedom.
g. Find the mean square within groups.
h. Test the hypothesis that the mean job satisfaction values are the same for all three work schedules. Use .05 for the level of significance.

Four-Day	Five-Day	Flextime
4	1	9
9	0	6
6	3	7
1	4	10

 22. A market study was done at Upton Company to examine the differences in hourly sales (in units) for a product that was priced at three different levels $1, $2, and $3. Random samples of unit sales were taken under the three prices. The sample sizes were $n_1 = 2$, $n_2 = 3$, and $n_3 = 2$ for the three prices, and other factors that influence sales were essentially held constant. The results are presented in the accompanying table.

a. Find the grand mean $\overline{\overline{X}}$ for all of the sample data and the sample group means $\overline{X}_1$, $\overline{X}_2$, and $\overline{X}_3$.
b. Find the between-groups sum of squares SSB.
c. Find the value for the between-groups degrees of freedom.
d. Find the between-groups mean square MSB.
e. Find the within-groups sum of squares SSW.
f. Find the value for the within-groups degrees of freedom.
g. Find the within-groups mean square MSW.
h. Show the ANOVA table.
i. Test the hypothesis that the means are the same for all three prices. Use .05 for the level of significance, and assume normal distributions and equal variances.

$1 Price	$2 Price	$3 Price
9	4	1
7	5	3
	6	

 23. Three different automatic milling machines at Castmetal, Inc., were set up to mill the same type of part. Observations were taken at random (see the accompanying table) to find out how many parts were being produced per hour by each machine. Only four observations were taken on machine 3 since the inspector taking the observations became ill and had to go home before he could complete his work. Test the hypothesis of equal mean hourly output for each machine. Use $\alpha = .01$.

Machine 1	Machine 2	Machine 3
105	100	104
105	98	106
110	112	99
107	114	109
102	100	

24. Several entrepreneurs who started a business that has now developed into a profitable corporation are interested in a partially tax-free reorganization. The type of reorganization selected will influence the amount of property that does not qualify for tax-free treatment assigned to each entrepreneur. Nonqualified property is known as "boot." The types of reorganization that are being considered are type A, merger, type B, stock for stock, type C, assets for stock, and type D, spin-off. Consulting accountants cannot determine the effects on the boot of the type of reorganization for all of the entrepreneurs for a reasonable fee. Thus random samples of entrepreneurs were taken under each type of reorganization, and the results for the boot are presented in the accompanying table. The sample sizes were different for the four types of reorganization.

    a. Find the grand mean $\overline{\overline{X}}$ for all of the sample data and the sample group means $\overline{X}_1$, $\overline{X}_2$, $\overline{X}_3$, and $\overline{X}_4$.
    b. Find the between-groups sum of squares SSB.
    c. Find the value for the between-groups degrees of freedom.
    d. Find the between-groups mean square MSB.
    e. Find the within-groups sum of squares SSW.
    f. Find the value for the within-groups degrees of freedom.
    g. Find the within-groups mean square MSW.
    h. Show the ANOVA table.
    i. Test the hypothesis that the means are the same for all four groups. Use .05 for the level of significance, and assume normal distributions and equal variances.

Type A	Type B	Type C	Type D
5	9	13	10
9	5	11	16
	7	15	

25. The accompanying table gives the piece rates of pay for samples of workers in three different apparel sewing shops owned by House of Lauren. Test the hypothesis H_0: $\mu_1 = \mu_2 = \mu_3$. Let $\alpha = .01$.

Shop 1	Shop 2	Shop 3
2.30	2.00	1.65
2.35	1.80	1.90
2.25	2.15	1.85
2.00	2.10	1.80
1.95	1.90	1.75
2.10	1.75	1.95
2.40		2.00
		1.90

26. In a study of the effects of different pressures on the yield of a dye at Burleson Industries textile mill, five containers of dye were sampled for each of three pressures. The results are given in the accompanying table. Test the hypothesis that the mean yields are the same for all three pressure settings. Let the chance of a type I error be .05. An SPSS printout for this problem is presented in Computer Exhibit 10.3.

200 mm	500 mm	800 mm
32.4	37.8	30.3
32.6	38.2	30.5
32.1	37.9	30.0
32.4	38.0	30.1
32.3	37.8	29.7

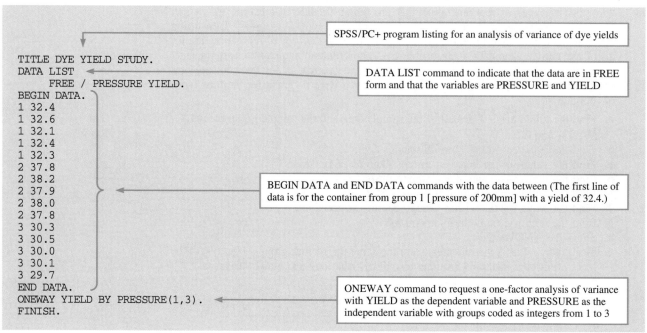

SPSS/PC+　　　**COMPUTER EXHIBIT 10.3** Analysis of Variance for Dye Yields at Burleson Industries

SPSS/PC+ program listing for an analysis of variance of dye yields

```
TITLE DYE YIELD STUDY.
DATA LIST
    FREE / PRESSURE YIELD.
BEGIN DATA.
1 32.4
1 32.6
1 32.1
1 32.4
1 32.3
2 37.8
2 38.2
2 37.9
2 38.0
2 37.8
3 30.3
3 30.5
3 30.0
3 30.1
3 29.7
END DATA.
ONEWAY YIELD BY PRESSURE(1,3).
FINISH.
```

DATA LIST command to indicate that the data are in FREE form and that the variables are PRESSURE and YIELD

BEGIN DATA and END DATA commands with the data between (The first line of data is for the container from group 1 [pressure of 200mm] with a yield of 32.4.)

ONEWAY command to request a one-factor analysis of variance with YIELD as the dependent variable and PRESSURE as the independent variable with groups coded as integers from 1 to 3

(continues)

 27. An individual investor is interested in examining the differences in portfolio management with respect to the average maturity (in days) for three tax-exempt money market funds. The funds are tax-free income, tax-exempt, and municipal bond. Random samples were taken from different times, and the maturities are presented in the accompanying table.

   a. Show the ANOVA table.
   b. Test the hypothesis that the means are the same for all three groups. Use .05 for the level of significance.

Fund 1	Fund 2	Fund 3
56	65	61
70	59	67
65	57	64
80	67	64
61	55	55
76	66	65
74	62	67
67	64	79
73	58	
54	63	

 28. An accountant with Arthur Anderson was interested in the impact of price changes over time on a firm's return on assets (ROA). ROA is an accounting measure of performance. Three methods of accounting are (1) historical cost, (2) current cost, and (3) constant-dollar, and the two latter methods attempt to adjust for changing prices. A sample was taken and firms were assigned at random to use one of the accounting

---

COMPUTER EXHIBIT 10.3 Analysis of Variance for Dye Yields at Burleson Industries
(Continued)

SPSS/PC+

```
- - - - - - - - - - O N E W A Y - - - - - - - - - -

     Variable  YIELD

    By Variable  PRESSURE

                        Analysis of Variance

                        Sum of      Mean        F       F
        Source       D.F.  Squares   Squares   Ratio   Prob.

Between Groups        2   162.1773   81.0887  1589.9739  .0000

Within Groups        12     .6120      .0510

Total                14   162.7893
```

ANOVA table

---

methods to compute *ROA*. The results are presented by group according to the accompanying table. The firms were all essentially the same, except for the accounting method they used. Notice that the data are presented in a form that gives the *ROA* and the accounting method as two items of data for each firm. Some data analysis packages require this form, and other packages use a unique command to perform analysis of variance if the data are entered in this form.

a. Show the ANOVA table.
b. Test the hypothesis that the means are the same for all three groups. Use .05 for the level of significance.

ROA	Group	ROA	Group	ROA	Group
12	1	10	2	4	3
13	1	10	2	6	3
11	1	11	2	6	3
12	1	7	2	10	3
11	1	7	2	7	3
12	1	9	2	3	3
12	1	8	2	5	3
13	1	11	2	1	3
12	1	5	2	8	3
9	1	7	2	4	3

29. The data in Table 10.8 give profitability measures in the form of five-year average returns on equity for U. S. banks with Northeast, Southeast, and West locations. Assume that the listings are independent random samples. Test the hypothesis that banks in these regions had equal mean profitabilities. Use .05 for the level of significance.

30. Samples of start-up owners' equity (in hundreds of thousands of dollars) for small business firms that within the first year of operation have resulted in Chapter 7 bankruptcy or Chapter 11 bankruptcy and firms that are solvent gave the results in the accompanying table. The samples are random and independent and the distributions are normal with equal standard deviations. Test to see whether the means of start-up owners' equity are equal for small business firms that are in Chapter 7 or Chapter 11 bankruptcy or that are solvent. Use .05 for the level of significance. You may want to

TABLE 10.8   Return-on-Equity Data for Banks

Northeast	Return on Equity (%)	Southeast	Return on Equity (%)	West	Return on Equity (%)
State Street Boston	21.3	First Wachovia	18.5	Security Pacific	16.4
Midlantic Banks	21.0	SunTrust Banks	18.5	Valley National	14.1
Republic New York	18.9	Barnett Banks Fla	18.1	Rainier Bancorp	13.9
Fleet Financial	18.7	Citizens & Southern	16.7	Wells Fargo	13.7
Hartford National	17.2	First Union	16.6	First Interstate Bncp	13.7
Bank of New England	16.2	NCNB	16.5	US Bancorp	13.4
First Fidelity Bncp	16.1	Third National	16.2	United Banks of Colo	12.9
Bank of New York	15.8	Florida Natl Banks	15.9	First Security	10.5
Norstar Bancorp	15.7	First Tennessee Natl	14.8	California First Bank	8.2
BayBanks	15.3	Southeast Banking	13.3		
KeyCorp	15.2				
Shawmut	13.4				
Marine Midland Banks	10.4				

Source: *Forbes*, January 12, 1987, pp. 74–77.

compute sample means, sums of squared deviations, and mean squares for this problem by using hand calculations.

Chapter 7	Chapter 11	Solvent
3	2	12
8	1	9
5	4	10
0	5	13

 31. Accounting fraud can result in misstatements in balance sheets. For example, accounts receivable can be misstated if the receivables are nonexistent. Auditors for James, Judd, and Stephens took samples of accounts receivable misstatement risk, and the results are given in the accompanying table. The audits were completed with exceptions noted, completed with no exceptions, or not completed.

a. Provide an ANOVA table summarizing your results.

b. Test to see whether the means of misstatement risk are equal for the different audit outcomes. Use .05 for the level of significance.

c. If it is appropriate, use multiple pairwise comparisons to conclude which audit result has the lowest accounts receivable misstatement risk. Use 95% for the family confidence level.

Completed With Exceptions	Completed Without Exceptions	Not Completed
7	1	5
5	9	15
4	5	
1		
8		

Reference: Brown, C. E., and I. Soloman (1991), "Configural Information Processing in Auditing," *The Accounting Review,* 66: 100–19.

Refer to the 141 companies listed in the Ratio Data Set in Appendix A.   **Ratio Data Set**

Questions

32. Locate the data for the current assets–sales ratio (*CA/S*). Some companies in the file are in the (1) transportation and utilities industry, some are in the (2) retail industry, and some are in the (3) services industry. Use analysis of variance to test whether the companies in the three industries have equal means for this ratio. Use a significance level of .05.

33. Locate the data for the quick assets–sales ratio (*QA/S*). Some companies in the file are in the (1) transportation and utilities industry, some are in the (2) retail industry, and some are in the (3) services industry. Use analysis of variance to test whether the companies in the three industries have equal means for this ratio. Use a significance level of .05.

34. Locate the data for the current assets–current liabilities ratio (*CA/CL*). Some companies in the file are in the (1) transportation and utilities industry, some are in the (2) retail industry and some are in the (3) services industry. Use analysis of variance to test whether the companies in the three industries have equal means for this ratio. Use a significance level of .05.

35. Locate the data for the net income–total assets ratio (*NI/TA*). Some companies in the file are in the (1) transportation and utilities industry, some are in the (2) retail industry, and some are in the (3) services industry. Use analysis of variance to test whether the companies in the three industries have equal means for this ratio. Use a significance level of .05.

Refer to the 113 applicants for credit listed in the Credit Data Set in Appendix A.   **Credit Data Set**

Questions

36. Use analysis of variance to test the hypothesis that the population means for JOBINC are the same for those who were granted credit and those who were denied credit. Omit those applicants who did not list a JOBINC value. Let $\alpha = .05$.

37. Compare the results you obtained for the first Credit Data Set question of Chapter 9 with the results you obtained in Question 36.

TESTING THE EFFECTIVENESS OF CONTRIBUTION-SOLICITING METHODS   **CASE 10.1**

The Metropolitan Charities United Association raises money for charitable causes in a large eastern urban area. The director of the association, Mr. William P. Dankful, and his staff are interested in determining what contribution-soliciting methods produce the best results.

The association has decided that it will test three solicitation methods: direct mail, telephone calls, and personal visits. In the test, three solicitors were used to make the telephone calls and the personal visits. All three made 15 calls and 15 visits each in order to eliminate any bias that might be introduced by the differing sales abilities of the three solicitors. Of course, none of the solicitors was involved in the direct-mail solicitations, since these were nonpersonal contacts.

The results of the solicitations are presented in Table 1. The figures represent the amount of money donated to the association on each contact.

TABLE 1   Solicitation Results for Metropolitan Charities

Solicitor	Telephone Call	Personal Visit	Solicitor	Telephone Call	Personal Visit	Solicitor	Telephone Call	Personal Visit
1	$ .00	$ 2.00	2	$ 3.00	$ 7.00	3	$ .00	$ .00
1	.00	.00	2	.00	10.00	3	15.00	.00
1	5.00	10.00	2	.00	8.00	3	5.00	.00
1	10.00	.00	2	20.00	10.00	3	.00	6.00
1	.00	.00	2	5.00	.00	3	.00	4.00
1	8.00	2.00	2	.00	.00	3	3.00	.00
1	.00	1.00	2	.00	2.00	3	2.50	3.50
1	15.00	125.00	2	4.00	5.50	3	.50	110.00
1	.00	5.00	2	5.00	.00	3	.00	.00
1	.00	.00	2	.00	1.50	3	100.00	9.50
1	.00	12.00	2	.00	30.00	3	.00	20.00
1	10.00	.00	2	.00	20.00	3	.00	.00
1	.00	.00	2	15.00	.00	3	5.00	40.00
1	.00	50.00	2	25.00	10.00	3	3.00	.00
1	5.00	.00	2	.00	.00	3	.00	5.00

The responses to 50 direct-mail solicitations were as follows:

$ .00	$ .00	$10.00	$ .00	$ .00	$ .00	$ .00	$ .00	$ .50	$ .00
.00	.00	.00	.00	.00	.00	.00	.00	.00	7.50
.00	.00	2.00	.00	.00	.00	.00	2.50	.00	.00
5.00	.00	5.50	.00	.00	.00	.00	.00	75.00	.00
.00	.00	.00	.00	50.00	.00	.00	.00	.00	.00

The costs of each contact differ. A direct-mail contact is estimated to cost approximately $.20 for printing and mailing. The telephone contacts cost only the labor of the caller, telephone rental, and office space costs. These are estimated to be about $.35 per contact. (Calls are made until the potential donor is reached.) The personal visits cost the most. They involve labor costs of the visitor and mileage that must be paid for the visitor's automobile. These costs run about $3.00 per contact.

a. Determine the hypothesis you would need to test the statistical significance of the methods' results given that a contribution is made.

b. Use analysis of variance to test the hypothesis that all three methods produce the same mean size of contribution given that a contribution is made.

REFERENCES

Cochran, W. G., and Gertrude M. Cox. 1957. *Experimental Designs,* 2nd ed. New York: Wiley.

Hocking, R. R. 1985. *The Analysis of Linear Models.* Monterey, Calif.: Brooks/Cole.

Kempthorne, Oscar. 1952. *The Design and Analysis of Experiments.* New York: Wiley.

Miller, R. G. 1981. *Simultaneous Statistical Inference,* 2nd ed. New York: Springer Verlag.

Neter, John, William Wasserman, and M. H. Kutner. 1990. *Applied Linear Statistical Models,* 3rd ed. Homewood, Ill.: Irwin.

Ostle, Bernard. 1963. *Statistics in Research,* 2nd ed. Ames: Iowa State University Press.

Snedecor, George W., and William G. Cochran. 1989. *Statistical Methods,* 8th ed. Ames: Iowa State University Press.

Scheffé, H. 1959. *Analysis of Variance.* New York: Wiley.

Optional Topics

Analysis of Variance and Two-Sample Tests
Introduction To Experimental Designs
Two-Factor Analysis of Variance and a Randomized Complete Block Design
Two-Factor Analysis of Variance and a Factorial Design

CHAPTER 10 SUPPLEMENT

In the chapter introduction we indicated that analysis of variance is an extension of the two-sample test of hypothesis for continuous populations with common variances in Chapter 9. We use the t test of Chapter 9 to test the hypothesis H_0: $\mu_1 = \mu_2$ and the analysis of variance method with the F value to test the hypothesis that three or more population means are equal to one another.

From this relationship one might assume that the two-sample t test is related to analysis of variance for the special case of testing H_0: $\mu_1 = \cdots = \mu_k$, where $k = 2$. This assumption is correct and is demonstrated in the following problem.

10.7
ANALYSIS OF VARIANCE AND TWO-SAMPLE TESTS

PRACTICE PROBLEM 10.3

The Federated Department Store chain operates two outlets of the store in the same city. In an effort to determine whether the outlets are serving different types of customers, the marketing manager of the chain took samples of the charge account customers at each outlet. She found the mean monthly billing for each customer in the two samples. The sample results follow.

$$n_1 = 31 \qquad \overline{X}_1 = \$42.16 \qquad s_1 = \$8.93$$
$$n_2 = 31 \qquad \overline{X}_2 = \$35.55 \qquad s_2 = \$9.78$$

Test the hypothesis that there is no difference between the mean monthly billings of the two outlets' charge account customers. Let the alternative hypothesis be that the two outlets' population means are not the same. Use an α value of .05.

Solution Introduction This problem is going to be solved by using the two-sample test of Chapter 9 when $\sigma_1 = \sigma_2$ *and* using analysis of variance. The hypotheses are

$$H_0: \mu_1 = \mu_2 \qquad H_a: \mu_1 \neq \mu_2$$

Solution Using Two-Sample Methods of Chapter 9 The t test with equal standard deviations of Chapter 9 requires that we pool the sample variances. The best estimate of that common variance is

$$s^2_{\text{pooled}} = \frac{(n_1 - 1)s_1^2 + (n_2 - 1)s_2^2}{n_1 + n_2 - 2}$$
$$= \frac{(30)(8.93)^2 + (30)(9.78)^2}{31 + 31 - 2} = \frac{5261.80}{60} = 87.70$$

We can now use this value to obtain an estimated standard deviation of $\overline{X}_1 - \overline{X}_2$:

$$s_{\overline{X}_1-\overline{X}_2} = \sqrt{s^2_{pooled}\left(\frac{1}{n_1} + \frac{1}{n_2}\right)} = \sqrt{87.70\left(\frac{1}{31} + \frac{1}{31}\right)} = \sqrt{5.66} = 2.38$$

Then the hypothesis is tested by using the t ratio:

$$t = \frac{\overline{X}_1 - \overline{X}_2}{s_{\overline{X}_1-\overline{X}_2}} = \frac{42.16 - 35.55}{2.38} = \frac{6.61}{2.38} = 2.78$$

Since the alternative hypothesis is two-tailed and since α is .05, we compare the calculated $t = 2.78$ with the Table VI value of $t_{.025, 60} = 2.000$. Since the calculated t value exceeds the table t value, we reject the null hypothesis.

Solution Using Analysis of Variance To use analysis of variance, we need two estimates of the population variance. The first is found by using Equation (10.4). But that formula involves the grand mean $\overline{X}$. Since the two sample sizes were the same, the grand mean is merely the mean of the means:

$$\overline{\overline{X}} = \frac{42.16 + 35.55}{2} = 38.86$$

Now Equation (10.4) gives us

$$s^2_B = \frac{31[(42.16 - 38.86)^2 + (35.55 - 38.86)^2]}{(2 - 1)}$$

$$= \frac{31(21.85)}{1} = 677.23$$

The s^2_W estimate of the population variance is given by Equation (10.5). However, this value was computed in the first-solution approach as the pooled estimate of variance $s^2_{pooled} = 87.70$. Thus

$$s^2_W = 87.70$$

The F ratio for this problem is then

$$F = \frac{s^2_B}{s^2_W} = \frac{677.23}{87.70} = 7.72$$

The numerator of this ratio has one degree of freedom. The denominator has $(n_1 + n_2 - 2) = 60$ degrees of freedom. In the 5% section of Table VII we obtain $F_{.05, 1, 60} = 4.0012$, which is less than our calculated F value of 7.72. Therefore we reject the hypothesis of equal mean monthly billing in using the analysis of variance approach.

The two approaches arrived at the same conclusion—that we should reject the null hypothesis—but they have more in common than the conclusion. The parallel results of these two approaches are presented in the following summary.

Test Name	Calculated Statistic	Table Value	α	Conclusion
Two-sample t	$t = 2.78$	$t_{.025, 60} = 2.000$	.05	Reject H_0
Analysis of variance	$F = 7.72$	$F_{.05, 1, 60} = 4.0012$	.05	Reject H_0

Note in the second column that $(2.78)^2 \approx 7.72$ and that in the third column $(2.000)^2 \approx 4.0012$, with the exception of rounding differences. Thus the two tests not only give the same conclusion but give virtually identical mathematical results.

To demonstrate the connection between analysis of variance and the t test, we must examine the following relationship between the t and the F distributions:

$$t^2_{\alpha/2,\nu} = F_{\alpha,1,\nu} \tag{10.25}$$

Thus we can square the t values in the $\alpha/2$ column of the t table (Table VI) and we will obtain the F values found in the first column (one degree of freedom in the numerator) and α section of the F table (Table VII). Conversely, the positive square root of the F values in column 1 of the α section of Table VII gives the t values in the $\alpha/2$ column of Table VI. (The reader may select a few values from each table to test this relationship.)

Therefore a t test of two means where the area in a single tail is $\alpha/2$ corresponds to an ANOVA F test of two means where $\nu_1 = 1$ and this area is α. For the data in our example both the calculated and the table value of $t^2_{\alpha/2,\nu}$ equal the corresponding $F_{\alpha,1,\nu}$ values and verify the relationship of Equation (10.25). The reader will have additional opportunities to verify this relationship in the problems that follow.

Problems: Section 10.7

38. The manager of quality control at a sugar refinery was worried that two packaging production lines might be filling the packages to different weights. Thus samples of size 16 were taken from each of the production lines and the contents of 10-pound packages carefully weighed. The following sample results seem to indicate that the mean weights of the 10-pound packages from the two lines are the same, but there appears to be much more variation in the weights of the packages coming off the second line. Test the hypothesis that the two lines have the same variation in weights by testing $H_0: \sigma_1^2 = \sigma_2^2$, using $\alpha = .05$. Use $H_a: \sigma_1^2 \neq \sigma_2^2$. If the following data do not cause rejection of the null hypothesis, what must we do in order to test $H_0: \mu_1 = \mu_2$?

$$n_1 = 16 \qquad n_2 = 16$$
$$\overline{X}_1 = 10.15 \qquad \overline{X}_2 = 10.16$$
$$s_1 = .07 \qquad s_2 = .16$$

39. The vice president for maintenance at a Rent-a-Wreck auto rental firm is concerned about the uniformity of wear given by tires put on the company's rental cars. If the tires on a car wear uniformly, then they can all be replaced at one time—when they all get to roughly the same level of tread wear. However, if they wear in a non-uniform manner, the car must be taken into the shop two, three, or four different times to change the one or two tires with the greatest wear. This latter situation is much more expensive in terms of maintenance costs. Thus the vice president had the maintenance shops conduct a study in which they equipped 50 cars with Northwest tires and 50 cars with Eastern tires. Logs were kept on each of the cars, and when each car had traveled 20,000 miles from the time of the tire installation, the tread wear on each car's left rear tire was measured. Owing to lost records, blowouts, and the sale of some cars, there were only 31 Northwest and 41 Eastern tire sets left at the end of the test. Problems 39–41 refer to the data given here.

Tire Type	Sample Size	Sample Mean	Sample Variance
Northwest	31	2.46	.65
Eastern	41	2.66	.98

What are the values of k and all the n_j's in this problem? Test the hypothesis that the mean tread wear is the same for both brands of tire, using analysis of variance and an α value of .05.

40. Test the hypothesis that the mean tread wear in Problem 39 is the same for both brands of tire, using the two-sample t test of Section 9.8 with $\alpha = .05$. (Assume H_a: $\mu_1 \neq \mu_2$.)

41. Show that both the calculated and theoretical t and F values you used in Problems 39 and 40 are related. That is, verify that $t_{\alpha/2,\nu}^2 = F_{\alpha,1,\nu}$ for the values you used.

42. The following data are yields in bushels per acre for two oat varieties used at Oklahoma Land & Agronomics, Inc. Each was tried on eight different plots. Can we conclude that the yields are the same for the two varieties, A and B? Or is the alternative hypothesis more reasonable: H_a: $\mu_A \neq \mu_B$ if we use $\alpha = .10$? Use the two-sample t test when $\sigma_1 = \sigma_2$.

A	81.2	72.6	56.8	76.9	42.5	49.6	62.8	48.2
B	56.6	58.6	45.4	39.1	42.8	65.2	40.7	49.9

43. Use the data in Problem 42 to test the hypothesis H_0: $\mu_1 = \mu_2$, using analysis of variance with $k = 2$ and $\alpha = .05$.

44. Verify that the t values of Problem 42 are related to the F values of Problem 43 by the relationship $t^2 = F$.

10.8
INTRODUCTION TO EXPERIMENTAL DESIGNS

Analysis of variance is often applied to situations in which experiments have been conducted and the results are being evaluated. For instance, a chemical company may wish to conduct an experiment in which four different types of fertilizer are tested to determine whether they produce differences in crop yields. If the company has 100 acres on which it might run tests, it could plant four 25-acre plots and treat each plot with a different fertilizer. Thus each plot could be viewed as a sample of $n_j = 25$ one-acre sample elements. The results would be expressed in terms of four different means where $\overline{X}_j$ is the mean crop yield per acre in the jth plot, $j = 1, \ldots, 4$.

As another example of experimentation, a company might wish to test the effectiveness of three different sales presentations. The sales manager could train three teams of salespeople in these presentation methods and then let them call on customers for a period of time. At the end of that time the sales manager could examine the results of the test by comparing the means produced by the three sales teams where $\overline{X}_j$ is the mean number of units purchased per customer exposed to the jth sales presentation.

In both of these examples we have situations where test results are produced by applying k different *treatments* (such as fertilizers or sales presentations) to different groups of **experimental items.**

Definition

An **experimental item,** or **experimental unit,** is a physical item that is included in the sample and from which the data measurements are taken.

An experimental item is often a person, a machine, a company, a store, or a city. There are n_j experimental items that receive each treatment in both of the examples discussed above.*

> **Definition**
>
> When the selection of a treatment for each experimental item is made randomly, then the experiment uses a **completely randomized design.**
>
> Variables used to define treatments are called **factors.**
>
> The categories defined by a factor are referred to as the **levels** of the factor.

The examples presented above use a single factor to define the treatments. In Section 10.10 we present an experimental design that uses two factors to define the treatments.

In the Sierra Pacific filter example that was first introduced in Section 10.1, an experimental item is a container of water, the data measurement is the pollution count, the filters constitute the factor, and the factor has four levels, one for each type of filter. If there is a single factor, as in the filter example, then the treatments correspond to the factor levels. The completely randomized design with k groups or k factor levels corresponds to the completely randomized experiment for two groups (Section 9.8).

The ANOVA approach can be used just as it was described in Sections 10.1 and 10.4 to test the hypothesis that the treatment groups in single-factor experiments have equal means. The s_B^2, or MSB, value in experimental situations is often called the treatments variance estimate, or $MSTR$; and the s_W^2, or MSW, value is called the error variance estimate, or MSE. The reason for calling s_W^2 the error variance estimate is that it measures the variations of measurements for all items treated alike (all the items in the same group). Thus even if the different treatments produce no impact on the measurements, we would expect to see a variation in measurements as large as s_W^2 simply because of natural random variation within the experimental items and of errors in our measurements. That is, s_W^2 measures random error or variation in our measurements.

The analysis of variance methods that we have considered to this point have included only one treatment factor or grouping variable. Experiments sometimes have more than one factor or grouping variable, and in this section we consider the **randomized complete block design,** an experimental design with a two-way classification scheme that includes a treatment factor and a grouping variable.

Experimental items are said to be **homogeneous** if they can be expected to respond similarly when subjected to an experimental treatment. When all of the experimental items are not homogeneous, we can sometimes sort the items into groups, called **blocks,** such that the items are homogeneous within each block, and the items are heterogeneous when we compare one block with another.

10.9
TWO-FACTOR ANALYSIS OF VARIANCE AND A RANDOMIZED COMPLETE BLOCK DESIGN

---

*In some designs, experimental items and experimental units are not equivalent. For example, a new training program may be assigned to a department (an experimental unit), and data values may be obtained from individuals in the department (items).

> **Definition**
>
> When the items included in an experiment can be separated into n homogeneous groups, or blocks, and each experimental item included in a block is assigned randomly to one of the k levels of a treatment factor, we have a **randomized complete block design.**

In this section we consider the case where exactly *one* experimental item is included in each different combination of a block and a treatment (or cell), so overall there will be kn experimental items. With this design the data can be arranged in a two-way table consisting of blocks and treatments with exactly one experimental item in each cell of the table. We will arrange the blocks as rows and the treatments as columns in the table. Because there is one experimental item in every cell of the table, the term *complete* is used in describing the design.

The purpose of a randomized complete block design is to improve the precision of the experiment by decreasing the error variance (estimated previously by s_w^2 or MSW in Sections 10.3 and 10.4), without increasing the size of the experiment. The error variance may be decreased by removing variation attributable to differences among blocks, if the blocking is effective, resulting in a more powerful hypothesis test of the equality of the treatment group means. The blocking is effective if the values of the dependent variable have more variation between blocks than they have within the blocks.

=== **EXAMPLE 10.9**   A new experiment was conducted at Sierra Pacific Electric Company to study the water pollution counts for containers of water (the experimental items) randomly sampled from a production cooling system when it is fitted with three different types of filters (treatments). In this example, the water comes from four different sources (blocks): (1) culinary water, (2) a recycling plant, (3) a stream, and (4) a pond. Logically, the pollution counts for the containers of water will be dependent on the water sources. One would expect a large variation in pollution counts from one water source (block), such as culinary water, to another, such as water from a pond. A single container of water from each type of water source (block) must be taken randomly from each type of filter (treatment or factor level) for this randomized complete block design. The results for this experiment, in the form of water pollution counts, are presented in Table 10.9. We will use this data for analysis of variance computations.

We want to test the hypothesis that the mean pollution count is the same for all three filters. That is, where $\mu_{.j}$ is the population mean of the dependent variable for the jth treatment (column), we wish to test H_0: $\mu_{.1} = \mu_{.2} = \mu_{.3}$ against the alternative hypothesis that one or more of the filters produce a different mean pollution count.   ===

We will account for the blocking variable while testing the hypothesis of this example by using the methods of **two-factor analysis of variance.** The blocking variable in

		Treatment		
TABLE 10.9	*Block*	*Filter 1*	*Filter 2*	*Filter 3*
Pollution Counts at Sierra	*1. Culinary Water*	6	4	2
Pacific: Three Filters	*2. Recycled Water*	11	7	3
(Treatments) and Four	*3. Stream Water*	15	11	7
Water Sources (Blocks)	*4. Pond Water*	24	22	20

the example stems from a characteristic of the experimental materials (the different water sources). In other studies—in a production setting, for example—different production machines or machine operators may serve as blocks. If the experimental item is a person, then sex, age, income, or education if often used for blocking purposes.

Because a treatment factor and a blocking variable are both included in a randomized complete block design, we require a two-factor analysis of variance. For this design the **total sum of squares** (*SST*) is partitioned into three parts: the **sum of squares between treatments** (*SSTR*), the **sum of squares between blocks** (*SSBL*), and the **error sum of squares** (*SSE*).

Equation for Partitioning Sum of Squares

$$SST = SSTR + SSBL + SSE \tag{10.26}$$

The following notation will be used during the computation of the various sums of squares:

X_{ij} = value of dependent variable for the experimental
item in ith block and jth treatment

k = number of treatments

n = number of blocks

n_T = total number of experimental items sampled = kn

$$\overline{X}_{.j} = \text{sample mean for } j\text{th treatment (column)} = \frac{\sum\limits_{i=1}^{n} X_{ij}}{n} \tag{10.27}$$

$$\overline{X}_{i.} = \text{sample mean for } i\text{th block (row)} = \frac{\sum\limits_{j=1}^{k} X_{ij}}{k} \tag{10.28}$$

$\overline{\overline{X}}$ = *grand mean* = sample mean for all experimental items

$$= \frac{\sum\limits_{i=1}^{n}\sum\limits_{j=1}^{k} X_{ij}}{n_T} \tag{10.29}$$

Definitional Formulas

The formulas that *define* the four different sums of squares in Equation (10.26) are as follows:

$$SST = \sum_{i=1}^{n}\sum_{j=1}^{k}(X_{ij} - \overline{\overline{X}})^2 \tag{10.30}$$

$$SSTR = n\sum_{j=1}^{k}(\overline{X}_{.j} - \overline{\overline{X}})^2 \tag{10.31}$$

$$SSBL = k\sum_{i=1}^{n}(\overline{X}_{i.} - \overline{\overline{X}})^2 \tag{10.32}$$

If *SST*, *SSTR*, and *SSBL* have been computed previously, then

$$SSE = SST - SSTR - SSBL \tag{10.33}$$

or, by definition, *SSE* can be computed by using

$$SSE = \sum_{i=1}^{n} \sum_{j=1}^{k} (X_{ij} - \overline{X}_{i.} - \overline{X}_{.j} + \overline{\overline{X}})^2 \tag{10.34}$$

We will use the basic definitional formulas (10.29–10.34) in the practice problem so that we can emphasize the meaning of the various sums of squares.

The value for the degrees of freedom of *SST* is $n_T - 1$; *SSTR* has $k - 1$ degrees of freedom: *SSBL* has $n - 1$ degrees of freedom; and *SSE* has $(k - 1)(n - 1)$ degrees of freedom. The degrees of freedom are additive, as shown by the following equation:

$$n_T - 1 = (k - 1) + (n - 1) + (k - 1)(n - 1) \tag{10.35}$$

Mean squares are computed by dividing sums of squares by their associated degrees of freedom in formulas for computing the **between-treatments mean square** (*MSTR*), the **between-blocks mean square** (*MSBL*), and the **error mean square** (*MSE*).

Equations for Mean Squares

$$MSTR = \frac{SSTR}{k - 1} \tag{10.36}$$

$$MSBL = \frac{SSBL}{n - 1} \tag{10.37}$$

$$MSE = \frac{SSE}{(k - 1)(n - 1)} \tag{10.38}$$

If kn independent samples of size one item each are taken from kn normal populations that have equal variances and for which the null hypothesis $H_0: \mu_{.1} = \mu_{.2} = \cdots = \mu_{.k}$ is true, then the ratio

$$F = \frac{MSTR}{MSE} \tag{10.39}$$

follows the *F* distribution with the following degrees of freedom: $\nu_1 = k - 1$ and $\nu_2 = (k - 1)(n - 1)$.

A large *F* value occurs in Equation (10.39) only if *MSTR* is large relative to *MSE*. But this condition exists only if the sample means for the treatments (columns), the $\overline{X}_{.j}$ values, differ greatly from one another. And this condition usually exists only if the population treatment means, the $\mu_{.j}$, are truly different, meaning that we should reject the null hypothesis. In sum, then, if the *F* value computed from Equation (10.39) is larger than the

Source	df	SS	MS	F	**TABLE 10.10**
Between treatments	$k - 1$	SSTR	MSTR	$F = MSTR/MSE$	**Two-Factor ANOVA Table**
Between blocks	$n - 1$	SSBL	MSBL	$F = MSBL/MSE$	**for a Randomized**
Error	$(k - 1)(n - 1)$	SSE	MSE		**Complete Block Design**
Total	$n_\text{T} - 1$	SST			

F_{α, ν_1, ν_2} value from Table VII, where α is the significance level, $\nu_1 = k - 1$, and $\nu_2 = (k - 1)(n - 1)$, then we reject the null hypothesis $H_0\colon \mu_{.1} = \mu_{.2} = \cdots = \mu_{.k}$.

We can also test the null hypothesis of equal means for the blocks, $H_0\colon \mu_{1.} = \mu_{2.} = \cdots = \mu_{n.}$, against the alternative hypothesis that one or more of the blocks has a different mean. The null hypothesis is rejected if the ratio

$$F = \frac{MSBL}{MSE} \qquad (10.40)$$

is greater than F_{α, ν_1, ν_2} from Table VII, where $\nu_1 = n - 1$ and $\nu_2 = (k - 1)(n - 1)$. However, usually the results with respect to the treatments are our main interest, and the blocks are used to reduce experimental error, so most of the time this hypothesis test is not of great interest.

The results of a two-factor analysis of variance for a randomized, complete block design are usually presented in a table. A summary of the quantities included in a two-factor analysis of variance (ANOVA) table for a randomized complete block design is presented in Table 10.10.

 PRACTICE PROBLEM 10.4

We now reconsider the experimental situation at Sierra Pacific that was introduced in Example 10.9 in this section. In this example we studied the water pollution counts for containers of water that were randomly sampled from a production cooling system when it was fitted with three different types of filters (treatments) and the water came from four different sources (blocks). The water pollution counts, the sample treatment, (column) means, the sample block (row) means, and the grand mean are presented in Table 10.11.

We want to test the null hypothesis that the mean pollution counts are equal for all three filters (treatments), $H_0\colon \mu_{.1} = \mu_{.2} = \mu_{.3}$, against the alternative hypothesis that one or more of the filters produce a different mean pollution count. Let $\alpha = .05$.

		Treatment			
Block	Filter 1	Filter 2	Filter 3	Block (Row) Mean	**TABLE 10.11**
1. Culinary Water	6	4	2	$\overline{X}_{1.} = 4$	**Pollution Counts and**
2. Recycled Water	11	7	3	$\overline{X}_{2.} = 7$	**Sample Means at Sierra**
3. Stream Water	15	11	7	$\overline{X}_{3.} = 11$	**Pacific: Three Filters**
4. Pond Water	24	22	20	$\overline{X}_{4.} = 22$	**(Treatments) and Four**
Treatment (Column) Mean	$\overline{X}_{.1} = 14$	$\overline{X}_{.2} = 11$	$\overline{X}_{.3} = 8$	Grand mean, $\overline{\overline{X}} = 11$	**Water Sources (Blocks)**

Solution We will need the sums of squares, the degrees of freedom, the mean squares, the *F* statistic, and F_{α, ν_1, ν_2} (from Table VII). The sums of squares [using Equations (10.30–10.34)] are as follows:

$$SST = \sum_{i=1}^{n} \sum_{j=1}^{k} (X_{ij} - \overline{\overline{X}})^2$$

$$= (6 - 11)^2 + (4 - 11)^2 + (2 - 11)^2$$
$$+ (11 - 11)^2 + (7 - 11)^2 + (3 - 11)^2$$
$$+ (15 - 11)^2 + (11 - 11)^2 + (7 - 11)^2$$
$$+ (24 - 11)^2 + (22 - 11)^2 + (20 - 11)^2 = 638$$

$$SSTR = n \sum_{j=1}^{k} (\overline{X}_{\cdot j} - \overline{\overline{X}})^2$$

$$= 4[(14 - 11)^2 + (11 - 11)^2 + (8 - 11)^2] = 4(18) = 72$$

$$SSBL = k \sum_{i=1}^{n} (\overline{X}_{i \cdot} - \overline{\overline{X}})^2$$

$$= 3[(4 - 11)^2 + (7 - 11)^2 + (11 - 11)^2 + (22 - 11)^2]$$
$$= 3(186) = 558$$

$$SSE = SST - SSTR - SSBL = 638 - 72 - 558 = 8$$

Or using the definitional formula, we have

$$SSE = \sum_{i=1}^{n} \sum_{j=1}^{k} (X_{ij} - \overline{X}_{i \cdot} - \overline{X}_{\cdot j} + \overline{\overline{X}})^2$$

$$= (6 - 4 - 14 + 11)^2 + (4 - 4 - 11 + 11)^2$$
$$+ (2 - 4 - 8 - 11)^2 + (11 - 7 - 14 + 11)^2$$
$$+ (7 - 7 - 11 + 11)^2 + (3 - 7 - 8 + 11)^2$$
$$+ (15 - 11 - 14 + 11)^2 + (11 - 11 - 11 + 11)^2$$
$$+ (7 - 11 - 8 + 11)^2 + (24 - 22 - 14 + 11)^2$$
$$+ (22 - 22 - 11 + 11)^2 + (20 - 22 - 8 + 11)^2 = 8$$

The two methods for computing *SSE* both result in the value 8, providing a check for the computations.

The value for the degrees of freedom associated with *SST* is $n_T - 1 = 12 - 1 = 11$; *SSTR* has $k - 1 = 3 - 1 = 2$ degrees of freedom; *SSBL* has $n - 1 = 4 - 1 = 3$ degrees of freedom; and *SSE* has $(k - 1)(n - 1) = (3 - 1)(4 - 1) = (2)(3) = 6$ degrees of freedom.

The mean squares [using Equations (10.36–10.38)] are as follows:

$$MSTR = \frac{SSTR}{k - 1} = \frac{72}{3 - 1} = 36$$

$$MSBL = \frac{SSBL}{n - 1} = \frac{558}{4 - 1} = 186$$

$$MSE = \frac{SSE}{(k - 1)(n - 1)} = \frac{8}{(3 - 1)(4 - 1)} = 1.333$$

Source	df	SS	MS	F	**TABLE 10.12**
Between treatments	2	72	36	27	**Two-Factor Analysis of**
Between blocks	3	558	186	139.5	**Variance Table for a**
Error	6	8	1.333		**Randomized Complete**
Total	11	638			**Block Design at**
					Sierra Pacific

The results of a two-factor analysis of variance for a randomized complete block design are usually presented in a table like Table 10.10. Table 10.12 shows the sources of variation, the values for the degrees of freedom, the sums of squares, and the mean squares for this water pollution problem.

To test the null hypothesis that the mean pollution count is the same for all three filters, H_0: $\mu_{.1} = \mu_{.2} = \mu_{.3}$, we use Equation (10.39) and compute

$$F = \frac{MSTR}{MSE} = \frac{36}{1.333} = 27$$

For this test we find F_{α,ν_1,ν_2}, where $\alpha = .05$, $\nu_1 = k - 1 = 3 - 1 = 2$, and $\nu_2 = (k - 1)(n - 1) = (3 - 1)(4 - 1) = 6$, in Table VII. The value from the table is $F_{.05,2,6} = 5.1433$. Our calculated value, $F = 27$, exceeds the value from the table. Therefore the computed value is in the rejection region, and we reject the null hypothesis. We conclude that one or more of the filters produce a pollution count different from the others.

In this problem we do not test the hypothesis of equal block means because we are primarily interested in the differences in means owing to treatments. ▬

A computer was used to analyze the Sierra Pacific pollution count data for the randomized block design, and the results are presented in Computer Exhibits 10.4A, 10.4B, and 10.4C.

We will now consider **multiple pairwise comparisons of treatment means.** For the randomized complete block design we have been discussing, if the null hypothesis of equal treatment means, H_0: $\mu_{.1} = \mu_{.2} = \cdots = \mu_{.k}$, is rejected, then m multiple pairwise comparisons of means, as discussed in Section 10.5, can be used with a family confidence level of $100(1 - \alpha)\%$ to determine which pairs of treatment means differ. Three adjustments to Equation (10.23) must be made in order to find the confidence intervals that are used to make the pairwise comparisons. First, the difference in means is $\overline{X}_{.j} - \overline{X}_{.j'}$. Second, MSE replaces MSW. Third, the degrees of freedom change to $(k - 1)(n - 1)$. Thus, the formula for computing confidence intervals that are used to test the null hypotheses of equality of pairwise treatment means, H_0: $\mu_{.j} = \mu_{.j'}$, where j and j' denote a pair of treatment groups, is as follows.

Equations for Confidence Limits for Multiple Pairwise Comparisons of Treatment Means

$$L = (\overline{X}_{.j} - \overline{X}_{.j'}) - t_{\alpha/(2m),(k-1)(n-1)} \sqrt{MSE\left(\frac{1}{n} + \frac{1}{n}\right)}$$

$$R = (\overline{X}_{.j} - \overline{X}_{.j'}) + t_{\alpha/(2m),(k-1)(n-1)} \sqrt{MSE\left(\frac{1}{n} + \frac{1}{n}\right)}$$

(10.41)

COMPUTER EXHIBIT 10.4A Analysis of Variance for the Randomized Block Design
with Sierra Pacific Pollution Count Data

MINITAB

```
MTB > read c1   c2   c3   ◄────    Command to read the pollution count data, water type and filter type into c1, c2 and c3
DATA >      6    1    1
DATA >      4    1    2
DATA >      2    1    3
DATA >     11    2    1
DATA >      7    2    2
DATA >      3    2    3
DATA >     15    3    1
DATA >     11    3    2
DATA >      7    3    3
DATA >     24    4    1
DATA >     22    4    2
DATA >     20    4    3
DATA > end

MTB > name c1 'Polcount'  c2 'Water'  c3 'Filter'   ◄──  Command name to name c1 Polcount c2 Water and c3 Filter

MTB > anova c1 = c2 c3   ◄────
                                    Command anova for a two-way analysis of variance of
                                    c1 as the dependent variable and c2 and c3 as factors
Factor     Type Levels Values
Water      fixed     4    1    2    3    4
Filter     fixed     3    1    2    3
```

As in Section 10.5, if a confidence interval does *not* contain the value 0, then we reject the corresponding null hypothesis and conclude that the two population treatment means are different.

 PRACTICE PROBLEM 10.5

The null hypothesis of equal treatment mean water pollution counts at Sierra Pacific was rejected in Practice Problem 10.4. Therefore the problem we now face is to decide which filter (or filters) results in lower mean pollution counts. In other words, we want to use multiple pairwise comparisons to determine which pairs of treatment means are significantly different by testing the null hypotheses H_0: $\mu_{.j} = \mu_{.j'}$, where j and j' denote the treatments, or filters. Let the family confidence level be at least 95%.

Solution To determine which filter (or filters) results in lower mean water pollution counts, we can make $m = 3$ multiple comparisons of means by testing the following three null hypotheses:

$$H_{0_1}: \mu_{.1} = \mu_{.2} \qquad H_{0_2}: \mu_{.1} = \mu_{.3} \qquad H_{0_3}: \mu_{.2} = \mu_{.3}$$

The alternative hypothesis in each case is that the two means are not equal.

The sample mean water pollution counts for the three different types of filters were presented in Table 10.11, and they are

$$\overline{X}_{.1} = 14 \qquad \overline{X}_{.2} = 11 \qquad \overline{X}_{.3} = 8$$

The sample size for each of the treatment groups is $n = 4$. The value for *MSE* from Table 10.12 is 1.333, and the value for the degrees of freedom associated with *MSE* is $(k - 1)(n - 1) = (3 - 1)(4 - 1) = 6$. Because the family confidence level must be at least 95%, we will let $\alpha = .05$. Each of the $m = 3$ individual confidence intervals will be

COMPUTER EXHIBIT 10.4A Analysis of Variance for the Randomized Block Design
with Sierra Pacific Pollution Count Data (Continued)   MINITAB

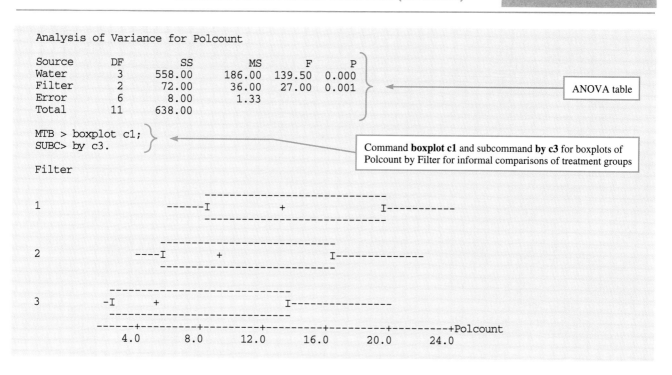

Analysis of Variance for Polcount

Source	DF	SS	MS	F	P
Water	3	558.00	186.00	139.50	0.000
Filter	2	72.00	36.00	27.00	0.001
Error	6	8.00	1.33		
Total	11	638.00			

ANOVA table

MTB > boxplot c1;
SUBC> by c3.

Command **boxplot c1** and subcommand **by c3** for boxplots of Polcount by Filter for informal comparisons of treatment groups

found by letting $\alpha/m = .05/3 = .017$, so the area we place in a single tail to determine the value of t for computing a confidence interval is $\alpha/(2m) = .05/[(2)\,(3)] = .008333$. Thus the value of t we will use in Equation (10.41) to determine the confidence limits is

$$t_{\alpha/(2m),(k-1)(n-1)} = t_{.008333,6} = 3.2875$$

Now to compare the pair of means for filter 1 and filter 2, we test the null hypothesis $H_{0_1}: \mu_{.1} = \mu_{.2}$. For this hypothesis $\overline{X}_{.1} - \overline{X}_{.2} = 14 - 11 = 3$, and the confidence limits that we find by using Equation (10.41) are as follows:

$$L = (\overline{X}_{.1} - \overline{X}_{.2}) - t_{\alpha/(2m),(k-1)(n-1)} \sqrt{MSE\left(\frac{1}{n} + \frac{1}{n}\right)}$$

$$= 3 - 3.2875 \sqrt{1.333\left(\frac{1}{4} + \frac{1}{4}\right)} = 3 - 2.68 = .32$$

$$R = (\overline{X}_{.1} - \overline{X}_{.2}) + t_{\alpha/(2m),(k-1)(n-1)} \sqrt{MSE\left(\frac{1}{n} + \frac{1}{n}\right)}$$

$$= 3 + (3.2875) \sqrt{1.333\left(\frac{1}{4} + \frac{1}{4}\right)} = 3 + 2.68 = 5.68.$$

We will use the format of the following expression to present the confidence intervals:

$$L \le \mu_{.j} - \mu_{.j'} \le R \qquad\qquad (10.42)$$

SAS

COMPUTER EXHIBIT 10.4B Analysis of Variance for the Randomized Block Design with Sierra Pacific Pollution Count Data

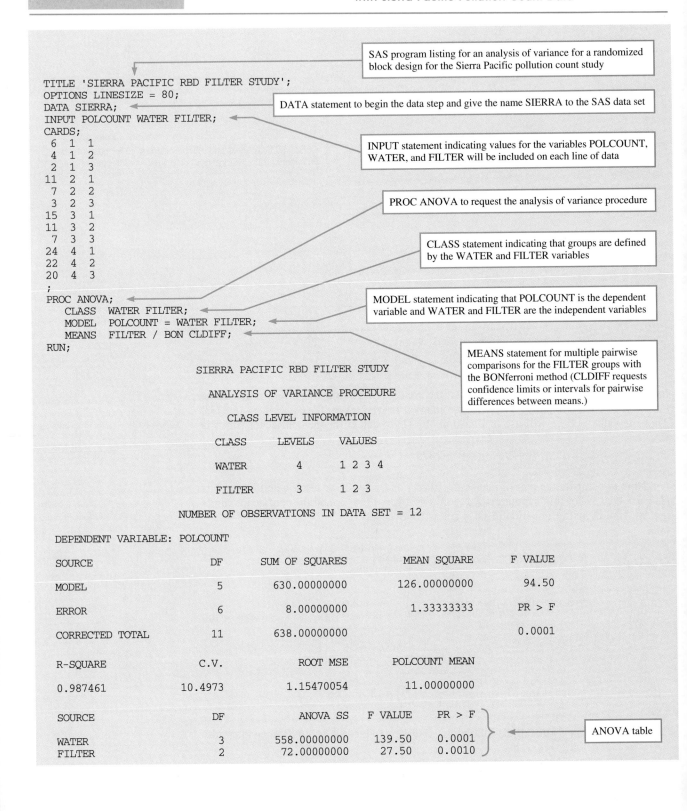

SAS program listing for an analysis of variance for a randomized block design for the Sierra Pacific pollution count study

```
TITLE 'SIERRA PACIFIC RBD FILTER STUDY';
OPTIONS LINESIZE = 80;
DATA SIERRA;
INPUT POLCOUNT WATER FILTER;
CARDS;
  6  1  1
  4  1  2
  2  1  3
 11  2  1
  7  2  2
  3  2  3
 15  3  1
 11  3  2
  7  3  3
 24  4  1
 22  4  2
 20  4  3
;
PROC ANOVA;
    CLASS  WATER FILTER;
    MODEL  POLCOUNT = WATER FILTER;
    MEANS  FILTER / BON CLDIFF;
RUN;
```

DATA statement to begin the data step and give the name SIERRA to the SAS data set

INPUT statement indicating values for the variables POLCOUNT, WATER, and FILTER will be included on each line of data

PROC ANOVA to request the analysis of variance procedure

CLASS statement indicating that groups are defined by the WATER and FILTER variables

MODEL statement indicating that POLCOUNT is the dependent variable and WATER and FILTER are the independent variables

MEANS statement for multiple pairwise comparisons for the FILTER groups with the BONferroni method (CLDIFF requests confidence limits or intervals for pairwise differences between means.)

SIERRA PACIFIC RBD FILTER STUDY

ANALYSIS OF VARIANCE PROCEDURE

CLASS LEVEL INFORMATION

CLASS	LEVELS	VALUES
WATER	4	1 2 3 4
FILTER	3	1 2 3

NUMBER OF OBSERVATIONS IN DATA SET = 12

DEPENDENT VARIABLE: POLCOUNT

SOURCE	DF	SUM OF SQUARES	MEAN SQUARE	F VALUE
MODEL	5	630.00000000	126.00000000	94.50
ERROR	6	8.00000000	1.33333333	PR > F
CORRECTED TOTAL	11	638.00000000		0.0001

R-SQUARE	C.V.	ROOT MSE	POLCOUNT MEAN
0.987461	10.4973	1.15470054	11.00000000

SOURCE	DF	ANOVA SS	F VALUE	PR > F
WATER	3	558.00000000	139.50	0.0001
FILTER	2	72.00000000	27.50	0.0010

ANOVA table

COMPUTER EXHIBIT 10.4B Analysis of Variance for the Randomized Block Design with Sierra Pacific Pollution Count Data (Continued)   **SAS**

To test H_0: $\mu_{.j} = \mu_{.j'}$ with SAS, use the following rule.

Rejection rule: If $|\overline{X}_{.j} - \overline{X}_{.j'}| > t_{\alpha/(2m),\,(k-1)(n-1)}\sqrt{MSW(1/n + 1/n)}$, then reject H_0.

For the Sierra Pacific data,

Rejection rule: If $|\overline{X}_{.j} - \overline{X}_{.j'}| > 2.6842$, then reject H_0.

```
BONFERRONI (DUNN) T TESTS FOR VARIABLE: POLCOUNT
NOTE: THIS TEST CONTROLS THE TYPE I EXPERIMENTWISE ERROR RATE
      BUT GENERALLY HAS A HIGHER TYPE II ERROR RATE THAN TUKEY'S

      ALPHA=0.05  CONFIDENCE=0.95  DF=6  MSE=1.33333
      CRITICAL VALUE OF T=3.28746
      MINIMUM SIGNIFICANT DIFFERENCE=2.6842  ◄

COMPARISONS SIGNIFICANT AT THE 0.05 LEVEL ARE INDICATED BY '***'

                     SIMULTANEOUS              SIMULTANEOUS
                        LOWER      DIFFERENCE     UPPER
          FILTER     CONFIDENCE     BETWEEN    CONFIDENCE
        COMPARISON     LIMIT        MEANS        LIMIT

        1   - 2        0.3158       3.0000       5.6842     ***
        1   - 3        3.3158       6.0000       8.6842     ***

        2   - 1       -5.6842      -3.0000      -0.3158     ***
        2   - 3        0.3158       3.0000       5.6842     ***

        3   - 1       -8.6842      -6.0000      -3.3158     ***
        3   - 2       -5.6842      -3.0000      -0.3158     ***
```

Confidence intervals for multiple pairwise comparisons of treatment means using the Bonferroni method

Note: ROOT MSE = $\sqrt{MSE}$

The interval for comparing filter 1 and filter 2 is thus

$$.32 \le \mu_{.1} - \mu_{.2} \le 5.68$$

Because the interval does not include 0, we reject the null hypothesis H_{0_1}: $\mu_{.1} = \mu_{.2}$. Our conclusion is that the mean water pollution counts that result from the use of filter 1 and filter 2 are different.

Confidence intervals for testing the null hypotheses H_{0_2} and H_{0_3} were also computed, and they are presented below by using the format of Equation (10.42):

$$3.32 \le \mu_{.1} - \mu_{.3} \le 8.68 \qquad \text{(Does not include 0; reject } H_{0_2}\text{: } \mu_{.1} = \mu_{.3}.)$$

$$.32 \le \mu_{.2} - \mu_{.3} \le 5.68 \qquad \text{(Does not include 0; reject } H_{0_3}\text{: } \mu_{.2} = \mu_{.3}.)$$

The results for the three pairwise comparisons enable us to conclude that $\mu_{.1}$, $\mu_{.2}$, and $\mu_{.3}$ are different from each other on a pairwise basis. From the family of confidence intervals we conclude that the mean pollution count for filter 1 is higher than the mean pollution count for filter 2 by an amount somewhere between .32 and 5.68. We also conclude that the mean pollution count for filter 1 is higher than the mean pollution count for

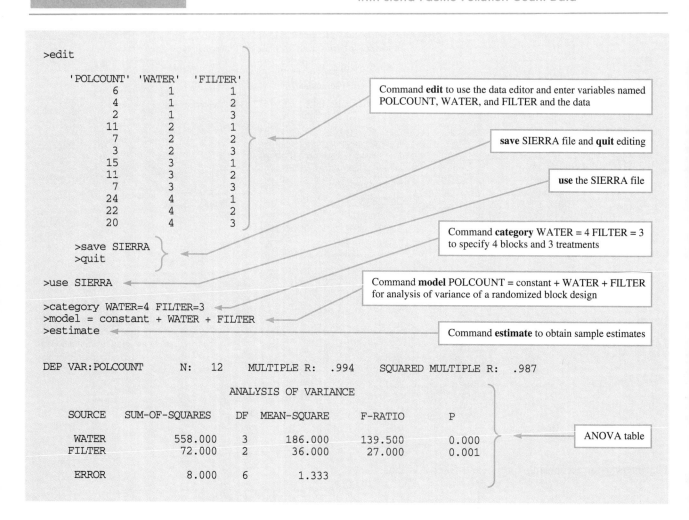

| MYSTAT | COMPUTER EXHIBIT 10.4C Analysis of Variance for the Randomized Block Design with Sierra Pacific Pollution Count Data |

```
>edit

    'POLCOUNT' 'WATER'  'FILTER'
          6        1        1
          4        1        2
          2        1        3
         11        2        1
          7        2        2
          3        2        3
         15        3        1
         11        3        2
          7        3        3
         24        4        1
         22        4        2
         20        4        3

    >save SIERRA
    >quit

>use SIERRA

>category WATER=4 FILTER=3
>model = constant + WATER + FILTER
>estimate


DEP VAR:POLCOUNT       N:  12    MULTIPLE R:  .994    SQUARED MULTIPLE R:  .987

                          ANALYSIS OF VARIANCE

      SOURCE   SUM-OF-SQUARES   DF  MEAN-SQUARE   F-RATIO       P

      WATER         558.000      3     186.000    139.500    0.000
      FILTER         72.000      2      36.000     27.000    0.001

      ERROR           8.000      6       1.333
```

Command **edit** to use the data editor and enter variables named POLCOUNT, WATER, and FILTER and the data

save SIERRA file and **quit** editing

use the SIERRA file

Command **category** WATER = 4 FILTER = 3 to specify 4 blocks and 3 treatments

Command **model** POLCOUNT = constant + WATER + FILTER for analysis of variance of a randomized block design

Command **estimate** to obtain sample estimates

ANOVA table

filter 3, and that the mean pollution count for filter 2 is higher than the mean pollution-count for filter 3. Thus our decision is to use filter 3 in order to achieve the *lowest* pollution count.

The randomized complete block design with one observation in each block and treatment combination (or cell) that was presented in this section assumes that there is no *interaction* between treatments and blocks. For our example this assumption means that there are no differential influences of the filters on the pollution counts that depend on a water source. We also note that the test of the equality for two means where the items were paired (presented in Section 9.13) corresponds to a randomized complete block experiment with $n = 2$ blocks.

If data sets are large, or if the computations are burdensome, and a computer is not available, then the following computational formulas can be used to determine the values of the various sums of squares; although the general availability of computers tends to prevail over the use of computational formulas.

Computational Formulas for a Randomized Complete Block Design

For T = the sum of the values of the dependent variable X_{ij} for all experimental items, or

$$T = \sum_i \sum_j X_{ij}$$

for $T_{i.}$ = the sum of the values of the dependent variable for the experimental items in block i, or

$$T_{i.} = \sum_j X_{ij}$$

and for $T_{.j}$ = the sum of the values of the dependent variable for the experimental items in treatment j, or

$$T_{.j} = \sum_i X_{ij}$$

the *computational formulas* are as follows:

$$SST = \sum_i \sum_j X_{ij}^2 - \frac{T^2}{n_T} \tag{10.43}$$

$$SSTR = \frac{\sum_j T_{.j}^2}{n} - \frac{T^2}{n_T} \tag{10.44}$$

$$SSBL = \frac{\sum_i T_{i.}^2}{k} - \frac{T^2}{n_T} \tag{10.45}$$

$$SSE = SST - SSTR - SSBL \tag{10.46}$$

Problems: Section 10.9

45. Rothamstet agricultural experiment station conducted a randomized complete block experiment in order to make comparisons of yields among three varieties of barley. Six different parcels of land (blocks) were used with each variety assigned at random to a segment of a parcel. The yields (in bushels per acre) are given in the accompanying table.

Block	*Barley*		
	Variety 1	*Variety 2*	*Variety 3*
Parcel 1	45	40	30
Parcel 2	40	42	37
Parcel 3	46	38	26
Parcel 4	38	42	25
Parcel 5	35	45	27
Parcel 6	43	48	35

a. Construct the analysis of variance table.
b. Test the hypothesis of equal means for the three varieties. Use a significance level of .05.

46. ACME Mfg., Inc., conducted an experiment to determine whether employee productivity (units produced per day) is different among employees that are paid by three different methods: piece rate, hourly rate, salary. Nine employees were sorted into three equal-sized groups on the basis of each employee's degree of organizational commitment (high, medium, low). Employees within the high-, medium-, and low-organizational-commitment groups were assigned randomly to a method of payment. The results are given in the accompanying table.

a. Construct the analysis of variance table.
b. Test the hypothesis of equal mean productivities for the different methods of paying the employees. Use a significance level of .05.
c. If appropriate, use multiple pairwise comparisons and a family confidence level of at least 95% to determine which payment method results in the highest mean productivity. [*Hint:* $t_{.008333,4} = 3.9608$.]

		Payment Method		
	Block	*Piece Rate*	*Hourly Rate*	*Salary*
1.	*High*	12	9	10
2.	*Medium*	9	6	4
3.	*Low*	8	4	3

47. An experiment was conducted to see whether financial analysts could detect a change in the financial condition of a firm more effectively by using financial accounting information in the form of tables of financial ratios, by using multi-dimensional computer graphics based on the same ratios, or by using both the tables and the graphics. Of the nine analysts three were experts, three had some experience, and three were inexperienced. The expert, experienced, and inexperienced analysts were assigned at random to receive one of the forms of information. The results, in the form of scores for an index of effectiveness, are given in the accompanying table. An edited printout from MINITAB for the analysis of variance is presented in Computer Exhibit 10.5.

a. Construct the analysis of variance table.
b. Test the hypothesis of equal mean effectiveness for analysts using the three forms of information. Use a significance level of .05.

MINITAB

COMPUTER EXHIBIT 10.5 Analysis of Variance for the Randomized Block Design with Financial Condition Data

```
Analysis of Variance for Effindex

Source    DF        SS        MS      F      P
Block      2    70.222    35.111  37.18  0.003
Info       2    49.556    24.778  26.24  0.005
Error      4     3.778     0.944
Total      8   123.556
```

	Information		
Block	Tables	Graphics	Tables and Graphics
1. Expert	10	15	16
2. Experienced	8	12	15
3. Inexperienced	5	7	9

REFERENCE: Stock, D., and C. J. Watson (1984), "Human Judgment Accuracy, Multidimensional Graphics, and Humans versus Models," *Journal of Accounting Research,* 22 (Spring): 192–206.

48. Pete, Marvin, and Mainman, a CPA accounting firm, conducted an experiment to study the effectiveness of three approaches to electronic data-processing (EDP) auditing: parallel simulation, test of transactions, and test deck application. The nine auditors who participated in the experiment were sorted into three categories with respect to experience. Three of the auditors were managers, three were staff II accountants, and three were staff I accountants. The auditors within each experience category were assigned randomly to use one of the three auditing approaches during a test audit. The effectiveness ratings are given in the accompanying table. An edited BMDP printout for the analysis of variance is given in Computer Exhibit 10.6.

   a. Construct the analysis of variance table.
   b. Test the hypothesis of equal mean effectiveness for the three EDP-auditing approaches. Use a significance level of .05.

	EDP Audit Approach		
Block	Parallel Simulation	Transactions Test	Test Deck
1. Manager	18	14	17
2. Staff II	12	7	11
3. Staff I	8	5	9

49. National Supermarkets, Inc., has performed a market study to examine the sales of a product under three different package designs. The nine supermarkets included in the study were sorted into three categories. Three were located in regional shopping malls, three in smaller nonregional malls, and three were not located in malls. The similar stores within each of these categories were each randomly assigned one of the package designs. The sales (in dollars per day) for the product in the nine supermarkets are given in the accompanying table.

   a. Construct the analysis of variance table.
   b. Test the hypothesis of equal mean sales for the three package designs. Use a significance level of .05.
   c. If appropriate, use multiple pairwise comparisons and a family confidence level of at least 95% to determine which type of package results in the highest mean amount of sales. [*Hint:* $t_{.008333,4} = 3.9608$.]

	Package		
Block	1	2	3
1. Regional Mall	100	85	80
2. Nonregional Mall	80	67	65
3. Nonmall	70	56	50

BMDP

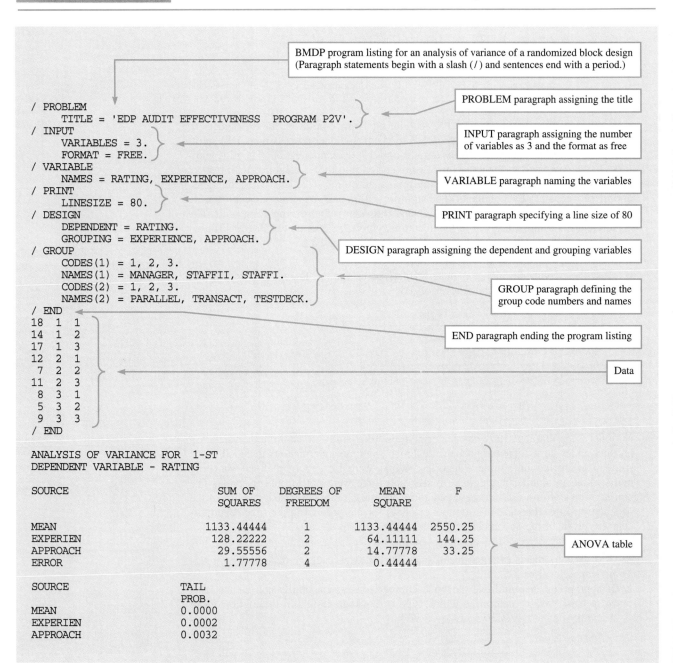

BMDP program listing for an analysis of variance of a randomized block design (Paragraph statements begin with a slash (/) and sentences end with a period.)

```
/ PROBLEM
    TITLE = 'EDP AUDIT EFFECTIVENESS  PROGRAM P2V'.
/ INPUT
    VARIABLES = 3.
    FORMAT = FREE.
/ VARIABLE
    NAMES = RATING, EXPERIENCE, APPROACH.
/ PRINT
    LINESIZE = 80.
/ DESIGN
    DEPENDENT = RATING.
    GROUPING = EXPERIENCE, APPROACH.
/ GROUP
    CODES(1) = 1, 2, 3.
    NAMES(1) = MANAGER, STAFFII, STAFFI.
    CODES(2) = 1, 2, 3.
    NAMES(2) = PARALLEL, TRANSACT, TESTDECK.
/ END
18   1   1
14   1   2
17   1   3
12   2   1
 7   2   2
11   2   3
 8   3   1
 5   3   2
 9   3   3
/ END
```

PROBLEM paragraph assigning the title

INPUT paragraph assigning the number of variables as 3 and the format as free

VARIABLE paragraph naming the variables

PRINT paragraph specifying a line size of 80

DESIGN paragraph assigning the dependent and grouping variables

GROUP paragraph defining the group code numbers and names

END paragraph ending the program listing

Data

```
ANALYSIS OF VARIANCE FOR  1-ST
DEPENDENT VARIABLE - RATING
```

SOURCE	SUM OF SQUARES	DEGREES OF FREEDOM	MEAN SQUARE	F
MEAN	1133.44444	1	1133.44444	2550.25
EXPERIEN	128.22222	2	64.11111	144.25
APPROACH	29.55556	2	14.77778	33.25
ERROR	1.77778	4	0.44444	

ANOVA table

SOURCE	TAIL PROB.
MEAN	0.0000
EXPERIEN	0.0002
APPROACH	0.0032

50. Three different numerically controlled drilling machines are used to drill holes in iron castings at Prefab Metalworks, Inc. An experiment was conducted using two people (blocks) who program the machines. Both programmers wrote a program for each machine (the machines require their own programs); the program will give the same pattern of holes in the castings. The order in which the programs were

written was randomly selected. The results, in terms of the number of castings drilled per hour for each programmer-machine combination, are presented in the accompanying table.

a. Construct the analysis of variance table.

b. Test the hypothesis of the equality of the mean number of castings produced per hour by the three machines. Use a significance level of .05.

Block	Machine		
	1	2	3
Programmer 1	15	12	11
Programmer 2	10	8	6

The experimental designs considered up to this point have used only one factor, or independent variable, to define the treatments. Experiments can have more than one factor, and in this section we consider experiments that use two factors (denoted factor A and factor B) to define the experimental treatments. An experiment that uses two or more factors to define the experimental treatments has a **factorial design.** Factorial designs can have more than two factors, but those designs are beyond the scope of this text.

> **Definition**
>
> Experiments having a **factorial design** use two or more factors to define the experimental treatments.

If there are a levels of factor A and b levels of factor B, then there will be ab combinations of a level of factor A and a level of factor B. Each combination of a level of factor A and a level of factor B is a **treatment** in the factorial design. We will require that exactly n different experimental items receive each treatment, and n *must be greater than 1.* The experimental items must be assigned to a treatment randomly, so this experiment is another example of a completely randomized design.

A factorial design with a levels of factor A and b levels of factor B is referred to as an $a \times b$ **factorial design.** For example, if factor A has two levels and factor B has two levels, then we have a 2×2 factorial design. With a factorial design the data can be arranged in a two-way table consisting of the levels of factor A and the levels of factor B. Each cell of the table corresponds to a treatment, so there are n experimental items in each cell. We will arrange the data in a table by letting the levels of factor A be *rows* and the levels of factor B be *columns.*

The factorial design that we consider is "complete" because there are experimental items that receive each treatment. It is *balanced* (sometimes called orthogonal) because each treatment is applied to an equal number (n) of experimental items. The number of experimental items receiving each treatment is often referred to as the number of **replications.**

One purpose of an experiment with a factorial design is to allow the study of the effects of two or more factors on the dependent variable simultaneously. An advantage of the factorial design is that to conduct one experiment and simultaneously study the effects of two factors is more efficient than to conduct two separate experiments to study the effects of the two factors.

	Factor A,	Factor B, Selling Price	
TABLE 10.13	Package Design	Higher	Lower
Sales (in Dollars): Three	Package 1	14	17
Package Designs		12	13
(Factor A) and Two Selling	Package 2	37	35
Prices (Factor B) at		41	41
Federated	Package 3	12	11
		16	9

=== **EXAMPLE 10.10**   Federated Department Stores, Inc., owns a chain of very similar retail stores and conducted an experiment by using a factorial design to study the sales of a product, in dollars (the dependent variable), for different package designs (factor A) and different selling prices (factor B). There were three different package designs (levels 1, 2, and 3 for factor A), so $a = 3$. There were two selling prices, and they were set at 99 cents for the higher price and 95 cents for the lower price (levels 1 and 2 for factor B), so $b = 2$. Thus there are $ab = (3)(2) = 6$ combinations of a package design and a selling price, or six treatments in a 3×2 factorial design.

A sample of 12 stores (the experimental items) was taken, and $n = 2$ separate stores were assigned at random to receive one of the six treatments. That is, each store used a single combination of a package design and a selling price, with two stores each receiving one of the six separate combinations of a package design and a selling price. The results for this experiment, in the form of sales (in dollars), are presented in Table 10.13. We will use this data for analysis of variance computations.

We want to test three separate hypotheses. First, we test the hypothesis that there are *no interactions* for any treatment (combination of a package design and a selling price). In a test for interactions the number of experimental items in each cell, or the number of replications n, must be greater than 1. The term *interaction* for this example means that there is a differential influence of a package design on the amount of sales, depending on the selling price. That is, the difference between the population means for any two levels of factor B is not the same for each of the levels of factor A. We will explain interactions more fully in the discussion following this example.

Second, we test the hypothesis that the mean population amount of sales is the same for all three package designs. That is, where $\mu_{i.}$ is the population mean of the dependent variable for the ith level (row) of factor A, we wish to test H_0: $\mu_{1.} = \mu_{2.} = \mu_{3.}$ against the alternative hypothesis that one or more of the levels of factor A produce a different mean for the dependent variable.

Third, we test the hypothesis that the mean population sales amount is the same for the two selling prices. That is, where $\mu_{.j}$ is the population mean of the dependent variable for the jth level (column) of factor B, we wish to test H_0: $\mu_{.1} = \mu_{.2}$ against the alternative hypothesis that the two levels of factor B produce different means for the dependent variable. ===

Now let us explore what is meant by the term *interaction*. An **interaction** is a joint effect of two factors. An interaction that results in a large mean value for sales in our example would be called *synergism*.

To illustrate the meaning of interactions between two factors, we consider the following new example that involves 2×2 factorial designs. Notice that this example is completely separate from Example 10.10, and it is used only to explain interactions. We do *not* use the data from the following example in later analysis of variance computations.

=== **EXAMPLE 10.11**   An American Stores retail outlet studied the effects on sales (in dollars) of displaying products on high or low shelf positions and with two different

TABLE 10.14

Population Treatment Means Sales, μ_{ij}, for Products 1, 2, and 3

(a) Product 1, Showing No Interactions

Factor A, Shelf Position	Factor B, Package	
	Package 1	Package 2
Low	$\mu_{11} = 10$	$\mu_{12} = 20$
High	$\mu_{21} = 15$	$\mu_{22} = 25$

(b) Product 2, with Interactions Present

Factor A, Shelf Position	Factor B, Package	
	Package 1	Package 2
Low	$\mu_{11} = 10$	$\mu_{12} = 20$
High	$\mu_{21} = 15$	$\mu_{22} = 30$

(c) Product 3, with Interactions Present

Factor A, Shelf Position	Factor B, Package	
	Package 1	Package 2
Low	$\mu_{11} = 10$	$\mu_{12} = 30$
High	$\mu_{21} = 20$	$\mu_{22} = 12$

types of packages. Three separate 2×2 factorial experiments were conducted using three different products. Factor A in each experiment is the shelf position, and the two levels for this factor are a low shelf position and a high shelf position. Factor B is the type of package, and the two levels for this factor are package 1 and package 2. A treatment in this example is a combination of a shelf position and a type of package.

Let the population mean for the treatment defined by the ith level of factor A and the jth level of factor B be denoted μ_{ij}. Now suppose that these means are known. The values of the population treatment means for the three separate experiments that correspond to the three different products are presented in Tables 10.14(a), 10.14(b), and 10.14(c).

Now let us plot the treatment mean sales values for product 1 against the levels of, say, factor B (package 1 and package 2) on the horizontal axis. We also draw lines that connect the two treatment means that correspond to each level of factor A (low and high shelf positions). The resulting plot is presented as Figure 10.7(a). If the lines that connect the treatment means are *parallel,* then there are *no interactions* present. The reason that the lines are parallel for product 1 is as follows: For each of the levels of factor B (package 1 and package 2), the differences between the treatment means for the levels of factor A (low and high shelf positions) are the same, and the magnitudes of these differences are both equal to 5. Also, and equivalently, for each of the levels of factor A the differences between the treatment means for the levels of factor B are the same, and the magnitudes of these differences are both equal to 10. For product 1, then, there are no interactions between shelf position and type of package.

Next, for product 2 the treatment means that are presented in Table 10.14(b) are plotted in Figure 10.7(b). For product 2 the lines that connect the means that correspond to the levels of factor A are *not parallel.* Thus there are interactions present. For package 1 notice that the difference between the treatment means for the low and high shelf positions is $15 - 10 = 5$, but for package 2 the difference between the treatment means for the low and high shelf positions is $30 - 20 = 10$. Thus these differences are not equal when there are interactions present. The plot shows that the mean sales value is uncharacteristically large when a high shelf position is combined with package 2; synergism results from the combination of a high shelf position and package 2. The plot also shows, for a given package, that the high shelf position

FIGURE 10.7

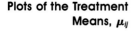

Plots of the Treatment Means, μ_{ij}

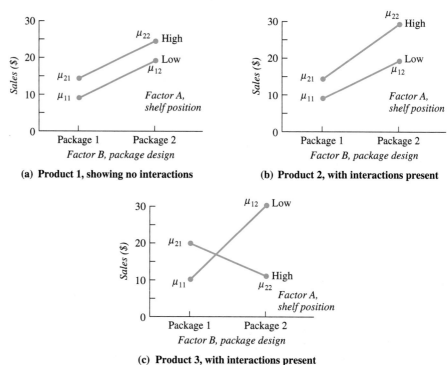

(a) **Product 1, showing no interactions**

(b) **Product 2, with interactions present**

(c) **Product 3, with interactions present**

always results in higher mean sales than does the low shelf position; furthermore, for a given shelf position package 2 always results in higher mean sales than does package 1. Hence interaction makes the interpretation of the effects of the treatments more complex, although in this case the interpretation is still not too difficult.

For product 3 the treatment means that are presented in Table 10.14(c) are plotted in Figure 10.6(c). For this situation the lines that connect the means that correspond to the high and low shelf positions are *not parallel*, and, in fact, they cross. Thus there are interactions present. We see from Figure 10.6(c) that for a given package the high shelf position does not always result in higher mean sales; furthermore, for a given shelf position package 2 does not always result in higher mean sales. Hence in this case the interpretation of the effects of the treatments is more difficult.

Notice that we have illustrated just two types of interactions here. Many other types of interactions are also possible.

Because two factors are included in a factorial design, we require a two-factor analysis of variance. For this design the **total sum of squares** (*SST*) is partitioned into four parts: the **sum of squares between levels of factor** *A* (*SSA*), the **sum of squares between levels of factor** *B* (*SSB*), the **sum of squares owing to interactions** (*SSAB*), and the **error sum of squares** (*SSE*). Notice that *AB* is used to designate interactions (not *A* multiplied times *B*).

> **Equation for Partitioning Sum of Squares**
>
> $$SST = SSA + SSB + SSAB + SSE \qquad (10.47)$$

The following notation will be used for the computations of the various sums of squares:

X_{ijk} = value of dependent variable for kth experimental item in ith level of factor A and jth level of factor B

a = number of levels of factor A

b = number of levels of factor B

n = number of experimental items that receive treatment defined by ith level of factor A and jth level of factor B, or number of *replicates*

n_T = total number of experimental items sampled = nab

$\overline{X}_{ij}$ = sample mean for treatment corresponding to ith level of factor A and jth level of factor B (or sample mean for cell defined by row i and column j)

$$= \frac{\sum\limits_{k=1}^{n} X_{ijk}}{n} \qquad (10.48)$$

$\overline{X}_{.j}$ = sample mean for jth level (column) of factor B

$$= \frac{\sum\limits_{i=1}^{a} \sum\limits_{k=1}^{n} X_{ijk}}{na} \qquad (10.49)$$

$\overline{X}_{i.}$ = sample mean for ith level (row) of factor A

$$= \frac{\sum\limits_{j=1}^{b} \sum\limits_{k=1}^{n} X_{ijk}}{nb} \qquad (10.50)$$

$\overline{\overline{X}}$ = grand mean, defined as sample mean for all experimental items

$$= \frac{\sum\limits_{i=1}^{a} \sum\limits_{j=1}^{b} \sum\limits_{k=1}^{n} X_{ijk}}{n_T} \qquad (10.51)$$

Definitional Formulas

The formulas that *define* the five different sums of squares in Equation are as follows:

$$SST = \sum_{i=1}^{a} \sum_{j=1}^{b} \sum_{k=1}^{n} (X_{ijk} - \overline{\overline{X}})^2 \tag{10.52}$$

$$SSA = nb \sum_{i=1}^{a} (\overline{X}_{i.} - \overline{\overline{X}})^2 \tag{10.53}$$

$$SSB = na \sum_{j=1}^{b} (\overline{X}_{.j} - \overline{\overline{X}})^2 \tag{10.54}$$

$$SSAB = n \sum_{i=1}^{a} \sum_{j=1}^{b} (\overline{X}_{ij} - \overline{X}_{i.} - \overline{X}_{.j} + \overline{\overline{X}})^2 \tag{10.55}$$

And if *SST, SSA, SSB,* and *SSAB* have been computed previously, then

$$SSE = SST - SSA - SSB - SSAB \tag{10.56}$$

or, by definition,

$$SSE = \sum_{i=1}^{a} \sum_{j=1}^{b} \sum_{k=1}^{n} (X_{ijk} - \overline{X}_{ij})^2 \tag{10.57}$$

We will use the basic definitional formulas (10.52–10.57) in the practice problem so that we can emphasize the meanings of the various sums of squares.

The value for the degrees of freedom associated with *SST* is $n_T - 1$; *SSA* has $a - 1$ degrees of freedom; *SSB* has $b - 1$ degrees of freedom; *SSAB* has $(a - 1)(b - 1)$ degrees of freedom; and *SSE* has $n_T - ab$ degrees of freedom. The degrees of freedom are additive, as shown in the following equation:

$$n_T - 1 = (a - 1) + (b - 1) + (a - 1)(b - 1) + (n_T - ab) \tag{10.58}$$

Since mean squares are computed by dividing sums of squares by their associated degrees of freedom, the **mean square for factor A (MSA)**, the **mean square for factor B (MSB)**, the **mean square due to interaction ($MSAB$)**, and the **error mean square (MSE)** are computed according to the following formulas.

Equations for Mean Squares

$$MSA = \frac{SSA}{a - 1} \tag{10.59}$$

$$MSB = \frac{SSB}{b - 1} \tag{10.60}$$

$$MSAB = \frac{SSAB}{(a - 1)(b - 1)} \tag{10.61}$$

$$MSE = \frac{SSE}{n_T - ab} \tag{10.62}$$

Source	df	SS	MS	F	
Factor A	$a - 1$	SSA	MSA	$F = MSA/MSE$	**TABLE 10.15**
Factor B	$b - 1$	SSB	MSB	$F = MSB/MSE$	**Two-Factor ANOVA Table**
Interactions (AB)	$(a - 1)(b - 1)$	$SSAB$	$MSAB$	$F = MSAB/MSE$	**for a Factorial Design**
Error	$n_T - ab$	SSE	MSE		
Total	$n_T - 1$	SST			

If ab independent samples of size n items each are taken from ab normal populations that have equal variances, and if the corresponding null hypothesis concerning factor A, factor B, or the interactions is true, then the ratio

$$F = \frac{MSA}{MSE} \tag{10.63}$$

follows the F distribution with degrees of freedom of $\nu_1 = a - 1$ and $\nu_2 = n_T - ab$; the ratio

$$F = \frac{MSB}{MSE} \tag{10.64}$$

follows the F distribution with degrees of freedom of $\nu_1 = b - 1$ and $\nu_2 = n_T - ab$; and the ratio

$$F = \frac{MSAB}{MSE} \tag{10.65}$$

follows the F distribution with degrees of freedom of $\nu_1 = (a - 1)(b - 1)$ and $\nu_2 = n_T - ab$.

The F value computed in each of the three formulas (10.63–10.65) is relatively large if the numerator mean square is large in comparison with MSE. The respective numerator mean squares will be large if the sample means for the respective factors are quite different (the $\overline{X}_{i.}$ values are quite different or the $\overline{X}_{.j}$ values are quite different) or if there are appreciable interactions in the sample. These results tend to occur if the population mean values are different or if the population interactions are not zero, respectively. Thus if the F value computed in any of the formulas (10.63–10.65) is larger than the corresponding F_{α, ν_1, ν_2} values from Table VII, then we reject the matching null hypothesis. Notice that the formulas for computing the values for the degrees of freedom for the numerators ν_1 are different for each of these F ratios.

The results of the analysis of variance for a factorial design are often presented in a table. A summary of the quantities included in a two-factor ANOVA table for a factorial design is presented in Table 10.15.

Note that if the hypothesis of no interactions is rejected and the interactions are considered to be important, then the interpretation of the results of the hypothesis tests concerning factors A and B may be difficult. Thus the test for interactions is usually the first test that is undertaken. The existence of interactions is often a finding that is important, and if significant interactions exist, the results of the hypothesis tests for factors A and B may not have important meanings. A plot like those presented in Figure 10.7, except that sample treatment means are plotted rather than population treatment means, is often effective for examining or displaying significant interactions when there are two factors.

PRACTICE PROBLEM 10.6

We now take up the Federated Department Stores experimental example that we introduced in this section. In this example we studied the sales (in dollars) of a product at 12

TABLE 10.16	Factor A, Package Design	Factor B, Selling Price		Sample Means for Packages (Rows)
		Higher	Lower	
Sales (in Dollars) and Sample Means: Three Package Designs (Factor A) and Two Selling Prices (Factor B) at Federated	Package 1	14 12 $\overline{X}_{11} = 13$	17 13 $\overline{X}_{12} = 15$	$\overline{X}_{1.} = 14$
	Package 2	37 41 $\overline{X}_{21} = 39$	35 41 $\overline{X}_{22} = 38$	$\overline{X}_{2.} = 38.5$
	Package 3	12 16 $\overline{X}_{31} = 14$	11 9 $\overline{X}_{32} = 10$	$\overline{X}_{3.} = 12$
Sample Means for Prices (Columns)		$\overline{X}_{.1} = 22$	$\overline{X}_{.2} = 21$	Grand mean, $\overline{\overline{X}} = 21.5$

retail stores. The experiment had a factorial design with three package designs (the three levels of factor A) and two selling prices (the two levels of factor B). Two stores were assigned randomly to receive each treatment (combination of a package design and a selling price). The sales amounts, the sample means for the treatments (cells), the sample means for the three packages (rows), the sample means for the two selling prices (columns), and the grand mean are presented in Table 10.16.

We want to test three separate hypotheses. First, we test the hypothesis that there are *no interactions* for any of the combinations of a package design and a selling price, against the alternative that one or more of the combinations of a package design and a selling price produce an interaction. Second, we test the hypothesis that the population mean sales amount is the same for all package designs (rows), or $H_0: \mu_{1.} = \mu_{2.} = \mu_{3.}$, against the alternative that one or more of the package designs produce a different population mean sales amount. Third, we test the hypothesis that the population mean sales amount is the same for the two selling prices (columns), or $H_0: \mu_{.1} = \mu_{.2}$, against the alternative that the two selling prices produce different population mean sales amounts. Let $\alpha = .05$.

Solution We will need the sums of squares, the values for the various degrees of freedom, the mean squares, the F statistics, and the values of F_{α, ν_1, ν_2} (using the appropriate values for the various degrees of freedom) from Table VII.

The sums of squares [using Equations (10.52–10.57)] are as follows:

$$SST = \sum_{i=1}^{a} \sum_{j=1}^{b} \sum_{k=1}^{n} (X_{ijk} - \overline{\overline{X}})^2$$

$$= (14 - 21.5)^2 + (12 - 21.5)^2 + (17 - 21.5)^2 + (13 - 21.5)^2$$
$$+ (37 - 21.5)^2 + (41 - 21.5)^2 + (35 - 21.5)^2 + (41 - 21.5)^2$$
$$+ (12 - 21.5)^2 + (16 - 21.5)^2 + (11 - 21.5)^2 + (9 - 21.5)^2$$

$$= 1809$$

$$SSA = nb \sum_{i=1}^{a} (\overline{X}_{i.} - \overline{\overline{X}})^2$$

$$= (2)(2)[(14 - 21.5)^2 + (38.5 - 21.5)^2 + (12 - 21.5)^2]$$

$$= 4(435.5) = 1742$$

$$SSB = na \sum_{j=1}^{b} (\overline{X}_{.j} - \overline{\overline{X}})^2$$

$$= (2)(3)[(22 - 21.5)^2 + (21 - 21.5)^2] = 6(.5) = 3$$

$$SSAB = n \sum_{i=1}^{a} \sum_{j=1}^{b} (\overline{X}_{ij} - \overline{X}_{i.} - \overline{X}_{.j} + \overline{\overline{X}})^2$$

$$= (2)[(13 - 14 - 22 + 21.5)^2 + (15 - 14 - 21 + 21.5)^2$$
$$+ (39 - 38.5 - 22 + 21.5)^2 + (38 - 38.5 - 21 + 21.5)^2$$
$$+ (14 - 12 - 22 + 21.5)^2 + (10 - 12 - 21 + 21.5)^2]$$

$$= (2)(9) = 18$$

$$SSE = SST - SSA - SSB - SSAB = 1809 - 1742 - 3 - 18 = 46$$

Or using the definitional formula, we have

$$SSE = \sum_{i=1}^{a} \sum_{j=1}^{b} \sum_{k=1}^{n} (X_{ijk} - \overline{X}_{ij})^2$$

$$= (14 - 13)^2 + (12 - 13)^2 + (17 - 15)^2 + (13 - 15)^2$$
$$+ (37 - 39)^2 + (41 - 39)^2 + (35 - 38)^2 + (41 - 38)^2$$
$$+ (12 - 14)^2 + (16 - 14)^2 + (11 - 10)^2 + (9 - 10)^2 = 46$$

The two methods of computing SSE both result in the value 46, providing a check for the computations.

The value for the degrees of freedom associated with SST is $n_T - 1 = 12 - 1 = 11$; SSA has $a - 1 = 3 - 1 = 2$ degrees of freedom; SSB has $b - 1 = 2 - 1 = 1$ degree of freedom; $SSAB$ has $(a - 1)(b - 1) = (3 - 1)(2 - 1) = (2)(1) = 2$ degrees of freedom; and SSE has $= (3 - 1)(4 - 1) = (2)(3) = 6$ degrees of freedom.

The mean squares [using Equations (10.59–10.62)] are as follows:

$$MSA = \frac{SSA}{a - 1} = \frac{1742}{3 - 1} = 871$$

$$MSB = \frac{SSB}{b - 1} = \frac{3}{2 - 1} = 3$$

$$MSAB = \frac{SSAB}{(a - 1)(b - 1)} = \frac{18}{(3 - 1)(2 - 1)} = 9$$

$$MSE = \frac{SSE}{n_T - ab} = \frac{46}{12 - (3)(2)} = 7.667$$

The sources of variation, the degrees of freedom, the sums of squares, and the mean squares for the sales example are presented in Table 10.17.

To test the first hypothesis that there are no interactions in the population, against the alternative that there are interactions in the population, we use Equation (10.65) and calculate

$$F = \frac{MSAB}{MSE} = \frac{9}{7.667} = 1.174$$

TABLE 10.17	*Source*	*df*	*SS*	*MS*	*F*
Two-Factor Analysis of	Factor A,	2	1742	871	113.609
Variance Table for the	package design				
Federated Sales Example	Factor B,	1	3	3	.391
	selling price				
	Interactions (AB)	2	18	9	1.174
	Error	6	46	7.667	
	Total	11	1809		

For this test we find F_{α,ν_1,ν_2}, where $\alpha = .05$, $\nu_1 = (a - 1)(b - 1) = (3 - 1) \times (2 - 1) = 2$, and $\nu_2 = n_T - ab = 12 - (3)(2) = 6$, in Table VII. The value from the table is $F_{.05,2,6} = 5.1433$. Our calculated value, $F = 1.174$, is smaller than the value from the table. Therefore we do not reject the null hypothesis. We conclude that there are no interactions in the population, and we continue to test the other hypotheses.

To test the second hypothesis that the population mean sales amount is the same for all three package designs (rows), or $H_0: \mu_{1.} = \mu_{2.} = \mu_{3.}$, against the alternative that one or more of the packages produce a different population mean sales amount, we use Equation (10.63) and calculate

$$F = \frac{MSA}{MSE} = \frac{871}{7.667} = 113.609$$

For this test we find F_{α,ν_1,ν_2} in Table VII, where $\alpha = .05$, $\nu_1 = a - 1 = 3 - 1 = 2$, and $\nu_2 = n_T - ab = 12 - (3)(2) = 6$. The value from the table is $F_{.05,2,6} = 5.1433$. Our calculated value, $F = 113.609$, exceeds the value from the table. Therefore the calculated value is in the rejection region, and we reject the null hypothesis. We conclude that one or more of the package designs produce a mean amount of sales different from the others.

To test the third hypothesis that the population mean sales amount is the same for the two selling prices (columns), or $H_0: \mu_{.1} = \mu_{.2}$, against the alternative that the two prices produce different mean sales amounts, we use Equation (10.64) and calculate

$$F = \frac{MSB}{MSE} = \frac{3}{7.667} = .391$$

For this test we find F_{α,ν_1,ν_2} in Table VII, where $\alpha = .05$, $\nu_1 = b - 1 = 2 - 1 = 1$, and $\nu_2 = n_T - ab = 12 - (3)(2) = 6$. The value from the table is $F_{.05,1,6} = 5.9874$. Our calculated value, $F = .391$, is smaller than the value from the table. Therefore we do not reject the null hypothesis. We conclude that the population mean sales values for the two selling prices are not different. Notice that the two selling prices are similar (99 cents and 95 cents).

A computer was used to analyze Federated Department Stores sales data for a 3×2 factorial design, and the results are presented in Computer Exhibits 10.7A, 10.7B, and 10.7C.

We will now consider **multiple pairwise comparisons of treatment means** or **factor level means.** For the factorial design we have been discussing, if the null hypothesis of no interactions is rejected, then multiple pairwise comparisons of treatment means μ_{ij} can be made in order to conclude which pairs of means are different. Likewise, if the hypothesis of no interactions is *not* rejected, and the null hypothesis of equal means for the levels of factor A is rejected, the null hypothesis of equal means for the levels of factor B is rejected, or if both of these null hypotheses are rejected, then multiple pairwise comparisons of factor level means can be made in order to conclude which pairs of level means for factor A and which pairs of level means for factor B are different.

COMPUTER EXHIBIT 10.7A Analysis of Variance for a 3 × 2 Factorial Design with
Federated Department Stores Sales Data

MINITAB

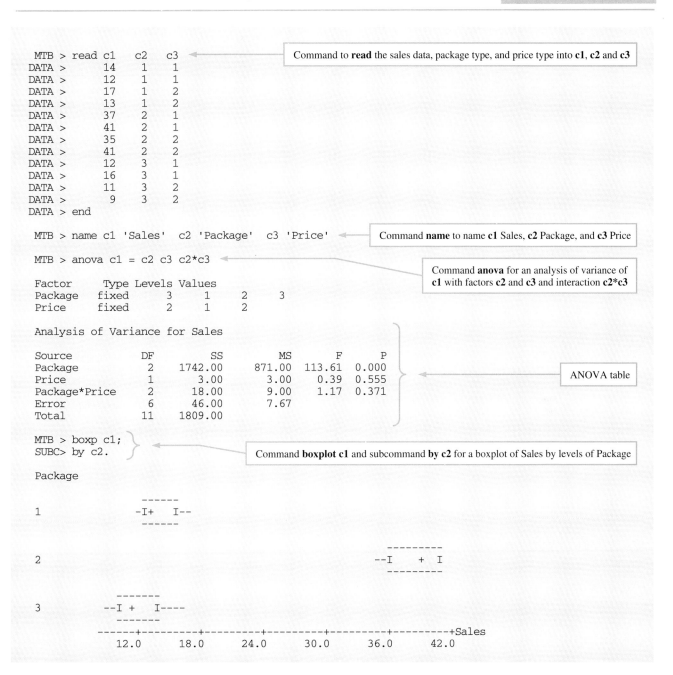

```
MTB > read c1   c2   c3          ←     Command to read the sales data, package type, and price type into c1, c2 and c3
DATA >      14   1   1
DATA >      12   1   1
DATA >      17   1   2
DATA >      13   1   2
DATA >      37   2   1
DATA >      41   2   1
DATA >      35   2   2
DATA >      41   2   2
DATA >      12   3   1
DATA >      16   3   1
DATA >      11   3   2
DATA >       9   3   2
DATA > end

MTB > name c1 'Sales'  c2 'Package'  c3 'Price'     ←     Command name to name c1 Sales, c2 Package, and c3 Price

MTB > anova c1 = c2 c3 c2*c3      ←     Command anova for an analysis of variance of
                                        c1 with factors c2 and c3 and interaction c2*c3
Factor     Type Levels Values
Package    fixed     3     1    2    3
Price      fixed     2     1    2

Analysis of Variance for Sales

Source          DF         SS        MS        F      P
Package          2    1742.00    871.00   113.61  0.000          ←     ANOVA table
Price            1       3.00      3.00     0.39  0.555
Package*Price    2      18.00      9.00     1.17  0.371
Error            6      46.00      7.67
Total           11    1809.00

MTB > boxp c1;
SUBC> by c2.              ←     Command boxplot c1 and subcommand by c2 for a boxplot of Sales by levels of Package

Package

1              ------
             -I+   I--
              ------

                                                    ---------
2                                                   --I   +  I
                                                    ---------

              -------
3            --I +   I----
              -------
        ------+---------+---------+---------+---------+---------+Sales
            12.0      18.0      24.0      30.0      36.0      42.0
```

If the null hypothesis of *no interactions* is rejected, then m multiple pairwise comparisons of treatment means μ_{ij}, as discussed in Section 10.5, can be used with a family confidence level of at least $100(1 - \alpha)\%$ to determine which pairs of treatment means differ. Three adjustments to Equation (10.41) must be made in order to find the confi-

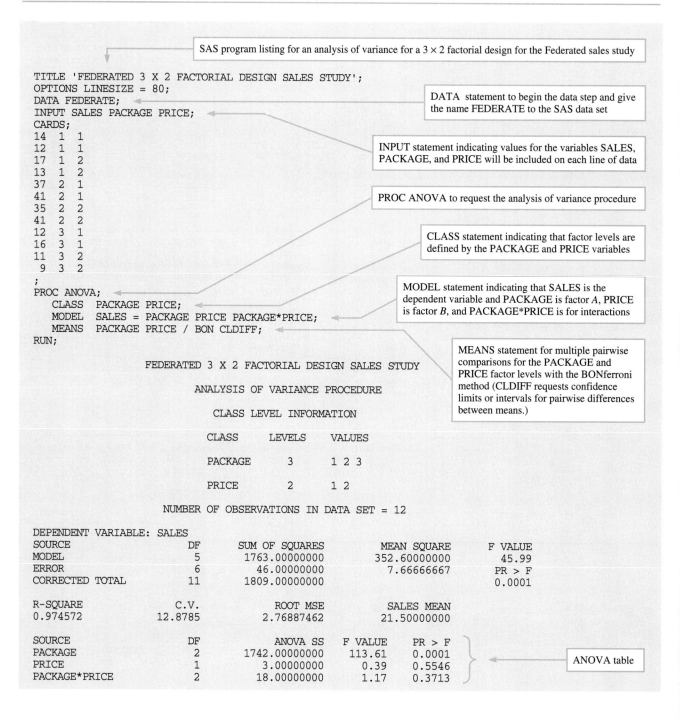

SAS

COMPUTER EXHIBIT 10.7B Analysis of Variance for a 3 × 2 Factorial Design with Federated Department Stores Sales Data

SAS program listing for an analysis of variance for a 3 × 2 factorial design for the Federated sales study

```
TITLE 'FEDERATED 3 X 2 FACTORIAL DESIGN SALES STUDY';
OPTIONS LINESIZE = 80;
DATA FEDERATE;
INPUT SALES PACKAGE PRICE;
CARDS;
14  1  1
12  1  1
17  1  2
13  1  2
37  2  1
41  2  1
35  2  2
41  2  2
12  3  1
16  3  1
11  3  2
 9  3  2
;
PROC ANOVA;
   CLASS   PACKAGE PRICE;
   MODEL   SALES = PACKAGE PRICE PACKAGE*PRICE;
   MEANS   PACKAGE PRICE / BON CLDIFF;
RUN;
```

DATA statement to begin the data step and give the name FEDERATE to the SAS data set

INPUT statement indicating values for the variables SALES, PACKAGE, and PRICE will be included on each line of data

PROC ANOVA to request the analysis of variance procedure

CLASS statement indicating that factor levels are defined by the PACKAGE and PRICE variables

MODEL statement indicating that SALES is the dependent variable and PACKAGE is factor A, PRICE is factor B, and PACKAGE*PRICE is for interactions

MEANS statement for multiple pairwise comparisons for the PACKAGE and PRICE factor levels with the BONferroni method (CLDIFF requests confidence limits or intervals for pairwise differences between means.)

```
             FEDERATED 3 X 2 FACTORIAL DESIGN SALES STUDY

                     ANALYSIS OF VARIANCE PROCEDURE

                        CLASS LEVEL INFORMATION

                  CLASS      LEVELS     VALUES

                  PACKAGE       3       1 2 3

                  PRICE         2       1 2

           NUMBER OF OBSERVATIONS IN DATA SET = 12
```

```
DEPENDENT VARIABLE: SALES
SOURCE                  DF      SUM OF SQUARES          MEAN SQUARE       F VALUE
MODEL                    5      1763.00000000        352.60000000         45.99
ERROR                    6        46.00000000          7.66666667        PR > F
CORRECTED TOTAL         11      1809.00000000                            0.0001

R-SQUARE              C.V.            ROOT MSE          SALES MEAN
0.974572          12.8785          2.76887462          21.50000000

SOURCE                  DF            ANOVA SS     F VALUE      PR > F
PACKAGE                  2      1742.00000000      113.61      0.0001
PRICE                    1         3.00000000        0.39      0.5546
PACKAGE*PRICE            2        18.00000000        1.17      0.3713
```

ANOVA table

dence intervals that are used to make the pairwise comparisons. First, the difference between treatment means is $\overline{X}_{ij} - \overline{X}_{i'j'}$. Second, the value for the degrees of freedom changes to $n_T - ab$. Third, the sample sizes for the experimental items that receive the various treatments are all equal to n.

COMPUTER EXHIBIT 10.7B Analysis of Variance for a 3 × 2 Factorial Design with Federated Department Stores Sales Data (Continued)

SAS

Since equality of PACKAGE means was rejected,

to test H_0: $\mu_{i.} = \mu_{i'.}$ with SAS, use the following rejection rule.

Rejection rule: If $|\bar{X}_{i.} - \bar{X}_{i'.}| > t_{\alpha/(2m),\,(n_T - na)}\sqrt{MSW[1/(na) + 1/(na)]}$, then reject H_0.

For the Federated sales data,

Rejection rule: If $|\bar{X}_{i.} - \bar{X}_{i'.}| > 6.4365$, then reject H_0.

```
BONFERRONI (DUNN) T TESTS FOR VARIABLE: SALES
NOTE: THIS TEST CONTROLS THE TYPE I EXPERIMENTWISE ERROR RATE
      BUT GENERALLY HAS A HIGHER TYPE II ERROR RATE THAN TUKEY'S

      ALPHA=0.05  CONFIDENCE=0.95  DF=6  MSE=7.66667
      CRITICAL VALUE OF T=3.28746
      MINIMUM SIGNIFICANT DIFFERENCE=6.4365   ◄────

COMPARISONS SIGNIFICANT AT THE 0.05 LEVEL ARE INDICATED BY '***'

                    SIMULTANEOUS              SIMULTANEOUS
                       LOWER     DIFFERENCE      UPPER
          PACKAGE    CONFIDENCE   BETWEEN     CONFIDENCE
         COMPARISON    LIMIT       MEANS        LIMIT

          2  - 1      18.064      24.500       30.936    ***
          2  - 3      20.064      26.500       32.936    ***

          1  - 2     -30.936     -24.500      -18.064    ***
          1  - 3      -4.436       2.000        8.436

          3  - 2     -32.936     -26.500      -20.064    ***
          3  - 1      -8.436      -2.000        4.436
```

Confidence intervals for multiple pairwise comparisons of factor level means using the Bonferroni method

```
BONFERRONI (DUNN) T TESTS FOR VARIABLE: SALES
NOTE: THIS TEST CONTROLS THE TYPE I EXPERIMENTWISE ERROR RATE
      BUT GENERALLY HAS A HIGHER TYPE II ERROR RATE THAN TUKEY'S

      ALPHA=0.05  CONFIDENCE=0.95  DF=6  MSE=7.66667
      CRITICAL VALUE OF T=2.44691
      MINIMUM SIGNIFICANT DIFFERENCE=3.9117

COMPARISONS SIGNIFICANT AT THE 0.05 LEVEL ARE INDICATED BY '***'

                    SIMULTANEOUS              SIMULTANEOUS
                       LOWER     DIFFERENCE      UPPER
           PRICE     CONFIDENCE   BETWEEN     CONFIDENCE
         COMPARISON    LIMIT       MEANS        LIMIT

          1  - 2      -2.912       1.000        4.912

          2  - 1      -4.912      -1.000        2.912

                         MEANS

         PACKAGE    PRICE      N        SALES

            1         1        2     13.0000000
            1         2        2     15.0000000
            2         1        2     39.0000000
            2         2        2     38.0000000
            3         1        2     14.0000000
            3         2        2     10.0000000
```

Note: ROOT MSE = $\sqrt{MSE}$

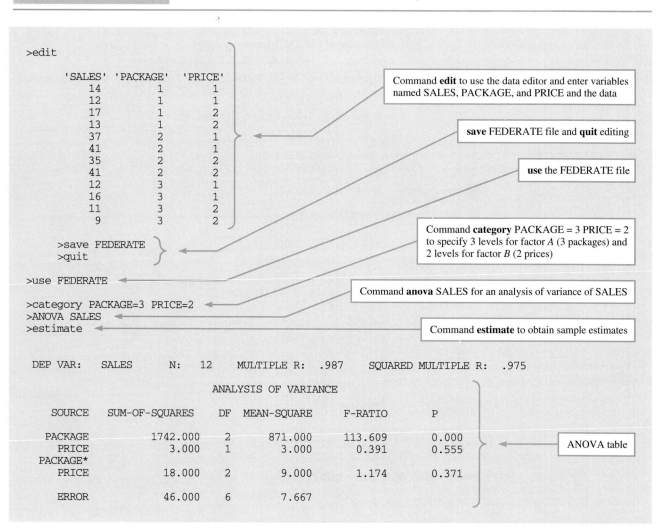

COMPUTER EXHIBIT 10.7C Analysis of Variance for a 3 × 2 Factorial Design with Federated Department Stores Sales Data

```
>edit

      'SALES'  'PACKAGE'  'PRICE'
         14        1         1
         12        1         1
         17        1         2
         13        1         2
         37        2         1
         41        2         1
         35        2         2
         41        2         2
         12        3         1
         16        3         1
         11        3         2
          9        3         2

    >save FEDERATE
    >quit

>use FEDERATE

>category PACKAGE=3 PRICE=2
>ANOVA SALES
>estimate
```

Command **edit** to use the data editor and enter variables named SALES, PACKAGE, and PRICE and the data

save FEDERATE file and **quit** editing

use the FEDERATE file

Command **category** PACKAGE = 3 PRICE = 2 to specify 3 levels for factor A (3 packages) and 2 levels for factor B (2 prices)

Command **anova** SALES for an analysis of variance of SALES

Command **estimate** to obtain sample estimates

DEP VAR: SALES   N: 12   MULTIPLE R: .987   SQUARED MULTIPLE R: .975

ANALYSIS OF VARIANCE

SOURCE	SUM-OF-SQUARES	DF	MEAN-SQUARE	F-RATIO	P
PACKAGE	1742.000	2	871.000	113.609	0.000
PRICE	3.000	1	3.000	0.391	0.555
PACKAGE* PRICE	18.000	2	9.000	1.174	0.371
ERROR	46.000	6	7.667		

ANOVA table

Thus for computing confidence intervals to test the null hypotheses of the **equality of pairwise treatment means,** H_0: $\mu_{ij} = \mu_{i'j'}$, where ij and $i'j'$ denote a pair of treatments, the following equations are used.

Equations for Confidence Limits for Multiple Pairwise Comparisons of Treatment Means

$$L = (\overline{X}_{ij} - \overline{X}_{i'j'}) - t_{\alpha/(2m),(n_T - ab)} \sqrt{MSE\left(\frac{1}{n} + \frac{1}{n}\right)}$$

$$R = (\overline{X}_{ij} - \overline{X}_{i'j'}) + t_{\alpha/(2m),(n_T - ab)} \sqrt{MSE\left(\frac{1}{n} + \frac{1}{n}\right)}$$

(10.66)

If an *interval does not* contain the value 0, then we reject the corresponding null hypothesis and conclude that the two treatment means differ.

Now consider the case when the hypothesis of no interactions is *not* rejected. If the null hypothesis of equal means for the levels of factor A, H_0: $\mu_{1.} = \mu_{2.} = \cdots = \mu_{a.}$ is rejected, if the null hypothesis of equal means for the levels of factor B, H_0: $\mu_{.1} = \mu_{.2} = \cdots = \mu_{.b}$ is rejected, or if both are rejected, then m multiple pairwise comparisons of means, as discussed in Section 10.5, can be used with a family confidence level of at least $100(1 - \alpha)\%$ to determine which pairs of factor level means differ. Two adjustments to Equation (10.66) must be made in order to find the confidence intervals that are used to make the pairwise comparisons. First, for pairwise comparisons between the means for the levels of factor A, the difference between means is $\overline{X}_{i.} - \overline{X}_{i'.}$; and for pairwise comparisons between the means for the levels of factor B, the difference between means is $\overline{X}_{.j} - \overline{X}_{.j'}$. Second, the sample sizes for the levels of factor A are all equal to nb, and the sample sizes for the levels of factor B are all equal to na.

Thus for computing confidence intervals to test the null hypotheses of the **equality of pairwise means for the levels of factor A,** H_0: $\mu_{i.} = \mu_{i'.}$, where i and i' denote a pair of levels, the following formulas are used.

Equations for Confidence Limits for Multiple Pairwise Comparisons of Means of Levels of Factor A

$$L = (\overline{X}_{i.} - \overline{X}_{i'.}) - t_{\alpha/(2m),(n_T - ab)} \sqrt{MSE\left(\frac{1}{nb} + \frac{1}{nb}\right)}$$

$$R = (\overline{X}_{i.} - \overline{X}_{i'.}) + t_{\alpha/(2m),(n_T - ab)} \sqrt{MSE\left(\frac{1}{nb} + \frac{1}{nb}\right)}$$

(10.67)

In addition, the formulas for computing confidence intervals to test the null hypotheses of the **equality of pairwise means for the levels of factor B,** H_0: $\mu_{.j} = \mu_{.j'}$, where j and j' denote a pair of levels, is as follows.

Equations for Confidence Limits for Multiple Pairwise Comparisons of Means of Levels of Factor B

$$L = (\overline{X}_{.j} - \overline{X}_{.j'}) - t_{\alpha/(2m),(n_T - ab)} \sqrt{MSE\left(\frac{1}{na} + \frac{1}{na}\right)}$$

$$R = (\overline{X}_{.j} - \overline{X}_{.j'}) + t_{\alpha/(2m),(n_T - ab)} \sqrt{MSE\left(\frac{1}{na} + \frac{1}{na}\right)}$$

(10.68)

As in Section 10.5, if an interval does *not* contain the value 0, then we reject the corresponding null hypothesis and conclude that the two means are different.

 PRACTICE PROBLEM 10.7

Because we rejected the null hypothesis of the equality of the means for the levels of factor A (package designs) in the Federated Department Stores analysis of variance of Practice Problem 10.6, the problem we now face is to decide which package design (or designs) results in higher mean sales values. In other words, we want to use multiple pairwise comparisons to determine which pairs of means for all of the levels of factor A are significantly different by testing the null hypotheses H_0: $\mu_{i.} = \mu_{i'.}$, where i and i' denote the levels. Let the family confidence level be at least 95%.

Solution To determine which package design (or designs) results in higher mean sales at Federated, we can make $m = 3$ multiple comparisons of means by testing the following three null hypotheses:

$$H_{0_1}: \mu_{1.} = \mu_{2.} \qquad H_{0_2}: \mu_{1.} = \mu_{3.} \qquad H_{0_3}: \mu_{2.} = \mu_{3.}$$

The alternative hypothesis in each case is that the two means are not equal.

The sample mean sales values for the three different types of packages were presented in Table 10.14, and they are as follows:

$$\overline{X}_{1.} = 14 \qquad \overline{X}_{2.} = 38.5 \qquad \overline{X}_{3.} = 12$$

The sample sizes for each of the treatments (cells) is $n = 2$, and the number of levels for factor B is $b = 2$, so the number of experimental items in each level of factor A is $nb = (2)(2) = 4$. From Table 10.17 the value for MSE is 7.667, and the value for the degrees of freedom associated with MSE is $n_T - ab = 12 - (3)(2) = 6$. Because the family confidence level must be at least 95%, we will let $\alpha = .05$. Each of the $m = 3$ individual confidence intervals will be found by letting the area we place in a single tail of the t distribution be $\alpha/(2m) = .05/[(2)(3)] = .008333$. Thus the value of t we will use in Equation (10.67) to determine the confidence limits is

$$t_{\alpha/(2m),(k-1)(n-1)} = t_{.008333,6} = 3.2875$$

Now to compare the pair of means for package 1 and package 2, we test the null hypothesis H_{0_1}: $\mu_{1.} = \mu_{2.}$. For this hypothesis $\overline{X}_{1.} - \overline{X}_{2.} = 14 - 38.5 = -24.5$, and the confidence limits that we find by using Equation (10.67) are as follows:

$$L = (\overline{X}_{1.} - \overline{X}_{2.}) - t_{\alpha/(2m),(n_T-ab)} \sqrt{MSE\left(\frac{1}{nb} + \frac{1}{nb}\right)}$$

$$= -24.5 - (3.2875) \sqrt{7.667\left[\frac{1}{(2)(2)} + \frac{1}{(2)(2)}\right]} = -24.5 - 6.44$$

$$= -30.94$$

$$R = (\overline{X}_{1.} - \overline{X}_{2.}) + t_{\alpha/(2m),(n_T-ab)} \sqrt{MSE\left(\frac{1}{nb} + \frac{1}{nb}\right)}$$

$$= -24.5 + (3.2875) \sqrt{7.667\left[\frac{1}{(2)(2)} + \frac{1}{(2)(2)}\right]} = -24.5 + 6.44$$

$$= -18.06$$

We will use the format of the following expression to present the confidence intervals:

$$L \leq \mu_{i.} - \mu_{i'.} \leq R \tag{10.69}$$

The interval for comparing package 1 and package 2 is thus

$$-30.94 \leq \mu_{1.} - \mu_{2.} \leq -18.06$$

Because the interval does not include 0, we reject the null hypothesis H_{0_1}: $\mu_{1.} = \mu_{2.}$. Our conclusion is that the mean sales values that result from the use of package 1 and package 2 are different.

Confidence intervals for testing the null hypotheses H_{0_2} and H_{0_3} were also computed, and they are presented by using the format of Equation (10.69):

$$-4.44 \leq \mu_{1.} - \mu_{3.} \leq 8.44 \qquad \text{(Includes 0; do not reject } H_{0_2}: \mu_{1.} = \mu_{3.}\text{)}$$

$$20.06 \leq \mu_{2.} - \mu_{3.} \leq 32.94 \qquad \text{(Does not include 0; reject } H_{0_3}: \mu_{2.} = \mu_{3.}\text{)}$$

The results for the three pairwise comparisons enable us to conclude that $\mu_{1.}$ and $\mu_{3.}$ are different from $\mu_{2.}$. But $\mu_{1.}$ and $\mu_{3.}$ are not different from each other. From the family of confidence intervals we conclude, with at least 95% confidence, that the mean sales for package 2 is higher than the mean sales for package 1 by an amount somewhere between 18.06 and 30.94, and also that the mean sales for package 2 is higher than the mean sales for package 3 by an amount somewhere between 20.06 and 32.94. Thus our decision is to use package 2 in order to achieve the highest sales. ▬▬

If data sets are large, or if the computations are burdensome, and a computer is not available, then the following computational formulas can be used to determine the values of the various sums of squares; although the general availability of computers and data analysis software packages tend to prevail over the use of computational formulas.

Computational Formulas for an $a \times b$ Factorial Design

For $T =$ the sum of the values of the dependent variable X_{ijk} for all experimental items, or

$$T = \sum_i \sum_j \sum_k X_{ijk}$$

$T_{i.} =$ the sum of the values of the dependent variable for the experimental items in level i of factor A, or

$$T_{i.} = \sum_j \sum_k X_{ijk}$$

$T_{.j} =$ the sum of the values of the dependent variable for the experimental items in level j of factor B, or

$$T_{.j} = \sum_i \sum_k X_{ijk}$$

and $T_{ij} =$ the sum of the values of the dependent variable for the experimental items in level i of factor A and level j of factor B, or

$$T_{ij} = \sum_k X_{ijk}$$

the *computational formulas* are as follows:

$$SST = \sum_i \sum_j \sum_k X_{ijk}^2 - \frac{T^2}{n_T} \qquad (10.70)$$

$$SSA = \frac{\sum_i T_{i.}^2}{nb} - \frac{T^2}{n_T} \qquad (10.71)$$

$$SSB = \frac{\sum_j T_{.j}^2}{na} - \frac{T^2}{n_T} \qquad (10.72)$$

$$SSE = \sum_i \sum_j \sum_k X_{ijk}^2 - \frac{\sum_i \sum_j T_{ij}^2}{n} \qquad (10.73)$$

$$SSAB = SST - SSA - SSB - SSE \qquad (10.74)$$

Problems: Section 10.10

51. Three manufacturers of flexible-disk drives for personal computers wish to sell their brands of disk drives to a company that assembles personal computers of two types, one with an 80386 microprocessor and one with an 80486 microprocessor. The assembling firm is concerned about the operating time prior to failure that the different brands achieve on the two types of computers. Twelve disk drives are selected randomly such that two of each brand are used with computers that have the 80386 microprocessor and two are used with computers that have the 80486 microprocessor. The disk drives are then operated until they fail. The time (in thousands of hours) are given in the accompanying table. Use a signficance level of .05.

    a. Construct the analysis of variance table.
    b. Test the hypothesis of no interactions.
    c. Test the hypothesis of equal means for the three brands of disk drives.
    d. Test the hypothesis of equal means for the two types of computers.

Brand of the Flexible-Disk Drive	Microprocessor	
	80386	80486
Brand A	6.4	4.3
	5.8	4.8
Brand B	8.2	8.2
	9.4	8.8
Brand C	1.8	1.5
	2.2	4.3

52. A company that has a chain of very similar retail stores conducted an experiment by using a factorial design to study the sales of a product (in dollars) when it was positioned at different levels on the shelves (high, middle, low) and located at different locations in the stores (front and back). A sample of 12 stores was taken, and 2 separate stores were assigned at random to use a single combination of a shelf position and a location in the store for displaying the product. The results for this experiment, in the form of sales of the product (in dollars), are presented in the accompanying table. Use a significance level of .05.

    a. Construct the analysis of variance table.
    b. Test the hypothesis of no interactions.
    c. Test the hypothesis of equal means for the three shelf positions.
    d. Test the hypothesis of equal means for the two locations in the store.
    e. If appropriate, use multiple pairwise comparisons and a family confidence level of at least 95% to determine which shelf position results in the highest mean amount of sales. [*Hint:* $t_{.008333, 6} = 3.2875$.]

Shelf Position	Location in the Store	
	Front	Back
High	9	12
	7	8
Middle	32	30
	36	36
Low	7	6
	11	4

COMPUTER EXHIBIT 10.8 Analysis of Variance for a Factorial Design with Corn Yield Data

---

```
Analysis of Variance for Yield

Source         DF        SS         MS       F      P
Fertiliz        1     80.08      80.08   45.76  0.001
Corn            2    612.50     306.25  175.00  0.000
Fertiliz*Corn   2     11.17       5.58    3.19  0.114
Error           6     10.50       1.75
Total          11    714.25
```

53. An agricultural experiment station conducted an experiment with a factorial design in order to make comparisons of yields among three varieties of corn when they were treated with two amounts of fertilizer, a normal amount and a heavy amount. Twelve similar parcels of land were used with two randomly selected parcels, each receiving one of the varieties and one of the amounts of fertilizer. The yields (in bushels per acre) are given in the accompanying table. Use a significance level of .05. An edited print-out from MINITAB for the analysis of variance is given in Computer Exhibit 10.8.

   a. Construct the analysis of variance table.
   b. Test the hypothesis of no interactions.
   c. Test the hypothesis of equal means for the two amounts of fertilizer.
   d. Test the hypothesis of equal means for the three varieties.
   e. If appropriate, use multiple pairwise comparisons and a family confidence level of at least 95% to determine which variety (or varieties) of corn results in the highest mean yield. [*Hint:* $t_{.008333, 6} = 3.2875$.]

	Corn		
Fertilizer	*Variety 1*	*Variety 2*	*Variety 3*
Heavy	47	42	28
	45	44	27
Normal	40	38	26
	38	36	24

54. Crust, Inc., conducted a field experiment with a factorial design in order to study the surface amount of new tooth decay (square millimeters, over a period of one year) for adolescents who were assigned randomly to use one of two different toothpastes and one of two different mouthwashes. The results are presented in the accompanying table. Use a signficance level of .05.

   a. Test for no significant effects due to interactions.
   b. Test for no significant effects due to toothpaste.
   c. Test for no significant effects due to mouthwash.

	Toothpaste	
Mouthwash	*1*	*2*
Mouthwash 1	1	2
	2	3
Mouthwash 2	2	3
	1	3

55. A researcher designed a factorial study to examine the rates of return on assets (*ROAs*) as percentages of companies that (a) pursued different diversification strategies known as (1) single-business, (2) dominant-business, or (3) conglomerate, and (b) were in industries that were mainly (1) service-oriented or (2) manufacturing-oriented. The data are presented in the accompanying table. Use a significance level of .05. An edited SPSS printout for the analysis of variance is given in Computer Exhibit 10.9.

   a. Use the ANOVA methods for a factorial design to test the hypothesis of no interactions.

SPSS/PC+   **COMPUTER EXHIBIT 10.9** Analysis of Variance for a 2 × 2 Factorial Design of Return on Assets

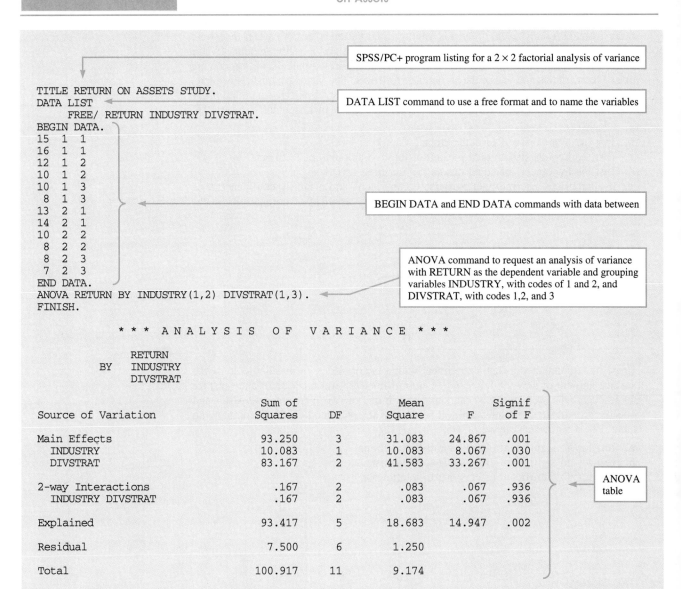

SPSS/PC+ program listing for a 2 × 2 factorial analysis of variance

```
TITLE RETURN ON ASSETS STUDY.
DATA LIST
      FREE/ RETURN INDUSTRY DIVSTRAT.
BEGIN DATA.
15  1  1
16  1  1
12  1  2
10  1  2
10  1  3
 8  1  3
13  2  1
14  2  1
10  2  2
 8  2  2
 8  2  3
 7  2  3
END DATA.
ANOVA RETURN BY INDUSTRY(1,2) DIVSTRAT(1,3).
FINISH.
```

DATA LIST command to use a free format and to name the variables

BEGIN DATA and END DATA commands with data between

ANOVA command to request an analysis of variance with RETURN as the dependent variable and grouping variables INDUSTRY, with codes of 1 and 2, and DIVSTRAT, with codes 1,2, and 3

```
* * *   A N A L Y S I S   O F   V A R I A N C E   * * *
```

              RETURN
      BY   INDUSTRY
              DIVSTRAT

Source of Variation	Sum of Squares	DF	Mean Square	F	Signif of F
Main Effects	93.250	3	31.083	24.867	.001
INDUSTRY	10.083	1	10.083	8.067	.030
DIVSTRAT	83.167	2	41.583	33.267	.001
2-way Interactions	.167	2	.083	.067	.936
INDUSTRY DIVSTRAT	.167	2	.083	.067	.936
Explained	93.417	5	18.683	14.947	.002
Residual	7.500	6	1.250		
Total	100.917	11	9.174		

ANOVA table

b. Use the ANOVA methods for a factorial design to test the hypothesis of equal mean *ROA*s for the diversification strategies.

c. Use the ANOVA methods for a factorial design to test the hypothesis of equal mean *ROA*s for the service and manufacturing industries.

d. If appropriate, use multiple pairwise comparisons and a family confidence level of at least 95% to determine which diversification strategy results in the highest mean *ROA*. [*Hint:* $t_{.008333, 6} = 3.2875.$]

	Diversification Strategy		
Industry	Single Business	Dominant Business	Conglomerate
Service	15	12	10
	16	10	8
Manufacturing	13	10	8
	14	8	7

Tests Using
Categorical Data

11

In Chapter 9, we considered a test of the equality of proportions for two binomial populations, H_0: $p_1 = p_2$. In this chapter, we extend the hypothesis test of two proportions to a test of several, say k, proportions.

A hypothesis test for several proportions is called a multinomial test. The statistic that we use with a multinomial test requires us to introduce a distribution known as the *chi-square distribution*.

We also use chi-square distributions in this chapter for *contingency table tests*. We classify items in contingency tables by using two categorical variables, and we use contingency tables to test the hypothesis of *independence* of the categorical variables or to test the hypothesis of *homogeneity* of k proportions H_0: $p_1 = p_2 = \cdots = p_k$.

11.1
MULTINOMIAL DISTRIBUTIONS

In earlier chapters, we discussed the binomial distribution in some detail. Recall that the binomial distribution is the appropriate model when we are sampling from a population in which the items belong to one of two possible classes and the sample is taken in such a manner that the probability of obtaining an element from a given class remains constant—that is, it is not affected by the sampling process. Not all actual statistical problems, of course, conform to this model. Frequently, analysts must consider the more general case where the elements of the population are classified as belonging to one of k classes, where $k \geq 2$. We refer to a population with a categorical variable that has k classes as a *multinomial population,* and if the proportions of the elements belonging to each class are not changed by the selection of the sample, the appropriate model is a **multinomial distribution.** This condition typically exists when we are sampling from an infinite population that has multiple characteristics or from a similar finite population with replacement. If we are sampling from a large finite multinomial population without replacement and the sample size n is small compared to the population size N, then the distribution for the sample outcomes is approximately multinomially distributed.

> **Definition**
>
> A **multinomial distribution** is the *joint* distribution for the random variables, $Y_1, Y_2, \ldots, Y_k$, the numbers of elements in a random sample of size n that belong to each of the k classes of the multinomial population, Y_j being the number that belong to the jth class.

The corresponding sample proportions, or fractions, Y_j/n we denote by $\hat{p}_j$. These quantities satisfy the conditions

$$\sum_{j=1}^{k} Y_j = n \tag{11.1}$$

$$\sum_{j=1}^{k} \hat{p}_j = 1 \tag{11.2}$$

The parameters of the multinomial distribution are the sample size n and the proportions $p_1, p_2, \ldots, p_k$ of the elements in the population that belong to each of the k classes, p_j being the proportion that belongs to the jth class.* Since the k classes contain all of the elements,

$$\sum_{j=1}^{k} p_j = 1 \tag{11.3}$$

As one might expect, each p_j is estimated unbiasedly by the corresponding sample proportion

$$\hat{p}_j = \frac{Y_j}{n}$$

---

*The multinomial probability distribution function or the probability of obtaining $Y_1, Y_2, \ldots,$ and Y_k numbers of elements in the k classes is $P(Y_1, Y_2, \ldots, Y_k) = [n!/(Y_1! \, Y_2! \cdots Y_k!)] p_1^{Y_1} p_2^{Y_2} \cdots p_k^{Y_k}$ where $Y_j = 0, 1, \ldots, n$.

The basic hypothesis to be considered here is that the proportions belonging to the k classes of a categorical variable are equal to a set of specified values. Thus, H_0: $p_1 = p_{1_0}$, $p_2 = p_{2_0}, \ldots, p_k = p_{k_0}$. An exact test for this hypothesis would be difficult to apply, particularly for large sample sizes; therefore it is usually tested by a procedure involving approximations.

If the hypothesis is true, the mean, or *expected,* number of elements of the jth class in a sample of size n is

$$E_j = np_{j_0}$$

The *observed* number of the jth class is Y_j. If we calculate for each class the quantity

$$\frac{(Y_j - E_j)^2}{E_j}$$

we have the squared difference, or squared residual, between the observed and expected numbers in that class $(Y_j - E_j)^2$ relative to its expected number E_j. The sum of these squared residuals relative to expected frequencies over all classes, denoted by χ^2 (Greek chi, squared), is

$$\chi^2 = \frac{(Y_1 - E_1)^2}{E_1} + \frac{(Y_2 - E_2)^2}{E_2} + \cdots + \frac{(Y_j - E_j)^2}{E_j}$$

This equation can be expressed as follows.

Equation for Chi-Square

$$\chi^2 = \sum_{j=1}^{k} \frac{(Y_j - E_j)^2}{E_j} \qquad\qquad (11.4)$$

The χ^2 statistic is a measure of the lack of agreement between the data and the hypothesis. The idea is that if the null hypothesis is true, then the observed frequencies

$$Y_1, Y_2, \ldots, Y_j$$

ought not deviate too much from their respective expected values

$$E_1, E_2, \ldots, E_j$$

The χ^2 statistic gathers together the squared residuals between Y_j and E_j relative to E_j for all the values of j. To test the null hypothesis, we ask whether the statistic in Equation (11.4) has a larger value than can reasonably be accounted for by the workings of chance.

For sufficiently large sample sizes the distribution of the statistic of Equation (11.4) can be approximated by a χ^2 distribution with $\nu = k - 1$ degrees of freedom. Note that the degrees of freedom are equal to one less than the number of *classes* and are not related to the size of the sample.

FIGURE 11.1

The χ^2 Distributions with 2, 4, and 10 Degrees of Freedom

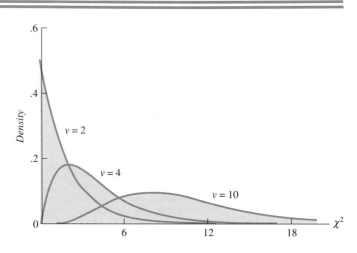

A χ^2 distribution, or **chi-square distribution,** is a continuous distribution ordinarily derived as the sampling distribution of a sum of squares of independent standard normal variables. It is a skewed distribution and only nonnegative values of the variable χ^2 are possible. A χ^2 distribution has a single parameter, the degrees of freedom ν. The χ^2 distributions for degrees of freedom equal to 2, 4, and 10 are shown in Figure 11.1. The figure shows that the skewness decreases as the degrees of freedom increase. In fact, as the degrees of freedom increase without limit, χ^2 distributions approach a normal distribution.

Percentage points of χ^2 distributions are given in Table VIII of Appendix C and are defined by

$$P(\chi_\nu^2 \geq \chi_{\alpha,\nu}^2) = \alpha$$

That is, $\chi_{\alpha,\nu}^2$ is that value for the χ^2 distribution with ν degrees of freedom such that the area to the *right*, the probability of a *larger* value, is equal to α. For example, the upper 5% point $\chi_{.05,20}^2$ for χ^2 with 20 degrees of freedom is 31.41 (see Figure 11.2).

To test the hypothesis H_0 that $p_j = p_{j_0}$ for all j, we calculate the multinomial χ^2 of Equation (11.4) and compare the calculated value with the percentage points of the χ^2 distribution with $k - 1$ degrees of freedom. Since good agreement between the observed and expected numbers will result in a small χ^2 value and since perfect agreement will give a value of 0 for χ^2, we are justified in rejecting the hypothesis only when χ^2 is large. Hence this test is always a one-tailed test on the upper tail of the χ^2 distribution. The rejection rule is as follows:

Rejection rule: If $\chi^2 > \chi_{\alpha,k-1}^2$, then reject H_0.

One note of caution should be observed. In order for the statistic defined in Equation (11.4) to approximate the chi-square distribution, the expected frequencies, the E_j values, should all be 5 or greater. If very small expected frequencies were allowed, then a small disagreement between Y_j and E_j could appear to be abnormally large when the squared difference $(Y_j - E_j)^2$ is divided by a small E_j value. If one category does not have an expected frequency of 5 or more, we can sometimes combine two or more categories.

FIGURE 11.2

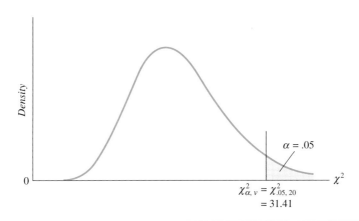

The procedure for testing that the proportions belonging to the k classes are equal to a set of specific proportions is summarized as follows.

> **Hypotheses and Rejection Rule for Testing k Proportions**
>
> H_0: $p_1 = p_{1_0}, p_2 = p_{2_0}, \ldots, p_k = p_{k_0}$.
>
> H_a: One or more of the proportions are not equal to the hypothesized values.
>
> Rejection rule: If $\chi^2 > \chi^2_{\alpha,\nu}$, then reject H_0;
>
> where: $\chi^2 = \sum \dfrac{(Y_j - E_j)^2}{E_j}$
>
> $\nu = k - 1$
>
> and E_j are all greater than or equal to 5.

 PRACTICE PROBLEM 11.1

In the Business Credit Institution industry the accounts receivable for companies are classified as being "current," "moderately late," "very late," and "uncollectable." Industry figures show that the ratio of these four classes is 9:3:3:1. Pratt Associates has 800 accounts receivable, with 439, 168, 133, and 60 falling in each class. Are these numbers in agreement with the industry ratio? Let $\alpha = .05$.

Solution The hypothesis test is as follows.

Step One H_0: $p_1 = 9/16$, $p_2 = 3/16$, $p_3 = 3/16$, and $p_4 = 1/16$.

Step Two H_a: One or more of the proportions are not equal to the proportions given in the null hypothesis.

TABLE 11.1	Class	Observed Y_j	Expected $E_j = np_{j_0}$	$Y_j - E_j$	$(Y_j - E_j)^2$	$(Y_j - E_j)^2/E_j$
Calculations for Pratt Associates	Current	439	450	−11	121	0.27
	Moderately late	168	150	18	324	2.16
	Very late	133	150	−17	289	1.93
	Uncollectable	60	50	10	100	2.00
					Total	6.36

Step Three n = 800, α = .05.

Step Four The value of the degrees of freedom is $\nu = k - 1 = 4 - 1 = 3$, and the critical value from the chi-square table is

$$\chi^2_{.05,3} = 7.81$$

Thus, the rejection rule is

Rejection rule: If $\chi^2 > 7.81$, then reject H_0

Step Five The test statistic is χ^2. The calculations may be conveniently arranged in tabular form, as shown in Table 11.1. All of the expected frequency E_j values (third column) are greater than 5, so the test statistic closely approximates the chi-square distribution. The χ^2 value from the calculations shown in Table 11.1 is

$$\chi^2 = 6.36$$

Step Six Since 6.36 is not greater than 7.81 (see Figure 11.3), we do not reject H_0. The data do not differ sufficiently from what we would expect under the null hypothesis to reject those proportions.

FIGURE 11.3

Hypothesis Test for Pratt Associates

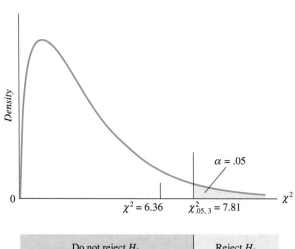

Class	Observed Y_j	Expected $E_j = np_{j_0}$	$Y_j - E_j$	$(Y_j - E_j)^2$	$(Y_j - E_j)^2/E_j$	TABLE 11.2
Red	60	40	20	400	10	Calculations for Plastics
Blue	20	40	−20	400	10	Products
Yellow	40	40	0	0	0	
				Total	20	

In this problem we may be interested in testing the hypothesis that some, but not all, of the proportions are equal to specified values. For instance, we might be interested in testing whether the proportions in the first two classes are 9/16 and 3/16, respectively, regardless of the proportions in the last two. In such cases the number of classes is reduced to conform with the hypothesis by combining the unspecified classes into one. For our accounts receivable here the classes would be "current," "moderately late," and "other." If we are interested in only one class, the problem falls into the framework of the binomial test, for we would have two classes: the class of interest and the class consisting of "all other classes."

 PRACTICE PROBLEM 11.2

Plastics Products, Inc., sells its product in three primary colors: red, blue, and yellow. The marketing manager feels that customers have no color preference for the product. Thus, her null hypothesis is $H_0: p_1 = p_2 = p_3 = 1/3$. That is, each of the three colors (1 = red, 2 = blue, 3 = yellow) is preferred by one-third of the purchasers. To test this hypothesis, the manager set up a test in which 120 purchasers were given equal opportunity to buy the product in each of the three colors. The results were that 60 bought red, 20 bought blue, and 40 bought yellow. Test the marketing manager's null hypothesis, using $\alpha = .05$.

Solution To calculate the χ^2 statistic of Equation (11.4), we need to find the expected frequencies, the E_j values. Since the hypothesis states that one-third of the people prefer each of the colors, then $E_1 = E_2 = E_3 = 120(1/3) = 40$. The calculation of χ^2 is facilitated by Table 11.2. Note, again, that all the values in the E_j column exceed 5.

Note that the sum of the figures in the last column of the table is the χ^2 value for this problem. This figure is compared with $\chi^2_{.05,2}$ since $\alpha = .05$ and there are $\nu = 3 - 1 = 2$ degrees of freedom. Table VIII yields a $\chi^2_{.05,2}$ value of 5.99. Since the calculated χ^2 value is 20 and exceeds 5.99, we should reject the hypothesis that people have no color preference with this product. It appears that red is the most popular color and blue is the least popular.

Problems: Section 11.2

Answers to odd-numbered problems are given in the back of the text.

1. The bank credit card operation of National Bancorporation knows from credit applications that the education level of its card holders has the following distribution:

$$
\begin{array}{rl}
\text{Some high school} & .02 = p_1 \\
\text{High school complete} & .15 = p_2 \\
\text{Some college} & .25 = p_3 \\
\text{College complete} & \underline{.58} = p_4 \\
& 1.00
\end{array}
$$

Of the 500 card holders whose cards have been called in for failure to pay their charges or for other abuses, the education distribution is as follows:

Some high school	50
High school complete	100
Some college	190
College complete	160
	500

Test the hypothesis that credit reliability is independent of education. That is, compare the distribution of the education of the 500 abusers with what would be expected for an education distribution in 500 random card holders. Let $\alpha = .01$.

2. United Labs testing agency set up six color TV sets in the lobby of a Chicago hotel. The brand names of the sets were covered, the positions of the sets were changed each day, and the sets were adjusted daily by an independent TV repairman. Over a period of several weeks 2700 people were asked: "Please look carefully at each TV screen and then indicate which color TV has the best picture." The number of people selecting each of the brands is given in the accompanying table. Can we reject the hypothesis that all the brands have pretty much the same picture quality? Let $\alpha = .005$.

Brand	Number Selecting Brand As Best
U	1350
V	567
W	243
X	240
Y	162
Z	138
	2700

3. Galstone Polling Agency reported that in a sample of 156 persons 81 were Democrats, 52 were Republicans, and 23 were independent or belonged to other parties. From these results, can one conclude that these political groups are equally strong in this area? That is, test the hypothesis H_0: $p_1 = p_2 = p_3$, using an α value of .05.

4. Rating Sciences, Inc., a television program-rating service, surveyed 600 families where the television was turned on during prime time on weeknights. They found the following numbers of people tuned to the various networks:

NBC⎤		210
CBS ⎬Commercial	170
ABC⎦		165
PBS		55
		600

   a. Test the hypothesis that all four networks have the same proportion of viewers during this prime-time period. Use $\alpha = .05$.
   b. Eliminate the results for PBS and repeat the test of hypothesis for the three commercial networks, using $\alpha = .05$ again.

5. Use the data in Problem 4 to test the hypothesis that each of the three major networks has 30% of the weeknight prime-time market and PBS has 10%. That is, test H_0: $p_1 = .30$, $p_2 = .30$, $p_3 = .30$, $p_4 = .10$, and let $\alpha = .005$. Would you reject the hypothesis if α were .05?

6. Banks, thrifts, credit unions, and armored-car companies lost $68 million from crimes in 1990, according to reports from the Federal Bureau of Investigation (FBI). The high

times for bank crimes of robbery, burglary, and larceny by day of week are given by the following frequency counts:

Day	Mon.	Tues.	Wed.	Thurs.	Fri.
Crimes	1,559	1,518	1,306	1,375	1,946

a. Test the hypothesis that bank crimes are uniformly distributed by day of the work week. Use a significance level of .05.
b. When should the bank manager beef up security at the bank?

REFERENCE: *American Banker* (1991), "Bank crimes rose 16% in '90, FBI Reports" (July 8): 5.

7. A study was done by a hospital's quality control administrators to determine what factors influence labor induction rates for maternity patients. Hospital administrators wanted to decrease hospitalization costs and morbidity for health care consumers and increase profitability at the same time. One factor that was studied was the procedure of inducing labor. Oxytocin, a hormone released by the pituitary gland, stimulates contractions during labor, and health care providers use synthetic hormone to induce and accelerate labor. Some induced labors are necessary for the well-being of the patients. But some labors are induced to provide a more dependable schedule for the health care provider, even though too rapid a labor can result in complications for patients, such as maternal injuries and fetal distress. The numbers of labor inductions by times of the day are given by the following frequency counts:

Time	9 a.m.–12 n.	12–3 p.m.	3–6 p.m.	6–9 p.m.	9 p.m.–12 m.
Inductions	80	220	300	50	25

a. Test the hypothesis that labor inductions are uniformly distributed by time of the work day. Use a significance level of 05.
b. Does it appear that labor inductions occur randomly for the well-being of patients?

REFERENCE: D. Scully (1980), *Men Who Control Women's Health* (Boston: Houghton Mifflin).

In this section, we extend the topic of testing the equality of two binomial proportions to a *contingency table test*. Many sets of countable data can be classified according to two or more categorical variables or criteria of classification, and often we would like to know whether or not these various criteria are independent of one another. For instance, we might like to know whether a person's color preference for a product is independent of gender. In this case we could take a sample of people, record their gender and their favorite color, and classify their responses by gender and by color preference. In a similar manner, we might want to test to see whether brand preference and income level are independent.

If the data from a random sample taken from a large or an infinite population are to be classified by two categorical variables or criteria, the case we treat here, the classes based on one of the criteria can be represented by the rows in a two-way table, and the classes based on the other can be represented by the columns. A cell of the table is formed by the intersection of a row and column. We may then count the number of elements in the sample that belong to each cell of the table. The general two-way **contingency table** or **cross tabulation** with r rows and c columns is presented in Table 11.3.

The notation used in Table 11.3 is as follows:

11.3
CONTINGENCY TABLES

		Columns					Row
Rows	1	2	3	$\cdots$	c		Totals
1	Y_{11}	Y_{12}	Y_{13}	$\cdots$	Y_{1c}		R_1
2	Y_{21}	Y_{22}	Y_{23}	$\cdots$	Y_{2c}		R_2
3	Y_{31}	Y_{32}	Y_{33}	$\cdots$	Y_{3c}		R_3
$\vdots$	$\vdots$	$\vdots$	$\vdots$		$\vdots$		$\vdots$
r	Y_{r1}	Y_{r2}	Y_{r3}	$\cdots$	Y_{rc}		R_r
Column Totals	C_1	C_2	C_3	$\cdots$	C_c		n

The $r \times c$ Contingency Table

Y_{ij} = number in the random sample observed to belong to ith row and jth column; first subscript, i, denotes row; second, j, column

R_i = total observed number in ith row; found by adding across row

C_j = total observed number in jth column; found by adding down column

$$n = \text{sample size} = \sum_{i=1}^{r} R_i = \sum_{j=1}^{c} C_j$$

To test the hypothesis that the *categorical variables* or *criteria of classification in the rows and columns are independent,* we compute an expected number of sample elements for each cell E_{ij} and employ a χ^2 statistic that approximately follows the chi-square distribution.

By *independence* in our hypothesis we mean that the proportion of each row total belonging in the jth column is the same for all rows (or that the proportion of each column total belonging in the ith row is the same for all columns). We express this condition mathematically by saying that the probability that a random element will belong in the (i, j)th cell is equal to the product of the probability that it belongs to the ith row times the probability that it belongs to the jth column. Symbolically, our hypothesis is

$$H_0: \ p_{ij} = p_{i.} \, p_{.j} \qquad i = 1, 2, \ldots, r \qquad j = 1, 2, \ldots, c$$

where: p_{ij} = probability of ith row *and* jth column
$p_{i.}$ = probability of ith row ignoring columns
$p_{.j}$ = probability of jth column ignoring rows

The null and alternative hypotheses are sometimes more understandable when they are stated as follows:

H_0: The row and column classification criteria are independent.

H_a: The row and column classification criteria are dependent.

The marginal probability $p_{i.}$ is estimated by the observed fraction in the ith row, R_i/n, and $p_{.j}$ by the observed fraction in the jth column, C_j/n. Under the hypothesis of independence the value of p_{ij}, or of $p_{i.} \, p_{.j}$, is estimated by the product of R_i/n and C_j/n. The expected number is the sample size n multiplied by the estimated probability, $(R_i/n)(C_j/n)$. Therefore, the estimate of the expected frequency, E_{ij}, in the ijth cell of an $r \times c$ contingency table is found by multiplying the total frequency of its row by the total frequency of its column and dividing by n, as given in the following equation:

$$E_{ij} = \frac{(row\ total)(column\ total)}{sample\ size} = \frac{R_i C_j}{n} \qquad (11.5)$$

We can now find the expected number for each cell in the table.

To perform the test, we find the contribution of each cell to the χ^2. The contribution of the (i, j)th cell is $(observed\ cell\ frequency - expected\ cell\ frequency)^2/(expected\ cell\ frequency)$ or $(Y_{ij} - E_{ij})^2/E_{ij}$. There are a total of $r \times c$ such contributions, and the calculated χ^2 is their sum, as given in Equation (11.6).

Equation for Chi-Square for a Contingency Table

$$\chi^2 = \sum_{i=1}^{r} \sum_{j=1}^{c} \frac{(Y_{ij} - E_{ij})^2}{E_{ij}} \qquad (11.6)$$

This χ^2 has $(r - 1)(c - 1)$ degrees of freedom, the number of rows less one multiplied by the number of columns less one since both ΣR_i and ΣC_j must be equal to n.

If the calculated χ^2 exceeds the tabular value for probability α, we reject the hypothesis and conclude that rows and columns do not represent independent classifications. If the calculated χ^2 does not exceed the tabular value, we cannot reject the hypothesis. Since good agreement between the observed and expected numbers will result in a small χ^2 value, and perfect agreement will give a value of 0 for χ^2, we are justified in rejecting the hypothesis only when χ^2 is large. Hence, this test is always a one-tailed test on the upper tail of the χ^2 distribution, and we have the following rejection rule, where $\nu = (r - 1)(c - 1)$.

Rejection rule: If $\chi^2 > \chi^2_{\alpha,\nu}$, then reject H_0.

One note of caution should be observed: In order for the statistic defined in Equation (11.6) to closely approximate the chi-square distribution, the expected frequencies, the E_{ij} values, should all be 5 or greater. If very small expected frequencies were allowed, then a small disagreement between Y_{ij} and E_{ij} could appear to be abnormally large when the squared difference $(Y_{ij} - E_{ij})^2$ is divided by a small E_{ij} value. If one category does not have an expected frequency of 5 or more, we can sometimes combine two or more categories.

The procedure for **testing the independence of row and column criteria of classification** is summarized as follows.

Hypotheses and Rejection Rule for a Contingency Table Test

H_0: The row and column classification criteria are independent.

H_a: The row and column classification criteria are dependent.

Rejection rule: If $\chi^2 > \chi^2_{\alpha,\nu}$, then reject H_0;

where: $\chi^2 = \sum_{i=1}^{r} \sum_{j=1}^{c} \frac{(Y_{ij} - E_{ij})^2}{E_{ij}}$

Y_{ij} = number of sample elements observed to belong to the ith row and jth column

R_i = total observed number in the ith row

C_j = total observed number in the jth column

$E_{ij} = \frac{R_i C_j}{n}$ and $E_{ij} \geq 5$

and there are $\nu = (r - 1)(c - 1)$ degrees of freedom.

The following example illustrates the computational procedure.

EXAMPLE 11.1   A human resources manager at Litton Field Enterprises was interested in knowing whether the voluntary absence behavior of the firm's employees was independent of marital status. The employee files contained data on marital status—with categories of married, divorced, widowed, and single—and on voluntary absenteeism behavior—with categories of often absent, seldom absent, and never absent. Table 11.4 gives the results for a random sample of 500 employees in the form of the numbers of employees in each cell of a two-way contingency table. Test the hypothesis that voluntary absence behavior is independent of marital status for this firm. Let the level of significance be .05.

For the data given in Table 11.4 the expected numbers are as follows:

$$E_{11} = \frac{(100)(150)}{500} = 30 \qquad E_{12} = \frac{(100)(100)}{500} = 20$$

$$E_{13} = \frac{(100)(50)}{500} = 10 \qquad E_{14} = \frac{(100)(200)}{500} = 40$$

$$E_{21} = \frac{(200)(150)}{500} = 60 \qquad E_{22} = \frac{(200)(100)}{500} = 40$$

$$E_{23} = \frac{(200)(50)}{500} = 20 \qquad E_{24} = \frac{(200)(200)}{500} = 80$$

$$E_{31} = \frac{(200)(150)}{500} = 60 \qquad E_{32} = \frac{(200)(100)}{500} = 40$$

$$E_{33} = \frac{(200)(50)}{500} = 20 \qquad E_{34} = \frac{(200)(200)}{500} = 80$$

TABLE 11.4

Contingency Table for Employee Voluntary Absence Behavior and Marital Status

Absence Behavior	Marital Status				Row Totals
	1. Married	2. Divorced	3. Widowed	4. Single	
1. Often Absent	36	16	14	34	100
2. Seldom Absent	64	34	20	82	200
3. Never Absent	50	50	16	84	200
Column Totals	150	100	50	200	500

Since all of the expected frequencies exceed 5, the test statistic closely approximates a chi-square distribution. The steps of the hypothesis test are as follows.

Step One H_0: The criteria of classification are independent, or voluntary absence behavior is independent of marital status.

Step Two H_a: The criteria of classification are dependent.

Step Three $n = 500$, $\alpha = .05$.

Step Four The value of the degrees of freedom is

$$\nu = (r - 1)(c - 1) = (3 - 1)(4 - 1) = 6$$

and the critical value from the chi-square table is

$$\chi^2_{.05,6} = 12.59$$

Thus, the rejection rule is as follows.

Rejection rule: If $\chi^2 > 12.59$ then reject H_0

Step Five The test statistic is χ^2. The calculated value is

$$\chi^2 = \frac{(36 - 30)^2}{30} + \frac{(16 - 20)^2}{20} + \frac{(14 - 10)^2}{10} + \frac{(34 - 40)^2}{40}$$
$$+ \frac{(64 - 60)^2}{60} + \frac{(34 - 40)^2}{40} + \frac{(20 - 20)^2}{20} + \frac{(82 - 80)^2}{80}$$
$$+ \frac{(50 - 60)^2}{60} + \frac{(50 - 40)^2}{40} + \frac{(16 - 20)^2}{20} + \frac{(84 - 80)^2}{80}$$
$$= 10.88$$

Step Six Since 10.88 is not greater than 12.59 (see Figure 11.4), we do not re-

FIGURE 11.4

Contingency Table Test of Litton Field Enterprises

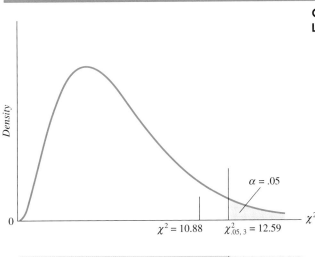

ject H_0. The data do not differ sufficiently from what we would expect under the null hypothesis to reject the hypothesis that voluntary absence behavior is independent of marital status. ═══

A computer was used to obtain contingency table analyses for the absenteeism and marital status data in Example 11.1. The results are presented in Computer Exhibits 11.1A, 11.1B, and 11.1C. The results given in the printouts are comparable to the results we obtained in the example.

Note that in Example 11.1 there is *one* population (the firm's employees) from which a sample of 500 employees is taken. The column totals 150, 100, 50, and 200 in Table 11.4 are random; they were not set in advance of sampling. In Example 11.1, we want to know whether marital status is independent of absence behavior. The test is for **independence** of two criteria of classification in the population.

In other cases there are two or more populations, and a random sample is taken independently from each population. There is nothing random about the column totals here; the randomness is in the way these totals split into the rows. We want to determine whether the populations have the same proportions for the categories included in the problem. The test is for **homogeneity** or equality of proportions for the populations. Despite this difference in point of view the computations for the test procedures are the same in the two cases.

MINITAB	**COMPUTER EXHIBIT 11.1A** Contingency Table Analysis of the Absenteeism Behavior and Marital Status Data

---

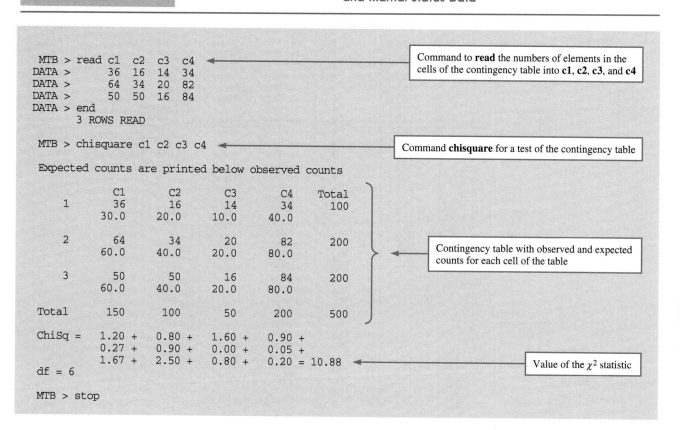

```
MTB > read c1  c2  c3  c4  ←─────          Command to read the numbers of elements in the
DATA >      36  16  14  34                  cells of the contingency table into c1, c2, c3, and c4
DATA >      64  34  20  82
DATA >      50  50  16  84
DATA > end
       3 ROWS READ

MTB > chisquare c1 c2 c3 c4  ←─────         Command chisquare for a test of the contingency table

Expected counts are printed below observed counts

              C1       C2       C3       C4    Total
      1       36       16       14       34      100
            30.0     20.0     10.0     40.0

      2       64       34       20       82      200
            60.0     40.0     20.0     80.0              Contingency table with observed and expected
                                                         counts for each cell of the table
      3       50       50       16       84      200
            60.0     40.0     20.0     80.0

  Total      150      100       50      200      500

ChiSq =    1.20 +   0.80 +   1.60 +   0.90 +
           0.27 +   0.90 +   0.00 +   0.05 +
           1.67 +   2.50 +   0.80 +   0.20 = 10.88  ←─────   Value of the $\chi^2$ statistic
  df = 6

MTB > stop
```

COMPUTER EXHIBIT 11.1B Contingency Table Analysis of the Absenteeism Behavior and Marital Status Data

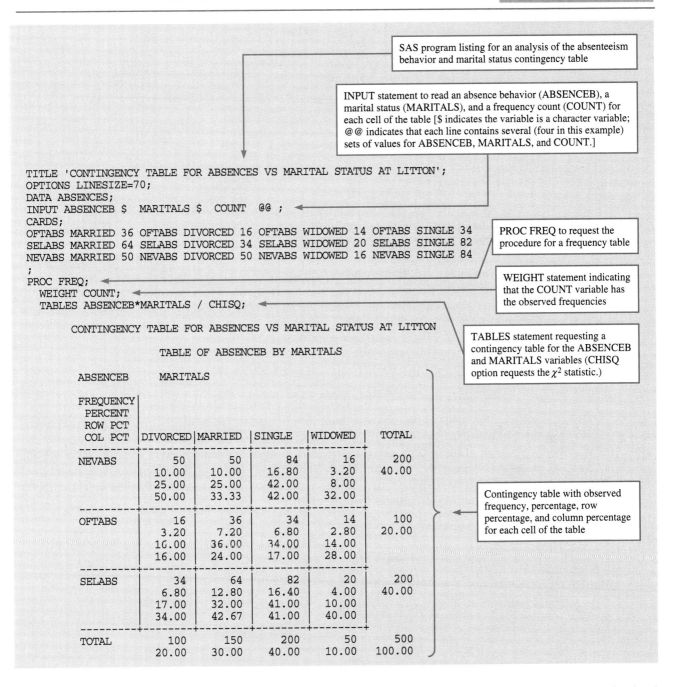

SAS program listing for an analysis of the absenteeism behavior and marital status contingency table

INPUT statement to read an absence behavior (ABSENCEB), a marital status (MARITALS), and a frequency count (COUNT) for each cell of the table [$ indicates the variable is a character variable; @@ indicates that each line contains several (four in this example) sets of values for ABSENCEB, MARITALS, and COUNT.]

```
TITLE 'CONTINGENCY TABLE FOR ABSENCES VS MARITAL STATUS AT LITTON';
OPTIONS LINESIZE=70;
DATA ABSENCES;
INPUT ABSENCEB $ MARITALS $ COUNT @@ ;
CARDS;
OFTABS MARRIED 36 OFTABS DIVORCED 16 OFTABS WIDOWED 14 OFTABS SINGLE 34
SELABS MARRIED 64 SELABS DIVORCED 34 SELABS WIDOWED 20 SELABS SINGLE 82
NEVABS MARRIED 50 NEVABS DIVORCED 50 NEVABS WIDOWED 16 NEVABS SINGLE 84
;
PROC FREQ;
   WEIGHT COUNT;
   TABLES ABSENCEB*MARITALS / CHISQ;
```

PROC FREQ to request the procedure for a frequency table

WEIGHT statement indicating that the COUNT variable has the observed frequencies

TABLES statement requesting a contingency table for the ABSENCEB and MARITALS variables (CHISQ option requests the χ^2 statistic.)

CONTINGENCY TABLE FOR ABSENCES VS MARITAL STATUS AT LITTON

TABLE OF ABSENCEB BY MARITALS

ABSENCEB   MARITALS

FREQUENCY PERCENT ROW PCT COL PCT	DIVORCED	MARRIED	SINGLE	WIDOWED	TOTAL
NEVABS	50 10.00 25.00 50.00	50 10.00 25.00 33.33	84 16.80 42.00 42.00	16 3.20 8.00 32.00	200 40.00
OFTABS	16 3.20 16.00 16.00	36 7.20 36.00 24.00	34 6.80 34.00 17.00	14 2.80 14.00 28.00	100 20.00
SELABS	34 6.80 17.00 34.00	64 12.80 32.00 42.67	82 16.40 41.00 41.00	20 4.00 10.00 40.00	200 40.00
TOTAL	100 20.00	150 30.00	200 40.00	50 10.00	500 100.00

Contingency table with observed frequency, percentage, row percentage, and column percentage for each cell of the table

(continues)

The simplest contingency table, the 2 × 2, can be used to test the hypothesis that two binomial populations have the same relative frequency of successes. Let p_1 be the proportion in the first population and p_2 the proportion in the second. Using

SAS

COMPUTER EXHIBIT 11.1B Contingency Table Analysis of the Absenteeism Behavior and Marital Status Data (Continued)

```
STATISTICS FOR TABLE OF ABSENCEB BY MARITALS

STATISTIC                          DF      VALUE       PROB
---------------------------------------------------------
CHI-SQUARE                          6     10.883       0.092    ←  Value and P-value for the χ² statistic
LIKELIHOOD RATIO CHI-SQUARE         6     10.754       0.096
MANTEL-HAENSZEL CHI-SQUARE          1      1.438       0.231
PHI                                        0.148
CONTINGENCY COEFFICIENT                    0.146
CRAMER'S V                                 0.104

SAMPLE SIZE = 500
```

the information contained in two independent random samples, one from each population, we can test the hypothesis $H_0: p_1 = p_2$.

The results of the two samples may be summarized in a two-way table as shown in Table 11.5. In the table n_1 and n_2 are the two sample sizes, and Y_1 and Y_2 are the observed number of successes in the first and second samples, respectively.

We test the hypothesis that the proportions of successes and failures are the same in the two populations. This result could be true only if $p_1 = p_2$. The χ^2 variable for the 2×2 table will have one degree of freedom, since $\nu = (r - 1)(c - 1) = (2 - 1)(2 - 1) = 1$.

 PRACTICE PROBLEM 11.3

For a test of the puncture resistance of a new type of off-the-road-vehicle tire, a random sample of 120 of the new type of tire and an independent random sample of 180 of the old type of tire were driven over the same special hazardous course for 24 hours by National Tire & Rubber, Inc. Just 6 of the new tires were punctured, whereas 18 of the old tires had punctures. Can we conclude that the new tire resists punctures differently than the old tire?

Solution Here

$$Y_1 = 6 \qquad\qquad Y_2 = 18$$

$$n_1 = 120 \qquad\qquad n_2 = 180$$

$$n_1 - Y_1 = 114 \qquad n_2 - Y_2 = 162$$

		Population		
	Outcome	*Population 1*	*Population 2*	*Total*
Success		Y_1	Y_2	$Y_1 + Y_2$
Failure		$n_1 - Y_1$	$n_2 - Y_2$	$n_1 + n_2 - Y_1 - Y_2$
Sample Size		n_1	n_2	$n_1 + n_2$

TABLE 11.5

2 × 2 Contingency Table for Testing $H_0: p_1 = p_2$

COMPUTER EXHIBIT 11.1C Contingency Table Analysis of the Absenteeism Behavior and Marital Status Data

MYSTAT

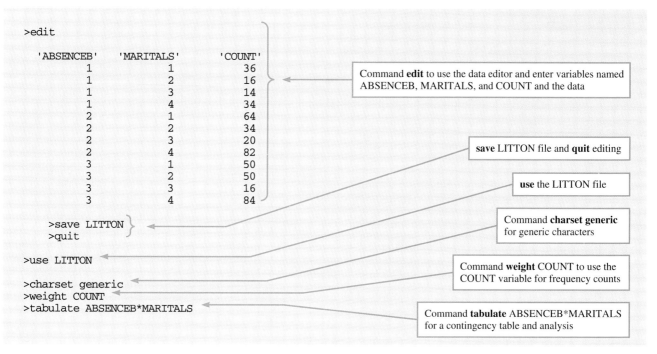

```
>edit

   'ABSENCEB'    'MARITALS'        'COUNT'
        1             1              36
        1             2              16
        1             3              14
        1             4              34
        2             1              64
        2             2              34
        2             3              20
        2             4              82
        3             1              50
        3             2              50
        3             3              16
        3             4              84

       >save LITTON
       >quit

   >use LITTON

   >charset generic
   >weight COUNT
   >tabulate ABSENCEB*MARITALS
```

Command **edit** to use the data editor and enter variables named ABSENCEB, MARITALS, and COUNT and the data

save LITTON file and **quit** editing

use the LITTON file

Command **charset generic** for generic characters

Command **weight** COUNT to use the COUNT variable for frequency counts

Command **tabulate** ABSENCEB*MARITALS for a contingency table and analysis

(continues)

The two-way contingency table is shown in Table 11.6.

The expected frequencies are calculated as shown next:

$$E_{11} = (24)(120)/300 = 9.6 \qquad E_{12} = (24)(180)/300 = 14.4$$
$$E_{21} = (276)(120)/300 = 110.4 \qquad E_{22} = (276)(180)/300 = 165.6$$

The calculated χ^2 is

$$\chi^2 = \frac{(6 - 9.6)^2}{9.6} + \frac{(18 - 14.4)^2}{14.4} + \frac{(114 - 110.4)^2}{110.4}$$
$$+ \frac{(162 - 165.6)^2}{165.6} = 2.45$$

Since the calculated value, 2.45, is less than the tabular value for $\chi^2_{.05,1}$, which is 3.84, we have no reason, at the .05 level of significance, to reject the hypothesis that the puncture rates for the two tires are equal.

	Population			
Outcome	*New Tire*	*Old Tire*	*Total*	**TABLE 11.6**
Punctured	6	18	24	**Two-Way Contingency**
Not Punctured	114	162	276	**Table for National Tire**
Sample Size	120	180	300	

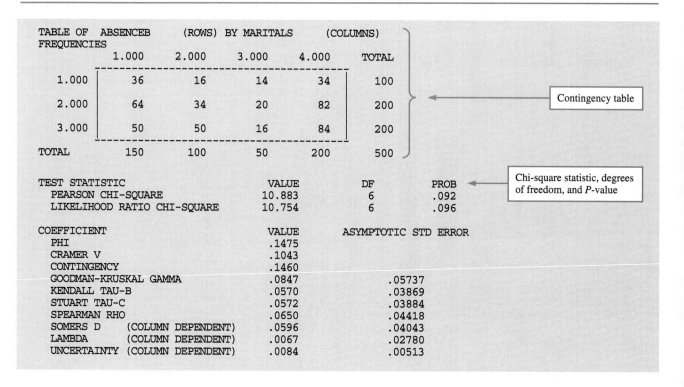

MYSTAT

COMPUTER EXHIBIT 11.1C Contingency Table Analysis of the Absenteeism Behavior and Marital Status Data (Continued)

```
TABLE OF  ABSENCEB    (ROWS) BY MARITALS    (COLUMNS)
FREQUENCIES
             1.000    2.000    3.000    4.000    TOTAL
         ----------------------------------------
  1.000  |   36       16       14       34    |   100
         |                                    |
  2.000  |   64       34       20       82    |   200
         |                                    |
  3.000  |   50       50       16       84    |   200
         ----------------------------------------
TOTAL       150      100       50      200       500
```

Contingency table

```
TEST STATISTIC                       VALUE        DF       PROB
  PEARSON CHI-SQUARE                 10.883        6        .092
  LIKELIHOOD RATIO CHI-SQUARE        10.754        6        .096
```

Chi-square statistic, degrees of freedom, and *P*-value

```
COEFFICIENT                          VALUE      ASYMPTOTIC STD ERROR
  PHI                                .1475
  CRAMER V                           .1043
  CONTINGENCY                        .1460
  GOODMAN-KRUSKAL GAMMA              .0847         .05737
  KENDALL TAU-B                      .0570         .03869
  STUART TAU-C                       .0572         .03884
  SPEARMAN RHO                       .0650         .04418
  SOMERS D      (COLUMN DEPENDENT)   .0596         .04043
  LAMBDA        (COLUMN DEPENDENT)   .0067         .02780
  UNCERTAINTY   (COLUMN DEPENDENT)   .0084         .00513
```

Problems: Section 11.3

8. Consider the accompanying table, which classifies a stock's price change as up, down, or no change for both today's and yesterday's prices. Price changes were examined for 100 days. If stock markets are "efficient" or if stock prices follow a *random walk*— or, in other words, if the current price of a stock fully reflects available information concerning a stock's expected future returns—then, among other things, the price change today for a stock must be independent of yesterday's price change. Test the hypothesis that daily stock price changes for this stock are independent. Let $\alpha = .05$.

	Price Change Yesterday		
Price Change Today	Up Yesterday	No Change	Down Yesterday
Up Today	18	6	16
No Change	8	6	6
Down Today	16	10	14

9. A large Acme Precision Products metal fabrication shop operates three shifts. The accompanying data give the distribution of accidents among the three shifts by type of accident. Test the hypothesis that shift and accident type are unrelated. Let $\alpha = .01$.

	Shift		
Accident Type	Day	Swing	Night
Minor	130	95	48
Serious	40	35	12

10. A random sample of 450 heads of households was taken in the trading area for a large financial institution. These people were asked to classify their own attitudes and their parents' attitudes toward borrowing money as follows:

      A: Borrow only for large purchases or emergencies.
      B: Borrow whenever you want something and can't pay for it now.
      C: Never borrow.

The heads of households were then classified as being *BA*, *CB*, and so on. For example, a *BA* classification meant that the respondent had a *B* attitude toward debt while his or her parents had an *A* attitude. The 450 respondents' classifications are presented in the accompanying table. Test the hypothesis that there is no relationship between debt attitudes of the respondents and how they perceive the debt attitudes of their parents. Let $\alpha = .05$.

	Parent		
Respondent	*A*	*B*	*C*
A	120	20	10
B	90	60	20
C	90	20	20

11. An appliance manufacturer offers a product in four models: standard, deluxe, super deluxe, and majestic. The Rand Department Store has compiled data (see the accompanying table) on the purchasers of this appliance relating the model of appliance purchased with the charge account balance of the purchaser at the time of purchase. Is there evidence of a relationship between the charge account balance and the model of appliance purchased? Let $\alpha = .025$.

Account Balance	*Model Purchased*			
	Standard	*Deluxe*	*Super Deluxe*	*Majestic*
Under $50	40	16	5	10
$50–$100	24	12	15	8
Over $100	16	12	30	16

12. Two chemical treatments were applied to random samples of seeds planted at Five Lakes Land & Agriculture, Inc. After treatment, germination tests were conducted, with the results as given in the accompanying table. Do these data indicate that the chemicals differ in their effects on germination? Let the chance of a type I error be .05.

	Treatment	
Seed	*Chemical 1*	*Chemical 2*
Germinated	120	30
Did Not Germinate	110	15

13. The only randomized study of the effectiveness of cancer screening by mammography and clinical examination in the United States was undertaken by the Health Insurance Plan (HIP) of Greater New York. The study was designed to test whether annual mammography and clinical examination decrease the mortality rate for breast cancer. Of the patients 40 to 49 years of age at the start of the study, one group of randomly selected patients was offered annual screening by mammography and clinical examination for cancer, whereas a second group of randomly selected patients was not offered screening. The accompanying table shows the frequency counts of the subjects that had died from cancer or not, after 13 years, for the screened and nonscreened

groups. Do the sample results support the hypothesis that mortality rates are homogeneous for a program offering annual cancer screening by mammography and clinical examination and for a nonscreened group for those 40 to 49 years of age? Use a .05 level of significance.

Survival	Screen	Frequency
Alive	Screened	14,804
Alive	Not screened	14,852
Dead	Screened	45
Dead	Not screened	59

REFERENCE: Moskowitz, M. (1983), "Minimal Breast Cancer Redux," *Radiologic Clinics of North America*, 21: 93–113.

14. A study was done by a hospital's quality control administrators to determine what factors influence infection rates for surgical incisions. Hospital administrators wanted to decrease hospitalization costs and morbidity for health care consumers and increase profitability at the same time. One factor that was studied was the "routine" procedure of shaving the operation site, which is explained at times by health care providers as a procedure to decrease infection rates. However, shaving can result in skin nicks that are favorable for or increase infection. Of the patients where the operation site was shaved and of the patients where the operation site was not shaved, the accompanying table shows the number of patients in each group having infected or noninfected incisions. Do the sample results differ sufficiently from those expected under the hypothesis that infection rates are the same for shaving and for not shaving of surgical incisions to reject that hypothesis? Use a .05 level of significance.

Shaving	Infection	Count
Shaved	Not infected	977
Shaved	Infected	23
Not shaved	Not infected	991
Not shaved	Infected	9

REFERENCE: Cruse, P., and R. Foord (1973), "A Five-Year Prospective Study of 23,649 Surgical Wounds," *Archives of Surgery*, 107: 206–10.

11.4
SUMMARY

In this chapter we dealt with tests using frequency counts for categorical data. The tests involve chi-square computations and chi-square distributions. The first test involves the multinomial $H_0: p_1 = p_2 = \cdots = p_k$, which is an extension of the two-sample test of $H_0: p_1 = p_2$.

The chi-square distribution was also used to test the hypothesis that items classified according to two different characteristics exhibit *independence* in those characteristics or *homogeneity* for the proportions. The table showing the classifications of the items by two characteristics is a contingency table, and the chi-square test here checks for dependence or homogeneity. Table 11.7 summarizes tests using the chi-square distribution.

The sample size for a contingency table test should be large enough so the expected frequencies are greater than or equal to 5. Furthermore, it is sometimes difficult to ascertain whether independence or homogeneity or the lack thereof occur uniformly among all categories or only among some categories when a categorical variable has three or more categories.

TABLE 11.7 Summary of the Uses of the Chi-Square Distribution

Type of Problem	Null Hypothesis	Test Statistic
Multinomial hypothesis	$H_0: p_1 = p_2 = \cdots = p_k$ or $H_0: p_1 = p_{1_0}, p_2 = p_{2_0}, \cdots, p_k = p_{k_0}$	$\chi^2 = \sum_j \dfrac{(Y_j - E_j)^2}{E_j}$ where: $E_j = np_{j_0}$

Comments:
— The E_j values should all be 5 or more. If they are not, two or more classes can be combined.
— Note that this test can be used to test the hypothesis that several p values are equal to a predetermined set of values, not all necessarily equal.
— Degrees of freedom are $k - 1$.

Contingency table analysis	$H_0:$ Row and column classifications are independent or $H_0:$ Homogeneity of proportions	$\chi^2 = \sum_{i=1}^{r} \sum_{j=1}^{c} \dfrac{(Y_{ij} - E_{ij})^2}{E_{ij}}$

Comments:
— The E_{ij} values should all be 5 or more. If they are not, some rows and/or columns can be combined.
— Each E_{ij} value is computed as $R_i C_j / n$ where $R_i =$ total in row i, and $C_j =$ total in column j.
— Degrees of freedom are $(r - 1)(c - 1)$, where r is the number of rows in the table and c is the number of columns.
— The hypothesis $H_0: p_1 = p_2$ can be tested by using a 2×2 contingency table.

Answers to odd-numbered problems are given in the back of the text.

REVIEW PROBLEMS

15. The grades presented in the accompanying table were given to a class of 100 students. The expected numbers are those corresponding to the proportions for a grading "curve." Can we conclude from the expected and observed or given frequencies that the instructor used the curve? Use $\alpha = .05$.

Grade	Number Given, Y_j	Number Expected, E_j
A	10	7
B	30	24
C	40	38
D	16	24
F	4	7

16. An automobile manufacturer ships one model of its economy car in just four colors. The manufacturer's contract with its dealers states that the colors red, yellow, green, and blue are in the proportions 8:7:3:2. A sample of 400 contains 180 red, 120 yellow, 40 green, and 60 blue cars.

   a. Estimate the proportion of each color in the manufacturer's shipments.
   b. Test the hypothesis that one-tenth of the cars are blue. Let $\alpha = .01$.
   c. Test the hypothesis that the proportions are 8:7:3:2. Let $\alpha = .005$.

17. A sample of 300 peak electrical loads for rural consumers of Southern California Pacific Power are distributed among three periods as follows:

2:00 a.m.–10:00 a.m.   80
10:00 a.m.–6:00 p.m.   90
6:00 p.m.–2:00 a.m.   130

   a. Estimate the proportions of peak loads that occur in each period.
   b. Test the hypothesis that the loads are uniformly distributed among the periods. That is, test H_0: $p_1 = p_2 = p_3 = 1/3$. Let $\alpha = .05$.

18. The probabilities for the various outcomes when a pair of dice are tossed are as shown in the accompanying table, with the frequencies observed when a pair of dice were tossed 1800 times.

   a. What is the estimated probability of throwing a 7 with these dice? Of throwing a 12?
   b. At the 1% level of significance, are the observed frequencies in agreement with the theoretical values?

Total on Dice	Probability	Frequency	Total on Dice	Probability	Frequency
2	1/36	40	7	6/36	330
3	2/36	108	8	5/36	223
4	3/36	175	9	4/36	228
5	4/36	184	10	3/36	128
6	5/36	225	11	2/36	87
			12	1/36	72

19. Fire department records in Suburbia show that the last 350 fires were distributed according to day of the week as follows:

Monday	35	Friday	65
Tuesday	35	Saturday	75
Wednesday	45	Sunday	55
Thursday	40	Total	350

   Test the hypothesis that fires are uniformly distributed throughout the week. That is, test H_0: $p_1 = p_2 = p_3 = p_4 = p_5 = p_6 = p_7 = 1/7$. Use $\alpha = .01$.

20. American Apparel, Inc., has always produced three sizes of sports shirts: small, medium, and large. The output of its shop has always been 25% small size, 50% medium size, and 25% large size. This breakdown was determined several years ago on the basis of a guess about the proportion of small, medium, and large men in the population. The guess was made by the company president. Recently, the new company president questioned these proportions, so a random sample of 200 men was selected and measured for sport shirt size. In this sample 35 were small, 90 were medium, and 75 were large. Test the hypothesis that the proportions are $p_1 = .25$, $p_2 = .50$, and $p_3 = .25$ in the population, using an α of .01.

21. Banks, thrifts, credit unions, and armored-car companies lost $68 million from crimes in 1990, according to reports from the Federal Bureau of Investigation (FBI). The high times for bank crimes of robbery, burglary, and larceny by times of the working day are given by the following frequency counts:

Time	9–11 a.m.	11 a.m.–1 p.m.	1–3 p.m.	3–6 p.m.
Crimes	1,926	1,939	2,215	1,542

   a. Test the hypothesis that bank crimes are distributed proportionally by hour for time periods of the day—that is, test H_0: $p_1 = p_2 = p_3 = 2/9$ and $p_4 = 3/9$. Use a significance level of .05.

b. When should the bank manager beef up security at the bank?

REFERENCE: *American Banker* (1991), "Bank crimes rose 16% in '90, FBI Reports" (July 8): 5.

22. A study was done by a hospital's quality control administrators to determine what factors influence "emergency" cesarean surgical rates for maternity patients. Hospital administrators wanted to decrease hospitalization costs and morbidity for health care consumers and increase profitability at the same time. One factor that was studied was the procedure of cesarean delivery. Some cesarean surgical procedures are necessary for the well-being of the patients. But some are justified by providers in terms of the learning experience when performed by residents, the greater financial gain for the providers, or the more dependable schedules for the provider, even though cesarean surgeries result in morbidity for the patient. The numbers of cesarean surgeries by times of the day are given by the following frequency counts:

Time	9 a.m.–12 n.	12–3 p.m.	3–6 p.m.	6–9 p.m.	9 p.m.–12 m.
Cesareans	280	430	510	120	60

a. Test the hypothesis that cesarean surgeries are uniformly distributed by time of the work day. Use a significance level of .05.
b. Does it appear that cesarean surgeries occur randomly for the well-being of patients?

REFERENCE: Scully, D. (1980), *Men Who Control Women's Health* (Boston: Houghton Mifflin).

23. A sample of the employees of Federated Airlines was asked to indicate a preference for one of three pension plans. The results are given in the accompanying table. Is there reason to believe that their preferences are dependent on job classification? Let $\alpha = .01$.

	Pension Plan		
Job Class	Plan 1	Plan 2	Plan 3
Supervisory	29	13	10
Clerical	19	80	19
Labor	22	57	81

24. Questionnaires were mailed to graduates with degrees in accounting from Iowa State University. The numbers who returned the questionnaires are given in the accompanying table. Test the hypothesis that returning or not returning the questionnaire is independent of the level of degree earned. Let $\alpha = .05$.

	Degree		
Result	B.S.	M.S.	Ph.D.
Returned	47	42	46
Not Returned	13	13	8

25. Do the accompanying data give us reason to believe that the fouling of boat bottoms can be reduced by the application of an antifouling paint at the Thor Corporation shipyard? Let $\alpha = .01$.

	Paint	
Treatment Result	Antifouling	Standard
No Fouling	40	25
Some Fouling	50	50
Much Fouling	10	45

26. Two independent polls were taken to investigate public opinion with regard to a proposed recreational facility. One was conducted by a polling agency, the other by a local newspaper, see the accompanying table. Are the two polls homogeneous with respect to division of opinion? Let $\alpha = .025$, and construct a three-row, two-column contingency table, where observed frequencies are found by multiplying sample sizes by the reported percentages divided by 100. Discuss anything unusual you see in the data.

	Polling Organization	
Opinion	Polling Agency	Newspaper
Favor	52.5%	47.5%
Oppose	35.0%	37.5%
Undecided	12.5%	15.0%
Sample Size	300	250

27. A recruiter for a health maintenance organization needed to hire a physician to practice in an outpatient clinic located in a rural town. The recruiter hypothesized that physicians' specialties were independent of the location of physicians' practices. A random sample of physicians was selected from a reference listing, and the numbers practicing in specialty and location classifications are given in the accompanying table. Test the recruiter's hypothesis, and let $\alpha = .05$.

	Specialty		
Location	1. Family Practice	2. Surgery	3. Internist
1. Rural	30	4	10
2. Suburban	20	40	20
3. Urban	5	30	15

REFERENCES: Watson, C. J. (1980), "An Empirical Model of Physician Practice Location Decisions," *Computers and Biomedical Research*, 13: 363–81; and Watson, C. J., and D. J. Croft (1978), "A Multiple Discriminant Analysis of Physician Specialty Choice," *Computers and Biomedical Research*, 11: 405–21.

28. An internal auditor at American Department Stores sampled accounts receivable at three retail stores and categorized each account as overdue or not overdue, as shown in the accompanying table. Test the hypothesis that the proportions of overdue accounts are homogeneous at the three stores. Use .05 for the significance level.

	Retail Store		
Account	Store 1	Store 2	Store 3
Overdue	20	30	25
Not Overdue	80	65	70

29. During a survey of patrons of shopping centers people were asked their main reason for shopping at two different types of shopping centers. Frequencies of responses for the categories are listed in the accompanying table. Test to see whether the type of shopping center is independent of the reason for shopping at different centers. Use .05 for the level of significance.

	Shopping Center	
Survey Response	Mall	Strip Mall
Convenience	15	48
Variety	38	25

30. The only randomized study of the effectiveness of cancer screening by mammography and clinical examination in the United States was undertaken by the Health Insurance Plan (HIP) of Greater New York. The study was designed to test whether annual mammography and clinical examination decrease the mortality rate for breast cancer. Of the patients 50 years or older at the beginning of the study, one group of randomly selected patients was offered annual screening by mammography and clinical examination for cancer, whereas a second group of randomly selected patients was not offered screening. The accompanying table shows the frequency counts of the subjects that had died from cancer or not, after 13 years, for the screened and nonscreened groups. Do the sample results support the hypothesis that mortality rates are homogeneous for a program offering annual cancer screening by mammography and clinical examination and for a nonscreened group for those 50 or more years of age? Use a .05 level of significance. Moskowitz (1980) wrote that "HIP proved unequivocally that screening can benefit women older than 50." Based on the hypothesis test results, do you agree with Moskowitz (1980)?

Survival	Screen	Frequency
Alive	Screened	16,080
Alive	Not screened	16,000
Dead	Screened	71
Dead	Not screened	89

REFERENCES: Moskowitz, M. (1980), "How Can We Decrease Breast Cancer Mortality?" *CA*, 30: 272–75; and Moskowitz, M. (1983), "Minimal Breast Cancer Redux," *Radiologic Clinics of North America*, 21: 93–113.

31. A study in Denmark examined whether there is a relationship between social class indices for fathers and social class indices for sons or whether they are independent. Inherited wealth, parental influence, or access to education may result in sons' having a greater propensity to achieve greater social class levels when their fathers are in higher social classes. Social class indices (measured from 1 for lower to 5 for higher social class) and frequency counts for the number of elements in each father and son social class combination are given in the accompanying table. Test the hypothesis that there is no relationship between social classes for fathers and sons. Use a significance level of .05.

Father	Son	Count	Father	Son	Count	Father	Son	Count
1	1	18	5	2	8	4	4	348
2	1	24	1	3	16	5	4	201
3	1	23	2	3	109	1	5	2
4	1	8	3	3	289	2	5	21
5	1	6	4	3	175	3	5	95
1	2	17	5	3	69	4	5	198
2	2	105	1	4	4	5	5	246
3	2	84	2	4	59			
4	2	49	3	4	217			

REFERENCE: Svalastoga, K. (1959), *Prestige, Class and Mobility* (London: Heineman).

Refer to the 141 companies listed in the Ratio Data Set in Appendix A.

Ratio Data Set

Questions

32. Locate the data for the current assets–sales (*CA/S*) ratio and the Standard Industry Classification Code for classifying the companies by industry.

    a. Use a computer to obtain a two-row by two-column crosstabulation for the companies that have *CA* greater than or equal to *S* (thus the ratio is greater than or equal to one) and those that do not, and the companies that are manufacturing (SIC codes 2000–3999) or nonmanufacturing companies.

   b. Test the hypothesis that the two categorical variables are independent. Use a significance level of .05.

33. Locate the data for the quick assets–sales (*QA/S*) ratio and the Standard Industry Classification Code for classifying the companies by industry.

   a. Use a computer to obtain a two-row by two-column crosstabulation for the companies that have *QA* greater than or equal to *S* (thus the ratio is greater than or equal to one) and those that do not, and the companies that are manufacturing (SIC codes 2000–3999) or nonmanufacturing companies.

   b. Test the hypothesis that the two categorical variables are independent. Use a significance level of .05.

   c. Is the chi-square approximation a close approximation in this problem? Explain.

34. Locate the data for the current assets–current liabilities (*CA/CL*) ratio and the Standard Industry Classification Code for classifying the companies by industry.

   a. Use a computer to obtain a two-row by two-column crosstabulation for the companies that have *CA* greater than or equal to *CL* (thus the ratio is greater than or equal to one) and those that do not, and the companies that are manufacturing (SIC codes 2000–3999) or nonmanufacturing companies.

   b. Test the hypothesis that liquidity and being a manufacturing or nonmanufacturing company, as measured by the two categorical variables, are not related. Use a significance level of .05.

35. Locate the data for the net income–total assets (*NI/TA*) ratio as a measure of profitability and the Standard Industry Classification Code for classifying the companies by industry.

   a. Use a computer to obtain a two-row by two-column crosstabulation for the companies that have not suffered losses (zero or positive values) and those that have suffered losses (negative values for the *NI/TA* variable) and the companies that are manufacturing (SIC codes 2000–3999) or nonmanufacturing companies.

   b. Test the hypothesis that having a loss or not is independent of being a manufacturing company or not. Use a significance level of .05.

Credit Data Set   *Refer to the 113 applicants for credit listed in the Credit Data Set in Appendix A.*

Questions   36. Use the chi-square distribution and a 2 × 2 contingency table to test the hypothesis that the proportion of women is the same for the populations of people who were granted and denied credit. Use an α value of .10.

37. Compare the results you obtained for the second Credit Data Set question in Chapter 9 with the results you obtained in Question 36.

CASE 11.1   EVALUATING THE MALDISTRIBUTION OF HEALTH CARE PROVIDERS

"We all are aware of the serious maldistribution of physicians in this country today," began Dr. William Samuelson as he opened the meeting of the Midnation Medical School Admissions Committee. "Physicians are badly needed in our rural areas, but they concentrate in the urban areas. We need thousands more general practitioners, but most of our students seem to specialize. We have 20 times more applicants for positions in each freshman class than we can accept. At least 80% of those applicants are more qualified than you or I were when we were accepted to medical school years ago."

Midnation Medical School faces a problem that confronts medical schools across the country today. It has more qualified applicants than it needs. Now it is looking for entrance criteria other than past scholastic achievement, national test scores, and letters of

Physician's Practice Town	Physician's Hometown			Total	
	Rural	*Small*	*Urban*	*Total*	**TABLE 1**
Rural	62	28	37	127	**Physician's Contingency**
Small	25	73	61	159	**Table**
Urban	18	64	182	264	
Total	105	165	280	550	

recommendation. That is, Midnation Medical School is now interested in producing the "right" kind of physicians—those who will help fill the gaps in medical personnel of this country.

Dr. Samuelson proceeded to show the members of the Admissions Committee the results of a study that he had one of his graduate assistants do for him. The study seemed to show that there is a relationship between the type of city in which a medical student grew up and the type of city in which he eventually established a practice.

Dr. Samuelson's study involved 550 physicians who were graduates of Midnation Medical School. The hometowns of the physicians were classified as being rural, small, or urban in the following way:

> Rural: population under 5000
> Small: population between 5000 and 100,000
> Urban: population over 100,000

The cities in which the physicians set up practice were classified in the same way. Table 1 shows this classification.

Some of the Admissions Committee members told Dr. Samuelson that he wasn't telling them anything they didn't know or couldn't have guessed. "You're just saying that small-town applicants tend to go back to the small towns," interjected Dr. Bills, a committee member. "Hold on just a minute," responded Dr. Samuelson. "I have a few things you probably haven't seen before. Table 2 shows the relationship between a physician's practice location and the type of hometown the physician's spouse came from."

"Thus it appears, gentlemen," Dr. Samuelson continued, "that we ought to begin asking our applicants about the hometowns their spouses came from. It appears that the spouse's hometown may be just as significant in identifying would-be rural practitioners as the applicant's hometown."

"In terms of identifying those applicants who are likely to go into general practice," Dr. Bills asked, "how will this help us?"

"The tables I have shown you so far won't help in that regard, but our study did uncover some interesting relationships between the MCAT scores and the practice preference of physicians." The MCAT scores to which Dr. Samuelson referred are the Medical College Admissions Test scores. Most medical schools require applicants to take the MCAT

Physician's Practice Town	Spouse's Hometown			Total	
	Rural	*Small*	*Urban*	*Total*	**TABLE 2**
Rural	264	29	33	126	**Spouse's Contingency**
Small	20	70	60	150	**Table**
Urban	5	53	184	242	
Total	89	152	277	518	

| | MCAT | Type of Practice | | | |
TABLE 3	Score	General	Specialized	Researcher-Academic	Total
	1st Fourth	33	69	30	132
	2nd Fourth	33	57	22	112
	3rd Fourth	46	92	20	158
	4th Fourth	35	80	14	129
	Total	147	298	86	531

Contingency Table for MCAT Score and Type of Practice

as part of the admissions procedure. Dr. Samuelson then showed the committee Table 3. Physicians were classified according to their primary vocational interests as general practitioners (GPs), specialists, or researcher-academics. They also were classified by the way in which they scored on the MCAT when they were admitted to medical school. The first fourth refers to those who scored in the top 25% of those taking the test that year and who were admitted to medical school.

"This table indicates that if we want to turn out researcher-academics, we ought to admit only students who score very high on the MCAT. On the other hand, if we want to produce GPs, we ought to begin admitting more students in the lower fourths of the MCAT."

This suggestion brought vigorous discussion from the Admissions Committee members. "That's outlandish," Dr. Bills blurted out. "We don't want to turn this place into a mediocre medical school with policies like that!" Dr. Horne, chairman of Community and Family Medicine, was cool toward the idea of admitting students with lower MCAT scores, but he did soften his stand by saying, "Its not as though we'd be letting in a bunch of dummies. Even the applicants in the lower fourths these days are smarter than we were when we were in the upper. When you think about it, we've been admitting on the basis of grade point average for too long. We've been admitting scholars to medical school. Do scholars make good bedside docs?"

The strongest opposition came from Dr. Wilder. He dealt very little with the students, except on research projects. He was known for his ability to secure government and foundation grants for medical research work. Dr. Wilder suggested, "You've got to do more homework before you can convince me of the validity of any of this. I'm not for admitting a bunch of country bumpkins to this school unless you can prove to me that your tables about hometowns are statistically significant. And by the way, how come you started out with 550 in your study and get only 518 and 531 in your next two tables? Are you sure you followed good research techniques? I want to turn out a few more GPs like the rest of you, but your table about MCAT scores scares me. In the first place, I'm not sure there's any real relationship there, and in the second place, if there is, I'd just as soon we admit plenty of those high scorers who might be able to help me in my work."

Case Assignment

   a. Is there a relationship between the size of a physician's hometown and the size of the town in which a physician practices?

   b. Is there a relationship between the size of a physician's spouse's hometown and the size of the town in which a physician practices?

   c. Is there a relationship between a physician's MCAT scores and the type of practice a physician selects upon completing medical training?

REFERENCES: Watson, C. J. (1980), "An Empirical Model of Physician Practice Location Decisions," *Computers and Biomedical Research,* 13: 363–81; and Watson, C. J., and D. J. Croft (1978), "A Multiple Discriminant Analysis of Physician Specialty Choice," *Computers and Biomedical Research,* 11: 405–21.

Bishop, Y., S. Fienberg, and P. Holland. 1975. *Discrete Multivariate Analysis.* Cambridge: MIT Press.

Conover, W. J. 1980. *Practical Nonparametric Statistics,* 2nd ed. New York: Wiley.

Gibbons, J. D. 1985. *Nonparametric Statistical Inference,* 2nd ed. New York: Marcel Dekker.

Siegel, S. 1956. *Nonparametric Statistics for the Social Sciences.* New York: McGraw-Hill.

Snedecor, George W., and William G. Cochran. 1989. *Statistical Methods,* 8th ed. Ames: Iowa State University Press.

Optional Topics

Testing a Binomial Proportion with Chi-Square
Testing for Goodness of Fit

In Chapter 9 we studied tests concerning the proportion p in a binomial population. The two-sided test considered in that chapter is a special instance of the χ^2 test for the case where k is 2. It is instructive to trace the connection.

We relabel the success category in the population as class 1 and the failure category as class 2. If p_0 is the proportion of successes in the population, then class 1 has proportion $p_{1_0} = p_0$ and class 2 has proportion $p_{2_0} = 1 - p_0$. And now the null hypothesis

$$H_0: \ p_1 = p_0 \quad \text{and} \quad p_2 = 1 - p_0$$

is exactly the same thing as the null hypothesis

$$H_0: \ p = p_0$$

considered in Chapter 9.

The number Y_1 of observations from a sample of size n falling in class 1 is the number of successes X, and the number Y_2 of observations falling in class 2 is the number of failures $n - X$. Now the χ^2 value of Equation (11.4) for testing H_0 is

$$
\begin{aligned}
\chi^2 &= \frac{(Y_1 - np_{1_0})^2}{np_{1_0}} + \frac{(Y_2 - np_{2_0})^2}{np_{2_0}} \\
&= \frac{(X - np_0)^2}{np_0} + \frac{[(n - X) - n(1 - p_0)]^2}{n(1 - p_0)}
\end{aligned}
\tag{11.7}
$$

Algebra reduces this expression to

$$
\chi^2 = \left[\frac{X - np_0}{\sqrt{np_0(1 - p_0)}} \right]^2
\tag{11.8}
$$

which is the square of the Z statistic used to test the hypothesis $H_0: p = p_0$ in Chapter 9.

 PRACTICE PROBLEM 11.4

At one time the gender ratio in an occupation was 8 women to every man. Suppose that in a recent survey we found that a random sample of 450 contained 68 men. Would we be justified in concluding that the ratio has changed?

Solution If the ratio is 8 to 1, 1 out of every 9 persons is a male. Therefore if the ratio has not changed, the proportion p of males is $1/9$. We test the hypothesis $H_0: p = 1/9$ against the two-sided alternative $H_a: p \neq 1/9$ at the 1% level of significance. Let Y_1 be the number of men in the sample; then

$$Y_1 = 68 \qquad Y_2 = 450 - 68 = 382$$

$$E_1 = 450(1/9) = 50 \qquad E_2 = 450(8/9) = 400$$

$$\chi^2 = \frac{(68 - 50)^2}{50} + \frac{(382 - 400)^2}{400} = 7.29$$

From Table VIII of Appendix C, $\chi^2_{.01,1}$ is found to be 6.63. The calculated χ^2 is greater than the tabular value; therefore we reject the null hypothesis and conclude that the ratio has changed.

In this calculation we have used Equation (11.7). Using Equation (11.8) instead gives

$$\chi^2 = \left[\frac{68 - (450)(1/9)}{\sqrt{(450)(1/9)(8/9)}} \right]^2 = 2.7^2 = 7.29$$

which checks. The value 2.7 to be squared here is the value of the Z statistic used in Chapter 9 for a two-sided test $H_0: p = p_0$. At the 1% level we are to reject if the statistic has absolute value exceeding $Z_{.005}$, or 2.576. Our value 2.7 exceeds the table value, so again we reject H_0. The two procedures necessarily lead to the same conclusion because 6.63, the cutoff point for the χ^2 test, is 2.576^2.   ▬

Since, in general $\chi^2_{\alpha,1}$ is equal to $Z^2_{\alpha/2}$, a χ^2 test based on Equation (11.7) is always the same as the two-sided test of $H_0: p = p_0$. Note that the χ^2 test is always one-sided, while the corresponding Z test is two-sided.

Problems: Section 11.5

38. A random sample of 200 drivers contained 62 who had been involved in one or more accidents. Would we reject the hypothesis that the proportion of accident-free drivers is equal to .75? That is, test $H_0: p = .75$ against the alternative $H_a: p \neq .75$, using $\alpha = .10$.

     a. First perform the test by using the method discussed in Chapter 9.

     b. Repeat the test by using the χ^2 test shown in this section.

39. A market survey taken by American Office Computers, Inc., just prior to the introduction of a new product indicated that 5% of those who were offered the product would buy it. After the product had been on the market for three months, it had been offered to 6000 potential purchasers and 300 had bought it.

     a. Calculate the χ^2 value for this problem.

     b. Are the survey and sales results consistent with this χ^2 value?

11.6
TEST FOR GOODNESS OF FIT

In tests of hypotheses in the preceding chapters, we assumed that the population was normal and tested the hypothesis $\mu = \mu_0$. But what if we want to check on the assumption of normality itself? The multinomial χ^2 **goodness-of-fit test** can be applied.

The null hypothesis for a goodness-of-fit test is that the distribution of the population from which the sample is taken *is* the one specified. The alternative hypothesis is that the actual distribution is *not* the specified distribution. Generally, a researcher specifies only

	Null Hypotheses	Parameter(s) to Be Estimated	Degrees of Freedom Lost	**TABLE 11.11**
H_0:	Population is normal	μ, σ	2	**Hypothesis Tests for**
H_0:	Population is normal with mean $\mu = \mu_0$	σ	1	**Goodness of Fit and**
H_0:	Population is normal with $\sigma = \sigma_0$	μ	1	**Degrees of Freedom**
H_0:	Population is normal with $\mu = \mu_0$ and $\sigma = \sigma_0$	None	0	
H_0:	Population is Poisson	λ	1	
H_0:	Population is Poisson with $\lambda = \lambda_0$	None	0	

the name of the distribution and uses the sample data to estimate the particular parameters of the distribution. In this situation one degree of freedom is lost for each parameter that has to be estimated. However, if the researcher completely specifies the distribution including parameter values, then no additional degrees of freedom are lost. Table 11.11 illustrates some possibilities.

EXAMPLE 11.2 Suppose Western Electronics wanted to test, at the 1% level, the hypothesis that the 106 daily absence values at Western Electronics of Table 2.1 come from a normal population. If we combine the first two categories and the last two, the data reduce to Table 11.12. The reason for combining the first few and last few classes is to ensure that the expected frequency for each class is at least 5. If this requirement is not met, then a small difference in Y_j and E_j could give an abnormally large figure when $(Y_j - E_j)^2$ is divided by a very small E_j figure.

Now we must check whether the observed frequencies Y_j in the table agree well with a normal distribution having *some* mean and standard deviation—a mean and a standard deviation unspecified in advance. Our first step is to estimate the mean and standard deviation from the sample by $\overline{X}$ and s, which can be calculated from the *grouped* data of Table 2.4 as

$$\overline{X} = 141.68 \quad \text{and} \quad s = 5.56$$

The next step is to compute what probability a normally distributed random variable X with mean 141.68 and standard deviation 5.56 has of falling in each of the four classes represented in Table 11.12. To do so, we standardize the three class boundaries between the four classes of Table 11.12 as follows:

$$Z = (132.5 - 141.68)/5.56 = -1.65$$

$$Z = (139.5 - 141.68)/5.56 = -.39$$

$$Z = (146.5 - 141.68)/5.56 = .87$$

Class	Absences	No. Observations, Y_j	**TABLE 11.12**
1	Less than 132.5	5	**Distribution of Absences**
2	132.5 to 139.5	26	**in 106 Days at Western**
3	139.5 to 146.5	59	**Electronics**
4	More than 146.5	16	
		$n = 106$	

FIGURE 11.5

Areas for Classes Used for
Goodness-of-Fit Test of
Absences at Western
Electronics

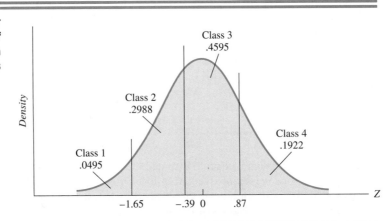

From the standard normal table the probability that a normally distributed X falls in the first class, for example, is

$$P(X \le 132.5) = P(Z \le -1.65) = .5000 - .4505 = .0495$$

The probabilities for the other three classes were determined by similar methods, and the resulting probabilities are shown in Figure 11.5.

And now we compare the observed frequencies in Table 11.12 against these probabilities. Since the probability of class 2 is .2988 and the total number of observations is 106, we expect, on the average, $106 \times .2298$, or 31.67, observations to fall in class 2. Table 11.13 shows the remaining calculations.

The number of degrees of freedom, ν, is found by determining the number of *usable* categories, k, minus the number of parameters that are estimated from the sample data, m, minus 1, which results in $k - m - 1$ degrees of freedom. The number of usable categories refers to the number of categories remaining after some cells are combined or pooled in order to meet the guideline that the expected frequency for each class be at least 5. Thus since we estimated the mean with $\overline{X}$ and the standard deviation with s, the number of estimated parameters in this example is $m = 2$, and the number of degrees of freedom is $k - m - 1 = 4 - 2 - 1 = 1$. If we had hypothesized some specific values for μ and σ, rather than using the sample data to estimate these parameters, then we would have had $k - 1 = 4 - 1 = 3$ degrees of freedom. The rejection rule follows.

Rejection rule: If $\chi^2 > \chi^2_{\alpha,\nu}$, then reject the null hypothesis.

	Class	Probability	E_j	Y_j	$Y_j - E_j$	$(Y_j - E_j)^2/E_j$
TABLE 11.13						
Calculation of the χ^2	1	.0495	5.25	5	−0.25	0.01
Goodness-of-Fit Test for	2	.2988	31.67	26	−5.67	1.02
the Western Electronics	3	.4595	48.71	59	10.29	2.17
Data	4	.1922	20.37	16	−4.37	0.94
	Total	1.0000	106.00	106	0.00	$\chi^2 = 4.14$

We are to test, at the 1% level, the hypothesis that the data in Table 11.12 fit a normal curve. The 1% point on a χ^2 distribution with 1 degree of freedom is $\chi^2_{.01,1}$, or 6.67. The calculated value, 4.14, is less than 6.67; consequently, we do not reject the hypothesis that absences are normally distributed.

The multinomial χ^2 may be used to test the goodness of fit of data with other distributions as well. We need only remember that each expected number should be at least 5 so that the calculated χ^2 statistic closely follows the χ^2 distribution, and that we lose one additional degree of freedom for every parameter estimated from the sample.

Problems: Section 11.6

40. For the accompanying distribution of current values (as percentages) of initial investments on the Southwest Regional Stock Exchange, test the hypothesis that the data came from a normal distribution. Let $\alpha = .05$, and use the procedure of this section. [*Hint:* Let the first class be "under 50" and the last class "125 or more."]

Class	Observed Frequency
$25 \le X < 50$	15
$50 \le X < 75$	25
$75 \le X < 100$	30
$100 \le X < 125$	20
$125 \le X < 150$	10
	100

41. The accompanying table gives the observed frequencies for the number of defective video recorders found in a sample of 150 lots of recorders at Electronic Components, Inc. We wanted to test the hypothesis that the number of defective recorders per lot has a Poisson distribution, and we proceeded as follows: First, we found the estimated value of the mean by taking the weighted average of the number of defectives with the observed frequencies as the weights. The estimated value is 2, and it estimates the parameter λ for the Poisson distribution:

$$P(X = x) = \frac{e^{-\lambda}\lambda^x}{x!} \qquad x = 0, 1, 2, \ldots$$

Second, we obtained the probabilities for each value of the number of defectives from tables of Poisson probabilities (Table IV in Appendix C):

$$P(X = 0) = .135$$
$$P(X = 1) = .270$$
$$P(X = 2) = .270$$
$$P(X = 3) = .180$$
$$P(X = 4) = .090$$

and so on. Third, we computed the expected frequency corresponding to each possible number of defectives by taking $nP(X = x)$, and these expected frequencies are also given in the accompanying table. Fourth, we combined the last four observed and expected frequencies so that no expected frequency is less than 5. Use a test for goodness of fit and a significance level of .05 to test the hypothesis that the number of defectives has a Poisson distribution. The information given in the accompanying table can be used to calculate χ^2. [*Hint:* There are 4 degrees of freedom since we had to estimate λ.]

Number Defective	Observed Frequency	Expected Frequency (Rounded)
0	23	20
1	39	41
2	43	41
3	23	27
4	10	14
5	7 ⎫	
6	4 ⎬ 12	⎫ 7
7	1 ⎭	

42. The accompanying table gives the observed number of customers arriving at teller windows each minute at Mortgage Bank, Inc. Using the procedure outlined in Problem 41, test the hypothesis that the data have a Poisson distribution. Let $\alpha = .025$. [*Hint:* The estimated value of λ can be shown to be 1.3. Also, note that two or three classes might need to be combined.]

Arrivals	0	1	2	3	4	5
Frequency	28	39	15	12	5	1

43. The following distribution gives the number of westbound cars arriving at an intersection on Route 66 in 60-second intervals. Can you conclude that a Poisson distribution provides a suitable model? Let $\alpha = .025$. (Show that the estimate of λ is 3.5.)

Number of Cars	0	1	2	3	4	5	6	7	8	9	10
Observed Frequency	8	23	39	53	36	30	15	7	5	3	1

44. Given the following distribution for the yields (in grams) obtained from 100 hills of corn at Great Basin Land & Farms, Inc., test that a normal distribution is an appropriate model. Let $\alpha = .05$, and use the procedure of Section 11.4. [*Hint:* Let the first class be "under 300" and the last class "1500 or more."]

Yield (grams)	Number of Hills
$100 \leq X < 300$	3
$300 \leq X < 500$	7
$500 \leq X < 700$	15
$700 \leq X < 900$	26
$900 \leq X < 1100$	22
$1100 \leq X < 1300$	13
$1300 \leq X < 1500$	9
$1500 \leq X < 1700$	5

Regression and Correlation

12

In our previous discussion of statistical inference, we have been concerned with estimating and testing hypotheses about a single variable X. In many applications, however, we are interested in examining relationships between two variables Y and X. For example, a banker might be interested in the relationship between a bank's rate of savings withdrawals Y and the current yield on corporate bonds X. Our objective in this chapter is to study the relationship between two variables Y and X.

One way to study the relationship between two variables is by means of regression. **Regression analysis** is the process of estimating a functional relationship between a random variable Y and a variable X—that is, we estimate the parameters of a linear regression *equation*. A regression equation is often used to predict a value of Y for a given value of X.

Another means of studying the relationship between two variables is called **correlation analysis;** it involves measuring the direction and strength of a linear relationship between two random variables. The numerical value that is used to measure the direction and strength of a linear relationship between two random variables is called the *correlation coefficient*.

In many problems, both regression analysis and correlation analysis are used. We obtain an equation relating Y with X, and we measure the direction and strength of the linear relationship between the two variables. The problem of relating Y to two or more X variables, known as multiple regression analysis, is the topic of Chapter 13.

12.1
SIMPLE LINEAR REGRESSION

To use a regression analysis, we must know or assume the functional form of the relationship between the variables. This relationship is expressed in the form of a mathematical function in which Y, the dependent variable, is set equal to some expression that depends only on X, the independent variable, and on certain constants or parameters. For instance, the simplest form of the relationship between two variables suggests that one variable is a constant multiple of the other. That is,

$$Y = B_1 X \tag{12.1}$$

In this form the value of Y is always B_1 times X. A one-unit change in X is associated with a B_1-units change in Y. This form is a special case of the general linear equation

$$Y = B_0 + B_1 X \tag{12.2}$$

where B_0 is called the *intercept* and B_1 is called the *slope* of the equation. This equation implies that Y has a base value of B_0 when $X = 0$ and that the value of Y changes B_1 units for every unit change in X.

=== **EXAMPLE 12.1**   Consider the pricing formula for a company's product. The company's base price is $2.00 per case, but the price drops 2¢ per case for every added case purchased up to 40 cases. The pricing formula is represented by

$$Y = 2.02 - .02X \quad \text{for} \quad 0 \le X \le 40$$

where Y is the price per case and X is the number of cases ordered.

Note that the intercept of this equation must be 2.02 (the price of $X = 0$ cases), so the price per case for $X = 1$ case is $Y = 2.02 - (.02)(1) = 2.02 - .02 = 2.00$. The slope of the equation is $-.02$, which indicates that for every added case ordered (one-unit increase in X), the price drops (since the sign is negative) 2¢. With an order of $X = 40$ cases, the price per case is $Y = 2.02 - (.02)(40) = 1.22$. The total cost is $(1.22)(40) = \$48.80$. ===

Some variables are not related by the simple linear form just discussed. Two variables might be related according to a formula such as

$$Y = B_0 + B_1 X + B_2 X^2 + 2^X$$

This equation is a rather complex relationship and not one we would expect to see often. Yet it raises the following question: If we are studying the relationship between two variables, how do we know whether the form of the relationship is similar to that in Equations (12.1) and (12.2) or some more complicated expression such as the previous equation? The answer is that in a given situation we may arrive at the functional form by either of two methods: (1) from analytical or theoretical considerations or (2) from studying scatter diagrams.

Theoretical considerations, for instance, suggest that over a relatively narrow range of prices the amount of a commodity supplied will have a linear relationship like that shown in Equation (12.2). That is,

$$P = B_0 + B_1 Q$$

is suggested by economists as the relationship between price P and quantity supplied Q in a perfectly competitive market.

Notice that the relationships that have been presented in this section so far could be termed *mathematical relationships* or *models.* They would not be termed *statistical relationships* or *models* because they do not include any variation or probability distributions.

The second means of obtaining the functional form of the relationship between two variables involves the use of scatter diagrams (sometimes called *scattergrams*). If we take a sample of elements from a population and we obtain data values on two variables, Y and X, for each element, then for the ith element we have the pair of values (Y_i, X_i). A sample of size n consists of the n pairs of values as follows:

$$\frac{Y \quad X}{\begin{array}{l} Y_1, X_1 \\ Y_2, X_2 \\ \quad \cdot \\ \quad \cdot \\ \quad \cdot \\ Y_n, X_n \end{array}}$$

We draw a **scatter diagram** by plotting the pairs of (Y, X) values as points in a plane, where Y is measured along the vertical axis and X along the horizontal axis. Examples of scatter diagrams are presented in Figure 12.1. Lines that fit the points are also shown on the scatter diagrams.

After the points have been plotted, observation of the diagram may reveal a pattern to the points that indicates what functional form may be used for the purposes of the analysis. We are concerned in this chapter only with simple *linear* relationships, in which the points in the scatter diagram appear to lie along a straight line or in which theoretical reasoning leads us to conclude that two variables are linearly related. Thus, this chapter deals only with relationships like those shown in Figures 12.1(a) and 12.1(b). Linear relationships have the general linear form given by Equation (12.2).

▬▬ **EXAMPLE 12.2** The United Park City Properties real estate investment firm is considering the construction of a condominium complex in a city. It wishes to price its units in line with what is being charged by other builders in the area. Thus the firm may wish to study the relationship between condominium selling

Scatter Diagrams   FIGURE 12.1

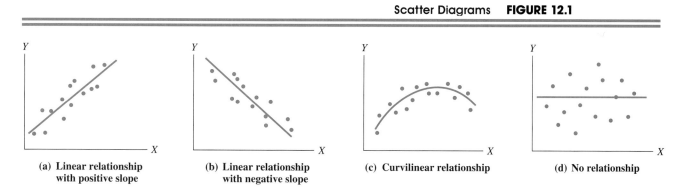

(a) Linear relationship with positive slope

(b) Linear relationship with negative slope

(c) Curvilinear relationship

(d) No relationship

prices and living area (measured in squared feet) for the condominiums in this city. One would expect that the larger condominiums would sell at higher prices, and the relationship between area X and price Y might be like that depicted in Figure 12.1(a).   ═══

In Example 12.2 no one would expect that there is a perfect relationship between price and area and that knowing area would allow us to predict a condominium's price exactly. Because of sampling variation and variations in price owing to other factors such as land costs, view, and access to shopping, the observed values or points will not all lie on the line but will be scattered to some degree about the line.

To model a linear relationship that allows variation in Y for a given X, we can use a regression model. The **simple linear regression model** follows.

Simple Linear Regression Model

$$Y_i = \beta_0 + \beta_1 X_i + \epsilon_i \tag{12.3}$$

where: Y_i = the value of the *dependent* variable for the ith observation

X_i = the value of the *independent* variable for the ith observation

ϵ_i = the value for a random fluctuation or *error* for the ith observation

β_0 = a parameter that represents the population regression *intercept,* or β_0 is the mean of Y when the value of X is equal to zero

β_1 = the parameter that represents the *slope* of the population regression line, or β_1 indicates the change in the mean value of the dependent variable Y for a unit increase in the independent variable X

Regression model (12.3) is *simple* because it only includes one independent or X variable.

In our condominium example, the model $Y_i = \beta_0 + \beta_1 X_i + \epsilon_i$ states that a condominium's price Y_i is some base or constant value β_0, plus an average of β_1 additional dollars for every additional square foot of area, plus some error value (which can be positive or negative) that cannot be predicted. The equation shows how area is associated with price. Note that the previous sentence said "is associated with" rather than "causes." That is, just because there is a relationship between two variables, we cannot assume that two variables are related by cause and effect. In the case of some physical relationships, however, the variables are causally related. For example, increases in temperature cause increases in the volume of a gas that is held under constant pressure. However, the price of a condominium Y may be only partially related to the area X of the unit through cause and effect. Perhaps the larger units also have many more frills that make them more expensive to purchase. Thus if we were to find, for instance, that $\beta_1 = \$40$, to say that an additional square foot of area "causes" the price to rise by an average of $40 would be incorrect, because the price rise may be caused by numerous factors such as the frills put in the larger units as well as the added area, or by chance variation. For this reason, in regression analysis we are careful to say that an additional unit of X "is associated with" an average of β_1 additional units of Y.

To complete the simple linear regression model, we specify three **assumptions** that must be met by the population data if we are to make inferences about the population:*

1. The X values are known, that is, not random.
2. For each value of X, Y is normally and independently distributed with mean $\mu_{Y|X}$ equal to $\beta_0 + \beta_1 X$ and variance $\sigma^2_{Y|X}$, where β_0, β_1, and $\sigma^2_{Y|X}$ are unknown parameters.
3. For each X the variance of Y is the same; that is, $\sigma^2_{Y|X} = \sigma^2$ for all X.

Under these assumptions the equation for **population simple regression** is as follows.

Population Simple Linear Regression Equation

$$\mu_{Y|X} = \beta_0 + \beta_1 X \tag{12.4}$$

That is, the mean of Y for a specific value of X, $\mu_{Y|X}$, is equal to $\beta_0 + \beta_1 X$, where the constants β_0 and β_1 are the population intercept and slope. The methods of analysis developed in the following sections enable us to use a random sample of data to estimate the parameters β_0, β_1, and $\sigma^2_{Y|X}$.

We can estimate the parameters of the population simple linear regression, Equation (12.4), by using sample data. The result is the equation for **sample,** or **estimated, simple linear regression.**

Sample (or Estimated) Simple Linear Regression Equation

$$\hat{Y} = b_0 + b_1 X \tag{12.5}$$

In Equation (12.5) the value of $\hat{Y}$ is the estimate of the mean value of Y given X, or $\hat{Y}$ estimates $\mu_{Y|X}$. The value of b_0 that is determined from the sample data is the estimate of the population regression intercept β_0; and the value of b_1 that is determined from the sample data is the estimate of the population regression slope β_1.

12.2
FINDING THE SLOPE AND THE INTERCEPT OF A REGRESSION LINE

Up to this point we have discussed only the form that regression equations might take. In this chapter we will deal only with simple linear regression equations of the form $\mu_{Y|X} = \beta_0 + \beta_1 X$. In this section we deal with the problem of how we use sample data to make estimates of the regression line's intercept β_0 and its slope β_1.

EXAMPLE 12.3 United Park City Properties real estate investment firm mentioned in the previous section took a random sample of $n = 5$ condominium

---

*Alternatively, the assumptions of the regression model are that the X value is known, and ϵ_i are independent and normally distributed with mean 0 and variance σ^2.

units that recently sold in the city. The sales prices Y (in thousands of dollars) and the areas X (in hundreds of square feet) for each unit are as follows:

Y = Sales Price ($\times$ $1000)	X = Area (square feet) ($\times$ 100)
36	9
80	15
44	10
55	11
35	10

This table shows that the first condominium in the sample had an area of 900 square feet and sold for $36,000. Figure 12.2 shows the scatter diagram for this sample. The figure shows that there is a likely positive slope to the points.

The real estate investment firm would like to know $\mu_{Y|X} = \beta_0 + \beta_1 X$, the regression equation for the line that would pass through a scatter diagram of *all* the condominium sales in the city. That is, the firm would like to know the population regression line. However, there are sales prices and areas for only a small sample of five condominium sales. Using these data, the firm will be able to *estimate* the population regression equation $\mu_{Y|X} = \beta_0 + \beta_1 X$. To distinguish between a regression equation calculated from population data and one calculated from sample data, we indicated that the sample regression equation is

$$\hat{Y} = b_0 + b_1 X$$

To obtain values b_0 and b_1 from the sample data of Example 12.3, we could draw a straight line through the points in Figure 12.2 in such a way that it passes close to all the points in the scatter diagram. Then with a ruler we could measure the intercept b_0 and the slope b_1. However, this method is very inexact, and two people drawing lines through the same set of points might obtain quite different estimates for the intercept and the slope of the line.

FIGURE 12.2

Scatter Diagrams for Five Condominium Sales

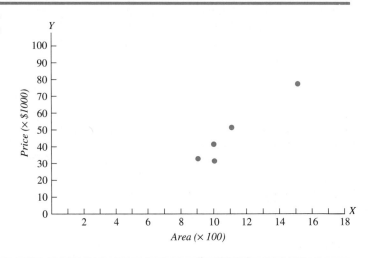

FIGURE 12.3

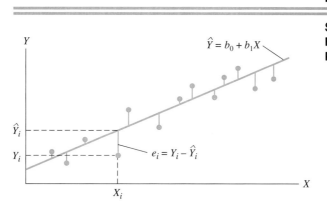

The method used by statisticians to find the intercept and the slope of the regression line passing through a set of points is called the **least squares method.** The principle of least squares is illustrated in Figure 12.3. For every observed Y_i in a sample of points, there is a corresponding predicted value $\hat{Y}_i$, equal to $b_0 + b_1 X_i$ as given by Equation (12.5). The sample deviation of the observed value Y_i from the predicted $\hat{Y}_i$ is $e_i = Y_i - \hat{Y}_i$, equal to $Y_i - b_0 - b_1 X_i$. The sample error or **residual** e_i parallels the random population error ϵ_i. The sum of squares of these deviations from the fitted line is

$$\sum e_i^2 = \sum (Y_i - \hat{Y}_i)^2 = \sum (Y_i - b_0 - b_1 X_i)^2 \tag{12.6}$$

where the estimators b_0 and b_1 are the functions of the sample values that make this sum of squares a minimum, or *least,* value.

Equation (12.6) is minimum when the values for b_0 and b_1 are as given in the following equations.

Equations for Slope b_1 and Intercept b_0

$$b_1 = \frac{SCP}{SSX} \tag{12.7}$$

$$b_0 = \overline{Y} - b_1 \overline{X} \tag{12.8}$$

where: SCP = the **sum of cross products of deviations**

$$SCP = \sum (X_i - \overline{X})(Y_i - \overline{Y}) = \sum X_i Y_i - \frac{(\sum X_i)(\sum Y_i)}{n} \tag{12.9}$$

SSX = the **sum of squared deviations** for X

$$SSX = \sum (X_i - \overline{X})^2 = \sum X_i^2 - \frac{(\sum X_i)^2}{n} \tag{12.10}$$

Notice that Equation (12.8) implies that the regression line goes through the mean of the X's and the mean of the Y's or through the point $(\overline{X}, \overline{Y})$. The slope b_1 and intercept b_0 of Equations (12.7) and (12.8) that are found by the method of least squares are minimum-variance unbiased estimators of the population slope β_1 and intercept β_0.* These properties for estimators were discussed in Section 8.1.

We will find the slope b_1 and intercept b_0 of the sample regression line $\hat{Y} = b_0 + b_1 X$ for the United Park City Properties data from the five condominiums by using data analysis software. Then we will use Equations (12.7) and (12.8) to work out the calculations in Example 12.4. After we have found the slope b_1 and intercept b_0, we will discuss the meaning of the sample regression equation.

A computer and data analysis software were used to analyze the United Park City Properties condominium data of Example 12.3 and Table 12.1 (see Example 12.4), and the printouts are presented in Computer Exhibits 12.1A, 12.1B, and 12.1C. The estimated equation as presented in the printouts is $\hat{Y} = -34.5 + 7.68X$. The printouts also give a great deal of information that we will discuss in the remainder of this chapter. You may want to compare the results on the printouts to the results in the remaining sections as we progress through the chapter (some slight differences between our results in the chapter and the results in the printouts are due to rounding). It is usually quite simple to adapt from one printout to the next, and printouts from other data analysis software are given with some of the supplemental problems for this chapter.

*Calculus is usually used if one wants to derive Equations (12.7) and (12.8) from Equation (12.6). For those with a calculus background, the derivation of the equations for estimating the slope and intercept follows. The sum of squared deviations that must be minimized is the function $F(b_0, b_1) = \Sigma(Y - \hat{Y})^2$ of the unknown quantities b_0 and b_1. Substituting $b_0 + b_1 X$ for $\hat{Y}$ into this equation gives $F(b_0, b_1) = \Sigma(Y - b_0 - b_1 X)^2$. To find the minimum value of the function, we take the partial derivatives with respect to b_0 and b_1 and set them equal to 0. Thus, $\partial F(b_0, b_1)/\partial b_0 = -2\Sigma(Y - b_0 - b_1 X) = 0$ and $\partial F(b_0, b_1)/\partial b_1 = -2\Sigma(Y - b_0 - b_1 X)(X) = 0$. Simplifying gives the next two equations, known as the normal equations: $\Sigma Y = nb_0 + b_1 \Sigma X$ and $\Sigma XY = b_0 \Sigma X + b_1 \Sigma X^2$, from which the equations for the slope and intercept can be derived. Since the second partial derivatives of $F(b_0, b_1)$ are positive, we have found the minimum.

		Price, Y	Area, X	Y^2	X^2	XY
TABLE 12.1						
United Park City		36	9	1,296	81	324
Properties Condominium		80	15	6,400	225	1200
Data and Calculations		44	10	1,936	100	440
		55	11	3,025	121	605
		35	10	1,225	100	350
	Total	250	55	13,882	627	2919

$$\overline{Y} = \frac{\Sigma Y}{n} = \frac{250}{5} = 50$$

$$\overline{X} = \frac{\Sigma X}{n} = \frac{55}{5} = 11$$

$$SCP = \Sigma XY - \frac{\Sigma X \Sigma Y}{n} = 2919 - \frac{(55)(250)}{5} = 169$$

$$SSX = \Sigma X^2 - \frac{(\Sigma X)^2}{n} = 627 - \frac{(55)^2}{5} = 22$$

COMPUTER EXHIBIT 12.1A Regression of Sales Price and Area for United Park City
Properties Condominium Data

MINITAB

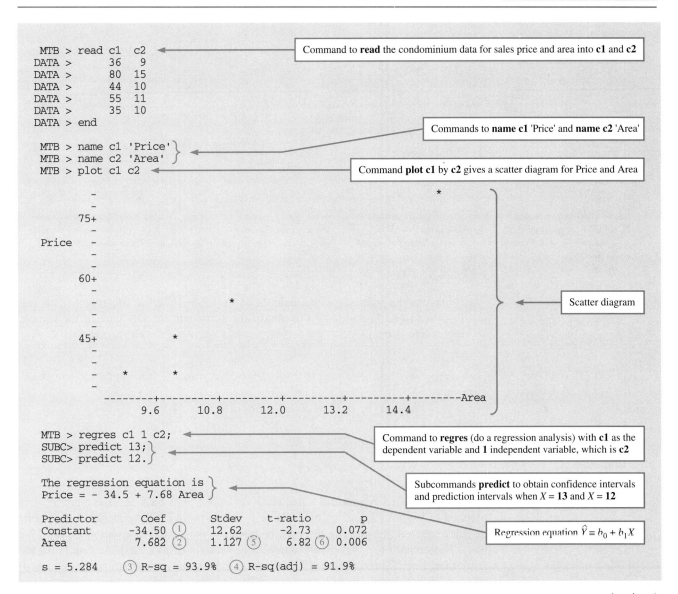

```
MTB > read c1  c2          ←——  Command to read the condominium data for sales price and area into c1 and c2
DATA >      36   9
DATA >      80  15
DATA >      44  10
DATA >      55  11
DATA >      35  10
DATA > end
                                Commands to name c1 'Price' and name c2 'Area'
MTB > name c1 'Price' ⎫
MTB > name c2 'Area'  ⎭  ←——
MTB > plot c1 c2          ←——  Command plot c1 by c2 gives a scatter diagram for Price and Area


       -
       -
  75+
       -
Price  -
       -
       -
  60+
       -
       -                                                             *
       -                                                                   Scatter diagram
       -                          *
  45+
       -                    *
       -
       -         *          *
       -
       --------+---------+---------+---------+---------+--------Area
              9.6       10.8      12.0      13.2      14.4


MTB > regres c1 1 c2;      ←——  Command to regres (do a regression analysis) with c1 as the
SUBC> predict 13; ⎫             dependent variable and 1 independent variable, which is c2
SUBC> predict 12. ⎭

The regression equation is ⎫   Subcommands predict to obtain confidence intervals
Price = - 34.5 + 7.68 Area ⎭   and prediction intervals when X = 13 and X = 12

Predictor      Coef        Stdev    t-ratio        p
Constant     -34.50 ①      12.62      -2.73    0.072          Regression equation Ŷ = b₀ + b₁X
Area           7.682 ②      1.127 ⑤     6.82 ⑥ 0.006

s = 5.284    ③ R-sq = 93.9%  ④ R-sq(adj) = 91.9%
```

The regression equation is $\hat{Y} = b_0 + b_1 X$

(continues)

EXAMPLE 12.4   Table 12.1 has been prepared from the United Park City Properties sample real estate data in Example 12.3. Table 12.1 includes columns for Y, X, Y^2, X^2, and XY and the sums of these columns. Computations for the sample means $\overline{Y}$ and $\overline{X}$, the sum of cross product deviations SCP, and the sum of squared deviations for X or SSX are also included in Table 12.1. The mean price of the five units $\overline{Y}$ was found to be 50, or $50,000. The mean area $\overline{X}$ was found to be 11, or 1100 square feet. The sum of cross products deviations SCP was found to be 169, and the sum of squares of the X deviations SSX was found to be 22. The column headed Y^2 and its total will be used in subsequent sections.

MINITAB	COMPUTER EXHIBIT 12.1A Regression of Sales Price and Area for United Park City Properties Condominium Data (Continued)

```
Analysis of Variance

SOURCE      DF         SS        MS       F         p
Regression   1     1298.2    1298.2   46.49     0.006
Error        3       83.8      27.9
Total        4     1382.0

    Fit   Stdev.Fit           95% C.I.  ⑨     95% P.I.
   65.36 ⑦     3.27 ⑧  (  54.97,   75.76)   (  45.59,   85.13)

   57.68        2.62   (  49.35,   66.01)   (  38.91,   76.45) ⑩

MTB > stop
```

ANOVA table

Note: Some values on the printout will be explained in subsequent sections.
① b_0, estimated regression intercept; ② b_1, estimated regression slope; ③ $s_{Y|X}$, standard error of the estimate; ④ r^2, sample coefficient of determination (as a percentage); ⑤ s_{b_1}, standard error of the sample regression slope; ⑥ t, t statistic for testing H_0: $\beta_1 = 0$; ⑦ $\hat{Y}$, estimated value of $\mu_{Y|X_0}$ or pre- dicted value of Y given $X = 13$; ⑧ $s_{\hat{Y}_0}$, standard error of the sample regression line or; standard deviation of a fitted Y value, $\hat{Y}$; ⑨ 95% confidence interval for $\mu_{Y|X_0}$ given $X = 13$; ⑩ 95% prediction interval for an individual value Y_0 given $X = 12$.

From the values that we calculated in Table 12.1, we are prepared to use Equations (12.7) and (12.8) to obtain the slope and the intercept of the sample regression line:

$$b_1 = \frac{SCP}{SSX} = \frac{169}{22} = 7.68$$

$$b_0 = \overline{Y} - b_1\overline{X} = 50 - (7.68)(11) = -34.5$$

Thus, the sample regression line is

$$\hat{Y} = -34.5 + 7.68X$$

The estimated regression equation $\hat{Y} = -34.5 + 7.68X$ for Example 12.4 is graphed in Figure 12.4, along with the $n = 5$ data points. This equation tells us that the least squares line passing through the points in the scatter diagram of Figure 12.2 has a Y intercept of -34.5. This value can be thought of as a "starting point" for the regression line. (Some people might wish to attach the interpretation that a condo- minium with $X = 0$—that is, zero square feet of area, which is not possible—would sell for $-\$34,500$.) The slope of 7.68 means that for condominiums in which the floor space ranges from 900 to 1500 square feet, each additional unit of X (100 square feet of area) is associated with 7.68 additional units of Y (in thousands of dol- lars). That is, 100 added square feet is associated with an added \$7680 in price, on the average. The same result suggests that the average price increase is \$76.80 per square foot.

Note that this interpretation of the slope of the regression line is valid only for condominiums with areas in the range of 900 to 1500 square feet, which is the range of sizes for condominiums in the data set used to construct the line. We cannot ex- trapolate outside this range of X values and suggest that the average price increase is

---

COMPUTER EXHIBIT 12.1B Regression of Sales Price and Area for United Park City Properties Condominium Data

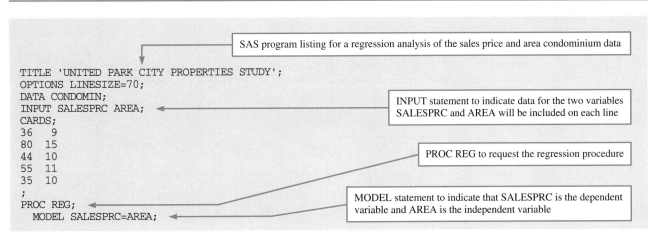

SAS program listing for a regression analysis of the sales price and area condominium data

```
TITLE 'UNITED PARK CITY PROPERTIES STUDY';
OPTIONS LINESIZE=70;
DATA CONDOMIN;
INPUT SALESPRC AREA;
CARDS;
36   9
80  15
44  10
55  11
35  10
;
PROC REG;
  MODEL SALESPRC=AREA;
```

INPUT statement to indicate data for the two variables SALESPRC and AREA will be included on each line

PROC REG to request the regression procedure

MODEL statement to indicate that SALESPRC is the dependent variable and AREA is the independent variable

(continues)

$76.80 per square foot for condominiums larger or smaller than those in our data set. The price change per square foot might be either greater or smaller than this figure, especially if the relationship between area and price is nonlinear outside the range of 900 to 1500 square feet.

After we have found numerical values for the estimators b_0 and b_1 by using Equations (12.7) and (12.8), we have the regression equation that can be used to obtain **predictions.** For instance, if the real estate investment firm is planning to build condominiums that have $X = 13$ (or 1300 square feet), it would predict a selling

FIGURE 12.4

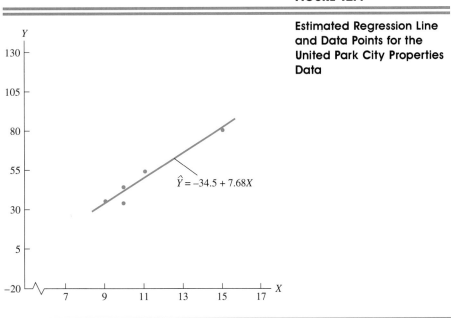

Estimated Regression Line and Data Points for the United Park City Properties Data

$\hat{Y} = -34.5 + 7.68X$

SAS	**COMPUTER EXHIBIT 12.1B** Regression of Sales Price and Area for United Park City Properties Condominium Data (Continued)

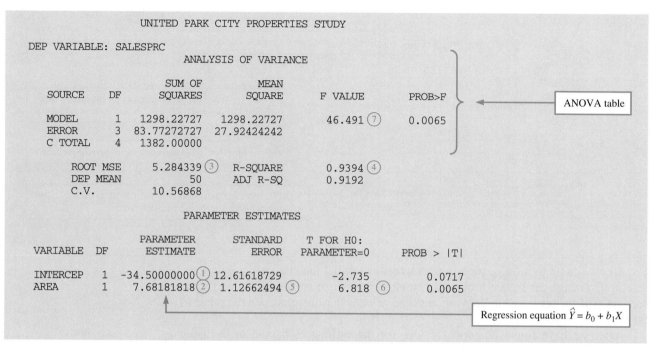

UNITED PARK CITY PROPERTIES STUDY

DEP VARIABLE: SALESPRC

ANALYSIS OF VARIANCE

SOURCE	DF	SUM OF SQUARES	MEAN SQUARE	F VALUE	PROB>F
MODEL	1	1298.22727	1298.22727	46.491 ⑦	0.0065
ERROR	3	83.77272727	27.92424242		
C TOTAL	4	1382.00000			

ANOVA table

ROOT MSE	5.284339 ③	R-SQUARE	0.9394 ④
DEP MEAN	50	ADJ R-SQ	0.9192
C.V.	10.56868		

PARAMETER ESTIMATES

VARIABLE	DF	PARAMETER ESTIMATE	STANDARD ERROR	T FOR H0: PARAMETER=0	PROB > |T|
INTERCEP	1	-34.50000000 ①	12.61618729	-2.735	0.0717
AREA	1	7.68181818 ②	1.12662494 ⑤	6.818 ⑥	0.0065

Regression equation $\hat{Y} = b_0 + b_1 X$

Note: Additional information will be explained in subsequent sections. ① b_0, estimated regression intercept; ② b_1, estimated regression slope; ③ $s_{Y|X}$, standard error of the estimate; $s_{Y|X} = \sqrt{MSE} = $ ROOT MSE; ④ r^2, sample coefficient of determination; ⑤ s_{b_1}, standard error of the sample regression slope; ⑥ t, t statistic for testing H_0: $\beta_1 = 0$; ⑦ F, F statistic for testing H_0: $\beta_1 = 0$.

price of $\hat{Y} = -34.5 + 7.68(13) = 65.36$. Thus a selling price of \$65,360 would be consistent with the prices obtained for the five condominiums in the sample. This result assumes, of course, that the firm's condominiums would have approximately the same type of amenities possessed by the condominiums used to construct the regression line.

MYSTAT	**COMPUTER EXHIBIT 12.1C** Regression of Sales Price and Area for United Park City Properties Condominium Data

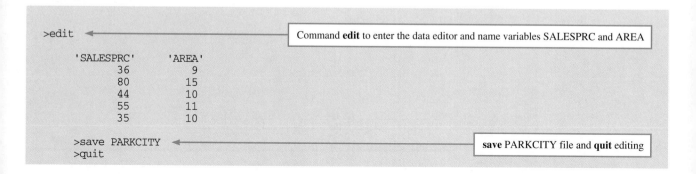

```
>edit                              Command edit to enter the data editor and name variables SALESPRC and AREA

      'SALESPRC'      'AREA'
             36           9
             80          15
             44          10
             55          11
             35          10

>save PARKCITY                     save PARKCITY file and quit editing
>quit
```

COMPUTER EXHIBIT 12.1C Regression of Sales Price and Area for United Park City Properties Condominium Data (Continued)   **MYSTAT**

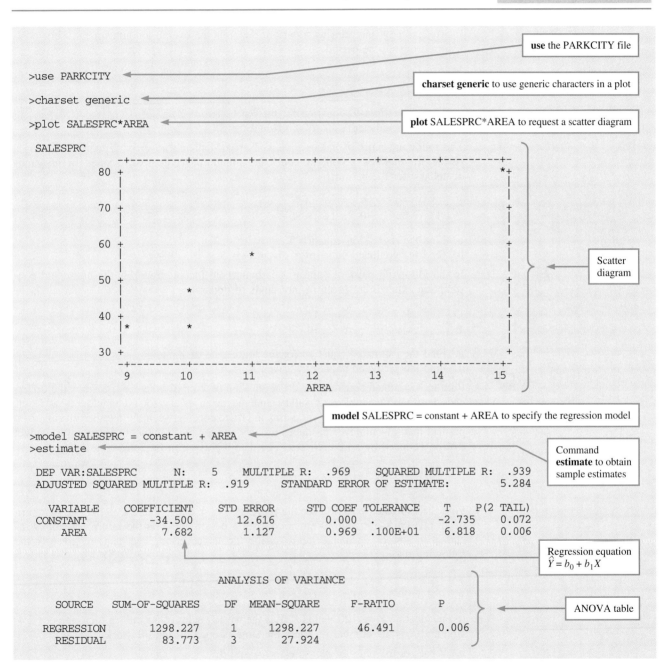

use the PARKCITY file

```
>use PARKCITY
```
charset **generic** to use generic characters in a plot
```
>charset generic
```
plot SALESPRC*AREA to request a scatter diagram
```
>plot SALESPRC*AREA
```

Scatter diagram

model SALESPRC = constant + AREA to specify the regression model

```
>model SALESPRC = constant + AREA
>estimate
```

Command **estimate** to obtain sample estimates

```
DEP VAR:SALESPRC     N:    5   MULTIPLE R:  .969    SQUARED MULTIPLE R:  .939
ADJUSTED SQUARED MULTIPLE R:  .919     STANDARD ERROR OF ESTIMATE:      5.284
```

VARIABLE	COEFFICIENT	STD ERROR	STD COEF	TOLERANCE	T	P(2 TAIL)
CONSTANT	-34.500	12.616	0.000	.	-2.735	0.072
AREA	7.682	1.127	0.969	.100E+01	6.818	0.006

Regression equation $\hat{Y} = b_0 + b_1 X$

ANALYSIS OF VARIANCE

SOURCE	SUM-OF-SQUARES	DF	MEAN-SQUARE	F-RATIO	P
REGRESSION	1298.227	1	1298.227	46.491	0.006
RESIDUAL	83.773	3	27.924		

ANOVA table

Note that the line $\hat{Y} = -34.5 + 7.68X$ passes through the points in Figure 12.4 in such a way that it is "close" to the points. Closeness is defined in a rather explicit way, however. That is, there is no other line that could be fitted to those points for which the sum of squares of vertical deviations from the line would be smaller. For this reason, the line is called the *least squares line*.

Problems: Sections 12.1–12.2

Answers to odd-numbered problems are given in the back of the text.

  1. For the following pairs of values, X is the number of units of a product produced at Intertechnology, Inc., during a period, and Y is the overall amount for the labor costs incurred during the period:

X	0	1	2	3	4	5	6
Y	1	2	3	5	8	11	12

   a. Find the estimated equation for the regression of Y on X.
   b. Find the predicted value of Y given $X = 8$.

  2. The following pairs of values give X, Narco Medical's advertising expense during a period (in hundreds of dollars), and Y, the amount of the company's sales (in thousands of dollars) for the period:

X	8	9	7	6	5	1
Y	15	11	10	11	8	5

   a. Find the regression equation $\hat{Y} = b_0 + b_1 X$.
   b. Find the predicted value of Y given $X = 5$.

3. In the following pairs of values, X is household income and Y is the household consumption expenditure (both in thousands of dollars):

X	1	5	9	13	17
Y	7	6	9	8	10

   a. Find the estimated equation for the regression of Y on X.
   b. Find the predicted value of Y given $X = 6$.

4. Find the estimated equation for the regression of Y on X, where X measures the difference between an employee's salary and the mean salary for all employees at Compugraph Corporation, and Y measures the difference between the number of units of a product that the employee produces per day and the mean number of units of a product that is produced by all employees. The data are given in the following listing:

X	−10	−6	−2	2	6	10	14	16	20	23
Y	−8	−5	−4	−3	0	4	7	8	11	10

5. Find the estimated equation for the regression of Y on X, where Y is the rate of return of an investor's portfolio of stocks during a period and X is the return for stocks included in a stock exchange index during the same period. The data are given in the following listing:

X	12	13	14	15	16	17	18
Y	−8	−6	−4	0	6	12	14

6. In the equation $\hat{Y} = 120 + .32X$, Y is the weekly sales of a store (in thousands of dollars) and X is the number of hours the store is open during the week.

   a. What are the units of measure for b_0 and b_1?
   b. If the store stays open 60 hours next week, what is its predicted sales level?
   c. If the store doesn't open at all next week, what does the equation predict for its sales level?
   d. If the store doesn't open at all next week, what would you predict for its sales level?
   e. Why is the answer in part c not valid?

The previous sections of this chapter have discussed the concept of regression analysis and methods to find the intercept and slope of the sample regression line. Several topics in regression analysis that will be discussed in following sections are related to this section's topic of how the variation in a regression problem's dependent variable Y can be partitioned into two parts. Many of the computations in sections that follow this one are simplified by using the partitioning of the sums of squares that will be found in this section.

For a single variable Y the variation in Y is measured by the sum of squared deviations of the Y values from their mean $\overline{Y}$, or the sum of squares, all of which can be considered as being due to random or unexplained variation. Hence the estimated variance of Y is based on the total sum of squares. In the regression situation, however, some of the observed variation among the sample Y's is associated with the relationship between Y and X. Figure 12.5(a) shows one observed point (X_i, Y_i), a fitted point $(X_i, \hat{Y}_i)$, and the point on the fitted line whose coordinates are the means $(\overline{X}, \overline{Y})$.

12.3
PARTITIONING THE SUM OF SQUARES IN SIMPLE REGRESSION

Total, Explained, and Unexplained Deviation and Variation   **FIGURE 12.5**

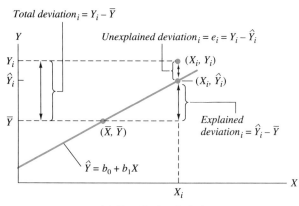

(a) Partitioning deviation

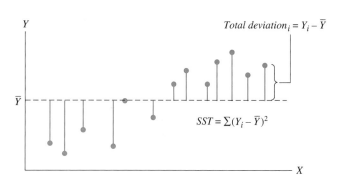

(b) Total variation

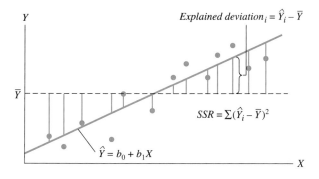

(c) Explained variation

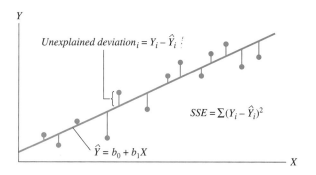

(d) Unexplained variation

We see that the *total deviation* from the mean $Y_i - \overline{Y}$ is the sum of the deviation *explained* by regression $\hat{Y}_i - \overline{Y}$ and the residual, or *unexplained* deviation $Y_i - \hat{Y}_i$. That is,

$$(Y_i - \overline{Y}) = (\hat{Y}_i - \overline{Y}) + (Y_i - \hat{Y}_i) \tag{12.11}$$

$$\begin{array}{ccc} \text{Total} & = & \text{Explained} \\ \text{deviation} & & \text{deviation} \end{array} + \begin{array}{c} \text{Unexplained} \\ \text{deviation} \end{array}$$

The explained deviation from $\hat{Y}_i$ to $\overline{Y}$, the first term on the right of the equation, is associated with the relationship between Y and X. Thus, this portion of the total deviation of Y_i from the mean $\overline{Y}$ may be said to be accounted for by the regression of Y on X, or the explained deviation is due to the regression of Y on X. The second term on the right of the equation, the deviation of Y_i from $\hat{Y}_i$, is the unexplained deviation, or the residual. If we square both sides of Equation (12.11) and sum over all n values, we obtain the **sum of squares total SST**, the **sum of squares explained by regression SSR**, and the **sum of squares due to error SSE** in the following equation which **partitions the sum of squares.**

Equation for Partitioning Sum of Squares

$$\sum (Y_i - \overline{Y})^2 = \sum (\hat{Y}_i - \overline{Y})^2 + \sum (Y_i - \hat{Y}_i)^2 \tag{12.12}$$

$$SST \quad = \quad SSR \quad + \quad SSE$$

$$\begin{array}{ccc} \text{Total} & = & \text{Explained} \\ \text{variation} & & \text{variation} \end{array} + \begin{array}{c} \text{Unexplained} \\ \text{variation} \end{array}$$

The sums of squared deviations—*SST, SSR,* and *SSE*—are known as *total variation, explained variation* and *unexplained variation.* Figures 12.5(b–d) depict the total variation *SST,* the explained variation *SSR,* and the unexplained variation *SSE* for a set of data values.

The various sums of squares may be found more simply by using the following formulas, if we use a sufficient number of significant digits to avoid rounding errors.

Equations for Sums of Squares

$$SST = \sum (Y_i - \overline{Y})^2 = \sum Y^2 - \frac{(\sum Y)^2}{n} \tag{12.13}$$

$$SSR = \sum (\hat{Y}_i - \overline{Y})^2 = (b_1)(SCP) \tag{12.14}$$

$$SSE = \sum (Y_i - \hat{Y}_i)^2 = SST - (b_1)(SCP) \tag{12.15}$$

The partitioning we have discussed is conveniently summarized in Table 12.2, which is known as an **analysis of variance table.** Where k is the number of indepen-

Source of Variation	Degrees of Freedom	Sum of Squares	Mean Square	F	**TABLE 12.2**
Regression	1	$SSR = (b_1)(SCP)$	$MSR = \dfrac{SSR}{1}$	$\dfrac{MSR}{MSE}$	**Analysis of Variance Table for Simple Regression**
Error	$n - 2$	$SSE = SST - (b_1)(SCP)$	$MSE = \dfrac{SSE}{n - 2}$		
Total	$n - 1$	$SST = \Sigma Y_2 - \dfrac{(\Sigma Y)^2}{n}$			

dent or X variables included in the equation, the sum of squares for regression SSR has k degrees of freedom, the sum of squares of error deviations SSE has $n - k - 1$ degrees of freedom, and the total sum of squares SST has, as usual, $n - 1$ degrees of freedom. The degrees of freedom, as well as the sum of squares, are partitioned, or the degrees of freedom are additive. In simple regression, we have one X variable, so $k = 1$, and the next equation shows the partitioning of the *degrees of freedom:*

$$(n - 1) = (k) + (n - k - 1)$$

or (12.16)

$$(n - 1) = (1) + (n - 2)$$

Recall from the analysis of variance that **mean squares** are sums of squares divided by their associated degrees of freedom. Two mean squares that are important in regression are the **mean square due to regression MSR** and the **mean square due to error MSE,** which are defined as follows for simple regression analysis.

> **Mean Square Due to Regression MSR and Mean Square Due to Error MSE for Simple Regression**
>
> $$MSR = \frac{SSR}{1} \qquad (12.17)$$
>
> $$MSE = \frac{SSE}{n - 2} \qquad (12.18)$$

The degrees of freedom and mean squares are also summarized in Table 12.2. The sums of squares and the degrees of freedom are additive, whereas the mean squares are not. The column for F will be discussed in subsequent examples.

The analysis of variance table for the United Park City Properties condominium data of Example 12.4, including the degrees of freedom, sums of squares, and mean squares, is given in the printouts of Computer Exhibit 12.1. In Example 12.5, we will find these values by using the equations that are summarized in Table 12.2.

EXAMPLE 12.5   For the United Park City Properties condominium data of Example 12.3 and using the calculations of Table 12.1, the total variation of the Y's about their mean $\overline{Y}$ is measured by the total sum of squares SST with degrees of freedom $n - 1 = 5 - 1 = 4$:

$$SST = \sum(Y_i - \overline{Y})^2 = \sum Y^2 - \frac{(\sum Y)^2}{n} = 13{,}882 - \frac{(250)^2}{5} = 1382$$

The portion of this variation that is associated with the regression is the sum of squares SSR with 1 degree of freedom:

$$SSR = (b_1)(SCP) = (7.682)(169) = 1298.2$$

The remaining unexplained or random error variation is the variation of the observed Y's about the estimated line. It is measured by the sum of squares of deviations SSE. This sum of squares has $n - 2 = 5 - 2 = 3$ degrees of freedom and is found by subtraction:

$$SSE = SST - (b_1)(SCP) = 1382 - (7.682)(169) = 83.8$$

The mean square due to regression MSR and the mean square due to error MSE from Equations (12.17) and (12.18) are

$$MSR = \frac{SSR}{1} = \frac{1298.2}{1} = 1298.2$$

$$MSE = \frac{SSE}{n - 2} = \frac{83.8}{5 - 2} = 27.9$$

The summary is given in Table 12.3, which is also known as an *analysis of variance table.*

The sums of squares and the degrees of freedom are additive in Table 12.3, just as Equations (12.12) and (12.16) say they must be. Rounding the values during these computations can result in errors in the sums of squares, so be sure to carry sufficient significant digits during these calculations. The sums of squares and mean squares in Table 12.3 will be used extensively to simplify the computations required in the following sections. You may want to compare the results given in Table 12.3 to the analysis of variance tables in the printouts of Computer Exhibit 12.1. The values in the tables are not exactly equal due to rounding.

	Source of Variation	Degrees of Freedom	Sum of Squares	Mean Square	F
TABLE 12.3					
Analysis of Variance	Regression	1	1298.2	1298.2	46.5
Table for the	Error	3	83.8	27.9	
Condominium Problem	Total	4	1382		

Problems: Section 12.3

7. Use the Intertechnology, Inc., data of Problem 1.

   a. Find *SST*, *SSR*, and *SSE*.
   b. Find the degrees of freedom associated with the sums of squares.
   c. Find *MSR* and *MSE*.
   d. Present the results in a summary table.

8. Refer to the Narco Medical Company data of Problem 2.

   a. Find *SST*, *SSR*, and *SSE*.
   b. Find the degrees of freedom associated with the sums of squares in part a.
   c. Find *MSR* and *MSE*.
   d. Present your results for parts a and b in a summary table.

9. Use the income and consumption expenditure data of Problem 3.

   a. Find *SST*, *SSR*, and *SSE*.
   b. Find the degrees of freedom associated with the sums of squares in part a.
   c. Find *MSR* and *MSE*.
   d. Present your results for parts a and b in a summary table.

12.4
STANDARD ERROR OF THE ESTIMATE $s_{Y|X}$ AND COEFFICIENT OF DETERMINATION r^2

In many regression problems, the major reason for constructing the regression equation is to obtain a tool that is useful in predicting the value of Y, the dependent variable, from some known value of X, the associated independent variable. Thus, we often wish to assess the accuracy of the regression line in predicting the Y values. In Section 12.1, we noted that the error in predicting a particular Y_i value, when the intercept β_0 and the slope β_1 are known, is a term designated ϵ_i. Thus we can measure the accuracy of prediction for a regression equation by examining these ϵ_i values. We mentioned earlier that the error terms average to zero: $E(\epsilon_i) = 0$. But the variance of these error terms, called the *mean squared deviation* around the regression line, is used to measure the dispersion of the points around the line. This value, denoted $\sigma^2_{Y|X}$, is simply the variance of the Y_i around the regression line.

In the typical regression problem, we do not know the population of all Y_i values. We usually have a sample, and the **sample mean squared error** is as follows:

$$s^2_{Y|X} = \frac{1}{n-2} \sum e_i^2 \qquad (12.19)$$

Or since the sample error or residual is $e_i = Y_i - \hat{Y}_i$, we have the following equation.

Equation for Sample Mean Squared Error

$$s^2_{Y|X} = \frac{1}{n-2} \sum (Y_i - \hat{Y}_i)^2 = \frac{1}{n-2}(SSE) = MSE \qquad (12.20)$$

For the regression model, $s^2_{Y|X}$ is an unbiased estimator of $\sigma^2_{Y|X}$.

Equation (12.19) shows why $s^2_{Y|X}$ is a good estimate of the accuracy of predictions using the sample regression equation. If the residuals, the e_i values, are large, then the $s_{Y|X}^2$ figure will be large. If all the Y_i values lie right on the regression line, however, then each $e_i = 0$ and $s_{Y|X}^2 = 0$. The notation $s_{Y|X}^2$ refers to the estimated variance of the population of all Y_i values around their mean $\mu_{Y|X}$ for a given value of X. This notation is used to distinguish the sample variance around the regression line, $s_{Y|X}^2$, from the sample variance defined in Equation (3.10), which specifies

$$s_Y^2 = \frac{1}{n-1} \sum (Y_i - \overline{Y})^2 = \frac{1}{n-1}[SST]$$

Note the similarities in Equation (12.20) and Equation (3.10). This comparison shows that $s_{Y|X}^2$ is the sample variance of the individual Y_i values with respect to the regression line's predicted values, $\hat{Y}_i$. That is, $s_{Y|X}^2$ is the variance of the deviations shown in Figure 12.3. The sum of squared deviations around the regression line is divided by $n - 2$, the number of degrees of freedom in the sum of squares. Two degrees of freedom were lost from the data set through the calculation of b_0 and b_1 for the regression line. The s_Y^2 figure is similar to $s_{Y|X}^2$ except that its reference point is $\overline{Y}$ rather than the regression line.

Thus s_Y^2 can be thought of as the average squared deviation of the Y_i values from the mean, while $s_{Y|X}^2$ can be thought of as the average squared deviation of the Y_i values from the sample regression line's predictions.

The square root of $s_{Y|X}^2$ is the standard deviation of the Y_i values around the sample regression line. It is known as the **standard error of the estimate** and is given in the following equation.

Equation for Standard Error of the Estimate $s_{Y|X}$

$$s_{Y|X} = \sqrt{s_{Y|X}^2} = \sqrt{MSE} \qquad\qquad (12.21)$$

The value of the standard error of the estimate $s_{Y|X}$ for the United Park City Properties data of Example 12.3 is presented in Computer Exhibit 12.1A as $s = 5.284$, in Computer Exhibit 12.1B as ROOT MSE 5.284339, and in Computer Exhibit 12.1C as STANDARD ERROR OF ESTIMATE: 5.284.

We will use Equation (12.21) to find the value of $s_{Y|X}$ in the following example.

EXAMPLE 12.6　If we use the United Park City Properties data of Example 12.5 in the previous section, we find that

$$s_Y^2 = \frac{1}{n-1} \sum (Y_i - \overline{Y})^2 = \frac{1}{n-1}(SST)$$

$$= \frac{1}{4}(1382) = 345.50 \text{ (thousand dollars)}^2$$

and

$$s_{Y|X}^2 = \frac{1}{n-2} \sum (Y_i - \hat{Y})^2 = \frac{1}{n-2}(SSE)$$

$$= \frac{1}{3}(83.8) = 27.9 \text{ (thousand dollars)}^2$$

The standard error of the estimate is

$$s_{Y|X} = \sqrt{s_{Y|X}^2} = \sqrt{27.9} = 5.28$$

Thus the variance of the condominium prices around the mean price, $\overline{Y}$, is 345.5 (thousand dollars)^2, while the variance of the condominium prices around the regression line's predicted values, $\hat{Y}_i$, is only 27.9 (thousand dollars)^2. The variance around the regression line is quite a bit smaller than the variance around the mean. That is, 27.9 is very much smaller than 345.50. The standard error of the estimate $s_{Y|X}$ is used for statistical inference in regression analysis. ≡

The result in Example 12.6 is usually the case in the regression equation, since *the variation of Y around the regression line*, $\Sigma(Y_i - \hat{Y}_i)^2$, *is less than or equal to the variation around the mean*, $\Sigma(Y_i - \overline{Y})^2$.

This condition suggests a way of measuring the accuracy of a regression line's predictions. The accuracy has been defined in terms of the squared error of prediction. The most commonly used measure of this accuracy is called the sample **coefficient of determination r^2**, and it is defined as follows.

Definition and Equation for the Coefficient of Determination

The **coefficient of determination r^2** is the proportion of the total variation of Y that is explained by a linear relationship between Y and X and it is given by the following equation

$$r^2 = 1 - \frac{\Sigma(Y_i - \hat{Y}_i)^2}{\Sigma(Y_i - \overline{Y})^2} = 1 - \frac{SSE}{SST} = \frac{SSR}{SST} \qquad (12.22)$$

This formula needs a little explanation. The numerator of the fraction SSE/SST measures the squared error that remains in the predictions of Y when we use the sample regression equation. The denominator is the sum of squared deviations around the mean $\overline{Y}$ and can be thought of as the sum of the squared errors in predicting the Y_i values if we were to use $\overline{Y}$ rather than the regression equation as our predictor. Thus the fraction SSE/SST is the proportion of the squared errors of prediction around the mean that remains *unexplained* when we use the regression equation as a predictor. The r^2 value is 1.0 minus the proportion of *unexplained* squared error; thus, r^2 is the *proportion of squared error that the regression equation can explain, or eliminate*, when we use it as the predictor rather than use $\overline{Y}$ as a predictor.

The value of r^2 for the United Park City Properties condominium data is given in Computer Exhibit 12.1 as 0.939. We will use Equation (12.22) to find r^2 in the following example.

≡ **EXAMPLE 12.7** For the United Park City Properties example on condominium prices we find that

$$r^2 = \frac{SSR}{SST} = \frac{1298.2}{1382} = .939$$

The value of the sample coefficient of determination r^2 can be interpreted in the following way. One person looking at the condominium price data in Example 12.5 might forecast prices in this way: Every time someone asks the potential selling price of a condominium, he or she could say, "I suppose that $50,000 will be the selling price since that is the average selling price of the condominiums in my sample." Another person looking at the same data might forecast prices in this way: Every time someone asks the potential selling price of a condominium, he or she could say, "Tell me the number of square feet in the condominium, and I will plug that figure into the regression equation $\hat{Y} = -34.5 + 7.68X$ to get your predicted price." This second set of predictions would eliminate $r^2 = .939$ of the squared errors, or variation, in the first set of predictions. Thus, 93.9% of the squared errors, or variation, of prediction can be eliminated by using the regression equation to predict prices rather than using $\overline{Y}$ as the predictor.

The r^2 figure is a value that ranges between zero and unity. That is, if the regression equation is no better at predicting Y than the mean $\overline{Y}$, then no variation or error has been eliminated, and $r^2 = 0$. However, if the regression equation is a perfect predictor of the Y_i values, then all the Y_i values in the scatter diagram will lie right on the regression line, and all of the squared error of prediction has been eliminated. Thus, $r^2 = 1$. Values of r^2 between zero and unity indicate the relative strength of the relationship between the X variable and the Y variable in the regression equation. The r^2 values that are close to unity indicate that there is a strong relationship and that the regression equation will give relatively accurate predictions of the Y_i figures once the associated X value is known and entered into the regression equation.

Since r^2 is the fraction of the total variation in Y that is accounted for by the regression, $1 - r^2$ must be the fraction of the variation in Y that is unaccounted for: the fraction associated with the errors of prediction. The latter quantity is sometimes called the *coefficient of alienation*.

The r^2 value associated with a regression equation is generally used as a measure of the equation's accuracy regardless of the context of the problem.*

In the example presented to this point, the slope of the regression line has been positive. That is, large values of X have been associated with large values of Y, and the regression line's graph has sloped from lower left to upper right in scatter diagrams like Figure 12.3. However, many regression lines have a negative slope, like the one shown in Figure 12.6. This line slopes from upper left to lower right and indicates that large values of X are associated with small values of Y and vice versa. For example, in a particular industry high inventory levels might be associated with

---

*Numerous computer programs that perform simple regression show an output called *"Adjusted r^2."* This figure is computed by the following formula:

$$Adjusted\ r^2 = \frac{s_Y^2 - s_{Y|X}^2}{s_Y^2} = 1 - \frac{s_{Y|X}^2}{s_Y^2} = 1 - \frac{SSE/(n-2)}{SST/(n-1)}$$

In our condominium example, the *Adjusted r^2* value is

$$Adjusted\ r^2 = 1 - \frac{84.08/(5-2)}{1382/(5-1)} = .919$$

The adjustment between the r^2 and the *Adjusted r^2* is in the division of *SST* by $n-1$ and *SSE* by $n-2$. That is, we adjust for the degrees of freedom for *SSE* and *SST*. When n is large, the two r^2 values are very close.

FIGURE 12.6

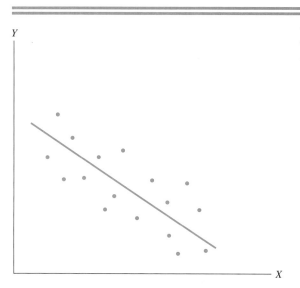

low profitability. In certain job classifications high pay might be associated with low employee turnover.

Problems: Section 12.4

10. Use the Intertechnology, Inc., data of Problem 1.
    a. Find $s_{Y|X}$.
    b. Find the coefficient of determination r^2.
    c. Explain the meaning of the coefficient of determination.
11. Refer to the Narco Medical Company data of Problem 2.
    a. Find $s_{Y|X}^2$.
    b. Find the coefficient of determination r^2.
12. Refer to the income and consumption expenditure data of Problem 3.
    a. Find $s_{Y|X}^2$.
    b. Find the coefficient of determination r^2.
13. In the equation $\hat{Y} = b_0 + b_1 X$, Y is the amount of production material used (in square feet). The X variable is the number of hours the production line is run. What are the units associated with b_0? With b_1? With $s_{Y|X}$?

12.5
INFERENCE AND PREDICTION IN REGRESSION ANALYSIS

In Example 12.3, we noted that the real estate investment firm would like to know the regression line showing the relationship between price and area for *all* the condominiums sold in the city. That is, the firm would like to know $\mu_{Y|X} = \beta_0 + \beta_1 X$, the population regression line. However, since the firm only has a sample taken from the population, we found the sample regression line $\hat{Y} = b_0 + b_1 X$. We also found the predicted price at which a condominium with $X = 13$ (1300 square feet of area) would sell: $65,360. However, this estimate was made by using the *sample* regression line, not the population regression line.

In Chapter 8, we discussed the topic of statistical inference and showed how sample results could be used to construct confidence intervals within which we think the population parameters will lie. Under the assumptions for the simple regression model, we can use our sample regression information to make inferences about what the population regression parameters will be. That is, we can answer the following questions:

1. What is the slope β_1 of the population regression line?
2. What will be the *mean* selling price $\mu_{Y|X_0}$ for condominiums that have an area of X_0?
3. What will be the selling price Y_0 for an *individual* condominium that has an area of X_0?

We discuss methods to answer these three questions about inference and prediction in the next three subsections. For statistical inference purposes the three assumptions listed in Section 12.1 must be met by the population data.

12.5.1 Hypotheses and Inferences Concerning β_1

To test hypotheses or make inferences concerning β_1, we must know the sampling distribution of b_1. The term *sampling distribution of b_1* refers to the different values of b_1 that would be computed from many different samples of Y values with fixed values for X. We shall present several different sample regression lines with different slopes (Figure 12.9). Under the assumptions of the regression model, the sampling distribution of b_1 is normal with mean $E(b_1) = \beta_1$ and standard deviation σ_{b_1}, which can be estimated by the **standard error of the sample regression slope** s_{b_1}. The value of s_{b_1} can be found by using the following equation.

Equation for Standard Error of Sample Regression Slope

$$s_{b_1} = \frac{s_{Y|X}}{\sqrt{SSX}} \qquad\qquad (12.23)$$

Recall that the numerator of Equation (12.23), or $s_{Y|X}$, is the standard error of the estimate.

If we wish to *test a hypothesis concerning the slope of the population's regression line,* we proceed as follows. Under the null hypothesis that the slope is equal to a given value β_{1_0}, that is, H_0: $\beta_1 = \beta_{1_0}$, the quantity

$$t = \frac{b_1 - \beta_{1_0}}{s_{b_1}} \qquad\qquad (12.24)$$

has a t distribution with $n - 2$ degrees of freedom.

A common hypothesis is that $\beta_1 = 0$. We want to know whether or not there is a linear association between the variables. If there is not, there is nothing to be gained by using the X's, since they would contribute nothing to the analysis or prediction of the Y's. Under this hypothesis, H_0: $\beta_1 = 0$, we have the following test.

Hypothesis Test of H_0: $\beta_1 = 0$ with a t Statistic

H_0: $\beta_1 = 0$

H_a: $\beta_1 \neq 0$

Rejection rule: If $t < -t_{\alpha/2, n-2}$ or $t > t_{\alpha/2, n-2}$, then reject H_0;

where: $t = \dfrac{b_1}{s_{b_1}}$ (12.25)

The quantity b_1/s_{b_1} is distributed as t with $n-2$ degrees of freedom only when β_1 is zero. In this situation the alternative hypothesis is usually that β_1 is not equal to zero. There are, of course, situations in which one is interested only in knowing whether or not the slope is greater than or equal to β_{1_0} or is less than or equal to β_{1_0}. In these circumstances we would use Equation (12.24) to compute the value of t, and the rejection rule for the single-sided alternative would be defined by using $t_{\alpha, n-2}$ or $-t_{\alpha, n-2}$ rather than $t_{\alpha/2, n-2}$.

EXAMPLE 12.8 To test the null hypothesis H_0: $\beta_1 = 0$ against the alternative H_a: $\beta_1 \neq 0$ for the United Park City Properties condominium problem of Example 12.2 with $\alpha = .05$, we recall from Example 12.4 that $SSX = 22$ and from Example 12.6 that $s_{Y|X} = 5.28$. Thus we calculate s_{b_1} as follows:

$$s_{b_1} = \frac{s_{Y|X}}{\sqrt{SSX}} = \frac{5.28}{\sqrt{22}} = 1.13$$

In the t table we find that $t_{\alpha/2, n-2} = t_{.05/2, 5-2} = t_{.025, 3} = 3.182$, and we have the following rejection rule:

Rejection rule: If $t < -3.182$ or $t > 3.182$, then reject H_0.

Now $b_1 = 7.68$, so we compute t as follows:

$$t = \frac{b_1}{s_{b_1}} = \frac{7.68}{1.13} = 6.8$$

Since $t = 6.8$ is greater than 3.182, we reject the hypothesis that β_1 is zero at the 5% level of significance (see Figure 12.7). We conclude that there is an underlying linear relationship.

The values for s_{b_1} and t can also be found on the printouts in Computer Exhibit 12.1.

An alternative way to test the hypothesis that two variables have no relationship to one another is to examine the mean square figures from the regression summary table. Table 12.2 showed that the mean square due to regression MSR was 1298.2 and had 1 degree of freedom, and the mean square due to error MSE was 27.9 and had 3 degrees of freedom. In the analysis of variance chapter, we found that we could test

FIGURE 12.7

t Test of H_0: $\beta_1 = 0$ Against H_a: $\beta_1 \neq 0$ for United Park City Properties

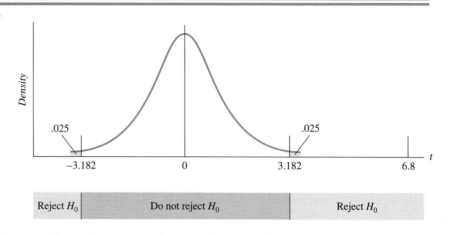

the significance of two mean squares by taking their ratio and comparing that with the *F* distribution. In the regression problem, the situation where there is no relationship between the variables being considered is characterized by a regression line of $\hat{Y} = \overline{Y}$. In that case, the mean square due to regression *MSR* is zero since the sum of squares of the regression line values about the mean is zero, because the regression line and the mean line are the same line. The larger the mean square due to regression *MSR*, the more likely is a significant relationship between the two variables, so $\beta_1 \neq 0$. The ratio of the mean square due to regression *MSR* to the mean square due to error *MSE* is an *F*-distributed variable with $\nu_1 = 1$ and $\nu_2 = n - 2$ under the hypothesis of no relationship. Thus, the *F* value computed from the data is

$$F = \frac{MSR}{MSE}$$

If this value exceeds $F_{\alpha, 1, n-2}$, we reject the hypothesis H_0: $\beta_1 = 0$ and conclude that there is a significant linear relationship between *Y* and *X*, as summarized below. The *F* and *t* tests of H_0: $\beta_1 = 0$ against H_a: $\beta_1 \neq 0$ are equivalent in simple regression analysis. However, the *F* and *t* tests are used for different purposes in multiple regression in Chapter 13.

The procedure for testing H_0: $\beta_1 = 0$ against H_a: $\beta_1 \neq 0$ with an *F* statistic is summarized as follows.

> **Hypothesis Test of H_0: $\beta_1 = 0$ Against H_a: $\beta_1 \neq 0$ with an F Statistic**
>
> $$H_0: \beta_1 = 0$$
> $$H_a: \beta_1 \neq 0$$
>
> Rejection rule: If $F > F_{\alpha, 1, n-2}$, then reject H_0;
>
> where: $F = \dfrac{MSR}{MSE}$ (12.26)

FIGURE 12.8

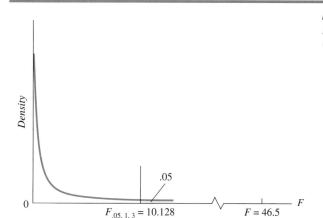

F Test of H_0: $\beta_1 = 0$ Against H_a: $\beta_1 \neq 0$ for United Park City Properties

EXAMPLE 12.9   Use the F statistic to test the null hypothesis H_0: $\beta_1 = 0$ against the alternative H_a: $\beta_1 \neq 0$ for the United Park City Properties condominium problem of Example 12.2 with $\alpha = .05$.

Now, $\nu_1 = 1$ and $\nu_2 = n - 2 = 5 - 2 = 3$, so in the F table we find that $F_{\alpha, 1, n-2} = F_{.05, 1, 3} = 10.128$, and we have the following rejection rule.

Rejection rule: If $F > 10.128$, then reject H_0.

To compute F, we find the mean squares in summary Table 12.3 and calculate

$$F = \frac{MSR}{MSE} = \frac{1298.2}{27.9} = 46.5$$

Therefore since $F = 46.5$ exceeds 10.128, we reject the hypothesis that β_1 is zero at the 5% level of significance (see Figure 12.8). We conclude that there is an underlying linear relationship, just as we did by using the t test.

Notice that the value for F can be found on the printouts in Computer Exhibit 12.1.

If we want a **confidence interval estimate for the slope** of the population regression equation β_1 with a confidence level of $100(1 - \alpha)\%$, we form the interval defined by the limits of Equation (12.27).

Equations for Limits of the Confidence Interval Estimate for the Slope β_1

$$L = b_1 - t_{\alpha/2, n-2} s_{b_1}$$
$$R = b_1 + t_{\alpha/2, n-2} s_{b_1}$$

(12.27)

12.5.2 Confidence Interval for a Mean Value of Y Given a Value X_0

Let us now turn our attention to the next question to be examined in this section: What will be the mean value of Y for some specific estimation using X_0? Rephrased in terms of our initial example, this question becomes, What will be the mean selling price $\mu_{Y|X_0}$ for a condominium that has area X_0? The *point estimate* of the mean value of Y for a condominium that has area X_0, that is, $\mu_{Y|X_0}$, is $\hat{Y}_0 = b_0 + b_1 X_0$. We find the **confidence interval** for the mean value of Y given a value X_0 by using the following usual form.

Expression for Limits of the Confidence Interval for Mean Value of Y Given X_0

$$\hat{Y}_0 \pm t_{\alpha/2,\, n-2} s_{\hat{Y}_0} \tag{12.28}$$

The **standard error of $\hat{Y}_0$**, $s_{\hat{Y}_0}$, has a rather complicated formula.

Equation for Standard Error of $\hat{Y}_0$

$$s_{\hat{Y}_0} = s_{Y|X} \sqrt{\frac{1}{n} + \frac{(X_0 - \overline{X})^2}{SSX}} \tag{12.29}$$

The formula for $s_{\hat{Y}_0}$ is an estimate of how the regression line's estimate, $\hat{Y}_0$, would be scattered around the population value, $\mu_{Y|X_0}$, if many sample regression lines were constructed. The variation in $\hat{Y}_0$ from the true $\mu_{Y|X_0}$ at X_0 for several different sample regression lines is depicted in Figure 12.9.

The formula for $s_{\hat{Y}_0}$ has two terms in it. If we move $s_{Y|X}$ under the square root sign, the formula becomes

$$s_{\hat{Y}_0} = \sqrt{\frac{s^2_{Y|X}}{n} + \frac{s^2_{Y|X}(X_0 - \overline{X})^2}{SSX}} \tag{12.30}$$

The term on the far right under the square root sign indicates the error in estimation that could be introduced by the possibility of the sample regression line's slope being in error. Note that the $(X_0 - \overline{X})^2$ factor in this term implies that the error in estimation gets larger for X values that are further from the mean $\overline{X}$. That is, if the sample regression line's slope b_1 is different from the population slope β_1, the estimated values out toward the ends of the sample regression line get "whipped" far from the actual values that would be given by the population regression line $\mu_{Y|X} = \beta_0 + \beta_1 X$, if we knew that line's equation.* Figure 12.10 illustrates this point by showing that

---

*Technically, the regression line has no "ends," but in practical regression problems the highest and lowest values of X in the sample data form a relevant range for X and indicate approximately where we would locate "ends" of the regression line. Extrapolating beyond the relevant range for X for prediction can be problematic.

FIGURE 12.9

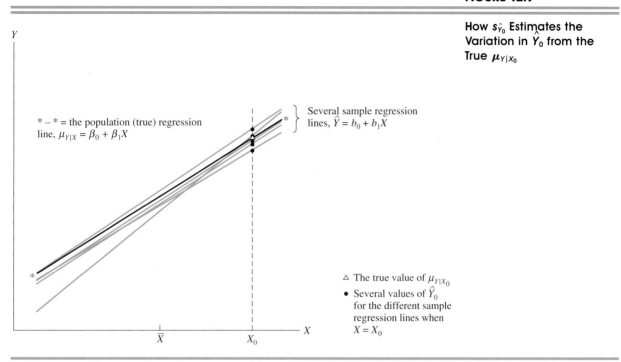

How $s_{\hat{Y}_0}$ Estimates the Variation in $\hat{Y}_0$ from the True $\mu_{Y|X_0}$

the width of the confidence interval defined in Equation (12.28) is narrower for X values close to $\overline{X}$ and wider at values of X far from $\overline{X}$ (the values are for the condominium example, as described in Practice Problem 12.1).

The first term under the square root sign in Equation (12.30) indicates the prediction error that could be introduced by the possibility of the sample regression line's being located at the wrong mean height along the Y axis, even if the slope were correct. That is, even if the sample slope b_1 were exactly equal to the population slope β_1, the intercept value might be in error. The $s_{Y|X}^2/n$ term is a measure of that error's contribution to the error in predicting $\mu_{Y|X_0}$, using the sample regression line with some specific X_0 value substituted to give $\hat{Y}_0 = b_0 + b_1 X_0$.

 PRACTICE PROBLEM 12.1

The United Park City Properties real estate investment firm described in Example 12.4 plans to build condominiums with $X_0 = 13$, or area of 1300 square feet. Find a 95% confidence interval for the mean selling price of 1300-square-foot condominiums in this city.

Solution In Example 12.4 we found that the sample regression equation predicts a selling price of $\hat{Y} = -34.5 + (7.68)(13) = 65.36$, or \$65,360. We can find the standard error of this estimate by using Equation (12.21) and values that can be obtained from the original example:

$$n = 5 \qquad \overline{X} = 11 \qquad SSX = 22$$
$$s_{Y|X} = \sqrt{MSE} = \sqrt{27.9} = 5.28$$

FIGURE 12.10

Confidence Intervals for a
Mean Value of Y Given
$X = X_0$

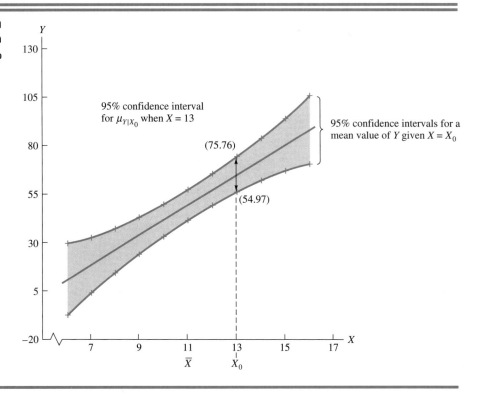

So

$$s_{\hat{Y}_0} = s_{Y|X}\sqrt{\frac{1}{n} + \frac{(X_0 - \overline{X})^2}{SSX}} = 5.28\sqrt{\frac{1}{5} + \frac{(13 - 11)^2}{22}} = 3.27$$

Now we can find the 95% confidence interval for the mean selling price by using Expression (12.28) and $t_{.025, 3} = 3.182$:

$$\hat{Y}_0 \pm t_{\alpha/2, n-2}s_{\hat{Y}_0} \text{ is } 65.36 \pm (3.182)(3.27)$$

or

54.97   to   75.76

This interval is depicted in Figure 12.10, and you can find this interval on the printout in Computer Exhibit 12.1A.

Thus, the real estate investment firm can be 95% sure that the mean selling price for condominiums with 1300 square feet in area will lie between \$54,970 and \$75,760. This interval is quite wide and does not provide a very precise range, but part of the problem lies in the facts that (1) the sample size of $n = 5$ is rather small and (2) only one variable, area of the condominium, was used to predict the price.

12.5.3 Prediction Interval for a New Individual Value Y_0 Given a Value X_0

The final question to be examined in this section is, What will be the *individual* value Y_0 for some specific prediction for a new and given X_0? Rephrased in terms of our initial example, this question becomes, What will be the selling price Y_0 of another independently selected individual condominium that has area X_0? Notice that this problem differs from the problem considered in the previous subsection, because we are now interested in the selling price of an additional *individual* (or single) condominium with area X_0, not the *mean* selling price of all condominiums with area X_0.

The **prediction interval** for an individual value Y_0 for a new and given value X_0 is found by using the following familiar form.

Expression for Limits of the Prediction Interval for a New Individual Value Y_0 Given X_0

$$\hat{Y}_0 \pm t_{\alpha/2,\, n-2} s_{Y_0} \tag{12.31}$$

where: $\hat{Y}_0 = b_0 + b_1 X_0$

The sample standard deviation s_{Y_0} when predicting an individual value Y_0 given X_0 is as follows.

Equation for Sample Standard Deviation for Predicting a New Individual Value Y_0 Given X_0

$$s_{Y_0} = s_{Y|X} \sqrt{1 + \frac{1}{n} + \frac{(X_0 - \overline{X})^2}{SSX}} \tag{12.32}$$

The variance for predicting an individual Y_0 must include the variance of the individual points about the population regression line in addition to the variation due to the many possible sample regression lines. Hence the additional term, which is equal to 1, is included beneath the square root for the standard deviation given in Equation (12.32).

Since the standard deviation for predicting an individual value Y_0 is larger than the standard deviation for estimating a mean value of Y for a given X_0, the prediction interval for an individual Y_0 is wider than the confidence interval for a mean value of Y given the same value X_0. Prediction intervals for an individual value Y_0 and confidence intervals for a mean value of Y for our condominium example are depicted in Figure 12.11.

 PRACTICE PROBLEM 12.2

A buyer plans to purchase an *individual* condominium with $X_0 = 12$, or area of 1200 square feet, from the real estate firm described in Example 12.4. Find a 95% pre-

FIGURE 12.11

Prediction Intervals for an
Individual Value of Y Given
$X = X_0$ and Confidence Intervals for a Mean Value of
Y Given $X = X_0$

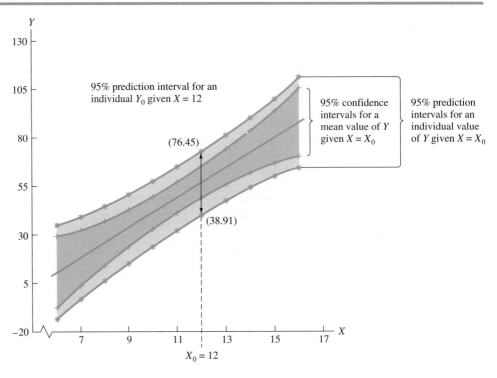

diction interval for the selling price of an individual 1200-square-foot condominium
in this city.

Solution In Example 12.4, we found the sample regression equation for predicting
selling price to be

$$\hat{Y} = -34.5 + 7.68X$$

So for the given area $X_0 = 12$,

$$\hat{Y}_0 = -34.5 + (7.68)(12) = 57.68$$

or $57,680.

We can find the sample standard deviation for the individual selling price by using
Equation (12.32) and values from the original example:

$$n = 5 \qquad \overline{X} = 11 \qquad SSX = 22 \qquad s_{Y|X} = 5.28$$

So

$$s_{Y_0} = s_{Y|X} \sqrt{1 + \frac{1}{n} + \frac{(X_0 - \overline{X})^2}{SSX}}$$

$$= 5.28 \sqrt{1 + \frac{1}{5} + \frac{(12 - 11)^2}{22}} = 5.90$$

Now we can find the 95% prediction interval for the selling price of an individual condominium by using Expression (12.31) and $t_{.025, 3} = 3.182$:

$$\hat{Y}_0 \pm t_{\alpha/2, n-2} s_{Y_0} \text{ is } 57.68 \pm (3.182)(5.90)$$

or

38.91 to 76.45

This interval is depicted in Figure 12.11, and it is given on the printout in Computer Exhibit 12.1A. Thus the buyer can be 95% confident that the selling price of an individual condominium with a 1200-square-foot area will be between $38,910 and $76,450.

Remember that the confidence and prediction intervals and hypothesis tests demonstrated in this section are valid only if the variables involved exhibit the three characteristics assumed at the beginning of the section. That is, the X values must be nonrandom, the Y values must be normally and independently distributed around the regression line, and the variance around the regression line must be the same regardless of the X values. If these conditions are not met, the least squares regression line may still be a somewhat accurate predictor of the Y_i values, but the confidence intervals and hypothesis tests described in this section may not be valid.

Problems: Section 12.5

14. Use the Intertechnology, Inc., data given for Problem 1.
    a. Find the standard deviation of the regression line's slope, s_{b_1}.
    b. Test the hypothesis $H_0: \beta_1 = 0$ against $H_a: \beta_1 \neq 0$ by using a t statistic. Let $\alpha = .05$.

15. Use the Intertechnology, Inc., data for Problem 1.
    a. Find $s_{\hat{Y}_0}$ when $X_0 = 8$.
    b. Find the 95% confidence interval estimate for the mean value of Y given $X_0 = 8$.
    c. Find s_{Y_0} when $X_0 = 8$.
    d. Find the 95% prediction interval for the individual value Y_0 given that $X_0 = 8$.

16. Use the Narco Medical Company data given for Problem 2.
    a. Find the standard deviation of the regression line's slope, s_{b_1}.
    b. Test the hypothesis $H_0: \beta_1 = 0$ against $H_a: \beta_1 \neq 0$ by using a t statistic. Let $\alpha = .05$.
    c. Test the hypothesis $H_0: \beta_1 = 0$ against $H_a: \beta_1 \neq 0$ by using an F statistic. Let $\alpha = .05$.

17. Use the income and consumption expenditure data given in Problem 3.
    a. Find the standard deviation of the regression line's slope, s_{b_1}.
    b. Test the hypothesis $H_0: \beta_1 = 0$ against $H_a: \beta_1 \neq 0$ by using a t statistic. Let $\alpha = .05$.
    c. Find $s_{\hat{Y}_0}$ when $X_0 = 6$.
    d. Find the 95% confidence interval estimate for the mean value of Y given $X_0 = 6$.
    e. Find s_{Y_0} when $X_0 = 6$.
    f. Find the 95% prediction interval for the individual value Y_0 given that $X_0 = 6$.

18. List the assumptions needed for hypothesis tests and confidence interval estimation in simple regression analysis.

19. In a regression problem, $n = 30$, $\Sigma X_i = 15$, $\Sigma Y_i = 30$, $SCP = -30$, $SSX = 10$, and $SST = 160$.

    a. Find the regression equation $\hat{Y} = b_0 + b_1 X$.
    b. Estimate the variance $\sigma_{Y|X}^2$.
    c. Test H_0: $\beta_1 = 0$ against H_a: $\beta_1 \neq 0$. Let $\alpha = .05$.

20. In a regression analysis, $n = 25$, $\Sigma X_i = 75$, $\Sigma Y_i = 50$, $\Sigma X_i^2 = 625$, $\Sigma X_i Y_i = 30$, and $\Sigma Y_i^2 = 228$.

    a. Find the regression equation.
    b. Find $s_{Y|X}^2$ and $s_{b_1}^2$.
    c. Test that $\beta_1 = 0$. Let $\alpha = .01$. Use H_a: $\beta_1 \neq 0$.

21. Use $n = 41$, $\overline{Y} = 10$, $\overline{X} = 12$, $SSX = 400$, $SCP = 100$, and $SST = 64$.

    a. Find the predicted value of Y given $X = 8$.
    b. Find a 99% confidence interval for a mean observation on Y given $X_0 = 8$.
    c. Find a 99% prediction interval for an individual value Y_0 given $X_0 = 8$.

12.6

CORRELATION

In the preceding sections of this chapter we have been concerned with the use of regression techniques to estimate the parameters of an assumed linear relation between X and the mean of Y given X. We assumed that the values of X were known, and we allowed them to be selected and controlled by the experimenter; that is, we did not assume that X was a random variable.

We now consider methods that are appropriate when we assume that X and Y are both random variables and have a joint distribution. We want to make inferences about the degree of linear relationship between them without estimating the regression line.

One of the measures of a joint distribution of X and Y is the *product moment correlation coefficient* or simply the **correlation coefficient ρ** (Greek rho).

Definition

The **correlation coefficient ρ** is a measure of the direction and the strength of linear association between two random variables. It is dimensionless, and it may take any value between -1 and 1, inclusive.

If ρ is either -1 or 1, the variables have a perfect linear relationship in that all of the points lie exactly on a straight line. If ρ is near -1 or 1, there is a high degree of linear association.

A **positive correlation** means that as one variable increases, the other likewise increases. A **negative correlation** means that as one variable increases, the other decreases. Heights and weights of humans are positively correlated, but the age of a car and its trade-in value are negatively correlated. If ρ is equal to zero for two variables, then we say that the variables are *uncorrelated* and that there is no linear association between them. Bear in mind that ρ measures only linear relationship. The variables may be perfectly correlated in a curvilinear relationship, but ρ can still equal zero.

The joint distribution for two random variables that is considered in this section is the **bivariate normal distribution.** The population correlation coefficient ρ is a parameter of this distribution, as are the means μ_X and μ_Y and the standard deviations σ_X and σ_Y. Since the two random variables X and Y are included in a bivariate normal distribution, the density $f(X, Y)$ forms a third dimension.*

A bivariate normal distribution with $\rho = .8$, $\mu_X = 0$, $\mu_Y = 0$, $\sigma_X = 1$, and $\sigma_Y = .75$ is shown as a surface in Figure 12.12. For the joint distribution of X and Y to be bivariate normal, then, among other things, the marginal or univariate distribution for X must be normal and the marginal or univariate distribution for Y must be normal. The normal distributions for X and Y are shown in the $(X, Density)$ and $(Y, Density)$ planes in Figure 12.12. If a plane that is parallel to the (X, Y) plane intersects the bivariate normal surface, then the intersection is an ellipse. The ellipse can be projected onto the (X, Y) surface and is called a *concentration ellipse,* or an *isodensity ellipse.* The size of the ellipse can be selected so that the ellipse will be expected to encompass a given percentage of the points (X_i, Y_i) in random samples and larger ellipses will encompass a larger percentage of the points. Four isodensity ellipses are shown on the (X, Y) plane in Figure 12.12.

*The bivariate normal density function, which has similarities with the univariate normal density function, is

$$f(X, Y) = \frac{1}{2\pi\sigma_X\sigma_Y\sqrt{1 - \rho^2}} \exp\left\{ -\frac{1}{2(1 - \rho^2)} \left[\left(\frac{X - \mu_X}{\sigma_X} \right)^2 + \left(\frac{Y - \mu_Y}{\sigma_Y} \right)^2 - 2\rho \left(\frac{X - \mu_X}{\sigma_X} \right) \left(\frac{Y - \mu_Y}{\sigma_Y} \right) \right] \right\}$$

FIGURE 12.12

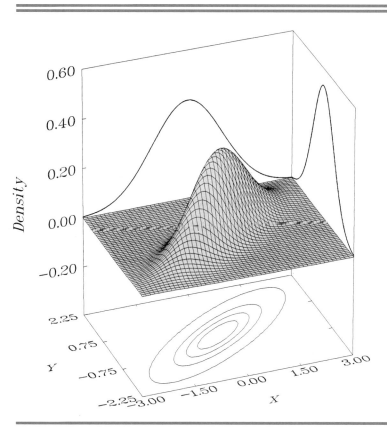

Bivariate Normal Distribution, Univariate Normal Distributions for X and Y, and Ellipses with $\rho = .8$, $\mu_X = 0$, $\mu_Y = 0$, $\sigma_X = 1$, and $\sigma_Y = .75$

FIGURE 12.13   Isodensity Ellipses and Samples of Points for Nine Bivariate Normal Distributions

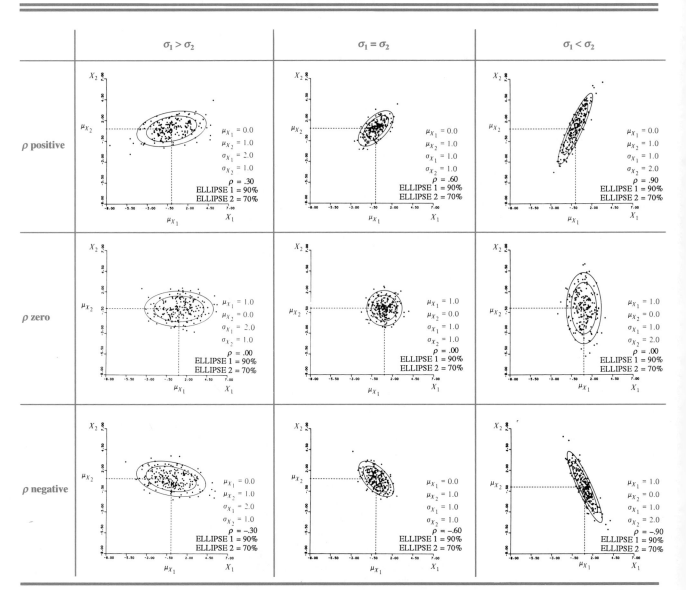

Since the value of ρ can be negative, zero, or positive, and since σ_X can be less than, equal to, or greater than σ_Y, Figure 12.13 shows nine forms for bivariate normal distributions or for scatter diagrams of points (X_i, Y_i) sampled from bivariate normal distributions, with the two random variables labeled X_1 and X_2, and a sample of 200 points (X_{i1}, X_{i2}) from each distribution. The largest ellipse in each plot is expected to encompass about 90% of samples of points, for example, so that about 20 of the 200 sample points should fall *outside* this ellipse.

The means μ_{X_1} and μ_{X_2} locate the centroids of the distributions (or ellipses). The orientations and spreads of the distributions are determined by the correlation coefficient ρ and the standard deviations σ_{X_1} and σ_{X_2}.

The value of ρ increases from .3, to .6, to .9 in the top row of plots in Figure 12.13, and it changes from $-.3$, to $-.6$, to $-.9$ in the bottom row of plots. The figure shows that as the correlation coefficient ρ increases (in absolute value), the points get closer to a line, and the *degree* of the linear association between the two random variables X_1 and X_2 increases. The positive or negative sign of the correlation coefficient indicates the *direction* of the relationship between the two variables.

The population correlation coefficient ρ, which measures the correlation between two random variables X and Y, can be obtained through a value called the **covariance** of X and Y. The population covariance of N ordered pairs of data (X_i, Y_i) is

$$\sigma_{XY} = \text{Cov}(X, Y) = \frac{\sum_{i=1}^{N} (X_i - \mu_X)(Y_i - \mu_Y)}{N} \qquad (12.33)$$

If the two variables move together, then the covariance is positive. However, if the two variables move in directly opposite directions, then the covariance is a negative value. Note that the population covariance is computed somewhat like the variance, which is denoted by σ^2. Unlike the variance, however, a covariance can be negative. If the two variables are totally uncorrelated, then the covariance is zero.

Equation (12.33) defines the population covariance and assumes that the population is composed of N pairs that we have under examination. However, when the pairs we are examining form a sample, then we estimate the population covariance with the **sample covariance** s_{XY}:

$$s_{XY} = \widehat{\text{Cov}}(X, Y) = \frac{\sum_{i=1}^{n} (X_i - \overline{X})(Y_i - \overline{Y})}{n - 1} = \frac{SCP}{n - 1} \qquad (12.34)$$

where $\widehat{\text{Cov}}(X, Y)$ denotes the estimated covariance of X and Y. The divisor $n - 1$ in Equation (12.34) is the number of degrees of freedom.

The covariance of pairs of values is a measure of how much they vary together in that a "big" positive covariance indicates that the variables move together, a "big" negative covariance indicates that they move inversely, and a "small" covariance indicates that they are uncorrelated with one another. However, the covariance's size depends on the units in which the variables X and Y are measured. In the real estate example of Section 12.2, the units of the covariance are square feet ($\times$ 100)–dollars ($\times$ 1000), since X was measured in terms of square feet ($\times$ 100) and Y was price in dollars ($\times$ 1000). We could change the size of the covariance by expressing X in terms of square yards and Y in terms of cents. The covariance needs to be adjusted to measure the degree of correlation between two variables. If we divide the covariance by the standard deviations of X and Y, then we obtain the dimensionless population correlation coefficient ρ as

$$\rho = \frac{\sigma_{XY}}{\sigma_X \sigma_Y} \qquad (12.35)$$

The **sample correlation coefficent** r is an estimator for ρ.

Equation for Sample Correlation Coefficient

$$r = \frac{s_{XY}}{s_X s_Y} = \frac{SCP}{\sqrt{(SSX)(SSY)}} \qquad (12.36)$$

The notation SSY in Equation (12.36) is the same as SST defined in Equation (12.13); thus SSY can be computed as

$$SSY = SST = \sum(Y_i - \overline{Y})^2 = \sum Y_i^2 - \frac{(\sum Y_i)^2}{n}$$

The r^2 value defined in Equation (12.22) is algebraically the square of the r value defined in Equation (12.36).

▆▆▆ PRACTICE PROBLEM 12.3

An economist was interested in studying the relationship between the way families spend tax refunds and the size of their incomes. Thus he took a sample of six families and looked at $X =$ annual family income and $Y =$ the percentage of the tax refund spent within three months of receipt. Table 12.4 shows the actual data for the families and the statistics used in a correlation problem. Find the correlation coefficient by using both forms of Equation (12.36).

Solution Equation (12.36) requires that we find the covariance of X and Y and the variance of X and of Y. Using the sample data, we obtain

$$s_{XY} = \frac{SCP}{n-1} = \frac{-1280}{6-1} = -256$$

$$s_X^2 = \frac{SSX}{n-1} = \frac{456}{6-1} = 91.2$$

$$s_Y^2 = \frac{SST}{n-1} = \frac{5150}{6-1} = 1030$$

	Percentage of Refund Spent, Y	Family Income (× $1000), X	Y^2	X^2	XY
TABLE 12.4					
Income and Tax Refund	70	13	4,900	169	910
Expenditure Data and	55	18	3,025	324	990
Calculations	100	9	10,000	81	900
	40	25	1,600	625	1000
	15	36	225	1296	540
	20	19	400	361	380
Total	300	120	20,150	2856	4720

$$\overline{Y} = \frac{300}{6} = 50 \qquad \overline{X} = \frac{120}{6} = 20$$

$$SCP = \sum XY - \frac{\sum X \sum Y}{n} = 4720 - \frac{(120)(300)}{6} = -1280$$

$$SSX = \sum X^2 - \frac{(\sum X)^2}{n} = 2856 - \frac{(120)^2}{6} = 456$$

$$SSY = SST = \sum Y^2 - \frac{(\sum Y)^2}{n} = 20150 - \frac{(300)^2}{6} = 5150$$

Applying Equation (12.36), we find the value

$$r = \frac{s_{XY}}{\sqrt{s_X^2 s_Y^2}} = \frac{-256}{\sqrt{(91.2)(1030)}} = -.835$$

We can apply the rightmost form of Equation (12.36) by using the data directly from Table 12.4 without going through the covariance and variances first:

$$r = \frac{SCP}{\sqrt{(SSX)(SSY)}} = \frac{-1280}{\sqrt{(456)(5150)}} = -.835$$

Thus, regardless of which formula is used, $r = -.835$. So the two variables, family income and percent of tax refund spent in the first three months, have a rather strong negative correlation. That is, higher-income families tend to spend a lower percentage of their refund in the first three months than low-income families do. This negative correlation is one of the reasons that economists recommend that tax cuts be aimed at lower-income groups—they spend a higher percentage of the tax savings and give the economy a substantial boost. ▬▬

Simple regression techniques and correlation methods are related. In correlation, r is an estimator for the population correlation coefficient ρ. In regression, r^2 is simply a measure of closeness of fit.* Thus the sample correlation coefficient r is used to estimate the direction and the strength of the relationship between two random variables, whereas the coefficient of determination r^2 is the proportion of the squared error that the regression equation can explain when we use the regression equation rather than the sample mean as a predictor.

If we want to test the hypothesis that the variables are not linearly related, that is, that ρ is 0, we may use an ordinary t test, where

$$t = \frac{r}{\sqrt{(1 - r^2)/(n - 2)}} \tag{12.37}$$

When ρ is 0, the t statistic has the t distribution with $n - 2$ degrees of freedom. To test $H_0: \rho = 0$, in a two-sided test we merely evaluate t by Equation (12.37) and compare it with the tabular t value for the given probability of type I error as summarized next.

Hypothesis Test of H_0: $\rho = 0$ Against H_2: $\rho \neq 0$

H_0: $\rho = 0$

H_a: $\rho \neq 0$

Rejection rule: If $t < -t_{\alpha/2, n-2}$ or $t > t_{\alpha/2, n-2}$, then reject H_0;

where: $t = \dfrac{r}{\sqrt{(1 - r^2)/(n - 2)}}$ $\qquad$ (12.37)

---

*If Y and X are both random variables, then the regression analysis results discussed previously are applicable if the distribution for Y_i given X_i is normal with mean $\beta_0 + \beta_1 X_i$ and variance $\sigma_{Y|X}^2$ and if the probability distribution of X_i is independent from β_0, β_1 and $\sigma_{Y|X}^2$. If Y and X are bivariate normally distributed, then the regression results still apply; however, there are two regressions, one for Y regressed on X and one for X regressed on Y.

PRACTICE PROBLEM 12.4

Use the data in Practice Problem 12.3 to test the hypothesis that there is no correlation between the size of a family's income and the percentage of a tax refund spent in the first three months after receiving it. Use, as the alternative hypothesis, that there is no correlation between these two variables. Let $\alpha = .05$.

Solution The hypothesis test is as follows.

Step One $H_0: \rho = 0$.

Step Two $H_a: \rho \neq 0$.

Step Three $n = 5$, $\alpha = .05$.

Step Four The value of the degrees of freedom is $n - 2 = 6 - 2 = 4$, and the critical value of t from the table is

$$t_{.025, 4} = 2.776$$

Thus, the rejection rule is

Rejection rule: If $t < -2.776$ or $t > 2.776$, then reject H_0.

Step Five Using Equation (12.37) to calculate t, we obtain

$$t = \frac{-.835}{\sqrt{(1 - (-.835)^2)/(6 - 2)}} = \frac{-.835}{.2751} = -3.03$$

Step Six Since -3.03 is less than -2.776 (see Figure 12.14), we reject H_0 and conclude that there is a negative correlation between the size of a family's income and the percentage of a tax refund spent in the first three months after receiving it.

The method just illustrated and Equation (12.37) cannot be used if the hypothesis is that the population correlation coefficient is something other than zero. That is,

FIGURE 12.14

t Test of $H_0: \rho = 0$ Against $H_a: \rho \neq 0$ for Income and Tax Refund Expenditure

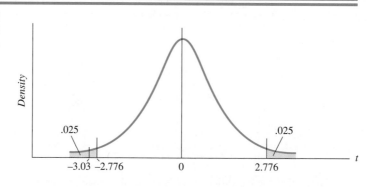

other methods are needed to test a null hypothesis such as H_0: $\rho = .6$. These methods will not be treated here.

Problems: Section 12.6

22. For the following set of quantities, find the sample correlation coefficient. Test the hypothesis that $\rho = 0$. Let $\alpha = .05$. Use H_a: $\rho \neq 0$.

$$n = 11 \qquad SSY = 400 \qquad SCP = 400 \qquad SSX = 625$$

23. From the following information, find the sample correlation coefficient. Test the hypothesis that $\rho = 0$. Let $\alpha = .01$ and H_a: $\rho \neq 0$.

$$n = 29 \qquad SSY = 64 \qquad SCP = 40 \qquad SSX = 100$$

24. Records were kept at Carter Wallace, Inc., on the scores received by 14 job applicants on a manual-dexterity test and their production output after one week on the job.

   a. Use the accompanying data to estimate the correlation between the test scores and production output.
   b. Is there a significant correlation at the .05 level?

Score	Output	Score	Output
124	120	66	84
84	103	31	30
13	16	43	62
13	20	19	26
48	86	117	121
61	36	50	93
112	153	72	83

25. Data were collected for six cities on their populations (in millions) and the market value of the cities' real estate (in billions of dollars), as given in the following listing:

Population	4	6	5	5	6	4
Real Estate Value	1	5	4	4	7	3

   a. Find the sample correlation between population and real estate value.
   b. Test the significance of the correlation at the $\alpha = .05$ level.

26. A certified public accountant with Coopers & Lybrand wonders whether percentages of long-term assets and long-term debt from common-size balance sheets are associated for corporations. A sample of data was collected for 12 corporations and is given in the accompanying table.

   a. Find the sample correlation between long-term asset and long-term debt balance sheet percentages.
   b. Test the significance of the correlation at the $\alpha = .05$ level.

Long-Term Assets	Long-Term Debt	Long-Term Assets	Long-Term Debt
54	28	47	30
47	26	69	38
60	39	62	43
56	43	45	24
64	24	48	36
26	16	39	20

REFERENCE: Stowe, John, Collin J. Watson, and Terry D. Robertson (1980), "Relationships Between the Two Sides of the Balance Sheet: A Canonical Correlation Analysis," *Journal of Finance,* 35 (September): 973–80.

FIGURE 12.15

Isodensity Ellipses and
Sample Points

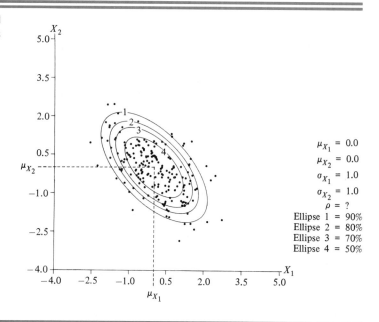

27. Figure 12.15 shows isodensity ellipses and a sample of points for a bivariate normal distribution. What is your estimate of the correlation coefficient for this distribution?

12.7
SUMMARY

We have studied the relationship between the two variables Y and X in this chapter. *Regression analysis* was used to obtain the simple linear regression equation $\hat{Y} = a + bX$ for a sample of data. Partitioning the sum of squares in regression provided us with a summary table that gives information used to measure the accuracy of prediction, to test hypotheses, and to perform estimation in regression analysis.

Correlation analysis was used to measure the direction and strength of the relationship between two random variables. This measurement for a sample of data gave us a numerical form called the *correlation coefficient r*. Correlation problems are often concerned simply with determining whether there is a relationship and not with obtaining a prediction equation. In many problems both regression analysis and correlation analysis are used. That is, an equation relating two variables is obtained, and the strength of the relationship between the variables is measured.

Numerous formulas have been presented in this chapter. Some of the more important ones are summarized in Table 12.5. Note that the square of r from Equation (12.36) is mathematically equivalent to the r^2 of Equation (12.22). But the interpretation and uses of r^2 in a regression problem are somewhat different from the interpretation of r in a correlation problem. The value of r^2 computed from Equation (12.22) in a regression problem is used to measure the goodness of fit of the data points around a regression line. In a correlation problem, however, a regression line is usually not even constructed. Thus, the r value computed from Equation (12.36) in a correlation problem is used to measure the direction and strength of the linear relationship between two variables.

TABLE 12.5 Summary of Formulas for Regression and Correlation

Formula		Comments
$b_1 = \dfrac{SCP}{SSX}$	(12.7)	—b_1 and b_0 are the *slope* and *intercept* of the least squares sample regression line $\hat{Y} = b_0 + b_1 X$.
$b_0 = \overline{Y} - b_1 \overline{X}$	(12.8)	—*SCP* is the *sum of cross products of deviations* for X and Y, and *SSX* is the *sum of squared deviations* for X.
where: $SCP = \sum XY - \dfrac{\sum X \sum Y}{n}$	(12.9)	
$SSX = \sum X^2 - \dfrac{(\sum X)^2}{n}$	(12.10)	
$SST = \sum Y^2 - \dfrac{(\sum Y)^2}{n}$	(12.13)	—*SST* is the *total sum of squares*, *SSR* is the *sum of squares* due to or explained by *regression*, and *SSE* is the *sum of squares* due to *error*.
$SSR = (b_1)(SCP)$	(12.14)	—*SST* has $n - 1$ degrees of freedom, *SSR* has 1 degree of freedom, and *SSE* has $n - 2$ degrees of freedom in simple regression.
$SSE = SST - (b_1)(SCP)$	(12.15)	
$MSR = \dfrac{SSR}{1}$	(12.17)	
$MSE = \dfrac{SSE}{n-2}$	(12.18)	
$s_{Y\mid X}^2 = \dfrac{1}{n-2}\sum(Y_i - \hat{Y}_i)^2$		—This value measures the dispersion of the Y_i values around the regression line. The s_Y^2 value measures the dispersion of the Y_i values around the mean $\overline{Y}$.
$\qquad = \dfrac{1}{n-2}(SSE) = MSE$	(12.20)	
$s_{Y\mid X} = \sqrt{s_{Y\mid X}^2} = \sqrt{MSE}$	(12.21)	—$s_{Y\mid X}$ is the *standard error of the estimate* or the standard deviation of the Y_i values around the regression line.
$r^2 = 1 - \dfrac{\sum(Y_i - \hat{Y}_i)^2}{\sum(Y_i - \overline{Y})^2}$		—r^2 varies between 0 and 1 and measures the relative strength of the relationship in X and Y.
$\qquad = 1 - \dfrac{SSE}{SST} = \dfrac{SSR}{SST}$	(12.22)	—The correlation coefficient r is equal to $\sqrt{r^2}$ with the same algebraic sign as the slope b_1.

(continues)

The calculations in regression analysis and correlation problems can become very tedious, especially if n, the number of pairs of Y and X, is large. Thus computer programs have been prepared that will aid the user in problems of this type. Edited computer printouts for several regression analysis programs, along with labels that identify some of the values that are on the printouts, are presented with the example

TABLE 12.5   Summary of Formulas for Regression and Correlation (Continued)

Formula		Comments
$s_{b_1} = \dfrac{s_{Y\|X}}{\sqrt{SSX}}$	(12.23)	—s_{b_1} measures the variation of the sample regression slope b_1 around the population regression slope β_1 under repeated sampling. —s_{b_1} is used in confidence intervals for β_1 and hypothesis tests about β_1. See Equations (12.23), (12.25), and (12.27).
$t = \dfrac{b_1}{s_{b_1}}$	(12.25)	—$t = b_1/s_{b_1}$ follows the t distribution with $\nu = n - 2$ degrees of freedom under the assumptions of the regression model and is used to test H_0: $\beta_1 = 0$ against H_a: $\beta_1 \neq 0$.
$F = \dfrac{MSR}{MSE}$	(12.26)	—$F = MSR/MSE$ follows the F distribution with $\nu_1 = 1$ and $\nu_2 = n - 2$ degrees of freedom under the assumptions of the regression model and if H_0 is true and is used to test H_0: $\beta_1 = 0$ against H_a: $\beta_1 \neq 0$.
$s_{\hat{Y}_0} = s_{Y\|X}\sqrt{\dfrac{1}{n} + \dfrac{(X_0 - \overline{X})^2}{SSX}}$	(12.29)	—This value estimates the variation of the $\hat{Y}$ values around the population regression line at the point $X = X_0$. The $s_{\hat{Y}_0}$ figure is used in construction of *confidence intervals* or hypothesis tests concerning the *mean* value of Y that might be associated with $X = X_0$.
$s_{Y_0} = s_{Y\|X}\sqrt{1 + \dfrac{1}{n} + \dfrac{(X_0 - \overline{X})^2}{SSX}}$	(12.32)	—This value estimates the variation of the prediction error at point $X = X_0$. —The s_{Y_0} value is used in construction of *prediction intervals* for an *individual value* of Y that might be associated with $X = X_0$.
$s_{XY} = \widehat{\text{Cov}}(X, Y) = \dfrac{SCP}{n - 1}$	(12.34)	—The sample covariance can be positive or negative. —Covariance is an intermediate figure on the way to finding the correlation coefficient.
$r = \dfrac{s_{XY}}{\sqrt{s_X s_Y}} = \dfrac{SCP}{\sqrt{(SSX)(SSY)}}$	(12.36)	—This formula is the equivalent of the formula for $r = \sqrt{r^2}$ in Equation (12.22). Although the numerical computation is the same, the interpretation of r here is a measure of *joint* variation between X and Y when they are both random variables.

in the chapter and with a few of the problems for this chapter. Generally, you can adapt to the printouts of other programs after studying those presented here, and we recommend that you work some regression and correlation problems with the aid of a computer.

28. Voluntary employee absenteeism is an important organizational problem. Voluntary absences are under the worker's control, and they are generally short-term and not allowed by personnel policies. Frequency absence, the total number of absence occurrences over a time period, is a popular measure of voluntary absence. Voluntary absences are thought to be related to an employee's job tenure. Measures of frequency absence Y and job tenure X for a sample of employees at Electronics Assemblies, Inc., are given in the following listing:

Frequency	7	0	5	9	9	18
Tenure	8	9	7	6	5	1

   a. Find the regression equation $\hat{Y} = b_0 + b_1 X$.
   b. Find the predicted value of Y given $X = 5$.
   c. Find *SST*, *SSR*, and *SSE*.
   d. Find the degrees of freedom associated with the sums of squares in part c.
   e. Find *MSR*.
   f. Find *MSE*.
   g. Present your results for parts c through f in a summary table.
   h. Find $s_{\hat{Y}|X}^2$.
   i. Find the standard error of the estimate $s_{Y|X}$.
   j. Find the coefficient of determination r^2.
   k. Find the standard deviation of the regression line's slope s_{b_1}.
   l. Test the hypothesis that Y and X are not related. That is, test H_0: $\beta_1 = 0$ against H_a: $\beta_1 \neq 0$ by using a t statistic. Let $\alpha = .05$.
   m. Test the hypothesis H_0: $\beta_1 = 0$ against H_a: $\beta_1 \neq 0$ by using an F statistic. Let $\alpha = .05$.
   n. Find $s_{\hat{Y}_0}$ when $X_0 = 5$.
   o. Find the 95% confidence interval estimate for the mean value of Y given $X_0 = 5$.
   p. Find s_{Y_0} when $X_0 = 5$.
   q. Find the 95% prediction interval for the individual value Y_0 given that $X_0 = 5$.

   REFERENCE: Watson, Collin J. (1981), "An Evaluation of Some Aspects of the Steers and Rhodes Model of Employee Attendance," *Journal of Applied Psychology,* 66: 385–89.

29. The relationship between stock prices and dividends is presently unknown. However, empirical studies suggest that stock prices Y in the current time period are related to dividends paid X in the period. Stock prices and dividends for the current period are given in the accompanying listing. You may wish to use deviations from the sample means during the analysis.

Stock Price Y	14	12	12	12	10
Dividend X	11	9	10	11	9

   a. Find the regression equation $\hat{Y} = b_0 + b_1 X$.
   b. Find the predicted value of Y given $X = 10$.
   c. Find *SST*, *SSR*, and *SSE*.
   d. Find the degrees of freedom associated with the sums of squares in part c.
   e. Find *MSR*.
   f. Find *MSE*.
   g. Present your results for parts c through f in an ANOVA summary table.
   h. Find $s_{\hat{Y}|X}^2$.
   i. Find the standard error of the estimate $s_{Y|X}$.
   j. Find the coefficient of determination r^2.

    k. Find the standard deviation of the regression line's slope s_{b_1}.

    l. Test the hypothesis that Y and X are not related. That is, test H_0: $\beta_1 = 0$ against H_a: $\beta_1 \neq 0$ by using a t statistic. Let $\alpha = .05$.

    m. Test the hypothesis H_0: $\beta_1 = 0$ against H_a: $\beta_1 \neq 0$ by using an F statistic. Let $\alpha = .05$.

    n. Find $s_{\hat{Y}_0}$ when $X_0 = 5$.

    o. Find the 95% confidence interval estimate for the mean value of Y given $X_0 = 10$.

    p. Find s_{Y_0} when $X_0 = 10$.

    q. Find the 95% prediction interval for the individual value Y_0 given that $X_0 = 10$.

REFERENCE: Brennan, M. J. (1991), "A Perspective on Accounting and Stock Prices," *The Accounting Review*, 66: 67–79.

30. In the development of a performance measure for financial investments, a characteristic line is often useful. A characteristic line corresponds to a regression equation between the rate of return for an investment Y and the rate of return for a market index X. The higher the slope of the characteristic line, the greater is the sensitivity of the investment to changes in rates of return for the market. Rates of return for a mutual fund and for a market index are given in the following listing:

Mutual Fund	10	7	5	12	7	7
Market Index	4	6	3	8	5	4

    a. Find the characteristic line $\hat{Y} = b_0 + b_1 X$.

    b. Find the predicted value of Y given $X = 5$.

    c. Find *SST*, *SSR*, and *SSE*.

    d. Find the degrees of freedom associated with the sums of squares in part c.

    e. Find *MSR*.

    f. Find *MSE*.

    g. Present your results for parts c through f in a summary table.

    h. Find $s_{Y|X}^2$.

    i. Find the standard error of the estimate $s_{Y|X}$.

    j. Find the coefficient of determination r^2.

31. The following questions refer to the characteristic line of the mutual fund that was found for the data of Problem 30.

    a. Find the standard deviation of the regression line's slope s_{b_1}.

    b. Test the hypothesis that the rate of return for the mutual fund is insensitive to changes in the rates of return for the market. That is, test H_0: $\beta_1 = 0$ against H_a: $\beta_1 \neq 0$ by using a t statistic. Let $\alpha = .05$.

    c. Test the hypothesis H_0: $\beta_1 = 0$ against H_a: $\beta_1 \neq 0$ by using an F statistic. Let $\alpha = .05$.

    d. Find $s_{\hat{Y}_0}$ when $X_0 = 5$.

    e. Find the 95% confidence interval estimate for the mean value of Y given $X_0 = 5$.

    f. Find s_{Y_0} when $X_0 = 5$.

    g. Find the 95% prediction interval for the individual value Y_0 given that $X_0 = 5$.

32. Hexcell Corporation has figured the standard number of workers that should be assigned to a work crew on the production line. In an experiment to verify the validity of the standard, the size of the work crew was varied from the standard. In the accompanying data, X represents the number of people above or below standard size of the crew. Crew sizes from three people below standard ($X = -3$) to three people above standard ($X = 3$) were used. Variable Y represents the number of defects produced in a shift with a crew of the corresponding size X.

X	-3	-2	-1	0	1	2	3
Y	36	33	24	15	9	6	3

a. Find the regression equation $\hat{Y} = b_0 + b_1 X$.
b. Find the predicted value of Y given $X = -1$.
c. Explain the meaning of $\hat{Y}$ for $X = 0$.
d. Explain the meaning of the slope b_1.

33. Use the data given for Problem 32.

   a. Find the standard error of estimate $s_{Y|X}$.
   b. Find the coefficient of determination.

34. Use the data given for Problem 32.

   a. Find s_{b_1}.
   b. Test the hypothesis that $\beta_1 = 0$ against the alternative hypothesis that $\beta_1 \neq 0$. Let $\alpha = .10$.

35. Use the data given for Problem 32, and assume that for tomorrow's shift the work crew will operate with one more person than standard.

   a. Find a 95% confidence interval for the mean number of defects that will be produced by the crew on tomorrow's shift.
   b. Find a 95% prediction interval for the individual number of defects that will be produced by the crew on tomorrow's shift.

36. A sample of 32 graduates with B.S. degrees in business administration reported their starting salaries in their first professional positions. The estimated average annual starting salary by year X is given by

$$\hat{Y} = 7130 + 700(X - 1964)$$

   a. Does a *linear* equation seem valid as a predictor of starting salary?
   b. How do you interpret the various components of this equation?
   c. What is the estimated mean annual starting salary for the year 1992?

37. In the equation $\hat{Y} = b_0 + b_1 X$, where X is pounds and Y is dollars, what units are associated with b_0? With b_1? With r^2?

38. In a regression analysis $n = 38$, $\overline{Y} = 20$, $\overline{X} = 7$, $SST = 900$, $SSX = 60$, and $SCP = 180$.

   a. Find the regression equation.
   b. Find a 90% confidence interval for β_1.
   c. What fraction of the squared error of prediction in Y can be eliminated by using the regression line as a predictor rather than $\overline{Y}$?

39. In a regression analysis $n = 18$, $\overline{X} = 6$, $\overline{Y} = 20$, $SSX = 100$, $SSY = 400$, and $SCP = -120$.

   a. Find the equation for the regression of Y on X.
   b. Find the estimated variance $s_{Y|X}^2$.
   c. Test the hypothesis $\beta_1 = 0$ against H_a: $\beta_1 \neq 0$, using $\alpha = .01$.
   d. Find and interpret the coefficient of determination r^2.

40. The accompanying table gives annual salary (in thousands of dollars), the number of years with the company, and the proportion of sales quota met last year for 16 district sales managers for Digital, Inc., an office equipment manufacturer. An edited printout, where Y is salary and X is years, is presented in Computer Exhibit 12.2.

   a. Find the equation for the regression of salary on years of employment.
   b. Find the equation for the regression of salary on proportion of quota met.
   c. Is the regression of salary on proportion of quota met significant? That is, is β_1 significantly different from zero? Use H_a: $\beta_1 \neq 0$ and $\alpha = .01$.
   d. Which variable, years of employment or proportion of quota met, is the better predictor of salary? [*Hint:* Compare the coefficients of determination.]

SAS	COMPUTER EXHIBIT 12.2 Printout for Digital, Inc.

```
DEP VARIABLE: SALARY
                        SUM OF          MEAN
SOURCE         DF       SQUARES         SQUARE        F VALUE          PROB>F

MODEL          1        581.254         581.254       16.667  (7)      0.0011
ERROR          14       488.245         34.874675
C TOTAL        15       1069.499

        ROOT MSE        5.905478  (3)   R-SQUARE      0.5435  (4)
        DEP MEAN        33.556250       ADJ R-SQ      0.5109
        C.V.            17.59874

                        PARAMETER       STANDARD      T FOR H0:
VARIABLE       DF       ESTIMATE        ERROR         PARAMETER=0      PROB>|T|

INTERCEP       1        17.248644 (1)   4.258602      4.050            0.0012
YEARS          1        1.196888  (2)   0.293174 (5)  4.083  (6)       0.0011
```

Note: (1) b_0, estimated regression intercept; (2) b_1, estimated regression slope; (3) $s_{Y|X}$, estimated standard deviation of Y_i values around regression line, or standard error of estimate; $s_{Y|X} = \sqrt{MSE}$ = ROOT MSE; (4) r^2, sample coefficient of determination; (5) s_{b_1}, standard error of sample regression slope; (6) t, t statistic for testing H_0: $\beta_1 = 0$; (7) F, F statistic for testing H_0: $\beta_1 = 0$.

Salary	Years	Quota	Salary	Years	Quota
33.0	23	0.90	30.0	9	0.98
34.1	12	0.92	33.1	10	1.90
21.0	9	0.89	38.1	17	0.88
29.5	13	0.59	46.6	17	0.97
38.4	12	1.24	21.1	8	0.78
25.3	10	0.57	38.9	15	0.94
21.4	8	0.77	38.7	14	0.80
47.7	26	0.91	40.0	15	0.99

41. The accompanying data show the sales produced (in thousands of dollars) by five new salespeople in their second, fourth, sixth, and eighth months on the job at Lear Siegler, Inc. Note that there are 20 observations, each value of X (time) occurring with 5 different Y's (sales figures). Thus, the value in the lower right corner of the table represents an (X_i, Y_i) point of $(8, 26.3)$. An edited printout of a regression analysis for the data of this problem is presented in Computer Exhibit 12.3.

a. Find the regression equation.
b. Test the hypothesis that $\beta_1 \geq 1.0$. Use H_a: $\beta_1 < 1.0$ and let $\alpha = .025$.
c. What fraction of the variation in Y is accounted for by time?
d. Estimate the mean sales given that $X = 5$.
e. Find a 95% confidence interval for the mean sales given that $X = 5$.
f. Find a 95% prediction interval for an individual sales amount given that $X = 5$.

X = Time (months)	2	4	6	8
Y = Sales (× $000)	21.3	23.2	25.5	25.9
	21.7	23.4	23.6	25.2
	21.4	23.3	24.7	27.6
	22.1	23.7	26.0	27.1
	20.7	23.0	24.3	26.3

COMPUTER EXHIBIT 12.3 Regression of Sales and Time at Lear Siegler   MINITAB

```
MTB > read    c1     c2  ◄──────────────────── Command read to enter the data into c1 and c2
DATA >      21.3    2
DATA >      21.7    2
DATA >      21.4    2
DATA >      22.1    2
DATA >      20.7    2
DATA >      23.2    4
DATA >      23.4    4
DATA >      23.3    4
DATA >      23.7    4
DATA >      23.0    4
DATA >      25.5    6
DATA >      23.6    6
DATA >      24.7    6
DATA >      26.0    6
DATA >      24.3    6
DATA >      25.9    8
DATA >      25.2    8
DATA >      27.6    8
DATA >      27.1    8
DATA >      26.3    8
DATA > end
                                                                    name the variables
MTB > name c1 'SALES' c2 'TIME'  ◄─────────────────────

MTB > regres c1 1 c2  ◄──────── Command regres for c1 as the dependent variable with 1 independent variable c2

The regression equation is  ⎫
SALES = 19.9 + 0.822 TIME    ⎬ ◄──────────────────────  Regression equation

Predictor      Coef      Stdev     t-ratio         p
Constant     19.8900 ①  0.3819      52.08      0.000
TIME          0.82200 ②  0.06973 ⑤  11.79  ⑥ 0.000

s = 0.6973 ③  R-sq = 88.5% ④  R-sq(adj) = 87.9%

Analysis of Variance

SOURCE       DF         SS          MS          F         p
Regression    1       67.568      67.568      138.97    0.000
Error        18        8.752       0.486
Total        19       76.320
```

42. The accompanying data represent the number of hours since rush hour and the carbon monoxide (CO) counts (in parts per million) on an August day in the heart of a large western city. An edited printout from a regression program for the data of this problem is presented in Computer Exhibit 12.4. The circled numbers correspond to the identifying labels that are also given in the figure.

   a. Fit the regression of CO count on time.

BMDP　　　　　　**COMPUTER EXHIBIT 12.4** Printout for Regression

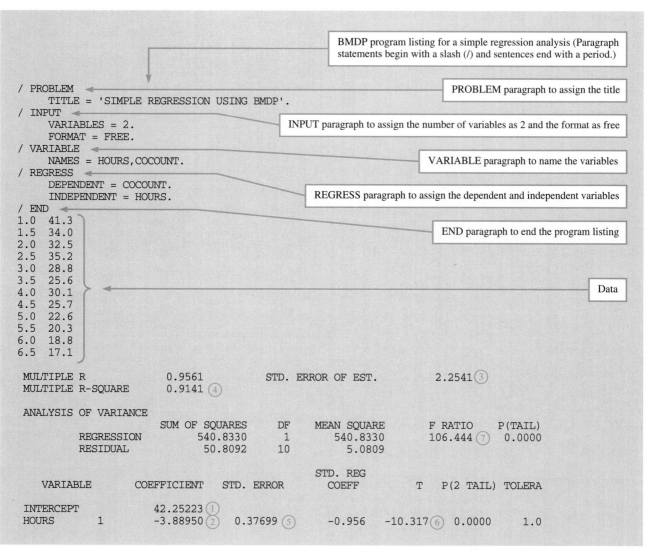

BMDP program listing for a simple regression analysis (Paragraph statements begin with a slash (/) and sentences end with a period.)

```
/ PROBLEM
     TITLE = 'SIMPLE REGRESSION USING BMDP'.
/ INPUT
     VARIABLES = 2.
     FORMAT = FREE.
/ VARIABLE
     NAMES = HOURS,COCOUNT.
/ REGRESS
     DEPENDENT = COCOUNT.
     INDEPENDENT = HOURS.
/ END
1.0    41.3
1.5    34.0
2.0    32.5
2.5    35.2
3.0    28.8
3.5    25.6
4.0    30.1
4.5    25.7
5.0    22.6
5.5    20.3
6.0    18.8
6.5    17.1
```

PROBLEM paragraph to assign the title

INPUT paragraph to assign the number of variables as 2 and the format as free

VARIABLE paragraph to name the variables

REGRESS paragraph to assign the dependent and independent variables

END paragraph to end the program listing

Data

```
MULTIPLE R                0.9561        STD. ERROR OF EST.           2.2541 ③
MULTIPLE R-SQUARE         0.9141 ④

ANALYSIS OF VARIANCE
                    SUM OF SQUARES    DF    MEAN SQUARE      F RATIO     P(TAIL)
        REGRESSION      540.8330       1      540.8330      106.444 ⑦   0.0000
        RESIDUAL         50.8092      10        5.0809

                                             STD. REG
    VARIABLE       COEFFICIENT   STD. ERROR    COEFF          T    P(2 TAIL) TOLERA

INTERCEPT            42.25223 ①
HOURS        1       -3.88950 ②   0.37699 ⑤   -0.956    -10.317 ⑥  0.0000     1.0
```

Note: ① b_0, estimated regression intercept; ② b_1, estimated regression slope; ③ $s_{Y|X}$, standard error of estimate; ④ r^2, sample coefficient of determination; ⑤ s_{b_1}, standard error of sample regression slope; ⑥ t, t statistic for testing H_0: $\beta_1 = 0$; ⑦ F, F statistic for testing H_0: $\beta_1 = 0$.

b. Find a 98% confidence interval for β_1.

c. Find and interpret the coefficient of determination.

Hours	CO Count	Hours	CO Count
1.0	41.3	4.0	30.1
1.5	34.0	4.5	25.7
2.0	32.5	5.0	22.6
2.5	35.2	5.5	20.3
3.0	28.8	6.0	18.8
3.5	25.6	6.5	17.1

43. The accompanying data are the ultimate loads resulting from tests at Weyerhauser Company of joints made with different sizes of common nails. Note that the first row indicates (X_i, Y_i) points of $(30, 11.03)$, $(50, 11.97)$, and $(60, 13.69)$. Each following row represents similar pairs, where the first element of the pair is 30, 50, or 60.

a. Find the equation for the regression of ultimate load on nail size.
b. Find a 95% confidence interval for the slope of the line.
c. Is b_1 significantly different from zero? Use $H_a: \beta_1 \neq 0$ and $\alpha = .05$.
d. Find a 95% confidence interval for the mean ultimate load of joints of this type made with 30d nails.
e. Find a 95% prediction interval for an individual value for the ultimate load of a joint of this type made with 30d nails.

Nail Size	30d	50d	60d
Ultimate Load	11.03	11.97	13.69
	10.64	13.63	14.82
	10.48	13.56	15.23
	10.02	15.20	15.45
	10.31	14.84	16.28

44. The accompanying data are the average incomes per acre, Y, produced by a commercial system for taking game animals over a ten-year period.

a. Find the regression equation predicting income per acre from time in years.
b. Find a 99% confidence interval for β_1.
c. What fraction of the variation in income is accounted for by the relationship?
d. What is the predicted income per acre for this year? Why is this prediction likely to involve a great deal of error?

Year	Income per Acre	Year	Income per Acre
1956	.51	1961	.89
1957	.59	1962	1.01
1958	.64	1963	1.07
1959	.74	1964	1.10
1960	.78	1965	1.18

45. A newly developed low-pressure snow tire has been tested to see how it wears under normal, dry-weather conditions. Twenty of the new tires were mounted on the right front wheels of 20 standard passenger cars. These cars were then driven at high speeds on a dry test track for varying lengths of time. The tread wear (in millimeters) was then recorded for each tire (see the accompanying table).

a. Estimate the relationship between hours driven X and tread wear Y.
b. Find a 95% confidence interval for the slope of the line.

Hours	Wear	Hours	Wear
13	.1	62	.4
25	.2	105	.7
27	.2	88	.6
46	.3	63	.4
18	.1	77	.5
31	.2	109	.7
46	.3	117	.8
57	.4	35	.2
75	.5	98	.6
87	.6	121	.8

46. Monthly percentage rates of return for a company's stock (Y) and for a stock exchange index (X) that represents a market portfolio are given in the accompanying table. The estimated regression slope for Y regressed on X is called the stock's "beta" by security analysts. If the slope is greater than 1, then this company's stock is called aggressive, because the stock's rate of return increases (or decreases) more sharply than does that of the market. That is, the stock's rate of return is more volatile than that of the market. If the slope is less than 1, then the stock is called defensive.

  a. Find the estimated regression equation.
  b. Is the stock agressive or defensive?
  c. Find the value of r^2.
  d. Test the hypothesis $H_0: \beta_1 = 0$ against $H_a: \beta_1 \neq 0$, using a significance level of .05.

Y	X	Y	X
0.17	1.67	-0.87	-1.07
-2.67	-1.06	9.38	7.55
7.67	4.15	10.19	4.73
8.54	2.46	-4.70	-1.99
0.97	5.78	2.54	-0.94
-13.99	-9.41	-8.44	-4.56

47. The executives at a national weight loss clinic feel that there is a relationship between the weight of a person entering their program and the number of pounds lost. To confirm this suspicion, they took a sample from their records. The sample data are as follows:

X = *Beginning Weight*	205	165	289	154	142	306	261	177
Y = *Weight Loss*	25	15	36	12	15	146	73	50

The regression information for this relationship is as follows:

$$\hat{Y} = -67.78 + .54X \qquad r^2 = .583 \qquad s_{Y|X} = 31.60 \qquad s_{b_1} = .186$$

$$SSX = 28{,}911.87 \qquad SSY = SST = 14{,}362.00$$

$$SCP = 15{,}557.48$$

  a. Test the hypothesis that there is no relationship between beginning weight and weight loss, using the t test on the slope of the regression line. Let $\alpha = .05$.
  b. Repeat the test in part a, using the test on r, the correlation coefficient.
  c. Repeat the test in part a, using analysis of variance on the partitioned sums of squares for regression.

48. The accompanying data show the gas mileages obtained (in miles per gallon), Y, for a compact car in driving at a constant speed for 3 hours with various amounts X of a gasoline additive mixed in a full tank of gas.

  a. Plot the points on graph paper.
  b. Find and plot the point $(\bar{X}, \bar{Y})$.
  c. Find b_0 and b_1 and, hence, the regression equation $\hat{Y} = b_0 + b_1 X$. Draw this line on the graph.
  d. Find the deviation of each Y from the corresponding $\hat{Y}$. Square and sum these deviations.
  e. Find the sum of squares for deviations from regression or error by the formula.

$$\sum (Y_i - \hat{Y}_i)^2 = SST - (b_1)(SCP)$$

How do you account for the difference between this figure and that obtained in part d?

f. Construct a summary table showing the partitioning of the sum of squares and degrees of freedom.

g. Find the estimated variance $s^2_{Y|X}$ and the variance and standard deviation of b_1.

h. Find the 95% confidence interval for β_1.

i. Test the hypothesis that $\beta_1 = 0$ against $H_a: \beta_1 \neq 0$, using a t test and $\alpha = .05$.

j. Test the hypothesis that $\rho = 0$ against $H_a: \rho \neq 0$, using a t test and $\alpha = .05$.

k. Test the hypothesis that there is no correlation between mileage and the amount of additive, using an F test and the data developed in part f. Use $\alpha = .05$.

Y	X	Y	X	Y	X	Y	X
26.8	45	29.7	68	28.1	54	26.4	44
28.1	59	27.5	49	29.2	63	27.7	56
28.5	66	29.0	57	26.9	48	28.2	62
27.0	47	30.2	67	28.6	55	26.7	50
28.8	61	26.3	43	31.0	67	30.8	70

49. Seven executives working for the same firm have expense accounts. The personnel people in the firm feel that there is a relationship between the executives' annual salaries and the amount they claim in expenses each year. Thus, they ran the following regression analysis relating the variables:

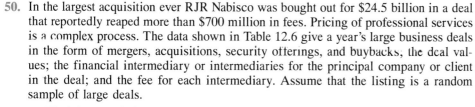

X = Salary ($\times$ 1000)	25	30	35	40	40	38	22
Y = Expenses ($\times$ 1000)	.9	1.2	2.2	1.1	.8	1.6	.6

The regression results are as follows:

$$\hat{Y} = .227 + .0296X \qquad r = .398 \qquad s_{Y|X} = .5475 \qquad s_{b_1} = .0306$$

$$SSY = SST = 1.78 \qquad SSX = 320.86 \qquad SCP = 9.50$$

a. Test the hypothesis that there is no relationship between annual salary and expenses claimed, using the t test on the regression coefficient b_1. Use $\alpha = .05$.

b. Repeat the test of part a, using the t test on r.

c. Repeat the test of part a, using analysis of variance on the table of partitioned sums of squares.

50. In the largest acquisition ever RJR Nabisco was bought out for $24.5 billion in a deal that reportedly reaped more than $700 million in fees. Pricing of professional services is a complex process. The data shown in Table 12.6 give a year's large business deals in the form of mergers, acquisitions, security offerings, and buybacks; the deal values; the financial intermediary or intermediaries for the principal company or client in the deal; and the fee for each intermediary. Assume that the listing is a random sample of large deals.

a. Find the correlation coefficient between the value of the deal and the fee for the intermediary (the total fee if there are two intermediaries).

b. Is there a significant linear relationship between deal values and intermediary fees? Use a significance level of .05.

c. Find a linear equation that could possibly be used to estimate intermediary fees from deal values.

d. Is it advisable to use a linear relationship between deal values and intermediary fees to predict intermediary fees? Explain.

51. For the following set of quantities,

$$n = 18 \qquad SSY = 100 \qquad SCP = 36 \qquad SSX = 36$$

find the sample correlation coefficient. Test the hypothesis that $\rho = 0$. Let $\alpha = .05$. Use $H_a: \rho \neq 0$.

TABLE 12.6   Business Deal Data

Deal	Value ($000)	Transaction	Financial Intermediaries (Client)	Fee ($000)	Fee (as % of Deal)
GENERAL ELECTRIC (diversified manufacturing) acquires RCA (consumer electronics, entertainment)	$6,406,000	Acquisition for cash, completed June 9	GOLDMAN SACHS (General Electric)	$15,000	0.23
BCI HOLDINGS acquires BEATRICE COS. (food, consumer products)	$6,216,000	Leveraged buy-out for cash and preferred stock, April 16	KOHLBERG KRAVIS ROBERTS (BCI holdings)	$40,000	0.64
BURROUGHS (computers, electronics) merges with SPERRY (computers, electronics) to form UNISYS	$4,753,700	Purchase of 51% of Sperry common stock through cash tender offer, followed by purchase of remaining shares for cash and preferred stock, September 16	LAZARD FRERES JAMES D. WOLFENSOHN (Burroughs)	$7,500 $5,000	0.16 0.11
SAFEWAY STORES HOLDINGS acquires SAFEWAY STORES (supermarkets)	$4,284,311	Leveraged buy-out for cash, debentures, and common stock warrants, November 24	KOHLBERG KRAVIS ROBERTS MORGAN STANLEY (Safeway Holdings)	$45,000 $10,000	1.05 0.23
MACY MERGER acquires R.H. MACY (department stores)	$3,564,992	Leveraged buy-out for cash, July 15	GOLDMAN SACHS (Macy Merger)	$30,250	0.85
CAPITAL CITIES COMMUNICATIONS acquires ABC (broadcasting)	$3,509,000	Acquisition for cash, June 6	FIRST BOSTON (ABC)	$7,500	0.21
CAMPEAU (real estate developer) acquires ALLIED STORES	$3,505,716	Acquisition for cash, December 31	FIRST BOSTON (Campeau)	$7,000	0.20
UNION CARBIDE (chemicals) repurchases 55% of its common stock	$3,332,853	Tender offer to exchange cash, notes and debentures for 38.8 million shares, to fend off hostile takeover bid by GAF, January 15	MORGAN STANLEY	$35,000	1.05
CLEVELAND ELECTRIC (public utility) merges with TOLEDO EDISON (public utility) to form CENTERIOR ENERGY	$3,140,493	Merger through change of common stock, April 29	MORGAN STANLEY (Cleveland Electric)	$3,794	0.12
USX (steel, oil, and gas) acquires TEXAS OIL & GAS	$2,996,646	Acquistion for common stock, February 11	GOLDMAN SACHS (USX)	$7,000	0.23

52. From the following information,

$$n = 18 \qquad SSY = 25 \qquad SCP = 54 \qquad SSX = 144$$

find the sample correlation coefficient. Test the hypothesis that $\rho = 0$. Let $\alpha = .01$ and H_a: $\rho \neq 0$.

Business Deal Data (Continued)   **TABLE 12.6**

Deal	Value ($000)	Transaction	Financial Intermediaries (Client)	Fee ($000)	Fee (as % of Deal)
UNION CARBIDE (chemicals) repurchases debt issued to finance stock buyback	$2,736,000	Cash tender offer for $13\frac{1}{4}\%$ notes due 1993, $14\frac{1}{4}\%$ notes due 1996, and 15% debentures due 2006 to reduce interest costs and eliminate financial restrictions, December 5	FIRST BOSTON	$1,725	0.06
OCCIDENTAL PETROLEUM acquires MIDCON (gas pipelines)	$2,576,188	Purchase of 53% of MidCon common stock through cash tender offer, followed by exchange of Occidental shares for remaining stock, April 1	FIRST BOSTON (Occidental)	$7,000	0.27
MAY DEPARTMENT STORES acquires ASSOCIATED DRY GOODS (department stores)	$2,380,082	Acquisition for common stock, October 6	MORGAN STANLEY (May)	$6,875	0.29
BCI HOLDINGS sells bonds	$1,750,000	Offering of $800 million of $12\frac{1}{2}\%$ debentures due 1988 and $950 million of $12\frac{3}{4}\%$ debentures due 2001, to help finance BCI's acquisition of Beatrice Cos., April 10	DREXEL BURNHAM LAMBERT	$62,188	3.55
Consortium of communications companies acquires GROUP W CABLE from WESTINGHOUSE ELECTRIC	$1,730,000	Acquisition for cash, June 19	FIRST BOSTON SHEARSON LEHMAN BROTHERS (Westinghouse)	$4,875 $4,875	0.28 0.28

53. Find the sample correlation coefficient and test H_0: $\rho = 0$ against H_a: $\rho \neq 0$ with $\alpha = .05$ for the following information:

$$n = 25 \qquad \sum Y = 50 \qquad \sum X = 75$$
$$\sum Y^2 = 228 \qquad \sum X^2 = 625 \qquad \sum XY = 30$$

54. The relationship between earnings and stock prices is presently unknown. However, empirical studies suggest that stock returns in the current time period are related to earnings from a previous period. Stock returns for the current period and earnings from a previous period are given in the following listing:

Stock Return	4	−1	0	2	0	1
Earnings	6	−1	−6	6	0	7

a. Find the sample correlation for the variables.
b. Test the significance of the correlation at the $\alpha = .05$ level?

REFERENCE: Lipe, R. (1989), "The Relations Between Stock Returns and Accounting Earnings Given Alternative Information," *The Accounting Review,* 65: 49–71.

 55. Data were collected from five employees on their commitment to the organization and on their job satisfaction. These two attitudinal variables are measured by indexes, and the results are presented in the following listing:

Commitment	−2	−4	2	0	4
Satisfaction	−4	−2	0	2	4

a. Find the sample correlation between commitment and satisfaction.
b. Test the significance of the correlation at the $\alpha = .05$ level.

REFERENCE: Watson, Collin J., K. D. Watson, and J. Stowe (1985), "Univariate and Multivariate Distributions of the Job Descriptive Index's Measures of Job Satisfaction," *Organizational Behavior and Human Decision Processes,* 35: 241–51.

56. F. Galton was interested in the association between the heights of fathers and the heights of their sons. Data were collected from 12 father-and-son pairs, and their height (in feet) are presented in the following table:

Father	5.9	5.7	6.0	5.9	6.1	5.3	5.7	6.2	6.0	5.7	5.8	5.6
Son	6.0	5.9	6.3	6.3	5.9	5.6	6.0	6.3	6.4	5.8	6.1	5.7

a. Find the sample correlation between the heights of the fathers and sons.
b. Test the significance of the correlation at the $\alpha = .05$ level.

REFERENCE: Galton, F. (1890), "Kinship and Correlation," *North American Review,* 150: 419–31.

57. The net interest sensitivity of banks, called the "gap" by bankers, is the difference between interest-sensitive asset and interest-sensitive liability accounts. An analyst with First Chicago Corporation is concerned about the association between balance sheet percentages of interest-sensitive loans and core deposits (with maturities longer than one year). Data, collected from 12 large banks, are presented in the accompanying table.

a. Find the sample correlation between interest-sensitive loans and core deposits.
b. Test the significance of the correlation at the $\alpha = .05$ level.

Interest-Sensitive Loans	Core Deposits	Interest-Sensitive Loans	Core Deposits
17	39	38	34
38	45	59	22
38	25	32	35
24	43	29	36
33	32	31	35
33	46	48	18

REFERENCE: Simonson, D., J. Stowe, and Collin J. Watson (1983), "A Canonical Correlation Analysis of Commercial Bank Asset/Liability Structures," *Journal of Financial and Quantitative Analysis,* 18: 125–40.

 58. A human resources manager with Mario Laboratories believes that there is an association between employee absenteeism and job tenure. A sample of data was collected for 12 employees and is given in the following listing:

Absences	4	8	3	7	8	5	5	9	5	5	8	8
Job Tenure	5	4	2	6	6	4	5	5	3	4	5	5

a. Find the sample correlation between absences and job tenure.
b. Test the significance of the correlation at the $\alpha = .05$ level.

REFERENCE: Watson, Collin J. (1981), "An Evaluation of Some Aspects of the Steers and Rhodes Model of Employee Attendance," *Journal of Applied Psychology,* 66 (June): 385–89.

Refer to the 141 companies listed in the Ratio Data Set in Appendix A.

Ratio Data Set

Questions

59. Locate the data for the net income–total assets (*NI/TA*) ratio as a measure of profitability and the current assets–sales (*CA/S*) ratio.

   a. Use a computer to obtain the sample regression equation for estimating *NI/TA* from *CA/S*.
   b. Test whether the linear relationship between the two variables is statistically significant. Use a significance level of .05.

60. Locate the data for the net income–total assets (*NI/TA*) ratio as a measure of profitability and the quick assets–sales (*QA/S*) ratio.

   a. Use a computer to obtain the sample regression equation for estimating *NI/TA* from *QA/S*.
   b. Test whether the linear relationship between the two variables is statistically significant. Use a significance level of .05.

61. Locate the data for the net income–total assets (*NI/TA*) ratio as a measure of profitability and the current assets–current liabilities (*CA/CL*) ratio.

   a. Use a computer to obtain the sample regression equation for estimating *NI/TA* from *CA/CL*.
   b. Test whether the linear relationship between the two variables is statistically significant. Use a significance level of .05.

62. Locate the data for the current assets–sales (*CA/S*) ratio and the quick assets–sales (*QA/S*) ratio.

   a. Use a computer to obtain the sample correlation coefficient for these two variables.
   b. Test whether the linear relationship between the two variables is statistically significant. Use a significance level of .05.

63. Locate the data for the current assets–sales (*CA/S*) ratio and the current assets–current liabilities (*CA/CL*) ratio.

   a. Use a computer to obtain the sample correlation coefficient for these two variables.
   b. Test whether the linear relationship between the two variables is statistically significant. Use a significance level of .05.

Refer to the 113 applicants for credit listed in the Credit Data Set in Appendix A.

Credit Data Set

Questions

64. Find the regression line that would allow us to predict JOBINC for the group granted credit as a linear function of AGE.

65. Use the results obtained in Question 64 to test the hypothesis $H_0: \beta_1 = 0$ against the alternative hypothesis $H_a: \beta_1 \neq 0$, using $\alpha = .10$.

66. Find the coefficient of correlation between TOTBAL and TOTPAY for the applicants who were denied credit.

CASE 12.1

PREDICTING CATASTROPHIC FAILURE OF THE SPACE SHUTTLE *CHALLENGER*

The space shuttle *Challenger* was in launch position the morning of January 28, 1986, with seven astronauts and passengers on board. Just prior to the launch, sheets of ice clung hauntingly to the fuselage. Moments later, with national television coverage as it blasted into orbit, the shuttle disintegrated in a catastrophic explosion. The remains of the astronauts and passengers were never recovered.

Thiokol Corporation manufactures the two solid-fuel rocket motors that propelled the shuttle into space. The night before the catastrophe, executives of Thiokol and the National Aeronautics and Space Administration debated whether they should launch the shuttle according to schedule or postpone the mission. The weather report called for a temperature of 31° F at blast off. At the conclusion of the debate, Thiokol executives recommended that the shuttle be launched on schedule because they felt that they did not have conclusive evidence that the low temperature would influence the capability of the solid-fuel rocket motors to thrust their payload into orbit.

From April 12, 1981, to January 12, 1986, prior to the catastrophe, the space shuttle had flown 24 successful missions. Six primary O-rings were used to seal the sections of the two solid-fuel rocket motors that were used to thrust the shuttle into space. On several flights, the motors had experienced O-ring erosion or gas blow-by incidents. O-ring incidents are extremely dangerous because a failed O-ring can allow super-hot gases to escape the solid-propellant motors and ignite the liquid hydrogen fuel tank. On one of the 24 flights, the motors were not recovered. The number of erosion or blow-by incidents, and the temperature of the rocket joints for 23 successful flights prior to the catastrophe are given in the accompanying listing.

Mission	O-Ring Incidents	Temperature (°F)	Mission	O-Ring Incidents	Temperature (°F)
1	0	66	13	0	67
2	1	70	14	2	53
3	0	69	15	0	67
4	0	68	16	0	75
5	0	67	17	0	70
6	0	72	18	0	81
7	0	73	19	0	76
8	0	70	20	0	79
9	1	57	21	2	75
10	1	63	22	0	76
11	1	70	23	1	58
12	0	78			

To predict the number of erosion or blow-by O-ring incidents using joint temperature as the independent variable, we use linear regression analysis. The results of a regression analysis of the number of O-ring incidents as the dependent variable Y on the joint temperature as the independent variable X are shown in Figure 1. The regression equation is

$$\hat{Incidents} = 3.70 - .0475(Temperature)$$

The prediction for the number of O-ring incidents when $X = 31°F$ is

$$\hat{Incidents} = 3.70 - .0475(31) = 2.25$$

FIGURE 1

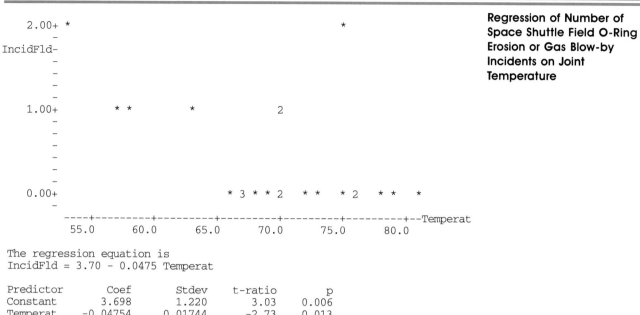

Regression of Number of Space Shuttle Field O-Ring Erosion or Gas Blow-by Incidents on Joint Temperature

```
2.00+  *                                          *
       -
IncidFld-
       -
       -
       -
       -
1.00+          *  *              *                2
       -
       -
       -
       -
       -
0.00+                        *  3  *  *  2    *  *    *  2    *  *     *
       -
       ----+---------+---------+---------+---------+---------+--Temperat
         55.0      60.0      65.0      70.0      75.0      80.0
```

The regression equation is
IncidFld = 3.70 - 0.0475 Temperat

Predictor	Coef	Stdev	t-ratio	p
Constant	3.698	1.220	3.03	0.006
Temperat	-0.04754	0.01744	-2.73	0.013

s = 0.5774   R-sq = 26.1%   R-sq(adj) = 22.6%

Analysis of Variance

SOURCE	DF	SS	MS	F	p
Regression	1	2.4762	2.4762	7.43	0.013
Error	21	7.0021	0.3334		
Total	22	9.4783			

Unusual Observations

Obs.	Temperat	IncidFld	Fit	Stdev.Fit	Residual	St.Resid
14	53.0	2.000	1.179	0.313	0.821	1.69 X
21	75.0	2.000	0.133	0.153	1.867	3.35R

R denotes an obs. with a large st. resid.
X denotes an obs. whose X value gives it large influence.

Fit	Stdev.Fit	95% C.I.	95% P.I.
2.225	0.683	(0.803, 3.646)	(0.364, 4.086) XX

X denotes a row with X values away from the center
XX denotes a row with very extreme X values

The regression results suggest that there was a relationship between joint temperature and O-ring incidents. However, the normality assumption is questionable so the probability values given in the results may not be realistic; furthermore the 31°F temperature forces an extrapolation beyond the temperatures experienced in previous missions. Nevertheless, it would seem prudent, if the executives had been given the prediction from regression analysis, to have postponed the launch.

Case Assignment

  a. The regression results indicate that mission 21, with a temperature of 75° and two O-ring incidents, is an influential observation. Omit the data for observation 21, and find the regression equation for the remaining 22 observations.

  b. Predict the number of O-ring incidents for a temperature of 31°F based on the 22 remaining observations.

REFERENCES: Dalal, S, E. Fowlkes, and B. Hoadley (1989), "Risk Analysis of the Space Shuttle: Pre-*Challenger* Prediction of Failure," *Journal of the American Statistical Association*, 84: 945–57; and Presidential Commission on the Space Shuttle *Challenger* Accident (1986), *Report of the Presidential Commission on the Space Shuttle Challenger Accident*, vols. 1 and 2 (Washington, D.C.: U.S. Government Printing Office).

CASE 12.2

PREDICTING THE VALUE OF REAL ESTATE

Bill McFarland is a real estate broker who specializes in selling farmland in a large western state. Since Bill advises many of his clients about pricing of their land, he is interested in developing a pricing formula of some type. He feels he could increase his business greatly if he could accurately determine the value of a farmer's land.

    Bill's friend, Mr. Martin Alluvial, is a geologist and has told Bill that the soil and rock characteristics in most of the area where Bill sells do not vary much. Thus according to Mr. Alluvial, the price of the land should depend solely on the acreage, and any variation from a fixed price per acre is due only to psychological factors in the buyer and seller and not to the value or worth of the land.

    Mr. McFarland has sold 35 plots of land in the past year at the selling prices and acreages listed in the following table:

Selling Price ($000)	Acreage	Selling Price ($000)	Acreage	Selling Price ($000)	Acreage
60	20.0	382	133.0	69	22.0
130	40.5	5	2.0	220	81.5
25	10.2	42	13.0	235	78.0
300	100.0	60	21.6	50	16.0
85	30.0	20	6.5	25	10.0
182	56.5	145	45.0	290	100.0
115	41.0	61	19.2	118	41.0
24	10.0	20	6.0	17	6.0
60	18.5	15	3.1	10	3.0
92	30.0	485	210.1	41	14.0
77	25.6	892	305.2	200	70.0
122	42.0	46	14.0		

    Mr. McFarland would like to convince Melvin Curtis to sell his land. Mr. Curtis owns 250 acres of land in the western portion of what Mr. McFarland considers his "territory." However, Mr. Curtis is unwilling to sell until he can get an accurate feeling for the worth of his land. Bill got him to say that he would be willing to sell for approximately the same price per acre that people in the area were selling for last year "plus about 10% for inflation this year." However, Mr. Curtis said that he'd have to see some firm data on what land was going for last year before he would sell.

In the meantime, Bill McFarland needs to give Orin West an estimate of what he might get from his 35-acre plot. Mr. West wants his estimate early next week.

Case Assignment

a. Find the simple regression line that could be used to predict selling price as a function of acreage.

b. Find a measure of how well the regression line predicts selling price.

c. Find an interval within which you can be 95% sure that (1) the mean selling price of 250-acre plots will lie and (2) the mean selling price of 35-acre plots will lie.

REDUCING INVESTMENT RISK WITH PORTFOLIO THEORY

CASE 12.3

Financial analysts often use the standard deviation, or variance of the probability distribution of expected future returns, as a measure of the riskiness of an investment. Some investors prefer to invest in U.S. Treasury bills or insured certificates of deposit, which have low risk and low returns. Other investors who seek higher returns prefer common stocks, even though stocks are riskier.

Portfolios of investments include different combinations of securities. Diversifying a portfolio means selecting a variety of securities for the portfolio or not putting all of one's eggs in a single basket, and some investors seek to decrease riskiness for a given level of return through diversification. Portfolio theory applies principles of statistics to show investors the benefits of diversification.

The expected value of a sum of random variables is the sum of the expected values, and the expected value of a constant multiplied by a random variable is the constant multiplied by the expected value. Consequently, the expected or mean return of a portfolio is the weighted mean return of the individual securities. The mean returns for securities are generally not known; therefore, the parameters of investment return distributions are typically estimated by using a sample of empirical returns. For a portfolio consisting of two securities with returns X_1 and X_2, the estimate of the portfolio mean return is

$$\overline{X}_{\text{portfolio}_{1,2}} = w_1 \overline{X}_1 + w_2 \overline{X}_2$$

where w_j is the proportion of the value of the portfolio invested in security i, $\overline{X}_j$ is the sample mean return for security j, and $j = 1, 2$.

The variance of a random variable multiplied by a constant is the constant squared multiplied by the variance of the variable. Furthermore, the variance of a sum of *independent* random variables is the sum of the variances. However, the returns of securities are generally not independent; security returns tend to move together or they experience co-movements. The direction and degree of linear co-movement of security returns X_1 and X_2 is measured by their covariance $\text{Cov}(X_1, X_2)$ or by their correlation ρ, since $\rho = \text{Cov}(X_1, X_2)/(\sigma_{X_1} \sigma_{X_2})$. The standard deviation of the return of a two-security portfolio that accounts for the co-movement (covariance or correlation) of the returns of the securities is

$$\sigma_{\text{portfolio}_{1,2}} = \sqrt{w_1^2 \sigma_{X_1}^2 + w_2^2 \sigma_{X_2}^2 + 2w_1 w_2 \, \text{Cov}(X_1, X_2)}$$

$$= \sqrt{w_1^2 \sigma_{X_1}^2 + w_2^2 \sigma_{X_2}^2 + 2w_1 w_2 \rho \sigma_{X_1} \sigma_{X_2}}$$

The estimate of the standard deviation of the return of a two-security portfolio using empirical return data is

$$s_{\text{portfolio}_{1,2}} = \sqrt{w_1^2 s_{X_1}^2 + w_2^2 s_{X_2}^2 + 2w_1 w_2 \, \widehat{\text{Cov}}(X_1, X_2)}$$

$$= \sqrt{w_1^2 s_{X_1}^2 + w_2^2 s_{X_2}^2 + 2w_1 w_2 r s_{X_1} s_{X_2}}$$

Samples of percentage returns for stocks of (1) GAuto X_1, (2) Xoff X_2, and (3) GZE X_3 are shown in the following table:

X_1	X_2	X_3
0	-2	7
-2	0	3
4	2	1
2	4	0
6	6	-1

An investor wants to put 50% of the value of an investment portfolio in GAuto and the other 50% in either Xoff or GZE. Furthermore, the investor wishes to minimize risk for a given level of return. Should the investor select a portfolio of 50% (1) GAuto and 50% (2) Xoff, or 50% (1) GAuto and 50% (3) GZE?

Descriptive statistics for the stock returns of the three corporations are shown in Table 1.

For a portfolio consisting of 50% (1) GAuto and 50% (2) Xoff the estimated mean and standard deviation of portfolio returns are

$$\overline{X}_{\text{portfolio}_{1,2}} = w_1\overline{X}_1 + w_2\overline{X}_2 = .5(2) + .5(2) = 2$$

$$s_{\text{portfolio}_{1,2}} = \sqrt{w_1^2 s_{X_1}^2 + w_2^2 s_{X_2}^2 + 2w_1 w_2 r s_{X_1} s_{X_2}}$$

$$= \sqrt{.5^2(10) + .5^2(10) + 2(.5)(.5)(.8)(3.162)(3.162)} = 3$$

For a portfolio of 50% (1) GAuto and 50% (3) GZE the estimated mean and standard deviation of portfolio returns are

$$\overline{X}_{\text{portfolio}_{1,3}} = w_1\overline{X}_1 + w_3\overline{X}_3 = .5(2) + .5(2) = 2$$

$$s_{\text{portfolio}_{1,3}} = \sqrt{w_1^2 s_{X_1}^2 + w_3^2 s_{X_3}^2 + 2w_1 w_3 r s_{X_1} s_{X_3}}$$

$$= \sqrt{.5^2(10) + .5^2(10) + 2(.5)(.5)(-.7)(3.162)(3.162)}$$

$$= 1.225$$

Each of the two investment portfolios has a sample mean return of 2%. However, the portfolio consisting of (1) GAuto and (2) Xoff has a standard deviation of 3%; whereas, the portfolio of (1) GAuto and (3) GZE has a standard deviation of only 1.225%. The standard deviations are different because the correlation between the returns of GAuto and Xoff is positive (.8), whereas the correlation between the returns of GAuto and GZE is negative ($-.7$). Thus, to minimize risk for a given level of return, the investor should select GAuto and GZE for the portfolio. Accordingly, the investor can use principles of statistics to decrease the riskiness of a portfolio by diversification. Minimizing risk through diversification requires a judicious selection of stocks.

Case Assignment   For a portfolio of 50% (2) Xoff and 50% (3) GZE, find

a. the estimated mean portfolio returns.
b. the estimated standard deviation of portfolio returns.

REFERENCES: Jacob, N., and R. Pettit (1989), *Investments*, 2nd ed. (Homewood, Ill: Irwin); Markowitz, H. (1952), Portfolio Selection, *J. of Finance*, 6: 77–91; and Tobin, J. (1958), "Liquidity Preference as a Behavior Toward Risk," *Review of Economic Studies*, 25: 65–86.

Row	X_1	X_2	X_3	$(X_1)^2$	$(X_2)^2$	$(X_3)^2$	$X_1 X_2$	$X_1 X_3$	$X_2 X_3$
1	0	−2	7	0	4	49	0	0	−14
2	−2	0	3	4	0	9	0	−6	0
3	4	2	1	16	4	1	8	4	2
4	2	4	0	4	16	0	8	0	0
5	6	6	−1	36	36	1	36	−6	−6
Total	10	10	10	60	60	60	52	−8	−18

TABLE 1

Stock Return Summary

$$\overline{X}_1 = \frac{\Sigma X_1}{n} = \frac{10}{5} = 2 \qquad \overline{X}_2 = \frac{\Sigma X_2}{n} = \frac{10}{5} = 2 \qquad \overline{X}_3 = \frac{\Sigma X_3}{n} = \frac{10}{5} = 2$$

$$s_{X_1}^2 = \frac{SSX_1}{n-1} = \frac{\Sigma X_1^2 - (\Sigma X_1)^2/n}{n-1} = \frac{60 - 10^2/5}{5-1} = \frac{40}{4} = 10$$

$$s_{X_1} = \sqrt{10} = 3.162$$

$$s_{X_2}^2 = \frac{SSX_2}{n-1} = \frac{\Sigma X_2^2 - (\Sigma X_2)^2/n}{n-1} = \frac{60 - 10^2/5}{5-1} = \frac{40}{4} = 10$$

$$s_{X_2} = \sqrt{10} = 3.162$$

$$s_{X_3}^2 = \frac{SSX_3}{n-1} = \frac{\Sigma X_3^2 - (\Sigma X_3)^2/n}{n-1} = \frac{60 - 10^2/5}{5-1} = \frac{40}{4} = 10$$

$$s_{X_3} = \sqrt{10} = 3.162$$

$$SCP_{X_1 X_2} = \Sigma X_1 X_2 - \frac{\Sigma X_1 \Sigma X_2}{n} = 52 - \frac{(10)(10)}{5} = 32$$

$$SCP_{X_1 X_3} = \Sigma X_1 X_3 - \frac{\Sigma X_1 \Sigma X_3}{n} = -8 - \frac{(10)(10)}{5} = -28$$

$$SCP_{X_2 X_3} = \Sigma X_2 X_3 - \frac{\Sigma X_2 \Sigma X_3}{n} = -18 - \frac{(10)(10)}{5} = -38$$

$$r_{X_1 X_2} = \frac{SCP_{X_1 X_2}}{\sqrt{SSX_1 SSX_2}} = \frac{32}{\sqrt{(40)(40)}} = .8$$

$$r_{X_1 X_3} = \frac{SCP_{X_1 X_3}}{\sqrt{SSX_1 SSX_3}} = \frac{-28}{\sqrt{(40)(40)}} = -.7$$

$$r_{X_2 X_3} = \frac{SCP_{X_2 X_3}}{\sqrt{SSX_2 SSX_3}} = \frac{-38}{\sqrt{(40)(40)}} = -.95$$

REFERENCES

Atkinson, A. C. 1985. *Plots, Transformations, and Regression.* Oxford: Clarendon Press.

Belsley, D., E. Kuh, and R. Welsh. 1980. *Regression Diagnostics: Identifying Influential Data and Sources of Collinearity.* New York: Wiley.

Bowerman, B., R. O'Connell, and D. Dickey. 1986. *Linear Statistical Models: An Applied Approach.* Boston: Duxbury Press.

Draper, N. R., and H. Smith. 1981. *Applied Regression Analysis,* 2nd ed. New York: Wiley.

Neter, J., W. Wasserman, and M. Kutner. 1989. *Applied Linear Regression Models,* 2nd ed. Homewood, Ill.: Irwin.

Snedecor, George W., and William G. Cochran. 1989. *Statistical Methods,* 8th ed. Ames: Iowa State University Press.

Younger, M. S. 1985. *A Handbook for Linear Regression,* 2nd ed. Boston: Duxbury Press.

For statistical inference and prediction in regression analysis, the regression model discussed in Section 12.1 included three conditions or assumptions that must be met by the population data.

**12.8
RESIDUAL ANALYSIS**

1. The X values are known—that is, the X values are not random.
2. For each value of X, Y is normally and independently distributed with mean $\mu_{Y|X}$ equal to $\beta_0 + \beta_1 X$ and variance $\sigma^2_{Y|X}$, where β_0, β_1, and $\sigma^2_{Y|X}$ are unknown parameters.
3. For each X, the variance of Y is the same—that is, $\sigma^2_{Y|X} = \sigma^2$ for all X.

The first condition implies that if we are to make predictions by using a regression equation, we must know the value of X to use in the regression equation—it will not materialize simultaneously with the Y value. The examples we have used satisfy this condition.

The second condition implies that for each value of X, the Y values will be normally distributed with mean $\mu_{Y|X}$ equal to $\beta_0 + \beta_1 X$. Figure 12.16 illustrates the meaning of this condition. It shows that for a fixed value of X, the distribution of Y_i values around the regression line follows the normal distribution. Each distribution has a mean $\mu_{Y|X}$ that falls on the linear regression line $\beta_0 + \beta_1 X$. The second condition also requires that the Y's should be *independently* distributed. Figure 12.17 shows an example where the Y values are not independently distributed. If we know the value of Y at a particular value of X, then we have a good idea of what the value of Y will be at the next value of X. For independence to exist, the Y values must be unrelated to one another at the different levels of X.

The third condition implies that the disturbances around the regression line must be the same for small values of X as they are for large values of X. The variance $\sigma^2_{Y|X}$ of each distribution shown on Figure 12.16 is the same, or each variance is a constant σ^2. The $\sigma^2_{Y|X}$ value is the population's equivalent of the sample's $s^2_{Y|X}$, which is the variance of the Y_i values around the sample regression line.

To check whether the data values used to estimate the model parameters conform to the assumptions of regression analysis, we use scatter diagrams and **residual analysis.** In residual analysis, we examine whether the residual values $e_i = Y_i - \hat{Y}_i$ are reasonably consistent with the model assumptions. Figure 12.18 shows scatter diagrams and **residual plots,** or plots of e_i against X (we could also plot e_i against $\hat{Y}_i$), to illustrate a case where the data are reasonably consistent with the assumptions and two cases in which the assumptions are not met. Figure 12.18(a) shows a set of data values that are consistent with

FIGURE 12.16

Linear Regression Model, with Normal Distribution of Y_i Values Around a Regression Line at $X = X_0$

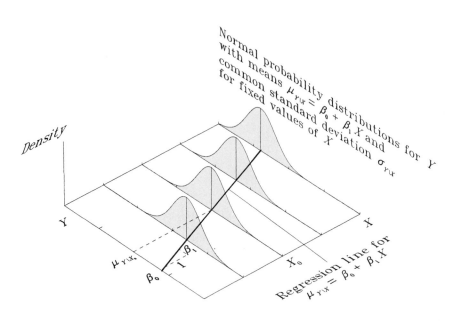

the assumption that the means of the distributions for Y for each value of X, $\mu_{Y|X}$, fall on the linear regression line $\beta_0 + \beta_1 X$. The residuals fall in a seemingly random pattern around the zero line. A nonlinear pattern of residuals and methods to adjust for nonlinearity are discussed in the next chapter.

The random pattern of residuals in Figure 12.18(a) also conforms reasonably with the requirement of constant variance, since the variation of the residuals is similar for each different value of X. In this situation the data are *homoscedastic*. In contrast,

FIGURE 12.17

Linear Regression Model, with Dependent Y Values

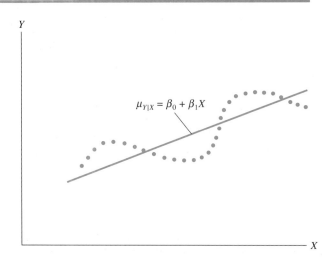

Linear Patterns of Residuals with Constant Variance and Nonconstant Variance **FIGURE 12.18**

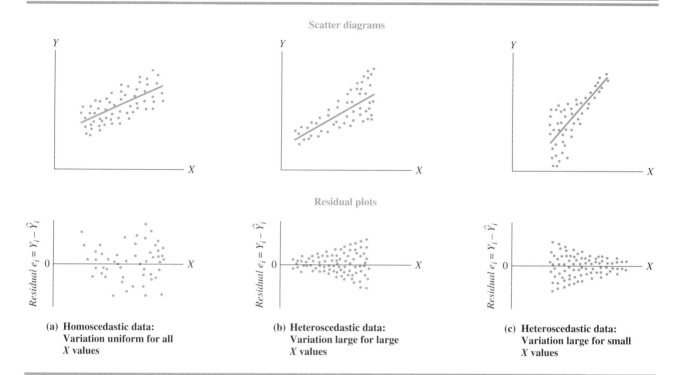

(a) **Homoscedastic data: Variation uniform for all X values**

(b) **Heteroscedastic data: Variation large for large X values**

(c) **Heteroscedastic data: Variation large for small X values**

Figures 12.18(b) and (c) show residual plots that have patterns in the residual plots that suggest that the variance of the residuals is not constant for different X values. In these cases we say that the data are *heteroscedastic*. The references at the end of the chapter discuss methods for transforming the data or changing the analytical methods to adjust for nonconstant variances, which is sometimes a problem for cross-sectional data, or data that has *not* been collected over time.

To informally check whether each value of X, Y is normally distributed, we can use a histogram to examine the distributional pattern of the residuals, or we can examine whether the residuals conform to the empirical rule for bell-shaped data (see Chapter 3). Computer Exhibit 12.5 shows a set of residuals from a regression analysis and the corresponding histogram that is reasonably consistent with the normality assumption. Formal tests of normality can also be applied to the residuals to check the normality assumption.

When the Y values for the regression model are obtained periodically over time and the Y_i are dependent, the error terms are termed *autocorrelated*. A time series plot of the residuals and a test for autocorrelation (the Durbin-Watson test) can be used to check time series residuals for consistency with the assumption of independence.

Some analysts prefer to plot standardized residuals or studentized deleted residuals against X (or against $\hat{Y}_i$). **Standardized residuals** (or studentized residuals) are sometimes defined as

$$\frac{e_i}{\sqrt{MSE - s_{\hat{Y}_i}^2}}$$

If the standardized residual for observation i is greater in absolute value than, say, 2, then observation i can be an *influential* observation on the parameter estimates or an *outlier*, and it may be omitted if the data are in error or it deserves further investigation.

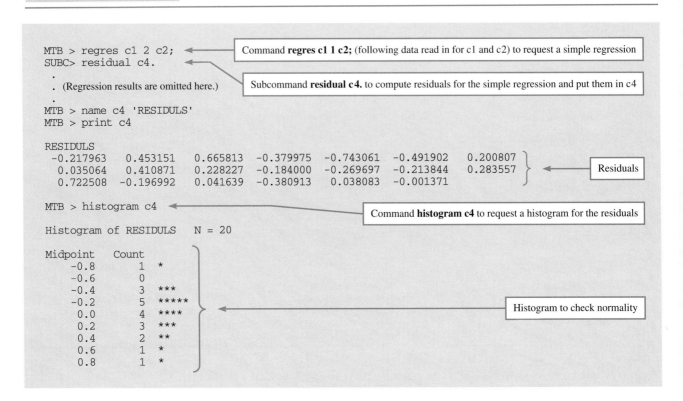

An alternative expression for the denominator of the standardized residual is

$$\sqrt{MSE(1 - h_i)}$$

where:
$$h_i = \frac{1}{n} + \frac{(X_i - \overline{X})^2}{SSX}$$

Here, h_i is referred to as a **leverage** measure, and if h_i is greater than $2(k + 1)/n$, then the observation may be *influential* on the parameter estimates or X_i may be an *outlier* compared to the other X data values.

Studentized deleted residuals are defined as

$$\frac{e_{(i)}}{\sqrt{MSE_{(i)} - s_{\hat{Y}_{(i)}}^2}}$$

where the notation (i) indicates that the data for observation i is removed during the regression analysis. If the studentized deleted residual for observation i is greater in absolute value than $t_{\alpha, n-k-2}$ when α is, say, .05 or .10, then observation i can be an *influential* observation. However, when a data set contains several influential observations, the influence of one observation can be **masked** by the influence of another observation so it can be difficult to locate all of the influential observations by these methods.

EXAMPLE 12.10 Standardized and studentized deleted residuals for the United Park City Properties condominium problem of Example 12.2 were found with the aid of a computer, and the results are presented in Computer Exhibits 12.6A and 12.6B. The largest standardized residual in absolute value is 1.59, so none of the observations are questionable by this method. The largest studentized deleted residual in absolute value is 3.33 for the fifth observation. Since $t_{.05,5-1-2} = 2.920$, this observation may be somewhat influential. The data values for the fifth observation were found to be valid so we will not discard the observation.

COMPUTER EXHIBIT 12.6A Standardized and Studentized Deleted Residuals for the United Park City Properties Data

MINITAB

```
MTB > read c1  c2
DATA >     36   9
DATA >     80  15
DATA >     44  10
DATA >     55  11
DATA >     35  10
DATA > end

MTB > name c1 'Price' c2 'Area' c3 'StdzdRes' c4 'StuDelRs'

MTB > regres c1 1 c2 c3;
SUBC> tresiduals c4.
```

Command **regres c1 1 c2 c3;** for a regression with standardized residuals saved in c3

Subcommand **tresiduals c4.** to save studentized deleted residuals in c4

```
The regression equation is
Price = - 34.5 + 7.68 Area

Predictor       Coef      Stdev    t-ratio        p
Constant      -34.50      12.62      -2.73    0.072
Area           7.682      1.127       6.82    0.006

s = 5.284       R-sq = 93.9%    R-sq(adj) = 91.9%

Analysis of Variance

SOURCE         DF         SS         MS        F        p
Regression      1     1298.2     1298.2    46.49    0.006
Error           3       83.8       27.9
Total           4     1382.0

Obs.    Area     Price      Fit Stdev.Fit  Residual   St.Resid
  1      9.0     36.00    34.64     3.27      1.36       0.33
  2     15.0     80.00    80.73     5.09     -0.73      -0.51
  3     10.0     44.00    42.32     2.62      1.68       0.37
  4     11.0     55.00    50.00     2.36      5.00       1.06
  5     10.0     35.00    42.32     2.62     -7.32      -1.59

MTB > print c3 c4

 ROW   StdzdRes    StuDelRs
   1    0.32821     0.27293
   2   -0.51034    -0.43605
   3    0.36639     0.30608
   4    1.05787     1.09085
   5   -1.59430    -3.33082
```

Standardized residuals and studentized deleted residuals

MYSTAT   **COMPUTER EXHIBIT 12.6B** Standardized and Studentized Deleted Residuals for the
United Park City Properties Data

```
>edit

    'SALESPRC'      'AREA'
          36           9
          80          15
          44          10
          55          11
          35          10

    >save PARKCITY
    >quit

>use PARKCITY
>save RESIDS    ◄─────────────────────────────────────  Command save RESIDS to save the residuals
>model SALESPRC = constant + AREA
>estimate

DEP VAR:SALESPRC      N:    5    MULTIPLE R:   .969    SQUARED MULTIPLE R:   .939
ADJUSTED SQUARED MULTIPLE R:   .919      STANDARD ERROR OF ESTIMATE:      5.284

    VARIABLE     COEFFICIENT     STD ERROR     STD COEF  TOLERANCE     T    P(2 TAIL)
    CONSTANT       -34.500         12.616        0.000       .       -2.735   0.072
        AREA         7.682          1.127        0.969    .100E+01    6.818   0.006

                       ANALYSIS OF VARIANCE

    SOURCE    SUM-OF-SQUARES     DF    MEAN-SQUARE      F-RATIO        P

    REGRESSION     1298.227       1      1298.227        46.491      0.006          Studentized deleted
    RESIDUAL         83.773       3        27.924                                   residual flagged by
                                                                                    MYSTAT
WARNING: Case    5 is an outlier     (STUDENTIZED RESIDUAL =        -3.331)
```

Problems: Section 12.8

67. Monthly percentage rates of return for a stock Y and the returns for an index of stocks X were given in Problem 46.

    a. Find the residuals when Y is regressed on X.
    b. Provide a check on the linearity assumption with a scatter plot and a residual plot.

68. Using the residuals found in Problem 67, provide a check on the normality regression assumption with a histogram of the residuals.

69. Use the monthly percentage rates of return for a stock Y and the returns for an index of stocks X that were given in Problem 46.

    a. Find the standardized (studentized) residuals.
    b. Find influential observations that have standardized residuals greater than 2 in absolute value.

70. Use the monthly percentage rates of return for a stock Y and the returns for an index of stocks X that were given in Problem 46.

    a. Find the studentized deleted residuals.
    b. Find influential observations that have studentized deleted residuals above 2.92 in absolute value.

Multiple Regression

In the previous chapter, we examined the relationship between a dependent variable Y and an independent variable X by using simple regression and correlation. In many business and economics problems, however, we are interested in the relationship between a dependent variable Y and several independent variables $X_1, X_2, \ldots, X_k$. For example, the amount of gas pumped at a service station Y might be related to the location of the station X_1 as well as the number of pumps X_2. Our objective in this chapter is to study the relationship between a dependent variable Y and several independent variables $X_1, X_2, \ldots, X_k$.

One way to study the relationship between a dependent variable Y and two or more X variables is **multiple regression.** Multiple regression equations are used when economic theory or reasoning suggest that the prediction of our dependent variable can be made more accurate by using more than one independent variable.

In this chapter, we present the multiple regression model and the procedures used in multiple regression analysis. We use the computer for calculations in multiple regression analysis; otherwise, the calculations are generally too burdensome for practical purposes. We will use regression for time series analysis in the following chapter.

13.1
MULTIPLE LINEAR REGRESSION

In this section, we present the multiple linear regression model in the context of the United Park City Properties real estate example that was introduced in simple regression. The real estate example is extended to include k different independent or X variables in this chapter.

To examine the relationship between a dependent variable Y and two independent variables X_1 and X_2, we obtain from the ith element in our sample the set of numbers (Y_i, X_{i1}, X_{i2}). If there are n elements in the sample, we have a sample of size n that consists of the n sets of values for Y, X_1 and X_2 as follows:

Y	X_1	X_2
$Y_1,$	$X_{11},$	X_{12}
$Y_2,$	$X_{21},$	X_{22}
.		
.		
.		
$Y_n,$	$X_{n1},$	X_{n2}

A sample of data for a dependent variable and k independent variables can be used to estimate the parameters of a multiple regression model.

EXAMPLE 13.1   Let us continue with the real estate example presented in Chapter 12. The United Park City Properties real estate investment firm is interested in estimating the price of condominiums by using a sample of five recently sold condominium units in a city. In addition to knowing the data on sales price Y and area X_1 for each condominium that was presented in Section 12.2, the firm also knows the land costs X_2 that can be assigned to each of the condominium units. In addition, it knows several other variables, including the number of bedrooms in each condominium. The multiple linear regression model would be used for this example, which has k different independent or X variables.

With k independent variables, $X_1, X_2, \ldots, X_k$, the **multiple linear regression model** is as follows.

Multiple Linear Regression Model

$$Y_i = \beta_0 + \beta_1 X_{i1} + \beta_2 X_{i2} + \cdots + \beta_k X_{ik} + \epsilon_i \qquad (13.1)$$

where: Y_i = the value of the *dependent* variable for the ith observation

X_{ij} = the value of the jth *independent* variable, for $j = 1, 2, \ldots, k$, for the ith observation

ϵ_i = the value of a *random fluctuation* or *error* for the ith observation

β_0 = a parameter that represents the population regression *intercept,* or β_0 is the mean of Y when the values for X_1, $X_2, \ldots, X_k$ are all equal to zero

β_j = the parameter that represents the *slope* of the population regression surface with respect to the jth independent variable, or β_j indicates the change in the mean value of Y for a unit increase in the independent variable X_j, when all other independent variables in the model are held constant

---

We will make the following **assumptions** for the multiple linear regression model.*

1. The X values for each independent variable are known—that is, are not random.
2. For each set of values for $X_1, X_2, \ldots, X_k$, Y is normally and independently distributed with mean $\mu_{Y|X_1, X_2, \ldots, X_k}$ equal to $\beta_0 + \beta_1 X_1 + \beta_2 X_2 + \cdots + \beta_k X_k$ and variance $\sigma^2_{Y|X_1, X_2, \ldots, X_k}$ where $\beta_0, \beta_1, \beta_2, \ldots, \beta_k$ and $\sigma^2_{Y|X_1, X_2, \ldots, X_k}$ are unknown parameters.
3. For each set of values for $X_1, X_2, \ldots, X_k$, the variance of Y given $X_1, X_2, \ldots, X_k$ is the same—that is, $\sigma^2_{Y|X_1, X_2, \ldots, X_k} = \sigma^2$ for all sets of $X_1, X_2, \ldots, X_k$ values.

Under these assumptions the **population multiple linear regression equation** is as follows.

Population Multiple Linear Regression Equation

$$\mu_{Y|X_1, X_2, \ldots, X_k} = \beta_0 + \beta_1 X_1 + \beta_2 X_2 + \cdots + \beta_k X_k \qquad (13.2)$$

That is, the mean of Y for any set of fixed $X_1, X_2, \ldots, X_k$ is equal to $\beta_0 + \beta_1 X_1 + \beta_2 X_2 + \cdots + \beta_k X_k$, where $\beta_0, \beta_1, \beta_2, \ldots, \beta_k$ are the population intercept and slopes, respectively.

The methods of analysis developed in the following sections enable us to use a random sample of data to estimate the parameters $\beta_0, \beta_1, \beta_2, \ldots, \beta_k$, and $\sigma^2_{Y|X_1, X_2, \ldots, X_k}$.

The parameters of the population multiple linear regression Equation (13.2) can be estimated by using sample data to obtain the **sample, or estimated, multiple linear regression equation.**

Sample (or Estimated) Multiple Linear Regression Equation

$$\hat{Y} = b_0 + b_1 X_1 + b_2 X_2 + \cdots + b_k X_k \qquad (13.3)$$

In Equation (13.3) the value of $\hat{Y}$ is the estimate of the mean value of Y given $X_1, X_2, \ldots, X_k$, or $\hat{Y}$ estimates $\mu_{Y|X_1, X_2, \ldots, X_k}$; the value of b_0 that is determined from the sample data is the estimate of the population regression intercept β_0; and the values of the b_j's that are determined from the sample data are the estimates of the population regression slopes, the β_j's.

In this chapter, as in Chapter 12, we will assume that the regression equation desired in each case has the linear form of Equation (13.1). That is, we will deal with the case in which the independent variables, the X_j's, in the equation have an additive impact on Y. In this section, we will deal with the problem of how we use sample data to make estimates of the regression equation's coefficients.

13.2

FINDING THE COEFFICIENTS FOR A MULTIPLE REGRESSION EQUATION BY USING A COMPUTER

*Alternatively, the assumptions of the multiple regression model are that the value of X for each independent variable is known, and ϵ_i are independent and normally distributed with mean 0 and variance σ^2.

EXAMPLE 13.2   The United Park City Properties real estate investment firm described in Example 13.1 is now interested in estimating the price of condominiums Y by using just two independent variables: the area of the condominium X_1 and the land cost that can be assigned to each condominium X_2. (To find this cost, take the total cost of the land and divide it by the number of units in the condominium complex.) The land costs would likely reflect the desirability of the condominium units' location; and the more desirable locations, all other things being equal, would likely contain the higher-priced units.

Consider the data in Table 13.1, which are the data presented in Section 12.2 with the land costs added. This table of numbers indicates that the first condominium sold for \$36,000, contained 900 square feet of area, and was built on land that cost \$8000 per unit.

Our objective is to find the values of b_0, b_1, and b_2, which can be used as coefficients in the regression equation represented by the following equation:

$$\hat{Y} = b_0 + b_1 X_1 + b_2 X_2 \tag{13.4}$$

Equation (13.4) in multiple regression, like corresponding Equation (12.5) in simple regression, is a linear equation. The multiple regression equation (13.4) cannot be graphed in two dimensions. Three dimensions would be needed, one for each of three variables Y, X_1, and X_2. The three-dimensional scatter diagram would appear as a series of points scattered around a plane rather than a line, the plane being all the points that satisfy the equation. Figure 13.1 shows a colored surface as an example of such a plane.

EXAMPLE 13.3   The value of b_0 is the intercept of the sample regression plane with the Y-axis, and it forms the beginning point for the sample regression equation. The value of b_1 is an estimate of the average price change (in thousands of dollars) associated with a one-unit change (100 square feet) in X_1, assuming X_2 is held constant. Likewise, b_2 estimates the average change in price (in thousands of dollars) associated with a one-unit change (\$1000) in land costs, assuming X_1 is held constant.

To determine numerical values for b_0, b_1, and b_2, we again use the least squares method introduced in Section 12.2. However, in this case we find the values of b_0, b_1, and b_2 that minimize the sum of the squared errors $SSE = \Sigma (Y_i - \hat{Y}_i)^2$ or the sum of the squared vertical distances between the (Y, X_1, X_2) points in the three-dimensional space and the plane determined by the sample regression equation (13.4).

TABLE 13.1	$Y = Sales\ Price$ (× \$1000)	$X_1 = Area\ (square\ feet)$ (× \$100)	$X_2 = Land\ Costs$ (× \$1000)
Condominium Data for United Park City Properties	36	9	8
	80	15	7
	44	10	9
	55	11	10
	35	10	6

FIGURE 13.1

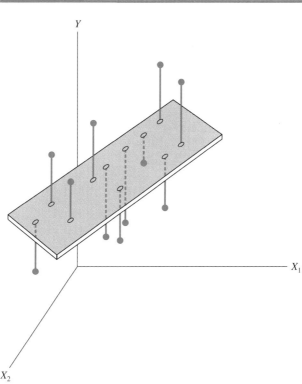

Regression Plane with One Dependent Variable, *Y*, and Two Independent Variables, *X*₁ and *X*₂

The coefficients of Equation (13.4), with Y as the dependent variable and X_1 and X_2 as the independent variables, can be found by solving a system of three simultaneous linear equations in which b_0, b_1, and b_2 are the unknowns.* However, a computer is usually used to find b_0, b_1, and b_2. The b_0, b_1, and b_2 that are derived by the method of least squares are minimum-variance unbiased estimators of the unknown parameters β_0, β_1, and β_2 for the multiple regression model.

EXAMPLE 13.4   The most common method for finding the estimated regression equation (13.4), and the other regression results that we will discuss in Sections 13.3 through 13.5, is by using data analysis software, such as MYSTAT, MINITAB, SAS, SPSS, BMDP, and so forth. In Computer Exhibits 13.1A, 13.1B, and 13.1C, we present MINITAB, SAS, and MYSTAT printouts for the United Park City Properties condominium problem that we have been studying, where Y is the sales price (in thousands of dollars), X_1 is the area (in hundreds of square feet), and X_2 is the land cost (in thousands of dollars).

*These equations are sometimes referred to as the **normal equations** and are as follows:

$$\sum Y = b_0 n + b_1 \sum X_1 + b_2 \sum X_2$$
$$\sum X_1 Y = b_0 \sum X_1 + b_1 \sum X_1^2 + b_2 \sum X_1 X_2$$
$$\sum X_2 Y = b_0 \sum X_2 + b_1 \sum X_1 X_2 + b_2 \sum X_2^2$$

Thus, the estimated multiple linear regression equation, with the intercept and slope coefficients expressed with two significant digits to the right of the decimal point, for the condominium example is

$$\hat{Y} = -61.29 + 8.07X_1 + 2.82X_2$$

The intercept value is $b_0 = -61.29$ (meaning that a condominium with no area, $X_1 = 0$, and built on land that cost nothing, $X_2 = 0$, which is impossible, would sell for $-\$61,290$). The 8.07 coefficient on the X_1 term indicates that an additional 100 square feet in area is associated with an additional \$8070 in price, as-

MINITAB	**COMPUTER EXHIBIT 13.1A** Multiple Regression for the United Park City Properties Condominium Data

```
MTB > read c1 c2 c3          ◄──── Command read to enter the condominium data into c1 (sales price), c2 (area), and c3 (land cost)
DATA > 36  9  8
DATA > 80 15  7
DATA > 44 10  9
DATA > 55 11 10
DATA > 35 10  6
DATA > end
       5 ROWS READ

MTB > name c1 'SALESPRC' c2 'AREA' c3 'LANDCOST'

MTB > regres c1 2 c2 c3;     ◄──── Command regres for c1 as the dependent variable with 2 independent variables c2 and c3
SUBC> predict 10 9.

The regression equation is                         Subcommand predict to obtain a confidence interval
SALESPRC = - 61.3 + 8.07 AREA + 2.82 LANDCOST       and a prediction interval for X_1 = 10 and X_2 = 9

Predictor      Coef       Stdev      t-ratio        p
Constant     -61.289 ①    7.531       -8.14      0.015
AREA           8.0664 ②   0.4218 ⑥    19.12 ⑧   0.003
LANDCOST       2.8199 ③   0.6256 ⑦     4.51 ⑨   0.046

S = 1937 ④      R-sq = 99.5% ⑤   R-sq(adj) = 98.9%

Analysis of Variance

SOURCE        DF          SS          MS         F         p
Regression     2      1374.49      687.25    183.09    0.005          ◄──── ANOVA table
Error          2         7.51        3.75
Total          4      1382.00

SOURCE        DF      SEQ SS
AREA           1     1298.23
LANDCOST       1       76.27

⑩   Fit  Stdev.Fit       95% C.I. ⑪        95% P.I. ⑫
    44.754     1.101   ( 40.014, 49.493)  ( 35.165, 54.343)
```

Note: ① b_0, estimated regression intercept; ② b_1, estimated regression coefficient for X_1; ③ b_2, estimated regression coefficient for X_2; ④ $S_{Y|12}$, estimated standard deviation of Y_1 values around regression surface; $S_{Y|12} = \sqrt{MSE}$; ⑤ R^2, coefficient of multiple determination (as a percentage); ⑥ s_{b_1}, standard error of coefficient b_1; ⑦ s_{b_2}, standard error of coefficient b_2; ⑧ t, t statistic for testing H_0: $\beta_1 = 0$; ⑨ t, t statistic for testing H_0: $\beta_2 = 0$; ⑩ $\hat{Y}$ for $X_1 = 10$ and $X_2 = 9$; ⑪ 95% confidence interval for $\mu_{Y|X_1, X_2}$ given $X_1 = 10$ and $X_2 = 9$; ⑫ 95% prediction interval for an individual value Y_0 given $X_1 = 10$ and $X_2 = 9$.

COMPUTER EXHIBIT 13.1B Multiple Regression for the United Park City Properties Condominium Data   **SAS**

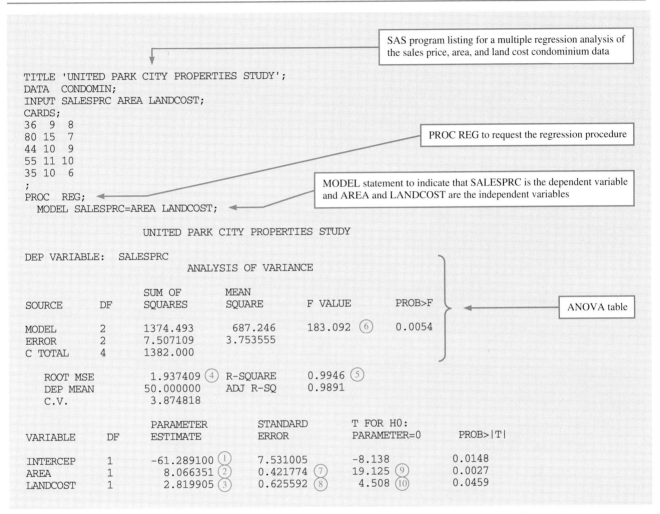

SAS program listing for a multiple regression analysis of the sales price, area, and land cost condominium data

```
TITLE 'UNITED PARK CITY PROPERTIES STUDY';
DATA   CONDOMIN;
INPUT SALESPRC AREA LANDCOST;
CARDS;
36  9  8
80 15  7
44 10  9
55 11 10
35 10  6
;
PROC  REG;
    MODEL SALESPRC=AREA LANDCOST;
```

PROC REG to request the regression procedure

MODEL statement to indicate that SALESPRC is the dependent variable and AREA and LANDCOST are the independent variables

UNITED PARK CITY PROPERTIES STUDY

DEP VARIABLE: SALESPRC

ANALYSIS OF VARIANCE

SOURCE	DF	SUM OF SQUARES	MEAN SQUARE	F VALUE	PROB>F
MODEL	2	1374.493	687.246	183.092 ⑥	0.0054
ERROR	2	7.507109	3.753555		
C TOTAL	4	1382.000			

ANOVA table

ROOT MSE	1.937409 ④	R-SQUARE	0.9946 ⑤
DEP MEAN	50.000000	ADJ R-SQ	0.9891
C.V.	3.874818		

VARIABLE	DF	PARAMETER ESTIMATE	STANDARD ERROR	T FOR H0: PARAMETER=0	PROB>|T|
INTERCEP	1	-61.289100 ①	7.531005	-8.138	0.0148
AREA	1	8.066351 ②	0.421774 ⑦	19.125 ⑨	0.0027
LANDCOST	1	2.819905 ③	0.625592 ⑧	4.508 ⑩	0.0459

Note: ① b_0, estimated regression intercept; ② b_1, estimated regression coefficient for X_1; ③ b_2, estimated regression coefficient for X_2; ④ $S_{Y|12}$, estimated standard deviation of Y_i values around regression surface; $S_{Y|12} = \sqrt{MSE} = $ ROOT MSE; ⑤ R^2, coefficient of multiple determination; ⑥ F, F statistic for testing H_0: $\beta_1 = \beta_2 = 0$; ⑦ s_{b_1}, standard error of coefficient b_1; ⑧ s_{b_2}, standard error of coefficient b_2; ⑨ t, t statistic for testing H_0: $\beta_1 = 0$; ⑩ t, t statistic for testing H_0: $\beta_2 = 0$.

suming X_2 is held constant. The 2.82 coefficient on the X_2 term indicates that an additional \$1000 in land cost is associated with an additional \$2820 in price, assuming X_1 is held constant.

Notice that many other results for the regression analysis are also presented in Computer Exhibit 13.1. We will discuss these results in Sections 13.3 through 13.5. Compare the values in the printout with the various values we will compute by using the formulas we present in Sections 13.3 through 13.5, and note that they are the same (except for rounding). If you have access to a multiple regression computer program that is different from those we have used, you can generally adapt to the printouts from other programs after studying the results presented here.

MYSTAT	**COMPUTER EXHIBIT 13.1C** Multiple Regression for the United Park City Properties Condominium Data

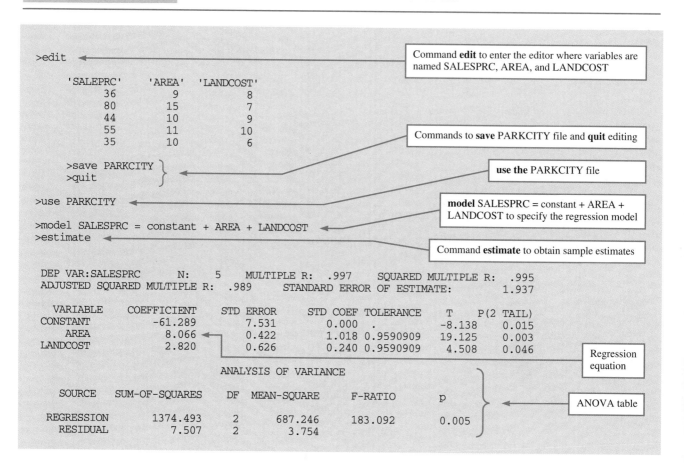

```
>edit  ◄──────────────────────────────────     Command edit to enter the editor where variables are
                                                named SALESPRC, AREA, and LANDCOST

        'SALEPRC'    'AREA'   'LANDCOST'
            36          9          8
            80         15          7
            44         10          9         Commands to save PARKCITY file and quit editing
            55         11         10
            35         10          6

        >save PARKCITY ⎫                          use the PARKCITY file
        >quit          ⎬ ◄──────────────
                       ⎭
>use PARKCITY  ◄─────────────────────────      model SALESPRC = constant + AREA +
                                               LANDCOST to specify the regression model
>model SALESPRC = constant + AREA + LANDCOST  ◄──
>estimate  ◄───────────────────────────────    Command estimate to obtain sample estimates


DEP VAR:SALESPRC      N:    5   MULTIPLE R:   .997   SQUARED MULTIPLE R:   .995
ADJUSTED SQUARED MULTIPLE R:   .989     STANDARD ERROR OF ESTIMATE:        1.937

    VARIABLE    COEFFICIENT   STD ERROR    STD COEF  TOLERANCE     T    P(2 TAIL)
    CONSTANT       -61.289       7.531       0.000       .       -8.138   0.015
        AREA         8.066 ◄     0.422       1.018  0.9590909    19.125   0.003
    LANDCOST         2.820       0.626       0.240  0.9590909     4.508   0.046

                          ANALYSIS OF VARIANCE

     SOURCE     SUM-OF-SQUARES   DF   MEAN-SQUARE     F-RATIO       p

  REGRESSION       1374.493       2     687.246      183.092     0.005
    RESIDUAL          7.507       2       3.754
```

Regression equation

ANOVA table

The examples in the next three sections involve only two independent variables, X_1 and X_2. But all the results discussed in Sections 13.3 through 13.5 can be extended to problems involving any number of independent variables.

Problems: Section 13.2

Answers to odd-numbered problems are given in the back of the text.

1. A regression equation was found to be $\hat{Y} = 20 + 14X_1 - 7X_2$. Which of the following statements are correct?

   a. A 1-unit increase in X_1 causes Y to increase by 14 units.
   b. Variable Y is more highly correlated with X_1 than with X_2 since the coefficient on X_1 is positive.
   c. If the value of X_2 is large enough, one can obtain negative predictions of Y.

2. An analyst for Goldman-Sachs found that the closing prices of six stocks on the last day of last month seem to be quite highly correlated with their latest reported earnings-per-share figures and with the percentage of earnings growth they experienced in the

past year. The figures are given in the accompanying table. Use a computer to show that the regression equation is approximately $\hat{Y} = .72 + 5.94X_1 + 1.08X_2$.

Y, Closing Price per Share	X₁, Latest Earnings per Share	X₂, Percentage Earnings Growth
$10	$1.50	3
18	2.00	10
22	2.00	8
30	2.50	6
30	3.00	10
40	7.00	−1

3. For the regression equation developed in Problem 2, give the units and physical interpretation of each coefficient: b_0, b_1 and b_2.

4. The tax legislation committee of a state's legislature is considering a proposal to give residents of the state a tax rebate. The committee feels that the rebate would be spent in the state and thus boost the state's economy. The last time the legislature gave the residents of the state a tax rebate was in 1982. To estimate the impact of a rebate on spending in the state, the chairman of the committee asked a legislative analyst to gather data concerning the ways families spent rebates in 1982. The analyst selected five 1982 rebate recipients (in a realistic problem, many more would be selected) and determined how they spent their rebates, using extensive interviews. The accompanying table is one of the tables from the report. Construct the regression equation relating these three variables, letting Y = percentage of rebate spent, X_1 = family income (in thousands of dollars), and X_2 = number of dependents in the family, and then show that the result is $\hat{Y} = 19 - 1.0X_1 + 12.3X_2$. Interpret the meaning of the constant and the two regression coefficients in this equation.

Percentage of Rebate Spent	Family Income (× $1000)	Number of Dependents
10	27	1
35	30	4
50	13	4
55	11	5
100	9	6

13.3
PARTITIONING THE SUM OF SQUARES IN MULTIPLE REGRESSION

In previous sections of this chapter, we have discussed the concept of multiple regression analysis and methods to find the intercept and slopes of the sample regression surface. Several topics in multiple regression that will be discussed in sections that follow are related to how the variation in the dependent variable Y of a multiple regression problem can be partitioned into two parts. One reason for partitioning the sum of squares in multiple regression is that many of the computations in the following sections are facilitated by use of the partitioning that we present in this section.

In multiple regression analysis, as well as in simple regression analysis [see Section (12.3)], the **total sum of squared deviations** of the dependent variable Y from its mean $\overline{Y}$ is labeled *SST* and can be partitioned into the **sum of squares due to re-**

gression, labeled *SSR*, and the **sum of squares of residuals or of deviations due to error,** labeled *SSE,* giving Equation (13.5).

Equation for Partitioning Sum of Squares in Multiple Regression

$$\sum(Y_i - \overline{Y})^2 = \sum(\hat{Y}_i - \overline{Y})^2 + \sum(Y_i - \hat{Y}_i)^2$$
$$SST \quad = \quad SSR \quad + \quad SSE$$

(13.5)

Thus, the same type of partitioning of the dependent variable Y can be accomplished in multiple regression. However, there are no simple, short computational formulas in multiple regression for finding the separate terms for SSR and SSE of Equation (13.5). We must calculate the individual $\hat{Y}_i$ values to achieve the partitioning of Equation (13.5), or we can allow the computer to find the sums of squares.

In multiple regression, the sum of squares for regression SSR has k degrees of freedom; the sum of squares due to error SSE has $n - k - 1$ degrees of freedom; and the total sum of squares SST has, as usual, $n - 1$ degrees of freedom. Equation (13.6) shows that the **degrees of freedom** are additive:

$$(n - 1) = (k) + (n - k - 1)$$

(13.6)

Recall from the analysis of variance that **mean squares** are sums of squares divided by their associated degrees of freedom. Two mean squares that are important in multiple regression are the mean square due to regression MSR and the mean square due to error MSE, which are defined as follows.

Mean Square Due to Regression *MSR* and Mean Square Due to Error *MSE* for Multiple Regression

$$MSR = \frac{SSR}{k}$$

(13.7)

$$MSE = \frac{SSE}{n - k - 1}$$

(13.8)

The equations for sums of squares, degrees of freedom, and mean squares are summarized in an analysis of variance table in Table 13.2, which corresponds to Table 12.2 in simple regression. The sums of squares and the degrees of freedom are additive, whereas the mean squares are not.

The analysis of variance table in multiple regression is usually found by using a computer. The printouts in Computer Exhibit 13.1 give the analysis of variance table for the United Park City Properties condominium data. In addition, we use the terms of Equation (13.5) to find the analysis of variance table for the condominium data in Example 13.5.

Source of Variation	Degrees of Freedom	Sum of Squares	Mean Square	F	**TABLE 13.2**
Regression	k	$SSR = \Sigma(\hat{Y}_i - \overline{Y})^2$	$MSR = \dfrac{SSR}{k}$	$F = \dfrac{MSR}{MSE}$	**Analysis of Variance Table for Multiple Regression**
Error	$n - k - 1$	$SSE = \Sigma(Y_i - \hat{Y}_i)^2$	$MSE = \dfrac{SSE}{n - k - 1}$		
Total	$n - 1$	$SST = \Sigma(Y_i - \overline{Y})^2$ $= \Sigma Y^2 - \dfrac{(\Sigma Y)^2}{n}$			

EXAMPLE 13.5   Table 13.3 shows the calculations of the terms in partitioning of prices in our condominium sales example. With the exception of rounding differences Equation (13.5) holds since $\Sigma(Y_i - \hat{Y}_i)^2 + \Sigma(\hat{Y}_i - \overline{Y})^2 = 7.5 + 1374.5$ is equal to the value of $\Sigma(Y_i - \overline{Y})^2 = 1382$.

The value for degrees of freedom associated with SST is $n - 1 = 5 - 1 = 4$; SSR has $k = 2$ degrees of freedom, and SSE has $n - k - 1 = 5 - 2 - 1 = 2$ degrees of freedom. Notice that the degrees of freedom are additive since the degrees of freedom for SST is 4, which is the sum of the 2 degrees of freedom for SSR and the 2 degrees of freedom for SSE.

The MS values were obtained for the sample data by using Equations (13.7) and (13.8), and the values are

$$MSR = \frac{SSR}{2} = \frac{1374.5}{2} = 687.25$$

$$MSE = \frac{SSE}{n - k - 1} = \frac{7.5}{5 - 2 - 1} = 3.75$$

Table 13.4 is the analysis of variance table that summarizes the values for the degrees of freedom, sums of squares, and mean squares for the condominium sales example. We will use these values in Sections 13.4 and 13.5 to measure how well the data fit the model and for purposes of inference in multiple regression. The minor differences between the values in Table 13.4 and those given in the printouts of Computer Exhibit 13.1 are due to rounding.

Calculations for Sums of Squares for United Park City Properties   **TABLE 13.3**

Item	Y_i	$\hat{Y}_i$	$\hat{Y}_i - \overline{Y}$	$(\hat{Y}_i - \overline{Y})^2$	$Y_i - \hat{Y}_i$	$(Y_i - \hat{Y}_i)^2$	$Y_i - \overline{Y}$	$(Y_i - \overline{Y})^2$
1	36	33.8673	−16.1327	260.264	2.13270	4.54842	−14	196
2	80	79.4455	29.4455	867.037	0.55450	0.30748	30	900
3	44	44.7536	−5.2464	27.525	−0.75356	0.56785	−6	36
4	55	55.6398	5.6398	31.807	−0.63981	0.40936	5	25
5	35	36.2938	−13.7062	187.859	−1.29384	1.67402	−15	225
Total	250		0.0000	1374.492	0.0000	7.50711	0	1382
	$\overline{Y} = 250/5 = 50$			$SSR = 1374.492$		$SSE = 7.50711$		$SST = 1382$

TABLE 13.4	Source of Variation	Degrees of Freedom	Sum of Squares	Mean Square	F
Analysis of Variance Table for the Condominium Example, Multiple Regression	Regression	2	1374.5	687.25	183.1
	Error	2	7.5	3.75	
	Total	4	1382.0		

Problems: Section 13.3

5. A multiple regression analysis, with sales (in dollars) of a product as the dependent variable Y and advertising expenditure X_1 and market share X_2 as the independent variables, was done with a sample of $n = 10$ stores. The values for the sum of squares due to regression and the sum of squares total were found to be $SSR = 15$ and $SST = 24$.
   a. Find SSE.
   b. Find the values for the degrees of freedom associated with SSR, the degrees of freedom associated with the sum of squares due to error SSE, and the degrees of freedom associated with SST.
   c. Find MSR and MSE.
   d. Present your results for parts a, b, and c in a summary table.

6. Refer to the Goldman-Sachs analyst's data for Problem 2 for the multiple regression analysis of stock price with earnings per share and growth in earnings.
   a. Find SSR, SSE, and SST.
   b. Find the value for the degrees of freedom associated with each of the sums of squares in part a.
   c. Find MSR and MSE.
   d. Present your results in an analysis of variance table.

7. Use the data of Problem 4 for the multiple regression analysis of rebate spent with income and number of dependents.
   a. Find SSR, SSE, and SST.
   b. Find the value for the degrees of freedom associated with each of the sums of squares in part a.
   c. Find MSR and MSE.
   d. Present your results in an analysis of variance table.

13.4
STANDARD ERROR OF THE ESTIMATE AND COEFFICIENT OF MULTIPLE DETERMINATION R^2

In Section 12.4, we learned that the accuracy of prediction of a simple regression line could be measured in two ways. First, it was measured by the variance of the errors, $s_{Y|X}^2$. This value measures the dispersion of the Y_i values around the *simple regression line* $\hat{Y} = b_0 + b_1 X$. In *multiple regression* there is a similar term that measures the dispersion of the Y_i values around the regression surface. The dispersion of the residuals for a multiple regression equation with k independent variables, $\hat{Y} = b_0 + b_1 X_1 + b_2 X_2 + \cdots + b_k X_k$, is expressed as the following variance.

> **Equation for Sample Mean Squared Error *MSE* in Multiple Regression**
>
> $$S_{Y|12\ldots k}^2 = \frac{1}{n-k-1}\sum(Y_i - \hat{Y}_i)^2 = \frac{1}{n-k-1}(SSE) = MSE$$
>
> (13.9)

The symbol $S^2_{Y|12\ldots k}$ represents the variance of Y around the regression surface obtained for the variables $X_1, X_2, \ldots, X_k$. For the multiple regression model, $S^2_{Y|12\ldots k}$ is an unbiased estimator of $\sigma^2_{Y|X_1, X_2, \ldots, X_k}$.

The square root of $S^2_{Y|12\ldots k}$ is the standard deviation of the Y_i values around the regression surface and is known as the **standard error of the estimate.**

> **Equation for Standard Error of the Estimate $S_{Y|12\ldots k}$ in Multiple Regression**
>
> $$S_{Y|12\ldots k} = \sqrt{S^2_{Y|12\ldots k}} = \sqrt{MSE} \qquad (13.10)$$

In Section 12.4, we also learned that in regression problems the accuracy of prediction is evaluated in terms of the squared errors. The coefficient of determination r^2, which we developed to measure this accuracy in simple regression problems, has a direct parallel in multiple regression known as the **coefficient of multiple determination R^2** and is given next.

> **Definition and Equation for Coefficient of Multiple Determination**
>
> The **coefficient of multiple determination R^2** is the proportion of the total variation of Y that is explained by the relationship between Y and the X variables and is given by the following equation:
>
> $$R^2 = 1 - \frac{\sum(Y_i - \hat{Y}_i)^2}{\sum(Y_i - \overline{Y})^2} = 1 - \frac{SSE}{SST} = \frac{SSR}{SST} \qquad (13.11)$$

The interpretation given to r^2 in simple regression is essentially equivalent to the interpretation of its counterpart, R^2.* The coefficient of multiple determination R^2 is the proportion of squared error in estimating the Y_i values that can be eliminated by using the regression equation $\hat{Y} = b_0 + b_1 X_1 + b_2 X_2 + \cdots + b_k X_k$ as the estimator rather than $\overline{Y}$. The multiple coefficient of determination R^2, like r^2, is a unitless number between 0 and 1. If there is no linear relationship whatever between the dependent variable Y and the independent variables used in the regression equation, $R^2 = 0$. However, if the regression equation fits the data perfectly and can be used to make exact estimates or predictions of the Y_i values, then $R^2 = 1$.

══ **EXAMPLE 13.6** The standard error of the estimate given in the printouts of Computer Exhibit 13.1 is $S_{Y|12} = 1.937$, and the coefficient of multiple determination is $R^2 = .995$, or 99.5%. We will use Equations (13.10) and (13.11) to find these values for the United Park City Properties condominium data.

---

*We use the lowercase $s_{Y|X}$ and r^2 figures when referring to the sample standard error of estimate and the coefficient of determination in *simple,* single-independent-variable regression problems. The uppercase $S_{Y|12\ldots k}$ and R^2 will refer to the corresponding measures in *multiple* regression problems.

We use the United Park City Properties regression equation relating condominium prices to area and land costs to find the standard error of the estimate $S_{Y|12}$ and the coefficient of multiple determination R^2. In Example 13.5 the regression equation was found to be $\hat{Y} = -61.29 + 8.07X_1 + 2.82X_2$. For the five condominiums in the sample problem the areas and land costs were X_1 and X_2, and the sums of squares were found to be $SST = 1382$, $SSR = 1374.5$, and $SSE = 7.5$. Then,

$$S_{Y|12}^2 = \frac{1}{n - k - 1}(SSE) = \frac{1}{5 - 2 - 1}(7.5) = 3.75$$

$$S_{Y|12} = \sqrt{S_{Y|12}^2} = \sqrt{MSE} = \sqrt{3.75} = 1.94$$

The multiple regression equation's standard error of estimate is $S_{Y|12} = \$1.94$ (thousand). If the condominium prices are normally distributed around the regression plane, *about* two-thirds of the prices should fall within $\pm(1)S_{Y|12} = \pm1.94$ of the value estimated by the regression equation.

The regression equation is a great improvement over the mean condominium price, $\overline{Y} = \$50,000$, as a predictor of a condominium's price. This improvement is indicated by the R^2 value, calculated as follows:

$$R^2 = 1 - \frac{SSE}{SST} = \frac{SSR}{SST} = \frac{1374.5}{1382} = .995$$

This value of R^2 implies that 99.5% of the squared error, or variation, in estimating condominium prices can be eliminated by using the regression equation to predict prices rather than saying each condominium will probably sell for the average price of $\overline{Y} = \$50,000$.

The coefficient of multiple determination R^2 is an important summary statistic that is used to help evaluate how well the multiple regression model fits the data. However, it is all too easy to allow a high R^2 to become an objective of multiple regression model building, rather than to use R^2 as an indicator of how well a model that is developed according to business or economic theory fits the data. Artificially high R^2 values can be obtained if the sample includes just a few observations. Indeed, most examples included in textbooks use a reduced number of observations to relieve the student from more computations than are necessary to illustrate the procedures. In actuality, many multiple regression equations are developed by using a large sample size.

Also, the R^2 value will generally increase as more independent variables are included in a multiple regression equation, given a fixed number of observations. The reason is that as additional independent variables X_j's are included in a regression equation, the value of SST does not change, but SSR generally increases; so SSE generally decreases, and therefore R^2 generally increases. The additional independent variables may not contribute *significantly* to the explanation of the dependent variable Y, but they do increase R^2. Adding more independent variables in the regression equation for the purpose of increasing R^2 often results in "overfitting" and may result in worse models rather than better ones.

To help prevent overfitting in regression analysis, we can use an *Adjusted R^2* value as the measure of how well the model fits the data. The **Adjusted R^2** value incorporates the effect of including additional independent variables in a multiple regression equation. Indeed, the term *adjusted* means "adjusted for degrees of freedom."

Many multiple regression computer programs print a value for the *Adjusted* R^2. This value is computed by the following formula:

$$Adjusted\ R^2 = 1 - \frac{\sum(Y_i - \hat{Y}_i)^2/(n - k - 1)}{\sum(Y_i - \overline{Y})^2/(n - 1)}$$

$$= 1 - \frac{(SSE)/(n - k - 1)}{(SST)/(n - 1)} \qquad (13.12)$$

In our condominium example the *Adjusted* R^2 value is

$$Adjusted\ R^2 = 1 - \frac{7.5/(5 - 2 - 1)}{1382/(5 - 1)} = .989$$

When n is large, the two R^2 values are very close. The *Adjusted* R^2 value may get smaller when an additional independent variable is included in a regression equation, because even though the unexplained variation, *SSE*, almost always decreases, this decrease may be offset since the associated degrees of freedom, $(n - k - 1)$, also decreases. Consequently, *Adjusted* R^2's are often used to compare the explanatory abilities of different multiple regression equations.

Problems: Section 13.4

8. A multiple regression analysis with sales (in dollars) of a product as the dependent variable Y, and advertising expenditure X_1 and market share X_2 as the independent variables, was done with a sample of $n = 10$ Nordstock, Inc., retail outlets. The values for the sum of squares due to regression and the sum of squares total were found to be $SSR = 15$ and $SST = 24$.

    a. Find $S_{Y|12}^2$.
    b. Find the standard error of the estimate $S_{Y|12}$.
    c. Find the coefficient of multiple determination R^2.

9. Refer to Problem 2 for the Goldman-Sachs analyst's multiple regression analysis of stock price with earnings per share and growth in earnings.

    a. Find $S_{Y|12}^2$.
    b. Find the standard error of the estimate $S_{Y|12}$.
    c. Find the coefficient of multiple determination R^2.

10. Refer to Problem 4 for the multiple regression analysis of rebate spent with income and number of dependents.

    a. Find $S_{Y|12}^2$.
    b. Find the standard error of the estimate $S_{Y|12}$.
    c. Find the coefficient of multiple determination R^2.

In Section 12.5, we discussed the topic of statistical inference in simple regression analysis. Under the assumptions of the multiple regression model given in Section 13.1, we can use our sample information to make statistical inferences in multiple regression analysis. We can answer questions such as the following:

1. Is the overall multiple regression relationship significant?
2. Are the individual regression coefficients β_j's significant? What are the confidence interval estimates for the individual regression coefficients β_j's?

13.5

STATISTICAL INFERENCE AND PREDICTION IN MULTIPLE REGRESSION ANALYSIS

3. What will be the *mean* selling price $\mu_{Y \mid X_{1_0}, X_{2_0}}$ for condominiums that have an area X_{1_0} and a land cost X_{2_0}? What will be the selling price Y_0 for an *individual* condominium that has an area X_{1_0} and a land cost of X_{2_0}?

For statistical inference purposes in multiple regression analysis the three conditions listed in Section 13.1 must be met by the population data.

13.5.1 Testing the Significance of the Multiple Regression Relationship

In Section 12.5, we showed how a simple regression relationship can be tested for significance by using the ratio of mean squares $F = MSR/MSE$ to test the hypothesis H_0: $\beta = 0$. If the value of the F ratio is large, then the amount of variation in Y that is explained by the regression equation (as measured by MSR in the numerator of the ratio) is large relative to the amount of variation that is left unexplained or is due to error (as measured by MSE in the denominator). Using the same logic, and under the assumptions listed in Section 13.1, we can use the ratio

$$F = \frac{MSR}{MSE} \tag{13.13}$$

to test the *significance of a multiple regression relationship,* where MSR has $\nu_1 = k$ degrees of freedom and MSE has $\nu_2 = n - k - 1$ degrees of freedom in multiple regression analysis. That is, $F = MSR/MSE$ follows the F distribution with $\nu_1 = k$ and $\nu_2 = n - k - 1$ degrees of freedom if the null hypothesis is true and under the assumptions listed in Section 13.1; F can be used to test the following hypotheses:

H_0: $\beta_1 = \beta_2 = \cdots = \beta_k = 0$.

H_a: One or more of the β_j values are not equal to zero.

The F ratio tests the null hypothesis H_0: $\beta_1 = \beta_2 = \cdots = \beta_k = 0$ that there is *no* linear relationship between the X_j variables and the dependent variable Y. The alternative hypothesis is true if one or more of the X_j independent variables are related to the dependent variable Y. A large value for F indicates that we should reject the null hypothesis that there is no multiple regression relationship, giving the following hypothesis test.

Hypothesis Test of the Significance of a Multiple Regression Relationship

H_0: $\beta_1 = \beta_2 = \cdots = \beta_k = 0$.

H_a: One or more of the β_j values are not equal to zero.

Rejection rule: If $F > F_{\alpha, k, n-k-1}$, then reject H_0;

where: $F = MSR/MSE$

=== **EXAMPLE 13.7**   We want to test the significance of the multiple regression relationship for United Park City Properties and the condominium data of Example 13.2, where Y is the sales price, X_1 is the area, and X_2 is the cost of the land for the condominiums. That is, we want to test the following hypotheses:

H_0: $\beta_1 = \beta_2 = 0$.

H_a: One or more of the β_j values are not equal to zero.

The significance level is .05. We computed the mean squares previously, and we find in Table 13.4 that $MSR = 687.25$ and that $MSE = 3.75$. Now, $\nu_1 = k = 2$ and $\nu_2 = n - k - 1 = 5 - 2 - 1 = 2$. Thus, from Table VII we find $F_{\alpha, k, n-k-1} = F_{.05, 2, 2} = 19.000$, and our rejection rule is as follows:

Rejection rule: If $F > 19.000$, then reject H_0.

The value of the F ratio for the sample data is

$$F = \frac{MSR}{MSE} = \frac{687.25}{3.75} = 183$$

Since the value computed for the sample $F = 183$ is greater than $F_{.05, 2, 2} = 19.000$, we reject the null hypothesis (see Figure 13.2). We conclude that area X_1 and land costs X_2 are related to the prices of condominiums Y.

13.5.2 Hypotheses and Inferences Concerning β_j's

In Section 12.5.1, we discussed methods to test hypotheses or to construct confidence intervals for the slope β of a population simple linear regression equation. To test hypotheses or to construct a confidence interval in multiple regression for the coefficient β_j of independent variable X_j, say, coefficient β_2 for variable X_2, we must know the sampling distribution of the sample coefficient b_j. Under the assumptions of the multiple regression model, the sampling distribution of each b_j is normal with mean $E(b_j) = \beta_j$ and standard deviation σ_{b_j}, which can be estimated by the **stan-**

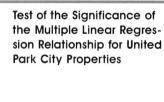

FIGURE 13.2

Test of the Significance of the Multiple Linear Regression Relationship for United Park City Properties

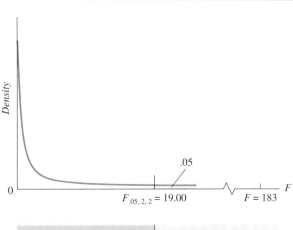

dard error of the sample regression coefficient s_{b_j}. The value for any s_{b_j} can be obtained from the multiple regression output from computers.

If we wish to test the hypothesis H_0: $\beta_j = 0$, we find the value for

$$t = \frac{b_j}{s_{b_j}} \tag{13.14}$$

Under the null hypothesis H_0: $\beta_j = 0$ and the assumptions from the multiple regression model, b_j/s_{b_j} follows the t distribution with $\nu = n - k - 1$ degrees of freedom. We must note, however, that $t = b_j/s_{b_j}$ is valid only if the null hypothesis is that β_j is zero. If the null hypothesis is that β_j equals some specific value, say, β_{j_0}, then the numerator in Equation (13.14) is replaced with $b_j - \beta_{j_0}$. The procedure for testing H_0: $\beta_j = 0$ is summarized as follows.

Hypothesis Test of H_0: $\beta_j = 0$ Against H_a: $\beta_j \neq 0$

H_0: $\beta_j = 0$

H_a: $\beta_j \neq 0$

Rejection rule: If $t < -t_{\alpha/2, n-k-1}$ or $t > t_{\alpha/2, n-k-1}$, then reject H_0;

where: $t = \dfrac{b_j}{s_{b_j}}$ $\tag{13.15}$

The ratio b_j/s_{b_j} in Equation (13.15) has $\nu = n - k - 1$ degrees of freedom since s_{b_j} is a function of $S^2_{Y|12}$, which has $\nu = n - k - 1$ degrees of freedom.

Although you will probably want to use a computer to find the s_{b_j} values, the values for s_{b_1} and s_{b_2} in a multiple regression problem with two independent variables X_1 and X_2 can be calculated by using Equation (13.16) and Equation (13.17), respectively. We present these equations for the insight that you can obtain by studying them:

$$s_{b_1} = \frac{S_{Y|12}}{\sqrt{SSX_1(1 - r^2_{12})}} \tag{13.16}$$

$$s_{b_2} = \frac{S_{Y|12}}{\sqrt{SSX_2(1 - r^2_{12})}} \tag{13.17}$$

where: $S_{Y|12} = \sqrt{MSE}$ is the sample standard error of the estimate

r^2_{12} is the coefficient of determination between X_1 and X_2

$SSX_1 = \sum (X_{i1} - \overline{X}_1)^2$

$SSX_2 = \sum (X_{i2} - \overline{X}_2)^2$

If X_1 and X_2 are correlated, then r^2_{12} is greater than zero. The standard deviations of the estimated slopes will be large if X_1 and X_2 are highly correlated.

The hypothesis test H_0: $\beta_j = 0$ is a test of the assertion that variable X_j adds no **additional** or **marginal** explanatory power to the regression equation over and above that which the other independent or X variables that are included in the regression equation have already provided. Thus failure to reject this hypothesis would not allow us to conclude that X_j is unrelated to Y. It would allow us to conclude only that there is not enough evidence to suggest that X_j is related to the *prediction errors* in Y after the contribution of the other independent variables to the predictions is considered. Conversely, rejecting the hypothesis H_0: $\beta_j = 0$ implies that X_j is related to the variation in Y not explained by the other independent variables.

The quantity $t = b_j / s_{b_j}$ in Equation (13.15) follows the t distribution with $\nu = n - k - 1$ degrees of freedom only when β_j is zero. In this situation the alternative hypothesis is usually that β_j is not equal to zero. In the situation where we need a one-tailed test, we can use Equation (13.15) to compute the value of $t = b_j / s_{b_j}$. For the alternative hypothesis H_a: $\beta_j > 0$, the rejection rule is as follows:

> Rejection rule: If $t > t_{\alpha, n-k-1}$, then reject H_0.

For H_a: $\beta_j < 0$, the rejection rule is as follows:

> Rejection rule: If $t < -t_{\alpha, n-k-1}$, then reject H_0.

If we want a **confidence interval estimate for slope β_j** of the population multiple regression equation with a confidence level of $100(1 - \alpha)\%$, we form the interval defined by Equation (13.18).

Equations for Limits of the Confidence Interval for β_j

$$L = b_j - t_{\alpha/2, n-k-1} s_{b_j}$$
$$R = b_j + t_{\alpha/2, n-k-1} s_{b_j}$$

(13.18)

EXAMPLE 13.8   To test the null hypothesis that condominium land costs X_2 contribute no explanatory power to the United Park City Properties multiple regression on condominium price Y over and above that which is provided by condominium area X_1, we test H_0: $\beta_2 = 0$ against the alternative H_a: $\beta_2 \neq 0$. The level of significance is .05. Recall that the sample multiple regression line was determined in Example 13.4 to be $\hat{Y} = -61.29 + 8.07X_1 + 2.82X_2$.

The hypothesis test will require us to find the standard error of b_2 on the printouts of Computer Exhibit 13.1 or to determine the value of s_{b_2} by using Equation (13.17). The standard error of b_2 as given in Computer Exhibit 13.1 is $s_{b_2} = .6256$, which rounds to .63. The preliminary calculations in Table 13.5 will help us find the value for s_{b_2} by using Equation (13.17).

In Example 13.6 we found that the standard error of the estimate $S_{Y|12} = 1.94$. Using the square of Equation (12.36), we find that the coefficient of determination for X_1 and X_2 is

$$r_{12}^2 = \frac{(SCP_{X_1 X_2})^2}{(SSX_1)(SSX_2)} = \frac{(-3)^2}{(22)(10)} = .04$$

Now we are prepared to calculate s_{b_2} as follows:

$$s_{b_2} = \frac{S_{Y|12}}{\sqrt{SSX_2(1 - r_{12}^2)}} = \frac{1.94}{\sqrt{(10)(1 - .04)}} = .63$$

TABLE 13.5	Area X_1	Land Cost, X_2	X_1^2	X_2^2	$X_1 X_2$
	9	8	81	64	72
	15	7	225	49	105
Condominium Data and	10	9	100	81	90
Calculations for United	11	10	121	100	110
Park City Properties	10	6	100	36	60
Total	55	40	627	330	437

$$\overline{X}_1 = \frac{\Sigma X_1}{n} = \frac{55}{5} = 11$$

$$\overline{X}_2 = \frac{\Sigma X_2}{n} = \frac{40}{5} = 8$$

The sum of cross product deviations for X_1 and X_2, $SCP_{X_1 X_2}$, as determined by using Equation (12.9), is

$$SCP_{X_1 X_2} = \Sigma(X_1 - \overline{X}_1)(X_2 - \overline{X}_2) = \Sigma X_1 X_2 - \frac{\Sigma X_1 \Sigma X_2}{n}$$

$$= 437 - \frac{(55)(40)}{5} = -3$$

and

$$SSX_1 = \Sigma(X_1 - \overline{X}_1)^2 = \Sigma X_1^2 - \frac{(\Sigma X_1)^2}{n}$$

$$= 627 - \frac{(55)^2}{5} = 22$$

$$SSX_2 = \Sigma(X_2 - \overline{X}_2)^2 = \Sigma X_2^2 - \frac{(\Sigma X_2)^2}{n}$$

$$= 330 - \frac{(40)^2}{5} = 10$$

With these preliminary results we are now prepared to test our null hypothesis $H_0: \beta_2 = 0$ against the alternative $H_a: \beta_2 \neq 0$. The significance level is .05, and there are $\nu = n - k - 1 = 5 - 2 - 1 = 2$ degrees of freedom. In Table VI, we find $t_{\alpha/2, n-k-1} = t_{.025, 2} = 4.303$, and we have the following rejection rule.

Rejection rule: If $t < -4.303$ or $t > 4.303$, then reject H_0.

We found that $b_2 = 2.82$ and $s_{b_2} = .63$, so Equation (13.17) gives

$$t = \frac{b_2}{s_{b_2}} = \frac{2.82}{.63} = 4.5$$

Notice that $t = 4.51$ can be found on the printouts of Computer Exhibit 13.1.

Since $t = 4.5$ exceeds 4.303, we reject H_0 and conclude that β_2 is not equal to zero, given that X_1 is also used to predict Y. In other words, we conclude that the land cost variable X_2 contributes additional explanatory power in predicting condominium price Y over and above that which is provided by area X_1. ▬▬

PRACTICE PROBLEM 13.1

In the United Park City Properties condominium example, find the 95% confidence interval estimate for β_1.

Solution To construct the confidence interval, we must find s_{b_1} on the printouts of Computer Exhibit 13.1 or use Equation (13.16). The standard error of b_1 on the printouts is $s_{b_1} = .4218$, which rounds to .42. From the value for the standard error of the estimate $S_{Y|12} = 1.94$ that was found in Example 13.6, the preliminary calculations that were made in Table 13.5, and the coefficient of determination for X_1 and X_2 of $r_{12}^2 = .04$ that was computed in Example 13.8, we are now prepared to calculate s_{b_1} by using Equation (13.16) as follows:

$$s_{b_1} = \frac{S_{Y|12}}{\sqrt{SSX_1(1 - r_{12}^2)}} = \frac{1.94}{\sqrt{(22)(1 - .04)}} = .42$$

With these preliminary results, we are now prepared to answer the question posed in this problem. The regression equation was determined in Section 13.2 to be $\hat{Y} = -61.29 + 8.07X_1 + 2.82X_2$. Thus, we find the 95% confidence interval for β_1 by using Equation (13.18):

$$L = b_1 - t_{\alpha/2, n-k-1}s_{b_1} = 8.07 - (4.303)(.42) = 6.26$$
$$R = b_1 + t_{\alpha/2, n-k-1}s_{b_1} = 8.07 + (4.303)(.42) = 9.88$$

Practice Problem 13.1 introduces an issue we must discuss: In problems where X_1 and X_2 are highly correlated (as would be indicated by a high r_{12}^2 value), we must be cautious about the interpretation we place on confidence intervals or hypothesis tests about β_1 and β_2. Under these circumstances a confidence interval for β_1 gives a range of values within which we are confident that β_1 lies, assuming that both X_1 and X_2 are in the regression equation. That is, the confidence interval applies only to the coefficient of X_1 in the population regression equation $\mu_{Y|X_1, X_2} = \beta_0 + \beta_1 X_1 + \beta_2 X_2$. The value of β_1 in the population equation relating only X_1 to Y (that is, in $\mu_{Y|X_1} = \beta_0 + \beta_1 X$) *might be rather different*. The condition where X_1 and X_2 are highly correlated to one another is called **multicollinearity.** This condition and the problems it presents will be discussed more fully in Section 13.8.

The same caution applies if there are several independent variables X_j's in a multiple regression analysis and if pairs of independent variables or combinations of independent variables are highly correlated. Notice that multicollinearity is not a problem for our condominium example since $r_{12}^2 = .04$ is close to zero for condominium areas X_1 and condominium land costs X_2.

13.5.3 Confidence Interval for a Mean Value of *Y* and Prediction Interval for a New Individual Value

The final questions to be examined in this section are

What will be the **mean** of *Y* for some specific estimation using new values X_{1_0} and X_{2_0}, $\mu_{Y|X_{1_0}, X_{2_0}}$?

What will be the **individual** value Y_0 for some specific prediction for a given X_{1_0} and X_{2_0}?

The methods used to answer these questions in multiple regression are analogous to the methods used in simple regression in Sections 12.5.2 and 12.5.3.

The *point estimate* of $\mu_{Y|X_{1_0}, X_{2_0}}$, the mean value of Y given X_{1_0} and X_{2_0}, is $\hat{Y}_0 = b_0 + b_1 X_{1_0} + b_2 X_{2_0}$. A **confidence interval** for the *mean* value is found by the usual form.*

Expression for Limits of the Confidence Interval for Mean Value of Y Given X_{1_0} and X_{2_0}

$$\hat{Y}_0 \pm t_{\alpha/2, n-k-1} S_{\hat{Y}_0}$$

(13.19)

The **prediction interval** for a new *individual* value Y_0 given X_{1_0} and X_{2_0} is also found by the following familiar form.

Expression for Limits of the Prediction Interval for a New Individual Value of Y_0 Given X_{1_0} and X_{2_0}

$$\hat{Y}_0 \pm t_{\alpha/2, n-k-1} S_{Y_0}$$

(13.20)

 PRACTICE PROBLEM 13.2

A buyer plans to purchase a condominium with $X_{1_0} = 10$, or an area of 1000 square feet, and $X_{2_0} = 9$, or a land cost of \$9000 per unit, from United Park City Properties. Find a 95% *confidence interval* for the mean selling price of condominiums with 1000 square feet and a land cost of \$9000 in this city. United Park City Properties would like to know this interval because it sells many such condominiums. Also, find a 95% *prediction interval* for the selling price of an individual condominium with 1000 square feet and a land cost of \$9000 in this city for a condominium purchaser.

Solution The sample multiple regression equation for estimating the mean selling price and for predicting an individual selling price where $X_{1_0} = 10$ and $X_{2_0} = 9$ gives

$$\hat{Y} = -61.29 + 8.07(10) + 2.82(9) = 44.8$$

or \$44,800.

To ease the computational burden, we use the computer printout in Computer Exhibit 13.1A to find the 95% confidence interval and the 95% prediction interval.

The 95% confidence interval for the mean selling price given on the printout is 40.014 to 49.493. Thus, United can be 95% confident that the mean selling price for

---

*The sample multiple regression surface for X_1 and X_2 will be different for different samples and has a standard error $S_{\hat{Y}_0}$ with a rather involved formula:

$$S_{\hat{Y}_0} = S_{Y|X_1, X_2} \sqrt{\frac{1}{n} + \frac{(X_{1_0} - \overline{X}_1)^2}{SSX_1(1 - r_{12}^2)} + \frac{(X_{2_0} - \overline{X}_2)^2}{SSX_2(1 - r_{12}^2)} - \frac{2(X_{1_0} - \overline{X}_1)(X_{2_0} - \overline{X}_2)(SCP_{X_1X_2})}{SSX_1(SSX_2)(1 - r_{12}^2)}}$$

To find the sample standard deviation S_{Y_0} for a *prediction interval*, add a term with the value 1 beneath the square root.

condominiums with 1000 square feet in area and with land costs of $9000 per unit will lie between $40,014 and $49,493.

The 95% prediction interval for an individual selling price given on the printout is 35.165 to 54.343. Thus, the purchaser can be 95% confident that the selling price for an individual condominium with 1000 square feet in area and with land costs of $9000 per unit will lie between $35,165 and $54,343.

Problems: Section 13.5

11. A multiple regression analysis with sales (in thousands of dollars) of a product as the dependent variable Y, and advertising expenditure X_1 (in thousands of dollars) and market share X_2 as the independent variables, was done by Goldblatt Brothers, Inc., with a sample of $n = 10$ retail outlets. The analysis of variance table for the multiple regression is presented in the accompanying table. The sample multiple regression equation is $\hat{Y} = 10 + 3.2X_1 + 6.3X_2$, $SSX_1 = 2.26$, $SSX_2 = .8$, and $r_{12}^2 = .6$.

    a. Test the overall significance of the multiple regression relationship. That is, test $H_0: \beta_1 = \beta_2 = 0$ against H_a: One or more of the β_j values is not equal to zero. Let the level of significance be .05.
    b. Test the hypothesis that X_2 adds no explanatory power to the multiple regression equation over and above that which is provided by X_1. That is, test $H_0: \beta_2 = 0$ against $H_a: \beta_2 \neq 0$, and let the level of significance be .05.

Source of Variation	Degrees of Freedom	Sum of Squares	Mean Square
Regression	2	15	7.5
Error	7	9	1.29
Total	9	24	

12. Refer to the Goldman-Sachs regression equation found for Problem 2.

    a. Test the hypothesis $H_0: \beta_2 = 0$ against $H_a: \beta_2 \neq 0$. Use a significance level of .05. [*Hint:* $r_{12}^2 = .45$.]
    b. What do you have to assume about the distribution of stock prices to conduct the hypothesis test?

13. For the Goldman-Sachs regression equation found in Problem 2, find an interval within which you can be 95% confident that β_1 lies for all stocks of this type in the population. [*Hint:* $r_{12}^2 = .45$.]

14. For the regression equation found in Problem 4, where Y is percentage of rebate spent, X_1 is family income, and X_2 is number of dependents, test the hypothesis $H_0: \beta_1 \geq 0$ against the alternative $H_a: \beta_1 < 0$. Let the level of significance be .05. [*Hint:* $r_{12}^2 = .51$.]

15. Find a 95% confidence interval estimate for β_2 for the population of tax rebate recipients considered in Problem 4. [*Hint:* $r_{12}^2 = .51$.]

13.6 INDICATOR VARIABLES

In the previous sections, we have utilized quantitative independent variables, X_j's, in our regression equations. **Quantitative variables** take on values that are measured on a scale. For example, we used the area in square feet and the land costs in dollars of condominiums as independent variables in a multiple regression problem. Frequently, one may wish to include **qualitative** or **categorical** independent variables in a regression problem. Examples of qualitative variables and some possible categories are sex—male or female; education—college degree or no college degree; and marital status—married or not married. *Qualitative variables* correspond to attributes of the observations that are included in our samples.

EXAMPLE 13.9   Southern Textiles, Inc., wants to study the relationship between an employee's wages Y and his or her experience X_1. But some of the employees being considered have college degrees and some do not have college degrees, and Southern wishes to include the qualitative variable of having a college degree or not having a college degree in the study of wages. It seems reasonable to expect that having a college degree or not having a college degree would be associated with having a difference in wages.

If qualitative variables are to be included in a regression equation, they must be quantified; in other words, they must be assigned numerical values. Quantification can be accomplished by using **indicator variables,** which are also termed *dummy variables* or *binary variables*. Indicator variables are assigned the values 0 or 1.

EXAMPLE 13.10   For our example the qualitative variable "education" has two categories (having a college degree or not having a college degree), so we can include variable X_2 as an indicator $(0, 1)$ variable. The value 0 is assigned to the indicator variable X_2 for one category (not having a college degree), and the value 1 is assigned to the indicator variable X_2 for the other category (having a college degree). That is,

$$X_2 = \begin{cases} 0 \text{ if the employee has } no \text{ college degree} \\ 1 \text{ if the employee has a college degree} \end{cases}$$

If we take a sample of employees in order to predict wages Y from experience on the job X_1 and whether or not the employee has a college degree X_2, then our estimated regression equation will be

$$\hat{Y} = b_0 + b_1 X_1 + b_2 X_2 \tag{13.21}$$

Now if an employee does *not* have a college degree, so that $X_2 = 0$, then the estimated regression equation (13.21) becomes

$$\hat{Y} = b_0 + b_1 X_1 + b_2(0) \quad \text{or} \quad \hat{Y} = b_0 + b_1 X_1 \tag{13.22}$$

The estimated regression equation (13.22) graphs as a straight line with intercept b_0 and slope b_1.

If the employee *has* a college degree, so that $X_2 = 1$, then the estimated regression equation (13.21) becomes

$$\hat{Y} = b_0 + b_1 X_1 + b_2(1) \quad \text{or} \quad \hat{Y} = (b_0 + b_2) + b_1 X_1 \tag{13.23}$$

The estimated regression equation (13.23) is a straight line with intercept $(b_0 + b_2)$ and slope b_1.

EXAMPLE 13.11   The two equations (13.22) and (13.23), for the employees who do not have a college degree and who do have a college degree, have the same slope b_1. But the predicted wages $\hat{Y}$ for employees who have a college degree (where $X_2 = 1$) are always different by the amount b_2 than the predicted wages of those who do not have a college degree. Thus, b_2 in this example is the estimated difference in wages that is associated with having a college degree.

In general, the estimated coefficient for the indicator variable estimates the *differential influence on Y* of having an attribute, when *not* having an attribute is assigned the value 0 and having an attribute is assigned the value 1 under the coding method of indicator variables $(0, 1)$.

 PRACTICE PROBLEM 13.3

For the Southern Textiles example we have been discussing, we wish to predict employee wages $\hat{Y}$ by using the employee's experience X_1 and the employee's education X_2. Employees are categorized as having a college degree or not having a college degree in their personnel files, so the variable "education" is a qualitative variable. Thus, X_2 will be an indicator $(0, 1)$ variable.

Using the sample of data for 20 employees presented in Table 13.6, find the estimated regression equation, and delineate the estimated difference in wages that is associated with having a college degree. Also, test the hypothesis that the differential influence of having a college degree is zero, given that experience X_1 is also used to predict wages Y. That is, test $H_0: \beta_2 = 0$ against $H_a: \beta_2 \neq 0$, and use a significance level of .05.

Solution A data analysis software package was used to compute the estimated multiple regression equation, and an edited printout of the results is presented in Computer Exhibit 13.2.

The estimated regression equation is

$$\hat{Y} = .008 + .485X_1 + 5.532X_2$$

Since $b_2 = 5.532$, the estimated difference in wages that is associated with having a college degree is $5532.

Wages Y ($\times$ $1000)	Experience X_1 (months)	Education, X_2 (College Degree = 1, No College Degree = 0)	
			TABLE 13.6
27.1	47.2	1	**Data for Southern Textiles**
20.1	40.1	0	
25.1	37.1	1	
22.3	44.7	0	
25.2	41.9	1	
27.4	46.1	1	
13.8	17.0	1	
11.0	29.2	0	
22.4	30.7	1	
30.3	59.8	0	
28.5	48.0	1	
26.7	55.3	0	
21.9	42.9	0	
22.1	47.2	0	
18.7	40.1	0	
21.8	36.5	1	
11.8	20.0	0	
14.1	30.7	0	
23.1	36.8	1	
30.8	49.9	1	

| | MINITAB | | COMPUTER EXHIBIT 13.2 Multiple Regression with an Indicator Variable for Southern Textiles, Inc. |

```
MTB > read    c1      c2    c3
DATA >        27.1   47.2   1
DATA >        20.1   40.2   0
DATA >        25.1   37.1   1
DATA >        22.3   44.7   0
DATA >        25.2   41.9   1
DATA >        27.4   46.1   1
DATA >        13.8   17.0   1
DATA >        11.0   29.2   0
DATA >        22.4   30.7   1
DATA >        30.3   59.8   0
DATA >        28.5   48.0   1
DATA >        26.7   55.3   0
DATA >        21.9   42.9   0
DATA >        22.1   47.2   0
DATA >        18.7   40.1   0
DATA >        21.8   36.5   1
DATA >        11.8   20.0   0
DATA >        14.1   30.7   0
DATA >        23.1   36.8   1
DATA >        30.8   49.9   1
DATA > end
       20 ROWS READ
 MTB > name c1 'WAGES'
 MTB > name c2 'EXPERNCE'
 MTB > name c3 'EDUC'

 MTB > regres c1 2 c2 c3   ◄──────
```

Command to **regres c1** (for wages Y) on the **2** variables **c2** (for experience X_1) and **c3** (for the indicator variable of having a college degree, coded 1, or *not* having a college degree, coded 0)

```
The regression equation is
WAGES = 0.01 + 0.485 EXPERNCE + 5.53 EDUC

Predictor      Coef       Stdev     t-ratio        p
Constant       0.007      1.279        0.01     0.995
EXPERNCE       0.48518    0.02933     16.54     0.000
EDUC           5.5321     0.6170       8.97     0.000

s = 1.374      R-sq = 95.1%     R-sq(adj) = 94.5%

Analysis of Variance

SOURCE        DF          SS          MS         F        p
Regression     2       623.42      311.71    165.07    0.000
Error         17        32.10        1.89
Total         19       655.52

SOURCE        DF       SEQ SS
EXPERNCE       1       471.61
EDUC           1       151.80
```

We will reject the null hypothesis $H_0: \beta_2 = 0$ if $t = b_2/s_{b_2}$ is greater than $t_{\alpha/2,\,n-k-1}$ or less than $-t_{\alpha/2,\,n-k-1}$. For this data set,

$$t = \frac{5.532}{.617} = 8.97$$

Now $t = 8.97$ is greater than $t_{.025,17} = 2.110$, so we reject H_0, concluding that there is a significant difference in employees' wages associated with having a college degree.

The estimated regression equation for those not having a college degree (where $X_2 = 0$) is

$$\hat{Y} = .008 + .485X_1 + 5.532(0) \quad \text{or} \quad \hat{Y} = .008 + .485X_1$$

The estimated regression equation for those having a college degree (where $X_2 = 1$) is

$$\hat{Y} = .008 + .485X_1 + 5.532(1) = (.008 + 5.532) + .485X_1$$

or

$$\hat{Y} = 5.540 + .485X_1$$

The data for this example and the two regression lines, one for employees who do not have a college degree and one for those who do have a college degree are plotted in Figure 13.3. The estimated difference in wages that is associated with having a college degree, $b_2 = 5.532$, is represented by the vertical distance between the two lines, or by the difference between the intercepts.

We have illustrated methods to quantify a qualitative variable that has two categories. When the qualitative variable had $c = 2$ categories, we used $c - 1 = 2 - 1 = 1$ indicator $(0, 1)$ variable. If we need to quantify a qualitative variable with c categories, where c is greater than 2, say, $c = 3$ categories, then we use $c - 1 = 3 - 1 = 2$ indicator $(0, 1)$ variables.

EXAMPLE 13.12   If our education variable had three categories (no college degree, undergraduate degree, or graduate degree), we would require $c - 1 = 3 - 1 = 2$ indicator $(0, 1)$ variables, and our coding method would be as presented in Table 13.7.

FIGURE 13.3

Regression Lines with an Indicator (0, 1) Variable for Southern Textiles, Inc.

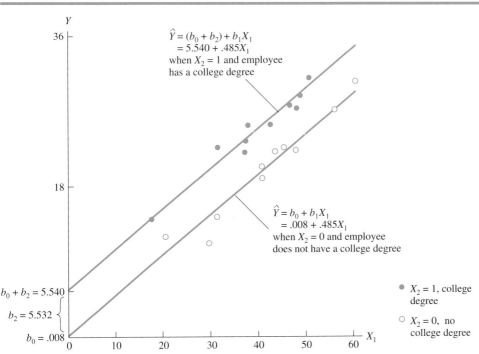

TABLE 13.7	*Qualitative Variable*	X_2	X_3
Two Indicator (0, 1) Variables to Quantify a Qualitative Variable with c = 3 Categories	No college degree	0	0
	Undergraduate degree	1	0
	Graduate degree	0	1

For the coding method presented in Table 13.7, the "no college degree" category is assigned the value 0 for both indicator variables X_2 and X_3, and this category is known as the *reference category*. The indicator variable X_2 is assigned the value 1 when the employee has a college degree, and the estimated differential influence on wages of having a "college degree" relative to the "no college degree" or reference category is the value of b_2. The indicator variable X_3 is assigned the value 1 when the employee has a graduate degree, and the estimated differential influence on wages of having a "graduate degree" relative to the "no college degree" or reference category is the value of b_3.

In general, the reference category is assigned the value 0 for all $c - 1$ indicator variables. Now, let a given category be assigned the value 1 for indicator variable X_j. The estimated coefficient b_j is the estimate for the differential influence on the independent variable Y of being in this particular category, relative to being in the reference category.

Problems: Section 13.6

16. An economist with Third Bank System Incorporated wants to develop a multiple regression model to predict individual consumption expenditures Y from income X_1 and the sex X_2 of the individual (female = 1, male = 0). Data were collected for 20 people and are presented in the accompanying table.

   a. Find the estimated multiple regression equation.
   b. Explain the meaning of the value of b_2.
   c. Test the hypothesis $H_0: \beta_2 = 0$ against $H_a: \beta_2 \neq 0$. Use a significance level of .05.

Consumption Expenditures (× $1000)	*Income* (× $1000)	*Sex*	*Consumption Expenditures* (× $1000)	*Income* (× $1000)	*Sex*
12.8	22.5	0	11.6	17.8	0
13.8	21.4	1	10.3	17.3	0
12.6	17.9	1	14.3	21.2	0
10.9	16.6	0	14.8	24.0	0
11.2	15.2	1	9.0	14.8	0
14.6	20.9	1	16.1	25.9	0
13.3	13.9	1	15.4	25.0	0
14.6	21.8	1	12.6	15.8	1
17.0	19.9	1	10.9	18.7	0
16.7	19.6	1	15.2	23.6	0

17. A prospective home buyer engaged the services of a consultant to develop a multiple regression model to predict mortgage interest rates Y by using the yield on long-term Treasury bonds X_1 and the type of the mortgage X_2. There were two types of mortgages that the buyer was considering: variable-rate (coded 1) and fixed-rate

(coded 0). All were 30-year mortgages. Data were collected for 20 mortgages and are presented in the accompanying table.

a. Find the estimated multiple regression equation.

b. What is the estimated influence of the variable-rate feature of mortgages on the mortgage interest rate?

c. Test the hypothesis $H_0: \beta_2 = 0$ against $H_a: \beta_2 \neq 0$. Use a significance level of .05.

Mortgage Rate (%)	Treasury Bond Yield (%)	Mortgage Type	Mortgage Rate (%)	Treasury Bond Yield (%)	Mortgage Type
13.8	11.7	1	13.9	11.9	1
14.2	12.1	1	16.5	13.8	1
15.0	12.1	1	12.6	12.5	1
14.7	10.7	0	13.3	10.8	0
12.6	7.9	0	14.1	10.6	0
16.4	11.8	0	14.1	15.2	1
15.2	12.4	1	14.3	14.1	1
14.5	12.8	1	12.7	10.1	1
18.3	12.9	0	13.2	11.3	0
9.9	8.6	1	17.8	14.5	0

18. A production manager with Atlantic Rivet & Machining Company wants to develop a multiple regression model to estimate the chatter Y (in vibrations per second) of the metal machining lathe's cutting tool Y from the turning speed X_1 (in revolutions per second) and from the type of cutting tool X_2. The cutting tools that are used on the machine are of two types. One is made of high-speed cutting steel (coded 0), and one has a carbide tip (coded 1). Data were collected for 20 machined parts and are presented in the accompanying table. An edited printout from a multiple regression program for this problem is presented in Computer Exhibit 13.3.

a. Find the estimated multiple regression equation.

b. What is the meaning of b_2?

c. Test the hypothesis $H_0: \beta_2 = 0$ against $H_a: \beta_2 \neq 0$. Use a significance level of .05.

Chatter	Turning Speed	Tool Type	Chatter	Turning Speed	Tool Type
94.2	20.7	1	79.9	16.7	1
117.1	24.1	0	80.3	10.0	0
88.6	16.8	0	95.8	18.9	0
98.4	22.6	0	78.5	8.5	0
100.6	24.0	1	94.6	25.5	1
96.4	16.6	0	76.5	16.8	1
99.8	14.8	0	83.6	16.8	0
85.2	16.9	1	67.5	16.5	1
117.1	32.4	1	117.8	22.6	0
93.9	25.5	1	89.9	16.7	0

In Chapter 12, we introduced the simple linear regression model with one independent X variable. In this chapter, we have introduced the multiple linear regression model where we have k independent variables $X_1, X_2, \ldots, X_k$. Thus, the estimated multiple regression equation is

$$\hat{Y} = b_0 + b_1 X_1 + b_2 X_2 + \cdots + b_k X_k$$

where the value of $\hat{Y}$ is the estimate of the mean value of Y given $X_1, X_2, \ldots, X_k$. A computer was used to find the sample regression equation.

13.7

SUMMARY

MINITAB   **COMPUTER EXHIBIT 13.3** Multiple Regression for Atlantic Rivet & Machining

```
MTB > read     c1     c2    c3
DATA >       94.2   20.7    1
DATA >      117.1   24.1    0
DATA >       88.6   16.8    0
DATA >       98.4   22.6    0
DATA >      100.6   24.0    1
DATA >       96.4   16.6    0
DATA >       99.8   14.8    0
DATA >       85.2   16.9    1
DATA >      117.1   32.4    1
DATA >       93.9   25.5    1
DATA >       79.9   16.7    1
DATA >       80.3   10.0    0
DATA >       95.8   18.9    0
DATA >       78.5    8.5    0
DATA >       94.6   25.5    1
DATA >       76.5   16.8    1
DATA >       83.6   16.8    0
DATA >       67.5   16.5    1
DATA >      117.8   22.6    0
DATA >       89.9   16.7    0
DATA > end
      20 ROWS READ
MTB > name c1 'CHATTER'
MTB > name c2 'TURNSPD'
MTB > name c3 'TOOL'

MTB > regres c1 2 c2 c3  ◄─────
```

> Command to **regres** (for multiple regression) **c1** (lathe chatter) on the **2** independent variables **c1** (turning speed) and **c2** (an indicator variable for the cutting tool tip coded carbide = 1, steel = 0)

```
The regression equation is
CHATTER = 55.4 + 2.32 TURNSPD - 15.7 TOOL

Predictor       Coef      Stdev     t-ratio         p
Constant      55.358      5.674        9.76     0.000
TURNSPD       2.3209     0.3086        7.52     0.000
TOOL         -15.700      3.380       -4.65     0.000

s = 6.844       R-sq = 77.7%     R-sq(adj) = 75.1%

Analysis of Variance

SOURCE        DF          SS          MS         F        p
Regression     2      2780.7      1390.4     29.69    0.000
Error         17       796.2        46.8
Total         19      3576.9

SOURCE        DF      SEQ SS
TURNSPD        1      1770.2
TOOL           1      1010.5
```

In this chapter, we discussed the partitioning of the sum of squares total *SST* into the sum of squares due to regression *SSR* and the sum of squares due to error *SSE*. The standard error of the estimate $S_{Y|12\ldots k}$ and the coefficient of multiple determination R^2 were given as measures of the accuracy of predictions in multiple regression. We also studied a test for the overall significance of a multiple regression relationship and tests of significance on the individual coefficients.

We introduced indicator $(0, 1)$ variables and showed how qualitative variables can be used as independent variables in a regression equation. Several problems, warnings, and limitations of regression analysis were all discussed in the chapter.

Computers are usually used for multiple regression analysis, and we presented printouts from widely used packages with our example in Section 13.2. We also examined some computational formulas for the case of two independent variables, because they provide insight to multiple regression analysis. The results using the computational formulas can be compared with those given on the printouts. Printouts from several different data analysis software packages are given with the problems for this chapter. It is usually quite easy to adapt from one printout to the next.

Some important formulas and hypothesis-testing procedures in multiple regression are summarized in Table 13.8.

TABLE 13.8   Summary of Formulas and Hypothesis Tests for Multiple Regression

Description	*Formula*	
Sum of squares due to regression	$SSR = \sum(\hat{Y}_i - \overline{Y})^2$	(13.5)
Sum of squares due to error	$SSE = \sum(Y_i - \hat{Y}_i)^2$	(13.5)
Sum of squares total	$SST = \sum(Y_i - \overline{Y})^2$	(13.5)
Standard error of the estimate	$S_{Y|12\ldots k} = \sqrt{\dfrac{1}{n-k-1}\sum(Y_i - \hat{Y}_i)^2}$ $= \sqrt{\dfrac{1}{n-k-1}[SSE]} = \sqrt{MSE}$	(13.10)
Coefficient of multiple determination	$R^2 = 1 - \dfrac{\sum(Y_i - \hat{Y}_i)^2}{\sum(Y_i - \overline{Y})^2} = 1 - \dfrac{SSE}{SST} = \dfrac{SSR}{SST}$	(13.11)
Test of the overall significance of the multiple regression relationship, or $H_0: \beta_1 = \beta_2 = \cdots = \beta_k = 0$	Rejection rule: If $F > F_{\alpha,k,n-k-1}$, then reject H_0. where: $F = \dfrac{SSR/k}{SSE/(n-k-1)} = \dfrac{MSR}{MSE}$	(13.13)
Test of marginal significance of individual regression coefficients, or $H_0: \beta_j = 0$ against $H_a: \beta_j \neq 0$	Rejection rule: If $t < -t_{\alpha/2,n-k-1}$, or $t > t_{\alpha/2,n-k-1}$, then reject H_0. where: $t = b_j/s_{b_j}$	(13.15)

Answers to odd-numbered problems are given in the back of the text.

REVIEW PROBLEMS

19. Some of the differences between simple and multiple regression are (you may check as many responses as you want) as follows.

   a. Simple regression uses one independent variable, and multiple regression uses two or more.

b. Simple regression equations represent lines, geometrically, and multiple regression equations represent curves.

c. Multiple regression equations usually give more accurate predictions of Y, as long as there are two or more independent variables associated with Y.

20. A financial analyst for Smith Brookfield wants to develop a multiple regression model to predict the average annual rate of return on stocks Y from the price-earnings ratio X_1 and a measure of risk known as a stock's beta X_2. Data were collected for ten stocks and are presented in the accompanying table.

   a. Find the estimated multiple regression equation.
   b. Find the predicted value for Y given that $X_1 = 12.0$ and $X_2 = 1.2$.
   c. Determine the coefficient of multiple determination, and explain its meaning.
   d. Test the hypothesis $H_0: \beta_1 = \beta_2 = 0$. Use a significance level of .05.
   e. Test the hypothesis $H_0: \beta_1 = 0$ against $H_a: \beta_1 \neq 0$. Use a significance level of .05.

Rate of Return (%)	Price-Earnings Ratio	Beta
16.3	8.7	1.3
12.4	11.2	1.0
14.5	8.0	1.1
9.1	9.7	1.0
18.2	8.6	1.3
13.6	12.6	1.0
10.4	7.4	1.1
12.8	12.7	1.1
14.3	11.3	1.1
15.6	9.3	1.2

21. A human resources manager with Analog Electronics Corporation engaged the services of a consultant to develop a multiple regression model to predict employee time lost to absenteeism Y from the employee's tenure on the job X_1 and the employee's job satisfaction X_2. Data were collected for ten employees and are presented in the accompanying table. Absenteeism is measured in annual total days absent, tenure is in months, and job satisfaction is measured by an index.

   a. Find the estimated multiple regression equation, and explain the meaning of b_1.
   b. Find the predicted value for Y given that $X_1 = 40$ and $X_2 = 50$.
   c. Determine the coefficient of multiple determination.
   d. Test the hypothesis $H_0: \beta_1 = \beta_2 = 0$. Use a significance level of .05.
   e. Test the hypothesis $H_0: \beta_2 = 0$ against $H_a: \beta_2 \neq 0$. Use a significance level of .05.

Absence	Tenure	Job Satisfaction
22.1	54.3	43.3
19.9	50.3	49.6
22.2	47.9	53.4
22.6	49.7	54.0
9.0	61.2	74.3
29.6	32.9	27.8
30.1	36.9	41.7
22.1	56.5	48.6
21.8	46.0	62.5
25.2	40.9	59.4

22. BZ Mining, Inc., has five large pumps that pump water from the mines it operates. Management has been concerned lately about the amount of money required to repair

malfunctioning pumps. These costs are sums over and above the amounts spent for routine maintenance. To get a better idea of the relationship between these costs and other factors connected with the pumps, the operations superintendent ran a regression relating the variables shown in the accompanying table. Show that the regression equation is $\hat{Y} = 314.7 + .2X_1 + 4.7X_2$. Find the predicted mean monthly repair cost for a pump of this type that averaged 20 hours of operation per week and was 10 months old at the first of the year.

Y, Mean Monthly Repair Costs over Past Year	X_1, Mean Weekly Hours of Operation over Past Year	X_2, Age of Pump at First of Year (months)
$643	28	80
613	26	48
494	15	27
250	15	2
400	16	13

23. Show that $S^2_{Y|12} \approx 8100$ and $R^2 \approx .84$ for the BZ Mining regression equation found in Problem 22. What is the meaning of $R^2 = .84$?

24. Upon seeing the results of the BZ Mining regression analysis in Problem 22, the operations superintendent said, "There certainly isn't a very significant relationship between the variables we used in the regression equation and the repair costs." Verify his statement.

   a. Test the hypothesis $H_0: \beta_1 \le 0$ versus $H_a: \beta_1 > 0$ with $\alpha = .10$. [Hint: $r^2_{12} = .84$. That $R^2 = .84$ also is only coincidence in this problem.]
   b. Test the hypothesis $H_0: \beta_2 \le 0$ versus $H_a: \beta_2 > 0$ with $\alpha = .10$.
   c. Test the hypothesis $H_0: \beta_1 = \beta_2 = 0$, using the F test and $\alpha = .05$.

25. A marketing manager with Nexclor Industries, Inc., engaged the services of a consultant to develop a multiple regression model to predict sales Y from advertising expenditures X_1 and market share X_2. Data were collected for ten periods and are presented in the accompanying table. Sales are measured in thousands of dollars, as are advertising expenditures, and market shares are given as percentages. An edited printout from a data analysis package's multiple regression program for the data of this problem is presented in Computer Exhibit 13.4.

   a. Find the estimated multiple regression equation.
   b. Find the predicted value for Y given that $X_1 = 5$ and $X_2 = 25$.
   c. Determine the coefficient of multiple determination, and explain its meaning.
   d. Test the hypothesis $H_0: \beta_1 = \beta_2 = 0$. Use a significance level of .05.
   e. Test the hypothesis $H_0: \beta_2 = 0$ against $H_a: \beta_2 \neq 0$. Use a significance level of .05.

Sales	Advertising Expenditure	Market Share
10.8	4.3	26.2
12.6	4.0	32.2
8.3	4.1	17.3
9.2	4.6	16.7
11.1	5.5	18.9
10.9	4.5	13.2
7.9	4.3	14.4
11.6	2.8	27.1
8.2	2.6	20.8
9.0	3.1	22.0

MYSTAT

COMPUTER EXHIBIT 13.4 Multiple Regression of Sales Y on Advertising Expenditure X_1 and Market Share X_2 for Nexclor

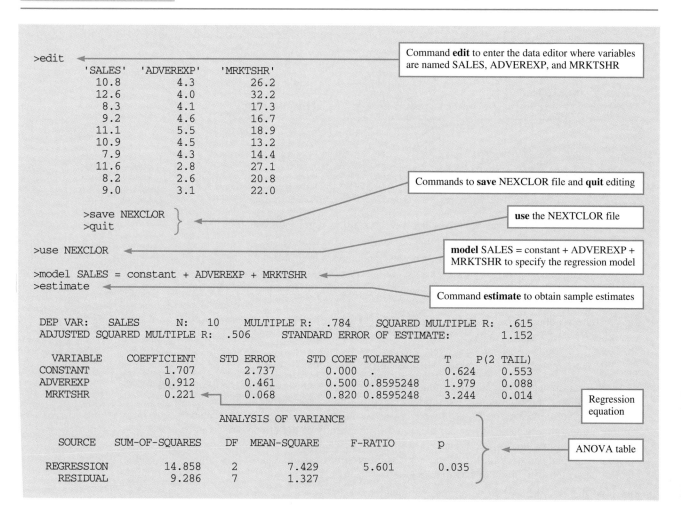

```
>edit
        'SALES'   'ADVEREXP'    'MRKTSHR'
         10.8        4.3          26.2
         12.6        4.0          32.2
          8.3        4.1          17.3
          9.2        4.6          16.7
         11.1        5.5          18.9
         10.9        4.5          13.2
          7.9        4.3          14.4
         11.6        2.8          27.1
          8.2        2.6          20.8
          9.0        3.1          22.0

       >save NEXCLOR
       >quit

>use NEXCLOR

>model SALES = constant + ADVEREXP + MRKTSHR
>estimate
```

Command **edit** to enter the data editor where variables are named SALES, ADVEREXP, and MRKTSHR

Commands to **save** NEXCLOR file and **quit** editing

use the NEXTCLOR file

model SALES = constant + ADVEREXP + MRKTSHR to specify the regression model

Command **estimate** to obtain sample estimates

```
DEP VAR:    SALES     N:   10    MULTIPLE R:  .784    SQUARED MULTIPLE R:   .615
ADJUSTED SQUARED MULTIPLE R:   .506    STANDARD ERROR OF ESTIMATE:       1.152

   VARIABLE    COEFFICIENT   STD ERROR    STD COEF TOLERANCE    T    P(2 TAIL)
   CONSTANT       1.707        2.737        0.000    .         0.624   0.553
   ADVEREXP       0.912        0.461        0.500  0.8595248   1.979   0.088
   MRKTSHR        0.221        0.068        0.820  0.8595248   3.244   0.014

                 ANALYSIS OF VARIANCE

   SOURCE    SUM-OF-SQUARES   DF   MEAN-SQUARE    F-RATIO      p

  REGRESSION      14.858       2      7.429        5.601     0.035
   RESIDUAL        9.286       7      1.327
```

Regression equation

ANOVA table

    26. Plager Dillon accounting firm developed a multiple regression model to predict CPA exam scores Y from grade point averages X_1 and months of accounting experience X_2. Data were collected for ten accountants and are presented in the accompanying table.

CPA Exam Score (%)	GPA	Experience (months)
74.2	3.0	8.5
68.1	2.7	7.1
50.2	2.6	5.9
68.3	3.2	8.6
71.6	2.4	13.3
68.9	3.2	7.4
59.6	3.1	6.8
39.8	2.4	3.8
86.6	3.5	9.1
71.2	2.8	12.6

COMPUTER EXHIBIT 13.5 Multiple Regression of CPA Exam Score Y on Grade Point
Average X_1 and Experience X_2 **BMDP**

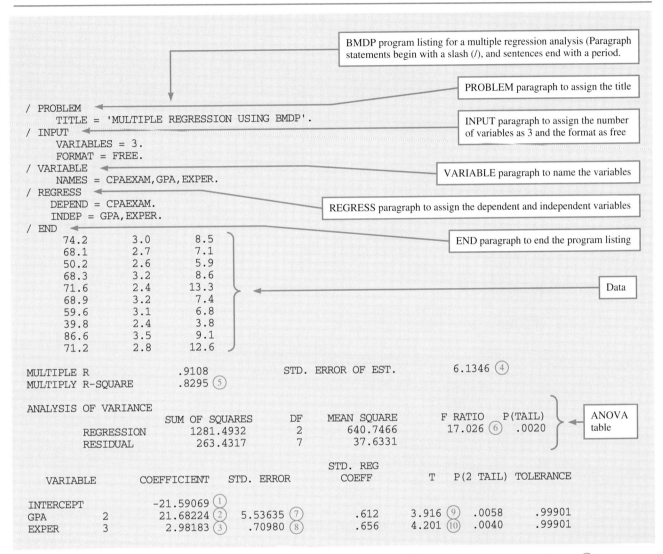

BMDP program listing for a multiple regression analysis (Paragraph statements begin with a slash (/), and sentences end with a period.

PROBLEM paragraph to assign the title

```
/ PROBLEM
     TITLE = 'MULTIPLE REGRESSION USING BMDP'.
/ INPUT
     VARIABLES = 3.
     FORMAT = FREE.
/ VARIABLE
     NAMES = CPAEXAM,GPA,EXPER.
/ REGRESS
     DEPEND = CPAEXAM.
     INDEP = GPA,EXPER.
/ END
     74.2        3.0        8.5
     68.1        2.7        7.1
     50.2        2.6        5.9
     68.3        3.2        8.6
     71.6        2.4       13.3
     68.9        3.2        7.4
     59.6        3.1        6.8
     39.8        2.4        3.8
     86.6        3.5        9.1
     71.2        2.8       12.6
```

INPUT paragraph to assign the number of variables as 3 and the format as free

VARIABLE paragraph to name the variables

REGRESS paragraph to assign the dependent and independent variables

END paragraph to end the program listing

Data

MULTIPLE R		.9108	STD. ERROR OF EST.		6.1346 ④	
MULTIPLY R-SQUARE		.8295 ⑤				

ANALYSIS OF VARIANCE

	SUM OF SQUARES	DF	MEAN SQUARE	F RATIO	P(TAIL)
REGRESSION	1281.4932	2	640.7466	17.026 ⑥	.0020
RESIDUAL	263.4317	7	37.6331		

ANOVA table

VARIABLE		COEFFICIENT	STD. ERROR	STD. REG COEFF	T	P(2 TAIL)	TOLERANCE
INTERCEPT		-21.59069 ①					
GPA	2	21.68224 ②	5.53635 ⑦	.612	3.916 ⑨	.0058	.99901
EXPER	3	2.98183 ③	.70980 ⑧	.656	4.201 ⑩	.0040	.99901

Note: ① b_0, estimated regression intercept; ② b_1, estimated regression coefficient for X_1; ③ b_2, estimated regression coefficient for X_2; ④ $S_{Y|12}$, standard error of the estimate; $S_{Y|12} = \sqrt{MSE}$; ⑤ R^2, coefficient of multiple determination; ⑥ F, F statistic for testing H_0: $\beta_1 = \beta_2 = 0$; ⑦ s_{b_1}, standard error of coefficient b_1; ⑧ s_{b_2}, standard error of coefficient b_2; ⑨ t, t statistic for testing H_0: $\beta_1 = 0$; ⑩ t, t statistic for testing H_0: $\beta_2 = 0$.

An edited printout from a BMDP regression program for the data of this problem is presented in Computer Exhibit 13.5.

a. Find the estimated multiple regression equation.
b. Find the predicted value for Y given that $X_1 = 3.0$ and $X_2 = 9.0$.
c. Determine the coefficient of multiple determination.
d. Test the hypothesis H_0: $\beta_1 = \beta_2 = 0$. Use a significance level of .05.
e. Test the hypothesis that X_2 adds no additional explanatory power to the equation over that which X_1 provides. Use a significance level of .05.

27. A real estate broker with Berg Land & Development engaged the services of a consultant to develop a multiple regression model to predict the sales price of condominiums Y from the area of the floor space X_1 (in square feet) and whether or not the condominium has access to a swimming pool X_2. Data were collected for 20 condominiums and are presented in the accompanying table. Having access to a pool was coded with the value 1, and not having a pool was coded 0.

a. Find the estimated multiple regression equation.
b. Find the predicted value for Y given that $X_1 = 12$ and the condominium has a pool.
c. What is the estimated influence of having a swimming pool on the sales price of a condominium?
d. Test the hypothesis $H_0: \beta_2 = 0$ against $H_a: \beta_2 \neq 0$. Use a significance level of .05.

Sales Price (× 1000)	Area (square feet) (× 100)	Pool	Sales Price (× 1000)	Area (square feet) (× 100)	Pool
67.9	12.0	1	59.1	11.1	0
69.5	13.8	1	59.3	10.5	0
67.2	14.6	0	64.3	11.9	1
61.1	12.6	0	56.6	9.7	0
68.3	12.1	1	50.3	9.2	1
43.5	8.0	0	63.7	12.5	0
59.2	11.0	1	65.5	13.4	1
52.6	9.6	0	61.4	12.9	0
56.2	9.9	1	63.2	13.3	0
56.1	10.6	0	64.8	13.5	1

28. A retail store manager with Petrie Stores, Inc., wants to develop a multiple regression model to predict the amount of sales of a product per month Y from the monthly advertising expenditures X_1 and whether the month was December (coded 1) or another month (coded 0), X_2. Data were collected for 20 months and are presented in the accompanying table.

a. Find the estimated multiple regression equation.
b. Find the predicted value for Y given that $X_1 = 2$ and the month is July.
c. What is the estimated incremental amount of sales of this product during the month of December?
d. Test the hypothesis $H_0: \beta_2 = 0$ against $H_a: \beta_2 \neq 0$. Use a significance level of .05.

Sales (× $1000)	Advertising (× $1000)	Month	Sales (× $1000)	Advertising (× $1000)	Month
14.8	2.2	1	14.4	2.4	1
17.3	3.0	1	5.8	1.3	0
8.2	2.4	0	12.2	1.3	1
12.3	2.0	1	15.3	2.0	1
11.2	2.1	0	12.6	2.0	1
16.0	2.4	1	12.0	1.3	1
11.1	2.7	0	9.6	2.0	0
10.0	2.2	0	11.4	2.2	0
14.5	1.1	1	10.1	2.1	1
11.9	2.1	1	17.6	2.8	1

29. A marketing manager in charge of a profit center for Max Department Stores Company is interested in developing a multiple regression model to predict the marketing budget Y (in tens of thousands of dollars) allowed for a product from previous productsales X_1 (in tens of thousands of dollars) and the stage of the life cycle for the product X_2. The life cycle of the product is coded as growth = 0 and maturity = 1. Data were collected for 20 products and are presented in the accompanying table.

a. Find the multiple regression equation for Y regressed on X_1 and X_2.
b. Find the predicted value for Y given $X_1 = 50$ and $X_2 = 1$.
c. Find SST, SSR, SSE, k, $n - k - 1$, MSR, and MSE, and present the results in an analysis of variance table.
d. Find the value of the standard error of the estimate $S_{Y|12}$.
e. Find the coefficient of multiple determination R^2.
f. Test the overall significance of the regression relationship; that is, test H_0: $\beta_1 = \beta_2 = 0$. Use a significance level of .05.
g. Test each independent variable separately to see whether it contributes explanatory power to the regression equation over and above that provided by the other independent variable; that is, test H_0: $\beta_j = 0$ against H_a: $\beta_j \neq 0$. Use .05 for the level of significance.

Budget	Sales	Life Cycle	Budget	Sales	Life Cycle
9	36	1	10	51	1
10	53	1	9	58	1
4	29	0	7	46	0
4	22	0	9	50	1
3	28	0	9	63	1
9	45	1	9	43	1
6	52	0	6	46	0
8	46	1	10	50	1
7	72	0	6	61	0
9	54	1	12	69	1

30. A human resources manager with Pueblo Federated Corporation is interested in developing a multiple regression model to estimate the salary Y (in thousands of dollars) for employees from experience X_1 (in years) with the firm and from performance X_2 (as measured by an index). Data were collected for 10 employees and are presented in the accompanying table.

a. Find the multiple regression equation for Y regressed on X_1 and X_2.
b. Find the predicted value for Y given $X_1 = 10$ and $X_2 = 60$.
c. Find SST, SSR, SSE, k, $n - k - 1$, MSR, and MSE, and present the results in an analysis of variance table.
d. Find the value of the standard error of the estimate $S_{Y|12}$.
e. Find the coefficient of multiple determination R^2.
f. Test the overall significance of the regression relationship; that is, test H_0: $\beta_1 = \beta_2 = 0$. Use a significance level of .05.
g. Test each independent variable separately to see whether it contributes explanatory power to the regression equation over and above that provided by the other independent variable; that is, test H_0: $\beta_j = 0$ against H_a: $\beta_j \neq 0$. Use .05 for the level of significance.

Salary	Experience	Performance
44	11	73
44	10	67
43	10	55
37	9	56
43	12	71
42	11	68
44	12	68
37	10	70
42	12	80
55	13	85

31. A manager of industrial sales territories with Rhode Island Energy Resources Company wants to develop a multiple regression model to predict a sales representative's annual sales Y (in hundreds of thousands of dollars) from the time that the representative has been assigned to the sales territory X_1 (in months) and the average level of customer satisfaction in the territory X_2 (as measured by an index). Data were collected for ten representatives and are presented in the accompanying table.

   a. Find the multiple regression equation for Y regressed on X_1 and X_2.
   b. Find the predicted value for Y given $X_1 = 20$ and $X_2 = 70$.
   c. Find SST, SSR, SSE, k, $n - k - 1$, MSR, and MSE, and present the results in an analysis of variance table.
   d. Find the value of the standard error of the estimate $S_{Y|12}$.
   e. Find the coefficient of multiple determination R^2.
   f. Test the overall significance of the regression relationship; that is, test H_0: $\beta_1 = \beta_2 = 0$. Use a significance level of .05.
   g. Test each independent variable separately to see whether it contributes explanatory power to the regression equation over and above that provided by the other independent variable; that is, test H_0: $\beta_j = 0$ against H_a: $\beta_j \neq 0$. Use .05 for the level of significance.

Sales	Time	Satisfaction
54	29	77
43	21	68
36	22	78
56	24	64
66	30	75
61	29	73
50	24	70
44	22	70
31	12	54
49	28	82

32. **Bank Balance Sheets.** A financial analyst for the Federal Deposit Insurance Corporation wants to examine a relationship between the two sides of banks' balance sheets. The variables of interest are (1) core deposits (percentage of liability and capital), (2) interest-sensitive loans (percentage of assets), and (3) nonsensitive loans (percentage of assets). A sample of data was obtained from large banks, and the data are presented in the accompanying table.

   a. Use a computer to obtain the sample regression equation for estimating core deposits from interest-sensitive and nonsensitive loans.
   b. Find the coefficient of multiple determination.

Core Deposits	Sensitive Loans	Nonsensitive Loans
18.	28.	15.
36.	37.	32.
19.	29.	16.
27.	37.	25.
35.	40.	28.
25.	33.	19.
31.	25.	27.
32.	50.	26.
30.	32.	26.
25.	27.	24.

c. Test whether the overall linear relationship between the variables is statistically significant. Use a significance level of .05.

d. Test each independent variable separately to see if it adds additional explanatory power to the regression equation over that provided by the other independent variable. Use .05 for the level of significance for each test.

REFERENCE: Simonson, Don, John Stowe, and Collin J. Watson (1983), "A Canonical Correlation Analysis of Commercial Bank Asset/Liability Structures," *Journal of Financial and Quantitative Analysis,* 18 (March): 125–40.

33. **Bank Balance Sheets.** A financial analyst for the Federal Deposit Insurance Corporation wants to examine a relationship between the two sides of banks' balance sheets. The variables of interest are (1) purchased funds (percentage of liabilities and capital), (2) interest-sensitive loans (percentage of assets), and (3) nonsensitive loans (percentage of assets). A sample of data was obtained from large banks, and the data are presented in the accompanying table.

a. Use a computer to obtain the sample regression equation for estimating purchased funds from interest-sensitive and nonsensitive loans.

b. Find the coefficient of multiple determination.

c. Test whether the overall linear relationship between the variables is statistically significant. Use a significance level of .05.

d. Test each independent variable separately to see if it adds additional explanatory power to the regression equation over that provided by the other independent variable. Use .05 for the level of significance for each test.

Purchased Funds	Sensitive Loans	Nonsensitive Loans
22.	28.	15.
35.	37.	32.
29.	29.	16.
33.	37.	25.
33.	40.	28.
30.	33.	19.
21.	25.	27.
43.	50.	26.
32.	32.	26.
29.	27.	24.

REFERENCE: Simonson, Don, John Stowe, and Collin J. Watson (1983), "A Canonical Correlation Analysis of Commercial Bank Asset/Liability Structures," *Journal of Financial and Quantitative Analysis,* 18 (March): 125–40.

34. **Relationships Between Personal Characteristics and Leisure Activities.** The busi ness manager for Southbuys wants to examine the relationships between age, education, income, and leisure activity in the form of attendance at art exhibits. The variables of interest are (1) age (years), (2) education (years), (3) income (thousands of dollars), and (4) art exhibits (annual attendance). A sample of data was obtained from consumers of leisure activities, and the data are presented in the accompanying table.

a. Use a computer to obtain the sample regression equation for estimating annual attendance at art exhibits from age, education, and income.

b. Find the coefficient of multiple determination.

c. Test whether the overall linear relationship between the variables is statistically significant. Use a significance level of .05.

d. Test each independent variable separately to see if it adds additional explanatory power to the regression equation over that provided by the other independent variables. Use .05 for the level of significance for each test.

Age	Education	Income	Art Exhibits
18.	11.	35.	11.
35.	13.	38.	10.
21.	14.	35.	15.
35.	16.	50.	22.
25.	14.	36.	13.
21.	13.	39.	14.
39.	13.	37.	13.
31.	12.	34.	7.
20.	14.	41.	15.
40.	12.	29.	12.

 35. Relationships Between Personal Characteristics and Leisure Activities. The business manager for Paramour Communications wants to examine the relationship between age, education, income, and leisure activity in the form of attendance at movies. The variables of interest are (1) age (years), (2) education (years), (3) income (thousands of dollars), and (4) movies (annual attendance). A sample of data was obtained from consumers of leisure activities and the data are presented in the accompanying listing.

a. Use a computer to obtain the sample regression equation for estimating annual attendance at movies from age, education, and income.
b. Find the coefficient of multiple determination.
c. Test whether the overall linear relationship between the variables is statistically significant. Use a significance level of .05.
d. Test each independent variable separately to see if it adds additional explanatory power to the regression equation over that provided by the other independent variables. Use .05 for the level of significance for each test.

Age	Education	Income	Movies
18.	11.	35.	25.
35.	13.	38.	12.
21.	14.	35.	21.
35.	16.	50.	9.
25.	14.	36.	18.
21.	13.	39.	27.
39.	13.	37.	4.
31.	12.	34.	17.
20.	14.	41.	17.
40.	12.	29.	7.

Ratio Data Set   *Refer to the 141 companies listed in the Ratio Data Set in Appendix A.*

Questions   36. Locate the data for the net income–total assets (NI/TA) ratio as a measure of profitability, the current assets–sales (CA/S) ratio, the quick assets–sales ratio (QA/S), and the current assets–current liabilities (CA/CL) ratio.

a. Use a computer to obtain the sample regression equation for estimating NI/TA from CA/S, QA/S, and CA/CL.
b. Find the coefficient of multiple determination.
c. Test whether the overall linear relationship between the variables is statistically significant. Use a significance level of .05.
d. Test each independent variable separately to see if it adds additional explanatory power to the regression equation over that provided by the other independent variables. Use .05 for the level of significance for each test.

37. Locate the data set for the net income–total assets (NI/TA) ratio as a measure of profitability, the current assets–sales (CA/S) ratio, and the current assets–current liabilities (CA/CL) ratio.

   a. Use a computer to obtain the sample regression equation for estimating NI/TA from CA/S and CA/CL.
   b. Find the coefficient of multiple determination.
   c. Test whether the overall linear relationship between the variables is statistically significant. Use a significance level of .05.
   d. Test each independent variable separately to see if it adds additional explanatory power to the regression equation over that provided by the other independent variable. Use .05 for the level of significance for each test.

38. Locate the data for the net income–total assets (NI/TA) ratio as a measure of profitability, the current assets–sales (CA/S) ratio, and the quick assets–sales ratio (QA/S).

   a. Use a computer to obtain the sample regression equation for estimating NI/TA from CA/S and QA/S.
   b. Find the coefficient of multiple determination.
   c. Test whether the overall linear relationship between the variables is statistically significant. Use a significance level of .05.
   d. Test each independent variable separately to see if it adds additional explanatory power to the regression equation over that provided by the other independent variable. Use .05 for the level of significance for each test.

39. Locate the data for the current assets–sales (CA/S) ratio, the quick assets–sales ratio (QA/S), and the current assets–current liabilities (CA/CL) ratio.

   a. Use a computer to obtain the sample regression equation for estimating CA/S from QA/S and CA/CL.
   b. Find the coefficient of multiple determination.
   c. Test whether the overall linear relationship between the variables is statistically significant. Use a significance level of .05.
   d. Test each independent variable separately to see if it adds additional explanatory power to the regression equation over that provided by the other independent variables. Use .05 for the level of significance for each test.

Refer to the 113 applicants for credit listed in the Credit Data Set in Appendix A.

Credit Data Set

Questions

40. Use a computer program to obtain the multiple regression equation in which JOBINC is the dependent variable and the independent variables are SEX, JOBYRS, TOTBAL, and MSTATUS. Use only the applicants who were granted credit and who listed a JOBINC figure.

41. Test two separate hypotheses. First, test the hypothesis that JOBYRS adds no additional explanatory power to the prediction of JOBINC, given that all of the other independent variables are also included in the regression equation. Then test the hypothesis that TOTBAL adds no additional explanatory power to the prediction of JOBINC, given that all of the other independent variables are also included in the regression equation. Use H_a: $\beta_j \neq 0$ and $\alpha = .05$ in these two tests.

PREDICTING THE MARGIN OF VICTORY IN COLLEGE FOOTBALL GAMES   **CASE 13.1**

Managers in the broadcasting and entertainment industries and administrators of college athletic programs are concerned about the success of college football teams. Top-20 college football teams generate a great deal of revenue for college athletic departments. Advertisers are also interested in the outcomes of football games.

To examine whether variables such as the amount of rushing yardage, the amount of passing yardage, the number of turnovers, the time of possession, and the home field advantage influence the final score in a college football game, an analyst developed a multiple regression model. The dependent variable for the multiple regression model was the margin of a victory in college football games. The margin of victory is the winning team's final score minus the score for the losing team.

The analyst wanted to examine whether the margin of victory could be explained by the following variables:

1. The difference in rushing yardage (rushing yardage for the winning team minus rushing yardage for the losing team).
2. The difference in passing yardage.
3. The difference in the number of turnovers made.
4. The difference in ball possession times.
5. An indicator variable for the home field advantage (home team coded 1; visiting team coded 0).

Ninety games for top-20 football teams in Division I of the National Collegiate Athletic Association were selected randomly for the 1985 college football season. The data for the dependent and independent variables were collected, a multiple regression analysis was conducted, and the results for the multiple regression equation are given in the following table:

Predictor	Coefficient	t ratio
Intercept	3.22	2.06*
Difference in rushing yardage	.11	12.50*
Difference in passing yardage	.09	10.19*
Difference in turnovers	−2.80	−5.75*
Difference in time of possession	−.01	−3.94*
Indicator for home field advantage	3.04	1.68

Dependent Variable: Margin of victory

$R^2_{adjusted} = .72$

*P-value $< .05$.

The regression analysis indicates that all the independent variables, except for the home field advantage, are marginally significant in the model, at the .05 level of significance. Furthermore, about 72% of the variation in margin of victory is explained by the predictor variables.

The coefficients in the regression equation generally agree, holding the other variables constant, with what fans would expect. For example, as the rushing yardage increases, the regression equation indicates that the margin of victory also tends to increase. An additional yard gained by rushing over and above the rushing yardage of an opponent is associated with a .11 point increase in the margin of victory. An additional passing yard over and above passing yardage of an opponent is associated with a .09 increase in the margin of victory.

The sample regression coefficient for the difference in number of turnovers made indicates that turnovers are very important in explaining the margin of victory. An increase of one turnover above the turnovers of an opponent is associated with a *decrease* of 2.80 points in the margin of victory, holding the other variables constant. The coefficient for the difference in time of possession is −.01, indicating that an increase in time of possession of one second over an opponent's time of possession is associated with a decrease in the margin of victory.

The regression results indicate that the home field advantage is not statistically significant in explaining the margin of victory in college football games for top-20 teams, given the other variables in the model.

Although there are many variables that are reported in the media about college football games, it appears that analysts, coaches, and fans may wish to pay attention to variables such as rushing and passing yardage and the number of turnovers in explaining the final point spreads in top-20 college football games.

a. List the multiple regression equation found for the college football games.
b. A football announcer states: "Dad-gum-it, time of possession favors that hoss of a visiting team." Does the announcer's statement suggest that the visiting team should be favored to win the game in light of the multiple regression results, presuming that other variables are the same?

Case Assignment

REFERENCE: Wagner, G. (1987), "College and Professional Football Scores: A Multiple Regression Analysis," *American Economist,* XXXI: 33–37.

PREDICTING THE NUMBER OF BIDDERS FOR THE OIL LEASE LOTTERY

CASE 13.2

Refer to Case 5.2 for a description of Western Petroleum and its Plan B investment in oil lease packages.

Mr. Malcolm Maxfield and Saul Stewart worked together to develop a report on the probability of winning no leases and on the expected returns from a Western Petroleum plan B. Mr. Maxfield did most of the computation and writing work, with some technical aid from Mr. Stewart, who was more current on the theory needed in these problems. Reviewing the necessary probability theory, making the computations, and writing up the results took Maxfield approximately four days.

During this time Mr. Maxfield was impressed with the fact that the probabilities and expected returns were rather sensitive to the number of applications he assumed would be made for each parcel. For instance, he found that the probability of obtaining one or more parcels in a plan B ranged from .993 (if the fewest people bid) down to .658 (if many people bid on each parcel). Under the optimistic assumption that few people would bid on each parcel, the expected number of parcels won by the holder of a plan B was found to be 4.727. But under the pessimistic assumption that many people bid on each parcel, that expected number of parcels went down to 1.063.

Since the results of his computations were so sensitive to the number of people bidding on the parcels, Mr. Maxfield determined that he would continue his study of the situation in an attempt to find a means of estimating or forecasting how many people would be bidding for a parcel. He commented, "I know that class I parcels are attracting more bids, but there is even a wide variation in the number of bidders on those parcels. If I could get a list of the characteristics of these parcels and how many people bid, maybe I could develop some kind of a forecasting equation based on the characteristics of the property."

With this objective in mind, Maxfield began gathering data on the characteristics and number of bidders on various parcels of land in the BLM lotteries during the past year. Since so many parcels of land had been involved, Maxfield could not find information on all of them. Thus he took a random sample of several parcels and arranged their data by class (see Table 1).

Each row in the table represents a parcel of land that the BLM put up for bid in the lotteries last year. The first column shows how many people bid on that parcel. The second column shows the parcel's classification. These classifications (I through III) are determined by the BLM. The third column in the table shows the number of miles (as the crow flies) from the center of the parcel to the nearest producing oil well. The output of the nearest producing well is listed in the fourth column. This output is measured in terms of barrels per month. The fifth column contains the subjective probability estimate that oil could be found in geological formations like those known to be on the parcel in question.

TABLE 1	Number Bidding	Class	Miles to Nearest Producer	Output of Nearest Producer	Geologist's Likelihood Estimate	Parcel Size
Characteristics and	300	I	8.0	5000	.40	3.5
Number of Bidders in	375	I	3.2	5000	.40	4.0
BLM Lotteries	200	I	8.0	4500	.15	10.2
	531	I	.8	6500	.50	1.3
	115	I	12.0	3600	.08	5.6
	629	I	1.6	7800	.65	2.9
	152	I	5.9	4200	.05	6.5
	131	I	11.8	3600	.05	9.8
	192	I	4.0	2600	.20	6.3
	427	I	2.6	7800	.35	10.1
	93	II	25.0	7100	.20	3.8
	120	II	36.0	1500	.25	4.2
	46	II	17.8	3500	.15	2.1
	38	II	49.6	3800	.15	1.6
	29	II	48.3	5100	.15	.8
	77	II	18.5	4200	.20	2.8
	102	II	17.1	5300	.33	3.5
	82	II	16.5	3100	.25	3.0
	53	II	26.1	4100	.18	3.0
	48	II	18.1	3400	.15	2.3
	10	III	135.1	3600	.02	4.6
	21	III	151.5	4500	.04	10.1
	16	III	80.5	3200	.03	8.6
	35	III	52.1	3900	.05	16.1
	40	III	51.6	4100	.04	21.7
	12	III	124.1	3600	.01	2.2
	15	III	103.1	4700	.01	3.2
	11	III	176.1	4300	.02	2.0
	19	III	142.6	3600	.02	3.6
	37	III	81.5	3100	.04	4.2

These estimates were made by a geologist who served as a consultant to Western Petroleum. The final column in the table shows the acreage on the parcels. Thus the first parcel listed in the table was one on which 300 people bid. It was a class I parcel as defined by the BLM, and it was 8 miles from a well that produced 5000 barrels of oil per month. The geologist estimated that 40% of the formations like those found on this parcel yielded producing wells, and the parcel had an area of 3.5 acres.

Mr. Maxfield showed his table to Saul Stewart and asked him for suggestions on how it might be used to develop an estimating procedure for the number of bids to expect on each parcel. Mr. Stewart replied that multiple linear regression seemed like the technique that would help Maxfield. Stewart wrote down an equation for Maxfield,

$$\hat{Y} = b_0 + b_1 X_1 + b_2 X_2 + b_3 X_3 + b_4 X_4 + b_5 X_5$$

and stated that $\hat{Y}$ would be the estimated number of bidders for a parcel and that X_1 through X_5 would be the five characteristics of the parcels Maxfield had listed in his table. In addition, Stewart offered to have an assistant at S&W enter Maxfield's data into a standard computer program to produce the regression equation. Malcolm was very pleased with the offer and turned his data over to Stewart.

On Wednesday of the next week Maxfield and Stewart met for lunch to discuss the computer output Mr. Stewart's assistant had produced. Saul Stewart explained the meaning of various figures in the computer output, and Maxfield felt that he understood the rough idea of how the output could be interpreted.

However, for over a week Malcolm was unable to get back to the computer output and study it owing to the press of other business. When he did, he found that he was a little confused on some of the figures. In particular, he was quite upset about the fact that the variable X_2, miles to nearest producer, seemed to have a positive coefficient in his prediction equation. This indicated to him that a parcel far from a producing well would have more bidders than a parcel near a producing well. To resolve the confusion in his mind. Maxfield called Stewart with the idea of asking him for additional explanations and about the possibility of obtaining new computer runs of the data. Mr. Stewart, however, was out of town for the next two weeks.

Case Assignment

a. What is the prediction (regression) equation?
b. Which variables are significantly related to the number of bidders for each parcel, and how can you determine "significance?"
c. How good a job does the regression equation do in predicting the number of bidders on a parcel? How much better is the equation in predicting than just using the mean number of bidders as your estimate?
d. Look at the algebraic signs on the coefficients in the regression equation. Do they make sense?

REFERENCES

Belsley, D., E. Kuh, and R. Welsch. 1980. *Regression Diagnostics: Identifying Influential Data and Sources of Collinearity.* New York: Wiley.

BMDP Statistical Software. 1985. *BMDP Statistical Software Manual.* Los Angeles: University of California Press.

Bowerman, B., R. O'Connell, and D. Dickey. 1986. *Linear Statistical Models: An Applied Approach.* Boston: Duxbury Press.

Draper, N. R., and H. Smith. 1981. *Applied Regression Analysis,* 2nd ed. New York: Wiley.

Neter, J., W. Wasserman, and M. Kutner. 1989. *Applied Linear Regression Models,* 2nd ed. Homewood, Ill.: Irwin.

Neter, J., W. Wasserman, and M. Kutner. 1990. *Applied Linear Statistical Models,* 3rd ed. Homewood, Ill.: Irwin.

Ryan, T., B. Joiner, and B. Ryan. 1985. *Minitab Handbook.* Boston: Duxbury Press.

SAS Institute, Inc. 1985. *SAS User's Guide: Statistics, Version 5 Edition.* Cary, N.C.: SAS Institute, Inc.

SPSS, Inc. 1988. *SPSS/PC+* Chicago: SPSS, Inc.

Younger, M. S. 1985. *A First Course in Linear Regression,* 2nd ed. Boston: Duxbury Press.

Optional Topics

Residual Analysis, Multicollinearity, and Special Problems in Multiple Regression
Stepwise Multiple Regression
Curvilinear Regression
Testing Several Coefficients of a Multiple Regression Equation
Standardized Regression Coefficients

CHAPTER 13 SUPPLEMENT

Computers make the solution to multiple regression problems rather simple to obtain. However, the ease of obtaining a mathematical solution does not relieve the program user from the responsibility of being cautious in using the results. Several special problems can plague the user if he or she is not careful. Some of the more common problems are discussed briefly in this section.

Violating the basic assumptions of multiple regression can lead us to incorrect conclusions. Recall that these assumptions are as follows:

1. The independent variables are nonrandom.
2. The dependent-variable values are normally and independently distributed around the regression surface, defined by $\mu_{Y|X_1, X_2, \ldots, X_k} = \beta_0 + \beta_1 X_1 + \cdots + \beta_k X_k$.
3. The variance of the dependent variable around the regression plane is constant.

One of the best ways to determine whether some of the basic assumptions of the regression procedure are met is to graph the **prediction errors,** or **residuals** $e_i = Y_i - \hat{Y}_i$, against $\hat{Y}$, or against the independent variables after the regression equation has been constructed. Figure 13.4(a) shows a situation that suggests that the variance around the regression plane is not constant; the variance of the residuals seems to be larger for larger values of the independent variable X_j. The pattern of the residual plot in Figure 13.4(b) suggests that the regression relationship is not linear, because the residuals are not scattered randomly above and below the line where $e_i = 0$.

A second problem that can create difficulty in multiple regression is that of **multicollinearity.** This problem exists when pairs or combinations of the independent variables are highly correlated with one another. For example, we may have a data set in which people's ability to perform a certain type of heavy work is the dependent variable Y, which we would like to estimate. Two logical independent variables are the people's heights and weights. However, the heights and weights of people are usually very highly correlated. Thus, this data set would be said to exhibit the characteristic of multicollinearity.

A specific problem introduced by the presence of multicollinearity in the data set involves the reliability of the regression equation's coefficients, the b_j's. This problem can be demonstrated in the two-independent-variable case by the value

$$s_{b_1} = \frac{S_{Y|12}}{\sqrt{SSX_1(1 - r_{12}^2)}}$$

The r_{12}^2 term in this equation is the coefficient of determination between the two independent variables X_1 and X_2. If these two variables are highly correlated (so that we have multi-

13.8
RESIDUAL ANALYSIS, MULTICOLLINEARITY, AND SPECIAL PROBLEMS IN MULTIPLE REGRESSION

FIGURE 13.4

Residual Plots

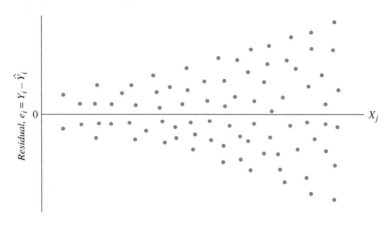

(a) Variance of the residuals not constant with respect
     to the independent variable X_j

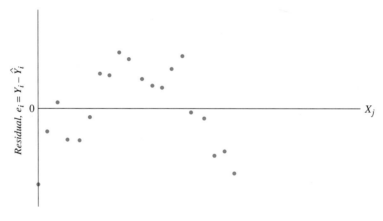

(b) Residual plot suggesting that the relationship
     between Y and X_j is not linear

collinearity), then r_{12}^2 is close to 1. This condition causes s_{b_1} to be large. In fact, the use of collinear independent variables can cause s_{b_j} values to be inflated. This large value, in turn, may cause the regression results user to fail in detecting significant relationships between the dependent variable and some independent variables. Furthermore, if there is perfect multicollinearity among independent variables where, say, r_{12}^2 is 1, then a unique, unbiased least squares regression equation does not exist; and if there is a high degree of multicollinearity, then computers may encounter difficulties in finding a solution.

Also, when there is multicollinearity among independent variables, as is often the case in business and economic regression problems, the value for b_1 in the sample regression equation $\hat{Y} = b_0 + b_1 X_1$ can be rather different from the value for b_1 in the sample multiple regression equation $\hat{Y} = b_0 + b_1 X_1 + b_2 X_2$. If we *omit an important independent X variable* that is correlated with other independent variables that are included in a regression model, then the sample intercept and slope coefficients are biased due to **misspecification** and tests of the effects of the included variables can be misleading. It is wise for practitioners of multiple regression analysis to be cautious while interpreting regression coefficients under the condition of multicollinearity. However, predictions of Y are not hampered under the condition of multicollinearity.

⎯ **EXAMPLE 13.13** For the United Park City Properties simple regression of the price of condominiums Y against the area of condominiums in Section 12.2, we determined the sample regression equation to be $\hat{Y} = -34.5 + 7.68X$, where X denoted area. For the multiple regression of price Y, area X_1, and land costs X_2 in Section 13.2 for the same condominiums, we determined the sample multiple regression equation to be $\hat{Y} = -61.29 + 8.07X_1 + 2.82X_2$. Notice that the regression coefficients for condominium areas are not the same (7.68 in the simple regression and 8.07 in the multiple regression), but they are quite close. The two coefficients are not exactly equal in this problem, but they are quite similar. The degree of multicollinearity is very low, as indicated by the value $r_{12}^2 = .04$. If the value of r_{12}^2 were equal to zero, then we would have no multicollinearity in this problem, and the sample regression coefficients for area would be equal in the two equations. If there is multicollinearity among independent variables, then sample regression coefficients for the variables can be different from one equation to another. ⎯

Another special problem encountered by many regression program users is that of **overfitting.** This problem occurs when we use too many independent variables and the program produces a long regression equation containing numerous independent variables. The R^2 measure for such an equation may be quite high, but the increase in the R^2 value produced by the addition of the last several variables may be insignificant, in the statistical sense.

This problem can often be detected by noting that the coefficients, the b_j values, on several of the independent variables are not significantly different from zero. That is, for several variables the hypothesis $H_0: \beta_j = 0$ cannot be rejected.

An additional problem involving regression analysis is that of **extrapolation.** The regression equation constructed from a particular data set is valid for making predictions of the dependent variable Y *only if* the independent variables used in the predictions, the X_j's are similar in value to those in the original data set.

⎯ **EXAMPLE 13.14** In the condominium sales example presented in Section 13.2, the condominiums used to develop the regression equation ranged in area from 900 to 1500 square feet. If we were to attempt to predict the price of a condominium containing 2500 square feet, we would be guilty of extrapolating beyond the range of the independent variable's observed data values. Similarly, if we use any values of the independent variables that are outside the range of the values used in constructing the regression equation, we are extrapolating and cannot place much confidence in our predicted value. ⎯

There are situations, however, in which using values of the independent variables that are outside the ranges of the observed data is unavoidable. In these cases those doing the extrapolation must be careful not to take their estimates of Y too seriously.

The final caution we will give concerns the **assumption of causation** many users of regression inappropriately read into their results. The correlation coefficient is often subject to misinterpretation. The false assumption is made at times that because two variables are related, a change in one *causes* a change in the other. If one has a point to prove, one can easily succumb to this fallacy and use a correlation coefficient to "prove" a cause-and-effect relationship that may not exist.

⎯ **EXAMPLE 13.15** Research has shown that there is a negative correlation between smoking and grades. If one objects to smoking, one seizes upon this fact as objective evidence that smoking is harmful, that smoking causes low grades. This statement may be true, but one could also argue that low grades result in increased nervous tension, which, in turn, causes the individual to smoke more. In our gas station example in the chapter introduction, we found that number of pumps and gasoline sales were positively related. But from our regression equation it is impossible to say if more pumps cause higher sales or if higher sales have caused the installation of more pumps or if both variables are related to some other factor, such as location of the station.

This last suggestion illustrates another danger in interpretation of the results of a correlation analysis. Frequently, two variables may appear to be highly correlated when one does not cause the other but both are caused by a third variable.

We must also be very careful in our interpretation of the coefficients in a regression equation in this situation.

EXAMPLE 13.16   We found in the previous section that the coefficient on X_1, area of the condominium (in hundreds of square feet), was 8.07, meaning $8070 for each additional 100 square feet of area. This value does *not* mean that 100 additional square feet in area in a condominium will *cause* the price to rise $8525.10. It merely indicates that there is an **association** between additional area and price within the data set and that good estimates of price can be obtained if "area" is weighted by $8070. However, in this example it makes much more sense to conclude that there is a cause-and-effect relationship between price and area since additional area requires additional materials and labor to be expended during construction. These added costs would naturally force up the price. However, the $8070 figure may not measure the direct effect of those increased costs.

The problems outlined in this section should convince the reader that the application of multiple regression to a problem requires judgment and good sense. When it is applied with care by competent individuals, multiple regression analysis is one of the most powerful tools available to managers and economists today.

Problems: Section 13.8

42. A simple regression analysis for Y regressed against X_1 at WatStat, Inc. resulted in the residual plot presented in Figure 13.5. Discuss the problem suggested by the plot.

FIGURE 13.5

Residual Plot for WatStat, Inc.

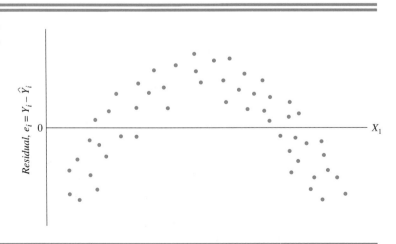

43. A Big Five accounting firm is concerned about the risk of incurring liabilities due to lawsuits stemming from the opinions that are given in its audit practice. The firm wants to develop a model to predict its annual liability Y (in tens of thousands of dollars) from its clients' average annual earnings X_1 (in millions of dollars) and from the firm size X_2 (in tens of thousands of billable hours). Data were collected for 10 years and are presented in the accompanying table. A regression of Y on X_1 gave the equation $\hat{Y} = -2.25 + 5.12X_1$, and a regression of Y on both X_1 and X_2 gave the equation $\hat{Y} = -3.91 + 4.98X_1 + .062X_2$.

a. Find the coefficient of determination for X_1 and X_2; that is, find r_{12}^2.

b. Why does the coefficient for X_1 change from 5.12 in the simple regression equation to 4.98 in the multiple regression equation?

Liability	Earnings	Billable Hours
54	12	45
43	9	51
36	9	52
56	10	49
66	12	62
61	12	53
50	10	45
44	9	51
31	6	37
49	11	57

44. Discuss the problem of multicollinearity in the BZ Mining data for Problem 22 and the difficulties it introduces.

13.9
STEPWISE MULTIPLE REGRESSION

In Section 13.2, we discussed how a computer can be used to obtain the coefficients in the sample regression line $\hat{Y} = b_0 + b_1X_1 + b_2X_2$. Sometimes, however, the person constructing a multiple regression equation does not wish to determine the regression equation between the dependent variable Y and a set of independent variables (the X_j's) in this manner. For instance, we may have a set of 15 or 20 independent variables that may be associated with the dependent variable whose value we would like to predict. However, to construct a regression equation using all 15 or 20 variables would not be parsimonious.

We may wish to construct a regression equation by using a subset of original independent variables in order to obtain an equation of a manageable size. However, this desired method raises the problem of which subset of the X_j's would do the best job in predicting the Y_i values. Stepwise multiple regression is one procedure commonly used to address this problem.

13.9.1 Stepwise Multiple Regression Concepts

In **stepwise multiple regression** the regression equation is constructed one variable at a time. That is, first the equation $\hat{Y} = b_0 + b_jX_j$ is constructed by using the independent variable X_j that is *most highly correlated* with Y. At the second step, a new variable is brought into the equation to produce $\hat{Y} = b_0 + b_jX_j + b_{j'}X_{j'}$, where $X_{j'}$ is the variable that does the best job of explaining the variation in Y *that has not already been explained* by X_j. That is, $X_{j'}$ is the variable that does the best job in eliminating the prediction errors that were present when using the first equation. Note that the b_0 and b_j values of the second equation may be somewhat different from their values in the first equation. In the third step of a stepwise regression problem, a new variable is brought into the equation to form

$\hat{Y} = b_0 + b_j X_j + b_{j'} X_{j'} + b_{j''} X_{j''}$. Again, $X_{j''}$ is the variable that does the best job in eliminating the prediction errors of the second equation. A variable may also be removed from the equation. This process is usually continued until we feel that there are enough variables in the equation.

Other stepwise procedures are possible, including *backward elimination,* where the regression practitioner begins with all of the X_j variables in the equation and then proceeds to eliminate variables from the equation.

Stepwise multiple regression requires extensive computations. For this reason, stepwise regression problems are solved by using a computer.

13.9.2 Stepwise Regression with a Computer

In previous sections, we used a simple example wherein we constructed regression equations to be used in predicting the sales price of condominiums at United Park City Properties. An actual study involving the objectives outlined in this example would differ from what has been presented in previous sections in three important aspects:

1. More condominium sales would be used in the data set. That is, n would be larger than the value of 5 used in previous examples.
2. More variables would be used to describe each condominium. That is, the number k of independent variables would be larger than the one or two used in previous examples.
3. A computer and data analysis software would be used to make the computations discussed in the previous sections. The meanings of the calculated measures would be the same, but the computer would do the tedious computational work.

In this section, we will examine a more realistic example, using stepwise regression to predict condominium sales prices and explain the output of a computer program used to solve it. We will use a data set that has $n = 20$ observations of condominium sales and $k = 5$ independent variables. The data we will use in this expanded example are presented in Table 13.9. The dependent variable that we are interested in predicting is still condominium sales price, as indicated in the second column of the table. The five independent variables used to predict sales price describe each condominium's area, land cost (just as in previous examples), distance from the center of town (in miles), number of bedrooms, and parking features.

The last two independent variables deserve special attention. They are not continuous variables like price, area, land cost, and distance. They are discrete. The number of bedrooms is an integer and takes on values of 1, 2, 3, or 4. The parking variable is used to indicate whether the condominium unit has a covered parking area. A value of 1 indicates that covered parking is available, and a 0 indicates that it is not. Thus the parking variable is an indicator variable.

Since the data presented in Table 13.9 are so voluminous, applying regression techniques would be very tedious without the aid of a computer program. Also, since not all of the five independent variables are likely to add significantly to the accuracy of sales price predictions, applying a computerized stepwise regression program to these data is appropriate.

There are many preprogrammed stepwise regression software packages that could be applied to the data in Table 13.9. One of the most commonly available is the stepwise regression program that is part of the Statistical Package for the Social Sciences (SPSS). This program requires that the user know very little about computers. However, the user must be able to interpret the computer output. The remainder of this section is devoted to presenting the major outputs of the SPSS stepwise regression program using the data in Table 13.9. The output explained in the following sections is very similar to the output from many multiple regression computer packages.

The first pieces of information printed by the program are the means, standard deviations, and correlations of all the variables used in the regression problem. These descriptive statistics, shown in Computer Exhibit 13.6, reveal that for the enlarged data set the

i, Condominium Number	Y, Sales Price	X_1, Unit Area	X_2, Land Cost	X_3, Distance to City Center	X_4, Number of Bedrooms	X_5, Covered Parking	**TABLE 13.9**
1	36	9	8	4	1	0	**Expanded Data for**
2	80	15	7	12	4	1	**Condominium Prices at**
3	44	10	9	6	2	1	**United Park City**
4	55	11	10	0	2	1	**Properties**
5	35	10	6	16	2	0	
6	62	12	6	10	3	1	
7	42	11	11	20	2	0	
8	77	16	12	1	3	1	
9	32	7	6	17	2	0	
10	46	9	10	6	3	1	
11	60	10	11	5	3	1	
12	36	10	4	11	2	1	
13	100	18	14	6	4	1	
14	45	10	5	2	2	0	
15	46	11	6	13	3	1	
16	35	8	8	3	1	0	
17	25	7	4	1	1	0	
18	48	12	8	16	3	1	
19	40	11	9	8	2	1	
20	58	13	10	2	3	1	

Note: As an example, condominium 8 sold for $77,000. It contained 1600 square feet and bore $12,000 in land costs. It was located 1 mile from the center of the city and had 3 bedrooms and a covered parking area. Condominiums 1–5 are the same condominiums used in a previous example, but additional information about each unit has been added. For example, the first condominium was located 4 miles from the city center and had 1 bedroom and no covered parking.

mean condominium price is $50,100. The large standard deviation of $18,515.70 indicates that there is quite a large variation in the prices. The other means and standard deviations are self-explanatory, but the parking variable mean deserves attention. Since parking is a dummy variable that takes on only 0 (for no covered parking) and 1 (for covered parking), the mean of .65 indicates that 65% of the condominiums in the example have covered parking available.

The correlation matrix gives the correlation coefficients measuring the degree of association between all possible pairs of variables in this problem. Several things should be noted about this matrix. First, all the figures on the main (upper left to lower right) diagonal equal unity, because these values show the correlations of the variables with themselves, and every variable is perfectly correlated ($r = 1$) with itself. Also, the values in the correlation matrix are symmetrical around the main diagonal. That is, the figure in row j and column j' is the same as the figure in row j' and column j: $r_{jj'} = r_{j'j}$. This result is obvious since the correlation of variable X_j with variable $X_{j'}$ should be the same as the correlation of $X_{j'}$ with X_j. The order in which we mention the variables is irrelevant.

The first column (and also the first row) of the correlation matrix shows us that the condominium sales prices are most highly correlated with X_1 = area since $r_{YX_1} = .925$. The variable next most highly correlated with price is number of bedrooms X_4, since $r_{YX_4} = .838$. The correlation of price and distance is interesting in that it is negative: $r_{YX_3} = -.170$. This value indicates that there is a slight tendency for the condominiums located away from the city center to be priced lower than those close to the city.

The next important output of the SPSS stepwise regression program is step 1 of the regression procedure. In this step, the variable most highly correlated with price is used to

SPSS/PC+   **COMPUTER EXHIBIT 13.6** Means, Standard Deviations, and Correlation Matrix from
                      Stepwise Regression for United Park City Properties

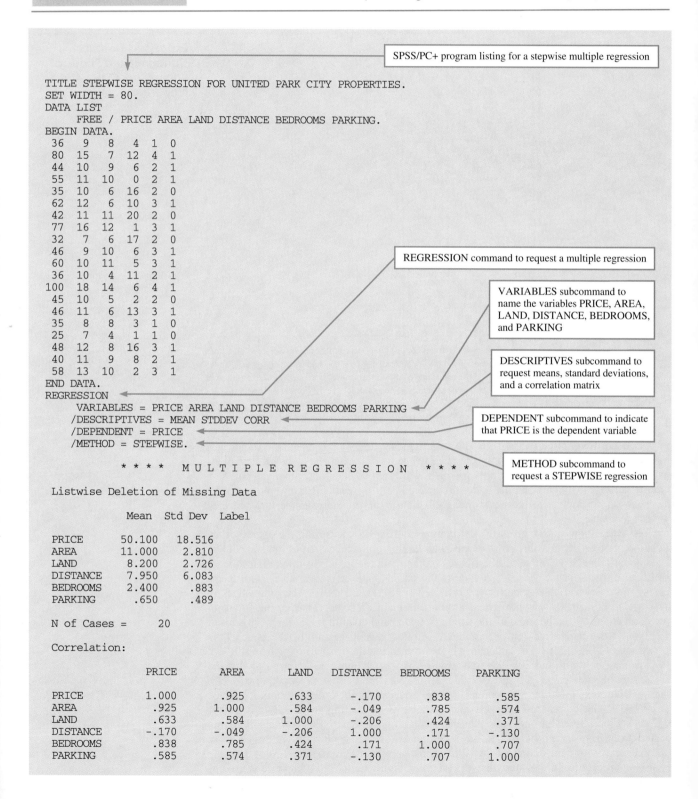

SPSS/PC+ program listing for a stepwise multiple regression

```
TITLE STEPWISE REGRESSION FOR UNITED PARK CITY PROPERTIES.
SET WIDTH = 80.
DATA LIST
      FREE / PRICE AREA LAND DISTANCE BEDROOMS PARKING.
BEGIN DATA.
 36   9   8   4   1   0
 80  15   7  12   4   1
 44  10   9   6   2   1
 55  11  10   0   2   1
 35  10   6  16   2   0
 62  12   6  10   3   1
 42  11  11  20   2   0
 77  16  12   1   3   1
 32   7   6  17   2   0
 46   9  10   6   3   1
 60  10  11   5   3   1
 36  10   4  11   2   1
100  18  14   6   4   1
 45  10   5   2   2   0
 46  11   6  13   3   1
 35   8   8   3   1   0
 25   7   4   1   1   0
 48  12   8  16   3   1
 40  11   9   8   2   1
 58  13  10   2   3   1
END DATA.
REGRESSION
      VARIABLES = PRICE AREA LAND DISTANCE BEDROOMS PARKING
      /DESCRIPTIVES = MEAN STDDEV CORR
      /DEPENDENT = PRICE
      /METHOD = STEPWISE.
```

REGRESSION command to request a multiple regression

VARIABLES subcommand to name the variables PRICE, AREA, LAND, DISTANCE, BEDROOMS, and PARKING

DESCRIPTIVES subcommand to request means, standard deviations, and a correlation matrix

DEPENDENT subcommand to indicate that PRICE is the dependent variable

METHOD subcommand to request a STEPWISE regression

```
          * * * *   M U L T I P L E   R E G R E S S I O N   * * * *

Listwise Deletion of Missing Data

            Mean   Std Dev   Label

PRICE      50.100   18.516
AREA       11.000    2.810
LAND        8.200    2.726
DISTANCE    7.950    6.083
BEDROOMS    2.400     .883
PARKING      .650     .489

N of Cases =      20

Correlation:

              PRICE     AREA      LAND    DISTANCE   BEDROOMS   PARKING

PRICE        1.000     .925      .633     -.170       .838       .585
AREA          .925    1.000      .584     -.049       .785       .574
LAND          .633     .584     1.000     -.206       .424       .371
DISTANCE     -.170    -.049     -.206     1.000       .171      -.130
BEDROOMS      .838     .785      .424      .171      1.000       .707
PARKING       .585     .574      .371     -.130       .707      1.000
```

obtain a simple regression equation $\hat{Y} = b_0 + b_1 X$. We saw from the correlation matrix that $X_1 =$ area is the most highly correlated with price. Thus, the program prints the results shown in Computer Exhibit 13.7. This output indicates that the dependent variable is price and that the first variable entered into the equation is area. For the moment we will ignore the next four figures and go directly to the output entitled Variables in the Equation. The first two columns indicate that the initial regression equation is

$$\hat{Y} = -16.93 + 6.09X_1$$

The column headed by Beta indicates that the standardized coefficient for X_1 is $Beta_1 = .924663$. In other words, a one–standard deviation change in X_1 is associated with a .924663–standard deviation change in price (see Section 13.12). The value .591449, which is labeled SEB, is s_{b_1}.

We noted in Equation (13.15) that the hypothesis H_0: $\beta_1 = 0$ could be tested by calculating

$$t = \frac{b_1}{s_{b_1}}$$

The output gives

$$t = \frac{b_1}{s_{b_1}} = \frac{6.09333}{.591449} = 10.302$$

COMPUTER EXHIBIT 13.7 Stepwise Regression at First Step for United Park City Properties

SPSS/PC+

```
Variable(s) Entered on Step Number
   1..     AREA

Multiple R            .92466
R Square              .85500
Adjusted R Square     .84695
Standard Error        7.24375

Analysis of Variance
                    DF      Sum of Squares      Mean Square
Regression           1         5569.30667       5569.30667
Residual            18          944.49333         52.47185

F =      106.13894        Signif F =  .0000
------------------ Variables in the Equation ------------------

Variable            B         SE B        Beta        T    Sig T

AREA          6.093333     .591449     .924663    10.302   .0000
(Constant)  -16.926667    6.704542                -2.525   .0212
------------- Variables not in the Equation -------------

Variable     Beta In  Partial  Min Toler       T   Sig T

LAND         .140321   .299117    .658876    1.292   .2135
DISTANCE    -.124333  -.326120    .997572   -1.422   .1730
BEDROOMS     .291937   .474673    .383333    2.224   .0400
PARKING      .080588   .173274    .670330     .725   .4781
```

with a P-value labeled as Sig T of .0000. This value can be used to test the hypothesis H_0: $\beta_1 = 0$ against the alternative H_a: $\beta_1 \neq 0$. Since the P-value is .0000, we can reject the hypothesis that $\beta_1 = 0$.

Let us now return to the four figures we passed over. These values involve the R, R^2, and $S_{Y|1}$ figures associated with this first equation. (Since this first equation is a *simple* regression equation, the exact notation might better be r, r^2, and $s_{Y|1}$ since lowercase letters are reserved to indicate simple correlation and regression results.) The Multiple R and R Square figures are computed from Equation (13.11). Thus we can say that 85.5% of the square error of prediction can be eliminated by using the first regression equation to predict sales prices rather than just using $\overline{Y} = \$50,100$ as our predictor.

The Standard Error value is $S_{Y|1}$. It is the standard deviation of the actual prices around the regression line. If the prices are normally distributed about the line, then approximately two-thirds of the condominiums' prices will lie within $\pm\$7243.75$ of the regression equation's predicted values.

The next major output produced by the SPSS program is an Analysis of Variance table, as shown in Computer Exhibit 13.7. We can see that this table produces an F value of

$$F = \frac{5569.30667}{52.47185} = 106.13894$$

with a P-value or Signif F of .0000. With the F test we also reject H_0: $\beta_1 = 0$.

At each step, the SPSS program also presents information concerning the independent variables that are not included in the regression equation yet, as shown in Computer Exhibit 13.7. For each of the variables not in the equation, we have listed first Beta In. This value is the $Beta_j$ that will be associated with each of the X_j values *if* they are chosen to enter the regression equation at the next step.

The values in the Partial column are **partial correlation coefficients.** They show the correlation between each of the independent variables, the X_j's and the dependent variable Y after the influence of X_1, area, has been removed from them. That is, the Partial correlation figure indicates which variable will be the most effective in improving the regression equation's accuracy if it is entered at the next step. For our example we can see that the number of bedrooms X_4 is the variable with the highest Partial value, .474673, and thus the variable that will be entered into the regression equation at the next step. The square of the jth Partial value can be interpreted as the proportion of the *currently unexplained variation* in Y, sales price, that can be eliminated if the variable X_j is entered in the regression equation at the next step. Since R Square = .85500,* the unexplained variation in Y at this point is $1.0 - .85500 = .14500$, or 14.5%. The square of the Partial correlation value for number of bedrooms is $(.474673)^2 = .2253$. Thus, 22.5% of the unexplained 14.5% can be eliminated by including X_4 as the next variable in the equation. So the new R Square at the second step should be

Second-step R Square = $.85500 + (.2253)(.14500) = .88767$

We will see in step 2 that this value is indeed the result.

The column labeled Min Toler in Computer Exhibit 13.7 is for minimum tolerance, and it provides information about the tolerance of a variable. The **tolerance** of an independent variable in a regression equation is the proportion of variation for the variable that is not eliminated or explained by the other independent variables in the equation. The minimum tolerance for an X_j that is currently not in the equation is the smallest tolerance that would be attained by any independent variable in the equation if X_j were included in the

---

*In the SPSS program the partial figure is related to the R Square value and not to the Adjusted R Square value.

equation. A tolerance value will be in the interval from 0 to 1. If X_j is entered into a regression equation, then its tolerance would be

$$\text{Tolerance}(X_j) = 1 - R_j^2$$

where R_j^2 is the coefficient of determination between X_j and the other independent variables that are currently in the equation.

In Computer Exhibit 13.7 the variable in the equation at step 1 is area X_1. Since the correlation coefficient between distance X_3 and area X_1 given in the correlation matrix of Computer Exhibit 13.7 is $-.049$, R_3^2 at step 1 is simply r_{13}^2, or $(-.049)^2$, and the tolerance for distance X_3, if it were entered in the equation, would be

$$\text{Tolerance}(X_3) = 1 - (-.049)^2 = .997572$$

The Min Toler is also .997572, since 1 minus the coefficient of determination for distance X_3 with area X_1 is .997572, and 1 minus the coefficient of determination for area X_1 with distance X_3 is also .997572. Thus, 99.8% of the variance in distance X_3 is not explained by area X_1 (or equivalently, 99.8% of the variance in area X_1 is not explained by distance X_3). A tolerance close to 1, such as .997572, indicates that there is essentially no multicollinearity among the independent variables; whereas a tolerance close to 0 indicates a high degree of multicollinearity. If tolerance or minimum tolerance for a variable is below a default level (such as .01) within the program, or, optionally, below a value specified by the program user, then the variable will not be entered into the equation.*

Finally, the T column for the variables not in the equation indicates $t = b_j / s_{b_j}$ values that would be assigned to each variable *if* that variable were to be introduced into the equation at the next step. We see that number of bedrooms X_4 has the highest t value for the variables not in the equation, and in the next step we will see that the t associated with b_4 is 2.224. The Sig T column shows the associated *P*-values.

The SPSS output associated with step 2 of the stepwise regression program is shown in Computer Exhibit 13.8. Several things can be noted about the results in the second step.

First, the variable "number of bedrooms," X_4, was entered into the equation as was predicted in the last step. Second, its Beta and t values are just what the previous step indicated. The R Square value of .88767 at this step is just what was calculated in step 1. The inclusion of X_4 in the equation reduced the Standard Error from $7243.75 to $6560.51.

If we compare the first and second equations, we find

$$\hat{Y} = -16.93 + 6.09X_1$$

and

$$\hat{Y} = -15.01 + 4.58X_1 + 6.12X_4$$

With the introduction of X_4 the intercept and b_1 changed values. That is, in the new equation the predictive impact of the number of bedrooms in each condominium is now included in the b_4 coefficient. In the previous $b_1 = 6.09$ coefficient, the impact of both area and number of bedrooms was included. In the new $b_1 = 4.58$ coefficient, the impact of the number of bedrooms has been isolated. The new equation indicates that the predicted price of a condominium increases $4583 for each additional 100 square feet in area, assuming the number of bedrooms is held constant. The price increases $6125 for each additional bedroom, assuming area is held constant.

---

*Rather than using tolerance, some computers compute a **variance inflation factor** (*VIF*) that is related to tolerance as follows:

$$VIF(X_j) = 1/\text{Tolerance}(X_j)$$

COMPUTER EXHIBIT 13.8 Stepwise Regression at Second Step for United Park City Properties

```
Variable(s) Entered on Step Number
   2..     BEDROOMS

Multiple R               .94216
R Square                 .88767
Adjusted R Square        .87446
Standard Error           6.56051

Analysis of Variance
                   DF       Sum of Squares       Mean Square
Regression          2          5782.11469        2891.05734
Residual           17           731.68531          43.04031

F =       67.17092        Signif F =  .0000
------------------ Variables in the Equation ------------------

Variable                B          SE B          Beta           T   Sig T

AREA              4.582609      .865175       .695410        5.297  .0001
BEDROOMS          6.124559     2.754346       .291937        2.224  .0400
(Constant)      -15.007638     6.133190                     -2.447  .0256
-------------- Variables not in the Equation -------------

Variable      Beta In  Partial  Min Toler         T  Sig T

LAND          .156232  .377499   .306615      1.631  .1225
DISTANCE     -.209586 -.587724   .339427     -2.906  .0103
PARKING       .041418 -.087340   .285637      -.351  .7304
```

The two Beta values indicate that price changes are more sensitive to changes in area ($Beta_1 = .69541$) than to changes in number of bedrooms ($Beta_4 = .29194$) (see Section 13.12).

Now we note that the output at step 2 gives us three different t values to interpret. The two t values associated with b_1 and b_4, the regression coefficients for area and bedrooms, measure the marginal significance of the two variables individually. That is, the t associated with area X_1 tests the hypothesis that area does not make a significant contribution to price predictions over and above that made by bedrooms X_4. The t value associated with bedrooms tests the hypothesis that the number of bedrooms X_4 does not make a significant marginal contribution to price predictions that area has not already made. The null hypotheses can be stated as follows:

H_0: $\beta_1 = 0$ when bedrooms X_4 is in the equation and held constant.

H_0: $\beta_4 = 0$ when area X_1 is in the equation and held constant.

The computer results can be used to show that each of these hypotheses is rejected since the t values for bedrooms X_4 and area X_1 are 2.224 and 5.297 and the associated P-values, .0400 and .0001, are less than .05. Thus, we can conclude that β_1 and β_4 differ from zero when the other variable is in the equation and held constant.

The F value that appears in the table for summarizing the partitioning of the sum of squares measures the overall significance of the relationship between the dependent vari-

---

COMPUTER EXHIBIT 13.9 Stepwise Regression at Third Step for United Park City Properties

---

```
Variable(s) Entered on Step Number
   3..    DISTANCE

Multiple R           .96253
R Square             .92647
Adjusted R Square    .91269
Standard Error      5.47121

Analysis of Variance
                   DF     Sum of Squares      Mean Square
Regression          3        6034.85321       2011.61774
Residual           16         478.94679         29.93417

F =      67.20138        Signif F =  .0000
------------------- Variables in the Equation ------------------

Variable               B        SE B        Beta          T   Sig T

AREA            3.922424     .756449     .595227      5.185   .0001
BEDROOMS        8.525096    2.441067     .406362      3.492   .0030
DISTANCE        -.637994     .219566    -.209586     -2.906   .0103
(Constant)     -8.434838    5.592710                 -1.508   .1510
-------------- Variables not in the Equation -------------

Variable     Beta In  Partial  Min Toler        T   Sig T

LAND         .110448  .322650   .295620      1.320   .2066
PARKING     -.164418 -.399236   .222966     -1.686   .1124

End Block Number   1   PIN =     .050 Limits reached.
```

able Y and the independent variables area X_1 and bedrooms X_4 *combined*. The value of this F can be taken as a measure to test the joint hypothesis $H_0: \beta_1 = \beta_4 = 0$ against the alternative H_a: One or more of the independent variables is significantly related to Y. The calculated value of this F is 67.17, and the P-value is .0000. Thus we can confidently reject the hypothesis that the combined predictive power of area X_1 and bedrooms X_4 is zero.

An analysis of the Variables not in the Equation output shows that the next variable to enter the equation is distance, because it has the largest (negative) Partial figure. That is, distance can explain $(-.587724)^2 = .345$ of the variation in price that has not already been explained by area and bedrooms. The Beta, Min Toler, and t values that will be associated with X_3, distance, at step 3 are also shown.

The output of step 3 in the SPSS stepwise regression program is presented in Computer Exhibit 13.9.

$$\hat{Y} = -8.43 + 3.92X_1 - .64X_3 + 8.53X_4$$

Note that the predicted sales price drops approximately \$640 for each mile the condominium is located from the center of the city. Note also that the predicted price is most sensitive to changes in area ($Beta_1 = .595227$), followed by bedrooms ($Beta_2 = .406362$) and distance ($Beta_3 = -.209586$). All of the variables' marginal coefficients are individually different from zero at the 5% level of significance since the t statistics, 5.185, 3.492, and -2.906, all have P-values less than .05.

The overall significance of the regression equation's prediction is measured by the $F = 67.20138$ in the table partitioning the sum of squares. The P-value is .0000 and is less than .05, so we conclude that variables area X_1, distance X_3, and bedrooms X_4 combined are significantly related to price Y.

This program was terminated at the end of step 3 since the P-values for the t statistics for the variables not in the equation are all above .05 (the default value for PIN required to bring a variable into the equation).

At this point the variables X_2, land, and X_5, parking, are not in the equation. It also appears that were they to be included, they would have regression coefficients not significantly different from zero. This result is indicated by the t values of 1.320 and -1.686 associated with these two variables. Both of them have P-values above .05, so we would not reject the hypothesis H_0: $\beta_j = 0$ at the next step. Thus stopping at step 3 appears to be a good decision. This observation does not mean that the land and parking variables do not have any association with price or any predictive power. It means that they have no additional significant association with price *over and above* that which is contained in area, distance, and bedrooms. These three independent variables together can account for 92.647% (R Square) of the squared variation in price.

Practitioners of regression analysis should be cautious in the use of stepwise regression, because this method is too often used as a "fishing expedition" to find regression equations that provide high R^2's, especially when the independent variables are quite highly correlated. Although stepwise regression can be a very useful technique, it can also be easily abused. Stepwise regression does not tell us whether variables that are not included in the equation are insignificant due to multicollinearity or because they are associated with Y in a nonlinear way. Business and economic theory, thoughtful conceptualization of the problem, and experience should guide the development of regression models, not fishing expeditions with the aid of stepwise regression.

Problems: Section 13.9

45. Bender International, Inc., has been experiencing problems recently with turnover among employees in a department. The work done in the department is largely unskilled, but the department figures that it must have an employee remain on the job at least one year before the hiring and training costs have been recovered. To determine whether any of the preemployment information gathered on applicants for jobs in the department is useful in predicting the length of time they will spend on the job, the personnel manager of the company took a random sample of 25 people who recently quit. The manager listed their ages at the date of job application, sex (0 = male, 1 = female), number of weeks worked in the department before quitting, and the score obtained by the applicant in a test administered to determine how well suited the person is for the job. The data are presented in the accompanying table. Computer Exhibit 13.10 presents the stepwise SPSS multiple regression output for this problem. The questions can be answered by using information contained in the output.

   a. On the average, how many years are people staying on the job in this department?
   b. Explain the meaning of the mean value of .6000 for SEX.
   c. Which two independent variables are most highly correlated with one another, and what is their r^2 value?
   d. Write the regression equation produced by the SPSS program. Interpret the meaning of the coefficients in the equation.
   e. Test the hypothesis H_0: $\beta_3 = 0$ against H_a: $\beta_3 \neq 0$, using $\alpha = .05$. (Remember, $X_3 = $ TEST.)

Y = JOBTENUR, Length of Stay on Job	X₁ = AGE, Age at Time of Application	X₂ = SEX	X₃ = TEST	Y = JOBTENUR, Length of Stay on Job	X₁ = AGE, Age at Time of Application	X₂ = SEX	X₃ = TEST
45	21	0	67	29	41	1	55
3	18	0	12	125	30	0	92
15	34	1	72	432	45	0	99
85	26	1	19	16	19	1	20
12	20	1	43	99	26	0	50
92	42	1	97	582	32	1	89
1	18	0	60	376	30	0	91
5	21	0	73	191	21	1	62
81	51	1	78	43	20	1	19
10	20	1	50	23	20	0	29
13	23	1	60	84	29	1	36
12	22	1	40	1	19	0	40
30	35	1	68				

46. A stock market analyst with Soloman Brothers wants to use stepwise regression to develop a model to predict the annual percentage rate of return of stocks Y from the risk of the stock as measured by the variability in security analyst's forecasts of the performance for the individual stocks X_1 (the standard deviation of the forecasted performances) and the percentage rate of inflation X_2. Data were collected for 10 years and are presented in the accompanying table.

   a. What variable enters the model at the first step?
   b. What is the regression equation after the first step?
   c. What is r^2 after the first step?
   d. What is the regression equation after the second step?
   e. What is R^2 after the second step?
   f. Do the individual independent variables add additional explanatory power to the regression equation over and above that contributed by the other variable after the second step? That is, test each coefficient for significance after the second step by using $t = b_j/s_{b_j}$. Use .05 for the level of significance in each test.

Stock Rate of Return	Variability	Inflation
10	9	8
7	10	6
−4	8	3
8	11	6
1	10	3
−3	8	4
24	15	9
11	10	8
10	11	6
4	12	3

47. A Big Five accounting firm is concerned about the risk of incurring liabilities owing to lawsuits stemming from the opinions that are given in its audit practice. The firm wants to use stepwise regression to develop a model to predict its annual liability Y (in tens of thousands of dollars) from its clients' average annual earnings X_1 (in millions of dollars) and from the firm size X_2 (in tens of thousands of billable hours).

SPSS/PC+

COMPUTER EXHIBIT 13.10 Stepwise Regression for Job Tenure Y, with Age X_1, Sex X_2, and Test Score X_3

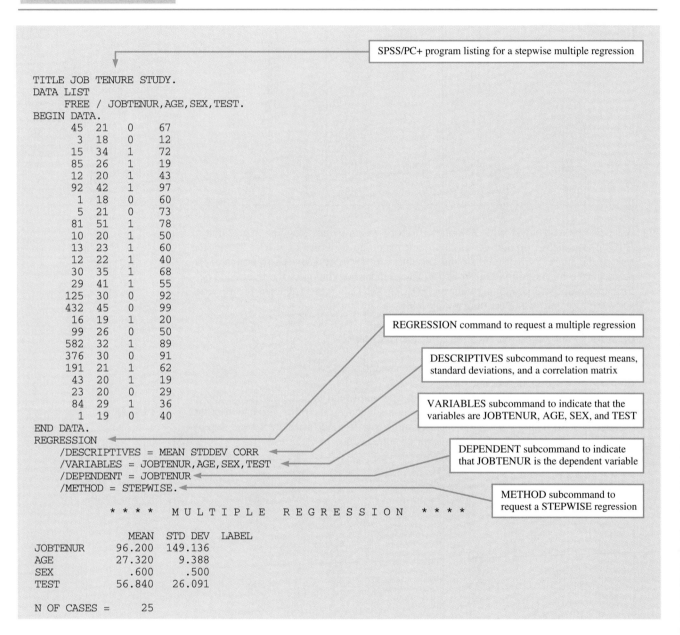

SPSS/PC+ program listing for a stepwise multiple regression

```
TITLE JOB TENURE STUDY.
DATA LIST
     FREE / JOBTENUR,AGE,SEX,TEST.
BEGIN DATA.
       45   21   0    67
        3   18   0    12
       15   34   1    72
       85   26   1    19
       12   20   1    43
       92   42   1    97
        1   18   0    60
        5   21   0    73
       81   51   1    78
       10   20   1    50
       13   23   1    60
       12   22   1    40
       30   35   1    68
       29   41   1    55
      125   30   0    92
      432   45   0    99
       16   19   1    20
       99   26   0    50
      582   32   1    89
      376   30   0    91
      191   21   1    62
       43   20   1    19
       23   20   0    29
       84   29   1    36
        1   19   0    40
END DATA.
REGRESSION
     /DESCRIPTIVES = MEAN STDDEV CORR
     /VARIABLES = JOBTENUR,AGE,SEX,TEST
     /DEPENDENT = JOBTENUR
     /METHOD = STEPWISE.
```

REGRESSION command to request a multiple regression

DESCRIPTIVES subcommand to request means, standard deviations, and a correlation matrix

VARIABLES subcommand to indicate that the variables are JOBTENUR, AGE, SEX, and TEST

DEPENDENT subcommand to indicate that JOBTENUR is the dependent variable

METHOD subcommand to request a STEPWISE regression

```
          * * * *   M U L T I P L E   R E G R E S S I O N   * * * *

                   MEAN     STD DEV   LABEL
JOBTENUR         96.200     149.136
AGE              27.320       9.388
SEX                .600        .500
TEST             56.840      26.091

N OF CASES =        25
```

Data were collected for 10 years and are presented in the accompanying table. If it is necessary to do so, then force both X_1 and X_2 to enter the equation.

a. What variable enters the model at the first step?
b. What is the regression equation after the first step?
c. What is r^2 after the first step?
d. What is the regression equation after the second step?
e. What is R^2 after the second step?

COMPUTER EXHIBIT 13.10 Stepwise Regression for Job Tenure Y, with Age X_1, Sex X_2, and Test Score X_3 (Continued)

SPSS/PC+

```
CORRELATION:

               JOBTENUR        AGE         SEX        TEST

JOBTENUR        1.000         .391       -.083        .563
AGE              .391        1.000        .224        .630
SEX             -.083         .224       1.000       -.142
TEST             .563         .630       -.142       1.000


            * * * *   M U L T I P L E   R E G R E S S I O N   * * * *

VARIABLE(S) ENTERED ON STEP NUMBER
   1..      TEST

MULTIPLE R            .56289
R SQUARE             .31685
ADJUSTED R SQUARE    .28715
STANDARD ERROR     125.91620

ANALYSIS OF VARIANCE
                     DF      SUM OF SQUARES        MEAN SQUARE
REGRESSION            1         169131.56897      169131.56897
RESIDUAL             23         364662.43103       15854.88831

F =      10.66747        SIGNIF F =   .0034
```

(continues)

Liability	Earnings	Billable Hours
54	12	45
43	9	51
36	9	52
56	10	49
66	12	62
61	12	53
50	10	45
44	9	51
31	6	37
49	11	57

48. The program in Problem 45 for Bender International, Inc., stopped after step 1 of the stepwise regression. Comment on the impact of multicollinearity and the fact that AGE is not included in the final regression equation.

49. Problem 22 involved the BZ Mining firm that was trying to construct a meaningful regression equation associating mean monthly repair costs over the past year Y with mean weekly hours of operation of the pumps X_1 and age of the pump at the first of the year X_2. That regression equation did not show very significant results. Thus the management of the company contacted other mine operators who owned similar pumps and expanded the data set by using the information on an additional 13 pumps. Also, management added a new variable, X_3, showing the frequency of routine maintenance performed throughout the year. The data for all 18 pumps are presented in the

SPSS/PC+	**COMPUTER EXHIBIT 13.10** Stepwise Regression for Job Tenure Y, with Age X_1, Sex X_2, and Test Score X_3 (Continued)

```
----------------- VARIABLES IN THE EQUATION ------------------

VARIABLE              B        SE B       BETA         T   SIG T

TEST             3.21752     .98512     .56289     3.266  .0034
(CONSTANT)     -86.68390   61.39683               -1.412  .1714

------------ VARIABLES NOT IN THE EQUATION -------------

VARIABLE      BETA IN  PARTIAL  MIN TOLER       T  SIG T

AGE            .06018   .05656    .60340     .266  .7929
SEX         -2.566E-03  -.00307   .97971    -.014  .9886

FOR BLOCK NUMBER  1   PIN =  .050 LIMITS REACHED. (*)
```

(*) PIN in the last line refers to the *P*-value (labeled SIG T here) required to bring a variable into the equation. If any variable not in the equation has *P*-value to enter the equation (SIG T) that is less than .05 (the default value for PIN), then the variable with the smallest *P*-value (SIG T) will be entered. Otherwise the PIN = .050 LIMITS REACHED message is given, and the stepwise regression is stopped.

accompanying table. The SPSS stepwise multiple regression program was applied to the data in the table. The preliminary output of the program is presented in the top section of Computer Exhibit 13.11(a), and the correlation matrix is presented in the bottom section of Computer Exhibit 13.11(a).

a. On the average, how many years old are the pumps?

b. On the average, how much time passes between performances of routine maintenance? Note that in the program the labels given the variables were $Y =$ RPAIRCST, $X_1 =$ HOURSOP, $X_2 =$ AGE, and $X_3 =$ FREQ.

$Y = RPAIRCST$, Mean Monthly Repair Costs During Past Year	$X_1 = HOURSOP$, Mean Weekly Hours of Operation	$X_2 = AGE$, Age of Pump at First of Year	$X_3 = FREQ$, Frequency of Routine Maintenance
643	28	80	36
613	26	48	12
494	15	27	18
250	15	2	52
400	16	13	26
791	20	70	6
836	31	68	6
124	15	14	39
359	15	29	26
500	15	76	6
611	35	52	13
492	28	61	18
250	19	11	52
612	35	34	39
317	41	16	39
199	19	1	52
216	26	17	52
321	30	35	39

COMPUTER EXHIBIT 13.11 Stepwise Regression for Mining Firm Pump Data

---

(a) Means, Standard Deviations, and Correlations

```
* * * *   M U L T I P L E   R E G R E S S I O N   * * * *

             MEAN   STD DEV   LABEL

RPAIRCST   446.000   208.117
HOURSOP     23.833     8.368
AGE         36.333    26.183
FREQ        29.500    17.068

N OF CASES =     18

CORRELATION:

             RPAIRCST      HOURSOP        AGE         FREQ

RPAIRCST      1.000          .338         .822        -.778
HOURSOP        .338         1.000         .250        -.032
AGE            .822          .250        1.000        -.754
FREQ          -.778         -.032        -.754         1.000
```

(continues)

50. Refer to BZ Mining's successive outputs from the stepwise program presented in (a) through (c) of Computer Exhibit 13.11.

   a. Which of the four variables has the greatest *relative* dispersion? [*Hint:* Find the coefficient of variation for each variable.]
   b. Which variable is most highly correlated with the dependent variable, Y? What is their r^2 value?
   c. Which two independent variables are most strongly correlated with one another? What is their r^2 value?
   d. Which two independent variables are least strongly correlated with one another? What is their r^2 value?

51. Step 1 of the stepwise multiple regression output for the data in Problem 49 is presented in Computer Exhibit 13.11(b).

   a. Write the regression equation after step 1 has been performed.
   b. What is R^2 at this point?
   c. Test the significance of β_2. That is, test $H_0: \beta_2 = 0$ against the alternative H_a: $\beta_2 \neq 0$, using $\alpha = .01$. (Remember that $X_2 = $ AGE.)

52. Step 2 of the stepwise regression output for the data in Problem 49 is presented in Computer Exhibit 13.11(c).

   a. Write the regression equation at this point.
   b. What is $S_{Y|123}^2$?
   c. What is R^2?
   d. Test the hypothesis $H_0: \beta_3 = 0$ against $H_a: \beta_3 \neq 0$, using $\alpha = .05$. (Remember that $X_3 = $ FREQ.)
   e. What interpretation can be given to the constant term 420.66755? Explain the meaning of the coefficients on AGE and FREQ.
   f. Test the hypothesis $H_0: \beta_2 = \beta_3 = 0$, using $\alpha = .05$.
   g. If the program is allowed to run one more step, what will be the T value (for $t = b_j / s_{b_j}$) associated with X_1 in the new equation?

SPSS/PC+

COMPUTER EXHIBIT 13.11 Stepwise Regression for Mining Firm Pump Data (Continued)

(b) Step 1

```
VARIABLE(S) ENTERED ON STEP NUMBER
    1..     AGE

MULTIPLE R              .82155
R SQUARE               .67494
ADJUSTED R SQUARE      .65462
STANDARD ERROR      122.30798

ANALYSIS OF VARIANCE
                        DF       SUM OF SQUARES        MEAN SQUARE
REGRESSION               1         496968.13189      496968.13189
RESIDUAL                16         239347.86811       14959.24176

F =       33.22148        SIGNIF F =  .0000

----------------- VARIABLES IN THE EQUATION ------------------

VARIABLE                 B          SE B         BETA        T   SIG T

AGE              6.53020      1.13297      .82155      5.764   .0000
(CONSTANT)     208.73591     50.25517                  4.154   .0007

------------- VARIABLES NOT IN THE EQUATION -------------

VARIBLE        BETA IN   PARTIAL   MIN TOLER        T   SIG T

HOURSOP         .14138    .24010     .93753      .958   .3533
FREQ           -.36701   -.42287     .43153    -1.807   .0908
```

53. The program described in Problems 51 and 52 was not allowed to run through step 3.
    a. Show that $X_1 =$ HOURSOP was not significantly enough correlated with the un-explained variation in Y to be included in the equation. [*Hint:* Test H_0: $\beta_1 = 0$, using $\alpha = .05$ and the T value in the VARIABLES NOT IN THE EQUATION section of step 2.]
    b. Where should the program be stopped, after step 1 or after step 2? Why?

54. Discuss the extent of the collinearity between AGE and FREQ in Problems 51 through 53. Which values in the output of step 2 might be influenced by this collinearity?

55. The National Football League examines passing data and rates quarterbacks by using a point system. A sports consultant with an MBA degree was retained by CBS, Inc., to use stepwise regression to develop a model to predict rating points for quarterbacks in the American Conference. Data were collected on percentage of passes completed, average yards gained, touchdown percentage, interception percentage, and rating points for 16 quarterbacks and are presented in the accompanying table.
    a. In what order do the variables enter the stepwise regression model?
    b. What is the regression equation after the first step?
    c. What is r^2 after the first step?
    d. What is the regression equation after the fourth step?
    e. What is R^2 after the fourth step?
    f. Do the individual independent variables add additional explanatory power to the regression equation over and above that contributed by the other variables after the

COMPUTER EXHIBIT 13.11 Stepwise Regression for Mining Firm Pump Data (Continued)

SPSS/PC+

(c) **Step 2**

```
VARIABLE(S) ENTERED ON STEP NUMBER
  2..    FREQ

MULTIPLE R          .85619
R SQUARE            .73307
ADJUSTED R SQUARE   .69747
STANDARD ERROR    114.46924

ANALYSIS OF VARIANCE
                DF      SUM OF SQUARES      MEAN SQUARE
REGRESSION       2        539767.89963     269883.94982
RESIDUAL        15        196548.10037      13103.20669

F =      20.59679       SIGNIF F =  .0000

------------------ VARIABLES IN THE EQUATION ------------------

VARIABLE             B          SE B        BETA        T     SIG T

AGE            4.33067      1.61415      .54483     2.683    .0170
FREQ          -4.47510      2.47611     -.36701    -1.807    .0908
(CONSTANT)   420.66755    126.34485                 3.330    .0046
```

(continues)

final step? That is, test each coefficient for significance after the final step by using $t = b_j/s_{b_j}$. Use .05 for the level of significance in each test.

| | Passing (Minimum of 180 Attempts) | | | | |
Quarterback	% Completed	Avg Yd Gained	Pct TD	Pct Int	Points
Marino, Miami	60.8	7.67	7.0	3.6	93.1
O'Brien, New York	63.1	7.88	5.5	3.7	90.2
Eason, New England	61.5	7.45	4.1	2.3	88.5
Krieg, Seattle	59.3	7.64	5.4	3.1	88.3
Kelly, Buffalo	59.2	7.56	4.8	3.6	83.6
Esiason, Cincinnati	56.9	8.05	4.3	3.6	82.3
Kosar, Cleveland	57.5	7.15	3.0	2.0	81.4
Plunkett, Los Angeles	51.9	7.78	5.5	3.8	80.2
Elway, Denver	56.3	7.06	3.9	2.8	79.7
Kenney, Kansas City	52.3	6.23	4.6	3.5	72.2
Fouts, Kansas City	58.8	6.94	3.5	5.5	68.8
Blackledge, Kansas City	45.7	5.71	4.8	2.9	68.0
Wilson, Los Angeles	53.8	7.17	5.0	6.3	67.4
Malone, Pittsburgh	50.8	5.48	3.9	4.5	61.8
Moon, Houston	52.7	7.15	2.8	5.7	61.6
Trudeau, Indianapolis	48.9	5.34	1.9	4.3	53.5

SOURCE: *The Sporting News*, December 22, 1986, p. 23. Reprinted by permission of *The Sporting News*.

COMPUTER EXHIBIT 13.11 Stepwise Regression for Mining Firm Pump Data
                                    (Continued)

```
------------- VARIABLES NOT IN THE EQUATION -------------

VARIBLE          BETA IN   PARTIAL   MIN TOLER        T   SIG T

HOURSOP           .21573    .39185     .38046      1.594   .1333

FOR BLOCK NUMBER  1   PIN =   .100 LIMITS REACHED. (*)
```

(*) PIN = .100 LIMITS REACHED in the last line indicates that the *P*-value of .1333 (or SIG T of .1333) for the
variable (HOURSOP) that is not in the equation is greater than .100 so the program stops after Step 2. PIN = .100 was
assigned in the program statements.

13.10
CURVILINEAR
REGRESSION

The sample regression equations that we have considered in the previous sections have
represented straight lines or planes—in other words, they have been *linear*. At times, however, theory, experience, or a scatter diagram may suggest a *curvilinear relationship* between the dependent variable Y and the independent variable X. In this section, we
consider two methods for estimating curvilinear relationships by using linear regression.
First, we discuss methods of estimating *polynomial relationships*. Second, we consider
some *transformations* that allow curvilinear relationships to be transformed into linear
relationships. The linear relationships can then be estimated by using linear regression.

EXAMPLE 13.17   Oklahoma Land & Agronomics, Inc., is interested in increasing
crop production and has installed a center-pivot, mobile sprinkling system to water an
alfalfa crop. Oklahoma has collected data for the weekly growth of alfalfa Y (in
inches) and for the water flow setting X_1 on the mobile sprinkling system (higher settings apply greater amounts of water). Other conditions are held constant. These data
are presented in Table 13.10 and are plotted as a scatter diagram in Figure 13.6. The
scatter diagram indicates that the relationship between alfalfa growth and the water
flow setting on the sprinkling system is curvilinear. That is, continually increasing
the amount of water applied by the sprinkling system does not always result in increased alfalfa growth.

One way to study the relationship between the dependent variable Y and the independent variable X_1 when the relationship between the two is curvilinear is to use a **polyno-**

TABLE 13.10	Growth, Y	Water Flow Setting, X_1	Growth, Y	Water Flow Setting, X_1
Data for Alfalfa Growth	1.6	0.0	6.2	3.2
Y and Water Flow Setting	2.9	0.3	6.3	3.5
X_1 at Oklahoma Land &	3.7	0.6	6.5	3.8
Agronomics	3.2	0.9	7.1	4.1
	3.5	1.3	7.6	4.4
	4.2	1.6	6.7	4.7
	5.3	1.9	6.9	5.1
	5.5	2.2	6.4	5.4
	6.2	2.5	6.7	5.7
	6.3	2.8	6.5	6.0

FIGURE 13.6

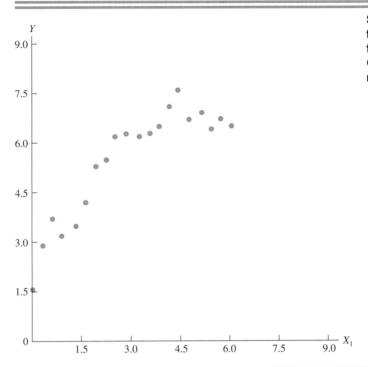

Scatter Diagram for the Alfalfa Growth Y and the Water Flow Setting X_1 for the Oklahoma Land & Agronomics Data

mial regression model. Polynomial regression models can have one or more independent variables, and the independent variables can be raised to various powers. However, we will restrict our discussion to a polynomial regression model with one independent variable. Furthermore, the independent variable will be raised to the first power in one term of the regression model and to the second power in another term of the regression model. This particular polynomial regression model that has the independent variable X_1 raised to the second power is a **quadratic regression model,** and its regression equation can be expressed as follows.

Equation for Quadratic Regression Model

$$Y_i = \beta_0 + \beta_1 X_{i1} + \beta_2 X_{i1}^2 + \epsilon_i \qquad (13.24)$$

To estimate the quadratic regression model's parameters, β_0, β_1, and β_2, we simply let

$$X_2 = X_1^2 \qquad (13.25)$$

and use the ordinary least squares method of multiple linear regression to find the values for the estimators b_0, b_1, and b_2 in the following equation:

$$\hat{Y} = b_0 + b_1 X_1 + b_2 X_2 \qquad (13.26)$$

Most data analysis packages allow the user to create new variables from variables that already exist. That is, X_2 can be created by the computer in this case by letting $X_2 = X_1^2$.

[To reduce multicollinearity, data analysts at times compute deviational scores of the form $X_1 - \overline{X}_1$ and then substitute $X_1 - \overline{X}$ for X_1 in Equations (13.24), (13.25), and (13.26).]

 PRACTICE PROBLEM 13.4

For the Oklahoma Land & Agronomics data on alfalfa growth Y and water flow setting X_1 presented in Table 13.10, find the estimated curvilinear (quadratic) regression equation

$$\hat{Y} = b_0 + b_1 X_1 + b_2 X_1^2$$

Use multiple regression, and let $X_2 = X_1^2$, as illustrated in Equation (13.26). Also, by using a graph of the estimated quadratic regression equation, determine the approximate value for the water setting X_1 that results in the highest estimated growth per week of alfalfa $\hat{Y}$.

Solution A multiple regression program was used to create X_2, by letting $X_2 = X_1^2$, and to find the estimated equation. Some of the results are presented in Computer Exhibit 13.12. The estimated curvilinear regression equation is

$$\hat{Y} = 1.82 + 2.17 X_1 - .23 X_1^2$$

MINITAB   **COMPUTER EXHIBIT 13.12** Quadratic Regression for Oklahoma Land & Agronomics

---

```
MTB > set c1                                              Command set to enter the
DATA >    1.6     2.9     3.7     3.2     3.5     4.2     5.3     5.5     6.2     6.3      GROWTH data in c1
DATA >    6.2     6.3     6.5     7.1     7.6     6.7     6.9     6.4     6.7     6.5
DATA > end

MTB > set c2                                              Command set to enter the
DATA >    0.0     0.3     0.6     0.9     1.3     1.6     1.9     2.2     2.5     2.8      WATFLO data in c2
DATA >    3.2     3.5     3.8     4.1     4.4     4.7     5.1     5.4     5.7     6.0
DATA > end

MTB > let c3 = c2**2                                      Command let c3 = c2**2 to create the water flow
                                                          squared variable as (water flow)² and put it in c3

MTB > name c1 'GROWTH' c2 'WATFLO' c3 'WATFLOSQ'
                                                          name the c1, c2, and c3 variables
                                                          GROWTH, WATFLO, and WATFLOSQ
MTB > regress c1 2 c2 c3

The regression equation is
GROWTH = 1.82 + 2.17 WATFLO - 0.231 WATFLOSQ             Command regres c1 2 c2 c3 for a regression of
                                                          GROWTH as the dependent variable with 2 indepen-
                                                          dent variables WATFLO and WATFLOSQ
Predictor        Coef         Stdev       t-ratio         p
Constant        1.8180       0.2466        7.37       0.000
WATFLO          2.1655       0.1904       11.37       0.000
WATFLOSQ       -0.23083      0.03063      -7.54       0.000

s = 0.4087      R-sq = 94.7%      R-sq(adj) = 94.0%

Analysis of Variance

SOURCE          DF          SS           MS           F          p
Regression       2        50.247       25.123      150.44     0.000
Error           17         2.839        0.167
Total           19        53.085

SOURCE          DF        SEQ SS
WATFLO           1        40.761
WATFLOSQ         1         9.485
```

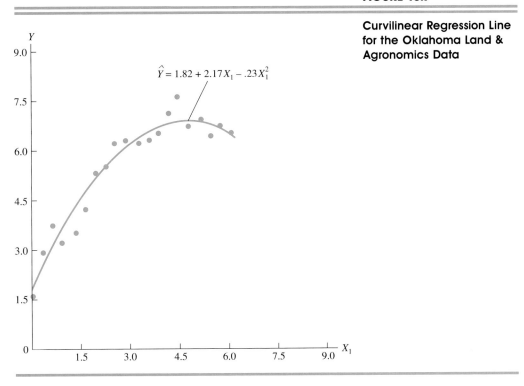

A plot of this curvilinear equation is presented in Figure 13.7. The figure shows that the highest amount of estimated growth per week $\hat{Y}$ is about 6.9 inches and is associated with a water flow setting X_1 of about 4.7.

By comparing the results for the analysis of the curvilinear (quadratic) regression equation (13.24) and the results for a simple linear regression equation,

$$\hat{Y} = b_0 + b_1 X_1 \tag{13.27}$$

when both are found by using the data in Table 13.10, we can obtain insight about which equation is most consistent with the data. An edited computer printout of the analysis for both the simple and the quadratic regression equations is presented in Computer Exhibit 13.13. The prediction errors, or *residuals*, $e_i = Y_i - \hat{Y}_i$, for the simple regression were computed and are also presented in Computer Exhibit 13.13.

COMPUTER EXHIBIT 13.13 Residuals for Simple Regression and Residuals for Quadratic Regression for Oklahoma Land & Agronomics

MINITAB

```
MTB > set c1                                                   Command set to enter the
DATA >    1.6    2.9    3.7    3.2    3.5    4.2    5.3    5.5    6.2    6.3     GROWTH data in c1
DATA >    6.2    6.3    6.5    7.1    7.6    6.7    6.9    6.4    6.7    6.5
DATA > end

MTB > set c2                                                   Command set to enter the
DATA >    0.0    0.3    0.6    0.9    1.3    1.6    1.9    2.2    2.5    2.8     WATFLO data in c2
DATA >    3.2    3.5    3.8    4.1    4.4    4.7    5.1    5.4    5.7    6.0
DATA > end
```

(continues)

MINITAB

COMPUTER EXHIBIT 13.13 Residuals for Simple Regression and Residuals for Quadratic Regression for Oklahoma Land & Agronomics (Continued)

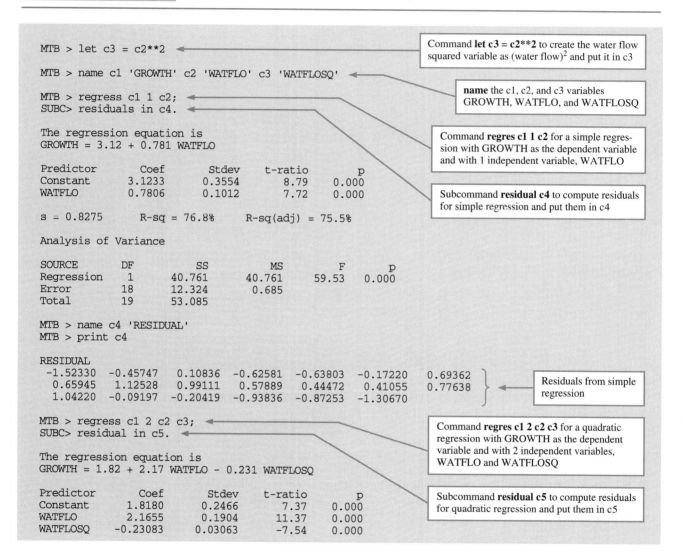

For a significance level of .05, the results for the simple regression equation (13.27) indicate that Y and X_1 are significantly linearly related, because $F = 59.53$ is greater than $F_{.05, 1, 18} = 4.4139$. The coefficient of determination for the simple regression is $r^2 = .77$. However, when the curvilinear regression equation (13.24) was used, the coefficient of multiple determination was $R^2 = .95$. In addition, the results for the *Adjusted* R^2 were similar to the results for R^2. Therefore, when we compare the simple regression equation (13.27) with the curvilinear regression equation (13.24), the curvilinear regression results in a superior fit for this data set.

A plot of the estimated simple linear regression equation (13.27), using the data of Table 13.10, is presented in Figure 13.8. A plot of the sample residuals, the e_i's, versus the water flow setting, independent variable X_1, for the simple regression is presented in Figure 13.9.

The pattern of the computed residual values that were listed in Computer Exhibit 13.13 suggests that the simple linear regression equation (13.27) tends to overpredict

COMPUTER EXHIBIT 13.13 Residuals for Simple Regression and Residuals for Quadratic Regression for Oklahoma Land & Agronomics (Continued)

MINITAB

```
s = 0.4087       R-sq = 94.7%      R-sq(adj) = 94.0%

Analysis of Variance

SOURCE        DF         SS        MS       F       p
Regression    2      50.247    25.123   150.44   0.000
Error        17       2.839     0.167
Total        19      53.085

SOURCE        DF      SEQ SS
WATFLO        1      40.761
WATFLOSQ      1       9.485
```

(continues)

alfalfa growth at both the lower and the upper values of the water flow setting X_1. Notice that these residuals are mainly negative as depicted in Figure 13.9. For the middle water flow values the simple linear regression equation (13.27) is underpredicting alfalfa growth Y. Notice that these residuals are positive.

The pattern of the residual plot (Figure 13.9) also suggests that a curvilinear regression would be more appropriate than the simple linear regression, because the residuals are not randomly scattered above and below the line where $e_i = 0$. The assumptions for the regression analysis model indicate that the residuals should be scattered above and below

FIGURE 13.8

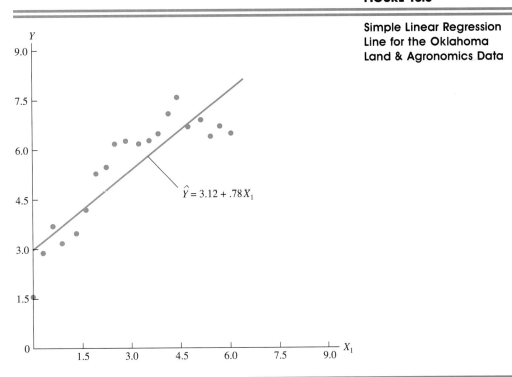

Simple Linear Regression Line for the Oklahoma Land & Agronomics Data

$\hat{Y} = 3.12 + .78X_1$

MINITAB

COMPUTER EXHIBIT 13.13 Residuals for Simple Regression and Residuals for Quadratic Regression for Oklahoma Land & Agronomics (Continued)

```
MTB > name c5 'RESIDULS'
SUBC> print c5

RESIDULS
 -0.217963    0.453151    0.665813   -0.379975   -0.743061   -0.491902    0.200807
  0.035064    0.410871    0.228227   -0.184000   -0.269697   -0.213844    0.283557
  0.722508   -0.196992    0.041639   -0.380913    0.038083   -0.001371

MTB > histogram c5

Histogram of RESIDULS    N = 20

Midpoint   Count
   -0.8       1     *
   -0.6       0
   -0.4       3     ***
   -0.2       5     *****
    0.0       4     ****
    0.2       3     ***
    0.4       2     **
    0.6       1     *
    0.8       1     *
```

> Residuals from quadratic regression

> Command **histogram c5** to request a histogram for the residuals

the estimated regression line for the values of X_1. In fact, if we had estimated the simple regression equation (13.27) prior to estimating the curvilinear regression equation (13.24), then the pattern of residuals in the simple regression residual plot (Figure 13.9) would have suggested that we should use a curvilinear regression equation.

The sample residuals for the curvilinear regression equation (13.24) are presented in Computer Exhibit 13.13. A histogram of the residuals is given in Computer Exhibit 13.13, and the residuals are plotted against X_1 in Figure 13.10. The pattern of residuals for the curvilinear regression (13.24) appears to be consistent with the assumption that they be scattered in a random fashion. The histogram of the residuals given in Computer Ex-

FIGURE 13.9

Residual Plot for the Simple Regression for Oklahoma Land & Agronomics

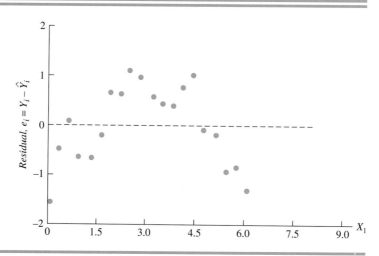

FIGURE 13.10

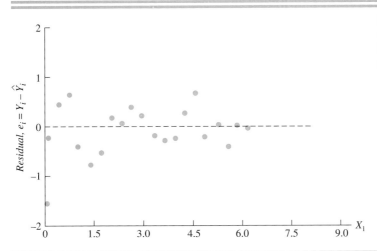

hibit 13.13 seems consistent with the assumption of normality. Thus the curvilinear regression equation appears to fit the data more appropriately than does the simple linear regression equation.

In addition to the use of quadratic regression equations to estimate curvilinear relationships, **transformations,** such as a logarithmic transformation, may be applied to some curvilinear regression equations in order to make them linear. For example, consider the following nonlinear population regression model:

$$Y = \beta_0 \beta_1^{X_1} \epsilon$$

If we take the logarithm to the base 10, say, of both sides of this equation, we get the linear regression model

$$\log(Y) = \log(\beta_0) + \log(\beta_1)X_1 + \log(\epsilon)$$

For $Y' = \log(Y)$, $\beta_0' = \log(\beta_0)$, $\beta_1' = \log(\beta_1)$, and $\epsilon' = \log(\epsilon)$, this linear regression model can be written as

$$Y' = \beta_0' + \beta_1' X_1 + \epsilon'$$

We can now find estimates for the slope and the intercept of this linear regression model by following the procedure of taking a sample of data for Y and X_1, transforming Y by letting $Y' = \log(Y)$, and using simple linear regression with the Y' and X_1 values to find the estimated regression equation

$$\hat{Y}' = b_0' + b_1' X_1$$

To predict Y for a given value of X_1, we first find $\hat{Y}'$ by using this simple regression equation. Then we take the antilogarithm of $\hat{Y}'$ in order to obtain the predicted value $\hat{Y}$.

Some other useful transformations are, for example, to let $Y' = 1/Y$ and then find the estimated regression equation $\hat{Y}' = b_0 + b_1 X$; or let $X' = \log(X)$ or $X' = 1/X$ and then find the estimated regression equation $\hat{Y} = b_0 + b_1 X'$. The references at the end of this chapter present examples of many types of transformations that can be used to transform curvilinear equations into linear equations. The linear equations then conform to the methods of linear regression.

Problems: Section 13.10

56. An agricultural economist at Dairy Products, Inc., wants to develop a regression model to predict the weekly growth of alfalfa Y (in inches) from the amount of fertilizer applied to a plot of land X_1 (in units of 100 pounds per acre). Data were collected for 15 plots when the amount of fertilizer was varied and are presented in the accompanying table.
    a. Find the estimated curvilinear regression equation $\hat{Y} = b_0 + b_1 X_1 + b_2 X_1^2$.
    b. Find the predicted value for Y given that $X_1 = 5$.
    c. Test the hypothesis $H_0: \beta_1 = \beta_2 = 0$. Use a significance level of .05.
    d. Find the estimated simple linear regression equation $\hat{Y} = b_0 + b_1 X_1$.
    e. Which equation fits the data better?

Alfalfa Growth	Fertilizer	Alfalfa Growth	Fertilizer
2.5	0.0	8.0	3.4
4.2	0.4	8.0	3.9
5.1	0.9	7.7	4.3
4.6	1.3	7.7	4.7
5.0	1.7	7.8	5.1
5.8	2.1	8.2	5.6
7.1	2.6	8.5	6.0
7.3	3.0		

57. A young urban professional wants to develop a regression model to predict individual caviar consumption Y (in ounces per year) from the consumer's annual income X_1 (in hundred thousands of dollars). Data were collected for 15 individual consumers of caviar and are presented in the accompanying table.
    a. Find the estimated curvilinear (quadratic) regression equation $\hat{Y} = b_0 + b_1 X_1 + b_2 X_1^2$.
    b. Find the predicted value for Y given that $X_1 = 4$.
    c. Find the estimated simple linear regression equation $\hat{Y} = b_0 + b_1 X_1$.
    d. Find the estimated simple linear regression equation $\hat{Y}' = b_0 + b_1 X_1$, where $Y' = \log_{10}(Y)$.
    e. Find the estimated simple linear regression equation $\hat{Y}' = b_0 + b_1 X_1'$, where $Y' = \log_{10}(Y)$ and $X_1' = \log_{10}(X_1 + 1)$.
    f. Which equation fits the data better?

Caviar	Income	Caviar	Income
0.2	0.0	11.0	2.9
2.6	0.4	12.6	3.2
3.6	0.7	13.8	3.6
0.8	1.1	16.1	3.9
1.3	1.4	19.2	4.3
3.1	1.8	23.9	4.6
6.8	2.1	28.6	5.0
8.0	2.5		

58. The chief executive officer for Floor Covering Mills, Inc., wants to develop a regression model to predict the monthly profit of the company Y (in thousands of dollars) from the units of product X_1 that the firm produces (in thousands of units). Data were collected for 15 months when the production level was varied and are presented in the accompanying table. Apparently, if the production level is too high, the amount of inventory increases and the profits decrease.

a. Find the estimated curvilinear regression equation $\hat{Y} = b_0 + b_1 X_1 + b_2 X_1^2$.
b. Find the predicted value for Y given that $X_1 = 5$.
c. Plot the estimated profit equation, and from the plot, determine the production level that maximizes profit.

Profit	Production	Profit	Production
3.5	0.0	64.8	4.0
24.4	0.5	63.4	4.5
36.1	1.0	57.8	5.0
26.5	1.5	55.2	5.5
30.6	2.0	54.9	6.0
39.3	2.5	59.6	6.5
55.6	3.0	62.7	7.0
56.6	3.5		

59. A North Seas Energy, Inc., oil refinery outputs two types of gasoline, regular and lead-free. Because both products are refined at the same facility, increasing the output of one product results in a decrease in the output of the other. A production manager wants to develop a regression model to predict the weekly output of regular gasoline Y (in thousands of barrels) from the output of lead-free gasoline X_1 (in thousands of barrels). Data were collected for 15 weeks and are presented in the accompanying table.

a. Find the estimated curvilinear regression equation $\hat{Y} = b_0 + b_1 X_1 + b_2 X_1^2$.
b. Find the predicted value for Y given that $X_1 = 4$.
c. Find the estimated simple linear regression equation $\hat{Y} = b_0 + b_1 X_1$.
d. Which equation fits the data better?

Regular	Lead-Free	Regular	Lead-Free
19	0	15	6
21	1	13	6
21	1	9	7
17	2	7	8
16	3	5	9
16	4	4	9
17	4	2	10
15	5		

13.11
TESTING SEVERAL COEFFICIENTS OF A MULTIPLE REGRESSION EQUATION

In Section 13.5, we considered a test of the hypothesis $H_0: \beta_j = 0$. This hypothesis test is used to test the assertion that a single independent variable X_j adds no additional explanatory power to a multiple regression equation over and above that which is provided by the other independent variables that are included in the equation. In this section, we want to extend our hypothesis-testing capability by learning how to test the assertion that *two or more* independent variables add no additional explanatory power to a multiple regression equation over and above that which is provided by the other independent variables that are included in the model. In other words, we want to test the null hypothesis that the population regression coefficients for two or more variables are equal to zero. As an example, if a population multiple regression equation has three independent variables, X_1, X_2, and X_3, so that $\mu_{Y|123} = \beta_0 + \beta_1 X_1 + \beta_2 X_2 + \beta_3 X_3$, then we may want to test the null hypothesis $H_0: \beta_2 = \beta_3 = 0$ against the alternative hypothesis H_a: One or more of these coefficients do not equal zero.

EXAMPLE 13.18   A human resources manager for National Business Forms, Inc., is concerned about the relationship between employees' absenteeism Y and their job tenure X_1, job satisfaction X_2, and commitment to the organization X_3. The last two variables, job satisfaction X_2 and organizational commitment X_3, are employee attitudes that are measured by indexes. Scores for the indexes are determined by the employees' responses to questions that are included on questionnaires. Higher scores for the indexes are associated with more job satisfaction or greater degrees of organizational commitment. The manager wants to know if employees' attitudes about their jobs are related to absenteeism such that the attitudes add no additional explanatory power to a multiple regression equation over and above that which job tenure X_1 provides. That is, the manager wants to test the null hypothesis

$$H_0: \beta_2 = \beta_3 = 0$$

against the alternative hypothesis

$$H_a: \text{One or more of these coefficients do not equal zero.}$$

The population regression coefficients β_2 and β_3 correspond to the attitudinal variables of job satisfaction X_2 and organizational commitment X_3, respectively.

The estimated multiple regression equation that includes all of the independent variables for the absenteeism example is termed the **full estimated equation.**

Full Estimated Equation

$$\hat{Y} = b_0 + b_1 X_1 + b_2 X_2 + b_3 X_3 \tag{13.28}$$

A segment of a multiple regression equation—in other words, a null hypothesis such as the one described above—can be tested by using an approach known as a **general linear test.** This approach uses a full estimated regression equation and a reduced estimated regression equation that is based on the null hypothesis. The reduced estimated regression equation omits the independent variables whose coefficients are being tested to see whether they are equal to zero in the null hypothesis. Equation (13.28) is the full estimated regression equation for the absenteeism example. The population multiple regression coefficients that are being tested to see whether they are equal to zero in the absenteeism example are β_2 and β_3, so the **reduced estimated regression equation** is obtained by omitting X_2 and X_3 from the full regression equation.

Reduced Estimated Equation

$$\hat{Y} = b_0 + b_1 X_1 \tag{13.29}$$

A value that is termed the **extra sum of squares** serves as an indicator of the additional explanatory power that is gained by including X_2 and X_3 in the full regression equation over and above that which X_1 provides in the reduced regression equation. The value of the extra sum of squares is found by the following formula:

$$\textit{Extra sum of squares} = SSE_{\text{reduced}} - SSE_{\text{full}} \tag{13.30}$$

In Equation (13.30) SSE_{full} is the sum of squared deviations from regression (errors or residuals) for the full estimated regression equation (13.28), and $SSE_{reduced}$ is the sum of squared deviations from regression for the reduced estimated regression equation (13.29).

The value of $SSE_{reduced}$ is almost always larger than the value of SSE_{full} because the extra independent variables that are included in the full estimated regression equation generally explain at least some of the sample variation in Y that is not explained by the independent variable(s) in the reduced estimated regression equation. The value of the difference between $SSE_{reduced}$ and SSE_{full} is termed the *extra sum of squares* because it is the value of the additional sample variation in Y that is explained by adding the extra independent variables in the regression equation. In other words, with the use of the full estimated regression equation (13.28) rather than the reduced estimated regression equation (13.29) for predictive purposes, the additional sample variation in Y that is explained is equal to the value known as the extra sum of squares, and this value is found by using Equation (13.30).

A computer is usually used to obtain the results for both the full regression equation and the reduced regression equation. The same sample of data is used to find both of the equations, but some of the independent variables are omitted when the reduced regression equation is computed. The values of SSE_{full} and $SSE_{reduced}$ are given in the partitioned sums of squares tables of the two computer printouts. Then one simply computes the extra sum of squares by using Equation (13.30).

The general linear test approach uses an F statistic to determine whether the value of the extra sum of squares is so large that we should reject H_0. If the value of the extra sum of squares due to the additional variables included in the full estimated multiple regression equation is large enough, then we reject the null hypothesis (H_0: $\beta_2 = \beta_3 = 0$ in the absenteeism example) and conclude that one or more of the coefficients of the additional variables do not equal zero. In other words, if the null hypothesis is rejected, then the conclusion is that one or more of the additional variables that are included in the full model, but not in the reduced model, provide additional explanatory power over and above the independent variables that are included in the reduced model.

The F statistic is obtained by the following formula.

Equation for *F* Statistic for General Linear Test

$$F = \frac{(SSE_{reduced} - SSE_{full})/(k - q)}{SSE_{full}/(n - k - 1)} \qquad (13.31)$$

In Equation (13.31), k is the number of independent variables (X_j's) in the full regression equation, q is the number of independent variables (X_j's) in the reduced regression equation, and n is the sample size.

The value of F that is obtained for the sample by using Equation (13.31) is compared with F_{α, ν_1, ν_2}, where α is the level of significance, $\nu_1 = k - q$, and $\nu_2 = n - k - 1$; and we have the following rejection rule:

Rejection rule: If $F > F_{\alpha, k-q, n-k-1}$, then reject H_0.

PRACTICE PROBLEM 13.5

For the National Business Forms, Inc., absenteeism example we have been discussing in this section, we wish to test the null hypothesis that employee attitudes about job satisfaction X_2 and organizational commitment X_3 add no additional explanatory power to a multiple regression equation over and above that which is provided by job tenure X_1. The regression equation is used to predict the annual time lost to absences Y by individual employees. In other words, for the population regression equation $\mu_{Y|123} = \beta_0 + \beta_1 X_1 + \beta_2 X_2 + \beta_3 X_3$, we wish to test the null hypothesis

$$H_0: \beta_2 = \beta_3 = 0$$

against the alternative hypothesis

H_a: One or more of these coefficients do not equal zero.

Let the significance level be equal to .05. A sample of data for 15 employees was collected by the manager and is presented in Table 13.11.

Solution A computer was used to find the full estimated regression equation (13.28) and the reduced estimated regression equation (13.29) for this set of data. The printout is presented in Computer Exhibit 13.14. The values of SSE_{full} and SSE_{reduced} are labeled on the printouts.

The F statistic for this sample is computed by using Equation (13.31), where the number of independent variables in the full regression equation is $k = 3$, the number of independent variables in the reduced regression equation is $q = 1$, and the sample size is $n = 15$. The F statistic is

$$F = \frac{(341.54 - 34.62)/(3 - 1)}{34.62/(15 - 3 - 1)} = 48.76$$

The values for the degrees of freedom are $\nu_1 = k - q = 3 - 1 = 2$ and $\nu_2 = n - k - 1 = 15 - 3 - 1 = 11$. We find from Table VIII the value $F_{.05, 2, 11} = 3.9823$. The computed $F = 48.76$ is greater than $F_{.05, 2, 11} = 3.9823$, so we reject the null hypothesis and conclude that one or both of the two employee attitudes do contribute additional explanatory power to the multiple regression equation over and above that which job tenure provides. Thus the human resources manager will use the full estimated regression equation to predict absences rather than the reduced regression equation.   ▬

For the absenteeism example there is a single independent variable (X_1) in the reduced regression equation, so $q = 1$. And there are three independent variables $(X_1, X_2,$

	Absences (days), Y	Tenure (months), X_1	Satisfaction, X_2	Organizational Commitment, X_3
TABLE 13.11				
	22.5	36.8	53.7	58.1
National Business Forms,	26.2	52.8	57.5	64.4
Inc., Absenteeism Data	28.7	29.9	54.1	75.6
	18.4	39.9	58.2	44.6
	27.1	58.7	50.1	56.5
	18.6	49.4	37.6	51.1
	23.5	52.7	55.1	54.9
	11.5	36.2	29.9	52.4
	25.2	36.9	57.7	58.7
	23.1	42.0	44.8	58.2
	10.5	22.0	45.8	44.5
	17.4	38.9	44.8	50.3
	23.6	35.1	51.9	69.4
	19.5	49.7	48.0	47.3
	17.0	43.8	41.2	54.4

COMPUTER EXHIBIT 13.14 Full and Reduced Regression Equations for National
Business Forms, Inc., Absence Data

MINITAB

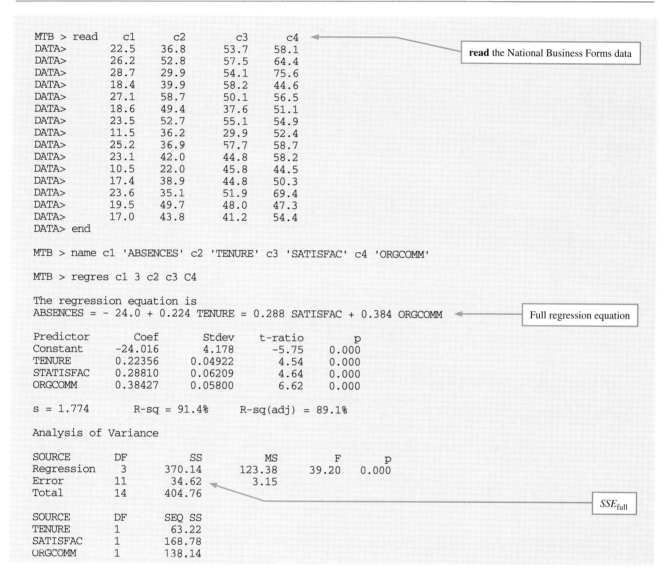

```
MTB > read    c1      c2      c3      c4          ◄─────        read the National Business Forms data
DATA>    22.5   36.8    53.7    58.1
DATA>    26.2   52.8    57.5    64.4
DATA>    28.7   29.9    54.1    75.6
DATA>    18.4   39.9    58.2    44.6
DATA>    27.1   58.7    50.1    56.5
DATA>    18.6   49.4    37.6    51.1
DATA>    23.5   52.7    55.1    54.9
DATA>    11.5   36.2    29.9    52.4
DATA>    25.2   36.9    57.7    58.7
DATA>    23.1   42.0    44.8    58.2
DATA>    10.5   22.0    45.8    44.5
DATA>    17.4   38.9    44.8    50.3
DATA>    23.6   35.1    51.9    69.4
DATA>    19.5   49.7    48.0    47.3
DATA>    17.0   43.8    41.2    54.4
DATA> end

MTB > name c1 'ABSENCES' c2 'TENURE' c3 'SATISFAC' c4 'ORGCOMM'

MTB > regres c1 3 c2 c3 C4

The regression equation is
ABSENCES = - 24.0 + 0.224 TENURE = 0.288 SATISFAC + 0.384 ORGCOMM    ◄─────    Full regression equation

Predictor       Coef        Stdev      t-ratio        p
Constant      -24.016       4.178       -5.75      0.000
TENURE         0.22356      0.04922      4.54      0.000
STATISFAC      0.28810      0.06209      4.64      0.000
ORGCOMM        0.38427      0.05800      6.62      0.000

s = 1.774      R-sq = 91.4%     R-sq(adj) = 89.1%

Analysis of Variance

SOURCE        DF          SS         MS          F         p
Regression     3       370.14      123.38      39.20     0.000
Error         11        34.62        3.15
Total         14       404.76                                    ◄─────    SSE_full

SOURCE        DF       SEQ SS
TENURE         1        63.22
SATISFAC       1       168.78
ORGCOMM        1       138.14
```

(continues)

and X_3) in the full regression equation, so $k = 3$. Of course, q and k will take on different
values in other problems, and q must always be less than k. If the number of variables
omitted from the full regression equation is 1, so that $q = k - 1$, then the hypothesis test
that is based on the extra sum of squares that was described in this section is equivalent to
the hypothesis test, or t test, for the marginal or additional significance of a regression coefficient that was presented in Section 13.5. Recall that the null hypothesis for testing the
marginal significance of a regression coefficient in Section 13.5 was H_0: $\beta_j = 0$. These
two tests are equivalent since $t^2_{\alpha/2, \nu} = F_{\alpha, 1, \nu}$. Thus, the t test in Section 13.5 tests the marginal or additional significance of a single regression coefficient given that the other independent variables are also included in the regression equation.

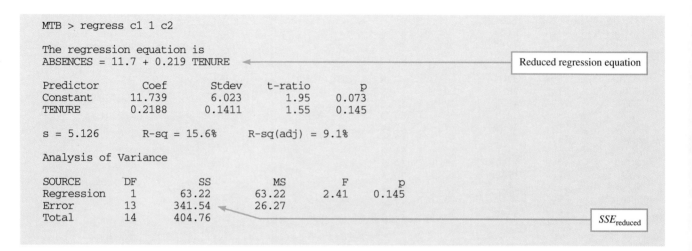

MINITAB

COMPUTER EXHIBIT 13.14 Full and Reduced Regression Equations for National Business Forms, Inc., Absence Data (Continued)

```
MTB > regress c1 1 c2

The regression equation is
ABSENCES = 11.7 + 0.219 TENURE          ←————————————————  Reduced regression equation

Predictor      Coef       Stdev      t-ratio       p
Constant      11.739      6.023        1.95      0.073
TENURE         0.2188     0.1411       1.55      0.145

s = 5.126        R-sq = 15.6%      R-sq(adj) = 9.1%

Analysis of Variance

SOURCE        DF        SS         MS        F        p
Regression     1      63.22      63.22     2.41    0.145
Error         13     341.54      26.27
Total         14     404.76    ←———————————————————————  SSE_reduced
```

Problems: Section 13.11

60. A real estate developer wants Northwest Realty Trust, Inc., to develop a regression model to predict the number of housing starts Y in his region from the interest rate on long-term Treasury bills X_1, the federal funds short-term interest rate X_2, and the short-term interest rate on money market deposits X_3. The developer is particularly interested in knowing if short-term interest rates add no additional explanatory power in a multiple regression equation that is used to predict the number of housing starts over and above that provided by long-term interest rates. Data were collected for 15 periods and are presented in the accompanying table.

   a. Find the full estimated regression equation $\hat{Y} = b_0 + b_1 X_1 + b_2 X_2 + b_3 X_3$.
   b. Predict the number of housing starts if $X_1 = 13$, $X_2 = 10$, and $X_3 = 10$.
   c. Find the reduced estimated regression equation $\hat{Y} = b_0 + b_1 X_1$.

Housing Starts (× 100)	Treasury Bill Rate (%)	Federal Funds Rate (%)	Money Market Rate (%)
10.8	12.9	10.3	9.5
12.1	14.3	11.4	12.2
12.9	12.6	8.3	8.4
9.5	12.9	12.3	10.8
12.4	14.8	10.3	9.4
9.5	13.7	9.4	8.7
11.2	14.2	11.6	11.3
7.2	12.3	8.7	9.0
11.7	13.0	10.4	9.1
11.0	13.3	9.0	8.0
6.8	11.1	10.6	9.4
9.1	12.8	10.1	9.2
11.2	12.8	9.6	10.4
9.8	13.8	11.1	9.9
9.0	13.2	10.0	10.2

d. Predict the number of housing starts if $X_1 = 13$.

e. Find the value of the extra sum of squares.

f. Test the null hypothesis $H_0: \beta_2 = \beta_3 = 0$. Use a significance level of .05.

61. A personnel manager at Equipment Rental & Leasing, Inc., wants to develop a regression model to predict salaries Y of the company's salespeople from their number of years of higher education X_1, a job performance index X_2, and the sales experience of the employee X_3. The manager is mainly concerned about knowing whether the employees' job situations, as measured by performance and experience, add no additional explanatory power to a multiple regression equation that is used to predict salaries over and above that provided by education. Data were collected for 15 salespeople and are presented in the accompanying table.

a. Find the full estimated regression equation $\hat{Y} = b_0 + b_1 X_1 + b_2 X_2 + b_3 X_3$.

b. Find SSE_{full}.

c. Find the reduced estimated regression equation $\hat{Y} = b_0 + b_1 X_1$.

d. Find $SSE_{reduced}$.

e. Test the null hypothesis $H_0: \beta_2 = \beta_3 = 0$. Use a significance level of .05.

f. Does an employee's job situation add additional explanatory power over and above that which is provided by his or her educational background in predicting salary, according to your conclusion in part e?

Salary (× $1000)	Education (years)	Performance (%)	Experience (months)
21.7	2.9	81.8	39.7
24.1	6.3	90.9	55.6
25.8	0.4	75.7	49.1
18.9	4.2	91.3	33.3
24.7	7.6	83.5	39.3
19.1	6.5	69.0	34.1
22.3	6.6	89.5	44.6
14.3	4.2	58.2	37.3
23.4	2.6	85.3	37.8
22.1	4.1	72.3	35.2
13.6	0.9	73.1	29.2
18.3	4.1	74.6	33.6
22.4	2.3	78.1	52.8
19.7	6.4	81.7	34.0
18.0	5.3	72.3	41.9

62. A corporate finance officer with Securities & Commodities, Inc., wants to develop a regression model to predict the percentage of long-term debt Y on a corporation's balance sheet from the percentage of long-term assets X_1, the percentage of cash X_2, and the percentage of inventory X_3. The financial officer is particularly interested in knowing whether short-term accounts such as cash and inventory percentages add no additional explanatory power in a multiple regression equation that is used to predict the percentage of long-term debt over and above that provided by the percentage of long-term assets. Data were collected for 15 corporations and are presented in the accompanying table.

a. Find the full estimated regression equation $\hat{Y} = b_0 + b_1 X_1 + b_2 X_2 + b_3 X_3$.

b. Find SSE_{full}.

c. Find the reduced estimated regression equation $\hat{Y} = b_0 + b_1 X_1$.

d. Find $SSE_{reduced}$.

e. Test the null hypothesis $H_0: \beta_2 = \beta_3 = 0$. Use a significance level of .05.

Long-Term Debt (%)	Long-Term Assets (%)	Cash (%)	Inventory (%)
35.0	46.4	9.1	18.6
42.4	72.5	10.1	37.7
47.3	41.6	2.9	11.7
26.8	47.5	15.4	26.7
44.2	81.7	6.6	9.4
27.2	61.3	6.6	12.6
37.0	70.1	11.3	28.8
13.0	36.5	7.1	25.9
40.3	48.8	8.8	12.2
36.3	54.4	5.0	5.1
10.9	15.3	13.6	27.1
24.8	45.3	9.6	18.1
37.2	44.9	7.0	30.6
29.1	62.6	10.9	19.0
23.9	51.9	8.9	28.7

63. A manager for Coastal Apparel, Inc., wants to develop a regression model to predict employee organizational commitment Y as measured on an index from the employee's age X_1, a role conflict index X_2, and a role ambiguity index X_3. The manager is mainly concerned about knowing whether the employee's attitudes concerning their roles in the organization as measured by role conflict and role ambiguity add no additional explanatory power in a multiple regression equation that is used to predict organizational commitment over and above that provided by age. Data were collected for 15 employees and are presented in the accompanying table.

   a. Find the full estimated regression equation $\hat{Y} = b_0 + b_1 X_1 + b_2 X_2 + b_3 X_3$.
   b. Find SSE_{full}.
   c. Find the reduced estimated regression equation $\hat{Y} = b_0 + b_1 X_1$.
   d. Find $SSE_{reduced}$.
   e. Test the null hypothesis $H_0: \beta_2 = \beta_3 = 0$. Use a significance level of .05.
   f. Does an employee's perceptions about his or her role in the organization add additional explanatory power in predicting organizational commitment over and above that provided by age, according to your conclusion in part e?

Organizational Commitment	Age (years)	Role Conflict	Role Ambiguity
64.2	35.2	31.6	42.1
70.4	42.7	39.9	59.3
74.5	36.0	4.8	33.8
57.3	34.1	52.1	46.6
71.8	45.2	37.0	43.0
57.6	37.5	43.8	52.0
65.8	41.2	46.5	53.9
45.8	29.3	46.2	64.7
68.6	36.6	26.7	34.0
65.2	37.3	27.5	38.6
44.1	23.8	48.2	50.7
55.7	33.3	42.8	48.6
66.0	35.2	24.8	52.5
59.2	38.1	50.1	50.1
54.9	34.7	46.3	61.1

64. A management information systems specialist at Computer Graphic Systems, Inc., wants to develop a regression model to predict the amount of time that it takes to

write the code for a computer program Y from the number of lines of code that make up the program X_1, the mean number of statements included in a program module X_2, and the percentage of code X_3 that is written by following structured techniques. The manager is mainly concerned about knowing whether the variables about modular and structured programming add no additional explanatory power in a multiple regression equation that is used to predict coding time over and above that provided by the number of lines of code. Data were collected for 15 computer programs and are presented in the accompanying table.

a. Find the full estimated regression equation $\hat{Y} = b_0 + b_1 X_1 + b_2 X_2 + b_3 X_3$.
b. Find SSE_{full}.
c. Find the reduced estimated regression equation $\hat{Y} = b_0 + b_1 X_1$.
d. Find $SSE_{reduced}$.
e. Test the null hypothesis $H_0: \beta_2 = \beta_3 = 0$. Use a significance level of .05.

Coding Time (hours)	Lines of Code (× 100)	Mean Number of Statements per Module	Structured Techniques (%)
108.4	10.1	48.1	65.2
120.7	13.8	55.0	92.8
128.9	10.5	36.8	58.6
94.6	9.6	58.9	73.7
123.7	15.1	50.8	64.2
95.3	11.3	46.4	58.1
111.6	13.1	56.5	81.8
71.6	7.1	42.2	62.7
117.2	10.8	48.1	60.7
110.4	11.2	42.6	51.3
68.2	4.4	49.3	62.1
91.3	9.1	48.4	61.3
112.0	10.1	44.3	77.3
98.5	11.6	54.1	67.4
89.9	9.9	48.6	73.0

In Example 13.4 for the United Park City Properties condominium data our regression equation is $\hat{Y} = -61.29 + 8.07X_1 + 2.82X_2$, and the coefficients on X_1 and X_2 imply the following: A 100-square-foot change in condominium area is associated with an $8070 change in price, and a $1000 change in land cost is associated with a $2820 change in price. One might wish to ask, however, Which variable is condominium price more sensitive to, area or land cost? The $b_1 = 8.07$ and $b_2 = 2.82$ cannot be used to answer this question because they are expressed in different units. To be exact, the units on b_1 and b_2 are

$$b_1 = \$8.07 \text{ thousand per 100 square feet of area}$$

$$b_2 = \$2.82 \text{ thousand per \$1000 of land cost}$$

One way to overcome this problem of differing units is to convert b_1 and b_2 to **standardized regression coefficients** or **beta coefficients**. These are defined as follows:

$$Beta_1 = b_1 \sqrt{\frac{SSX_1}{SSY}}$$

$$Beta_2 = b_2 \sqrt{\frac{SSX_2}{SSY}}$$

(13.32)

13.12
STANDARDIZED REGRESSION COEFFICIENTS

The beta coefficients are the regression equation's coefficients expressed in terms of the standard deviations of the variables involved. Coefficient $Beta_1$ is the number of standard deviations that Y changes for a one–standard deviation change in X_1. Coefficient $Beta_2$ is the number of standard deviations Y will change for a one–standard deviation change in X_2. The standardized regression coefficient $Beta_j$ should not be confused with the regression parameter β_j ($Beta_j$ and β_j are different).

EXAMPLE 13.19   In our United Park City Properties condominium example we found $SSY = 1382$ (in Section 12.3). Thus

$$Beta_1 = (8.07)\sqrt{\frac{22}{1382}} = 1.02$$

$$Beta_2 = (2.82)\sqrt{\frac{10}{1382}} = .24$$

These figures indicate that a one–standard deviation change in X_1 (area) is associated with a 1.02–standard deviation change in price. But a one–standard deviation change in X_2 (land cost) is associated with a change of only .24 standard deviation in price. These beta coefficients are dimensionless numbers that can answer our original question about the sensitivity of price to changes in area and land costs. Apparently, in our sample data price is far more sensitive to changes in area ($Beta_1 = 1.02$) than it is to changes in land cost ($Beta_2 = .24$). In problems that have several independent variables, the variables are sometimes ranked by the absolute values of their beta coefficients to give the relative sensitivity of Y to each of the X_i's.

In general, we should be cautious and *not* use the ranking described in Example 13.19 to indicate the relative importance of variables when multicollinearity is present, because the values of the regression coefficients are influenced by the other independent variables that are included in the equation.

Problems: Section 13.12

65. For the regression equation found in Problem 28, find the *Beta* coefficients associated with X_1 and X_2. To which variable are the estimates of Y more sensitive, X_1 or X_2?

66. Find the *Beta* coefficients associated with X_1 and X_1^2 in the regression equation found in Problem 57.

Time Series Analysis

In this chapter we introduce methods used to analyze data that are collected over time. The purpose of this chapter is to present statistical methods that managers can use for forecasting.

We examine the notion of a *time series of data* in the first section. Second, we introduce the components of a time series and discuss reasons for their existence. Methods that can be used to smooth large variations in a time series are presented next.

Multiplicative and additive models of time series are also covered in this chapter. The use of regression analysis as a method of measuring and representing trend in a time series is presented. Various methods for measuring seasonal variation and a discussion of the analysis of cycles are given. How the information that we know about the components in a time series can be used for forecasting is discussed.

14.1
TIME SERIES

Historical data, mentioned several times in previous chapters, can be used in a number of ways: to estimate probabilities, in regression and correlation problems, and in tests of hypotheses when we want to determine whether a particular population currently has the same characteristics it has had in the past.

Much of the historical data available to businesspeople and economists has been recorded at regular time intervals to form a **time series.**

> **Definition**
>
> A **time series** is a sequence of data values for a variable $Y_1, Y_2, \ldots, Y_n$ that are separated by equal time intervals. Y_t is the notation for the value of the time series at time t.

For instance, gross national product figures are easily obtained by the calendar quarter and many companies keep track of revenue and expense figures by the month. Thus, a sequence of quarterly gross national product figures from the first quarter of 1983 through the last quarter of 1992 form a time series consisting of 4×10, or 40, values. Monthly sales of a company from January of 1978 through December of 1992 form a time series of 12×15, or 180, values.

EXAMPLE 14.1   Table 14.1 shows the time series values for monthly employment at International Systems Corporation for 20 quarters. Sometimes, to see very much in a set of numbers is difficult, so time series are often presented in graphs and charts. The data in Table 14.1 are graphed in Figure 14.1. Time series data are most often graphed by using a **time series plot** like the one in Figure 14.1. The pattern present in the data is quite evident when the time series plot is used. Apparently, the general level of employment was rising slightly over the 20-quarter period, but there were also marked seasonal fluctuations in employment. The summer months had especially high employment, but the winter months were characterized by low employment. This pattern is typical of companies in the construction or recreation industries.

Observing patterns such as these can be useful to management in planning for the future. The regular fluctuation suggested by Figure 14.1 can be used in forecasting almost any figure connected to the company's level of employment. For instance, the company's treasurer may be interested in budgeting wage and salary figures by quarter for next year. Thus, an estimate of what the employment level will be is needed. If all employees are covered by accident insurance,

				Year		
TABLE 14.1	*Quarter*	*Five Years Ago*	*Four Years Ago*	*Three Years Ago*	*Two Years Ago*	*Last Year*
Number of People	1st	150	160	175	180	200
Employed on Last Day of	2nd	175	190	200	220	250
Quarter at International	3rd	450	460	480	495	520
Systems	4th	140	145	160	180	190

FIGURE 14.1

Time Series Plot for the
Quarterly Employment Fig-
ures at International Systems Corporation

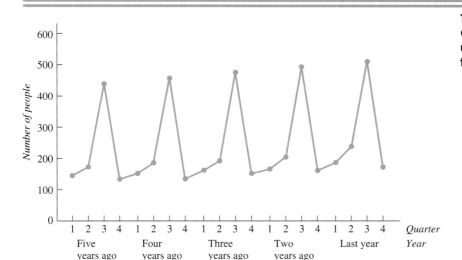

to know how many employees the company will have will also be helpful for purposes of budgeting insurance premium costs. There are many other tasks in which a forecast of the company employment level might be useful — from determining how many company uniforms to order, to planning an expansion of the company cafeteria.

But before one can use the patterns in a set of data as an aid in forecasting, these patterns must be identified and described in a concrete way. For instance, to say that "employment for the company whose data are graphed in Figure 14.1 is rising slightly, is high in the summer, and is low in the winter" is correct but rather vague. It would be more concrete and useful in forecasting to say that "employment seems to be rising at an average rate of about 6% per year, and third-quarter employment averages about 2.8 times as much as first-quarter employment." **Time series analysis** consists of breaking down a time series into its components and analyzing these components so that concrete, quantitative statements about the patterns the data reveal can be made. It is these concrete statements that will be useful in forecasting future values of the time series. The sections that follow in this chapter discuss the components of time series and methods for describing each component in a fashion that will be useful for forecasting purposes.

Problems: Section 14.1

Answers to odd-numbered problems are given in the back of the text.

1. Consider a time series whose first value was recorded in January of 1960. The last period for which there are records is September of 1992.

   a. How many full months of data are available?
   b. How many full quarters of data are available?
   c. How many full years of data are available?

2. Space Vehicle Corporation started keeping weekly sales records on Monday in the first week of July of 1964. How many weeks of data will they have collected by next Friday?

14.2
COMPONENTS OF TIME SERIES

Analysts usually divide time series into four components: long-term trend, seasonal variations, cyclical variations, and irregular variations.

Trends in time series are the long-term movements of the series that can be characterized by steady or only slightly variable rates of change. Trends can be represented by straight lines (when the data have a steady rate of change) or by smooth curves (when the rate of change is slightly variable). The time series graphed in Figure 14.1 has a slight upward trend that could be described with a straight line. The time series graphed in Figure 14.2 has a downward trend that can be represented with a smooth curve. The basic economic forces that produce long-term trend movements in time series are population changes, inflation, business competition, and technological changes.

Fuel oil consumption rises in the winter and falls in the summer. But gasoline consumption has an opposite seasonal variation, rising in the summer and falling in the winter. As their name indicates, **seasonal variations** in a time series are those variations that occur rather predictably at a particular time each year. Quite obviously, seasonal variations can be found only in data recorded at intervals of less than a year; quarterly, monthly, or weekly data might well indicate variations of this type. The quarterly time series in Figure 14.1 shows marked seasonal variation in addition to a slight trend.

Cyclical variations are movements in a time series that, like seasonal variations, are recurrent but that, unlike seasonal variations, occur in cycles of longer than one year. Cyclical variations can be found in time series recorded on an annual basis as well as in data recorded at more frequent intervals. The time series in Figure 14.3 is an annual series that shows cyclical variation in the number of housing starts in the United States from 1959 through 1991. Cycles are not constant in amplitude and duration, and this lack of regular pattern makes their future occurrence difficult to predict. Cyclical variations are often caused by general economic conditions, government policy changes, or shifts in consumer tastes and purchasing habits.

Irregular variations constitute the class of time series movements that do not fit into the other three categories previously discussed. Essentially, they are the movements left in a time series when trend, seasonal variations, and cyclical varia-

FIGURE 14.2

Data with a Curved Downward Trend

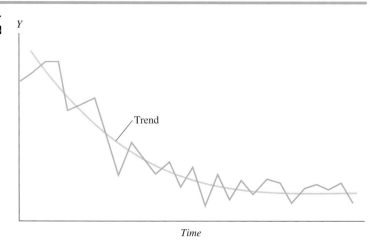

FIGURE 14.3

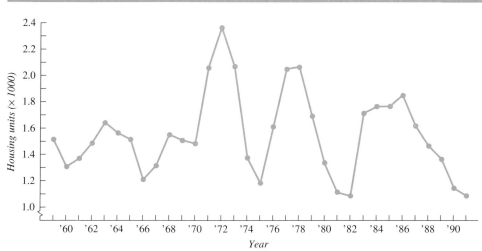

Private and Public Housing Starts, 1959–1991 (SOURCE: Data from the *Statistical Abstract of the United States*)

tions have been identified. Irregular variations cannot be predicted by using historical data, and they are not periodic in nature. They are caused by such factors as changes in the weather, wars, strikes, government legislation, and elections. For instance, a particularly harsh winter will cause a variation in crop production and fuel oil use. A strike in the automotive industry will result in a drop in production. A strike in the steel industry is typically preceded by unusually high production, as steel users stockpile material in anticipation of short supplies. The passage of antipollution legislation in the late 1970s and early 1980s caused a sharp increase in the sales of firms producing air and water purification equipment. All of these unusual changes would be called irregular variations. One can often determine the causes of irregular variations that are large. Figure 14.4 shows the time series of United States imports of handguns for private use from 1963 to 1992. The high figures in the 1966–1968 period correspond to a period of urban violence and political assassination.

FIGURE 14.4

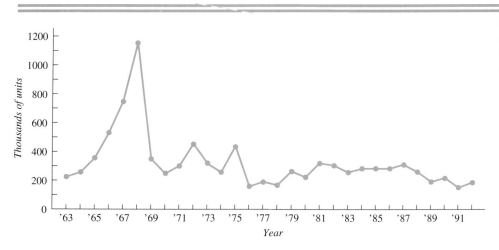

Imports of Handguns for Private Use (SOURCE: Data from the *Statistical Abstract of the United States*)

3. Give an example of a time series you think would have the following trends.

   a. A moderately increasing linear trend.
   b. A decreasing linear trend.
   c. A curvilinear trend.

4. Give an example of a time series you think would have the following indexes.

   a. High seasonal indexes in the first half of the year and low seasonal indexes in the second half of the year.
   b. High seasonal indexes in winter and summer and low seasonal indexes in spring and fall.

5. The accompanying data indicate the number of mergers that took place in an industry over a 19-year period.

   a. Plot these data on a time series chart.
   b. What type of trend (linear or curved) might best be fit to this time series?
   c. Is there evidence of seasonal variation in this series?

Year	Mergers	Year	Mergers	Year	Mergers
1	23	8	64	14	150
2	23	9	47	15	165
3	31	10	96	16	192
4	23	11	125	17	210
5	32	12	140	18	250
6	32	13	160	19	300
7	42				

14.3
SMOOTHING

Sometimes, the seasonal and irregular variations in time series are so large that determining whether trend and cyclical components exist is difficult. Large variations can be smoothed from a series by using a number of smoothing techniques. Two such techniques that we will discuss are **moving averages** and *exponential smoothing*.

Definition

A **moving average** of the values in a time series is found by averaging two or more consecutive values in the series and letting the computed value replace one of the values averaged.

For instance, a three-month average can be constructed by using the average of values for January, February, and March as the smoothed figure for February, and the average of data for February, March, and April as the March smoothed figure, and so on. The average is *moving* in the sense that as the average for each new month is calculated, a more recent value of the original time series is brought into the calculation of the average and an earlier one is dropped out.

Time Period	Actual Number of Units Sold	Five-Month Moving Average	Exponentially Smoothed Average with $\alpha = .5$	TABLE 14.2
Two Years Ago				**Actual and Smoothed Sales of Electronic Calculators at Mammoth Department Store**
1 January	21	—	21.0	
2 February	20	—	20.5	
3 March	19	18.4	19.8	
4 April	18	17.2	18.9	
5 May	14	17.6	16.5	
6 June	15	19.4	15.8	
7 July	22	20.8	18.9	
8 August	28	23.0	23.5	
9 September	25	25.0	24.3	
10 October	25	24.6	24.7	
11 November	25	24.0	24.9	
12 December	20	24.0	22.5	
Last Year				
13 January	25	23.8	23.8	
14 February	25	24.4	24.4	
15 March	24	27.6	24.2	
16 April	28	29.0	26.1	
17 May	36	29.0	31.0	
18 June	32	28.8	31.5	
19 July	25	27.6	28.3	
20 August	23	25.0	25.7	
21 September	22	23.0	23.8	
22 October	23	23.4	23.4	
23 November	22	—	22.7	
24 December	27	—	24.9	

 PRACTICE PROBLEM 14.1

Table 14.2 shows the 24-month sales record for electronic calculators in the business equipment department of the Mammoth Department Store. Find the 5-month moving average time series.

Solution The general formula for finding the 5-month moving average in month t is

$$\overline{Y}_t = \frac{Y_{t-2} + Y_{t-1} + Y_t + Y_{t+1} + Y_{t+2}}{5}$$

For instance, the moving average value for time period 11 from November of 2 years ago is

$$\overline{Y}_{11} = \frac{25 + 25 + 25 + 20 + 25}{5} = 24.0$$

The second column of figures in Table 14.2 shows the moving average values for this series, and the actual and 5-month moving average figures are graphed in Figure 14.5.

FIGURE 14.5

Sales Record for Electronic
Calculators in the Mam-
moth Department Store

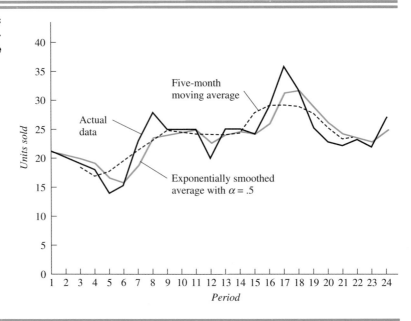

Two important points are demonstrated in Practice Problem 14.1. The first is that a moving average cannot be calculated for the first two and last two periods in the time series. The calculations for the moving average in January of two years ago would require the sales figures for December and November of three years ago, and these figures are not given. Similarly, the moving average for November of last year would require the sales figures for both December of last year and January of this year, and this latter figure is not available. In general, the longer the span covered by the moving average, the more periods are lost at the beginning and the end of the time series. Thus, a six-month moving average loses three months on each end, a nine-month moving average does not have values for the first four months or the last four months, and so forth.

The second point is that the smoothed values contain less variation than the original values. This observation is always true, and the longer the span covered by the moving average, the less variable will be the smoothed series. Thus a six-month moving average will contain less variation than the five-month series. Of course, if too many periods are used in the calculation of the moving average, the smoothed series becomes nearly a straight line, and rather than being exposed, any trend or cyclical variations that may exist are smoothed out entirely.

A second way of smoothing data is called *exponential smoothing.* An exponentially smoothed value can be obtained for each time period. Let S_t be the smoothed value for period t. We begin by setting S_1 equal to Y_1, the actual value of the time series in the first period. Then,

$$S_2 = \alpha Y_2 + (1 - \alpha)S_1$$

where α is a number between zero and one and is called the **smoothing constant.** For period t we have the following **exponential smoothing** formula.

> **Equation for Exponentially Smoothed Value for Period t**
>
> $$S_t = \alpha Y_t + (1 - \alpha)S_{t-1} \qquad 0 \le \alpha \le 1 \qquad\qquad (14.1)$$

Equation (14.1) shows that this period's smoothed value is a weighted average of two figures: the actual value in the time series for this period and the previous smoothed value. But the previous smoothed value is the weighted average of the actual time series value for the previous period and the smoothed value the period before than. If we were to proceed backward in time, we would find that

$$S_t = \alpha Y_t + (1 - \alpha)\alpha Y_{t-1} + (1 - \alpha)^2\alpha Y_{t-2}$$
$$+ (1 - \alpha)^3\alpha Y_{t-3} + \cdots + (1 - \alpha)^{t-1}Y_1 \qquad\qquad (14.2)$$

Equation (14.2) shows that S_t is a weighted sum of all the previous values in the time series back to time $t = 1$. But the values further and further in the past have lighter and lighter weights since α and $1 - \alpha$ are fractional values that become smaller and smaller as we multiply them together and raise them to powers as in the formula.

The analyst smoothing the data can set the value of α anywhere between zero and one, thereby controlling how much weight current and past values in the time series will receive. If α is large, say .95, then S will be composed primarily of the current value of the time series Y_t, and only 5% of S_t will be dependent on historical values of the series. However, if α is small, say .10, then S_t will be composed primarily of the weighted historical values in the time series, and only 10% of S_t will be dependent on the current value Y_t. If the analyst wants the smoothed series to follow the actual series quite closely — that is, little smoothing is desired—he or she should use a large smoothing constant α. The resulting smoothed series will be very sensitive to changes in the current values of the time series. But if a smoothed series that has most of the random, volatile variation taken out of it is desired, the analyst should choose a small smoothing constant. The resulting series will be quite smooth and reflect historical values of the time series far more than current values. Forecasters commonly use a value between .1 and .3 for α.

PRACTICE PROBLEM 14.2

Find the exponential average values for the Mammoth Department Store's calculator sales, using a smoothing constant of $\alpha = .5$. These values are presented in the third column of numbers in Table 14.2. They are also plotted on the graph in Figure 14.5.

Solution The calculations for the first few figures are as follows:

$$S_1 = Y_1 = 21.0$$
$$S_2 = \alpha Y_2 + (1 - \alpha)S_1 = .5(20) + .5(21.0) = 20.5$$
$$S_3 = \alpha Y_3 + (1 - \alpha)S_2 = .5(19) + .5(20.5) = 19.8$$
$$S_4 = \alpha Y_4 + (1 - \alpha)S_3 = .5(18) + .5(19.8) = 18.9$$

To illustrate that Equation (14.1) and Equation (14.2) are equivalent, we compute S_3 again using Equation (14.2) as follows:

$$S_3 = \alpha Y_3 + (1 - \alpha)S_2$$
$$= \alpha Y_3 + (1 - \alpha)[\alpha Y_2 + (1 - \alpha)S_1]$$
$$= \alpha Y_3 + (1 - \alpha)(\alpha)Y_2 + (1 - \alpha)^2 Y_1$$
$$= (.5)(19) + (.5)(.5)(20) + (.5)^2(21) = 19.8$$

Computers are often used to make time series computations less burdensome. Data analysis software packages were used to find the 5-month moving averages and the exponentially smoothed values for the Mammoth Department Store calculator data, and the results are presented in Computer Exhibits 14.1A and 14.1B.

Problems: Section 14.3

6. When a four-month moving average is found for a time series, how many months do not have averages associated with them at the following points?
   a. At the first of the time series.   **b.** At the end of the time series.
7. Find the three-year moving average values for the merger time series in Problem 5.
8. Find a four-year moving average series for the merger data in Problem 5. Center the average *on* the years.
9. Find the exponentially smoothed series for the past 10 years in Problem 5:
   a. By using $\alpha = .2$ **b.** By using $\alpha = .8$.
10. Judging from your results in Problem 9, which value of α should be used if the person smoothing the merger data wants the smoothed series to be highly responsive to sudden changes in the merger rate?

14.4
MEASURING THE COMPONENTS OF A TIME SERIES

When the original data of a time series have been smoothed, the results often uncover evidence that the time series contains elements of trend, seasonal, and/or cyclical variations. The data analyst may well be interested in obtaining reliable means of measuring these components to use in a model or equation to forecast future values of the time series.

The classical equation that is used to represent a time series is defined as follows.

> **Classical Multiplicative Equation to Represent Time Series**
>
> $$Actual\ value = T \times S \times C \times I \tag{14.3}$$

In Equation (14.3) the trend value T is measured in the same units as the times series, and S, C, and I are indexes of seasonal, cyclical, and irregular variation, respectively.

If we wish to **forecast** a future period's value of the time series, we can find the forecast by doing the following steps.

Step One   Project a trend value.

Step Two   Multiply the trend value by a seasonal index number that measures the typical deviation of the time series from the trend for that season.

COMPUTER EXHIBIT 14.1A Time Series Computations: Five-Month Moving Average
of Calculator Sales

MINITAB

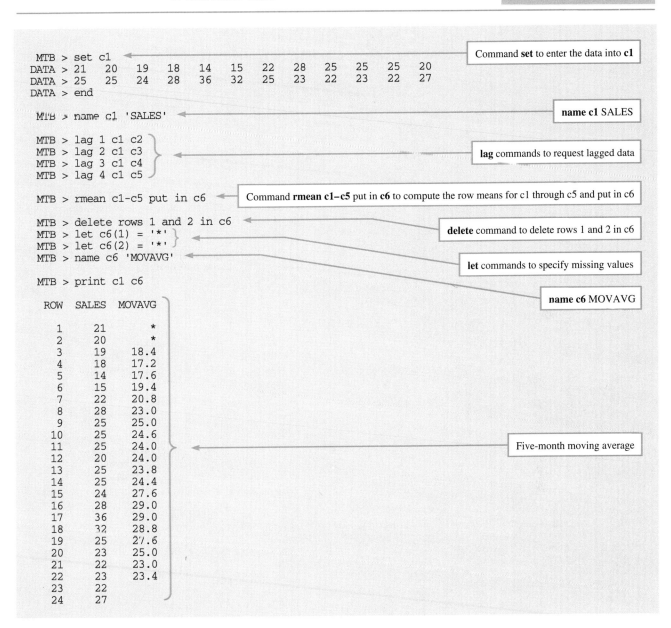

```
MTB > set c1                                                    Command set to enter the data into c1
DATA > 21    20    19    18    14    15    22    28    25    25    25    20
DATA > 25    25    24    28    36    32    25    23    22    23    22    27
DATA > end
MTB > name c1 'SALES'                                           name c1 SALES

MTB > lag 1 c1 c2
MTB > lag 2 c1 c3                                               lag commands to request lagged data
MTB > lag 3 c1 c4
MTB > lag 4 c1 c5

MTB > rmean c1-c5 put in c6      Command rmean c1–c5 put in c6 to compute the row means for c1 through c5 and put in c6

MTB > delete rows 1 and 2 in c6
MTB > let c6(1) = '*'                                           delete command to delete rows 1 and 2 in c6
MTB > let c6(2) = '*'
MTB > name c6 'MOVAVG'                                          let commands to specify missing values

MTB > print c1 c6                                               name c6 MOVAVG

ROW   SALES   MOVAVG
  1     21      *
  2     20      *
  3     19     18.4
  4     18     17.2
  5     14     17.6
  6     15     19.4
  7     22     20.8
  8     28     23.0
  9     25     25.0
 10     25     24.6
 11     25     24.0                                             Five-month moving average
 12     20     24.0
 13     25     23.8
 14     25     24.4
 15     24     27.6
 16     28     29.0
 17     36     29.0
 18     32     28.8
 19     25     27.6
 20     23     25.0
 21     22     23.0
 22     23     23.4
 23     22
 24     27
```

Step Three   Multiply the value in Step Two by an index that measures the cyclical variation expected for that period.

Since irregular variations are, by their very nature, unpredictable, the forecast is completed after the third step. That is,

$$\text{Forecasted value} = T \times S \times C \qquad (14.4)$$

MYSTAT   **COMPUTER EXHIBIT 14.1B** Time Series Computations: Exponentially Smoothed
                    Calculator Sales

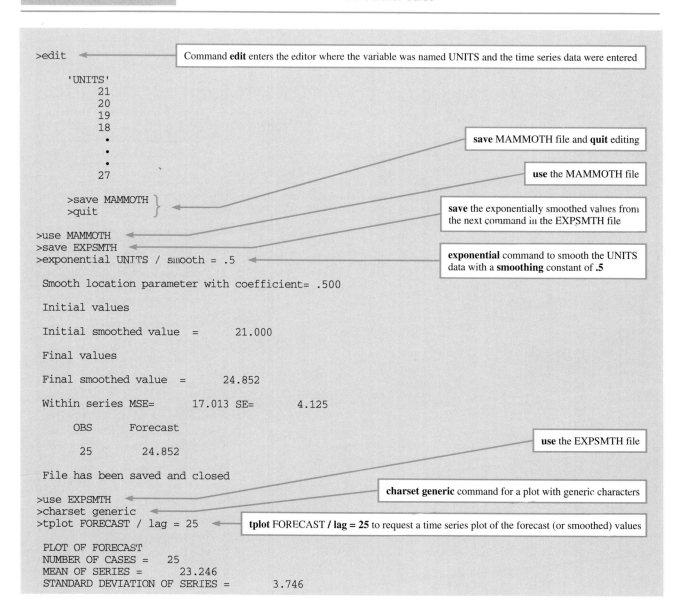

```
>edit ◄─────────────────   Command edit enters the editor where the variable was named UNITS and the time series data were entered

     'UNITS'
        21
        20
        19
        18                                                              save MAMMOTH file and quit editing
         •
         •
         •                                                              use the MAMMOTH file
        27
                                                                        save the exponentially smoothed values from
    >save MAMMOTH                                                       the next command in the EXPSMTH file
    >quit
>use MAMMOTH ◄─────────
>save EXPSMTH                                                           exponential command to smooth the UNITS
>exponential UNITS / smooth = .5 ◄───                                   data with a smoothing constant of .5

Smooth location parameter with coefficient= .500

Initial values

Initial smoothed value   =      21.000

Final values

Final smoothed value  =      24.852

Within series MSE=       17.013 SE=        4.125

     OBS        Forecast
                                                                        use the EXPSMTH file
     25         24.852

File has been saved and closed
                                                                        charset generic command for a plot with generic characters
>use EXPSMTH ◄──────
>charset generic ◄──────
>tplot FORECAST / lag = 25 ◄───    tplot FORECAST / lag = 25 to request a time series plot of the forecast (or smoothed) values

PLOT OF FORECAST
NUMBER OF CASES =     25
MEAN OF SERIES =        23.246
STANDARD DEVIATION OF SERIES =        3.746
```

Thus the forecaster would benefit from having some numerical measures of T, S, and C to use in Equation (14.4). Section 14.5 contains an explanation of how the trend T can be measured in a time series. Sections 14.6 and 14.7 show various means of measuring the seasonal variation S. Section 14.8 discusses the analysis of cycles C, and Section 14.9 deals with how the knowledge of the components in a time series can be used in both short- and long-term forecasting.

COMPUTER EXHIBIT 14.1B Time Series Computations: Exponentially Smoothed
Calculator Sales (Continued)

MYSTAT

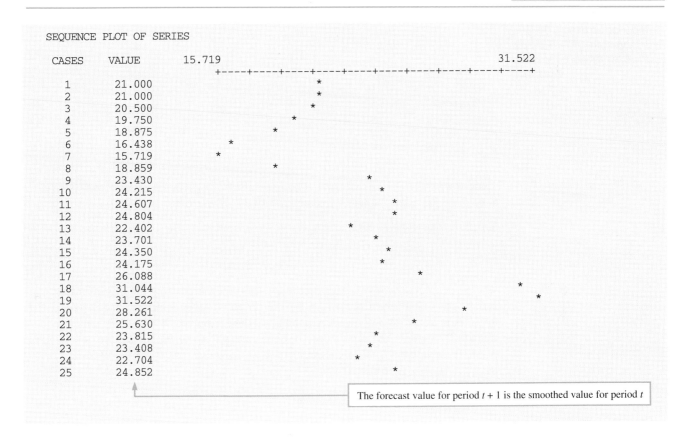

SEQUENCE PLOT OF SERIES

```
   CASES    VALUE    15.719                                        31.522
                     +----+----+----+----+----+----+----+----+----+----+
     1      21.000                        *
     2      21.000                        *
     3      20.500                      *
     4      19.750                    *
     5      18.875                 *
     6      16.438            *
     7      15.719         *
     8      18.859                 *
     9      23.430                       *
    10      24.215                        *
    11      24.607                          *
    12      24.804                          *
    13      22.402                     *
    14      23.701                       *
    15      24.350                        *
    16      24.175                       *
    17      26.088                           *
    18      31.044                                    *
    19      31.522                                      *
    20      28.261                               *
    21      25.630                          *
    22      23.815                      *
    23      23.408                     *
    24      22.704                    *
    25      24.852                        *
```

The forecast value for period $t + 1$ is the smoothed value for period t

14.5
TREND ANALYSIS

When the original data or the smoothed data in a time series show evidence of trend, the statistician may well be interested in finding a suitable way of measuring and representing the trend. If the time series shows a steady upward or downward movement, a straight line can be used to represent the trend. If the overall rise or fall occurs at an increasing or decreasing rate, a curved line should be used. In either case the most common method for finding a **trend line** is to fit a least squares line to the data. The procedure is the same as for fitting regression least squares lines to pairs of observations. In the present case, the pairs of points are the values of the time series data paired with the dates of the observations.

PRACTICE PROBLEM 14.3

The data in Table 14.3 represent the closing price on the Northwest Regional Stock Exchange of a speculative stock over the past nine quarters of trading. Monthly, weekly, or daily data have not been presented since the stock's price fluctuates so

TABLE 14.3	Quarter	Price of Stock on Last Day of Quarter
Closing Prices for a Speculative Stock on the Northwest Regional Stock Exchange over Nine Quarters	1	$37.75
	2	$44.00
	3	$46.25
	4	$44.00
	5	$51.75
	6	$49.00
	7	$58.50
	8	$58.00
	9	$70.25

much over short periods of time. The general rise in price is evident from the quarterly data. Fit a least squares linear trend line to these data.

Solution The general form of the linear trend line is

$$\hat{Y} = b_0 + b_1 t \tag{14.5}$$

where t is a numerical value for the time period measured from the origin of the time series. The selection of the origin for the series is rather arbitrary. For instance, the quarters in Table 14.3 could have been numbered from 0 to 8 instead of 1 to 9. Equations (12.7) and (12.8) can be used to obtain the values for b_0 and b_1, the coefficients of the least squares line. In this case,

$$b_0 = 34.26$$
$$b_1 = 3.36$$

So the least squares trend line becomes

$$\hat{Y} = 34.26 + 3.36t$$

These coefficients can be interpreted as meaning that in time period $t = 0$ the trend value of the stock would be $34.26 per share, and that over the nine-quarter period the price of the stock rose an average of $3.36 per share per quarter. The actual values of the time series and the least squares trend line are plotted in Figure 14.6.

Practice Problem 14.3 illustrates two important points about the least squares line. The first is that the least squares method of fitting a line to a set of points gives heavy weight to the larger fluctuations. Large deviations result in very large squared deviations, which strongly influence the values of b_0 and b_1 calculated by Equations (12.7) and (12.8). Thus, a few large fluctuations like the rapid price rises in quarters 7 and 9 can unduly influence the slope of the least squares line. Note that in Figure 14.7 the seventh- and ninth-quarter price rises have pulled the line upward so that the price fluctuations in the first six quarters are not *around* the line but near or below it.

The second point illustrated by the problem concerns the shape of the trend line. There is a slight indication that the general rate of price increase for the stock is in-

FIGURE 14.6

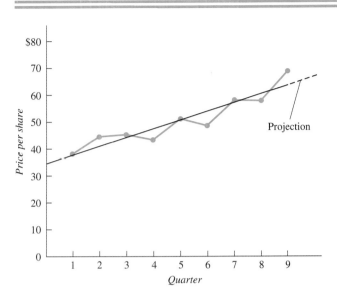

creasing. Thus, a trend curve might better describe the movement of this time series. A least squares curve could be fitted to the time series by using the multiple regression techniques discussed in Chapter 13.

 PRACTICE PROBLEM 14.4

Fit a polynomial trend line of the form $\hat{Y} = b_0 + b_1 t + b_2 t^2$ to the Northwest Regional Stock Exchange stock price data in Table 14.3.

Solution To perform the multiple regression required for the solution to this problem, we must compute the second independent variable needed in the equation, t^2. The following table shows the data from Table 14.3 rearranged in the format used for the problems in Chapter 13. Note that the third column is just the square of the second.

Y, Closing Prices of Stock	t, Quarter	t^2, (Quarter)^2
37.75	1	1
44.00	2	4
46.25	3	9
44.00	4	16
51.75	5	25
49.00	6	36
58.50	7	49
58.00	8	64
70.25	9	81

With the data in this form, we used a computer to solve for b_0, b_1, and b_2 just as though we were finding the regression equation coefficients for any two-independent-variable equation $\hat{Y} = b_0 + b_1 X_1 + b_2 X_2$. The only difference here is that $X_1 = t$ and $X_2 = t^2$. Its results gave

$$\hat{Y} = 39.90 + .28t + .31t^2$$

This equation does a slightly better job in fitting the data points than does the one produced in Practice Problem 14.3. That is, R^2 for this equation is .91, and the r^2 for the simple regression equation of Practice Problem 14.3 is .87. In this particular problem, the indications of curved trend are so slight that it would probably be best to wait a few more quarters before concluding that the trend is not linear. ▬

Problems: Section 14.5

11. Fit a least squares trend line to the merger data in Problem 5.

12. The accompanying quarterly data show the number of appliances (in thousands) returned to National Appliances, Inc., for warranty service over the past five years.

    a. Plot a frequency histogram of this time series.

    b. Find the equation of the least squares linear trend line that fits this time series. Let $t = 1$ be the first quarter five years ago.

    c. What would be the trend line value for the second quarter of the current year, that is, two periods beyond the end of the actual data?

	Quarter			
Year	*First*	*Second*	*Third*	*Fourth*
5 years ago	1.2	0.8	0.6	1.1
4 years ago	1.7	1.2	1.0	1.5
3 years ago	3.1	3.5	3.5	3.2
2 years ago	2.6	2.2	1.9	2.5
1 year ago	2.9	2.5	2.2	3.0

14.6
SEASONAL INDEXES

Next, we turn our attention to measuring the seasonal variation in a time series S by using seasonal indexes. **Seasonal indexes** are numbers that vary from a base of 100. When a month (week, quarter, or other seasonal period) is assigned an index value of 100, this value indicates that that month has no seasonal variation. Two methods of obtaining seasonal indexes are presented here. The first involves finding a seasonal index that compares the seasonal mean value with a grand mean value. This method is most appropriate for time series that do not have strong trends or cyclical variations. The second method, known as a **ratio-to-moving-average** method, involves comparing each season's actual value with a yearly moving average to obtain an index. The index arrived at are then averaged over all the time periods in the series. This method is much more widely used since it can give meaningful seasonal indexes for data with strong trends and cyclical variations.

The practice problems in this section consist of problems in which seasonal indexes are desired for each of the four quarters in a year. If indexes were to be calculated for each month of the year, techniques employed would be the same. We obtain quarterly indexes here merely to demonstrate the method without using excessive amounts of space.

▬ PRACTICE PROBLEM 14.5

Table 14.4 presents ten years of quarterly data for the sales of XB–7, a product used in combating lawn larvae. The manufacturer recommends that XB–7 be applied in the spring and fall. Find the seasonal sales indexes for the four quarters of the year.

Year	First	Second	Third	F
1	257	288	263	
2	291	368	341	3
3	319	485	325	3
4	305	364	336	38
5	332	435	410	449
6	368	520	415	444
7	332	464	405	468
8	351	440	411	668
9	355	504	449	527
10	408	490	740	649
Total	3318	4358	4095	4688
Grand total				16,459
Quarterly mean*	331.8	435.8	409.5	468.8
Grand mean†				411.48
Quarterly index‡	80.7	105.9	99.5	113.9
Sum of indexes				400.0

TABLE 14.4

Calculation of Quarterly Sales Index for Product XB-7 (in tons)

*The total for the quarter divided by 10.
†(Sum of quarterly means)/4, or 16,459/40.
‡(Quarterly mean/grand mean) × 100.

Solution The indexes have been found in Table 14.4 by using the first method of calculating seasonal indexes. The mean sales value for each quarter is taken over the ten-year period and then is expressed as a percentage of the grand mean over all quarters in the ten-year period. The grand mean is just the mean of the four quarterly means. Thus the seasonal index for the first quarter is the mean for the quarter divided by the grand mean times 100:

$$\text{Seasonal index for first quarter} = \frac{331.8}{411.48} \times 100 = 80.7$$

Thus, on the average, the first quarter's sales are only about 80.7% of the average quarter's sales. In contrast, the fourth quarter's sales are about 13.9% above the average quarter's sales. Since the seasonable indexes vary from a base of 100, the four quarterly indexes add to 400. When monthly indexes are calculated, their total should be 1200, averaging a value of 100 for each month.

 PRACTICE PROBLEM 14.6

Table 14.5 presents in somewhat different form the ten years of quarterly data for the sales of product XB–7. In this problem, seasonal indexes are to be computed by using a ratio-to-moving-average method of comparing actual sales values with a four-quarter moving average.

Solution There are five steps in this process. The first three are shown in Table 14.5 and the fourth and fifth in Table 14.6.

Step One Determine a four-quarter moving average. This average is found by summing four adjacent values and dividing the total by 4. Now, to which quarter should this moving average be assigned? In Practice

TABLE 14.5

Calculations for Steps 1–3 in Finding Seasonal Sales Indexes of Product XB–? (in ton...

	Column 1: Actual Data	Column 2: Four-Quarter Moving Average of Column 1	Column 3: Two-Quarter Moving Average of Column 2	Column 4: Actual Data as Percentage of Column 3
First	257	—	—	—
Second	288	—	—	—
Third	263	279.75	284.0	92.6
Fourth	311	288.25	298.2	104.3
2 — First	291	308.25	318.0	91.5
2 — Second	368	327.75	339.9	108.3
2 — Third	341	352.00	355.5	95.9
2 — Fourth	408	359.00	373.6	109.2
3 — First	319	388.25	386.2	82.6
3 — Second	485	384.25	380.9	127.3
3 — Third	325	377.50	375.8	86.5
3 — Fourth	381	374.00	358.9	106.2
4 — First	305	343.75	345.1	88.4
4 — Second	364	346.50	346.8	105.0
4 — Third	336	347.00	350.4	95.9
4 — Fourth	383	353.75	362.6	105.6
5 — First	332	371.50	380.8	87.2
5 — Second	435	390.00	398.2	109.2
5 — Third	410	406.50	411.0	99.8
5 — Fourth	449	415.50	426.1	105.4
6 — First	368	436.75	437.4	84.1
6 — Second	520	438.00	437.4	118.9
6 — Third	415	436.75	432.2	96.0
6 — Fourth	444	427.75	420.8	105.5
7 — First	332	413.75	412.5	80.5
7 — Second	464	411.25	414.2	112.0
7 — Third	405	417.25	419.6	96.5
7 — Fourth	468	422.00	419.0	111.7
8 — First	351	416.00	416.8	84.2
8 — Second	440	417.50	442.5	99.4
8 — Third	411	467.50	468.0	87.8
8 — Fourth	668	468.50	476.5	140.2
9 — First	355	484.50	489.2	72.6
9 — Second	504	494.00	476.4	105.8
9 — Third	449	458.75	465.4	96.5
9 — Fourth	527	472.00	470.2	112.1
10 — First	408	468.50	504.9	80.8
10 — Second	490	541.25	556.5	88.1
10 — Third	740	571.75	—	—
10 — Fourth	649	—	—	—

Problem 14.1, where a five-month moving average was calculated, this question did not arise. With an odd number of periods (five) involved in the moving average, each of the moving average values could be assigned to the middle (third) period of the five-period unit. However, when the number of periods is even (like four), the moving average

	Quarter				
Year	First	Second	Third	Fourth	**TABLE 14.6**
1	—	—	92.6	104.3	**Calculations for Steps 4**
2	91.5	108.3	95.9	109.2	**and 5 in Finding Seasonal**
3	82.6	127.3	86.5	106.2	**Sales Indexes of Product**
4	88.4	105.0	95.9	105.6	**XB-7 (in tons)**
5	87.2	109.2	99.8	105.4	
6	84.1	118.9	96.0	105.5	
7	80.5	112.0	96.5	111.7	
8	84.2	99.4	87.8	140.2	
9	72.6	105.8	96.5	112.1	
10	80.8	88.1	—	—	
Mean percentage	83.5	108.2	94.1	111.1	
Total*				396.9	
Seasonal index	84.2	109.0	94.8	112.0	
Total†				400.0	

*Total of percentages.
†Total of seasonal indexes.

cannot be assigned to a middle period since one does not exist. The middle of four quarters is between the second and third quarter. Thus the moving averages in the second column of Table 14.5 are centered *between* quarters. To find a moving average that can be assigned to a particular quarter, we must perform Step 2.

Step Two Form a two-quarter moving average from the four-quarter moving average column. This figure, column 3 of the table, is now centered *on* a quarter instead of *between* quarters. This figure is also a four-quarter moving average that is centered on a particular quarter. For instance, the first moving average is centered on the third quarter of the first year. It is actually a weighted average of quarterly sales data with the following weights:

Year	Quarter	Weight
1	First	1
1	Second	2
1	Third	2
1	Fourth	2
2	First	1
	Total	8

Similarly, each quarter's moving average consists of data from five quarters, where the first and last quarter are weighted once and the intervening quarters are weighted twice.

Step Three Express each quarter's actual figure as a percentage of its moving average. The fourth column of Table 14.5 contains these percentages.

Step Four For each quarter (first, second, third and fourth), find the mean percentage value from Step 3 over all the years. This step is done in

Table 14.6. For instance, the mean percentage value for the first quarter is 83.5. This value indicates that on the average the sales figure for the first quarter was 83.5% of the four-quarter moving average centered on the first quarter. But these mean percentages cannot be used as seasonal indexes themselves, because seasonal indexes vary from a base of 100. The four mean percentages do not add to 400—they add to 396.9. Thus a final calculation is needed to obtain a seasonal index.

Step Five Adjust the mean percentages so that they add to 400. To accomplish this step, each mean percentage figure is multiplied by an adjustment factor of 400/396.9, or 1.008. The final result is a new percentage figure that varies from a base of 100 and represents the seasonal index.

The steps outlined in Practice Problem 14.6 are very similar to those that would be followed in calculating monthly seasonal indexes. The minor modifications that would be made can be seen by comparing the steps of the problem with the following procedure.

Step One Determine a 12-month moving average.
Obtain a 2-month moving average of the figures found in Step One.

Step Two This result is a modified 12-month moving average.

Step Three Express each month's actual figure as a percentage of the moving average in Step Three.

Step Four For each month (January, February, . . . , December), find the mean percentage value from Step Three over all years.

Step Five Adjust the mean percentages so that they total 1200.

The seasonal indexes calculated in Practice Problems 14.5 and 14.6 do not differ a great deal from one another. They both show that sales are high in the spring and fall and low in the winter and summer. However, the indexes developed in Practice Problem 14.6 are preferable to those from Practice Problem 14.5. They are preferable because the data over the ten-year period used in the examples show a rather strong trend, and the second of our two methods for developing seasonal indexes is most appropriate for data with strong trend.

The difficulty of calculating seasonal indexes by the second method is usually not a serious factor, since in most problems where much data is involved, computer programs are written to perform the calculations.

Problems: Section 14.6

13. Determine the quarterly seasonal indexes for the National Appliances warranty service time series in Problem 12 by using the method described in Practice Problem 14.5. Then find values for the same indexes by using the ratio-to-moving-average method. If you ignore the differences in the calculations required, which of the two methods of finding seasonal indexes is preferable for calculating seasonal indexes for this series? Why?

14. The accompanying time series shows the number of building permits issued by month for Redrock County.

   a. Use the method of Practice Problem 14.5 to find monthly seasonal construction indexes.

b. Use a ratio-to-moving-average method to find monthly seasonal construction indexes.

c. Discuss the advantages of the first method over the ratio-to-moving-average method for this time series.

| | | *Year* | |
Month	Year 1	Year 2	Year 3
January	1015	664	1106
February	901	743	1022
March	1319	1147	1679
April	1590	1284	2011
May	1555	1250	1985
June	1473	1352	1938
July	1252	1408	1943
August	1249	1287	2045
September	1293	1309	1738
October	1234	1409	1797
November	946	1269	1722
December	841	1214	1496

15. The values in the accompanying table show the percentages of students dropping a particular class over the past three years at a school using the three-quarter academic year.

a. Use the method of Practice Problem 14.5 to find the indexes of drops for the academic quarters.

b. Use a ratio-to-moving-average method to find the indexes of drops for the academic quarters. (*Note:* Step 2 in the process can be skipped, because Step 2 is only used to center the moving average on a period in the data set. Since this data set has only three periods per year, the three-quarter moving average computed in Step 1 will already be centered.)

| | | *Quarter* | |
Year	Fall	Winter	Spring
Two years ago	5	4	8
Last year	6	4	10
This year	9	8	12

In the previous section, we found that the second, more complex method of finding seasonal indexes is the appropriate method for data containing a strong trend. However, we can combine our knowledge of regression and time series in a method that accounts for trends in the data. One need only have access to a multiple regression package and a computer.

The method proceeds as follows. First, we must select one season as the *base season*. If the data are quarterly, we may wish to select the first calendar quarter, winter, as the base season. All other **seasonal indexes** will be measured with respect to this base season. There are no absolute rules governing the selection of the base season.

Next, the time series data are recorded over the historical time period, using dummy variables to represent those seasons other than the base season.

14.7

FINDING SEASONAL MEASURES USING MULTIPLE REGRESSION

EXAMPLE 14.2   One year of quarterly sales data for a company might be $50, $125, $105, and $60 thousand. If winter were selected as the base season, then these data could be recorded with the use of indicator variables, as follows.

		Season		
Y, Sales	t, Time	S_2, Spring	S_3, Summer	S_4, Fall
50	1	0	0	0
125	2	1	0	0
105	3	0	1	0
60	4	0	0	1

Note that the seasonal variables—spring, summer, and fall—are all zero in the time periods that do not correspond to their season. This notation allows us to represent all four seasons with only three variables. For instance, the winter period is represented by three zeros in the seasonal variables $(0, 0, 0)$. Spring is represented by $(1, 0, 0)$, summer by $(0, 1, 0)$, and fall by $(0, 0, 1)$.   ▄▄

With the data represented in this fashion, we can construct a regression equation of the form

$$\hat{Y} = b_0 + b_1 t + b_2 S_2 + b_3 S_3 + b_4 S_4 \tag{14.6}$$

where $\hat{Y}$ represents the predicted sales figure, t is the time period, and the S_j's are indicator variables representing spring, summer, and fall. To obtain a forecast using this regression equation, we must determine the time period t and the season S_j for which the forecast is desired. If we desire a forecast of sales for the summer quarter following the last data point, then $t = 7$ and the seasonal indications are $(0, 1, 0)$. The forecast would then be

$$\hat{Y} = b_0 + b_1(7) + b_2(0) + b_3(1) + b_4(0) = b_0 + b_1(7) + b_3$$

In this result, we can see that the forecast consists of the trend value $b_0 + b_1(7)$ plus the amount b_3. This b_3 is the seasonal adjustment figure for summer sales. The same line of reasoning shows that b_2 is the seasonal adjustment figure for spring and that b_4 gives this adjustment for fall. Since winter is the base quarter, no seasonal adjustment figure is added to the trend. In realistic problems, of course, more than four data points are used to obtain the regression equation.

 PRACTICE PROBLEM 14.7

The data in the accompanying table represent the quarterly fuel consumption (in thousands of gallons) of Long-Haul Trucking, Inc., over the past five years. Find the seasonal adjustments for fuel consumption by multiple regression, using indicator seasonal variables. Predict fuel consumption for fall of this year.

		Season		
Year	Winter	Spring	Summer	Fall
5 years ago	80	100	150	105
4 years ago	95	115	170	105
3 years ago	110	150	190	145
2 years ago	115	175	210	180
Last year	120	185	220	190

Year	Quarter	t, Time	Y, Actual Data	S_2, Spring	S_3, Summer	S_4, Fall	$\hat{Y}$, Predicted Values
Five years ago	Winter	1	80	0	0	0	66.2
	Spring	2	100	1	0	0	107.2
	Summer	3	150	0	1	0	150.2
	Fall	4	105	0	0	1	107.2
Four years ago	Winter	5	95	0	0	0	85.1
	Spring	6	115	1	0	0	126.1
	Summer	7	170	0	1	0	169.1
	Fall	8	105	0	0	1	126.1
Three years ago	Winter	9	110	0	0	0	104.0
	Spring	10	150	1	0	0	145.0
	Summer	11	190	0	1	0	188.0
	Fall	12	145	0	0	1	145.0
Two years ago	Winter	13	115	0	0	0	122.9
	Spring	14	175	1	0	0	163.9
	Summer	15	210	0	1	0	206.9
	Fall	16	180	0	0	1	163.9
Last year	Winter	17	120	0	0	0	141.8
	Spring	18	185	1	0	0	182.8
	Summer	19	220	0	1	0	225.8
	Fall	20	190	0	0	1	182.8

TABLE 14.7

Data for Fuel Consumption at Long-Haul Trucking with Time and Indicator Variables

Solution The first step in the solution is to select a base season. For convenience, we will use winter as that base. Winter seems to be the lowest fuel consumption season and is perhaps appropriate as a base. Next, we record the data over the past five years, using indicator variables to represent the spring, summer, and fall seasons. Table 14.7 shows how these data are recorded. Two data analysis software packages were used to obtain the multiple regression equation, and the results are presented in Computer Exhibits 14.2A and 14.2B. The regression equation follows:

$$\hat{Y} = 61.5 + 4.7t + 36.3S_2 + 74.6S_3 + 26.8S_4$$

This regression line indicates that the forecast is $\hat{Y} = 61.5 + 4.7(0) = 61.5$ at time $t = 0$. That is, the constant term can be interpreted as the equation's estimate of the fuel consumption trend for fall of six years ago, before seasonal adjustments are made. The trend coefficient, 4.7, suggests that, all other things being constant, the fuel consumption has been rising at a rate of 4700 gallons per quarter. The seasonal adjustment figures are 36.3, 74.6, and 26.8 for spring, summer, and fall, respectively. There is no seasonal adjustment for winter. Thus, winter forecasts are equal to the trend. However, spring's consumption is 36,300 gallons above the trend, summer's is 74,600 above trend, and fall's is 26,800 above trend.

The forecast for fall of this year is made by using $t = 24$ and a set of dummy variables for fall, $(0, 0, 1)$. Thus,

$$\hat{Y} = 61.5 + 4.7(24) + 36.3(0) + 74.6(0) + 26.8(1)$$

$$= 61.5 + 112.8 + 26.8 = 201.1 \text{ or } 201,100 \text{ gallons}$$

MINITAB   COMPUTER EXHIBIT 14.2A Multiple Regression for Seasonal Adjustments at Long-Haul Trucking

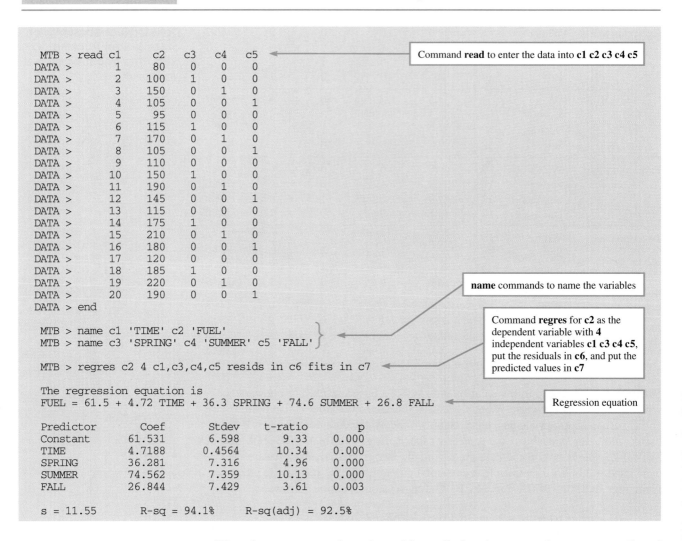

```
MTB > read c1      c2    c3   c4    c5    ◄────    Command read to enter the data into c1 c2 c3 c4 c5
DATA >      1      80    0    0     0
DATA >      2     100    1    0     0
DATA >      3     150    0    1     0
DATA >      4     105    0    0     1
DATA >      5      95    0    0     0
DATA >      6     115    1    0     0
DATA >      7     170    0    1     0
DATA >      8     105    0    0     1
DATA >      9     110    0    0     0
DATA >     10     150    1    0     0
DATA >     11     190    0    1     0
DATA >     12     145    0    0     1
DATA >     13     115    0    0     0
DATA >     14     175    1    0     0
DATA >     15     210    0    1     0
DATA >     16     180    0    0     1
DATA >     17     120    0    0     0
DATA >     18     185    1    0     0
DATA >     19     220    0    1     0
DATA >     20     190    0    0     1
DATA > end

MTB > name c1 'TIME' c2 'FUEL'                        ◄──  name commands to name the variables
MTB > name c3 'SPRING' c4 'SUMMER' c5 'FALL'

MTB > regres c2 4 c1,c3,c4,c5 resids in c6 fits in c7   ◄──  Command regres for c2 as the
                                                             dependent variable with 4
                                                             independent variables c1 c3 c4 c5,
The regression equation is                                   put the residuals in c6, and put the
FUEL = 61.5 + 4.72 TIME + 36.3 SPRING + 74.6 SUMMER + 26.8 FALL  ◄── predicted values in c7
                                                                        Regression equation
Predictor       Coef       Stdev     t-ratio        p
Constant      61.531       6.598        9.33    0.000
TIME           4.7188      0.4564      10.34    0.000
SPRING        36.281       7.316        4.96    0.000
SUMMER        74.562       7.359       10.13    0.000
FALL          26.844       7.429        3.61    0.003

s = 11.55       R-sq = 94.1%     R-sq(adj) = 92.5%
```

There is some uncertainty about this prediction, however, since we are guilty of extrapolating our forecast beyond the range of the independent variable t. That is, $t = 24$ is beyond the range of t values used to develop the regression equation. Unfortunately, when one is forecasting by using a time series regression equation, there is no way to avoid such extrapolations. The predicted consumption values for the five-year period are shown in the last column of Table 14.7. These figures will be used in the discussion for Section 14.8. Note that the R^2 figure for the regression equation developed in this problem is .94. This value suggests that the regression equation does a rather good job of fitting the predicted to the actual sales values.

In Section 14.4 we mentioned that Equation (14.3) represents the classical time series model, where

$$Actual\ value = T \times S \times C \times I$$

COMPUTER EXHIBIT 14.2A Multiple Regression for Seasonal Adjustments at Long-Haul Trucking (Continued)

MINITAB

```
Analysis of Variance

SOURCE        DF          SS          MS        F        p
Regression     4     31895.6      7973.9    59.82    0.000
Error         15      1999.4       133.3
Total         19     33895.0

SOURCE        DF      SEQ SS
TIME           1     17743.2
SPRING         1        29.0
SUMMER         1     12383.2
FALL           1      1740.3

MTB > name c7 'YHAT'
MTB > print c7

YHAT
    66.250   107.250   150.250   107.250    85.125   126.125   169.125
   126.125   104.000   145.000   188.000   145.000   122.875   163.875
   206.875   163.875   141.750   182.750   225.750   182.750
```
← Predicted values

This model is sometimes referred to as the *multiplicative model* since all the components are multiplied. This name also helps to distinguish the model from the **additive time series model,** which follows.

Additive Time Series Model

$$Actual\ value = T + S + C + I \tag{14.7}$$

When we use the regression Equation (14.6) to forecast a future value of a time series, we are using a short form of this additive model. That is, use of the dummy variables to represent the seasons always results in a forecast of the form

$$\hat{Y} = T + S_j \tag{14.8}$$

where S_j is the amount to be *added* to the trend for the jth season. An even better forecast could be obtained if we had some measure of the likely cyclical and irregular variations that could be added to Equation (14.8) to predict the time series value more accurately.

One may wonder which of the two methods of representing time series values is more appropriate.

$$Actual\ value = T \times S \times C \times I$$

or

$$Actual\ value = T + S + C + I$$

The answer, of course, depends on the nature of the time series involved. On the whole, however, the first method seems to be preferred by most time series analysts.

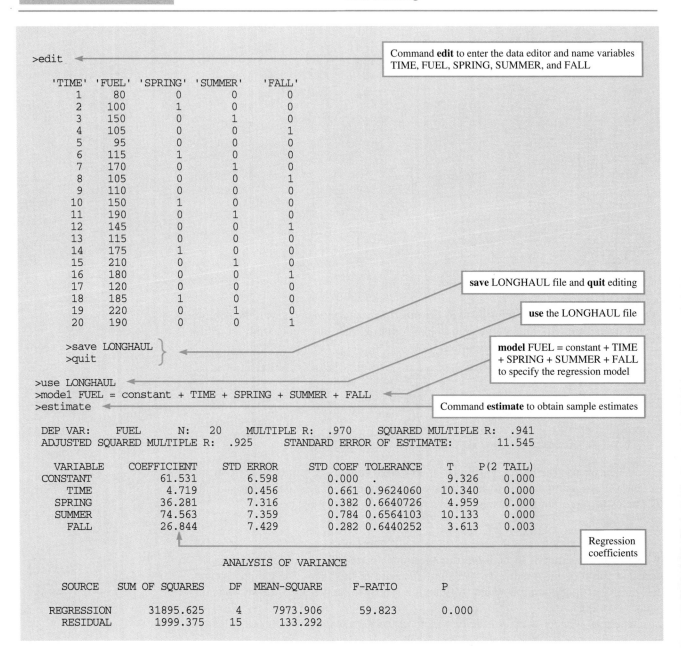

| MYSTAT | COMPUTER EXHIBIT 14.2B Multiple Regression for Seasonal Adjustments at Long-Haul Trucking |

```
>edit  ◄──────────────────────────          Command edit to enter the data editor and name variables
                                            TIME, FUEL, SPRING, SUMMER, and FALL

    'TIME'  'FUEL'  'SPRING'  'SUMMER'  'FALL'
       1      80       0         0         0
       2     100       1         0         0
       3     150       0         1         0
       4     105       0         0         1
       5      95       0         0         0
       6     115       1         0         0
       7     170       0         1         0
       8     105       0         0         1
       9     110       0         0         0
      10     150       1         0         0
      11     190       0         1         0
      12     145       0         0         1
      13     115       0         0         0
      14     175       1         0         0
      15     210       0         1         0
      16     180       0         0         1
      17     120       0         0         0
      18     185       1         0         0
      19     220       0         1         0
      20     190       0         0         1
    >save LONGHAUL  ⎫
    >quit           ⎭
>use LONGHAUL  ◄─────────────────
>model FUEL = constant + TIME + SPRING + SUMMER + FALL  ◄──────
>estimate  ◄────────────────

DEP VAR:    FUEL     N:   20    MULTIPLE R:   .970    SQUARED MULTIPLE R:   .941
ADJUSTED SQUARED MULTIPLE R:   .925    STANDARD ERROR OF ESTIMATE:      11.545

    VARIABLE    COEFFICIENT   STD ERROR   STD COEF TOLERANCE     T     P(2 TAIL)
    CONSTANT        61.531       6.598      0.000   .           9.326    0.000
        TIME         4.719       0.456      0.661 0.9624060    10.340    0.000
      SPRING        36.281       7.316      0.382 0.6640726     4.959    0.000
      SUMMER        74.563       7.359      0.784 0.6564103    10.133    0.000
        FALL        26.844       7.429      0.282 0.6440252     3.613    0.003

                        ANALYSIS OF VARIANCE

    SOURCE    SUM OF SQUARES   DF   MEAN-SQUARE    F-RATIO      P

    REGRESSION    31895.625     4     7973.906     59.823     0.000
    RESIDUAL       1999.375    15      133.292
```

Command **edit** to enter the data editor and name variables TIME, FUEL, SPRING, SUMMER, and FALL

save LONGHAUL file and **quit** editing

use the LONGHAUL file

model FUEL = constant + TIME + SPRING + SUMMER + FALL to specify the regression model

Command **estimate** to obtain sample estimates

Regression coefficients

EXAMPLE 14.3   To understand the preference for multiplicative time series models, we can examine the forecast of fuel consumption at Long-Haul Trucking, Inc., produced in Practice Problem 14.7. That forecast consisted of the trend value $T = 61.5 + 4.7(24) = 174.3$ thousand gallons plus 26.8 thousand gallons for the fall seasonal figure. Thus, the total forecast was 201.1 thousand gallons. One might say that the seasonal variation for fall is $(26.8/174.3) \times$

$100 = 15.4\%$ above the trend. One might also say that the seasonal variation for fall is 26.8 thousand gallons above the trend. For this prediction, either statement is correct, and they both say the same thing and result in the same forecast.

But for a future fall forecast, say next year, the two views of seasonal variation (multiplicative and additive) give different forecasts. In that future year, the additive model requires that we still add 26.8 thousand gallons to the trend figure to obtain our forecast. The multiplicative approach would involve raising the forecast by a figure similar to the 15.4%. If there is a strong upward trend in a time series, the 15.4% will be quite a bit larger than 26.8 thousand gallons.

The basic difference between the multiplicative and additive models is that the first assumes that seasonal variations produce the same *percentage* change each time a season arrives, and the second assumes that the seasonal variations produce the same *amount* of change. Most analysts agree that typical time series behave in the former manner—seasonal variations seem to be constant percentages through the years rather than constant amounts. Thus, the multiplicative model is more generally used in forecasting except in short-term-forecasting situations or where the time series exhibits little trend. In these cases, the percentage change and the additive change are very nearly the same.

Problems: Section 14.7

16. Use the method described in this section to obtain seasonal figures for the National Appliances data given in Problem 12. [*Hint:* This method requires a least squares trend line. A multiple regression computer program must be applied to the data, using $t = 1$ in the first quarter five years ago and using three dummy variables. Let the first quarter be the base quarter.]

17. Compare the National Appliances data three years ago in Problem 12 with the other years. Give a one- or two-sentence discussion of how that year of data might affect the trend line constructed in Problem 16.

18. Use the method described in this section to find seasonal indexes for the Redrock County building permits data given in Problem 14. [*Hint:* A computer program that performs multiple regressions should be used along with 11 dummy variables. Let January be the base.]

14.8 CYCLICAL VARIATION

Cyclical variations are movements in a time series that recur over periods longer than a year. These variations have no particular patterns of the sort we see in seasonal variations. Since cycles differ from one another in their amplitudes and durations, one cannot reasonably develop indexes to measure their average strength or amplitude. One can, however, analyze historical data and identify cyclical variations. One method for doing this analysis is outlined here.

Throughout this analysis of cyclical variation we will represent the time series values by using the multiplicative model:

$$Actual\ value = T \times S \times C \times I$$

where the trend value T is measured in the same units as the time series, and S, C, and I are indexes of seasonal, cyclical, and irregular variation, respectively. If the data for the time series are annual, then there can be no seasonal variation in the time series and

$$Actual\ value = T \times C \times I \tag{14.9}$$

Generally, moving averages are considered to contain only trend and cyclical components. If the moving average is a *12-month moving average,* it does not contain seasonal variations and the irregular variations have been averaged out. Thus,

$$12\text{-}month\ moving\ average = T \times C \tag{14.10}$$

For quarterly data,

$$Four\text{-}quarter\ moving\ average = T \times C \tag{14.11}$$

A *12-month moving average* can be obtained for each month in a time series (except the first six and last six), or a *four-quarter moving average* can be obtained for each quarter (except the first two and last two), using the method illustrated in column 3 of Table 14.5. If a least squares trend equation has been constructed for the time series, then a trend value can also be obtained for each month or quarter. If the trend value is divided into the moving average, the resulting ratio is a measure of the cyclical component for that period. That is,

$$\frac{Moving\ average}{Trend} = \frac{T \times C}{T} = C \tag{14.12}$$

This ratio is usually expressed as a percentage. A value of $C = 100$ would indicate that there is no cyclical variation in the time series for that period. If this percentage is determined for each period in the time series, its movements can serve to indicate the cyclical variations of the series.

There is another method of arriving at a measure of cyclical variation. It involves dividing each value in the time series by the trend. If the actual time series data are annual, then

$$\frac{Actual\ value}{Trend\ value} = \frac{T \times C \times I}{T} = C \times I \tag{14.13}$$

If the actual data are monthly or quarterly, then the actual value can be divided by the trend value adjusted for seasonal variation. For instance, the trend value may be 60 for a particular month, and the seasonal index for that month may be 105. Then the trend value adjusted for seasonal variation would be 60×1.05, or 63. When the actual figures are divided by this trend adjusted for seasonal variation, the result is

$$\frac{Actual\ value}{Adjusted\ trend} = \frac{T \times S \times C \times I}{T \times S} = C \times I \tag{14.14}$$

The results in Equations (14.13) and (14.14) are both proportions that are usually converted to percentages. Both proportions contain irregular variation in addition to cyclical variation. A moving average of these ratios can be taken to average out the irregular variations. The final result, a set of percentages for each period in the time series, offers a measure of the cyclical variation in the series.

These percentage variations are not very useful in forecasting changes in time series values since cylces are by their nature somewhat irregular in duration and amplitude. However, the percentages can be used to identify turning points in cycles for historical time series. They can also be used to adjust historical data to remove cyclical components in much the same way as seasonal indexes are used to adjust historical data values to deseasonalize them. (Methods of deseasonalization are discussed in Section 14.9.)

 PRACTICE PROBLEM 14.8

In Section 14.6 data on sales of a lawn treatment product, XB–7, were used to develop seasonal indexes. The ratio-to-moving-average method yielded quarterly seasonal indexes of 84.2, 109.9, 94.8, and 112.0. The simple linear trend line for the ten years of data used in that example is

$$\hat{Y} = 278.9 + 6.46t$$

Use this trend line and the seasonal indexes to remove the trend and seasonal variation from the last year of sales data.

Solution   As a first step in the solution, we must use the trend line to obtain trend values for the last four quarters:

$$\hat{Y}_{37} = 278.9 + (6.46)(37) = 517.9$$

$$\hat{Y}_{38} = 278.9 + (6.46)(38) = 524.4$$

$$\hat{Y}_{39} = 278.9 + (6.46)(39) = 530.8$$

$$\hat{Y}_{40} = 278.9 + (6.46)(40) = 537.3$$

Next, we adjust these trend values for seasonal variation, using the seasonal indexes to obtain $T \times S$ for each quarter as follows:

$$(517.9)(.842) = 436.1$$

$$(524.4)(1.090) = 571.6$$

$$(530.8)(.948) = 503.2$$

$$(537.3)(1.120) = 601.8$$

Note that in these calculations the seasonal indexes were expressed as proportions, not as percentages.

The resulting $T \times S$ values contain trend and seasonal variation. If we divide the $T \times S$ figures into the *Actual values* $= T \times S \times C \times I$, we obtain just $C \times I = (T \times S \times C \times I)/(T \times S)$, which measures the cyclical and irregular components of the time series. Thus, the $C \times I$ values are as follows:

$$(408/436.1) \times 100 = 93.6$$

$$(490/571.6) \times 100 = 85.7$$

$$(740/503.2) \times 100 = 147.1$$

$$(649/601.8) \times 100 = 107.8$$

These four values contain only cyclical and irregular variation. Since the third quarter's value of 147.1 is so much larger than the others, it appears that there was some irregular influence in the sales during that period (perhaps a large promotion of the product), which may even have carried over into the fourth quarter. ▬▬

Problems: Section 14.8

19. The trend line for the National Appliances warranty data in Problem 12 is $\hat{Y} = 1.0 + .1t$. Use this trend line and the indexes found in the first part of Problem 13 to express the last two years of original data as $(C \times I) \times 100$. Discuss why the first method of finding seasonal indexes in Problem 13 may be appropriate.

20. For the enrollment decrease data in Problem 15 the trend line is $\hat{Y} = 3.50 + .77t$, where $t = 1$ in the fall quarter, two years ago. Use the seasonal indexes found in Problem 15a and this trend line to remove the trend and seasonal variation in the data and express the results as $(C \times I) \times 100$. Why would it be impossible for these data to show marked cyclical variation?

14.9
USING TIME SERIES RESULTS AND FORECASTING

In the previous sections of this chapter, we have discussed methods of analyzing historical time series data in order to separate the series into its components of trend, seasonal, cyclical, and irregular variation. Although the analysis of a time series to identify its components is interesting, the work involved can be justified only if the results can somehow be used. This section explains some of the ways in which results from time series analysis can be used in **adjusting** data and in **forecasting.**

One of the most common uses of seasonal indexes is *adjustment of actual data to account for seasonal variation.*

▬▬ **EXAMPLE 14.4**   The company that sells product XB–7, which was described in Practice Problems 14.5 and 14.6, might have a goal of selling 2350 tons of the product next year. This goal averages to 587.5 tons per quarter. If it sells 500 tons the first quarter, at first glance the firm seems to be below its target sales level. However, due to seasonal variations in sales, the first quarter is always slow. If we use the seasonal index found in Practice Problem 14.6 for the first quarter, we know from historical data that the first quarter represents about 84.2% of the average quarterly sales for the year. Taking this seasonal variation into account, we would expect sales of only .842 × 587.5, or 494.7 tons, in the first quarter.

In the same way, the .842 figure can be used to adjust the actual sales figure to a deseasonalized value. That is, if seasonal variation is taken into account, the actual sales figure of 500 tons is equivalent to an adjusted sales value of 500/ .842, or 593.8 tons. This figure can then be compared with the average quarterly sales target of 587.5 tons needed to attain the yearly goal. Either way the result is expressed, the company seems to be ahead of schedule in meeting its goal.▬▬

Most of the economic time series statistics we see in publications read by the general public are "seasonally adjusted." This term means that the actual figure has been **deseasonalized** by dividing it by the appropriate seasonal index expressed as a proportion rather than a percentage. The deseasonalized figures show trends and cyclical variations much more clearly. The irregular variations are still present, however, as can be seen in the following equation:

$$\frac{T \times S \times C \times I}{S} = T \times C \times I \qquad (14.15)$$

The results of time series analysis can be used in several different **forecasting** situations. Some of these are presented here in the form of problems and examples. Practice Problem 14.9 demonstrates how a yearly forecast can be broken down into forecasts for each season of the year by using seasonal indexes. Example 14.5 demonstrates how a forecast can be developed for a particular month or quarter by using a trend equation and a seasonal index. Practice Problem 14.10 shows the next period's (month or quarter) forecast can be obtained by using only this period's data and seasonal indexes.

 PRACTICE PROBLEM 14.9

The Specialty Building Products Company would like to develop a forecast of monthly sales for the coming year. An industry sales forecast financed by the industrywide trade association and Specialty's knowledge of its historical percentage of industry sales have enabled the company to determine that its annual sales for the coming year will be about $15 million. Use the monthly sales indexes in Table 14.8 to break down the annual forecast into monthly forecasts.

Solution Since the total of all the seasonal indexes is 1200, the monthly forecasts can be expressed as

$$\frac{Seasonal\ index}{1200} \times Annual\ forecast$$

Thus, the forecast figures are calculated as follows:

January: $(86/1200) \times \$15$ million $= \$1.075$ million
February: $(88/1200) \times \$15$ million $= \$1.100$ million

and so on. The entire set of 12 monthly forecasts is as follows:

January	$1.075 million	July	$ 1.488 million
February	1.100 million	August	1.325 million
March	1.150 million	September	1.250 million
April	1.250 million	October	1.225 million
May	1.300 million	November	1.212 million
June	1.425 million	December	1.200 million
		Total	$15.000 million

EXAMPLE 14.5 In January, the executive assistant to the governor of a state was given the task of forecasting state revenues from sales taxes during each quarter of the current year. For this task, the assistant used a quarterly trend

Month	Index	Month	Index	**TABLE 14.8**
January	86	July	119	**Monthly Sales Indexes**
February	88	August	106	**for Specialty Building**
March	92	September	100	
April	100	October	98	
May	104	November	97	
June	114	December	96	

equation and seasonal indexes supplied by the state treasurer's office. The trend equation and indexes had been developed by using sales tax collection data from the past 25 years. The data were adjusted to eliminate differences in the state sales tax rate over the 25-year period. The trend equation was

$$\hat{Y} = \$151 + \$3.1t$$

where all figures are in millions of dollars and $t = 1$ occurred in the first quarter of the first year in the 25-year period. Since the first quarter of the first year in the 25-year period was $t = 1$, the first quarter of the current year is $t = 101$, the second quarter is $t = 102$, and so forth. The quarterly sales tax collection *indexes* were as follows:

First quarter	80
Second quarter	95
Third quarter	104
Fourth quarter	121

The governor's assistant proceeded to develop the forecast by using the formula

$$Forecast = T \times \frac{S}{100} \qquad (14.16)$$

Recall that forecasts of cyclical and irregular variations cannot be developed from historical data by using techniques presented in this chapter since they do not follow predictable patterns. Thus, the projection was based solely on the trend and seasonal data. Any deviations from the forecast would then be due mainly to business cycle fluctuations and irregular variations.

Combining the forecasting Equation (14.16) with the trend equation, the assistant arrived at a general forecasting formula:

$$\hat{Y}_{Qtr} = (151 + 3.1t)\frac{S}{100}$$

The forecasted revenues by quarter are as follows:

$$\hat{Y}_1 = [151 + 3.1(101)](80/100) \ = 464.1 \times .80 = 371.3$$
$$\hat{Y}_2 = [151 + 3.1(102)](95/100) \ = 467.2 \times .95 = 443.8$$
$$\hat{Y}_3 = [151 + 3.1(103)](104/100) = 470.3 \times 1.04 = 489.1$$
$$\hat{Y}_4 = [151 + 3.1(104)](121/100) = 473.4 \times 1.21 = \underline{572.8}$$

$$\text{Total} \quad 1877.0$$

The sum of the quarterly forecasts gave an annual forecast of $1.877 billion in sales tax revenues.

The method of forecasting demonstrated in Example 14.5 consists of using the trend line and seasonal indexes. This method could be used to make projections to the year 2000 and beyond. Remember, however, that projections into the future can become very unreliable beyond one or two years. Projections five to ten years in the future can be made with confidence only for those time series proven to be highly stable and regular in the past.

 PRACTICE PROBLEM 14.10

The purchasing department of the Sky Kitchen Catering Service is planning its ordering for the current month. Sky Kitchen prepares meals for airlines to serve to passengers on flights departing from a city in the Midwest. It has found that the number of meals it is asked to prepare each month follows the same seasonal pattern as airline traffic. Thus the service can use the local airport's airline traffic seasonal indexes to help it project the number of meals it must prepare. Last month, it prepared 9232 meals. The purchasing agent would like to know how many meals to plan for this month.

Solution The last month was June, with a seasonal airline traffic index of 118; and the current month is July, with a seasonal index of 124. The projection for this month's meals can be found by using the following relationship:

$$\frac{Meals\ prepared\ in\ July}{Meals\ prepared\ in\ June} = \frac{Seasonal\ index\ for\ July}{Seasonal\ index\ for\ June}$$

That is, the ratio of actual meals prepared in July to those prepared in June should be about the same as the ratio of the airline traffic seasonal index for July to the index for June. All of the values in the formula are known but one—the meals to be prepared in July. We have

$$\frac{July\ meals}{9232} = \frac{124}{118}$$

or

$$July\ meals = 9232 \times (124/118) = 9701$$

The purchasing agent can plan for about 9700 meals.   ▬

The method of forecasting just described can be used to obtain only short-range forecasts. The number of meals the catering service should plan on serving in August could also be calculated by using the ratio of the August index to the June index. In fact, projections could be made through the end of the year. However, projections beyond 60 or 90 days using the method may be in error. A 60-day projection may be needed for August if the final figures on the number of meals served in July are not available on the first of August. This situation is common, where the last period's data are not available in time to use in the next period's forecast. The forecast must be based on data from two periods back. In Practice Problem 14.10 the August forecast might be

$$August\ meals = Meals\ prepared\ in\ June$$
$$\times \frac{Seasonal\ index\ for\ August}{Seasonal\ index\ for\ June}$$

But once again, the greater the time between the present and the forecast period, the less confidence we have in the forecast.

Problems: Section 14.9

21. Actual billings for the investment consulting firm of John & Stew Associates were $79,852 in March, and the March seasonal index for this firm's billings is 107. What is the seasonally adjusted March billing figure? What would be the expected annual billings based on the March figure?

22. The U.S. Department of Commerce reports seasonally adjusted personal income for the first quarter of the year to be $412.5 billion. If the quarterly seasonal index of personal income is 96, what was the actual personal-income figure for the quarter?

23. The accompanying time series represents the number of patients received in an Allied Healthcare, Inc., hospital emergency room. The seasonal indexes for each quarter are also given. Find the seasonally adjusted figures for the time series. Will these seasonal indexes tell the emergency room manager how much staffing and what supplies to order for each quarter?

	Quarter			
	First	*Second*	*Third*	*Fourth*
Patient Visits	16,440	12,295	10,667	13,379
Seasonal Index	118	75	90	117

24. If the manager of the Allied Healthcare emergency room in Problem 23 has forecasted 60,000 patient visits to the emergency room next year, how should he break down his forecast by quarter?

25. The Peterson Products Company has determined the seasonal indexes in the accompanying table for its expenditures on materials for production.

    a. Do these indexes vary from a base of 100? How can you tell?
    b. If the total annual expenditure for materials at Peterson Products next year is expected to be $2.4 million, find the figures that would be budgeted for each month.

Month	*Index*	*Month*	*Index*
January	50	July	90
February	90	August	130
March	140	September	170
April	65	October	140
May	105	November	90
June	90	December	40

26. The Peterson Products Company of Problem 25 spends $122,000 for materials in January and $197,500 in February next year.

    a. What would be their two seasonally adjusted figures?
    b. What would be their expected annual material expenditure on the basis of these first two months?

14.10
SUMMARY

In this chapter, we considered the notion of a time series of data. The methods in this chapter are important forecasting tools for managers. We examined the components of a time series—trend and seasonal, cyclical, and irregular variations—and the reasons for their existence. Moving average and exponential-smoothing methods that can be used to smooth large variations in a time series were covered in Section 14.3.

Multiplicative and additive models of time series were introduced in this chapter. Regression analysis was employed to measure and represent trend in a time se-

TABLE 14.9 Summary of Time Series Formulas

Description	Formula	
n-period moving average for time t	$$\overline{Y}_t = \frac{\sum(n \text{ consecutive values around } t)}{n}$$	
Exponential smoothing	$S_t = \alpha Y_t + (1 - \alpha)S_{t-1} \qquad 0 \le \alpha \le 1$	(14.1)
Multiplicative model	$Actual\ value = T \times S \times C \times I$	(14.3)
Additive model	$Actual\ value = T + S + C + I$	(14.7)
Linear trend line	$\hat{Y} = b_0 + b_1 t$	(14.5)
Quadratic trend line	$\hat{Y} = b_0 + b_1 t + b_2 t^2$	

ries. Various methods for measuring seasonal variation and a discussion of the analysis of cycles were given. A section of the chapter also dealt with how the components in a time series can be used for forecasting.

Table 14.9 summarizes some important methods covered in this chapter.

Time series analysis and forecasting are important topics in statistics. This chapter serves as an introduction to these topics. To cover these methods in depth would require a complete book. In this chapter, we have introduced the concept of an index number. In a subsequent chapter, we will extend the concept of index numbers by introducing price and quantity indexes.

Short answers are given in the margin; for complete answers and solutions, see the Instructor's Manual.

Answers to odd-number problems are given in the back of the text.

REVIEW PROBLEMS

27. Give an example of a time series that would show a large irregular variation (up or down) like the one in Figure 14.4. What is the cause of this variation?

28. The accompanying data show the percentage of new people hired each year in a company, broken down by the quarter in which the new employees actually joined the company.

   a. Construct a time series plot or line chart from these data.
   b. Does there seem to be any seasonal variation in the percentage of new people joining the company?
   c. There does not seem to be any trend in these data. However, the personnel director of the company says, "We are hiring more people every year." Why does the trend in hiring fail to show up?

	Quarter			
Year	First	Second	Third	Fourth
Three years ago	24.7	26.5	22.8	26.0
Two years ago	24.5	25.7	25.0	24.8
One year ago	26.6	24.0	26.6	22.8

29. When a 12-month average is found for a time series that begins in January of one year and ends in December of another year, which month has the following?

    a. The first moving average value.   **b.** The last moving average value.

30. What would an exponentially smoothed time series look like if the value $\alpha = 1.0$ were used?

31. The accompanying time series shows the number of firms in the advertising agency industry over a 25-year period.

    a. Find the five-year moving averages for this series.
    b. Find the exponentially smoothed averages for this series with $\alpha = .6$.
    c. Plot the two averages on the same graph.

Year	Firms	Year	Firms	Year	Firms
1	437	10	906	18	1071
2	467	11	941	19	1067
3	526	12	968	20	1049
4	683	13	981	21	1098
5	739	14	1011	22	1114
6	772	15	1051	23	1151
7	804	16	1056	24	1163
8	841	17	1063	25	1202
9	873				

32. Find the intercept and slope for a trend line fit to the advertising agency time series in Problem 31.

33. The Urban Transit Authority bus company has supplied the accompanying data showing the number of accidents involving its buses over the past five years.

    a. Use the method of Practice Problem 14.5 to find the four seasonal indexes for accidents.
    b. Use a multiple regression method to find the four seasonal figures. [*Hint:* Use winter of five years ago as $t = 1$, and let winter be the base quarter.]

	Season			
Year	Winter	Spring	Summer	Fall
Five years ago	15	10	9	12
Four years ago	15	14	11	15
Three years ago	22	21	12	19
Two years ago	45	22	14	20
Last year	32	25	18	24

34. Use a ratio-to-moving-average method to find the four seasonal indexes for the Urban Transit Authority data in Problem 33.

35. In Problem 33, note the value for winter two years ago. Does it appear that this value is out of line with the others owing to cyclical or irregular variation? What might have caused this unusually high value?

36. Sales of the Leasing Company have a trend line that is represented by the equation $\hat{Y}_t = \$55,000 + \$1250t$, where $t = 1$ in the first quarter of 1967. The four quarterly seasonal indexes of sales are 107, 100, 82, and 111.

    a. Approximately how much have sales been increasing each quarter in the past?
    b. What was the approximate level of sales in the first quarter of 1967?
    c. Forecast quarterly sales for the year 1999.

37. The dollar volume of small-business loans made by the Midstate Valley Bank and Trust Company has followed a trend line of $\hat{Y}_t = \$152,000 + \$950t$, where the time period involved is months and $t = 1$ in July of 1973. Since most of these small-business loans are used to finance inventories in the later half of the year, the seasonal indexes on these loans are larger in the second half than in the first half. The monthly seasonal indexes are as follows:

January	56	July	100
February	64	August	110
March	62	September	137
April	99	October	167
May	83	November	191
June	67	December	64
		Total	1200

   a. What was the approximate dollar volume of loans in July of 1973?
   b. Forecast the loan volume for March of the year 2008.
   c. Forecast the loan volume for May of 1996.
   d. Which of the two forecasts above is likely to be more accurate? Why?
   e. Forecast the monthly loan volumes for the year 1999.

38. Central Airlines generated 172,889 passenger miles last month. Its seasonal index for passenger miles for last month is 110, but the index for this month is only 92. Using these pieces of information, forecast this month's figure for number of passenger miles flown.

39. If Central (Problem 38) has a seasonal index of passenger miles of 98 for next month, what would be the forecast for next month's figure?

40. On February 1 the president of Midstate Valley Bank and Trust Company (Problem 37) received a report showing that the bank had loaned $238,560.00 to small businesses in January. Use this figure and the January and February seasonal indexes to forecast the loan volume for February.

41. The quarterly indexes for housing starts in a certain state are winter, 60; spring, 170; summer, 100; and fall, 70.

   a. If the state's Department of Housing expects 140,000 housing starts this year, how should it divide its annual forecast between the four quarters?
   b. Assume that during the winter of this year the state issued permits for 20,000 housing starts. What is the seasonally adjusted figure?
   c. On the basis of winter's 20,000 housing starts, how should the Department of Housing revise its annual forecast? Give a new numerical forecast for the year.

42. Assume the Department of Housing mentioned in Problem 41 developed a least squares regression equation relating housing starts Y to time t. That equation has the form

$$\hat{Y} = 10,000 + 500t + 12.5t^2$$

where the winter of 1973 is the quarter in which $t = 1$.

   a. What would the Department of Housing forecast for the spring of 1999? (Recall the note of caution about the accuracy of predictions using extrapolation, however.)
   b. Which of the following statements best describes the way in which housing starts have been changing over time?
   (i) They have been growing at a declining rate.
   (ii) They have been falling at an increasing rate.
   (iii) They have been growing at an increasing rate.
   (iv) They have been growing at a constant rate.
   (v) They have not been growing or declining.

Ratio Data Set   *Refer to the 141 companies listed in the Ratio Data Set in Appendix A.*

Questions   43. Locate the data for the current assets–sales (*CA/S*) ratio, the quick assets–sales ratio (*QA/S*), the current assets–current liabilities (*CA/CL*) ratio, and the net income–total assets (*NI/TA*) ratio. Are any of the variables in the ratio data set time series variables?

44. If a sample of data is taken from several elements at one point in time, then the data are said to be cross-sectional data. Are any of the variables in the Ratio Data Set cross-sectional variables?

45. Explain what would constitute a time series for the *NI/TA* variable in the Ratio Data Set.

46. Explain how a time series for the *NI/TA* variable in the Ratio Data Set would be related to a process.

Credit Data Set   *Refer to the 113 applicants for credit listed in the Credit Data Set in Appendix A.*

Questions   47. Are any of the variables listed in the Credit Data Set time series variables?

48. List some variables that the department store's credit office would probably track on a time series basis.

49. In which months would the seasonal indexes for "number of credit applications made" likely be high for this department store?

CASE 14.1	**ADJUSTING TIME SERIES DATA AT THE BUREAU OF LABOR STATISTICS**

Seasonal variation in a time series is often measured by using seasonal indexes. A ratio-to-moving-average method for obtaining seasonal indexes was discussed in the text in the seasonal indexes section. Seasonal adjustment of time series data is standard practice at the Bureau of Labor Statistics (BLS). For example, the BLS collects time series data for retail sales and employment in the United States by month.

An economic time series may be affected by regular intra-yearly (seasonal) movements that result from climatic conditions, model changeovers, vacation practices, holidays, and similar factors. Often such effects are large enough to mask the short-term, underlying movements of the series. If the effect of such intra-yearly repetitive movements can be isolated and removed, the evaluation of a series may be made more perceptive. For example, retail sales are seasonally adjusted to account for increased sales during holiday seasons and employment series are adjusted to account for decreased employment during winter months.

Seasonal movements are found in almost all economic time series. They may be regular, yet they do show variation from year to year and are subject to changes in pattern over time. Because these intra-early patterns are combined with the underlying growth or decline and cyclical movements of the series (trend-cycle) and also random irregularities, it is difficult to estimate the pattern with exactness.

The Bureau of Labor Statistics began work on seasonal factor methods in 1959. Prior to that time, when additional data became available and seasonal factors were generated from the lengthened series, the new factors sometimes differed markedly from the corresponding factors based on the shorter series. This difference could affect any portion of the series. It was difficult to accept a process by which the addition of recent information could affect significantly the seasonal factors for periods as much as 15 years earlier, especially since this meant that factors could never become final. The first BLS method, introduced in 1960, had two goals: First, to stabilize the seasonal factors for the earlier part of the series, and second, to minimize the revisions in the factors for the recent period.

Since 1960, the bureau has made numerous changes and improvements in its techniques and in methods of applying them. Thus far, all the changes have been made within the scope of the ratio-to-moving-average and difference-from-moving-average types of approaches. The BLS 1960 method, entitled "The BLS Seasonal Factor Method," was further refined, with the final version being introduced in 1966. It was in continuous use for many bureau series (especially employment series based on the establishment data) until 1980. In 1967 the Bureau of the Census introduced "The X–11 Variant of the Census Method II Seasonal Adjustment Program," better known as simply X–11. The X–11 provided some useful analytical measures along with many more options than the BLS method. Taking advantage of the X–11's additional flexibility, BLS began making increasing use of the X–11 method in the early 1970s, especially for seasonal adjustment of the labor force data based on the household survey.

Later, in the 1970s, Statistics Canada, the Canadian national statistical agency, developed an extension of the X–11 called "The X–11 ARIMA Seasonal Adjustment Method." The X–11 ARIMA provided the option of using ARIMA (autoregressive integrated moving average) modeling and forecasting techniques to extrapolate some extra data at the end of a time series to be seasonally adjusted. The extrapolated data help alleviate the effects of the inherent limitations of the moving average techniques at the ends of series. After extensive testing and research showed that use of X–11 ARIMA would help to further minimize revisions in factors for recent periods, BLS began using the X–11 ARIMA procedure in 1980 for most of its official seasonal adjustment.

The time series and seasonal indexes for the percentage rate of return on a short-term financial instrument are given in the following table:

	Quarter			
	First	*Second*	*Third*	*Fourth*
Rate of Return	6.52	7.30	9.25	10.04
Seasonal Index	122.9	85.9	124.9	66.3

The seasonally adjusted rates of return are

$$(6.52/122.9)(100) = 5.31$$

$$(7.30/85.9)(100) = 8.50$$

$$(9.25/124.9)(100) = 7.41$$

$$(10.04/66.3)(100) - 15.14$$

The seasonal adjustment shows an appreciable increase in the rate of return. Failure to account for the seasonal variations in rates of return make it more difficult to detect the increase in returns.

The percentage rate of return on a long-term bond calculated for four quarters were 9.95, 8.23, 8.54, and 8.31. Quarterly indexes for long-term bond returns were 97.9, 89.1, 104.1, and 108.9. Find the seasonally adjusted rates of return.

Case Assignment

REFERENCE: U.S. Department of Labor, Bureau of Labor Statistics (1988), *BLS Handbook of Methods,* Bulletin 2285 (Washington, D.C.: U.S. Government Printing Office).

REFERENCES

U.S. Department of Labor, Bureau of Labor Statistics. 1988. *BLS Handbook of Methods,* Bulletin 2285. Washington, D.C.: U.S. Government Printing Office.

Box, G. E. P., and G. Jenkins. 1976. *Time Series Analysis, Forecasting and Control,* 2nd ed. San Francisco: Holden-Day.

Brown, R. G. 1963. *Smoothing, Forecasting, and Prediction of Time Series.* Englewood Cliffs, N.J.: Prentice-Hall.

Durbin, J., and G. S. Watson 1951. "Testing for Serial Correlation in Least Squares Regression. II." *Biometrika,* 38: 159–78.

Hanke, J., and A. Reitsch. 1989. *Business Forecasting,* 3rd ed. Newton, Mass.: Allyn and Bacon.

Kmenta, J. 1986. *Elements of Econometrics,* 2nd ed. New York: Macmillan.

Montgomery, D., L. Johnson, and J. Gardiner. 1990. *Forecasting and Time Series Analysis,* 2nd ed. New York: McGraw-Hill.

Nelson, C. R. 1973. *Applied Time Series for Managerial Forecasting.* San Francisco: Holden-Day.

Neter, J., W. Wasserman, and M. Kutner. 1989. *Applied Linear Regression Models,* 2nd ed. Homewood, Ill.: Irwin.

SAS Institute, Inc. 1984. *SAS/ETS User's Guide,* Version 5 ed. Cary, N.C.: SAS Institute, Inc.

The simple and multiple regression models that we have examined all assumed that the observations were independent of one another. When time series data are used in regression analysis, the effect of the error or residual in one period often carries over to a subsequent period. When error terms in different periods, say, ϵ_t and ϵ_{t-1}, are correlated, they are said to be **autocorrelated** or **serially correlated.**

Autocorrelation between errors is similar to the sound effect of striking a piano key. The sound from striking a piano key is reverberant when the key is depressed, but the sound may linger and have an effect on a subsequent tone when another key is struck. The result is analogous to autocorrelation among the error terms wherein the random effect from one period persists to subsequent periods.

In business and economic regressions that involve time series data, ϵ_t, the error for period t, is often correlated with the error for the preceding period, ϵ_{t-1}. Correlation of the errors in successive periods is known as *first-order autocorrelation*. Thus, first-order autocorrelation is defined by the following equation:

$$\rho = \frac{\text{Covariance}(\epsilon_t, \epsilon_{t-1})}{\sqrt{\text{Variance}(\epsilon_t)\text{Variance}(\epsilon_{t-1})}}$$

When the assumption of independence between observations is violated in least squares estimation, the effects of autocorrelation are as follows:

1. The least squares estimators of the regression coefficients are unbiased, but they do not retain the minimum variance property.
2. The estimators of the variances using the least squares estimators, such as *MSE* and $s_{b_j}^2$, are biased estimators, and the confidence intervals and tests of significance that include the estimates are not rigorously applicable.

A test of the significance of first-order autocorrelation in a regression model, when least squares estimation is used with time series data, is the Durbin–Watson test. In typical business applications, the hypotheses for the Durbin–Watson test are

$$H_0: \rho \leq 0 \quad \text{and} \quad H_a: \rho > 0$$

The null hypothesis indicates that the errors from one period to the next do not exhibit positive autocorrelation, whereas the alternative hypothesis implies positive autocorrelation.

The Durbin–Watson test statistic d is given by

$$d = \frac{\sum\limits_{t=2}^{n}(e_t - e_{t-1})^2}{\sum\limits_{t-1}^{n}e_t^2}$$

where $e_t = Y_t - \hat{Y}_t$ is the residual for period t.

Values of d tend to be small when e_t and e_{t-1} take on similar values and therefore are correlated. Thus, small values of d are observed when the data support the alternate hypothesis and large values of d are observed when the data better support the null hypothesis. The rejection rule is as follows.

Rejection rule: If $d < d_L$, then reject H_0.

Here d_L is the lower Durbin–Watson bound. Values of d_L (and d_U) were tabulated by Durbin and Watson (1951) for different α values, different numbers of observations n, and different numbers of independent variables in the regression equation k. If $d > d_U$, then we do not reject H_0, and if $d_L \le d \le d_U$, then the test is inconclusive. The bounds d_L and d_U are referred to as Durbin–Watson test bounds. Table IX in the appendix presents the values of d_L and d_U. The Durbin–Watson test is appropriate for a regression model with fixed independent variables, an intercept, and time series data.

The **Durbin–Watson test for autocorrelation** can be summarized as follows.

Hypotheses and Rejection Rule for Durbin-Watson Test for Autocorrelation

H_0: $\rho \le 0$

H_a: $\rho > 0$

Rejection rule: If $d < d_L$, then reject H_0;

where: $d = \dfrac{\sum\limits_{t=2}^{n}(e_t - e_{t-1})^2}{\sum\limits_{t=1}^{n}e_t^2}$

$e_t = Y_t - \hat{Y}_t$

and d_L is the lower Durbin–Watson test bound for α from Table IX.

 PRACTICE PROBLEM 14.11

A financial analyst with First Metro Bancshares wishes to test the significance of the relationship between the yield on 90-day Treasury bills Y_t and the discount rate at the New York Federal Reserve Bank X_t. The yield and discount rate data were collected monthly for two years as percentages, and the results are presented in the accompanying listing. The analyst is interested in the effect of first-order autocorrelation on the variance esti-

	Year 1				Year 2		
Month	Period, t	Yield (%)	Discount (%)	Month	Period, t	Yield (%)	Discount (%)
1	1	7.76	8.00	1	13	7.04	7.50
2	2	8.22	8.00	2	14	7.03	7.50
3	3	8.57	8.00	3	15	6.59	7.10
4	4	8.00	8.00	4	16	6.06	6.83
5	5	7.56	7.81	5	17	6.12	6.50
6	6	7.01	7.50	6	18	6.21	6.50
7	7	7.05	7.50	7	19	5.84	6.16
8	8	7.18	7.50	8	20	5.57	5.82
9	9	7.08	7.50	9	21	5.19	5.50
10	10	7.17	7.50	10	22	5.18	5.50
11	11	7.20	7.50	11	23	5.35	5.50
12	12	7.07	7.50	12	24	5.49	5.50

SOURCE: U. S. Department of Labor, Bureau of Labor Statistics (1986), *Business Statistics, 1986* (Washington D. C.: U. S. Government Printing Office).

mates that are used to test the significance of the relationship between yield and the discount rate. We wish to use the Durbin–Watson d statistic to test the significance of a positive, first-order autocorrelation for the errors when Y_t is used in a linear regression with X_t. Use a significance level of .05.

Solution A data analysis software package gave the following regression results:

$$\hat{Y}_t = -.557 + 1.04 X_t$$

$$MSE = .073 \qquad s_{b_1} = .06283$$

The hypothesis test is as follows.

Step One $H_0: \rho \leq 0.$

Step Two $H_a: \rho > 0.$

Step Three $n = 24$, and the significance level is .05.

Step Four The Durbin–Watson test bounds with $n = 24$, the number of independent variables $k = 1$, and $\alpha = .05$, as determined from Table IX, are $d_L = 1.27$ and $d_U = 1.45$. Consequently, our rejection rule is as follows.

Rejection rule: If $d < 1.27$, then reject H_0.

Step Five The value of d is computed from the data as shown in Table 14.10, and the result is

$$d = \frac{\sum_{t=2}^{24}(e_t - e_{t-1})^2}{\sum_{t=1}^{24} e_t^2} = \frac{1.1164}{1.6100} = .69$$

TABLE 14.10	t	Yield, Y_t	Discount, X_t	$\hat{Y}_t$	$e_t = Y_t - \hat{Y}_t$	$e_t - e_{t-1}$	$(e_t - e_{t-1})^2$	e_t^2
Data for First Metro Banc-	1	7.76	8.00	7.761075	−.001075	—	—	.000001
shares and Computations	2	8.22	8.00	7.761075	.458925	.460000	.211600	.210612
for the Durbin–Watson	3	8.57	8.00	7.761075	.808924	.349999	.122500	.654358
Test for Autocorrelation	4	8.00	8.00	7.761075	.238925	−.570000	.324900	.057085
	5	7.56	7.81	7.563519	−.003519	−.242443	.058779	.000012
	6	7.01	7.50	7.241189	−.231188	−.227670	.051833	.053448
	7	7.05	7.50	7.241189	−.191188	.040000	.001600	.036553
	8	7.18	7.50	7.241189	−.061189	.130000	.016900	.003744
	9	7.08	7.50	7.241189	−.161189	−.100000	.010000	.025982
	10	7.17	7.50	7.241189	−.071188	.090000	.008100	.005068
	11	7.20	7.50	7.241189	−.041189	.030000	.000900	.001697
	12	7.07	7.50	7.241189	−.171188	−.130000	.016900	.029305
	13	7.04	7.50	7.241189	−.201189	−.030000	.000900	.040477
	14	7.03	7.50	7.241189	−.211188	−.010000	.000100	.044601
	15	6.59	7.10	6.825279	−.235279	−.024091	.000580	.055356
	16	6.06	6.83	6.544540	−.484540	−.249261	.062131	.234779
	17	6.12	6.50	6.201416	−.081416	.403125	.162510	.006629
	18	6.21	6.50	6.201416	.008584	.090000	.008100	.000074
	19	5.84	6.16	5.847892	−.007892	−.016477	.000271	.000062
	20	5.57	5.82	5.494370	.075630	.083522	.006976	.005720
	21	5.19	5.50	5.161642	.028358	−.047272	.002235	.000804
	22	5.18	5.50	5.161642	.018358	−.010000	.000100	.000337
	23	5.35	5.50	5.161642	.188358	.170000	.028900	.035479
	24	5.49	5.50	5.161642	.328358	.140000	.019600	.107819
					0.00000		1.1164	1.6100

$$\hat{Y}_t = -.557 + 1.04X_t \qquad MSE = .073 \qquad s_{b_1} = .06283$$

$$d = \frac{\sum\limits_{t=2}^{24}(e_t - e_{t-1})^2}{\sum\limits_{t=1}^{24}e_t^2} = \frac{1.1164}{1.6100} = .69$$

Step Six Since $d = .69$ is less than 1.27, we reject H_0. The data support the hypothesis that the time series has positive autocorrelation. Thus, the tests of significance for the regression are not strictly applicable.

The computation of the Durbin–Watson test statistic d is burdensome; hence, the test is almost always done with a computer. A computer was used to calculate the Durbin–Watson test statistic d and to plot the residuals against time for the First Metro Bancshares example, and the results are presented in Computer Exhibit 14.3. The plot depicts autocorrelation by showing that the residuals do not follow a random pattern over time.

Two corrective measures when the error terms exhibit autocorrelation are to include more independent variables in the equation or to transform the variables. Kmenta (1986) and Neter, Wasserman, and Kutner (1989) provide extended discussions of corrective measures for autocorrelated errors.

COMPUTER EXHIBIT 14.3 Durbin–Watson Test for Autocorrelation for First Metro Bancshares

```
MTB > set c1              ◄───────────────────────   Command set to enter the YIELD data into c1
DATA >  7.76   8.22   8.57   8.00   7.56   7.01
DATA >  7.05   7.18   7.08   7.17   7.20   7.07
DATA >  7.04   7.03   6.59   6.06   6.12   6.21
DATA >  5.84   5.57   5.19   5.18   5.35   5.49
DATA > end

MTD > set c2              ◄───────────────────────   Command set to enter the DISCOUNT data into c1
DATA >  8.00   8.00   8.00   8.00   7.81   7.50
DATA >  7.50   7.50   7.50   7.50   7.50   7.50
DATA >  7.50   7.50   7.10   6.83   6.50   6.50
DATA >  6.16   5.82   5.50   5.50   5.50   5.50
DATA > end

MTB > name c1 'YIELD'     }◄───────────────────   name c1 YIELD and name c2 DISCOUNT
MTB > name c2 'DISCOUNT'  }

MTB > regres c1 1 c2;     }      Command regres c1 1 c2; and subcommand residuals c3; and subcommand dw. for a regression
SUBC> residuals c3;       }◄──   of YIELD on DISCOUNT with residuals put into c3 and a Durbin–Watson test statistic
SUBC> dw.                 }

The regression equation is
YIELD = - 0.557 + 1.04 DISCOUNT

Predictor       Coef      Stdev     t-ratio         p
Constant     -0.5571     0.4438       -1.26     0.223
DISCOUNT      1.03977    0.06283      16.55     0.000

s = 0.2705      R-sq = 92.6%      R-sq(adj) = 92.2%

Analysis of Variance

SOURCE       DF        SS          MS         F        p
Regression    1    20.045      20.045    273.91    0.000
Error        22     1.610       0.073
Total        23    21.655
```

(continues)

Problems: Section 14.11

50. A financial analyst with Second Metro Bancshares wishes to test the significance of the relationship between the yields on 90-day Treasury bills Y_t and the prime rate of interest charged by banks on short-term business loans X_t. The yield and prime rate data were collected monthly for two years, and the results are presented in the accompanying listing. The analyst is interested in the effect of autocorrelation on the variance estimates that are used to test the significance of the relationship between yields and the prime rate. Use the Durbin–Watson d statistic to test the significance of a positive, first-order autocorrelation for the errors when Y_t is used in a linear regression with X_t. Use a significance level of .05.

MINITAB

COMPUTER EXHIBIT 14.3 Durbin–Watson Test for Autocorrelation for First Metro Bancshares (Continued)

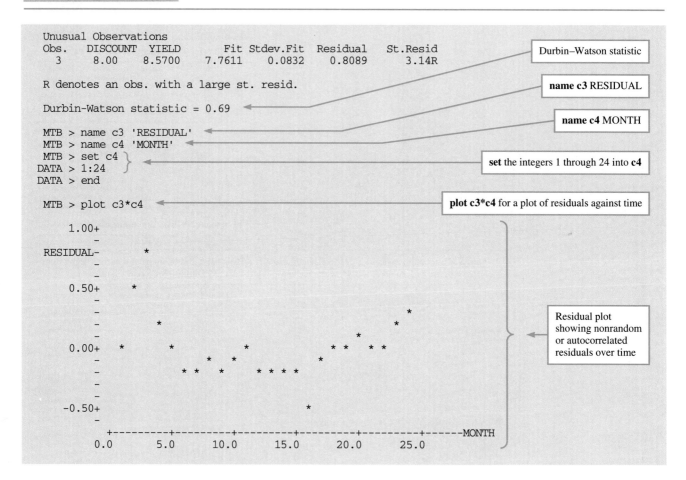

```
Unusual Observations
Obs.   DISCOUNT  YIELD      Fit Stdev.Fit  Residual    St.Resid         ┌─────────────────────┐
  3      8.00   8.5700   7.7611   0.0832    0.8089        3.14R          │ Durbin–Watson statistic │
                                                                        └─────────────────────┘
R denotes an obs. with a large st. resid.                               ┌─────────────────────┐
                                                                        │ name c3 RESIDUAL    │
Durbin-Watson statistic = 0.69                                          └─────────────────────┘
                                                                        ┌─────────────────────┐
MTB > name c3 'RESIDUAL'                                                │ name c4 MONTH       │
MTB > name c4 'MONTH'                                                   └─────────────────────┘
MTB > set c4  ⎫
DATA > 1:24   ⎬                                                  set the integers 1 through 24 into c4
DATA > end    ⎭

MTB > plot c3*c4                                          plot c3*c4 for a plot of residuals against time

     1.00+
         -
RESIDUAL-          *
         -
     0.50+      *
         -
         -
         -             *                                       *
         -                                                   *
     0.00+   *      *        *             *  *   *  *
         -                *     *                *
         -             *  *   *        *  *  *  *
         -
         -
    -0.50+                         *
         -
         +---------+---------+---------+---------+---------+------MONTH
        0.0       5.0      10.0      15.0      20.0      25.0
```

┌─────────────────────┐
│ Residual plot   │
│ showing nonrandom   │
│ or autocorrelated   │
│ residuals over time │
└─────────────────────┘

| | Year 1 | | | | Year 2 | | |
Month	t	Yield	Prime	Month	t	Yield	Prime
1	1	7.76	10.61	1	13	7.04	9.50
2	2	8.22	10.50	2	14	7.03	9.50
3	3	8.57	10.50	3	15	6.59	9.10
4	4	8.00	10.50	4	16	6.06	8.83
5	5	7.56	10.31	5	17	6.12	8.50
6	6	7.01	9.78	6	18	6.21	8.50
7	7	7.05	9.50	7	19	5.84	8.16
8	8	7.18	9.50	8	20	5.57	7.90
9	9	7.08	9.50	9	21	5.19	7.50
10	10	7.17	9.50	10	22	5.18	7.50
11	11	7.20	9.50	11	23	5.35	7.50
12	12	7.07	9.50	12	24	5.49	7.50

SOURCE: U.S. Department of Labor, Bureau of Labor Statistics (1986), *Business Statistics, 1986* (Washington, D.C.: U.S. Government Printing Office).

51. The following quarterly data show the number of appliances (in thousands) returned to National Appliances, Inc., for warranty service over the past five years. Let $t = 1$ be the first quarter five years ago. Test for a significant positive autocorrelation for the errors when Y_t is used in a simple linear regression with t. Use a significance level of .05.

Year	1st Quarter	2nd Quarter	3rd Quarter	4th Quarter
5 years ago	1.2	0.8	0.6	1.1
4 years ago	1.7	1.2	1.0	1.5
3 years ago	3.1	3.5	3.5	3.2
2 years ago	2.6	2.2	1.9	2.5
1 year ago	2.9	2.5	2.2	3.0

52. Reconsider the quarterly data given for Problem 51, which show the number of appliances (in thousands) returned to National Appliances, Inc., for warranty service over the past five years. For an additive seasonal regression model (use Y_t in a multiple regression with t and three indicator variables for the second, third, and fourth quarters), test the significance of a positive, first-order autocorrelation for the errors. Let $t = 1$ in the first quarter five years ago, and use a significance level of .05.

53. The accompanying time series shows the number of building permits issued by month for Redrock County. For an additive seasonal regression model (use Y_t in a regression with t and 11 indicator variables for the months of February through December), test the significance of a positive, first-order autocorrelation for the errors. Let $t = 1$ in January of the first year, and use a significance level of .05.

Month	Year 1	Year 2	Year 3
January	1015	664	1106
February	901	743	1022
March	1319	1147	1679
April	1590	1284	2011
May	1555	1250	1985
June	1473	1352	1938
July	1252	1408	1943
August	1249	1287	2045
September	1293	1309	1738
October	1234	1409	1797
November	946	1269	1722
December	841	1214	1496

54. Test for a significant positive autocorrelation for the errors in the trend line for Problem 31. Use a significance level of .05.

55. Test for a significant positive autocorrelation for the regression errors of Problem 33 part b. Use a significance level of .05.

Time series of n successive observations $Y_1, Y_2, \ldots, Y_n$, are encountered often in management, business, and economics. For example, quarterly employment and monthly sales time series are often examined by business analysts. Time series generally cannot be predicted with certainty since each value is generated by an underlying process and the outcome for the process is uncertain. Consequently, the observed value of a time series in time t is associated with a random variable Y_t. A time series variable in one period Y_t is often associated with, or is often influenced by, a variable in a preceding period. The cor-

14.12
AUTOCORRELATION AND AUTOREGRESSIVE MODELS

relation between time series variables in successive periods, Y_{t-1} and Y_t, is known as **auto-correlation** at **lag** 1 because Y_{t-1} and Y_t differ by one period. A time series that is generated with the same mean, variance, and correlations at lag k over time is known as a **stationary** time series; that is, the time series is generated by a *stationary process*.

Regression models with Y_t as the dependent variable and one or more variables of lag k, Y_{t-k}, as independent variables are known as **autoregressive models.** Autoregressive models can be used to take advantage of dependencies between Y_t and Y_{t-k}, say, for purposes of forecasting. We discuss autocorrelation and autoregressive models for stationary time series in the following subsections.

14.12.1 Autocorrelation

To examine a time series for autocorrelation, we often begin with a time series plot, as illustrated in the next example.

EXAMPLE 14.7   Internal auditors at National Transportation, Inc. audit the company's accounting procedures. The auditors are now examining the accuracy of the accounts receivable ledger. Over the past 24 years, in sequence, the percentage error amounts in the accounts receivable ledger at National (from left to right, top row to bottom row) are provided in the following listing:

1.78971	1.50243	0.70456	−0.38646	−1.00450	−0.21128
0.60950	0.29844	−0.66711	−0.00756	0.10621	0.52230
1.47254	0.62074	−0.35636	−1.72158	−2.04567	−0.51280
0.10631	−0.36978	−0.09626	−0.05865	−0.15951	0.05144

A time series plot for the amount of error in the accounts receivable ledger is shown in Figure 14.7. The observed values of the time series tend to wander or meander upward and downward over time. Furthermore, error amounts tend to stay close

FIGURE 14.7

Time Series Plot of the Accounts Receivable Error Amounts at National Transportation

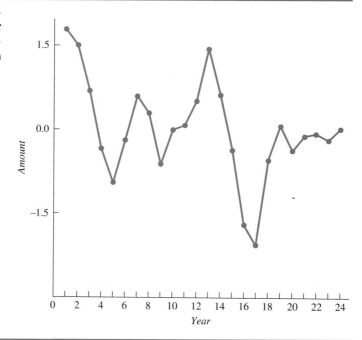

together over short periods of time—that is, positive error amounts tend to be followed closely by positive error amounts and negative error amounts tend to be followed closely by negative error amounts. Time series that exhibit these patterns suggest that the data are autocorrelated.

Correlation between time series variables that differ by k periods, Y_{t-k} and Y_t, for time series that are stationary is known as **autocorrelation** at **lag k**. Autocorrelation at lag k for stationary series is measured by the **autocorrelation coefficient ρ_k,** which is defined as follows.

Definition

The **autocorrelation coefficient ρ_k** measures the degree and direction of linear association between the stationary time series random variables Y_{t-k} and Y_t separated by constant *lag k*. The autocorrelation coefficient is defined by the following equation:

$$\rho_k = \frac{\text{Covariance}(Y_t, Y_{t-k})}{\sqrt{\text{Variance}(Y_t)\,\text{Variance}(Y_{t-k})}}$$

The variables Y_{t-k} and Y_t are said to be *autocorrelated* because the time series is correlated with *itself* at lag k time periods.

Figure 14.8(a) allows us to visualize distributions for time series variables Y_{t-k} and Y_t where, say, $k = 1$. The surface plotted in the figure is the joint distribution for Y_{t-1} and Y_t.

Autocorrelation for a Time Series   **FIGURE 14.8**

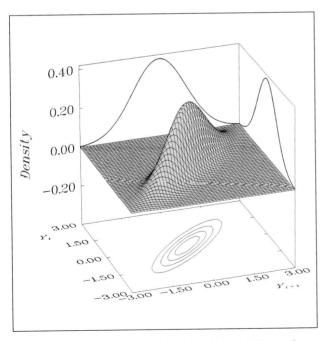

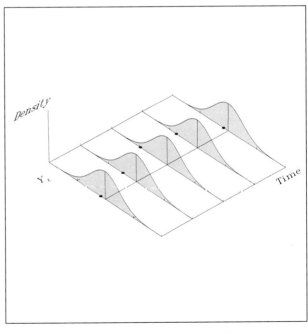

(a) Autocorrelation depicted by the surface and ellipses of the joint distribution for Y_{t-1} and Y_t

(b) Autocorrelation depicted by the meandering pattern of observed time series values when Y_t is stationary

In this case, the joint distribution is a bivariate normal distribution. The variables Y_{t-1} and Y_t are autocorrelated, as depicted by the surface and by the ellipses projected on the (Y_{t-1}, Y_t) plane.

The (marginal) distributions for Y_{t-1} and Y_t shown in Figure 14.8(a) have the same mean, the same standard deviation, and the same functional form. The common mean is $\mu = 0$, the standard deviation is $\sigma = 1$, and the probability distributions are both normal.

Another way to visualize a stationary time series with *autocorrelation* is illustrated in Figure 14.8(b). In this case, realized or observed time series values are plotted against time as points or dots; furthermore, identical distributions are displayed for Y_t across time. The observed values of the time series tend to wander or meander first upward and then downward over time. When a time series value is observed above the mean, the subsequent value also tends to be above the mean owing to autocorrelation. If any values were below the mean then subsequent values would also tend to be below the mean. Consequently, the time series exhibits positive autocorrelation.

The surface for the joint distribution of Y_{t-1} and Y_t and the ellipses in Figure 14.8(a) depict a statistical model of autocorrelation; whereas, Figure 14.8(b) shows identical distributions and observed time series values for a series that is autocorrelated.

The autocorrelation of a stationary time series at lag k, ρ_k, is estimated by the **sample autocorrelation coefficient at lag k, r_k,** as follows.

Definition and Equation for the Sample Autocorrelation Coefficient at lag k

The **sample autocorrelation coefficient at lag k, r_k,** is a measure of the direction and degree of linear association between an observed time series and the time series lagged by k periods. The sample autocorrelation coefficient can be computed according to the following equation:

$$r_k = \frac{SCP_{Y_t, Y_{t-k}}}{SSY_t} = \frac{\sum_{t=k+1}^{n} (Y_t - \overline{Y})(Y_{t-k} - \overline{Y})}{\sum_{t=1}^{n}(Y_t - \overline{Y})^2}$$

where: $\overline{Y} = (1/n)\sum_{t=1}^{n} Y_t$

The sample autocorrelations r_k for lags $k = 1, 2, \ldots$ are referred to as the **sample autocorrelation function (ACF).** The computation for a sample autocorrelation coefficient at lag 1, r_1, is illustrated in the following example.

EXAMPLE 14.8   The percentage error amounts in the accounts receivable ledger over the past 24 years at National Transportation were given in the previous example. In this example, we find the lag 1 time series, or the Y_{t-1} series. Next, we construct a scatter diagram for Y_{t-1} and Y_t. The scatter diagram helps us see the linear association between Y_{t-1} and Y_t for the sample data. Finally, we compute the lag 1 sample autocorrelation coefficient r_1.

The accounts receivable ledger error amounts at National Transportation over 24 years are displayed in Table 14.11. The series lagged by one year, Y_{t-1}, is simply the

Year, t	Error Amount, Y_t	Lag 1 Year, Y_{t-1}	$(Y_t - \overline{Y})$	$(Y_{t-1} - \overline{Y})$	$(Y_t - \overline{Y}) \times (Y_{t-1} - \overline{Y})$	$(Y_t - \overline{Y})^2$	**TABLE 14.11**
1	1.78971	—	1.78193	—	—	3.17527	**Computation of the**
2	1.50243	1.78971	1.49466	1.78193	2.66337	2.23400	**Sample Autocorrelation**
3	0.70456	1.50243	0.69678	1.49466	1.04145	0.48551	**Coefficient at Lag 1, r_1, for**
4	−0.38646	0.70456	−0.39424	0.69678	−0.27470	0.15543	**Accounts Receivable**
5	−1.00450	−0.38646	−1.01228	−0.39424	0.39908	1.02471	**Ledger Errors at National**
6	−0.21128	−1.00450	−0.21906	−1.01228	0.22175	0.04799	**Transportation**
7	0.60950	−0.21128	0.60172	−0.21906	−0.13181	0.36207	
8	0.29844	0.60950	0.29066	0.60172	0.17490	0.08448	
9	−0.66711	0.29844	−0.67489	0.29066	−0.19617	0.45548	
10	−0.00756	−0.66711	−0.01533	−0.67489	0.01035	0.00024	
11	0.10621	−0.00756	0.09843	−0.01533	−0.00151	0.00969	
12	0.52230	0.10621	0.51452	0.09843	0.05064	0.26473	
13	1.47254	0.52230	1.46476	0.51452	0.75365	2.14552	
14	0.62074	1.47254	0.61296	1.46476	0.89784	0.37572	
15	−0.35636	0.62074	−0.36413	0.61296	−0.22320	0.13259	
16	−1.72158	−0.35636	−1.72935	−0.36413	0.62971	2.99067	
17	−2.04567	−1.72158	−2.05345	−1.72935	3.55114	4.21666	
18	−0.51280	−2.04567	−0.52057	−2.05345	1.06897	0.27100	
19	0.10631	−0.51280	0.09854	−0.52057	−0.05130	0.00971	
20	−0.36978	0.10631	−0.37756	0.09854	−0.03720	0.14255	
21	−0.09626	−0.36978	−0.10404	−0.37756	0.03928	0.01082	
22	−0.05865	−0.09626	−0.06642	−0.10404	0.00691	0.00441	
23	−0.15951	−0.05865	−0.16728	−0.06642	0.01111	0.02798	
24	0.05144	−0.15951	0.04367	−0.16728	−0.00730	0.00191	
	0.18667				10.5970	18.6291	

$$\overline{Y} = (1/n) \sum_{t=1}^{n} Y_t = (1/24) \sum_{t=1}^{24} Y_t = (1/24)(.18667) = .0077778$$

$$r_1 = \frac{SCP_{Y_t, Y_{t-1}}}{SSY_t} = \frac{\sum_{t=2}^{24}(Y_t - \overline{Y})(Y_{t-1} - \overline{Y})}{\sum_{t=1}^{24}(Y_t - \overline{Y})^2} = \frac{10.5970}{18.6291} = .5688$$

column of data for Y_t shifted down by one year. Notice that lagging a series by one period results in a *missing* value at the beginning of the Y_{t-1} series. Shifting the first value Y_1 down one year during the lagging procedure creates the missing value for Y_{t-1} at $t = 1$; we also omit one value at the end of the series. Lagging a series by k periods creates k missing values at the beginning of the series since Y_1 is shifted by k periods.

A scatter diagram of Y_{t-1} and Y_t for the ledger error amounts is shown in Figure 14.9. The pattern of data in the scatter diagram suggests that the time series has a positive autocorrelation at lag 1.

We compute the value of r_1 in Table 14.11 and the result is

$$r_1 = \frac{SCP_{Y_t, Y_{t-1}}}{SSY_t} = \frac{\sum_{t=2}^{24}(Y_t - \overline{Y})(Y_{t-1} - \overline{Y})}{\sum_{t=1}^{24}(Y_t - \overline{Y})^2} = \frac{10.5970}{18.6291} = .5688$$

FIGURE 14.9

Scatter Diagram Suggest-
ing Positive Sample Auto-
correlation for Lag 1 Error
Amounts Y_{t-1} and Error
Amounts Y_t at National
Transportation

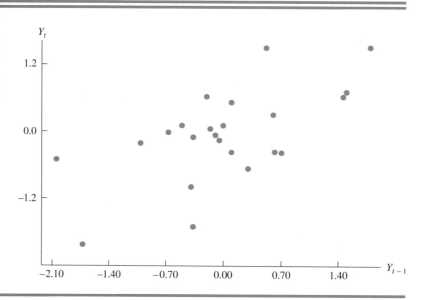

The sample autocorrelation at lag 1 indicates that the time series is associated with itself at lag 1. Attempts to predict or forecast future values of the time series should try to take advantage of the dependence between the successive values of the series. We also found the sample autocorrelations at lags 2, 3, 4, and 5 and the r_k values at lags 1 through 5 are

$$r_1 = .568839 \qquad r_2 = -.031487 \qquad r_3 = -.328528$$

$$r_4 = -.287153 \qquad r_5 = -.004555$$

14.12.2 Autoregressive Models

To predict or forecast future values of time series that are autocorrelated, we endeavor to take advantage of the dependence between the successive time series values as we predict future values of Y_t. A regression model with Y_t as the dependent variable and with lag 1 values of the series Y_{t-1} as the independent variable is known as a **first-order autoregressive AR(1) model,** as given next.

First-Order Autoregressive AR(1) model

$$Y_t = \beta_0 + \beta_1 Y_{t-1} + \epsilon_t$$

where: Y_t is the dependent variable
Y_{t-1} is the independent variable
β_0 and β_1 are parameters that specify the dependence of Y_t on Y_{t-1}
ϵ_t is error in time t that is independent of $Y_{t-1}, Y_{t-2}, \ldots$

The first order autoregressive model can be extended to an *autoregressive model of order p,* an AR(p) model, as follows:

$$Y_t = \beta_0 + \beta_1 Y_{t-1} + \beta_2 Y_{t-2} + \cdots + \beta_p Y_{t-p} + \epsilon_t$$

One way we can estimate the parameters of autoregressive models is by using the now familiar least squares method with the observed data values from the time series. The estimated first-order autoregressive model is

$$\hat{Y}_t = b_0 + b_1 Y_{t-1}$$

and is usually obtained from a computer.

To predict or forecast one period beyond period n with an AR(1) model, the forecasted value $\hat{Y}_{n+1}$ is

$$\hat{Y}_{n+1} = b_0 + b_1 Y_n$$

and the forecasted value j periods beyond period n, $\hat{Y}_{n+j}$, is

$$\hat{Y}_{n+j} = b_0 + b_1 \hat{Y}_{n+j-1} \qquad j \geq 2$$

To predict or forecast j periods beyond period n with an AR(p) model, the forecasted value $\hat{Y}_{n+j}$ is

$$\hat{Y}_{n+j} = b_0 + b_1 \hat{Y}_{n+j-1} + b_2 \hat{Y}_{n+j-2} + \cdots + b_p \hat{Y}_{n+j-p}$$

where $\hat{Y}_{n+m}$ is the forecast of Y_{n+m} when $m \geq 1$, and $\hat{Y}_{n+m}$ is the observed value Y_{n+m} when $m \leq 0$.

EXAMPLE 14.9 The percentage error amounts in the accounts receivable ledger over the past 24 years at National Transportation were given in the data listing in Example 14.7. We previously found that the autocorrelation at lag 1 for the time series of the error amounts was $r_1 - .5688$, suggesting that an AR(1) model may be appropriate for forecasting purposes. Thus, to forecast the amount of error in the accounts receivable ledger in the year following the end of the series ($t = 25$), we first use National's time series error amounts to estimate a first order autoregressive AR(1) model. We then use the estimated model to obtain the forecast.

A computer was used to find and plot the sample autocorrelation function (ACF) and to estimate the parameters for the AR(1) model for the error amounts and the results are presented in Computer Exhibit 14.4. Notice that the sample autocorrelation function gives the same autocorrelations that we found previously. The estimated AR(1) model is

$$\hat{Y}_t = -.073 + .569 Y_{t-1}$$

We now compute the forecasted amount of error in accounts receivable in the year following the end of the series. The forecasted value with $j = 1$ and $n = 24$ at time $t = n + j = 24 + 1 = 25$, is

$$\hat{Y}_{n+j} = \hat{Y}_{24+1} = \hat{Y}_{25}$$
$$= -.073 + .569 Y_{n+j-1} = -.073 + .569 Y_{24+1-1}$$
$$= -.073 + .569 Y_{24} = -.073 + .569(0.05144) = -.0437$$

The forecasted percentage amount of error ($-.0437$) can now be applied to National's typical ledger amounts to find the predicted error in the ledger. For example, if National typically has \$100,000,000 in accounts receivables, the predicted error in dollars is $(-.0437/100)(\$100,000,000) = \$43,700$. Internal auditors can use the

MINITAB

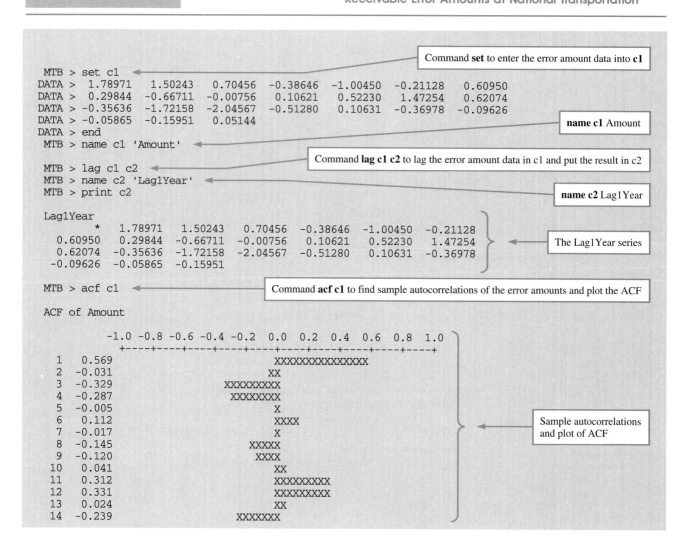

forecasted amount of error in the accounts receivables ledger to decide whether the accounting procedures need to be changed.

Autoregressive models are included as a special case of a broader class of time series models known as **autoregressive integrated moving average (ARIMA) models.** The ARIMA models include models that can be represented as follows:

$$Y_t = \underbrace{\beta_0}_{\text{Constant}} + \underbrace{\beta_1 Y_{t-1} + \beta_2 Y_{t-2} + \cdots + \beta_p Y_{t-p}}_{\text{Autoregressive terms}}$$

$$+ \underbrace{\epsilon_t}_{\text{Error}} \underbrace{- \theta_1 \epsilon_{t-1} - \theta_2 \epsilon_{t-2} - \cdots - \theta_q \epsilon_{t-q}}_{\text{Moving average terms}}$$

where ϵ_t are independent.

COMPUTER EXHIBIT 14.4 Sample Autocorrelations, Sample Autocorrelation Function (ACF), First-Order Autoregressive AR(1) Model Estimates, and ARIMA (1, 0, 0) Estimates for Accounts Receivable Error Amounts at National Transportation (Continued)

MINITAB

```
MTB > regres c1 1 c2
```
← Command **regres c1 1 c2** for a regression of the error amount on Lag1Year

```
The regression equation is
Amount = - 0.073 + 0.569 Lag1Year
```
← Regression equation estimates for the AR(1) model

```
23 cases used 1 cases contain missing values

Predictor      Coef       Stdev    t-ratio       p
Constant     -0.0730      0.1387     -0.53     0.604
Lag1Year      0.5687      0.1541      3.69     0.001

s = 0.6652      R-sq = 39.3%      R-sq(adj) = 36.4%

Analysis of Variance
```
← Command **arima 1 0 0 c1** to compute estimates for an ARIMA(1,0,0) model of the error amounts in c1

```
SOURCE       DF         SS          MS         F        p
Regression    1      6.0248      6.0248     13.62    0.001
Error        21      9.2910      0.4424
Total        22     15.3158

Unusual Observations
Obs.Lag1Year      Amount       Fit  Stdev.Fit   Residual   St.Resid
 16    -0.36      -1.722     -0.276     0.150     -1.446      -2.23R
 18    -2.05      -0.513     -1.236     0.345      0.724       1.27 X

R denotes an obs. with a large st. resid.
X denotes an obs. whose X value gives it large influence.

MTB > arima 1 0 0 c1

Estimates at each iteration
Iteration          SSE    Parameters
      0         16.8276    0.100    0.097
      1         14.3347    0.250    0.068
      2         12.6058    0.400    0.044
      3         11.6127    0.550    0.029
      4         11.3615    0.633    0.031
      5         11.3181    0.663    0.039
      6         11.3089    0.677    0.043
      7         11.3068    0.684    0.045
      8         11.3063    0.687    0.046
      9         11.3062    0.688    0.046
     10         11.3062    0.689    0.047
     11         11.3062    0.689    0.047
     12         11.3062    0.690    0.047
Unable to reduce sum of squares any further

Final Estimates of Parameters
Type       Estimate    St. Dev.   t-ratio
AR   1       0.6896      0.1548      4.46
Constant     0.0466      0.1426      0.33
Mean         0.1503      0.4596
```
← Estimates for the ARIMA(1,0,0) model

```
No. of obs.:  24
Residuals:    SS = 10.6359  (backforecasts excluded)
              MS =  0.4834  DF = 22
```

Note: ACF = auto correlation function

The ARIMA models that we consider are referred to as ARIMA(p, d, q) models. The notation p refers to the number of autoregressive terms in the model. The ARIMA model that we have given is for stationary time series, which we noted are the only time series we consider in this section. If a time series is not stationary, then *changes* in the successive time series values or the first *differences,* $Y_t - Y_{t-1}$, are often stationary. If the first differences are not stationary, then the differences of the differences, or the second differences, or the differences of logarithms of the series values are often stationary. The notation d refers to the degree of differencing specified to obtain a stationary series. The notation q refers to the number of lagged error terms that make up the *moving average* part of the model.

ARIMA models are often estimated by using an iterative nonlinear least squares estimation procedure. Data analysis software can provide us with the estimates and interested readers are referred to the references at the end of the chapter for a complete discussion of the estimation procedure.

═══ EXAMPLE 14.10   The percentage error amounts in the accounts receivable ledger over the past 24 years at National Transportation were given in the data listing in Example 14.7. The first order ($p = 1$) autoregressive model for the data was estimated by the method of least squares in Example 14.9. The error amounts appear to be stationary, so $d = 0$, and we do not postulate any moving average terms so $q = 0$. Thus, we estimate an ARIMA($1, 0, 0$) model.

To compare the ARIMA($1, 0, 0$) estimates and the AR(1) least squares estimates for the ledger error data, we used a computer to find estimates for the ARIMA($1, 0, 0$) model and the results are presented in Computer Exhibit 14.4.

The estimated ARIMA($1, 0, 0$) model is

$$\hat{Y}_t = .0466 + .6896Y_{t-1}$$

The ARIMA($1, 0, 0$) and AR(1) estimated models are similar and provide similar forecasts for the ledger error amounts. The estimates differ somewhat owing to the differences in the estimation procedures.   **══**

We have compared the results of our estimates of an AR(1) model and an ARIMA($1, 0, 0$) model simply to introduce the general class of autoregressive integrated moving average models. A complete explanation of ARIMA models is beyond the scope of our discussion. However, we hope to develop parsimonious models with few autoregressive or moving average terms by identifying, estimating, and applying diagnostic checks to our postulated models. Readers interested in expanded discussions are referred to Box and Jenkins (1976), who are credited with the development of ARIMA models, or to the references at the end of the time series chapter.

Problems: Section 14.12

56. Internal auditors at National Transportation, Inc., audit the company's accounting procedures. The auditors are now examining the accuracy of the total assets account on the balance sheet. Over the past 24 years, in sequence, the percentage error amounts in the assets account at National (from left to right, top row to bottom row) are provided in the accompanying listing.

   a. Construct a time series plot for the data.
   b. Does the time series appear to be stationary?
   c. Does the time series appear to be autocorrelated?

−2.55	−1.54	−0.90	−1.82	−1.72	−3.38	−3.21	−2.53
−2.68	−1.05	−1.55	−1.27	−0.51	0.58	−1.01	−0.76
−0.38	−0.24	−0.16	1.36	2.76	1.84	1.81	0.69

57. Use the asset error time series provided in Problem 56.

    a. Find the lag 1 series Y_{t-1}.

    b. Construct a scatter diagram for Y_{t-1} and Y_t.

    c. Does the time series appear to be autocorrelated at lag 1?

58. For the asset error time series provided in Problem 56, find the sample autocorrelation coefficient at lag 1.

59. Use the asset error time series provided in Problem 56.

    a. Estimate a first-order autoregressive model for the data.

    b. Find the forecast for Y_t when $t = 25$.

Statistical Process Control

15

Process control is an important area of statistical practice in business, industry, engineering, and science. To continuously *improve* quality and productivity during the production of goods and services, managers must control and reduce variation in processes. We discuss concepts of managing processes and give a definition of *quality* in this chapter. The importance of quality system management standards (ISO 9000 standards) is introduced.

A *process* is defined, and the *statistical control* of variables associated with outputs of processes is discussed. We provide three-dimensional graphics that allow readers to visualize distributions for outputs of processes that are *in control* and processes that are *out of control*. We also provide graphics that allow readers to visualize the effects of *tampering* with a stable process.

We define *statistical process control* in this chapter and describe methods of statistical process control, which are employed to monitor characteristics of processes over time. We introduce *control charts,* such as $\overline{X}$-charts, R-charts, p-charts, and I-charts, that are used to monitor process means, process variation, proportions of nonconforming items, and changes in processes. Statistical process control is the basis for an effective quality management system.

15.1
QUALITY IMPROVEMENT

To compete effectively in the production of goods and services, managers are constantly attempting to improve quality and productivity. Although quality of a product or service has been variously defined from the perspective of managers, employees, products, services, customers, or investors, **quality** for an organization committed to a quality management program can be defined as follows.

Definition

Quality of a product, service, or organization is measured by how well the product, service, or organization meets or exceeds the expectations of its users or customers.

A product is more than a good provided by an organization. The concept of a quality product or a quality service must be extended to encompass the needs of suppliers, owners, employees, customers, and society. Quality exists when customers come back but products do not. Organizations committed to quality are now considering *customers as partners,* and they seek to enroll suppliers and customers as partners in the pursuit of quality.

EXAMPLE 15.1 A successful customer partnership was developed between Minitab, Inc., a developer of statistical software, and 3M, Inc. MINITAB product planners and representatives of 3M concluded that they needed to encourage the use of statistical software among intermittent users at 3M while sustaining its regular use among data analysts. To do so, 3M needed a product with a powerful *command language* for advanced users and a simplified *menu* for neophytes. The customer-as-partner process changed MINITAB's product development efforts and ultimately led to the development of MINITAB's user interface, which now includes both a command language and a menu system (Ryan, 1991).

REFERENCE: Ryan, B. (1991), "Minitab Embraces TQM, Improves Customer/Supplier Relations," *Keeping TAB Newsletter* (December): 2.

Many companies that are committed to quality improvement and to developing partnerships with both customers and suppliers are adopting **quality system management** programs. A set of international standards for quality systems, known as the *ISO 9000 series of standards,* are used for total quality management (TQM) purposes. Some of the ISO 9000 standards are for external quality assurance purposes and are used in contractual situations. Others contain guidance on factors affecting the quality of products and services. Companies can be audited and certified under the ISO 9000 series of standards. The quality certification can be used to assure potential customers of a company's commitment to quality. One advantage for quality system certification is that it is not necessary for potential customers who are entering contracts with certified companies to conduct their own separate quality audits. Furthermore, by adopting the ISO 9000 standards, companies can consistently provide high-quality products with less rework, fewer returns, and fewer customer problems. Consequently, they can increase productivity and profitability.

A philosophy of quality management that stresses continuous improvement of quality and productivity has been recommended by Deming (1986). The Deming quality improvement position is summarized in 14 points as follows: (1) Create constancy of purpose for improvement of product and service; (2) adopt the new philosophy of constantly attempting to improve quality; (3) cease dependence on inspection to achieve quality; (4) end the practice of awarding business on the basis of price tag alone, and instead, minimize total cost by working with a single supplier; (5) improve constantly and forever every process for planning, production, and service; (6) institute training on the job; (7) adopt and institute leadership; (8) drive out fear; (9) break down barriers between staff areas; (10) eliminate slogans, exhortations, and targets for the work force; (11) eliminate numerical quotas for the work force and numerical goals for management; (12) remove barriers that rob people of pride of workmanship, and eliminate the annual rating or merit system; (13) institute a vigorous program of education and self-improvement for everyone; and (14) put everybody in the company to work to accomplish the transformation. An in-depth elaboration on these points is provided by Deming (1986) and Gitlow et al. (1989).

15.2
PROCESS CONTROL

Goods and services are often produced by using continuing processes, such as those involved in the ongoing manufacture of memory chips by electronics companies or the provision of auditing services over time by accounting firms. A **process** is defined as follows.

> **Definition**
>
> A **process** is a set of activities that transforms inputs into outputs.

When a process transforms inputs into outputs the transformation usually changes the inputs or it adds value to the inputs. For example, the ongoing manufacture of memory chips at Digital Industries, Inc., converts raw material inputs such as silicon, oxide film, and doping chemicals into memory chips. The memory chips use thousands of transistors to store information in a digital format and are invaluable in management information systems. The process of auditing financial statements and reports by public accounting firms adds value by certifying that the information is in accordance with generally accepted accounting principles.

The outputs of processes have characteristics of interest, which are the variables associated with the outputs. For example, memory chips are made up of transistors that have conducting zones, and the electrical charge associated with a conducting zone is a variable that can be measured after the chip is produced. Generally, a variable that is associated with an output from a process depends on the outcome of the process, and the outcome cannot be predicted with certainty. Consequently, a variable associated with an output from a process, or an output variable, is a random variable whose numerical value depends on the outcome of the process.

In our discussion of random sampling, we noted that a random sample from a *process* is a collection of n random variables $X_1, X_2, \ldots, X_n$ that are generated from the same probability distribution and that are statistically independent of each other. In short, the X_i are independent and identically distributed. The same ideas give us the definition of a **process** that is in **statistical control** as follows.

> **Definition**
>
> A **process** is in **statistical control** if the random variables associated with the outputs from the process over time are generated from the same probability distribution and are statistically independent of one another.

FIGURE 15.1   Process Distributions

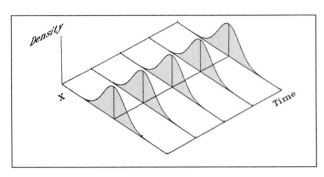

(a) Process in control

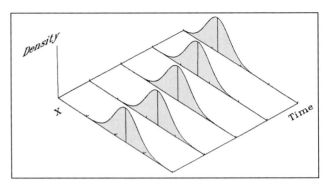

(b) Process out of control due to changing means

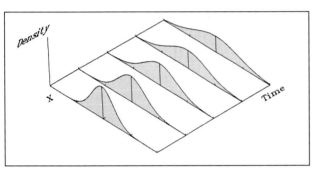

**(c) Process out of control due to changing
standard deviations**

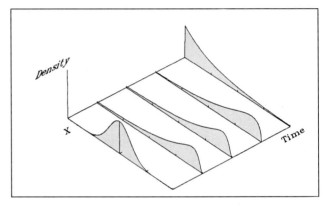

(d) Process out of control due to changing distributions

A process that is in statistical control is also referred to as a **stable process** or a **random process.**

EXAMPLE 15.2   We consider concepts of process control in this example by using visualization of probability distributions. The distributions shown in Figure 15.1(a) were generated by a process that was *in control*. The distributions had the same mean, standard deviation, and functional form, and the random variables were independent over time. In contrast, the distributions shown in Figure 15.1(b) were generated by a process that was *out of control* due to an increasing mean. Processes could also be out of control due to means that are decreasing or moving in a cyclical or seesaw pattern. The distributions shown in Figure 15.1(c) were generated by a process that was *out of control* due to increasing process variation. The distributions shown in Figure 15.1(d) were generated by a process that was *out of control* due to changing functional forms. The distribution was symmetrical, then it became positively skewed, finally it was negatively skewed.

The distribution shown in Figure 15.1(e) was generated by a process that was *in control*. The surface in the figure is the bivariate normal distribution for output variables X_1 and X_2. The surface and the contours [projected on the

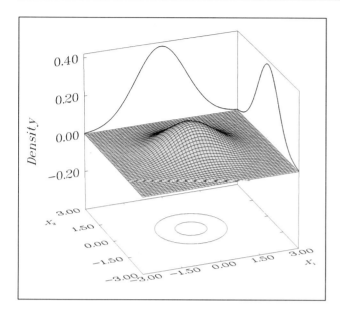

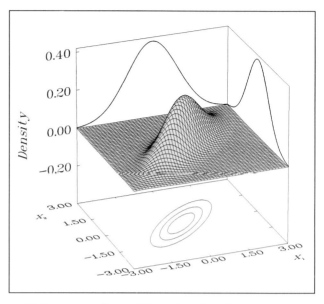

(e) **Process in control (Output variables at X_1 and X_2 are independent and have the same mean, standard deviation, and functional form.)**

(f) **Process out of control due to dependence between output variables over time**

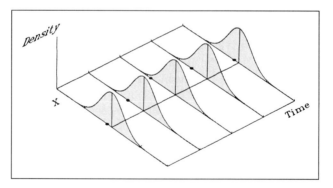

(g) **Process out of control due to dependence between output variables as suggested by pattern in observed values over time**

(X_1, X_2) plane] show that the variables were uncorrelated. Bivariate normally distributed random variables that are uncorrelated are also independent. Thus, X_1 and X_2 are independent. The marginal distributions for X_1 and X_2 [depicted on the $(X_1, Density)$ and $(X_2, Density)$ planes] had the same mean $\mu_1 = \mu_2 = 0$, the same standard deviation $\sigma_1 = \sigma_2 = 1$, and they were both normal. Thus, X_1 and X_2 were identically distributed. Consequently, X_1 and X_2 were independent and identically distributed and the process that generated the variables was in control.

In contrast, the distribution shown in Figure 15.1(f) was generated by a process that was *out of control* due to dependence. The output distributions were identical; however, the variables were correlated, as shown by the surface. Variables that are bivariate normally distributed with nonzero correlation are dependent. Consequently, X_1 and X_2 were identically distributed, but dependent, and the process that generated the variables was out of control due to dependence.

A process that generates output variables with the same mean, variance, and correlations at time lag k (correlations between X_{t-k} and X_t for $k = 1, 2, \ldots$) is known as a **stationary process.** Thus, another way to say that a process is in control is to say that the process is stationary and that its output variables are independent. Correlations between X_{t-k} and X_t are known as **autocorrelations.** The distribution shown in Figure 15.1(f) allows us to visualize positive *autocorrelation* for variables X_1 and X_2. If output variables X_{t-k} and X_t are bivariate normally distributed with nonzero correlations then the variables are dependent. Thus, the process that generated the distribution shown in Figure 15.1(f) was stationary because the variables had the same mean, variance, and correlations; but the process was out of control due to dependence.

Figure 15.1(g) also allows us to visualize distributions and observed data values generated by a process that was *out of control* due to dependence. The observed data values are shown as points. The observed values meandered upward and downward. When a data value was above the mean, the next data value also tended to be above the mean due to autocorrelation. The process generating the output variables was stationary, but it was out of control due to dependence.*

A process that is out of control may have an **assignable** or **special cause** that produces a change in the distribution. Assignable causes need to be investigated and eliminated if they are detrimental to the process. Assignable causes of variation can be incorporated in the process if they are advantageous. For example, a new supplier may be able to provide better quality components, resulting in a beneficial change in the process. Assignable causes may be attributable to such factors as differences among raw materials, employees, or machines. Assignable causes of variation must be removed to bring a process in control.

Processes that are in statistical control still retain variation [see Figure 15.1(a)]. The variation inherent in a process that is in control is chance or **common cause variation.** Common cause variation in a process is the result of the design of the process and may be attributable to the inputs, activities, machines, people, or other components that make up the process. To improve a process, we reduce common cause variation. For example, investing in new machinery or in personnel development training programs can improve processes by reducing common cause variation.

To gain control of processes and improve quality, managers use methods of **statistical process control,** which we define as follows.

> **Definition**
>
> **Statistical process control** is monitoring, isolating, and removing variation to gain control or to improve a process.

---

*Alwan and Roberts (1988) recommended control charts that account for autocorrelated output variables.

Isolating and acting on *assignable causes* of variation are often responsibilities of those who observe or measure the output variable. In contrast, isolating and removing *common causes* of variation are generally responsibilities of managers.

━━ **EXAMPLE 15.3** In this example, we use graphs of distributions of an output variable to visualize the idea of **improving a process.** We improve a process that is in control by reducing common cause variation. The first two distributions shown in Figure 15.2 have relatively high common cause variation; improvements in the process then reduce the variation, thereby improving the process.[*]
━━

Although processes must be controlled and improved to increase quality and productivity, managers should not overadjust processes for phantom special causes of variation. *Overadjusting* or *tampering* with processes that are in control can only make performance worse by increasing variation or by creating an unstable process. We consider an experiment in the next example that demonstrates the effects of overadjusting a process that is in control.

━━ **EXAMPLE 15.4** In this example, we consider the results of an experiment, known as the *funnel experiment,* to demonstrate the effects of overadjusting a process when it is in control.[†] To perform the funnel experiment, we repeatedly drop a ball bearing through a funnel onto a target. Since the ball bearing ricochets through the funnel, we cannot expect to hit the center of the bull's eye each time. Rather, we expect error as a result of chance variation. If the funnel is *fixed* above the center of the target, then the process of dropping the ball bearing through the funnel and observing points where the ball bearing hits represents a process that is in control. We now run simulations of the experiment under three rules. First, the results with *Rule 1, keep the funnel fixed,* are shown in Figure 15.3(a) with successive points where the ball bearing hit the target connected by lines. With the funnel fixed, we have a stable distribution of points centered on the bull's eye with variation due to common cause or chance. Second, the results with *Rule 2, move the funnel in the opposite direction from the hit point*

---

[*]Although we have depicted processes that are in control by using normally distributed output variables, other processes might generate distributions that are not normal.

[†]The idea of the funnel experiment is attributed to L. S. Nelson by Deming (1986).

FIGURE 15.2

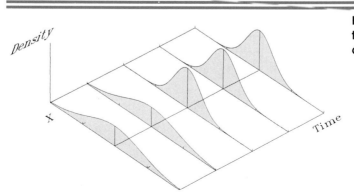

Improving a Stable Process to Reduce Common Cause or Chance Variation

FIGURE 15.3   Funnel Experiment: A Ball Bearing Dropped Through a Funnel onto a Target with a
Bull's Eye

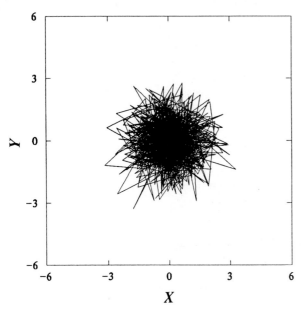

(a) Rule 1: The funnel is fixed.

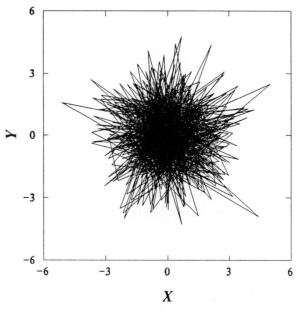

(b) Rule 2: The funnel is moved in the opposite direction
from the hit point by the amount of the error.

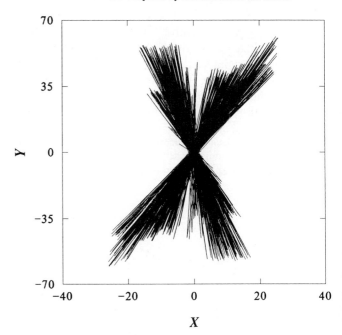

(c) Rule 3: The funnel is moved in the opposite direction
from the bull's eye by the amount of the error.

by the amount of the error, are shown in Figure 15.3(b). With the funnel moved in the opposite direction from the previous hit point, we have a stable distribution of points centered on the bull's eye, but we can see that the chance variation has increased markedly. Third, the results with *Rule 3, move the funnel in the opposite direction from the bull's eye by the amount of the error,* are shown in Figure 15.3(c). With the funnel moved in the opposite direction from the bull's eye, we see a pattern that looks like a bow tie; furthermore, we see that the variation increased tremendously (it increased so much that we changed the scales of the graph) and the process was out of control.

The results of the funnel experiment show us that if we overadjust or tamper with a process when it is in control, we only make things worse. Tampering with a process that is in control can only increase the process variation or drive the process out of control.

Managers that adopt a quality improvement view need to monitor processes, and they need to isolate and eliminate variation from processes by using methods of statistical process control. Many of the methods that we have discussed in previous chapters, such as time series plots, histograms, box plots, probability distributions, sampling, analysis of variance, and regression analysis can assist managers in understanding, monitoring, isolating, and removing variation from processes. The *control charts* that we discuss in the next section are the fundamental method that is used for purposes of statistical process control.

Problems: Sections 15.1–15.2

1. Distributions of the dollar value of new accounts opened each week that were generated by a new account acquistion process over five weeks at First Jeopardy Bank are shown in Figure 15.4(a). Was the process in control or out of control? Why?

2. Distributions of the dollar value of mistakes on bank wires each week that were generated by an electronic money transfer process over five weeks and observed data values at Zero Certainty Bank are shown in Figure 15.4(b). Was the process in control or out of control? Why?

Distributions for Banking Processes   FIGURE 15.4

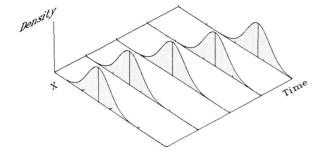
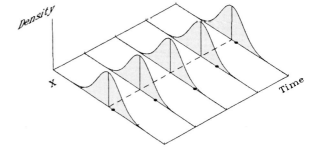

15.3
CONTROL CHARTS

Managers use control charts to monitor and improve processes during the production of goods and services. Control charts were popularized by Shewhart (1931) and advanced more recently by Deming (1986) and others who follow the quality management system philosophy with ongoing quality improvements. Process characteristics—such as the mean or variability of an output variable, or the fraction of defective or nonconforming items produced by a process—are often monitored with **control charts.**

> **Definition**
>
> **Control charts** are sequence plots of descriptive measures or measurements that are used to monitor, control, and improve processes.

In addition to descriptive measures or measurements for a process characteristic, control charts generally include a center line, as well as upper and lower control limits. The **center line** is plotted at a central value of the characteristic. The **upper control limit** and the **lower control limit** are boundary values for an interval that is expected to encompass essentially all of the sample results when a process is in control.

The control charts that we consider are for processes where standards for the process are not known and where data are collected over time from the process. Control charts based on data collected over time from the process are generally constructed using a standard procedure. First, we periodically take samples and calculate descriptive measures for an output variable, or we take individual measurements for an output variable. Next, we find a central value for the characteristic, and we find lower and upper control limits (*LCL* and *UCL*). Control limits are generally the central value plus or minus three standard deviations as follows:

$$LCL = Central\ value - 3(Standard\ deviation)$$

$$UCL = Central\ value + 3(Standard\ deviation)$$

If the process is in control and the central value and standard deviation are known, the plus or minus three standard deviation limits encompass essentially 99.7% of the values for a normally distributed random variable. Thus, if a process is in control, the results should only very infrequently fall outside of the control limits.

Control charts are used as a basis for action. If one or more of the sample results fall outside of the control limits, or if patterns in the results suggest the behavior of the process is not random, then we say the *process is out of control,* and we attempt to bring the process in control.* To bring a process in control, we isolate and remove special causes of variation if they are detrimental to the process. If the sample results are all contained between the control limits, then we say that the *process is in control.*

*Rules for patterns that indicate a process might be out of control include the following: several consecutive sample results, say, 8 points, on one side of the central value; a few consecutive results, say, 3 points, falling within one standard deviation inside a control limit; and several consecutive results, say, 6 points, that increase or decrease in value. Readers interested in extended discussions of rules for patterns that indicate a process might be out of control may wish to consult the chapter references.

If a process that is in control has too much variation due to common causes, then we reduce chance variation by making fundamental improvements in the process.

In statistical process control, samples are also known as *subgroups*. Besides taking samples periodically, say, hourly or daily, we can also select subgroups on a rational basis to isolate special causes of variation. For example, if we suspect special causes of variation due to a difference in raw materials, we may select samples at times when we are using the different materials.

Control charts are used to monitor a process, to warn that corrective action should be taken to remove a special cause of variation, or to indicate that a process should be changed to achieve a more favorable process mean or to reduce process variability. We discuss control charts for process means, ranges, fractions of defective or nonconforming items, and individual data values in this section. The charts are generally known as $\overline{X}$-charts (or *XBAR*-charts), *R*-charts, *p*-charts and *I*-charts (or *X*-charts). Many other types of control charts are also used to monitor processes.

15.3.1 Control Chart for a Process Mean: $\overline{X}$-Chart

A control chart for a process mean, or an $\overline{X}$-chart, is used to identify a change in a process or to identify fluctuations in a process when the characteristic being examined is a quantitative variable.

=== **EXAMPLE 15.5**   Digital Industries, Inc., manufactures memory chips for computers in an ongoing process. To store information correctly in a digital format, the electrical charge for a chip's conducting zone must not change or fluctuate extensively. If the charge fluctuates extensively or does not fluctuate in a random pattern, then the chip will not operate properly. Digital uses $\overline{X}$-charts to monitor the mean electrical charge for the chips produced by the process. Digital also uses $\overline{X}$-charts in contractual situations to show potential new customers that the chip manufacturing process is in control and is constantly being monitored and improved. Digital's customers also use the control charts so they know that the chip manufacturing process is in control, and they do not have to rely on costly, redundant, or out-dated inspection plans.   ===

To construct an $\overline{X}$-chart for a process characteristic, random samples are taken periodically from the process. Samples of constant size n are taken fairly frequently from a process in a typical application of control charts for monitoring a process mean.

Consider k samples of constant size (n is often 4 or 5) taken periodically from a process. Let X_{ij} be the value of the process characteristic for the jth item in the jth sample. We can find the mean for the jth sample or subgroup as

$$\overline{X}_j = \frac{\sum_{i=1}^{n} X_{ij}}{n} \quad \text{for} \quad j = 1, 2, \ldots, k$$

The sample variance for the jth subgroup or sample is

$$s_j^2 = \frac{\sum_{i=1}^{n} (X_{ij} - \overline{X}_j)^2}{n - 1} \quad \text{for} \quad j = 1, 2, \ldots, k$$

The process mean estimated from the data is the mean of the subgroup means $\overline{\overline{X}}$, as follows:

$$\overline{\overline{X}} = \frac{\sum\limits_{j=1}^{k} \overline{X}_j}{k}$$

The standard deviation of the sampling distribution for the subgroup means can be estimated as follows:*

$$\hat{\sigma}_{\overline{X}} = \sqrt{\frac{\hat{\sigma}_X^2}{n}} = \sqrt{\frac{(s_1^2 + s_2^2 + \cdots + s_k^2)/k}{n}} = \frac{\hat{\sigma}_X}{\sqrt{n}}$$

The estimated standard deviation of the output variable or the characteristic of interest is denoted $\hat{\sigma}_X$.

The central line and the lower and upper control limits for the $\overline{X}$-chart are given as follows.†

Central Line and Control Limits for $\overline{X}$-Chart

Central line: $\overline{\overline{X}} = \dfrac{\sum\limits_{j=1}^{k} \overline{X}_j}{k}$

Control limits: $LCL = \overline{\overline{X}} - 3\hat{\sigma}_{\overline{X}}$ and $UCL = \overline{\overline{X}} + 3\hat{\sigma}_{\overline{X}}$

The $\overline{X}$-chart is constructed by plotting the subgroup means, the central line, and the lower and upper control limits over time.

=== **EXAMPLE 15.6**   Digital Industries, Inc., randomly selected $n = 4$ memory chips hourly for $k = 12$ consecutive hours. The measurements for the electrical charges (in microvolts) for a conducting zone are given in Table 15.1. The subgroup means $\overline{X}_j$'s and sample variances s_j^2's were computed and are also given in the table. We want to construct the $\overline{X}$-chart for the electrical charges to monitor the process.

---

*The estimate of the standard deviation of the sampling distribution of the sample mean is straightforward and was used by Shewhart (1931); however, there are several alternative methods for estimating the standard deviation (Minitab, 1991; SAS, 1986). Shewhart recommended 25 or more subgroups for control charts.

†An alternative method for finding control limits for $\overline{X}$-charts when n is small or a computer is unavailable is

$$\overline{\overline{X}} \pm A_2 \overline{R}$$

where A_2 is a constant for a given sample size, as shown in the accompanying listing, and $\overline{R}$ is defined in the next section.

n	2	3	4	5	6	7	8	9	10
A_2	1.88	1.02	.73	.58	.48	.42	.37	.34	.31

Electrical Charges for Memory Chips' Conducting Zones to Construct $\overline{X}$-Chart   TABLE 15.1

Item	Subgroup (hour)											
	1	*2*	*3*	*4*	*5*	*6*	*7*	*8*	*9*	*10*	*11*	*12*
1	9.9	9.1	9.6	11.4	9.9	10.3	11.2	10.5	8.5	11.0	10.5	7.8
2	9.9	9.8	9.4	9.4	10.6	9.8	11.1	9.5	10.3	9.8	12.2	10.6
3	11.0	9.9	10.7	9.2	9.6	9.7	9.6	9.6	8.3	10.4	10.3	10.1
4	9.8	11.2	9.9	7.9	10.0	9.9	10.0	10.6	10.2	8.9	13.3	10.1
Sample Mean, $\overline{X}_j$	10.15	10.00	9.90	9.47	10.03	9.93	10.47	10.05	9.33	10.03	11.57	9.65
Sample Variance, s_j^2	0.32	0.77	0.33	2.09	0.18	0.07	0.64	0.34	1.15	0.80	2.05	1.58

The mean of the subgroup means, $\overline{\overline{X}}$, which is the center line of the control chart, is found as follows:

$$\overline{\overline{X}} = \frac{\sum_{j=1}^{12} \overline{X}_j}{k} = \frac{10.15 + 10.00 + \cdots + 9.65}{12} = 10.05$$

The standard deviation of the sampling distribution for the subgroup means can be found as

$$\hat{\sigma}_{\overline{X}} = \sqrt{\frac{(s_1^2 + s_2^2 + \cdots + s_k^2)/k}{n}} = \sqrt{\frac{(.32 + .77 + \cdots + 1.58)/12}{4}}$$

$$= .46$$

The lower and upper control limits are

$$LCL = \overline{\overline{X}} - 3\hat{\sigma}_{\overline{X}} = 10.05 - 3(.46) = 10.05 - 1.38 = 8.67$$

$$UCL = \overline{\overline{X}} + 3\hat{\sigma}_{\overline{X}} = 10.05 + 3(.46) = 10.05 + 1.38 = 11.43$$

The $\overline{X}$-chart was constructed by plotting the subgroup means $\overline{X}_j$'s, the central line at $\overline{\overline{X}} = 10.05$, and the lower and upper control limits, 8.67 and 11.43, for the 12 hours, and the results are presented in Figure 15.5. Figure 15.5 shows that the sample mean for the eleventh hour, $\overline{X}_{11} = 11.57$, is outside the upper control limit, so the process is said to be out of control. If the sample mean for the charge on a conducting zone is as high as 11.57 microvolts, then some of the memory chips can be expected to malfunction. ==

Recall that a process that is out of control may have an *assignable* or *special cause* that generates a point outside the control limits. Assignable causes need to be investigated and variation due to assignable causes must be removed. *Chance* or *common variation* can be managed or perhaps reduced by improving the production system.

== **EXAMPLE 15.7**   Digital Industries, Inc., examined the production process for memory chips and found that a new batch of etching acid was introduced during

FIGURE 15.5

Control Chart for Process Mean: $\overline{X}$-Chart for Electrical Charge for Memory Chips at Digital Industries, Inc.

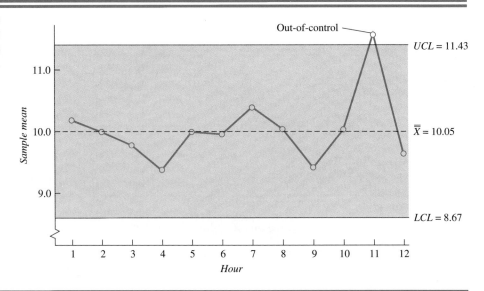

the eleventh hour of production. The acid had been in storage and was not mixed thoroughly before it was used to etch the chips. Some of the acid was too strong and resulted in etchings that were too deep, thereby increasing the electrical charge on the conducting zone and causing the process to go out of control, as shown on the $\overline{X}$-chart in Figure 15.5. Operations managers then required that all of the chips produced during the eleventh hour undergo quality inspections, and they instituted a policy that requires that acids be mixed thoroughly before the etching process is started.

If a process is not in control due to changes in the variability of a process characteristic, then the control limits for an $\overline{X}$-chart may not be appropriate. For this reason, a control chart for process variability, such as a range or R-chart, is used in conjunction with an $\overline{X}$-chart to monitor process variability. R-charts are discussed next.

15.3.2 Control Chart for Process Variation: *R*-Chart

A control chart for process variation, as measured by sample or subgroup ranges, or an **R-chart,** is a practical tool that is used to identify changes in the variation of a process characteristic. The process characteristic is a quantitative variable, and an R-chart often accompanies an $\overline{X}$-chart.

EXAMPLE 15.8   Digital Industries, Inc., the manufacturer of memory chips in an ongoing process, uses R-charts to monitor the variability of the electrical charge for a conducting zone as they examine the quality of the chip manufacturing process. Digital's customers require assurance that the chip manufacturing process is in control such that the variability for the chips' conducting zones is not so high as to affect the chips' ability to store digital information accurately.

If X_{ij} is the value of the process characteristic for the ith item in the jth sample, we can find the range for the jth subgroup, R_j, as the difference between the largest and smallest data values for the subgroup as follows:

$$R_j = X_{ij\text{largest}} - X_{ij\text{smallest}} \quad \text{for} \quad j = 1, 2, \ldots, k$$

The mean of the ranges is

$$\overline{R} = \frac{\sum\limits_{j=1}^{k} R_j}{k}$$

The standard deviation of the sampling distribution of the ranges is denoted by $\hat{\sigma}_R$.

The central line and the upper and lower control limits for the R-chart are found by using the following equations.

Central Line and Control Limits for R-Chart

Central line: $\quad \overline{R} = \dfrac{\sum\limits_{j=1}^{k} R_j}{k}$

Control limits: $\quad LCL = \overline{R} - 3\hat{\sigma}_R = D_3 \overline{R}$

$$UCL = \overline{R} + 3\hat{\sigma}_R = D_4 \overline{R}$$

where D_3 and D_4 are constants for samples of fixed sizes, as given in Table 15.2.

An R-chart is constructed by plotting the subgroup ranges, the central line, and the lower and upper control limits over time.

EXAMPLE 15.9 Digital Industries, Inc., randomly selected $n = 4$ memory chips hourly for $k = 12$ consecutive hours, and the measurements for the electri-

Sample size, n	Constant		Sample size, n	Constant		
	D_3	D_4		D_3	D_4	**TABLE 15.2**
2	0.00	3.27	12	0.28	1.72	**Constants for Finding**
3	0.00	2.57	13	0.31	1.69	**Control Limits for R-Charts**
4	0.00	2.28	14	0.33	1.67	
5	0.00	2.11	15	0.35	1.65	
6	0.00	2.00	16	0.36	1.64	
7	0.08	1.92	17	0.38	1.62	
8	0.14	1.86	18	0.39	1.61	
9	0.18	1.82	19	0.40	1.60	
10	0.22	1.78	20	0.41	1.59	
11	0.26	1.74				

TABLE 15.3   Electrical Charges Data to Construct R-Chart

| | Subgroup (hour) | | | | | | | | | | | |
Item	1	2	3	4	5	6	7	8	9	10	11	12
1	9.9	9.1	9.6	11.4	9.9	10.3	11.2	10.5	8.5	11.0	10.5	7.8
2	9.9	9.8	9.4	9.4	10.6	9.8	11.1	9.5	10.3	9.8	12.2	10.6
3	11.0	9.9	10.7	9.2	9.6	9.7	9.6	9.6	8.3	10.4	10.3	10.1
4	9.8	11.2	9.9	7.9	10.0	9.9	10.0	10.6	10.2	8.9	13.3	10.1
Sample Range, R_j	1.2	2.1	1.3	3.5	1.0	0.6	1.6	1.1	2.0	2.1	3.0	2.8

cal charges for a conducting zone are given in Table 15.3. The subgroup ranges R_j's were computed and are also given in the table. We want to construct the R-chart for the electrical charges. The R-chart is used to monitor variability for the process.

To construct the R-chart, we compute the central line for the ranges or the varilability as the mean of the subgroup ranges $\overline{R}$, as follows:

$$\overline{R} = \frac{\sum\limits_{j=1}^{12} R_j}{k} = \frac{1.2 + 2.1 + 1.3 + \cdots + 2.8}{12} = 1.86$$

The constants D_3 and D_4 for finding control limits with samples of size $n = 4$ from Table 15.9 are $D_3 = .00$ and $D_4 = 2.28$. Thus, the lower and upper control limits are

$$LCL = \overline{R} - 3\hat{\sigma}_R = D_3\overline{R} = (.00)(1.86) = .00$$

$$UCL = \overline{R} + 3\hat{\sigma}_R = D_4\overline{R} = (2.28)(1.86) = 4.24$$

An R-chart was constructed by plotting the subgroup ranges R_j's, the central line at $\overline{R} = 1.86$, and the lower and upper control limits, .00 and 4.24, for the 12 hours, and the results are presented in Figure 15.6. Figure 15.6 shows that the subgroup ranges R_j are all encompassed between the control limits and that the variability is not changing in any predictable pattern. The R-chart indicates that the process variability is in control.

Computers are often used to create control charts. Software packages were used to construct $\overline{X}$- and R-charts for the Digital Industries data, and the results are presented in Computer Exhibits 15.1A and 15.1B.

The process characteristic that we have monitored with $\overline{X}$- and R-charts was measured on a quantitative scale. Control charts can also be used to monitor process characteristics that are attributes or categorical measures. To monitor the proportion or fraction of defective or nonconforming items coming from a process, we use a p-chart, as discussed next.

15.3.3 Control Chart for Process Fraction Nonconforming: p-Chart

Control charts are often used to monitor process characteristics that are attributes, such as defective or nondefective items produced by a manufacturer. To monitor the

FIGURE 15.6

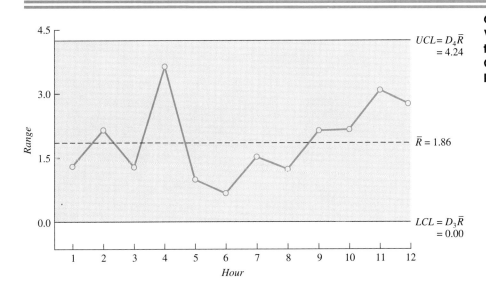

proportion or fraction of defective or nonconforming items coming from a process, we use a ***p*-chart.**

EXAMPLE 15.10   Digital Industries, Inc., the manufacturer of memory chips for computers in an ongoing process, uses control charts to monitor the fraction of memory chips that are defective or nonconforming. Customers who receive nonconforming memory chips may assemble the chips into devices that will not operate properly. To continually improve customer service, Digital monitors the fraction of nonconforming memory chips by using *p*-charts.

To construct a control chart for the fraction nonconforming, we take random samples periodically from the process. Samples of size 100 or more are taken fairly frequently from a process in a typical application of control charts for monitoring a process fraction nonconforming.

If X_j is the number of nonconforming or defective items in a sample of size n_j for the *j*th subgroup, we can find the fraction or proportion of nonconforming items for the *j*th subgroup $\hat{p}_j$ as

$$\hat{p}_j = \frac{X_j}{n_j}$$

The process fraction nonconforming $\bar{p}$ for all of the data can be determined by pooling the number of nonconforming items and dividing by the total sample size from all of the subgroups, as follows:

$$\bar{p} = \frac{\sum\limits_{j=1}^{k} X_j}{\sum\limits_{j=1}^{k} n_j}$$

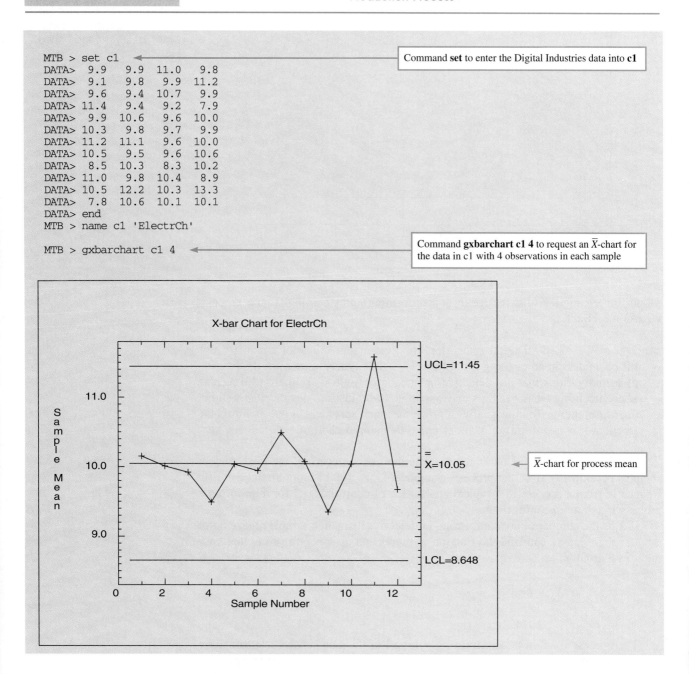

MINITAB

COMPUTER EXHIBIT 15.1A $\bar{X}$- and R-Charts for the Digital Industries Memory Chip Production Process

```
MTB > set c1                          ◄───────────  Command set to enter the Digital Industries data into c1
DATA>  9.9    9.9   11.0    9.8
DATA>  9.1    9.8    9.9   11.2
DATA>  9.6    9.4   10.7    9.9
DATA> 11.4    9.4    9.2    7.9
DATA>  9.9   10.6    9.6   10.0
DATA> 10.3    9.8    9.7    9.9
DATA> 11.2   11.1    9.6   10.0
DATA> 10.5    9.5    9.6   10.6
DATA>  8.5   10.3    8.3   10.2
DATA> 11.0    9.8   10.4    8.9
DATA> 10.5   12.2   10.3   13.3
DATA>  7.8   10.6   10.1   10.1
DATA> end
MTB > name c1 'ElectrCh'

MTB > gxbarchart c1 4     ◄───────  Command gxbarchart c1 4 to request an X̄-chart for
                                    the data in c1 with 4 observations in each sample
```

X-bar Chart for ElectrCh

UCL=11.45

$\bar{\bar{X}}$=10.05   ◄─── $\bar{X}$-chart for process mean

LCL=8.648

The standard error of the proportion or fraction of successes for a binomial distribution was defined in Chapter 5 as

$$\sigma_{\hat{p}} = \sqrt{p(1 - p)/n}$$

COMPUTER EXHIBIT 15.1A $\overline{X}$- and R-Charts for the Digital Industries Memory Chip Production Process (Continued)

MINITAB

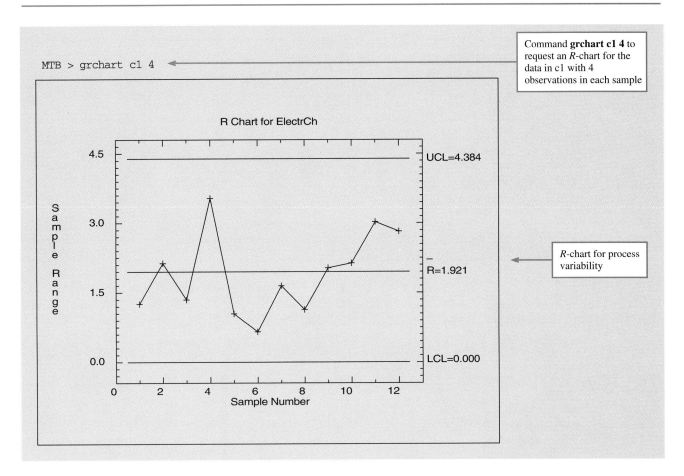

```
MTB > grchart c1 4
```

Command **grchart c1 4** to request an R-chart for the data in c1 with 4 observations in each sample

R Chart for ElectrCh

UCL=4.384

$\overline{R}$=1.921

LCL=0.000

R-chart for process variability

The central line and the lower and upper control limits for a p-chart are found by using the following equations.

Central Line and Control Limits for p-Chart

Central line: $\overline{p} = \dfrac{\sum\limits_{j=1}^{k} X_j}{\sum\limits_{j=1}^{k} n_j}$

Control limits: $LCL = \overline{p} - 3\sqrt{\overline{p}(1 - \overline{p})/n_j}$

$UCL = \overline{p} + 3\sqrt{\overline{p}(1 - \overline{p})/n_j}$

The p-chart is constructed by plotting each subgroup fraction nonconforming, the central line, and the lower and upper control limits over time. If the sample sizes

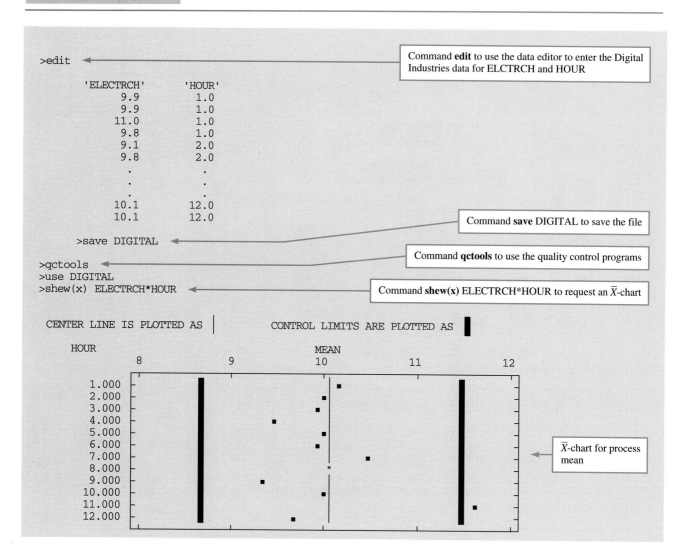

MYSTAT

COMPUTER EXHIBIT 15.1B $\overline{X}$- and R-Charts for the Digital Industries Memory Chip Production Process

```
>edit  ◄─────────────────────────────    Command edit to use the data editor to enter the Digital
                                          Industries data for ELCTRCH and HOUR
        'ELECTRCH'      'HOUR'
           9.9          1.0
           9.9          1.0
          11.0          1.0
           9.8          1.0
           9.1          2.0
           9.8          2.0
            .            .
            .            .
            .            .
          10.1         12.0
          10.1         12.0                 Command save DIGITAL to save the file
     >save DIGITAL  ◄──────────────
>qctools  ◄─────────────────────────      Command qctools to use the quality control programs
>use DIGITAL
>shew(x) ELECTRCH*HOUR  ◄─────────        Command shew(x) ELECTRCH*HOUR to request an X̄-chart

CENTER LINE IS PLOTTED AS │         CONTROL LIMITS ARE PLOTTED AS  ▐

     HOUR                          MEAN
             8           9          10          11          12
     1.000  ┌─        ▐            ▪                         ▐  ─┐
     2.000  ┤         ▐           ▪                          ▐   │
     3.000  ┤         ▐          ▪                           ▐   │
     4.000  ┤         ▐     ▪                                ▐   │
     5.000  ┤         ▐          ▪                           ▐   │
     6.000  ┤         ▐         ▪                            ▐   │
     7.000  ┤         ▐                  ▪                   ▐   │      ◄──  X̄-chart for process
     8.000  ┤         ▐            "                         ▐   │           mean
     9.000  ┤         ▐    ▪                                 ▐   │
    10.000  ┤         ▐          ▪                           ▐   │
    11.000  ┤         ▐                              ▪       ▐   │
    12.000  └─        ▐      ▪                               ▐  ─┘
```

for the subgroups are the same, then the control limits are constant. Otherwise, the control limits vary from one subgroup to the next as the sample sizes change.

EXAMPLE 15.11   Digital Industries, Inc., randomly selected $n = 200$ memory chips hourly for $k = 12$ consecutive hours, and the number of nonconforming memory chips for each hour is given in Table 15.4. The subgroup fractions nonconforming were computed and are also given in the table. We want to construct a p-chart to monitor the fraction nonconforming for the chip-manufacturing process.

To construct the p-chart, we find the central line for the fraction nonconforming by pooling the number of nonconforming items and dividing by the to-

COMPUTER EXHIBIT 15.1B $\overline{X}$- and R-Charts for the Digital Industries Memory Chip Production Process (Continued)

MYSTAT

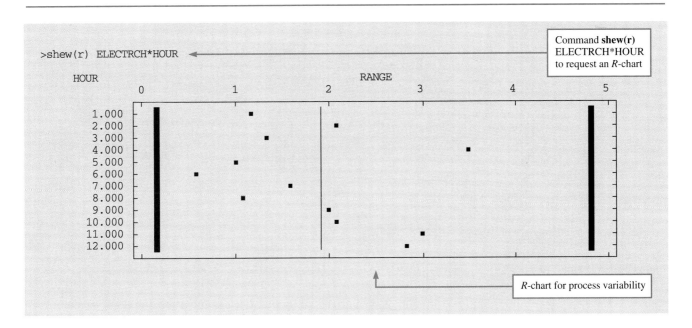

```
>shew(r) ELECTRCH*HOUR
```

Command **shew(r)** ELECTRCH*HOUR to request an *R*-chart

R-chart for process variability

tal of the sample sizes in the subgroups, as follows:

$$\overline{p} = \frac{\sum_{j=1}^{k} X_j}{\sum_{j=1}^{k} n_j} = \frac{12 + 15 + 13 + \cdots + 7}{200 + 200 + 200 + \cdots + 200} = \frac{133}{2400} = .055$$

The lower and upper control limits for the *p* chart are

$$LCL = \overline{p} - 3\sqrt{\overline{p}(1 - \overline{p})/n_j} = .055 - 3\sqrt{.055(1 - .055)/200}$$

$$= .055 - .048 = .007$$

$$UCL = \overline{p} + 3\sqrt{\overline{p}(1 - \overline{p})n_j} = .055 + 3\sqrt{.055(1 - .055)/200}$$

$$= .055 + .048 = .103$$

Number of Nonconforming Chips for Constructing *p*-Chart   TABLE 15.4

		Subgroup (hour)										
Item	*1*	*2*	*3*	*4*	*5*	*6*	*7*	*8*	*9*	*10*	*11*	*12*
Sample Size, n_j	200	200	200	200	200	200	200	200	200	200	200	200
Number Nonconforming, X_j	12	15	13	7	14	6	13	9	12	13	12	7
Fraction Nonconforming, $\hat{p}_j$	.060	.075	.065	.035	.070	.030	.065	.045	.060	.065	.060	.035

FIGURE 15.7

Control Chart for Process
Fraction Nonconforming:
p-Chart for Fraction of
Nonconforming Memory
Chips at Digital Industries,
Inc.

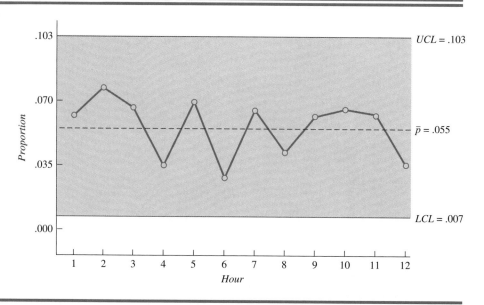

A *p*-chart was constructed by plotting the subgroup fractions nonconforming $\hat{p}_j$, the central line at $\bar{p} = .055$, and the lower and upper control limits, .007 and .103, for the 12 hours, and the results are presented in Figure 15.7. Figure 15.7 shows that the fraction nonconforming for all of the subgroups are encompassed between the control limits and that the fractions do not appear to be changing in any predictable pattern. The *p*-chart indicates that the fraction of nonconforming items produced by the process is in control.

Computers are often used to generate control charts. Data analysis software was used to construct a *p*-chart for the Digital Industries example, and the results are presented in Computer Exhibit 15.2.

15.3.4 Control Chart for Individual Data Values: *I*-Chart

The control charts that we have discussed required us to take samples of data values periodically from a process. However, we often take just one individual data value periodically from a process. A control chart for individual data values, or an *I*-chart (or *X*-chart), is used to identify a change in a process or to identify fluctuations in a process. Individual data values are often used when the process provides outputs infrequently, when measurements require destruction of a product, or when sampling is too expensive.

EXAMPLE 15.12 National Illumination, Inc., manufactures high-intensity laser lamps that are used in surveying equipment. The length of the useful life of the laser lamps is important for the equipment to operate accurately. National recommends that the lamps be replaced after 24 hours of continuous use. National uses control charts to monitor the quality of the lamp manufacturing process. Testing requires destruction of a lamp since the lamp is illuminated continuously to failure. Consequently, National tests individual lamps. Since data values

COMPUTER EXHIBIT 15.2 *p*-Chart for the Digital Industries Memory Chip Production Process

MINITAB

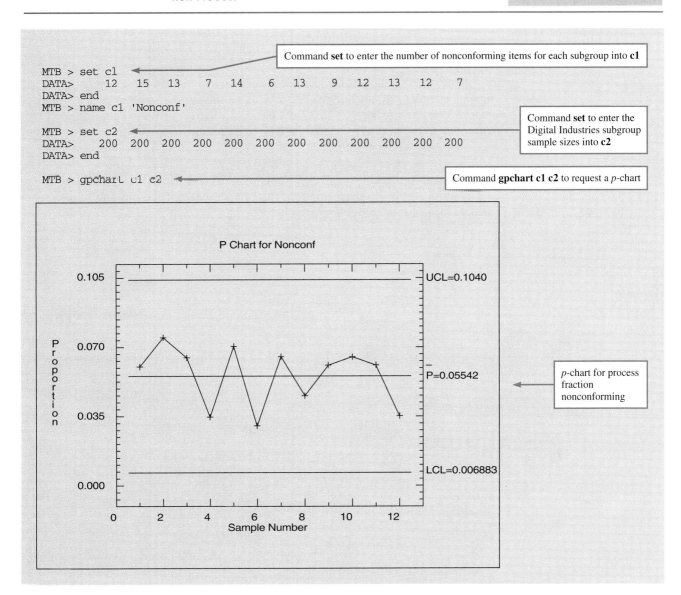

Command **set** to enter the number of nonconforming items for each subgroup into **c1**

```
MTB > set c1
DATA>     12    15    13    7    14    6    13    9    12    13    12    7
DATA> end
MTB > name c1 'Nonconf'

MTB > set c2
DATA>    200   200   200   200   200   200   200   200   200   200   200   200
DATA> end

MTB > gpchart c1 c2
```

Command **set** to enter the Digital Industries subgroup sample sizes into **c2**

Command **gpchart c1 c2** to request a *p*-chart

P Chart for Nonconf

UCL=0.1040

P=0.05542

LCL=0.006883

p-chart for process fraction nonconforming

are obtained for individual lamps, National uses *I*-charts to monitor the process or to identify fluctuations in the process. National uses *I*-charts in contractual situations to show potential new customers that the lamp manufacturing process is in control and is constantly being monitored and improved.

To construct an *I*-chart for a process characteristic, individual measurements are taken periodically from the process. Samples of size $n = 50$ or more individual data values are often taken from a process in a typical application of *I*-charts for monitoring the process.

Let X_i be the value of the process characteristic for the ith item in the sequence. We can find the sample mean as

$$\overline{X} = \frac{\sum\limits_{i=1}^{n} X_i}{n}$$

Because the sample data value for the ith item is an individual measurement, we can not estimate the standard deviation by using subgroups. To estimate the variation in a process, we use the variation from one data value to the next. The variation from one value to the next is measured by the moving range R_i, as follows:

$$R_i = |X_i - X_{i-1}| \qquad i = 2, 3, \ldots, n$$

The control limits are then based on the sample mean of the moving ranges. The sample mean for the moving ranges is

$$\overline{R} = \frac{\sum\limits_{i=2}^{n} R_i}{n - 1}$$

The central line and the upper and lower control limits for the I-chart are given as follows.

Central Line and Control Limits for I-Chart

Central line: $\overline{X} = \sum\limits_{i=1}^{n} X_i / n$

Control limits: $LCL = \overline{X} - 3\hat{\sigma}_R = \overline{X} - 2.66\overline{R}$

$UCL = \overline{X} + 3\hat{\sigma}_R = \overline{X} + 2.66\overline{R}$

where the moving range is

$$R_i = |X_i - X_{i-1}| \qquad i = 2, 3, \ldots, n$$

and the sample mean moving range is

$$\overline{R} = \frac{\sum\limits_{i=2}^{n} R_i}{n - 1}$$

The I-chart is constructed by plotting the X_i data values, the central line, and the upper and lower control limits over time.

━━ **EXAMPLE 15.13**   National Illumination, Inc., randomly selected $n = 50$ lamps over 50 days and the measurements for the lifetimes in hours of continuous op-

		Lifetime (hours)					*Moving Range*			**TABLE 15.5**
25.6	23.9	23.6	26.9	27.7	—	1.7	0.3	3.3	0.8	**Lifetimes and Moving**
25.6	24.5	20.8	27.4	24.1	2.1	1.1	3.7	6.6	3.3	**Ranges for Constructing**
23.7	21.3	22.9	23.0	21.0	0.4	2.4	1.6	0.1	2.0	**I-Chart**
22.8	22.6	27.1	24.8	19.0	1.8	0.2	4.5	2.3	5.8	
23.7	24.2	20.9	25.1	23.6	4.7	0.5	3.3	4.2	1.5	
25.6	21.4	24.6	27.9	20.8	2.0	4.2	3.2	3.3	7.1	
22.9	21.9	22.9	23.7	24.5	2.1	1.0	1.0	0.8	0.8	
28.3	23.2	23.1	23.3	23.3	3.8	5.1	0.1	0.2	0.0	
23.0	28.8	21.0	27.3	26.0	0.3	5.8	7.8	6.3	1.3	
23.6	23.1	23.6	21.1	35.0	2.4	0.5	0.5	2.5	13.9	

eration are given in Table 15.5. The moving ranges $R_i = |X_i - X_{i-1}|$ were computed and are also given in the table. We want to construct the *I*-chart for the lifetimes of the laser lamps to monitor the process.

The mean of the individual data values $\overline{X}$, which is the center line of the control chart, is found as follows:

$$\overline{X} = \frac{\sum_{i=1}^{50} X_i}{n} = \frac{25.6 + 23.9 + \cdots + 35.0}{50} = \frac{1205.7}{50} = 24.114$$

The moving range when $i = 2$ is

$$R_2 = |X_i - X_{i-1}| = |X_2 - X_{2-1}| = |X_2 - X_1| = |23.9 - 25.6| = 1.7$$

The remaining moving range values were found by the same method. Notice that the moving range is missing when $i = 1$.

The sample mean for the moving range values is

$$\overline{R} = \frac{\sum_{i=2}^{n} R_i}{n-1} = \frac{1.7 + 0.3 + \cdots + 13.9}{50 - 1} = \frac{134.211}{49} = 2.739$$

The lower and upper control limits are

$$LCL - \overline{X} \quad 2.66\overline{R} = 24.114 - 2.66(2.739) = 24.114 - 7.286$$

$$= 16.83$$

$$UCL = \overline{X} + 2.66\overline{R} = 24.114 + 2.66(2.739) = 24.114 + 7.286$$

$$= 31.40$$

The *I*-chart was constructed by plotting the individual X_i values, the central line at $\overline{X} = 24.114$, and the upper and lower control limits, 16.83 and 31.40. The results are presented in Figure 15.8.

Figure 15.8 shows that $X_{50} = 35.0$, the individual value when $i = 50$, is outside of the upper control limit. Consequently, we say that the process is out of

FIGURE 15.8

Control Chart for Individual
Data Values: *I*-Chart for
Lifetimes of Laser Lamps at
National Illumination, Inc.

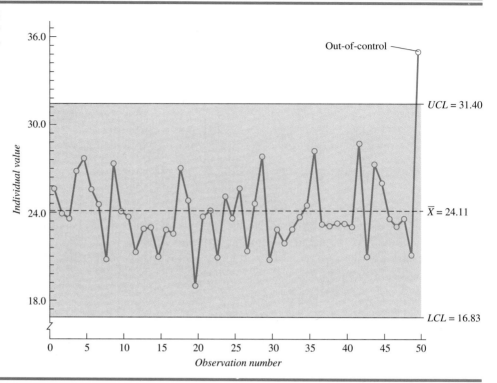

control. If the lifetime is as high as 35 hours, then some of the laser lamps will last an exceedingly long time and National should search for an assignable cause that may be influencing the process. Perhaps National can improve the quality of the lamps if they can determine why the lifetime increased.

A computer was used to construct an *I*-chart for the National Illumination example, and the results are presented in Computer Exhibit 15.3.

If a process is not in control due to changes in the variability of a process characteristic, then the control limits for an *I*-chart may not be appropriate. For this reason, an *R*-chart, which we discussed previously, is usually used in conjunction with an *I*-chart to monitor process variability.

We have monitored the memory chip manufacturing process with $\overline{X}$-, and *R*-, and *p*-charts. However, many other control charts, such as *I*-charts, are available for monitoring processes during the production of goods and services. Readers interested in extended discussions may wish to consult the chapter references.

Problems: Section 15.3

3. Southwest Airways uses Basin International Airport as a hub for its flight operations. To examine passenger check-in times at its check-in counter, Southwest randomly selected $n = 4$ passengers daily for $k = 7$ consecutive days from Monday through Sunday for its 4:00 p.m. flight to Los Angeles. The check-in times are given in the accompanying table. The subgroup means $\overline{X}_j$'s and sample variances s_j^2's were com-

COMPUTER EXHIBIT 15.3 *I*-Chart for the National Illumination Laser Lamp Manufacturing Process

MINITAB

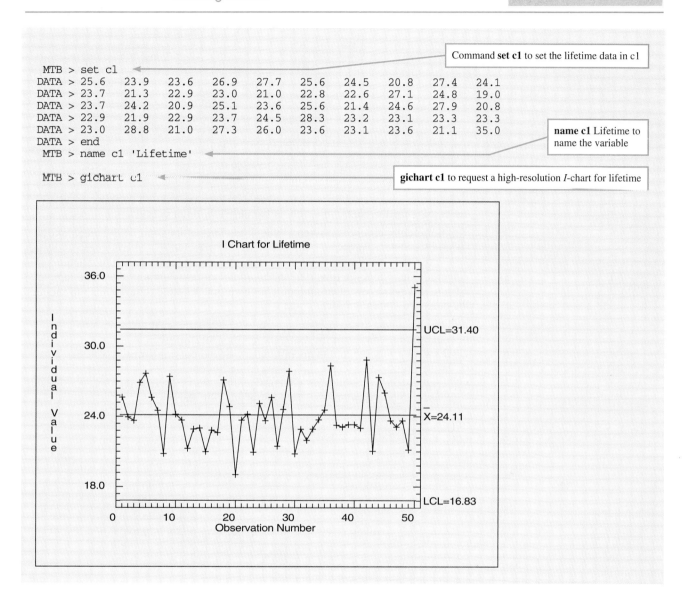

Command **set c1** to set the lifetime data in c1

```
MTB > set c1
DATA > 25.6   23.9   23.6   26.9   27.7   25.6   24.5   20.8   27.4   24.1
DATA > 23.7   21.3   22.9   23.0   21.0   22.8   22.6   27.1   24.8   19.0
DATA > 23.7   24.2   20.9   25.1   23.6   25.6   21.4   24.6   27.9   20.8
DATA > 22.9   21.9   22.9   23.7   24.5   28.3   23.2   23.1   23.3   23.3
DATA > 23.0   28.8   21.0   27.3   26.0   23.6   23.1   23.6   21.1   35.0
DATA > end
MTB > name c1 'Lifetime'
MTB > gichart c1
```

name c1 Lifetime to name the variable

gichart c1 to request a high-resolution *I*-chart for lifetime

I Chart for Lifetime

puted and are also given in the table. Use the data to monitor the process with an $\overline{X}$-chart.

a. Find the center line.
b. Find the lower control limit and the upper control limit.
c. Do any of the sample means fall outside of the control limits?
d. Is the process considered to be in control?
e. If the process is not considered to be in control, then what assignable cause can you think of that may be influencing the process?

			Subgroup (day)			
	M	T	W	Th	F	S	Su
Item	1	2	3	4	5	6	7
1	7.1	10.3	6.7	15.3	8.5	12.1	20.5
2	9.6	12.1	8.8	11.1	13.4	10.2	21.0
3	6.9	11.6	7.6	5.6	9.5	12.1	21.2
4	10.4	10.8	11.0	13.2	8.5	9.3	29.5
Sample Mean, $\overline{X}_j$							
	8.500	11.200	8.525	11.300	9.975	10.925	23.050
Sample Variance, s_j^2							
	3.113	0.647	3.463	17.380	5.436	1.976	18.577

4. Refer to the Southwest Airways data given with Problem 1. Use a computer to construct an $\overline{X}$-chart for the check-in times. Is the check-in process said to be under control?

5. To examine variability in passenger check-in times at its Basin International Airport check-in counter, Southwest Airways randomly selected $n = 4$ passengers daily for $k = 7$ consecutive days from Monday through Sunday for its 4:00 p.m. flight to Los Angeles. The check-in times are given in the accompanying table. The subgroup ranges, R_j's were computed and are also given in the table. Use the data to monitor the process with an R-chart.

   a. Find the center line.
   b. Find the lower control limit and the upper control limit.
   c. Do any of the sample ranges fall outside of the control limits?
   d. Is the process variability considered to be in control?
   e. If the process variability is not considered to be in control, then what assignable cause can you think of that may be influencing the process?

			Subgroup (day)			
	M	T	W	Th	F	S	Su
Item	1	2	3	4	5	6	7
1	7.1	10.3	6.7	15.3	8.5	12.1	20.5
2	9.6	12.1	8.8	11.1	13.4	10.2	21.0
3	6.9	11.6	7.6	5.6	9.5	12.1	21.2
4	10.4	10.8	11.0	13.2	8.5	9.3	29.5
Sample Range, R_j							
	3.5	1.8	4.3	9.7	4.9	2.8	9.0

6. Refer to the Southwest Airways data given with Problem 5. Use a computer to construct an R-chart for the check-in times. Is the check-in process considered to be under control?

7. To examine its baggage handling operations, Southwest Airways randomly selected $n = 100$ checked bags daily for $k = 7$ consecutive days from Monday through Sunday for its 4:00 p.m. flight to Los Angeles. The numbers of lost or mishandled bags for the samples taken each day are given in the accompanying table. Use the data to monitor the process with a p-chart.

   a. Find the center line.
   b. Find the lower control limit and the upper control limit.
   c. Do any of the sample statistics fall outside of the control limits?

    d. Is the process considered to be in control?

    e. If the process is not considered to be in control, then what assignable cause can you think of that may be influencing the process?

			Subgroup (day)			
M	T	W	Th	F	S	Su
1	2	3	4	5	6	7

			Number Nonconforming, X_j			
9	11	7	3	6	7	6

			Sample Size, n_j			
100	100	100	100	100	100	100

8. Refer to the Southwest Airways data given with Problem 7. Use a computer to construct a *p*-chart for the lost or mishandled bags. Is the check-in process considered to be under control?

9. National Illumination, Inc., randomly selected $n = 50$ lamps over 50 days. The measurements for the illumination power of individual lamps are given in the accompanying listing. Use the data to monitor the process with an *I*-chart.

    a. Find the center line.

    b. Find the lower control limit and the upper control limit.

    c. Do any of the individual data values fall outside of the control limits?

    d. Is the process considered to be in control?

    e. If the process is not considered to be in control, then what assignable cause can you think of that may be influencing the process?

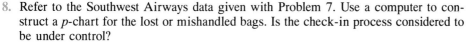
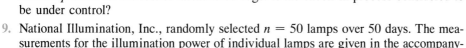

9.1	11.3	11.3	9.0	8.0	9.8	11.8	9.6	10.2	11.7
11.4	9.8	10.6	10.4	9.5	10.7	8.2	10.8	10.1	10.8
9.0	11.2	7.6	10.2	11.3	11.2	8.8	10.5	10.7	8.5
10.4	10.7	11.6	9.1	10.1	9.3	9.7	9.4	9.8	11.2
9.6	9.4	10.2	10.8	10.1	10.5	10.5	11.4	10.8	9.6

10. Refer to the National Illumination, Inc., data given with Problem 9. Use a computer to construct an *I*-chart for the illuminating power of the lamps manufactured by the company. Is the manufacturing process considered to be under control?

We satisfy designers and consumers of products and services when our products and services conform to *specification limits.* For example, the memory chip manufactured by Digital Industries, Inc., is designed to store digital information accurately if the electrical charge for a conducting zone is between 7 and 13 microvolts. The values that bound the interval are known as **specification limits,** which are defined as follows.

15.4
PROCESS CAPABILITY

> **Definition**
>
> **Specification limits** are boundary values for an interval of values that conform to the requirements of designers or consumers for a characteristic of an output of a process.

An **upper specification limit, USL,** is the largest value that satisfies the specifications and a **lower specification limit, LSL,** is the smallest value that satisfies the specifications.

We discussed control charts that are used to monitor processes in the previous section. Processes that are in control still retain variation and control limits are dependent on the variation inherent in the process. Specification limits and control limits are generally *not* the same values.

To provide outputs that satisfy users of products and services, processes must have the capability to transform inputs into outputs that conform to specifications. Investigation of how well the outputs of a process that is in control conform to specifications is known as **process capability analysis.**

Definition

Process capability analysis uses data taken from a process that is in control to determine how well the process meets specifications for a product or service.

In this section, we discuss process capability analysis of the *proportion of process outputs that conform to specifications*. In addition, we discuss two *process capability indexes* denoted as C_p and C_{pk}.

If the output variable for a process is normally distributed with the mean and standard deviation estimated from the sample data, then the proportion of process outputs that conform to specifications can be estimated by using the standard normal distribution. When we have k subgroups with n data values in each subgroup, the *standardized values,* or Z values, for the specification limits are

$$\frac{LSL - \overline{\overline{X}}}{\hat{\sigma}_X} \quad \text{and} \quad \frac{USL - \overline{\overline{X}}}{\hat{\sigma}_X}$$

where $\overline{\overline{X}}$ is the estimated mean and $\hat{\sigma}_X$ is the estimated standard deviation of the output variable, as defined in the section on control charts. The proportion of outputs that conform to specifications can then be estimated as

$$P\left(\frac{LSL - \overline{\overline{X}}}{\hat{\sigma}_X} \leq Z \leq \frac{USL - \overline{\overline{X}}}{\hat{\sigma}_X}\right)$$

The proportion is found by using areas of the standard normal distribution for the Z values. A diagram showing a histogram for the data values, a superimposed normal distribution, specification limits, and the estimated mean plus or minus three standard deviations is often used to display the results.

EXAMPLE 15.14 Engineers at Digital Industries, Inc., design memory chips to store information accurately in a digital format so long as the electrical charge for a conducting zone is between 7 and 13 microvolts. Measurements for $n = 4$ memory chips selected randomly by hour for $k = 12$ hours were given in Table 15.1. To analyze the capability of the manufacturing process to produce chips that conform to specifications in this example, we find an estimate for the proportion of outputs that fall between the specification limits.

In Example 15.6, we used the electrical charge data to find the estimated mean,

$$\overline{\overline{X}} = \frac{\sum \overline{X}_j}{k} = \frac{10.15 + 10.00 + \cdots + 9.65}{12} = 10.05$$

and the estimated standard deviation of the sampling distribution of the subgroup means,

$$\hat{\sigma}_{\overline{X}} = \frac{\hat{\sigma}_X}{\sqrt{n}} = .46$$

Thus, the estimated standard deviation of the output distribution $\hat{\sigma}_X$ is

$$\hat{\sigma}_X = (\hat{\sigma}_{\overline{X}})(\sqrt{n}) = (.46)(\sqrt{4}) = .92$$

The standardized values or Z values for the specification limits are

$$Z = \frac{LSL - \overline{\overline{X}}}{\hat{\sigma}_X} = \frac{7 - 10.05}{.92} = -3.32$$

and

$$Z = \frac{USL - \overline{\overline{X}}}{\hat{\sigma}_X} = \frac{13 - 10.05}{.92} = 3.21$$

From the standard normal table, the area between $Z = -3.32$ and $Z = 0$ is .4995, and the area between $Z = 0$ and $Z = 3.21$ is .4993. Consequently, the proportion of chips that conform to specification limits is estimated, based on a normal distribution for the electrical charges, to be

$$P\left(\frac{LSL - \overline{\overline{X}}}{\hat{\sigma}_X} \leq Z \leq \frac{USL - \overline{\overline{X}}}{\hat{\sigma}_X}\right) = P(-3.32 \leq Z \leq 3.21)$$
$$= .4995 + .4993 = .9988$$

Thus, we estimate that a high proportion of Digital's memory chips conforms to specifications. Nevertheless, if the production process is improved, then Digital could obtain an even higher level of product quality. Recall that Digital has instituted a change in the process to ensure that the etching acid is thoroughly mixed prior to the etching operation, based on the $\overline{X}$-chart.

Figure 15.9 shows a histogram for Digital's chip data values, a superimposed normal distribution, specification limits, and the estimated mean plus or minus three standard deviations. The area under the normal curve between the specification limits is .9988.

We measure the range of the specification limits relative to the estimated variability of the process by using two **process capability indexes, C_p and C_{pk},** as defined next.

FIGURE 15.9

Histogram, Normal Curve, and Specification Limits for Electrical Charge on Memory Chips at Digital Industries, Inc.

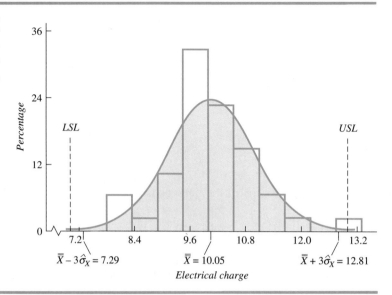

Definition

Process capability indexes, C_p and C_{pk}, are measures of the range of the specification limits relative to the range for the characteristic of the output of a process as estimated from the process variation. The C_p index is defined as

$$C_p = \frac{USL - LSL}{6\hat{\sigma}_X}$$

The process capability for the lower specification limit (*CPL*) is

$$CPL = \frac{\overline{\overline{X}} - LSL}{3\hat{\sigma}_X}$$

and the process capability for the upper specification limit (*CPU*) is

$$CPU = \frac{USL - \overline{\overline{X}}}{3\hat{\sigma}_X}$$

The C_{pk} index is equal to *CPL* or *CPU*, whichever is smaller.

    If the output variable for a process is normally distributed, then $\pm 3\hat{\sigma}_X$, or $6\hat{\sigma}_X$, will encompass about 99.7% or essentially all of the output from the process. Consequently, C_p and C_{pk} must be greater than or equal to 1 if the process is capable of producing outputs that are essentially all contained between the specification limits. In fact, the capability of the process to provide outputs that conform to specifications is even more satisfactory if both indexes are greater than or equal to, say, 1.33. The C_p index is limited because it does not account for the mean of the process. Consequently, a process with a mean that is not centered between the specification limits may have a value greater than or equal to 1 for the C_p index even though the process is not capable of producing outputs that conform to specifications. The C_{pk} index is preferable, in this situation, because C_{pk} accounts for the estimated mean of the process. If the estimated mean $\overline{\overline{X}}$ falls exactly between *LSL* and *USL*, then C_p will be equal to C_{pk}.

EXAMPLE 15.15 Digital Industries, Inc., memory chips are designed to store information accurately in a digital format so long as the electrical charge for a conducting zone is between 7 and 13 microvolts. To analyze the capability of the manufacturing process to produce chips that conform to specifications, we find process capability indexes C_p and C_{pk}.

The estimated mean, which we used previously, is

$$\overline{\overline{X}} = 10.05$$

and the estimated standard deviation of the output distribution is

$$\hat{\sigma}_X = .92$$

Thus, the C_p index is

$$C_p = \frac{USL - LSL}{6\hat{\sigma}_X} = \frac{13 - 7}{6(.92)} = \frac{6}{5.52} = 1.09$$

The process capability for the lower specification limit (*CPL*) is

$$CPL = \frac{\overline{\overline{X}} - LSL}{3\hat{\sigma}_X} = \frac{10.05 - 7}{3(.92)} = \frac{3.05}{2.76} = 1.11$$

and the process capability for the upper specification limit (*CPU*) is

$$CPU = \frac{USL - \overline{\overline{X}}}{3\hat{\sigma}_X} = \frac{13 - 10.05}{3(.92)} = \frac{2.95}{2.76} = 1.07$$

Since C_{pk} is equal to *CPL* = 1.11 or *CPU* = 1.07, whichever is smaller, we have

$$C_{pk} = 1.07$$

The process capability indexes C_p and C_{pk} for Digital's electrical charge data are both greater than 1. Thus, the data indicate that the process is capable of producing essentially all chips that conform to specifications. Nevertheless, the process would be considered even more satisfactory if it could be improved and the indexes could be increased to, say, 1.33. Since *CPU* = 1.07 is somewhat smaller than *CPL* = 1.11, the process mean is estimated to be somewhat closer to the upper specification limit than it is to the lower specification limit. Production managers at Digital should continue to monitor the process and may need to adjust the process to center the mean between the two specification limits. Recall that

Digital has decided to change the process, based on the $\overline{X}$-chart, to ensure that the etching acid is thoroughly mixed prior to the etching operation.   ═══

Problems: Section 15.4

11. Southwest Airways uses Basin International Airport as a hub for its flight operations. To examine passenger check-in times at its check-in counter, Southwest randomly selected $n = 4$ passengers daily for $k = 7$ consecutive days from Monday through Sunday for its 4:00 p.m. flight to Los Angeles. The check-in times are given in the accompanying table. The subgroup means $\overline{X}_j$'s and sample variances s_j^2's were computed and are also given in the table. Southwest wants to get passengers through the check-in procedures in from 7 to 17 minutes. Use the data to examine the process capability.

    a. Find the proportion of outputs that conform to specifications.
    b. Find the process capability index C_p.
    c. Find the process capability index C_{pk}.
    d. Is the check-in process capable of providing service that conforms to specifications for all of the customers?

| | *Subgroup (day)* | | | | | | |
Item	*M* 1	*T* 2	*W* 3	*Th* 4	*F* 5	*S* 6	*Su* 7
1	7.1	10.3	6.7	15.3	8.5	12.1	20.5
2	9.6	12.1	8.8	11.1	13.4	10.2	21.0
3	6.9	11.6	7.6	5.6	9.5	12.1	21.2
4	10.4	10.8	11.0	13.2	8.5	9.3	29.5
Sample Mean, $\overline{X}_j$							
	8.500	11.200	8.525	11.300	9.975	10.925	23.050
Sample Variance, s_j^2							
	3.113	0.647	3.463	17.380	5.436	1.976	18.577

 12. Use a computer to analyze the process capability. Is the check-in process capable of meeting the specifications for all of the customers?

15.5
SUMMARY

Statistical process control was introduced in this chapter. To improve quality and productivity during the production of goods and services, managers must monitor, control, and improve processes. We discussed concepts of managing processes and gave a definition of *quality*. Quality system management standards (ISO 9000 standards) were introduced.

In this chapter, a *process* was defined, and the variables associated with output of processes were shown in figures so that processes in control and processes out of control could be visualized. *Statistical process control* was defined. We introduced *control charts*—such as $\overline{X}$-charts, R-charts, p-charts, and I-charts—that are used to monitor processes. We also discussed methods that are used to analyze the capability of a process to provide outputs that conform to specifications. Table 15.6 summarizes formulas that can be used to construct control charts and to analyze process capability.

TABLE 15.6 **Summary of Formulas**

Control Chart	Formula	Comment
$\overline{X}$-chart for process mean	Central line: $$\overline{\overline{X}} = \frac{\sum_{j=1}^{k} \overline{X}_j}{k}$$ Control limits; $$LCL = \overline{\overline{X}} - 3\hat{\sigma}_{\overline{X}}$$ $$UCL = \overline{\overline{X}} + 3\hat{\sigma}_{\overline{X}}$$	k samples, each of size n, are taken periodically from a process, and X_{ij} is the value of the process characteristic for the ith item in the jth sample. The mean for the jth subgroup is $$\overline{X}_j = \frac{\sum_{i=1}^{n} X_{ij}}{n} \quad \text{for} \quad j = 1, 2, \ldots, k$$ The sample variance for the jth subgroup or sample is $$s_j^2 = \frac{\sum_{i=1}^{n} (X_{ij} - \overline{X}_j)^2}{n - 1}$$ and the standard deviation of the sampling distribution for the subgroup means can be estimated as $$\hat{\sigma}_{\overline{X}} = \sqrt{\frac{(s_1^2 + s_2^2 + \cdots + s_k^2)/k}{n}}$$
R-chart for process variability	Central line: $$\overline{R} = \frac{\sum_{j=1}^{k} R_j}{k}$$ Control limits: $$LCL = \overline{R} - 3\hat{\sigma}_R = D_3\overline{R}$$ $$UCL = \overline{R} + 3\hat{\sigma}_R = D_4\overline{R}$$	Where there are k samples or subgroups, the range for the jth subgroup, R_j, is the difference between the maximum and minimum values for the subgroup, and D_3 and D_4 are constants for samples of fixed sizes, as given in Table 15.2
p-chart for process fraction nonconforming	Central line: $$\overline{p} = \frac{\sum_{j=1}^{k} X_j}{\sum_{j=1}^{k} n_j}$$ Control limits: $$LCL = \overline{p} - 3\sqrt{\overline{p}(1 - \overline{p})/n_j}$$ $$UCL = \overline{p} + 3\sqrt{\overline{p}(1 - \overline{p})/n_j}$$	X_j is the number of nonconforming or defective items in a sample of size n_j for the jth subgroup, and the fraction or proportion of nonconforming items for the jth subgroup $\hat{p}_j$ is $$\hat{p}_j = \frac{X_j}{n_j}$$

(continues)

TABLE 15.6   Summary of Formulas (Continued)

Control Chart	*Formula*	*Comment*
I-chart for individual values	Central line: $$\overline{X} = \sum_{i=1}^{n} X_i/n$$ Control limits: $$LCL = \overline{X} - 2.66\overline{R}$$ $$UCL = \overline{X} + 2.66\overline{R}$$	Where X_i is the ith individual data value in the sequence of n values, the moving range is $$R_i = |X_i - X_{i-1}| \quad i = 2, 3, \ldots, n$$ and the sample mean moving range is $$\overline{R} = \frac{\sum_{i=2}^{n} R_i}{n-1}$$

Process Capability: *Proportion of Outputs That Conform to Specifications*	*Process Capability Indexes C_p and C_{pk}*

With k subgroups and n data values in each subgroup, the standardized values, or Z values, for the lower specification limit (LSL) and the upper specification limit (USL) are

$$\frac{LSL - \overline{\overline{X}}}{\hat{\sigma}_X} \quad \text{and} \quad \frac{USL - \overline{\overline{X}}}{\hat{\sigma}_X}$$

where:

$$\overline{\overline{X}} = \frac{\sum_{j=1}^{k} \overline{X}_j}{k} \text{ is the estimated mean}$$

and

$$\hat{\sigma}_X = (\hat{\sigma}_{\overline{X}})(\sqrt{n})$$

is the estimated standard deviation of the output variable.
The proportion of outputs that conform to specifications is estimated, based on a normal distribution for the variable, as

$$P\left(\frac{LSL - \overline{\overline{X}}}{\hat{\sigma}_X} \le Z \le \frac{USL - \overline{\overline{X}}}{\hat{\sigma}_X}\right)$$

$$C_p = \frac{USL - LSL}{6\hat{\sigma}_X}$$

C_{pk} index is equal to *CPL* or *CPU*, whichever is smaller,

where:

$$CPL = \frac{\overline{\overline{X}} - LSL}{3\hat{\sigma}_X}$$

and

$$CPU = \frac{USL - \overline{\overline{X}}}{3\hat{\sigma}_X}$$

REVIEW PROBLEMS

13. A financial analyst with First Metro Bancshares wishes to examine the yields on 90-day Treasury bills. The information will be used to examine the changes over time of short-term interest rates. The yield data were collected monthly for two years, and the results are presented in order (left to right from first row to last row) in the accompanying listing.

  a. Construct a time series plot for the data. A time series plot is a plot of data values for a variable over time.
  b. Do the yields appear to be generated from a stable process?

c. Would the central tendency for the yields be a good approximation for the short term interest rate in the month following the data collection period?

7.76	8.22	8.57	8.00	7.56	7.01	7.05	7.18
7.08	7.17	7.20	7.07	7.04	7.03	6.59	6.06
6.12	6.21	5.84	5.57	5.19	5.18	5.35	5.49

SOURCE: U.S. Department of Labor, Bureau of Labor Statistics, (1986), *Business Statistics, 1986* (Washington, D.C.: U.S. Government Printing Office).

14. An economist with First Metro Bancshares wishes to examine the discount rate at the New York Federal Reserve Bank. The information will be used to examine the pattern of interest rates for banks borrowing funds from the Federal Reserve. The discount data were collected monthly for two years, and the results are presented in order (left to right from first row to last row) in the accompanying listing.

a. Construct a time series plot for the data. A time series plot is a plot of data values for a variable over time.
b. Do the rates appear to be generated from a stable process?
c. Would the central tendency for the rates be a good approximation for the interest rate for bank borrowing one month after the data were collected?

8.00	8.00	8.00	8.00	7.81	7.50	7.50	7.50
7.50	7.50	7.50	7.50	7.50	7.50	7.10	6.83
6.50	6.50	6.16	5.82	5.50	5.50	5.50	5.50

SOURCE: U.S. Department of Labor, Bureau of Labor Statistics, (1986), *Business Statistics, 1986* (Washington, D.C.: U.S. Government Printing Office).

15. Precision Autotech, Inc., produces valves for automobile engines by using a hot metal–forging manufacturing process. To examine the diameters of the valves prior to machining and grinding, Precision randomly selected $n = 4$ valves hourly for $k = 8$ consecutive hours during a production shift. The diameters (in centimeters) are given in the accompanying table. The subgroup means $\overline{X}_j$'s and sample variances s_j^2's were computed and are also given in the table. Use the data to monitor the process with an $\overline{X}$-chart.

a. Find the center line.
b. Find the lower control limit and the upper control limit.
c. Do any of the sample means fall outside of the control limits?
d. Is the process considered to be in control?
e. If the process is not considered to be in control, then what assignable cause can you think of that may be influencing the process?

	Subgroup (hour)							
Item	*1*	*2*	*3*	*4*	*5*	*6*	*7*	*8*
1	5.0	5.2	4.6	4.9	4.8	5.4	5.7	5.9
2	5.0	5.0	5.0	5.2	5.0	5.0	5.5	5.8
3	5.0	5.4	4.7	5.2	5.2	5.3	6.5	6.0
4	4.9	4.9	5.3	5.0	5.2	4.4	5.7	5.9
Sample Mean, $\overline{X}_j$								
	4.975	5.125	4.900	5.075	5.050	5.025	5.850	5.900
Sample Variance, s_j^2								
	.0025	.0492	.1000	.0225	.0367	.2025	.1967	.0067

16. Refer to the Precision Autotech data given with Problem 15. Use a computer to construct an $\overline{X}$-chart for the diameters. Is the process considered to be under control?

17. Precision Autotech, Inc., produces valves for automobile engines by using a hot metal–forging manufacturing process. To examine the variation in the diameters of the valves, Precision randomly selected $n = 4$ valves hourly for $k = 8$ consecutive hours during a production shift. The diameters (in centimeters) are given in the accompanying table. The subgroup ranges, R_j's were computed and are also given in the table. Use the data to monitor the process with an R-chart.

   a. Find the center line.
   b. Find the lower control limit and the upper control limit.
   c. Do any of the sample statistics fall outside of the control limits?
   d. Is the process considered to be in control?
   e. If the process is not considered to be in control, then what assignable cause can you think of that may be influencing the process?

				Subgroup (hour)				
Item	1	2	3	4	5	6	7	8
1	5.0	5.2	4.6	4.9	4.8	5.4	5.7	5.9
2	5.0	5.0	5.0	5.2	5.0	5.0	5.5	5.8
3	5.0	5.4	4.7	5.2	5.2	5.3	6.5	6.0
4	4.9	4.9	5.3	5.0	5.2	4.4	5.7	5.9
				Sample Range, R_j				
	0.1	0.5	0.7	0.3	0.4	1.0	1.0	0.2

 18. Refer to the Precision Autotech data given with Problem 17. Use a computer to construct an R-chart for the diameters. Is the process considered to be under control?

19. Precision Autotech, Inc., produces valves for automobile engines by using a hot metal–forging manufacturing process. To examine the manufacturing process, Precision randomly selected $n = 100$ valves hourly for $k = 8$ hours during a production shift, and the number of nonconforming valves for the sample taken each hour is given in the accompanying table. Use the data to monitor the process with a p-chart.

   a. Find the center line.
   b. Find the lower control limit and the upper control limit.
   c. Do any of the sample statistics fall outside of the control limits?
   d. Is the process considered to be in control?
   e. If the process is not considered to be in control, then what assignable cause can you think of that may be influencing the process?

			Subgroup (hour)				
1	2	3	4	5	6	7	8
			Number Nonconforming, X_j				
7	8	6	3	5	7	16	18
			Sample Size, n_j				
100	100	100	100	100	100	100	100

 20. Refer to the Precision Autotech data given with Problem 19. Use a computer to construct a p-chart for the valves. Is the process considered to be under control?

21. Electrical Motors, Inc., produces drive shafts for small electrical motors by using a numerically controlled milling machine. To monitor the manufacturing process, EM

randomly selected $n = 4$ shafts daily for two working weeks, so $k = 10$. The lengths (in centimeters) are given in the accompanying table. The subgroup means $\bar{X}_j$'s and sample variances s_j^2's were computed and are also given in the table. Use the data to monitor the process with an $\bar{X}$-chart.

a. Find the center line.
b. Find the lower control limit and the upper control limit.
c. Do any of the sample means fall outside of the control limits?
d. Is the process considered to be in control?
e. If the process is not considered to be in control, then what assignable cause can you think of that may be influencing the process?

					Subgroup (day)					
Item	1	2	3	4	5	6	7	8	9	10
1	9.5	10.4	11.5	10.5	9.1	9.7	10.2	10.6	10.6	9.1
2	8.9	9.6	9.1	9.0	9.9	10.0	10.8	9.6	8.9	10.9
3	9.2	9.7	10.3	10.0	9.9	9.3	9.0	9.6	9.1	11.1
4	10.2	9.8	9.1	9.6	9.5	10.9	9.3	11.4	9.3	10.4
Sample Mean, $\bar{X}_j$										
	9.450	9.875	10.000	9.775	9.600	9.975	9.825	10.300	9.475	10.375
Sample Variance, s_j^2										
	.31000	.12917	1.32000	.40250	.14667	.46250	.68250	.76000	.58917	.80917

22. Refer to the Electrical Motors data given with Problem 21. Use a computer to construct an $\bar{X}$-chart for the lengths. Is the process considered to be under control?

23. Electrical Motors, Inc., produces drive shafts for small electrical motors by using a numerically controlled milling machine. To monitor the manufacturing process, EM randomly selected $n = 4$ shafts daily for two working weeks, so $k = 10$ days. The lengths (in centimeters) are given in the accompanying table. The subgroup ranges, R_j's were computed and are also given in the table. Use the data to monitor the variability of the process with an R-chart.

a. Find the center line.
b. Find the lower control limit and the upper control limit.
c. Do any of the sample statistics fall outside of the control limits?
d. Is the process considered to be in control?
e. If the process is not considered to be in control, then what assignable cause can you think of that may be influencing the process?

					Subgroup (day)					
Item	1	2	3	4	5	6	7	8	9	10
1	9.5	10.4	11.5	10.5	9.1	9.7	10.2	10.6	10.6	9.1
2	8.9	9.6	9.1	9.0	9.9	10.0	10.8	9.6	8.9	10.9
3	9.2	9.7	10.3	10.0	9.9	9.3	9.0	9.6	9.1	11.1
4	10.2	9.8	9.1	9.6	9.5	10.9	9.3	11.4	9.3	10.4
Sample Range, R_j										
	1.3	0.8	2.4	1.5	0.8	1.6	1.8	1.8	1.7	2.0

24. Refer to the Electrical Motors data given with Problem 23. Use a computer to construct an R-chart for the lengths. Is the process considered to be under control?

25. Electrical Motors, Inc., produces drive shafts for small electrical motors by using a numerically controlled milling machine. To monitor the manufacturing process, EM randomly selected $n = 100$ shafts daily for two work weeks, so $k = 10$ days. The number of nonconforming shafts are given in the accompanying table. Use the data to monitor the process with a p-chart.

a. Find the center line.
b. Find the lower control limit and the upper control limit.
c. Do any of the sample statistics fall outside of the control limits?
d. Is the process considered to be in control?
e. If the process is not considered to be in control, then what assignable cause can you think of that may be influencing the process?

Subgroup (day)									
1	2	3	4	5	6	7	8	9	10
Number Nonconforming, X_j									
10	11	10	6	15	11	6	11	10	11
Sample Size, n_j									
100	100	100	100	100	100	100	100	100	100

26. Refer to the Electrical Motors data given with Problem 25. Use a computer to construct a p-chart for the nonconforming shafts. Is the process considered to be under control?

27. National Illumination, Inc., selected $n = 50$ lamps in sequence, and the measurements (in centimeters) for the diameters of individual lamps are given in the accompanying listing (in left to right sequence from top row to bottom row). Use the data to monitor the process with an I-chart.

a. Find the center line.
b. Find the lower control limit and the upper control limit.
c. Do any of the individual data values fall outside of the control limits?
d. Is the process considered to be in control?
e. If the process is not considered to be in control, then what assignable cause can you think of that may be influencing the process?

1.99	1.99	2.02	1.99	2.02	1.99	1.99	2.02	2.01	2.01
1.99	2.00	2.02	2.00	2.01	1.98	2.00	2.01	2.01	2.01
1.99	2.01	2.00	2.01	1.99	2.02	2.00	1.98	2.01	2.00
2.00	2.01	1.97	1.99	2.01	1.99	1.99	2.00	2.00	1.99
2.00	2.00	2.00	2.00	2.01	1.99	2.01	2.00	2.01	1.99

28. Refer to the National Illumination data given with Problem 27. Use a computer to construct an I-chart for the diameter of the lamps manufactured by the company. Is the manufacturing process considered to be under control?

29. Precision Autotech, Inc., produces valves for automobile engines by using a hot metal–forging manufacturing process. To examine the diameters of the valves prior to machining and grinding, Precision randomly selected $n = 4$ valves hourly for $k = 8$ consecutive hours during a production shift. The diameters (in centimeters) are given in the accompanying table. The subgroup means $\overline{X}_j$'s and sample variances s_j^2's were computed and are also given in the table. Precision engineers have designed the forged valves with specification limits of 4.6 and 5.8 centimeters. Use the data to examine the process capability.

a. Find the proportion of outputs that conform to specifications.
b. Find the process capability index C_p.

c. Find the process capability index C_{pk}.
d. Is the process capable of providing outputs that conform to the specifications for all of its products?

Item	Subgroup (day)							
	1	*2*	*3*	*4*	*5*	*6*	*7*	*8*
1	5.0	5.2	4.6	4.9	4.8	5.4	5.7	5.9
2	5.0	5.0	5.0	5.2	5.0	5.0	5.5	5.8
3	5.0	5.4	4.7	5.2	5.2	5.3	6.5	6.0
4	4.9	4.9	5.3	5.0	5.2	4.4	5.7	5.9
	Sample Mean, $\overline{X}_j$							
	4.975	5.125	4.900	5.075	5.050	5.025	5.850	5.900
	Sample Variance, s_j^2							
	.0025	.0492	.1000	.0225	.0367	.2025	.1967	.0067

30. Refer to the Precision Autotech data given with Problem 29. Use a computer to analyze the process capability. Is the process capable of meeting the specifications for all of its products?

31. A **Pareto diagram** is a bar chart for a categorical variable that results from a quality improvement investigation. Each category of the variable represents a factor that results in a nonconformity or a problem for a product or service. The lengths of the bars on the chart are equal to the frequencies (or relative or percentage frequencies) for the categories. Categories are ordered according to their frequency counts, and it often turns out that the majority of nonconformities or difficulties can be traced to a few factors. Quality improvement efforts can then be directed to the more important factors. The diagram is extended at times to a combination bar and line chart where the line shows the cumulative frequency (cumulative relative frequency or cumulative percentage) over the ordered categories.

The typographical errors (nonconformities) in a CVS policy and procedures manual were traced to the following sources: (1) typesetting, 10 errors; (2) text editing, 3 errors; (3) original manuscript, 2 errors; (4) carryover from previous manual, 1 error.

a. Construct a Pareto diagram by using a bar chart.
b. Based on frequencies of the factors, where should quality improvement efforts be directed?

CONTROLLING THE QUALITY OF GOODS AND SERVICES

CASE 15.1

The statistical control of quality is one use of statistical methods in the production of goods and services. A manufacturer of small motors had a quality control problem that resulted in too many malfunctioning motors. Some of the steel rods used in the motors were loose in their bearings. The diameters of the loose rods were smaller than the lower limit called for in the specifications. During an inspection a quality control analyst collected measurements of the diameters of 500 rods. The distribution of the diameters of the rods obtained by the analyst is shown in Figure 1.

The gap in the histogram of Figure 1 for the third class interval (rod diameters of .9985 to .9995 centimeter) and the peak for the fourth interval seemed strange to the analyst. The specification limit for the rod diameters called for the rejection of rods if their diameters were less than .9995 centimeter. Upon further investigation the analyst deter-

FIGURE 1

Distribution of Diameters of
Steel Rods

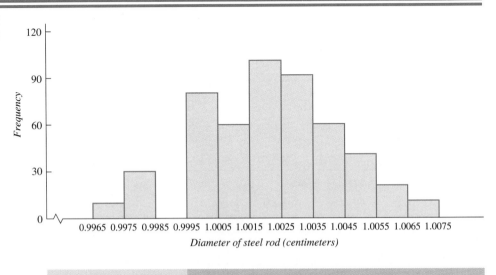

Diameter of steel rod (centimeters)

| Rods rejected | Rods accepted |

mined that the inspectors were accepting parts that were barely defective and were recording the diameters of these rods at just above .9995 centimeter. The result was that none of
the rods were tallied in the third class, creating a gap in the histogram, and too many rods
were tallied in the fourth class, creating a peak.

When the inspectors were made aware of the trouble that a rod with an undersized
diameter would cause for the motors, the inspections were carried out properly and the
problem was corrected. In this case, product quality was improved by using a histogram
in a statistical analysis of the rod diameters, and the number of malfunctioning motors
was decreased dramatically.

Case Assignment

a. Comment on an estimated value for the central tendency of the rod diameters.
b. Comment on how spread out the rod diameters are, or the dispersion of the rod
   diameters.
c. Comment on the symmetry of the histogram for the rod diameters compared with
   the symmetry you would expect for these diameters.
d. Estimate the number of rods rejected by the inspectors from the 500 steel rods
   whose diameters are depicted in the histogram.

REFERENCE: Deming, W. E. (1976), "Making Things Right. What Is the Statistical Control of Quality?"
in *Statistics: A Guide to Business and Economics,* ed. by J. M. Tanur et al. (San Francisco: Holden-Day).

REFERENCES

Alwan, L., and H. Roberts. 1988. "Time-Series Modeling for Statistical Process Control."
    Journal of Business and Economic Statistics, 6: 87–95.
American Society for Testing and Materials. 1976. *ASTM Manual of Presentation of
    Data and Control Chart Analysis.* Philadelphia: ASTM.
Deming, W. Edwards. 1986. *Out of the Crisis.* Cambridge, Mass.: MIT Press.

Duncan, A. 1986. *Quality Control and Industrial Statistics,* 5th ed. Homewood, Ill.: Irwin.

Dzus, G. 1991. "Planning a Successful ISO 9000 Assessment." *Quality Progress,* 24: 43–48.

Gitlow, H., S. Gitlow, A. Oppenheim, and R. Oppenheim. 1989. *Tools and Methods for the Improvement of Quality.* Homewood, Ill.: Irwin.

Grant, E., and R. Leavenworth. 1988. *Statistical Quality Control,* 6th ed. New York: McGraw-Hill.

Minitab, Inc. 1991. *MINITAB QC Supplement: Quality Control and Improvement.* State College, Pa.: Minitab, Inc.

National Standards Authority of Ireland. 1987. *Notes for Guidance on ISO 9000.* Merrimack, N.H.: NSAI.

Ryan, B. 1991. "Minitab Embraces TQM, Improves Customer/Supplier Relations." *Keeping TAB Newsletter* (December): 2.

Ryan, T. 1989. *Statistical Methods for Quality Improvement.* New York: Wiley.

SAS Institute, Inc. 1986. *SAS/QC User's Guide,* Version 5 ed. Cary, N.C.: SAS Institute, Inc.

Shewhart, W. 1931. *Economic Control of Quality of Manufactured Products.* New York: Van Nostrand.

Western Electric, Inc. *Statistical Quality Control Handbook.* 1956. Western Electric, Inc.: Mack Printing, Easton, PA.

Decision Analysis

16

I n previous chapters, we have dealt with subjects that involve *decision making*. For instance, we have discussed the subject of probability with the aim of *deciding* what the chances are for the occurrence of certain possible events (such as drawing an ace from a deck of cards). We have studied the subject of statistical estimation with the aim of *deciding* what the mean of a population might be (such as the mean daily sales of a product), given only sample information from which to work. In both cases, and in most of the others we have discussed, the decisions involved have been only preliminary. That is, they are decisions that are usually made prior to taking some decisive action (such as marketing a new product).

Before decision makers take an *action,* they usually consider what the *costs* or *payoffs* may be. That is, they consider the economic consequences of their actions. Up to this point, we have ignored economic consequences even in hypothesis testing and other types of problems where decisive action was implied. In this chapter, we introduce the *economic consequences* of taking various actions.

There is a general class of problems where a decision maker must choose between two or more courses of action. The following list gives examples of the types of choices that have to be made by decision makers every day:

Market or not market a new product.

Build a new plant in England, Germany, or the United States.

Invest in stocks, bonds, or mutual funds.

Sue a competitor or try out-of-court settlement.

Use high-, low-, or moderate-quality materials.

All decision problems have characteristics in common. In each case, the best course of action depends on the economic consequences of the alternative actions and the *probabilities* that these economic consequences will be realized. The objective of this chapter is to present ways in which the economic consequences and the probabilities of realizing them can be combined to determine which of several courses of action is "best" for the decision maker.

16.1
PAYOFF TABLE

The first step a manager takes when a decision must be made is to define exactly what alternative actions are available and what the economic consequences or payoffs are for choosing each of these alternatives. The payoffs for an action may be different depending on what future event occurs. A payoff table can be used to present information about alternative actions, events, and economic consequences.

A payoff table is a rectangular array of numbers. Usually, it is drawn so that the *columns* of the array represent each of the alternative **actions** A_j from which the decision maker must choose. The *rows* of the array represent **events** E_i that might take place, or current "states of nature" that will influence the consequences of the decision maker's actions. The *numbers* or *values* in the array at any particular row or column, X_{ij}, represent the **economic consequences** that will accrue to the decision maker if he or she takes the action represented by column j and if the event represented by row i occurs.

EXAMPLE 16.1   Consider the following bets for a coin-tossing game. The player tosses a coin three times or until a head appears, whichever comes first. If a head appears on the first toss, the player receives $1; if the first head appears on the second toss, the player receives $2; if it appears on the third toss, the player receives $4; and if no head appears in three tosses, the player must pay $8. Someone presented with the opportunity of playing this game must decide whether or not to play.

The payoff table faced by such a person is presented in Table 16.1. This payoff table has two *columns,* since the person making the decision has two possible *actions*—to play or not to play. The table has four *rows,* each representing an *event* or state of nature that might take place. The *payoffs* in the table represent what the decision maker would receive under the various combinations of decisions and events. The negative value indicates that there is to be a pay *out* (cost), not a payoff (profit).

EXAMPLE 16.2   Consider the decision faced by the owner of Equipment Leasing Services. The owner has been offered $80,000 for this business if she sells it this month. However, she is tempted to hold on to the business for a year since the business may be awarded a contract that, according to her best calculations, will yield a profit of $30,000 in current dollars over its one-year life. One year of operation without the contract will not yield any profit since she is operating just at the break-even point.

With these thoughts in mind the owner approached the potential buyer of her business and asked if the buyer would consider waiting a year. The purchaser responded positively but added that the purchase price would be $70,000 in current dollars in that case and would have to be agreed upon now.

		Action	
TABLE 16.1	*Event*	A_1, *Play*	A_2, *Not Play*
Payoff Table for	E_1, *head on first toss*	1	0
Coin-Tossing Game	E_2, *first head on second toss*	2	0
	E_3, *first head on third toss*	4	0
	E_4, *no head appears*	−8	0

| | Action | |
Event	A_1, Sell Now	A_2, Hold a Year
E_1, contract awarded	80	100
E_2, contract not awarded	80	70

TABLE 16.2

Payoff Table for Sale of Equipment Leasing Services (Payoffs in Thousands of Dollars)

The owner of Equipment Leasing faces the payoff situation shown in Table 16.2. The owner must decide between selling now or holding her business for one year. If she sells now, she will receive $80,000 for sure. Her payoff one year from now is $70,000 plus whatever is realized from the contract bid.

Payoff tables can also be used in problems where all the consequences are costs. Rather than indicating all the costs as negative payoffs, users of such tables simply enter the costs as positive figures and then keep in mind the fact that the values are costs, not profits. Payoff tables can also be used in problems where the results of an action are not expressed in terms of monetary values. For instance, a payoff table could be constructed showing the production output associated with several different plant layouts (different actions) under various possible demand situations (different events).

The payoff tables discussed in the examples of this section are helpful in laying out a statement of the problem. However, the payoff table by itself does not usually indicate which course of action should be taken. Only under special circumstances can one make a decision from the payoff table alone. Consider the decision of an investor who needs to decide whether to invest in stocks (action 1), in bonds (action 2), or in mutual funds (action 3). The events for this problem correspond to possible states of nature for the economy. During the investment period, the economy may be in conditions of inflation (event 1), in disinflation (event 2), or in stagnation (event 3). The actions, events, and payoffs for the investor are presented in Table 16.3.

From this table we see that action 3 is the preferred action since its consequences are better than those for the other two actions, *regardless of what event occurs.* Action 3 (invest in mutual funds) is said to *dominate* the other actions in this case. One action **dominates** another when all of its payoffs are equal to or better than the other's and one or more of its payoffs is better than the other's regardless of what event occurs. Actions that are dominated are *inadmissable* and are eliminated from further consideration. In Examples 16.1 and 16.2 there are no dominant actions. The decision makers must rely on other information to help them make their decisions.

In the examples that have been presented so far in this section, the payoffs that appear in the cells of the payoff tables have been given in the problem statements. However, often the decision maker must calculate these payoffs by using information about the decision-making situation. The following problem presents such a situation.

| | Action | | |
Event	A_1, Stocks	A_2, Bonds	A_3, Mutual Funds
E_1, inflation	−5	2	8
E_2, disinflation	7	−3	9
E_3, stagnation	4	6	7

TABLE 16.3

An Investor's Payoff Table with a Dominant Action

PRACTICE PROBLEM 16.1

The manager of Pacific Concrete, Inc., has an opportunity to buy up to three car-loads of cement (which is used as an ingredient in concrete). The Northwestern Construction Company is willing to buy some of the cement from him, but it hasn't told him how many carloads it will take. He must make his decision today. Any cement not bought by Northwestern can be sold to a broker. Construct Pacific's payoff table, using the following facts:

Cost to Pacific	$65/ton
Tons per carload	120
Price to Northwestern	$80/ton
Price to broker if Northwestern doesn't buy	$60/ton

Solution Since the manager for Pacific must buy in carload lots, his possible actions are buy 0, buy 1, buy 2, or buy 3 carloads. Northwestern may order any number of carloads, but Pacific's manager would be concerned only with its orders up to 3 since this is the maximum number of carloads he could sell. Thus, the events are order 0, order 1, order 2, or order 3 or more carloads.

We can find each cell value in the payoff table by computing total revenue less total cost:

$$Profit = Total\ revenue - Total\ cost$$

$$= (Revenue\ from\ Northwestern\ orders + Revenue\ from\ broker)$$

$$- (Cost\ of\ cement\ ordered)$$

Since all the events and actions are expressed in terms of carloads, the costs and profits must also be so expressed. Thus the cost of one carload is ($65 per ton) × (120 tons per carload) = $7800 per carload. The revenue from Northwestern on a carload is ($80 per ton) × (120 tons per carload) = $9600 per carload. The revenue from the broker for a purchased carload that Northwestern does not order is ($60 per ton) × (120 tons per carload) = $7200 per carload. The payoff values for the "order 2" event row can be found as shown in Table 16.4.

The payoffs from each row can be computed in the same fashion. The entire table is presented in Table 16.5, and the reader can verify the values in each cell. This table has no dominant actions, and how many carloads Pacific's manager should buy is not obvious. The manager needs more information before he can determine his best action. The type of information he needs is discussed in the next section.

		Carloads Purchased by Pacific			
TABLE 16.4	*Pacific's Revenues and Costs*	*Buy 0*	*Buy 1*	*Buy 2*	*Buy 3*
Calculating Payoffs to	Revenue from Northwestern	0	9600	19,200	19,200
Pacific Concrete	Revenue from broker	0	0	0	7,200
Assuming That North-	Cost of cement	0	(7800)	(15,600)	(23,400)
western Orders 2 Carloads	Payoffs to Pacific	0	1800	3,600	3,000

	Pacific's Action			
Event: Northwestern's Order	Buy 0	Buy 1	Buy 2	Buy 3
Order 0	0	−600	−1200	−1800
Order 1	0	1800	1200	600
Order 2	0	1800	3600	3000
Order 3 +	0	1800	3600	5400

TABLE 16.5

Payoff Table for Pacific Concrete

16.2
PRIOR PROBABILITIES

Decision makers faced with a payoff table would have no problem choosing an action if they knew what event was going to occur. They would simply examine the row associated with that event and choose the action giving the best payoff in that row. For instance, if the business owner in Example 16.2 *knew* the contract was not going to be awarded, she would sell her business now. But she does not know what event is going to take place. Thus, she is trying to make her decision under uncertainty. It is uncertainty that makes the decision-making process difficult.

Decision makers can more sharply state their feelings about the uncertainty they face if they assign probabilities to each of the events in their payoff tables.

EXAMPLE 16.3 For the betting game in Example 16.1, assigning probabilities is a relatively easy matter involving probability theory. The events of Example 16.1 were, again,

E_1: *Head on the first toss* E_3: *First head on the third toss*

E_2: *First head on the second toss* E_4: *No head appears*

The probability of a head on the first toss of the coin is $P(E_1) = 1/2$. If the first head (H) is to appear on the second toss, the first toss must be a tail (T). Using the notation of Chapter 4, we have $P(E_2) = P(TH)$. Since tosses of a coin are independent, we can multiply the probabilities of a tail and a head:

$$P(E_2) = P(TH) = (1/2)(1/2) = 1/4 = .250$$

In a similar manner,

$$P(E_3) = P(TTH) = (1/2)(1/2)(1/2) = 1/8 = .125$$

and

$$P(E_4) = P(TTT) = (1/2)(1/2)(1/2) = 1/8 = .125$$

Now a probability column can be added to the payoff table of Example 16.1. The new table is presented as Table 16.6.

EXAMPLE 16.4 The task of adding probabilities to the payoff table of Example 16.2 for the owner of Equipment Leasing Services is not so simple. The events for the owner of Equipment Leasing of Example 16.2 were, again, as follows:

E_1: *Contract awarded* E_2: *Contract not awarded*

			Action	
TABLE 16.6	*Event*	*Probability*	A_1, *Play*	A_2, *Not Play*
Payoff Table 16.1 with Probabilities	E_1, *head on first toss*	.500	1	0
	E_2, *first head on second toss*	.250	2	0
	E_3, *first head on third toss*	.125	4	0
	E_4, *no head appears*	.125	-8	0

In the coin-tossing game we could calculate the *exact* probabilities of the events from our knowledge of the physical structure of a coin and our knowledge of probabilities. In this new situation, we have little to go on except the business owner's subjective feelings about the relative likelihood of the two events. These feelings are based on her knowledge of the contract and the other bidders and on her experience in bidding on similar, if not identical, contracts. It would help her in evaluating the two possible actions—sell now versus hold the business for one year—if she could make an explicit statement about the subjective probabilities of the two events. Suppose that after some consideration the business owner states that there is a .20 chance that she will win the contract (and thus a .80 chance that she will not). The payoff table then becomes Table 16.7.

Subjective probability estimates were introduced in Section 4.1. Some people claim that subjective probabilities like those just discussed are meaningless. They feel that subjective probabilities are only guesses and thus very likely to be in error. Of course, people's guesses about the probabilities of various events can be very far from correct. Yet if the person making the guesses is close to the problem and has had a great deal of experience in similar situations, guesses may be quite good even though they may not be exact or correct in all cases. Furthermore, few decision makers can really avoid being influenced in their choice of actions by their subjective feelings concerning the likelihood of each event. In this case, argue those who favor using subjective probabilities in decision making, isn't it best if decision makers make an explicit statement of these feelings in the form of subjective probabilities? They are going to be influenced by their subjective probabilities in any case, so they might as well get them out in the open where they can see them and use them. The material presented in the remainder of this book assumes that subjective probabilities are legitimate and can be entered into the probability columns of payoff tables. The probabilities entered into a payoff table are called **prior probabilities.**

			Action	
TABLE 16.7	*Event*	*Probability*	A_1, *Sell Now*	A_2, *Hold a Year*
Payoff Table for the Owner of Equipment Leasing with Probabilities (Payoffs in Thousands of Dollars)	E_1, *contract awarded*	.20	80	100
	E_2, *contract not awarded*	.80	80	70

Problems: Sections 16.1–16.2

Answers to odd-numbered problems are given in the back of the text.

1. The Semiconductor Devices Company is trying to decide whether it should market a
   new product. The company's profit (or loss) on the product depends on the percentage
   of the market it can capture. The break-even market share is 10%. The management
   feels that it will make $40,000 for every percentage point above 10% it is able to cap-
   ture. But if the company fails to achieve a 10% market share, management has decided
   that losses will amount to

   $$\$20,000 + \$30,000(10 - p)$$

   where p is the market share expressed as a percentage (that is, not as a decimal but as
   a number between 0 and 100). The $20,000 cost is not incurred once the company
   achieves a 10% market share; it applies only if the company fails to meet the break-
   even point. Semiconductor Devices feels that its chances at various possible market
   shares are as given in the accompanying table.

   a. What are the actions open to the company?
   b. Construct a payoff table for the company?

Market Share (%)	Probability
7	.05
8	.10
9	.12
10	.18
11	.25
12	.20
13	.10
	1.00

2. The Atlantic Tool Company is planning its first (and probably last) production run of
   a new, special-purpose machine. It already has five orders for this machine. The machine sells for $8500 and has variable costs of $2600 per machine. Fixed costs total
   $30,000 and are made up of new-equipment costs and a manufacturing right the com-
   pany has purchased. The entire fixed cost is to be charged to this production run.

   The company is uncertain about how many of these machines will eventually be
   ordered and thus how many it should produce. The company must make the production
   run this month in order to meet the delivery date on the first few machines. Uncertainty
   about the eventual total sales of the machine is expressed in the following probability
   distribution:

Sales	5	6	7	8
P(Sales)	.25	.35	.30	.10

   a. What are the actions the company could take?
   b. What are the events that might occur?
   c. Construct the payoff table.

3. Consider the accompanying payoff table, which presents the *costs* for an International
   Building, Ltd., construction project (in thousands of dollars) where there are four alternative methods of construction ($A_1, A_2, A_3,$ and A_4) and three types of weather condi-

tions that might be encountered, snow, rain, or clear (E_1, E_2, and E_3). Eliminate any dominated actions in this problem.

		Action			
Event	Probability	A_1	A_2	A_3	A_4
E_1	.20	20	20	26	30
E_2	.30	25	30	35	30
E_3	.50	20	18	10	22

16.3
EXPECTED MONETARY VALUE

Entering the probabilities of the events in the payoff table helps to sharpen the statement of the problem facing the decision maker. But it still does not indicate directly which action should be taken. The appropriateness of an action in any problem depends on the combination of the possible payoffs together with the probabilities of realizing those payoffs. Thus, we need a method of combining an action's payoffs with the event's probabilities. This method was presented in Section 5.3, where the expected value of random variables was discussed.

If X is a random variable and each value that the random variable may take is the lowercase x_i, where there are k distinct values ($i = 1, 2, 3, \ldots, k$), the expected value of X, denoted $E(X)$, is the weighted mean of the possible values that X might take. The weights are the probabilities associated with these values. Equation (5.1) for the expected value of X is repeated here:

$$E(X) = \mu = \sum_{i=1}^{k} x_i P(X = x_i) \tag{16.1}$$

If the decision maker chooses a particular action A_j, the payoff is a random variable that may assume any of the values X listed in the jth column of the payoff table. If the value X_{ij} is the payoff associated with the ith event's occurrence ($i = 1, 2, \ldots, m$) after the jth action ($j = 1, 2, \ldots, n$) has been taken, then we can define the **expected monetary value** (**EMV**) of A_j according to the following formula.

> ### Equation for Expected Monetary Value for the *j*th Action
>
> $$EMV_j = \sum_{i=1}^{m} X_{ij} P(E_i) \qquad j = 1, 2, \ldots, n \tag{16.2}$$

The summation in Equation (16.2) is over m events, and there are n actions from which to choose. The equation tells us that to compute an expected monetary value EMV for an action, we must select a column in the payoff table (since actions are arranged in columns), multiply each payoff in the column by its corresponding probability, and then add the products. We then repeat the procedure for each column or action.

There is an expected monetary value for every action in a payoff table, and this figure is a weighted mean of the payoffs associated with the action. It is a value combining the probabilities of the events with the payoffs in the table. Thus, the decision maker can choose the action with the highest weighted payoff, the one with the *largest* expected monetary value, which we denote EMV^*.

			Action		
Event	Probability	A_1, Play	A_2, Not Play	**TABLE 16.8**	
E_1, head on first toss	.500	1	0	**Payoff Table for Coin-**	
E_2, first head on second toss	.250	2	0	**Tossing Game (Payoffs in**	
E_3, first head on third toss	.125	4	0	**Thousands of Dollars)**	
E_4, no head appears	.125	−8	0		

 PRACTICE PROBLEM 16.2

Consider the coin-tossing game of Example 16.1 again. The payoff table for this game is presented again as Table 16.8. Find the *EMV* for each action.

Solution The *EMV* of each of the actions is computed as follows:

$$EMV_1 = \$1(.500) + \$2(.250) + \$4(.125) - \$8(.125)$$

$$= \$.50 + \$.50 + \$.50 - \$1.00 = \$.50$$

$$EMV_2 = \$0(.500) + \$0(.250) + \$0(.125) + \$0(.125) = \$0$$

The decision maker would play the game (choose A_1) if he desires the action with the highest expected monetary value since $EMV^* = EMV_1 = \$.50$. Of course, he cannot *expect* a $.50 gain if he plays; he either gains $1, $2, or $4 or loses $8 on each play of the game. The EMV^* value of $.50 means that if the decision maker plays the game repeatedly, he will gain $.50 per play on the average. If he were to play the game 1000 times and were to add up his winnings and subtract his losses at the end of that period, his net gain would be about $500, or $.50 per day.

 PRACTICE PROBLEM 16.3

Calculate the expected monetary values of the actions confronted by the Equipment Leasing Services business owner who wants to sell her business. Her payoff table is repeated as Table 16.9.

Solution The *EMV*'s of the actions are

$$EMV_1 = \$80(.20) + \$80(.80) = \$16 + \$64 = \$80 \text{ (thousand)}$$

$$EMV_2 = \$100(.20) + \$70(.80) = \$20 + \$56 = \$76 \text{ (thousand)}$$

Apparently, the best decision based on expected monetary value is to sell the business now since $EMV^* = \$80$ (thousand). The figure of $80,000 is the actual

			Action		
Event	Probability	A_1, Sell Now	A_2, Hold a Year	**TABLE 16.9**	
E_1, contract awarded	.20	80	100	**Payoff Table for Sale of**	
E_2, contract not awarded	.80	80	70	**Equipment Leasing**	
				Services (Payoffs in	
				Thousands of Dollars)	

price the owner will receive for her business; it is not difficult to interpret. But what is the meaning of $EMV_2 = \$76,000$? In a sense, it means that if the owner were to face this decision many times in a row and were to choose to hold her business for one year each time, then some of the time she would get $100,000 and some of the time she would get $70,000, and her gains each time she made this decision would average out to $76,000 per sale of the business. ▬

In Practice Problem 16.3, to suggest that the owner of Equipment Leasing Services would face the task of selling her business several times in the future is somewhat ridiculous. Her decision is a one-shot decision. Interpreting expected monetary value as the long-run average payoff when there is no long run is somewhat misleading. In one-shot decision problems like those in Practice Problem 16.3, we would do better to view the expected monetary value as a number that shows us the weighted average of the payoffs associated with an action. Then we choose the action that has the highest weighted average of payoffs.

Problems: Section 16.3

4. Given the Semiconductor Devices Company scenario of Problem 1, which action has the highest EMV?
5. Refer to the Atlantic Tool Company scenario of Problem 2.
   a. Find the EMV for each action.
   b. What action should the company take on the basis of the EMV? Why?
6. Refer to the International Building, Ltd., construction project scenario in Problem 3.
   a. Find the EMV for each action (you may omit any dominated actions).
   b. Select the best action on the basis of the EMV.

16.4
DECISION TREES

Payoff tables are useful for structuring decision-making problems so that the decision maker can understand the alternative actions that will be considered, the events that might occur, and the possible payoffs. A second method for structuring a decision-making problem uses a **decision tree.** A decision problem that is structured as a payoff table can also be shown as a decision tree, as the following example illustrates.

▬ **EXAMPLE 16.5**   The decision tree in Figure 16.1 represents the problem faced by the owner of the Equipment Leasing Services business who is trying to decide whether or not to sell her business. A decision tree like this one is read from left to right. First, the decision tree shows that the owner must select between the actions of selling now or holding her business for one year, as indicated by the branches that represent actions. Second, the decision tree shows that the owner finds out whether the contract will be awarded, as indicated by the branches that represent events. Probabilities of events are given on the event branches. The payoffs or net monetary flows that the owner may receive, depending on the action taken and the event that occurs, are given at the endpoints of the branches at the right of the tree.

The decision tree for Equipment Leasing Services is evaluated like a payoff table. The probabilities on the event branches are multiplied by the corresponding payoffs at the ends of the branches, and these products are then summed to arrive at expected values. The expected values are then recorded in the circles at the nodes for the event branches. The owner then chooses between the actions

FIGURE 16.1

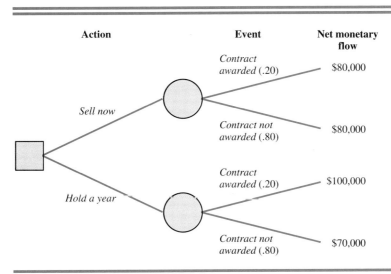

Decision Tree for the Sale of Equipment Leasing Services

and selects the action that gives the highest expected monetary value. The highest expected value is recorded in the squares at the nodes for the action branches. To keep track of the selected action at a decision node, the decision maker crosses out the other actions. The evaluated decision tree is given in Figure 16.2 and shows that the best decision, the one with the highest expected monetary value, is for the owner to sell her business now.

A few conventions are usually followed to add clarity to decision trees. *Branches* on the tree representing **actions** to be chosen by the decision maker are drawn from *squares* that are known as **decision nodes.** *Branches* on the tree representing **events** that might occur are drawn from *circles* that are called **event nodes.** Decisions are placed on the tree according to the order in which they must be made,

FIGURE 16.2

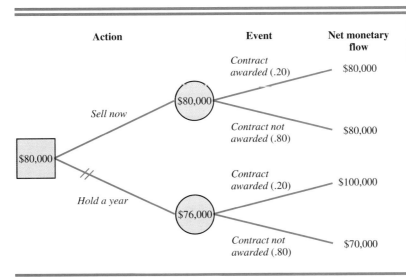

Evaluated Decision Tree for the Sale of Equipment Leasing Services

and events are placed on the tree according to the order in which the decision maker will learn about their occurrence. **Probabilities** are recorded on event branches, and the probabilities on the branches emanating from each event node must sum to *one*.

In complex decision trees decision makers often keep track of *partial monetary flows* by recording them on the branches of the tree where they take place. The partial monetary flows are summed along each path of branches to obtain a **net monetary flow** or **payoff** for each branch that ends on the right of the tree. The **payoffs** are placed at the ends of the branches, and they represent the net gains or losses if the sequence of decisions and events eventually leads along a path of nodes and branches to that endpoint.

To evaluate a decision tree, one multiplies the probabilities on the event branches for each node at the right of the tree by the corresponding payoffs at the ends of the branches and then sums them to give **expected monetary values.** If *partial monetary flows* are recorded on the tree, they are *ignored* during the evaluation of the tree. The expected values are recorded in the circles for the event nodes. Decision makers choose from among the action branches emanating from decision nodes and *select the best action,* the action that has the highest expected monetary value. They then *cross out* the actions not selected for that decision node and record the highest expected value in the square.

Decision trees are always evaluated from right to left or *backward,* with expected values always being computed for event nodes and choices among actions always being made for decision nodes. The actions that the decision maker should take are those that are *not* crossed out at the end of the evaluation process. The evaluation of decision trees from right-to-left is based on the principle of **backward induction,** which indicates that the best act can be selected at a decision node only if the best acts at decision nodes to the right have already been selected.

We usually do not represent a payoff table problem in the form of a decision tree for the simple reason that a payoff table is more compact. A payoff table is appropriate when the decision maker must select one action from several alternative actions and the payoff is dependent on the occurrence of an event. A payoff table with ten actions and ten events could easily fit on this page, but a readable decision tree describing the same problem would probably not fit the page.

If a decision analysis problem is **sequential,** in that it calls for a sequence of decisions to be made, or if there are sequences of events that determine the net monetary flows, then the problem can *only* be adequately represented with a decision tree, not with a payoff table. Consider the following problem.

▀▀▀ PRACTICE PROBLEM 16.4

ACME Manufacturing, an equipment manufacturer, operates two plants, one quite large and the other small. The smaller plant will run out of work next month and will have a one-week period when none of the currently scheduled projects can be worked on. Management is considering closing the small plant, and according to labor contracts, it must make any closure announcement now. However, it fears that if an announcement is made, employees at the larger plant will stage a protest sympathy strike. Management feels that there is about an .85 chance this event will happen. Information from the large plant indicates that there is about a .75 chance the strike will last one day and a .25 chance it will last two days. A strike at the large plant would cost the company about $450,000 per day.

If the smaller plant is left open during the slack week (five working days), there will no longer be any question of a strike, but it will cost the company $100,000 per

FIGURE 16.3

Decision Tree for ACME Manufacturing

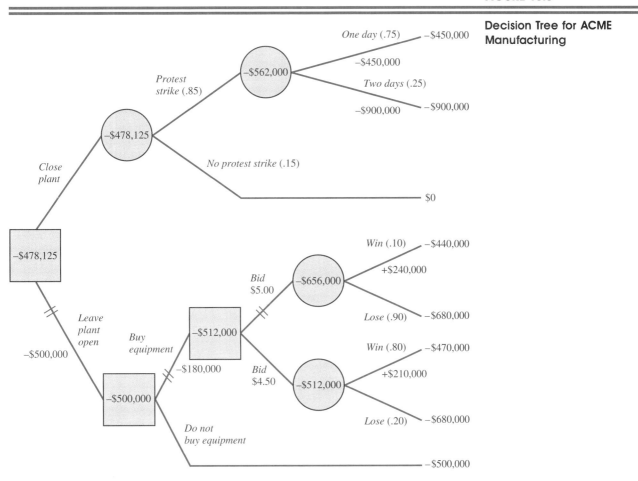

day more than if the plant were shut down. To cover part of these costs, the company might submit a bid to a local customer for manufacturing 60,000 units of a special product the company has never made before. To be eligible for this bid, the company would have to buy and install equipment costing $180,000 to show that it has the capability of doing the job, prior to entering a bid. The entire cost of the equipment would be charged against the project since there seems little likelihood the equipment would be used again in the near future. The management of the small plant cannot decide whether a bid of $5.00 per unit or $4.50 per unit would be best (bids are being accepted only in increments of $.50). It believes that there is a .10 chance it will win with a $5.00 bid and an .80 chance it will win with a $4.50 bid. It will not consider lower bids. Materials cost about $1.00 per unit. Should management announce a plant closure? If not, should it bid on the additional work? If so, what should be the bid?

Solution The decision tree for this problem is presented in Figure 16.3. The net monetary flows at the ends of the branches in this example are the net losses to the company if the sequence of decisions and events leads to that point. The partial monetary flows on the branches are the losses and gains shown at the point where they occur; they are netted out to the end and can be ignored in the evaluation of the decision tree.

The gains shown if the bid were won are calculated as contribution to fixed cost and profit; that is, (revenue − materials costs) × 60,000 units. The tree has been evaluated from *right to left,* beginning at the net dollar losses.

Figure 16.3 shows that the company should go ahead and close the small plant since that action has the lowest expected *cost,* $478,125. Actually, the real costs will be either $450,000, $900,000, or nothing; but the $478,124 figure is a weighted average of these costs. It was found by weighting the $562,500 figure in the upper circle by .85 and the zero cost of no strike by .15. The $562,500 figure, in turn, was obtained by finding the expected cost of the one- and two-day strikes.   ▬

The ACME Manufacturing decision problem illustrates that the use of a decision tree or payoff table cannot be merely a mechanical exercise. For instance, the difference between the expected monetary value of closing the plant and leaving it open is slightly under $22,000, which is not a very large amount relative to the rest of the figures in the problem. Thus, the company might want to take a more careful look at its decision. If it does, it will find that the decision is quite sensitive to the probabilities on which it bases its calculations. If the probability of a strike after the plant closure were much higher than .85, the best decision would shift to leaving the plant open. In fact, if the probability of a strike rises to .89, then leaving the plant open becomes the best decision. The new *EMV,*

$$EMV = -\$562,500(.89) + \$0(.11) = -\$500,625$$

is higher by $625 than the cost of leaving the plant open for five days.

Similar analysis of the effects of changing other probabilities can be made. Since the probabilities used in this problem were all subjective and thus subject to error, to make such a **sensitivity analysis** would be wise. The company should also investigate further to see whether it can get more reliable probability estimates. Without more reliable estimates the alternatives of closing the plant and keeping it open seem to have very close to the same desirability. If company officials do not refine their probability estimates, they might wish to employ one of the alternative decision criteria mentioned in Section 16.7 to help decide between keeping the plant open and closing it. The company also might wish to consider its social responsibility in this problem and could decide in favor of keeping the plant open since the economic factors in favor of closure do not appear to be compelling.

When the time period covered by a decision tree is a year or more, the monetary values in the problem should be discounted to present value by using an appropriate discount rate.

Problems: Section 16.4

7. Solve Problem 1 for Semiconductor Devices by using a decision tree instead of a payoff table.

8. A manager for Forest Products, Inc., needs to decide whether to postpone a burn (a prescribed fire) that has been planned for the next day on one of the company's forestry plots or to commit resources to the burn. If the burn is postponed, then the cost to Forest Products of notifying the burn team of the postponement of the burn and of placing the fire management equipment back in storage is $400.

If the resources are committed to the burn, then the burn will be carried out if tomorrow's weather is good. The cost to Forest of carrying out the burn is $3000. The local weather forecast indicates that the probability of good weather tomorrow is .5. If tomorrow's weather is bad, then the manager will cancel the burn. But a cancellation of the burn tomorrow will result in a cost of $1000 to Forest due to the mobilization of the burn team.

Vegetation characteristics and topography, among other factors, have an influence on whether or not the burn will be successful. Sixty percent of all prescribed burns by Forest in similar locations have been successful in the past. Prescribed burns that are successful result in reduced fire hazards, in enhanced control of diseases and insects, and in stimulation of new growth of the timber; the value of these benefits to Forest is $7000. Prescribed burns that are not successful result in benefits of $4000 to Forest.

a. Draw a decision tree to represent the decision facing the manager for Forest Products, Inc.
b. What should the manager do on the basis of expected monetary value?
c. What is the *EMV* for Forest?
d. If the local weather forecast reduces the probability of good weather tomorrow to .2, then what should the manager do, and what is the *EMV*? Comment on the sensitivity of the manager's decision to the probability of good weather.

9. The Board Aluminum Plating Company has been troubled recently by a process in which it does aluminum plating on an airplane engine part. The problem is due to irregular flows of electricity during the plating process. From historical data the figures show that about 80% of the time they get a uniform flow of electricity, and each batch of parts (there are 1000 in a batch) comes out 90% good. The other 20% of the time they seem to get an irregular flow of electricity during the plating process and end up with only 75% good parts.

Board engineers are currently examining three alternatives for solving this problem. The first is to do what they do now. They send the batches on to assembly where touch-up work on the defectives is done. Each touch-up costs an added $12 per part. The second alternative is to set up a special new rework area where defectives could be reworked immediately after the plating process but before they go to assembly. All items coming out of such a rework area would be good ones. However, the added fixed and variable costs of the rework area figure out to be roughly $10 per part for the first 150 parts that had to be reworked and $15 per part for every part over 150. The extra cost is due to overtime pay required after 150 parts. The third alternative would be to install a voltage regulator on the plating process. The cost accountants have spread the cost of such a device and its maintenance over the useful life of the regulator and the plating process and have determined that it costs about $2200 per batch. Even with the regulator, however, only 90% of the parts would come out as good, regardless of the uniformity of the electricity flow.

Set up Board's payoff table and determine the *EMV*'s for each alternative.

Corresponding to every payoff table is another table called the **opportunity loss table.** A decision maker might construct an opportunity loss table rather than a payoff table as an aid to choosing the best course of action. However, there is usually less work in constructing a payoff table first and then converting it to an opportunity loss table, as the following example will illustrate.

16.5
OPPORTUNITY LOSS TABLE

══ **EXAMPLE 16.6** International Container, Inc., must decide where it will locate a new plant to produce a new packing product. The plant is to be constructed in London, Beijing, or New York (A_1, A_2, and A_3), and the major market for the

	Event, Major Market	Probability	Action (Plant Location)		
TABLE 16.10			A_1, London	A_2, Beijing	A_3, New York
Payoff Table for International Container	E_1, Europe	.45	40	12	−5
	E_2, Far East	.30	−10	50	18
	E_3, United States	.25	28	−2	20

product may be in Europe, the Far East, or the United States (E_1, E_2, and E_3). The profits (in millions of dollars) for the new product are given as payoffs in Table 16.10.

To construct an opportunity loss table, we proceed as follows. We assume that one of the events has already taken place. Then for each of the possible actions we ask ourselves the question, How much better off could we have been if we had chosen one of the other actions open to us? For example, assume that event 1 has occurred. Then assume that action 1 was the chosen action and a payoff of 40 has been received. We now ask ourselves, How much better off could we have been if we had chosen either of the other two actions, action 2 or action 3? Since their payoffs are 12 and −5, respectively, we could not have put ourselves in a better situation by choosing either of them. Thus, action 1 has zero opportunity loss associated with it when event 1 occurs; we did not have the opportunity to do better than a payoff of 40. In other words, building the plant in London has zero opportunity loss associated with it when the major market for the product turns out to be Europe; we did not have the opportunity to do better than a payoff of 40 even if we had built the plant in Beijing or New York.

Next, we assume that instead of action 1 we had chosen action 2 and that event 1 occurred. Under these conditions our payoff would be 12. We ask ourselves again. How much better off could we have been by choosing some other action? If we had chosen action 3, we would have been worse off since the payoff there is −5. But if we had chosen action 1 instead of action 2, we would have gained 40 and thus could have been better off by 40 − 12, or 28. By choosing action 2 when event 1 occurs, we would *lose the opportunity* to gain an additional 28 units of payoff. Thus, the opportunity loss associated with event 1 and action 2 is 28. In other words, we have forgone profits of 28 by building our plant in Beijing rather than in London when the major market for the product happens to be Europe, because we would have made profits of 40 if we had built the plant in London. However, now we will make only 12 since we built the plant in Beijing.

To obtain the opportunity loss for event 1 and action 3, we proceed in the same way. In the case that event 1 occurs, either action 1 or action 2 will yield a better payoff than the payoff value of −5 associated with action 3. Action 1, with payoff 40, again is the best action, so choosing action 3 when event 1 occurs has an opportunity loss of 40 − (−5), or 45. That is, we would be better off by 45 payoff units if action 1 were chosen instead of action 3 when event 1 occurs. In other words, we would be better off by 45 payoff units if we had built the plant in London instead of in New York when the major market for the product happens to be Europe.

The opportunity loss table is sometimes called the *regret table*. The values in this table contain measures of how badly we might feel if we chose the wrong action. That is, these values measure the feelings of regret we would experience.

In general, the opportunity loss for any action in a particular row of a payoff table is computed as the difference between that action's payoff and the best payoff in that row. If we let X_i^* be the best payoff in the ith row, then the **opportunity loss** (OL_{ij}) for action j in the face of event i is as given by the following formula.

Equation for Opportunity Loss

$$OL_{ij} = |X_i^* - X_{ij}| \qquad (16.3)$$

We use the absolute value sign in Equation (16.3) for the following reason. If the payoff table being used contains *profits* as entries, then X_i^* will be the *largest* value in the ith row, and OL_{ij} will be a positive number. But if the payoff table contains *costs*, then X_i^* is the *smallest* value in the ith row; and if we do *not* take the absolute value of $X_i^* - X_{ij}$, then OL_{ij} will be negative. Since opportunity loss figures are customarily recorded as positive numbers, the absolute value sign in Equation (16.3) allows us to record the *difference* between X_i^* and X_{ij} as a positive value in the cases where the payoff table lists costs.

EXAMPLE 16.7 The computations for the opportunity losses associated with the second and third events of Table 16.10 for the International Container plant location problem are as follows:

$$OL_{21} = |50 - (-10)| = 60 \qquad OL_{22} = |50 - 50| = 0$$

$$OL_{23} = |50 - 18| = 32 \qquad OL_{31} = |28 - 28| = 0$$

$$OL_{32} = |28 - (-2)| = 30 \qquad OL_{33} = |28 - 20| = 8$$

Table 16.11 is the entire opportunity loss table corresponding to the plant location payoff Table 16.10.

Several observations should be made about Table 16.11. First, it contains all positive or zero values. Even though the entries in the table are considered losses, they are made positive, and the user of the table simply remembers that he or she is

Event, Major Market	Probability	Action (Plant Location)			TABLE 16.11
		A_1, London	A_2, Beijing	A_3, New York	
E_1, Europe	.45	0	28	45	**Opportunity Loss Table for the International Container Plant Location Problem**
E_2, Far East	.30	60	0	32	
E_3, United States	.25	0	30	8	

dealing with a table of losses. Use of the absolute value sign in Equation (16.3) assures us that we will always have positive entries. Second, there is one zero value in each row of the table. There will always be at least one zero value in each row of an opportunity loss table. The zero value appears at the position occupied by the best payoff value for each row of the corresponding payoff table. Finally, the entries in the opportunity loss table are measures of the cost to the decision maker of not knowing which event is going to occur. If the decision maker knew in advance what event was to take place, he would always choose the best action and would never incur an opportunity loss. The positive opportunity losses in the table, then, are measures of the losses he may incur simply through not knowing in advance what event will occur. They are measures of the cost of the uncertainty he faces.

The calculation of opportunity losses associated with an event in a payoff table of *costs* is demonstrated in the next example.

■■ EXAMPLE 16.8   Assume that event i has the following *costs* associated with it:

Event	Act 1	Act 2	Act 3
Event i	5000	3500	6500

We can find the opportunity loss associated with the first act by using Equation (16.3). In this case, $X_i^* = 3500$ since it is the *lowest cost* associated with event i. Thus,

$$OL_{i1} = |3500 - 5000| = 1500$$

$$OL_{i2} = |3500 - 3500| = 0$$

$$OL_{i3} = |3500 - 6500| = 3000$$

That is, the opportunity loss associated with action 1 and event i is 1500, because when event i occurs and we have chosen action 1, the cost is 5000. But we could have been 1500 *better off* had we chosen action 2 and incurred a cost of only 3500. The opportunity loss associated with action 2 is zero since action 2 is the lowest-cost act when event i occurs. Finally, action 3 has an opportunity loss of 3000 since we could have saved this much by choosing action 2 at a cost of 3500 rather than action 3 at a cost of 6500.   **■■**

How does a measure like opportunity loss help determine the best course of action? In the previous section we approached the problem of selecting the best action by calculating the expected monetary value of each of the possible actions and choosing the action with the highest *EMV*. Let us look at an example.

■■ EXAMPLE 16.9   The calculation of the *EMV* for the International Container plant location payoff table, Table 16.10, yields the following results:

$$EMV_1 = 40(.45) - 10(.30) + 28(.25) = 18.0 - 3.0 + 7.0 = 22.0$$

$$EMV_2 = 12(.45) + 50(.30) - 2(.25) = 5.4 + 15.0 - .5 = 19.9$$

$$EMV_3 = -5(.45) + 18(.30) + 20(.25) = -2.25 + 5.4 + 5.0 = 8.15$$

By the criterion of selecting the action with the highest *EMV*, action 1, with $EMV^* = 22$, appears to be the best.   **■■**

An alternative criterion for Example 16.9 is suggested by the measures of the cost of our uncertainty about which event will occur. That is, we may want to find the **expected opportunity loss for the jth action** (EOL_j) and then select the action that has the *minimum* expected opportunity loss, which we denote EOL^*, rather than the action with the highest expected monetary value EMV^*. The expected opportunity loss EOL for each action is calculated the same way as the EMV value, except that we work with the opportunity losses in the opportunity loss table, as follows.

Equation for the Expected Opportunity Loss for the jth Action

$$EOL_j = \sum_{i=1}^{m} OL_{ij}P(E_i) \qquad j = 1, 2, \ldots, n \qquad (16.4)$$

The summation is over m events, and there are n actions to choose from. Equation (16.4) simply tells us that to compute an expected opportunity loss EOL for an action, we must select a column in the opportunity loss table (since actions are arranged in columns), multiply each opportunity loss in the column by its corresponding probability, and then add the products. We then repeat the procedure for each column or action.

The calculations for the expected opportunity losses for the International Container plant location problem, using the opportunity loss table, Table 16.11, are as follows:

$$EOL_1 = 0(.45) + 60(.30) + 0(.25) = 0 + 18.0 + 0 = 18.0$$

$$EOL_2 = 28(.45) + 0(.30) + 30(.25) = 12.6 + 0 + 7.5 = 20.1$$

$$EOL_3 = 45(.45) + 32(.30) + 8(.25) = 20.25 + 9.6 + 2.0 = 31.85$$

So action 1 is also the most desirable action if we choose to minimize the expected opportunity loss, since $EOL^* = EOL_1 = 18.0$ is the lowest of the three figures.

It is no coincidence that action 1 has both the lowest EOL and the highest EMV. It is always true that the action with the best expected monetary value computed from the payoff table also has the lowest expected opportunity loss computed from the opportunity loss table.

Problems: Section 16.5

10. Refer to the Semiconductor Devices scenario of Problem 1.
    a. Find the opportunity loss table.
    b. Find the EOL for each action, and give the action with the minimum expected opportunity loss EOL^*.

11. Refer to the Atlantic Tool Company scenario of Problem 2.
    a. Find the opportunity loss table.
    b. What action should Atlantic take on the basis of the EOL? Why?

12. Refer to the International Building, Ltd., construction project scenario in Problem 3.
    a. Find the EOL for each action (you may omit any dominated actions).
    b. Select the best action on the basis of the EOL.

16.6
EXPECTED VALUE OF PERFECT INFORMATION

We can find the best decision for problems discussed in this chapter by using the payoff table or the opportunity loss table. These tables simply present two views of the same problem. However, there is one advantage to using the opportunity loss table. Since the entries in this table represent the costs of not knowing what event is going to take place, the expected opportunity loss of the best action, denoted *EOL**, represents the expected cost of uncertainty for the best action. That is,

$$EOL^* = Expected\ cost\ of\ uncertainty \tag{16.5}$$

This relationship is important because it gives an upper limit on how much decision makers should be willing to spend to eliminate the uncertainty facing them. The most anyone would spend to remove uncertainty is the amount that the uncertainty costs. Thus, decision makers should be willing to pay an amount equal to or less than the expected cost of uncertainty for a perfect predictor of which event will take place. This amount is called the **expected value of perfect information** (*EVPI*):

$$EVPI = EOL^* \tag{16.6}$$

According to this relationship, the person faced with the payoff table (Table 16.10) should be willing to pay a maximum of 18.0 units for information about which event is going to occur. Actually, the person would pay substantially less if the information were only a forecast and not a guaranteed accurate prediction.

▰▰ PRACTICE PROBLEM 16.5

Consider the coin-tossing game discussed in Section 16.1. The opportunity loss table for that game is presented in Table 16.12. Calculate the expected opportunity losses.

Solution The *EOL*'s are as follows:

$$EOL_1 = 0(.500) + 0(.250) + 0(.125) + 8(.125) = 0 + 0 + 0 + 1.0$$
$$= 1.0$$
$$EOL_2 = 1(.500) + 2(.250) + 4(.125) + 0(.125) = .5 + .5 + .5 + 0$$
$$= 1.5$$

So *EOL** = 1.0, and action 1 is the preferred action. Also, *EVPI* = *EOL** = 1.0. Thus a player would be willing to pay up to $1 per play to learn in advance which event was going to occur.

			Action	
TABLE 16.12	*Event*	*Probability*	*A₁, Play*	*A₂, Not Play*
Opportunity Loss Table	E_1, *head on first toss*	.500	0	1
for Coin-Tossing Game	E_2, *first head on second toss*	.250	0	2
	E_3, *first head on third toss*	.125	0	4
	E_4, *no head appears*	.125	8	0

| Event | Probability | Action | | TABLE 16.13 |
		A_1, Sell Now	A_2, Hold a Year	
E_1, contract awarded	.20	20	0	**Opportunity Loss Table**
E_2, contract not awarded	.80	0	10	**for Sale of Equipment**
				Leasing Services (Payoffs
				in Thousands of Dollars)

 PRACTICE PROBLEM 16.6

The Equipment Leasing Services business owner who has the opportunity to sell her business now or hold it for one year may construct from the payoff table (Table 16.7) the opportunity loss table presented as Table 16.13. From the tabulated values, calculate the expected opportunity losses.

Solution The *EOL*'s are as follows:

$$EOL_1 = 20(.20) + 0(.80) = 4.0 + 0 = 4.0$$

$$EOL_2 = 0(.20) + 10(.80) = 0 + 8.0 = 8.0$$

So *EOL** is $4000, and the action "sell now" is the best action. But the owner should be willing to spend up to $4000 to gain more information about her chances of winning the contract, since $EOL^* = EVPI$.

There is a second way of calculating the expected value of perfect information that may help to clarify its meaning.

EXAMPLE 16.10 Once again, consider the coin-tossing game. The payoff table for this game is reproduced again as Table 16.14. Let us assume for a moment that the decision maker is offered the chance to play this game many times in a row. If he knew in advance that event 1 were going to occur—that is, if he had a perfect predictor—he would surely play since he would gain $1. Also, if he knew that E_2 or E_3 were going to occur, he would play and receive $2 or $4. But if he knew E_4 were going to occur, he would surely decline the offer to play and end up gaining and losing nothing. Thus a perfect predictor would allow the decision maker to earn $1 on 50% of the plays, $2 on 25% of the plays, $4 on 12.5% of the plays, and $0 on 12.5% of the plays.

The average payoff per play using a perfect predictor is called the **expected payoff under certainty** (*EPUC*). For the betting game this value is

$$EPUC = 1(.500) + 2(.250) + 4(.125) + 0(.125)$$

$$= .50 + .50 + .50 + 0 = 1.50$$

| Event | Probability | Action | | TABLE 16.14 |
		A_1, Play	A_2, Not Play	
E_1, head on first toss	.500	1	0	**Payoff Table for**
E_2, first head on second toss	.250	2	0	**Coin-Tossing Game**
E_3, first head on third toss	.125	4	0	
E_4, no head appears	.125	−8	0	

Using a perfect predictor, the decision maker could average $1.50 per play. In Section 16.3 we found that the expected monetary value of the decision to play the game every time was $.50. If the decision maker can average $.50 per play without perfect information and $1.50 per play with perfect information, he knows that the per-play value of the perfect information is $1.50 − $.50, or $1.00.

If **expected value of perfect information *EVPI*** is the amount by which the expected payoff under certainty differs from the *largest* expected monetary value, denoted *EMV**. Or

$$EVPI = |EPUC - EMV^*| \qquad (16.7)$$

Since according to Equation (16.6) the expected value of perfect information also equals the expected opportunity loss of the best action, we have the following equation.

Equation for Expected Value of Perfect Information

$$EVPI = |EPUC - EMV^*| = EOL^* \qquad (16.8)$$

In Practice Problem 16.5 the *EOL* of the best action (to play the game) was calculated to be $1, so the relationship checks out.

 PRACTICE PROBLEM 16.7

In Practice Problem 16.6 the *EVPI* for the Equipment Leasing Services business owner facing the decision about selling her business was calculated to be $4000. Does this result coincide with the value of the *EVPI* calculated by the alternative method?

Solution If the owner had a perfect predictor and knew for sure which event was going to take place, then when event 1 was imminent, she would choose action 2, to hold, and gain 100. This information can be seen from Table 16.9. When event 2 was imminent, she would choose action 1, to sell, and gain 80. These two possibilities are presented in Table 16.15.

Since there is only a 20% chance that a perfect predictor would forecast E_1 and an 80% chance it would forecast E_2, the *EPUC* is

$$EPUC = 100(.20) + 80(.80) = 20 + 64 = 84$$

In Section 16.3 the *EMV** was found to be $80,000, so by Equation (16.7) the expected value of perfect information is found to be

$$EVPI = |84 - 80| = 4$$

or since all figures are in thousands of dollars, $4000.

Event Known to Be Imminent	Best Action	Payoff for Best Action	**TABLE 16.15**
E_1, contract awarded	A_2, hold	100	**Best Payoffs under Certainty for Sale of Equipment Leasing Services**
E_2, contract not awarded	A_1, sell	80	

Problems: Section 16.6

13. Refer to the Semiconductor Devices scenario of Problem 1.
    a. Find the *EPUC*.
    b. A market research firm will do a study for the company to determine just what market share it can capture. The survey will cost $72,000 and is guaranteed to be accurate. Should the company have the survey done? Why?

14. Refer to the Atlantic Tool Company scenario of Problem 2.
    a. Find the *EPUC*.   b. Find the *EVPI*.
    c. Compare the value of *EVPI* to the value of *EOL**.
    d. What is the maximum amount a decision maker faced with this table would be willing to pay for information about which event is going to occur?

15. Refer to the International Building, Ltd., construction project scenario in Problem 3.
    a. Find the *EPUC*.   b. Find the *EVPI*.

16.7
DECISION CRITERIA

By using expected monetary value as the basis for making decisions, the decision maker will end up doing well on the average or in the long run. But if someone is very worried about the short run or doesn't believe that the laws of probability hold in his case, then he might do better by abandoning the use of expected monetary value in evaluating alternative courses of action. Several other methods for selecting a course of action are presented in this section. Their use will be demonstrated for the payoff table given as Table 16.16.

The **optimist's criterion** is sometimes called the *maximax criterion*. Simply stated, it suggests that the decision maker should select that action with the best payoff. Since 20 is the highest payoff in Table 16.16, the optimist decision maker would select the action associated with that payoff, A_4. This decision ignores completely the fact that event 3, which must occur if the decision maker is to realize a payoff of 20, is the least likely of the three events. Even if E_3 had a probability of .001, the optimist decision maker would make the same choice. The optimist's criterion ignores event probabilities and all payoffs in the table except the best one. Thus even though A_4 is associated not only with the highest but also with some of the lowest payoffs in the table, the optimist's decision is still A_4.

Event	Probability	Action

Event	Probability	A_1	A_2	A_3	A_4	**TABLE 16.16**
E_1	.45	8	−6	12	−4	**Demonstration Payoff Table**
E_2	.30	10	12	15	−8	
E_3	.25	6	18	−10	20	

One can see why this criterion is called the optimist's criterion. A person who is truly optimistic (and only optimistic) *would* ignore the probabilities of events and possible adverse payoffs in order to obtain a large payoff. The reason for the name *maximax criterion* is not so clear. It can be explained in the following way. Suppose that the decision maker examined each action separately and chose the maximum payoff associated with each. For the situation represented in Table 16.16 he would find that the four actions had maximum payoffs of 10, 18, 15, and 20, respectively. Then he would choose the action that produced the *maximum* of these *maximums*. By choosing the largest payoff in the table, the maximax decision maker behaves as though he were selecting the action that maximizes the maximum payoff for each action. This two-step maximization process leads to the term *maximax*.

However, this term cannot be applied to the process that would be employed by an optimist approaching a payoff table of costs. The optimist would look at each action and list the *minimum* cost associated with each action and then choose the *minimum* of these minimum costs. Thus, the optimist's criterion would have to be called the *minimin criterion* when applied to payoff tables filled with costs. To avoid this confusion, we will use only one term: the *optimist's criterion*.

Given the obvious drawbacks of the optimist's criterion—put simply, that it ignores everything in the payoff table except the best payoff—one might wonder why anyone would consider it as an aid to decision making. There are very few circumstances where this method of selecting an action is valid. A manager whose business was in danger of bankruptcy and who needed a large payoff in order to save it from ruin might rationally choose the largest payoff in the table as a guide. For instance, a manager who needs 19 units or more to save a business might use the optimist's criterion for payoffs in Table 16.16, since only the optimistic decision will give the amount needed to remedy the situation. In this case, that decision seems reasonable.

The **pessimist's criterion** is, in some respects, the opposite of the optimist's criterion. A pessimist assumes that no matter what action is taken, the worst possible event will take place. Thus, the pessimistic decision maker feels that if she takes action 1 in Table 16.16, event 3 is bound to happen since it yields the lowest payoff for A_1, a payoff of 6. And if a pessimist were to take action 2, she feels that event 1 is sure to take place since that would yield the worst possible payoff for A_2, a value of -6. In short, no matter which action she takes, she is bound to end up with the minimum payoff for that action: 6, -6, -10, or -8. But even a pessimist will make the best of a bad situation. Thus, the pessimistic decision maker selects the action that will *maximize* these *minimum* outcomes. This action is action 1, with a guaranteed payoff of at least 6. (This two-step process leads some people to call the pessimist's criterion the *maximin criterion*.)

The pessimistic decision maker ignores the probabilities in the payoff table and concentrates only on the adverse outcomes of each action. This method of selecting an action guarantees against large losses and establishes a floor below which the payoff cannot fall. However, it also leads to decisions where large payoffs may not be realized even though they may be quite probable.

The pessimist's decision criterion is appropriate in situations where an organization or person is in financial difficulty and cannot afford a large loss. It may be used to select those actions that ensure against the chances of disastrous results at the expense of passing up possible good results. This approach to decision making is very conservative and is appropriate only in cases where it is imperative that adverse results be avoided.

The **maximum-likelihood criterion** does not ignore probabilities. In fact, the first step in making a maximum-likelihood decision is to determine which event is most likely to occur. In Table 16.16 that event is event 1, which has probability .45 of occurring. The second step in applying the maximum-likelihood criterion involves listing the payoffs associated with that most probable event. For our example the payoffs associated with E_1 are 8, −6, 12, and −4. The final step is selecting the action with the best payoff among these. Thus the maximum-likelihood decision maker would select action 3, since the payoff of 12 is the maximum value for the payoffs of the most likely event.

This criterion ignores the probabilities and payoffs of all the events in the table except that one event with the greatest probability of occurrence. This approach would not be very appropriate if the probabilities of the events were, say, $P(E_1) = .34$, $P(E_2) = .33$, and $P(E_3) = .33$. In this case event 1 is the most probable again, but to concentrate all attention on E_1 and ignore the other two possible events would seem inadvisable.

There are two situations when the maximum-likelihood criterion would seem appropriate. The first is the case in which one event is so much more probable than the others that it seems foolish to consider anything but that event. Thus, if $P(E_1) = .98$, $P(E_2) = .01$, and $P(E_3) = .01$, the maximum-likelihood criterion seems appropriate. We all use this decision rule when we cross a street. Although there is a small chance that we will be hit by a car while crossing, the probability of our making it across safely is so high that we usually ignore the consequences involved with an accident's occurring even though these consequences might be very serious. A second situation where the maximum-likelihood criterion is appropriate is the one in which choosing a course of action involves a great deal of preparation for carrying out the action even before the event is known. If the decision maker has time or resources to prepare for only one of the possible events, which one should she prepare for? The one that has the highest probability of occurring is the logical choice.

The **equal-likelihood criterion** is used when the decision maker considers each of the events in the payoff table to be equally likely. For instance, when the decision maker has absolutely no idea about the relative probabilities of the events or he has *insufficient reason* to assign unequal probabilities to the events, he may have to assume that each is as likely as any other. So the decision maker assigns equal probabilities to each of the events and proceeds as though he were calculating the expected monetary value of the actions. For Table 16.16 the equal-likelihood decision maker would assume $P(E_1) = 1/3$, $P(E_2) = 1/3$, and $P(E_3) = 1/3$. Then he would calculate the EMV of each action by using these probabilities:

$$EMV_1 = 8(1/3) + 10(1/3) + 6(1/3) = 24/3 = 8.00$$

$$EMV_2 = -6(1/3) + 12(1/3) + 18(1/3) = 24/3 = 8.00$$

$$EMV_3 = 12(1/3) + 15(1/3) - 10(1/3) = 17/3 = 5.67$$

$$EMV_4 = -4(1/3) - 8(1/3) + 20(1/3) = 8/3 = 2.67$$

Thus the equal-likelihood decision maker would be indifferent between actions 1 and 2 since they both have EMV's of 8.00 when the events are considered equally likely.

The preceding calculation illustrates that there is a somewhat faster way of determining the EMV of an action when all the probabilities are equal. This method involves summing all the payoffs in one column of the payoff table and then dividing the sum by the number of possible events. For A_1 the sum of the payoffs in Table 16.16 is 24, and EMV_1 is 24/3, or 8.00.

The equal-likelihood criterion gives equal weight to every event and thus equal weight to every payoff in the table. It takes into account all the possible payoffs but does not allow for the fact that events might have unequal probabilities of occurring. There are likely to be few situations in which a decision maker cannot distinguish any difference between the relative probabilities of the events in his payoff table.

The **expected monetary value criterion** is sometimes called **Bayes' criterion** after the Reverend Thomas Bayes, an eighteenth-century English minister and mathematician. This criterion was fully explained in Section 16.3. It has the advantage of combining both the probabilities of the events and the payoffs of the actions in a logical way. As we have learned, the Bayesian decision maker makes good decisions in the long run, or on the average. Only when some of the situations discussed previously arise should the decision maker revert to some of the other decision-making criteria.

The Bayesian decision for Table 16.16 is found by using the usual *EMV* calculations:

$$EMV_1 = 8(.45) + 10(.30) + 6(.25) = 3.6 + 3.0 + 1.5 = 8.1$$

$$EMV_2 = -6(.45) + 12(.30) + 18(.25) = -2.7 + 3.6 + 4.5 = 5.4$$

$$EMV_3 = 12(.45) + 15(.30) - 10(.25) = 5.4 + 4.5 - 2.5 = 7.4$$

$$EMV_4 = -4(.45) - 8(.30) + 20(.25) = -1.8 - 2.4 + 5.0 = .8$$

The action with the highest *EMV* (and also the lowest *EOL*) is action 1: $EMV_1 = EMV^* = 8.1$.

Most decision makers would be well off if they used only the criterion of expected monetary value in making their decisions, valuing the other criteria chiefly as tiebreakers. If two or more actions have the same expected monetary values, decision makers can employ one of the other criteria to help them choose between the tied actions. If they desire to insure themselves against large losses, they may revert to the pessimist's criterion to break the tie, for instance.

Problems: Section 16.7

16. Several people looked at the same payoff table and made the comments listed in parts a through h. Examine their comments and determine which of the following decision-making criteria the different speakers would likely use:

> Optimist's
> Pessimist's
> Maximum likelihood
> Equal likelihood
> Expected monetary value

a. "No matter what I do, the worst always happens. I'm going to protect myself against the inevitable disaster."
b. "I play the averages. That way I come out best in the long run."
c. "There won't be any 'long run' for me if I don't strike it rich on this deal. I've got to have that big profit or I'm sunk."
d. "None of those probabilities look right to me. I just don't think you can tell what's going to happen."

e. "All of the probabilities in that table are too small to consider except for event 2."
f. "No matter what I do, I come out smelling like a rose. I might as well go for broke."
g. "I'm going to make my choice in such a way that I give weight to both the payoffs and the probabilities."
h. "I've only got time to prepare for one of those events. I just can't consider more than one at a time."

17. Refer to the International Building, Ltd., payoff table in Problem 3.

   a. Find the optimist's decision.
   b. Find the pessimist's decision.
   c. What is the maximum-likelihood decision?
   d. Find the best decision assuming equal likelihood for all three events.

18. Two professional football teams are playing in the Super Bowl. Assume that there are only 48 seconds left in the game and team A is leading team B by the score of 27 to 21.

   a. Team B has fourth down and six on their own 33 yard line. The quarterback decides to go for a touchdown and throws a long, deep pass. What decision criterion is he using?
   b. Assume that the pass in part a was incomplete and team A takes over with 43 seconds left in the game. Team A then runs out the clock by running four straight plays in which the quarterback merely falls on the ball after taking it from the center. What decision criterion is team A using here?

In all the examples in this chapter we have assumed that a rational decision maker would choose the action with the highest expected monetary value. That is, we have assumed that expected monetary value is the best decision-making criterion. This assumption seems logical in many instances. However, consider the following situation.

16.8
UTILITY

EXAMPLE 16.11 A casino in Monte Carlo offers you two alternatives. The casino will pay you $10,000 for sure, *or* it will allow you to engage in this game: You toss a coin; if it comes up heads you win $200,000, and if it comes up tails you must pay $160,000. Your payoff table is presented in Table 16.17.

Suppose you decide to approach this situation by figuring the expected monetary value for each of the two actions. Thus,

$$EMV_1 = 10,000(.5) + 10,000(.5) = 5000 + 5000 = 10,000$$

$$EMV_2 = 200,000(.5) - 160,000(.5) = 100,000 - 80,000 = 20,000$$

The decision to toss the coin has an expected monetary value of $20,000 and thus appears to be better than the decision simply to accept the $10,000. Yet

		Action		
Event	*Probability*	A_1, *Sure Thing*	A_2, *Toss Coin*	**TABLE 16.17**
E_1, *heads*	.5	10,000	200,000	**Payoff Table for**
E_2, *tails*	.5	10,000	-160,000	**the Gamble**

you might justifiably feel that you would rather have the $10,000 straight out than toss the coin and run the risk of having to pay the gambler $160,000. That is, you may prefer a sure $10,000 to a gamble with an *EMV* of $20,000. This very natural feeling is based on the fact that people make decisions in terms of the *utility* involved in the problem. The reason that many people would take $10,000 for sure instead of tossing the coin is that their utility for the sure $10,000 is higher than their utility for the gambling situation—even though the gambling situation has a higher expected monetary value.   ▬▬

A person's **utility** for an amount of money, a gamble, or even a physical object is, in a sense, the amount of well-being these things produce in the person. Utility is subjective and varies among people. One person's utility for an amount of money may be very different from another person's utility for the same amount. A million-aire might easily choose to engage in the gamble of tossing a coin for a $200,000 gain or a $160,000 loss. A millionaire could afford the loss of $160,000 without much difficulty, and the possible gain of $10,000 is peanuts.

Since sheer amount of money does not always measure desirability or utility to a decision maker, the payoff tables discussed in the previous sections should be filled in with utility figures rather than dollar amounts. If they were, decision makers would choose the action with the highest expected utility value rather than the highest expected monetary value. The problem with using utility figures is that it is very difficult to measure a person's utility. This section will not deal with actual nu-merical measures of utility. Only cases where monetary value serves as a good mea-sure or where different techniques can be used to express utility will be discussed.

▬▬ **EXAMPLE 16.12**   Let us reconsider the gambling situation presented in Ex-ample 16.11 and change the scale of the problem. Suppose you are offered the chance to take $.10 for sure *or* toss a coin and receive $2.00 for a head and pay $1.60 for a tail. This is the same problem as the one in Example 16.11 except that the consequences are 100,000 times smaller.

The expected value of the sure thing is $.10, and the expected value of toss-ing the coin is $.20. Many people who would not have tossed the coin when the payoffs were $200,000 and −$160,000 would not hesitate to do so with the smaller payoffs. In general, for small amounts of money the monetary value is a good measure of utility. In cases involving relatively small dollar amounts, the action with the highest expected monetary value is also the one with the highest expected utility.

But what does *relatively small* mean? It means small relative to the total wealth of the person or organization making the decision. Thus a millionaire would likely be willing to toss the coin for the large stakes, but a college student would not. However, the student might change his or her mind once the payoffs had been reduced by a factor of 100,000. That is, $20,000 seems a small amount to some very wealthy people, just as $.20 seems a small amount to most college students.   ▬▬

Thus expected monetary value seems to be a good decision-making criterion when the payoffs involved are small relative to the overall wealth, assets, or budget of the decision maker. But when the payoffs are large, the utility criterion must be con-sidered. How do decision makers take utility into account in the situation involving relatively large payoffs? They might consider using an alternative decision-making criterion like one of those discussed in the previous section.

The *decision analysis* approach to decision making was presented in this chapter. The decision in this type of problem depends on alternative actions and events and their probabilities and on the associated economic consequences.

In this chapter, *payoff tables, decision trees,* and *opportunity loss tables* were introduced to allow us to organize actions and events and their associated payoffs, costs, or opportunity losses. The criteria of selecting the best action on the basis of the *highest expected monetary value* or the *lowest expected opportunity loss* were presented.

Decision trees and a right-to-left solution procedure were introduced for selecting the best set of actions in problems involving a sequence of decisions or events.

The concept of the *expected value of perfect information* was also discussed. Equations that are important in decision analysis are summarized in Table 16.18. Decision criteria that do not account for probabilities, including the optimist's, pessimist's, maximum-likelihood, and equal-likelihood criteria, were also presented.

TABLE 16.18 Summary of Equations for Decision Analysis

Description	Equation
Expected monetary value of the jth action	$$EMV_j = \sum_{i=1}^{m} X_{ij}P(E_i) \quad \text{for} \quad j = 1, 2, \ldots, n \qquad \textbf{(16.2)}$$ where: X_{ij} = the payoff for the ith event and the jth action $\quad\quad P(E_i)$ = the probability of the ith event
Opportunity loss for the ith event and jth action	$$OL_{ij} = |X_i^* - X_{ij}| \qquad \textbf{(16.3)}$$ where: X_i^* = the best payoff in the ith row
Expected opportunity loss of jth action	$$EOL_j = \sum_{i=1}^{m} OL_{ij}P(E_i) \quad \text{for} \quad j = 1, 2, \ldots, n \qquad \textbf{(16.4)}$$
Expected value of perfect information	$$EVPI = |EPUC - EMV^*| = EOL^* \qquad \textbf{(16.8)}$$ where: $EPUC$ = the expected payoff under certainty $\quad\quad EMV^*$ = the highest (best) expected monetary value $\quad\quad EOL^*$ = the lowest (best) expected opportunity loss

Answers to odd-numbered problems are given in the back of the text.

REVIEW PROBLEMS

19. The accompanying payoff table represents the cost (in thousands of dollars) to a county of building a section of new road under three alternative proposals. The cost depends not only on the nature of the proposal but also on the weather conditions that might exist during the winter.

   The county has received a bid on the work indicating that the job could be done by a contractor for a flat $120,000. Alternatively, the contractor would be willing to begin the work in the fall as the first stage of a two-stage contract. Payment for the second portion of the work would depend on the severity of the winter. The costs of the two-stage contract are presented in the payoff table. Finally, the county could do the work itself, and past experience and cost estimates indicate that the costs depend to a great extent on the weather. The cost estimates for this proposal are also presented in the table.

a. Find the action with the best *EMV*.
b. If the county could pay for a meteorological study of the likely winter conditions in this area, what is the maximum amount it should be willing to pay for it?
c. What would the optimist's decision be?
d. What would the pessimist's decision be?
e. Find the decision assuming equally likely events.
f. What is the maximum-likelihood decision?

		Action		
Event	*Probability*	A_1, *Flat-Fee Contract*	A_2, *Have County Do It*	A_3, *Two-Stage Contract*
E_1, *light winter*	.25	120	70	100
E_2, *moderate winter*	.50	120	105	120
E_3, *bad winter*	.25	120	150	130

20. The accompanying payoff table for Agriculture Chemicals, Inc., presents the profits (in thousands of dollars) for manufacturing a product where good-, moderate-, or poor-quality materials can be selected (A_1, A_2, and A_3) to produce a product, and the demand for the product might be high or low (E_1 and E_2).

a. Find the action with the highest *EMV*.
b. How much can the probabilities change, and in what direction, before the preferred action will change?
c. Under what probability combinations for E_1 and E_2 will action A_2 become the action with the highest *EMV*?
d. Assuming the original probabilities hold, how much can the profit $X_{23} = 400$ fall before A_3 is no longer a desirable action?

		Action (Quality)		
Event	*Probability*	A_1, *Good*	A_2, *Moderate*	A_3, *Poor*
E_1, *high demand*	.70	100	60	50
E_2, *low demand*	.30	70	200	400

21. Mr. Ernest Morney speculates in real estate. Currently, he has an option to buy a piece of property. The value of the property depends on whether the City Planning Commission will rezone it to the classification Mr. Morney desires. The commission will not meet until late next week, at which time it will decide to take one of three steps: rezone the property as Mr. Morney desires, deny the rezoning application outright, or table the request until the next meeting some six weeks away. However, Mr. Morney is in tight financial circumstances. He has to complete the deal on this piece of property within three weeks to free up his resources for another transaction at that time. The problem is compounded by the fact that his option to buy this property expires tomorrow, and the property owner wants him to revew it or buy the property outright today. The chances associated with the requested rezoning change seem to be about .15 for approval and about twice that for denial. The monetary values involved are the following:

Cost of the new option $10,000	Cost of land purchased now $150,000
Sales price if rezoning denied $135,000	Sales price if rezoned $250,000
Cost of land under new option $155,000	Sales price if tabled $170,000

a. Construct the payoff table facing Mr. Morney. Assume that if it is not profitable to purchase the land under a set of circumstances, he will not do so. [*Hint:* Make sure that one of the possible actions you consider is "do nothing".]

b. Are any of the actions dominated by the others? If so, which one(s)?

c. Find the *EMV* of each action.

d. What is the maximum amount Mr. Morney would be willing to spend in finding out what the commission is likely to do?

22. The accompanying payoff table presents the profits for an AAA Industries, Ltd., manufacturing operation (in thousands of dollars), where there are three alternative machines that can be selected (A_1, A_2, and A_3) to produce a product, and the demand for the product might be high, moderate, or low (E_1, E_2, and E_3).

a. Find the *EMV* for each action.   **b.** Find the *EMV**.

c. Find the opportunity loss table.   **d.** Find the *EOL* for each action.

e. Find the *EOL**.   **f.** Find the *EPUC*.   **g.** Find the *EVPI*.

		Action		
Event	*Probability*	A_1, *Machine 1*	A_2, *Machine 2*	A_3, *Machine 3*
E_1, *high demand*	.20	7	28	−9
E_2, *moderate demand*	.35	10	−20	1
E_3, *low demand*	.45	−8	13	15

23. Use the payoff table in Problem 20.

a. Find the optimist's decision.   **b.** Find the pessimist's decision.

c. Find the maximum-likelihood decision.

d. Find the equal-likelihood decision.

e. Find the *EPUC*.

24. Mary Hoffma, a financial analyst for International Electrical Controls, has just completed a month-long study to determine the potential profits from a three-part contract that is to be awarded by a computer manufacturer six months from now. The profit potentials depend on how many parts of the contract International gets (it has bid on all three). Ms. Hoffma has considered the fact that her company will enjoy the economies of lengthy production runs, but only up to a certain point in volume. She has also considered the special equipment whose costs will have to be charged to the contract.

The potential profits of the various-sized contracts depend on whether the company decides to do all the work itself, subcontract half of it to J.J. Electronics, or subcontract all of it to J.J. Electronics, the only subcontractor International would trust with work of this nature. Ms. Hoffma has worked with J.J.'s financial vice president in determining her figures associated with the subcontracting. The vice president has told her that J.J. would not consider taking any of the work unless International is willing to make a commitment right now about how much of the work J.J. will get, assuming Hoffma's company wins some work, because of J.J.'s other possible opportunities and its need to prepare for such a large volume of work. In addition, Ms. Hoffma is informed, J.J. will charge International a flat $3000 "spadework" fee in the event that having been promised all or part of the work now, they end up getting none owing to International's winning none of the contract parts.

Ms. Hoffma's payoffs are given in the accompanying table (all figures are in thousands of dollars). She feels there is a .35 probability that International will get one or more parts of the contract. The probability that they get only one is estimated to be .20; that they get two is .10.

a. Fill in the first row and the probability column of the table.
b. Find the action with the highest *EMV*.
c. Construct an opportunity loss table and find the action with lowest *EOL*.
d. What is the *EVPI* for this problem?
e. Show that $EMV_j + EOL_j = EPUC$ for all actions A_j, where $j = 1, 2, 3$.

		Action		
Event	*Probability*	*Do It Alone*	*Subcontract Everything*	*Go 50–50 with J. J.*
Zero parts				
One part		20	5	10
Two parts		25	15	20
Three parts		−14	18	16

25. The Overthrust Petroleum Analysis Company (OPAC) has leased a tract of land located in the overthrust belt and needs to decide whether to drill for oil on the tract or to sell the lease. From a great deal of experience the manager of the oil company believes that the probability of obtaining a productive well on the site during a drilling operation is .20. The cost of drilling at the site is $1 million. The company will receive $5 million in cash if it drills the well and the well is productive. Prior to drilling, the lease can be sold to major oil company for $650 thousand.

a. Draw a decision tree to represent the decision facing OPAC.
b. What should OPAC do, on the basis of expected monetary value?
c. What is the *EMV* for OPAC?

26. Fabricated Moldings, Inc., a manufacturing company, faces a serious problem. Its chief raw material is copper. However, there is a national copper strike in progress. The company's supplier of copper can no longer ship copper to the company, but the company could buy what it needs for a large contract from a reputable foreign supplier. Thus the company's major problem is deciding whether it will wait out the copper strike or whether it will buy from the foreign supplier in order to satisfy the contract it is currently working on.

If the company decides to wait out the strike, then the president of the United States may invoke the Taft-Hartley Act, forcing the strikers back to work. There seems to be about a 60% chance he will do so. If he does, then labor officials indicate there is about a 50–50 chance that the workers will obey the order. If they obey the order, the contract will be completed since copper will become available, and the company will make $400,000 on the contract. However, if the contract is not completed (owing to the workers' not going back to work or for any other reason), the company will lose $200,000. If the president does not invoke the Taft-Hartley Act, there is still about a 10% chance that the strike will be settled in time for the company to finish the contract. But, of course, there has to be a 90% chance that the contract will not be finished under these circumstances, and the company will lose the $200,000.

If the company decides to purchase copper from the foreign supplier, there is some question about whether the supplier will be able to make delivery in time for the contract to be completed. The purchasing manager has stated, "I think there's an 80% chance that it can get the stuff here on time." Even if it delivers on time, however, there is only a .5 chance that the company's own employees would not be willing to use foreign materials bought in order to help break another domestic union's strike. The ramifications of buying the foreign copper are also financial. Since the copper is a special type that cannot be used on other orders in the foreseeable future. The extra money spent on the foreign material would be lost if the contract is not finished. The extra cost (from shipping and high prices brought on by the copper shortage) is $100,000. Thus if the company uses foreign copper on the contract, its profits will be reduced to $300,000. The loss on the contract also goes to $300,000 if the foreign material is purchased but the contract can't be completed through late delivery or the company's own worker boycott of the material.

Construct the decision tree outlining the problem facing the company. Determine the best course of action, using the *EMV* criterion.

27. The Pemform Company has 35 small retail stores located in the western United States. The San Jose store is across the street from a very large vacant tract of land. Two parties are currently seeking to buy the land. One would build a large shopping mall on the land; the other would use it to put up an amusement park.

The Pemform Company retained a consultant who specializes in evaluating the effects that changing neighborhoods and traffic patterns have on businesses' revenues. According to the consultant, a shopping mall on the vacant land will mean an added $50,000 per year in revenues for Pemform. However, the amusement park would change the surrounding area in such a way that the company could expect a $20,000-per-year drop in revenues.

The Pemform management is friendly with the real estate agent handling the negotiations for the land. At this time the real estate agent says there is a 40% chance that the shopping mall purchasers will obtain the land, a 30% chance that the amusement park people will win out, and a 30% chance that everything will fall through and the land will remain vacant.

The person who owns the building where Pemform's San Jose store is located has heard that the land across the street is going to be developed. On this basis the owner is demanding a long-term lease at a rent increase amounting to an additional $7000 per year beginning this month.

The Pemform store could be moved to another nearby location where the rent and operating costs would be approximately the same as are currently being paid. The location consultant estimates that revenues in the new location would be approximately $10,000 per year lower than at present and would be unaffected by the development of the vacant land. The incremental costs of the move would be nearly negligible in the eyes of the management.

a. Draw a decision tree showing Pemform's problem.
b. Calculate the *EMV* values for each decision. (Ignore discounting of annuities.)
c. What would be the optimist's decision for Pemform?
d. What would be the maximum-likelihood decision for Pemform?
e. Could this problem be presented in a payoff table?

28. Mr. J. B. Hight owns several metal fabricating firms in the Midwest. He has been quite successful in recent years and is now considering a major expansion. He plans to expand by purchasing 100% of the stock in either AAA Fabricators, Inc. or BZ Best, Inc.

AAA has 1 million shares of stock outstanding, and it could be acquired for $6 million. In calculating the desirability of the purchase, Mr. Hight would charge the purchase price of the stock against the stock's first three years' earnings. The annual earnings per share of AAA are currently $2.50 per share. However, the future earnings depend on antipollution legislation of the state where AAA is located. J. B.'s accountants figure that if Bill No. 1 is passed, earnings per share will drop to $2.25 for the next three years. If Bill No. 2 is passed, the per-share earnings are figured at $2.00 for three years. Mr. Hight's executive vice president has investigated the chances of passage for these bills and says they are as follows:

No legislation passed	20%
Bill No. 1	30% chance of passage
Bill No. 2	50% chance of passage

BZ Best can be purchased for $5 million. Its earnings per share are currently $3, and there are 800,000 shares outstanding. The future earnings of the company depend on the outcome of a suit that has been filed against it by two competing firms. If BZ Best loses the court fight, it will have to pay damages and fines, which are expected to reduce earnings per share to $1 for the next few years. Since Mr. Hight calculates the desirability of his investments on the basis of results in the first three years, the amount of time it will take to settle the suit affects his evaluation. J. B.'s lawyer estimates that there is a .4 probability that the suit will be ruled on one year from the purchase date, and a .6 probability that it will be ruled on two years after the purchase. She also estimates that there is a .5 probability of BZ Best's winning or losing the suit regardless of the date of its settlement.

a. Construct Mr. Hight's decision tree. At the ends of the branches, indicate the annual cash flows over the three-year period.
b. Determine the *EMV*'s of each potential purchase, ignoring discounting.
c. Determine the *EMV*'s of each potential purchase, using discounted cash flows and a discount rate of 10%.

29. The Martvig Construction Company is trying to decide which of two construction jobs it should bid on. The first is a job for constructing a bridge over a canyon. The second is the construction of a small highway interchange. Mr. Martvig estimates that there is a 50–50 chance he could win the bridge contract and a 60% chance he could win the interchange contract. However, he has determined that he will bid on only one of the two jobs. He plans that his bid (if he makes it) on the highway interchange job will be for $2.0 million. He estimates his costs will run about $1.8 million. His bid for the bridge (if he makes it) will be $1.5 million. However, he is not sure what his costs will be. The costs depend on how much drilling will have to be done at the site where the bases to the bridge will be anchored on either side of the canyon. If the amount of drilling required is moderate, then the cost of meeting the contract will be about $1.0 million. However, the amount of drilling required may be extensive. Then the costs depend on who will do the drilling.

Mr. Martvig has three drilling companies that he subcontracts from time to time. The companies are Miller Drilling, Bentler Properties, and Gadin Probers. If Miller does the job, it will charge $800,000 more than if the moderate drilling is required. Bentler will charge $900,000 more than if moderate drilling is required, and Gadin will charge $1 million more than if moderate drilling is required. One would think that Mr. Martvig would naturally want to work with Miller Drilling. However, because of the work schedules of the three companies, Mr. Martvig says there is only a .3 chance that he can get Miller Drilling to do the extensive drilling—if it is required. Mr. Martvig thinks there is a .4 chance that he will have to go with Bentler and a .3 chance that he will end up with the high-priced Gadin's doing the work. Mr. Martvig says that if the amount of drilling required is moderate, his own drilling people can handle the drilling job, and their cost has been built into the $1.0 million cost of meeting the contract just mentioned.

Since the cost picture on the bridge contract is so heavily influenced by the type of drilling that might be required, Mr. Martvig asked a geologist to look at the situation. The geologist's evaluation of the situation was summarized in the last statement of the report she submitted: "So many factors will determine what type of drilling will be required that we can only estimate the probabilities of what you will find once construction begins. We estimate that there is a 30% chance that extensive drilling will be required. Thus there is a 70% chance that moderate drilling will suffice."

Draw the decision tree facing Martvig Construction Company. At the ends of the branches, indicate the profit that will result if the company ends up at that point. Then indicate which bid the company should enter on the basis of the *EMV*.

Refer to the 113 applications for credit listed in the Credit Data Set in Appendix A.

30. Consider the accompanying payoff table.
    a. Discuss the payoffs that should be entered in the four cells of this table. What cost and profit items should be considered in arriving at these payoffs?
    b. Discuss the probabilities that should be entered in this table. Do they differ from one applicant to the next? Could they be estimated for "the average applicant" on the basis of data likely to be available in the department store's credit office?

		Payoffs for Actions	
Event	Probability	Grant Credit	Deny Credit
Good credit risk			
Poor credit risk			

31. The credit manager of the department store has said, "We use the expected monetary value criterion in determining whether to grant credit to individuals, where possible. We feel this criterion is more appropriate than any of the other commonly used criteria." Discuss the validity of this statement.

MAKING DECISIONS WITH THE AID OF A COMPUTER

CASE 16.1

Realistic decision problems that include sequential decisions and sequences of events can generate rather complex decision trees. The evaluation of each branch of these trees can be especially difficult when event probabilities must be estimated or when it is desirable to evaluate the tree with several different probability values. For this reason computer software programs are available to help evaluate complex decision trees.

One software package that allows the decision maker to use a personal computer to graphically depict and evaluate decision trees is known as ARBORIST. The program allows the user to build a decision tree for a decision problem, to browse through the tree, to compute expected values, to generate probability distributions of payoffs, and to conduct sensitivity analyses, among other things.

A facsimile of a display from ARBORIST for the decision that the owner of the Equipment Leasing Services business must make concerning selling her business is presented in Figure 1.

Software for graphically depicting and analyzing decision trees does not relieve the user of the responsibility of constructing a decision tree, which is often a very difficult part of the problem. Since the structure of the decision tree depends completely on the problem, there can be few general rules governing decision-tree drawing. But these programs will do the tedious work of evaluating the tree and ultimately listing the most desired courses of action.

The use of computers is especially valuable in doing *sensitivity* analysis on complex decision trees. Changing the probabilities of various events or the monetary flows to see whether the changes will alter the expected values of the actions enough to cause the decision maker to select a new set of actions can be done easily with the aid of a personal computer.

Use a computer to evaluate the decision tree for Fabricated Moldings, Inc., as described in Problem 26.

Case Assignment

FIGURE 1   Display for the Small-Business Owner's Decision Tree, Using ARBORIST Software

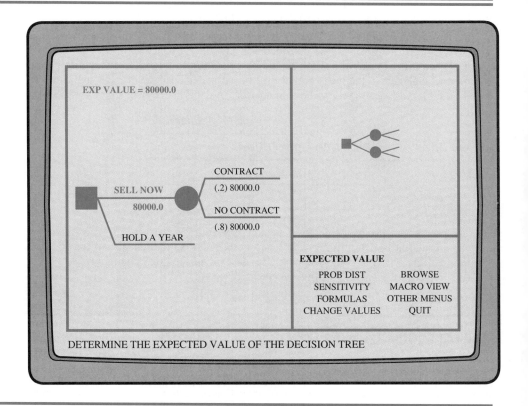

EXP VALUE = 80000.0

SELL NOW
80000.0

CONTRACT
(.2) 80000.0

NO CONTRACT
(.8) 80000.0

HOLD A YEAR

EXPECTED VALUE

PROB DIST	BROWSE
SENSITIVITY	MACRO VIEW
FORMULAS	OTHER MENUS
CHANGE VALUES	QUIT

DETERMINE THE EXPECTED VALUE OF THE DECISION TREE

CASE 16.2   MAKING CAPITAL INVESTMENT DECISIONS

Frances Phillips has worked for Farnsworth County Building Department since she was in high school. She worked for the department part time during the school year and each summer, right through her four years of study at the local university. After graduation, she began working full time and eventually worked into her current job, which involves a combination of personnel functions and purchasing.

Frances is faced with a decision that has been bothering her for a week. She must authorize purchase of some new welding equipment for the department, and she has to determine which company will be given the order for the equipment. The problem is compounded by the fact that the companies from which she might buy are putting on pressure to make a decision right now—before the final specifications for the equipment have been drawn up by the Building Engineering Division at the county's headquarters. These specifications were originally due to be completed six weeks ago, but higher priority projects have been occupying Building Engineering. Frances feels she has three alternatives:

1. Order the Felt Company equipment now for a cost of $15,000.
2. Order the Norton Company equipment at the specially discounted price of $10,000.
3. Delay her decision until after the specifications have been completed by Building Engineering.

Frances was at first doubtful about her authority to purchase prior to the completion of the specifications. But she checked with the County Attorney's Office and was informed that she is fully empowered to authorize any purchases she deems "in the county's best interests" and that the specifications are meant to help her in making her purchasing decision, but not to prevent her from making a prudent purchase. For instance, if the specifications are slow in being completed and a good purchase seems possible, she can act now. However, the county attorney advised Frances that any items that are bought before specifications are completed and that are later found not meeting the specification should be returned to the vendor or modified to meet the specifications. On the basis of this opinion, Frances feels that it would be permissible for her to make an early purchase of the welding equipment if she can document her case for not waiting for completion of the specification.

The only problem with delaying the decision is that the future price will likely be greater for both companies' equipment. The Felt people are planning a 10% price increase on the first of next month—well before the specifications are due to be completed. Also, the Norton salesman will give the $10,000 price now but says that the price will go back up to its normal $14,000 level after the first of the month. He doesn't think there is a very good chance that Phillips could talk the factory into the $10,000 price after that data. He said, "There's only about a 10% chance that you could get that price later, but you can take the chance if you want to. My commission would be larger on the higher price."

It would appear the Norton equipment is preferable due to its lower price tag, but there is some question as to whether it will perform up to the specifications that are still being written. There is no doubt that the Felt welding equipment will fill the bill, but Frances has indicated to her associates that she is only "80% sure" that the Norton equipment can meet specifications. If it doesn't, the county could buy the Felt equipment or stay with Norton since the Norton people say there is a 50–50 chance they could modify their equipment so it will do the job specified. Norton would attempt the modifications only if the county buys the machine first. These modifications would, of course, add to the cost of the machine. The range of this added cost is "from $1500 to $8000," according to the Norton Company salesman. This range of prices holds now or at any purchase date in the foreseeable future. The cost of modifications would not have to be paid by the county if Norton is unsuccessful in an attempt to make the machine meet the county's specifications.

In view of these complications, Frances Phillips has decided that she will have to approach the decision from a rational point of view and document carefully the reasons for her decision.

Case Assignment

a. Draw a decision tree Phillips could use to make her decision.
b. Determine the course of action with the best expected monetary value.
c. How high must the probability of the event "Norton Doesn't Meet Specifications" rise before buying the Felt equipment now becomes the best decision according to the expected monetary value criterion?

REFERENCE: Baird, B. (1989), *Managerial Decisions Under Uncertainty* (New York: Wiley).

REFERENCES

Baird, Bruce F. 1989. *Managerial Decisions Under Uncertainty.* New York: Wiley.
Bierman, Harold, Jr., Charles P. Bonini, and Warren H. Hausman. 1991. *Quantitative Analysis for Business Decisions,* 8th ed. Homewood, Ill.: Irwin.
Hammond, John S. 1967. "Better Decisions with Preference Theory." *Harvard Business Review,* November-December.

Holloway, C. 1979. *Decision Making Under Uncertainty.* Englewood Cliffs, N.J.: Prentice Hall.

McNamee, P., and J. Celona. 1987. *Decision Analysis for the Professional.* Redwood City, Calif.: Scientific Press.

Samson, D. 1988. *Managerial Decision Analysis.* Homewood, Ill.: Irwin.

Schlaifer, Robert. 1969. *Analysis of Decisions under Uncertainty:* New York: McGraw-Hill.

Scientific Press. 1987. *Supertree Software.* Redwood City, Calif.: Scientific Press.

Texas Instruments, Inc. 1986. *ARBORIST Decision Tree Software.* Austin, Tex.: Texas Instruments, Inc.

Decision Analysis with Sample Information

17

The decision problems of Chapter 16 had three common elements: (1) they had alternatives from which the decision maker had to choose, (2) they had uncertain states of nature or events that might take place, and (3) they had variable payoffs that were conditional upon what action was taken and what event occurred. Without each of these elements the decision makers would have no problem. If they had no alternatives, they would face no decision; if they knew what event was going to occur, their decision would be trivial; and if all the payoffs were the same rather than being conditional upon the events and actions, then what the decision maker decided to do would not matter. In this chapter, we concentrate on the second element of these problems—the uncertainty of the events and how this uncertainty can be measured and expressed.

In some types of betting games, we can calculate the exact probabilities associated with each of the game's events. But managers of public and private enterprises do not deal with betting games; they deal with the real world. Thus, the problem of assigning probabilities to the events facing them is not a simple task. Sometimes, managers can estimate probabilities of events by looking at historical data and determining how often each of the events has occurred in the past. In the absence of historical data, a manager may have only subjective estimates of the probabilities involved. In either case, managers make decisions under conditions of uncertainty.

One of the actions often open to decision makers is to delay the decision and obtain information allowing them to specify more accurately the event probabilities of the problem. They usually do so by undertaking an objective survey, sample, or experiment. In this chapter, we deal with methods of combining subjective or historical probabilities with the results of current, objective information to obtain better event probabilities; and guidelines for determining just when objective, experimental information should be sought.

17.1
SUBJECTIVE PROBABILITIES

When decision makers have no historical information to guide their assessment of the event probabilities, they have no alternative but to use **subjective probabilities.** When there are only a few events that might occur, estimating subjective probabilities may not be too difficult a task.

EXAMPLE 17.1 The HC&B Company is expecting a shipment of parts to arrive either this week or next week. Shipment receivers feel that there is a .4 chance that the shipment will arrive this week and a .6 chance that it will arrive next week. Also, they feel that there is about a .25 chance that the shipment may arrive damaged. The set of events faced by HC&B is shown in Figure 17.1 (which might be a set of event forks on a large decision tree). Factors that might influence HC&B's estimate of the arrival probabilities include: its experience with the supplier, the last information it had concerning the location of the shipment, the delivery records of the railroads and trucking companies that might handle the shipment, and the season of the year. The factors influencing its subjective estimates of damage probabilities include: its record of past experience with damaged goods, its knowledge of shipping facilities at the supplier's location, the reputation for damaged goods of the railroads and trucking companies handling the shipment, and its own record for damaging goods as they are received.

The more knowledge the company has about factors such as these, the more likely its subjective probabilities will be accurate. But even then, the probabilities are only guesses. The person making the subjective probability estimates does have one way of checking himself for realistic estimates. He can use probabilities like the values of .6, .4, and .25 arrived at by the HC&B Company to compute the probabilities of *compound,* or *joint, events* and see whether these probabilities seem realistic. For instance, if the arrival time and the condition of the shipment upon its arrival are considered independent, then the joint proba-

FIGURE 17.1

HC&B Company's Shipment Problem

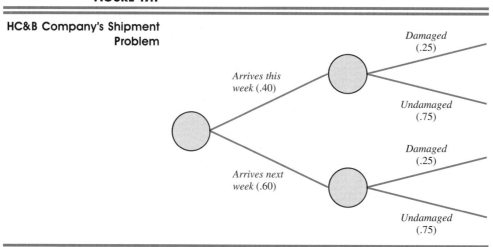

bilities of arrival time and shipment conditions can be found by multiplying probabilities:

$$P(Arrives\ this\ week\ damaged) = (.4)(.25) = .10$$

$$P(Arrives\ this\ week\ undamaged) = (.4)(.75) = .30$$

$$P(Arrives\ next\ week\ damaged) = (.6)(.25) = .15$$

$$P(Arrives\ next\ week\ undamaged) = (.6)(.75) = .45$$

If the probabilities of the joint events seem unrealistic to the person making the estimates, he knows he must change the estimates of the individual events. If on examining the joint probabilities, he realizes that there is actually about a two-thirds chance that the shipment will arrive next week undamaged, then he knows that he must either change that two-thirds estimate or change the probabilities on arrivals and/or damage. If he holds to his estimate that $P(Arrives\ next\ week\ undamaged) = 2/3$, then he must make an upward adjustment in either $P(Arrives\ next\ week)$ or $P(Undamaged)$ or both. ═══

Checking the consistency of subjective probabilities of simple events by calculating the joint probabilities of compound events is a fairly easy task when there are only a few events involved. However, many decision situations involve the possibility of numerous events. The task of finding subjective probabilities for all these events is more difficult.

═══ **EXAMPLE 17.2** Consider the problem faced by a manufacturing plant using a chemical in its operations. Production planners feel that use of this chemical is most likely to fall in the range of 50,000 to 80,000 gallons. Perhaps they feel that there is quite a large chance—say, as much as 50% chance—that use will lie outside this range and a very small chance that use will be very high—as much as 120,000 gallons. They are reasonably sure that the use will definitely not fall below 35,000 gallons. In this case there are 120,000 − 34,999, or 85,001, possible events if the amount of chemical used in the process is to be measured to the nearest gallon.

To try to assign probabilities to every possible amount of chemical measured to the nearest gallon would be unrealistic. Our planners would do better to treat the amount of chemical used in the process as a continuous variable and try to put some of their subjective feelings about use to work to obtain a smooth, use probability curve like the one in Figure 17.2. This curve fits the description of the use probabilities given previously. About 50% of the area under the curve is between 50,000 and 80,000 gallons. The curve terminates at 35,000 and 120,000 gallons and is slightly skewed to the upper side. The probability that chemical use will be between any two figures can be taken from this curve by measuring the area under the curve between those two points and expressing that area as a proportion of the total area under the curve. The areas can be approximated by using a series of rectangles (such as those in the figure) having a total area similar to the area under the curve. To calculate the areas in the rectangles is a simple matter, and these areas are good approximations to the areas under the curve. ═══

FIGURE 17.2

Probability of Volume of
Chemical Used in a Pro-
cess, Showing Rectangular
Approximation of Areas

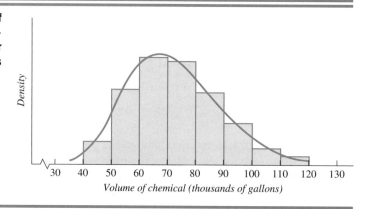

Obtaining a curve like the one in Figure 17.2 can be quite difficult. Usually, constructing a cumulative probability curve is easier. The cumulative curve corresponding to the curve in Figure 17.2 is shown in Figure 17.3. This cumulative probability curve can be obtained in the following way: The person constructing the curve determines the chemical use figure that she believes divides all possible use values into two equally likely halves. This value determines where the .5 probability point will go on the cumulative curve. For Example 17.2, assume that the company determines that volumes of chemical used in the process above or below 75,000 gallons are equally likely. Thus, the first point on the cumulative curve is (75, .5). Next, the two extreme points of the curve are plotted. Since the company is sure that use will lie between 35,000 and 120,000 gallons, the cumulative probability that use is 120,000 or less is 1.0, and the probability that it is 35,000 or less is 0. Thus, the next two points on the curve are (35, 0) and (120, 1.0).

This process can be continued by determining the amount of chemical that the company feels will be exceeded with only 25% probability. This value may be

FIGURE 17.3

Cumulative Probability
Curve for Volume of
Chemical Used in a Process

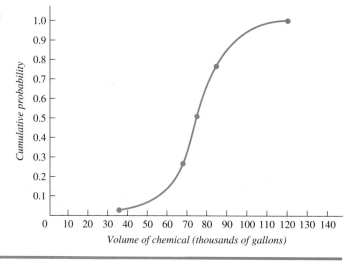

85,000 gallons and thus gives the cumulative curve point (85, .75). Once again, the company will determine an amount of the chemical. This figure should be chosen so that there is only a 25% chance that the amount of chemical will fall below it. Assume that this value is 68,000 gallons. This value gives the cumulative probability curve the point (68, .25). The cumulative curve in Figure 17.3 was obtained by fitting a smooth curve through the five points developed previously. One can continue the process outlined and develop many more points for the curve, but the increased work may not be justified by the small amount of added accuracy.

The cumulative curve can be used to find the probabilities that the amount of chemical will be between any two values, and rectangular approximation of areas is not needed. Assume, for instance, that the company's planners desire to know the probability that the chemical volume will lie between 80,000 and 90,000 gallons. The cumulative probability figure for 80,000 gallons is read from Figure 17.3 as .66, and the figure for 90,000 gallons is .84. Thus the probability that the amount of chemical used in the process will lie in that 10,000-gallon range is .84 − .66, or .18.

━━ **EXAMPLE 17.3** Suppose that the company has a policy of ordering the chemical in 10,000-gallon lots in order to obtain the best possible bulk price. The chemical is known to deteriorate or spoil if it is not used over a specified period, so the planners want to order the correct amount. They can construct a payoff table showing the alternatives of ordering 40,000 gallons, 50,000 gallons, and so on, up to 120,000 gallons. The events in this table will be intervals for the amounts of chemical as follow:

E_1: 39,999 gal or less $\qquad$ E_6: 80,000 to 89,999 gal

E_2: 40,000 to 49,999 gal $\qquad$ E_7: 90,000 to 99,999 gal

E_3: 50,000 to 59,999 gal $\qquad$ E_8: 100,000 to 109,999 gal

E_4: 60,000 to 69,999 gal $\qquad$ E_9: 110,000 to 120,000 gal

E_5: 70,000 to 79,999 gal

The probabilities associated with these events can be calculated from the cumulative probability curve in Figure 17.3. A precise reading of the curve gives the following calculations:

$P(E_1) = .02 - .00 = .02 \qquad P(E_6) = .84 - .66 = .18$

$P(E_2) = .05 - .02 = .03 \qquad P(E_7) = .94 - .84 = .10$

$P(E_3) = .12 - .05 = .07 \qquad P(E_8) = .98 - .94 = .04$

$P(E_4) = .32 - .12 = .20 \qquad P(E_9) = 1.00 - .98 = \underline{.02}$

$P(E_5) = .66 - .32 = .34 \qquad\qquad$ Total $\qquad$ 1.00

In each case the probability listed first is the cumulative probability that the amount of chemical will be *equal to* or *less than* the value at the upper end of the range specified by the event, and the probability that is subtracted is the probability that chemical use will be *less than* the value at the lower end of the range specified by the event. The difference is then the probability that use will lie *in* the range specified by the event.

These nine probabilities can now be entered into the probability column of a payoff table. The payoff table will be nine rows (events, or actual amount of the chemical used) by nine columns (actions, or amount ordered) and will contain 81 entries of costs X_{ij} associated with the chemical order sizes and potential chemical volume consumed over the next ordering period.   =

This payoff table is not constructed here since we are concentrating only on the probabilities of the events. What we should note is that we have been able to obtain nine rather precise and reasonable subjective probabilities for the volume of chemical used in the process from the cumulative probability curve in Figure 17.3. To estimate these probabilities by using direct guesses or rectangular approximations from the curve in Figure 17.2 would have been much more difficult. The problem remains, however, that these values are subjective probabilities and are not based on historical data or objective experimentation. The following section discusses methods by which subjective probabilities can be revised by using Bayes' theorem.

Problems: Section 17.1

Answers to odd-numbered problems are given in the back of the text.

1. A large nonprofit organization is self-insured. A benefits officer is trying to establish probabilities associated with various levels of claims on the organization's health insurance next year. She feels there is a 50–50 chance that claims will be more than or less than $225,000. She feels that the chance of having claims of $350,000 or more is .25 and the chance that they will be $200,000 or less is about .4. There is virtually no chance that claims will be less than $150,000 or more than $.5 million.
   a. Construct a cumulative probability curve for the benefits officer.
   b. Find the probability that health insurance claims will exceed $.25 million next year.
   c. Find the probability that claims will be between $250,000 and $400,000.

2. A salesman for Multisys, Inc., is trying to construct a probability curve that will indicate the likelihood of his selling various amounts of the company product next month. He feels there is a 50–50 chance of his selling more than or fewer than 2000 cases. He feels there is only a 25% chance that he can sell more than 2800 cases and only a 25% chance that he will sell less than 1600. He has never sold more than 3200 nor fewer than 1300 cases in one month.
   a. Construct a cumulative probability curve for this salesman.
   b. Use the curve to find the probability that next month's sales will be above 1800 cases.
   c. What is the probability that this salesman will sell between 2200 and 3000 cases?

3. A student feels there is a .5 chance he will get a score of 80 or better on the next 100-point midterm exam. He feels there is only a .25 chance that he will get 90 or better, and he feels there is a .25 chance that he will receive a score of 55 or worse.
   a. Construct a cumulative probability curve for the student's exam score.
   b. Grades on the next midterm are to be given as follows:

Test Score Ranges	100–90	75–89	65–74	50–64	0–49
Grade	A	B	C	D	F

What are the subjective probabilities the student should assign to getting each of the five grades?

We have analyzed several decision problems by using expected monetary values, and many probabilities have been subjective estimates. It is often advisable to obtain additional information in the form of samples, surveys, tests, or experiments that can be used to revise probabilities.

The revision of prior probabilities by employing sample information is done by using Bayes' theorem from Chapter 4. The sample information provides conditional probabilities, and Bayes' theorem combines the *prior probabilities* and the *conditional probabilities* to give *revised* or *posterior probabilities* that reflect the sample information. The revised probabilities can enrich the analysis and lead to improved decisions.

17.2
DECISION ANALYSIS WITH SAMPLE INFORMATION

EXAMPLE 17.4   Earth Structures, Inc., has been told that it will be awarded a contract to build a large flood control system, contingent upon the state legislature's voting funds for the project. The construction company feels that there is a .6 chance that the bill approving funds for the project will pass and a .4 chance that it will not. These values are subjective probabilities, based on the company president's "feel" for the mood of the legislature and some informal discussions with two golf partners who serve in the legislature.

Since the subjective probabilities are not at all lopsided in favor of the bill's passage, the company's managers are unsure whether they should hire new personnel, and purchase new equipment for the project. Company officers realize that they might supplement their own subjective probabilities about the bill's passage with objective information published in a local newspaper. The paper regularly predicts the fate of pending legislation on the basis of its political editor's interviews with key legislators. The paper lists its record of accuracy in the following way: For bills that eventually passed, the paper has correctly predicted passage 80% of the time and incorrectly predicted failure 20% of the time. For bills that eventually failed, the paper has correctly predicted failure 95% of the time and incorrectly predicted passage 5% of the time.

If we let *Pass* denote that a bill passes, *Fail* denote that it does not pass, *Passage predicted* denote that passage was predicted, and *Failure predicted* denote that failure was predicted, then the paper's record of accuracy can be presented as *conditional probabilities:*

$$P(Passage\ predicted\,|\,Pass) = .80 \qquad P(Failure\ predicted\,|\,Pass) = .20$$

$$P(Passage\ predicted\,|\,Fail) = .05 \qquad P(Failure\ predicted\,|\,Fail) = .95$$

The first probability, for example, is the probability that the paper will predict passage, given that the bill will pass.

Let us assume for a moment that the paper predicts passage for the flood control project bill. The company will be interested in the probabilities in the reverse order of those given previously—that is, the probability that the bill will pass, given that it is predicted to pass, $P(Pass\,|\,Passage\ predicted)$, and the probability that it will fail, given that it is predicted to pass, $P(Fail\,|\,Passage\ predicted)$. It has the published accuracy record and the *prior probabilities* of the company president, $P(Pass) = .6$ and $P(Fail) = .4$, with which to work. These probabilities are called prior probabilities because they are the probabilities of passage for the bill *prior* to performing any experiment or obtaining any objective information.

A formula for combining the known prior and conditional probabilities in order to obtain $P(Pass \mid Passage\ predicted)$ and $P(Fail \mid Passage\ predicted)$ was presented as Bayes' theorem in Chapter 4. Using the notation developed in Example 17.4, Bayes' theorem gives the following revised probabilities:

$$P(Pass \mid Passage\ predicted)$$

$$= \frac{P(Pass) \times P(Passage\ predicted \mid Pass)}{P(Passage\ predicted)}$$

$$= \frac{P(Pass) \times P(Passage\ predicted \mid Pass)}{\begin{array}{c} P(Pass) \times P(Passage\ predicted \mid Pass) \\ + P(Fail) \times P(Passage\ predicted \mid Fail) \end{array}} \qquad (17.1)$$

and

$$P(Fail \mid Passage\ predicted)$$

$$= \frac{P(Fail) \times P(Passage\ predicted \mid Fail)}{P(Passage\ predicted)}$$

$$= \frac{P(Fail) \times P(Passage\ predicted \mid Fail)}{\begin{array}{c} P(Fail) \times P(Passage\ predicted \mid Fail) \\ + P(Pass) \times P(Passage\ predicted \mid Pass) \end{array}} \qquad (17.2)$$

In Section 4.5 we noted that expressions like (17.1) and (17.2) are somewhat formidable in appearance. However, probability diagrams, which are similar to decision trees, can be used to demonstrate the meaning of the formulas. The example of determining whether the bill authorizing funds for construction of the flood control project will pass is diagramed in Figure 17.4. Note that the events *bill will pass* and *bill will fail* are assigned the company president's subjective probabilities of .6 and .4, respectively. The second set of forks in the tree presents the newspaper's record

FIGURE 17.4

Probability Tree for the Passage or Nonpassage of a Bill

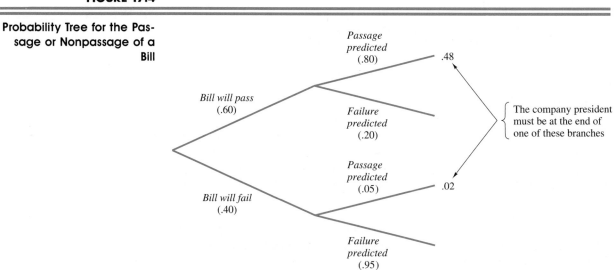

for accuracy. The problem we face is how to revise the .6 and .4 probabilities to reflect the newspaper's record for accuracy.

EXAMPLE 17.5 We assumed previously that the paper had predicted passage for the bill. Thus the company president must be at the end of the top branch in the tree or at the end of the third branch down, because the president *knows* that the paper has predicted passage of the bill and he is no longer concerned with the event that failure might be predicted. The president just doesn't know whether it has predicted passage for a bill that will eventually pass or not pass.

The joint probability that the bill will pass and that the paper will predict passage is $(.6)(.80) = .48$, which is listed at the end of the top branch. The joint probability that the bill will fail but that the paper will predict passage is $(.4)(.05) = .02$, which is listed at the end of the third branch. The chance of the company president's being at the end of either the first or the third branch is the sum of the two probabilities, $.48 + .02 = .50$. The fact remains, however, that the president *is* at the end of one of those two branches since the paper predicted the bill would pass. So in determining the probabilities that the bill will pass or not pass, the president should allocate probabilities to passage and nonpassage that are proportional to the contribution that the "bill will pass" and "bill will not pass" branches make to the total probability of .50. That is,

$$P(Pass \,|\, Passage\ predicted) = \frac{.48}{.48 + .02} = .96$$

$$P(Fail \,|\, Passage\ predicted) = \frac{.02}{.48 + .02} = .04$$

In this process we have divided the .50 chance that the paper predicts passage into its parts and have assigned relative weights of .96 and .04 to the chances that the prediction was made on a bill that will pass or a bill that will not pass. Again, this technique is possible since the president is no longer concerned with the event that the paper predicts failure.

This division into parts is exactly what Equation (17.1) accomplishes. If we examine the numerator, we find

$$P(Pass) \times P(Passage\ predicted \,|\, Pass) = (.6)(.80) = .48$$

In the denominator we find the sum of two expressions:

$$P(Pass) \times P(Passage\ predicted \,|\, Pass)$$
$$+ P(Fail) \times P(Passage\ predicted \,|\, Fail)$$
$$= (.6)(.80) + (.4)(.05) = .48 + .02$$

which is the denominator used previously in our more intuitive approach. By a similar analysis, we can show that Equation (17.2) also gives

$$P(Fail \,|\, Passage\ predicted) = .04$$

The .96 and .04 figures are the **revised probabilities** of passage and failure of the bill based on both the prior probabilities of the company president and on the newspaper's prediction of passage. We can see that the chance of passage now seems

	(1)	(2)	(3)	(4) Joint = Prior ×	(5)
TABLE 17.1	*Possible* *Event*	*Prior*	*Conditional*	*Conditional*	*Revised*
Revision of Probabilities	*Pass*	0.6	.80	.48	0.96 = *P*(*Pass* | *Pass predicted*)
for Bill Passage Given a	*Fail*	0.4	.05	.02	0.04 = *P*(*Fail* | *Pass predicted*)
Prediction of Passage		1.0		*Marginal* = .50	1.00

very high, because the newspaper predicts passage, and its record for accuracy is quite good.

Some people find that the computation of revised probabilities is more easily accomplished by using a table. Table 17.1 shows the calculation of the revised probabilities in tabular form. The first column of the table contains the events that might occur—passage or failure of the project's funding bill. The second column contains the prior probabilities concerning these events—$P(Pass)$ and $P(Fail)$. The third column is the critical column. It contains the conditional probabilities. The conditional probabilities are always the probabilities of observing the objective information obtained, given that the event specified in the row is true. Thus, the .80 value in the first row of the third column is the conditional probability that the bill's passage will be predicted (the observed objective information), given that the bill will eventually pass (the possible event in the first row).

The fourth column of Table 17.1 is the joint probability column. It is found by multiplying the prior column, term for term, by the conditional column. The .48 figure is the joint probability that the bill will pass *and* that success will be predicted for it, $.6 \times .80 = .48$. This value is also the figure that corresponds to the numerator in Equation (17.1). The .02 figure is the numerator of Equation (17.2) and is the joint probability that the bill will fail *and* that passage will be predicted for it, $.4 \times .05 = .02$. The sum of the probabilities in the fourth column is the marginal probability. The fifth column contains the revised probabilities of passage and failure of the bill, $P(Pass \mid Passage\ predicted)$ and $P(Fail \mid Passage\ predicted)$, and these are found by dividing the marginal probability into each of its joint components: $.48/.50 = .96$ and $.02/.50 = .04$.

The revised probability that the bill will pass is now .96 rather than the prior value of .6. This result is consistent with what one would expect, since the objective information, the newspaper's prediction, favors the passage of the bill. Since the bill's passage now seems quite sure, the company can proceed with more confidence in making its plans for the future. It could use the revised probabilities in any payoff table where the events listed in the table were passage or failure of this particular bill.

EXAMPLE 17.6   As a further illustration of the method of revising probabilities, Table 17.2 shows how the prior probabilities would have been revised *if the newspaper had predicted the bill would fail.* The conditional column in Table 17.2 shows the probabilities that the prediction of failure would be made for a bill that eventually will pass and fail, respectively. That is, the conditional probabilities are $P(Failure\ predicted \mid Pass)$ and $P(Failure\ predicted \mid Fail)$ and were given in the newspaper's record of accuracy. The joint probabilities are found, once again, by multiplying the prior column by the conditional column. The revised probabilities are then merely the individual joint probabilities

Possible Event	Prior	Conditional	Joint = Prior × Conditional	Revised	
Pass	0.6	.20	.12	0.24 = P(Pass | Fail predicted)	**TABLE 17.2**
Fail	0.4	.95	.38	0.76 = P(Fail | Fail predicted)	**Revision of Probabilities**
	1.0		Marginal = .50	1.00	**for Bill Passage Given a Prediction of Failure**

divided by the marginal: $P(Pass \mid Failure\ predicted) = .12/.50 = .24$ and $P(Fail \mid Failure\ predicted) = .38/.50 = .76$.

Thus if the newspaper had predicted failure for the bill, the company would have ended up feeling quite sure (probability .76) that the bill would, in fact, fail to pass the legislature. This result is, of course, contrary to the subjective feeling of the company president, but the newspaper's record for accuracy for predicting bill failures in the past has been so outstanding that were it to predict failure, the revised probabilities heavily favor failure. ═══

 PRACTICE PROBLEM 17.1

Pollution Control Industries, Inc., is preparing to market a new product. The marketing vice president has had a great deal of experience with new products and feels in a position to predict that the new product may capture 10%, 20%, or 30% of the market during its first year. Her prior probabilities about the product's ability to capture those market shares are .25, .55, and .20, respectively. (Actually, the marketing vice president probably realizes that the product might capture some intervening value like 18% but has used 10% to mean "between 5 and 15" and 20% to mean "between 15 and 25," and so forth.) The table of events and prior probabilities is as follows:

Event	Prior Probability
.10 Market share	P(Market share = .10) = 0.25
.20 Market share	P(Market share = .20) = 0.55
.30 Market share	P(Market share = .30) = 0.20
	1.00

The head of market research wishes to take a survey and use its results to check and, if necessary, modify these priors. In a market survey of 25 randomly selected customers, he finds only 3, or 12%, who say they would buy the product. Intuitively, he suspects that this poor showing of the product's saleability will change the prior probabilities about the market shares in the direction of placing heavier weight on the chance that there is only a 10% market awaiting the new product.

Solution Table 17.3 shows the procedure for revising the prior probabilities. The table has the same structure as Tables 17.1 and 17.2. The only difference is in the

Possible Event	Prior	Conditional	Joint = Prior × Conditional	Revised	
.10 Market share	0.25	.2265	.05663	0.41	**TABLE 17.3**
.20 Market share	0.55	.1358	.07469	0.55	**Revisions of Probabilities**
.30 Market share	0.20	.0242	.00484	0.04	**for New-Product Market**
	1.00		Marginal = .13616	1.00	**Share at Pollution Control**

way the conditional probabilities were obtained. In the construction company examples the conditional probabilities were obtained from the newspaper's record for accuracy in predicting passage and failure of pending legislation. But when the events are *proportions,* as they are in this problem, then the conditional probabilities can be obtained by using the binomial formula or the binomial table (Table III in Appendix C). The conditional probabilities, remember, are the probabilities of observing the objective information obtained, assuming each of the events is true.

Thus, the conditional probabilities in Table 17.3 are the binomial probabilities of obtaining 3 successes in 25 trials from populations that contain 10%, 20%, and 30% successes, respectively. In the notation of Chapter 5 these probabilities are as follows:

$$P(3) = \binom{25}{3}(.10)^3(.90)^{22} = .2265 \qquad \left\{ \begin{array}{l} \text{Assuming the true} \\ \text{market share is 10\%} \end{array} \right.$$

$$P(3) = \binom{25}{3}(.20)^3(.80)^{22} = .1358 \qquad \left\{ \begin{array}{l} \text{Assuming the true} \\ \text{market share is 20\%} \end{array} \right.$$

$$P(3) = \binom{25}{3}(.30)^3(.70)^{22} = .0242 \qquad \left\{ \begin{array}{l} \text{Assuming the true} \\ \text{market share is 30\%} \end{array} \right.$$

The joint probabilities are, once again, merely the product of the prior and the conditional probabilities for each row of the table. And the *revised probabilities* are the joint probabilities divided by the marginal. Just as the head of market research suspected, the revised probabilities differ from the prior probabilities in that the chances of achieving a 10% market share now appear to be much greater than before and the chances of achieving a 30% market share now appear negligible. ▬▬▬

Note that when sample, experimental, or objective information is used to revise a set of prior probabilities, all the prior values usually change. The fact that both the prior and the revised probabilities of a 20% market share were .55 is only a coincidence of our example.

One may ask why we even bother to use the subjective prior probabilities in any of these problems where we have experimental, objective information upon which to base probability statements about the events. Why didn't we, for instance, simply use the methods discussed in Chapter 8 to construct a confidence interval for p, the potential market share for the new product, based on the value of $\hat{p}$, the proportion of people in the sample saying they would buy the product? Some statisticians, those who have very little faith in the accuracy of the subjective probabilities that are revised by using the methods described in this chapter, would argue that such a confidence interval based on only objective information would be far more trustworthy than a probability forecast incorporating guess-type subjective information. But the people who feel that the revision of probabilities using experimental, objective information is a useful process support their view by saying that we all have hunches, insights, and intuitive feelings about the probabilities that certain events are going to take place. They suggest that we use these subjective feelings in our decision making anyhow and that their approach to the revision of event probabilities provides a framework within which to combine subjective feelings with objective information. Also, they claim, a person who is familiar with a situation can often come up with subjective event probabilities that are very accurate, and we would be foolish to throw out this valuable information and proceed only on the basis of the objective data.

Both positions, of course, are correct to a degree. When there is no basis upon which to arrive at prior judgments about the event that is going to take place, then we have no information with which to construct prior probabilities. In this case, to use only the objective information (perhaps in the form of a confidence interval) to estimate event probabilities would probably be better than to assume that all possible events are equally likely and then revise this uniform prior distribution. But if there is a great deal of historical information about which event might take place, or if the decision maker feels quite sure about subjective prior probabilities, including this information in the decision-making process by constructing a prior probability distribution and then revising it by using objective information would be best.

Problems: Section 17.2

4. Rework the example in Table 17.1, assuming that the prior probabilities for passage and failure are .20 and .80, respectively.

5. Historical records for the production output of a machine show the proportion of defectives produced by the machine in the accompanying table. A sample of 20 pieces has just been taken from a new batch of the machine's output, and one was found to be defective. Construct a revised distribution for the proportion of defectives.

Proportion p of Defectives	Percentage of Batches with Proportion p Defectives
.01	.7
.05	.2
.10	.1

6. The Matthew Tennis Ball Company has been having trouble controlling the quality of its output. It manufactures tennis balls in lots containing 1000 cans, and there are three balls per can. The last 50 lots have had the proportions of defects in the form of broken pressure seals on the cans as given in the accompanying table. A new lot is produced, and a sample of 25 randomly selected cans is inspected. Five of these are found to have broken pressure seals. Find a revised probability distribution for the proportion defective in the new lot.

Proportion Defective	Number of Lots
.01	10
.05	25
.10	10
.20	5

7. An Illinois construction firm has bid on a state highway job in a neighboring state. The company felt that it had a .35 chance of getting the job. However, it needs to revise the probability of getting the job on the basis of some new information. This new information is that the highway department in the state where the bid was made has never awarded a highway construction contract to a company outside its own state.

   a. What are the events that might take place?
   b. What are the prior probabilities for these events?
   c. What are the revised probabilities for the events?

17.3
WHEN TO SAMPLE

One final question has to be answered before we leave the topic of experimentation and revision of probabilities: How does one decide when it is appropriate to use prior probabilities in the decision tables presented in Chapter 16 and when it is best to revise these prior probabilities by using a sample or a survey or a test? There is no simple rule that can answer this question. Some mathematical methods can be employed in certain instances, but they are beyond the scope of this book. However, there are a few general guidelines that can be adopted.

When the expected value of perfect information calculated from a payoff table is very high and the cost of the experiment is very low, then one should usually perform the experiment and revise the payoff table's probability column. This situation may occur in many problems. For instance, gathering more information often involves no more than making a few telephone calls. In the construction company example of Section 17.2, the company desired to know whether a certain bill was going to pass the state legislature. It was able to revise its probabilities concerning the bill's passage by merely obtaining the local newspaper's forecast of the bill's fate and the newspaper's record for accurate predictions. This type of information gathering can usually be done very cheaply. In contrast, a survey to determine consumer buying habits may cost several thousands of dollars and would not be a worthwhile experiment if it were undertaken in order to revise the probability distribution in a problem where the *EVPI* is $600.

More than the cost of the experiment is involved. Only experiments that promise to provide accurate information should be undertaken. For instance, it would not have been wise to incorporate the newspaper's prediction about the passage of the flood control project's funding bill if the newspaper had been successful in its predictions only 35% or 40% of the time. But it may pay to perform a survey even though its cost is high and close to the *EVPI* of a problem if the information obtained is expected to be highly accurate. That is, we might be willing to spend almost as much as the expected value of perfect information if our information is nearly perfect in resolving the uncertainty we face.

Problems: Section 17.3

8. The *EVPI* for a decision problem is $9700. The probabilities for the events in this problem's payoff table can be revised by using an experiment that involves a sample. The setup cost of the experiment is $5200, and individual items in the sample cost $2 each for the first 1000 items. Then the cost drops to $1.50 on the remainder, but the first 1000 still cost $2 each. What is the maximum-size sample that should be taken to revise the payoff table probabilities?

9. How high would the setup costs for the sample in Problem 8 have to be for you to be absolutely sure that the sample should not be taken?

10. In Problems 8 and 9, what other piece of information would you want before you decided to take the sample?

17.4
SUMMARY

We addressed the uncertainty element of decision theory problems in this chapter. Managers often specify event probabilities, or prior probabilities, by studying historical data or by making subjective assessments. We examined methods that can be used to ensure that subjectively determined probabilities are essentially reliable.

When managers want to specify event probabilities more accurately, they can undertake a survey, a sample, or an experiment. Revised probabilities were obtained in this chapter by using Bayes' rule to combine prior probabilities with conditional probabilities that resulted from current, objective information. The revised probabilities were then used to solve decision theory problems. Finally, we considered guidelines for determining when objective, experimental information should be obtained.

Answers to odd-numbered problems are given in the back of the text.

11. Determine whether the following probability statements are consistent with one another.

   a. There is a .4 chance the stock will drop. There is a .2 chance that Sue Bradle will sell the stock if it drops. There is about an 8% chance that the stock will drop *and* Sue Bradle will sell it.

   b. There is a .5 chance that the economy will be up next year. There is a .8 chance that our sales will go up when the economy is up. Thus, there is a .4 chance that our sales will be up next year.

   c. There is a .5 chance we will get the contract and a .1 chance that Smith will quit if we do. There is a .6 chance Smith will quit if we lose the contract. There is a .25 chance Smith will quit.

   d. There is a .7 chance that management will decide to market the QB–7. The chance of high, moderate, or low sales for the QB–7 are .4, .3, and .3, respectively. The probability that the product will be marketed *and* will achieve moderate sales or worse is .56.

12. Use the method outlined in Problem 3 to determine a cumulative probability distribution for your scores on your next 100-point midterm exam. Then use the grading schedule in Problem 3 to assign probabilities to your possible grades.

13. A newspaper reporter does not know if she should go to the hotel where union and management contract negotiations are taking place. She feels that there is a .2 chance a contract agreement will be reached today and a .8 chance that there will be continued negotiations. Before she makes up her mind, she calls the hotel's night manager and asks how late negotiations lasted the past evening. The night manager informs her that the negotiations lasted all night. From her years of experience the reporter knows that all-night negotiations have preceded 80% of the contract agreement announcements. But she also knows that 30% of the nonproductive bargaining sessions go all night and produce no agreement. Construct a table like Table 17.1 and find the prior and revised probabilities for a contract agreement and continued negotiations.

14. A company buys several shipments of the R28 part each year. It buys from two manufacturers, A and B, and the parts from the two manufacturers are identical in external appearance. A shipment of parts is taken from inventory and found to be 90% defective. The company is not sure to which manufacturer the shipment should be returned. Manufacturer A usually supplies one-third of the shipments, and manufacturer B supplies two-thirds. Before taking the chance of shipping the defective R28 parts back to manufacturer B, the most probable supplier, the production manager finds that this particular shipment arrived in June of last year. Company records show that last year 40% of A's shipments arrived in June and 10% of B's arrived in that month.

   a. What are the prior probabilities for this problem?

   b. What are the conditional probabilities?

   c. Find the revised probabilities that the bad shipment came from manufacturers A and B.

15. Suppose that an experiment has been used many times in the past to predict whether event 1 or event 2 is going to take place. Past experience has shown that the experiment has been wrong in its prediction every time. Should this experiment be discontinued?

16. Colorectal cancer is a leading cause of death from cancer in the United States. A fecal occult–blood test has been developed by companies in the pharmaceutical industry for screening for early detection of this cancer, and the American College of Physicians recommends annual screening for persons over 50 years of age. Experimental studies have not shown the effectiveness of screening for the cancer; however, mortality studies are currently under way. The annual incidence rate of this cancer is somewhat rare and has been reported to be about .39 in 1000 asymptotic patients 50 years of age. The probability that a patient 50 years of age selected randomly for screening has this type of cancer is thus $P(Cancer) = .00039$. Unfortunately, screening tests for cancer are not perfect. The hemoglobin activity that is screened by the fecal occult–blood test may require 20 ml of blood to give a positive result. The probability of a positive test given that there is cancer has been estimated to be about .50. Hemoglobin activity from, say, the ingestion of vegetables can result in a positive test when there is no cancer. The probability of a positive test given that the individual does not have cancer has been estimated to be .025.

   a. Find the probability of having this cancer given a positive test result.
   b. What is the absolute change from the prior probability for the cancer to the revised probability given a positive test?
   c. What is the relative change from the prior probability for the cancer to the revised probability given a positive test? What is the percentage change?
   d. How might a pharmaceutical company advertise the effectiveness of the test if the company wants to increase its sales?
   e. Why would the American College of Physicians recommend screening for the cancer when experimental studies have not shown the effectiveness of screening?
   f. Find the probability of *not* having cancer given a negative test.
   g. If you were to be screened for this cancer and you received a negative test, would you feel better about being free of cancer?

   REFERENCES: Ransohoff, D., and C. Lang (1991), "Screening for Colorectal Cancer," *The New England Journal of Medicine*, 325: 37–41; and Brett, A. (1989), "Treating Hypercholesterolemia: How Should Practicing Physicians Interpret the Published Data for Patients?" *The New England Journal of Medicine*, 321: 676–79.

17. A production process at the Alpha-Omega Corporation is set so that its usual output is about 5% defective. On about 12% of the production runs, poor material is used, which causes the defective rate to jump to 10%. The process can be adjusted to produce 5% defectives even with the poor material, but this adjustment costs $400. Defective items cost $2 each to replace. If the adjusted process is run using good material, the defective rate is still 5%.

   a. Construct a payoff table showing the events and the alternative actions for this problem. Assume a run of 15,000 items is about to be made.
   b. What is the *EVPI* for this problem?

18. Assume that the process in Problem 17 can be run for a very short time to produce a sample that will indicate the approximate defective rate being turned out. Obtaining a sample of 15 items costs $30 in setup costs and $2.50 per item sampled. If one or more items are found defective in the sample of 15, the adjustment process will be performed for $400.

   a. What is the probability that the sample will cause the adjustment to be made when it does not need to be made?
   b. What is the probability that the adjustment will not be made after the sample when it really should be made?
   c. What is the cost of an unneeded adjustment?
   d. What is the cost of not adjusting when an adjustment is needed?

19. Draw a decision tree showing the problem faced by the Alpha-Omega Company in Problems 17 and 18. They must choose between A_1 = adjustment now, A_2 = no ad-

justment, and A_3 = sample and adjust if one or more defectives are found. Include all probabilities and all costs or losses from making wrong decisions at the ends of the branches on the tree. Then determine which action has the lowest expected cost.

20. A number of bankers are worried about actions the Federal Reserve Bank may take next week. After the meeting of the Federal Reserve, they feel that monetary policies will be either tightened up, making credit harder to grant, or loosened up, making it easier to loan money. Ms. Trudy Campo, president of Union County Bank, feels that there is a 40% chance that credit policies will be tightened. She says, "There is no chance that credit will remain the same after next week's meeting. It's going to be tighter or looser, I'm sure of that." To get more information concerning the chances that credit will be tighter or looser after next week, Ms. Campo contacted a respected economist, Dr. James Block. Dr. Block predicted that the Federal Reserve would lossen credit, but he gave this caution to Ms. Campo: "Remember, Trudy, I've only predicted 80% of the cases where credit was loosened. And in 15% of the cases where credit was tightened, I was caught predicting a loosening." Revise Ms. Campo's prior probabilities on the basis of Dr. Block's prediction.

21. Mr. T. Ramson does business with Union County Bank, discussed in Problem 20. Mr. Ramson needs to borrow $100,000, which he will use for one year. At the end of the year he will repay the lender the loan plus interest in a single payment. That is, there are no monthly or quarterly payments due on this loan. However, he has to decide if he will borrow the money from the bank or from a family trust that is willing to loan him the money. The loan from the trust will be at a 10% rate of interest, but the trust also would require some expenses for setting up the loan. They would be $800. The conditions of the loan at the bank are up in the air, and Mr. Ramson has to apply for the loan right now.

    If credit is loosened by the Federal Reserve, as discussed in Problem 20, the interest rate on Ramson's loan will be 8%—if the loan is granted. There is about a 90% chance that it will be granted under the looser rules. If the loan is granted, it will be granted for the full amount. If the loan is denied, Ramson will have to go to the family trust for the money. However, if credit is tightened by the Federal Reserve, the cost on Ramson's bank loan would be 10% per year. Also, under the tighter rules there is only a 75% chance that the Ramson loan will be approved by the bank. Even if the loan is approved, it may only be for half the $100,000 requested. There seems to be about a 70% chance that Ramson will get the full amount requested and a 30% chance that he will get only $50,000 under these tighter conditions. The remaining $50,000 he would then borrow from the family trust at the 10% interest rate and with the same $800 initial fee. The initial loan-processing fee at the bank is $300 plus one-fourth of 1% of the amount loaned. The processing fee is not charged if the loan is denied.

    a. Draw the decision tree facing Ramson. At the ends of the branches, enter the total cost of the loan, including interest and fees.
    b. Determine the expected monetary value of each action, using the prior probabilities of Ms. Campo in Problem 20.
    c. Repeat part b, using the revised probabilities found in Problem 20.
    d. Does the revision of the probabilities change the preferred action?

22. Kelso Chemical Company (KCC) has just completed production of a batch of RB–7, a chemical product used by one of Kelso's customers, Benson Builders (BB). Just as the batch is about to be shipped to BB, KCC's production manager learns that one of the constituent chemicals of RB–7 may have contained impurities that might prevent the RB–7 from performing satisfactorily for BB.

    If the product is shipped to BB and does not perform satisfactorily, BB's production run incorporating the RB–7 will be ruined and KCC will have to replace BB's RB–7 at a cost of $2000 and pay BB a penalty fee for its ruined production run. According to KCC's contract with BB, the penalty is $1000 if the ruined run is a type I run and $400 if it is a type II run. The production manager knows that BB's runs are approximately 50% of each type.

If the questionable batch of RB–7 is scrapped, a new, guaranteed-good batch will cost an additional $2000. Rather than ship or scrap the present batch immediately, the production manager can subject it to a test to determine its quality. The cost of the test, which must be done on an overtime basis, is about $500. If the test shows the batch to be bad, the production manager will scrap the batch and run a new one despite his feeling that the test is only 90% accurate. That is, 10% of the batches the test identifies as good (or bad) are really bad (or good). He feels that the probability of the test's showing the batch to be good is .7. From past experience he feels there are about two chances in three that the batch is good.

a. Draw a decision tree for the production manager's problem. Label all branches clearly, and enter the monetary consequences. Assign probabilities to the event forks.

b. Use the decision tree as an aid in calculating the expected cost of each choice facing KCC—that is, fill in the remainder of your decision tree. What should be done, and what is the expected cost of that action?

Credit Data Set   *Refer to the 113 applicants for credit listed in the Credit Data Set in Appendix A.*

Questions   23. The credit manager of the department store received an application from a woman who was 21 years of age. This woman had no job, listed no additional sources of income, owed nothing to creditors, and was single. From the data in Appendix A, give a subjective probability that the credit manager would grant this woman credit.

24. Assume that just prior to his decision on the woman mentioned in the previous question, the credit manager noted that the woman's address was in a very exclusive part of town. Upon further investigation the credit manager found that this woman was the daughter of the president of the city's second largest bank. Discuss the factors that would cause you to revise the probability you listed in the previous question. What is your subjective probability now?

CASE 17.1   **MAKING THE DECISION TO DRILL FOR OIL**

The Overthrust Petroleum Exploration Company (OPEC) has leased a tract of land located in the overthrust belt and needs to decide whether to drill for oil on the tract or sell the lease. Before drilling or selling the lease, OPEC may decide to contract with International Seismographers, Inc., to conduct seismographic readings at the site. The seismographic readings will be used by OPEC's geologist as an indicator of whether or not there is a geologic structure beneath the tract.

From a great deal of experience the manager of the oil company believes that the probability of obtaining a productive well on the site during a drilling operation is .20. The cost of drilling at the site is $1,000,000. OPEC will receive $5,000,000 if it drills the well and the well is productive. A major oil company has offered $600,000 to purchase OPEC's lease, but the offer will be withdrawn after 24 hours.

International Seismographers, Inc., will take seismographic readings at the lease site for a fee of $200,000, but International needs several days to conduct the readings. If the seismographic readings indicate that there is a geologic structure beneath the tract, then OPEC believes that it will be able to sell the lease to a major oil company for $1,200,000. However, if the readings indicate that a geologic structure does not exist beneath the tract, then the value of the lease will be $500,000.

OPEC's geologist feels that the probability of the seismographic readings' indicating a geologic structure beneath the tract of land, given that the tract will result in a productive oil well, is .90. The probability of the seismographic readings' indicating a geologic structure, given that the tract will not result in a productive well, is .20.

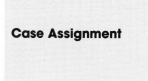

Case Assignment

  a. Draw a decision tree to represent the decisions facing OPEC.
  b. What is the probability that an oil well drilled on the site will be productive given that the readings indicate that there is a geologic structure?
  c. What is the probability that an oil well drilled on the site will be productive given that the readings do not indicate that there is a geologic structure?
  d. What should OPEC's strategy be on the basis of expected monetary value?
  e. What is the *EMV* for OPEC?

REFERENCES

Baird, Bruce F. 1989. *Managerial Decisions Under Uncertainty.* New York: Wiley.

Bierman, Harold, Jr., Charles P. Bonini, and Warren H. Hausman. 1991. *Quantitative Analysis for Business Decisions,* 8th ed. Homewood, Ill.: Irwin.

Hammond, John S. 1967. "Better Decisions with Preference Theory," *Harvard Business Review,* November-December.

Holloway, C. 1979. *Decision Making under Uncertainty.* Englewood Cliffs, N.J.: Prentice-Hall.

McNamee, P., and J. Celona. 1987. *Decision Analysis for the Professional.* Redwood City, Calif.: Scientific Press.

Schlaifer, Robert. 1969. *Analysis of Decisions under Uncertainty.* New York: McGraw-Hill.

Scientific Press. 1987. *Supertree Software.* Redwood City, Calif.: Scientific Press.

Texas Instruments, Inc. 1986. *ARBORIST Decision Tree Software.* Austin, Tex.: Texas Instruments, Inc.

Nonparametric Methods

18

In previous chapters, we often required samples taken from specific population distributions. For example, in testing hypotheses, we often required samples from populations that were normally distributed. But we also need methods of statistical inference that do not rely on specific population distributions. The need for techniques that do not rely on specific population distributions motivated the development of nonparametric methods. *Nonparametric methods* do not depend on specific population distributions; in particular, they do not require samples from normally distributed populations. Consequently, nonparametric methods are also known as *distribution-free methods*. Nonparametric methods are often based on the *ranks* or the order of quantitative data values. In contrast, parametric methods are based on parameters, such as means and standard deviations, and their estimators.

Because nonparametric methods do not require specific population distributions, they have the advantage of generality; they are inherently resistant to outliers and skewness; and they can be used with categorical variables, ranks, and frequency counts. Parametric methods can be more powerful than nonparametric methods, when the assumptions underlying the parametric methods hold; otherwise, power comparisons can be favorable for the nonparametric methods. The random samples that we used for purposes of inference in previous chapters are still required for nonparametric methods.

In this chapter, we consider the *Wilcoxon signed rank test* and the *sign test* for testing hypotheses about medians. We discuss the *Mann–Whitney rank sum test* for testing hypotheses about medians of two populations. We introduce *Spearman's rank correlation coefficient* as a nonparametric measure of correlation.

18.1
WILCOXON SIGNED RANK TEST OF A MEDIAN

The **median** of a continuous population is a number η (Greek eta) such that an observation X selected randomly from the population has probability .5 of being less than η and probability .5 of being greater. In symbols,

$$P(X < \eta) = P(X > \eta) = .5 \qquad (18.1)$$

The population is *symmetric* about its median η if for each positive a the observation X has the same probability of being less than $\eta - a$ as of being greater than $\eta + a$. In symbols,

$$P(X < \eta - a) = P(X > \eta + a) \qquad (18.2)$$

Geometrically, this expression means that the shaded areas in Figure 18.1 are equal for each a, so the frequency curve looks the same if viewed with a mirror. A normal curve with mean μ has median $\eta = \mu$ and is symmetric about η, but a nonnormal curve like the one in Figure 18.1 can also have the property of symmetry.

We tested hypotheses about population means in Chapter 9. But we also need to test hypotheses about population medians when populations are symmetric but not normally distributed. We can use the Wilcoxon signed rank test to test hypotheses about medians for symmetric populations.

Consider the null hypothesis that the population median is less than or equal to η_0 and the population is symmetric, together with the alternative hypothesis that the population median exceeds η_0 and the population is symmetric:

$$H_0: \ \eta \leq \eta_0, \text{ symmetric}$$
$$H_a: \ \eta > \eta_0, \text{ symmetric} \qquad (18.3)$$

If the population is normal, a t test applies to this problem. If the population is nonnormal, it may not be appropriate to use a t test, and another test is needed.

Let $X_1, X_2, \ldots, X_n$ be an independent sample from a nonnormal population. The **Wilcoxon statistic W** is a sum of signed ranks and is computed in four steps.

Step One   Subtract the hypothesized median η_0 from each observation to get

$$Y_i = X_i - \eta_0 \qquad (18.4)$$

If any Y_i values are 0, then omit them and reduce n accordingly.

FIGURE 18.1

Distribution Symmetric about Its Median η

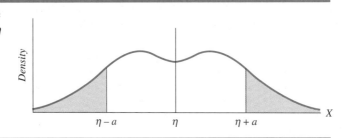

Step Two Arrange the observations $Y_i, Y_2, \ldots, Y_n$ in order of increasing *absolute value.*

Step Three Write the rank numbers $1, 2, \ldots, n$ in order, and attach the algebraic sign of the observation in the corresponding position in the arrangement of Step Two. The ranks with the appropriate signs attached are the signed ranks, R_i.

Step Four Compute the sum of the signed ranks, W, as*

$$W = \sum R_i \tag{18.5}$$

EXAMPLE 18.1 The annual returns for stocks on the Midwest Regional Exchange, after adjusting for inflation, are distributed symmetrically. A random sample of $n = 4$ annual, inflation-adjusted returns gave values of

+5.4   −7.3   −3.8   +9.5

We now compute the sum of signed ranks W for testing the hypothesis that the median inflation-adjusted return is less than or equal to 0, H_0: $\eta \le 0$, against the alternative H_a: $\eta > 0$.

The first step of the computation gives $Y_i = X_i - 0$ values since the value of the median given in the hypothesis is 0. Consequently, the Y_i values are the same as the X_i values in this example, and we do not need to omit any zero values. Next the Y_i values are arranged in order of increasing absolute value as follows:

−3.8   +5.4   −7.3   +9.5

The numbers 1, 2, 3, and 4, or ranks, with the corresponding signs attached R_i, are

−1   +2   −3   +4

The sum of the signed ranks, W, is

$$W = -1 + 2 - 3 + 4 = 2$$

In testing the hypotheses in (18.3), we reject H_0 in favor of H_a if W is excessively large.

It is reasonable to reject the null hypothesis H_0: $\eta \le 0$ when W is excessively large, just as it is reasonable to reject the null hypothesis H_0: $\mu \le 0$ when $\overline{X}$ is too large because W/n, which is proportional to W, is similar to $\overline{X}$. For the

---

*We use the sum of the signed ranks W for the Wilcoxon test, but some analysts use the sum of the positive signed ranks W^+. Tests using W or W^+ are equivalent since the two sums are related linearly as follows: The ranks for n different data values are $1, 2, \ldots, n$, and the sum of the ranks without signs is $n(n + 1)/2$. To find the sum of the negative ranks with signs, we negate the difference between the sum of the ranks without signs and the sum of the positive ranks, or we find $(-1)[n(n + 1)/2 - W^+]$. Therefore, the sum of the signed ranks is the sum of the positive ranks plus the sum of the negative ranks with signs, or $W = W^+ + (-1)[n(n + 1)/2 - W^+]$.

four inflation-adjusted stock return X_i values and the corresponding signed rank R_i values, we have

$$\overline{X} = \frac{+5.4 - 7.3 - 3.8 + 9.5}{4} = .95$$

and

$$\frac{W}{n} = \frac{-1 + 2 - 3 + 4}{4} = .5.$$

The sample mean $\overline{X} = .95$ and $W/n = .5$ are the balance points for the X_i and R_i values represented by weights in Figure 18.2. In Figure 18.2(a), the balance point $\overline{X}$ exceeds 0 because the positive X_i values overbalance the negative values. In Figure 18.2(b), the balance point W/n exceeds 0 because the positive signed rank R_i values overbalance the negative values. Consequently, W/n is a non-parametric point of balance, and it is reasonable to reject H_0: $\eta \leq 0$ if W/n is excessively above 0 or, since W is proportional to W/n, if W is excessively large. Changing from $\overline{X}$ to W enables us to test the location of a nonnormal distribution.

For fixed n the distribution of W is the same for every population satisfying the null hypothesis H_0. There are published tables of this distribution for small values of n (Gibbons, 1985). In the examples here, we will use the normal approximation, which is accurate for an n of 10 or more, as is the case in many practical applications. The mean and standard deviation of W are

$$\mu_W = 0 \qquad\qquad\qquad\qquad\qquad\qquad (18.6)$$

and

$$\sigma_W = \sqrt{\sum R_i^2} \qquad\qquad\qquad\qquad\qquad (18.7)$$

FIGURE 18.2

Parametric and Non-
parametric Balance Points

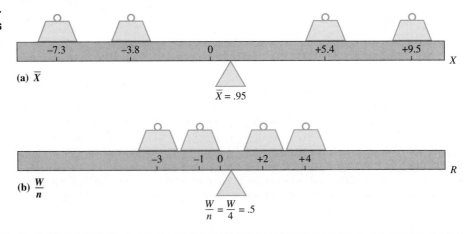

(a) $\overline{X}$

(b) $\dfrac{W}{n}$

Thus, the standardized variable is W/σ_W. To test at the level α, we reject H_0 if W/σ_W exceeds the corresponding percentage cutoff point Z_α on the normal curve.

The following procedure summarizes the **Wilcoxon signed rank test.**

Hypotheses and Rejection Rule for One-Sided Wilcoxon Signed Rank Test

H_0: $\eta \leq \eta_0$, symmetric

H_a: $\eta > \eta_0$, symmetric

Rejection rule: If $W/\sigma_W > Z_\alpha$, then reject H_0;

where: W is the sum of the signed ranks, or

$$W = \sum R_i \qquad \sigma_W = \sqrt{\sum R_i^2}$$

and n is greater than or equal to 10.

The same procedure is applicable for H_a: $\eta < \eta_0$ except that the rejection rule is as follows.

Rejection rule: If $W/\sigma_W < -Z_\alpha$, then reject H_0.

And for the hypotheses H_0: $\eta = \eta_0$ and H_a: $\eta \neq \eta_0$, the rejection rule is as follows.

Rejection rule: If $W/\sigma_W < -Z_{\alpha/2}$ or $W/\sigma_W > Z_{\alpha/2}$, then reject H_0.

In using the Wilcoxon signed rank test and other tests described in this chapter, we sometimes have to deal with **tied observations** or ranks. When two or more ranks are tied, we merely average the ranks and assign all the tied observations the average rank. For instance, in the five observations 3, 5, 7, 7, and 9, we see that the third and fourth observations are tied. The ranks to assign to the five observations are 1, 2, 3.5, 3.5, and 5 where 3.5 is the average of ranks 3 and 4 or $(3 + 4)/2 = 3.5$. The method of using the average rank for tied observations will be used throughout this chapter. The two-sided Wilcoxon signed rank test and the use of tied ranks are demonstrated in the next problem.

  PRACTICE PROBLEM 18.1

A market researcher with Catalog Apparel, Inc., constructed a questionnaire seeking consumer responses to a product. She hypothesized that people were equally split on whether they liked the product's package design. To test this hypothesis, she asked the following question: "What is your feeling about the product's package design?" The respondents could then check a response on a seven-point scale that ranged from 1 = strongly dislike to 7 = strongly favor. The response in the middle was 4 = indifferent. In a pilot test of her questionnaire she obtained the following responses to this question from 12 people:

7   3   5   4   7   1   2   2   5   7   6   5

Use the Wilcoxon W to test the hypothesis that the median response to the package design is 4. Let $\alpha = .05$.

Solution The steps of the hypothesis test are as follow.

Step One H_0: $\eta = 4$, symmetric.

Step Two H_a: $\eta \neq 4$, symmetric.

Step Three   The sample size is $n = 12$, and $\alpha = .05$.

Step Four   Since $Z_{.025} = 1.96$, we have the following rejection rule.

Rejection rule: If $W/\sigma_W < -1.96$ or $W/\sigma_W > 1.96$, then
reject H_0.

Step Five   If we subtract the hypothesized median of 4 from each response, we obtain the $Y_i = X_i - 4$ values as follow:

$$3 \quad -1 \quad 1 \quad 0 \quad 3 \quad -3 \quad -2 \quad -2 \quad 1 \quad 3 \quad 2 \quad 1$$

Next, we omit the 0 and arrange these values in increasing order of absolute value, to give

$$-1 \quad 1 \quad 1 \quad 1 \quad -2 \quad -2 \quad 2 \quad 3 \quad 3 \quad -3 \quad 3$$

The ranks can now be determined, but we note that there are several ties. Thus we find the signed ranks R_i are

$$-2.5 \quad 2.5 \quad 2.5 \quad 2.5 \quad -6 \quad -6 \quad 6 \quad 9.5 \quad 9.5 \quad -9.5 \quad 9.5$$

The value of W is found by summing the signed ranks, or

$$W = \sum R_i = -2.5 + 2.5 + 2.5 + \cdots + 9.5 = 18$$

The standard deviation of W is

$$\sigma_W = \sqrt{\sum R_i^2} = \sqrt{(-2.5)^2 + 2.5^2 + 2.5^2 + \cdots + 9.5^2}$$
$$= \sqrt{494} = 22.23$$

FIGURE 18.3

Test of Package Design at
Catalog Apparel, Inc.

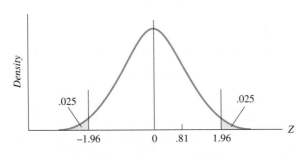

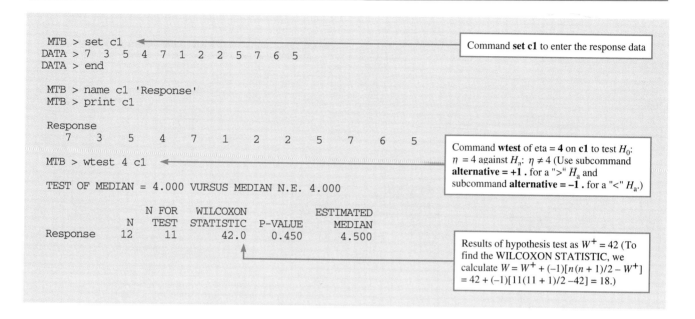

COMPUTER EXHIBIT 18.1A Wilcoxon Signed Rank Test for Catalog Apparel, Inc.,
Questionnaire Response Data

MINITAB

```
MTB > set c1
DATA > 7  3  5  4  7  1  2  2  5  7  6  5
DATA > end

MTB > name c1 'Response'
MTB > print c1

Response
  7    3    5    4    7    1    2    2    5    7    6    5

MTB > wtest 4 c1

TEST OF MEDIAN = 4.000 VURSUS MEDIAN N.E. 4.000

                  N FOR   WILCOXON              ESTIMATED
             N    TEST    STATISTIC  P-VALUE    MEDIAN
Response    12    11        42.0      0.450      4.500
```

Command **set c1** to enter the response data

Command **wtest** of eta = **4** on **c1** to test H_0: $\eta = 4$ against H_a: $\eta \neq 4$ (Use subcommand **alternative = +1**. for a ">" H_a and subcommand **alternative = –1**. for a "<" H_a.)

Results of hypothesis test as $W^+ = 42$ (To find the WILCOXON STATISTIC, we calculate $W = W^+ + (-1)[n(n + 1)/2 – W^+]$ $= 42 + (-1)[11(11 + 1)/2 –42] = 18$.)

Thus, the standardized value of W is

$$\frac{W}{\sigma_W} = \frac{18}{22.23} = .81$$

Step Six We see that .81 is less than 1.96 (see Figure 18.3), so we do not reject H_0. We do not conclude that the question responses have a population median different from 4, "indifferent."

A computer was also used to test the hypothesis of Catalog Apparel, Inc. Computer Exhibits 18.1A and 18.1B give the results for the Wilcoxon signed rank test.

Problems: Section 18.1

Answers to odd-numbered problems are given in the back of the text.

1. A financial analyst at Transpacific Corporation maintains the historical notion that the median equity–debt ratio for firms in an eastern region that are going concerns has been greater than or equal to 3. A sample of $n = 10$ going concerns is taken, and the resulting equity–debt ratios are presented in the accompanying listing. Do the sample results contradict the historical notion maintained by the analyst to such a degree that an investor should reject the notion? Assume symmetry and nonnormality, and use a significance level of .05.

   8 0 0 0 4 1 1 2 2 4

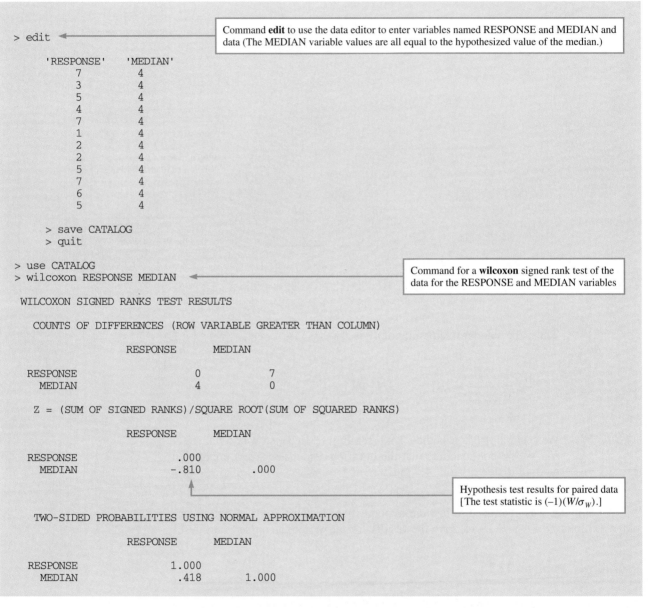

```
> edit  ←                 Command edit to use the data editor to enter variables named RESPONSE and MEDIAN and
                          data (The MEDIAN variable values are all equal to the hypothesized value of the median.)

       'RESPONSE'   'MEDIAN'
           7           4
           3           4
           5           4
           4           4
           7           4
           1           4
           2           4
           2           4
           5           4
           7           4
           6           4
           5           4

     > save CATALOG
     > quit

> use CATALOG
> wilcoxon RESPONSE MEDIAN  ←       Command for a wilcoxon signed rank test of the
                                    data for the RESPONSE and MEDIAN variables

WILCOXON SIGNED RANKS TEST RESULTS

   COUNTS OF DIFFERENCES (ROW VARIABLE GREATER THAN COLUMN)

                   RESPONSE        MEDIAN

   RESPONSE            0              7
    MEDIAN             4              0

   Z = (SUM OF SIGNED RANKS)/SQUARE ROOT(SUM OF SQUARED RANKS)

                   RESPONSE        MEDIAN

   RESPONSE          .000
    MEDIAN          -.810          .000
                         ↑
                                    Hypothesis test results for paired data
                                    [The test statistic is (−1)(W/σ_W).]

   TWO-SIDED PROBABILITIES USING NORMAL APPROXIMATION

                   RESPONSE        MEDIAN

   RESPONSE         1.000
    MEDIAN           .418         1.000
```

2. The accompanying measurements are a random sample of changes in annual stock prices for 16 stocks traded on the over-the-counter market. The changes are symmetric around their median η. Test H_0: $\eta \geq 0$ against H_a: $\eta < 0$ at the 2.5% level.

−12.6	−20.5	8.1	−27.3	−21.2	−29.4	15.5	−25.3
15.9	20.3	−11.7	−9.3	−28.8	32.7	19.2	−27.1

3. The instructions attached to a standard examination administered to employees by a personnel manager read: "The median score for people across the country on this examination is 75." The personnel manager just finished giving this examination to 11

people in one department. Test the hypothesis that the median score of all people hired into this department is 75. Use $\alpha = .10$ and an alternative hypothesis H_a: $\eta \neq 75$. Use the Wilcoxon signed rank test and the following data:

80 45 50 34 78 92 98 23 34 54 55

Consider the null hypothesis that the population median is less than or equal to η_0, together with the alternative hypothesis that the population median exceeds η_0:

$$H_0:\ \eta \leq \eta_0$$

$$H_a:\ \eta > \eta_0$$

If the population is normal, the t test applies to this problem. If the population is non-normal, the t test cannot be used in the hypothesis test, and another test is needed.

The Wilcoxon signed rank test cannot be used unless the population is symmetric about its median. In the absence of symmetry, we can use the **sign test** instead. The sign test, being less sensitive than the Wilcoxon signed rank test, should be used only if symmetry cannot be assumed.

On the basis of an independent sample $X_1, X_2, \ldots, X_n$, we are to test the null hypothesis that the population median η is η_0 against the alternative that it exceeds η_0. To apply the sign test, we count the number y of observations that exceed η_0 among $X_1, X_2, \ldots, X_n$. Because of Equation (18.1) for the median, the distribution of y under H_0 is binomial with $p \leq .5$, while under H_a it is binomial with $p > .5$. Thus, if n is large, say greater than or equal to 20, as is the case in many practical applications, we can use the normal approximation of the binomial, with

$$Z = \frac{y - \mu_y}{\sigma_y}$$

where: $\mu_y = np$

$\sigma_y = \sqrt{np(1 - p)}$

The procedure for the **sign test** is summarized as follows.

Hypotheses and Rejection Rule for One-Sided Sign Test

$$H_0:\ \eta \leq \eta_0$$

$$H_a:\ \eta > \eta_0$$

Rejection rule: If $Z > Z_\alpha$, then reject H_0;

where: $Z = \dfrac{y - \mu_y}{\sigma_y}$ $\mu_y = np$ $\sigma_y = \sqrt{np(1 - p)}$ (18.8)

and n is large, say greater than or equal to 20.

For H_a: $\eta < \eta_0$, we reject H_0 if $Z < Z_\alpha$; and for H_a: $\eta \neq \eta_0$, we reject H_0 if $Z < -Z_{\alpha/2}$ or $Z > Z_{\alpha/2}$. If n is less than 20, then see Practice Problem 18.3.

PRACTICE PROBLEM 18.2

Twenty secretaries using a new type of electric typewriter at Temporary Secretaries, Inc., increased their typing speeds by the following amounts (in words per minute):

$$
\begin{array}{cccccccccc}
+7 & -6 & +3 & +1 & +6 & +4 & +9 & -5 & +9 & -7 \\
-3 & +7 & -9 & +8 & +6 & -4 & +4 & +9 & -6 & +1
\end{array}
$$

Test, at the 5% level, the null hypothesis that the median gain in typing speed is less than or equal to zero against the alternative that it is greater than zero or positive.

Solution The hypotheses for this problem are

$$
H_0: \eta \le 0 \qquad H_a: \eta > 0
$$

Now $Z_{.05} = 1.645$, so the rejection rule is as follows.

Rejection rule: If $Z > 1.645$, then reject H_0.

There are 13 positive observations among the 20 changes in typing speeds, so the number of observations y that exceed the hypothesized median of zero is 13. The mean and standard deviation of the number of observations that exceed the median y (the number of successes in the terms used with the binomial distribution) are

$$
\mu_y = np = (n)(.5) = (20)(.5) = 10
$$

and

$$
\sigma_y = \sqrt{n(p)(1 - p)} = \sqrt{n(.5)(.5)} = \sqrt{20(.5)(.5)} = \sqrt{5} = 2.24
$$

So the standardized value is

$$
Z = \frac{y - 10}{2.24} = \frac{13 - 10}{2.24} = 1.34
$$

Since $Z = 1.34$ is less than 1.645, we do not reject the null hypothesis. That is, we conclude that there was no gain in typing speed.

The following problem is presented to illustrate three points from the preceding discussion. First, it presents a situation where the hypothesized median is nonzero. Second, it shows that when one or more of the values is zero (which is neither positive nor negative), the zero value is dropped from consideration and the sample size is reduced by one. Third, it demonstrates how to perform the sign test for small sample sizes.

PRACTICE PROBLEM 18.3

A market researcher with Catalog Apparel, Inc., constructed a questionnaire seeking consumer responses to a particular product. She hypothesized that people were equally split on whether they liked the product's package design. To test this hypothesis, she asked the following question: "What is your feeling about the product's package design?" The respondents could then check a response on a seven-point scale that ranged from 1 = strongly dislike to 7 = strongly favor. The response in the middle was 4 = indifferent. In a pilot test of her questionnaire she obtained the following responses to this question from 12 people:

7   3   5   4   7   1   2   2   5   7   6   5

Test H_0: $\eta = 4$ against the alternative H_a: $\eta \neq 4$, using the sign test. Use $\alpha = .05$.

Solution If we examine the data, we find that the list of responses, after 4 has been subtracted from each is

3   −1   1   0   3   −3   −2   −2   1   3   2   1

Arranging these values in order gives

−3   −2   −2   −1   0   1   1   1   2   3   3   3

Four of these values are negative, 7 are positive, and 1 is zero. Since zero is neither positive nor negative, we eliminate it from consideration and proceed as though we had only 11 observations.

The number of positive responses is $y = 7$, and the mean is

$$\mu_y = np = (11)(.5) = 5.5$$

The standard deviation is

$$\sigma_y = \sqrt{(n)(p)(1 - p)} = \sqrt{(11)(.5)(.5)} = 1.66$$

and the standardized value is

$$Z = \frac{y - \mu_y}{\sigma_y} = \frac{7 - 5.5}{1.66} = .90$$

There is seldom a situation in which a standardized variable of .90 would cause rejection of the null hypothesis. However, since our sample size of $n = 11$ (after one observation was dropped) is small, the test statistic cannot be compared with the normal distribution's critical values. Thus, we first must use the exact binomial probability distribution to determine the probability of obtaining 7 or more positive responses when we expect 5.5. Thus, we seek

$$P(X \geq 7 \mid n = 11, p = .5) = \sum_{k=7}^{11} \frac{11!}{k!\,(11 - k)!} (.5)^k (.5)^{11-k} = .2744$$

The probability .2744 is a P-value if the test is one-sided. However, since we have H_a: $\eta \neq 4$ and a two-sided test, we must double the .2744 probability value to arrive at a P-value of .5488. Since our P-value of .5488 is greater than our significance level of .05, we do not reject the null hypothesis. ▬▬

A computer was also used to test the hypothesis of Catalog Apparel, Inc. Computer Exhibits 18.2A and 18.2B give the results for the sign test.

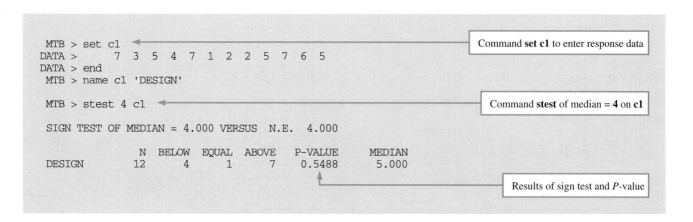

MINITAB   **COMPUTER EXHIBIT 18.2A** Sign Test for Catalog Apparel, Inc.

```
MTB > set c1  ◄─────────────────────────        Command set c1 to enter response data
DATA >      7  3  5  4  7  1  2  2  5  7  6  5
DATA > end
MTB > name c1 'DESIGN'

MTB > stest 4 c1  ◄─────────────────             Command stest of median = 4 on c1

SIGN TEST OF MEDIAN = 4.000 VERSUS  N.E.   4.000

                N  BELOW  EQUAL  ABOVE   P-VALUE     MEDIAN
DESIGN         12      4      1      7    0.5488      5.000
```
Results of sign test and P-value

Problems: Section 18.2

4. Twenty stock market analysts were asked to assess the probability that the Dow-Jones industrial average will rise 10 or more points during the next month. Their probability assessments are given in the accompanying listing. The distribution is asymmetric. Test the hypothesis H_0: $\eta \le .5$ against the alternative H_a: $\eta > 5$, using $\alpha = .05$. Which test seems appropriate?

.6	.7	.4	.9	.7	.8	.4	.3	.6	.7
.4	.9	.7	.6	.3	.6	.2	.4	.6	.9

5. The accompanying data are the times (in seconds) that it took a sample of employees to assemble a toy truck at a Cole Industries assembly plant. At the 5% level, can we conclude that the median assembly time for this toy truck is not equal to 3 minutes? Use H_a: $\eta \ne 180$ seconds as the alternative hypothesis.

190	198	181	181	208	198	187	189	191	178
199	176	174	183	188	165	166	177	182	191

 6. The instructions attached to a standard examination administered to employees by a personnel manager read: "The median score for people across the country on this examination is 75." The personnel manager just finished giving this examination to 11 people in one department. Test the hypothesis that the median score of all people hired into this department is 75. Use $\alpha = .10$ and an alternative hypothesis H_a: $\eta \ne 75$. Use the data that follow.

80	45	50	34	78	92	98	23	34	54	55

18.3
MANN–WHITNEY TEST OF TWO MEDIANS

The Mann–Whitney test is a nonparametric counterpart to the t tests of the equality of two means for normal distributions of Chapter 9. The Mann–Whitney test can be used to test whether two populations have the same medians, and it does not require the assumption of normality. The Mann–Whitney test is also known as the Wilcoxon rank sum test.

Let $X_{11}, X_{21}, \ldots, X_{n_1 1}$ be a random sample of size n_1 from population 1, and let $X_{12}, X_{22}, \ldots, X_{n_2 2}$ be an independent random sample of size n_2 from population 2. Let

COMPUTER EXHIBIT 18.2B Sign Test for Catalog Apparel, Inc.   **MYSTAT**

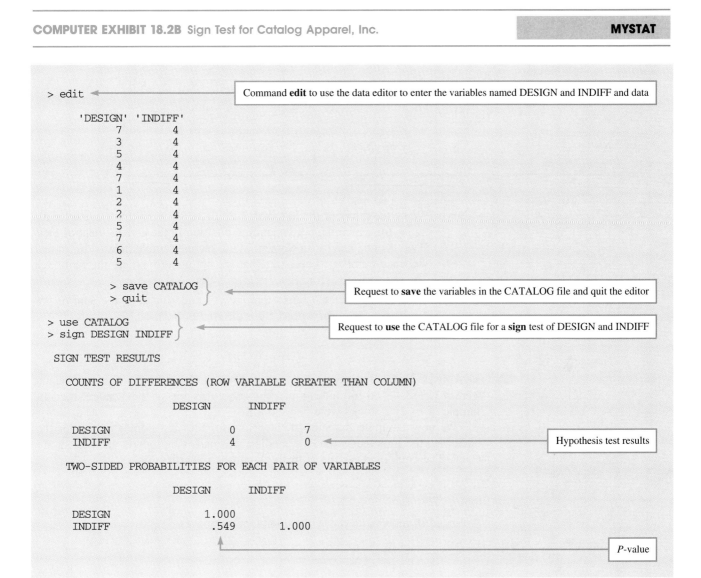

```
> edit  ←——————————  Command edit to use the data editor to enter the variables named DESIGN and INDIFF and data

      'DESIGN'  'INDIFF'
          7         4
          3         4
          5         4
          4         4
          7         4
          1         4
          2         4
          2         4
          5         4
          7         4
          6         4
          5         4

         > save CATALOG  ⎫  ←——  Request to save the variables in the CATALOG file and quit the editor
         > quit          ⎭

> use CATALOG        ⎫  ←——  Request to use the CATALOG file for a sign test of DESIGN and INDIFF
> sign DESIGN INDIFF ⎭

 SIGN TEST RESULTS

   COUNTS OF DIFFERENCES (ROW VARIABLE GREATER THAN COLUMN)

                    DESIGN      INDIFF

     DESIGN            0           7   ←——  Hypothesis test results
     INDIFF            4           0

   TWO-SIDED PROBABILITIES FOR EACH PAIR OF VARIABLES

                    DESIGN      INDIFF

     DESIGN         1.000
     INDIFF          .549       1.000
                      ↑  ←——————————————  P-value
```

H_0 be the null hypothesis that the two populations are the same, and let H_a be the hypothesis that the probability distribution for population 1 has the same shape as the probability distribution for population 2 but is shifted to the right with a larger median, as shown in Figure 18.4.

The statistic for this problem, the **rank sum** for the first population R_1, is computed in four steps.

Step One   Combine the two samples into one, underlining the observations in the second sample to keep track of them.

Step Two   Arrange the observations in the combined sample in order of increasing size, carrying along the lines under the observations in the second sample.

FIGURE 18.4

Distributions with the Median for Population 1, η_1,
Greater Than the Median
for Population 2, η_2

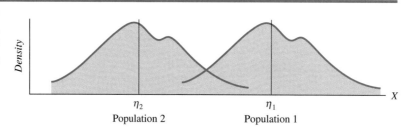

Step Three   Write the ranks or the numbers $1, 2, \ldots, n_1 + n_2$ in order, underlining
those for which the corresponding position in the arrangement of Step
Two is occupied by an element of the second sample (that is, by an underlined number). The ranks that are not underlined are the ranks for
the observations from the first population, R_{i1}.

Step Four   Find the sum of the ranks that are not underlined, or the sum of the
ranks for the observations from the first population, R_1, where

$$R_1 = \sum R_{i1}$$

EXAMPLE 18.2   The percentage returns for a random sample of $n_1 = 3$ bonds
were 3.1, -2.4, and .6, and the percentage returns for an independent random
sample of $n_2 = 2$ stocks were 7.7 and 2.8.

We now compute the sum of the ranks for the first population or for the
bonds for the hypothesis that the median returns for the bonds and stocks are
equal. The combined sample with stock returns underlined is

$$3.1 \quad -2.4 \quad .6 \quad \underline{7.7} \quad \underline{2.8}$$

Arranged in increasing order, the combined sample is

$$-2.4 \quad .6 \quad \underline{2.8} \quad 3.1 \quad \underline{7.7}$$

The numbers 1, 2, 3, 4, and 5 ($n_1 + n_2 = 5$), with underlines for stocks in
the same pattern, are

$$1 \quad 2 \quad \underline{3} \quad 4 \quad \underline{5}$$

The sum of the ranks for the first population or bonds is

$$R_1 = \sum R_{i1} = 1 + 2 + 4 = 7$$

If the medians for the two populations are equal, then the ranks for the two populations should be similar. If the ranks for the first population tend to be smaller or
larger than the ranks for the second population, so the sum of the ranks for the first
population is excessively small or excessively large, then the medians for the two
populations are likely to be different.

For each n_1 and n_2, the distribution of the sum of the ranks for the first population R_1 under H_0 has the desirable property that it does not depend on the shape of the density function common to the two populations. Tables are available for very small values of n_1 and n_2 (see Gibbons, 1985).

The mean and standard deviation of R_1 are

$$\mu_{R_1} = n_1(n_1 + n_2 + 1)/2 \tag{18.9}$$

and

$$\sigma_{R_1} = \sqrt{n_1 n_2(n_1 + n_2 + 1)/12} \tag{18.10}$$

For $n_1 + n_2$ greater than or equal to 12, as is the case in many practical applications, $(R_1 - \mu_{R_1})/\sigma_{R_1}$ is approximately normally distributed. We reject H_0: $\eta_1 = \eta_2$ in favor of H_a: $\eta_1 \neq \eta_2$ if $(R_1 - \mu_{R_1})/\sigma_{R_1}$ is less than $-Z_{\alpha/2}$ or greater than $Z_{\alpha/2}$. The procedure for a one-sided **Mann–Whitney test** is summarized as follows.

Hypotheses and Rejection Rule for a One-Sided Mann-Whitney Test of Two Medians

H_0: $\eta_1 \leq \eta_2$

H_a: $\eta_1 > \eta_2$

Rejection rule: If $(R_1 - \mu_{R_1})/\sigma_{R_1} > Z_\alpha$, then reject H_0;

where: $R_1 = \sum R_{i1}$
$\mu_{R_1} = n_1(n_1 + n_2 + 1)/2$
$\sigma_{R_1} = \sqrt{n_1 n_2(n_1 + n_2 + 1)/12}$

and $n_1 + n_2$ is greater than or equal to 12.

For the hypotheses H_0: $\eta_1 \geq \eta_2$ and H_a: $\eta_1 < \eta_2$, we reject the null hypothesis H_0 if $(R_1 - \mu_{R_1})/\sigma_{R_1} < Z_\alpha$. Other versions of rank sum tests are available, but the properties of the tests are equivalent to those for the test described above.

 PRACTICE PROBLEM 18.4

The cost of land per acre (in thousands of dollars) is presented in the accompanying table for 18 pieces of land in two different exclusive housing subdivisions. We want to test, at the 5% level, the null hypothesis of no difference in land prices between the two subdivisions, or the first subdivision's median price is greater than the second's. The alternative hypothesis is that the first subdivision's land prices are lower than the second's. Here n_1 is 10 and n_2 is 8.

Subdivision 1	40.1	37.5	46.3	44.2	35.1	48.2	41.7	37.3	50.3	43.9
Subdivision 2	49.4	41.3	45.8	39.7	44.4	43.1	48.8	47.5		

Solution The two samples merged into one and ordered are

35.1   37.3   37.5   39.7   40.1   41.3   41.7   43.1   43.9
44.2   44.4   45.8   46.3   47.5   48.2   48.8   49.4   50.3

The ranks, or the numbers $1, 2, \ldots, 18$ ($n_1 + n_2 = 18$), underlined in the same pattern, are

1   2   3   <u>4</u>   5   <u>6</u>   7   <u>8</u>   9   10   <u>11</u>   <u>12</u>   13   <u>14</u>   15   <u>16</u>   <u>17</u>   18

The sum of the ranks that are not underlined, the ranks for the first subdivision, are

$$R_1 = \sum R_{i1} = 1 + 2 + 3 + 5 + 7 + 9 + 10 + 13 + 15 + 18 = 83$$

The steps for the hypothesis test are as follow:

Step One H_0: $\eta_1 \geq \eta_2$.
Step Two H_a: $\eta_1 < \eta_2$.
Step Three $\alpha = .05$, $n_1 = 10$, $n_2 = 8$.
Step Four $-Z_{.05} = -1.645$, and the rejection rule is as follows.

> Rejection rule: If $(R_1 - \mu_{R_1})/\sigma_{R_1} < -1.645$, then reject H_0.

Step Five   The mean of R_1 is

$$\mu_{R_1} = n_1(n_1 + n_2 + 1)/2 = (10)(10 + 8 + 1)/2 = 95$$

and the standard deviation is

$$\sigma_{R_1} = \sqrt{n_1 n_2(n_1 + n_2 + 1)/12} = \sqrt{(10)(8)(10 + 8 + 1)/12}$$
$$= \sqrt{1520/12} = 11.25$$

Thus,

$$\frac{R_1 - \mu_{R_1}}{\sigma_{R_1}} = \frac{83 - 95}{11.25} = -1.07$$

Step Six   Since -1.07 is not less than -1.645, we do not reject H_0.   ▬

As in the previous section for this chapter, ties can be dealt with by averaging adjacent ranks. If ties are across populations, then the mean and the standard deviation for R_1 are approximations.

▬ PRACTICE PROBLEM 18.5

A group of salespeople at Industrial Structures, Inc., was given some sales training designed to boost their sales on a product. Another, independent group of salespeople was not given the training. The accompanying data show the sales (in thousands of dollars) for the two groups during the month after the training. Test the null hypothesis that the median sales for those (1) receiving the training is less than or equal to the median sales for those (2) not receiving the training against the alternative hypothesis that the median sales for those receiving the training is greater than the median sales for those not receiving the training. Let the level of significance be .05.

Salesperson	(1) Trained	(2) Untrained
1	4.3	2.6
2	1.2	1.5
3	4.6	4.2
4	2.8	2.5
5	4.6	3.0
6	2.0	1.8
7	2.9	2.6

Solution If we were solving this problem by using the techniques of Chapter 9, we would test the hypothesis H_0: $\mu_1 \leq \mu_2$ by using the two-sample techniques of Section 9.7. However, that technique requires us to make the rather strong assumption that the populations of sales for the untrained and trained are normally distributed. We will proceed instead by using the Mann–Whitney nonparametric test of H_0: $\eta_1 \leq \eta_2$.

In the Mann–Whitney test, we merge the sales values, underlining the values that are for salespeople not in the training program:

1.2   1.5   1.8   2.0   2.5   2.6   2.6   2.8   2.9   3.0   4.2   4.3   4.6   4.6

Next, the ranks, or the numbers $1, 2, \ldots, 14$, or the average ranks for ties, are placed in the same order, with appropriate values underlined to note the positions of the sales values from the second population:

1   2   3   4   5   6.5   6.5   8   9   10   11   12   13.5   13.5

Note that the sixth and seventh ordered values were tied, as were the thirteenth and fourteenth.

The sum of the ranks that are not underlined, the ranks for the first population, are

$$R_1 = \sum R_{i1} = 1 \mid 4 + 8 + 9 + 12 + 13.5 + 13.5 = 61$$

The steps for the hypothesis test are as follow.

Step One H_0: $\eta_1 \leq \eta_2$.

Step Two H_a: $\eta_1 > \eta_2$.

Step Three $\alpha = .05$, $n_1 = 7$, $n_2 = 7$.

Step Four $Z_{.05} = 1.645$, and the rejection rule is as follows.

Rejection rule: If $(R_1 - \mu_{R_1})/\sigma_{R_1} > 1.645$, then reject H_0.

MINITAB   **COMPUTER EXHIBIT 18.3** Mann-Whitney Test for Industrial Structures, Inc.

```
MTB > read    c1      c2  ◄────────     Command read to enter TRAINED1 and UNTRAIN2 data into c1 and c2
DATA >      4.3      2.6
DATA >      1.2      1.5
DATA >      4.6      4.2
DATA >      2.8      2.5
DATA >      4.6      3.0
DATA >      2.0      1.8
DATA >      2.9      2.6
DATA > end

MTB > name c1  'TRAINED1'
MTB > name c2  'UNTRAIN2'

MTB > mann-whitney alternative = 1 c1 c2  ◄──    Command mann-whitney alternative = 1 c1 c2
                                                 for a test of medians with $H_a$: $\eta_1 > \eta_2$
Mann-Whitney Confidence Interval and Test

 TRAINED1   N =   7      MEDIAN =      2.900
 UNTRAIN2   n =   7      MEDIAN =      2.600
 POINT ESTIMATE FOR ETA1-ETA2 IS      0.400
 95.9  PCT C.I. FOR ETA1-ETA2 IS (   -1.000,     2.000)
 W =     61.0  ◄───────                          Results of Mann–Whitney test
 TEST OF ETA1 = ETA2  VS.  ETA1 G.T. ETA2 IS SIGNIFICANT AT   0.1533

 CANNOT REJECT AT ALPHA = 0.05                              P-value
```

Step Five   The mean of R_1 is

$$\mu_{R_1} = n_1(n_1 + n_2 + 1)/2 = (7)(7 + 7 + 1)/2 = 52.5$$

and the standard deviation is

$$\sigma_{R_1} = \sqrt{n_1 n_2 (n_1 + n_2 + 1)/12} = \sqrt{(7)(7)(7 + 7 + 1)/12}$$
$$= \sqrt{735/12} = 7.83$$

Thus,

$$\frac{R_1 - \mu_{R_1}}{\sigma_{R_1}} = \frac{61 - 52.5}{7.83} = 1.09$$

Step Six   Since 1.09 is not greater than 1.645, we do not reject H_0. We do not
conclude that the sales training increases sales. ▬▬

A computer was also used to test the hypothesis of Industrial Structures, Inc.
Computer Exhibit 18.3 gives the results for the Mann–Whitney test.

Problems: Section 18.3

---

7. Independent random samples were taken of start-up owners' equity (in hundreds of
thousands of dollars) for small-business firms that within the first year of operation

have resulted in (1) nonbankruptcy or going concerns and (2) bankruptcy. The results are given in the accompanying table. Test the null hypothesis that the median start-up owners' equity for small-business firms that are not bankrupt within the first year is less than or equal to that for bankrupt firms. Use .10 for the level of significance.

| 1. Nonbankrupt Firms | 14 | 2 | 12 | 4 | 6 | 10 | 8 | 8 |
| 2. Bankrupt Firms | 10 | 10 | 0 | 0 | 2 | 2 | 4 | 4 |

8. Independent random samples of the amount of insider trading during mergers by employees of (1) investment banking firms and (2) brokerage houses gave the results listed in the accompanying table. Test the null hypothesis that the median amount of insider trading at investment banking firms is less than or equal to that at brokerage houses. Let the significance level be .10.

| 1. Investment Banking Firms | 10 | 5 | 5 | 5 | 8 | 8 | 8 |
| 2. Brokerage Houses | 6 | 0 | 5 | 2 | 2 | 3 | 3 |

9. The "triple witching hour" for the stock market occurs when futures and options for stocks expire around the same time. A future is a contract whereby a security is bought or sold with a specified price for delivery in the future. A stock option is a contract that gives the owner the right to buy or sell securities at a given price during a given period. Extreme changes in stock prices may accompany the triple witching hour. Independent random samples from nonnormal symmetric distributions of daily stock returns for a stock exchange at (1) the triple witching hour and (2) not at the triple witching hour gave the accompanying results. Test to see whether the median returns are equal or are not equal at the triple witching hour and at other times. Use a significance level of .05.

| 1. Triple Witching | 10 | 4 | 2 | 0 | 10 | 0 | 2 | 4 | 8 |
| 2. Not Triple Witching | 11 | 8 | 7 | 8 | 5 | 10 | 6 | 9 | 7 |

REFERENCE: Jacob, N., and R. Pettit (1990), *Investments,* 3rd ed. (New York: Wiley).

Correlation was discussed in Chapter 12 as a measure of the degree and direction of linear association between two variables that are normally distributed. If variables are measured by rankings, so that they are not normally distributed, then we need to use a measure of rank correlation.

18.4
SPEARMAN'S RANK CORRELATION

EXAMPLE 18.3 Suppose four students, Allen, Bill, Cole, and Darlene, are ranked in two subjects, mathematics and history, with these results:

Subject	Allen	Bill	Cole	Darlene
	Student			
Mathematics	4	2	3	1
History	3	1	4	2

That is, in mathematics Darlene is best (rank 1), Bill is next best (rank 2), Cole next (rank 3), and Allen is worst (rank 4); in history the ranking (best to worst) is Bill, Darlene, Allen, Cole. To check whether there is a connection between performance in the two subjects, we want a measure of the extent to which these two rankings agree.

This problem is similar to the problem in Section 12.6 of finding the correlation coefficient r between observations of two random variables. Recall that the Pearson correlation between two variables X and Y is

$$r = \frac{SCP}{\sqrt{(SSX)(SSY)}} = \frac{\sum(X_i - \overline{X})(Y_i - \overline{Y})}{\sqrt{\sum(X_i - \overline{X})^2 \sum(Y_i - \overline{Y})^2}}$$

However, the rankings of $n = 4$ students in mathematics and history do not follow the normal curve; they are uniformly distributed. Thus, we need a new method of finding correlations between two sets of *rankings*.

If we let R_{X_i} denote the ranks of the X_i values and R_{Y_i} denote the ranks of the Y_i values, then the correlation between the ranks is a nonparametric measure of the direction and degree of association between the two sets of rankings. The nonparametric measure of correlation is **Spearman's rank correlation r_s,** as given next.

Definition and Equation for Spearman's Rank Correlation

Spearman's rank correlation r_s is a measure of the direction and degree of the association between two sets of rankings, or

$$r_s = \frac{SCP_{R_X R_Y}}{\sqrt{(SSR_X)(SSR_Y)}} = \frac{\sum_{i=1}^{n}(R_{X_i} - \overline{R}_X)(R_{Y_i} - \overline{R}_Y)}{\sqrt{\sum_{i=1}^{n}(R_{X_i} - \overline{R}_X)^2 \sum_{i=1}^{n}(R_{Y_i} - \overline{R}_Y)^2}}$$

where R_{X_i} are the ranks of the X_i values and R_{Y_i} are the ranks of the Y_i values and the means of the ranks are

$$\overline{R}_X = \frac{\sum_{i=1}^{n} R_{X_i}}{n} \quad \text{and} \quad \overline{R}_Y = \frac{\sum_{i=1}^{n} R_{Y_i}}{n}$$

An equation that is often used to compute r_s that is equivalent to the definitional equation when there are no tied ranks is given next.

Equation for Spearman's Rank Correlation Measure

$$r_s = 1 - \frac{6\sum D_i^2}{n(n^2 - 1)} \tag{18.11}$$

where $D_i = R_{X_i} - R_{Y_i}$ is the difference between an observation's ranking on one variable and its ranking on the other.

Tied ranks are handled, as usual, by averaging adjacent ranks; however, r_s is then an approximate measure of association.

The sampling distribution of r_s has mean

$$\mu_{r_s} = 0$$

and standard deviation

$$\sigma_{r_s} = \sqrt{1/(n-1)}$$

When the sample size n is 10 or more, as is the case in many practical applications, the next statistic follows approximately the standard normal distribution:

$$Z = \frac{r_s}{\sqrt{1/(n-1)}} \tag{18.12}$$

Note that r_s is a value between -1 and 1. If the two sets of rankings agree completely, then all the D_i values are zero, and $r_s = 1$. This value, in turn, produces a Z value in Equation (18.12) that is large—and therefore significant. If the two sets of rankings are in perfectly reversed order, then the sum of the D_i^2 figures divided by $n(n^2 - 1)$ and multiplied by 6 always equals 2.0. Thus $r_s = -1$ and again produces a significant Z value that is negative. An r_s value of zero indicates that the rankings have no positive or negative correlation.

The procedure for **testing the significance of Spearman's rank correlation** is summarized as follows.

Hypotheses on Population Rank Correlation and Rejection Rule for Two-Sided Test

H_0: The rank correlation for the two variables is equal to zero.

H_a: The rank correlation for the two variables is *not* equal to zero.

Rejection rule: If $Z < -Z_{\alpha/2}$ or $Z > Z_{\alpha/2}$, then reject H_0;

where: $Z = \dfrac{r_s}{\sqrt{1/(n-1)}}$

and n is greater than or equal to 10.

For values of n less than 10, critical values of r_s can be found in special tables (Gibbons, 1985).

EXAMPLE 18.4   We can find Spearman's rank correlation r_s for the mathematics and history rankings of the four students described in Example 18.3 in the following way. The rankings for mathematics R_{X_i} and history R_{Y_i}, the differences between the rankings $D_i = R_{X_i} - R_{Y_i}$ and the squared differences D_i^2 are given in the following table:

Math, R_{X_i}	History, R_{Y_i}	$D_i = R_{X_i} - R_{Y_i}$	D_i^2
4	3	1	1
2	1	1	1
3	4	−1	1
1	2	−1	1
			$\Sigma D_i^2 = 4$

Then,

$$\sum_{i=1}^{n} D_i^2 = (1 + 1 + 1 + 1) = 4$$

and

$$r_s = 1 - \frac{(6)(4)}{4(4^2 - 1)} = .6$$

The following problem shows how r_s is calculated and tested for significance by using Equation (18.12).

 PRACTICE PROBLEM 18.6

A securities analyst for a brokerage house wanted to investigate the degree of association between a company's profitability and its liquidity. He ranked ten firms in one industry according to their profitability and liquidity, with the results as shown in the accompanying table. Use Spearman's rank correlation measure to test the null hypothesis H_0: There is no rank correlation between profitability and liquidity, against the alternative H_a: There is rank correlation (positive or negative) between profitability and liquidity. Let $\alpha = .05$.

Profitability	8	9	4	1	5	10	2	3	7	6
Liquidity	9	1	8	3	4	2	10	5	6	7

Solution The steps of the hypothesis test are as follow.

Step One H_0: *Rank correlation* $= 0$.

Step Two H_a: *Rank correlation* $\neq 0$.

Step Three The sample size is $n = 10$, and $\alpha = .05$.

Step Four The critical value of Z is $Z_{.025} = 1.96$. Consequently, the rejection rule is as follows.

Rejection rule: If $Z < -1.96$, or $Z > 1.96$ then reject H_0.

Step Five The differences in the profitability and liquidity rankings $D_i =

$R_{X_i} - R_{Y_i}$ are

$$-1 \quad 8 \quad -4 \quad -2 \quad 1 \quad 8 \quad -8 \quad -2 \quad 1 \quad -1$$

The ten D_i^2 values are listed and summed next:

$$\sum_{i=1}^{n} D_i^2 = (1 + 64 + 16 + 4 + 1 + 64 + 64 + 4 + 1 + 1)$$

$$= 220$$

Spearman's r_s value can be calculated from Equation (18.11) as

$$r_s = 1 - \frac{6(220)}{10(10^2 - 1)} = -.33$$

The test statistic is

$$Z = \frac{-.33}{\sqrt{1/(10 - 1)}} = \frac{-.33}{.333} = -.99$$

Step Six   The statistic $-.99$ is not less than -1.96, so these data do not lead us to reject the null hypothesis that profitability and liquidity rankings are not correlated.

Problems: Section 18.4

10. At the beginning of the basketball season local sportswriters met to rank the teams in the Out Back Conference. The preseason rankings and postseason standings for the teams are given in the accompanying table. Use Spearman's r_s to determine whether the preseason rankings have any meaning. That is, test the hypothesis that there is no rank correlation against the alternative that the rank correlation between preseason polls and postseason standings is positive. Let $\alpha = .05$. [*Hint:* For $n = 5$, r_s must exceed .9 for you to reject the null hypothesis.]

Team	Preseason Standing	Conference Finish
Cup Cake Junction	1	3
Tumbleweed Tech	2	4
Hoopla College	3	1
Frisbee State	4	2
Washout United	5	5

11. Two loan officers examined 11 loan applications and ranked them according to their desirability to the bank. See the accompanying table. Compute the Spearman rank correlation r_s and determine whether it is significant. Use the alternative that there is positive rank correlation, and let $\alpha = .05$.

Officer 1	1	7	4	2	3	6	5	9	10	8	11
Officer 2	1	6	5	2	3	4	7	11	8	10	9

12. The relationship between stock prices and dividends is presently unknown. However, empirical studies suggest that stock prices Y in the current time period are related to dividends paid X in the period. Stock prices and dividends for the current period are given in the accompanying listing.

Stock Price Y	14	12	12	12	10	18	10	8	9	8
Dividends X	11	9	10	11	9	13	12	11	9	7

a. Find the correlation between the rankings for X and Y.
b. Test the hypothesis that the rankings of Y and X are not related. Let $\alpha = .05$.

REFERENCE: Brennan, M. J. (1991), "A Perspective on Accounting and Stock Prices," *The Accounting Review,* 66: 67–79.

18.5
SUMMARY

In this chapter we examined the following nonparametric tests:

1. The *Wilcoxon signed rank test* is used to test the hypothesis that a population has a *median* η_0. It requires that we be able to assume the population has a symmetric shape. The Wilcoxon signed rank test is a nonparametric counterpart to the one-population t test of Chapter 9.

2. The *sign test* is also used to test the hypothesis that a population has a *median* η_0. It does not require the assumption of symmetry, but it is a somewhat weaker test. That is, it may not identify a population with a median different from η_0 as powerfully as the Wilcoxon signed rank test if the population is symmetric. The sign test is also a nonparametric counterpart to the one-population t test of Chapter 9.

3. The *Mann–Whitney test* is used to test the hypothesis that *two populations have the same median*. The only assumption required is that the shape of the two populations be the same (not necessarily symmetric). The assumption is a rather weak one since we are usually comparing populations of similar measures, which are likely to be quite similar in shape. The Mann–Whitney test is the nonparametric counterpart to the two-population t tests of Chapter 9.

4. *Spearman's rank correlation test,* with the statistic r_s, is used to test for correlation (positive or negative) between two sets of rankings of the same items. No assumptions are involved. Spearman's rank correlation r_s is a nonparametric counterpart of the correlation coefficient r of Chapter 12.

A summary of these methods is presented in Table 18.1.

REVIEW PROBLEMS

Answers to odd-numbered problems are given in the back of the text.

13. A financial analyst for James, Stephen & Associates maintains the historical view that takeover target firms have always experienced percentage sales growths with a median of 23 percent or less. A sample of $n = 10$ firms was taken from a western state and the sales growth percentages are given in the accompanying listing. Test the hypothesis that firms that are now takeover targets have increased their median percentage sales growth. Assume symmetry and nonnormality, and use a .05 level of significance.

    31   21   21   26   26   26   24   25   25   27

TABLE 18.1 **Summary of Four Nonparametric Tests**

Test Name	Null Hypothesis	Test Statistic
Wilcoxon signed rank	$H_0: \eta = \eta_0$, symmetric	W/σ_W where: $\sigma_W = \sqrt{\sum R_i^2}$

Comments:
  —W is the sum of signed ranks.
  —The population values must be symmetric around the median.
  —W is approximately normally distributed for n of 10 or more, as is the case in many practical applications.
  —For smaller n, tables of critical values for W are available.

Test Name	Null Hypothesis	Test Statistic
Sign test	$H_0: \eta = \eta_0$	y

Comments:
  —y is the number of values greater than (or less than) η_0.
  —No symmetry assumption is required.
  —For small n, y is binomially distributed with $p = .5$.
  —For large n, say n is greater than or equal to 20, as is the case in very many practical applications, y is approximately normal with mean $.5n$ and standard deviation $.5\sqrt{n}$ and $Z = (y - \mu_y)/\sigma_y$ follows the standard normal distribution.

Test Name	Null Hypothesis	Test Statistic
Mann–Whitney	$H_0: \eta_1 = \eta_2$	$\dfrac{R_1 - \mu_{R_1}}{\sigma_{R_1}}$ where: $\mu_{R_1} = \dfrac{(n_1)(n_1 + n_2 - 1)}{2}$ $\sigma_{R_1} = \sqrt{\dfrac{(n_1)(n_2)(n_1 + n_2 + 1)}{12}}$

Comments:
  —$R_1 = \sum R_{i1}$ is the sum of the first sample's ranks from the combined data set.
  —For small n_1 and n_2, tables of critical R_1 values are available.
  —For large n_1 and n_2, say, $n_1 + n_2 \geq 12$, as is the case in many practical applications, $(R_1 - \mu_{R_1})/\sigma_{R_1}$ is approximately normally distributed.

Test Name	Null Hypothesis	Test Statistic
Spearman's rank correlation	H_0: Rankings are not correlated	$r_s = 1 - \dfrac{6 \sum D_i^2}{n(n^2 - 1)}$

Comments:
  —D_i is the difference between an item's ranking in the two sets of rankings.
  —For $n < 10$, special tables of critical r_s values are available.
  —For $n \geq 10$, as is the case in many practical applications, $r_s/\sqrt{1/(n - 1)}$ approximately follows the standard normal distribution.

14. Vulvitzer Pharmaceutical, Inc., has developed 7-voricil, a new cauterizing solution for local application on the benign tumor condylomata acuminata. Ten milliliters of the currently used resin solution has provided curative therapy for a median of 5 square centimeters of tumor. The areas of tumors that receive adequate therapy are symmetric and nonnormal. A sample of clinical trials of 10 ml of 7-voricil resulted in

adequate therapy for the tumor areas given in the accompanying listing. Test to determine whether 7-voricil is more effective in cauterizing the benign tumors than is the current resin solution. Use a significance level of .05.

14   6   10   6   7   6   7   11   8   3

REFERENCE: Ferenczy, A. (1987), "Laser Treatment of Patients with Condylomata...," *CA*, 37: 334–47.

15. The following measurements represent the effective income tax rates for several young urban professionals after the most recent change in the tax code. Tax rates are not normally distributed. Use the sign test to test, at the 5% level, whether the measurements come from a population with median 10. Use H_a: $\eta \neq 10$.

| 13.6 | 8.0 | 15.8 | 12.2 | 3.5 | 17.3 | 9.5 | 11.4 | 12.8 | 10.7 |
| 17.1 | 12.7 | 13.9 | 11.4 | 7.5 | 16.3 | 13.2 | 17.0 | 12.1 | 26.3 |

16. The accompanying values are percentage rates of return for investments in real estate parcels in a major time-share resort development. Distributions of rates of return for real estate parcels may substantially differ from normality. A financial analyst wants to test whether the median return on real estate parcels is greater than or equal to zero, or if the median return is negative. Test H_0: $\eta \geq 0$ against H_a: $\eta < 0$ at the 2.5% level.

| −12.6 | −20.5 | 8.1 | −27.3 | −21.2 | −29.4 | 15.5 | −25.3 | −26.0 | −10.2 |
| 15.9 | 20.3 | −11.7 | −9.3 | −28.8 | 32.7 | 19.2 | −27.1 | −22.7 | −14.3 |

17. The following data are measurements for the number of customers serviced per shift at the drive-up windows of state-chartered banks (sample 1) and federal-chartered banks (sample 2). Use the Mann–Whitney test to check whether the two samples come from the same population. Use $\alpha = .10$.

Sample 1	120	136	107	109	129	117	125	110	124
Sample 2	131	144	116	111	103	122	141	139	130
	133	132	135	148

18. The daily catch of two fishing boats was recorded on a random basis. The results for two independent random samples are given in the accompanying table. Use the Mann–Whitney test to check whether the two samples come from the same distribution. Use a significance level of .02.

| *Boat 1* | 108 | 110 | 103 | 100 | 107 | 107 | 101 |
| *Boat 2* | 113 | 110 | 108 | 98 | 111 | 112 | 110 |

19. A garment manufacturer has a shop containing several hundred commercial sewing machines. The employees who operate the machines attended a program on the safe and efficient use of their machines. The times to machine breakdown for a sample of machines before the training are given in the accompanying table. The similar times to breakdown are listed for an independent sample of machines after the training. Use the Mann–Whitney test to determine whether the median time to breakdown differs before and after training. Use $\alpha = .10$; the alternative hypothesis is H_a: $\eta_1 > \eta_2$.

Machine	(1) Time Until Breakdown, After Training	(2) Time Until Breakdown, Before Training
1	42 weeks	23 weeks
2	39 weeks	32 weeks
3	24 weeks	44 weeks
4	26 weeks	12 weeks
5	48 weeks	55 weeks
6	15 weeks	6 weeks

20. Assume for the moment that a previous sample and test found that the hypothesis of normality for the two populations of times in Problem 19 could not be rejected. Use the methods of Section 9.8 to test the hypothesis of equal means for the data in Problem 19. Use H_a: $\mu_1 < \mu_2$ and $\alpha = .10$. Under the circumstances described here, which of the two approaches would be preferred?

21. Use the Mann–Whitney test to determine whether there is a difference between the shelf lives of a food product from two different firms. The shelf life figures (in days) are given in the accompanying table and were determined by an independent testing agency. Use H_0: $\eta_1 = \eta_2$ and H_a: $\eta_1 \neq \eta_2$, with $\alpha = .05$.

Firm 1	12	14	15	15	13	16	13	17	19	17	17
Firm 2	20	10	19	21	15	16	18	14	15		

22. The job performance of ten employees was ranked by two supervisors familiar with the work of all ten. Compute Spearman's rank correlation measure of the accompanying two sets of ratings. Is it significantly greater than 0 at the 2.5% level?

Supervisor 1	5	6	3	9	4	8	1	7	10	2
Supervisor 2	3	4	1	8	5	10	6	7	9	2

23. Two interviewers at the same company rated several job applicants. The ratings are given in the accompanying table. Use Spearman's rank correlation measure r_s to show whether the interviewers disagree on how they rank the applicants. Test the significance of r_s by using the alternative hypothesis H_a: There is negative rank correlation. Let $\alpha = .01$.

Interviewer 1	7	6	4	1	2	8	3	5	10	11	9
Interviewer 2	2	3	5	8	7	1	6	4	10	11	9

24. The second interviewer in Problem 23 complained that she thought the first interviewer was putting too much weight on an applicant's college GPA. To substantiate her point, she compared the applicant's rankings by GPA with the first interviewer's rankings. Calculate Spearman's rank correlation measure r_s, and test to see whether it is significantly different from zero at the $\alpha = .01$ level of significance. [*Hint:* Rank the GPA's first.]

Interviewer 1 Ranking	7	6	4	1	2	8	3	5	10	11	9
GPA	2.1	2.4	3.2	3.9	3.7	2.1	3.5	3.5	2.0	2.0	1.9

25. Two capital budgeting consultants were retained by Beard Enterprises to help the firm set its capital budgeting priorities for the coming year. The consultants ranked the projects as shown in the accompanying table. Do the rankings of the consultants differ significantly from one another? Use the alternative that there is positive rank correlation, and let $\alpha = .10$.

Consultant 1	7	9	12	10	1	8	5	11	2	4	3	6
Consultant 2	9	11	10	8	4	6	5	12	1	2	7	3

26. The relationship between earnings and stock prices is presently unknown. However, empirical studies suggest that stock returns in the current time period are related to earnings from a previous period. Daily stock returns (as percentages times 100) for the current period and earnings from a previous period are given in the following listing:

Stock Return	4	−1	0	2	0	1	5	3	6	10
Earnings	6	−1	−6	6	0	7	6	4	7	11

 a. Find the correlation between the rankings for X and Y.
 b. Test the hypothesis that the rankings of Y and X are not related. Let $\alpha = .05$.

REFERENCE: Lipe, R. (1989), "The Relations Between Stock Returns and Accounting Earnings Given Alternative Information," *The Accounting Review*, 65: 49–71.

Ratio Data Set   *Refer to the 141 companies listed in the Ratio Data Set in Appendix A.*

Questions   27. Locate the data for current assets–sales ratio (CA/S). The companies in the file are a random sample from the numerous companies in the COMPUSTAT files. A financial analyst maintains that companies in the population historically have had CA/S with a median of 1.40 or more. Has the ratio decreased on average? Use a significance level of .05.

28. Locate the data for the quick assets–sales ratio (QA/S). The companies in the file are a random sample from the numerous companies in the COMPUSTAT files. A financial analyst maintains that the population of companies historically have had QA/S with median equal to 1.40 or more. Has the ratio decreased overall? Use a significance level of .05.

29. Locate the data for the net income to total assets ratio (NI/TA). The companies on the list are a random sample of all of the numerous companies included in the COMPUSTAT data base. A stockbroker trying to sell stock for the companies maintains that companies have historically had NI/TA equal to a median of .04 or more. Has the ratio decreased? Use a significance level of .05.

30. Locate the data for the net income–total assets ratio (NI/TA). Some of the companies (1) suffered financial losses as reflected by negative values for NI/TA, whereas other companies (2) did not suffer losses. The data set also includes the current assets to current liabilities (CA/CL) ratio as a measure of liquidity. The companies in the file are a random sample of the numerous companies in the COMPUSTAT files. Test the hypothesis that the median liquidity is lower for companies that suffer losses. Use a significance level of .05, and assume symmetric distributions.

31. Locate the data for the net income–total assets (NI/TA) ratio as a measure of profitability and the current assets to sales (CA/S) ratio.

   a. Use a computer to obtain Spearman's rank correlation between NI/TA and CA/S.
   b. Test whether the association between the rankings for the two variables is statistically significant. Use a significance level of .05.

32. Locate the data for the net income–total assets (NI/TA) ratio as a measure of profitability and the quick assets to sales (QA/S) ratio.

   a. Use a computer to obtain Spearman's rank correlation between the two ratios.
   b. Test whether the association between the rankings for the two variables is statistically significant. Use a significance level of .05.

33. Locate the data for the current assets–sales (CA/S) ratio and the quick assets to sales (QA/S) ratio.

   a. Use a computer to obtain the sample correlation coefficient for these two variables.
   b. Test whether the linear relationship between the two variables is statistically significant. Use a significance level of .05.

Credit Data Set   *Refer to the 113 applicants for credit listed in the Credit Data Set in Appendix A.*

Questions   34. Use the Mann–Whitney test to test the hypothesis that the medians of JOBINC are equal for the groups who were granted and denied credit. Let $\alpha = .05$.

35. Rank applicants according to JOBINC and TOTBAL. Test the hypothesis that there is no rank correlation between income and debt balance. Let $\alpha = .05$.

COMPARING TEMPERATURE RANKINGS FOR THE SPACE SHUTTLE *CHALLENGER*

The space shuttle *Challenger* was launched on the morning of January 28, 1986, with seven astronauts and passengers on board. Just prior to the launch, sheets of ice clung hauntingly to the fuselage. Moments later, with national television coverage as it blasted to orbit, the shuttle disintegrated in an explosive fireball. The remains of the astronauts and passengers were never recovered.

Thiokol Corporation manufactures the two solid-fuel rocket motors that thrust the shuttle into space. The night before the catastrophe, executives of Thiokol and the National Aeronautics and Space Administration debated whether they should launch the shuttle according to schedule or postpone the mission. The weather report called for a temperature of 31°F at blast off. At the conclusion of the debate, Thiokol executives recommended that the shuttle be launched on schedule because they felt that they did not have conclusive evidence that the low temperature would influence the capability of the solid-fuel rocket motors to thrust their payload into orbit.

From April 12, 1981, to January 12, 1986, prior to the catastrophe, the space shuttle had flown 24 successful missions. Six primary O-rings were used to seal the sections of the two solid-fuel rocket motors that were used to thrust the shuttle into space. On several flights, the motors had experienced O-ring erosion or gas blow-by incidents. O-ring incidents are extremely dangerous because a failed O-ring can allow super-hot gases to escape the solid-fuel motors and ignite the liquid hydrogen fuel tank. On one flight, the motors were not recovered. The number of erosion or blow-by incidents and the temperature of the rocket joints for 23 successful flights prior to the catastrophe are given in the accompanying listing.

Mission	O-Ring Incidents	Temperature (°F)	Mission	O-Ring Incidents	Temperature (°F)
1	0	66	13	0	67
2	1	70	14	2	53
3	0	69	15	0	67
4	0	68	16	0	75
5	0	67	17	0	70
6	0	72	18	0	81
7	0	73	19	0	76
8	0	70	20	0	79
9	1	57	21	2	75
10	1	63	22	0	76
11	1	70	23	1	58
12	0	78			

To examine the differences in temperatures when the space shuttle experienced erosion or gas blow-by O-ring incidents compared to when there were no incidents, we use the Mann–Whitney nonparametric test of medians. Some of the temperature values seem quite low for the launch site, and the analysis based on ranks is resistant to the influence of surprisingly low temperature values.

Since higher temperatures should expand the O-rings and improve the rocket joint seals, we examine whether the data support the hypothesis that the median temperature η_1 was higher when there were no O-ring incidents than it was when there were O-ring incidents η_2 on missions prior to the catastrophe. The results of a Mann–Whitney test of the

FIGURE 1

Mann–Whitney Test, Using MINITAB, of Median Temperature for the Space Shuttle

```
NoIncid
   66    69    68    67    72    73    70    78    67    67    75    70
   81    76    79    76

Incident
   70    57    63    70    53    75    58

Mann-Whitney Confidence Interval and Test

NoIncid   N =  16     Median =       71.00
Incidnt   N =   7     Median =       63.00
Point estimate for ETA1-ETA2 is       9.00
95.1 pct c.i. for ETA1-ETA2 is (-0.01,15.00)
W = 223.5
Test of ETA1 = ETA2  vs.  ETA1 g.t. ETA2 is significant at 0.0192
The test is significant at 0.0188 (adjusted for ties)

Incid
                                    ------------------
  0                            ---I     +     I----------
                                    ------------------

                       -------------------------
1or2           ---------I         +         I----------
                       -------------------------
               ----+---------+---------+---------+---------+---------+---------+--Temperat
               55.0      60.0      65.0      70.0      75.0      80.0
```

median temperatures when there were one or more O-ring incidents and when there were no O-ring incidents are shown in Figure 1.

The steps of the nonparametric hypothesis test are as follow.

Step One H_0: $\eta_1 \le \eta_2$.

Step Two H_a: $\eta_1 > \eta_2$.

Step Three $\alpha = .05$, $n_1 = 16$, $n_2 = 7$.

Step Four   We have $Z_{.05} = 1.645$, and the rejection rule is as follows.

$$\text{Rejection rule:}\quad \text{If } (R_1 - \mu_{R_1})/\sigma_{R_1} > 1.645, \text{ then reject } H_0.$$

Step Five   The joint temperatures for the missions prior to the catastrophe are

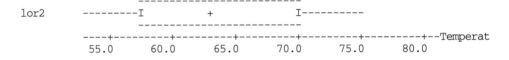

```
66   70   69   68   67   72   73   70   57   63
70   78   67   53   67   75   70   81   76   79
75   76   58
```

The rankings for the joint temperatures are

```
  5    12.5   10    9    7   15     16    12.5    2     4
12.5   21     7    1    7   17.5   12.5  23.0   19.5   22
17.5   19.5    3
```

and the corresponding number of O-ring incidents are

```
0  1  0  0  0  0  0  0  1  1
1  0  0  2  0  0  0  0  0  0
2  0  1
```

Then the sum of the ranks from population 1 when there were no O-ring incidents for a Mann–Whitney test is

$$R_1 = \sum R_{i1} = 5 + 10 + 9 + 7 + 15 + 16 + 12.5 + 21 + 7$$

$$+ 7 + 17.5 + 12.5 + 23 + 19.5 + 22 + 19.5$$

$$= 223.5$$

The mean of R_1 is

$$\mu_{R_1} = n_1(n_1 + n_2 + 1)/2 = (16)(16 + 7 + 1)/2 = 192$$

and the standard deviation is

$$\sigma_{R_1} = \sqrt{n_1 n_2 (n_1 + n_2 + 1)/12} = \sqrt{(16)(7)(16 + 7 + 1)/12}$$

$$= 14.97$$

The test statistic is

$$\frac{R_1 - \mu_{R_1}}{\sigma_{R_1}} = \frac{223.5 - 192}{14.97} = 2.10$$

Step Six   The statistic 2.10 is greater than 1.645 so we reject H_0. We conclude that the median temperature was higher when there were no O-ring incidents. The conclusion is paramount to the conclusion that the shuttle operated more safely prior to the catastrophe when the temperature was higher.

Case Assignment

a. The data indicate that mission 14 and mission 21 experienced 2 O-ring incidents. Omit the data for observations 14 and 21 and test whether the median temperature was higher when there were no O-ring incidents than it was when there was one incident. Use a significance level of .05.
b. What does your test indicate about the safety of the space shuttle prior to the catastrophe?

REFERENCES: Dalal, S., E. Fowlkes, and B. Hoadley (1989), "Risk Analysis of the Space Shuttle: Pre-*Challenger* Prediction of Failure," *Journal of the American Statistical Association*, 84: 945–57; Presidential Commission on the Space Shuttle *Challenger* Accident (1986), *Report of the Presidential Commission on the Space Shuttle Challenger Accident*, vols. 1 and 2 (Washington, D.C.: U.S. Government Printing Office).

CASE 18.2   **COMPARING CONSUMER RANKINGS OF ELECTRONIC CALCULATORS**

Bill McGraw and Leonard Alvey recently decided to make their break with Smithson Electronics, Inc. Both men had worked for Smithson for approximately fifteen years and had been satisfied with their treatment by the company for most of that time.

However, a year ago George Smithson, Jr., had taken over as vice-president of Research and Development. George Jr., son of the company's president, George Q. Smithson, knew little about research and development. McGraw and Alvey were development engineers under George Jr.'s direction. The two men soon began to grumble about young Smithson's conservative attitude toward the development of new products.

McGraw and Alvey reached their breaking point the next summer. Many electronics companies across the country were getting into the market for hand calculators. Smithson Electronics had the technology, manufacturing facilities, and marketing organization to produce and sell hand calculators at a low start-up cost. McGraw and Alvey could not convince George Smithson, Jr., that Smithson should get into the market. He felt that such a venture was "too risky."

It was this decision that brought Bill and Leonard to write their letters of resignation. They spent two months raising money from friends and relatives and securing a small loan from a local bank. They opened the doors of Blue Star Calculator Company in January. Blue Star was to assemble the Tecktron Calculator, a simple hand calculator designed by McGraw and Alvey. The company bought the components from several sources, one of which was their old employer, Smithson Electronics.

Sales of the Tecktron Calculator began slowly. It was priced to sell for $74.95, which was about in the middle of the price range for calculators with similar features. After one full year of operation Blue Star was not even close to profitability. Bill McGraw had taken a second mortgage on his home to pay off two of the smaller loans, and Leonard had borrowed enough from his father to pay off the bank. Both men were worried, but they hung their hopes on the Postal Service contract.

The Postal Service's Region Seven offices were in the market for hand calculators to be used by postal employees when making change and balancing their accounts at the end of the day. The purchasing officer was impressed with the Tecktron because of its simplicity. "Your machine has just the operations we need—no more and no less. If we buy your calculator, we won't be paying for a lot of fancy stuff our employees know nothing about and couldn't use on their jobs even if they did know," he told Leonard Alvey. Leonard had quoted the purchasing officer a very reasonable price and had high hopes of landing the contract.

That's when the bottom fell out from under Leonard and Bill's hopes. Just prior to the Postal Service's announcement as to which calculator would be bought, two different organizations published their evaluations of various simple hand calculators. The first evaluation appeared in *Appliance Buyer's Guide,* a publication that offered many consumer suggestions on the comparability of various appliances. The second evaluation was published in the monthly newsletter of Consumer Advisory Service, an organization with purposes similar to those of the *Appliance Buyer's Guide's* publisher. The Advisory Service, however, evaluated a much wider range of products. *Buyer's Guide* rated the Tecktron calculator quite low. The Advisory Service was not quite so negative.

It was not entirely clear what criteria the two organizations were using in rating the calculators they tested, but both placed heavy emphasis in their ratings on what they called "probable maintenance requirements." Both admitted that many of the calculators were so new to the market that they had not had time to develop firm statistics as to the maintenance records of the machines. However, both organizations developed their evaluations of probable maintenance records from design and quality of the calculators' components.

Buyer's Guide rated the Tecktron eighth among the 12 machines they examined for maintenance reliability, and the Advisory Service rated the Tecktron sixth among the same 12 machines. The persons rating the calculators also showed their evaluation of the

Calculator Brand	Ranks for Probable Maintenance Requirements		Rank for Ease of Use		Rank for Overall Adequacy		
	ABG	CAS	ABG	CAS	ABG	CAS	**TABLE 1**
Alpha	12	7	11	4	12	7	**Rankings of Calculators**
Cummins	1	11	3	5	2	10	
Detro	4	2	1	6	4	1	
Electra	5	12	4	10	6	12	
Felton	2	9	6	1	3	8	
Frazer	11	3	12	11	11	2	
Hulett	6	8	9	12	5	9	
Sharply	9	1	7	9	10	3	
TECKTRON	8	6	10	3	9	6	
Ventura	7	4	8	7	8	4	
Westlake	3	5	2	8	1	5	
Zermot	10	10	5	2	7	11	

12 machines in the areas of ease of use (how simple to follow the instructions were) and overall adequacy. The positions in which the evaluators ranked the various machines in three categories are presented in Table 1. In the table ABG refers to the rankings of *Appliance Buyer's Guide,* and CAS refers to those of the Consumer Advisory Service.

The Postal Service purchasing officer for Region Seven had read the results of the evaluations and called McGraw. "I still think your machine is the one we should buy. But I'm over a barrel now. I've got to justify my purchases to my superiors, and it would look as though I'm buying inferior stuff if they see these two reports."

McGraw tried to point out that the two people who rated the machines reached quite different conclusions on many of the machines. "Why look at how the Cummins machine turned out? *Buyer's Guide* rated it second overall, but the Advisory Service rated it tenth! The reverse was true for the Sharpy. *Buyer's Guide* rated it tenth, while the Advisory Service rated it third. It seems to me that these two groups must be using subjective criteria and judgments. Otherwise, they would have come in with ratings more consistent with one another. It also appears that there is an extremely high correlation between the ratings the machines were given in the maintenance reliability area and the overall rankings. If the evaluators were overly influenced by the maintenance ratings, then that's not giving the newer machines a fair shake. They admitted themselves that their maintenance data for the new machines was simply made up."

"What you say may be true," responded the purchasing officer, "but can you document some of what you say? I need some concrete statements to give my boss if he shows me these articles and asks why I didn't follow their evaluations."

Case Assignment

a. Find and test the statistical significance of Spearman's rank correlation measure for the ABG and CAS rankings of the following:

(1) Probable maintenance requirements
(2) Ease of use
(3) Overall adequacy

b. Use Spearman's rank correlation coefficient to compare the overall adequacy rankings with the probable maintenance requirements rankings for the following:

(1) The ABG rankings
(2) The CAS rankings

c. Write a report to the Postal Service purchasing officer discussing the results of your tests above, and note whether they confirm or contradict McGraw's allegations.

REFERENCES

Conover, W. J. 1980. *Practical Nonparametric Statistics,* 2nd ed. New York: Wiley.

Gibbons, J. D. 1985. *Nonparametric Statistical Inference,* 2nd ed. New York; Marcel Dekker.

Hogg, Robert V., and Allen T. Craig. 1978. *Introduction to Mathematical Statistics,* 4th ed. New York: Macmillan.

Kraft, Charles H., and Constance van Eeden. 1968. *A Nonparametric Introduction to Statistics.* New York: Macmillan.

Olds, E. G. 1938. "Distribution of Sums of Squares of Rank Differences for Small Samples." *Annals of Mathematical Statistics,* IX.

Siegel, S. 1956. *Nonparametric Statistics for the Social Sciences.* New York: McGraw-Hill.

Optional Topics

18.6 Wilcoxon Signed Rank Test with Paired Data
18.7 Runs Test for Randomness
18.8 Kruskal-Wallis Test of *k* Medians

**CHAPTER 18
SUPPLEMENT**

The Wilcoxon signed rank test can be used to test hypotheses about the median for the difference between paired data. We have n pairs of experimental units and two competing treatments (or a treatment and a nontreatment or control). Treatment 1 is given to a randomly selected element of pair i, which results in an observation, and treatment 2 is given to the other element of pair i, which results in a second observation. Under the null hypothesis that the two treatments have the same effect, the difference, which we denote X_i, has median zero and is symmetric about zero. Without the assumption of normality, which was required for the t test, we can still use the Wilcoxon signed rank test of Section 18.1.

**18.6
WILCOXON SIGNED
RANK TEST WITH
PAIRED DATA**

 PRACTICE PROBLEM 18.7

In Section 9.12 we used the t test with paired data on the null hypothesis that a particular heat treatment has no effect on the number of bacteria in skim milk at Kroft Foods, Inc., against the alternative that the treatment tends to reduce this number. Here $H_0: \eta \leq 0$ and $H_a: \eta > 0$. The X_i is the difference in the count $[\log(DMC)]$ for sample i before treatment and after treatment. The 12 differences are

.03   .14   1.17   .15 $-.02$ $-.04$ .44   .22 $-.01$ .13   .19   .70

Use the Wilcoxon signed rank test to test the hypothesis here.

Solution Since the hypothesized median is zero, we can omit Step 1 of the test and arrange the values in order of increasing absolute value; the Y_i values are

$-.01$ $-.02$ .03 $-.04$ .13   .14   .15   .19   .22   .44   .70   1.17

The numbers $1, 2, \ldots, 12$ with algebraic signs in the same pattern, the signed ranks R_i, are

-1 -2 $+3$ -4 $+5$ $+6$ $+7$ $+8$ $+9$ $+10$ $+11$ $+12$

The sum of the signed ranks is

$$W = \sum R_i = -1 - 2 + 3 - 4 + \cdots + 12 = 64$$

When there are no ties in the ranks, the standard deviation is

$$\sigma_W = \sqrt{\sum R_i^2} = \sqrt{n(n + 1)(2n + 1)/6}$$

MINITAB

COMPUTER EXHIBIT 18.4A Wilcoxon Signed Rank Test for Kroft Foods, Inc., with Paired Data

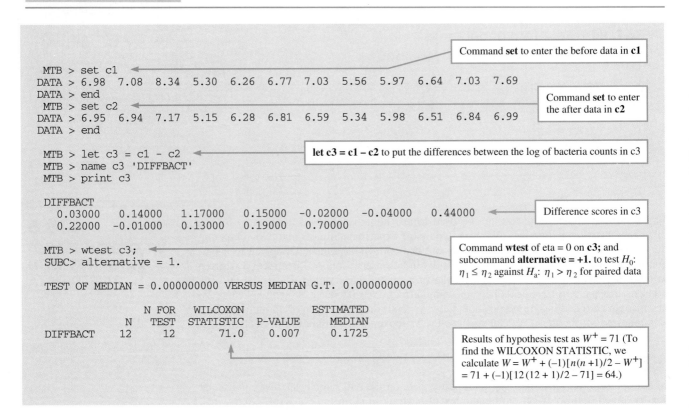

Command **set** to enter the before data in **c1**

```
MTB > set c1
DATA > 6.98   7.08   8.34   5.30   6.26   6.77   7.03   5.56   5.97   6.64   7.03   7.69
DATA > end
MTB > set c2
DATA > 6.95   6.94   7.17   5.15   6.28   6.81   6.59   5.34   5.98   6.51   6.84   6.99
DATA > end

MTB > let c3 = c1 - c2
MTB > name c3 'DIFFBACT'
MTB > print c3

DIFFBACT
   0.03000    0.14000    1.17000    0.15000   -0.02000   -0.04000    0.44000
   0.22000   -0.01000    0.13000    0.19000    0.70000

MTB > wtest c3;
SUBC> alternative = 1.

TEST OF MEDIAN = 0.000000000 VERSUS MEDIAN G.T.  0.000000000

                    N FOR   WILCOXON           ESTIMATED
              N     TEST    STATISTIC  P-VALUE  MEDIAN
DIFFBACT     12      12        71.0     0.007    0.1725
```

Command **set** to enter the after data in **c2**

let c3 = c1 – c2 to put the differences between the log of bacteria counts in c3

Difference scores in c3

Command **wtest** of eta = 0 on **c3**; and subcommand **alternative = +1.** to test H_0: $\eta_1 \le \eta_2$ against H_a: $\eta_1 > \eta_2$ for paired data

Results of hypothesis test as $W^+ = 71$ (To find the WILCOXON STATISTIC, we calculate $W = W^+ + (-1)[n(n+1)/2 - W^+]$ $= 71 + (-1)[12(12 + 1)/2 - 71] = 64$.)

so we have

$$\sigma_W = \sqrt{(12)(13)(25)/6} = \sqrt{650} = 25.5$$

So the standardized W value, W/σ_W, is $64/25.5$, or 2.51. The upper 5 percentage cutoff point $Z_{.05}$ on the normal curve being 1.645, we reject the null hypothesis of no effect.

A computer was also used to test the hypothesis of Kroft Foods, Inc. Computer Exhibits 18.4A and 18.4B give the results for the Wilcoxon signed rank test with paired data.

Problems: Section 18.6

36. A new weight reduction program that uses low-fat food supplements and peer group support is being marketed by Slim Is Us, Inc. Ten patients were selected at random to try the program with the results given in the accompanying listing. The weight difference distribution is symmetric and nonnormal. Should the Food and Drug Administration conclude that the program is effective? Use a significance level of .05.

| Weight Before | 123 | 98 | 135 | 198 | 265 | 350 | 240 | 198 | 167 | 223 |
| Weight After | 110 | 97 | 122 | 175 | 260 | 355 | 234 | 187 | 166 | 205 |

COMPUTER EXHIBIT 18.4B Wilcoxon Signed Rank Test for Kroft Foods, Inc., with Paired Data   **MYSTAT**

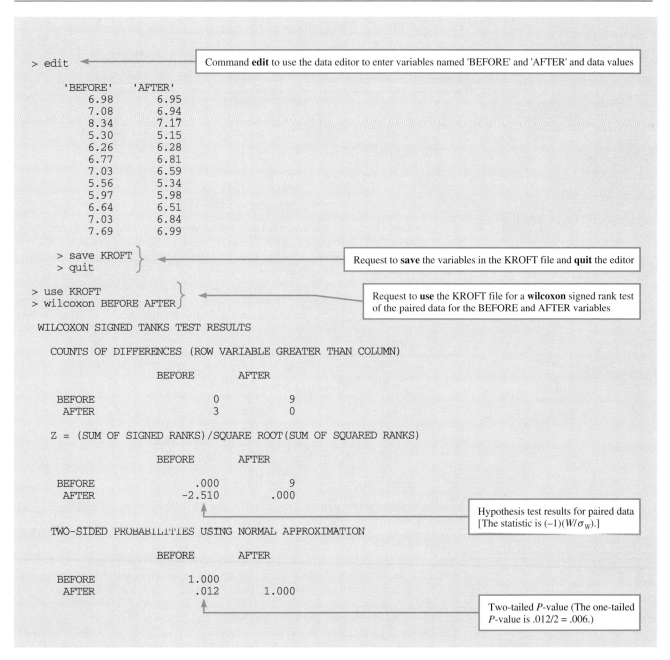

```
> edit                          Command edit to use the data editor to enter variables named 'BEFORE' and 'AFTER' and data values

    'BEFORE'    'AFTER'
       6.98       6.95
       7.08       6.94
       8.34       7.17
       5.30       5.15
       6.26       6.28
       6.77       6.81
       7.03       6.59
       5.56       5.34
       5.97       5.98
       6.64       6.51
       7.03       6.84
       7.69       6.99

  > save KROFT                   Request to save the variables in the KROFT file and quit the editor
  > quit

> use KROFT                      Request to use the KROFT file for a wilcoxon signed rank test
> wilcoxon BEFORE AFTER          of the paired data for the BEFORE and AFTER variables

WILCOXON SIGNED TANKS TEST RESULTS

  COUNTS OF DIFFERENCES (ROW VARIABLE GREATER THAN COLUMN)

                 BEFORE        AFTER

  BEFORE            0             9
  AFTER             3             0

  Z = (SUM OF SIGNED RANKS)/SQUARE ROOT(SUM OF SQUARED RANKS)

                 BEFORE        AFTER

  BEFORE          .000            9
  AFTER         -2.510          .000                          Hypothesis test results for paired data
                                                              [The statistic is (−1)(W/σ_W).]

TWO-SIDED PROBABILITIES USING NORMAL APPROXIMATION

                 BEFORE        AFTER

  BEFORE         1.000
  AFTER           .012         1.000
                                                              Two-tailed P-value (The one-tailed
                                                              P-value is .012/2 = .006.)
```

37. The management in a large assembly plant wishes to make changes in assembly techniques, but union officials are afraid that the changes will result in lower wages for their workers, who are paid on the basis of output. Thus, management set up a test, using a randomly selected group of 14 employees who had varying levels of experience in the assembly process. The assembly process was run by using the old assembly techniques, and then, after a suitable training period, it was run by using the new

techniques. The hourly wages earned by the 14 employees using the two techniques are given in the accompanying table. Is there sufficient evidence to indicate that the labor force will be able to make more money by using the new techniques? Test the hypothesis that the median wage change $\eta_1 - \eta_2$ is zero or less against the alternative hypothesis that the change is positive. Use a Wilcoxon signed rank test, and let the probability of a type I error be .10.

Employee	Technique (1) New	(2) Old	Employee	Technique (1) New	(2) Old
1	4.61	3.96	8	8.12	7.35
2	6.33	6.13	9	4.75	4.13
3	7.38	8.21	10	6.43	6.37
4	6.87	6.05	11	7.38	7.08
5	6.62	5.21	12	6.45	7.56
6	6.82	5.25	13	6.16	6.24
7	6.87	6.32	14	8.14	8.05

38. In a test of the effectiveness of two different sales approaches, each of 16 commercial cleaning compound salespeople used alternately both of the two approaches for the same period of time and the same number of sales contacts. The sales of the compound (in pounds) for each salesperson are given in the accompanying table. The distributions of sales are symmetric. Test the hypothesis that the median sales for approach 1 are less than or equal to median sales for approach 2 against the alternative that the median sales for approach 1 exceeds that for approach 2, using $\alpha = .05$.

Approach 1			Approach 2		
1	130	9	73	1	44	9	110
2	120	10	56	2	62	10	38
3	61	11	65	3	77	11	66
4	111	12	71	4	58	12	120
5	93	13	109	5	88	13	81
6	56	14	122	6	101	14	54
7	25	15	85	7	42	15	31
8	123	16	131	8	57	16	11

39. List two reasons why a paired comparison using the Wilcoxon signed rank test would not be appropriate for the data in Problem 21.

18.7
RUNS TEST FOR RANDOMNESS

In this section, we discuss a nonparametric test known as the **runs test for randomness.** Specifically, the test is for runs *above and below* a fixed value, such as the mean, for a sequence of data values.

We have required random samples in estimation and hypothesis tests, and we have assumed randomness for residuals in regression analysis. Sequences of data values examined in some time series analyses are assumed random and output variables from statistical processes are often random. Consequently, a test for randomness can be widely applicable.

The sample of n data values or outcomes for a runs test must be a sequence of values or outcomes arranged in the order of occurrence. Each data value or outcome is categorized into one of two categories, with n_1 falling in the first category and n_2 in the second. Some processes generate data values that naturally fall in one of two outcome categories and in other cases data values can be categorized as above or below a fixed value, such as the mean. For example, $n = 10$ sequential outputs from a process were observed over time and were categorized as non-defective (*S* for success) or defective (*F* for failure) as follows:

SFSFSSFFSS

For the sequence of successes and failures, $n_1 = 6$ successes and $n_2 = 4$ failures. If a sequence of quantitative data values are from a time series or from a process, the values can be categorized as being greater than or less than or equal to a fixed value, such as the mean or median of the sequence. If a data value is equal to the fixed value, we consider it as being below the fixed value when we find the number of runs.

The hypotheses to be tested for the process generating the sequence of data values are

H_0: The sequence is generated by a process that is random.

H_a: The sequence is generated by a process that is *not* random.

Recall that outcomes from a random process are independent and identically distributed.

The test for randomness is based on the number of **runs** in the sequence.

Definition

A **run** is a sequence of identical outcomes, preceded and succeeded by a different type of outcome or by no outcome at all.

A run at the beginning or end of a sequence is preceded or succeeded by no outcome at all.

The runs in the *SFSFSSFFSS* sequence of successes and failures for the non-defective and defective outputs from a process are

S F S F SS FF SS

so the total number of runs R is

$R = 7$

The runs test for randomness is based on the total number of runs in the sequence. If there too many runs or if there are too few runs in the sequence, then the sample data better support the alternative hypothesis and we reject H_0.

Under the null hypothesis, the distribution for the total number of runs R in a sequence has a mean of

$$\mu_R = 2n_1 n_2 / (n_1 + n_2) + 1$$

and a standard deviation of

$$\sigma_R = \sqrt{\frac{2n_1 n_2 (2n_1 n_2 - n_1 - n_2)}{(n_1 + n_2)^2 (n_1 + n_2 - 1)}}$$

For large samples, say n_1 and n_2 greater than or equal to 10, as is the case in many practical applications, the following statistic follows approximately the standard normal distribution:

$$Z = \frac{R - \mu_R}{\sigma_R}$$

We reject the null hypothesis of randomness according to the following rejection rule.

Rejection rule: If $Z < -Z_{\alpha/2}$ or $Z > Z_{\alpha/2}$, then reject H_0.

The procedure for a two-sided **runs test for randomness** is summarized as follows.

Hypotheses and Rejection Rule for a Two-Sided Runs Test for Randomness

H_0: The sequence is generated by a process that is random

H_a: The sequence is generated by a process that is *not* random

Rejection rule: If $Z < -Z_{\alpha/2}$ or $Z > Z_{\alpha/2}$ then reject H_0;

where: $Z = \dfrac{R - \mu_R}{\sigma_R}$

$R = \text{total number of runs}$

$\mu_R = 2n_1 n_2/(n_1 + n_2) + 1$

$\sigma_R = \sqrt{\dfrac{2n_1 n_2(2n_1 n_2 - n_1 - n_2)}{(n_1 + n_2)^2(n_1 + n_2 - 1)}}$

n is the sample size, the data values are arranged in order of occurrence with n_1 values in one category and n_2 in another, and $n_1 \geq 10$, $n_2 \geq 10$.

If we hypothesize, say, that a process is generating a sequence with a trend or a continually changing mean, such that the sequence is not random as a result of too few runs, then the test is one-sided and we reject H_0 if $Z < -Z_{\alpha/2}$; whereas if we hypothesize, say, that the process is generating a sequence with frequent changes or with a frequent cycle, such that the sequence is not random as a result of too many runs, then the test is one-sided and we reject H_0 if $Z > Z_{\alpha/2}$. If n_1 and n_2 are not greater than or equal to 10, then we can use a P-value from a computer for our test or we can use tables of critical values (Gibbons, 1985).

 PRACTICE PROBLEM 18.8

A financial analyst with First Metro Bancshares wished to examine the yields on 90-day Treasury bills. The information will be used to examine short-term interest rates over time. The yield data as percentages were collected monthly for two years (24 months), and the results are presented in order (left to right from top row to bottom row) in the accompanying listing. First, we want to construct a time series plot for the data. A time series plot is a plot of data values Y_t for a variable over time. We shall also include the mean yield for the sample data as a line on the plot. Finally, we want to test the hypothesis that the yields are generated from a process that is random. The runs test for randomness is based on runs above and below or equal to the sample mean. We will use a significance level of .05.

| 7.76 | 8.22 | 8.57 | 8.00 | 7.56 | 7.01 | 7.05 | 7.18 | 7.08 | 7.17 | 7.20 | 7.07 |
| 7.04 | 7.03 | 6.59 | 6.06 | 6.12 | 6.21 | 5.84 | 5.57 | 5.19 | 5.18 | 5.35 | 5.49 |

SOURCE: U.S. Department of Labor, Bureau of Labor Statistics (1986), *Business Statistics, 1986,* Washington D.C.: U.S. Government Printing Office.

FIGURE 18.5

Time Series Plot of Treasury Bill Yields and Mean Reference Line

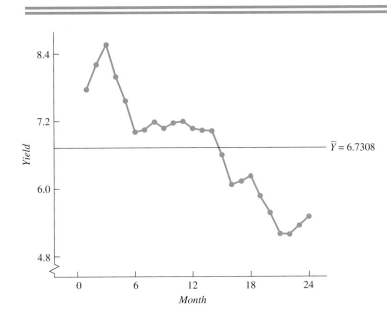

Solution A time series plot for the Treasury bill yield data and the sample mean as a line of reference is presented in Figure 18.5. The mean for the sample of treasury bill yields is $\bar{Y} = 6.7308$.

The time series plot displays the general downward trend in the Treasury bill yields during this period, suggesting that the short term interest rates were not generated by a random process. The runs test for randomness follows:

Step One H_0: The sequence is generated by a process that is random.

Step Two H_a: The sequence is generated by a process that is *not* random.

Step Three $n = 24$, $\alpha = .05$.

Step Four The time series plot shows $n_1 = 14$ data values above the mean and $n_2 = 10$ data values below the mean so we use the normal approximation. We have $\alpha/2 = .05/2 = .025$ and the standard normal table gives $Z_{\alpha/2} = Z_{.025} = 1.96$. Thus, the rejection rule is as follows.

Rejection rule: If $Z < -1.96$ or $Z > 1.96$ then reject H_0.

Step Five The time series plot shows only two runs, one above the mean and one below the mean, so

$$R = 2$$

The mean of R is

$$\mu_R = \frac{2n_1 n_2}{n_1 + n_2} + 1 = \frac{2(14)(10)}{14 + 10} + 1 = \frac{280}{24} + 1$$

$$= 11.67 + 1 = 12.67$$

FIGURE 18.6

Test of Randomness for
Treasury Bond Yields at First
Metro Bancshares

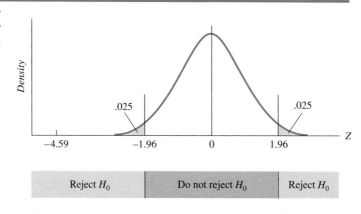

and the standard deviation is

$$
\sigma_R = \sqrt{\frac{2n_1 n_2 (2n_1 n_2 - n_1 - n_2)}{(n_1 + n_2)^2 (n_1 + n_2 - 1)}}
$$

$$
= \sqrt{\frac{2(14)(10)[2(14)(10) - 14 - 10]}{(14 + 10)^2 (14 + 10 - 1)}} = \sqrt{\frac{71680}{13248}} = 2.326
$$

The test statistic is

$$
Z = \frac{R - \mu_R}{\sigma_R} = \frac{2 - 12.67}{2.326} = \frac{-10.67}{2.326} = -4.59
$$

Step Six   The test statistic -4.59 is less than -1.96 (see Figure 18.6), so we reject H_0.

Short-term interest rates had a persistent downward trend during the period, so much so that we cannot consider the yields to be a random sample. The data do not support the notion that the yields are independent and identically distributed.

 PRACTICE PROBLEM 18.9

When a time series is generated by a process that is not random, it is often the situation with business, financial, and economic data that **changes** in the data values or **differences** between data values in the sequence are generated by a random process. Examining differences between time series data values is a method that is used often in time series analysis. The *first differences* of the time series $Y_1, Y_2, \ldots, Y_n$ are

$$
d_t = Y_t - Y_{t-1} \quad \text{for } t = 2, 3, \ldots, n
$$

First, we want to find the first differences for the First Metro Bancshares treasury bill yield data. Then we construct a time series plot for the differences. We shall also include the mean difference for the sample data as a line on the plot. Finally, we want to test the hypothesis that the yield differences are generated from a process that is random. The runs test for randomness is based on runs above and below the sample mean. We will use a significance level of .05.

FIGURE 18.7

Time Series Plot of Differences Between Treasury Bill Yields and Mean Reference Line

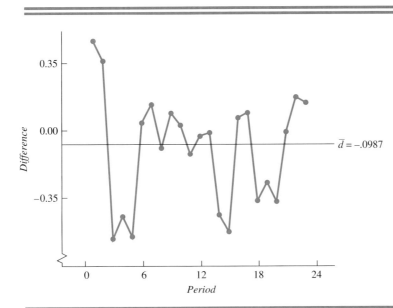

Solution The difference between the first two yields when $t = 2$ is

$$d_2 = Y_2 - Y_{2-1} = Y_2 - Y_1 = 8.22 - 7.76 = .46$$

The remaining differences were found in the same way and the 24 monthly data values resulted in 23 *differences* as follow:

—	.46	.35	−.57	−.44	−.55	.04	.13
−.10	.09	.03	−.13	−.03	−.01	−.44	−.53
.06	.09	−.37	−.27	−.38	−.01	.17	.14

A time series plot for the yield difference data and the sample mean difference is presented in Figure 18.7. The mean for the sample of differences is $\bar{d} = -.0987$. The time series plot displays a seemingly random appearance in the Treasury bill yield differences during this period, suggesting that the changes in short term interest rates may have been generated by a random process. The runs test for randomness follows.

Step One H_0: The sequence is generated by a process that is random.

Step Two H_a: The sequence is generated by a process that is *not* random.

Step Three $n = 23$, $\alpha = .05$.

Step Four   The time series plot shows $n_1 = 13$ data values above the mean and $n_2 = 10$ data values below the mean so we use the normal approximation. We have $\alpha/2 = .05/2 = .025$ and the standard normal table gives $Z_{\alpha/2} = Z_{.025} = 1.96$, and the rejection rule is as follows.

     Rejection rule: If $Z < -1.96$ or $Z > 1.96$ then reject H_0.

Step Five   The time series plot shows 11 runs above and below the mean, so

$$R = 11$$

The mean of R is

$$\mu_R = \frac{2n_1 n_2}{n_1 + n_2} + 1 = \frac{2(13)(10)}{13 + 10} + 1 = \frac{260}{23} + 1$$

$$= 11.30 + 1 = 12.30$$

and the standard deviation is

$$\sigma_R = \sqrt{\frac{2n_1 n_2 (2n_1 n_2 - n_1 - n_2)}{(n_1 + n_2)^2 (n_1 + n_2 - 1)}}$$

$$= \sqrt{\frac{2(13)(10)[2(13)(10) - 13 - 10]}{(13 + 10)^2 (13 + 10 - 1)}} = \sqrt{\frac{61620}{11638}} = 2.301$$

The test statistic is

$$Z = \frac{R - \mu_R}{\sigma_R} = \frac{11 - 12.30}{2.301} = \frac{-1.30}{2.301} = -.56$$

Step Six   The test statistic $-.56$ is not less than -1.96 so we do not reject H_0.

Differences in short term interest rates did not differ sufficiently from what we would expect under the randomness hypothesis to reject randomness. If first differences or changes in a sequence of data values are random, then the sequence may be following a *random walk*. If the differences have a mean that is not equal to zero, then the sequence may be following a random walk with *drift*. First differences are often used in the analysis of time series data.

A computer was used for the runs tests for randomness on the Treasury bill yield data and on the first differences for the yield data and the results are presented in Computer Exhibit 18.5.

Problems: Section 18.7

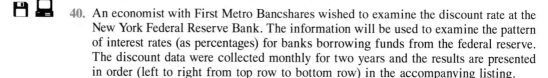

40. An economist with First Metro Bancshares wished to examine the discount rate at the New York Federal Reserve Bank. The information will be used to examine the pattern of interest rates (as percentages) for banks borrowing funds from the federal reserve. The discount data were collected monthly for two years and the results are presented in order (left to right from top row to bottom row) in the accompanying listing.

     a. Construct a time series plot for the data. A time series plot is a plot of data values for a variable over time. Include the mean rate for the sample data as a reference line on the plot. The sample mean is 7.009.

     b. Find the number of runs above and below or equal to the mean.

     c. Do the rates appear to be generated randomly?

8.00	8.00	8.00	8.00	7.81	7.50	7.50	7.50	7.50	7.50	7.50	7.50
7.50	7.50	7.10	6.83	6.50	6.50	6.16	5.82	5.50	5.50	5.50	5.50

Source: U.S. Department of Labor, Bureau of Labor Statistics (1986), *Business Statistics, 1986*, Washington D.C.: U.S. Government Printing Office.

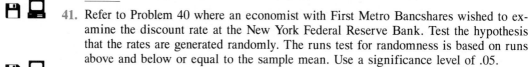

41. Refer to Problem 40 where an economist with First Metro Bancshares wished to examine the discount rate at the New York Federal Reserve Bank. Test the hypothesis that the rates are generated randomly. The runs test for randomness is based on runs above and below or equal to the sample mean. Use a significance level of .05.

42. Refer to Problem 40 where an economist with First Metro Bancshares wished to examine the discount rate at the New York Federal Reserve Bank. When a time series is

COMPUTER EXHIBIT 18.5 Test of Randomness for Treasury Bond Yields at First
Metro Bancshares

MINITAB

```
MTB > set c1
DATA. > 7.76   8.22   8.57   8.00   7.56   7.01   7.05   7.18   7.08   7.17   7.20   7.07
DATA > 7.04   7.03   6.59   6.06   6.12   6.21   5.84   5.57   5.19   5.18   5.35   5.49
DATA > end
MTB > Name c1 'Yield'

MTB > runs c1        ◄──────────────── runs c1 command for a runs test on yield

   Yield

   K =      6.7308

   THE OBSERVED NO. OF RUNS =    2
   THE EXPECTED NO. OF RUNS =  12.6667       ◄─── Results of the runs test for randomness on yield
   14 OBSERVATIONS ABOVE K    10 BELOW
          THE TEST IS SIGNIFICANT AT 0.0000

                                              difference c1 c2 command to find the first
MTB > difference c1 c2   ◄──────────────      differences of yield in c1 and put the results in c2
MTB > name c2 'DiffYld'
MTB > print c2

DiffYld
     *   0.46   0.35  -0.57  -0.44  -0.55   0.04   0.13  -0.10   0.09   0.03  -0.13      First differences
   -0.03 -0.01  -0.44  -0.53   0.06   0.09  -0.37  -0.27  -0.38  -0.01   0.17   0.14      between yields

MTB > copy c2 c3                              copy c2 c3; command and omit 1. subcommand to omit the '*'
SUBC> omit 1.            ◄──────────────      missing value runs c3 command for a runs test on the differences
MTB > name c3 'DiffYlds'
MTB > runs c3

   DiffYlds

   K =     -0.0987

   THE OBSERVED NO. OF RUNS =   11
   THE EXPECTED NO. OF RUNS =  12.3043
   13 OBSERVATIONS ABOVE K    10 BELOW       ◄─── Results of the runs test for randomness on differences
          THE TEST IS SIGNIFICANT AT   0.5709
          CANNOT REJECT AT ALPHA = 0.05
```

generated by a process that is not random, it is often the situation that *changes* in the data values or *differences* between data values in the sequence are generated by a random process.

a. Find the first differences for the discount rate data.
b. Construct a time series plot for the difference data. A time series plot is a plot of data values for a variable over time. Include the mean rate for the differences as a reference line on the plot. The sample mean is $-.1087$.
c. Find the number of runs above and below or equal to the mean.
d. Do the differences appear to be generated randomly?

43. Refer to Problem 42 where an economist with First Metro Bancshares wished to examine the discount rate differences at the New York Federal Reserve Bank. Test the hypothesis that the first differences between rates are generated randomly. The runs test for randomness is based on runs above and below or equal to the sample mean difference. Use a significance level of .05.

18.8
KRUSKAL–WALLIS TEST OF k MEDIANS

The Kruskal–Wallis test is a nonparametric counterpart to the test of the equality of k means that we discussed in the analysis of variance chapter. The F test of the equality of means was based on k normal distributions with common variance. The Kruskal–Wallis test is used to test whether k populations have the same median, and it does not require the assumption of normality. The Kruskal–Wallis test is equivalent to the Mann–Whitney test if $k = 2$ populations.

Let X_{ij} be the value of the dependent variable for the ith item in the jth group, where n_j is the sample size for the jth group and there are $j = 1, 2, \ldots, k$ groups. We take independent random samples $X_{11}, X_{21}, \ldots, X_{n_1 1}$ of size n_1 from population 1; $X_{12}, X_{22}, \ldots, X_{n_2 2}$ of size n_2 from population 2; and so forth, until we have $X_{1k}, X_{2k}, \ldots, X_{n_k k}$ as the random sample of size n_k from population k.

The hypotheses we test are

H_0: The k populations have the same distributions.

H_a: The k populations do not all have the same medians.

The procedure for the Kruskal–Wallis test is outlined below.

Step One Determine the number of groups, k; the number of observations for each group, n_j, and the total number of observations, $n_T = n_1 + n_2 + \cdots + n_k$.

Step Two Arrange the observations X_{ij} in the combined sample in order of increasing size, and assign ranks, R_{ij}, 1 to n_T, as in the Mann–Whitney rank sum test.

Step Three Compute the sum of the ranks R_j for each group as follows:

$$R_j = \sum_{i=1}^{n_j} R_{ij} \qquad j = 1, 2, \ldots, k$$

The average rank, $\overline{R}_j$, for the jth group is

$$\overline{R}_j = \frac{\sum_{i=1}^{n_j} R_{ij}}{n_j} \qquad j = 1, 2, \ldots, k$$

and the grand mean of the ranks for the combined sample is

$$\overline{\overline{R}} = \frac{\sum_{i=1}^{n_j} \sum_{j=1}^{k} R_{ij}}{n_T} = \frac{n_T(n_T + 1)}{2}$$

Step Four Compute the Kruskal–Wallis test statistic H as follows:

$$H = \frac{12}{n_T(n_T + 1)} \sum_{j=1}^{k} n_j(\overline{R}_j - \overline{\overline{R}})^2 = \frac{12}{n_T(n_T + 1)} \frac{\sum_{j=1}^{k} R_j^2}{n_j} - 3(n_T + 1)$$

Under the null hypothesis, the Kruskal–Wallis test statistic H is approximately chi square distributed with $k - 1$ degrees of freedom. The chi square approximation is reasonable if the sample sizes for each group are all greater than or equal to 5.

If the null hypothesis is true, then the sample mean ranks for the groups $\overline{R}_j$ are likely to be close to the sample grand mean rank $\overline{\overline{R}}$, and the Kruskal–Wallis test statistic H will likely be small. In contrast, if the medians for the k populations are not all equal, the sample mean ranks for some groups $\overline{R}_j$ are likely to be reasonably different from the sample grand mean rank $\overline{\overline{R}}$, and the Kruskal–Wallis test statistic H will likely be large. Consequently, we have the following rejection rule for the Kruskal–Wallis test.

Rejection rule: If $H > \chi^2_{\alpha,\,k-1}$, then reject H_0.

The **Kruskal–Wallis test** is summarized as follows.

Hypotheses and Rejection Rule for a Kruskal–Wallis Test of k Medians

H_0: The k populations have the same distributions

H_a: The k populations do not all have the same medians

Rejection rule: If $H > \chi^2_{\alpha,\,k-1}$, then reject H_0;

where: R_{ij} is the rank of X_{ij} in increasing size in the combined sample, the sum of the ranks for the jth group R_j is

$$R_j = \sum_{i=1}^{n_j} R_{ij} \qquad j = 1, 2, \ldots, k$$

the test statistic is

$$H = \frac{12}{n_T(n_T + 1)} \sum_{j=1}^{k} \frac{R_j^2}{n_j} - 3(n_T + 1)$$

and n_j is greater than 5.

The sample sizes are all greater than 5 in many practical business applications of the Kruskal–Wallis test. However, if any of the sample sizes are less than 5, then tables of critical values of the Kruskal–Wallis H statistic are available (Gibbons, 1985), or we can use a computer to find a P-value for the test. If there are tied observations or ranks, then we use the average tied ranks.*

 PRACTICE PROBLEM 18.9

A financial analyst with James, Stephens & Associates took independent random samples of start-up owners equity (in hundreds of thousands of dollars) for small business firms that within the first year of operation have resulted in (1) Chapter 7 bankruptcy, (2) Chapter 11 bankruptcy, and firms that are (3) solvent. The results are shown in the accompanying listing. Test whether the medians of start-up owners equity are different for

---

*If there are tied observations or ranks, then the Kruskal–Wallis H is sometimes adjusted as follows:

$$H_{\text{adjusted}} = \left\{ 1 / \left[1 - \sum_{i=1}^{q} (t_i^3 - t_i) / (n_T^3 - n_T) \right] \right\} H$$

where q is the number of distinct data values and t_i is the number of observations that are tied at the ith distinct value or t_i is 1 for a data value with no ties.

small business firms that are in Chapter 7 bankruptcy, Chapter 11 bankruptcy, or are solvent. Use .05 for the level of significance.

(1) Chapter 7	(2) Chapter 11	(3) Solvent
3	2	12
8	1	9
5	6	10
0	7	13
4	−1	11

Solution The ranks R_{ij} for the combined data, sums of ranks R_j for each group, and sample mean ranks $\overline{R}_j$ for each group, for the owners equity data are shown in the accompanying listing.

(1) Chapter 7	(2) Chapter 11	(3) Solvent
5	4	14
10	3	11
7	8	12
2	9	15
6	1	13
$R_1 = 30$	$R_2 = 25$	$R_3 = 65$
$\overline{R}_1 = \dfrac{30}{5} = 6$	$\overline{R}_2 = \dfrac{25}{5} = 5$	$\overline{R}_3 = \dfrac{65}{5} = 13$

The Kruskal–Wallis test of medians follows.

Step One H_0: The $k = 3$ populations have the same distributions.

Step Two H_a: The $k = 3$ populations do not all have the same medians.

Step Three $n_T = n_1 + n_2 + n_3 = 5 + 5 + 5 = 15;\ \alpha = .05.$

Step Four We have $k - 1 = 3 - 1 = 2$ degrees of freedom and the critical value from the chi square table is

$$\chi^2_{\alpha, k-1} = \chi^2_{.05, 2} = 5.99$$

The rejection rule is as follows.

Rejection rule: If $H > 5.99$, then reject H_0.

Step Five The Kruskal–Wallis H statistic for the data is

$$H = \frac{12}{n_T(n_T + 1)} \frac{\displaystyle\sum_{j=1}^{3} R_j^2}{n_j} - 3(n_T + 1)$$

$$= \frac{12}{(15)(15 + 1)}\left(\frac{30^2}{5} + \frac{25^2}{5} + \frac{65^2}{5}\right) - 3(15 + 1)$$

$$= (.05)(1150) - 48 = 9.5$$

Step Six The test statistic 9.5 is greater than 5.99 so we reject H_0.

The data better support the hypothesis that the median start-up owners equity values are not equal for small business firms that enter Chapter 7 or Chapter 11 bankruptcy or that are solvent after the first year of operation.

To informally examine which pairs of medians are not equal when we have rejected the null hypothesis, we often use box plots. For multiple pairwise comparisons of medians with an overall or family significance level ot at least α, we use the following expression to find the limits of confidence intervals. The confidence intervals are used to examine the differences between m pairs of populations denoted by j and j':

$$(\overline{R}_j - \overline{R}_{j'}) \pm Z_{\alpha/(2m)} \sqrt{\frac{n_T(n_T + 1)}{12}\left(\frac{1}{n_j} + \frac{1}{n_{j'}}\right)}$$

If the confidence interval does not contain the value 0, then we reject the hypothesis that populations j and j' have the same median. If we compare all possible pairs of medians, then we use $m = k(k - 1)/2$ confidence intervals.

 PRACTICE PROBLEM 18.10

We rejected the hypothesis of equal distributions for start-up owners equity (in hundreds of thousands of dollars) for small business firms that resulted in (1) Chapter 7 bankruptcy or (2) Chapter 11 bankruptcy and firms that are (3) solvent. We found the ranks, sums of ranks, and sample mean ranks for the equity data in the previous practice problem. We now use multiple pairwise comparisons to determine which pairs of medians of start-up owners equity for the groups are different. We use .05 for the overall level of significance α.

Solution We have $m = k(k - 1) = 3(3 - 1)/2 = 3$ pairwise comparisons to test whether the medians of the pairs of populations 1 and 2, 1 and 3, and 2 and 3 are significantly different. The $Z_{\alpha/(2m)}$ percentage point from the standard normal table is

$$Z_{.05/[(2)(3)]} = Z_{.00833} = 2.39$$

The limits of the confidence interval for testing the difference between the medians of populations $j = 1$ and $j' = 2$ are obtained from the following expression:

$$(\overline{R}_1 - \overline{R}_2) \pm Z_{\alpha/(2m)} \sqrt{\frac{n_T(n_T - 1)}{12}\left(\frac{1}{n_j} + \frac{1}{n_{j'}}\right)}$$

Thus, we have

$$(6 - 5) \pm 2.39 \sqrt{\frac{15(15 + 1)}{12}\left(\frac{1}{5} + \frac{1}{5}\right)}$$

which gives

$$1 \pm 2.39(2.828) \quad \text{or} \quad 1 \pm 6.76$$

The limits of the confidence interval are therefore -5.76 and 7.76. Since the interval contains 0, we do not reject the hypothesis that populations 1 and 2 have the same median.

The limits of the confidence interval for comparing the medians of populations 1 and 3 are obtained from the following expression:

$$(\overline{R}_1 - \overline{R}_3) \pm 2.39 \sqrt{\frac{15(15 + 1)}{12}\left(\frac{1}{5} + \frac{1}{5}\right)}$$

which gives

$$(6 - 13) \pm 6.76$$

The limits of the confidence interval are therefore -13.76 and $-.24$. Since the interval does not contain the value 0, we reject the hypothesis that the medians for populations 1 and 3 are equal.

The limits of the confidence interval for comparing the medians of populations 2 and 3 are obtained from the following expression

$$(\overline{R}_2 - \overline{R}_3) \pm 6.76$$

which gives

$$(5 - 13) \pm 6.76$$

The limits of the confidence interval are therefore -14.76 and -1.24. Since the interval does not contain the value 0, we reject the hypothesis that the medians for populations 2 and 3 are equal.

The multiple pairwise comparisons allow us to conclude that population 3, solvent companies, has a higher median start-up value for owners equity than do either of the bankrupt populations.

Problems: Section 18.8

44. Independent random samples of accounts receivable misstatement risk (owing to nonexistent accounts) during audits where the audits were (1) completed with exceptions noted, (2) completed with no exceptions, or (3) not completed, gave the accompanying results. Test whether the medians of the populations are different. Use .05 for the level of significance.

(1) Completed with Exceptions	(2) Completed Without Exceptions	(3) Not Completed
7	2	11
5	9	15
4	6	12
1	3	8.5
8	3.5	10

REFERENCE: Brown, C. E., and E. Soloman (1991), "Configural Information Processing in Auditing," *The Accounting Review,* 66: 100–19.

45. An accountant with Wheris, Waldo, and Bagady is interested in the ratio of debt–total assets (as a percentage) for corporations that have been assigned bond ratings of (1) *Baa,* (2) *AA* and (3) *AAA* by a ratings service. Random samples of debt to total assets (as percentages) for several corporations were taken from the *Baa, AA* and *AAA* ratings groups, respectively. The sample data are given in the accompanying listing. Test whether the medians of the populations are different. Use .05 for the level of significance.

(1) Baa	(2) AA	(3) AAA
41	25	15
31	22	18
38	17	14
31	21	13
26	23	12

REFERENCE: Stock, D., and Collin J. Watson (1984), "Human Judgment Accuracy, Multidimensional Graphics, and Humans Versus Models," *Journal of Accounting Research,* 22: 192–206.

46. The accountant with Wheris, Waldo, and Bagady is interested in the ratio of debt to total assets (as a percentage) for corporations that have been assigned bond ratings of (1) *Baa,* (2) *AA* and (3) *AAA* by a ratings service, as described in Problem 45. If it is appropriate to do so, use multiple pairwise comparisons to determine which pairs of population medians are different. Use .05 for the overall level of significance α.

Index Numbers

19

We introduced the concept of index numbers for time series analysis in Chapter 14 to measure changes across seasons (weeks, months, or quarters) from some base value (such as the mean weekly, monthly, or quarterly value) for the year. Accordingly, if the month of March has a seasonal index of 110, the March time series values are typically 10% above the value for the average month of the year.

Other types of index numbers show changes in time series by expressing values of the series as percentages of a base value. The base value is a fixed value selected from the series. For instance, if retail sales in dollars at Vogue Apparel, Inc., this year are 20% above what they were in year 1, the sales index for this year is 120. The base value is the fixed value of sales in the first year. In this chapter, we use *index numbers* to measure changes of a time series from a fixed value of the series. Index numbers give clear pictures of trends in sequences of data.

We introduce *price indexes* in this chapter to measure changes in prices of goods and services over time. Price indexes are important because prices of many goods and services have been increasing appreciably and because labor and pension-benefits contracts often contain cost-of-living escalation clauses that are tied to consumer price indexes.

19.1
INDEX NUMBERS

Index numbers are used in business and economics to show changes in a series of data values over time. For example, if the prices that a consumer pays for goods in year 5 were 25% above what they were in year 1, an index for the series for year 5 prices would be 125. **Index numbers** are defined as follows.

> **Definition**
>
> An **index number** measures the value of a time series in a period as a percentage of the time series in a base period.

The Bureau of Labor Statistics consumer price index (CPI) system includes some of the major indexes for the U.S. economy. Consumer price indexes measure changes in prices of goods and services and are usually called the cost-of-living indexes. Consumer price indexes are important because labor contracts and pension benefits often contain cost-of-living adjustment clauses. In addition, government fiscal and economic policies are influenced by movements in price indexes and long-term manufacturing and real-estate contracts are often tied to price indexes. Finally, changes in a consumer price index can impact consumer psychology. An index gives a clear picture of trend over time. The striking increases in prices that managers and consumers have experienced over the past several decades are depicted in the time series plot of a consumer price index in Figure 19.1.

We introduce index numbers in the following example that computes index numbers for a single quantity, the total electrical power generation of a utility company over three years.

FIGURE 19.1

Time Series Plot of a Consumer Price Index

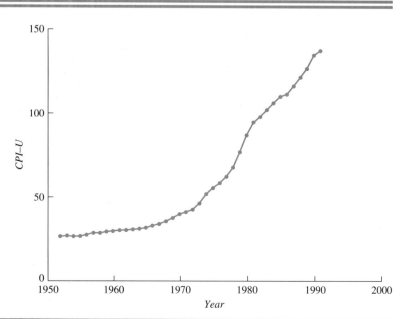

	1982	1985	1992	**TABLE 19.1**
Electric Power (billions of kilowatthours)	1329	1614	1916	**Electric Power Generation**
Generation Indexes				
Base year 1982	100.0	121.4	144.2	
Base year 1985	82.3	100.0	118.7	

EXAMPLE 19.1   Table 19.1 shows the total electrical power generation for the Atlantic Utility Company, Inc., for the years 1982, 1985, and 1992 in billions of kilowatthours. The ratio of the 1985 figure to that for 1982 is 1614/1329, or 1.214; multiplication by 100 gives 121.4, the second entry in the second row. We can say that in 1985 electrical power generation was 121.4% of what it was in 1982, or that generation was up 21.4%. The other two entries in the second row, 100.0 and 144.2, are computed in like manner: $100 \times (1329/1329)$ and $100 \times (1916/1329)$. The 1992 generation of power was 44.2% above the 1982 generation.

The indexes in the second row of Table 19.1 all have for their denominator 1329, the generation figure for 1982. In computing these index numbers, we have adopted the *base year* of 1982 as a standard of comparison. If we were to divide the four production figures by 1614, the figure for 1985, we would obtain the power generation indexes with 1985 as the base year. These indexes are given in the last row of the table. We see, for example, that power generation in 1982 was 82.3% of what it became in 1985.

The theory and computation of an index number are ordinarily more complicated than in our simple example, because an index number ordinarily represents the combined effect of changes in a number of quantities. In the next section, we find the combined effect of price changes in several items by using a simple aggregate index.

19.2
SIMPLE AGGREGATE INDEX

The prices of hundreds of items go into price indexes. A discussion of a price index, the CPI, is given later in this chapter, but to make its structure clear, we will construct a simple illustrative example involving the four items whose prices are listed in Table 19.2. We take 1982 as the base year, and the problem is to construct a sensible index for 1985 and 1992 compared with 1982. The first procedure that comes to

Item	1982 Price, p_{i0}	1985 Price, p_{i1}	1992 Price, p_{i2}	**TABLE 19.2**
1. Milk (dollars/quart)	$0.26	$0.31	$0.56	**Simple Aggregate Price Index**
2. Steak (dollars/pound)	1.05	1.17	2.15	
3. Butter (dollars/pound)	0.75	0.85	1.40	
4. Pepper (dollars/pound)	2.50	2.20	2.60	
Total	$4.56	$4.53	$6.71	
Simple Aggregate Price Index	100.0	99.3	147.1	

mind is simply to add the prices for a year and divide each sum by $4.56, the sum for 1982. This procedure gives the **simple aggregate indexes:**

$$100 \times \frac{4.56}{4.56} = 100.0 \qquad 100 \times \frac{4.53}{4.56} = 99.3 \qquad 100 \times \frac{6.71}{4.56} = 147.1$$

In other words, if p_{i0}, p_{i1}, and p_{i2} represent the respective prices of item i ($i = 1, 2, 3, 4$) for each of the three years, we have the following expressions.

Expressions for Simple Aggregate Index

$$100 \times \frac{\sum_i p_{i0}}{\sum_i p_{i0}} = 100 \qquad 100 \times \frac{\sum_i p_{i1}}{\sum_i p_{i0}} \qquad 100 \times \frac{\sum_i p_{i2}}{\sum_i p_{i0}} \qquad (19.1)$$

EXAMPLE 19.2   The index of 99.3 for 1985 makes it appear as if prices dropped from 1982 to 1985, a conclusion that defies reason for those who lived through that period of time. A glance at the table shows that although the prices of milk, steak, and butter rose considerably in that period, this increase was more than offset in our computation by the decrease in the price of pepper, which costs a great deal per pound. Since pepper is surely a small part of the average budget, we might get a better idea of food cost changes during this period if we throw out the figures for pepper. This change leads to new indexes:

	1982	1985	1992
Total Price (Pepper Omitted)	$2.06	$2.33	$4.11
Simple Aggregate Price Index	100.00	113.11	199.51

Now the indexes reflect an increase in prices. Acting on the observation that pepper played a disproportionate role in the first aggregate index seems to have put our computed figures more in line with what we know to have been the actual situation. But simply throwing out an item that does not seem important in the determination of an index is hardly a satisfactory way to determine the value of the index. Two people constructing an index number to measure the same basic price changes might disagree on which items should be thrown out. Or they might fail to delete from their working data an item that distorts the calculations in a less obvious way than the pepper in our example.

Considerations such as these suggest quite clearly that the prices in our data must somehow be weighted according to their importance in the food budget. The question of the units for which the prices are listed also suggests that prices should be weighted. The milk price is given in dollars per quart, but why not use gallons instead? Why is a quart of milk rather than a gallon comparable to a pound, the unit for the other three items? If milk cost is expressed in dollars per gallon, it becomes four times what it was and exerts a proportionally greater influence in the computation.

Item	1982 Price, p_{i0}	1985 Price, p_{i1}	1992 Price, p_{i2}	**TABLE 19.3**
1. Milk (dollars/gallon)	$1.04	$1.24	$2.24	**Simple Aggregate Price**
2. Steak (dollars/pound)	1.05	1.17	2.15	**Index Using Different**
3. Butter (dollars/pound)	0.75	0.85	1.40	**Purchasing Units**
Total	$2.84	$3.26	$5.79	
Simple Aggregate Price Index	100.0	114.8	203.9	

EXAMPLE 19.3   Table 19.3 shows how the simple aggregate price index would change if milk costs were expressed in dollars per gallon and pepper were left out of the computations. Note that when milk prices are expressed in dollars per gallon, the 1985 prices are 14.8% above the 1982 prices. But when they are expressed in dollars per quart, 1985 prices are only 13.1% above 1982 prices.

One might wonder which of these percentage increases is correct. The answer is that neither truly reflects the percentage price rise in food costs. Only two things can be said:

1. A quart of milk, a pound of steak, and a pound of butter cost 13.1% more in 1985 than they did in 1982.
2. A gallon of milk, a pound of steak, and a pound of butter cost 14.8% more in 1985 than they did in 1982.

With a change in the units of measure a simple aggregate index could be constructed to express almost any percentage price change desired.

In general, when an index number is constructed to reflect the combined effect of changes in a number of varying quantities, the simple aggregate index should *not* be used. A simple aggregate price index can be used legitimately only under the special conditions that all the prices are expressed in the same units (like dollars per pound) and equal amounts of each item are purchased.

Problems: Sections 19.1–19.2

Answers to odd-numbered problems are given in the back of the text.

1. Find the simple aggregate index of prices for the following data.

Item	1982 Unit Price	1982 Units Sold	1992 Unit Price	1992 Units Sold
A	$ 1.50	350	$ 3.00	600
B	15.00	100	18.50	125
C	8.50	200	15.00	300

2. An alloy is made up of 25% metal X, 35% metal Y, and 40% metal Z. During the past year, prices (per pound) of the metals have changed, as shown in the accompanying table. Find the simple aggregate index for the cost of the alloy.

Metal	Last Year	This Year
Metal X	$1.19	$2.50
Metal Y	6.72	6.89
Metal Z	3.50	2.90

3. For the accompanying retail sales volumes (sales in thousands of dollars), find the following sales indexes.

Year	Year 1	Year 2	Year 3	Year 4	Year 5
Sales	$64,978	$65,810	$68,352	$73,409	$78,992

a. With Year 1 as the base year.
b. With Year 5 as the base year.

19.3
PRICE INDEXES WITH QUANTITY WEIGHTS

We showed in the previous section that the index for changes in prices over a period of time depends on the quantities of the items that are purchased. Thus, a reasonable procedure is to weight changes in the prices of goods by the quantities of goods purchased in order to reflect the relative importance of each good or item in the total expenditures. This section presents various weighting schemes designed to accomplish this end.

EXAMPLE 19.4   Consider Table 19.4. This table shows the quantities for each of four items purchased by a typical family in 1982, 1985, and 1992. The four food items, together with the quantities in which they were purchased, are called a *market basket*. The market baskets are different for each of the years since different quantities of goods were purchased in each year. For any item i the 1982 price was p_{i0} and the 1982 quantity was q_{i0}. Hence $p_{i0}q_{i0}$ was the total amount spent on item i in the base year. For example, 728 quarts of milk at $.26 per quart (from Table 19.2) comes to .26 × 728, or $189.28, for 1982 expenditures for milk. Table 19.5 shows the amounts spent for each product in each year. In 1985 the typical family bought 735 quarts of milk (from Table 19.4) at $.31 per quart (from Table 19.2) and thus spent $227.85 on milk in 1985 (milk expenditure figure in Table 19.5). Table 19.5 also shows the total spent on each year's market basket of goods.

A set of index numbers known as a **weighted aggregate index** can be constructed from these total expenditure figures. Each weighted aggregate index number is a ratio of the total expenditure on the market basket for a particular year to the total expenditure for a base year.

		1982,	1985,	1992,
TABLE 19.4	*Item*	q_{i0}	q_{i1}	q_{i2}
Market Baskets for 1982, 1985, and 1992	1. Milk	728 qt	735 qt	737 qt
	2. Steak	312 lb	320 lb	350 lb
	3. Butter	55 lb	56 lb	56 lb
	4. Pepper	.3 lb	.3 lb	.3 lb

Item	1982 Expenditure, $p_{i0}q_{i0}$	1985 Expenditure, $p_{i1}q_{i1}$	1992 Expenditure, $p_{i2}q_{i2}$	
1. Milk	$189.28	$227.85	$ 412.72	**TABLE 19.5**
2. Steak	327.60	374.40	752.50	
3. Butter	41.25	47.60	78.40	**Total Expenditures for**
4. Pepper	0.75	0.66	0.78	**Yearly Market Baskets and**
Total	$558.88	$650.51	$1244.40	**Weighted Aggregate** **Index**
Weighted Aggregate *Index*	100.0	116.4	222.7	

Expressions for Weighted Aggregate Index

$$100 \times \frac{\sum_{i} p_{i0}q_{i0}}{\sum_{i} p_{i0}q_{i0}} = 100 \quad 100 \times \frac{\sum_{i} p_{i1}q_{i1}}{\sum_{i} p_{i0}q_{i0}} \quad 100 \times \frac{\sum_{i} p_{i2}q_{i2}}{\sum_{i} p_{i0}q_{i0}} \quad (19.2)$$

EXAMPLE 19.5 The last row in Table 19.5 consists of weighted aggregate indexes reflecting the change in the total amount of money spent on the four food items over the period 1982–1992. Since the quantities of items purchased changed, these indexes reflect changes in both price and amount purchased. We can say, then, that the *food bill* for these four items rose approximately 122.7% from 1982 to 1992. However, we cannot say that *prices* for these four items rose 122.7%, since the percentage increase reflects some increase in amount of goods purchased, too. Thus, the 122.7% figure reflects changes in price and standard of living.

What should we do if we desire to construct index numbers that reflect only changes in price over a period of time? We must hold the market basket constant in some way. There are two ways to do so. The first way results in a price index, known as a *Laspeyres index,* that uses quantities in the base period q_{i0} as the weights. The second way results in a price index, known as a *Paasche index,* that uses quantities in the nth period q_{in} as the weights. We compute Laspeyres price index numbers by using quantities in the base period q_{i0} as the weights in the following example.

EXAMPLE 19.6 We find Laspeyres price index numbers for our market basket of four items of food by assuming that the quantity of each item purchased in each year by a typical family does not change from the base year. The expenditures, assuming purchases of base year quantities q_{i0} have been calculated in Table 19.6. The 1982 expenditures in Table 19.6 are the same as the expenditures in Table 19.5. The expenditure columns for 1985 and 1992, however, were calculated by assuming that in 1985 and 1992 the typical family purchased the same quantity of each item as they purchased in the base year. Thus, the expenditure for milk in 1985 was computed to be $.31 per quart (the 1985 price) times

TABLE 19.6 Item	1982 Quantity, q_{i0}	1982 Expenditure, $p_{i0}q_{i0}$	1985 Expenditure, $p_{i1}q_{i0}$	1992 Expenditure, $p_{i2}q_{i0}$
1. Milk	728 qt	$189.28	$225.68	$ 407.68
2. Steak	312 lb	327.60	365.04	670.80
3. Butter	55 lb	41.25	46.75	77.00
4. Pepper	.3 lb	0.75	0.66	0.78
	Total	$558.88	$638.13	$1156.26
Weighted Price Index		100.0	114.2	206.9

Price Index with Base Year Quantity Weights (the Laspeyres Index Number)

728 quarts (the 1982 quantity), or $225.68. Notice that pepper plays a negligible role, as it should. The 1985 pepper expenditure was down $.09 from 1982, but this decrease hardly affects the total expenditure for the market basket because the family only purchases a very small amount of pepper.

The indexes in the last row of Table 19.6 were found by dividing the total expenditure in 1982, $558.88, into each of the other years' total expenditure figures. The resulting indexes, called the Laspeyres index numbers, measure the change in cost for a fixed buying pattern.

The **Laspeyres index number** for the nth period is obtained by using the following formula.

Equation for Laspeyres Index

$$L_{n:0} = 100 \times \frac{\sum_i p_{in}q_{i0}}{\sum_i p_{i0}q_{i0}}$$

(19.3)

The Laspeyres index has the algebraically equivalent form

$$L_{n:0} = 100 \times \frac{\sum_i p_{i0}q_{i0}(p_{in}/p_{i0})}{\sum_i p_{i0}q_{i0}}$$

In other words, $L_{n:0}$ is 100 times the weighted average of the price ratio p_{in}/p_{i0} for the various items, the weights being the amounts $p_{i0}q_{i0}$ spent in the base period.

An actual price index, such as the CPI, is computed as in our example, although it involves not 4 food items but around 400 items, including food, transportation, medical costs, and so on. The general description of a price index is as follows: Let the index i refer to the item; i ranges from 1 to 4 in our example and from 1 to about 400 for the CPI. Let the index n refer to the time period; n ranges over 0, 1, and 2 in our example, but any number of times can be considered and the period could be a month, say, rather than ten years. The price of item i in period n is p_{in}, and the quantity is q_{in}. The amount spent on item i in the base period is $p_{i0}q_{i0}$, for a grand total of $\sum_i p_{i0}q_{i0}$. If in period n the quantities were q_{i0} as in the base period, the total for item i would be $p_{in}q_{i0}$, and the grand total expended would be $\sum_i p_{in}q_{i0}$.

It is not mandatory that we use the base year quantities as weights for prices. Alternatively, we could use nth-period quantities q_{in} as weights. We use nth-period quantities as weights with both base period prices and nth-period prices to obtain the **Paasche index number** for the nth period according to the following formula.

Equation for Paasche Index

$$P_{n:0} = 100 \times \frac{\sum\limits_{i} p_{in} q_{in}}{\sum\limits_{i} p_{i0} q_{in}} = 100 \times \frac{\sum\limits_{i} p_{i0} q_{in}(p_{in}/p_{i0})}{\sum\limits_{i} p_{i0} q_{in}} \qquad (19.4)$$

The prices in both the numerator and the denominator of the Paasche index number are weighted by q_{in}. Thus, from one period to the next, a new $P_{n:0}$ index is calculated; and new weights, new q_{in} values, are needed for each period. The Laspeyres index is more convenient to use on a continuing basis, because the weights remain fixed from one period to the next. For this reason, the Laspeyres index is used more often than the Paasche index for measuring price changes.

There is one final index of price change that has been proposed over the years. It is called **Fisher's ideal index number,** *ideal* referring to its several subtle mathematical properties, which will not be discussed here. The Fisher index number is found by calculating the Laspeyres and the Paasche indexes first and then taking the square root of the product of these numbers.

Equation for Fisher's Ideal Index

$$F_{n:0} = \sqrt{L_{n:0} \times P_{n:0}} \qquad (19.5)$$

Since Fisher's index involves the Paasche index, it shares that index's disadvantage of requiring a new set of quantity weights q_{in} for each period. The ideal index number may have many ideal mathematical properties, but it is not so ideal from a practical point of view.

Index numbers can be used to measure more than just price changes. Sometimes, it is desirable to measure physical-volume changes. Index numbers that show changes in **physical volume** often measure production output. A weighted aggregate method can be used to construct an index to show the change from period 0 to period n.

Equation for Physical-Volume Index

$$I_{n:0} = \frac{\sum\limits_{i} p_{i0} q_{in}}{\sum\limits_{i} p_{i0} q_{i0}} \times 100 \qquad (19.6)$$

The quantities are weighted by the prices in the base period.

		Base Period 1982–1984				
		Mean Price per Unit,	*Mean Sales per Year,*	*1992 Quantity,*	*Base Period Value,*	*1992 Value,*
TABLE 19.7	*Item*	p_{i0}	q_{i0}	q_{in}	$p_{i0}q_{i0}$	$p_{i0}q_{in}$
Weighted Aggregate Physical-Volume Index	Hammers	$4.20	320	668	$1344.00	$2805.60
	Saws	8.60	292	591	2511.20	5082.60
	Nails	0.20/lb	2677 lb	5344 lb	535.40	1068.80
				Total	$4390.60	$8957.00
	Weighted Aggregate Physical-Volume Index				100.0	204.0

EXAMPLE 19.7   Table 19.7 shows the calculation of a physical-volume index for the sales output of three items carried by a hardware store. The sales output in 1992 is compared with the sales output in the 1982–1984 period in the index $I_{1992:0} = 204$, where in base period 0 (the 1982–1984 period) the value of the index is 100. The index was calculated by using Equation (19.6) and shows only the change in physical volume from the base period to 1992, since the base period prices were used as weights for the 1992 quantities.

Note that the prices and the quantities used for the three-year base period were the mean prices and mean quantities during those years. The value columns were found by multiplying the base period mean prices by the corresponding quantities in the base period and in 1992. Thus, the figure of $2805.60 for the 1992 value of hammers is the revenue that would have been generated in 1992 by selling the 1992 quantity of 668 hammers at 1982–1984 prices. The total value of $8957.00 is the value of the 1992 sales valued at the 1982–1984 prices. The index of $100 \times (\$8957.00/\$4390.60) = 204$ shows that physical volume of sales rose by 104% from 1982–1984 to 1992.

The reader may be confused at this point since so many ways of calculating price indexes are shown. The person may ask: What is the *true* price increase? The correct response to that question is another question: The *true* price increase of *what*? Each of the price indexes discussed in this section is a measure of price change. The indexes merely measure changes for different quantities of goods. The simple aggregate index measures the price change of one unit of each good. The weighted aggregate index measures the change in the cost of living from one period to the next. But this index is not a pure price index in that it reflects both price and quantity changes. The Laspeyres index measures only price changes. It measures the same set of items purchased in the given year as were purchased in the base year. And the Paasche index number measures only price changes, but it assumes that the purchaser bought in the base year exactly what was bought in the given year, which is seldom true. If the items in the market basket do not change and if the quantities purchased do not change, then the weighted aggregate, Laspeyres, and Paasche indexes all give the same value.

Problems: Section 19.3

4. Use the data in Problem 1 and find the weighted aggregate index number.
5. Use the data in Problem 1 and find the Laspeyres index number.

6. Use the data in Problem 1 and find the Paasche index number.

7. Use the data in Problem 1 and find Fisher's ideal index number.

8. Using the data in Problem 1, find a physical-volume index for 1992, weighting quantities with 1982 base year prices.

9. Use the data in Problem 2.

   a. Find the weighted aggregate index for the cost of the alloy.

   b. How can you explain the values of these index numbers in view of the fact that only two out of the three constituent metal prices rose?

Index number series are used primarily for comparative purposes. One might wonder, Why use 1967 as a base period? Why not use 1966 or 1982–1984? There are several criteria that should be considered when choosing a base period for an index.

 The first criterion is the time criterion: The base period should be fairly recent. Part of the justification for this criterion rests on the reasoning that an index number should help people quantitatively compare present conditions with past conditions. If the comparison is to be meaningful, the past (base period) should be recent enough that the person making the comparison can remember what conditions were like. It does not mean much to most people if they are told that prices now are 2000% above what they were in the Middle Ages. However, if they are told that prices are now 8% above what they were last year, they can remember last year and know that they have not changed their buying habits much since that time. The comparison tells them quite plainly that their standard of living this year is costing them 8% more than it did last year.

 A second criterion for selecting a base period is that it should be a period of normal activity for the series whose index is sought. If one were to construct an index of imports of handguns for private use into the United States, one would not be likely to use the years 1967–1968 for the base period, as the civil unrest experienced during these two years was cause for extremely high imports of handguns.

 A third criterion in selecting a base period is that of comparability. One often wishes to compare one index with another. For comparisons to be valid, the indexes should have the same base period. One company might find that an index of its raw materials costs is 105. Another might find that its index of costs is 145. We cannot safely conclude that costs have risen higher for the second company, however, without knowing the base from which the measurements were made. If the first company were using a 1982–1984 base period and the other were using a 1967 base period, then the indexes are not comparable. One says that costs have risen 5% from the 1982–1984 period, and the other says that costs have risen 45% from the 1967 period. The period 1982–1984 is now the base period for many general-purpose indexes prepared by government agencies.

 A final criterion for selecting a base period is that of the availability of data. The base period should be a period for which accurate and complete data are available. Thus, base periods back in the seventeenth century are likely to be unsatisfactory not only for having little relevance to us because they are not recent, but also for lacking accurate data with which to calculate the base values of the index. A good procedure may be to use census years as base years, since for these years complete data, rather than sample data, are available.

 One of the limitations of any price index is that it does not reflect changes in the quality of a product. For instance, a standard automobile with no options might have cost $2500 in 1960; the same model of car with no options might have cost $15,000 in 1992. It would be fallacious to conclude that the price index for this model of automobile had gone up to 480 based on the 1960 price at 100. The problem is that the 1992 automobile is not the same car as the 1960 version. The later model would have, as standard features, safety equipment, pollution control equipment, and luxury items not included in the 1960 model. These differences in quality affect the price of the automobile so that the $9500 price rise cannot be attributed solely to inflation. In order to eliminate as far as possible the differences in quality of products purchased from period to period, the items included in a price index are specified in enough detail to ensure that a similar item is priced for each period.

Problems: Section 19.4

10. Find the base period of the CPI.
11. Give examples of two goods or commodities for the following conditions.
    a. Quality has changed significantly in the past 25 years.
    b. Quality has remained unchanged for the past 25 years.

19.5
USES OF INDEXES

One of the most common uses of general-purpose price indexes is the adjusting of dollar values, often called **deflating**, or **reducing to constant dollars**. A consumer price index is often used to deflate wage and income figures.

EXAMPLE 19.8   An urban wage earner had a bonus of $6780 per year in 1986. By 1992 the worker's bonus was $7460, but the worker was unhappy. To see the reason for the worker's dissatisfaction, we must express the new bonus in terms of its purchasing power:

Year	CPI	Bonus	Constant-Dollar Bonus
1986	100	$6780	$6780/1.00 = $6780
1992	120	$7460	$7460/1.20 = $6217

In constant-dollar terms, our wage earner could purchase only $6217 worth of 1986 goods with the 1992 bonus of $7460. That is, the wage earner had actually lost ground in terms of buying power.

The reciprocal of a CPI is used to express the **purchasing power** of the dollar. In the preceding example a 1992 dollar would purchase only $1 × (100/120), or $.83 worth of 1986 goods. In other words, $1 in 1992 purchased only 83% of what it would purchase in 1986.

EXAMPLE 19.9   Interstate Fabricated Metal, Inc., sold $1,400,000 worth of goods in 1982 and $2,600,000 worth of goods in 1992. The sales volume has increased to 100 × (2,600,000/1,400,000), or 186% of the 1982 figure. However, part of that increase is due to inflation. A quick way to adjust for the effect of inflation is to reduce the 1992 sales figure to 1982 dollars. This reduction can be done with a 1992 producer price index, PPI, since the company involved is a manufacturer. The calculation is as follows:

Year	PPI	Actual Sales (millions)	Constant-Dollar Sales (millions)
1982	100.0	$1.4	$1.4/1.000 = $1.40
1992	303.1	$2.6	$2.6/3.031 = $0.86

The 1992 sales would have amounted to only about $860,000 at the 1982 prices. In this example a composite PPI was used, but in an actual example the manufacturer could use a subindex within the PPI group that pertained to its industry. If it had its own detailed price and sales quantity data convenient, it could compute a constant-dollar sales figure more accurately by using Equation (19.6). The deflating method demonstrated in our example results in only approximate constant-dollar figures, since the items whose prices are being deflated might not have experienced the average price changes reflected in the index. ▬▬

The most well-known economic series in the United States is the **gross national product,** GNP. It is usually deflated by an appropriate price index and expressed in terms of constant dollars. Thus changes in GNP reflect changes in national economic output rather than just price. The changes in constant-dollar GNP are called "real" GNP changes to distinguish them from changes caused by both price and output changes.

The **Dow–Jones Industrial Average** of stock prices is an index number of sorts. It is better called an "indicator" than an index number. It does not measure changes from some base period, but it is supposed to be the mean price of 30 specific industrial stocks. However, its value is not found by adding the prices of these stocks and dividing by 30, because over a period of time the 30 stocks in the industrial average have produced dividends and split. Also, there have been mergers of one company's stock with another; and periodically, some stocks are dropped from the list of the 30 industrials and new ones are added. When changes have occurred, adjustments have been made in the denominator used with this average.

▬▬ **EXAMPLE 19.10**   Suppose that we desire to construct a price indicator like the Dow–Jones industrial average, but we include only three stocks. On the first day that we publish our indicator, the prices of the stocks are $15, $20, and $25 per share. Then our mean price is

$$\frac{\$15 + \$20 + \$25}{3} = \$20$$

On the second day the second stock splits two for one. The price on the second stock should fall to somewhere in the neighborhood of $10 per share. If we ignored the split, we would find that with no real change in stock prices, our indicator would change. It would now be

$$\frac{\$15 + \$10 + \$25}{3} = \$16.67$$

This result would not be proper since the price indicator should not change if the prices do not change. The drop in the average was due only to a stock split. Thus, the denominator of our average must be adjusted to a new value D such that at the same prices for the stock ($15, $10 on the split stock, and $25) we still have an indicator average of $20. Thus, we seek a D value such that

$$\frac{\$15 + \$10 + \$25}{D} = \$20$$

The D value that satisfies this equation is 2.5. Now if the split stock's price rises to $11 per share after the split while the other prices remain unchanged, the new value of the indicator is

$$\frac{\$15 + \$11 + \$25}{2.5} = \$20.40$$

Over the years the many stock splits, stock dividends, and mergers that have occurred among the Dow–Jones 30 industrials have caused the denominator to drop far below the original value of 30, and it drops further each year. ═══

The Dow–Jones industrial average is as much a psychological indicator as a true index of general stock price movement on the New York Stock Exchange. The 30 stocks making up this average are not necessarily representative of typical or average stocks on the exchange. Sometimes, the price movements of the 30 stocks included in the index differ considerably from the movements in the general price level of stocks. When the Dow–Jones industrial average closed over the 3000 mark for the first time, many financial analysts correctly suggested that the stock market in general was not setting any new records. It was only the 30 industrials and a relatively few other stocks that were reaching new highs; most of the rest of the stocks were still not experiencing significant price rises. This lack of representativeness is one reason that the New York Stock Exchange Index was developed as an average price of *all* stocks on the New York Stock Exchange.

Problems: Section 19.5

12. If you earned $30,000 in 1986 when the CPI in your city was 110 and you earned $35,000 in 1992 when your city's CPI hit 140, express your 1992 purchasing power as a percentage of your 1986 purchasing power.

13. Assume the CPI for your city had the following values:

Year	1988	1989	1990	1991	1992
Value	109	110	115	120	125

   a. Find the purchasing power of the dollar in each of these years as a proportion of the 1982 dollar.
   b. Explain the meaning of these figures. [*Hint:* What percentage of 1982 goods could you buy in these years?]

14. Suppose you have a stock market indicator made up of five common stocks selling at the prices per share indicated in the following table:

Stock	Price per Share
1	$101.00
2	82.75
3	44.50
4	27.75
5	18.50

a. Find the market average.
b. Suppose that stock 1 split two for one and stock 4 split three for one. Find the new denominator D for your average.
c. After the splits, the prices settled to the values shown in the following table. Find the indicator's new value.

Stock	Price per Share
1	$51.00
2	88.75
3	48.50
4	9.50
5	18.25

19.6 SUMMARY

We have considered formulas for several price indexes in this chapter. Each index measures price changes in a slightly different way. Table 19.8 summarizes the formulas and what they measure.

The consumer price index (CPI) and the producer price index (PPI) are price indexes that are used extensively. The Dow–Jones industrial average of stock prices is an indicator of the mean price of a set of industrial stocks.

Price indexes, such as the CPI, are often used to adjust (or deflate) dollar measurements taken at different times to constant dollars. Also, the CPI is commonly

TABLE 19.8 Summary of Formulas

Index Name	Formula		Price Change from Period 0 to Period n For
Simple aggregate	$100 \times \dfrac{\sum_i p_{in}}{\sum_i p_{i0}}$	(19.1)	One unit of each item in the market basket
Weighted aggregate	$100 \times \dfrac{\sum_i p_{in} q_{in}}{\sum_i p_{i0} q_{i0}}$	(19.2)	Total amount spent on all items in the market basket
Laspeyres	$100 \times \dfrac{\sum_i p_{in} q_{i0}}{\sum_i p_{i0} q_{i0}}$	(19.3)	The market basket of goods purchased in the base period
Paasche	$100 \times \dfrac{\sum_i p_{in} q_{in}}{\sum_i p_{i0} q_{in}}$	(19.4)	The market basket of goods purchased in period n
Fisher's ideal	$\sqrt{L_{n:0} \times P_{n:0}}$	(19.5)	A market basket mixed between the periods

used as a measure of inflation, its reciprocal is used to express the purchasing power of the dollar, and cost-of-living wage adjustments are usually based on the CPI. Workers sometimes determine their real wages by using the CPI to deflate their current wages. Consequently, managers should understand price indexes.

REVIEW PROBLEMS

Answers to odd-numbered problems are given in the back of the text.

15. A company buys three major units from subcontractors. The company then assembles the three units and markets the total product. The prices charged by the subcontractors have changed over the past few years. The price and quantity data are given in the accompanying table.

 a. Find a price index reflecting the change in price of buying one of each item. What is the name of this index?
 b. Find and name the index reflecting the change in price of the components bought at the level of this year's purchase amounts.
 c. Find and name the price index reflecting the change in the total amount of money spent for subcontracted parts over the past five years.
 d. Find and name the price index reflecting the changes in price of the components bought at the level of use experienced five years ago.

| | Five Years Ago | | This Year | |
Item	Unit Price	Units Ordered	Unit Price	Units Ordered
A	$15.00	200	$20.00	250
B	8.00	150	9.00	100
C	12.00	300	16.00	500

16. A coal mining company is currently enjoying a boom in both price and demand for its product. It produces two types of coal: anthracite and bituminous. The accompanying data show the price per ton and demand (in thousands of tons) in three different years.

 a. Find the simple aggregate index between 1982 and 1987.
 b. Find the weighted aggregate index between 1982 and 1992.

| | 1982 | | 1987 | | 1992 | |
Type	Price	Quantity	Price	Quantity	Price	Quantity
Anthracite	$15	500	$23	1000	$42	2000
Bituminous	$12	1000	$18	2000	$34	3000

17. Use the data in Problem 16 to find the following Laspeyres index numbers:

 a. Between 1982 and 1987.   b. Between 1982 and 1992.

18. Use the data in Problem 16 to find the following Paasche index numbers:

 a. Between 1982 and 1987.   b. Between 1982 and 1992.

19. For the accompanying data for the retail price of selected electric appliances, find the Laspeyres retail price index for each year, using 1982 as the base.

	Average Unit Price			Thousands of
Appliance	1982	1985	1992	Units Sold, 1982
A	$295	$305	$415	6530
B	334	340	390	2050
C	250	261	360	1500
D	43	44	62	950

20. Identify which of the following index numbers could be found by using the data in Problem 19.

   a. Simple aggregate index.   **b.** Weighted aggregate index.
   c. Laspeyres index.   **d.** Paasche index.
   e. Fisher's ideal index.   **f.** Physical-volume index.

21. Using the data in Problem 15, find the physical-volume index for this year, weighting quantities by prices five years ago.

22. Using the data in Problem 16, find the physical-volume index for 1992. Use 1982 prices as weights and 1982 as the base year.

23. Using the data in Problem 16, find the physical-volume index for 1987. Use 1982 prices as weights and 1982 as the base year.

24. In 1991 the CPI was 135.5; in 1992 it was 140.0. If a worker in 1991 earned $12.54 per hour and in 1992 earned $13.00 per hour, did her purchasing power increase or decrease?

25. The Generous Company paid wages of $10,468,500 in June 1983 when the CPI was 100 and $12,450,000 for the same number of workers in fiscal 1992, when the CPI was 140. Did the workers receive a "real" wage increase in base period dollars?

26. For the Generous Company of Problem 25, what was the percentage change in real wages in base period dollars?

27. For the Generous Company of Problem 25, what were the 1992 wages in base period constant dollars?

28. For the Generous Company in Problem 25, what were the 1983 wages in 1992 constant dollars?

29. Why would an index series of transportation costs based in the year 1840 be useless today?

30. The price indicator in Problem 14 used weights of 1.0 for each stock when the indicator was first constructed. What other weighting factors might have been used?

31. On Wednesday afternoon Professor Ralph R. Bradshaw received a long-distance telephone call. The caller identified himself as the personnel manager of a large San Francisco company.

"I understand you're a good statistician, and I need help on a statistical-type problem," the caller began. Flattered, Professor Bradshaw offered to do what he could.

"We're transferring an employee from San Francisco to your city next month, and we need to adjust salary for the cost-of-living difference between the two locations. I can't find a cost-of-living index for your city. Are there any local organizations in your area that have an index I might use? You know, the Chamber of Commerce, state industrial promotion board, or utility companies?"

Professor Bradshaw responded that some indexes were available but that they would be useless for the purpose the personnel manager proposed. "What do you mean—useless?" asked the caller.

Discuss what Professor Bradshaw might tell his caller.

32. The GAPA company is concerned about the impact of changing prices on its financial statements. GAPA owned cash, inventory, and property at the end of year 5. However, the inventory was purchased at the end of year 3 and the property was purchased at the end of year 1. The historical or original dollar amounts for these asset accounts, the times that the assets were purchased, and the values of the CPI that correspond to these times are given in the accompanying table. Determine the value of total assets based on the historical amounts and based on constant-dollar amounts in terms of end-of-year-5 dollars.

Asset	Historical Dollars	Acquired At	Corresponding CPI
Cash	$100,000	End of year 5	125
Inventory	300,000	End of year 3	110
Property	800,000	End of year 1	105

Credit Data Set　*Refer to the 113 applicants for credit listed in the Credit Data Set in Appendix A.*

Questions　33. Is the cost of consumer credit (interest on consumer credit loans) part of the consumer price index?

34. As the consumer price index rises, what should be the impact of this rise on the credit-granting policies and procedures of the department store's credit office?

CASE 19.1　　　　**INDEXING CONSUMER PRICES**

The U.S. government publishes hundreds of indexes, and some of the most important indexes are the **consumer price indexes,** the **CPI**s. A consumer price index is a measure of the average changes in prices over time in a fixed market basket of goods and services. The Bureau of Labor Statistics publishes CPIs for two population groups: (1) a CPI for all urban consumers (CPI–U), which covers approximately 80% of the total population, and (2), a CPI for urban wage earners and clerical workers (CPI–W), which covers 32% of the total population. The CPI–U includes, in addition to wage earners and clerical workers, groups such as professional, managerial, and technical workers, the self-employed, short-term workers, the unemployed, and retirees and others not in the labor force.

A CPI is based on prices of food, clothing, shelter, and fuels, transportation fares, charges for doctors' and dentists' services, drugs, and the other goods and services that people buy for day-to-day living. Prices are collected in 85 urban areas across the country from about 57,000 housing units and approximately 19,000 retail establishments—department stores, supermarkets, hospitals, filling stations, and other types of stores and service establishments. All taxes directly associated with the purchase and use of items are included in the index. Prices of food, fuels, and a few other items are obtained every month in all 85 locations. Prices of most other commodities and services are collected every month in the five largest geographic areas and every other month in other areas. Prices of most goods and services are obtained by personal visits of the bureau's trained representatives. Some data, such as used-car prices, are obtained from secondary sources.

In the calculation of the index, price changes for the various items in each location are averaged together with weights that represent their importance in the spending of the appropriate population group. Local data are then combined to obtain a U.S. city average. Separate indexes are also published by size of city, by region of the country, for cross calculations of regions and population-size classes, and for 27 local areas. Area indexes do not measure differences in the level of prices among cities; they only measure the average change in prices for each area since the base period.

Year and Month	CPI–U, All Items	CPI–W, All Items	Year and Month	CPI–U, All Items	CPI–W, All Items	**TABLE 1**
1952, Dec.	26.7	26.9	1972, Dec.	42.5	42.7	Consumer Price Index for
1953, Dec.	26.9	27.0	1973, Dec.	46.2	46.5	All Urban Consumers
1954, Dec.	26.7	26.9	1974, Dec.	51.9	52.2	(CPI–U) and Urban Wage
1955, Dec.	26.8	27.0	1975, Dec.	55.5	55.8	Earners and Clerical
1956, Dec.	27.6	27.8	1976, Dec.	58.2	58.5	Workers (CPI–W): U.S. City
1957, Dec.	28.4	28.6	1977, Dec.	62.1	62.5	Average
1958, Dec.	28.9	29.1	1978, Dec.	67.7	68.1	
1959, Dec.	29.4	29.5	1979, Dec.	76.7	77.2	
1960, Dec.	29.8	30.0	1980, Dec.	86.3	86.9	
1961, Dec.	30.0	30.2	1981, Dec.	94.0	94.4	
1962, Dec.	30.4	30.6	1982, Dec.	97.6	98.0	
1963, Dec.	30.9	31.1	1983, Dec.	101.3	101.2	
1964, Dec.	31.2	31.4	1984, Dec.	105.3	104.8	
1965, Dec.	31.8	32.0	1985, Dec.	109.3	108.6	
1966, Dec.	32.9	33.1	1986, Dec.	110.5	109.3	
1967, Dec.	33.9	34.1	1987, Dec.	115.4	114.2	
1968, Dec.	35.5	35.7	1988, Dec.	120.5	119.2	
1969, Dec.	37.7	37.9	1989, Dec.	126.1	124.6	
1970, Dec.	39.8	40.0	1990, Dec.	133.8	132.2	
1971, Dec.	41.1	41.3	1991, Aug.	136.6	135.2	

The indexes measure price change from a designated reference date, 1982–1984, which equals 100.0. An increase of 7%, for example, is shown as 107.0. This change can also be expressed in dollars as follows: The price of a base period "market basket" of goods and services in the CPI has risen from $100 in 1982–1984 to $107. CPI–U and CPI–W index numbers for all items are given in Table 1.

Case Assignment

a. A wage earner has a current seasonal salary of $10,000. Find a recent listing of the CPI–U. Adjust the current salary to base period dollars by using the recent CPI–U index.

b. A wage earner has a current seasonal salary of $10,000. Find a recent listing of the CPI–W. Adjust the current salary to base period dollars by using the recent CPI–W index.

SOURCE: U.S. Department of Labor, Bureau of Labor Statistics (1991), *CPI Detailed Report, October 1991* (Washington, D.C.: U.S. Government Printing Office).

INDEXING PRODUCER PRICES

CASE 19.2

Dating back to 1890, **producer price indexes, PPI**s, are some of the oldest continuous time series published by the U.S. Bureau of Labor Statistics. A PPI was known as a wholesale price index, WPI, until 1978.

A producer price index measures average changes in selling prices received by domestic producers for their output. Most of the information used in calculating a producer price index is obtained through the systematic sampling of virtually every industry in the mining and manufacturing sectors of the economy. The PPI program (also known as the industrial price program) includes some data from other sectors as well—agriculture, fishing, forestry, services, and gas and electricity. Thus the title "Producer Price Index" refers to an entire "family" or system of indexes.

The PPI program contains the following:

1. Price indexes for nearly 500 mining and manufacturing industries, including approximately 8000 indexes for specific products and product categories.
2. Over 3000 commodity price indexes organized by type of product and end use.
3. Several major aggregate measures of price change organized by stage of processing.

Together, these elements constitute a system of price measures designed to meet the need for both aggregate information and detailed applications, such as following price trends in specific industries and products.

Measures of price change classified by industry, the most recent addition to the PPI system, now form the basis of the program. These indexes reflect the price trends of a constant set of goods and services that represent the total output of an industry. Industry index codes are based upon the standard industrial classification (SIC) system and provide comparability with a wide assortment of industry-based data for other economic phenomena, including productivity, production, employment, wages, and earnings.

Case Assignment

A manufacturer sold goods for $100,000 in the current period. Find a recent listing of a PPI. Adjust the current sales to base period dollars, using the recent PPI.

SOURCE: U.S. Department of Labor, Bureau of Labor Statistics (1988), *BLS Handbook of Methods*, Bulletin 2285 (Washington, D.C.: U.S. Government Printing Office).

REFERENCES

U.S. Department of Labor, Bureau of Labor Statistics. 1977. *The Consumer Price Index: Concepts and Content Over the Years.* Report 517. Washington, D.C.: U.S. Government Printing Office.

U.S. Department of Labor, Bureau of Labor Statistics. 1978. *The Consumer Price Index Revision—1978.* Washington, D.C.: U.S. Government Printing Office.

U.S. Department of Labor, Bureau of Labor Statistics. 1988. *BLS Handbook of Methods.* Bulletin 2285. Washington, D.C.: U.S. Government Printing Office.

U.S. Department of Labor, Bureau of Labor Statistics. 1991. *CPI Detailed Report, October 1991.* Washington, D.C.: U.S. Government Printing Office.

Data Sets

The data set presented on the following pages represents a random sample of 113 people who applied for charge account privileges at a well-known department store on the East Coast. Each line of data represents one applicant and gives ten pieces of information about that person. The nature and measurement of the ten variables are discussed in the following list.

CLASS indicates whether the department store granted credit to the individual. The value 1 indicates credit was granted, and the value 0 indicates it was not. The first 63 people in the list were granted credit, and the last 50 were not.

SEX Indicates whether the applicant was male (indicated by a 1) or female (indicated by a 0).

AGE Indicates the applicant's age, listed in years.

JOBYRS Indicates the number of years the applicant had held his or her current job. The value 99 indicates that the individual was not employed in an income-producing job at the time the application was made.

JOBINC Indicates the monthly income the applicant was receiving from his or her job at the time of application. A value of 9999 indicates that the applicant had no monthly income.

ADDINC Indicates the amount of additional income (over and above that received from a regular job) the applicant received each month. The figures listed here most often included income from commissions or rental property.

TOTBAL Indicates the total balance of debt owed by the applicant (exclusive of a home mortgage) at the time of application.

TOTPAY Indicates the total monthly payments the applicant was making on the debt balance listed above.

SPINC Indicates the applicant's spouse's monthly income. Some applicants listed the value 0, but many applicants merely left this item blank. A blank value is indicated in the data set by 9999.

MSTATUS Indicates the marital status of the applicant. Married applicants are indicated by a 1, and unmarried applicants (single, divorced, widowed) are indicated by a 0.

Questions concerning this sample data set are listed at the end of appropriate chapters in the text. To answer the questions at the end of a chapter, apply the concepts discussed in the chapter to these data. A computer is required for the solution of some of the more complex problems.

Credit Data Set

CLASS	SEX	AGE	JOBYRS	JOBINC	ADDINC	TOTBAL	TOTPAY	SPINC	MSTATUS
1	1	29	4	1200	200	5645	80	0	1
1	0	21	0	450	0	0	0	9999	0
1	1	23	1	700	0	1798	34	430	1
1	1	53	27	2000	0	0	0	9999	1
1	1	30	5	1200	0	3500	110	9999	1
1	1	25	3	925	0	828	103	500	1
1	1	47	20	1520	1100	0	0	9999	1
1	1	23	0	782	0	1626	79	9999	0
1	1	57	99	9999	880	0	0	850	1
1	1	34	2	2500	110	6000	70	9999	1
1	1	22	1	600	0	568	91	9999	1
1	1	44	8	1250	0	896	49	9999	1
1	0	53	9	600	755	0	0	9999	1
1	1	37	1	1200	0	0	176	9999	0
1	0	33	5	520	210	1000	28	9999	0
1	1	27	0	834	100	0	0	100	1
1	1	27	6	630	0	0	0	400	1
1	1	39	0	740	0	880	40	750	1
1	0	66	19	550	0	0	0	9999	1
1	1	35	3	1000	0	0	0	9999	0
1	1	37	0	1875	0	0	0	9999	1
1	1	40	11	2000	300	0	0	9999	1
1	1	24	4	1350	175	0	0	9999	0
1	1	60	99	9999	806	1740	36	9999	1
1	1	42	3	700	300	500	40	9999	0
1	1	48	7	4000	1000	18000	461	9999	0
1	1	31	4	800	0	0	0	500	1
1	0	29	3	600	150	0	0	9999	0
1	1	30	2	1000	0	1050	120	9999	0
1	1	78	30	1000	1620	0	0	9999	0
1	1	28	1	520	350	0	0	9999	1
1	1	22	1	650	0	0	0	250	1
1	1	39	5	800	0	0	0	9999	0
1	1	27	2	1100	0	800	55	9999	0
1	1	28	6	650	0	287	20	9999	1
1	1	65	99	9999	0	0	0	9999	1
1	1	56	25	2000	2000	0	0	9999	1
1	0	22	3	640	85	0	0	9999	0
1	1	22	6	750	0	0	0	9999	0
1	1	48	15	850	300	10800	163	150	1
1	1	63	6	1916	0	0	0	9999	1
1	1	28	0	9999	0	0	0	9999	1
1	1	32	1	2000	0	2800	0	9999	1
1	1	24	0	650	1000	0	0	9999	1
1	1	24	3	900	0	0	0	300	1
1	1	32	0	1450	0	2700	115	9999	1
1	1	70	99	9999	280	0	0	9999	1
1	1	35	0	700	100	700	60	9999	1
1	1	29	0	1060	0	457	51	9999	1
1	1	21	3	900	0	215	88	9999	0
1	1	28	8	1000	0	0	0	9999	1
1	1	60	6	2000	0	0	0	9999	0
1	1	27	2	1025	0	0	0	575	1

Credit Data Set (Continued)

CLASS	SEX	AGE	JOBYRS	JOBINC	ADDINC	TOTBAL	TOTPAY	SPINC	MSTATUS
1	1	29	0	1336	115	0	0	9999	1
1	1	21	2	900	0	200	43	9999	0
1	1	50	7	1500	0	0	0	9999	1
1	1	42	13	3000	0	0	0	9999	1
1	0	25	1	713	83	0	0	9999	0
1	1	44	99	9999	0	6132	130	1000	1
1	1	26	16	1000	0	0	0	9999	0
1	1	34	13	1374	0	0	0	9999	1
1	1	22	0	833	0	0	0	9999	1
1	1	36	2	2200	400	0	0	9999	1
0	0	34	4	400	0	0	0	1000	1
0	1	21	1	540	0	469	46	0	0
0	1	40	3	1500	0	360	37	300	1
0	1	25	3	865	0	1000	103	500	1
0	1	22	9	570	260	200	25	0	0
0	1	54	20	1000	0	2400	205	0	1
0	0	63	8	600	0	0	0	0	0
0	1	28	0	440	0	120	20	0	0
0	0	29	0	600	300	0	0	9999	0
0	0	22	2	350	85	820	57	9999	0
0	1	30	1	1000	0	5146	217	9999	1
0	1	30	9	600	0	5000	288	9999	0
0	1	45	11	2225	0	4000	78	0	1
0	1	26	2	950	0	500	95	9999	0
0	1	28	1	400	240	0	0	500	1
0	1	40	5	1300	0	9000	200	250	1
0	0	25	0	600	43	169	15	9999	0
0	1	21	0	400	0	0	0	9999	0
0	1	24	0	755	0	0	0	9999	1
0	1	21	1	645	0	0	0	300	1
0	1	39	1	1000	116	1356	117	9999	1
0	0	29	5	539	0	220	40	9999	0
0	1	24	1	400	0	0	0	9999	0
0	0	23	99	9999	0	0	0	9999	0
0	1	28	2	660	196	890	17	660	1
0	1	22	0	1265	0	250	57	9999	0
0	0	24	2	400	0	50	0	200	1
0	1	29	10	1200	0	150	30	9999	0
0	1	21	0	520	0	0	0	9999	0
0	1	28	1	300	0	0	0	9999	0
0	1	22	3	700	0	0	0	420	1
0	1	19	3	700	0	0	0	9999	0
0	1	52	28	755	0	0	0	9999	1
0	1	32	1	750	310	0	0	450	1
0	0	24	0	500	309	0	0	9999	0
0	1	20	1	900	0	0	0	9999	0
0	0	20	1	376	0	200	20	9999	0
0	1	22	1	450	0	3063	106	9999	1
0	1	23	3	800	0	0	0	9999	0
0	1	32	10	800	0	2800	104	9999	1
0	1	35	0	450	0	0	0	9999	0
0	1	20	2	750	0	0	0	9999	0

(continues)

Credit Data Set (Continued)

CLASS	SEX	AGE	JOBYRS	JOBINC	ADDINC	TOTBAL	TOTPAY	SPINC	MSTATUS
0	1	34	0	600	175	2709	52	9999	1
0	1	32	2	1800	400	90	460	9999	1
0	1	35	1	1600	0	3900	163	9999	1
0	1	36	5	1300	800	765	54	350	1
0	1	27	0	660	0	768	64	556	1
0	1	23	2	700	0	385	20	9999	0
0	1	28	2	700	0	297	28	9999	0
0	1	28	2	1200	0	0	0	500	1

RATIO DATA SET

The data set on the following pages represents a random sample of 141 companies from the COMPUSTAT files. Each line or record in the file represents one company and gives six items of data and the name of the company. The nature and measurement of the six variables are discussed in the following list.

CNUM Company identification number or CUSIP (Committee on Uniform Security Identification Procedures) number.

DNUM Standard Industry Classification (SIC) Code. SIC codes can be classified according to the accompanying listing.

Industry	SIC Code
Agriculture, forestry, and fishing	100–999
Mining	1000–1499
Construction	1500–1999
Manufacturing	2000–3999
Transportation and utilities	4000–4999
Wholesale	5000–5199
Retail	5200–5999
Financial	6000–6999
Services	7000–8999

CA/S Current assets divided by sales; a measure of inventory intensiveness.

QA/S Quick assets divided by sales; a measure of receivables intensiveness.

CA/CL Current assets divided by current liabilities; a measure of short-term liquidity.

NI/TA Net income divided by total assets; a measure of return on investment.

CONAME Company name

Ratio Data Set

CNUM	DNUM	CA/S	QA/S	CA/CL	NI/TA	CONAME
449842	800	0.854	0.854	3.611	0.123	IP TIMBERLANDS -LP
000776	1040	22.770	22.770	39.895	−0.005	ABM GOLD CORP -CL A
422704	1040	0.361	0.221	3.216	0.061	HECLA MINING CO
053435	1311	1.304	1.305	1.661	0.022	AVALON CORP
730448	1311	0.411	0.384	0.876	−0.001	POGO PRODUCING CO
739647	1311	1.248	1.248	3.512	0.074	PRAIRIE OIL ROYALTIES CO LTD
741016	1311	0.922	0.922	1.111	0.003	PRESIDIO OIL -CL B
909218	1381	0.395	0.338	1.344	−0.636	UNIT CORP
027420	1540	0.216	0.210	1.095	−0.258	AMERICAN MEDICAL BLDGS INC

Ratio Data Set (Continued)

CNUM	DNUM	CA/S	QA/S	CA/CL	NI/TA	CONAME
883556	1600	0.772	0.578	3.114	0.042	THERMO ELECTRON CORP
381370	1623	0.675	0.675	1.619	0.038	GOLDFIELD CORP
099599	2020	0.287	0.200	1.629	0.064	BORDEN INC
427866	2060	0.199	0.082	1.617	0.090	HERSHEY FOODS CORP
811850	2080	1.054	0.506	2.117	0.069	SEAGRAM CO LTD
411352	2111	0.750	0.626	2.407	0.090	HANSON PLC -ADR
346592	2200	0.582	0.235	5.490	−0.010	FORSTMANN & CO INC
851783	2211	0.385	0.266	3.019	0.051	SPRINGS INDUSTRIES INC
688222	2300	0.369	0.137	3.935	0.211	OSHKOSH B'GOSH INC -CL A
793897	2320	0.490	0.273	3.057	0.051	SALANT CORP
723886	2390	0.530	0.234	1.082	−0.192	PIONEER SYSTEMS INC
158496	2451	0.164	0.060	1.266	−0.052	CHAMPION ENTERPRISES INC
927457	2451	0.258	0.105	1.102	−0.135	VINTAGE ENTERPRISES INC
975628	2510	0.480	0.390	1.866	0.034	WINSTON FURNITURE COMPANY
257561	2621	0.285	0.161	1.901	0.056	DOMTAR INC
080555	2711	0.255	0.234	1.484	0.036	BELO (A.H.) CORP
579489	2711	0.170	0.146	1.399	0.074	MCCLATCHY NEWSPAPERS -CL A
650111	2711	0.189	0.168	1.027	0.094	NEW YORK TIMES CO -CL A
441560	2731	0.475	0.300	2.389	0.095	HOUGHTON MIFFLIN CO
680665	2800	0.352	0.211	1.683	0.046	OLIN CORP
453370	2833	0.509	0.272	2.952	0.041	INCSTAR CORP
525369	2834	0.605	0.310	1.718	0.056	LEINER (P) NUTRITIONAL PRODS
670100	2834	0.831	0.624	1.980	0.058	NOVO INDUSTRI A/S -ADR
832377	2834	0.542	0.407	1.300	0.128	SMITHKLINE BECKMAN CORP
479169	2844	0.464	0.312	2.111	0.025	JOHNSON PRODUCTS
127055	2890	0.394	0.274	1.345	0.027	CABOT CORP
520786	2890	0.692	0.528	4.353	0.152	LAWTER INTERNATIONAL INC
044540	2911	0.222	0.167	1.217	0.033	ASHLAND OIL INC
718507	2911	0.269	0.218	1.201	0.003	PHILLIPS PETROLEUM CO
929092	3069	0.923	0.838	4.615	−0.011	VULCAN CORP
115657	3140	0.300	0.107	2.684	0.069	BROWN GROUP INC
033038	3221	0.366	0.167	1.499	0.026	ANCHOR GLASS CONTAINER CORP
745075	3241	0.668	0.349	3.645	0.088	PUERTO RICAN CEMENT CO INC
686079	3310	0.240	0.116	1.545	0.096	OREGON STEEL MILLS INC
981811	3310	0.339	0.180	2.304	0.111	WORTHINGTON INDUSTRIES
047483	3312	0.523	0.188	2.164	−0.022	ATHLONE INDS
276317	3420	0.347	0.173	3.341	0.092	EASTERN CO
854616	3420	0.422	0.242	2.389	0.069	STANLEY WORKS
124800	3443	0.350	0.258	1.521	0.011	CBI INDUSTRIES INC
168088	3452	0.769	0.664	8.049	0.080	CHICAGO RIVET & MACHINE CO
303711	3452	0.664	0.449	2.476	0.024	FAIRCHILD INDUSTRIES INC
428399	3452	0.782	0.487	4.405	0.060	HI-SHEAR INDUSTRIES
282636	3460	2.608	1.543	0.534	0.005	EKCO GROUP INC
651192	3460	0.491	0.214	2.107	0.044	NEWELL COMPANIES
001909	3490	0.828	0.631	4.431	0.071	ARX INC
871565	3490	0.480	0.213	1.722	0.003	SYNALLOY CORP
489170	3540	0.473	0.213	2.564	0.053	KENNAMETAL INC
901476	3560	0.476	0.239	3.297	0.023	TWIN DISC INC
949391	3560	0.722	0.347	2.398	−0.021	WELDOTRON CORP
370838	3561	0.541	0.327	2.654	0.050	GENERAL SIGNAL CORP
816119	3567	0.631	0.517	1.281	0.022	SELAS CORP OF AMERICA
872479	3610	0.998	0.335	3.342	0.040	TII INDUSTRIES INC
878293	3622	0.515	0.342	1.905	0.067	TECH OPS SEVCON INC

(continues)

Ratio Data Set (Continued)

CNUM	DNUM	CA/S	QA/S	CA/CL	NI/TA	CONAME
422191	3630	0.520	0.330	2.727	0.016	HEALTH-MOR INC
723657	3651	0.573	0.416	1.836	0.035	PIONEER ELECTRONIC -ADR
002062	3661	19.488	19.332	24.194	−0.270	AT&E CORP
571263	3661	0.236	0.236	0.345	−0.670	MARLTON TECHNOLOGIES
034425	3663	0.595	0.380	3.889	0.018	ANDREW CORP
370442	3664	0.396	0.164	1.204	0.060	GENERAL MOTORS-CLASS H
849345	3670	0.432	0.199	2.303	0.008	SPRAGUE TECHNOLOGIES INC
584931	3679	0.738	0.496	2.271	0.048	MEDICORE INC
741555	3682	0.962	0.871	3.921	0.049	PRIME COMPUTER
875370	3682	0.664	0.575	3.584	0.109	TANDEM COMPUTERS INC
253651	3683	0.680	0.468	3.360	0.081	DIEBOLD INC
296656	3683	0.471	0.313	0.933	−0.133	ESPRIT SYSTEMS INC
253902	3687	0.802	0.703	3.692	0.137	DIGITAL COMMUNICATIONS ASSC
756231	3687	0.669	0.391	2.954	0.044	RECOGNITION EQUIPMENT INC
709352	3689	0.502	0.274	1.349	−0.041	PENRIL CORP
042167	3690	0.401	0.204	1.137	−0.186	ARMATRON INTERNATIONAL INC
269803	3714	0.444	0.314	2.104	0.080	EAGLE-PICHER INDS
317312	3714	0.535	0.280	1.700	0.110	FILTERTEK INC
883203	3720	0.404	0.179	1.689	0.049	TEXTRON INC
580169	3721	0.411	0.122	1.125	0.037	MCDONNELL DOUGLAS CORP
666807	3721	0.259	0.156	0.763	0.030	NORTHROP CORP
264147	3728	2.142	1.856	1.527	0.008	DUCOMMUN INC
867323	3728	0.541	0.216	2.078	0.023	SUNDSTRAND CORP
965010	3811	1.611	1.180	12.095	−0.086	WHITEHALL CORP
297425	3823	0.518	0.288	1.556	−0.158	ESTERLINE CORP
075887	3841	0.532	0.340	1.708	0.075	BECTON, DICKINSON & CO
912707	3841	0.392	0.252	2.894	0.082	U S SURGICAL CORP
009363	3842	0.416	0.237	2.568	0.032	AIRGAS INC
892027	3911	0.734	0.249	2.760	0.050	TOWN & COUNTRY CORP -CL A
292007	3944	0.826	0.687	1.596	−0.064	EMPIRE OF CAROLINA INC
631724	3944	0.392	0.268	1.404	0.069	NASTA INTERNATIONAL INC
544118	3960	0.414	0.270	1.148	−0.018	LORI CORP
690368	4411	0.478	0.478	3.445	0.028	OVERSEAS SHIPHOLDING GROUP
981904	4511	0.499	0.477	1.652	0.184	WORLDCORP INC
210902	4811	0.263	0.230	0.620	0.014	CONTEL CORP
670768	4811	0.309	0.277	1.289	0.056	NYNEX CORP
694890	4811	0.244	0.227	0.801	0.045	PACIFIC TELESIS GROUP
912889	4811	0.270	0.246	0.937	0.053	U S WEST INC
959807	4890	0.513	0.491	1.092	−0.048	WESTERN UNION CORP-NEW
276461	4924	0.339	0.282	1.716	0.047	EASTERN GAS & FUEL ASSOC
718009	4940	0.227	0.203	0.655	0.028	PHILADELPHIA SUBURBAN CORP
157829	4953	0.804	0.772	3.044	0.042	CHAMBERS DEVELOPMENT -CL A
617439	5030	0.248	0.082	2.978	0.110	MORGAN PRODUCTS LTD
042735	5065	0.462	0.239	3.610	−0.046	ARROW ELECTRONICS INC
477205	5094	0.512	0.191	1.814	−0.005	JEWELCOR INC
760930	5130	0.541	0.325	8.281	−0.059	RESOURCE RECOVERY TECH
059422	5161	0.286	0.203	1.507	0.059	BAMBERGER POLYMERS INC
853156	5200	0.347	0.116	2.883	0.034	STANDARD BRANDS PAINT CO
014752	5311	0.157	0.052	2.001	0.007	ALEXANDER'S INC
055270	5311	0.569	0.332	1.910	0.092	BAT INDS P.L.C. -ADR
030789	5331	0.266	0.034	2.056	0.039	AMES DEPT STORES INC
224174	5411	0.094	0.027	1.233	0.015	CRAIG CORP
781258	5411	0.145	0.052	1.637	0.047	RUDDICK CORP

Ratio Data Set (Continued)

CNUM	DNUM	CA/S	QA/S	CA/CL	NI/TA	CONAME
125820	5600	0.277	0.088	1.756	0.050	C M L GROUP
872539	5651	0.228	0.046	1.611	0.118	TJX COMPANIES INC
651576	5732	0.182	0.048	1.320	−0.026	NEWMARK & LEWIS INC
580400	5810	0.141	0.132	0.354	−0.302	MCFADDIN VENTURES INC
767754	5912	0.268	0.061	1.865	0.077	RITE AID CORP
931422	5912	0.192	0.030	1.769	0.076	WALGREEN CO
879496	6281	0.253	0.253	0.766	0.160	TELERATE INC
220291	6411	1.592	1.592	1.030	0.056	CORROON & BLACK CORP
363576	6411	1.398	1.398	1.028	0.064	GALLAGHER (ARTHUR J.) & CO
400095	6531	0.199	0.190	1.275	−0.018	GRUBB & ELLIS CO
577913	6552	3.264	3.264	0.963	0.001	MAXXAM INC
009293	7011	0.088	0.081	0.519	−0.058	AIRCOA HOTEL PARTNRS -LP-A
205477	7372	0.287	0.287	4.739	0.065	COMPUTER TASK GROUP INC
230032	7372	0.400	0.400	1.376	−0.317	CULLINET SOFTWARE INC
545700	7372	0.570	0.546	2.632	0.227	LOTUS DEVELOPMENT CORP
698631	7372	0.655	0.655	2.693	0.090	PANSOPHIC SYSTEMS INC
817615	7392	0.090	0.079	1.998	0.162	SERVICEMASTER -LP
835898	7399	1.440	1.426	1.181	0.078	SOTHEBY'S HOLDINGS -CL A
934436	7810	0.486	0.371	1.168	0.077	WARNER COMMUNICATIONS INC
800296	7813	0.597	0.597	2.418	−0.072	SANDY CORPORATION
847809	7814	0.580	0.458	2.975	0.161	SPELLING (AARON) PDS
798407	7948	1.754	1.754	6.162	0.068	SAN JUAN RACING ASSN
204015	8060	0.460	0.460	4.427	0.147	COMMUNITY PSYCHIATRIC CNTRS
126270	8911	0.302	0.302	1.421	−0.252	CRS SIRRINE INC
872625	8911	0.608	0.558	2.023	0.000	TRC COS INC
925297	8911	0.493	0.463	2.881	0.018	VERSAR INC

DATA DISK FILES

The data disk files documented in the accompanying listing are available on the data disk for *Statistics for Management and Economics*, Fifth Edition. The data disk is available from the publisher. The files on the data disk with a DAT file name extension are DOS files (ASCII or text files). Data values are separated by blanks.

Chapter/Problem	File Name	Description
1–19	CREDIT.DAT	Credit data for problems at end of most chapters (Missing value for JOBYRS is 99; JOBINC and SPINC is 9999.)
1–19	RATIO.DAT	Financial ratio data for problems at end of most chapters
1, 29	SALESEST.DAT	Year, sales, estimate data
Case 1.3	SPACESHT.DAT	Mission, incidents, temperature data for the space shuttle
2, 05	SALESCOM.DAT	Sales commissions data
2, 23	POLLUTNT.DAT	Pollutant data
2, 24	PRGEXP.DAT	Programming experience data
2, 27	BESTSTK.DAT	Price 1, price 2, dividends, return data
2, 28	BESTSTK.DAT	Price 1, price 2, dividends, return data
2, 29	BESTSTK.DAT	Price 1, price 2, dividends, return data

(continues)

**Data Disk Files
(Continued)**

Chapter/Problem	File name	Description
2, 30	COMPANY.DAT	Sales, profits, margin data (Missing value for MARGIN is 9999.)
2, 39	JAPANPE.DAT	Japan price earnings ratio data
2, 40	JAPANPE.DAT	Japan price earnings ratio data
2, 42	TBILL.DAT	Yield, discount rate, period data
2, 43	TBILL.DAT	Yield, discount rate, period data
2, 44	FEDDEBT.DAT	Total federal debt, government, public, reserve, GNP
2, 45	INCOME.DAT	Income, total, wages, supplements data
3, 21	MORTGAGE.DAT	Mortgage interest rate data
3, 22	WAITTIME.DAT	Waiting time data
3, 23	JOBSATIS.DAT	Job satisfaction data
3, 43	ENGINTHR.DAT	Engine thrust data
3, 44	BESTSTK.DAT	Price 1, price 2, dividends, return data
3, 45	COMPANY.DAT	Sales, profits, margin data (Missing value for MARGIN is 9999.)
3, 53	TBILL.DAT	Yield, discount rate, period data
3, 54	TBILL.DAT	Yield, discount rate, period data
3, 55	PERSINC.DAT	Income, tax, DPI, outlays, saving, saving percentage, constant DPI
Case 3.1	LOTTERY.DAT	Draft lottery data
Case 3.2	RATIO.DAT	Financial ratio data
4, 34	MRKTVAL.DAT	Market value, industry data
7, 37	DBTEQTY.DAT	Debt as a percentage of equity data
Case 7.1	RATIO.DAT	Financial ratio data
8, 32	PERATIO.DAT	Price earnings ratio data
8, 56	RANDSAMP.DAT	Random sample data
8, 66	SENSITIV.DAT	Sensitivity of tubes data
8, 106	GRNMAIL.DAT	Green-mail stock price decrease, group data
9, 11	MALFUNCT.DAT	Time to malfunction data
9, 29	EXCESSRT.DAT	Return, group data
9, 45	TESTMEAN.DAT	Normal data
9, 64	CEOHOURS.DAT	Hours, industry data
10, 27	MMFUND.DAT	Group, maturity data
10, 28	ROASSET.DAT	Return on assets, group data
10, 29	BANKROEQ.DAT	Group, return on equity data
11, 11	RANDDEPT.DAT	Account balance, model, frequency data
11, 31	SOCIALCL.DAT	Father's class, son's class, count data
12, 45	SNOWTIRE.DAT	Hours, wear data
12, 46	STKBETA.DAT	Return, index data
12, 50	DEAL.DAT	Deal value, fee and fee percentage data
Case 12.1	SPACESHT.DAT	Mission, incidents, temperature data for the space shuttle
13, 16	CONSUMP.DAT	Consumption, income, sex data
13, 18	CHATTER.DAT	Chatter, turn speed, tool type data
13, 20	RORPEBTA.DAT	Stock rate of return, price earnings, beta data
13, 21	ABSTENJS.DAT	Absence, tenure and job satisfaction data
13, 26	CPAEXAM.DAT	Exam score, grade point, experience data
13, 27	CONDOPRC.DAT	Condominium sales price, area, pool data
13, 28	SALESADV.DAT	Sales, advertising, month data
13, 32	BANKBALS.DAT	Core deposits, purchased funds, sensitive loans, nonsensitive loans data
13, 33	BANKBALS.DAT	Core deposits, purchased funds, sensitive loans, nonsensitive loans data

Chapter/Problem	File name	Description	Data Disk Files (Continued)
13, 34	LEISURE.DAT	Age, education, income, art exhibits, movies data	
13, 35	LEISURE.DAT	Age, education, income, art exhibits, movies data	
13, 45	JOBTENUR.DAT	Job tenure, age, sex, and test score data	
13, 49	RPAIRCST.DAT	Repair cost, operation hours, age, maintenance data	
13, 55	PASSING.DAT	Pass completion, yardage, touchdowns, interceptions, points data	
14, 12	APPLIANC.DAT	Appliance, time, indicator 2, indicator 3, indicator 4 data	
14, 13	APPLIANC.DAT	Appliance, time, indicator 2, indicator 3, indicator 4 data	
14, 56	ASSETERR.DAT	Asset errors data	
15, 13	TBILL.DAT	Yield, discount rate, period data	
15, 14	TBILL.DAT	Yield, discount rate, period data	
15, 16	VALVEDIA.DAT	Diameter, subgroup data	
15, 27	LAMPDIAM.DAT	Lamp diameter data	
17, 6	TENNISBL.DAT	Defective, number of lots data	
18, 16	TIMESHAR.DAT	Rate of return on time shares data	
18, 36	LOWFAT.DAT	Weights, group data	
18, 37	ASSEMBLY.DAT	Wages, technique data	
18, 38	SALESAPP.DAT	Sales, approach data	
18, 40	TBILL.DAT	Yield, discount rate, period data	
18, 41	TBILL.DAT	Yield, discount rate, period data	
18, 42	DISCNTRT.DAT	Discount rate, period data	
Case 18.1	SPACESHT.DAT	Mission, incidents, temperature data for the space shuttle	
19, 19	PRICEQNT.DAT	Price year one, price year two, price year three, quantity year one data	
Document	README5.DOC	Documents the files with corresponding problems	

Summation Notation

The study of statistics requires some familiarity with the mathematical shorthand used to express measures used in statistics. Because the operation of addition plays a large role in the study of statistics, we need a way to express sums in a compact and simple form. The summation notation meets this requirement and is defined and illustrated in this appendix for the student who needs a review of this topic.

If we have collected data that are the measurements of some characteristic of a number of individuals or items, such as the incomes of some group of persons, we designate the characteristic of interest by some letter or symbol, say, X. A second characteristic, such as educational level, is designated by another letter or symbol, say, Y. To differentiate between the same kind of measurements made on different items or individuals, or between similar repeated measurements made on the same element, we add a subscript to the corresponding symbol; thus X_1 stands for the income of the first person interviewed, X_2 for that of the second, and so on. In general, any arbitrary observed value is represented by X_i, where the subscript i is variable in the sense that it represents any one of the observed items and need only be replaced by the proper number in order to specify a particular observation. The income level and educational level of the ith, or general, individual are represented by X_i and Y_i, respectively.

Given a set of n observations, which we represent by $X_1, X_2, X_3, \ldots, X_n$, we can express their sum as

$$\sum_{i=1}^{n} X_i - X_1 + X_2 + X_3 + \cdots + X_n$$

where Σ (uppercase Greek sigma) means "the sum of," the subscript i is the index of summation, and the 1 and n that appear respectively below and above the operator Σ designate the range of the summation. The combined expression says, "Add all X's whose subscripts are between 1 and n, inclusive." In place of i we sometimes use j as the summation index; any letter will do.

If we want the sum of the squares of the n observations, we write

$$\sum_{i=1}^{n} X_i^2 = X_1^2 + X_2^2 + X_3^2 + \cdots + X_n^2$$

which says, "Add the squares of all observations whose subscripts are between 1 and n, inclusive." The sum of the products of two variables X and Y is written as

$$\sum_{i=1}^{n} X_i Y_i = X_1 Y_1 + X_2 Y_2 + X_3 Y_3 + \cdots + X_n Y_n$$

In those cases where the context makes it clear that *the sum is to be taken over all the data,* we sometimes simplify the summation notation by omitting the range of summation. Thus ΣX_i is the sum of all of the numbers, and ΣX_i^2 is the sum of the squares of all of the numbers.

We now look at the following algebraic rules that apply to summations.

Rule B.1

The summation of a sum (or difference) is the sum (or difference) of the summations:

$$\sum_{i=1}^{n} (X_i + Y_i - Z_i) = \sum_{i=1}^{n} X_i + \sum_{i=1}^{n} Y_i - \sum_{i=1}^{n} Z_i \tag{B.1}$$

Rule B.2

The summation of the product of a variable and a constant is the product of the constant and the summation of the variable:

$$\sum_{i=1}^{n} cY_i = cY_1 + cY_2 + \cdots + cY_n = c\sum_{i=1}^{n} Y_i \tag{B.2}$$

Rule B.3

The summation of a constant is the constant multiplied by the number of terms in the summation:

$$\sum_{i=1}^{n} c = c + c + \cdots + c = nc \qquad \text{where there are } n \text{ terms} \tag{B.3}$$

For example,

$$\sum_{i=1}^{6} c = c + c + c + c + c + c = 6c$$

where there are 6 terms.

PRACTICE PROBLEM B.1

Assume the following values are assigned to each of three variables X, Y, and Z:

$$X_1 = 2 \qquad Y_1 = -1 \qquad Z_1 = 2$$
$$X_2 = 0 \qquad Y_2 = 7 \qquad Z_2 = 3$$
$$X_3 = 5 \qquad Y_3 = 2 \qquad Z_3 = 6$$

Referring to Equation (B.1), show that Rule B.1 holds for these nine numbers.

Solution The left side of Equation (B.1) is

$$\sum_{i=1}^{3} (X_i + Y_i - Z_i) = [2 + (-1) - 2] + (0 + 7 - 3) + (5 + 2 - 6)$$

$$= (-1) + (4) + (1) = 4$$

The right side of Equation (B.1) is found by summing up each of the three columns (the X column, the Y column, and the Z column) and then adding the X and Y sums and subtracting the Z sum:

$$\sum_{i=1}^{3} X_i + \sum_{i=1}^{3} Y_i - \sum_{i=1}^{3} Z_i = (2 + 0 + 5) + (-1 + 7 + 2) - (2 + 3 + 6)$$

$$= (7) + (8) - (11) = 4$$

Two important but sometimes misread sums are the **sum of squared numbers,**

$$\sum_{i=1}^{n} X_i^2 = X_1^2 + X_2^2 + \cdots + X_n^2$$

and the **square of the sum of the numbers,**

$$\left(\sum_{i=1}^{n} X_i\right)^2 = (X_1 + X_2 + \cdots + X_n)^2$$

The former is found by squaring each X and then adding the squares; the latter by adding up the X's and then squaring the sum. These two expressions are usually unequal and often occur together in the same mathematical statement.

 PRACTICE PROBLEM B.4

Use the values of X given in Practice Problem B.1 to find both the sum of the squared numbers and the square of the sum of the numbers.

Solution The sum of the squared numbers is

$$\sum_{i=1}^{3} X_i^2 = (2)^2 + (0)^2 + (5)^2 = 4 + 0 + 25 = 29$$

The square of the sum of the numbers is

$$\left(\sum_{i=1}^{3} X_i\right)^2 = (2 + 0 + 5)^2 = (7)^2 = 49$$

Obviously, in this case the sum of squared numbers and the square of the sum of the numbers are not equal.

Summation notation like that introduced in this appendix is used throughout the text.

Problems: Appendix B

Answers to odd-numbered problems are given in the back of the text.

1. Write out the following summations as sums of the terms involved.

   a. $\displaystyle\sum_{i=1}^{5} i^2$ **b.** $\displaystyle\sum_{i=1}^{4} a^i Y_i$ **c.** $\displaystyle\sum_{i=2}^{5} Y_i^i$

   d. $\displaystyle\sum_{i=3}^{6} (Y_i - a^i)$ **e.** $\displaystyle\sum_{i=1}^{3} Y_i^2 - 3$ **f.** $\displaystyle\sum_{i=4}^{7} (-1)^{i+1} Y_i$

2. Given that

$$X_1 = 3 \qquad X_2 = 4 \qquad X_3 = 2 \qquad X_4 = -3$$
$$Y_1 = 2 \qquad Y_2 = 1 \qquad Y_3 = 5 \qquad Y_4 = 7$$

find the following:

a. $\sum_{i=1}^{4} X_i Y_i$

b. $\left(\sum_{i=1}^{4} X_i \right) \left(\sum_{i=1}^{4} Y_i \right)$

c. $\sum_{i=1}^{4} (X_i - Y_i)$

d. $\sum_{i=1}^{4} (X_i - 4)(Y_i - 5)$

e. $\sum_{i=1}^{4} X_i Y_i - \frac{1}{4} \left(\sum_{i=1}^{4} X_i \right) \left(\sum_{i=1}^{4} Y_i \right)$

f. $\sum_{i=1}^{4} X_i^2 Y_i$

3. Given that

$$X_1 = 3 \qquad X_4 = 6 \qquad X_7 = 3$$
$$X_2 = 2 \qquad X_5 = 1 \qquad X_8 = 5$$
$$X_3 = 4 \qquad X_6 = -2 \qquad X_9 = 1$$

find the following:

a. $\sum_{i=1}^{9} X_i$ **b.** $\sum_{i=1}^{9} X_i^2$ **c.** $\left(\sum_{i=1}^{9} X_i \right)^2$ **d.** $\sum_{i=4}^{9} X_i$ **e.** $\sum_{i=2}^{5} (X_i - 3)^2$

4. Use the data given for the X_1 through X_9 values in Problem 3 to find the following:

a. $\sum_{i=1}^{6} (X_i^2 + 2)$ **b.** $\sum_{i=1}^{5} X_i$ **c.** $\sum_{i=1}^{9} (X_i - 4)$

d. $\sum_{i=1}^{9} X_i - 4$ **e.** $\sum_{i=6}^{9} i X_i$ **f.** $\sum_{i=1}^{4} 8$

5. Show that the following equations hold.

a. $\sum_{i=1}^{n} (Y_i - c)^2 = \sum_{i=1}^{n} Y_i^2 - 2c \sum_{i=1}^{n} Y_i + nc^2$

[*Hint:* Expand the $(Y_i - c)^2$ term. Then apply the first, second, and third summation rules, in that order.]

b. $\sum_{i=1}^{n} (Y_i - \overline{Y}) = 0$, where $\overline{Y} = \frac{1}{n} \sum_{i=1}^{n} Y_i$

[*Hint:* Rearrange the expression on the right to read $n\overline{Y} = \Sigma Y$ and substitute for ΣY in the expression on the left.]

Tables

CONTENTS

TABLE I   Combinations of *n* Things Taken *x* at a Time

| | | | | | *Selected Values of x* | | | | | |
n	0	1	2	3	4	5	6	7	8	9	10
0	1										
1	1	1									
2	1	2	1								
3	1	3	3	1							
4	1	4	6	4	1						
5	1	5	10	10	5	1					
6	1	6	15	20	15	6	1				
7	1	7	21	35	35	21	7	1			
8	1	8	28	56	70	56	28	8	1		
9	1	9	36	84	126	126	84	36	9	1	
10	1	10	45	120	210	252	210	120	45	10	1
11	1	11	55	165	330	462	462	330	165	55	11
12	1	12	66	220	495	792	924	792	495	220	66
13	1	13	78	286	715	1287	1716	1716	1287	715	286
14	1	14	91	364	1001	2002	3003	3432	3003	2002	1001
15	1	15	105	455	1365	3003	5005	6435	6435	5005	3003
16	1	16	120	560	1820	4368	8008	11440	12870	11440	8008
17	1	17	136	680	2380	6188	12376	19448	24310	24310	19448
18	1	18	153	816	3060	8568	18564	31824	43758	48620	43758
19	1	19	171	969	3876	11628	27132	50388	75582	92378	92378
20	1	20	190	1140	4845	15504	38760	77520	125970	167960	184756

Note: These values are sometimes called the *binomial coefficients* since they are the coefficients in the terms of the expansion of $(a + b)^n$. Note the symmetry of each row. This is due to the fact that

$$_nC_x = \frac{n!}{x!\,(n - x)!} = \frac{n!}{(n - x)!\,x!} = _nC_{(n-x)}$$

Thus we can find $_{15}C_{12}$ even though there is no $x = 12$ column. We simply note that $_{15}C_{12} = _{15}C_3 = 455$, which is readily determined from Table I.

The value of $n!$ can be approximated by using *Stirling's formula* (with second-term approximation) as follows:

$$n! \approx (2\pi)^{1/2} e^{[-n + 1/(12n)]} n^{(n+.5)}$$

The percentage error for the approximation decreases as n increases and is less than .001% for $n \geq 10$. For example, for $n = 10$,

$$n! = 10! = (10)(9)(8)(7)(6)(5)(4)(3)(2)(1) = 3,628,800$$
$$\approx (2\pi)^{1/2} e^{\{-10 + 1/[(12)(10)]\}} 10^{(10+.5)} = 3,628,810$$

TABLE II Random Numbers

Line	1–5	6–10	11–15	16–20	21–25	26–30	31–35
1	39591	16834	74151	92027	24670	36665	00770
2	46304	00370	30420	03883	94648	89428	41583
3	99547	47887	81085	64933	66279	80432	65793
4	06743	50993	98603	38452	87890	94624	69721
5	69568	06483	28733	37867	07936	98710	98539
6	91240	18312	17441	01929	18163	69201	31211
7	97458	14229	12063	59611	32249	90466	33216
8	35249	38646	34475	72417	60514	69257	12489
9	38980	46600	11759	11900	46743	27860	77940
10	10750	52745	38749	87365	58959	53731	89295
11	36247	27850	73958	20673	37800	63835	71051
12	70994	66986	99744	72438	01174	42159	11392
13	99638	94702	11463	18148	81386	80431	90628
14	72055	15774	43857	99805	10419	76939	25993
15	24038	65541	85788	55835	38835	59399	13790
16	74976	14631	35908	28221	39470	91548	12854
17	35553	71628	70189	26436	63407	91178	90348
18	35676	12797	51434	82976	42010	26344	92920
19	74815	67523	72985	23183	02446	63594	98924
20	45246	88048	65173	50989	91060	89894	36036
21	76509	47069	86378	41797	11910	49672	88575
22	19689	90332	04315	21358	97248	11188	39062
23	42751	35318	97513	61537	54955	08159	00337
24	11946	22681	45045	13964	57517	59419	58045
25	96518	48688	20996	11090	48396	57177	83867
26	35726	58643	76869	84622	39098	36083	72505
27	39737	42750	48968	70536	84864	64952	38404
28	97025	66492	56177	04049	80312	48028	26408
29	62814	08075	09788	56350	76787	51591	54509
30	25578	22950	15227	83291	41737	59599	96191
31	68763	69576	88991	49662	46704	63362	56625
32	17900	00813	64361	60725	88974	61005	99709
33	71944	60227	63551	71109	05624	43836	58254
34	54684	93691	85132	64399	29182	44324	14491
35	25946	27623	11258	65204	52832	50880	22273
36	01353	39318	44961	44972	91766	90262	56073
37	99083	88191	27662	99113	57174	35571	99884
38	52021	45406	37945	75234	24327	86978	22644
39	78755	47744	43776	83098	03225	14281	83637
40	25282	69106	59180	16257	22810	43609	12224
41	11959	94202	02743	86847	79725	51811	12998
42	11644	13792	98190	01424	30078	28197	55583
43	06307	97912	68110	59812	95448	43244	31262
44	76285	75714	89585	99296	52640	46518	55486
45	55322	07598	39600	60866	63007	20007	66819
46	78017	90928	90220	92503	83375	26986	74399
47	44768	43342	20696	26331	43140	69744	82928
48	25100	19336	14605	86603	51680	97678	24261
49	83612	46623	62876	85197	07824	91392	58317
50	41347	81666	82961	60413	71020	83658	02415

Table II is abridged from *Table of 105,000 Random Decimal Digits,* Interstate Commerce Commission, Bureau of Transport Economics and Statistics, May 1949.

TABLE III　Binomial Probabilities (Cumulative)

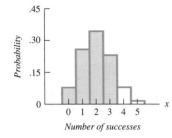

Number of successes

The values in the body of each table are the probabilities of $X \leq x$ successes for selected values of n and p, defined by the cumulative binomial probability function

$$P(X \leq x \mid n, p) = \sum_{k=0}^{x} \binom{n}{k} p^k (1 - p)^{n-k} \qquad x = 1, 2, \ldots, n - 1$$

where X is the number of successes in n trials and p is the probability of a success on each trial.

Example: $P(X \leq 3 \mid n = 5, p = .4) = .9130$

$n = 5$

x	.01	.05	.10	.20	.30	.40	.50	.60	.70	.80	.90	.95	.99	x
0	0.9510	0.7738	0.5905	0.3277	0.1681	0.0778	0.0313	0.0102	0.0024	0.0003	0.0000	0.0000	0.0000	0
1	0.9990	0.9774	0.9185	0.7373	0.5282	0.3370	0.1875	0.0870	0.0308	0.0067	0.0005	0.0000	0.0000	1
2	1.0000	0.9988	0.9914	0.9421	0.8369	0.6826	0.5000	0.3174	0.1631	0.0579	0.0086	0.0012	0.0000	2
3	1.0000	1.0000	0.9995	0.9933	0.9692	0.9130	0.8125	0.6630	0.4718	0.2627	0.0815	0.0226	0.0010	3
4	1.0000	1.0000	1.0000	0.9997	0.9976	0.9898	0.9688	0.9222	0.8319	0.6723	0.4095	0.2262	0.0490	4

$n = 10$

x	.01	.05	.10	.20	.30	.40	.50	.60	.70	.80	.90	.95	.99	x
0	0.9044	0.5987	0.3487	0.1074	0.0282	0.0060	0.0010	0.0001	0.0000	0.0000	0.0000	0.0000	0.0000	0
1	0.9957	0.9139	0.7361	0.3758	0.1493	0.0464	0.0107	0.0017	0.0001	0.0000	0.0000	0.0000	0.0000	1
2	0.9999	0.9885	0.9298	0.6778	0.3828	0.1673	0.0547	0.0123	0.0016	0.0001	0.0000	0.0000	0.0000	2
3	1.0000	0.9990	0.9872	0.8791	0.6496	0.3823	0.1719	0.0548	0.0106	0.0009	0.0000	0.0000	0.0000	3
4	1.0000	0.9999	0.9984	0.9672	0.8497	0.6331	0.3770	0.1662	0.0473	0.0064	0.0001	0.0000	0.0000	4
5	1.0000	1.0000	0.9999	0.9936	0.9527	0.8338	0.6230	0.3669	0.1503	0.0328	0.0016	0.0001	0.0000	5
6	1.0000	1.0000	1.0000	0.9991	0.9894	0.9452	0.8281	0.6177	0.3504	0.1209	0.0128	0.0010	0.0000	6
7	1.0000	1.0000	1.0000	0.9999	0.9984	0.9877	0.9453	0.8327	0.6172	0.3222	0.0702	0.0115	0.0001	7
8	1.0000	1.0000	1.0000	1.0000	0.9999	0.9983	0.9893	0.9536	0.8507	0.6242	0.2639	0.0861	0.0043	8
9	1.0000	1.0000	1.0000	1.0000	1.0000	0.9999	0.9990	0.9940	0.9718	0.8926	0.6513	0.4013	0.0956	9

$n = 15$

x	.01	.05	.10	.20	.30	.40	.50	.60	.70	.80	.90	.95	.99	x
0	0.8601	0.4633	0.2059	0.0352	0.0047	0.0005	0.0000	0.0000	0.0000	0.0000	0.0000	0.0000	0.0000	0
1	0.9904	0.8290	0.5490	0.1671	0.0353	0.0052	0.0005	0.0000	0.0000	0.0000	0.0000	0.0000	0.0000	1
2	0.9996	0.9638	0.8159	0.3980	0.1268	0.0271	0.0037	0.0003	0.0000	0.0000	0.0000	0.0000	0.0000	2
3	1.0000	0.9945	0.9444	0.6482	0.2969	0.0905	0.0176	0.0019	0.0001	0.0000	0.0000	0.0000	0.0000	3
4	1.0000	0.9994	0.9873	0.8358	0.5155	0.2173	0.0592	0.0093	0.0007	0.0000	0.0000	0.0000	0.0000	4
5	1.0000	0.9999	0.9978	0.9389	0.7216	0.4032	0.1509	0.0338	0.0037	0.0001	0.0000	0.0000	0.0000	5
6	1.0000	1.0000	0.9997	0.9819	0.8689	0.6098	0.3036	0.0950	0.0152	0.0008	0.0000	0.0000	0.0000	6
7	1.0000	1.0000	1.0000	0.9958	0.9500	0.7869	0.5000	0.2131	0.0500	0.0042	0.0000	0.0000	0.0000	7
8	1.0000	1.0000	1.0000	0.9992	0.9848	0.9050	0.6964	0.3902	0.1311	0.0181	0.0003	0.0000	0.0000	8
9	1.0000	1.0000	1.0000	0.9999	0.9963	0.9662	0.8491	0.5968	0.2784	0.0611	0.0022	0.0001	0.0000	9
10	1.0000	1.0000	1.0000	1.0000	0.9993	0.9907	0.9408	0.7827	0.4845	0.1642	0.0127	0.0006	0.0000	10
11	1.0000	1.0000	1.0000	1.0000	0.9999	0.9981	0.9824	0.9095	0.7031	0.3518	0.0556	0.0055	0.0000	11
12	1.0000	1.0000	1.0000	1.0000	1.0000	0.9997	0.9963	0.9729	0.8732	0.6020	0.1841	0.0362	0.0004	12
13	1.0000	1.0000	1.0000	1.0000	1.0000	1.0000	0.9995	0.9948	0.9647	0.8329	0.4510	0.1710	0.0096	13
14	1.0000	1.0000	1.0000	1.0000	1.0000	1.0000	1.0000	0.9995	0.9953	0.9648	0.7941	0.5367	0.1399	14

TABLE III   Binomial Probabilities (Cumulative) (Concluded)

n = 20

x	.01	.05	.10	.20	.30	.40	p .50	.60	.70	.80	.90	.95	.99	x
0	0.8179	0.3585	0.1216	0.0115	0.0008	0.0000	0.0000	0.0000	0.0000	0.0000	0.0000	0.0000	0.0000	0
1	0.9831	0.7358	0.3917	0.0692	0.0076	0.0005	0.0000	0.0000	0.0000	0.0000	0.0000	0.0000	0.0000	1
2	0.9990	0.9245	0.6769	0.2061	0.0355	0.0036	0.0002	0.0000	0.0000	0.0000	0.0000	0.0000	0.0000	2
3	1.0000	0.9841	0.8670	0.4114	0.1071	0.0160	0.0013	0.0000	0.0000	0.0000	0.0000	0.0000	0.0000	3
4	1.0000	0.9974	0.9568	0.6296	0.2375	0.0510	0.0059	0.0003	0.0000	0.0000	0.0000	0.0000	0.0000	4
5	1.0000	0.9997	0.9887	0.8042	0.4164	0.1256	0.0207	0.0016	0.0000	0.0000	0.0000	0.0000	0.0000	5
6	1.0000	1.0000	0.9976	0.9133	0.6080	0.2500	0.0577	0.0065	0.0003	0.0000	0.0000	0.0000	0.0000	6
7	1.0000	1.0000	0.9996	0.9679	0.7723	0.4159	0.1316	0.0210	0.0013	0.0000	0.0000	0.0000	0.0000	7
8	1.0000	1.0000	0.9999	0.9900	0.8867	0.5956	0.2517	0.0565	0.0051	0.0001	0.0000	0.0000	0.0000	8
9	1.0000	1.0000	1.0000	0.9974	0.9520	0.7553	0.4119	0.1275	0.0171	0.0006	0.0000	0.0000	0.0000	9
10	1.0000	1.0000	1.0000	0.9994	0.9829	0.8725	0.5881	0.2447	0.0480	0.0026	0.0000	0.0000	0.0000	10
11	1.0000	1.0000	1.0000	0.9999	0.9949	0.9435	0.7483	0.4044	0.1133	0.0100	0.0001	0.0000	0.0000	11
12	1.0000	1.0000	1.0000	1.0000	0.9987	0.9790	0.8684	0.5841	0.2277	0.0321	0.0004	0.0000	0.0000	12
13	1.0000	1.0000	1.0000	1.0000	0.9997	0.9935	0.9423	0.7500	0.3920	0.0867	0.0024	0.0000	0.0000	13
14	1.0000	1.0000	1.0000	1.0000	1.0000	0.9984	0.9793	0.8744	0.5836	0.1958	0.0113	0.0003	0.0000	14
15	1.0000	1.0000	1.0000	1.0000	1.0000	0.9997	0.9941	0.9490	0.7625	0.3704	0.0432	0.0026	0.0000	15
16	1.0000	1.0000	1.0000	1.0000	1.0000	1.0000	0.9987	0.9840	0.8929	0.5886	0.1330	0.0159	0.0000	16
17	1.0000	1.0000	1.0000	1.0000	1.0000	1.0000	0.9998	0.9964	0.9645	0.7939	0.3231	0.0755	0.0010	17
18	1.0000	1.0000	1.0000	1.0000	1.0000	1.0000	1.0000	0.9995	0.9924	0.9308	0.6083	0.2642	0.0169	18
19	1.0000	1.0000	1.0000	1.0000	1.0000	1.0000	1.0000	1.0000	0.9992	0.9885	0.8784	0.6415	0.1821	19

n = 25

x	.01	.05	.10	.20	.30	.40	p .50	.60	.70	.80	.90	.95	.99	x
0	0.7778	0.2774	0.0718	0.0038	0.0001	0.0000	0.0000	0.0000	0.0000	0.0000	0.0000	0.0000	0.0000	0
1	0.9742	0.6424	0.2712	0.0274	0.0016	0.0001	0.0000	0.0000	0.0000	0.0000	0.0000	0.0000	0.0000	1
2	0.9980	0.8729	0.5371	0.0982	0.0090	0.0004	0.0000	0.0000	0.0000	0.0000	0.0000	0.0000	0.0000	2
3	0.9999	0.9659	0.7636	0.2340	0.0332	0.0024	0.0001	0.0000	0.0000	0.0000	0.0000	0.0000	0.0000	3
4	1.0000	0.9928	0.9020	0.4207	0.0905	0.0095	0.0005	0.0000	0.0000	0.0000	0.0000	0.0000	0.0000	4
5	1.0000	0.9988	0.9666	0.6167	0.1935	0.0294	0.0020	0.0001	0.0000	0.0000	0.0000	0.0000	0.0000	5
6	1.0000	0.9998	0.9905	0.7800	0.3407	0.0736	0.0073	0.0003	0.0000	0.0000	0.0000	0.0000	0.0000	6
7	1.0000	1.0000	0.9977	0.8909	0.5118	0.1536	0.0216	0.0012	0.0000	0.0000	0.0000	0.0000	0.0000	7
8	1.0000	1.0000	0.9995	0.9532	0.6769	0.2735	0.0539	0.0043	0.0001	0.0000	0.0000	0.0000	0.0000	8
9	1.0000	1.0000	0.9999	0.9827	0.8106	0.4246	0.1148	0.0132	0.0005	0.0000	0.0000	0.0000	0.0000	9
10	1.0000	1.0000	1.0000	0.9944	0.9022	0.5858	0.2122	0.0344	0.0018	0.0000	0.0000	0.0000	0.0000	10
11	1.0000	1.0000	1.0000	0.9985	0.9558	0.7323	0.3450	0.0778	0.0060	0.0001	0.0000	0.0000	0.0000	11
12	1.0000	1.0000	1.0000	0.9996	0.9825	0.8462	0.5000	0.1538	0.0175	0.0004	0.0000	0.0000	0.0000	12
13	1.0000	1.0000	1.0000	0.9999	0.9940	0.9222	0.6550	0.2677	0.0442	0.0015	0.0000	0.0000	0.0000	13
14	1.0000	1.0000	1.0000	1.0000	0.9982	0.9656	0.7878	0.4142	0.0978	0.0056	0.0000	0.0000	0.0000	14
15	1.0000	1.0000	1.0000	1.0000	0.9995	0.9868	0.8852	0.5754	0.1894	0.0173	0.0001	0.0000	0.0000	15
16	1.0000	1.0000	1.0000	1.0000	0.9999	0.9957	0.9461	0.7265	0.3231	0.0468	0.0005	0.0000	0.0000	16
17	1.0000	1.0000	1.0000	1.0000	1.0000	0.9988	0.9784	0.8464	0.4882	0.1091	0.0023	0.0000	0.0000	17
18	1.0000	1.0000	1.0000	1.0000	1.0000	0.9997	0.9927	0.9264	0.6593	0.2200	0.0095	0.0002	0.0000	18
19	1.0000	1.0000	1.0000	1.0000	1.0000	0.9999	0.9980	0.9706	0.8065	0.3833	0.0334	0.0012	0.0000	19
20	1.0000	1.0000	1.0000	1.0000	1.0000	1.0000	0.9995	0.9905	0.9095	0.5793	0.0980	0.0072	0.0000	20
21	1.0000	1.0000	1.0000	1.0000	1.0000	1.0000	0.9999	0.9976	0.9668	0.7660	0.2364	0.0341	0.0001	21
22	1.0000	1.0000	1.0000	1.0000	1.0000	1.0000	1.0000	0.9996	0.9910	0.9018	0.4629	0.1271	0.0020	22
23	1.0000	1.0000	1.0000	1.0000	1.0000	1.0000	1.0000	0.9999	0.9984	0.9726	0.7288	0.3576	0.0258	23
24	1.0000	1.0000	1.0000	1.0000	1.0000	1.0000	1.0000	1.0000	0.9999	0.9962	0.9282	0.7226	0.2222	24

TABLE IV Poisson Probabilities

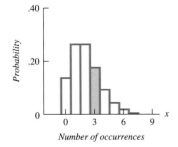

Number of occurrences

The values in the body of each table are the probabilities of exactly $X = x$ for selected values of λ, defined by the Poisson probability function

$$P(X = x) = \frac{e^{-\lambda}\lambda^x}{x!} \qquad x = 1, 2, \ldots$$

Examples: 1. If $\lambda = 2$, then $P(X = 3) = .1804$.
2. If $\lambda = 8.0$, then $P(X \leq 2) = .0003 + .0027 + .0107 = .0137$.

						λ				
x	.1	.2	.3	.4	.5	.6	.7	.8	.9	1.0
0	.9048	.8187	.7408	.6703	.6065	.5488	.4966	.4493	.4066	.3679
1	.0905	.1637	.2222	.2681	.3033	.3293	.3476	.3595	.3659	.3679
2	.0045	.0164	.0333	.0536	.0758	.0988	.1217	.1438	.1647	.1839
3	.0002	.0011	.0033	.0072	.0126	.0198	.0284	.0383	.0494	.0613
4	.0000	.0001	.0003	.0007	.0016	.0030	.0050	.0077	.0111	.0153
5	.0000	.0000	.0000	.0001	.0002	.0004	.0007	.0012	.0020	.0031
6	.0000	.0000	.0000	.0000	.0000	.0000	.0001	.0002	.0003	.0005
7	.0000	.0000	.0000	.0000	.0000	.0000	.0000	.0000	.0000	.0001
8	.0000	.0000	.0000	.0000	.0000	.0000	.0000	.0000	.0000	.0000

						λ				
x	1.1	1.2	1.3	1.4	1.5	1.6	1.7	1.8	1.9	2.0
0	.3329	.3012	.2725	.2466	.2231	.2019	.1827	.1653	.1496	.1353
1	.3662	.3614	.3543	.3452	.3347	.3230	.3106	.2975	.2842	.2707
2	.2014	.2169	.2303	.2417	.2510	.2584	.2640	.2678	.2700	.2707
3	.0738	.0867	.0998	.1128	.1255	.1378	.1496	.1607	.1710	.1804
4	.0203	.0260	.0324	.0395	.0471	.0551	.0636	.0723	.0812	.0902
5	.0045	.0062	.0084	.0111	.0141	.0176	.0216	.0260	.0309	.0361
6	.0008	.0012	.0018	.0026	.0035	.0047	.0061	.0078	.0098	.0120
7	.0001	.0002	.0003	.0005	.0008	.0011	.0015	.0020	.0027	.0034
8	.0000	.0000	.0001	.0001	.0001	.0002	.0003	.0005	.0006	.0009
9	.0000	.0000	.0000	.0000	.0000	.0000	.0001	.0001	.0001	.0002
10	.0000	.0000	.0000	.0000	.0000	.0000	.0000	.0000	.0000	.0000

						λ				
x	2.1	2.2	2.3	2.4	2.5	2.6	2.7	2.8	2.9	3.0
0	.1225	.1108	.1003	.0907	.0821	.0743	.0672	.0608	.0550	.0498
1	.2572	.2438	.2306	.2177	.2052	.1931	.1815	.1703	.1596	.1494
2	.2700	.2681	.2652	.2613	.2565	.2510	.2450	.2384	.2314	.2240
3	.1890	.1966	.2033	.2090	.2138	.2176	.2205	.2225	.2237	.2240
4	.0992	.1082	.1169	.1254	.1336	.1414	.1488	.1557	.1622	.1680
5	.0417	.0476	.0538	.0602	.0668	.0735	.0804	.0872	.0940	.1008
6	.0146	.0174	.0206	.0241	.0278	.0319	.0362	.0407	.0455	.0504
7	.0044	.0055	.0068	.0083	.0099	.0118	.0139	.0163	.0188	.0216
8	.0011	.0015	.0019	.0025	.0031	.0038	.0047	.0057	.0068	.0081
9	.0003	.0004	.0005	.0007	.0009	.0011	.0014	.0018	.0022	.0027
10	.0001	.0001	.0001	.0002	.0002	.0003	.0004	.0005	.0006	.0008
11	.0000	.0000	.0000	.0000	.0000	.0001	.0001	.0001	.0002	.0002
12	.0000	.0000	.0000	.0000	.0000	.0000	.0000	.0000	.0000	.0001
13	.0000	.0000	.0000	.0000	.0000	.0000	.0000	.0000	.0000	.0000

TABLE IV
Poisson Probabilities
(Continued)

x	3.1	3.2	3.3	3.4	3.5	3.6	3.7	3.8	3.9	4.0
0	.0450	.0408	.0369	.0334	.0302	.0273	.0247	.0224	.0202	.0183
1	.1397	.1304	.1217	.1135	.1057	.0984	.0915	.0850	.0789	.0733
2	.2165	.2087	.2008	.1929	.1850	.1771	.1692	.1615	.1539	.1465
3	.2237	.2226	.2209	.2186	.2158	.2125	.2087	.2046	.2001	.1954
4	.1733	.1781	.1823	.1858	.1888	.1912	.1931	.1944	.1951	.1954
5	.1075	.1140	.1203	.1264	.1322	.1377	.1429	.1477	.1522	.1563
6	.0555	.0608	.0662	.0716	.0771	.0826	.0881	.0936	.0989	.1042
7	.0246	.0278	.0312	.0348	.0385	.0425	.0466	.0508	.0551	.0595
8	.0095	.0111	.0129	.0148	.0169	.0191	.0215	.0241	.0269	.0298
9	.0033	.0040	.0047	.0056	.0066	.0076	.0089	.0102	.0116	.0132
10	.0010	.0013	.0016	.0019	.0023	.0028	.0033	.0039	.0045	.0053
11	.0003	.0004	.0005	.0006	.0007	.0009	.0011	.0013	.0016	.0019
12	.0001	.0001	.0001	.0002	.0002	.0003	.0003	.0004	.0005	.0006
13	.0000	.0000	.0000	.0000	.0001	.0001	.0001	.0001	.0002	.0002
14	.0000	.0000	.0000	.0000	.0000	.0000	.0000	.0000	.0000	.0001
15	.0000	.0000	.0000	.0000	.0000	.0000	.0000	.0000	.0000	.0000

x	4.1	4.2	4.3	4.4	4.5	4.6	4.7	4.8	4.9	5.0
0	.0166	.0150	.0136	.0123	.0111	.0101	.0091	.0082	.0074	.0067
1	.0679	.0630	.0583	.0540	.0500	.0462	.0427	.0395	.0365	.0337
2	.1393	.1323	.1254	.1188	.1125	.1063	.1005	.0948	.0894	.0842
3	.1904	.1852	.1798	.1743	.1687	.1631	.1574	.1517	.1460	.1404
4	.1951	.1944	.1933	.1917	.1898	.1875	.1849	.1820	.1789	.1755
5	.1600	.1633	.1662	.1687	.1708	.1725	.1738	.1747	.1753	.1755
6	.1093	.1143	.1191	.1237	.1281	.1323	.1362	.1398	.1432	.1462
7	.0640	.0686	.0732	.0778	.0824	.0869	.0914	.0959	.1002	.1044
8	.0328	.0360	.0393	.0428	.0463	.0500	.0537	.0575	.0614	.0653
9	.0150	.0168	.0188	.0209	.0232	.0255	.0281	.0307	.0334	.0363
10	.0061	.0071	.0081	.0092	.0104	.0118	.0132	.0147	.0164	.0181
11	.0023	.0027	.0032	.0037	.0043	.0049	.0056	.0064	.0073	.0082
12	.0008	.0009	.0011	.0013	.0016	.0019	.0022	.0026	.0030	.0034
13	.0002	.0003	.0004	.0005	.0006	.0007	.0008	.0009	.0011	.0013
14	.0001	.0001	.0001	.0001	.0002	.0002	.0003	.0003	.0004	.0005
15	.0000	.0000	.0000	.0000	.0001	.0001	.0001	.0001	.0001	.0002
16	.0000	.0000	.0000	.0000	.0000	.0000	.0000	.0000	.0000	.0000
17	.0000	.0000	.0000	.0000	.0000	.0000	.0000	.0000	.0000	.0000

x	5.1	5.2	5.3	5.4	5.5	5.6	5.7	5.8	5.9	6.0
0	.0061	.0055	.0050	.0045	.0041	.0037	.0033	.0030	.0027	.0025
1	.0311	.0287	.0265	.0244	.0225	.0207	.0191	.0176	.0162	.0149
2	.0793	.0746	.0701	.0659	.0618	.0580	.0544	.0509	.0477	.0446
3	.1348	.1293	.1239	.1185	.1133	.1082	.1033	.0985	.0938	.0892
4	.1719	.1681	.1641	.1600	.1558	.1515	.1472	.1428	.1383	.1339
5	.1753	.1748	.1740	.1728	.1714	.1697	.1678	.1656	.1632	.1606
6	.1490	.1515	.1537	.1555	.1571	.1584	.1594	.1601	.1605	.1606
7	.1086	.1125	.1163	.1200	.1234	.1267	.1298	.1326	.1353	.1377
8	.0692	.0731	.0771	.0810	.0849	.0887	.0925	.0962	.0998	.1033
9	.0392	.0423	.0454	.0486	.0519	.0552	.0586	.0620	.0654	.0688
10	.0200	.0220	.0241	.0262	.0285	.0309	.0334	.0359	.0386	.0413
11	.0093	.0104	.0116	.0129	.0143	.0157	.0173	.0190	.0207	.0225
12	.0039	.0045	.0051	.0058	.0065	.0073	.0082	.0092	.0102	.0113
13	.0015	.0018	.0021	.0024	.0028	.0032	.0036	.0041	.0046	.0052
14	.0006	.0007	.0008	.0009	.0011	.0013	.0015	.0017	.0019	.0022

(continues)

TABLE IV
Poisson Probabilities
(Continued)

x	5.1	5.2	5.3	5.4	5.5	5.6	5.7	5.8	5.9	6.0
15	.0002	.0002	.0003	.0003	.0004	.0005	.0006	.0007	.0008	.0009
16	.0001	.0001	.0001	.0001	.0001	.0002	.0002	.0002	.0003	.0003
17	.0000	.0000	.0000	.0000	.0000	.0001	.0001	.0001	.0001	.0001
18	.0000	.0000	.0000	.0000	.0000	.0000	.0000	.0000	.0000	.0000
19	.0000	.0000	.0000	.0000	.0000	.0000	.0000	.0000	.0000	.0000

x	6.1	6.2	6.3	6.4	6.5	6.6	6.7	6.8	6.9	7.0
0	.0022	.0020	.0018	.0017	.0015	.0014	.0012	.0011	.0010	.0009
1	.0137	.0126	.0116	.0106	.0098	.0090	.0082	.0076	.0070	.0064
2	.0417	.0390	.0364	.0340	.0318	.0296	.0276	.0258	.0240	.0223
3	.0848	.0806	.0765	.0726	.0688	.0652	.0617	.0584	.0552	.0521
4	.1294	.1249	.1205	.1162	.1118	.1076	.1034	.0992	.0952	.0912
5	.1579	.1549	.1519	.1487	.1454	.1420	.1385	.1349	.1314	.1277
6	.1605	.1601	.1595	.1586	.1575	.1562	.1546	.1529	.1511	.1490
7	.1399	.1418	.1435	.1450	.1462	.1472	.1480	.1486	.1489	.1490
8	.1066	.1099	.1130	.1160	.1188	.1215	.1240	.1263	.1284	.1304
9	.0723	.0757	.0791	.0825	.0858	.0891	.0923	.0954	.0985	.1014
10	.0441	.0469	.0498	.0528	.0558	.0588	.0618	.0649	.0679	.0710
11	.0244	.0265	.0285	.0307	.0330	.0353	.0377	.0401	.0426	.0452
12	.0124	.0137	.0150	.0164	.0179	.0194	.0210	.0227	.0245	.0263
13	.0058	.0065	.0073	.0081	.0089	.0090	.0108	.0119	.0130	.0142
14	.0025	.0029	.0033	.0037	.0041	.0046	.0052	.0058	.0064	.0071
15	.0010	.0012	.0014	.0016	.0018	.0020	.0023	.0026	.0029	.0033
16	.0004	.0005	.0005	.0006	.0007	.0008	.0010	.0011	.0013	.0014
17	.0001	.0002	.0002	.0002	.0003	.0003	.0004	.0004	.0005	.0006
18	.0000	.0001	.0001	.0001	.0001	.0001	.0001	.0002	.0002	.0002
19	.0000	.0000	.0000	.0000	.0000	.0000	.0001	.0001	.0001	.0001
20	.0000	.0000	.0000	.0000	.0000	.0000	.0000	.0000	.0000	.0000

x	7.1	7.2	7.3	7.4	7.5	7.6	7.7	7.8	7.9	8.0
0	.0008	.0007	.0007	.0006	.0006	.0005	.0005	.0004	.0004	.0003
1	.0059	.0054	.0049	.0045	.0041	.0038	.0035	.0032	.0029	.0027
2	.0208	.0194	.0180	.0167	.0156	.0145	.0134	.0125	.0116	.0107
3	.0492	.0464	.0438	.0413	.0389	.0366	.0345	.0324	.0305	.0286
4	.0874	.0836	.0799	.0764	.0729	.0696	.0663	.0632	.0602	.0573
5	.1241	.1204	.1167	.1130	.1094	.1057	.1021	.0986	.0951	.0916
6	.1468	.1445	.1420	.1394	.1367	.1339	.1311	.1282	.1252	.1221
7	.1489	.1486	.1481	.1474	.1465	.1454	.1442	.1428	.1413	.1396
8	.1321	.1337	.1351	.1363	.1373	.1381	.1388	.1392	.1395	.1396
9	.1042	.1070	.1096	.1121	.1144	.1167	.1187	.1207	.1224	.1241
10	.0740	.0770	.0800	.0829	.0858	.0887	.0914	.0941	.0967	.0993
11	.0478	.0504	.0531	.0558	.0585	.0613	.0640	.0667	.0695	.0722
12	.0283	.0303	.0323	.0344	.0366	.0388	.0411	.0434	.0457	.0481
13	.0154	.0168	.0181	.0196	.0211	.0227	.0243	.0260	.0278	.0296
14	.0078	.0086	.0095	.0104	.0113	.0123	.0134	.0145	.0157	.0169
15	.0037	.0041	.0046	.0051	.0057	.0062	.0069	.0075	.0083	.0090
16	.0016	.0019	.0021	.0024	.0026	.0030	.0033	.0037	.0041	.0045
17	.0007	.0008	.0009	.0010	.0012	.0013	.0015	.0017	.0019	.0021
18	.0003	.0003	.0004	.0004	.0005	.0006	.0006	.0007	.0008	.0009
19	.0001	.0001	.0001	.0002	.0002	.0002	.0003	.0003	.0003	.0004
20	.0000	.0000	.0001	.0001	.0001	.0001	.0001	.0001	.0001	.0002
21	.0000	.0000	.0000	.0000	.0000	.0000	.0000	.0000	.0001	.0001
22	.0000	.0000	.0000	.0000	.0000	.0000	.0000	.0000	.0000	.0000

TABLE IV
**Poisson Probabilities
(Concluded)**

					λ					
x	8.1	8.2	8.3	8.4	8.5	8.6	8.7	8.8	8.9	9.0
0	.0003	.0003	.0002	.0002	.0002	.0002	.0002	.0002	.0001	.0001
1	.0025	.0023	.0021	.0019	.0017	.0016	.0014	.0013	.0012	.0011
2	.0100	.0092	.0086	.0079	.0074	.0068	.0063	.0058	.0054	.0050
3	.0269	.0252	.0237	.0222	.0208	.0195	.0183	.0171	.0160	.0150
4	.0544	.0517	.0491	.0466	.0443	.0420	.0398	.0377	.0357	.0337
5	.0882	.0849	.0816	.0784	.0752	.0722	.0692	.0663	.0635	.0607
6	.1191	.1160	.1128	.1097	.1066	.1034	.1003	.0972	.0941	.0911
7	.1378	.1358	.1338	.1317	.1294	.1271	.1247	.1222	.1197	.1171
8	.1395	.1392	.1388	.1382	.1375	.1366	.1356	.1344	.1332	.1318
9	.1255	.1269	.1280	.1290	.1299	.1306	.1311	.1315	.1317	.1318
10	.1017	.1040	.1063	.1084	.1104	.1123	.1140	.1157	.1172	.1186
11	.0749	.0775	.0802	.0828	.0853	.0878	.0902	.0925	.0948	.0970
12	.0505	.0530	.0555	.0579	.0604	.0629	.0654	.0679	.0703	.0728
13	.0315	.0334	.0354	.0374	.0395	.0416	.0438	.0459	.0481	.0504
14	.0182	.0196	.0210	.0225	.0240	.0256	.0272	.0289	.0306	.0324
15	.0098	.0107	.0116	.0126	.0136	.0147	.0158	.0169	.0182	.0194
16	.0050	.0055	.0060	.0066	.0072	.0079	.0086	.0093	.0101	.0109
17	.0024	.0026	.0029	.0033	.0036	.0040	.0044	.0048	.0053	.0058
18	.0011	.0012	.0014	.0015	.0017	.0019	.0021	.0024	.0026	.0029
19	.0005	.0005	.0006	.0007	.0008	.0009	.0010	.0011	.0012	.0014
20	.0002	.0002	.0002	.0003	.0003	.0004	.0004	.0005	.0005	.0006
21	.0001	.0001	.0001	.0001	.0001	.0002	.0002	.0002	.0002	.0003
22	.0000	.0000	.0000	.0000	.0001	.0001	.0001	.0001	.0001	.0001
23	.0000	.0000	.0000	.0000	.0000	.0000	.0000	.0000	.0000	.0000
24	.0000	.0000	.0000	.0000	.0000	.0000	.0000	.0000	.0000	.0000

					λ					
x	9.1	9.2	9.3	9.4	9.5	9.6	9.7	9.8	9.9	10.0
0	.0001	.0001	.0001	.0001	.0001	.0001	.0001	.0001	.0001	.0000
1	.0010	.0009	.0009	.0008	.0007	.0007	.0006	.0005	.0005	.0005
2	.0046	.0043	.0040	.0037	.0034	.0031	.0029	.0027	.0025	.0023
3	.0140	.0131	.0123	.0115	.0107	.0100	.0093	.0087	.0081	.0076
4	.0319	.0302	.0285	.0269	.0254	.0240	.0226	.0213	.0201	.0189
5	.0581	.0555	.0530	.0506	.0483	.0460	.0439	.0418	.0398	.0378
6	.0881	.0851	.0822	.0793	.0764	.0736	.0709	.0682	.0656	.0631
7	.1145	.1118	.1091	.1064	.1037	.1010	.0982	.0955	.0928	.0901
8	.1302	.1286	.1269	.1251	.1232	.1212	.1191	.1170	.1148	.1126
9	.1317	.1315	.1311	.1306	.1300	.1293	.1284	.1274	.1263	.1251
10	.1198	.1209	.1219	.1228	.1235	.1241	.1245	.1249	.1250	.1251
11	.0991	.1012	.1031	.1049	.1067	.1083	.1098	.1112	.1125	.1137
12	.0752	.0776	.0799	.0822	.0844	.0866	.0888	.0908	.0928	.0948
13	.0526	.0549	.0572	.0594	.0617	.0640	.0662	.0685	.0707	.0729
14	.0342	.0361	.0380	.0399	.0419	.0439	.0459	.0479	.0500	.0521
15	.0208	.0221	.0235	.0250	.0265	.0281	.0297	.0313	.0330	.0347
16	.0118	.0127	.0137	.0147	.0157	.0168	.0180	.0192	.0204	.0217
17	.0063	.0069	.0075	.0081	.0088	.0095	.0103	.0111	.0119	.0128
18	.0032	.0035	.0039	.0042	.0046	.0051	.0055	.0060	.0065	.0071
19	.0015	.0017	.0019	.0021	.0023	.0026	.0028	.0031	.0034	.0037
20	.0007	.0008	.0009	.0010	.0011	.0012	.0014	.0015	.0017	.0019
21	.0003	.0003	.0004	.0004	.0005	.0006	.0006	.0007	.0008	.0009
22	.0001	.0001	.0002	.0002	.0002	.0002	.0003	.0003	.0004	.0004
23	.0000	.0001	.0001	.0001	.0001	.0001	.0001	.0001	.0002	.0002
24	.0000	.0000	.0000	.0000	.0000	.0000	.0000	.0001	.0001	.0001
25	.0000	.0000	.0000	.0000	.0000	.0000	.0000	.0000	.0000	.0000

TABLE V   Normal Probabilities: Areas of the Standard Normal Distribution

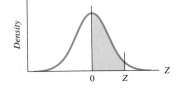

The values in the body of the table are the areas between the mean and the value of Z.

Z	.00	.01	.02	.03	.04	.05	.06	.07	.08	.09
.00	.0000	.0040	.0080	.0120	.0160	.0199	.0239	.0279	.0319	.0359
.10	.0398	.0438	.0478	.0517	.0557	.0596	.0636	.0675	.0714	.0753
.20	.0793	.0832	.0871	.0910	.0948	.0987	.1026	.1064	.1103	.1141
.30	.1179	.1217	.1255	.1293	.1331	.1368	.1406	.1443	.1480	.1517
.40	.1554	.1591	.1628	.1664	.1700	.1736	.1772	.1808	.1844	.1879
.50	.1915	.1950	.1985	.2019	.2054	.2088	.2123	.2157	.2190	.2224
.60	.2257	.2291	.2324	.2357	.2389	.2422	.2454	.2486	.2517	.2549
.70	.2580	.2611	.2642	.2673	.2703	.2734	.2764	.2793	.2823	.2852
.80	.2881	.2910	.2939	.2967	.2995	.3023	.3051	.3078	.3106	.3133
.90	.3159	.3186	.3212	.3238	.3264	.3289	.3315	.3340	.3365	.3389
1.00	.3413	.3438	.3461	.3485	.3508	.3531	.3554	.3577	.3599	.3621
1.10	.3643	.3665	.3686	.3708	.3729	.3749	.3770	.3790	.3810	.3830
1.20	.3849	.3869	.3888	.3907	.3925	.3944	.3962	.3980	.3997	.4015
1.30	.4032	.4049	.4066	.4082	.4099	.4115	.4131	.4147	.4162	.4177
1.40	.4192	.4207	.4222	.4236	.4251	.4265	.4279	.4292	.4306	.4319
1.50	.4332	.4345	.4357	.4370	.4382	.4394	.4406	.4418	.4429	.4441
1.60	.4452	.4463	.4474	.4484	.4495	.4505	.4515	.4525	.4535	.4545
1.70	.4554	.4564	.4573	.4582	.4591	.4599	.4608	.4616	.4625	.4633
1.80	.4641	.4649	.4656	.4664	.4671	.4678	.4686	.4693	.4699	.4706
1.90	.4713	.4719	.4726	.4732	.4738	.4744	.4750	.4756	.4761	.4767
2.00	.4772	.4778	.4783	.4788	.4793	.4798	.4803	.4808	.4812	.4817
2.10	.4821	.4826	.4830	.4834	.4838	.4842	.4846	.4850	.4854	.4857
2.20	.4861	.4864	.4868	.4871	.4875	.4878	.4881	.4884	.4887	.4890
2.30	.4893	.4896	.4898	.4901	.4904	.4906	.4909	.4911	.4913	.4916
2.40	.4918	.4920	.4922	.4925	.4927	.4929	.4931	.4932	.4934	.4936
2.50	.4938	.4940	.4941	.4943	.4945	.4946	.4948	.4949	.4951	.4952
2.60	.4953	.4955	.4956	.4957	.4959	.4960	.4961	.4962	.4963	.4964
2.70	.4965	.4966	.4967	.4968	.4969	.4970	.4971	.4972	.4973	.4974
2.80	.4974	.4975	.4976	.4977	.4977	.4978	.4979	.4979	.4980	.4981
2.90	.4981	.4982	.4982	.4983	.4984	.4984	.4985	.4985	.4986	.4986
3.00	.4987	.4987	.4987	.4988	.4988	.4989	.4989	.4989	.4990	.4990
3.10	.4990	.4991	.4991	.4991	.4992	.4992	.4992	.4992	.4993	.4993
3.20	.4993	.4993	.4994	.4994	.4994	.4994	.4994	.4995	.4995	.4995
3.30	.4995	.4995	.4995	.4996	.4996	.4996	.4996	.4996	.4996	.4997
3.40	.4997	.4997	.4997	.4997	.4997	.4997	.4997	.4997	.4997	.4998
3.50	.4998	.4998	.4998	.4998	.4998	.4998	.4998	.4998	.4998	.4998
3.60	.4998	.4998	.4999	.4999	.4999	.4999	.4999	.4999	.4999	.4999
3.70	.4999	.4999	.4999	.4999	.4999	.4999	.4999	.4999	.4999	.4999
3.80	.4999	.4999	.4999	.4999	.4999	.4999	.4999	.4999	.4999	.4999

Note: For example, if we want to find the area under the standard normal curve between $Z = 0$ and $Z = 1.96$, we find the $Z = 1.90$ row and .06 column (for $Z = 1.90 + .06 = 1.96$) and read .4750 at the intersection.

TABLE VI   *t* Distribution Values

Degrees of Freedom ν	$t_{.10}$	$t_{.05}$	$t_{.025}$	$t_{.01}$	$t_{.005}$
1	3.078	6.314	12.706	31.821	63.657
2	1.886	2.920	4.303	6.965	9.925
3	1.638	2.353	3.182	4.541	5.841
4	1.533	2.132	2.776	3.747	4.604
5	1.476	2.015	2.571	3.365	4.032
6	1.440	1.943	2.447	3.143	3.707
7	1.415	1.895	2.365	2.998	3.499
8	1.397	1.860	2.306	2.896	3.355
9	1.383	1.833	2.262	2.821	3.250
10	1.372	1.812	2.228	2.764	3.169
11	1.363	1.796	2.201	2.718	3.106
12	1.356	1.782	2.179	2.681	3.055
13	1.350	1.771	2.160	2.650	3.012
14	1.345	1.761	2.145	2.624	2.977
15	1.341	1.753	2.131	2.602	2.947
16	1.337	1.746	2.120	2.583	2.921
17	1.333	1.740	2.110	2.567	2.898
18	1.330	1.734	2.101	2.552	2.878
19	1.328	1.729	2.093	2.539	2.861
20	1.325	1.725	2.086	2.528	2.845
21	1.323	1.721	2.080	2.518	2.831
22	1.321	1.717	2.074	2.508	2.819
23	1.319	1.714	2.069	2.500	2.807
24	1.318	1.711	2.064	2.492	2.797
25	1.316	1.708	2.060	2.485	2.787
26	1.315	1.706	2.056	2.479	2.779
27	1.314	1.703	2.052	2.473	2.771
28	1.313	1.701	2.048	2.467	2.763
29	1.311	1.699	2.045	2.462	2.756
30	1.310	1.697	2.042	2.457	2.750
40	1.303	1.684	2.021	2.423	2.704
60	1.296	1.671	2.000	2.390	2.660
120	1.289	1.658	1.980	2.358	2.617
∞	1.282	1.645	1.960	2.326	2.576

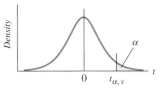

Note: For example, if $\alpha = .05$ and $\nu = 15$, then
$t_{\alpha, \nu} = t_{.05, 15} = 1.753$.

TABLE VII   *F* Distribution Values

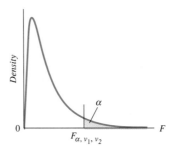

F Values When $\alpha = .05$

ν_1 ν_2	1	2	3	4	5	6	7	8	9
1	161.45	199.50	215.71	224.58	230.16	233.99	236.77	238.88	240.54
2	18.513	19.000	19.164	19.247	19.296	19.330	19.353	19.371	19.385
3	10.128	9.5521	9.2766	9.1172	9.0135	8.9406	8.8868	8.8452	8.8123
4	7.7086	6.9443	6.5914	6.3883	6.2560	6.1631	6.0942	6.0410	5.9988
5	6.6079	5.7861	5.4095	5.1922	5.0503	4.9503	4.8759	4.8183	4.7725
6	5.9874	5.1433	4.7571	4.5337	4.3874	4.2839	4.2066	4.1468	4.0990
7	5.5914	4.7374	4.3468	4.1203	3.9715	3.8660	3.7870	3.7257	3.6767
8	5.3177	4.4590	4.0662	3.8378	3.6875	3.5806	3.5005	3.4381	3.3881
9	5.1174	4.2565	3.8626	3.6331	3.4817	3.3738	3.2927	3.2296	3.1789
10	4.9646	4.1028	3.7083	3.4780	3.3258	3.2172	3.1355	3.0717	3.0204
11	4.8443	3.9823	3.5874	3.3567	3.2039	3.0946	3.0123	2.9480	2.8962
12	4.7472	3.8853	3.4903	3.2592	3.1059	2.9961	2.9134	2.8486	2.7964
13	4.6672	3.8056	3.4105	3.1791	3.0254	2.9153	2.8321	2.7669	2.7144
14	4.6001	3.7389	3.3439	3.1122	2.9582	2.8477	2.7642	2.6987	2.6458
15	4.5431	3.6823	3.2874	3.0556	2.9013	2.7905	2.7066	2.6408	2.5876
16	4.4940	3.6337	3.2389	3.0069	2.8524	2.7413	2.6572	2.5911	2.5377
17	4.4513	3.5915	3.1968	2.9647	2.8100	2.6987	2.6143	2.5480	2.4943
18	4.4139	3.5546	3.1599	2.9277	2.7729	2.6613	2.5767	2.5102	2.4563
19	4.3808	3.5219	3.1274	2.8951	2.7401	2.6283	2.5435	2.4768	2.4227
20	4.3513	3.4928	3.0984	2.8661	2.7109	2.5990	2.5140	2.4471	2.3928
21	4.3248	3.4668	3.0725	2.8401	2.6848	2.5757	2.4876	2.4205	2.3661
22	4.3009	3.4434	3.0491	2.8167	2.6613	2.5491	2.4638	2.3965	2.3419
23	4.2793	3.4221	3.0280	2.7955	2.6400	2.5277	2.4422	2.3748	2.3201
24	4.2597	3.4028	3.0088	2.7763	2.6207	2.5082	2.4226	2.3551	2.3002
25	4.2417	3.3852	2.9912	2.7587	2.6030	2.4904	2.4047	2.3371	2.2821
26	4.2252	3.3690	2.9751	2.7426	2.5868	2.4741	2.3883	2.3205	2.2655
27	4.2100	3.3541	2.9604	2.7278	2.5719	2.4591	2.3732	2.3053	2.2501
28	4.1960	3.3404	2.9467	2.7141	2.5581	2.4453	2.3593	2.2913	2.2360
29	4.1830	3.3277	2.9340	2.7014	2.5454	2.4324	2.3463	2.2782	2.2229
30	4.1709	3.3158	2.9223	2.6896	2.5336	2.4205	2.3343	2.2662	2.2107
40	4.0848	3.2317	2.8387	2.6060	2.4495	2.3359	2.2490	2.1802	2.1240
60	4.0012	3.1504	2.7581	2.5252	2.3683	2.2540	2.1665	2.0970	2.0401
120	3.9201	3.0718	2.6802	2.4472	2.2900	2.1750	2.0867	2.0164	1.9588
∞	3.8415	2.9957	2.6049	2.3719	2.2141	2.0986	2.0096	1.9384	1.8799

Note: For example, if $\alpha = .05$, $\nu_1 = 4$, and $\nu_2 = 7$, then $F_{\alpha, \nu_1, \nu_2} = F_{.05, 4, 7} = 4.1203$, where ν_1 is the numerator degrees of freedom and ν_2 is the denominator degrees of freedom.

TABLE VII   *F* **Distribution Values (Continued)**

F Values When α = .05

ν_1 / ν_2	10	12	15	20	24	30	40	60	120	∞
1	241.88	243.91	245.95	248.01	249.05	250.09	251.14	252.20	253.25	254.32
2	19.396	19.413	19.429	19.446	19.454	19.462	19.471	19.479	19.487	19.496
3	8.7855	8.7446	8.7029	8.6602	8.6385	8.6166	8.5944	8.5720	8.5494	8.5265
4	5.9644	5.9117	5.8578	5.8025	5.7744	5.7459	5.7170	5.6878	5.6581	5.6281
5	4.7351	4.6777	4.6188	4.5581	4.5272	4.4957	4.4638	4.4314	4.3984	4.3650
6	4.0600	3.9999	3.9381	3.8742	3.8415	3.8082	3.7743	3.7398	3.7047	3.6688
7	3.6365	3.5747	3.5108	3.4445	3.4105	3.3758	3.3404	3.3043	3.2674	3.2298
8	3.3472	3.2840	3.2184	3.1503	3.1152	3.0794	3.0428	3.0053	2.9669	2.9276
9	3.1373	3.0729	3.0061	2.9365	2.9005	2.8637	2.8259	2.7872	2.7475	2.7067
10	2.9782	2.9130	2.8450	2.7740	2.7372	2.6996	2.6609	2.6211	2.5801	2.5379
11	2.8536	2.7876	2.7186	2.6464	2.6090	2.5705	2.5309	2.4901	2.4480	2.4045
12	2.7534	2.6866	2.6169	2.5436	2.5055	2.4663	2.4259	2.3842	2.3410	2.2962
13	2.6710	2.6037	2.5331	2.4589	2.4202	2.3803	2.3392	2.2966	2.2524	2.2064
14	2.6021	2.5342	2.4630	2.3879	2.3487	2.3082	2.2664	2.2230	2.1778	2.1307
15	2.5437	2.4753	2.4035	2.3275	2.2878	2.2468	2.2043	2.1601	2.1141	2.0658
16	2.4935	2.4247	2.3522	2.2756	2.2354	2.1938	2.1507	2.1058	2.0589	2.0096
17	2.4499	2.3807	2.3077	2.2304	2.1898	2.1477	2.1040	2.0584	2.0107	1.9604
18	2.4117	2.3421	2.2686	2.1906	2.1497	2.1071	2.0629	2.0166	1.9681	1.9168
19	2.3779	2.3080	2.2341	2.1555	2.1141	2.0712	2.0264	1.9796	1.9302	1.8780
20	2.3479	2.2776	2.2033	2.1242	2.0825	2.0391	1.9938	1.9464	1.8963	1.8432
21	2.3210	2.2504	2.1757	2.0960	2.0540	2.0102	1.9645	1.9165	1.8657	1.8117
22	2.2967	2.2258	2.1508	2.0707	2.0283	1.9842	1.9380	1.8895	1.8380	1.7831
23	2.2747	2.2036	2.1282	2.0476	2.0050	1.9605	1.9139	1.8649	1.8128	1.7570
24	2.2547	2.1834	2.1077	2.0267	1.9838	1.9390	1.8920	1.8424	1.7897	1.7331
25	2.2365	2.1649	2.0889	2.0075	1.9643	1.9192	1.8718	1.8217	1.7684	1.7110
26	2.2197	2.1479	2.0716	1.9898	1.9464	1.9010	1.8533	1.8027	1.7488	1.6906
27	2.2043	2.1323	2.0558	1.9736	1.9299	1.8842	1.8361	1.7851	1.7307	1.6717
28	2.1900	2.1179	2.0411	1.9586	1.9147	1.8687	1.8203	1.7689	1.7138	1.6541
29	2.1768	2.1045	2.0275	1.9446	1.9005	1.8543	1.8055	1.7537	1.6981	1.6377
30	2.1646	2.0921	2.0148	1.9317	1.8874	1.8409	1.7918	1.7396	1.6835	1.6223
40	2.0772	2.0035	1.9245	1.8389	1.7929	1.7444	1.6928	1.6373	1.5766	1.5089
60	1.9926	1.9174	1.8364	1.7480	1.7001	1.6491	1.5943	1.5343	1.4673	1.3893
120	1.9105	1.8337	1.7505	1.6587	1.6084	1.5543	1.4952	1.4290	1.3519	1.2539
∞	1.8307	1.7522	1.6664	1.5705	1.5173	1.4591	1.3940	1.3180	1.2214	1.0000

(continues)

TABLE VII *F* Distribution Values (Continued)

F Values When α = .025

ν_2 \ ν_1	1	2	3	4	5	6	7	8	9
1	647.79	799.50	864.16	899.58	921.85	937.11	948.22	956.66	963.28
2	38.506	39.000	39.165	39.248	29.298	39.331	39.355	39.373	39.387
3	17.443	16.044	15.439	15.101	14.885	14.735	14.624	14.540	14.473
4	12.218	10.649	9.9792	9.6045	9.3645	9.1973	9.0741	8.9796	8.9047
5	10.007	8.4336	7.7636	7.3879	7.1464	6.9777	6.8531	6.7572	6.6810
6	8.8131	7.2598	6.5988	6.2272	5.9876	5.8197	5.6955	5.5996	5.5234
7	8.0727	6.5415	5.8898	5.5226	5.2852	5.1186	4.9949	4.8994	4.8232
8	7.5709	6.0595	5.4160	5.0526	4.8173	4.6517	4.5286	4.4332	4.3572
9	7.2093	5.7147	5.0781	4.7181	4.4844	4.3197	4.1971	4.1020	4.0260
10	6.9367	5.4564	4.8256	4.4683	4.2361	4.0721	3.9498	3.8549	3.7790
11	6.7241	5.2559	4.6300	4.2751	4.0440	3.8807	3.7586	3.6638	3.5879
12	6.5538	5.0959	4.4742	4.1212	3.8911	3.7283	3.6065	3.5118	3.4358
13	6.4143	4.9653	4.3472	3.9959	3.7667	3.6043	3.4827	3.3880	3.3120
14	6.2979	4.8567	4.2417	3.8919	3.6634	3.5014	3.3799	3.2853	3.2093
15	6.1995	4.7650	4.1528	3.8043	3.5764	3.4147	3.2934	3.1987	3.1227
16	6.1151	4.6867	4.0768	3.7294	3.5021	3.3406	3.2194	3.1248	3.0488
17	6.0420	4.6189	4.0112	3.6648	3.4379	3.2767	3.1556	3.0610	2.9849
18	5.9781	4.5597	3.9539	3.6083	3.3820	3.2209	3.0999	3.0053	2.9291
19	5.9216	4.5075	3.9034	3.5587	3.3327	3.1718	3.0509	2.9563	2.8800
20	5.8715	4.4613	3.8587	3.5147	3.2891	3.1283	3.0074	2.9128	2.8365
21	5.8266	4.4199	3.8188	3.4754	3.2501	3.0895	2.9686	2.8740	2.7977
22	5.7863	4.3828	3.7829	3.4401	3.2151	3.0546	2.9338	2.8392	2.7628
23	5.7498	4.3492	3.7505	3.4083	3.1835	3.0232	2.9024	2.8077	2.7313
24	5.7167	4.3187	3.7211	3.3794	3.1548	2.9946	2.8738	2.7791	2.7027
25	5.6864	4.2909	3.6943	3.3530	3.1287	2.9685	2.8478	2.7531	2.6766
26	5.6586	4.2655	3.6697	3.3289	3.1048	2.9447	2.8240	2.7293	2.6528
27	5.6331	4.2421	3.6472	3.3067	3.0828	2.9228	2.8021	2.7074	2.6309
28	5.6096	4.2205	3.6264	3.2863	3.0625	2.9027	2.7820	2.6872	2.6106
29	5.5878	4.2006	3.6072	3.2674	3.0438	2.8840	2.7633	2.6686	2.5919
30	5.5675	4.1821	3.5894	3.2499	3.0265	2.8667	2.7460	2.6513	2.5746
40	5.4239	4.0510	3.4633	3.1261	2.9037	2.7444	2.6238	2.5289	2.4519
60	5.2857	3.9253	3.3425	3.0077	2.7863	2.6274	2.5068	2.4117	2.3344
120	5.1524	3.8046	3.2270	2.8943	2.6740	2.5154	2.3948	2.2994	2.2217
∞	5.0239	3.6889	3.1161	2.7858	2.5665	2.4082	2.2875	2.1918	2.1136

TABLE VII *F* Distribution Values (Continued)

F Values When $\alpha = .025$

ν_2 \ ν_1	10	12	15	20	24	30	40	60	120	∞
1	968.63	976.71	984.87	993.10	997.25	1001.4	1005.6	1009.8	1014.0	1018.3
2	39.398	39.415	39.431	39.448	39.456	39.465	39.473	39.481	39.490	39.498
3	14.419	14.337	14.253	14.167	14.124	14.081	14.037	13.992	13.947	13.902
4	8.8439	8.7512	8.6565	8.5599	8.5109	8.4613	8.4111	8.3604	8.3092	8.2573
5	6.6192	6.5246	6.4277	6.3285	6.2780	6.2269	6.1751	6.1225	6.0693	6.0153
6	5.4613	5.3662	5.2687	5.1684	5.1172	5.0652	5.0125	4.9589	4.9045	4.8491
7	4.7611	4.6658	4.5678	4.4667	4.4150	4.3624	4.3089	4.2544	4.1989	4.1423
8	4.2951	4.1997	4.1012	3.9995	3.9472	3.8940	3.8398	3.7844	3.7279	3.6702
9	3.9639	3.8682	3.7694	3.6669	3.6142	3.5604	3.5005	3.4493	3.3918	3.3329
10	3.7168	3.6209	3.5217	3.4186	3.3654	3.3110	3.2554	3.1984	3.1399	3.0798
11	3.5257	3.4296	3.3299	3.2261	3.1725	3.1176	3.0613	3.0035	2.9441	2.8828
12	3.3736	3.2773	3.1772	3.0728	3.0187	2.9633	2.9063	2.8478	2.7874	2.7249
13	3.2497	3.1532	3.0527	2.9477	2.8932	2.8373	2.7797	2.7204	2.6590	2.5955
14	3.1469	3.0501	2.9493	2.8437	2.7888	2.7324	2.6742	2.6142	2.5519	2.4872
15	3.0602	2.9633	2.8621	2.7559	2.7006	2.6437	2.5850	2.5242	2.4611	2.3953
16	2.9862	2.8890	2.7875	2.6808	2.6252	2.5678	2.5085	2.4471	2.3831	2.3163
17	2.9222	2.8249	2.7230	2.6158	2.5598	2.5021	2.4422	2.3801	2.3153	2.2474
18	2.8664	2.7689	2.6667	2.5590	2.5027	2.4445	2.3842	2.3214	2.2558	2.1869
19	2.8173	2.7196	2.6171	2.5089	2.4523	2.3937	2.3329	2.2695	2.2032	2.1333
20	2.7737	2.6758	2.5731	2.4645	2.4076	2.3486	2.2873	2.2234	2.1562	2.0853
21	2.7348	2.6368	2.5338	2.4247	2.3675	2.3082	2.2465	2.1819	2.1141	2.0422
22	2.6998	2.6017	2.4984	2.3890	2.3315	2.2718	2.2097	2.1446	2.0760	2.0032
23	2.6682	2.5699	2.4665	2.3567	2.2989	2.2389	2.1763	2.1107	2.0415	1.9677
24	2.6396	2.5412	2.4374	2.3273	2.2693	2.2090	2.1460	2.0799	2.0099	1.9353
25	2.6135	2.5149	2.4110	2.3005	2.2422	2.1816	2.1183	2.0517	1.9811	1.9055
26	2.5895	2.4909	2.3867	2.2759	2.2174	2.1565	2.0928	2.0257	1.9545	1.8781
27	2.5676	2.4688	2.3644	2.2533	2.1946	2.1334	2.0693	2.0018	1.9299	1.8527
28	2.5473	2.4484	2.3438	2.2324	2.1735	2.1121	2.0477	1.9796	1.9072	1.8291
29	2.5286	2.4295	2.3248	2.2131	2.1540	2.0923	2.0276	1.9591	1.8861	1.8072
30	2.5112	2.4120	2.3072	2.1952	2.1359	2.0739	2.0089	1.9400	1.8664	1.7867
40	2.3882	2.2882	2.1819	2.0677	2.0069	1.9429	1.8752	1.8028	1.7242	1.6371
60	2.2702	2.1692	2.0613	1.9445	1.8817	1.8152	1.7440	1.6668	1.5810	1.4822
120	2.1570	2.0548	1.9450	1.8249	1.7597	1.6899	1.6141	1.5299	1.4327	1.3104
∞	2.0483	1.9447	1.8326	1.7085	1.6402	1.5660	1.4835	1.3883	1.2684	1.0000

(continues)

TABLE VII   *F* **Distribution Values (Continued)**

F Values When α = .01

ν_1 / ν_2	1	2	3	4	5	6	7	8	9
1	4052.2	4999.5	5403.3	5624.6	5763.7	5859.0	5928.3	5981.6	6022.5
2	98.503	99.000	99.166	99.249	99.299	99.332	99.356	99.374	99.388
3	34.116	30.817	29.457	28.710	28.237	27.911	27.672	27.489	27.345
4	21.198	18.000	16.694	15.977	15.522	15.207	14.976	14.799	14.659
5	16.258	13.274	12.060	11.392	10.967	10.672	10.456	10.289	10.158
6	13.745	10.925	9.7795	9.1483	8.7459	8.4661	8.2600	8.1016	7.9761
7	12.246	9.5466	8.4513	7.8467	7.4604	7.1914	6.9928	6.8401	6.7188
8	11.259	8.6491	7.5910	7.0060	6.6318	6.3707	6.1776	6.0289	5.9106
9	10.561	8.0215	6.9919	6.4221	6.0569	5.8018	5.6129	5.4671	5.3511
10	10.044	7.5594	6.5523	5.9943	5.6363	5.3858	5.2001	5.0567	4.9424
11	9.6460	7.2057	6.2167	5.6683	5.3160	5.0692	4.8861	4.7445	4.6315
12	9.3302	6.9266	5.9526	5.4119	5.0643	4.8206	4.6395	4.4994	4.3875
13	9.0738	6.7010	5.7394	5.2053	4.8616	4.6204	4.4410	4.3021	4.1911
14	8.8616	6.5149	5.5639	5.0354	4.6950	4.4558	4.2779	4.1399	4.0297
15	8.6831	6.3589	5.4170	4.8932	4.5556	4.3183	4.1415	4.0045	3.8948
16	8.5310	6.2262	5.2922	4.7726	4.4374	4.2016	4.0259	3.8896	3.7804
17	8.3997	6.1121	5.1850	4.6690	4.3359	4.1015	3.9267	3.7910	3.6822
18	8.2854	6.0129	5.0919	4.5790	4.2479	4.0146	3.8406	3.7054	3.5971
19	8.1850	5.9259	5.0103	4.5003	4.1708	3.9386	3.7653	3.6305	3.5225
20	8.0960	5.8489	4.9382	4.4307	4.1027	3.8714	3.6987	3.5644	3.4567
21	8.0166	5.7804	4.8740	4.3688	4.0421	3.8117	3.6396	3.5056	3.3981
22	7.9454	5.7190	4.8166	4.3134	3.9880	3.7583	3.5867	3.4530	3.3458
23	7.8811	5.6637	4.7649	4.2635	3.9392	3.7102	3.5390	3.4057	3.2986
24	7.8229	5.6136	4.7181	4.2184	3.8951	3.6667	3.4959	3.3629	3.2560
25	7.7698	5.5680	4.6755	4.1774	3.8550	3.6272	3.4568	3.3239	3.2172
26	7.7213	5.5263	4.6366	4.1400	3.8183	3.5911	3.4210	3.2884	3.1818
27	7.6767	5.4881	4.6009	4.1056	3.7848	3.5580	3.3882	3.2558	3.1494
28	7.6356	5.4529	4.5681	4.0740	3.7539	3.5276	3.3581	3.2259	3.1195
29	7.5976	5.4205	4.5378	4.0449	3.7254	3.4995	3.3302	3.1982	3.0920
30	7.5625	5.3904	4.5097	4.0179	3.6990	3.4735	3.3045	3.1726	3.0665
40	7.3141	5.1785	4.3126	3.8283	3.5138	3.2910	3.1238	2.9930	2.8876
60	7.0771	4.9774	4.1259	3.6491	3.3389	3.1187	2.9530	2.8233	2.7185
120	6.8510	4.7865	3.9493	3.4796	3.1735	2.9559	2.7918	2.6629	2.5586
∞	6.6349	4.6052	3.7816	3.3192	3.0173	2.8020	2.6393	2.5113	2.4073

TABLE VII   *F* Distribution Values (Continued)

F Values When α = .01

ν_1 / ν_2	10	12	15	20	24	30	40	60	120	∞
1	6055.8	6106.3	6157.3	6208.7	6234.6	6260.7	6286.8	6313.0	6339.4	6366.0
2	99.399	99.416	99.432	99.449	99.458	99.466	99.474	99.483	99.491	99.501
3	27.229	27.052	26.872	26.690	26.598	26.505	26.411	26.316	26.221	26.125
4	14.546	14.374	14.198	14.020	13.929	13.838	13.745	13.652	13.558	13.463
5	10.051	9.8883	9.7222	9.5527	9.4665	9.3793	9.2912	9.2020	9.1118	9.0204
6	7.8741	7.7183	7.5590	7.3958	7.3127	7.2285	7.1432	7.0568	6.9690	6.8801
7	6.6201	6.4691	6.3143	6.1554	6.0743	5.9921	5.9084	5.8236	5.7372	5.6495
8	5.8143	5.6668	5.5151	5.3591	5.2793	5.1981	5.1156	5.0316	4.9460	4.8588
9	5.2565	5.1114	4.9621	4.8080	4.7290	4.6486	4.5667	4.4831	4.3978	4.3105
10	4.8492	4.7059	4.5582	4.4054	4.3269	4.2469	4.1653	4.0819	3.9965	3.9090
11	4.5393	4.3974	4.2509	4.0990	4.0209	3.9411	3.8596	3.7761	3.6904	3.6025
12	4.2961	4.1553	4.0096	3.8584	3.7805	3.7008	3.6192	3.5355	3.4494	3.3608
13	4.1003	3.9603	3.8154	3.6646	3.5868	3.5070	3.4253	3.3413	3.2548	3.1654
14	3.9394	3.8001	3.6557	3.5052	3.4274	3.3476	3.2656	3.1813	3.0942	3.0040
15	3.8049	3.6662	3.5222	3.3719	3.2940	3.2141	3.1319	3.0471	2.9595	2.8684
16	3.6909	3.5527	3.4089	3.2588	3.1808	3.1007	3.0182	2.9330	2.8447	2.7528
17	3.5931	3.4552	3.3117	3.1615	3.0835	3.0032	2.9205	2.8348	2.7459	2.6530
18	3.5082	3.3706	3.2273	3.0771	2.9990	2.9185	2.8354	2.7493	2.6597	2.5660
19	3.4338	3.2965	3.1533	3.0031	2.9249	2.8422	2.7608	2.6742	2.5839	2.4893
20	3.3682	3.2311	3.0880	2.9377	2.8594	2.7785	2.6947	2.6077	2.5168	2.4212
21	3.3098	3.1729	3.0299	2.8796	2.8011	2.7200	2.6359	2.5484	2.4568	2.3603
22	3.2576	3.1209	2.9780	2.8274	2.7488	2.6675	2.5831	2.4951	2.4029	2.3055
23	3.2106	3.0740	2.9311	2.7805	2.7017	2.6202	2.5355	2.4471	2.3542	2.2559
24	3.1681	3.0316	2.8887	2.7380	2.6591	2.5773	2.4923	2.4035	2.3099	2.2107
25	3.1294	2.9931	2.8502	2.6993	2.6203	2.5383	2.4530	2.3637	2.2695	2.1694
26	3.0941	2.9579	2.8150	2.6640	2.5848	2.5026	2.4170	2.3273	2.2325	2.1315
27	3.0618	2.9256	2.7827	2.6316	2.5522	2.4699	2.3840	2.2938	2.1984	2.0965
28	3.0320	2.8959	2.7530	2.6017	2.5223	2.4397	2.3535	2.2629	2.1670	2.0642
29	3.0045	2.8685	2.7256	2.5742	2.4946	2.4118	2.3253	2.2344	2.1378	2.0342
30	2.9791	2.8431	2.7002	2.5487	2.4689	2.3860	2.2992	2.2079	2.1107	2.0062
40	2.8005	2.6648	2.5216	2.3689	2.2880	2.2034	2.1142	2.0194	1.9172	1.8047
60	2.6318	2.4961	2.3523	2.1978	2.1154	2.0285	1.9360	1.8363	1.7263	1.6006
120	2.4721	2.3363	2.1915	2.0346	1.9500	1.8600	1.7628	1.6557	1.5330	1.3805
∞	2.3209	2.1848	2.0385	1.8783	1.7908	1.6964	1.5923	1.4730	1.3246	1.0000

(continues)

TABLE VII   *F* **Distribution Values (Continued)**

F Values When α = .005

ν_2 \ ν_1	1	2	3	4	5	6	7	8	9
1	16211	20000	21615	22500	23056	23437	23715	23925	24091
2	198.50	199.00	199.17	199.25	199.30	199.33	199.36	199.37	199.39
3	55.552	49.799	47.467	46.195	45.392	44.838	44.434	44.126	43.882
4	31.333	26.284	24.259	23.155	22.456	21.975	21.622	21.352	21.139
5	22.785	18.314	16.530	15.556	14.940	14.513	14.200	13.961	13.772
6	18.635	14.544	12.917	12.028	11.464	11.073	10.786	10.566	10.391
7	16.236	12.404	10.882	10.050	9.5221	9.1554	8.8854	8.6781	8.5138
8	14.688	11.042	9.5965	8.8051	8.3018	7.9520	7.6942	7.4960	7.3386
9	13.614	10.107	8.7171	7.9559	7.4711	7.1338	6.8849	6.6933	6.5411
10	12.826	9.4270	8.0807	7.3428	6.8723	6.5446	6.3025	6.1159	5.9676
11	12.226	8.9122	7.6004	6.8809	6.4217	6.1015	5.8648	5.6821	5.5368
12	11.754	8.5096	7.2258	6.5211	6.0711	5.7570	5.5245	5.3451	5.2021
13	11.374	8.1865	6.9257	6.2335	5.7910	5.4819	5.2529	5.0761	4.9351
14	11.060	7.9217	6.6803	5.9984	5.5623	5.2574	5.0313	4.8566	4.7173
15	10.798	7.7008	6.4760	5.8029	5.3721	5.0708	4.8473	4.6743	4.5364
16	10.575	7.5138	6.3034	5.6378	5.2117	4.9134	4.6920	4.5207	4.3838
17	10.384	7.3536	6.1556	5.4967	5.0746	4.7789	4.5594	4.3893	4.2535
18	10.218	7.2148	6.0277	5.3746	4.9560	4.6627	4.4448	4.2759	4.1410
19	10.073	7.0935	5.9161	5.2681	4.8526	4.5614	4.3448	4.1770	4.0428
20	9.9439	6.9865	5.8177	5.1743	4.7616	4.4721	4.2569	4.0900	3.9564
21	9.8295	6.8914	5.7304	5.0911	4.6808	4.3931	4.1789	4.0128	3.8799
22	9.7271	6.8064	5.6524	5.0168	4.6088	4.3225	4.1094	3.9440	3.8116
23	9.6348	6.7300	5.5823	4.9500	4.5441	4.2591	4.0469	3.8822	3.7502
24	9.5513	6.6610	5.5190	4.8898	4.4857	4.2019	3.9905	3.8264	3.6949
25	9.4753	6.5982	5.4615	4.8351	4.4327	4.1500	3.9394	3.7758	3.6447
26	9.4059	6.5409	5.4091	4.7852	4.3844	4.1027	3.8928	3.7297	3.5989
27	9.3423	6.4885	5.3611	4.7396	4.3402	4.0594	3.8501	3.6875	3.5571
28	9.2838	6.4403	5.3170	4.6977	4.2996	4.0197	3.8110	3.6487	3.5186
29	9.2297	6.3958	5.2764	4.6591	4.2622	3.9830	3.7749	3.6130	3.4832
30	9.1797	6.3547	5.2388	4.6233	4.2276	3.9492	3.7416	3.5801	3.4505
40	8.8278	6.0664	4.9759	4.3738	3.9860	3.7129	3.5088	3.3498	3.2220
60	8.4946	5.7950	4.7290	4.1399	3.7600	3.4918	3.2911	3.1344	3.0083
120	8.1790	5.5393	4.4973	3.9207	3.5482	3.2849	3.0874	2.9330	2.8083
∞	7.8794	5.2983	4.2794	3.7151	3.3499	3.0913	2.8968	2.7444	2.6210

TABLE VII   *F* Distribution Values (Concluded)

F Values When α = .005

ν_1 / ν_2	10	12	15	20	24	30	40	60	120	∞
1	24224	24426	24630	24836	24940	25044	25148	25253	25359	25465
2	199.40	199.42	199.43	199.45	199.46	199.47	199.47	199.48	199.49	199.51
3	43.686	43.387	43.085	42.778	42.622	42.466	42.308	42.149	41.989	41.829
4	20.967	20.705	20.438	20.167	20.030	19.892	19.752	19.611	19.468	19.325
5	13.618	13.384	13.146	12.903	12.780	12.656	12.530	12.402	12.274	12.144
6	10.250	10.034	9.8140	9.5888	9.4741	9.3583	9.2408	9.1219	9.0015	8.8793
7	8.3803	8.1764	7.9678	7.7540	7.6450	7.5345	7.4225	7.3088	7.1933	7.0760
8	7.2107	7.0149	6.8143	6.6082	6.5029	6.3961	6.2875	6.1772	6.0649	5.9505
9	6.4171	6.2274	6.0325	5.8318	5.7292	5.6248	5.5186	5.4104	5.3001	5.1875
10	5.8467	5.6613	5.4707	5.2740	5.1732	5.0705	4.9659	4.8592	4.7501	4.6385
11	5.4182	5.2363	5.0489	4.8552	4.7557	4.6543	4.5508	4.4450	4.3367	4.2256
12	5.0855	4.9063	4.7214	4.5299	4.4315	4.3309	4.2282	4.1229	4.0149	3.9039
13	4.8199	4.6429	4.4600	4.2703	4.1726	4.0727	3.9704	3.8655	3.7577	3.6465
14	4.6034	4.4281	4.2468	4.0585	3.9614	3.8619	3.7600	3.6553	3.5473	3.4359
15	4.4236	4.2498	4.0698	3.8826	3.7859	3.6867	3.5850	3.4803	3.3722	3.2602
16	4.2719	4.0994	3.9205	3.7342	3.6378	3.5388	3.4372	3.3324	3.2240	3.1115
17	4.1423	3.9709	3.7929	3.6073	3.5112	3.4124	3.3107	3.2058	3.0971	2.9839
18	4.0305	3.8599	3.6827	3.4977	3.4017	3.3030	3.2014	3.0962	2.9871	2.8732
19	3.9329	3.7631	3.5866	3.4020	3.3062	3.2075	3.1058	3.0004	2.8908	2.7762
20	3.8470	3.6779	3.5020	3.3178	3.2220	3.1234	3.0215	2.9159	2.8058	2.6904
21	3.7709	3.6024	3.4270	3.2431	3.1474	3.0488	2.9467	2.8408	2.7302	2.6140
22	3.7030	3.5350	3.3600	3.1764	3.0807	2.9821	2.8799	2.7736	2.6625	2.5455
23	3.6420	3.4745	3.2999	3.1165	3.0208	2.9221	2.8198	2.7132	2.6016	2.4837
24	3.5870	3.4199	3.2456	3.0624	2.9667	2.8679	2.7654	2.6585	2.5463	2.4276
25	3.5370	3.3704	3.1963	3.0133	2.9176	2.8187	2.7160	2.6088	2.4960	2.3765
26	3.4916	3.3252	3.1515	2.9685	2.8728	2.7738	2.6709	2.5633	2.4501	2.3297
27	3.4499	3.2839	3.1104	2.9275	2.8318	2.7327	2.6296	2.5217	2.4078	2.2867
28	3.4117	3.2460	3.0727	2.8899	2.7941	2.6949	2.5916	2.4834	2.3689	2.2469
29	3.3765	3.2111	3.0379	2.8551	2.7594	2.6601	2.5565	2.4479	2.3330	2.2102
30	3.3440	3.1787	3.0057	2.8230	2.7272	2.6278	2.5241	2.4151	2.2997	2.1760
40	3.1167	2.9531	2.7811	2.5984	2.5020	2.4015	2.2958	2.1838	2.0635	1.9318
60	2.9042	2.7419	2.5705	2.3872	2.2898	2.1874	2.0789	1.9622	1.8341	1.6885
120	2.7052	2.5439	2.3727	2.1881	2.0890	1.9839	1.8709	1.7469	1.6055	1.4311
∞	2.5188	2.3583	2.1868	1.9998	1.8983	1.7891	1.6691	1.5325	1.3637	1.0000

TABLE VIII χ^2 Distribution Values

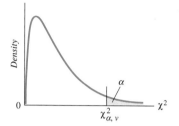

Degrees of Freedom ν	$\chi^2_{.100}$	$\chi^2_{.050}$	$\chi^2_{.025}$	$\chi^2_{.010}$	$\chi^2_{.005}$
1	2.71	3.84	5.02	6.63	7.88
2	4.61	5.99	7.38	9.21	10.60
3	6.25	7.81	9.35	11.34	12.84
4	7.78	9.49	11.14	13.28	14.86
5	9.24	11.07	12.83	15.09	16.75
6	10.64	12.59	14.45	16.81	18.55
7	12.02	14.07	16.01	18.48	20.28
8	13.36	15.51	17.53	20.09	21.96
9	14.68	16.92	19.02	21.67	23.59
10	15.99	18.31	20.48	23.21	25.19
11	17.28	19.68	21.92	24.72	26.76
12	18.55	21.03	23.34	26.22	28.30
13	19.81	22.36	24.74	27.69	29.82
14	21.06	23.68	26.12	29.14	31.32
15	22.31	25.00	27.49	30.58	32.80
16	23.54	26.30	28.85	32.00	34.27
17	24.77	27.59	30.19	33.41	35.72
18	25.99	28.87	31.53	34.81	37.16
19	27.20	30.14	32.85	36.19	38.58
20	28.41	31.41	34.17	37.57	40.00
21	29.62	32.67	35.48	38.93	41.40
22	30.81	33.92	36.78	40.29	42.80
23	32.01	35.17	38.08	41.64	44.18
24	33.20	36.42	39.36	42.98	45.56
25	34.38	37.65	40.65	44.31	46.93
26	35.56	38.89	41.92	45.64	48.29
27	36.74	40.11	43.19	46.96	49.64
28	37.92	41.34	44.46	48.28	50.99
29	39.09	42.56	45.72	49.59	52.34
30	40.26	43.77	46.98	50.89	53.67
40	51.81	55.76	59.34	63.69	66.77
50	63.17	67.50	71.42	76.15	79.49
60	74.40	79.08	83.30	88.38	91.95
70	85.53	90.53	95.02	100.43	104.22
80	96.58	101.88	106.63	112.33	116.32
90	107.60	113.14	118.14	124.12	128.30
100	118.50	124.34	129.56	135.81	140.17

Note: For example, if $\alpha = .05$ and $\nu = 20$, then
$\chi^2_{\alpha,\nu} = \chi^2_{.05, 20} = 31.41.$

Table VIII is abridged from Thompson, Catherine M.: "Table of Percentage Points of the χ^2 Distribution," *Biometrika,* Vol. 32 (1942), p. 187, by permission of *Biometrika* Trustees.

TABLE IX   Durbin-Watson Test Bounds

Level of Significance $\alpha = .05$

	Number of Independent Variables k									
	k = 1		k = 2		k = 3		k = 4		k = 5	
n	d_L	d_U	d_L	d_U	d_L	d_U	d_L	d_U	d_L	d_U
15	1.08	1.36	0.95	1.54	0.82	1.75	0.69	1.97	0.56	2.21
16	1.10	1.37	0.98	1.54	0.86	1.73	0.74	1.93	0.62	2.15
17	1.13	1.38	1.02	1.54	0.90	1.71	0.78	1.90	0.67	2.10
18	1.16	1.39	1.05	1.53	0.93	1.69	0.82	1.87	0.71	2.06
19	1.18	1.40	1.08	1.53	0.97	1.68	0.86	1.85	0.75	2.02
20	1.20	1.41	1.10	1.54	1.00	1.68	0.90	1.83	0.79	1.99
21	1.22	1.42	1.13	1.54	1.03	1.67	0.93	1.81	0.83	1.96
22	1.24	1.43	1.15	1.54	1.05	1.66	0.96	1.80	0.86	1.94
23	1.26	1.44	1.17	1.54	1.08	1.66	0.99	1.79	0.90	1.92
24	1.27	1.45	1.19	1.55	1.10	1.66	1.01	1.78	0.93	1.90
25	1.29	1.45	1.21	1.55	1.12	1.66	1.04	1.77	0.95	1.89
26	1.30	1.46	1.22	1.55	1.14	1.65	1.06	1.76	0.98	1.88
27	1.32	1.47	1.24	1.56	1.16	1.65	1.08	1.76	1.01	1.86
28	1.33	1.48	1.26	1.56	1.18	1.65	1.10	1.75	1.03	1.85
29	1.34	1.48	1.27	1.56	1.20	1.65	1.12	1.74	1.05	1.84
30	1.35	1.49	1.28	1.57	1.21	1.65	1.14	1.74	1.07	1.83
31	1.36	1.50	1.30	1.57	1.23	1.65	1.16	1.74	1.09	1.83
32	1.37	1.50	1.31	1.57	1.24	1.65	1.18	1.73	1.11	1.82
33	1.38	1.51	1.32	1.58	1.26	1.65	1.19	1.73	1.13	1.81
34	1.39	1.51	1.33	1.58	1.27	1.65	1.21	1.73	1.15	1.81
35	1.40	1.52	1.34	1.58	1.28	1.65	1.22	1.73	1.16	1.80
36	1.41	1.52	1.35	1.59	1.29	1.65	1.24	1.73	1.18	1.80
37	1.42	1.53	1.36	1.59	1.31	1.66	1.25	1.72	1.19	1.80
38	1.43	1.54	1.37	1.59	1.32	1.66	1.26	1.72	1.21	1.79
39	1.43	1.54	1.38	1.60	1.33	1.66	1.27	1.72	1.22	1.79
40	1.44	1.54	1.39	1.60	1.34	1.66	1.29	1.72	1.23	1.79
45	1.48	1.57	1.43	1.62	1.38	1.67	1.34	1.72	1.29	1.78
50	1.50	1.59	1.46	1.63	1.42	1.67	1.38	1.72	1.34	1.77
55	1.53	1.60	1.49	1.64	1.45	1.68	1.41	1.72	1.38	1.77
60	1.55	1.62	1.51	1.65	1.48	1.69	1.44	1.73	1.41	1.77
65	1.57	1.63	1.54	1.66	1.50	1.70	1.47	1.73	1.44	1.77
70	1.58	1.64	1.55	1.67	1.52	1.70	1.49	1.74	1.46	1.77
75	1.60	1.65	1.57	1.68	1.54	1.71	1.51	1.74	1.49	1.77
80	1.61	1.66	1.59	1.69	1.56	1.72	1.53	1.74	1.51	1.77
85	1.62	1.67	1.60	1.70	1.57	1.72	1.55	1.75	1.52	1.77
90	1.63	1.68	1.61	1.70	1.59	1.73	1.57	1.75	1.54	1.78
95	1.64	1.69	1.62	1.71	1.60	1.73	1.58	1.75	1.56	1.78
100	1.65	1.69	1.63	1.72	1.61	1.74	1.59	1.76	1.57	1.78

Note: For example, if $n = 20$, $\alpha = 0.05$, and $k = 2$ independent variables, $d_L = 1.10$ and $d_U = 1.54$.

(continues)

TABLE IX   Durbin-Watson Test Bounds (Concluded)

	Level of Significance $\alpha = .01$									
	Number of Independent Variables k									
	k = 1		k = 2		k = 3		k = 4		k = 5	
n	d_L	d_U	d_L	d_U	d_L	d_U	d_L	d_U	d_L	d_U
15	0.81	1.07	0.70	1.25	0.59	1.46	0.49	1.70	0.39	1.96
16	0.84	1.09	0.74	1.25	0.63	1.44	0.53	1.66	0.44	1.90
17	0.87	1.10	0.77	1.25	0.67	1.43	0.57	1.63	0.48	1.85
18	0.90	1.12	0.80	1.26	0.71	1.42	0.61	1.60	0.52	1.80
19	0.93	1.13	0.83	1.26	0.74	1.41	0.65	1.58	0.56	1.77
20	0.95	1.15	0.86	1.27	0.77	1.41	0.68	1.57	0.60	1.74
21	0.97	1.16	0.89	1.27	0.80	1.41	0.72	1.55	0.63	1.71
22	1.00	1.17	0.91	1.28	0.83	1.40	0.75	1.54	0.66	1.69
23	1.02	1.19	0.94	1.29	0.86	1.40	0.77	1.53	0.70	1.67
24	1.04	1.20	0.96	1.30	0.88	1.41	0.80	1.53	0.72	1.66
25	1.05	1.21	0.98	1.30	0.90	1.41	0.83	1.52	0.75	1.65
26	1.07	1.22	1.00	1.31	0.93	1.41	0.85	1.52	0.78	1.64
27	1.09	1.23	1.02	1.32	0.95	1.41	0.88	1.51	0.81	1.63
28	1.10	1.24	1.04	1.32	0.97	1.41	0.90	1.51	0.83	1.62
29	1.12	1.25	1.05	1.33	0.99	1.42	0.92	1.51	0.85	1.61
30	1.13	1.26	1.07	1.34	1.01	1.42	0.94	1.51	0.88	1.61
31	1.15	1.27	1.08	1.34	1.02	1.42	0.96	1.51	0.90	1.60
32	1.16	1.28	1.10	1.35	1.04	1.43	0.98	1.51	0.92	1.60
33	1.17	1.29	1.11	1.36	1.05	1.43	1.00	1.51	0.94	1.59
34	1.18	1.30	1.13	1.36	1.07	1.43	1.01	1.51	0.95	1.59
35	1.19	1.31	1.14	1.37	1.08	1.44	1.03	1.51	0.97	1.59
36	1.21	1.32	1.15	1.38	1.10	1.44	1.04	1.51	0.99	1.59
37	1.22	1.32	1.16	1.38	1.11	1.45	1.06	1.51	1.00	1.59
38	1.23	1.33	1.18	1.39	1.12	1.45	1.07	1.52	1.02	1.58
39	1.24	1.34	1.19	1.39	1.14	1.45	1.09	1.52	1.03	1.58
40	1.25	1.34	1.20	1.40	1.15	1.46	1.10	1.52	1.05	1.58
45	1.29	1.38	1.24	1.42	1.20	1.48	1.16	1.53	1.11	1.58
50	1.32	1.40	1.28	1.45	1.24	1.49	1.20	1.54	1.16	1.59
55	1.36	1.43	1.32	1.47	1.28	1.51	1.25	1.55	1.21	1.59
60	1.38	1.45	1.35	1.48	1.32	1.52	1.28	1.56	1.25	1.60
65	1.41	1.47	1.38	1.50	1.35	1.53	1.31	1.57	1.28	1.61
70	1.43	1.49	1.40	1.52	1.37	1.55	1.34	1.58	1.31	1.61
75	1.45	1.50	1.42	1.53	1.39	1.56	1.37	1.59	1.34	1.62
80	1.47	1.52	1.44	1.54	1.42	1.57	1.39	1.60	1.36	1.62
85	1.48	1.53	1.46	1.55	1.43	1.58	1.41	1.60	1.39	1.63
90	1.50	1.54	1.47	1.56	1.45	1.59	1.43	1.61	1.41	1.64
95	1.51	1.55	1.49	1.57	1.47	1.60	1.45	1.62	1.42	1.64
100	1.52	1.56	1.50	1.58	1.48	1.60	1.46	1.63	1.44	1.65

MINITAB Brief Reference

Starting MINITAB

To start MINITAB, at the DOS prompt, enter: MINITAB. After you clear the opening screen, you'll see the Session window.

Working with Menus

To open a menu, press ⟨Alt⟩ + the highlighted letter in the menu name.

To cancel a menu, press ⟨Esc⟩.

To choose a menu command, press the highlighted letter.

Stopping MINITAB

To stop MINITAB, choose File▶Exit:

Press ⟨Alt⟩ + ⟨F⟩ to open the File menu.

Press ⟨X⟩ to choose Exit.

Working with Dialog Boxes

To move forward, press ⟨Tab⟩.

To move backward, press ⟨Shift⟩ + ⟨Tab⟩.

To move among items within a group, press arrow keys.

To select an item in a dialog box, press ⟨Alt⟩ + the highlighted letter.

Press ⟨Tab⟩ to enter box, then type desired value.

Press ⟨↓⟩ to cycle through available choices.

To cancel, choose the Cancel button, or press ⟨Esc⟩.

To choose OK: Press ⟨Alt⟩ + ⟨O⟩, or move to the OK button and press ⟨Enter⟩.

To get Help, choose the ⟨?⟩ button, or press ⟨F1⟩.

To select a variable when you can select only one:

Make sure your cursor is in appropriate text box

Press ⟨F2⟩ to make variable list box active

Use arrow keys to highlight variable of interest

Press ⟨F2⟩ to select

To select a variable when you can select more than one:

Make sure cursor is in appropriate text box

Press ⟨F2⟩ to make variable list box active

Press ⟨Spacebar⟩ to highlight variables of interest

Press ⟨F2⟩ to select

Working with the File Dialog Box

To specify files or directories in the File dialog box:

Press ⟨Tab⟩ to move to the "Files" list box

Press an arrow key to highlight the desired file or directory name

Press ⟨Enter⟩ to select the highlighted file or directory

To type a file name in the "File name" text box:

Press ⟨Tab⟩ or ⟨Shift⟩ + ⟨Tab⟩ to move to the "File name" text box

Type the file name

Choose OK to process the file

Working with the Data Screen

To enter the Data screen	Choose Edit▶Data Screen
To open the menu	Press ⟨F10⟩
To close the menu	Press ⟨Esc⟩
To get help	Press ⟨F1⟩
To switch to the Session window	Choose Go to Minitab Session from the menu
Top of worksheet	⟨Ctrl⟩ + ⟨Home⟩
Bottom of worksheet	⟨Ctrl⟩ + ⟨End⟩
One screen to the right	⟨Ctrl⟩ + ⟨→⟩
One screen to the left	⟨Ctrl⟩ + ⟨←⟩
Delete cell	⟨F8⟩
Delete row	⟨Shift⟩ + ⟨F8⟩
Insert cell	⟨F7⟩
Insert row	⟨Shift⟩ + ⟨F7⟩
Edit a value within a cell	⟨F2⟩

This appendix presents MINITAB menu operations and commands in brief. More detail is given by using the HELP command in the program or the *Minitab Reference Manual*. Note that in this appendix angle brackets designate keys on the computer keyboard.

Notation

K   denotes a constant, which can be either a number such as
     8.3, or a stored constant such as K14.
C   denotes a column, which must be typed with a C directly
     in front, such as C12. Columns may be named.
E   denotes either a constant or a column.
'text' denotes text (either a filename or a column name) that is
     to be enclosed in single quote marks (apostrophes) when
     the command is entered.
[]   denotes an optional argument.

General Information

HELP	explains MINITAB commands
INFORMATION	gives status of worksheet
STOP	ends the current session

Input and Output of Data

NAME	for C is 'NAME', for C is 'NAME',..., for C is 'NAME'	
READ	the following data	into C,...,C
READ	data from 'FILENAME'	into C,...,C
SET	the following data	into C
SET	data from 'FILENAME'	into C
INSERT	the following data	between rows K and K of C,...,C
INSERT	data from 'FILENAME'	between rows K and K of C,...,C
INSERT	the following data	at the end of C,...,C
INSERT	data from 'FILENAME'	at the end of C,...,C
RETRIEVE	the MINITAB saved worksheet [in 'FILENAME']	
END	of data (optional)	
PRINT	the data in C,...,C	
PRINT	the data in K,...,K	
WRITE	[to 'FILENAME'] the data in C,...,C	
SAVE	[in 'FILENAME'] a copy of the worksheet	

Editing and Manipulating Data

COPY	C,...,C into C,...,C
USE	rows K,...,K
USE	rows where C = K,...,K
OMIT	rows K,...,K
OMIT	rows where C = K,...,K
DELETE	rows K,...,K of C,...,C

Arithmetic

LET = expression
    Expressions may use the arithmetic operators + − * / and
    ** (exponentiation) and any of the following:

    ABSOLUTE,   SQRT,   LOGTEN,   LOGE,   EXPO,
    ANTILOG, ROUND, SIN, COS, TAN, ASIN, ACOS,
    ATAN, SIGNS, COUNT, N, NMISS, SUM, MEAN,
    STDEV, MEDIAN, MIN, MAX, SSQ

    You can use subscripts to access individual numbers.
    Examples:

        LET C7 = 2*'INCOME' + 500
        LET C2 = SQRT(C1 − MIN(C1))
        LET C3(5) = 4.5

Plotting Data

High resolution commands begin with the letter G.
HISTOGRAM   of C,...,C
    INCREMENT = K
    START at K [end at K]
    BY C
    SAME scales for all columns
PLOT   y in C vs x in C
    Same subcommands as HISTOGRAM
MPLOT   C vs C and C vs C and ...C vs C
LPLOT   C vs C using tags in C
TSPLOT   of data in C
TPLOT   y in C vs x in C vs z in C
WIDTH   of all plots that follow is K spaces
HEIGHT   of all plots that follow is K lines

Basic Statistics

DESCRIBE C,...,C
TINTERVAL [with K% confidence] for data in C,...,C
TTEST [of mu = K] on data in C,...,C
    ALTERNATIVE = K
        Subcommand   ALTERNATIVE = −1   specifies   a
        "less than (<)" alternative hypothesis; subcommand
        ALTERNATIVE = +1 is used for a "greater than
        (>)" alternative hypothesis. The semicolon and alter-
        native subcommand are omitted for a "not equal to
        (≠)" alternative hypothesis.
TWOSAMPLE test and c.i. [K% confidence] on samples in C, C
    ALTERNATIVE = K
    POOLED procedure
TWOT test and c.i. [K% confidence] data in C, groups in C
    ALTERNATIVE = K
    POOLED procedure
CORRELATION between C,...,C [put into M]

Regression

REGRESS C on K predictors C,...,C [store standardized
    residuals in C [fits in C]]
    PREDICT for E,...,E
    NOCONSTANT in equation
    WEIGHTS are in C

MSE	put into K
COEFFICIENTS	put into C
XPXINV	put into M
HI	put into C (leverage)
RESIDUALS	put into C (observed − fit)
TRESID	put into C (studentized, deleted)
COOKD	put into C (Cook's distance)
DFITS	put into C
VIF	(variance inflation factors)
PURE	(pure error lack-of-fit test)
XLOF	(experimental lack-of-fit test)

STEPWISE regression of C on the predictors C,...,C

FENTER	= K	(default is 4)
FREMOVE	= K	(default is 4)
FORCE	C,...,C	
ENTER	C,...,C	
REMOVE	C,...,C	

BEST   K alternative predictors (default is 0)
STEPS   = K (default depends on output width)
BRIEF   output [using print code = K] from REGRESS

Analysis of Variance

AOVONEWAY   analysis of variance for samples in C, . . . , C
ONEWAY   analysis of variance, data in C, subscripts in C
             [store residuals in C [fits in C]]
TWOWAY   analysis of variance, data in C, subscripts in
             C, C [store residuals in C [fits in C]]
INDICATOR   variables for subscripts in C, put into
             C, . . . , C

Tables

CHISQUARE test on table stored in C, . . . , C
TABLE the data classified by C, . . . , C
         CHISQUARE analysis [output code = K]

Exploratory Data Analysis

STEM-AND-LEAF display of C, . . . , C
     TRIM of "outliers"
BOXPLOTS for C [levels in C]

Distributions

PDF for values in E [put into E]
         BINOMIAL   n = K p = K
         POISSON   mu = K
         NORMAL   mu = K sigma = K
         UNIFORM   a = K b = K
         T   v = K
         F   u = K v = K
         CHISQUARE   v = k
         EXPONENTIAL   b = K
CDF for values in E [put into E]
         same subcommands as PDF
INVCDF for values in E [put into E]
         same subcommands as PDF

Time Series

ACF   [up to K lags] for series in C [put into C]
PACF   [up to K lags] for series in C [put into C]
CCF   [up to K lags] between series in C and C
DIFFERENCES   [of lag K] for data in C, put into C
LAG   [by K] data in C, put into C
ARIMA   p = K, d = K, q = K, data in C [put residuals
                 in C [predicteds in C [coefficients in C]]]
ARIMA   p = K, d = K, q = K, P = K, D = K,
                 Q = K, S = K, data in C [put residuals in C
                 [put predicteds in C [put coefficients in C]]]
     CONSTANT   term in model
     NOCONSTANT term in model
     STARTING   values are in C
     FORECAST   [forecast origin = K] up to K leads ahead
                     [store forecasts in C [confidence limits in
                     C, C]]

Nonparametrics

RUNS test [above and below K] for C
STEST sign test [median = K] for C . . . C
     ALTERNATIVE = K
SINT   sign c.i. [K% confidence] for C . . . C
WTEST Wilcoxon one-sample rank test [median = K] for
         C . . . C
     ALTERNATIVE = K
WINT Wilcoxon c.i. [K% confidence] for C . . . C
MANN-WHITNEY test and c.i. [alt. = K] [K% confidence]
                     first sample in C, second sample in C
KRUSKAL-WALLIS test for data in C, subscripts into C
WALSH averages for C, put into C [indices in C and C]
WDIFF C and C, put into C [indices into C and C]
WSLOPE y in C, x in C, put slopes in C [indices into C, C]

Sorting

SORT C [carry along C . . . C] put into C [and C . . . C]
     BY   C . . . C
     DESCENDING C . . . C
RANK the values in C, put ranks into C

Statistical Process Control

High resolution commands begin with the letter G.
XBARCHART   for C . . . C, subgroups are in E
GXBARCHART for C . . . C, subgroups are in E
     MU   = K
     SIGMA   = K
     RSPAN   = K
     TEST   K · · · K
     SUBGROUP   size is E
     Subcommands used in all control charts
RCHART   for C . . . C, subgroups are in E
GRCHART for C . . . C, subgroups are in E
     SIGMA   = K
     SUBGROUP size is E
     Subcommands used in all control charts
PCHART number of nonconformities are in C . . . C, sample
             size = E
GPCHART number of nonconformities are in C . . . C, sample
             size = E
     P   = K
     TEST   K · · · K
     SUBGROUP size is E
     Subcommands used in all control charts

Miscellaneous

NOTE   comments may be put here
ERASE   E, . . . , E
RESTART   begin fresh MINITAB session
NEWPAGE   start next output on a new page
OW   output width = K spaces
OH   output height = K lines
IW   input width = K spaces
PAPER   direct output to a file to be printed
OUTFILE   'filename' (put all output in this file)
     OW = K
     OH = K

NOTERM
NOPAPER   output to terminal only

The symbol # anywhere on a line tells MINITAB to ignore everything after that on a line. To continue a command onto another line, end the first line with the symbol &.

Worksheet and Commands

MINITAB consists of a worksheet where data are stored and about 150 commands that operate on these data. The worksheet may contain columns of data denoted by C1, C2, C3, ..., stored constants denoted by K1, K2, K3, ..., and matrices denoted by M1, M2, M3, ...

A column may be given a name with the command NAME (Section 1). A name may be up to 8 characters long. Any characters except apostrophes may be used. Names may be used on any command in place of column numbers. When a name is used, the name must be enclosed in apostrophes (single quotes). For example, PLOT 'INCOME' vs 'AGE'.

Each command starts with a command word, which is usually followed by a list of arguments. An argument is a number, a column name or number, a stored constant, a matrix, or a file name. Only the command word and arguments are needed. All other text is solely for the readers' information.

Subcommands

Some MINITAB commands have subcommands, which are used for special options or to convey additional information. To use a subcommand, put a semicolon at the end of the main command line. Then type the subcommands. Start each subcommand on a new line; then end each subcommand line with a semicolon. When you are done, end the last subcommand with a period. If you forget the period, simply type the period on the next line.

If you discover an error on a subcommand, type ABORT as the next subcommand. This cancels the whole command, and you can then start over again with the main command.

Note: Subcommands are indented below commands in this brief reference.

Portions of this appendix are excerpts from the MINITAB *Quick Reference Card* and from MINITAB Release 8 *Reference Manual,* copyright 1991 by Minitab, Inc., reprinted with permission. All rights reserved.

Business
MYSTAT® Manual

CONTENTS

BUSINESS MYSTAT DOCUMENTATION

This is a real statistics program—it is not just a demonstration.
You can use Business MYSTAT to enter, transform, and save data,
and to perform a wide range of statistical evaluations.

Business MYSTAT is a subset of SYSTAT, our premier statistics package. We've geared Business MYSTAT especially for
teaching business statistics, with special statistical process control, forecasting and time series routines all in a single, easy-to-
use package.

Business MYSTAT provides descriptive statistics, cross-tabulation, Pearson and Spearman correlation coefficients, multiple regression, time series forecasting, and much more.

Business MYSTAT is available in Macintosh, IBM-PC/com-
patible, and VAX/VMS versions. Copies are available at a nominal cost.

For more information about SYSTAT, FASTAT, and our other
top-rated professional statistics and graphics packages, please call
or write.

Installation

Business MYSTAT requires 512K of RAM and a floppy or hard
disk drive. It can handle up to 50 variables and up to 32,000 cases.

Your Business MYSTAT disk contains three MYSTAT files:
MYSTAT.EXE (the program), MYSTAT.HLP (a file with information for on-line help), and DEMO.CMD (a demonstration that
creates a data file and demonstrates some of Business MYSTAT's
features) and QCTOOLS files.

Hard disk

Set up the CONFIG.SYS file   The CONFIG.SYS file in your
root directory must have a line that says FILES=20. You can modify your existing CONFIG.SYS file, or create one if you don't
have one, by typing the following lines from the DOS prompt (>).

```
COPY CONFIG.SYS + CON: CONFIG.SYS
FILES=20
```

To finish, press the F6 key and then press Enter or Return.

Set up the AUTOEXEC.BAT file   You may find it convenient to
install MYSTAT in its own directory and add that directory to the
PATH statement in your AUTOEXEC.BAT file. This file should
be located in the root directory (\) of your boot disk (the hard
disk). *If you do not already have an AUTOEXEC.BAT file, or if your
AUTOEXEC.BAT file does not have a PATH statement already,
please do the following:*

```
COPY AUTOEXEC.BAT + CON: AUTOEXEC.BAT
PATH=C:\;C:\MYSTAT
[F6, ENTER]
```

*If you already have an AUTOEXEC.BAT file that has a PATH state-
ment,* use a text editor to add the MYSTAT directory to the existing PATH statement.

Reboot your machine

Copy the files on the MYSTAT disk into a \SYSTAT directory
Now, make a \SYSTAT directory, insert the MYSTAT disk in

drive A, and copy the Business MYSTAT files into the directory.
(You must have the help file in the same directory as MY-
STAT.EXE. If you put MYSTAT.EXE in a directory other than
\SYSTAT, the help file MYSTATB.HLP must be either in that di-
rectory or the \SYSTAT directory.)

```
>MD \SYSTAT
>CD \SYSTAT
>COPY A:*.*
```

You are now ready to begin using Business MYSTAT. From now
on, all you need to do to get ready to use Business MYSTAT is
boot and move (CD) into the \SYSTAT directory. Save your MY-
STAT master disk as a backup copy.

Floppy disk

Boot your machine   Insert a "boot disk" into drive A. Close
the door of the disk drive and turn on the machine.

Set up a CONFIG.SYS file on your boot disk   The boot disk
must contain a file named CONFIG.SYS with a line FILES=20.
You can modify your existing CONFIG.SYS file, or create one if
you don't have one, by typing the following lines from the DOS
prompt (>).

```
COPY CONFIG.SYS + CON: CONFIG.SYS
FILES=20
```

To finish, press the F6 key and then press Enter or Return.

Set up an AUTOEXEC.BAT file on your boot disk   You must
be sure that you have a file called AUTOEXEC.BAT containing
the line PATH=A:\;B:\ on your boot disk. If you do not, type the
following lines from the DOS prompt (>).

```
COPY CON AUTOEXEC.BAT + CON: AUTOEXEC.BAT
PATH=A:\;B:\
[F6, Enter]
```

Reboot your machine

Make a copy of the Business MYSTAT disk   When you get a
DOS prompt (>), remove the boot disk. Put the Business MYSTAT
disk in drive A and a blank, formatted disk in drive B and type the
COPY command at the prompt:

```
>COPY A:*.* B:
```

Remove the master disk from drive A and store it. If anything happens to your working copy, use the master to make a new copy.

Use Business MYSTAT   Now, switch your working copy into
drive A. If necessary, make drive A the "logged" drive (the drive
your machine reads from) by issuing the command A: at the DOS
prompt (>).

Business MYSTAT reads and writes its temporary work files
to the currently logged drive. Since there is limited room on the
Business MYSTAT disk, you should read and write all your data,
output, and command files from a data disk in drive B.

From now on, all you need to do to use Business MYSTAT is
boot, insert your working copy, and log the A drive.

Getting Started

To start the program, type MYSTAT and press Enter.

```
>MYSTAT
```

When you see the MYSTAT logo, press Enter. You'll see a command menu listing the commands you can use in MYSTAT.

Command menu

The command menu shows a list of commands that are available for Business MYSTAT with QCTOOLS. The menu divides the commands into six groups: information, file handling, general, graphics, statistics, and time series analysis.

Business MYSTAT --- A Personal Version of SYSTAT					
DEMO	EDIT	MENU	PLOT	STATS	LOG
HELP		NAMES	BOX	TABULATE	MEAN
QCTOOLS	USE	LIST	HISTOGRAM	TTEST	SQUARE
	SAVE	FORMAT	STEM	CORRELATE	TREND
	PUT	NOTE			PCNTCHNG
	SUBMIT				DIFFRNCE
					INDEX
QUIT	OUTPUT	SORT	CHARSET	MODEL	SMOOTH
		RANK		CATEGORY	ADJSEAS
		WEIGHT		ANOVA	EXP
				ESTIMATE	TPLOT
					ACF,PACF
					CLEAR

As you become more experienced with MYSTAT, you might want to turn the menu off with the MENU command.

```
>MENU
```

Data Editor

MYSTAT has a built-in Data Editor with its own set of commands. To enter the Editor, use the EDIT command.

```
>EDIT
```

Inside the Editor, you can enter, view, edit, and transform data. When you are done with the Editor, type QUIT to get out of the Editor and back to MYSTAT, where you can do statistical and graphical analyses. To quit MYSTAT itself, type QUIT again.

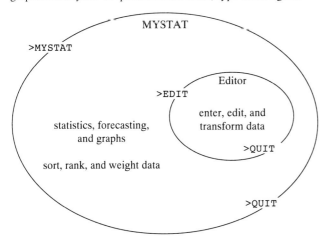

The Data Editor is an independent program inside MYSTAT and has its own commands. Five commands—USE, SAVE, HELP, QUIT, and FORMAT—can be used both in *and* out of the Editor.

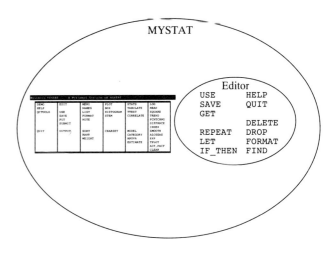

Demo

To see an on-line demonstration of MYSTAT, use the DEMO command. (Remember to press Enter after you type the command.)

```
>DEMO
```

When the demonstration is finished, Business MYSTAT returns you to the command menu. After you've seen the demo, you might want to remove the DEMO.CMD file and the CITIES.SYS file that it creates to save disk space.

Help

The HELP command provides instructions for any command—inside *or* outside the Editor. HELP lists and describes all the commands.

```
>HELP
```

You can get help for any specific command. For example:

```
>HELP EDIT
EDIT starts the MYSTAT full screen editor.

EDIT [filename]

EDIT (edit a new file)
EDIT CITIES (edit CITIES.SYS)

For further information, type EDIT [Enter],
[ESC], and then type HELP [Enter] inside the
data editor.
```

The second line shows a summary of the command. You see that any EDIT command must begin with the command word EDIT. The brackets indicate that specifying a file is optional. Anything in lowercase, like "filename," is just a placeholder—you should type a real file name (or a real variable name, or whatever).

Customizing DEMO and HELP   All help information is stored in the text file MYSTATB.HLP. You can use a text editor to customize your help information. Teachers can design special demonstrations by editing the file DEMO.CMD.

Data Editor

The Business MYSTAT Data Editor lets you enter and edit data, view data, and transform variables. First, we enter data; later, we show you how to use commands.

To use the Editor to create a new file, type EDIT and press Enter. If you already have a MYSTAT data file that you want to edit, specify a filename with the EDIT command.

```
>EDIT [<filename>]
```

If you do not specify a filename, you get an empty editor like the one above. MYSTAT stores data in a rectangular worksheet. *Variables* fill vertical columns and *cases* are in horizontal rows.

Entering data

To create a file, we first enter variable names in the top row. Variable names *must* be surrounded by single or double quotation marks, must begin with a letter, and can be no longer than 8 characters.

Character variables (those whose values are words and letters) must have names ending with a dollar sign ($). The quotation marks and dollar sign do not count toward the eight character limit.

Numeric variables (those whose values are numbers) can be named with subscripts; e.g., ITEM(3). Subscripts allow you to specify a range of variables for analyses. For example, STATS ITEM(1-3) does descriptive statistics on the first three ITEM*(i)* variables.

The cursor is already positioned in the first cell in the top row of the worksheet. Type 'CITY$' or "CITY$" and then press Enter.

Business MYSTAT uses variable names in upper-case whether you enter them in lower- or upper-case.

The cursor automatically moves to the second column. You are now ready to name the rest of the variables, pressing Enter to store each name in the worksheet.

```
'STATE$'
"POP"
'RAINFALL'
```

Now you can enter values. Move the cursor to the first blank cell under CITY$ by pressing Home. (On most machines, Home is the 7 key on the numeric keypad. If pressing the 7 key types a 7 rather than moving the cursor, press the NumLock key and try again.)

When the cell under CITY$ is selected, enter the first data value:

```
'New York'
```

When you press Enter, Business MYSTAT accepts the value and moves the cursor to the right. You can also use the cursor arrow keys to move the cursor.

Character values can be no longer than twelve characters. Like variable names, character values must be surrounded by single or double quotation marks. Unlike variable names, character values are case sensitive—upper-case is not the same as lower-case. For example, 'TREE' is not the same as 'tree' or 'Tree'. Enter a blank space surrounded by quotation marks for missing character values. To use single or double quotation marks as part of a value, surround the whole value with the opposite marks.

Numeric values can be up to 10^{35} in absolute magnitude. Scientific notation is used for long numbers; e.g., .000000000015 is equivalent to 1.5E-11. Enter a decimal (.) for missing numeric values.

To enter the first few cases: type a value, press Enter, and type the next value. The cursor automatically moves to the beginning of the next case when a row is filled. For example,

"New York"	"NY"	7164742	57.03
"Los Angeles"	"CA"	3096721	7.81
"Chicago"	"IL"	2992472	34

and so forth.

Case	CITY$	STATE$	POP	RAINFALL
1	New York	NY	7164742.000	57.030
2	Los Angeles	CA	3096721.000	7.810
3	Chicago	IL	2992472.000	34.000
4	Dallas	TX	974234.000	33.890
5	Phoenix	AZ	853266.000	14.910
6	Miami	FL	346865.000	60.020
7	Washington	DC	638432.000	37.730
8	Kansas City	MO	448159.000	38.770

Moving around

Use the cursor keys on the numeric keypad to move around in the Editor.

Esc		toggle between Editor and command prompt($>$)
Home	**7**	move to first cell in worksheet
$\uparrow$	**8**	move upward one cell
PgUp	**9**	scroll screen up
$\leftarrow$	**4**	move left one cell
$\rightarrow$	**6**	move right one cell
End	**1**	move to last case in worksheet
$\downarrow$	**2**	move down one cell
PgDn	**3**	scroll screen down

If these keys type numbers rather than move the cursor, press the NumLock key which toggles the keypad back and forth between typing numbers and performing the special functions. (If your computer does not have a NumLock key or something similar, consult your user's manual.)

Editing data

To change a value or variable name move to the cell you want, type the new value, and press Enter. Remember to enclose character values and variable names in quotation marks.

Data Editor Commands

When you have entered your data, press the Esc key to move the cursor to the prompt ($>$) below the worksheet. You can enter editor commands at this prompt. Commands can be typed in upper- or lower-case. Items in $<$angle brackets$>$ are placeholders; for instance, you should type a specific filename in place of $<$filename$>$.

DELETE and DROP

DELETE lets you remove an entire case (row) from the dataset in the Editor. You can specify a range or list of cases to be deleted. The following are valid DELETE commands.

```
DELETE 3            Deletes third case from dataset.
DELETE 3-10         Deletes cases 3 through 10.
DELETE 3, 5-8, 10   Deletes cases 3, 5, 6, 7, 8, and 10.
```

DROP removes variables from the dataset in the Editor. You can specify several variables or a range of subscripted variables.

```
DROP RAINFALL        Drops RAINFALL from dataset.
DROP X(1-3)          Drops subscripted variables X(1-3).
DROP X(1-3), GROUP$  Drops X(1-3) and GROUP$.
```

Saving files

The SAVE command saves the data in the Editor to a Business MYSTAT data file. You must save data in a data file before you can analyze them with statistical and graphic commands.

```
SAVE <filename>       Saves data in a MYSTAT file.
  /DOUBLE | SINGLE    Choose single or double precision.
```

Business MYSTAT filenames can be up to 8 characters long and must begin with a letter. Business MYSTAT adds a ".SYS" extension that labels the file as a Business MYSTAT data file. To specify a path name for a file, enclose the entire file name, including the file extension, in single or double quotation marks.

Business MYSTAT stores data in double-precision by default. You can choose single precision if you prefer: add /SINGLE to the end of the command. Always type a slash before command options.

```
SAVE CITIES/SINGLE          Saves CITIES.SYS in single
                            precision.
```

Single precision requires approximately half as much disk space as double precision and is accurate to about 9 decimal places. The storage option (single or double precision) does not affect computations, which always use double precision arithmetic (accurate to about 15 places).

```
SAVE CITIES             Creates data file CITIES.SYS.
SAVE b:new              Creates file NEW.SYS on
                        B drive.
SAVE 'C:\DATA\FIL.SYS'  Creates FIL.SYS in \DATA
                        directory of C.
```

You can save data in text files for exporting to other programs with the PUT command. PUT is not an Editor command, though. First QUIT the Editor, USE the data file, and PUT the data to a text file.

Reading files

USE reads a MYSTAT data file into the Editor.

```
USE [<filename>]      Reads data from MYSTAT data file.
```

Creating new data files

Use the NEW command to clear the worksheet and start editing a new data file.

Importing data from other programs

You can import data stored in a text (ASCII) file with the GET command. ASCII files contain only text and numbers—they have no special characters or formatting commands.

Save your data in an ASCII file according to the instructions given by any program's manual. The ASCII file must have a ".DAT" extension, data values must be separated by blanks or commas, and each case must begin on a new line.

Then, start Business MYSTAT. Use EDIT to get an empty worksheet, and enter variable names in the worksheet for each variable in the ASCII file. Next, toggle to the editor command line with the ESC key, and use GET to read the text file. Finally, SAVE the data.

```
GET [<filename>]      Reads data from ASCII text file.
```

Finding a case

FIND searches through the Editor starting from the current cursor position, and moves the cursor to the first value that meets the condition you specify. Try this:

```
>FIND POP>1000000
```

This moves the cursor to case 4. After Business MYSTAT finds a value, use the FIND command without an argument (that is, "FIND" is the entire command) to find the next case meeting the same condition. All functions, relations, and operators listed below are available.

Some valid FIND commands:

```
>FIND AGE>45 AND SEX$='MALE'
>FIND INCOME<10000 AND STATE$='NY'
>FIND (TEST1+TEST2+TEST3)>90
```

Decimal places in the Editor

The FORMAT command specifies the number (0–9) of decimal places to be shown in the Editor. The default is 3. Numbers are stored the way you enter them regardless of the FORMAT setting; FORMAT affects the Editor display only.

FORMAT=<#>	Sets number of decimal places to <#>.
/UNDERFLOW	Displays tiny numbers in scientific notation.

For example, to set a two-place display with scientific notation for tiny numbers use the command:

```
>FORMAT=2/UNDERFLOW
```

Transforming variables

Use LET and IF...THEN to transform variables or create new ones.

LET <var>=<exprn>	Transforms <var> according to <exprn>.
IF <exprn> THEN LET <var>=<exprn>	Transforms <var> conditionally according to <exprn>.

For example, we can use LET to create LOGPOP from POP.

```
>LET LOGPOP=LOG(POP)
```

```
MYSTAT Editor
Case   CITY$      STATE$      POP         RAINFALL    LOGPOP
  1    New York     NY     7164742.000     57.030     15.785
  2    Los Angeles  CA     3096721.000      7.810     14.946
  3    Chicago      IL     2992472.000     34.000     14.912
  4    Dallas       TX      974234.000     33.890     13.789
  5    Phoenix      AZ      853266.000     14.910     13.657
  6    Miami        FL      346865.000     60.020     12.757
  7    Washington   DC      638432.000     37.730     13.367
  8    Kansas City  MO      448159.000     38.770     13.013
  9
 10
 11
 12
 13
 14
 15
```

LET labels the last column of the worksheet LOGPOP and sets the values to the natural logs of the POP values. (If LOGPOP had already existed, its values would have been replaced.)

Use IF...THEN for *conditional* transformations. For example:

```
>IF POP>1000000 THEN LET SIZE$='BIG'
```

creates a new character variable, SIZE$, and assigns the value BIG for every city that has population greater than one million.

For both LET and IF...THEN, character values must be enclosed in quotation marks and are case sensitive (i.e., "MALE" is not the same as "male"). Use a period to indicate missing values.

Some valid LET and IF-THEN commands:

```
>LET ALPHA$='abcdef'
>LET LOGIT1=1/(1+EXP(A+B*X)
>LET TRENDY=INCOME>40000 AND CAR$='BMW'
>IF SEX$='Male' THEN LET GROUP=1
>IF group>2 THEN LET NEWGROUP=2
>IF A=-9 AND B<10 OR B>20 THEN LET
C=LOG(D)*SQR(E)
```

Functions, relations, and operators for LET and IF...THEN

+	addition	CASE	current case number
−	subtraction	INT	integer truncation
*	multiplication	URN	uniform random number
/	division	ZRN	normal random number
^	exponentiation		
<	less than	AND	logical and
<=	less than or equal to	OR	logical or
=	equal to	SQR	square root
<>	not equal to	LOG	natural log
>=	greater than or equal	EXP	exponential function
>	greater than	ABS	absolute value

Logical expressions   Logical expressions evaluate to one if true and to zero if false. For example, for LET CHILD=AGE<12, a variable CHILD would be filled with ones for those cases where AGE is less than 12 and zeros whenever AGE is 12 or greater.

Random data   You can generate random numbers using the REPEAT, LET and SAVE commands in the Editor. First, enter variable names. Press Esc to move the cursor to the command line. Then, use REPEAT to fill cases with missing values and LET to redefine the values.

REPEAT 20	Fills 20 cases with missing values.
LET A=URN	Fills A with uniform random data.
LET B=ZRN	Fills B with normal random data.
SAVE RANDOM	Saves data in file RANDOM.SYS.

Distribution functions   MYSTAT can compute the following distribution functions.

A *cumulative* distribution function computes the probability that a random value from the specified distribution falls below a given value; that is, it shows the proportion of the distribution below that value. *Inverse* distribution functions do the same thing backwards: you specify a cumulative probability or proportion between zero and one as an area, and MYSTAT shows the critical value for that cumulative proportion.

Distribution	*Cumulative*	*Inverse*
Normal	**ZCF(z)**	**ZIF(area)**
Exponential	**ECF(x)**	**EIF(area)**
Chi-square	**XCF(chisq,df)**	**XIF(area,df)**
t	**TCF(t,df)**	**TIF(area,df)**
F	**FCF(F,df1,df2)**	**FIF(area,df1,df2)**
Poisson	**PCF(k,lambda)**	**PIP(area,k)**
Binomial	**NCF(p,k,n)**	**NIF(area,k,n)**
Logistic	**LCF(x)**	**LIF(area)**
Studentized range	**SCF(x,k,df)**	**SIF(area,k,df)**
Weibull	**WCF(x,p,q)**	**WIF(area,p,q)**

The uniform distribution is uniformly distributed from zero to one. The normal distribution is a standard normal distribution with mean zero and standard deviation one. The exponential distribution has parameter one. The cumulative Poisson function calculates the probability of the number of random events between zero and k when the expected value is *lambda*. The binomial function provides the probability of k or more occurrences in n trials with the binomial probability p.

Use the cumulative distribution functions to obtain probabilities associated with observed sample statistics. Use the inverse distributions to determine critical values and to construct confidence intervals. Finally, to generate pseudo-random data for the functions, apply the appropriate inverse cumulative distribution function to uniform random data.

Leaving the Data Editor

Use the QUIT command to leave the Editor and return to the main Business MYSTAT menu.

```
QUIT
```

General MYSTAT Commands

Once your data are in a Business MYSTAT data file, you can use Business MYSTAT's statistical and graphics routines to examine them.

Using MYSTAT data files

Open a data file   To use a MYSTAT data file, first, you must open the file containing the data you want to analyze with the USE command:

```
USE <filename>       Reads the data in <filename>.
```

To analyze the data we entered earlier, type:

```
>USE CITIES
```

MYSTAT responds by listing the variables in the file.

```
VARIABLES IN MYSTAT FILE ARE:
     CITY$     STATE$     POP  RAINFALL LOGPOP
```

See variable names and data values   The NAMES command shows the variable names in the current file.

```
>NAMES
VARIABLES IN MYSTAT FILE ARE:
     CITY$     STATE$     POP  RAINFALL  LOGPOP
```

The LIST command displays the values of variables you specify. If you specify no variables, all variables are shown.

```
>LIST CITY$
                         CITY$

CASE     1          New York
CASE     2       Los Angeles
CASE     3           Chicago
CASE     4            Dallas
CASE     5           Phoenix
CASE     6             Miami
CASE     7        Washington
CASE     8       Kansas City

   8 CASES AND    5 VARIABLES PROCESSED
```

Decimal places   Use the FORMAT command to specify the number of digits to be displayed after the decimal in statistical output. This FORMAT command has the same syntax and works the same as in the Editor:

```
FORMAT=<#>         Sets number of decimal places to <#>.
   /UNDERFLOW      Uses scientific notation for tiny numbers.
```

Sorting and ranking data

SORT reorders the cases in a file in ascending order according to the variables you specify. You can specify up to ten numeric or char-

acter variables for nested sorts. Use a SAVE command after the USE command to save the sorted data into a MYSTAT file. Then, open the sorted file to do analysis.

```
>USE MYDATA
>SAVE SORTED
>SORT CITY$ POP
>USE SORTED
```

RANK replaces each value of a variable with its rank order within that variable. Specify an output file before ranking.

```
>USE MYDATA
>SAVE RANKED
>RANK RAINFALL
>USE RANKED
```

Weighting data

WEIGHT replicates cases according to the integer parts of the values of the weighting variable you specify.

```
WEIGHT <variable>       Weights according to variable
                        specified.
```

To turn weighting off, use WEIGHT without an argument.

```
>WEIGHT
```

Quitting

When you are done with your analyses, you can end your session with the QUIT command. Remember that the Data Editor also has a QUIT. To quit MYSTAT from the Editor, enter QUIT twice.

```
>QUIT
```

Notation used in command summaries

The box below describes the notation we use for command summaries in this manual.

Any item in angled brackets($<>$) is representative   insert an actual value or variable in its place. Replace <var> with a variable name, replace <#> with a number, <var$> with a character variable, and <gvar> with a numeric grouping variable.

Some commands have *options* you can use to change the type of output you get. Place a slash / before listing any options for your command. You only need one slash, no matter how many options.

A vertical line (|) means "or." Items in brackets ([ ]) are optional. Commas and spaces are interchangeable, except that *you must use a comma at the end of the line when a command continues to a second line.* You can abbreviate commands and options to the first two characters and use upper- and lower-case interchangeably.

Most commands allow you to specify particular variables. If you don't specify variables, MYSTAT uses its defaults (usually the first numeric variable or all numeric variables, depending on the command).

Statistics

Descriptive statistics

STATS produces basic descriptive statistics.

```
>STATS
TOTAL OBSERVATIONS: 8
```

	POP	RAINFALL	LOGPOP
N OF CASES	8	8	8
MINIMUM	346865.000	7.810	12.757
MAXIMUM	7164742.000	60.020	15.785
MEAN	2064361.375	35.520	14.028
STANDARD DEV	2335788.226	18.032	1.068

Here is a summary of the STATS command.

STATS <var1> <var2>...	Statistics for the variables specified.
MEAN SD SKEWNESS, KURTOSIS MINIMUM, MAXIMUM RANGE SUM, SEM	Choose which statistics you want.
/BY <gvar>	Statistics for each group defined by the grouping variable <gvar>. The data data must first be SORTed on the grouping variable.

For example, you can get the mean, standard deviation, and range for RAINFALL with the following command.

```
>STATS RAINFALL / MEAN SD RANGE
 TOTAL OBSERVATIONS:  8
```

	RAINFALL
N OF CASES	8
MEAN	35.520
STANDARD DEV	18.032
RANGE	52.210

Tabulation

TABULATE provides one-way and multi-way frequency tables. For two-way tables, MYSTAT provides the Pearson chi-square statistic. You can produce a table of frequencies, percentages, row percentages, or column percentages. You can tell MYSTAT to ignore missing data with the MISS option and suppress the chi-square statistic with NOSTAT.

TAB <var1>*<var2>...	Tabulates the variables you specify.
/LIST	Special list format table.
FREQUENCY PERCENT ROWPCT COLPCT MISS NOSTAT	Different types of tables.
TABULATE	Frequency tables of all numeric variables.
TABULATE AGE/LIST	Frequency table of AGE in list format.
TAB AGE*SEX	Two-way table with chi-square.
TAB AGE*SEX$*STATE$, /ROWPCT	Three-way row percentage table.

TAB A,AGE*SEX/FREQ, PERC	Two two-way frequency and cell percentage tables (A*SEX and AGE*SEX).
TAB AGE*SEX/MISS	Two-way table excluding missing values.
TAB AGE*SEX/NOSTAT	Two-way table excluding chi-square.

One-way frequency tables show the number of times a distinct value appears in a variable. Two-way and multi-way tables count the appearances of each unique combination of values. Multi-way tables count the appearances of a value in each subgroup. Percentage tables convert the frequencies to percentages of the total count; row percentage tables show percentages of the total for each row; and column percentage tables show percentages of the total for each column.

t-tests

TTEST does dependent and independent t-tests. A dependent (paired samples) t-test is used to test whether the means of two continuous variables differ. An independent t-test is used to test whether the means of two groups of a single variable differ.

To request an independent (two-sample) t-test, specify one or more continuous variables and one grouping variable. Separate the continuous variable(s) from the grouping variable with an asterisk. The grouping variable must have only two values.

To request a dependent (paired) t-test, specify two or more continuous variables. MYSTAT does separate dependent t-tests for each possible pairing of the variables.

TTEST <var1>...[*<gvar>]	Does t-test of the variables you specify.
TTEST A B	Dependent (paired) t-test of A and B.
TTEST A B C	Paired tests of A and B, A and C, B and C.
TTEST A*SEX$	Independent test.
TTEST A B C*SEX	Three independent tests.

You can also do a one-sample test by adding a variable to your data file that has a constant value corresponding to the population mean of your null hypothesis. Then do a dependent t-test on this variable and your data variable.

Correlation

PEARSON computes Pearson product moment correlation coefficients for the variables you specify (or all numerical variables). You can select pairwise or listwise deletion of missing data; pairwise is the default. RANK the variables before correlating to compute Spearman rank-order correlations. CORRELATE is a synonym for PEARSON.

PEARSON <var1>...	Pearson correlation matrix.
/PAIRWISE | LISTWISE	Pairwise or listwise deletion of missing values.
PEARSON	Correlation matrix of all numeric variables.
CORR HEIGHT IQ AGE	Matrix of three variables.
PEARSON /LISTWISE	Listwise deletion rather than pairwise.

Correlation measures the strength of linear association between two variables. A value of 1 or −1 indicates a perfect linear relationship; a value of 0 indicates that neither variable can be linearly predicted from the other.

Regression and ANOVA

Business MYSTAT computes simple and multiple regression and balanced or unbalanced ANOVA designs. For unbalanced designs, MYSTAT uses the method of weighted squares of means.

The MODEL and ESTIMATE commands provide linear regression. MODEL specifies the regression equation and ESTIMATE tells MYSTAT to start working. Your MODEL should almost always include a CONSTANT term.

```
>MODEL Y=CONSTANT+X          Simple linear regression.
>ESTIMATE

>MODEL Y=CONSTANT+X+Z        Multiple linear regression.
>ESTIMATE
```

Use CATEGORY and ANOVA commands for fully factorial ANOVA. CATEGORY specifies the number of categories (levels) for one or more variables used as categorical predictors (factors). ANOVA specifies the dependent variable and produces a fully-factorial design from the factors given by CATEGORY.

All CATEGORY variables must have integer values from 1 to *k*, where *k* is the number of categories.

```
>CATEGORY SEX=2          One-way design with independent
>ANOVA SALARY           variable SALARY and one factor
>ESTIMATE               (SEX) with 2 levels.

>CATEGORY A=2, B=3      Two-by-three ANOVA.
>ANOVA Y
>ESTIMATE
```

Saving residuals   Use a SAVE command before MODEL or ANOVA to save residuals in a file. MYSTAT saves model variables, estimated values, residuals, and standard errors of prediction as the variables ESTIMATE, RESIDUAL, and SEPRED. When you use SAVE with a linear model, MYSTAT lists cases with extreme studentized residuals or leverage values and prints the Durbin-Watson and autocorrelation statistics.

```
>SAVE RESIDS               >SAVE RESID2
>MODEL Y=CONSTANT+X+Z      >CATEGORY SEX-2
>ESTIMATE                  >ANOVA SALARY
                          >ESTIMATE
```

Using residuals   You can USE the residuals file to analyze your residuals with MYSTAT's statistical and graphic routines.

Graphics

Use the CHARSET command to choose the type of graphic characters to be used for printing and screen display. If you have IBM screen or printer graphic characters, use GRAPHICS; if not, use GENERIC. The GENERIC setting uses characters like +, −, and |.

```
CHARSET GRAPHICS      For IBM graphic characters.
CHARSET GENERIC       For any screen or printer.
```

Scatterplots

PLOT draws a two-way scatterplot of one or more Y variables on a vertical scale against an X variable on a horizontal scale. Use different plotting symbols to distinguish Y variables.

```
PLOT <yvar1>...*<xvar>      Plots <yvar(s)> against
                              <xvar>.
  /SYMBOL=<var$> |          Use character variable values
    '<char>'                   or character string as
                               plotting symbol.
  YMAX=<#>  YMIN=<#>        Specify range of X and Y
    XMAX=<#>  XMIN=<#>        values.
  LINES=<#>                 Specify number of screen
                               lines for graph.
PLOT A*B/SYMBOL='*'         Uses asterisk as plotting
                             symbol.
PLOT A*B/SYMBOL=SEX$        Uses SEX$ values for plotting
                             symbol.
PL Y1 Y2*X/SY='1','2'      Plot Y1 points as 1 and Y2
                             points as 2.
PLOT A*B/LINES=40          Limits graph size to 40 lines
                             on screen.
```

For example, we can plot LOGPOP against RAINFALL using the first letter of the values of CITY$ for plotting symbols.

```
>PLOT LOGPOP*RAINFALL/SYMBOL=CITY$
```

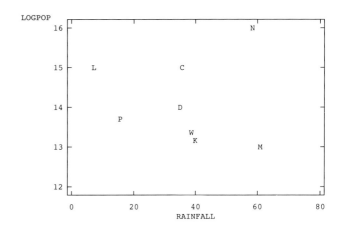

The SYMBOL option is powerful. If you are plotting several Y variables, you can label each variable by specifying its own plotting symbol.

```
>PLOT Y1 Y2*X/SYMBOL='1','2'
```

Or, you can name a character variable to plot each point with the first letter of the variable's value for the corresponding case:

```
>PLOT WEIGHT*AGE/SYMBOL=SEX$
```

Box plots

BOX produces box plots. Include an asterisk and a grouping variable for grouped box plots.

```
BOX <var1>... [*<gvar>]    [Grouped] box plots of the
                             variables.
  /GROUPS=<#>,             Show only the first <#>
                             groups.
  MIN=<#> MAX=<#>          Specify scale limits.
BOX                        Box plots of every numeric
                             variable.
BOX SALARY                 Box plot of SALARY only.
```

BOX SALARY*RANK   Grouped box plots of
                                           SALARY by RANK.
BOX INCOME*STATE$/GR=10   Box plots of first 10 groups
                                           only.

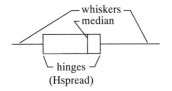

The center line of the box marks the *median*. The edges of the box show the upper and lower *hinges*. The median splits the ordered batch of numbers in half and the hinges split these halves in half again. The distance between the hinges is called the *Hspread*. The *whiskers* show the range of points within 1.5 Hspreads of the hinges. Points outside this range are marked by asterisks and those more than 3 Hspreads from the hinges are marked with zeros.

Histograms

HISTOGRAM displays histograms for one or more variables.

HISTOGRAM <var1>...   Histograms of variables
                                           specified.
   /BARS=<#>   Limits the number of bars used.
   SCALE,   Forces round cutpoints between
                                           bars.
   MIN=<#> MAX=<#>   Specifies scale limits.
HISTOGRAM   Histograms of every numeric
                                           variable.
HISTOGRAM A B/BARS=18   Forces 18 bars for histograms
                                           of A and B.
HIST A/MIN=0 MAX=10   Histogram of A with scale from
                                           0 to 10.

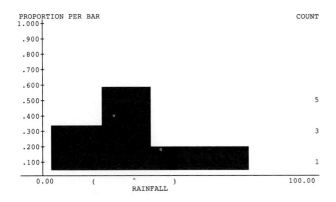

A histogram shows the *distribution* of a set of data values. The data are divided into equal-sized intervals along the horizontal axis. The number of values in each interval is represented by a vertical bar. The height of each bar can be measured in two ways: the axis on the right side shows the actual number of cases in the intervals, and the left axis shows the proportions of the data in the intervals. The mean is denoted with the ^ symbol and the mean plus or minus one standard deviation are denoted with parentheses.

Stem-and-leaf diagrams

STEM plots a stem-and-leaf diagram.

STEM <var1>...   Diagrams of variables specified.
   /LINES=<#>   Specify number of lines in diagram.
STEM   Stem-and-leaf of every numeric
                                           variable.
STEM TAX   Stem-and-leaf of TAX.
STEM TAX/LINES=20   Stem-and-leaf with 20 lines.

The numbers on the left side are *stems* (the most significant digits in which variation occurs). The *leaves* (the subsequent digits) are printed on the right. For example, in the following plot, the stems are 10's digits and the leaves are 1's digits.

```
STEM AND LEAF PLOT OF VARIABLE: RAINFALL,
   N = 48

MINIMUM IS:            7.000
LOWER HINGE IS:           26.500
MEDIAN IS:           36.000
UPPER HINGE IS:           43.000
MAXIMUM IS:           60.000

                   0   78
                   1   0114
                   1   55556
                   2
                   2 H 588
                   3   00113333
                   3 M 5557999
                   4 H 001123333
                   4   55567999
                   5   0
                   5   9
                   6   0
```

(We added 40 more cases to the dataset for this plot. You will get a different plot if you try this example with the CITIES.SYS data file.)

Hinges are defined under "Box plots," above.

Time Series Analysis

MYSTAT's time series analysis routines let you transform, smooth, and plot time series data. All of the routines operate in the computer's memory; that is, the results are saved in the working memory rather than being saved to a data file. When you execute a transformation or a smooth, the only results you see are messages like "Series is smoothed" or "Series is transformed."

Your usual strategy will be to plot the series with TPLOT, ACF, or PACF; do a transformation or smoothing operation; and then re-plot the data to see your progress. You can save a modified series in a new data file by preceding the transformation or smoothing command with a SAVE command. You can use the CLEAR command at any time to restore the variable's original values.

Usually, you will do your work in memory, and when you find some satisfactory results, CLEAR the series, and then repeat the successful transformation using the SAVE command. Suppose, for example, that you wanted to save a demeaned and then smoothed series:

```
>MEAN MYSERIES
>SAVE A:NEW
>SMOOTH MYSERIES
```

This saves the smoothed, demeaned series in the file NEW.SYS. If you also want to save the intermediate (demeaned but not smoothed) series, use two SAVE commands:

```
>SAVE A:MEAN
>MEAN MYSERIES
>SAVE A:FINAL
>SMOOTH MYSERIES
```

Transforming

LOG replaces each value in a series with its natural logarithm. Logging may remove nonstationary variability, such as increasing variances across time.

LOG <var>	Logs the variable specified.
LOG	Logs first numerical variable in file.
LOG SPEED	Logs SPEED.

MEAN subtracts the variable's mean from each value. Demeaning centers a series on zero.

MEAN <var>	Demeans the variable specified.
MEAN	Demeans the first numeric variable.
MEAN SUNSPOTS	Demeans SPEED.

SQUARE squares each value. Squaring can help normalize variance in a series.

SQUARE <var>	
SQUARE	
SQUARE MAGNITUD	

TREND removes linear trend to make a series "level."

TREND <var>	
TREND	
TREND WEATHER	

PCNTCHNG replaces each value with its percentage difference from the immediately preceding value.

PCNTCHNG <var>	
PCNTCHNG	
PCNTCHNG SERIES	

DIFFRNCE replaces each value with its difference from the immediately preceding value; you can specify a different lag to show differences from a greater number of positions previous. Differencing helps make some series "stationary" by removing the dependence of one point on the previous point.

DIFFRNCE /LAG=<#>]	Compute differences between values <#> cases apart.
DIFFRNCE	
DIFFRNCE STROKES/LAG=3	Difference interval of 3 values.

INDEX divides each value by the first value of the series, or by the base you specify.

INDEX <var>	
INDEX SERIES/BASE=12	Indexes series to its 12th value.

Smoothing

Smoothing removes local variations from a series and makes it easier to see the general shape of the series. Running smoothers replace each value with the mean or median of that value and its neighbors. The "window" is the number of values considered at a time; 3 is the default. You can optionally specify weights for each value, separated by commas.

SMOOTH <var1>...	Smooths the variables specified.
/MEAN=<#>|MEDIAN=<#> <#>.	Mean or median smoothing with window size of
WT=<#,#,...>	Specifies smoothing weights for each value.
SMOOTH SERIES	Running mean smoother, window of 3.
SMOOTH SERIES/MEAN=5	Running mean smoother, window of 5.
SMOOTH SERIES/MEDIAN=4	Running median smoother, window of 4.
SMOOTH SERIES, /WT=.1,.2,.1	Weighted mean smoother, window of 3.

Seasonal decomposition removes additive or multiplicative seasonal effects according to the period you specify; the default is 12.

ADJSEAS <var> /SEASON=<#>,	Smooths specified variable. Number of observations per period.
ADDI=<#>|MULT=<#>	Removes additive or multiplicative seasonal effects with weight <#>.
ADJSEAS	Adjusts first variable with period 12.
ADJSEAS SALES/SEASON=4	Adjusts SALES with period 4.

Exponential smoothing forecasts future observations as a weighted average (running smooth) of previous observations. You can optionally specify a linear or percentage growth component and an additive or multiplicative seasonal effects component. Specify the weights for each component and the number or range of cases to be forecast.

EXP <var>	Smooths specified variable.
/LINEAR=<#>|PERCENT=<#>,	Remove linear or percentage growth trend.
SEASON=<#>,	Number of observations per period.
ADDI=<#>|MULT=<#>,	Removes additive or multiplicative seasonal effects with weight <#>.
FORECAST=<#>|<#>-<#>,	Forecasts number or range of cases.
SMOOTH=<#>	Specifies weight for level component.
EXP/FORE=4	Default model with 4 forecasted cases.
EXP SALES/LINEAR ADD	Holt-Winter's model: linear trend, additive seasonals with period 12.
EXP SERIES/MULT SEAS=4	Multiplicative seasonality with period 4.

Plotting

TPLOT produces time series plots, which plot the variable against CASE (time). TPLOT's STANDARDIZE option removes the series mean from each value and divides each by the standard deviation. MYSTAT uses the first fifteen cases unless you specify otherwise with LAG.

By default, TPLOT fills the area from the left axis to the plotted point. Include the NOFILL option to specify that the plot not be filled. Use the CENTER option to fill the plot from the observed value to the mean of the series.

TPLOT <var>	Case plot of variable specified.
/LAG=<#>	Plots first <#> cases.
STANDARDIZE,	Standardizes before plotting.
CENTER|NOFILL,	Changes the way the plot is filled.
MIN=<#>,MAX=<#>	Sets scale limits.
TPLOT	Case plot of first numeric variable.
TPLOT PRICE/LAG=10	First 10 cases of PRICE.
TPLOT PRICE/STAN	Standardizes before plotting.

ACF plots show how closely the points in a series relate to the values immediately preceding them.

ACF <var>	Autocorrelation plot of specified variable.
/LAG=<#>	Plots first <#> cases.
ACF	Autocorrelation plot of first variable.
ACF PRICE/LAG=10	First 10 cases of PRICE.

PACF plots show the relationship of values in a series to preceding points after partialing out the influence of intervening points.

PACF <var>	Partial autocorrelation plot of variable.
/LAG=<#>43	Plots first <#> cases.
PACF	Partial autocorrelation plot of first variable.
PACF PRICE/LAG=10	First 10 cases of PRICE.

Clearing the series

CLEAR restores the initial values of the series, removing the effects of smoothings and transformations.

Submitting Files of Commands

You can operate MYSTAT in batch mode, where MYSTAT executes a series of commands from a file and you sit back and watch. (You've already seen a command file in action: the DEMO demonstration uses a file of commands, DEMO.CMD.)

The SUBMIT command reads commands from a file and executes the commands as though they were typed from the keyboard. Command files must have a ".CMD" file extension.

Use a word processor to create a file of commands—one command per line, with no extraneous characters. Save the file as a text (ASCII) file. (Use the command "COPY CON BATCH.CMD" and [F6] to type commands into a file if you have no word processor.)

SUBMIT <filename>	Submits file of commands.
SUBMIT COMMANDS	Reads commands from COMMANDS.CMD.
SUBMIT B:NEWJOB	Reads commands from file on drive B.

Redirecting Output

Ordinarily, Business MYSTAT sends its results to the screen. OUTPUT routes subsequent output to an ASCII file or a printer.

OUTPUT *	Sends output to the screen only.
OUTPUT @	Sends output to the screen and the printer.
OUTPUT <filename>	Sends output to the screen and a text file.

MYSTAT adds a .DAT suffix to ASCII files produced by OUTPUT. You must use OUTPUT * or QUIT the program to stop redirecting.

Printing and saving analysis results   To print analysis results, use the command OUTPUT @ before doing the analysis or analyses. Use OUTPUT <filename> to save analysis results in a file. Use OUTPUT * to turn saving or printing off when you are finished.

Printing data or variable names   You can print your data by using the command OUTPUT @ and then using the LIST command. Use LIST <var1>... to print only certain variables. Don't forget to turn printing off when you are done.

You can print your variable names by using OUTPUT @ and then NAMES. Don't forget to turn printing off when you are done.

Putting Comments in Output

NOTE allows you to write comments in your output. Surround each line with quotation marks, and issue another NOTE command for additional lines:

```
>NOTE 'Following are decriptive statistics
    for the POP'
>NOTE "and RAINFALL variables of the CITIES
    dataset."
>STATS POP RAINFALL
TOTAL OBSERVATIONS:  8

                    POP      RAINFALL

N OF CASES           8          8
MINIMUM         346865.000     7.810
MAXIMUM        7164742.000    60.020
MEAN           2064361.375    35.520
STANDARD DEV   2335788.226    18.032
>NOTE "Note that the average annual
    rainfall for these."
>NOTE 'cities is 35.52 inches.'
```

Saving Data in Text Files

You can save datasets to ASCII text files with the PUT command. PUT saves the current dataset in a plain text file suitable for use with most other programs. Text files have a .DAT extension.

USE <filename>	Opens the dataset to be exported.
PUT <filename>	Saves the dataset as a text file.

Note that the PUT command is not an Editor command. To save a newly created dataset in a plain text file, you must QUIT from the Editor, USE the datafile, and finally PUT the data in an ASCII file.

```
>EDIT
[editing session here]
>SAVE A:NEWSTUFF
>QUIT
>USE A:NEWSTUFF
>PUT A:NEWTEXT
```

The above commands would create a plain ASCII file called NEWTEXT.DAT on the A disk.

Index of Commands

This index lists all of MYSTAT's commands and the Editor commands, and it gives a brief description of each.

Editor commands

DELETE	delete a row (case)
DROP	drop a column (variable)
Esc key	toggle between command line and worksheet
FIND	find a particular data value
FORMAT	set number of decimal places in Editor
GET	read data from an ASCII file
HELP	get help for Editor commands
IF...THEN	conditionally transform or create a variable
LET	transform or create a variable
NEW	create a new data file
QUIT	quit the Editor and return to MYSTAT
REPEAT	fill cases with missing values
SAVE	save data in a data file
USE	read a data file into Editor

MYSTAT commands

ACF	autocorrelation plot
ADJSEAS	seasonal decomposition
ANOVA	analysis of variance
BOX	box-and-whisker plot
CATEGORY	specify factors for ANOVA
CHARSET	choose type of characters for graphs
CLEAR	restore original values of series
CORRELATE	Pearson correlation matrix
DEMO	demonstration of Business MYSTAT
DIFFRNCE	difference transformation
EDIT	edit a new or existing data file
ESTIMATE	start computations for regression
EXP	exponential smoothing
FORMAT	set number of decimal places in output
HELP	get help for MYSTAT commands
HISTOGRAM	draw histogram
INDEX	index transformation
LIST	display data values
LOG	log transformation
MEAN	demean a series
MENU	turn the command menu on/off
MODEL	specify a regression model
NAMES	display variable names
NOTE	put comment in output
OUTPUT	redirect output to printer or text file
PACF	partial autocorrelation plot
PCNTCHNG	percent transformation
PEARSON	Pearson correlation matrix
PLOT	scatterplot (X-Y plot)
PUT	save data in text file
QUIT	quit the MYSTAT program
RANK	rank data
SAVE	save results in a file

SMOOTH	smooth a series
SORT	sort data
SQUARE	square transformation
STATS	descriptive statistics
STEM	stem-and-leaf diagram
SUBMIT	submit a batch file of commands
SYSTAT	get information about SYSTAT
TABULATE	one-way or multi-way tables
TPLOT	case plot
TREND	remove trend from a series
TTEST	independent and dependent t-tests
USE	read a data file for analysis
WEIGHT	weight data

QCTOOLS DOCUMENTATION

Herbert H. Stenson

QCTOOLS is a set of quality control programs. It is designed to be used in conjunction with data files created by MYSTAT. The results of all analyses that are performed will be printed on your screen in either graphical or tabular form (or both, if you request it). You also have the option of saving the screen output into an ASCII file and/or saving numerical output into a MYSTAT file. You can switch to MYSTAT from QCTOOLS.

Installation

QCTOOLS requires a minimum of 256K of RAM and a floppy or hard disk drive. One of the disks that you received contains QCTOOLS.EXE (the program), QCTOOLS.HLP (a file with information for on-line help), and two sample data files (files with a .SYS suffix in their names).

If you have not yet installed MYSTAT, please do so now by following the directions given for MYSTAT installation. Note that you cannot use QCTOOLS to analyze data unless the data files have been created with MYSTAT.

Once you have MYSTAT installed, the procedure for installing QCTOOLS is very similar. Just follow the directions given for MYSTAT that correspond to the type of drive that you have, but substitute the word QCTOOLS for MYSTAT in these procedures. However, you will not have to set up or alter the AUTO-EXEC.BAT or CONFIG.SYS files again if you have already installed MYSTAT: just skip these steps. Be sure to make backup copies of both QCTOOLS and MYSTAT files.

If you are using a hard disk, the MYSTAT.EXE file and the MYSTATQ.HLP file must be in the same directory as the QC-TOOLS.EXE file and QCTOOLS.HLP. If you are using a system with no hard disk and a 720K or more floppy disk with both QC-TOOLS and MYSTAT on it, both programs must be in the same directory of that disk. If you are using a system with two 360K floppy disk drives then the QCTOOLS disk and the MYSTAT disk will both be used alternately in the A: drive. If these conditions are not met you will have trouble with the programs.

Getting Started

At your system prompt enter:

```
>QCTOOLS
```

after which you should press the Return or Enter key.

Command menu

The first thing that you will see on the screen is the QCTOOLS logo. Follow the instruction on the bottom of the screen and press your Enter key. Next you will see the QCTOOLS command menu. It shows all of the commands available in QCTOOLS. The commands are split into two major groups. General system commands are in the left half of the menu and commands to perform specific quality control procedures are in the right half of the menu. Here is a list of the commands you will see on the left half of the menu, and a brief description of what each command does. Asterisks indicate the default settings, if any.

DEMO	Run a tutorial demonstration of QCTOOLS.
HELP	Obtain on-line help for any command.
MENU	Turn the menu off if it is on*, or on if it is off.
MYSTAT	Switch to the MYSTAT program from within QCTOOLS.
QUIT	Leave QCTOOLS and return to DOS.
CHARSET	Set graphics for output to IBM graphics* or generic.
FORMAT	Set number of decimals to be used in output. (3*)
NAMES	List the names of the variables in a file that is open.
NOTE	Write a note or title on the output.
OUTPUT	Route output to a file, a printer, or to the screen* only.
PRINT	Print long or short* results.
SAVE	Save the output into a MYSTAT data file.
SUBMIT	Run a batch job from commands listed in a command file.
USE	Open a particular MYSTAT data file.
WEIGHT	Designate a particular variable as a weighting variable.

Except for the PRINT and MYSTAT commands, all of the commands listed above function in the same way as their counterparts in the MYSTAT program. The PRINT command allows you to see extended output for many QCTOOLS commands. Most often this extra output is a tabular listing of the numerical values that are plotted on a control chart. To see this output enter the command:

```
>PRINT LONG
```

prior to the command that produces the output. To switch back to the default output enter:

```
>PRINT SHORT
```

The PRINT command stays in effect until you change it or QUIT.

The command MYSTAT allows you to switch directly to MYSTAT from QCTOOLS. Any options that you have set within QCTOOLS (such as FORMAT or CHARSET) will be carried over to MYSTAT, as will any open files. Thus, if a data file is open (in use) in QCTOOLS it will still be open when you switch to MYSTAT. However, this command is sensitive to the installation of QCTOOLS and MYSTAT. Thus, you must follow the installation directions precisely for this command to work. The remainder of the commands in the list above are described in the text for the MYSTAT program and also on the HELP screens in QCTOOLS.

The WEIGHT command functions similarly to the WEIGHT command in MYSTAT, but needs some clarification. This command replicates cases according to a weighting variable that you specify, just as in MYSTAT. The values of the weighting variable are truncated to integers, and the sample size used in the entire analysis is the sum of these integers. Cases weighted less than one

are excluded from the analysis. EXCEPTION: Truncation to integers is not performed in the case of a Shewhart U plot where the "sample size" for a case is the number of sample units, which may be fractional. (See the discussion of Poisson charts below). The command syntax is:

```
>WEIGHT=<var>
```

The WEIGHT command must precede any analysis for which it is to be used. Type WEIGHT with nothing after it to cancel the WEIGHT command. You will use this command most often in QCTOOLS to indicate individual sample sizes if you are analyzing data that are already aggregated into a single statistic, such as the mean, by samples or subgroups. (See the later section on Input Data for more on this.)

The commands on the right half of the menu are to provide specific quality control analyses and charts. They are discussed in the paragraphs that follow.

If, after you have entered any QCTOOLS command, you wish to use the same command again, press the F9 key and the command will reappear. Then press the ENTER key to activate that command. You can also edit the previous command before pressing ENTER. QCTOOLS stores the 6 most recent commands, so you can press the F9 key repeatedly to retrieve them.

Control Charts

Eight Shewhart control charts can be produced with QCTOOLS. The charts are:

X	An X-bar chart of subgroup means
S2	A chart of subgroup variances
S	A chart of subgroup standard deviations
R	A chart of subgroup ranges
NP	Binomial count by subgroup
P	Binomial proportion by subgroup
C	Poisson count by subgroup
U	Poisson rate per sample unit

The generic form of the command to produce any of the Shewhart charts is:

```
SHEW(<type>) <yvar>*<xvar> [/ <option list>   ]
```

Replace <type> with one of the eight types listed above, and enclose it in parentheses. Replace <yvar> with the name of a file variable that contains the quality control data for each subgroup. Replace <xvar> with the name of a file variable that identifies the subgroup to which each data entry belongs. You must enter the asterisk between <yvar> and <xvar>. The <yvar> must be a numeric variable, but the <xvar> can be either character (ending with $) or numeric (not ending with $).

Options

Each chart has a similar list of options. Thus, the options are described only once. Options are described on the HELP screens and examples are given in the DEMO program. The options may be entered in any order after the / symbol. They must be separated by either a comma or a space. The options are:

CENTER=<#>	Place the center line of the chart at the number specified. The default center line is the estimated expected value of the statistic being plotted.

SLIMITS — Enter this as an option if you wish to plot sigma limits instead of probability limits.

LCL=<#> — Place the lower control limit at the number specified. If SLIMITS is used, then the value for LCL is interpreted as so-called "sigma" units. Otherwise the number is taken to be the actual value for the lower control limit. The default LCL is a probability limit corresponding to the value of ALPHA used, or is −3.00 if SLIMITS is used.

UCL=<#> — Same as the LCL, but now applying to the upper control limit instead of the lower. Default value is +3.00 if SLIMITS used.

SIGMA=<#> — Specify an a priori value for sigma, the population standard deviation. Note that this usually is not the standard deviation of the sampling distribution.

ALPHA=<#> — The proportion of the sampling distribution to be outside the LCL and/or UCL. If both control limits are to be used, then ALPHA/2 of the distribution is outside each control limit. If only one limit is used, then ALPHA of the distribution is outside that limit. Default value of ALPHA = .0027. This option is ignored if either LCL or UCL is specified in the option list.

UPPER | LOWER — Specify UPPER if you wish only an upper probability limit as the UCL. Specify LOWER to get only a lower probability limit. This option is ignored if either LCL or UCL is specified in the option list.

AGG — Indicates that the input data are already aggregated as the statistic that is to be plotted on the chart. You probably will want to use the WEIGHT command to specify individual sample (subgroup) sizes when you use the AGG option.

YMAX=<#> — Specify a maximum value for the Y-axis of the control chart.

YMIN=<#> — Specify a minimum value for the Y-axis of the control chart.

The last two options are useful for "zooming in" on a part of your data by rescaling the Y-axis of the control chart.

Input data

Each chart will accept either raw data or aggregated data as input. By raw data we mean that a value of <yvar> is provided for each and every individual case in each subgroup of data. Aggregated data consist of a statistic that has already been computed for each subgroup. For example, if you already have a file that contains the mean of some variable for each subgroup, then those means constitute aggregated input data for an X chart.

The easiest way to think about aggregated data is to consider the statistic that is being plotted for each type of Shewhart chart. That statistic is the definition of aggregated input data for the chart type in question. Thus, subgroup means are the aggregated input data for an X chart, subgroup variances are the aggregated data for an S2 chart, etc. If you use the AGG option, you also will have to use the WEIGHT command to inform QCTOOLS of the sample size for each subgroup whose aggregated data is entered. An example of this is given in the DEMO program.

Control limits

QCTOOLS computes probability limits for each chart. Measuring the upper and lower control limits are points on a statistical distribution that have a specific proportion of the distribution outside those limits. The proportion of the distribution outside the limits is termed ALPHA, and its default value is 0.0027. That is the value of ALPHA for control limits set at the mean plus or minus 3 standard errors of the mean for a normal distribution of means. If only one control limit is being used, then all of alpha is one tail of the distribution.

QCTOOLS computes probability limits by using the statistical distribution that is appropriate for the chart in question. Thus, a normal distribution is used for an X chart, a chi-square distribution is used for an S2 chart, the binomial distribution is used for an NP chart, etc. If, instead, you want to use "sigma limits" you may do so by using the SLIMITS option.

Sample size vs. sample units for Poisson charts

The theoretical sample size for a Poisson distribution is infinitely large. Thus, the usual concept of a sample size does not apply. Instead, we define "sample units"; an arbitrary unit of time, weight, area, volume, etc. A sample unit is assumed to have a very large number of opportunities for an event of interest to occur, with a small probability of occurrence. For example, the sample unit might be 100 square yards of cloth being examined for defects. Logical subgroups could contain different numbers of sample units. Then we might want to plot the average number of defects per sample unit (e.g. defects per 100 square yards). For example, one subgroup might contain, say, two 100-square-yard bolts, while another contained three 50-square-yard bolts. Then the former subgroup would contain 2 sample units, and the latter would contain 1.5 sample units (with a sample unit defined as 100 square yards). To reflect the number of sample units in a subgroup, use the WEIGHT command (along with the AGG option). WEIGHT can be used this way for Shewhart C or U charts only. THIS IS THE ONLY PLACE IN QCTOOLS WHERE THE WEIGHT VALUES WILL NOT BE ROUNDED DOWN TO INTEGERS AUTOMATICALLY.

UWMA control chart

The UWMA command is very much like the SHEW (X) command, except that it plots the unweighted moving average by subgroup, instead of the simple average for each subgroup. Its syntax is:

```
UWMA <yvar>*<xvar>
[/WIDTH=<#>,CENTER=<#>,LCL=<#>,UCL=<#>,
   SIGMA=<#>,ALPHA=<#>,UPPER|LOWER,
   AGG,YMAX=<#>,YMIN=<#>]
```

Replace <yvar> with the name of a variable containing the data for which the moving average is desired. Replace <xvar> with the name of a variable identifying the subgroup to which each <yvar> observation belongs.

All of the options except for WIDTH are the same as for the X chart. The WIDTH option allows you to select the number of subgroups over which the moving average is to be computed. For example, if you choose WIDTH=4, then the plotted value for each successive subgroup will be the average of all the data for the current subgroup along with the previous 3 subgroups. Notice that the very first subgroup has no subgroups previous to it, so the point plotted for it will just be the mean of that subgroup. The second subgroup has only one previous subgroup, so the moving average for it can contain only the data from the first and second subgroups, etc. The default value of WIDTH is 1, so that the default UWMA chart is identical to a Shewhart X chart. The default CENTER value is the mean of all observations being analyzed. The default probability limits are calculated using a normal distribution model, the same as for the X chart. However, the sample size on which the limits are based is the total number of observations encompassed by the subgroups selected by the WIDTH option. See the DEMO program for an example of the UWMA chart.

Other QCTOOLS Commands

The last column of the QCTOOLS menu contains a number of utility commands that are useful in analyzing quality control data.

FUNCTION command

This command will replace most of your statistical tables. It calculates the cumulative probability for a statistical function or the inverse function of a probability, plus some statistics for the range of a standard normal distribution. Its general syntax is:

```
FUNCTION <type>
```

The following table shows what to substitute for <type> for each of the available distributions. Lower case letters represent numbers.

Distribution	Cumulative Function	Inverse Function
Normal	ZCF(x)	ZIF(prob)
t	TCF(x,df)	TIF(prob,df)
F	FCF(x,df1,df2)	FIF(prob,df1,df2)
Chi-Square	XCF(x,df)	XIF(prob,df)
Exponential	ECF(x)	EIF(prob)
Binomial	NCF(x,n,p)	NIF(prob,n,p)
Poisson	PCF(x,mean)	PIF(prob,mean)
Relative Range	RCF(x,n)	RIF(prob,n)
Expected Relative Range	ER(n)	
SD of the Relative Range	SDR(n)	

In this notation, x always refers to the value of a variable, prob refers to a probability, df is degrees of freedom, n is sample size, and p is a binomial proportion. The Relative Range refers to the range of a standard normal distribution.

Examples of this command are:

```
FUNCTION ZCF(1.96)
     (The program will respond with P=0.975)
FUNCTION NIF(.75,2,.5)
     (The program will respond with X=1.000)
```

PARETO command

This command creates a chart showing frequencies of occurrence of an attribute sorted in descending order. They are plotted as a function of a variable that identifies each subgroup. Its syntax is:

```
PARETO <yvar>*<xvar> [/CUM,P,AGG]
```

The <yvar> must consist of zeros and ones if individual instances of an event are in the input file. The <xvar> identifies subgroups and may be either numeric or character. If <yvar> contains the number of instances of an event already aggregated by subgroup, then the program must be informed by using the AGG option. Note that the WEIGHT command simply replicates cases in the file and has nothing to do with sample sizes in this case. All sample sizes are assumed to be equal.

The default output shows frequencies. The CUM option will change this to cumulative frequencies. The P option will produce proportions (relative frequencies) instead. CUM and P used together will produce cumulative proportions. Use the SAVE command prior to PARETO to save the chart data into a file. Use PRINT LONG prior to PARETO in order to print the chart values on the screen after the plot.

Here are some examples of the command:

`PARETO DEFECTS*SAMPLE`	Input data are zeros and ones.
`PARETO ERRORS*DAY$ / AGG`	Input data are frequencies by sample.
`PARETO NUMBER*WEEK$ / CUM,P,AGG`	Cumulative proportions will be shown.

See the DEMO program for an example of the output for this command.

PCA command

The PCA command produces Process Control Indices for a variable that you name. Its syntax is:

```
PCA(<lsl,usl>) <var>
```

The numerical lower and upper specification limits for a process are the numbers that you enter in the parentheses in place of <lsl,usl>. You must type these limits in parentheses. Replace <var> with the name of a MYSTAT variable that contains the empirical data for which you have stated the upper and lower specification limits. You will get output showing basic statistics on the data in <var> along with the process control indices Cp and Cpk. (These indices are described in many quality control textbooks.) No chart will appear.

The WEIGHT command may be used to indicate replications of input data for this analysis, but the use of PRINT=LONG and SAVE will have no effect.

As an example, USE the sample file BOXES.SYS from your QCTOOLS disk, and then issue the following command:

```
PCA(15,25) OHMS
```

This will produce the process control indices and other statistics. This example assumes that the specification limits for OHMS are 15 and 25 ohms.

PLOT command

This command creates a scatterplot of a numeric variable, <yvar>, against an <xvar> that may be either numeric or character. Its syntax is:

```
PLOT <yvar>*<xvar> / XMIN=<#>,XMAX=<#>,
     YMIN=<#>,YMAX=<#>
```

If the type of the <xvar> is character, each successive value of the variable as it exists in your file will be plotted along with its corresponding <yvar> value. In this case, the use of XMIN and XMAX will be ignored. If the x-variable is numeric, a normal scatter-plot will be produced. In this case XMIN, XMAX, YMIN, and YMAX can be used to set limits on the axes of the plot, thus allowing you to 'zoom in' on parts of the graph by enlarging it.

Note that the Y axis of the plot is the horizontal dimension of your screen and the X axis is the vertical dimension.

The SAVE command is inoperative for the PLOT command, since the data that would be saved are already in a file.

PPLOT command

This command plots the values of a numeric variable against the corresponding percentage points of a normal distribution. Its syntax is:

```
PPLOT <var> / XMIN=<#>,XMAX=<#>,
  YMIN=<#>,YMAX=<#>
```

The program rank orders the variable given and then computes cumulative percentages of the data falling below each rank. These percentages are converted to expected standard normal deviates (Expected Z-values). The Expected Z-values are plotted against the ordered data values. If the data are approximately normally distributed a linear plot should be evident, as should points that do not conform to this pattern.

Your variable is the vertical axis of the plot, and Expected Z is the horizontal axis. The SAVE command is inoperative for this plot.

An example of the command is:

```
PPLOT EFFECTS / YMIN=-2,YMAX=2
```

It produces a PPLOT with Z values between -2 and 2.

Greek Alphabet

A	α	alpha		N	ν	nu
B	β	beta		Ξ	ξ	xi
Γ	γ	gamma		O	o	omicron
Δ	δ	delta		Π	π	pi
E	ϵ	epsilon		P	ρ	rho
Z	ζ	zeta		Σ	σ	sigma
H	η	eta		T	τ	tau
Θ	θ	theta		Υ	υ	upsilon
I	ι	iota		Φ	ϕ	phi
K	κ	kappa		X	χ	chi
Λ	λ	lambda		Ψ	ψ	psi
M	μ	mu		Ω	ω	omega

Answers to Odd-Numbered Problems

Chapter 1 Introduction to Statistics

1. The population is usually larger because the sample is a subset of the population.

3. This is a statistical inference problem.

5. The state is qualitative.

7. Line chart for sales:

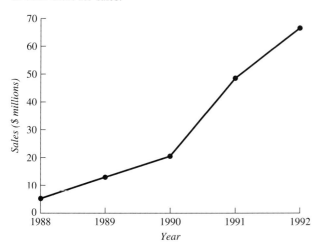

9. Bar (column) chart for number of restaurants:

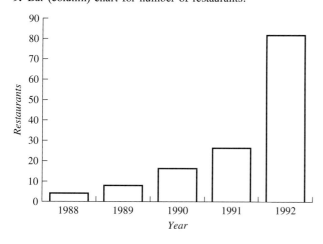

11. Combination chart for actual and estimated number of restaurants:

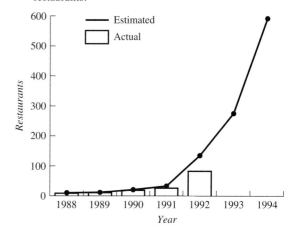

13. Rather than judging the heights of the barrel symbols, readers might judge the areas or volumes of the barrels.

15. This is a statistical inference problem.

17. The exact numbers available each year will change, but a recent edition of the *Statistical Abstract* showed that the three states with the lowest death rates were Alaska, 4.3 per thousand residents; Hawaii, 5.2 per thousand; and Utah, 5.6 per thousand.

19. Industry is qualitative; current assets/sales is quantitative; net income/total assets is quantitative.

21. a. Washington, D.C.   **b.** Washington, D.C.

23. a. 1911, New York City   **b.** About 30,000   **c.** 11

25. Bar chart for an individual investor's portfolio:

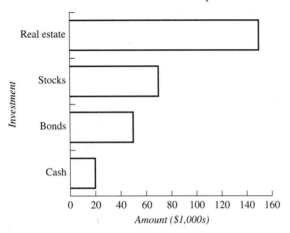

27. a. Grouped bar chart for classification proportions:

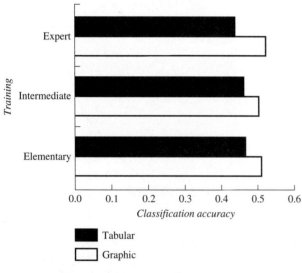

b. Graphics

29. Combination chart for actual and estimated sales:

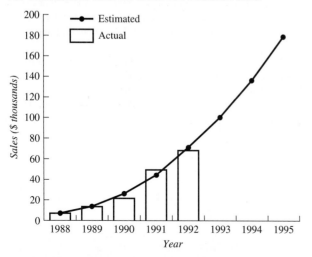

31. Pie charts for asset and liability structures of commercial banks:

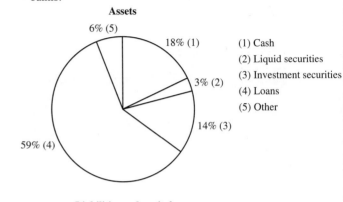

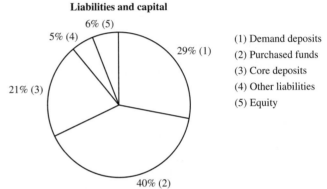

33. Rather than judging the heights of the computer symbols, the manager might judge the areas or volumes.

35. An item is a company.

37. There are seven variables (including the company name as an identifying variable).

39. Quantitative variables

47. a. The fleet may be traveling more miles; the type of driving may have changed.

b. The fleet's mean ton-miles per gallon

49. Inflation, taxes

51. a. First island   **b.** Second island

c. Stock more items for more mature people; stock fewer items for a youthful market.

Chapter 2 Data Distributions

1. a. 5   **b.** 6   **c.** 8   **d.** 14   **e.** 20

3. a. *Approximate number of classes* $= [(2)(250)]^{.3333} = 7.94$ so round upward and use 8 classes.

b. The approximate class interval or width is

$$\textit{Approximate class interval} = (79.9 - 8.8)/8$$
$$= 8.89$$

so the class interval can be 9.

c. The interval covered by the actual data is $79.9 - 8.8 = 71.1$, the interval covered by the classes is $(8)(9) = 72$, and the overlap is

$$\textit{Overlap} = 72 - 71.1 = .9$$

d. Half of the overlap is .9/2 = .45, so the lower boundary of the first class can be .45 units below the lowest value or

$$\textit{Lower boundary of the first class} = 8.8 - .45$$
$$= 8.35$$

The upper boundary of the first class is

$$\textit{Upper boundary of the first class} = 8.35 + 9$$
$$= 17.35$$

The class limits are the attainable values just within the boundaries of 8.35 and 17.35; thus, the class limits are 8.4 and 17.3.

5. a. 5   **b.** 28

c. Frequency distribution:

Class	Frequency
54.5 to 82.5	7
82.5 to 110.5	9
110.5 to 138.5	21
138.5 to 166.5	4
166.5 to 194.5	7

7. Cumulative distribution:

Less Than	Cumulative Frequency
54.5	0
82.5	7
110.5	16
138.5	37
166.5	41
194.5	48

9. a. 25

b. Boundaries: 24.5, 49.5, 74.5, 99.5, 124.5, 149.5

c. Histogram of ages of 100 accounts receivable:

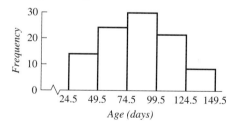

Age (days)

11. 7

13. a. $[(2)(40)]^{.3333} = 4.31$; 5 classes

b. The approximate class interval or width is

$$(7.4 - 0.1)/5 = 1.46$$

so the class interval can be 1.50.

c. Frequency distribution:

Class	Frequency
0.05 to 1.55	19
1.55 to 3.05	13
3.05 to 4.55	3
4.55 to 6.05	3
6.05 to 7.55	2

d. Histogram of service times:

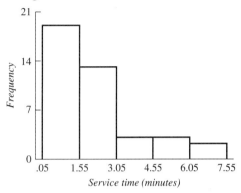

Service time (minutes)

15. a. Ogive of salaries (in dollars per month):

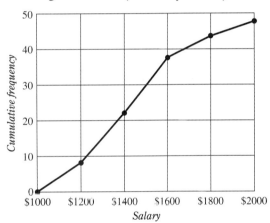

Salary

b. Approximately 1,450 and 1,670

17. a. Stem-and-left display:

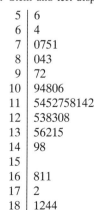

```
 5 | 6
 6 | 4
 7 | 0751
 8 | 043
 9 | 72
10 | 94806
11 | 5452758142
12 | 538308
13 | 56215
14 | 98
15 |
16 | 811
17 | 2
18 | 1244
19 | 3
```

b. The central tendency is about 120.

c. The data values vary from 56 to 193; quite a few of the data values are in the range from 110 to 129.

d. The data values are fairly symmetric around the central tendency.

e. No data values are far removed from the others.

f. There is a gap in the 150s.

g. There is a large concentration in the 110s and smaller concentrations in the 70s and 180s.

19. a. $[(2)(30)]^{.3333} = 3.91$; 4 classes

b. The approximate class interval or width is

$$(37 - 4)/4 = 8.25$$

so the class interval can be 9.

c. Frequency distribution:

Class	Frequency
2.5 to 11.5	2
11.5 to 20.5	6
20.5 to 29.5	14
29.5 to 38.5	8

d. Histogram of price–earnings ratios:

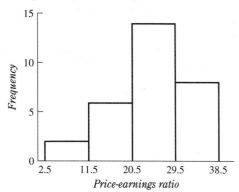

21. a. $[(2)(30)]^{.3333} = 3.91$; 4 classes

b. We use a class interval of .07.

c. A histogram for *NI/TA* follows:

Boundaries	Middle of Interval	Number of Observations	
−.145 to −.075	−0.1100	1	*
−.075 to −.005	−0.0400	0	
−.005 to .065	0.0300	11	***********
.065 to .135	0.1000	18	******************

d. About .07

e. High of about .12; low of about −.12

f. Stretched to the low side by the −.13 data value

g. The −.13 data value is removed from the others.

h. A gap exists between the −.13 and the next higher data value.

i. The data values are concentrated around .06 to .12.

23. a. $[(2)(80)]^{.3333} = 5.43$; so 6 classes. The approximate class interval or width is

$$(6.08 - 2.53)/6 = .592$$

so the class interval can be 0.6. The frequency distribution is

Class	Frequency
2.505 to 3.105	16
3.105 to 3.705	20
3.705 to 4.305	16
4.305 to 4.905	17
4.905 to 5.505	7
5.505 to 6.105	4
	80

b. Cumulative distribution:

Amount	Cumulative Frequency
Less than 2.505	0
Less than 3.105	16
Less than 3.705	36
Less than 4.305	52
Less than 4.905	69
Less than 5.505	76
Less than 6.105	80

c. Histogram of pollutant materials:

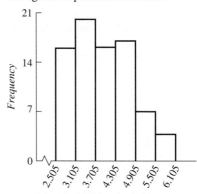

d. Stem-and-leaf display:

2. (50–74)	55 63 53
(75–99)	76 87 76 74 76 98 76 98
3. (00–24)	65 20 20 05 08 07 22 06
(25–49)	30 30 40 39 40 30 40 42
(50–74)	64 52 64 74 63 64 52 52 64 72 74 73 66
(75–99)	87 81 87 86 98 96 98
4. (00–24)	04 08 10 18
(25–49)	40 42 44 40 39 48 42 30
(50–74)	61 50 74 61 74 74 54 40 62 50
(75–99)	98
5. (00–24)	08 05 06
(25–49)	38 42
(50–74)	50 74 51
(75–99)	
6. (00–24)	06 08

The numbers in parentheses are not required on a completed display.

25. a. A, about 10 percent; B, about 20 percent
   b. B shows more variability than A.
   c. Neither is exactly symmetrical.
27. a. Histogram of stock returns:

MIDDLE OF INTERVAL	NUMBER OF OBSERVATIONS	
100	14	**************
150	29	*****************************
200	6	******
250	0	
300	0	
350	0	
400	0	
450	0	
500	1	*

   b. About 150%
   c. The data values have a high of about 500% and a low of about 100%; most of the values are between 100% to 200%.
   d. The data values are somewhat symmetric around 150%, but the one return stretches the histogram to the high side for returns.
   e. The highest return, 493%, is far removed to the high side.
   f. From the histogram, a gap exists from 250% to 450%.
   g. The data values are concentrated in the interval around 150%.
29. a. Histogram of percentage price increase ($n = 50$):

Midpoint	Count	
−50	1	*
0	0	
50	3	***
100	18	******************
150	23	***********************
200	4	****
250	0	
300	0	
350	0	
400	0	
450	0	
500	1	*

   b. About 130%
   c. The data values have a high of about 500% and low of about −50%; most of the values are between 50% to 200%.
   d. The data values are somewhat symmetric around 130%, but the one value stretches the histogram to the high side.
   e. The highest percentage, 493%, is far removed to the high side. The lowest percentage, −63%, is removed to the low side.
   f. A gap exists from about 225% to 490%.
   g. The data values are concentrated around 150%.

31. a. Stem-and-leaf of dividend ($n = 50$):

0	0000000000000000000000000000000000000000000000
0	6
1	0
1	
2	
2	
3	
3	
4	
4	
5	
5	
6	
6	9

   b. The central tendency is not well defined, but it is about 0%.
   c. The data values have a high of about $69 and a low of about $0; most of the values are close to $0.
   d. The data values are not symmetric mainly because the one $69.60 dividend value stretches the display to the high dividend side.
   e. The highest dividend, $69.60, is far removed to the high dividend side.
   f. A gap exists from about $10 to $69.
   g. The data values are concentrated around and just above $0.
33. a. A computer gave the following histogram of profits in millions of dollars:

Histogram of Profits	N = 55	
Midpoint	Count	
−1200	1	*
−800	0	
−400	2	**
0	45	***
400	4	****
800	1	*
1200	1	*
1600	1	*

   b. The central tendency is about $120 million.
   c. From the histogram, the data values have a high of about $1,600 million and a low of about −$1,200 million (−$1,167 million for USX).
   d. The data values are fairly symmetrical.
   e. The −$1,167 million value for USX is somewhat removed from the main concentration of data values on the low profits side.
   f. From the histogram, a gap exists between the lowest profit and the next lowest profit (−$1167 million for USX and −$384.7 million for Santa Fe Southern Pacific).
   g. The data values are concentrated around $0 million.
35. a. 4
   b. 10

c. A histogram from a computer follows:

Yen/$

Boundaries	MIDDLE OF INTERVAL	NUMBER OF OBSERVATIONS	
104.5 to 114.5	109.5	2	**
114.5 to 124.5	119.5	4	****
124.5 to 134.5	129.5	6	******
134.5 to 144.5	139.5	8	********

d. About 130

e. From the histogram, the data values have a high of about 144 and a low of about 105. There are more data values in classes with higher yen/$ values.

f. Skewed to the left

g. The distribution helps the investor understand the exchange rate risk. If the investor considers the exchange rate risk to be too high, she might liquidate some investments.

37. a. 4   b. 11

c. A histogram from a computer follows:

Spread

Boundaries	MIDDLE OF INTERVAL	NUMBER OF OBSERVATIONS	
78.5 to 89.5	84.0	4	****
89.5 to 100.5	95.0	9	********
100.5 to 111.5	106.0	5	*****
111.5 to 122.5	117.0	2	**

d. About 100

e. From the histogram, the data values have a high of about 122 and a low of about 79.

f. The data values are reasonably symmetrical; but not exactly symmetrical.

g. An investor could use the histogram to summarize the past spreads. The investor might compare the current spread of a long-term bond she owns to the past spreads. Thus she can assess the riskiness of her bond compared to the riskiness of the sample of bonds. If she thinks the spread is too high, she might sell her bond.

39. a. 5   b. 4

c. A histogram from a computer follows:

Japan_PE

Boundaries	MIDDLE OF INTERVAL	NUMBER OF OBSERVATIONS	
51.5 to 55.5	53.50	6	******
55.5 to 59.5	57.50	5	*****
59.5 to 63.5	61.50	15	***************
63.5 to 67.5	65.50	7	*******
67.5 to 71.5	69.50	2	**

d. About 60

e. From the histogram, the data values have a high of about 70 and a low of about 53. There are more data values in the third class.

41. a. Panel (b) is the histogram for the birth weight of babies since the distribution of birth weights of babies tends to be symmetrical and shaped like a bell.

b. Panel (a) is the histogram for waiting time since waiting time distributions tend to be asymmetrical and stretched to the right, high, or long waiting time side.

43. a. A computer gave the following histogram:

Histogram of Discount   N = 24

Midpoint	Count	
5.6	4	****
6.0	2	**
6.4	2	**
6.8	1	*
7.2	1	*
7.6	9	*********
8.0	5	*****

b. About 7.5 percent

c. A computer gave the following time series plot:

```
        8.00+ 1234
           -    5
     Discount-
           -         67890AB12
           -
        7.20+
           -                3
           -                4
           -
           -               56
        6.40+
           -                7
           -
           -
           -                8
        5.60+
           -                90AB
         +-----------+-----------+
         0          12          24
```

0 is month 10, A is month eleven, B is month 12

d. No

45. b. About 1242   d. No

Chapter 3 Descriptive Measures

1. a. 5   b. 4   c. 4
3. a. 4   b. 3   c. 3
5. a. 30.0   b. 27.0
7. 20.43
9. a. 5   b. 4   c. 2
11. a. 22.4   b. 58.78   c. 73.48   d. 7.67, 8.57
13. Panel (a) has more variation.
15. a. Both   b. .68   c. .95
    d. .025
17. 12.80%, 14.65%
19. 40%, 50%; B is larger.

21. Format may vary, but values are

Mean   =   12.08
Variance =   .085
StdDev   =   .29
Median   =   12.13
Minimum   =   11.55
Maximum   =   12.43
Range   =   .88
Sum   =   301.95
Cases   =   25

23. Format may vary, but values are

Mean   =   58.71
Variance =   317.98
StdDev   =   17.83
Median   =   58.3
Minimum   =   30.2
Maximum   =   87.8
Range   =   57.6
Sum   =   1467.7
Cases   =   25

A box plot provided by a computer follows:

```
BOX PLOT OF JOB SATISFACTION SCORES


                    ---------------------------
         -----------|I          +          I|----------
                    ---------------------------
        +---------+---------+---------+---------+---------+-----job_sat
        24        36        48        60        72        84
```

25. a. 4   **b.** 4   **c.** 3, 6   **d.** 5   **e.** 4   **f.** 2
27. a. 5   **b.** 5   **c.** 3, 5, 6   **d.** 6   **e.** 4   **f.** 2
29. a. 3   **b.** 3   **c.** No mode   **d.** 4   **e.** 2.5
   f. 1.58
31. a. 1   **b.** 1   **c.** −1, 1, 2   **d.** 6   **e.** 4   **f.** 2
33. a. 943.23   **b.** 941.5   **c.** 1392   **d.** 127,477.22
   e. 357.04   **f.** 123,227.98, 351.04
35. a. Population   **b.** 8.9, .13, .36, 8.80, 1.1
37. a. 5   **b.** 4   **c.** 2   **d.** 4   **e.** 5   **f.** −1 to 11
39. a. 2   **b.** 4   **c.** 2   **d.** 3   **e.** 5
41. a. At least .90   **b.** About .68
43. −1.25, 45,718.9, 213.82
45. b. Skewed to the right
   c. 0.4% to 16.70%
   d. −5.31% to 18.75%
   e. No
47. a. 130.60   **b.** 133   **c.** The median is resistant.
49. a. 98.85   **b.** 97.5   **c.** The median is resistant.
51. a. 60.743   **b.** 61
53. c. 3.821 to 9.641
55. a. 1768   **b.** 1256   **c.** Skewed   **d.** 4288.6
   e. 1795332   **f.** 1339.9
61. 2.532, 2, 1, 4.88, 2.21
63. 24.5, 35.12, 5.93, Yes

65. 13, 13
67. a. No   **b.** No   **c.** No   **d.** No   **e.** No
   f. Yes
69. a. 595.65, 650   **b.** 550.35; 42,744; 207

Chapter 4 Probability

1. a. About .10
   b. Relative frequency
3. a. Subjective   **b.** Relative frequency
5. Whether successive products that are coming off a production line that is operating as a stable process are defective or not are independent events is just one example.
7. .1034
9. a. .8464   **b.** .0064   **c.** .1472
11. 1/3
13. a. 8   **b.** 3   **c.** 4
15. a. 272   **b.** .29   **c.** .50
17. a. .025   **b.** .550   **c.** .682
19. a. .11   **b.** The revised probability is smaller.
21. a. .43   **b.** .57
23. 36 outcomes; 18 even, 30 greater than four
25. a. .5   **b.** .5   **c.** .25   **d.** .75
27. a. .48   **b.** .44   **c.** .08
29. a. 7/10   **b.** 7/10   **c.** 7/10   **d.** 49/100
   e. 49/100   **f.** Yes   **g.** Yes
31. a. .48   **b.** .06   **c.** .82   **d.** .30   **e.** .40
33. a. The four aces   **b.** 1/13
   c. The 26 cards from black suits   **d.** 1/2
   e. Ace of clubs, ace of spades   **f.** 2/52
   g. The 26 cards from black suits
   h. Ace of clubs, ace of spades   **i.** 1/13   **j.** 1/13
   k. Yes, since $P(A|B) = 1/13$ is equal to $P(A) = 1/13$
35. a. .83   **b.** .61   **c.** .60   **d.** .67
37. a. People often think it is more likely that the investor will increase her wealth.
   b. 1/2   **c.** 1/2
39. .0975
41. a. .9314   **b.** .0686
43. a. .02   **b.** .01   **c.** .0396
45. .47, .53
47. a. .00498   **b.** .9998
57. 120
59. 720
61. 504
63. Over 635 billion
65. a. 132   **b.** 55
67. 67,200
69. 210

Chapter 5 Discrete Probability Distributions

1. a. 1.00   **b.** .75   **c.** .25
3. a. H, T
   b. 1, 0
   c. .5, .5
5. a. 2   **b.** 1   **c.** 1

7. a. .3669   **b.** .0355
9. a. .2500   **b.** .0244
11. a. 2, 1.2   **b.** 2, 1.2
13. a. .0312   **b.** .9687   **c.** .3125
15. a. 4   **b.** $p_A < p_B < p_C$ **c.** $\mu_A < \mu_B < \mu_C$
    d. A is skewed right, B is symmetric, C is skewed left.
17. a. .3656   **b.** .0776
19. a. .0003   **b.** .4846   **c.** .9872
21. a. 2.25   **b.** .0507   **c.** .7042 (.7038 by interpolation)
23. a. .1042 by Poisson (exact answer is .1052)
25. a. .1311   **b.** .4606
27. a. 21, 2.5100   **b.** 20, 3.4641
29. a. Cross-hatch the bars with zero, one or two successes.
    b. .9298
31. a. .1216   **b.** .3917
33. a. .8042   **b.** .2053   **c.** .5886   **d.** 1.0000
35. a. .2916   **b.** .6561   **c.** .3439
37. a. .0205   **b.** .0000
39. a. $n_B > n_A$ **b.** $\mu_B > \mu_A$
    c. $\sigma_B^2 > \sigma_A^2$
    d. A and B are both skewed to the right (B is skewed less than A).
41. a. .3504   **b.** .0000   **c.** 7   **d.** 2.1
43. a. 7.5   **b.** .8491
    c. No, the probability of a stock-out seems too low (.1509).
45. a. .0018   **b.** .2921
47. a. .6703   **b.** .3297
49. a. .2240   **b.** .1606   **c.** .9502
51. .0680
53. a. .1296   **b.** .0614   **c.** .2995
63. a. .1143   **b.** .5143
65. a. .3263   **b.** .3916   **c.** 0
67. .580
69. .238
71. a. .5, .25   **b.** 1   **c.** .5   **d.** .707
73. a. 250 hours   **b.** 2500 hours squared   **c.** 50 hours

Chapter 6 Continuous Probability Distributions

1. a. 1   **b.** .5   **c.** .5
3. a. .64   **b.** .82
5. a. 5   **b.** .40
7. a. .3413   **b.** .1587
9. a. .9750   **b.** .0418
11. a. −1.90   **b.** Yes
13. a. .8413   **b.** .5670
15. a. .1056   **b.** .2643   **c.** Lower
17. .0062
19. .1056
21. a. .0548   **b.** .0232   **c.** .0000
23. a. .9713   **b.** .7062
25. .0104
27. a. .60   **b.** .15
29. a. 1.25   **b.** .60   **c.** .26
31. a. .25   **b.** .40

33. a. .6826   **b.** .8413
35. a. .2742   **b.** .5410
37. a. .7324   **b.** .2017
39. a. 1.70   **b.** −.50
41. a. −3.55   **b.** 1.79
43. a. .0062   **b.** .2670
45. .0279
47. 8856, 4650
49. a. .1359   **b.** .0228
51. 9.67
53. .9332
55. a. .0062   **b.** .0228
57. .8413
59. a. 40   **b.** .2327 (exact is .2374)
    c. .9545 (exact is .9509)
61. .0384 (exact is .0434)
63. a. .3228   **b.** Somewhat accurate
73. a. .6321   **b.** .0821
75. a. Exponential   **b.** .333   **c.** .9502
77. .0800
79. .6321

Chapter 7 Sampling Distributions

1. a. 30   **b.** 2
3. a. MT, MW, MTh, MF, TW, TTh, TF, WTh, WF, ThF
    b. Sample mean: 2 3 4 5 6 7 8
        Probability: .1 .1 .2 .2 .2 .1 .1
    c. 5.
5. a. .0228   **b.** .9544
7. a. .0668   **b.** .0919   **c.** .0002
9. a. .2207   **b.** .0001
11. a. .1587   **b.** CLT
13. .9987
15. a. Yes
    b. $E(\overline{X}) = \$100$, $\sigma_{\overline{x}} = 3$, $\overline{X}$ is approximately normally distributed due to the central limit theorem.
    c. .8413   **d.** .1587   **e.** No, .0228
17. a. .70, .0042   **b.** .0618
19. a. .0099   **b.** Yes   **c.** .5899
21. 50, 2.5
23. a. .0082   **b.** CLT
25. Exactly normal; approximately normal
27. .9544
29. a. .3446   **b.** 100, .1   **c.** Normal
    d. .0228   **e.** Individual part; sample mean
31. .9772
33. 2.945
35. a. .8413   **b.** 46.08 and 53.92   **c.** .0013
37. a. 152.66, 37.96   **b.** .0823, questionable   **c.** Yes
39. .4920
41. .0228
43. .9974
45. a. .0005   **b.** Yes
47. .3520
55. a. 32   **b.** 2.24   **c.** Exactly normal

57. .4176
59. .9786
61. .3936
63. .1230

Chapter 8 Estimation

1. a. 4 **b.** 36 **c.** 6 **d.** 2
3. a. 20 **b.** 100 **c.** 10 **d.** 2
5. a. 8 **b.** 16 **c.** 4 **d.** 1.33
7. a. 20 **b.** 4 **c.** 2 **d.** 1
9. 18.355 to 21.645
11. 14.977 to 16.623
13. 49
15. 2.064
17. 2.576
19. Yes
21. 2.154 to 5.846
23. 7.388 to 16.612
25. −1.208 to 17.208
27. 13.349 to 16.651
29. a. −58.26 to 55.76 **b.** No
31. a. 532.17 to 568.63 **b.** Yes
33. a. .2 **b.** 4 **c.** 2 **d.** .0064 **e.** .08
35. a. .98 **b.** .014 **c.** .95 to 1.01
37. .55 to .73
39. .036 to .064
41. .30 to .50
43. .03 to .21
45. 865
47. 68 more
49. 385
51. a. 50 **b.** 4 **c.** 2 **d.** 2/3
53. a. 4 **b.** 4 **c.** 2 **d.** .82
55. a. 6679 **b.** 4 **c.** 2 **d.** .82
57. 50
59. 17.869 to 22.131
61. −4.47 to 4.47
63. 52.71 to 59.29
65. −.31 to 4.31
67. 82.62 to 117.38
69. 2.454 to 5.546
71. 682.7 to 802.3
73. −44.57 to 51.07
75. a. .64 **b.** 23.04 **c.** 4.8 **d.** .002304
      e. .0480
77. .44 to .56
79. .75 to .85
81. .50 to .84
83. .77 to .91
85. .1342 to .2658
87. 67
89. 9604
99. a. 30, 10 **b.** 9.14 to 50.86
101. 1507.18 to 2992.82
103. 620.1 to 8879.9

105. a. 11.24 to 20.76 **b.** Yes
107. −.29 to 8.29
109. a. .28, .46, .18 **b.** .057 **c.** .047 to .313
111. .101 to .299
113. −.0016 to .2016

Chapter 9 Tests of Hypotheses

1. The employee is honest.
3. $H_0: p \leq .10$; $H_a: p > .10$
5. a. $H_0: \mu \leq 5$; $H_a: \mu > 5$
      b. .025
7. 13.333 is greater than 1.711, so reject H_0.
9. −2.5 is not less than −2.797, so do not reject H_0.
11. 3.586 is greater than 1.671, so reject H_0.
13. 1.852 is greater than 1.796, so reject H_0.
15. a. $H_0: \mu \geq 150$
      b. −2.556 is less than −1.740, so reject H_0.
17. 3 is greater than 2.896, so reject H_0.
19. a. 2 is greater than 1.28, so reject H_0.
      b. Z **c.** CLT
21. 1.057 is not greater than 1.645, so do not reject H_0.
23. −1.864 is not less than −1.96, so do not reject H_0.
25. 2 is greater than 1.761, so reject H_0.
27. −1.8438 is less than −1.734, so reject H_0.
29. 1.63 is not greater than 1.6526, so do not reject H_0.
31. 3.84 is greater than 2.228, so reject H_0.
33. a. $H_0: p_{\text{men}} \leq p_{\text{women}}$; $H_a: p_{\text{men}} > p_{\text{women}}$
      b. 1.591 is not greater than 1.645, so do not reject H_0.
35. −2.18 is less than −1.96, so reject H_0.
37. −.95 is not less than −1.645, so do not reject H_0.
39. 2.02 is not greater than 2.2693, so do not reject H_0.
41. 6.25 is greater than 3.2374, so reject H_0.
43. 2.3 is not greater than 6.9928, so do not reject H_0.
45. −7 is less than −1.6772, so reject H_0.
47. 1.5 is greater than 1.318, so reject H_0.
49. −2 is not less than −2.064, so do not reject H_0.
51. 8 is greater than 1.638, so reject H_0.
53. −1 is not less than −1.860, so do not reject H_0.
55. 1.717 is not greater than 2.145, so do not reject H_0.
57. −2 is less than −1.96, so reject H_0.
59. 3.677 is greater than 1.645, so reject H_0.
61. 42.7 is greater than 1.645, so reject H_0.
63. −2.5 is less than −1.714, so reject H_0.
65. 2 is greater than 1.761, so reject H_0.
67. 3.74 is greater than 1.782, so reject H_0.
69. 1.894 is greater than 1.761, so reject H_0.
71. .87 is not greater than 2.467, so do not reject H_0.
73. −1.366 is not less than −1.96, so do not reject H_0.
75. 6.32 is greater than 1.28, so reject H_0.
77. 2.49 is greater than 1.645, so reject H_0.
79. a. 6.25 is greater than 4.026, so reject H_0.
      b. No
81. 4 is greater than 3.797, so reject H_0.
91. a. Not rejecting that the pollution measurement is 132 or more when it is actually less than 132.
      b. .5000

93. a. .5000   **b.** .0500
95. 2 is not greater than 2.896, so do not reject H_0.
97. 1.816 is greater than 1.350, so reject H_0.
99. a. .9741 is not greater than 2.552, so do not reject H_0.
   b. Do not reject H_0 in either case.

Chapter 10 Analysis of Variance

1. a. 6, 4, 4, 10, 4, 7, 1   **b.** 36   **c.** 4
   d. 9 is greater than 5.1433, so reject H_0.
3. a. 3   **b.** 4
   c. 4.21 is not greater than 7.5910, so do not reject H_0.
5. a. 25.167   **b.** 297.33   **c.** 21.19
   d. 14.03 is greater than 4.2565 so reject H_0.
7. a. 8, 11, 5, 11, 5, 18, 4, 4, 8,   **b.** 30   **c.** 7
   d. 4.2857 is not greater than 4.7571, so do not reject H_0.
9. 97.22 is greater than 3.8853, so reject H_0.
11. a. Analysis of variance:

Source	df	SS	MS	F
Between	3	216	72	5.426
Within	11	146	13.27	
Total	14	362		

   b. 5.426 is greater than 3.5874, so reject H_0.
13. a. B rated bonds
   b. −5.37 to 5.37, −11.37 to −.63, −11.37 to −.63, B rated bonds
15. No
17. a. 5, 7, 6, 2, 1, 7, 3   **b.** 21   **c.** 22/6
   d. 5.727 is greater than 5.1433, so reject H_0.
19. a. 12.6   **b.** 201.45   **c.** 2.48
   d. 81.23 is greater than 4.7374, so reject H_0.
21. a. 5, 5, 2, 8   **b.** 72   **c.** 2   **d.** 36
   e. 54   **f.** 9   **g.** 6
   h. 6 is greater than 4.2565, so reject H_0.
23. .075 is not greater than 7.2057, so do not reject H_0.
25. 10.0421 is greater than 6.0129, so reject H_0.
27. a. Analysis of variance:

Source	df	SS	MS	F
Factor	2	182.4	91.2	1.99
Error	25	1144.3	45.8	
Total	27	1326.7		

   b. 1.99 is not greater than 3.3852 so do not reject H_0.
29. 6.83 is greater than 3.3277, so reject H_0.
31. a. Analysis of variance:

Source	df	SS	MS	F
Between	2	40	20	1.25
Within	7	112	16	
Total	9	152		

   b. 1.25 is not greater than 4.7374, so do not reject H_0.
   c. No

39. a. 2, 31, 41
   b. .842 is not greater than 3.9877, so do not reject H_0.
41. .84 = .84; 3.99 = 3.99
43. 3.589 is not greater than 4.6001, so do not reject H_0.
45. a.

Source	df	Sum of Squares	Mean Square	F-Ratio	P-Value
Treatments	2	565.444	282.722	16.3213	.00071
Blocks	5	105.111	21.0222	1.21360	.37020
Error	10	173.222	17.3222		
Total	17	843.778			

   b. 16.32 is greater than 4.1028, so reject H_0.
47. a.

Source	df	Sum of Squares	Mean Square	F-Ratio	P-Value
Treatments	2	49.5556	24.7778	26.2352	.00502
Blocks	2	70.222	35.1111	37.1764	.00261
Error	4	3.777779	.944447		
Total	8	123.556			

   b. 26.24 is greater than 6.9443, so reject H_0.
49. a.

Source	df	Sum of Squares	Mean Square	F-Ratio	P-Value
Treatments	2	550.889	275.445	120.930	.00026
Blocks	2	1336.22	668.111	293.326	.00005
Error	4	9.11084	2.27771		
Total	8	1896.22			

   b. 120.93 is greater than 6.9443, so reject H_0.
   c. 4.24 to 23.76, 8.57 to 28.09, −5.43 to 14.09; package 1
51. a.

Source	df	Sum of Squares	Mean Square	F-Ratio	P-Value
Disk Brand	2	77.0150	38.5075	44.3891	.00025
Microproc	1	.300842	.300842	.346792	.57742
Interactions	2	3.00166	1.50083	1.73006	.25513
Error	6	5.20500	.867500		
Total	11	85.5225			

   b. 1.73 is not greater than 5.1433, so do not reject H_0.
   c. 44.39 is greater than 5.1433, so reject H_0.
   d. .35 is not greater than 5.9874, so do not reject H_0.
53. a.

Source	df	Sum of Squares	Mean Square	F-Ratio	P-Value
Fertilizer	1	80.0833	80.0833	45.7619	.00051
Corn-Variety	2	612.500	306.250	175.000	.00000
Interactions	2	11.1667	5.58337	3.1905	.11381
Error	6	10.5000	1.75000		
Total	11	714.250			

b. 3.19 is not greater than 5.1433, so do not reject H_0.

c. 45.76 is greater than 5.9874, so reject H_0.

d. 175 is greater than 5.1433, so reject H_0.

e. $-.58$ to 5.58, 13.17 to 19.33, 10.67 to 16.83; variety 1 and variety 2

55. a. .067 is not greater than 5.1433, so do not reject H_0.

b. 33.267 is greater than 5.1433, so reject H_0.

c. 8.067 is greater than 5.9874, so reject H_0.

d. 1.90 to 7.10, 3.65 to 8.85, $-.85$ to 4.35; single business

Chapter 11 Tests Using Categorical Data

1. 260.41 is greater than 11.34, so reject H_0.

3. 32.346 is greater than 5.99, so reject H_0.

5. 7.22 is not greater than 12.84, so do not reject H_0; 7.22 is not greater than 7.81, so do not reject H_0.

7. a. 420.741 is greater than 9.49, so reject H_0.

b. No

9. 1.145 is not greater than 9.21, so do not reject H_0.

11. 30.275 is greater than 14.45, so reject H_0.

13. 1.833 is not greater than 3.84, so do not reject H_0.

15. 6.843 is not greater than 9.49, so do not reject H_0.

17. a. .267, .300, .433

b. 14 is greater than 5.99, so reject H_0.

19. 29 is greater than 16.81, so reject H_0.

21. a. 620.284 is greater than 7.81, so reject H_0.

b. 1 to 3 pm

23. 84.0395 is greater than 13.28, so reject H_0.

25. 24.11 is greater than 9.21, so reject H_0.

27. 43.84 is greater than 9.49, so reject H_0.

29. 17.228 is greater than 3.84, so reject H_0.

31. 754.104 is greater than 26.30, so reject H_0.

39. a. 0   **b.** Yes

41. 5.95 is not greater than 9.49, so do not reject H_0.

43. 3.87 is not greater than 16.01, so do not reject H_0.

Chapter 12 Regression and Correlation

1. a. 0, 2   **b.** 16

3. a. 6.2, .2   **b.** 7.4

5. -58, 4

7. a. 116, 112, 4   **b.** 6, 1, 5   **c.** 112, .8

d. 112, 4, 116; 1, 5, 6; 112, .8

9. a. 10, 6.4, 3.6   **b.** 4, 1, 3   **c.** 6.4, 1.2

d. 6.4, 3.6, 10.0; 1, 3, 4; 6.4, 1.2

11. a. 4   **b.** .7143

13. square feet, square feet/hour, square feet

15. a. .910   **b.** 13.66 to 18.34   **c.** 1.276

d. 12.72 to 19.28

17. a. .0866

b. 2.31 is not greater than 3.182, so do not reject H_0.

c. .554   **d.** 5.64 to 9.16   **e.** 1.227

f. 3.50 to 11.30

19. 2.5, -3 **b.** 2.5

c. -6 is less than -2.048, so reject H_0.

21. a. 9   **b.** 8.31 to 9.69   **c.** 6.21 to 11.79

23. .5; 3 is greater than 2.771, so reject H_0.

25. a. .8944

b. 4 is greater than 2.776, so reject H_0.

27. About $-.6$

29. a. 2, 1   **b.** 12   **c.** 8, 4, 4   **d.** 4, 1, 3   **e.** 4

f. 4/3   **g.** 4, 4, 8; 1, 3, 4; 4, 4/3   **h.** 4/3

i. 1.1547   **j.** .50   **k.** .57735

l. 1.732 is not greater than 3.182, so do not reject H_0.

m. 3 is not greater than 10.128, so do not reject H_0.

n. .5164   **o.** 10.357 to 13.643   **p.** 1.2649

q. 7.975 to 16.025

31. a. .5

b. 2 is not greater than 2.776, so do not reject H_0.

c. 4 is not greater than 7.7086, so do not reject H_0.

d. .816   **e.** 5.73 to 10.27   **f.** 2.160

g. 2.00 to 14.00

33. a. 2.683   **b.** .966

35. a. 9.08 to 14.92   **b.** 4.51 to 19.49

37. Dollars, dollars/pound, no dimensions

39. a. 27.2, -1.2 **b.** 16

c. -3 is less than -2.921, so reject H_0.

d. .36

41. a. 19.89, .822

b. -2.55 is less than -2.101, so reject H_0.

c. .885   **d.** 24.0   **e.** 23.67 to 24.33

f. 22.50 to 25.50

43. a. 5.9, .155

b. .114 to .196

c. 8.16 is greater than 2.160, so reject H_0.

d. 9.7 to 11.4

e. 8.4 to 12.7

45. a. .0028, .006572

b. .00627 to .00688

47. a. 2.90 is greater than 2.447, so reject H_0.

b. 2.90 is greater than 2.447, so reject H_0.

c. 8.385 is greater than 5.9874, so reject H_0.

49. a. .97 is not greater than 2.571, so do not reject H_0.

b. .97 is not greater than 2.571, so do not reject H_0.

c. .94 is not greater than 6.6079, so do not reject H_0.

51. 3 is greater than 2.120, so reject H_0.

53. -3.011 is less than -2.069, so reject H_0.

55. a. .8

b. 2.309 is not greater than 3.182, so do not reject H_0.

57. a. $-.682$

b. -2.95 is less than -2.228, so reject H_0.

67. a. $-2.05640, \ldots$

b. Linear

69. a. $-0.61136, \ldots$

b. Observation 5

Chapter 13 Multiple Regression

1. a. False   **b.** False   **c.** True

3. $b_0 = .72$ means $.72 is the predicted selling price of a stock with zero earnings per share and zero earnings growth. The units are dollars per share.

$b_1 = 5.94$ means a \$5.94 increase in price per share is associated with each one dollar increase in earnings per share, assuming X_2 is held constant. The coefficient has no units.

$b_2 = 1.08$ means a \$1.08 increase in price per share is associated with each 1% increase in earnings growth, assuming X_1 is held constant. The units are dollars per share per percentage point.

5. **a.** 9   **b.** 2, 7, 9   **c.** 7.5, 1.286
   d. 15, 9, 24; 2, 7, 9; 7.5, 1.286
7. **a.** 3807.8, 542.2, 4350   **b.** 2, 2, 4   **c.** 1903.9, 271.1
   d. 3807.8, 542.2, 4350; 2, 2, 4; 1903.9, 271.1
9. **a.** 34.37   **b.** 5.862   **c.** .815
11. **a.** 5.8 is greater than 4.7374, so reject H_0.
    b. 3.1 is greater than 2.365, so reject H_0.
13. .37 to 11.51
15. -14.72 to 39.32
17. **a.** Estimated regression coefficients are 5.977, .802, -1.930.
    b. $b_2 = -1.930$ means that variable rate loans have beginning interest rates that are estimated to be 1.93% below fixed rate loans while holding yield constant.
    c. -3.367 is less than -2.110, so reject H_0.
19. **a.** True   **b.** False   **c.** True
21. **a.** Estimated regression coefficients are 51.58, $-.40$, $-.19$. $b_1 = -.40$ means that a one month increase in tenure is associated with a decrease in absenteeism of .40 days per year, assuming X_2 is held constant.
    b. 26.1   **c.** .86
    d. 21.63 is greater than 4.7347, so reject H_0.
    e. -2.386 is less than -2.365, so reject H_0.
23. 8101.89; .84; 84% of the squared error in estimating repair expense by using the sample mean repair expense can be eliminated by estimating repair expense using the regression equation.
25. **a.** Estimated regression coefficients are 1.707, .912, .221.
    b. 11.792
    c. .615
    d. 5.60 is greater than 4.7374, so reject H_0.
    e. 3.24 is greater than 2.365, so reject H_0.
27. **a.** Estimated regression coefficients are 21.987, 3.205, 2.866.
    b. 63.313
    c. 2.866
    d. 2.233 is greater than 2.110, so reject H_0.
29. **a.** Estimated regression coefficients are 2.217, .071, 3.545.
    b. 9.312
    c. 105.2, 93.455, 11.745, 2, 17, 46.727, .691
    d. .831
    e. .888
    f. 67.6 is greater than 3.5915, so reject H_0.
    g. 4.67 is greater than 2.110, so reject H_0; 9.00 is greater than 2.110, so reject H_0.
31. **a.** Estimated regression coefficients are 52.92, 3.00, -1.07.
    b. 38.02

c. 1062, 1037.44, 24.56, 2, 7, 518.72, 3.51
d. 1.873
e. .977
f. 147.8 is greater than 4.7374, so reject H_0.
g. 15.75 is greater than 2.365, so reject H_0; -8.32 is less than -2.365, so reject H_0.
33. **a.** Estimated regression coefficients are 3.27, .736, .106.
    b. .84
    c. 18.41 is greater than 4.7374, so reject H_0.
    d. 5.29 is greater than 2.365, so reject H_0; .56 is not greater than 2.365, so do not reject H_0.
35. **a.** Estimated regression coefficients are 51.6, $-.810$, -1.19, .079.
    b. .867
    c. 13.00 is greater than 4.7571, so reject H_0.
    d. -5.93 is less than -2.447, so reject H_0; $-.90$ is not less than -2.447, so do not reject H_0; .23 is not greater than 2.447, so do not reject H_0.
43. **a.** .411
    b. Multicollinearity
45. **a.** 1.85 years
    b. 60% are female.
    c. AGE and TEST; $r^2 = .39660$
    d. Estimated regression coefficients are -86.68, 3.22.
    e. 3.27 is greater than 2.069, so reject H_0.
47. **a.** X_1
    b. Estimated regression coefficients are -2.250, 5.12.
    c. .7914
    d. Estimated regression coefficients are -3.911, 5.0, .06.
    e. .7923
49. **a.** 3.03
    b. One every .0339 year
51. **a.** Estimated regression coefficients are 208.74, 6.53.
    b. .67494
    c. 5.78 is greater than 2.921, so reject H_0.
53. **a.** 1.594 is not greater than 2.131, so do not reject H_0.
    b. After step 1
55. **a.** Percentage completions, percentage touchdowns, percentage interceptions, and average yardage gained.
    b. Estimated regression coefficients are -10.60, 3.11.
    c. .6358
    d. Estimated regression coefficients are 2.54, .827, 4.12, 3.34, -4.11.
    e. .9999
    f. 66.28, 54.77, and 91.53 are all greater than 2.201 and -122.57 is less than -2.201, so reject H_0 in each case.
57. **a.** Estimated regression coefficients are 1.32, $-.41$, 1.14.
    b. 17.92
    c. Estimated regression coefficients are -3.11, 5.29.
    d. Estimated regression coefficients are $-.12$, .34.
    e. Estimated regression coefficients are $-.42$, 2.37.
    f. Part a
59. **a.** Estimated regression coefficients are 19.92, $-.26$, $-.16$.
    b. 16.32

c. Estimated regression coefficients are 22.37, −1.85.
d. Part a
61. **a.** Estimated regression coefficients are −1.69, .02, .16, .24.
   b. 79.74
   c. Estimated regression coefficients are 19.83, .17.
   d. 178.29
   e. 6.80 is greater than 3.9823, so reject H_0.
   f. Yes
63. **a.** Estimated regression coefficients are 36.53, 1.22, −.33, −.13.
   b. 52.16
   c. Estimated regression coefficients are 11.27, 1.39.
   d. 406.26
   e. 37.34 is greater than 2.9823, so reject H_0.
   f. Yes
65. 1.14, .44

Chapter 14 Time Series Analysis

1. **a.** 393   **b.** 131   **c.** 32
3. Many examples are possible.
5. **b.** Curved
   c. No, seasonal variation takes place within one year; these data are annual.
7. Year 2—25.67, Year 3—25.67, . . . , Year 18—253.33
9. **a.** 96.00, 101.80, . . . , 201.23
   b. 96.00, 119.20, . . . , 288.21
11. $\hat{Y} = -34.77 + 14.56t$
13. First method: 109.0, 96.7, 87.2, 107.1
   Ratio-to-moving-average method: 121.2, 100.4, 80.3, 98.1
15. **a.** 90.9, 72.7, 136.4
   b. 99.0, 70.4, 130.6
17. The data for three years ago are not only larger than the other years, they are also in a different pattern. The data probably will increase the intercept of the trend line. They may increase the slope slightly also.
19. 103.7, 94.8, 87.2, 89.8; 98.5, 92.3, 87.0, 93.4
21. $74,628; $895,536
23. Adjusted visits: 13,932, 16,393, 11,852, 11,435
   No, the manager needs to plan for actual visits.
25. **a.** Yes, they sum to 1200.
   b. $100,000, 180,000, 280,000, . . . , $80,000
27. Stock market prices, among other examples.
29. **a.** July of year 1   **b.** June of year 2
31. **a.** 570.4, 637.4, . . . , 1145.6
   b. 437.0, 455.0, . . . , 1181.3
33. **a.** 137.6, 98.1, 68.3, 96.0
   b. 0.0, −8.35, −14.90, −10.65
35. Irregular variation; probably due to an unusually high number of winter snow and ice storms
37. **a.** $152,950   **b.** $339,853   **c.** $342,998
   d. The one for 1996, since 1996 is closer to the present.
   e. The forecasts for January through December are

   210,140, 240,768, 233,833, 374,319, 314,611.5, 254,600, 380,950, 420,090, 524,504.5, 640,946, 734,872.5, 246,848

39. 154,028
41. **a.** 21,000, 59,500, 35,000, 24,500
   b. 33,333   **c.** 133,333
51. .65 is less than 1.20, so reject H_0.
53. 0.31 is less than .90 (approximately), so reject H_0.
55. 1.90 is not less than .90, so do not reject H_0.
57. **a.** —, −2.55, . . . , 1.81   **c.** Yes
59. **a.** Estimated coefficients are .001, .830.
   b. .574

Chapter 15 Statistical Process Control

1. Out of control
3. **a.** 11.93   **b.** 7.91, 15.95   **c.** Yes   **d.** No
   e. Sunday business passengers
5. **a.** 5.14   **b.** 0.00, 11.72   **c.** No   **d.** Yes
   e. None
7. **a.** .07   **b.** 0.0, .147   **c.** No   **d.** Yes   **e.** None
9. **a.** 10.146   **b.** 6.895, 13.40   **c.** No   **d.** Yes
   e. None
11. **a.** .9370   **b.** .62   **c.** .61   **d.** No
13. **a.** The time series plot trends downward.
   b. No   **c.** No
15. **a.** 5.2375   **b.** 4.82, 5.65   **c.** Yes   **d.** No
   e. Employees may be tiring at the end of shifts.
17. **a.** .525   **b.** 0.00, 1.20   **c.** No   **d.** Yes
   e. None
19. **a.** .0875   **b.** .003, .172   **c.** Yes   **d.** No
   e. Employees may be getting tired toward the end of the shift.
21. **a.** 9.865   **b.** 8.74, 10.99   **c.** No   **d.** Yes
   e. None
23. **a.** 1.57   **b.** 0.00, 3.58   **c.** No   **d.** Yes
   e. None
25. **a.** .101   **b.** .011, .191   **c.** No   **d.** Yes
   e. None
27. **a.** 2.001   **b.** 1.962, 2.040   **c.** No   **d.** Yes
   e. None
29. **a.** .9681   **b.** .72   **c.** .68   **d.** No
31. **b.** Typesetting

Chapter 16 Decision Analysis

1. **a.** Market or not market the new product.
   b. Payoff table for Semiconductor:

		Action	
		A₁	*A₂*
Probability	*Event*	*Market*	*Not Market*
.05	7% Mkt Share	−110	0
.10	8% Mkt Share	−80	0
.12	9% Mkt Share	−50	0
.18	10% Mkt Share	0	0
.25	11% Mkt Share	40	0
.20	12% Mkt Share	80	0
.10	13% Mkt Share	120	0

3. A_1 dominates A_4, and A_2 dominates A_4.
5. a. -500, 3275, 4075, 2325   **b.** Make 7
7. Market, \$18,500
9. 1560, 1400, 3400
11. a. Table of opportunity losses (in dollars):

	Action			
Event	*Make 5*	*Make 6*	*Make 7*	*Make 8*
Sell 5	0	2600	5200	7800
Sell 6	5900	0	2600	5200
Sell 7	11800	5900	0	2600
Sell 8	17700	11800	5900	0

b. *Make* 7, 2800
13. a. 38,000   **b.** 19,500
15. a. 16,500   **b.** 4,200
17. a. A_3, 10,000   **b.** A_1, 25,000   **c.** A_3, 10,000
d. A_1, 21,667
19. a. Action 2, 107,500   **b.** 7,500   **c.** Action 2, 70,000
d. Action 1, 120,000   **e.** Action 2, 108,333
f. Action 2, 105,000
21. a. Payoff table:

		Action		
Event	*Prob.*	*A_1* *Do Nothing*	*A_2* *Renew*	*A_3* *Buy Outright*
Rezone	.15	0	85,000	100,000
Deny	.30	0	$-10,000$	$-15,000$
Table	.55	0	5,000	20,000

b. No   **c.** 0, 12,500, 21,500   **d.** 4,500
23. a. Action 3, 400   **b.** Action 1, 70
c. Action 1, 100   **d.** Action 3, 225   **e.** 190
25. b. Sell the lease.   **c.** 650,000
27. b. For remain at present location, $EMV = \$7000$; for
move to new location, $EMV = -\$10,000$.
c. Optimist's decision would be to remain at present
location.
d. Maximum likelihood decision would be to stay since
the most likely event is a shopping mall, which has the
best possible outcome.
e. Yes, there are no sequences of decisions or events.
29. Interchange, 120,000

Chapter 17 Decision Analysis with Sampling Information

1. b. Approximately .45
c. Approximately .30
3. b. .25, .35, .10, .07, .23 (approximately)
5. .5300, .3462, .1238
7. a. Win job and lose job
b. .35, .65
c. 0, 1

9. \$9,700
11. a. Consistent
b. Consistent, only if the company's sales cannot possibly
rise when the economy is down
c. Inconsistent
d. Inconsistent
13. .4, .6
15. No, do the opposite of what the experiment suggests.
17. Events are defective rates of .05 or .10. Actions are to
make an adjustment or not make an adjustment. Costs are
1900, 1500, 1900, 3000.
19. Do not adjust; 1680
21. b. Borrow from trust fund, \$10,800; borrow from bank,
\$9570.75.
c. With revised probabilities, borrow from bank,
\$8996.03.
d. Applying to the bank is still preferred.

Chapter 18 Nonparametric Methods

1. -1.29 is not less than -1.645, so do not reject H_0.
3. -1.87 is less than -1.645, so reject H_0.
5. 1.79 is not greater than 1.96, so do not reject H_0.
7. 1.79 is greater than 1.282, so reject H_0.
9. -1.86 is not less than -1.96, so do not reject H_0.
11. .88; 2.78 is greater than 1.645, so reject H_0.
13. 2.11 is greater than 1.645, so reject H_0.
15. 2.68 is greater than 1.96, so reject H_0.
17. -1.90 is less than -1.645, so reject H_0.
19. .4804 is not greater than 1.282, so do not reject H_0.
21. $-.99$ is not less than -1.96, so do not reject H_0.
23. .75 is not less than -2.33, so do not reject H_0.
25. 2.62 is greater than 1.282, so reject H_0.
37. 1.73 is greater than 1.282, so reject H_0.
39. The sample sizes are different and the data are not paired.
We have no reason to believe that the variation among any
proposed pairings would be greater than the variation between the units within the pairs.
41. -4.58 is less than -1.96, so reject H_0.
43. -4.46 is less than -1.96, so reject H_0.
45. 12.02 is greater than 5.99, so reject H_0.

Chapter 19 Index Numbers

1. 146.0
3. a. 100.0, 101.3, 105.2, 113.0, 121.6
b. 82.3, 83.3, 86.5, 92.9, 100.0
5. 158.4
7. 160.0
9. a. 103.6
b. Combined price changes resulted in an increase.
11. a. Most consumer goods such as automobiles, household
appliances, housing, and so forth
b. Most commodities such as coal, chemicals, metals, and
so forth

13. a. .92, .91, .87, .83, .80

   b. 92%, 91%, 87%, 83%, 80%

15. a. 128.6   **b.** 131.8   **c.** 178.2   **d.** 130.1

17. a. 151.3   **b.** 282.1

19. 100.0, 103.1, 135.7

21. 135.3

23. 200.0

25. No, −$1,575,643

27. $8,892,857

29. The year is not recent enough and quality has changed.

31. Comparisons of price indices between geographical locations is generally not appropriate.

Appendix B Summation Notation

1. a. $1^2 + 2^2 + 3^2 + 4^2 + 5^2$

   b. $aY_1 + a^2Y_2 + a^3Y_3 + a^4Y_4$

   c. $Y_2^2 + Y_3^3 + Y_4^4 + Y_5^5$

   d. $Y_3 - a^3 + Y_4 - a^4 + Y_5 - a^5 + Y_6 - a^6$

   e. $Y_1^2 + Y_2^2 + Y_3^2 - 3$

   f. $-Y_4 + Y_5 - Y_6 + Y_7$

3. a. 23   **b.** 105   **c.** 529   **d.** 14   **e.** 15

5. a. $\Sigma(Y_i - c)^2 = \Sigma(Y_i^2 - 2Y_ic + c^2)$

$$= \Sigma Y_i^2 - \Sigma 2Y_ic + \Sigma c^2$$

$$= \Sigma Y_i^2 - 2c\Sigma Y_i + nc^2$$

   b. $\Sigma(Y_i - \overline{Y}) = \Sigma Y_i - \Sigma \overline{Y} = \Sigma Y_i - n\overline{Y}$

$$= (n/n)\Sigma Y_i - n\overline{Y} = n\overline{Y} - n\overline{Y} = 0$$

Index

APPENDIX C, TABLE VII   F Distribution Values

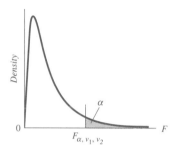

F Values When $\alpha = .05$

ν_1 / ν_2	1	2	3	4	5	6	7	8	9
1	161.45	199.50	215.71	224.58	230.16	233.99	236.77	238.88	240.54
2	18.513	19.000	19.164	19.247	19.296	19.330	19.353	19.371	19.385
3	10.128	9.5521	9.2766	9.1172	9.0135	8.9406	8.8868	8.8452	8.8123
4	7.7086	6.9443	6.5914	6.3883	6.2560	6.1631	6.0942	6.0410	5.9988
5	6.6079	5.7861	5.4095	5.1922	5.0503	4.9503	4.8759	4.8183	4.7725
6	5.9874	5.1433	4.7571	4.5337	4.3874	4.2839	4.2066	4.1468	4.0990
7	5.5914	4.7374	4.3468	4.1203	3.9715	3.8660	3.7870	3.7257	3.6767
8	5.3177	4.4590	4.0662	3.8378	3.6875	3.5806	3.5005	3.4381	3.3881
9	5.1174	4.2565	3.8626	3.6331	3.4817	3.3738	3.2927	3.2296	3.1789
10	4.9646	4.1028	3.7083	3.4780	3.3258	3.2172	3.1355	3.0717	3.0204
11	4.8443	3.9823	3.5874	3.3567	3.2039	3.0946	3.0123	2.9480	2.8962
12	4.7472	3.8853	3.4903	3.2592	3.1059	2.9961	2.9134	2.8486	2.7964
13	4.6672	3.8056	3.4105	3.1791	3.0254	2.9153	2.8321	2.7669	2.7144
14	4.6001	3.7389	3.3439	3.1122	2.9582	2.8477	2.7642	2.6987	2.6458
15	4.5431	3.6823	3.2874	3.0556	2.9013	2.7905	2.7066	2.6408	2.5876
16	4.4940	3.6337	3.2389	3.0069	2.8524	2.7413	2.6572	2.5911	2.5377
17	4.4513	3.5915	3.1968	2.9647	2.8100	2.6987	2.6143	2.5480	2.4943
18	4.4139	3.5546	3.1599	2.9277	2.7729	2.6613	2.5767	2.5102	2.4563
19	4.3808	3.5219	3.1274	2.8951	2.7401	2.6283	2.5435	2.4768	2.4227
20	4.3513	3.4928	3.0984	2.8661	2.7109	2.5990	2.5140	2.4471	2.3928
21	4.3248	3.4668	3.0725	2.8401	2.6848	2.5757	2.4876	2.4205	2.3661
22	4.3009	3.4434	3.0491	2.8167	2.6613	2.5491	2.4638	2.3965	2.3419
23	4.2793	3.4221	3.0280	2.7955	2.6400	2.5277	2.4422	2.3748	2.3201
24	4.2597	3.4028	3.0088	2.7763	2.6207	2.5082	2.4226	2.3551	2.3002
25	4.2417	3.3852	2.9912	2.7587	2.6030	2.4904	2.4047	2.3371	2.2821
26	4.2252	3.3690	2.9751	2.7426	2.5868	2.4741	2.3883	2.3205	2.2655
27	4.2100	3.3541	2.9604	2.7278	2.5719	2.4591	2.3732	2.3053	2.2501
28	4.1960	3.3404	2.9467	2.7141	2.5581	2.4453	2.3593	2.2913	2.2360
29	4.1830	3.3277	2.9340	2.7014	2.5454	2.4324	2.3463	2.2782	2.2229
30	4.1709	3.3158	2.9223	2.6896	2.5336	2.4205	2.3343	2.2662	2.2107
40	4.0848	3.2317	2.8387	2.6060	2.4495	2.3359	2.2490	2.1802	2.1240
60	4.0012	3.1504	2.7581	2.5252	2.3683	2.2540	2.1665	2.0970	2.0401
120	3.9201	3.0718	2.6802	2.4472	2.2900	2.1750	2.0867	2.0164	1.9588
∞	3.8415	2.9957	2.6049	2.3719	2.2141	2.0986	2.0096	1.9384	1.8799

Note: For example, if $\alpha = .05$, $\nu_1 = 4$, and $\nu_2 = 7$, then $F_{\alpha, \nu_1, \nu_2} = F_{.05, 4, 7} = 4.1203$, where ν_1 is the numerator degrees of freedom and ν_2 is the denominator degrees of freedom.

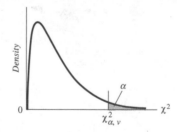

Degrees of Freedom ν	$\chi^2_{.100}$	$\chi^2_{.050}$	$\chi^2_{.025}$	$\chi^2_{.010}$	$\chi^2_{.005}$
1	2.71	3.84	5.02	6.63	7.88
2	4.61	5.99	7.38	9.21	10.60
3	6.25	7.81	9.35	11.34	12.84
4	7.78	9.49	11.14	13.28	14.86
5	9.24	11.07	12.83	15.09	16.75
6	10.64	12.59	14.45	16.81	18.55
7	12.02	14.07	16.01	18.48	20.28
8	13.36	15.51	17.53	20.09	21.96
9	14.68	16.92	19.02	21.67	23.59
10	15.99	18.31	20.48	23.21	25.19
11	17.28	19.68	21.92	24.72	26.76
12	18.55	21.03	23.34	26.22	28.30
13	19.81	22.36	24.74	27.69	29.82
14	21.06	23.68	26.12	29.14	31.32
15	22.31	25.00	27.49	30.58	32.80
16	23.54	26.30	28.85	32.00	34.27
17	24.77	27.59	30.19	33.41	35.72
18	25.99	28.87	31.53	34.81	37.16
19	27.20	30.14	32.85	36.19	38.58
20	28.41	31.41	34.17	37.57	40.00
21	29.62	32.67	35.48	38.93	41.40
22	30.81	33.92	36.78	40.29	42.80
23	32.01	35.17	38.08	41.64	44.18
24	33.20	36.42	39.36	42.98	45.56
25	34.38	37.65	40.65	44.31	46.93
26	35.56	38.89	41.92	45.64	48.29
27	36.74	40.11	43.19	46.96	49.64
28	37.92	41.34	44.46	48.28	50.99
29	39.09	42.56	45.72	49.59	52.34
30	40.26	43.77	46.98	50.89	53.67
40	51.81	55.76	59.34	63.69	66.77
50	63.17	67.50	71.42	76.15	79.49
60	74.40	79.08	83.30	88.38	91.95
70	85.53	90.53	95.02	100.43	104.22
80	96.58	101.88	106.63	112.33	116.32
90	107.60	113.14	118.14	124.12	128.30
100	118.50	124.34	129.56	135.81	140.17

Note: For example, if $\alpha = .05$ and $\nu = 20$, then
$\chi^2_{\alpha,\nu} = \chi^2_{.05,20} = 31.41$.

Table VIII is abridged from Thompson, Catherine M.: "Table of Percentage Points of the χ^2 Distribution," *Biometrika*, Vol. 32 (1942), p. 187, by permission of *Biometrika* Trustees.